Lecture Notes in Networks and Systems

Volume 247

The series "Lecture Notes in Networks and Systems" publishes the latest developments in Networks and Systems—quickly, informally and with high quality. Original research reported in proceedings and post-proceedings represents the core of LNNS.

Volumes published in LNNS embrace all aspects and subfields of, as well as new challenges in, Networks and Systems.

The series contains proceedings and edited volumes in systems and networks, spanning the areas of Cyber-Physical Systems, Autonomous Systems, Sensor Networks, Control Systems, Energy Systems, Automotive Systems, Biological Systems, Vehicular Networking and Connected Vehicles, Aerospace Systems, Automation, Manufacturing, Smart Grids, Nonlinear Systems, Power Systems, Robotics, Social Systems, Economic Systems and other. Of particular value to both the contributors and the readership are the short publication timeframe and the world-wide distribution and exposure which enable both a wide and rapid dissemination of research output.

The series covers the theory, applications, and perspectives on the state of the art and future developments relevant to systems and networks, decision making, control, complex processes and related areas, as embedded in the fields of interdisciplinary and applied sciences, engineering, computer science, physics, economics, social, and life sciences, as well as the paradigms and methodologies behind them.

Indexed by SCOPUS, INSPEC, WTI Frankfurt eG, zbMATH, SCImago.

All books published in the series are submitted for consideration in Web of Science.

More information about this series at https://link.springer.com/bookseries/15179

Alexey Beskopylny · Mark Shamtsyan
Editors

XIV International Scientific Conference "INTERAGROMASH 2021"

Precision Agriculture and Agricultural Machinery Industry, Volume 2

Springer

Editors
Alexey Beskopylny
Don State Technical University
Rostov- on-Don, Russia

Mark Shamtsyan
St. Petersburg State Institute of Technology
St. Petersburg, Russia

ISSN 2367-3370 ISSN 2367-3389 (electronic)
Lecture Notes in Networks and Systems
ISBN 978-3-030-80945-4 ISBN 978-3-030-80946-1 (eBook)
https://doi.org/10.1007/978-3-030-80946-1

This Springer imprint is published by the registered company Springer Nature Switzerland AG
The registered company address is: Gewerbestrasse 11, 6330 Cham, Switzerland

Contents

Agricultural Machinery and Agricultural Materials Science, Construction of Agricultural Facilities

High-Speed HFC-Boriding Kinetics

Alexey Ishkov and Vladimir Malikov

Abstract The paper presents studies of boride coatings, which were created using high-frequency treatment with currents. Chemical reactions characterized the boriding procedure at various temperatures are show. The choice of the main element of the reaction-boriding scheme is characterized, its role is revealed, and the probable processes leading to the realization of topochemical boriding initiated by HFC-heating are established. The kinetic curves received during the test at different temperatures are showed. Anamorphoses of the kinetic curves are showed, and the values of activation energy and parameters of the boriding process using HFC-heating are calculated on the basis of linearization of the Arrhenius equation.

Keywords Boriding procedure · Topochemical boriding · Arrhenius equation

1 Introduction

Obligating to the great properties (high yield strength, great strength to weight multiplier and good corrosion resistance) [1, 2], steel fusions are widely gone in aerospace and chemical industries. Nevertheless, low hardness and poor wear-friction restrict their use, especially in tribological applications [3, 4]. Most studies try to enhance their deterioration resistance using many surface hardening techniques [5]. Boriding, as one of the effective surface hardening treatments, can cause boron (B) atoms to diffuse through the metal substrate and effectively produce one or more hard boride layers, providing a significant enhancement in surface hardness and tribological properties [6, 7]. Boriding, owing to its simplicity, low cost, and flexibility properties, compared with another processes hardening, has been the most frequently used in hardening of steel. It can make hard (TiB_2 and TiB) boride layers on the superficies

A. Ishkov (✉)
Altay State Agricultural University, Krasnoarmeyskiy ave., 98, 656049 Barnaul, Altai region, Russia

V. Malikov
Altai State University, Lenina Avenue, 61, 656049 Barnaul, Altai region, Russia

A. Beskopylny and M. Shamtsyan (eds.), *XIV International Scientific Conference "INTERAGROMASH 2021"*, Lecture Notes in Networks and Systems 247,
https://doi.org/10.1007/978-3-030-80946-1_1

of steel alloys [8, 9], following in the enhancement of the properties (such as microhardness, compressive strength, and yield stress) and corrosion resistance for steel alloys [10, 11].

However, due to its relatively high treatment temperature and long treatment time, some defects (such as the microstructural changes and large deformation of base materials) are likely to be caused [12], largely affecting the mechanical properties and precision of borided samples. In order to deal with such prominent defects, during solid state boriding of steel alloy, rare earth (RE) elements can be used [13, 14]. It can not only accelerate the growth kinetics of boride layer [15] but also promote the efficiency of boriding process and save the treatment time. Thus, compared with the conventional solid state boriding of steel alloy, RE additions-solid state boriding (REs-boriding) can largely reduce the impact of high temperature and long treatment time on the mechanical properties and precision of base alloy, which can help to ensure the mechanical properties of base alloy as much as possible.

From the viewpoint of kinetics, the modelling for boriding kinetics can be regarded as a way to estimate the boride layer thicknesses to meet the various employment requirements of borided materials [16, 17]. Furthermore, there are a few studies exploring the modelling of the accelerated kinetics of boride layer growth for the solid state boriding of steel alloy. Recently, B. Sarma and K.S. Ravi Chandran established a diffusion model to estimate the effect of the boriding temperatures that are very close to the β-transus temperature on accelerating the kinetics of boriding of CP-Ti [18]. However, until now, no study is found to develop a kinetic model to quantitatively estimate the effect of RE on accelerating the kinetics of solid-state boriding.

It is necessary to carry out researches of SHS processes in a base model system and its modified versions in order to establish the rules governing the initiation and combustion of a charge of various compositions, as well as to control this process followed by a possibility of obtaining high-quality hardening boride coatings on the surface of a steel part with a thickness of at least 150–200 μm, and development of an active prototype technology.

The chosen area is relevant, since only a few works among the known ones are devoted to the use of HFC-heating in boriding, and the combination of various factors, stages and types of boriding in one process is one of the most widely studied areas in boriding (at least the diffusion one), and the use of a surface chemical (topochemical) reaction between iron and boron in order to obtain coatings, possible in the systems under study, is a new direction in boriding.

According to the data presented in paper [19], the first exothermic effects become apparent in oxides, mixtures of iron (II, III) oxide and iron spinel FeO and Fe_2O_3 with amorphous and crystalline boron under conditions of heating even at a temperature of 350 °C, which is explained by the reduction reaction of iron oxides with boron according to the following equations:

At higher temperature (500, 700 °C), additional exo-effects are observed caused by targeted reactions of the formation of FeB and Fe_2B borides and oxygen compounds of elements according to the following equations:

$$6FeO + 5B = 3FeB + Fe_3(BO_3)_2, \quad (1)$$

Then, extending the provisions of paper [20], the total process of boriding of the oxidized steel surface can be described by the following reaction:

$$2Fe_2O_3 + 8B = 4FeB + 2B_2O_3, \quad (2)$$

When the process proceeds not in a powder mixture, but on the surface of a compact material, as in case of boriding using HFC-heating, the FeB boride formed on the steel surface will further react with the base metal, propagating into its depth and forming Fe_2B boride according to the reaction:

$$FeB + Fe = Fe_2B, \quad (3)$$

and the final composition of the boride coating will also depend on Fe_2B disproportionation at higher temperatures.

Thus, the equilibrium state of the reversible reaction will definitely also have influence upon the chemical and phase composition and the structure of layers formed on the surface of steels during a topochemical reaction with boron at high temperatures (900 °C and more).

Additional acceleration of the topochemical process on a steel surface can also be achieved by using various deoxidizers present or specially introduced into the composition of fluxes used in HFC-surfacing and HFC-heating, thus simultaneously protecting the metal surface from oxidation and removing the process slags.

Taking into opinion the fact that HFC-heating allows to heat the steel surface to temperatures of 900–1000 °C, and all of the above reactions are exothermic, when applying HFC-heating for boriding, there is a real possibility not only to obtain boride coatings, but also to achieve their phase and structural homogenization due to remelting under the action of heat released in the surface layers during HFC-heating.

In paper [19], the thermodynamic calculations were also carried out and the Gibbs energy dependences were obtained for all observed and theoretically possible reactions, including those in the Fe-B system, and it was also shown that it has negative values (from −390 to −560 kJ/mole) across the entire studied temperature range and up to 1500 °C, which means that these boriding reactions are possible, and the type of the obtained products will depend mainly on the kinetic and equilibrium parameters of the boriding process.

The above summary theoretical equation for the topochemical boriding of an oxidized surface also indicates the need for a multiple (up to 100-fold) theoretical excess of amorphous boron in the mixture relative to oxidized forms of iron (as amorphous boron has a very low density), which, furthermore, will change permanently in the process of boriding, as some part of the element will be spent for surface deoxidation, and the other part will take part in the target reaction.

Gases evolving in the melt during combustion can lead to scattering (dispersion) of the melt, increase in porosity, expansion, decrease in density of SHS-products, and impairment of process initiation. However, in preliminary experiments and during combustion of cylindrical molded specimens, a slight increase in volume of SHS-process products was noted as compared with the initial substances.

Based on these considerations, boron should be replaced by its compounds capable to decompose and release an active element (B_4C, dissociating borides of *s*-metals and p-elements of the periodic system, ferroboron, etc.) under high temperatures and/or chemical influences.

The above circumstances determined the substance of the researches reflected in this paper, and their results are sequentially presented below. Initially, we've chosen the basic components of the reaction, established their role, and identified possible processes resulting in the topochemical boriding initiated by HFC-heating.

2 Material Choices and Design

To establish practically realizable reactions of surface chemical (topochemical) boriding and the accompanying high-temperature processes, a model mixture was prepared containing P-0.66 flux and boron carbide at the following ratio, wt.%: B_4C—84, P-0.66—16 [21, 22].

The main component of the P-0.66 flux is the eutectic mixture of B_2O_3 and $Na_4B_2O_7$ with a mass ratio of substances of 0.66, which melts when heated above 500–520 °C, forms complex mixtures of meta- and orthoborates, active boron in cationic (+) and anionic (−) forms, reacts with oxides on steel surface and other components of the mixture.

Calcium silicide ($CaSi_2$), included into the flux in an amount of up to 10–15 %wt, appears to be strong reducing agent, reactive compound, steel surface deoxidizing agent, and also as a high-temperature fluxer (due to released silicon) and others when starts thermally dissociate at 720–740 °C:

$$3Ca + B_2O_3 = 3CaO + 2B, \tag{4}$$

$$3Ca + Si + B_4C = Ca_3B_2 + SiC + 2B, \tag{5}$$

$$Ca + 2B_4C = CaC_2 + 8B, \tag{6}$$

To elucidate the nature and the most probable mechanism of boriding using HFC-heating, the process kinetics was studied, when it was conducted based on a mixture applied to 65G steel samples. The process kinetics was studied both when carrying out the borating process using HFC-heating in a protective medium (Ar gas supplied at a rate of 1.5–2 L/min into an aluminium cartridge with an inductor and a sample placed in it), and when carrying out the process in air under normal conditions.

Since reactions 1, 2 are heterogeneous and occur at the interface of several solid phases, namely a solid starting material (steel surface, boriding agent, flux, etc.)—a solid reaction product (FeB, Fe_2B), they can be considered as topochemical reactions (TCR).

Usually a topochemical reaction begins with the formation of seeds (nuclei) of crystallization of the solid reaction product on the surface of the initial solid. The centers of these nuclei are often associated with the presence of various types of defects on the surface: points of emergence of crystal lattice dislocations, vacancies (holes) or ions in interstices, crystal-crystallite interfaces in a polycrystalline material, domains, zones of nonstoichiometric composition, etc. Around these centers, nuclei, usually of a spherical type, start to grow. As they grow, the interface between the initial substance and the reaction product phases increases, and the reaction accelerates. When the nuclei fuse together and form a continuous reaction front, the phase interface and, accordingly, the reaction rate become maximum. Then, the process slows down due to the overlap of the nuclei themselves, which gradually disappear. From the presented phenomenological TCR model it can be seen that the rate of the entire process is determined by the rate of nuclei formation on the one hand, and by the rate of their growth and recombination on the other hand.

Like the rate of any chemical process, TCR depends on temperature and (to a lesser extent) pressure. Pressure dependence appears when gaseous components are the products or part of the starting materials of the TCR. At constant p and T, the TCR rate is a certain function of α, which must be determined from the experimental data.

Numerous experiments show [20] that the $\alpha = f(\tau)$ curve has a sigmoid form, and the $w = f(\tau)$ curve has its maximum at a certain value of τ_{max}.

This type of kinetic TCR curves in general form is explained as follows. At the initial time, the rate of appearance of nuclei and their growth is low (TCR induction period), afterwards their surface begins to grow rapidly, so that the rate of the process becomes maximum by w_{max} the time τ_{max}, and the degree of transformation α reaches ~ 0.5 for most known TCRs, that is, a half of the starting substances are consumed by this time as a rule. After that, no new nuclei are formed due to the absence of a free surface, and the overlap of the existing ones results in a decrease in the area of phase interface, and thus the reaction begins to slow down. The phase interface moves into the volume of the solid phase, and the TCR rate slows down even more.

Therefore, the actual TCR mechanism will be determined by the processes of nucleation and their recombination, which is probabilistic, and for kinetic calculations of such processes, two approaches to the rate of nucleus formation are applied.

According to the first of them, the nuclei are formed according to the exponential law:

$$N_t = N_0[1 - \exp(-kt)], \tag{7}$$

where N_0 and N_t are the numbers of nuclei at the initial and arbitrary points of time.

The second approach is based on the power law of nucleation:

$$N_t = K_n t^n \tag{8}$$

where K_n is the nuclei growth factor, and n is a whole or fractional number > 1.

In practice, the first mechanism is implemented very rarely, and the determining it TCR kinetic equations have a complex dependence $a = f(t)$. Therefore, to obtain the kinetic dependences of TCRs proceeding according to the second mechanism, the Erofeev-Kolmogorov equation is used:

$$\alpha = 1 - \exp(-kt^n), \tag{9}$$

where $n = \sigma + 3$, or $n = \sigma + m$ in general case.

σ is the number of units' steps of transformation of a new phase crystallization center into a growing nucleus,

m is the number of directions in which the nuclei grow, $m = 1, 2, 3$, and other.

Then, the TCR rate constant and the value of n can be easily found after double logarithmic transformation of dependence (12):

$$\ln[-\ln(1 - \alpha)] = \ln k + n \cdot \ln t, \tag{10}$$

which, when displaying experimental values in $\ln[-\ln(1 - \alpha)] = f(\ln t)$ coordinates, is a linear dependence, the parameters of which can be found graphically or by LSM.

The criterion for classifying the observed process as TCR will be a strong linear correlation of the experimental points of the linear dependence.

3 Experimental Results

In order to study the kinetics of boriding using HFC-heating, 65G steel specimens were boronized in a mixture for different periods of time. In our case, the degree of transformation can be calculated by the formula:

$$\alpha = \frac{h_{max} - h_t}{h_{max}}, \tag{11}$$

where h_{max} and h_t are the maximum and the current boride coat thickness, respectively.

The kinetic curves obtained in the experiment at temperatures of 900, 1000 and 1150 °C are shown in Fig. 1.

As can be seen from Fig. 1, the dependence of the thicknesses of boride coatings on steel 65G obtained at different temperatures of the process on the time of HFC-heating is sigmoid and has an induction period of ~ 30 s. The absence of noticeable processes in the system under study is obviously explained not so much by the

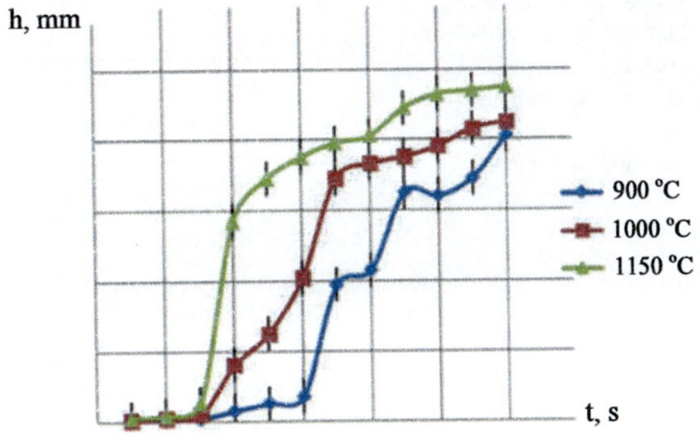

Fig. 1 Kinetic curves of 65G steel boriding from a mixture at temperatures of 900, 1000, and 1150 °C

accumulation of the number of nuclei in TCR, as by the release of a certain amount of active boron interacting with the steel surface according to reactions 4, 5, and 6, which depend little on temperature within the range we have studied.

Anamorphoses of kinetic curves of 65G steel boriding using HFC-heating in linear coordinates of the Erofeev-Kolmogorov equation are shown in Figs. 2, 3 and 4.

To determine the TCR activation energy, let's use the equation proposed by S. Arrhenius:

$$\frac{d\ln k}{dT} = \frac{E_\alpha}{RT^2}, \tag{12}$$

where k—reaction rate constant,

E_a—activation energy,

T—process absolute temperature,

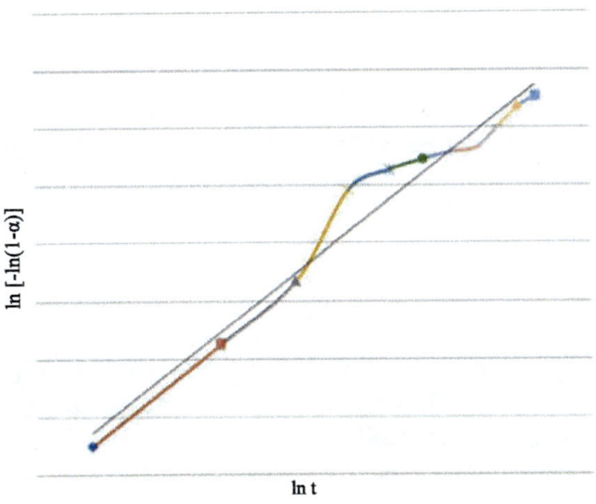

Fig. 2 Anamorphoses of kinetic curves of 65G steel boriding using HFC-heating in linear coordinates of the Erofeev-Kolmogorov equation at 1198 K

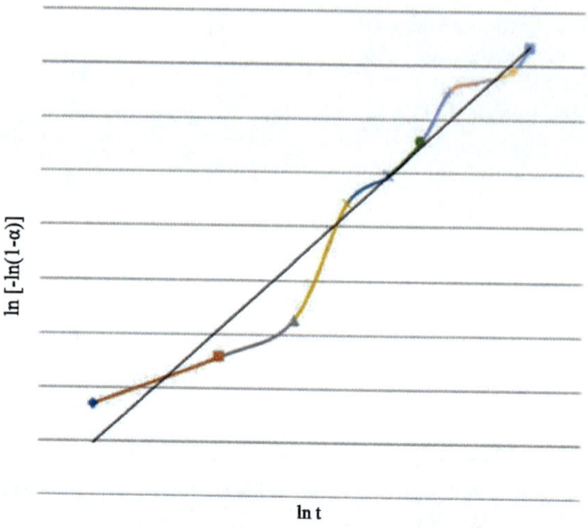

Fig. 3 Anamorphoses of kinetic curves of 65G steel boriding using HFC-heating in linear coordinates of the Erofeev-Kolmogorov equation at 1298 K

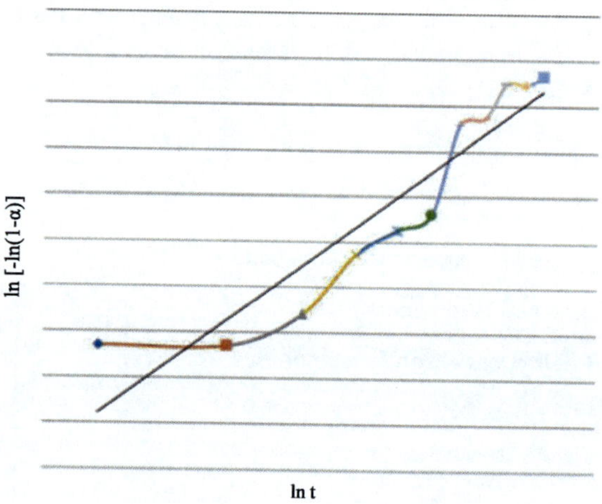

Fig. 4 Anamorphoses of kinetic curves of 65G steel boriding using HFC-heating in linear coordinates of the Erofeev-Kolmogorov equation at 1448 K

R—universal gas constant.

Arrhenius equation integration results in a dependence of reaction velocity (via the rate constant) on temperature:

$$\ln k = -\frac{E_\alpha}{RT} + \text{const.}, \quad (13)$$

Most generally, it turns out that the activation energy of the majority of real processes (diffusion, chemical, or mixed) depends on temperature. As a rule, this is due to their proceeding through several stages, the rates of which vary in different ways at different temperatures.

However, assuming that the formation of a coating layer in the process of boriding using HFC-heating is due to one, albeit hypothetical, lumped multistage topochemical reaction of type (2), the kinetics of which have been already studied and the thermodynamic potentials estimated, then the activation energy can be considered constant in the temperature range of practical implementation of boriding using HFC heating from 1198 to 1448 K.

When integrating the Arrhenius equation in a small T_1–T_2 temperature range, where it can be assumed that the chemical reaction E_a does not depend on temperature, the following expression is obtained:

$$\int_{T_1}^{T_2} \frac{d\ln k}{dT} = \ln KT_1 + \frac{E_a(T_2 - T_1)}{R \cdot T_1 \cdot T_2}, \tag{14}$$

from which it is easy to obtain an expression for the analytical determination of the activation energy, when a chemical reaction is carried out at two temperatures T_1 and T_2, its kinetics is studied, and the values of the chemical reaction rate constant at these temperatures kT_1 and kT_2 are found:

$$E_\alpha = \frac{R \cdot T_1 \cdot T_2}{T_2 - T_1}, \tag{15}$$

We have calculated the values of the activation energy of 65G steel boriding from a mixture using HFC-heating according to formula (15) and on the basis of linearization of the Arrhenius equation (Fig. 5).

The processing of linearized TCR kinetic curves by LSM made it possible to establish the parameters of the kinetic Eq. (9) for the process of steel boriding using HFC-heating (k, n) and calculate the apparent activation energy of TCR TPCHR (E_a*) according to the Arrhenius equation (Table 1).

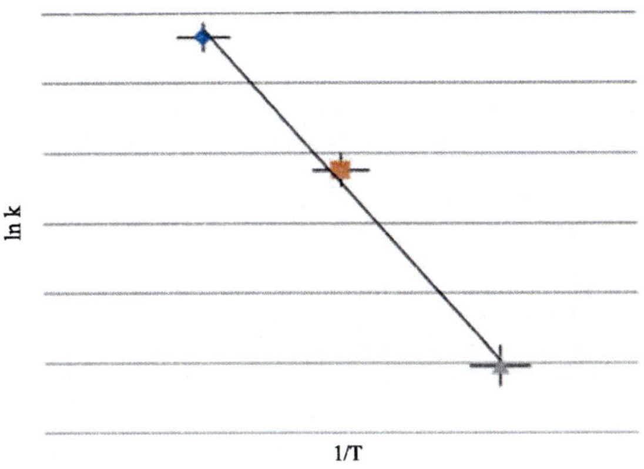

Fig. 5 Linearized Arrhenius equation

Table 1 Kinetic parameters of high-speed HFC-boriding of 65G steel using HFC-heating (n = 5; P = 0.95)

T, *K*	ln(*k*)	*N*	σ, m = const			E_a*, kJ/mole
			1	2	3	
1198	−10.35 ± 0.03	2.72 ± 0.02	1.72	0.72	–	365.8 ± 0.2
1298	−12.25 ± 0.02	3.32 ± 0.03	2.32	1.32	0.32	
1498	−15.05 ± 0.03	4.15 ± 0.03	3.15	2.15	1.15	

4 Conclusions

Kinetic curves of 65G steel boriding using HFC-heating at various temperatures were obtained. Anamorphoses of kinetic curves of boriding are plotted, and the values of activation energy and kinetic parameters of the borating process using HFC-heating are obtained on the basis of linearization of the Arrhenius equation. Following the conducted thermodynamic and kinetic studies, it was possible to confirm the possibility of TCR boriding using boronizing mixture and applying HFC-heating.

References

1. Hou Z, Zhao Y, Zeng W (2017) Effect of heat treatment on the microstructure development of TC21 alloy. Xiyou Jinshu Cailiao Yu Gongcheng/Rare Met Mater Eng 46(8):2087–2091
2. Zhu T, Li M (2010) Effect of hydrogen addition on the microstructure of TC21 alloy. Mater Sci Eng A 527(26):7080–7085. https://doi.org/10.1016/j.msea.2010.07.071
3. Zhu YS, Lu WZ, Zuo DW (2014) Microstructure and tribological properties of REs borided TC21 alloy. Surf Eng 30(8):612–618. https://doi.org/10.1179/1743294414Y.0000000292
4. Zhu YS, Lu WZ, Zuo DW, Sun YL, Wang H, Xu J (2013) Effects of rare earth additions during surface boriding on microstructure and properties of titanium alloy TC21. Mater Sci Technol 30(14):1765–1769. https://doi.org/10.1179/1743284713Y.0000000455
5. Zhang LC, Chen LY, Wang L (2019) Surface modification of titanium and titanium alloys: technologies, developments, and future interests. Adv Eng Mater 22. https://doi.org/10.1002/adem.201901258
6. Kul M, Oskay KO, Temizkan A (2019) Effect of boronizing composition on boride layer of boronized GGG-60 ductile cast iron. Vacuum 126:80–83. https://doi.org/10.1016/j.vacuum.2016.01.021
7. Kaouka A, Benarous K (2019) Electrochemical boriding of titanium alloy Ti-6Al-4V. J Market Res 8(6):6407–6412. https://doi.org/10.1016/j.jmrt.2019.10.024
8. Li P, Liu D, Bao W (2018) Surface characterization and diffusion model of pack borided TB2 titanium alloy. Ceram Int 44(15):18429–18437. https://doi.org/10.1016/j.ceramint.2018.07.060
9. Duan Y, Li P, Chen Z (2018) Surface evolution and growth kinetics of Ti6Al4V alloy in pack boriding. J Alloy Compd 742:690–701. https://doi.org/10.1016/j.jallcom.2018.01.383
10. Attar H, Bönisch M, Calin M (2014) Selective laser melting of in situ titanium-titanium boride composites: Processing, microstructure and mechanical properties. Acta Mater 76:13–22. https://doi.org/10.1016/j.actamat.2014.05.022

11. Chen Y, Zhang J (2017) Corrosion behaviour of selective laser melted ti-tib biocomposite in simulated body fluid. Electrochim Acta 232:89–97. https://doi.org/10.1016/j.electacta.2017.02.112
12. Duan Y, Wang X (2020) Characteristics, wear and corrosion properties of borided pure titanium by pack boriding near $\alpha \rightarrow \beta$ phase transition temperature. Ceram Int 46(10):16380–16387. https://doi.org/10.1016/j.ceramint.2020.03.197
13. Zhu YS, Lu WZ, Zuo DW (2018) A novel rare earth-salt bath nitriding of TC21-DT titanium alloy. Surf Eng 34(2):128–131. https://doi.org/10.1080/02670844.2016.1213035
14. Peng MJ, Duan YH (2018) Characteristics of surface layers on Ti_6Al_4V alloy borided with CeO_2 near the transition temperature. J Alloy Compd 769:1–9. https://doi.org/10.1016/j.jallcom.2018.07.365
15. Zhu YS, Lu WZ (2016) Effect of rare earth addition on diffusion kinetics of borided TC21-DT titanium alloy. Surf Eng 32(11):795–801. https://doi.org/10.1080/02670844.2016.1178413
16. Sarma B, Tikekar NM (2012) Kinetics of growth of superhard boride layers during solid state diffusion of boron into titanium. Ceram Int 38(8):6795–6805. https://doi.org/10.1016/j.ceramint.2012.05.077
17. Keddam M, Taktak S (2017) Characterization and diffusion model for the titanium boride layers formed on the Ti6Al4V alloy by plasma paste boriding. Appl Surf Sci 399:229–236. https://doi.org/10.1016/j.apsusc.2016.11.227
18. Sarma B, Ravi CKS (2011) Accelerated kinetics of surface hardening by diffusion near phase transition temperature: mechanism of growth of boride layers on titanium. Acta Mater 59(10):4216–4228. https://doi.org/10.1016/j.actamat.2011.03.046
19. Merzhanov AG, Borovinskaya IP, Volodin YuE (1972) On the mechanism of combustion of porous metal samples in nitrogen. Rep USSR Acad Sci 206(4):905–908
20. Itin VI, Nayborodenko YS (1989) High-temperature synthesis of intermetallic compounds. Tomsk State University, Tomsk, p 214
21. Ishkov AV, Malikov VN (2020) The study of the wear of the control knives of the straw chopper. J Phys: Conf Ser 1515(5):1–5. https://doi.org/10.1088/1742-6596/1515/5/052026
22. Dmitriev S, Malikov V, Ishkov A (2020) Investigation of thermal fields at phase boundaries in powder mixtures that are subject to melting and chemical transformation. Mater Sci Forum 992:1011–1015. https://doi.org/10.4028/www.scientific.net/MSF.992.1011

Optimal Heat Flux Reduction Inside Film Cooled Wall

Nicolay Kortikov and Vladimir Polishuk

Abstract This work is devoted to search for the global extreme of heat flux reduction into the film cooled wall. It was carried out by varying five main system parameters from a combination of the lateral average of both the adiabatic effectiveness and the heat transfer coefficient. The lateral average heat flux reduction is processed according to the *IOSO* technology. It yields a prediction of the heat transfer coefficient from the ejection position to far downstream, including effects of extreme blowing angles and hole spacing. Together with the calculation of the adiabatic effectiveness it provides an immediate determination of the stream wise heat flux reduction distribution of cylindrical hole film cooling configurations.

Keywords Optimization · Reduction heat flux · IOSO technology

1 Introduction

An increase in the initial temperature of the gas at the turbine inlet poses the problem of ensuring the operability of the elements of gas turbine units (GTU), which are exposed to high gas temperatures in combination with high external loads. The vanes and blades of the first stage of the high-temperature turbine [1] have a developed convective-film cooling scheme in which the cooler is fed through rows of perforations to the surface of the blade (Fig. 1a).

To create competitive samples in the field of gas turbine engineering [2], it is necessary to combine mathematical models [3, 4] and software systems [5–7] with search methods [8–10] for the most effective technical solutions within the framework of the optimization environment. Moreover, to solve optimization problems, it is necessary to solve the problem of integrating various programs within the framework of one project.

Figure 1b introduces the following notation: T_G, T_C—hot gas and coolant temperatures, respectively; $T_f = T_{AW}$—gas temperature equality T_f and adiabatic wall T_{AW}

N. Kortikov (✉) · V. Polishuk
Peter the Great St. Petersburg Polytechnic University, Polytechnicheskaya 29, St. Petersburg 195251, Russian Federation

A. Beskopylny and M. Shamtsyan (eds.), *XIV International Scientific Conference "INTERAGROMASH 2021"*, Lecture Notes in Networks and Systems 247,
https://doi.org/10.1007/978-3-030-80946-1_2

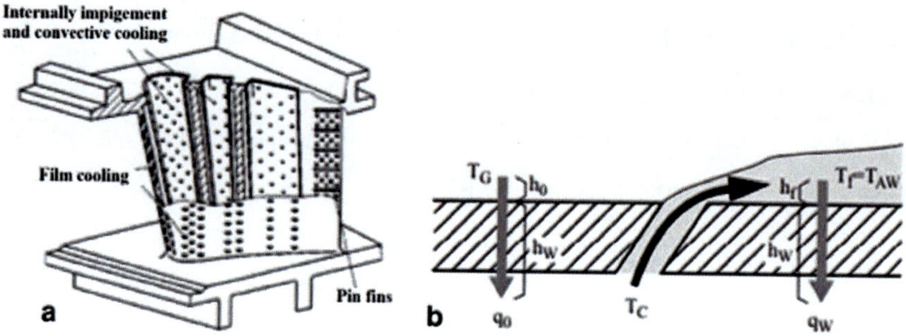

Fig. 1 Nozzle blade with convection-film cooling (**a**); heat transfer through the diabatic wall (**b**)

in the presence of film cooling; h_0 and h_f are heat transfer coefficients before and after the film on the upper surface of the plate; h_W—heat transfer coefficient on the bottom surface.

In papers [11, 12] introduced the concept of the net heat flux reduction (NHFR) to evaluate film cooling in terms of its wall heat flux effect

$$\eta_q = 1 - \frac{q_w}{q_0} \tag{1}$$

Best cooling is achieved for large values of NHFR, indicating maximum reduction of the heat flux into the film cooled wall q_w compared to that without cooling film q_0. Using the definitions of the heat flux into the cooled and the uncooled wall leads to

$$\eta_q = 1 - \frac{h_f}{h_0}(1 - \eta\theta) \tag{2}$$

Given the values of the heat transfer augmentation $\frac{h_f}{h_0}$ and the effectiveness $\eta = \frac{T_G - T_{AW}}{T_G - T_C}$, NHFR is obtained for a preset dimensionless wall temperature $\theta = \frac{T_G - T_C}{T_G - T_W}$:

$$\eta_q = 1 - \frac{2\left({}^{h_f}/_{h_0}\right)}{1 + \left({}^{h_f}/_{h_0}\right)}(1 - \eta) \tag{3}$$

The heat transfer measurements [13] indicate that configurations causing the highest heat transfer augmentation in general coincide with those configurations yielding the best overall effectiveness. Obviously, it is necessary to combine the measurements of both parameters to determine the wall temperature governing the heat flux reduction. Only a comparison based on the heat flux reduction allows for a final decision for one of several competing ejection configurations.

Currently, a large number of commercial software packages are known that declare the ability to organize a search for a global optimum for functions with a large number of variables. Among the most promising are the algorithms of *IOSO* technology [14]. The *IOSO NX GT 2.0* program is able to find a global extremum for functions with five of independent variables.

In this work, *IOSO NX GT 2.0* verification was carried out on the basis of integration with the adiabatic wall temperature and reduction of the heat flux calculation programs when the curtain was blown onto the plate through perforations (Fig. 1b) in order to increase the reduction of the heat flux of gas turbine blades.

2 Problem and Method

The appearance of the maximum value for the Stanton number $\overline{St}$ of cooling the plate with a gas curtain when blowing through perforations (Figs. 2, 3 and 4) is associated

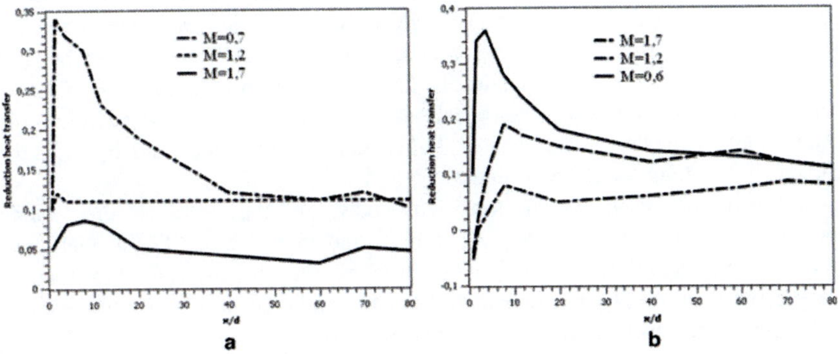

Fig. 2 Effect of the density ratio on the heat flux reduction: $\alpha = 30°$, $P/d = 3$, $DR = 1.2$ (**a**); $\alpha = 30°$, $P/d = 3$, $DR = 1.8$ (**b**)

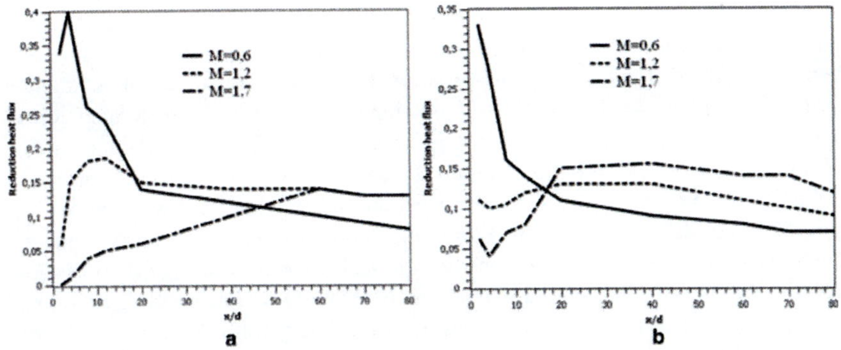

Fig. 3 Effect of the blowing angle on the heat flux reduction: $\alpha = 60°$, $P/d = 3$, $DR = 1.8$ (**a**); $\alpha = 90°$, $P/d = 3$, $DR = 1.8$ (**b**)

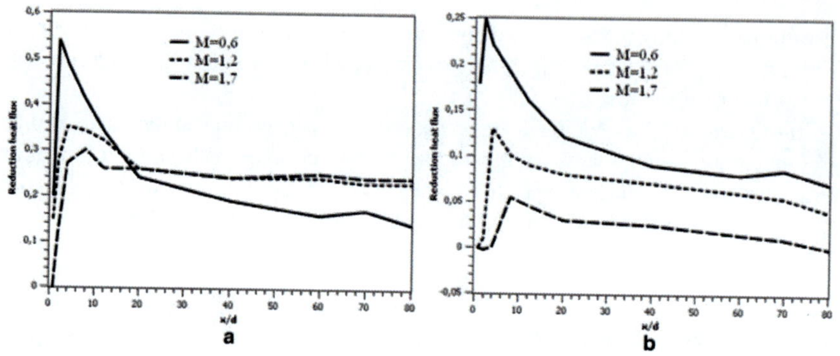

Fig. 4 Effect of the hole spacing on the heat flux reduction: $\alpha = 30°$, $P/d = 2$, $DR = 1.8$ (**a**); $\alpha = 30°$, $P/d = 5$, $DR = 1.8$ (**b**)

the Stanton number with various effects on the flow and heat transfer of the following factors (θ—dimensionless wall temperature, M is the injection parameter, DR is the ratio of the densities of the main and secondary flows, Tu_1 is the degree of turbulence of the main stream, α is the angle of blowing of the curtain, $\overline{P} = P/d$ is the relative step between the perforation holes (d is the diameter of the hole), δ_1/d—the dimensionless displacement thickness, L/d is relative cooling tube length, x/d is the dimensionless longitudinal coordinate).

Referring to a similarity and sensitivity analysis of the discussed film cooling situation (see [3]), this dependence can be formulated as a functional relation of similarity numbers. Employing the Stanton number as dimensionless heat transfer coefficient yields. In [4], to calculate the Stanton number $\overline{St}$ averaged over the width of the plate, the dependence (4) was proposed:

$$\overline{St} = \overline{St}\,(\theta, M, DR, Tu_1, \alpha, P/d, \delta_1/d, L/d, x/d) \tag{4}$$

In the presence of film cooling, the influence of the ejection on the heat transfer situation is of particular interest. Therefore, the ratio of the heat transfer coefficients on the surface with and without film ejection is regarded. Since the Stanton numbers of both cases should refer to the hot gas flow properties, the dimensionless augmentation ratio fulfils$h_f/h_0 = \overline{St}/\overline{St_0}$. It was obtained as a result of a generalization of experimental data, and its practical use is associated with the sequential calculation of thirty-one algebraic equations. Figure 2 compares heat flux reduction results of the typical geometry of shallow angle ejection and standard hole spacing for low (Fig. 2a) and high (Fig. 2b) density rations. Optimum overall heat flux reduction is obtained at $M = 0.6$ for low density ratio and $M = 0.85$ at high density ratio. The maximum heat flux reduction obtained is about 35% close to the ejection and 10–12% downstream, gradually increasing with density ratio. Blowing rates of $M = 1.7$ and beyond yield heat flux reductions in the order of 2–5%, indicating that a large amount of coolant is ejected without any beneficial effect.

Along the film cooled surface, the areas of dominating single jet in crossflow mixing and adjacent jet in crossflow interaction downstream can be clearly distinguished. For the steep angle ejection (Fig. 3a), the adverse effect of a rising blowing rate can be observed upstream $x/d = 30$ for lower blowing rates with a shift to $x/d = 50$ for high blowing rates.

For normal ejection, a distinct crossing point of all curves is present at $x/d = 18$. It obviously separates the single jet dominated region close to the ejection from the film flow dominated region downstream (Fig. 3b). The optimum is less pronounced than for shallow angle ejection.

Figure 4 shows the results of small pitch and large pitch ejection at typical shallow ejection angles and engine like high density ratios. At a hole spacing of $P/d = 2$ (Fig. 4a) the heat flux reduction is monotonically increasing with blowing rate over most of the regarded downstream length.

This shows the dominance of the adjacent jet interaction at small pitch resulting in stable cooling films. In contrast to the findings at larger hole spacings, the heat flux reduction is especially high at high blowing rates. The large pitch ejection (Fig. 4b) displays a completely different behavior. The almost purely jet in crossflow structured situation obviously exhibits optimum cooling conditions at a blowing rate of $M = 0.85$.

3 Results

The calculation results are presented in Table 1. The initial version of the film cooling system for the nozzle blade of the first stage of the turbine [2] is characterized by the parameters presented in the first line of Table 1. The search for the reduction heat flux was carried out on the basis of a mathematical model, including a Fortran program for sequential calculation of thirty-one algebraic relations and the *IOSO NX 2.0* package.

The calculation results indicate a small contribution of film cooling to a decrease in the maximum heat flux inside the wall for the initial version of the cooling system: $\eta_q = 0.0217$ at a relative coolant flow rate $\overline{G_c} = \dot{G}_c / \dot{G}_f = 5.8\%$, where $\dot{G}_c$ and $\dot{G}_f$—where mass flow rates of the cooler and hot gas flow.

Table 1 Parameters of the film cooling system of the gas turbine unit [12] before and after optimization

Parameter	P/d	α radian	*Tu* %	*M*	*DR*	$\overline{G_c}$ %	η_q
Before optimization	4.6	0.785	5	3.01	2.35	5.8	0.0217
After optimization: Pressure side of the nozzle blade (Fig. 1a)	3.0	1.5	9.9	1.682	1.5	5.0	0.0928

Optimization results obtained with the *IOSO NX 2.0* package indicate a possible increase in the maximum of reduction heat flux (up to $\eta_q = 0.0928$) with the parameters indicated in the second line of Table 1. These parameters correspond to the trends in the reduction heat flux shown in Figs. 2, 3 and 4.

The search for the maximum reduction heat flux was carried out by varying five independent parameters: $5.0\% \leq Tu_1 \leq 10\%$; $30° \leq \alpha \leq 90°$; $1.2 \leq DR \leq 2.5$; $3 \leq \overline{P} \leq 5$ and $0.5 \leq M \leq 3.0$.

These values correspond to the experimental data [3, 4], while the discrepancy of 15% can be explained by some difference in the initial data in the experiment and calculation.

4 Conclusions

It is shown that *IOSO NX GT 2.0* is a reliable tool for finding the optimal solution to one function of the target (the reduction heat flux). This algorithm can be successfully used in combination with various application programs. Only the decay of heat transfer and a simultaneous formation of a closed cooling film due to adjacent jet interaction assists heat flux reduction far downstream. As an example, a search was made for a global maximum of reduction heat flux and optimal parameter values for a film cooling system, as applied to operating conditions of gas turbine blades.

References

1. Krivonosova V, Lebedev A, Simin N, Zolotogorov M, Kortikov N (2011) Experimental and numerical analysis of high temperature gas turbine nozzle vane and film cooling effectiveness. In: Turbo Expo 2011. GT-45294, Proceedings of ASME. Vancouver, British Columbia, pp 1–9
2. Krivonosova VV, Kortikov NN (2018) Analysis and improvement of the cooling system of a gas turbine nozzle vane. St. Petersburg Polytechnic Univ J Eng Sci Technol 24(03):80–90. https://doi.org/10.18721/JEST.240307
3. Baldauf S, Scheurlen M, Schulz A, Wirtig S (2002) Correlation of film cooling effectiveness from thermographic measurements at engine like conditions. In: Proceeding of ASME. Turbo Expo 2002, Amsterdam, The Netherlands, GT-2002-30180, pp 1–14
4. Baldauf S, Scheurlen M, Schulz A, Wirtig S (2020) Heat flux reduction from film cooling and correlation of heat transfer coefficients from thermographic measurement at engine like conditions. In: Proceedings of ASME. Turbo Expo 2002, Amsterdam, The Netherlands, GT-2002-30181, pp 1–12
5. IOSO NS GT vers. 2.0. IOSO Technology Center. 2001–2003, Moscow. http://www.iosotech.com. Last accessed 21 Nov 2016
6. Egorov IN, Kretinin GV, Leshchenko IA, Kuptzov SV (2002) IOSO optimization toolkit—novel software to create better design. In: 9th AIAA/ISSMO symposium and exibit on multidisciplinary analysis and optimization. AIAA paper AIAA-5514, Atlanta, pp 1–12
7. Egorov IN, Kretinin GV, Fedechkin KS (2010) Multi-level robust design optimization fan. Workshop CEAS, Vrije Universiteit Brussels (VUB), Brussels, Belgium, pp 1–11

8. Lee KD, Kim SM, Kim KY (2011) Multi-objective optimization of film-cooling holes considering heat transfer and aerodynamic loss. In: Proceedings of ASME Turbo Expo 2011. GT-45402, Vancouver, British Columbia, Canada, pp 1–13
9. Lee KD, Choi DW, Kim KY (2013) Optimization of ejection angles of double-jet film-cooling holes using RBNN model. Int J Therm Sci 73:69–78
10. Lee KD, Kim SM, Kim KY (2012) Numerical analysis of film-cooling performance and optimization for a novel shaped film-cooling hole. In: ASME Paper GT-68529, pp 1–11
11. Sen B, Schmidt DL, Bogard DG (1996) Film cooling with compound angle holes: heat transfer. J Turbomach 118:800–806
12. Hay N, Lampard D, Saluja CL (1985) Effects of cooling films and the heat transfer coefficient on a flat plate with zero mainstream pressure gradient. J Eng Gas Turbines Power 107:105–110
13. Goldstein RJ, Jin P, Olson RL (1998) Film cooling effectiveness and mass/heat transfer downstream of one row of discrete holes, ASME Paper 98-GT-174, pp 1–12
14. Kortikov N (2021) Numerical optimization of film cooling system with injection through circular holes. In: International scientific conference energy management of municipal facilities and sustainable energy technologies EMMFT 2019. Advances in intelligent systems and computing, vol 1259, Springer, Cham. https://doi.org/10.1007/978-3-030-57453-6_51

Numerical Simulation of the Operation of a Chemical Reactor with an Open Cell Foam Catalyst

Sergei Solovev and Olga Soloveva

Abstract In this work, based on numerical modeling, we carried out a comparative analysis of the flow of chemically reactive gas flow through a catalyst using the example of the selective hydrogenation of acetylene. The catalyst models are built based on an open foam cell material and a traditional granular catalyst. We obtained the results of numerical studies in the form of fields of concentration of components of the gas mixture, vector fields of gas motion, and values of conversion and selectivity.

Keywords Numerical simulation · Open cell foam · Catalyst

1 Introduction

Most chemical industry reactions are exothermic; therefore, it is necessary to prevent overheating in the gas's contact zones and the surface of the catalyst granules [1–3]. A solution to this problem can be using an open cell foam material as a catalyst carrier. The authors of [4] investigated the flow in a layer consisting of cubes of an open cell foam material. Global and local porosity in this geometry increase convection by decreasing the superheat in the reactor. The use of such structures in a reactor is promising. Open cell foam materials are the subject of numerous studies due to their unique properties: large contact surface area and high porosity, which creates a low pressure drop and provides high energy efficiency [5–7]. Currently, metallic porous materials are widely used as multifunctional heat exchangers [8] and compact radiators for microelectronic devices [9] due to their heat transfer properties [10, 11]. Also, thermally conductive porous materials have been proposed as an effective solution for intensifying non-adiabatic catalytic processes in tubular reactors.

In chemical engineering, porous materials are widely used as catalyst carriers. During chemical reactions, the process parameters and geometric structure of fixed bed catalytic reactors, including porous materials, play a decisive role. For stable

S. Solovev (✉) · O. Soloveva
Kazan State Power Engineering University, Krasnoselskaja Street, 51, Kazan 420066, Russia
e-mail: solovev.sa@kgeu.ru

A. Beskopylny and M. Shamtsyan (eds.), *XIV International Scientific Conference "INTERAGROMASH 2021"*, Lecture Notes in Networks and Systems 247,
https://doi.org/10.1007/978-3-030-80946-1_3

and reliable reactors in industrial conditions, it is necessary to correctly set the operating parameters to control the reactor's coolant dynamics. The mode of operation of catalytic reactions associated with a chemical reaction or their field of application determines the characteristic requirements and limitations of reactors, including size, productivity, uniformity of heat transfer, pressure drop, and selectivity. Numerical simulations can characterize or optimize the performance of fixed bed catalytic reactors for industrial applications. For exothermic reactions, solid porous structures show good results. In [12], ceramic porous materials, honeycombs, and volumetric spheres were compared. The authors found that porous materials' main advantage is their radial mixing characteristics of gases and heat transfer from the reactor's pipe wall. In [13], porous metal granules with solid porous structures were compared in terms of pressure drop and heat transfer characteristics.

In this work, we performed a numerical simulation of acetylene's selective hydrogenation to ethylene when passing through an open cell foam catalyst. Ethylene is an essential component of the petrochemical industry. The main task in the production of ethylene is its purification from by-components, for example, acetylene. In this case, purification can be carried out by hydrogenation, when acetylene is converted to ethylene. However, the acetylene and the ethylene, which is hydrogenated to ethane, can be hydrogenated. Thus, the process should take place in the presence of a catalyst with selectivity parameters that allow acetylene's hydrogenation, but do not promote ethylene's hydrogenation.

2 Problem Formulation and Solution Method

Let us carry out a comparative analysis of the reactors' operation for acetylene's selective hydrogenation [14]. Experimental data are needed to build a model of the movement of a chemically reacting flow. Experimental studies were carried out for the reaction of selective hydrogenation of acetylene to ethylene. The hydrogenation of acetylene to ethylene occurs in a reactor with a granular bed of a palladium catalyst or a foam catalyst with palladium deposition (see Fig. 1).

The catalyst is tested in an isothermal cylindrical reactor. The reactor is a hollow steel tube 700 mm long and 20 mm in inner diameter. There is a 45 mm bed of granular catalyst mixed with quartz in the middle of the reactor or a sample of a foam catalyst 55 mm high. In front of the catalyst and on the catalyst, quartz is poured, a shapeless granule with a 3–5 mm diameter. Gas temperature 25–75 °C, pressure 1 atm. The gas used is a fraction of argon (24%), ethylene (75%), and acetylene (1%). Passing through the catalyst bed, acetylene is hydrogenated to ethylene

$$C_2H_2 + H_2 \rightarrow C_2H_4 + 176\ \text{kJ/mol}. \quad (1)$$

In addition to the primary reaction, undesirable side reactions can occur, for example, hydrogenation of ethylene to ethane's

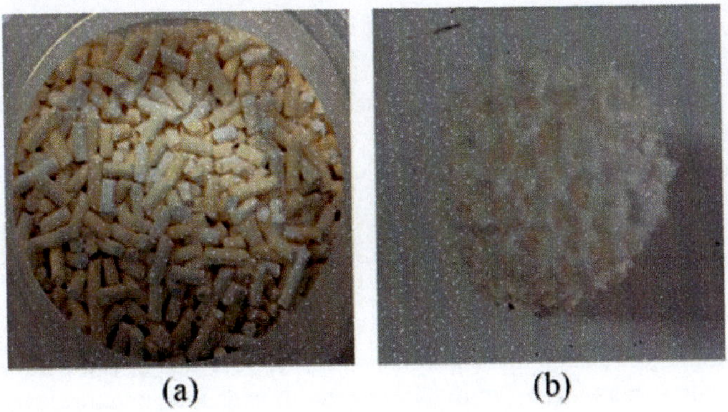

Fig. 1 Samples of catalyst: **a**—traditional granular, **b**—open cell foam

$$C_2H_4 + H_2 \rightarrow C_2H_6 + 138\ \text{kJ/mol}. \tag{2}$$

As the criteria for the catalyst operation, let us estimate the acetylene conversion and selectivity as

$$A = \frac{C_2H_2^{\text{inlet}} - C_2H_2^{\text{outlet}}}{C_2H_2^{\text{inlet}}} 100\%. \tag{3}$$

$$S = \frac{C_2H_2^{\text{inlet}} - C_2H_2^{\text{outlet}} - C_2H_6^{\text{outlet}}}{C_2H_2^{\text{inlet}}} 100\%. \tag{4}$$

We will carry out a detailed numerical simulation of reactors from experimental studies to analyze the effects that affect the reaction's activity and selectivity.

The solution to the problem of the movement of a chemically reacting gas flow is carried out by the finite volume method with the division of the considered region of the reactor into triangular elements. The considered flow is stationary. For a multi-component gas phase, the laws of conservation of mass, momentum, and energy are fulfilled.

Mass conservation equation

$$\nabla \cdot (\rho \vec{v}) = 0, \tag{5}$$

where ρ is the density, $\vec{v}$ is the velocity.

The momentum conservation equation

$$\nabla \cdot (\rho \vec{v}\vec{v}) = -\nabla p + \nabla \cdot \bar{\bar{\tau}} + \rho \vec{g}, \tag{6}$$

where p is the pressure, $\bar{\bar{\tau}}$ is the stress tensor. In Eq. (6), the stress tensor is

$$\bar{\bar{\tau}} = \mu\left(\nabla\vec{v} + \nabla\vec{v}^T\right) + \frac{2}{3}\mu\nabla\cdot\vec{v}\bar{\bar{I}}, \tag{7}$$

where μ is the viscosity, $\bar{\bar{I}}$ is unit tensor.

Mass conservation equation for the i-th component of the gas mixture

$$\nabla\cdot(\rho\vec{v}Y_i) = -\nabla\cdot\vec{J}_i + R_i, \tag{8}$$

where Y_i is the mass fraction of the i-th component of the gas mixture, R_i is the consumption of the i-th component of the chemical reaction, J_i is the diffusion flux of the i-th component arising from the concentration and temperature gradients.

$$\vec{J}_i = -\rho D_{m,i}\nabla Y_i - D_{T,i}\frac{\nabla T}{T}, \tag{9}$$

where T is the temperature, $D_{m,i}$ is the mass diffusion coefficient, $D_{T,i}$ is the temperature diffusion coefficient.

Energy conservation equation

$$\nabla\cdot(\rho\vec{v}h) + \nabla\cdot\left(\vec{J}_q\right) = \frac{\partial p}{\partial t} + \bar{\bar{\tau}} : \vec{v}, \tag{10}$$

where $h = \sum_{i=1}^{N} Y_i h_i$ is the enthalpy, and

$$\vec{J}_q = \lambda\nabla T + \sum_{i=1}^{N} h_i\vec{J}_i, \tag{11}$$

where λ is the thermal conductivity of the gas mixture, N is the number of components in the mixture.

For chemical reactions, the parameter R_i in Eq. (8) can be written as

$$R_i = M_{w,i}\sum_{r=1}^{N_R}\dot{R}_{i,r}, \tag{12}$$

where $M_{w,i}$ is the molecular weight of the i-th component of the mixture, N_R is the number of reactions involving the i-th component of the mixture.

$$\dot{R}_{i,r} = k_{f,r}\prod_{j=1}^{N}\left[C_{j,r}\right]_{wall}^{\eta'_{i,r}}, \tag{13}$$

where $C_{j,r}$ is the concentration of the *j*-th component of the mixture, $\eta'_{i,r}$ is the exponent for the reactant *j*-th component in the reaction, $k_{f,r}$ is the reaction rate constant.

$$k_{f,r} = A_r e^{-\frac{E_r}{RT}}, \tag{14}$$

where A_r is the pre-exponential factor, E_r is the activation energy.

We use ANSYS Fluent 19.2 software to solve the written system of equations. Many of the component properties parameters are taken from the database of the software used. We take the physicochemical properties of the components according to the polynomial dependence on temperature.

For a detailed simulation of the reactor, it is necessary to create a geometry consisting of randomly poured granules or representing a foam material with a random arrangement of connected cells. The primary methods for creating geometry with random elements in space are the Discrete Element Method (DEM). A dynamic process is considered, in which the speed of movement depends on the physical properties of a discrete system.

Then we will consider the constructed layer of granules cut from the reactor volume and the movement of the gas mixture in space. Next, the built area for gas movement will be subdivided into a grid. In this case, it is worth paying attention to the points of contact between the individual pellets and the reactor's walls. The constructed grid cells at such points have sharp bends, which leads to problems of solution convergence. There are four approaches to solving the convergence of the numerical solution at the points of contact: a decrease in the diameter of granules with the formation of gaps [15–17], an increase in the diameter of granules with the formation of intersections [18], a partial cutting of the surface of the granules at the points of contact [19], bringing bridges between granules at points of contact [20]. The authors of [21] analyzed the use of these approaches for fixed bed reactors, with particular attention to the parameters of pressure drop, porosity, and heat transfer. The authors showed that an adequate description of the parameters under study could be obtained only with the help of local modification of particles, for example, by cutting off part of the granules at the points of contact or by the bridge method. In [22], these two modifications of granules were studied with the numerical simulation of a chemical reaction in a fixed layer of spheres. The results obtained indicate similar values of the investigated parameters when leveling a part of the granules at the contact points and building bridges.

The acetylene hydrogenation reaction is very fast and is used in industry, for example, to purify the ethylene fraction from acetylene impurities. The advantage of using an open cell foam material as a catalyst carrier is a decrease in the resistance to gas flow as it passes through the catalyst bed and a developed catalyst surface area.

Let us carry out a comparative analysis of two types of catalysts: traditional granular and built based on a foam material. Let us create models that correspond to the experimental studies carried out. The constructed models are shown in Fig. 2. In the experiments, catalyst granules were mixed with quartz particles. When making

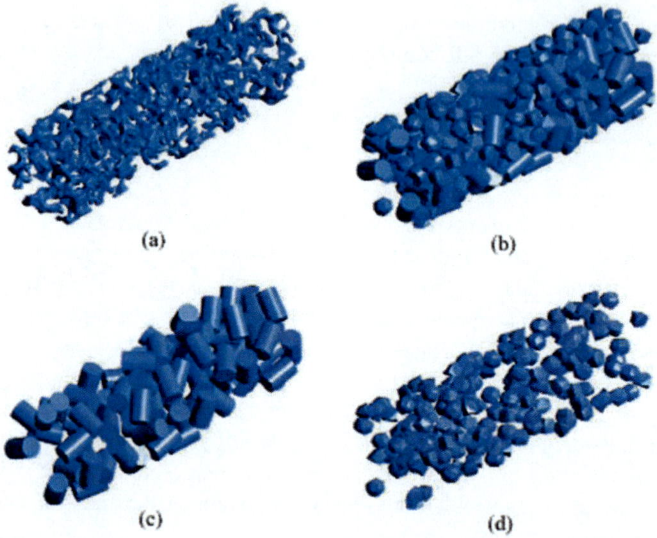

Fig. 2 Catalyst model: **a**—open cell foam material, **b**—catalyst granules with quartz, **c**—catalyst granules, **d**—quartz

a computational model, cylindrical catalyst granules and shapeless quartz particles were also selected.

To solve the problem numerically, we divide the entire area under consideration into finite elements of a triangular shape, the dimensions of which are sufficient to determine the specific factors of the phenomenon under study. We are using a uniform mesh with 10^{-4} m selected as the primary cell size. Condensation with a cell size of 2.5×10^{-5} m and 5×10^{-5} m was carried out near the granules' surface and at the contact points. This separation of the grid allows avoiding a sharp bending of the cells.

3 Results and Discussions

Experimental studies were carried out for a feed gas temperature range of 25 to 75 °C. Numerical calculations were also performed in this range of gas flow temperatures. The results of the calculated fields of the mass content of acetylene in the cross-section of the reactor are presented in Fig. 3 for a granular catalyst and in Fig. 4 for a catalyst based on an open cell foam material.

It can be seen from the pictures of the calculation results that the mass content of acetylene is significantly lower at the outlet of the reactor for a catalyst based

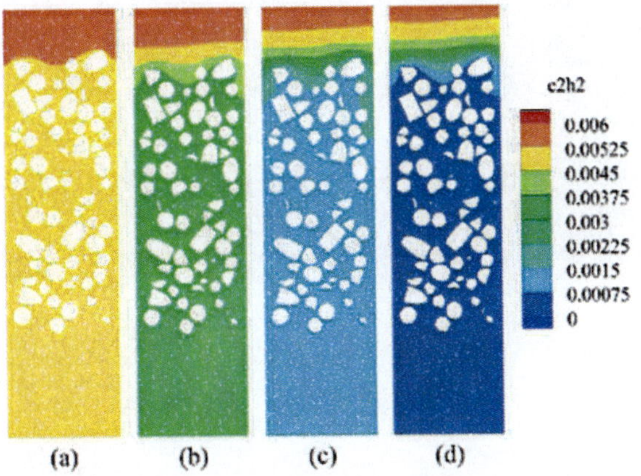

Fig. 3 Mass content of acetylene for a reactor with granular catalyst at different flow temperatures: **a**—30 °C, **b**—40 °C, **c**—50 °C, **d**—60 °C

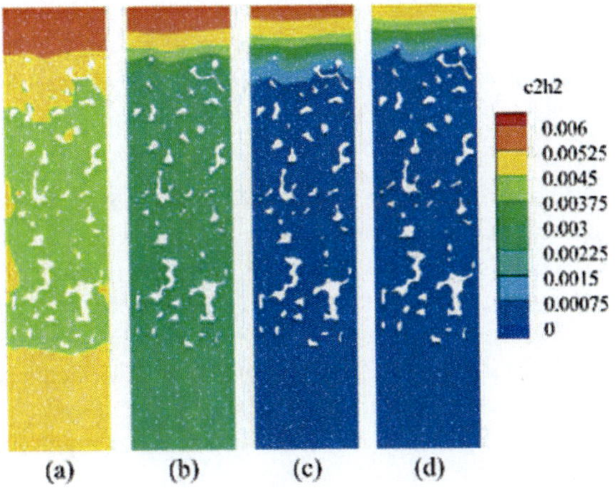

Fig. 4 Mass content of acetylene for a reactor with an open cell foam catalyst at different flow temperatures: **a**—30 °C, **b**—40 °C, **c**—50 °C, **d**—60 °C

on an open cell foam material than at the outlet of a reactor with a traditional granular catalyst. In this case, large values in acetylene conversion are observed in the calculations over the entire temperature range of the gas flow.

Next, we carry out a quantitative analysis of the conversion of acetylene when passing through the catalyst bed. Figure 5 shows the results of experimental studies and numerical calculations.

It can be seen from the figure that the results of numerical calculations show good agreement with the results of the experiments performed. In this case, the best agreement of the results is observed for an open cell foam catalyst. The most considerable deviations are observed for a granular catalyst at the smallest and largest gas temperatures in the selected range. In both experimental and computational experiments, catalysts based on an open cell foam catalyst show the best results in comparison with a traditional granular catalyst.

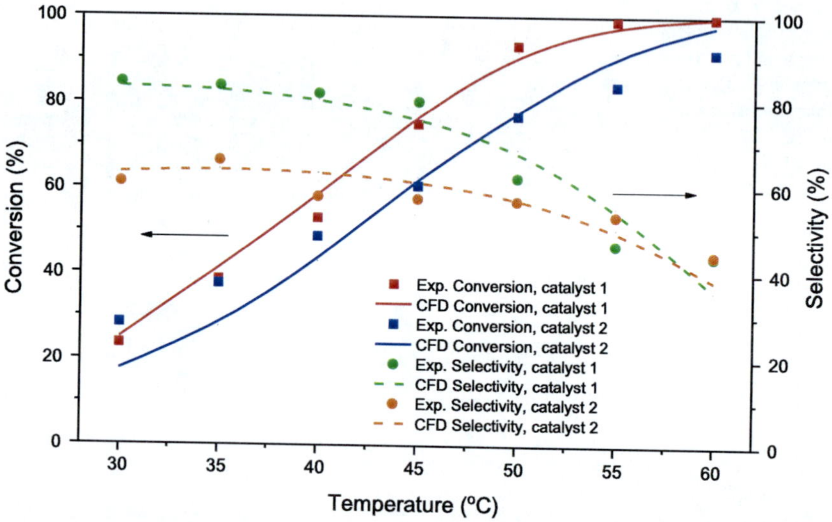

Fig. 5 Acetylene conversion and selectivity: catalyst 1 is open cell foam, catalyst 2 is traditional granular

Figure 6 shows the vector fields of gas movement through the catalyst bed for two studied samples. It can be seen that when passing through an open cell foam catalyst, the gas velocity field is more uniform. And the structural elements of the catalyst do not much disturb the flow. In this case, the picture for a granular catalyst shows both the formation of regions with a high velocity and stagnant circulation regions near the catalyst granules. The gas contact time with the catalyst decreases when driving at a high rate. And in stagnant circulation areas, the likelihood of side reactions increases. Both factors negatively affect the studied parameters of conversion and selectivity.

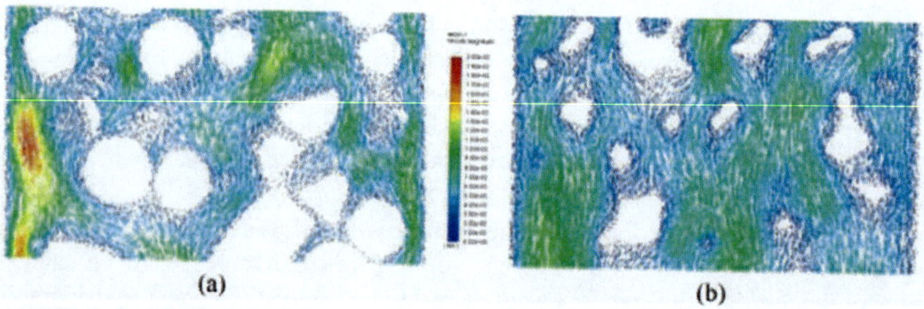

Fig. 6 Vector field of gas movement in the catalyst bed: **a**—traditional granular catalyst, **b**—open foam cell catalyst

4 Conclusion

We made models of an open cell foam catalyst and a traditional granular catalyst. A mathematical model of the process of selective catalytic hydrogenation of acetylene has been constructed. The calculations of the models of laboratory reactors of the catalyst are carried out, and the parameters of conversion and selectivity are compared with laboratory data. As a result of a comparative analysis of the gas movement fields directly near the catalyst surface, factors influencing the studied reaction were determined.

Acknowledgements The reported research was funded by Russian Foundation for Basic Research, grant №. 19-07-01188.

References

1. Soloveva OV, Solovyev SA (2016) Investigation of the influence of heated catalyst feeding system on the intensity of temperature-dependent chemical reaction in the fluidized bed apparatus. IOP Confers Ser: Mater Sci Eng 158(1):012086. https://doi.org/10.1088/1757-899x/158/1/012086
2. Solovev SA, Soloveva OV, Antipin AV, Shamsutdinov EV (2018) Investigation of internal elements impaction on particles circulation in a fluidized bed reactor. J Phys: Conf Ser 944:012114. https://doi.org/10.1088/1742-6596/944/1/012114
3. Solovev SA, Khusainov RR, Nasretdinova DR (2019) Numerical investigation of the granule size effect on the reaction product yield in a catalyst fixed bed. IOP Conf Ser: Mater Sci Eng 618(1):012096. https://doi.org/10.1088/1757-899x/618/1/012096
4. Das S, Deen NG, Kuipers JAM (2018) Multiscale modeling of fixed-bed reactors with porous (open-cell foam) non-spherical particles: hydrodynamics. Chem Eng J 334:741–759. https://doi.org/10.1016/j.cej.2017.10.047
5. Lu TJ, Stone HA, Ashby MF (1998) Heat transfer in open-cell metal foams. Acta Mater 46(10):3619–3635. https://doi.org/10.1016/S1359-6454(98)00031-7
6. Richardson JT, Peng Y, Remue D (2000) Properties of ceramic foam catalyst supports: pressure drop. Appl Cat A: Gen 204(1):19–32. https://doi.org/10.1016/S0926-860X(00)00508-1
7. Giani L, Groppi G, Tronconi E (2005) Mass-transfer characterization of metallic foams as supports for structured catalysts. Ind Eng Chem Res 44(14):4993–5002. https://doi.org/10.1021/ie0490886
8. Haack DP, Butcher KR, Kim T, Lu TJ (2001) Novel lightweight metal foam heat exchangers. ASME Congr Proc 1–7
9. Banhart J (2001) Manufacture, characterisation and application of cellular metals and metal foams. Progr Mater Sci 46(6):559–632. https://doi.org/10.1016/S0079-6425(00)00002-5
10. Coquard R, Loretz M, Baillis D (2008) Conductive heat transfer in metallic/ceramic open-cell foams. Adv Eng Mater 10(4):323–337. https://doi.org/10.1002/adem.200700331
11. Ghosh I (2009) Heat transfer correlation for high-porosity open-cell foam. Int J Heat Mass Transf 52(5–6):1488–1494. https://doi.org/10.1016/j.ijheatmasstransfer.2008.07.047
12. Patcas FC, Garrido GI, Kraushaar-Czarnetzki B (2007) CO oxidation over structured carriers: a comparison of ceramic foams, honeycombs and beads. Chem Eng Sci 62(15):3984–3990. https://doi.org/10.1016/j.ces.2007.04.039
13. Kolaczkowski ST, Awdry S, Smith T, Thomas D, Torkuhl L, Kolvenbach R (2016) Potential for metal foams to act as structured catalyst supports in fixed-bed reactors. Cat Today 273:221–233. https://doi.org/10.1016/j.cattod.2016.03.047

14. Khusainov RR, Solovev SA, Soloveva OV, Ilyasov IR (2020) Numerical simulation and experimental study of the acetylene hydrogenation reaction. IOP Conf Ser: Mater Sci Eng 734(1):012205. https://doi.org/10.1088/1757-899X/734/1/012205
15. Calis HPA, Nijenhuis J, Paikert BC, Dautzenberg FM, van den Bleek CM (2001) CFD modelling and experimental validation of pressure drop and flow profile in a novel structured catalytic reactor packing. Chem Eng Sci 56:1713–1720. https://doi.org/10.1016/s0009-2509(00)00400-0
16. Dixon AG, Nijemeisland M (2001) CFD as a design tool for fixed-bed reactors. Ind Eng Chem Res 40:5246–5254 (2001). https://doi.org/10.1021/ie001035a
17. Romkes SJP, Dautzenberg FM, van den Bleek CM, Calis HPA (2003) CFD modelling and experimental validation of particle-to-fluid mass and heat transfer in a packed bed at very low channel to particle diameter ratio. Chem Eng J 96:3–13. https://doi.org/10.1016/j.cej.2003.08.026
18. Guardo A, Coussirat M, Recasens F, Larrayoz MA, Escaler X (2006) CFD study on particle-to-fluid heat transfer in fixed bed reactors: convective heat transfer at low and high pressure. Chem Eng Sci 61:4341–4353. https://doi.org/10.1016/j.ces.2006.02.011
19. Eppinger T, Seidler K, Kraume M (2011) DEM–CFD simulations of fixed bed reactors with small tube to particle diameter ratios. Chem Eng J 166:324–331. https://doi.org/10.1016/j.cej.2010.10.053
20. Kuroki M, Ookawara S, Ogawa KA (2009) High-fidelity CFD model of methane steam reforming in a packed bed reactor. J Chem Eng Jpn 42(1):73–78. https://doi.org/10.1252/jcej.08we256
21. Dixon AG, Nijemeisland M, Stitt EH (2013) Systematic mesh development for 3D CFD simulation of fixed beds: contact points study. Com Chem Eng 48:135–153. https://doi.org/10.1016/j.compchemeng.2012.08.011
22. Dixon AG (2017) Local transport and reaction rates in a fixed bed reactor tube: endothermic steam methane reforming. Chem Eng Sci 168:156–177. https://doi.org/10.1016/j.ces.2017.04.039

The Cost of Flexible Elements of a Rectangular Profile

Vitaly Kuznetsov and Yulia Shaposhnikova

Abstract Hinge-supported bending reinforced concrete beams of rectangular profile, loaded with a uniformly distributed load, were considered. The main direction of the work is a theoretical study of the influence of the geometric dimensions of the section of beams on the consumption of materials and their cost in the product. The regulatory documents governing design and development activities in the Russian Federation, existing design solutions for bending elements, as well as actual works of Russian and foreign scientists corresponding to research in this area were used. Structural and analytical analysis was used. The normative, scientific and methodological materials were studied, which made it possible to establish the parameters and factors affecting the strength and cost of bent reinforced concrete elements. This is the basis for proposals for improving the calculation and reinforcement of bending elements. The main factors influencing the cost of concrete and reinforcement in a product are analyzed. Specific proposals for calculation and design were developed to determine the optimal cost structures. The proportions of the cross-sectional dimensions and the variable location of the neutral axis significantly affect the consumption of longitudinal reinforcement and the cost of the beam. The boundary ranges of the section dimensions, heights of the compressed zone and the degree of their influence on the cost of the element have been determined. Recommendations for design are given to ensure the construction of the minimum cost, but meet the requirements of strength and durability.

Keywords Material consumption · Product cost · Reinforcement · Reinforced concrete beams · Section dimensions

1 Introduction

When designing various structural elements, it is important, first of all, to ensure the subsequent safe operation and durability of structures. But in addition to this,

V. Kuznetsov · Y. Shaposhnikova (✉)
Moscow State University of Civil Engineering, Yaroslavskoe Shosse, 26, Moscow 129337, Russia

A. Beskopylny and M. Shamtsyan (eds.), *XIV International Scientific Conference "INTERAGROMASH 2021"*, Lecture Notes in Networks and Systems 247,
https://doi.org/10.1007/978-3-030-80946-1_4

the economic component is important—minimization of financial costs for the production of elements and structures.

Therefore, a theoretical study of the influence of the geometric dimensions of the section of beams on the consumption of materials and their cost in the product is an important and interesting engineering problem.

It is possible to obtain the calculated amount of reinforcement, providing the strength of the normal section, depending on the specified values of the geometric parameters of the beam and their relationship to each other. You can also set the cost of materials in the beam and in the product as a whole, which will allow you to design structures of a given strength, but the optimal cost. This is what this work is about.

2 Literature Review

Many Russian and foreign scientists dealt with the problems of optimal design of reinforced concrete beams, for example, Karpenko, Baikov, Skladnev, Alekseev, Jensen, Garstecki and many others.

The development of methods for calculating reinforced concrete structures was carried out by Baykov and Karpenko. In the works of Karpenko the general theory of deformation and destruction of reinforced concrete under various types of loading was formulated [1]. Baikov paid great attention to the development of calculation methods for precast concrete structures [2]. Skladnev was engaged in the problems of optimal design of reinforced concrete structures, taking into account reliability and efficiency [3]. Tamrazyan and Alekseytsev studied the problems of optimization of structures taking into account the ratio of production costs and risks of material losses in emergency situations [4, 5].

Chakrabarty studied the relationship between beam cost and unit cost of materials and beam sizes [6]. Jensen and Lapko investigated the design of shear reinforcement in reinforced concrete beams [7]. Coello et al. as well as Lee and Ahn, used genetic algorithms for optimal design beams [8, 9]. Jiin-Po, Guerra, Hare and others have studied the design optimization of reinforced concrete structures, including beams [10–12]. Also Garstecki et al. developed a software package for the optimal design of reinforced concrete beams and columns [13]. Demby addressed the problem of the optimal and safe design of reinforcement structures in reinforced concrete elements [14]. Nemirovsky considers critical characteristics of concrete failure, such as shrinkage, temperature sensitivity, and the influence of production technology [15].

Currently, reinforced concrete structures in the Russian Federation are calculated in accordance with [16, 17]. In Europe, a regulatory document is used [18], as well as various literature, for example [19, 20].

3 Methods

When designing bending elements, the strength of normal sections with single reinforcement is allowed to be determined at values $\xi \leq \xi_R$ ($\alpha_m \leq \alpha_R$). In this case, the limiting state occurs when the design resistance R_s in the tensile reinforcement reaches $\xi = \xi_R$ [13, 14]. However, the condition $\xi \leq \xi_R$ leaves the possibility of underutilizing the strength of concrete or reinforcement, which depends on the strength of concrete and reinforcement, their quantitative ratio, geometric characteristics of the cross-section and the actual location of the neutral axis.

The strength of an element in a normal section M_{sec} with values $\xi \leq \xi_R$ or $\alpha_m \leq \alpha_R$ can be determined equivalently by concrete (1) or reinforcement (2)

$$\mathrm{M} \leq \mathrm{M}_{sec} = \mathrm{R_b b x (h_0 - 0.5x)} \tag{1}$$

$$\mathrm{M} \leq \mathrm{M}_{sec} = \mathrm{A_S R_S (h_0 - 0.5x)} \tag{2}$$

The maximum value of the bearing capacity of an element with a single reinforcement, without the use of compressed reinforcement, $M_{sec}{}^{max}$ and with a coefficient of use of concrete and reinforcement equal to or close to unity, will be achieved at the height of the compressed zone, which is close to or equal to the boundary value, that is, at $\xi = \xi_R$ or $\alpha_m = \alpha_R$.

$$\mathrm{M_{sec}}^{\mathrm{max}} = \mathrm{R_b b x_R (h_0 - 0.5 x_R)} \tag{3}$$

Taking into account that $x_R = \xi_R h_0$

$$\mathrm{M_{sec}}^{\mathrm{max}} = \mathrm{R_b b \xi_R h_0 (h_0 - 0.5\, \xi_R h_0)} \tag{4}$$

or

$$M_{sec}{}^{max} = R_b b h_0^2 \xi_R (1 - 0.5\, \xi_R) \tag{5}$$

or

$$M_{sec}{}^{max} = R_b b h_0^2 \eta_R \tag{6}$$

where

$$\eta_R = \xi_R (1 - 0.5\, \xi_R) \tag{7}$$

The boundary value ξ_R in accordance with [13, 14] for elements without prestressing is allowed to be determined by Formula (8) or tables [13, 14]

$$\xi_R = 560/(700 + R_s). \tag{8}$$

Maximum moment taken by a normal section with full use of concrete strength

$$M_{sec}{}^{max} = \eta_b R_b b h_0^2 \tag{9}$$

Maximum moment taken by a normal section with full use of the strength of the reinforcement

$$\mathrm{M_{sec}}{}^{\mathrm{max}} = \eta_s \mathrm{A_s R_s h_0} \tag{10}$$

Full use of materials in the section is achieved at $\eta_b = \eta_s$.

The required amount of tensile reinforcement, determined in accordance with Formula (2)

$$A_s = M_{sec}{}^{max} / R\eta_s h_0 = \eta_b R_b b h_0 / \eta_s R_s \tag{11}$$

With reinforcement of class A400, the exhaustion of strength is achieved at $\xi_R = 0.531$

$$\eta_s = \xi_R(1 - 0.5\,\xi_R) = 0.531(1 - 0.5 \cdot 0.531) = 0.390 \tag{12}$$

Typically, the section height h depends on the design span *l*. Let's set the section height as $h = nl$, where the variable coefficient is $n = (0.05 \div 0.1)$. We take the section width *b* as the product of *m* by *h*, that is, $b = mh$, where the coefficient $m = (0.1 \div 0.5)$. We obtain an expression for the section width $b = mnl$. Then expression (11) can be transformed to form (13).

$$A_s = \eta_b R_b n^2 m l^2 / (\eta_s R_s) \tag{13}$$

In particular, for elements with A400 class reinforcement at $\eta_s = 0.390$, the expression for determining the area of longitudinal reinforcement in the design section will take the form

$$\mathrm{A_s} = \eta_b \mathrm{R_b n^2 m l^2} / (0.39\mathrm{R_s}) \tag{14}$$

Thus, by varying η_b, it is possible to obtain a calculated amount of reinforcement that ensures the strength of the normal section within the specified values of η_b. It is also possible to establish the cost of the materials of the beam and the product as a whole. This will make it possible to design structures of a given strength and optimal cost.

The object of the study was reinforced concrete beams without prestressing—B1, length $l = 3$ m; B2, $l = 6$ m; B3 $l = 9$ m. Rectangular profile beams, made of B20 concrete, reinforced with A400 class reinforcement bars. The section height was taken $h = nl$ for $n = (0.075; 0.1; 0.125)$, the section width $b = mh$ for the values $m = (0.1; 0.2; 0.3; 0.4; 0.5)$.

The influence of the section dimensions on the completeness of the use of materials at various values of the height of the compressed zone (coefficient $\eta_b = 0.05; 0.1; 0.15; 0.2; 0.3; 0.39$), as well as on the cost of materials and products in general, for the Moscow region at average prices for concrete 3500 rubles/m^3 (₽/m^3) and for reinforcement of the A400 class, with a diameter of 14–18 mm, 35,000 rubles/ton (₽/t). The use of other classes of reinforcement and concrete, of various strengths and costs, including in other countries, does not change the outlined approach to calculating the strength and price indicators of bent elements.

The investigated parameters were the costs of longitudinal reinforcement and concrete and their cost in the product with the following initial data. The height and width of the section are variable, the coefficient η_b is constant or the coefficient η_b is variable, the height of the section and the width of the section are constant.

4 Results and Discussion

The research results for the B2 beam are shown in Tables 1 and 2.

The table shows that with a fixed value of $\eta_b = 0.2$, the predominant component in the element price is the cost of reinforcement. The percentage of the cost of reinforcement and concrete does not depend on the height and width of the section. With the accepted class of concrete B20 and reinforcement A400, this is 76.9% for reinforcement and 23.1% for concrete. In Table 2 these data are highlighted in bold.

A graphical interpretation of cost indicators, under these conditions, is shown in Fig. 1.

Table 1 Absolute and relative cost of B2 beam materials with variable width and height of the section and a fixed value $\eta_b = 0.2$

Section width b (at m)	$h = 45$ cm ($n = 0.075$)			$h = 60$ cm ($n = 0.1$)			$h = 75$ cm ($n = 0.125$)		
	A_s	B	Σ	A_s	B	Σ	A_s	B	Σ
	₽/%	₽/%	₽/%	₽/%	₽/%	₽/%	₽/%	₽/%	₽/%
$b = mh$	2836.9	850.5	3687.4	5043.3	1512.0	6555.3	7880.2	2362.5	10,242.7
($m =$ 0.2)	76.9	23.1	100.0	76.9	23.1	100.0	76.9	23.1	100.0
$b = mh$	4255.3	1275.8	5531.1	7565.0	2268.0	9833.0	11,820.3	3543.8	15,364.0
($m =$ 0.3)	76.9	23.1	100.0	76.9	23.1	100.0	76.9	23.1	100.0
$b = mh$	5673.7	1701.0	7374.7	10,086.6	3024.0	13,110.6	15,760.4	4725.0	20,485.4
($m =$ 0.4)	76.9	23.1	100.0	76.9	23.1	100.0	76.9	23.1	100.0
$b = mh$	7092.2	2126.3	9218.4	12,608.3	3780.0	16,388.3	19,700.5	5906.3	25,606.7
($m =$ 0.5)	76.9	23.1	100.0	76.9	23.1	100.0	76.9	23.1	100.0

Table 2 The absolute and relative cost of materials in the B2 beam at constant width and height of the section ($m = 0.3$; $n = 0.1$) and variable values of η_b

η_b	₽			%		
	A_s	B	Σ	A_s	B	Σ
0.05	1891.2	2268.0	4159.2	45.5	54.5	100.0
0.1	3782.5	2268.0	6050.5	62.5	37.5	100.0
0.15	5673.7	2268.0	7941.7	71.4	28.6	100.0
0.2	**7565.0**	**2268.0**	**9833.0**	**76.9**	**23.1**	**100.0**
0.3	11,347.5	2268.0	13,615.5	83.3	16.7	100.0
0.39	14,751.7	2268.0	17,019.7	86.7	13.3	100.0

Fig. 1 The cost of materials for the B2 beam

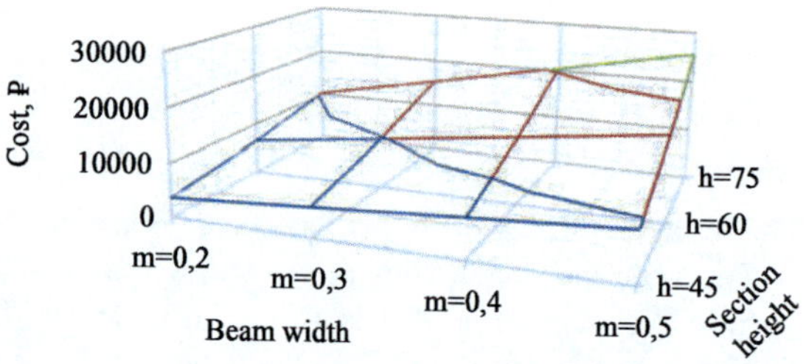

It should be borne in mind that the results obtained correspond to the full use of the strength of the reinforcement used. The influence of the variable height of the compressed zone, expressed through the coefficient η_b, is shown in Table 2.

The graphs presented in Fig. 2 illustrate the dynamics of changes in the cost of concrete and reinforcement in a product.

The cost indicators of the B2 beam were considered with constant width and height of the section ($m = 0.3$; $n = 0.1$) and variable values of η_b. It is significant

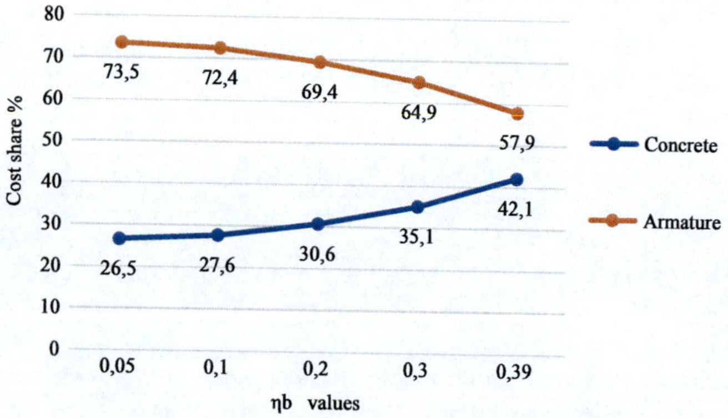

Fig. 2 The ratio of the cost of materials in the B2 beam (n = 0.1, m = 0.3)

that with an increase in η_b, the amount of reinforcement increases. When the strength of concrete and reinforcement is completely exhausted, that is, with $\eta_b = 0.39$, the amount of reinforcement becomes maximum. And this design situation leads to the creation of structures of maximum value.

However, in design practice, when determining the number of reinforcement, the value of the coefficient η_b (α_m) is often much less than the boundary value. This makes it possible to obtain elements of a given bearing capacity, but at a lower cost, when η_b (α_m) changes.

5 Conclusions

The conducted research allows designing flexible elements of a rectangular profile taking into account the real stress state and the cost of the materials used.

The applied technique is universal and applicable for other reinforced concrete structures used in construction practice in various countries.

References

1. Karpenko N (1996) General models of reinforced concrete mechanics. Stroyiszdat, Moscow
2. Baykov V, Sigalov A (2009) Reinforced concrete structures. General course. Stroyiszdat, Moscow
3. Skladnev N (1979) Optimal design of reinforced concrete structures taking into account the requirements of profitability, manufacturability, reliability, durability. Dissertation, Doctor of Technical Sciences, Moscow
4. Alekseytsev A, Tamrazyan A (2019) Optimal design of load-bearing structures of buildings taking into account the relative risk of accidents. Vestnik MGSU 14(7):819–830
5. Tamrazyan A, Alekseytsev A (2020) Modern methods for optimizing structural solutions for the bearing systems of buildings and structures. Vestnik MGSU 15(1):12–30
6. Chakrabarty BK (1992) Models for optimal design of reinforced concrete beams. Comput Struct 42(3):447–451
7. Jensen B, Lapko A (2009) On shear reinforcement design of structural concrete beams on the basis of theory of plasticity. J Civ Eng Manag 15(4):395–403
8. Coello C, Hernandez F, Farrera F (1997) Optimal design of reinforced concrete beams using genetic algorithms. Expert Syst Appl 12(1):101–108
9. Lee C, Ahn J (2003) Flexural design of reinforced concrete frames by genetic algorithm. J Struct Eng 129(6):762–774
10. Jiin-Po Y, Ren-Pei Y (2015) Optimal design of continuous reinforced concrete beams using neural networks. Trans Mach Learn Artif Intell 3(4). https://doi.org/10.14738/tmlai.34.1303
11. Guerra A, Kiousis P (2006) Design optimization of reinforced concrete structures. Comput Concr 3(5):313–334
12. Hare W, Nutini J, Tesfamariam S (2013) A survey of non-gradient optimization methods in structural engineering. Adv Eng Softw 59:19–28
13. Garstecki A, Glema A, Ścigałło J (1996) Optimal design of reinforced concrete beams and frames. Comput Assist Mech Eng Sci 3(3):223–231
14. Demby M, Scigallo J (2017) Design aspects of the safe structuring of reinforcement in reinforced concrete bending beams. Procedia Eng 172:211–217

15. Nemirovsky Y (2014) Problems and methods of structural design of reinforced concrete constructions. Izvestiya of Altai State University
16. SR 63.13330.2012 (2015) Concrete and reinforced concrete structures. The main provisions. Updated edition of SNiP 52-01-2003. Moscow
17. SR 52-103-2007 (2007) Monolithic reinforced concrete structures of buildings. Moscow. A guide for the design of concrete and reinforced concrete structures made of heavy concrete without prestressing reinforcement (to SR 52-101-2003). Moscow
18. EN 1992 Eurocode 2 (2004) Design of concrete structures
19. Seinturiere P (2006) Service limit state. IUT, Civil Engineering of Grenoble
20. Paille G (2013) Calculation of reinforced concrete structures. AFNOR, Paris

Biodiesel Production from Various Crops

Kirill Zhichkin, Vladimir Nosov, Lyudmila Zhichkina, Vyacheslav Zhenzhebir, Elena Artemova, and Nadezhda Sukhina

Abstract The article discusses the possibility of using various oilseeds for the production of biodiesel in the Samara region. Purpose of the work: to determine the economic efficiency of using various crops as the basis for the production of biodiesel, as well as to compare the options for the cultivation of spring rapeseed as the main crop for the production of biofuel. Within the framework of this, the following tasks were completed: a list of oilseeds was determined, the use of which is possible for the production of biodiesel in the Samara region conditions; analysis of the economic parameters of their cultivation; calculated the economic efficiency of various technologies for growing spring rape (6 options). It was revealed that crops such as spring rape, mustard and soybeans have the best economic parameters (cost, yield) in the Samara region. Among the technologies for the cultivation of spring rapeseed, the best results were shown by the following options: direct sowing with fertilizers and plowing without fertilizers (profitability 42.34% and 41.72%, respectively).

Keywords Biodiesel · Crop · Technological map

K. Zhichkin (✉) · L. Zhichkina
Samara State Agrarian University, Kinel 446442, Russia

V. Nosov
K.G. Razumovsky Moscow State University of Technologies and Management, Moscow 109004, Russia

V. Zhenzhebir
Plekhanov Russian University of Economics, Moscow 117997, Russia

E. Artemova
Kuban State Agrarian University Named after I.T. Trubilin, Krasnodar 350044, Russia

V. Nosov
Peoples' Friendship University, Moscow 117198, Russia

N. Sukhina
Kuban State Technological University, Krasnodar 350000, Russia

A. Beskopylny and M. Shamtsyan (eds.), *XIV International Scientific Conference "INTERAGROMASH 2021"*, Lecture Notes in Networks and Systems 247,
https://doi.org/10.1007/978-3-030-80946-1_5

1 Introduction

Currently, increasing attention is paid to the use of alternative fuels, due to the reduction in the worldwide supply of biogenic energy carriers, tightened exhaust emission standards, and limitation of carbon monoxide emission [1]. Its chemical composition allows it to be used in diesel engines without other substances that stimulate ignition. The following useful properties of biodiesel should also be noted:

- biodiesel undergoes almost complete biological decay: in the soil or in water, microorganisms recycle 99% of biodiesel in 28 days;
- less CO_2 emissions;
- low a number of components content in exhaust gases, such as carbon monoxide CO, unburned hydrocarbons, nitrogen oxides NO_X and soot;
- low sulfur content;
- good lubricating characteristics. An increase in the service life of the engine and fuel pump by an average of 60% is achieved [2].

In order to calculate the possibility of developing biofuel production in the Samara region and providing the region's agriculture with biofuel (biodiesel), it is necessary to choose a crop, determine the most profitable cultivation technology, select equipment (identify the number of complexes necessary to provide the region's agriculture with biofuel and calculate its payback), consider the possibility of state support for this project [3–8]. It is important to know how different technologies for the cultivation of oilseeds will affect the cost of the produced crop products and other resulting technical and economic indicators of the technology [9–12]. For this purpose, an economic assessment of various technologies for the cultivation of oilseeds, used in the arable fields of the Samara region, was carried out on the example of the cultivation of spring rape. A feature of spring rape as a crop is the possibility of using rape as a green manure fallow, and the economy can cultivate cash crops on the saved areas.

2 Methods and Materials

Agronomic, technical and economic information necessary for the effective implementation of crop production technology contains a technological map. For this calculation, the methodology for developing technological maps was used in accordance with standard OST 10 1.3 for each variant of the technology of rapeseed cultivation [13–16].

To solve the problem posed in the study, used specialized software for calculating technological maps in crop production, developed by scientists of the Samara State Agrarian University, was used (Fig. 1). With its help, the diesel

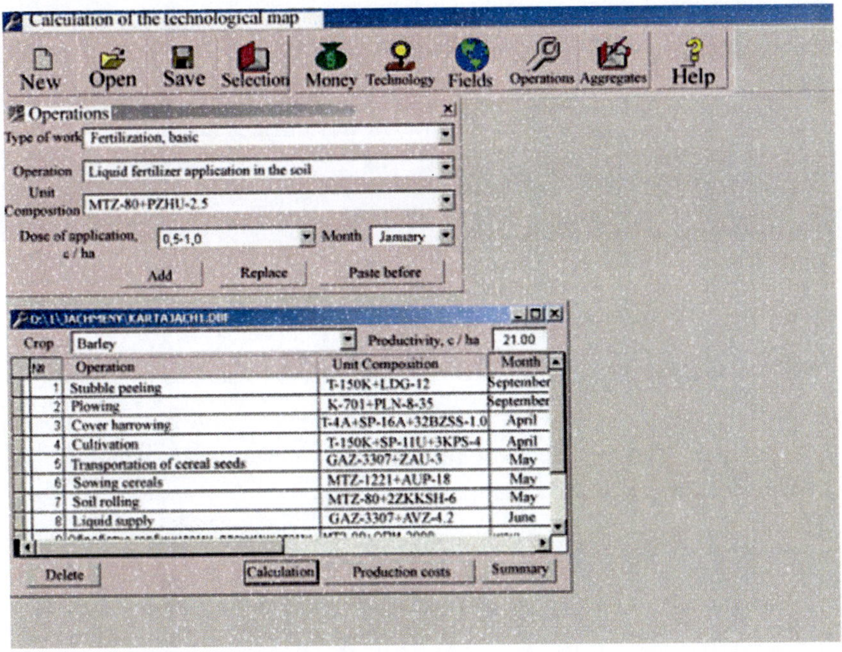

Fig. 1 An example of filling in the initial information for calculating the technological map in crop production [13]

fuel amount needed to complete the production program was determined. Subsequently, the monographic, abstract-logical method, situational and system analysis, economic-statistical methods, and the method of expert evaluations were used.

Considering that the total area of arable lands in the Samara region is 2871.2 thousand ha and that annually agricultural productions consume 60 kg of diesel fuel per 1 ha of arable land, the average consumption of diesel fuel is 172.2 thousand tons per year [17].

The results of the research are presented in tabular and graphical forms.

3 Results and Discussion

Table 1 shows a list of technological operations for each of the studied technologies for the cultivation of spring rapeseed. Cultivation technologies differed, first of all, by different intensity of soil cultivation: plowing by 25–27 cm, loosening by 10–12 cm, without mechanical tillage (direct sowing). Each variant of soil cultivation was studied at two levels of fertilization: without the use of fertilizers and against the background of the application of fertilizers at a dose of $N_{81}P_{38}K_{38}$, based on the calculation of the planned yield level of 15 cwt/ha.

Table 1 List of technological operations in the studied technologies of cultivation of spring rape

Rapeseed cultivation technologies based on		
Plowing	Shallow loosening	"Zero" tillage and direct seeding
1. Peeling 4–6 cm after harvesting the predecessor	1. Peeling 4–6 cm after harvesting the predecessor	Without autumn tillage
2. Application of complex mineral fertilizers randomly 1.5 c/ha diammofoska	2. Application of complex mineral fertilizers randomly 1.5 c/ha diammofoska	
3. Plowing by 25–27 cm with regrowth of weeds and fall of the predecessor	3. Loosening by 10–12 cm during the growth of weeds and fall of the predecessor	1. Application of the herbicide Hurricane 2 l/ha during the regrowth of perennial weeds and fall of the predecessor
4. Spring harrowing	4. Spring harrowing	Without spring tillage
5. Presowing cultivation by 3–4 cm	5. Presowing cultivation by 3–4 cm	
6. Sowing with a SZ-5.4 seeder with the simultaneous introduction of 0.6 c/ha of ammonium nitrate	6. Sowing with a SZ-5.4 seeder with the simultaneous introduction of 0.6 c/ha of ammonium nitrate	2. Sowing with the Amazone DMC seeder with the simultaneous introduction of 1.5 c/ha of diammofoska and 0.6 c/ha of ammonium nitrate
7. Rolling after sowing	7. Rolling after sowing	3. Rolling after sowing
8. Spraying with tank mix herbicide + insecticide + biostimulator	8. Spraying with tank mix herbicide + insecticide + biostimulator	4. Spraying with tank mix herbicide + insecticide + biostimulator
9. The introduction of ammonium sulfate by scattering into the feed 2.2 cwt/ha	9. The introduction of ammonium sulfate by scattering into the feed 2.2 cwt/ha	5. The introduction of ammonium sulfate by scattering into the feed 2.2 cwt/ha
10. Spraying with a tank mixture insecticide + biostimulator	10. Spraying with a tank mixture insecticide + biostimulator	6. Spraying with a tank mixture insecticide + biostimulator
11. Mowing into rolls	11. Mowing into rolls	7. Mowing into rolls
12. Selection and threshing of rolls	12. Selection and threshing of rolls	8. Selection and threshing of rolls
13. Oilseed transportation	13. Oilseed transportation	9. Oilseed transportation
14. Primary cleaning of oilseeds	14. Primary cleaning of oilseeds	10. Primary cleaning of oilseeds

Thus, six variants of technologies for cultivation of spring rapeseed, differing in the levels of costs for their implementation, have been put for study (Table 2).

The use of technologies based on minimum tillage and direct sowing can reduce the cost of fuels and lubricants on average by more than two times from 50 to 26% in the share of total costs. However, with direct sowing, the need for a double increase in the use of crop protection products increases—from 5.7% with traditional technology to 18.5% of the total costs with direct sowing.

Increasing the yield of rapeseed is a fundamental factor in obtaining gross harvest and sales proceeds. The high cost of fertilizers, and as a result, a large share of investments in the sector of variable costs is reflected in the profitability of the production of spring rapeseed with various cultivation technologies [18–24].

In the developed technology of direct sowing with the use of a full range of fertilizers, despite the high level of operating costs, the highest profitability is achieved—42.34%. This is primarily due to an increase in the yield of oilseeds of rape: from 16.4 c/ha for plowing to 17.0 c/ha for "zero" tillage and direct sowing due to the moisture retained by surface plant residues in the soil and full provision of plant nutrition due to the applied mineral fertilizers. The use of direct sowing technology can reduce production costs by more than 2.8 thousand rub/ha. This is due to a decrease in the cost of performing energy-intensive tillage operations—deep plowing and subsequent pre-sowing tillage [25–28].

Direct sowing technology allows to reduce fuel consumption per hectare by half compared to the technology taken for control (using plowing) from 55 to 23.6 kg/ha and reduce labor costs per unit of production to 0.87 man-h/t produced oilseeds.

Thus, on the basis of the performed calculations of economic efficiency in the production of oilseeds of spring rape, along with the traditional technology in the Samara region, it is advisable to use the technology of direct sowing with the introduction of a full range of fertilizers. This technology in the conditions of insufficient moisture supply of the growing season allows to obtain a higher yield and reduces production costs per unit of manufactured product—rapeseed oilseeds. The use of direct sowing technology can reduce the cost of fuels and lubricants on average by more than two times, compared to conventional technology with plowing and reduce the per hectare consumption of motor fuel and labor costs by reducing energy-intensive operations [29].

Previously, we identified agricultural crops cultivated in the Samara region and suitable for the manufacture of oil (Table 3). In this regard, the interest in considering the economics of cultivation of these oilseeds is growing, studying the features of their place in crop rotation and technology, and the influence of these factors on the cost. The final value of the cost of each oilseed crop will be the minimum value among the considered cultivation technologies.

An example of revealing the dependence of the cost of cultivating oilseeds on the technology of their cultivation is given on spring rape. Spring rapeseed was selected as a sample crop taking into account the highest oil yield (1000 kg of oil per hectare). Further crops are presented in descending order of the value of this indicator [30].

In the Samara region, a certain technology for growing sunflower has developed, which allows you to get high yields at an earlier date while reducing financial and

Table 2 Comparative economic efficiency of technologies for cultivation of spring rapeseed

Indicators	Using plowing		Shallow moldless processing		Direct seeding	
	Without the use of fertilizers	$N_{81}P_{38}K_{38}$	Without the use of fertilizers	$N_{81}P_{38}K_{38}$	Without the use of fertilizers	$N_{81}P_{38}K_{38}$
Seeds, rub/ha	1035	1035	1035	1035	1035	1035
Fertilizers, rub/ha	0	3204	0	3134	0	3204
Plant protection products, rub/ha	492	492	518	518	1078	1078
Fuels and lubricants, rub/ha	4271	4371	2850	2950	1530	1580
Repair of equipment, rub/ha	180	184	163	168	153	157
Road transport, rub/ha	28	39	20	36	18	42
Electricity, rub/ha	14	20	11	19	9	22
Wages, rub/ha	743	774	677	710	458	492
Total: variable costs, rub/ha	6763	10,119	5274	8570	4281	7610
Depreciation deductions, rub/ha	1802	1841	1630	1675	1527	1569
Total: fixed costs, rub/ha	1802	1841	1630	1675	1527	1569
Total costs, rub/ha	8565	11,960	6904	10,245	5808	9179
Total revenue, rub/ha	12,138	14,856	8165	12,755	6224	13,541
Profit, rub/ha	3573	2896	1261	2020	343	3886
Profitability, %	41.72	24.21	18.27	19.72	5.91	42.34
Cost of 1 ton, rub	5227	6010	6941	5614	7046	5399

Table 3 Economic efficiency of oilseeds cultivation

Indicators	Rape	Sunflower	False flax	Mustard	Pumpkin	Flax	Soy
Seeds, rub/ha	1035	602	72	1440	198	7500	3780
Fertilizers, rub/ha	3204	2350	654	235	153	752	2685
Plant protection products, rub/ha	1078	246	832	389	432	517	950
Fuels and lubricants, rub/ha	1580	1894	1450	1930	1320	3235	3430
Repair of equipment, rub/ha	157	128	143	164	115	174	197
Road transport, rub/ha	42	35	15	37	18	35	45
Electricity, rub/ha	22	18	7	15	6	19	25
Wages, rub/ha	492	762	420	754	940	754	774
Total: variable costs, rub/ha	7610	6035	3593	4964	3182	12,986	11,886
Depreciation deductions, rub/ha	1569	1640	1462	1846	1327	1795	1890
Total: fixed costs, rub/ha	1569	1640	1462	1846	1327	1795	1890
Total costs, rub/ha	9179	7675	5055	6810	4509	14,781	13,776
Total revenue, rub/ha	14,147	10,873	6572	8998	5475	19,889	18,365
Profit, rub/ha	4968	3198	1517	2188	966	5108	4589
Profitability, %	54.1	41.7	30.0	32.1	21.4	34.6	33.3
Cost of 1 ton, rub	5381	10,923	8128	6200	12,004	9073	6937

labor costs. Taking into account the use of new hybrids and varieties with improved characteristics, farmers are able to achieve excellent results in this important branch of agriculture.

Growing sunflower in accordance with the most promising technology allows you to get a good income from this industry, since sowing requires about 10 kg of seeds per 1 ha, and the yield per hectare can reach 25 cwt. Moreover, not only vegetable oil is obtained from the collected seeds, but also meal, husk, cake, which can become a tangible additional source of income.

Compliance with crop rotation when growing sunflower with the correct alternation of crops in the field is the key to a successful harvest. Sunflower seeds can be sown at the same place no earlier than after 6 years, otherwise the seeds of broomrape and pathogens will accumulate in the ground, which can have an extremely negative effect on the crop.

Mineral and organic fertilizers applied in sufficient quantities contribute to the increase in yield and acceleration of the development of sunflower. Throughout the growing season, sunflower needs phosphorus, nitrogen, potassium fertilizers, as well as trace elements such as boron, zinc and manganese.

False flax is a small-seeded oilseed crop of the cruciferous family. It is an annual plant with an erect, branched stem up to 100 cm tall.

Agricultural enterprises are interested in obtaining oil on unproductive lands, which will strengthen the economy of the economy (average yield of 10–13 cwt/ha). Its seeds contain over 40% oil and 30% crude protein.

The most promising technology is the cultivation of spring camelina using herbicidal fallow, the preparation of which is carried out using soil-protective moisture-saving technology.

Mustard is a triple industrial crop due to its widespread use. It is grown for high quality edible oil, mustard powder and green animal feed. In addition, mustard is widely used as green manure crops, since it has the unique property of absorbing difficult-to-reach forms of nutrients from the soil and transforming them into easily digestible forms.

The average pumpkin yield is relatively low—13.7 c/ha. On average, the mass of 1000 seeds in pumpkins does not exceed 420 g. The seeds remain viable for up to 6–8 years.

Pumpkin seeds are most responsive to the application of manure or other organic fertilizers in doses of 20–40 t/ha. These works are performed after the main plowing, presowing cultivation or cutting of irrigation furrows. With intensive technology, mineral fertilizers are applied simultaneously with sowing. The oil content in oil flax seeds reaches 32–48%. In 7–8-field crop rotations, flax should occupy no more than one field, and return it to its original place no earlier than 6–7 years later. In crop rotations, flax is placed after the best predecessors: perennial grasses, according to the seam turnover, cereals, legumes, winter crops, going in pairs or after grasses, as well as after row crops. Soybeans are the most widespread pulses of the world. Its seeds contain on average 36–42% of complete protein, consisting of globulins and a small amount of albumin, 19–22% of semi-drying oil and up to 30% of carbohydrates. The best predecessors of soybeans in many areas of its cultivation include green manure fallows, the layer and turnover of the layer of perennial grasses, cereals and spike crops that go in clean and busy pairs, as well as row crops (corn, potatoes, sugar beets, sweet potatoes, etc.) Soybeans are not grown after Sudanese grass and sunflower, it is not recommended to place it after corn, under which the herbicides simazine and atrazine were applied. The optimal saturation of crop rotations with soy is from 22 to 40%. To avoid the negative impact of repeated crops on the yield of soybeans, it is recommended to return it to the field no earlier than after 4 years. In the calculation, we neglect such crops as corn, oats and lupine—this is due to the low content of seed oil (less than 200 kg of oil per hectare). This indicator is two times lower in comparison with the oilseeds described above. Thus, we assume a decrease in the oil yield in these crops.

4 Conclusion

The values given in Table 3 are determined from the calculation of the most successful cultivation technologies for each crop. The most profitable oilseeds for cultivation are rapeseed and sunflower, it is interesting that these crops are widely used in the production of biofuel. The amount of diesel fuel used to cultivate the planted area of the Samara region is 77,062.9 tons of diesel fuel. Based on described calculations of comparative effectiveness of biofuel production using MIXER-2 11 AB with a production capacity of 5000 tons per year, we were able to determine that 15 units will be enough to provide agricultural producers of the Samara region with biofuel.

For greater convenience, 5 units should be located in every agricultural zone of the Samara region.

References

1. Zhichkin K, Nosov V, Zhichkina L, Panchenko V, Zueva E, Vorob'eva D (2020) Modelling of state support for biodiesel production. E3S Web Conf 203:05022
2. Morkovkin DE, Gibadullin AA, Romanova JuA, Erygin YuV, Ziadullaev UlS (2019) Formation of a national environmental strategy for the fuel and energy complex. IOP Conf Ser Mater Sci Eng 537:042064
3. Zhichkin K, Nosov V, Zhichkina L, Kuznecova E, Poletaeva L (2020) Economic aspects of the biodiesel use in agricultural production. E3S Web Conf 164:09026
4. Atadashi IM, Aroua MK, Abdul Aziz AR, Sulaiman NMN (2012) Production of biodiesel using high free fatty acid feedstocks. Renew Sustain Energy Rev 16(5):3275–3285. https://doi.org/10.1016/j.rser.2012.02.063
5. Ishak S, Kamari A (2019) A review of optimum conditions of transesterification process for biodiesel production from various feedstocks. Int J Environ Sci Technol 16(5):2481–2502
6. Wan Ghazali WNM, Mamat R, Masjuki HH, Najafi G (2015) Effects of biodiesel from different feedstocks on engine performance and emissions: a review. Renew Sustain Energy Rev 51:585–602. https://doi.org/10.1016/j.rser.2015.06.031
7. Borugadda VB, Goud VV (2012) Biodiesel production from renewable feedstocks: status and opportunities. Renew Sustain Energy Rev 16(7):4763–4784
8. Lakomiak A, Zhichkin KA (2020) Economic aspects of fruit production: a case study in Poland. BIO Web Conf 17:00236. https://doi.org/10.1051/bioconf/20201700236
9. Baskar G, Aiswarya R (2016) Trends in catalytic production of biodiesel from various feedstocks. Renew Sustain Energy Rev 57:496–504
10. Yuryeva AA, Kuternin MI, Gibadullin AA, Romanova JuA (2019) Formation of mechanisms for the development of innovative activity in the industrial production of the Russian Federation. J Phys Conf Ser 1399:033099
11. Canakci M, Sanli H (2008) Biodiesel production from various feedstocks and their effects on the fuel properties. J Ind Microbiol Biotechnol 35(5):431–441. https://doi.org/10.1007/s10295-008-0337-6
12. Lin C-Y, Lin H-A, Hung L-B (2006) Fuel structure and properties of biodiesel produced by the peroxidation process. Fuel 85(12–13):1743–1749
13. Zhichkin K, Nosov V, Zhichkina L, Zhenzhebir V, Rubtsova S (2020) The agricultural crops production profitability in modern conditions. E3S Web Conf 175:13008. https://doi.org/10.1051/e3sconf/202017513008

14. Zhichkin K, Nosov V, Zhichkina L (2019) Economic mechanism of the machine-tractor park updating in the Samara region. IOP Conf Ser Earth Environ Sci 403:012073. https://doi.org/10.1088/1755-1315/403/1/012073
15. Zhichkin K, Nosov V, Zhichkina L, Burlankov P, Ponomareva L, Ponomarev E (2020) Use of biodiesel in Russian agricultural production. E3S Web Conf 164:06031. https://doi.org/10.1051/e3sconf/202016406031
16. Abbaszaadeh A, Ghobadian B, Omidkhah MR, Najafi G (2012) Current biodiesel production technologies: a comparative review. Energy Convers Manage 63:138–148. https://doi.org/10.1016/j.enconman.2012.02.027
17. Demirbas A (2009) Progress and recent trends in biodiesel fuels. Energy Convers Manage 50(1):14–34. https://doi.org/10.1016/j.enconman.2008.09.001
18. Issariyakul T, Dalai AK (2014) Biodiesel from vegetable oils. Renew Sustain Energy Rev 31:446–471. https://doi.org/10.1016/j.rser.2013.11.001
19. Ambat I, Srivastava V, Sillanpää M (2018) Recent advancement in biodiesel production methodologies using various feedstock: a review. Renew Sustain Energy Rev 90:356–369. https://doi.org/10.1016/j.rser.2018.03.069
20. Lopatkin DS, Shushunova TN, Shaldina GE, Gibadullin AA, Smirnova IL (2019) Renewable and small energy development management. J Phys Conf Ser 1399:033061. https://doi.org/10.1088/1742-6596/1399/3/033061
21. Chuah LF, Klemeš JJ, Yusup S, Bokhari A, Akbar MM (2017) A review of cleaner intensification technologies in biodiesel production. J Clean Prod 146:181–193. https://doi.org/10.1016/j.jclepro.2016.05.017
22. Karmakar A, Karmakar S, Mukherjee S (2010) Properties of various plants and animals feedstocks for biodiesel production. Bioresour Technol 101(19):7201–7210
23. Aransiola EF, Ojumu TV, Oyekola OO, Madzimbamuto TF, Ikhu-Omoregbe DIO (2014) A review of current technology for biodiesel production: state of the art. Biomass Bioenergy 61:276–297. https://doi.org/10.1016/j.biombioe.2013.11.014
24. Hajjari M, Tabatabaei M, Aghbashlo M, Ghanavati H (2017) A review on the prospects of sustainable biodiesel production: a global scenario with an emphasis on waste-oil biodiesel utilization. Renew Sustain Energy Rev 72:445–464
25. Atabani AE, Silitonga AS, Badruddin IA, Mahlia TMI, Masjuki HH, Mekhilef S (2012) A comprehensive review on biodiesel as an alternative energy resource and its characteristics. Renew Sustain Energy Rev 16(4):2070–2093
26. Lin L, Cunshan Z, Vittayapadung S, Xiangqian S, Mingdong D (2011) Opportunities and challenges for biodiesel fuel. Appl Energy 88(4):1020–1031
27. Marchetti JM, Miguel VU, Errazu AF (2007) Possible methods for biodiesel production. Renew Sustain Energy Rev 11(6):1300–1311
28. Datta A, Mandal BK (2016) A comprehensive review of biodiesel as an alternative fuel for compression ignition engine. Renew Sustain Energy Rev 57:799–821. https://doi.org/10.1016/j.rser.2015.12.170
29. Fukuda H, Kondo A, Noda H (2001) Biodiesel fuel production by transesterification of oils. J Biosci Bioeng 92(5):405–416
30. Hoekman SK, Broch A, Robbins C, Ceniceros E, Natarajan M (2012) Review of biodiesel composition, properties, and specifications. Renew Sustain Energy Rev 16(1):143–169. https://doi.org/10.1016/j.rser.2011.07.143

Molecular Relaxation in a Liquid Crystal After Switching Off the Acoustic Action

Olga Denisova

Abstract Relaxation processes in thin (up to 125 μm) homeotropic layers of nematic liquid crystals (n-methoxybenzylidene-n-butylaniline), which were observed after switching off periodic shear oscillations with a low frequency (about 200 Hz), were experimentally studied. The foner method was used to excite vibrations in a liquid crystal sample. The amplitude of the oscillations was determined by inductive and optical methods. The cell with the sample was made in the form of a multi-layer structure. A liquid crystal was placed between the three glass plates. The middle plate floated in the liquid crystal and transmitted oscillatory movements to it. The dependences of the angle of deviation of the director from the normal to the cell on the time and relaxation times, the relaxation time on the shear amplitude and exposure time, as well as on the thickness of the liquid crystal layer; the time dependences of the optical signal at different exposure times and oscillation amplitudes; the dependences of the maximum values of the angle of rotation of the director in the *XY* plane on the shear amplitude and exposure time. It is found that after switching off the external perturbation, the relaxation processes of the liquid crystal director field depend on the exposure time, the thickness of the liquid crystal layer, and the oscillation amplitude. The relaxation dependences of the angle of inclination of the director after switching off the shift are described by an exponential dependence. Relaxation times are tens of seconds. The value of the relaxation time coincides with the relaxation time for the classical Fredericks effect and is about 30 s.

Keywords Liquid crystal · Relaxation · Acoustic effect · Orientation transition · Shear action

O. Denisova (✉)
Ufa State Petroleum Technological University, Kosmonavtov Str., 1, Ufa 450062, Republic of Bashkortostan, Russia
e-mail: denisovaolga@bk.ru

A. Beskopylny and M. Shamtsyan (eds.), *XIV International Scientific Conference "INTERAGROMASH 2021"*, Lecture Notes in Networks and Systems 247,
https://doi.org/10.1007/978-3-030-80946-1_6

1 Introduction

Despite the fairly wide application of liquid crystals (LC) in engineering, medicine, instrumentation, and design, the relevance of their research can hardly be overestimated, since they are opening up more and more new practical applications [1–12]. The authors of [1] experimentally studied the relaxation of a planar-oriented liquid crystal in the smectic A-phase, nematic and isotropic phases in the frequency range from 1 Hz to 5.5 MHz under the action of a direct current when the sample is heated. It is found that the relaxation processes are affected by a change in the direction of polarization associated with the movement of the ensemble of crystal molecules.

In [2], layered liquid crystal smectic phases and their response and relaxation under the action of a low-frequency alternating field (20 Hz–10 MHz) were considered. The data obtained indicate a non-Baian type of asymmetric dispersion, with peaks in loss and permittivity associated with time-scaled relaxations.

Liquid crystals are used to model molecular dynamics, which plays an important role in characterizing the structure of ribonucleic acid (RNA). The simulation complements the information obtained by experimental methods, such as nuclear magnetic resonance (NMR) [3].

The dielectric relaxation in the liquid crystal matrix was studied by spectroscopy to understand the role of its resistivity in the relaxation of colored structures. Two types of liquid crystal mixtures were considered: one had high ion contamination and low resistivity, and the other had high resistivity and contained a negligible amount of ions. The samples were doped with azo or anthraquinone dye. The use of both dyes in a liquid crystal enriched with ions allowed to increase the concentration of ions and slow down the relaxation processes [4].

The article [5] presents a study of the dynamic light scattering and the influence of the interface boundaries on the dynamics of the behavior of nematic liquid crystals filled with aerosil particles, which were placed in a matrix with cylindrical pores. A significant change in the relaxation properties of the LC was found in this situation. In the first case, the relaxation is associated with fluctuations in the orientation of the nematic director, similar to those in a bulk liquid crystal. In the second case, the relaxation proceeds more slowly, apparently, this is due to fluctuation changes in the layers bordering the solid pore wall or the surface of the aerosil particles.

Ferroelectric liquid crystals were studied in [6, 7]. A low-frequency mode of dielectric relaxation is experimentally detected. The dependencies of its behavior on the heating temperature of the sample, as well as on the applied displacement field, were studied. As the temperature and field strength increase, the relaxation frequency shifts to the high-frequency region.

Liquid crystal elastomers exhibit nontrivial mechanical properties [8]: elasticity, reversible deformation, and the tangent of the loss angle above the glass transition point. The behavior at temperatures below the glass transition temperature in the rubber–nematic phase and in the isotropic state–rubber phase was analyzed. It is found that when modeling the fluidity in the nematic–rubber phase, which was carried out using two characteristic relaxation times, they showed a decrease with increasing temperature. At a temperature of 62 °C in a rubber–nematic state characterized by a low threshold stress and a short relaxation time, a reversible reorientation of the domains between the cycles is detected and a stress–strain curve with a high hysteresis is obtained.

The paper [9] presents the results of numerical simulation of the effect of surface waves on photosensitive thin films of liquid crystals. The results obtained relate the wave characteristics (amplitude, frequency, and phase delay) to the viscoelastic reaction of the sample. Recommendations for creating surface waves with the specified parameters are offered.

Analyzing a wide range of problems related to liquid crystals considered in the scientific literature in recent years, it should be noted that the issues of relaxation of the liquid crystal director field under acoustic shear action are poorly reflected. Liquid crystal acoustics is a very promising field, including the study of optical effects observed in acoustic flows of liquid crystals. In scientific works, the high-frequency range up to 100 MHz is mainly studied, where the mechanisms that determine the relaxation processes are associated with the rotation of molecules around their axis [13]. In the case of a low-frequency shift, viscoelastic relaxation processes of the orientation order work, information about which is practically absent. In this regard, it is relevant and important to study the low-frequency acoustic relaxation of thin homeotropic nematic layers under the action of a low-frequency shift.

2 Materials and Methods

To conduct experimental studies of relaxation processes in nematic liquid crystals, an installation based on the foner method was assembled. A detailed description of the method used is described in [14–17]. The measurements were carried out in the sound frequency range of 0.02–20 kHz. Nematic layers with a thickness of up to 125 μm with a homeotropic orientation of liquid crystal molecules were studied. The glass substrates used were sprayed with metallic chrome, which created the desired orientation. The object of the study was MBBA (n-methoxybenzylidene-n-butylaniline), which has a nematic phase at room temperature. The cell consisted of two semi-transparent plates, between them was placed a third thinner plate with a thickness of 150 μm with a chromium coating sprayed on both sides, which oscillated in the horizontal direction. Gaskets were made of cover glass in the form of narrow strips. To create shear vibrations in the NLC sample, the membrane of the sound vibration source was connected using a glass waveguide. The oscillation amplitude of the thin cell substrate was measured by inductive method.

In [19, 20], the orientational and azimuthal instability of nematic liquid crystals (NLC) under the action of a low-frequency periodic shift was investigated. They manifested themselves as follows. When the threshold value of the amplitude of the external action $a_1 = 0.8$ μm is reached, the behavior of the NLC director is no longer affected by the movement of the vibrating plate, which is expressed in the appearance of a stationary angle of inclination of the LC molecules θ_c. With a further increase in the shift amplitude after reaching the threshold $a_2 = 1.4\ \mu m$, the director, turning, leaves the original plane of motion.

In the experimental situation under study, the director moves in the *ZY* plane, where *Y* coincides with the horizontal plane and the direction of oscillation of the plate ($Y \| \nu$), and *Z*—with the normal to the plane of the cell. This motion is characterized by an angle θ (the angle of deviation of the director from the equilibrium state), and the rotation of the director in the *XY* plane is characterized by an angle φ′.

3 Results and Discussion

To study the relaxation processes under the influence of periodic shear vibrations, a MBBA sample with a thickness of $h_c = 125$ mm and a shear frequency of f = 190 Hz was selected. The dependencies of the transmitted light intensity on the exposure amplitude I(a), the exposure time $I(t_n)$, and the time I(t) were analyzed, from which the dependencies of the mean square of the sine of the stationary angle of inclination of the director $<\sin 2\theta_c>$ on the shear amplitude at a fixed exposure time, on the exposure time at a fixed amplitude, and the relaxation dependencies $<\sin 2\theta_c>$ after switching off the shear effect were calculated. Figure 1a shows the relaxation dependencies $<\sin 2\theta_c>$ when the shift is switched off and the acoustic flow action time is different ($t_n = 5$–120 s), which monotonically fall to zero. In general, these dependencies are exponential. In this case, the relaxation time coincides with the relaxation time for the classical Fredericks effect and is equal to

$$\tau = \frac{\gamma_1 h_c^2}{\pi^2 K_{22}} \tag{1}$$

From the expression (1) for a given h_c thickness, the value $\tau \approx 30$ s follows (the data of the material constants $\gamma 1 = 77$ sP, $K_{22} = 4 \cdot 10^{-7}$ din are taken from [18]). The relaxation times were determined based on the relaxation dependencies of the mean square of the stationary angle of inclination of the director $<\sin 2\theta_c>$ after switching off the periodic shift. Figure 1b–d show the dependencies of the relaxation times τ on the shift amplitude a, the exposure time t_n, and the thickness of the LC layer h_c. From the dependencies τ(a) (Fig. 1b), it follows that the values of the relaxation times monotonically increase with the amplitude of the impact. For a fixed amplitude, with only the time of exposure to t_n changing, we get the dependence $\tau(t_n)$ (Fig. 1b). The value of τ first increases to a certain value and the time of action tn then it is saturated, which indicates the presence of long-term modes in the LC system, leading to the establishment of equilibrium under the influence of the shift. Since the data presented above related to the situation when the layer thickness $h_c = 125$ mm, it is quite natural to investigate the influence of the spatial dimensions of the LC system—the thickness of the nematic layer on the characteristic relaxation times.

Figure 1d shows the dependencies of the relaxation times on the thickness of the LC layer $\tau(h_c)$ at different shear amplitudes and a fixed exposure time, so that a

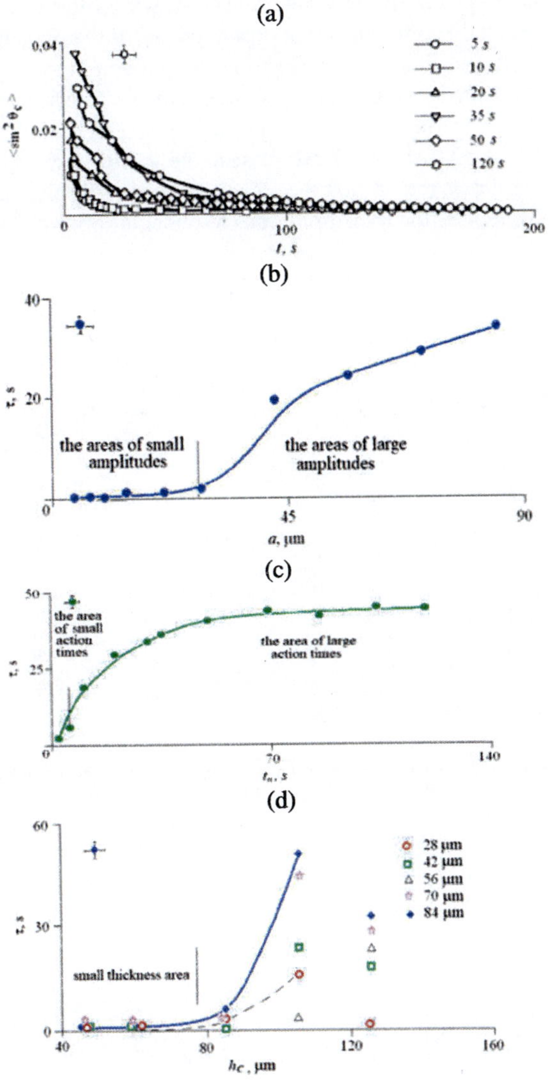

Fig. 1 Dependencies of **a** $<\sin^2\theta_c>$ on the time and relaxation times; **b** relaxation time on the shift amplitude; **c** relaxation time on the exposure time; **d** relaxation time on the thickness of the LC layer

family of curves of the type $\tau(h_c, a, t_n)$ is obtained, where the amplitude a and the exposure time tn are parameters. From the presented dependencies, it follows that the relaxation times τ with an increase in the thickness of the LC layer are well described by the analytical dependence (1) of the type $\tau = h_c^2$, which reflects the process of reorientation of the director and the transition to the equilibrium state of the LC system under the classical Fredericks effect. To study the relaxation processes under the effect of the director's exit from the oscillation plane to the third dimension, a cell with an LC layer with a thickness of $h_c = 105$ mm was selected, since this thickness is the beginning of the nonlinear growth of the threshold in thickness. To do this, we studied the relaxation dependencies of the optical signal after switching off the shift at different exposure times tn and amplitudes a (the Nichols are crossed, and the direction of the shift coincides with the polarizer) (Figs. 2 and 3). Figure 2 shows the relaxation dependencies of the optical signal I(t) at different exposure times.

Let us first consider the case of the shortest exposure time (Fig. 2a), i.e. at $t_n = 5$ s. It follows that at first the signal I(t) drops sharply, and then slowly relaxes to zero. Moreover, the transition times to the equilibrium homeotropic state are about 10 s. When the exposure time increases to 10 s, the nature of the optical signal changes.

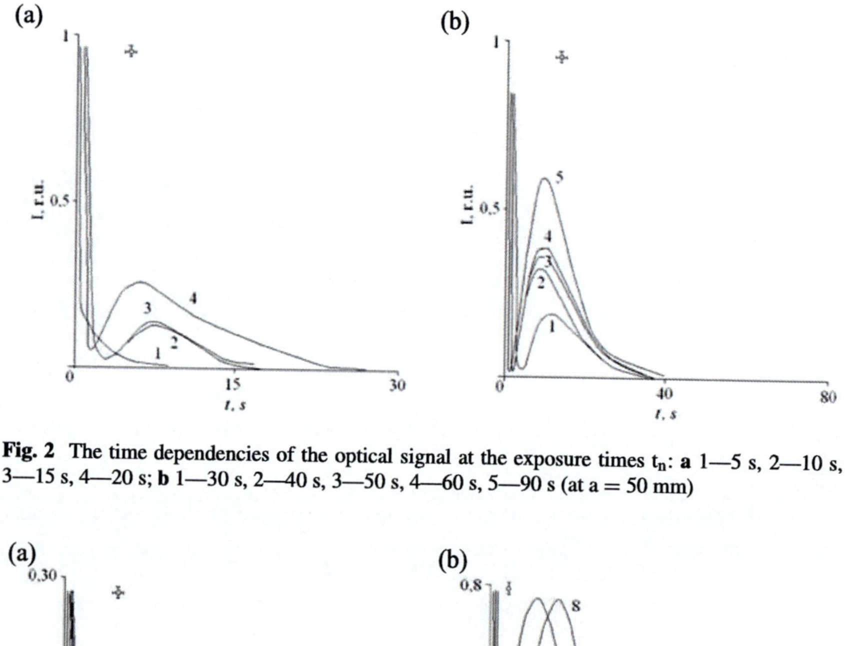

Fig. 2 The time dependencies of the optical signal at the exposure times t_n: **a** 1—5 s, 2—10 s, 3—15 s, 4—20 s; **b** 1—30 s, 2—40 s, 3—50 s, 4—60 s, 5—90 s (at a = 50 mm)

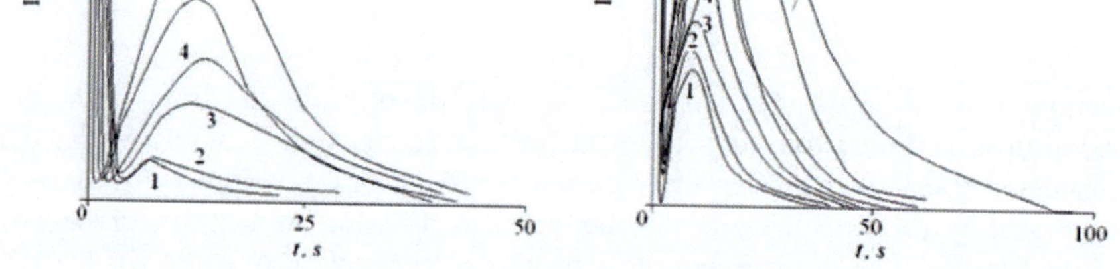

Fig. 3 Time dependencies of the optical signal at shift amplitudes a (at t = 50 s): **a** 1—19.4 mm, 2—21 mm, 3—22.4 mm, 4—25.2 mm, 5—26.6 mm, 6—28 mm; **b** 1—30.8 mm, 2—33.6 mm, 3—35 mm, 4—36.4 mm, 5—39.2 mm, 6—42 mm, 7—49 mm, 8—56 mm

On the one hand, the initial phase of a sharp decline is preserved, on the other hand, after reaching the minimum, the layer is again illuminated, and the position of the maximum intensity on the time axis does not depend on the time of exposure as a parameter, although the intensity of the maximum itself increases. The appearance of enlightenment is associated with the development of the twist mode, which affects the formation of reverse flows in the LC. In fact, the increase in the maximum amplitude also reflects the increase in the maximum angle of the director's exit from the initial plane of the ZY oscillation.

The relaxation dependencies of I(t) at different shift amplitudes are shown in Fig. 3. They have a character similar to the curves shown in Fig. 2, but unlike the first case, the position of the maximum is shifted on the time axis in the direction of large times. According to the time dependencies of the optical signal, the dependencies of the rotation angle of the director φ' in the XY plane on the shift amplitude, the exposure time tn, and the relaxation dependencies φ' after switching off the exposure were calculated.

Let us first consider how the maximum angle of rotation φ'_{max} depends on the amplitude of the impact (the time of the impact tn is fixed and is equal to 50 s) and on the time of the impact at a fixed amplitude a = 50 mm. This can be done by the maximum values of the layer illumination intensity. Figure 4a, b show the dependencies $\varphi'_{max}(a)$ and $\varphi'_{max}(t_n)$.

The angle of rotation of the director in the XY plane increases from the amplitude of the shift a and the time of the impact tn and tends to its maximum value in this range of amplitudes and times to $\varphi'_{max} \approx 45°$, otherwise, if it becomes more than 45°, we should observe a gradual damping of the layer at φ'_{max}, tending to 90°. However, this behaviour does not detect the intensity of the transmitted light. If we fix the amplitude of the shift a and change the time of the impact, we get the dependencies of the angle φ'_{max}, which characterizes the time process of reorientation of the director when the external influence is turned on. This dependence reveals the tendency of the director's reorientation to $\varphi'_{max} = 45°$.

Now we will consider the relaxation process at large times $t_n \approx 10$–20 s. It was shown above that in this case there is a maximum intensity (Figs. 2 and 3). According to them, the dependences of the angle of the maximum deviation of the director in

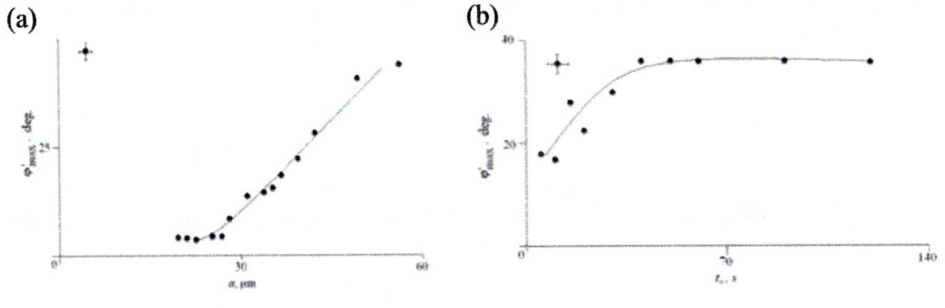

Fig. 4 The dependencies of the maximum values of the rotation angle of the director φ'_{max} in the XY plane **a** on the shear amplitude at $t_n = 50$ s; **b** the exposure time at a = 50 mm ($h_c = 105$ mm)

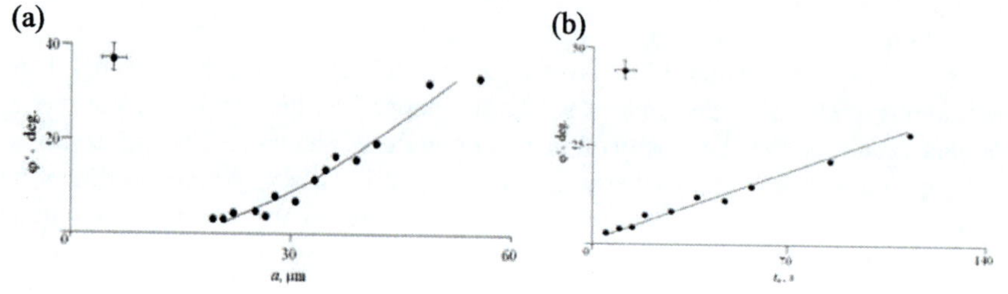

Fig. 5 The dependencies of the rotation angle of the director φ′ in the XY plane on **a** the shift amplitude at $t_n = 50$ s; **b** the time ($h_c = 105$ mm)

the XY plane were constructed, corresponding to the appearance of these maxima (Fig. 5a).

The behaviour of the angle φ′(a) at a fixed $t_n = 50$ s correlates with the behaviour φ'_{max}a), calculated from the initial phase of the intensity of the transmitted light after switching off the exposure (Fig. 4a). According to the data of the time dependencies I(t, a_i) for t_n = const, the dependencies φ′(t, a_i) were also constructed, some of which are shown in Fig. 5b. It follows from them that φ′(t) relaxes for a long time (20–50 s) to the equilibrium state. In essence, this process is analogous to the relaxation under twist deformation of a planar oriented NLC in a magnetic field and can be described by the equation [18]:

$$K_{22}\frac{\partial^2\theta}{\partial z^2} = \gamma_1\frac{\partial\theta}{\partial t}.$$

The solution of this equation (for any θ) gives an exponential decline of the angle θ with a set of constant times:

$$\tau_n^{-1} = \frac{K_{22}}{\gamma_1}\left[\frac{(2n+1)\pi}{d}\right]^2$$

The time estimates from this formula give $\tau_o \approx 10$ s (for n = 0) (data on material parameters are taken from [18]).

Using the data for the optical signal I(t) (Figs. 2 and 3), the dependences of the relaxation time on the shift amplitude a, the exposure time t_n, and the thickness of the LC layer h_c were calculated. It follows from Fig. 6a that the relaxation times τ depend linearly on the shear amplitude. Analyzing the dependence of the relaxation time τ on the time of exposure to t_n, we can conclude that this value τ first increases sharply, then goes to a plateau (Fig. 6b), and is similar to the behaviour of the relaxation time τ during the formation of a stationary angle of inclination of the director under the action of a shift.

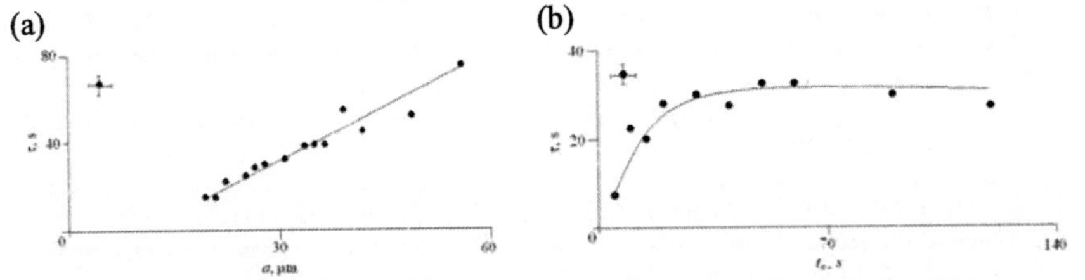

Fig. 6 Dependencies of the relaxation times τ on a the shear amplitude at $t_n = 50$ s; b the exposure time at a = 50 mm

4 Conclusions

Thus, we note that the relaxation processes of reorientation of the director after switching off the perturbation are described by an exponential time dependence, and the characteristic times τ are tens of seconds. Such long relaxation times can be associated with the formation of loops or domain walls in the LC volume, the relaxation of which takes a very long time. The dependencies of the relaxation times on the thickness of the LC layer at different amplitudes of the action are quadratic in thickness and are well described by the formula (1), applicable for the case of relaxation with the Fredericks effect in a magnetic field. The latter indicates a deep connection between these two phenomena. In addition, the dependence of the time of establishing equilibrium in the LC system on the time of exposure shows that there is a long-lived twist mode in the LC layer, which leads to the exit of the director from the original plane of oscillation to the third dimension and the formation of a twist deformation.

The experimental results obtained can be useful in the further development of the theory of nonlinear phenomena of liquid crystals. The results of the presented studies can be used in the development of seismic sensors, acoustic low-frequency converters and sensor devices, modulators and light deflectors.

References

1. Anoop KS, Jongyoon K, Sunggu Y (2017) Anomalously high dielectric strength and low frequency dielectric relaxation of a bent-core liquid crystal with a large kink angle. Curr Appl Phys 17:858
2. Ashok K, Chalapathy PV, Potukuchi DM (2019) Low-frequency dielectric spectroscopy and distinct *relaxation* modes in smectic phases of *liquid crystal* dimers. J Mol Liq 278:568
3. Bergonzo C, Grishaev A (2019) Accuracy of MD solvent models in RNA structure refinement assessed via *liquid-crystal* NMR and spin *relaxation* data. J Struct Biol 207:250
4. Rahman M, Hsieh C, Wang C, Lee W (2010) Dielectric relaxation dynamics in liquid crystal—dye composites. Dyes Pigm 84:128
5. Aliev F, Basu S (2006) Relaxation of director reorientations in nanoconfined liquid crystal: dynamic light scattering investigation. J Non-Cryst Solids 352:4983

6. Chandra A, Rakash J, Joshi T, Biradar A (2014) Probing the effect of temperature and electric field on the low frequency dielectric relaxation in a ferroelectric liquid crystal mesogen. J Mol Liq 198:280
7. Pandey S, Vimal T (2015) Analysis of physical parameters and collective dielectric relaxations in core/shell quantum dot ferroelectric liquid crystal composite. J Mol Liq 211:157
8. Merkel D, Shaha R, Yakacki C, Frick C (2019) Mechanical energy dissipation in polydomain nematic liquid crystal elastomers in response to oscillating loading. Polymer 166:148
9. Mehta K, Peeketi A (2020) Modeling of surface waves in photo-responsive viscoelastic liquid crystal thin films under a moving light source. Mech Mater 147:103388
10. Johri M, Tiwari A (2020) Study of the dielectric response and thermo-dynamical properties of pentyl cyano biphenyl (PCB) liquid crystal using microwave cavity spectrometer as a probe. Measurement 166:108156
11. Peibao J, Yu X, Li K (2020) Controllable vibration of liquid crystal elastomer beams under periodic illumination. Int J Mech Sci 170:105366
12. Rastogi A, Agrahari K, Srivastava A, Manohar R (2020) Effect of graphene oxide dispersion in antiferroelectric liquid crystal mixture in the verge of SmC* to SmCA* phase transition. Chin J Phys 67:91
13. Gunduz IE (2018) 3D printing of extremely viscous materials using ultrasonic vibrations. Addit Manuf 22:98
14. Denisova OA, Chuvyrov AN (2012) Structural transitions in liquid crystals. Influence of oscillating flows and electric fields. LAP LAMBERT, Germany
15. Denisova OA (2012) Nonequilibrium structural transformations of liquid crystals in electric fields and acoustic ceilings. UGUES, Ufa
16. Leonov VV, Denisova OA (2016) Electrodynamics of viscous flow of condensed media. High-Tech 17(2):37
17. Denisova OA (2013) One of the methods of experimental studies of liquid crystals. Electr Data Process Fac Syst 9:107
18. de Gennes PG, Prost J (1993) The physics of liquid crystals, 2nd edn. Clarendon Press, Oxford
19. Denisova OA, Chuvyrov AN (2011) Possible mechanism of the effect of azimuthal instability of liquid crystals. Chelyabinsk J Phys Math 15:42
20. Denisova OA, Chuvyrov AN (2013) Acoustic analog of the Fredericks transition in rapidly oscillating liquid crystal flows. Liq Cryst Appl 2:37

Microstructure and Composition Boriding Coating Based on Boron Carbide and Amorphous Boron

Alexey Ishkov and Vladimir Malikov

Abstract The paper shows studies boride coatings, which were created using treatment with currents. Chemical reactions characterized the boriding procedure at various temperatures are presented. To establish practically realizable reactions of surface chemical (topochemical) boriding and the accompanying high-temperature processes, a model mixture was prepared containing P-0.66 flux and boron carbide. A method of the experimental study of coatings applied on specimens by HFC-heating was developed. A study of the composition and properties of the products formed during the SHS process in the basic mixture on the surface of S355 steel parts during HFC-heating has been made, and it has been shown that aluminum oxide, aluminum nitride (by-product), and amorphous boron are formed under these conditions, and composite porous inhomogeneous coatings up to 250 μm thick, based on aluminum oxide, SHS products in the basic mixture and traces of high-speed HFC-boriding products are formed on the workpiece surface. The formed coatings are composite materials based on alumina, SHS products in the base mixture and the product of fast HFC-boriding of the surface of a steel part being formed during SHS-boriding, having the porous heterogeneous structure, insufficient thickness and low hardness.

Keywords SHS processes · HFC-heating · Hardening coatings · Coating powders

1 Introduction

The successful solution of production efficiency are closely connected with more rapid scientific and technical progress. For this purpose, it is necessary to create and implement fundamentally new types of technology and engineering connected with material science. This especially concerns parts subject to friction, upon which the safety and durability of modern machinery [1].

A. Ishkov (✉)
Altay State Agricultural University, Krasnoarmeyskiy Ave., 98, 656049 Barnaul, Russia

V. Malikov
Altai State University, Lenina Avenue, 61, 656049 Barnaul, Russia

A. Beskopylny and M. Shamtsyan (eds.), *XIV International Scientific Conference "INTERAGROMASH 2021"*, Lecture Notes in Networks and Systems 247,
https://doi.org/10.1007/978-3-030-80946-1_7

Nevertheless, low hardness and poor wear-friction restrict their use, especially in tribological applications [2–4].

Tribology is a universally faced problem in the mining and chemical industries. Various tribology processes occur in different industries, such as abrasion, corrosion, adhesion and erosion [5]. Erosion and corrosion predominantly affect the efficiency and life of machinery used in the chemical and mining industries [6].

Improving operational properties and wear resistance of essential machine parts is acute for modern machine engineering. These problems are solved both at the stage of manufacture by improving production technologies and application of modern materials and protective and wear resistant coatings [7].

Various types of coating powders are used to protect the surfaces of different components of machinery [8].

Among the works devoted to technological methods of improving operational properties the ones aimed at improving stability of cutting processes, materials and coatings should be mentioned at steel [9].

Owing to the combination of good mechanical properties and excellent corrosion resistance, steel is widely used in mining machinery [10, 11].

High corrosion and wear resistances are basic requirements of materials operating in aggressive environments such as those found in the petrochemical industry. Among the commercially available alloys, steels are well-established candidates for applications where good mechanical properties and high corrosion resistance are demanded. Although largely used, conventional steel grades are not generally regarded as wear-resistant materials, even though some of those alloys can be cold-worked (ferritic and austenitic) or heat-treated (martensitic) to improve their hardness.

An interesting approach proved to effectively increase the wear resistance of stainless steels is the use of minor additions of boron to form hard phases [12].

Arji et al. [13] performed an experimental study to investigate erosive wear boride coating sprayed on mild steel. They found that B had improved the wear of mild steel substrate. Karimi et al. [14] also investigated the porosity of sprayed boride coatings after the remelting processes.

The process of conventional diffusion boriding is a quite common method of steel and steel parts case-hardening [15]. When boriding, an obtained saturable surface often contains two-phases consisting of mixed borides, FeB and Fe_2B, and a transition zone represented by solid solution of boron and other elements contained in the steel in α-Fe form. After conventional boriding, the coating microstructure most often represents boride needles intergrown by their lower parts forming the layer of coating. The resulting internal tensile stresses in the borated coating significantly reduce its ductility, so stripping and chipping of such strengthening coating occurs, up to its complete destruction subject to relatively small bending, impact, or compressive stresses, and especially to alternate loads and vibrations.

All the above mentioned restricts the mass use of isothermal boriding for steel parts surfaces hardening [16].

However, in the case of boriding using HFC-heating, the rate of the formation of coating increased tenfold due to the enhancement of the heating rate, and there is no need in the protective atmosphere subject to addition of special fluxes, as the boriding time does not exceed several minutes. Besides, HFC-heating makes it possible to shorten the time of exposure of the base material to high temperatures, and easily combine the boriding process with the subsequent heat treatment. In the future, a short time of boriding when using HFC-heating can make the use of this method of boriding in processing lines, in mass production and with a large number of parts to be hardened.

The main prerequisite for developing the new boriding method in controlled topochemical reaction Fe + B was the results first published in the work [17], where the authors of the methods of differential thermal analysis (DTA) showed that chemical interactions between boron, iron and its oxides that are always on steel surface and lead to the formation of iron borides, begin at 500–700 °C in an inert, rarefied atmosphere and end after the formation of d-metals borides. Based on the data from the work, the total boriding process of steel oxidized surface can be described by the following reaction:

$$2Fe_2O_3 + 8B = 4FeB + 2B_2O_3. \quad (1)$$

The preliminary experiments that we have been conducted showed the possibility of the formation of boride coatings on the surface of steel parts subjected to HFC-heating. In the course of preliminary experiments, it has been revealed and confirmed that the rate of reaction and thus the formation of a coating increases by a factor of ten compared to the isothermal steel case-hardening in the furnace, when increasing the rate of heating, and also it has been found that the use of borate fluxes, for example, the original P-0.66 flux, is mandatory for obtaining high-quality extended coatings.

The research of SHS-processes in the system and its modified versions is necessary to establish the patterns of initiation and combustion of mixtures of various compositions, as well as to control this process with the possibility of subsequent production of high-quality hardening boride coatings on the surface of a steel part 150–200 μm thick, and the development of a prototype of an effective technology.

2 Materials and Methods

For studying the processes of boriding using HFC-heating of surfaces made of major brands of structural and alloyed steels used in machine building, and determination of the impact of carbon and allying elements contents in steel on the physical, physical and mechanical properties and wear resistance of boride coatings, the specimens of coatings on steels of St3, St45, 40KhN, 50G, 50KHGA, and 65 G grades have been obtained from the boriding mixture based on boron carbide and amorphous boron under conditions of HFC-heating. The boriding process temperature in all cases was 950–1250 °C, while the time of saturation was 40–180 s.

To form hardening coatings by borating, engineers use an ELSIT-100-40/70 (OOO ELSIT, Tomsk) high-frequency transistor converter (inverter). The ELSIT inverter consists of a frequency converter power block based on IGBT (Germany) units with a microprocessor control system and a transformer unit with resonance capacitors, which provide galvanic network isolation and load matching. The inverter power is 100 kW in the frequency range 40–70 kHz.

We used a laboratory SNOL PL-20/14N (ZAO Nakal) induction furnace for heat treatment of hardened parts (quenching, tempering). A SNOL PL furnace is intended for heating and performing heat treatment of various materials under laboratory conditions.

Templates and specimens for metallographic tests were prepared from steel plates with surface alterations showed themselves after SHS process under HFC-heating in the places of application of coatings. The following parameters were investigated and established: the structure of samples in the zones of surface alterations (of presumably formed coatings); the borated layer thickness; microhardness and its distribution over the height of the coating.

To prepare thin sections, the following equipment was used: Labotom-03 Struerus manual cutting machine for preliminary specimen cutting; CitoPress-1 Struerus automatic electro-hydraulic press for template pressing into plastic; LaboPol-5 Struerus grinding and polishing machine.

To identify the surface structure of the specimens, the surfaces of prepared microsections was treated with 4% Nitric Acid solution in ethanol for 5–7 s. After the sections were etched, images of specimens were taken using an OLYMPUS GX51 inverted metallographic microscope.

The microhardness was measured by Vickers method using the KMT-1 apparatus, the load F = 1 N, exposure time t = 10 s, magnification 370*.

Fig. 1 Microstructure of the coating obtained on steel S355 during the SHS-process occurring in the course of HFC-heating of specimen No. 1 (general view, magnification ×100)

Fig. 2 Microstructure of the coating obtained on steel S355 during the SHS-process occurring in the course of HFC-heating of specimen No. 1 (transition zone, magnification ×500)

Fig. 3 Microstructure of the coating obtained on steel S355 during the SHS-process occurring in the course of HFC-heating of specimen No. 1 (borated layer, magnification ×500)

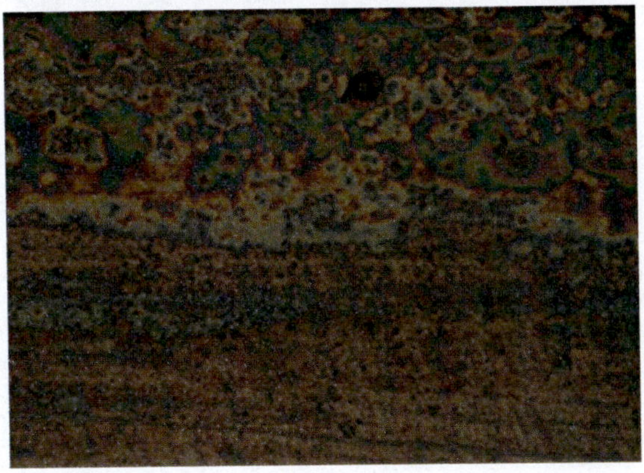

Fig. 4 Microstructure of the coating obtained on steel S355 during the SHS-process occurring in the course of HFC-heating of specimen No. 1 (borated layer boundary, magnification ×500)

Fig. 5 Microstructure of the coating obtained on steel S355 during the SHS-process occurring in the course of HFC-heating of specimen No. 2 (area 1, general view, magnification ×500)

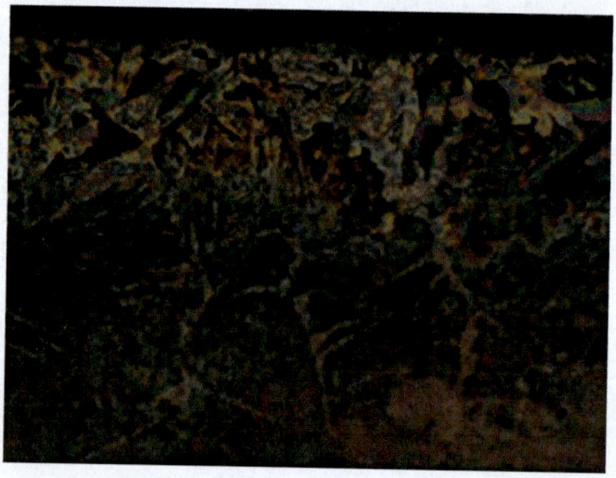

Fig. 6 Microstructure of the coating obtained on steel S355 during the SHS-process occurring in the course of HFC-heating of specimen No. 2 (area 1, general view, magnification ×1000)

3 Experimental Results

In Figs. 1, 2, 3, 4, 5 and 6 show the results of investigation of the coating structure obtained during SHS taking place in the process of HFC-heating of specimen No. 1 mounted on a S355 steel plate (Russian analogue is 17G1S steel) at various magnifications.

As it shown in Fig. 1, when the SHS process is taking place during HFC-heating of molded cylindrical sample No. 1 made of the basic mixture, the formation of a porous coating with a complex structure of a transition zone with several components in its composition was really observed at the place of its application to the surface of the S355 steel plate. The thickness of the resulting coating ranged from 150 to 250 μm.

The complex structure of the coating-metal matrix transition zone, the multi-component composition and porosity of the resulting coating are clearly visible on polarized-light images made with high magnification (see Figs. 2, 3 and 4).

Structures similar to those observed above were obtained both on steel S355 and during SHS-process occurring in the course of HFC-heating of specimen No. 3 applied to the surface of a steel plate with the difference that their thickness was 70–100 μm.

The decrease in the thickness of the coatings formed during the SHS process occurring in the course of HFC-heating of sample No. 3 is explained by its lesser height (volume, mass), and hence the smaller quantity of surface-active products formed during combustion.

Besides, a lesser amount of the initial composition in cylindrical specimen No. 3 forms a lesser quantity of heat necessary for melting the products and possible surface reaction of boriding.

The coatings obtained during the SHS process occurring in the course of HFC-heating of sample No. 2 (sheath) have a completely different structure.

As it seen from Figs. 5, 6, 7 and 8, two zones (areas) of coating with different thicknesses and structures are produced in a coating formed by the SHS-process occurring in the course of HFC-heating of the sheathing made of 2–3 mm thick basic composition applied to the surface of S355 steel.

The central zone of coating (section 1) has formed at the site of maximum high-frequency heating of the steel plate with a sheathing of basic composition applied to it, and the peripheral zone of coating (section 2) has formed along the borders of the coating, within the areas of minimum heating, maximum heat removal and long-term oxygen exposure in the course of formation of the coating.

The microstructure and composition of the peripheral zone coating are similar to those for the coating obtained during SHS-process occurring in the course of HFC-heating of specimens Nos. 1 and 3, the structure of the coating-base metal transition zone is also similar (see Figs. 7 and 8, area 2).

Fig. 7 Microstructure of the coating obtained on steel S355 during the SHS-process occurring in the course of HFC-heating of specimen No. 2 (area 2, general view, magnification ×500)

Fig. 8 Microstructure of the coating obtained on steel S355 during the SHS-process occurring in the course of HFC-heating of specimen No. 2 (area 2, general view, magnification ×1000)

The microstructure and coating composition of the central zone differs from those for previously obtained coatings (see Figs. 5 and 6, area 1). In the coating formed in the central zone, despite its insignificant average thickness (50–70 μm), the formation of a structure characteristic of boride phases is observed: closed dark carbide regions, light interlayers of iron-boride and iron-carbo-boride eutectics, inclusions of dark crystalline carbides and carboborides (see Figs. 5, 6 and 7).

Based on the microscopic images, the maximum thickness of the layer for each specimen. It was 163 μm—for the first specimen, 66 μm—for the second, and 77 μm—for the third ones.

4 Conclusions

Thus, in the course of the research, a method of the experimental study of coatings applied on specimens by HFC-heating was developed. It is clear that the formed coatings are composite materials based on alumina, SHS products in the base mixture and the product of fast HFC-boriding of the surface of a steel part being formed during SHS-boriding, but having the porous heterogeneous structure, insufficient thickness and low hardness due to the high content of alumina and side reactions (nitriding of aluminum, boron burnout, etc.).

It is possible to obtain more uniform, extended and harder coatings are only subject to modifying the base mixture: increase of aluminum content, addition of fluxes and adjustment of the particle-size and chemical composition of the powder material.

References

1. Zhu T, Li M (2010) Effect of hydrogen addition on the microstructure of TC21 alloy. Mater Sci Eng A 527(26):7080–7085
2. Zhu YS, Lu WZ, Zuo DW (2014) Microstructure and tribological properties of REs borided TC21 alloy. Surf Eng 30(8):612–618
3. Zhu YS, Lu WZ, Zuo DW, Sun YL, Wang H, Xu J (2013) Effects of rare earth additions during surface boriding on microstructure and properties of titanium alloy TC21. Mater Sci Technol 30(14):1765–1769
4. Patil CS, Ansari MI (2021) Influence of micro B4C ceramic particles addition on mechanical and wear behavior of aerospace grade Al-Li alloy composites. Sadhana Acad Proc Eng Sci 46(1):11
5. Ansari F, Ries MD, Pruitt L (2016) Effect of processing, sterilization and crosslinking on UHMWPE fatigue fracture and fatigue wear mechanisms in joint arthroplasty. J Mech Behav Biomed Mater 53:329–340
6. Chattopadhyay R (1993) High silt wear of hydroturbine runners. Wear 162:1040–1044
7. Krasnyy V (2019) The use of nanomaterials to improve the wear resistance of machine parts under fretting corrosion conditions. IOP Conf Ser Mater Sci Eng 560:012186
8. Singh J, Kumar S, Mohapatra S (2018) Erosion wear performance of Ni-Cr-O and NiCrBSiFe-WC(Co) composite coatings deposited by HVOF technique. Ind Lubr Tribol 71(4):610–619
9. Maksarov V, Khalimonenko A, Timofeev D (2014) Machining quality when lathing blanks with ceramic cutting tools. Agron Res 12(1):269–278
10. Chiu KY, Cheng FT, Man HC (2006) Corrosion behavior of AISI 316L stainless steel surface-modified with NiTi. Surf Coat Technol 200:6054–6061
11. Sun Y, Rana V (2011) Tribocorrosion behaviour of AISI 304 stainless steel in 0.5 M NaCl solution. Mater Chem Phys 129:138–147
12. Koga G, Zepon G (2019) Wear resistance of boron-modified supermartensitic stainless steel coatings produced by high-velocity oxygen fuel process. J Therm Spray Technol 28:2003–2014
13. Arji R, Dwivedi D, Gupta S (2009) Some studies on slurry erosion of flame sprayed Ni-Cr-Si-B coating. Ind Lubr Tribol 61(1):4–10
14. Karimi M, Salimijazi H, Golozar M (2016) Effects of remelting processes on porosity of NiCrBSi flame sprayed coatings. Surf Eng 32(3):238–245
15. Ishkov A, Malikov V (2019) Chemical reactions at high-speed HFC-boriding. J Phys Conf Ser 1399(4):044116
16. Dmitriev SF, Ishkov AV (2016) Subminiature eddy current transducers for studying boride coatings. J Phys Conf Ser 729(1):012018
17. Yankauskas V, Katinas E, Skirkus R, Aleknyavichene V (2014) The method of hardening soil rippers by surfacing and technical–economic assessment. J Frict Wear 35:270–277

The Stress State of Corner Cutout Area of the Model Boundary by Photoelastic Method

Lyudmila Frishter

Abstract The stress–strain state in the zone of the corner cutout of the area boundary is characterized by the singularity of the solution of the homogeneous boundary value problem and the complexity of its analysis. This article focuses on the stress state (SS) in the neighborhood of an irregular point of the area boundary on flat composite models with different cutout angles of the polymer material end. Forced deformations are created in one of the areas of the model, the forced deformations gap along the area contact boundary extends to the top of the corner cutout on the boundary of the model. The solution is obtained by means of experiments on optically sensitive material models by photoelastic method using the free temperature deformation defrosting property. A comparison of the SS obtained experimentally on the model and the parent distribution of SS in the zone adjacent to the irregular point of the area boundary is given. A conclusion is made on coincidence of the positions of the pure shear line obtained experimentally on the model and the neutral axis for the corresponding wedge under the action of a concentrated load. This paper is aimed at analyzing SS in the zone of the corner cutout of the boundary of the region and identifying the self-balanced SS corresponding to the solution of the homogeneous boundary value problem in the area with an irregular boundary point.

Keywords Stress state · Corner cutout of the area boundary · Photoelastic method · Singularity of the solution

1 Introduction

The stress–strain state (SSS) of composite structures is characterized by the concentration of stresses at the junctions of elements with different design of the boundary: special lines, points, such as the reentrant angle, etc. The complex SSS occurs in the stress concentration area, which is due to the shape of the boundary or "geometric

L. Frishter (✉)
National Research University Moscow State University of Civil Engineering, 26 Yaroslavskoye Shosse, 129337 Moscow, Russian Federation

A. Beskopylny and M. Shamtsyan (eds.), *XIV International Scientific Conference "INTERAGROMASH 2021"*, Lecture Notes in Networks and Systems 247,
https://doi.org/10.1007/978-3-030-80946-1_8

factor" and the finite discontinuity of the given forced strains, mechanical properties, going out to an irregular point of the area boundary. The complexity of the SSS analysis in the corner cutout zone of the area boundary is due to the singularity of the solution of the homogeneous boundary value problem of elasticity.

The solution of elliptic equations for areas with non-smooth boundaries is considered in papers of Williams, Kondratiev, Fufaev, Ufliand, Kalandiia, Cherepanov [1], Bodzhi, Aksentian, Aleksandrova, Chobanian. In papers [1–7], the stress nature in the corner zone of a plane wedge for different boundary conditions on its sides is determined. In the neighborhood of irregular boundary points, the solution of the elliptic boundary value problem is given in the form of asymptotic series and an infinitely differentiable function. The components of these series contain solutions of homogeneous boundary value problems for model areas: a cone or a wedge. These solutions depend on local characteristics: the values of the solid and plane angle and the type of boundary conditions [7–14]. The values of the solution expansion coefficients in the neighborhood of the singular point are unknown and depend on the problem as a whole.

The application of the well-known experimental photoelastic method [9, 15–20] makes it possible to obtain a solution of the elasticity problem on the model with a corner cutout of the boundary. The most developed and effective is the method for solving plane problems of deformable body mechanics [9, 20].

Experimentally, the stress state in the neighborhood of an irregular point of the boundary area is obtained by photoelastic method using the free temperature deformation defrosting property [12, 15–20] on flat composite polymer models with different cutout corners of the boundary. Forced deformations are created in one of the areas of the model, the forced deformations gap along the area contact boundary extends to the top of the corner cutout on the boundary of the model.

This paper is aimed at analyzing SS in the zone of the corner cutout of the boundary of the region and identifying the self-balanced SS corresponding to the solution of the homogeneous boundary value problem in the area with an irregular boundary point.

2 Materials and Methods

2.1 *Experimental Solution of the Boundary Corner Cutout Problem on a Flat Model by Photoelastic Method*

An experimental solution of a plane problem for an area with a cutout at the boundary, the geometry of which causes the occurrence of stress concentration, is considered.

The plane area consists of parts Ω_1 and Ω_2, with elasticity module E, Poisson's material ratio ν and linear expansion coefficient α. Temperature deformations $\alpha T \delta_{ij}$ are created in one of the areas—Ω_2, the area Ω_1 is not loaded. Along the area contact line $\Gamma = \Omega_1 \cap \Omega_2$, there is a deformation jump $\Delta\varepsilon_{ij} = \alpha T \delta_{ij}$, which goes to the body boundary S to the point $O(0, 0)$—the top of the corner cutout of the area boundary. For different angles of the corner cutout solution of the boundary at its top, there is a feature of the stress state, which is experimentally realized on the model made of optically sensitive material as a stress raiser (see Fig. 1).

The elastic problem under the action of temperature deformations is solved by the photoelastic method using the property of "defrosting" of free temperature deformations [12, 16, 20]. A model of homogeneous epoxy material with given dimensions l = 180 mm, h = 25(23) mm is glued from two elements. The Ω_2 area element is cut from a plate obtained from the longitudinal shear of a hollow cylinder frozen under the action of a longitudinal uniformly distributed compression load. The height of the cylinder before "freezing" is H = 309 mm, the outer diameter is D = 180 mm, the inner diameter is 150 mm. The ends of the cylinder are cut off after freezing and only the middle part of the cylinder, which is in a flat deformed state, is considered. The compression of the cylinder after applying the load and freezing is $\Delta H = 9$ mm, relative elongation along the longitudinal axis OZ of the cylinder is $\varepsilon_z = \frac{\Delta H}{H} = \frac{9}{300} \approx 0.03$. Compression stress is

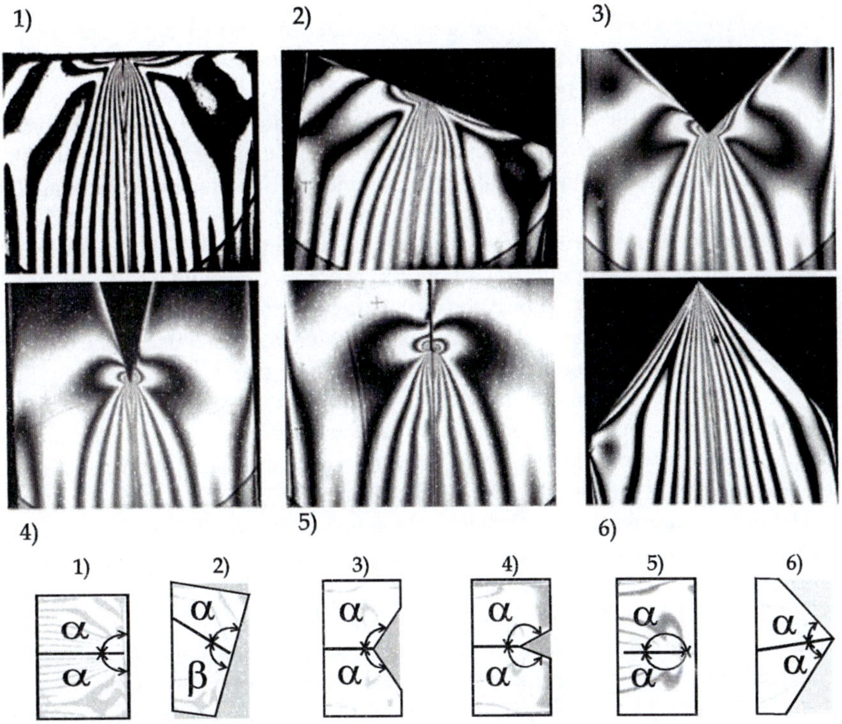

Fig. 1 Patterns of strips in models $t = 3$ mm thick for different cases of the model end solution: (1) $2\alpha = 180°$; (2) $\alpha + \beta = 180°$, $\alpha = 105°$, $\beta = 75°$; (3) $2\alpha = 270°$; (4) $2\alpha = 330°$; (5) $2\alpha = 360°$—narrow cutout—gap; (6) $2\alpha = 90°$

$\sigma_z = -\frac{P}{F} = -E\varepsilon_z = -200 \cdot 0.03 = -6$ kg/cm^2, $\sigma_z = \frac{\sigma^{1.0} m}{t} \Rightarrow m = \frac{\sigma_z \cdot t}{\sigma^{1.0}}$, where t is a shear thickness; $\sigma^{1.0}$ is a price of the model strip material; $\sigma^{1.0} = 0.341$. According to these data, in the shear $t = 0.3$ cm thick, the number of strips is $m = 5.3$.

The area Ω_1 is cut from free unloaded material. The areas Ω_1 and Ω_2 are glued together, then the model is "defrosted". The model generates SSS due to the action in the Ω_2 area of temperature deformations. $\varepsilon_0 = -\varepsilon_z = -\alpha T \delta_{ij}$.

The pattern of the strips corresponding to the stress–strain state from the action of temperature deformations for different shapes of the model end boundary is shown in Fig. 1. Model thickness (1)...(6) is the same and equals to $t = 3$ mm. In case (6), stress concentration at the top of point $O(0, 0)$ is not observed and stresses at the very top $O(0, 0)$ are zero.

2.2 *Analysis of the Stress State Obtained Experimentally on the Photoelastic Method Model*

The solution obtained experimentally on the composite model in the area of the corner cutout of the boundary under the action of discontinuous forced deformations is analyzed. An irregular point of the area boundary is considered as a top of the cutout of the end of the flat model lying on the contact boundary of the model parts.

We consider the neighborhood of an irregular point of the boundary adjacent to the singular point from which the irregular point is removed. Experimentally, the singular area on the model is defined as the neighborhood of the top of the corner cutout of the boundary, in which the interference fringes are blurred and the isochrome picture is not readable at any magnification of the neighborhood fragment.

The stress state in the neighborhood of the irregular point of the model boundary can be represented as Areas I, II, III according to Fig. 2a, where, due to the oblique symmetry of the stress diagrams, one of the areas Ω_2 is given.

Area I is an area adjacent to the symmetry axis of the model $\Gamma{:}\, y = 0$, where the main compressive stresses σ_2 considerably exceed the modulus of the main tensile stresses σ_1. As the point moves away from the top of the cutout in area I, the σ_1 stresses nature may change from tensile to compressive.

The main vector on the radial sites at the points of the area I (see Fig. 1a) has a normal component $\sigma_i = \sigma_r$ considerably superior in modulo to the tangential one σ_θ, which agrees with the ratio of the values of the principal stresses: $|\sigma_2| \gg \sigma_1$; $\sigma_1 \to 0$ at $(x, y) \to (x, 0)$; $\sigma_1|\Gamma = 0$; $\Gamma{:}\, y = 0$.

Area III is the area adjacent to the model cutout boundary, where the tensile principal stresses σ_1 considerably exceed the modulus of compressive stresses σ_2: $\sigma_1 \gg |\sigma_2|$; $\sigma_2 \to 0$ at $(x, y) \to S$; S—model boundary; $S{:}\, y = ax$, $a = tg\alpha$, 2α—angle of the model end opening; $\sigma_2|S = 0$.

Area II is a transitional area in which sharp angles, isochrome and significant gradients of the isocline parameter forming loops are observed. The largest isocline

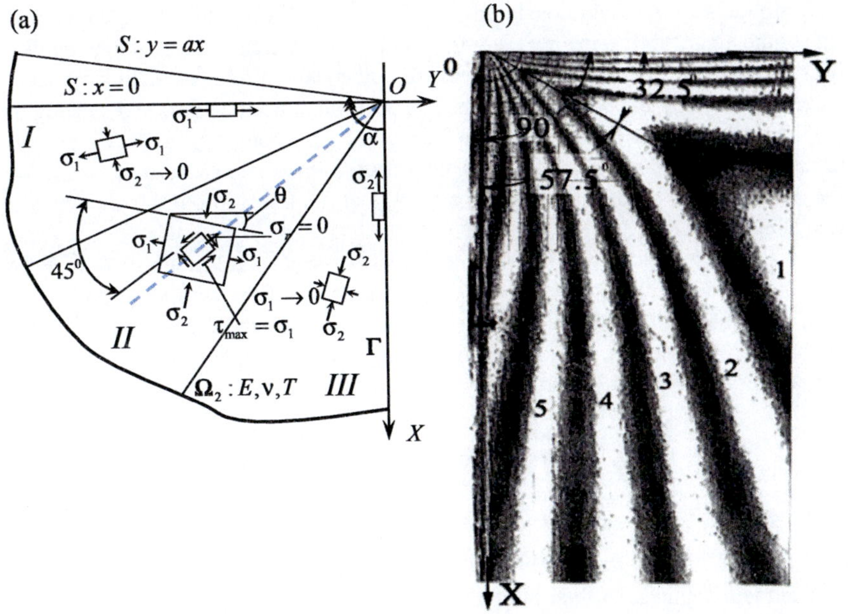

Fig. 2 SS scheme in the neighborhood of an irregular point of the boundary of a plane area. -----pure shear line

parameter is observed for the inner, nested in the remaining isocline loops, and varies from 45°...50° up to 70° when increasing the opening of the model end from $2\alpha = 180°$ to $2\alpha = 260°$.

There is a change in the direction of the main sites rotation in area II. If the main sites rotate clockwise in area III, then, passing area II, they change the direction of rotation to the opposite, so that the rotation of the main sites at the points of area I is observed in a clockwise direction as before.

In the transition area II (see Fig. 2a), a pure shear site is observed. The principal stresses at these points are equal in modulo and opposite in sign: $\sigma_1 = -\sigma_2$. The sites tilted at an angle of 45° to the main ones are in pure shear conditions, i.e. $\tau_{\max} = \frac{\sigma_1 - \sigma_2}{2} = \sigma_1$, and the normal stresses across these sites are zero: $\sigma_i = \frac{(\sigma_1 + \sigma_2)}{2} = 0$. According to the experimental data, the pure shear sites coincide with the radial ones. The radial stress as the normal stress over the pure shear site is zero: $\sigma_r = 0$, and shear stresses $\tau_{\max} = \frac{m\sigma^{1.0}}{2t} \neq 0$. Therefore, the isochrome orders m at the points of pure shear sites are not equal to zero. According to the experimental data, pure shear sites are observed at the tops of sharp angles of isochromes and their vicinities.

The stresses are continuous in the area $\Omega_2 : 0 \leq \theta \leq \alpha$, so there is a subarea in transition area II in which pure shear and a stress state close to it are observed. For this "pure shear" subarea, there are sites where the normal stresses coincide or are close to the radial stresses and are equal or close to zero: $\sigma_i \cong \sigma_r \cong 0$ (see Fig. 2).

Main stresses σ_2 continuously varying from the highest value along the contact line of the area Γ: y = 0 in area I, going to area II "pure shear" subarea, then quickly decrease to zero values in area III. The sign may also change for main stresses σ_2 when approaching the boundary of the area S. Similar, tensile stresses σ_1, changing continuously from the highest values at the model cutout boundary in area III, passing to the "pure shear" subarea II, then rapidly decreasing to zero values in area I. For the main stresses σ_1 the sign may also change at the transition to the area contact line $\Gamma = \Omega_1 \cap \Omega_2$, Γ: y = 0.

Radial stresses in area I: $\sigma_r \cong \sigma_2$ at $(x, y) \to (x, 0)$, and in area III: $\sigma_r \cong \sigma_1$ at $(x, y) \to (x, \alpha x)$, $S{:}\, y = ax$, $a = tg\alpha$. In the transition area II on pure shear sites and close to them $\sigma_r = 0$ or $\sigma_r \approx 0$. Such a distribution of radial stresses in the neighborhood of the top of the model end cutout corresponds to the theoretical radial stress distribution of the form:

$$\sigma_r = cf(r)f(\theta) = \frac{\alpha ET}{r^{1-\lambda}}(c_1 \cos\theta + c_2 \sin\theta),\ \sigma_\theta = \tau_{r\theta} = 0 \quad (1)$$

At the pure shear sites, at $\theta = \theta_0$, $r > r_0$, r_0—sufficiently small, $\sigma_r = 0$, $\sigma_r = \frac{\alpha ET}{r^{1-\lambda_0}}(c_1 \cos\theta_0 + c_2 \sin\theta_0) = 0 \Rightarrow c_1 = -c_2\left(\frac{\sin\theta_0}{\cos\theta_0}\right)$.

The theoretical radial distribution of stresses in the neighborhood of the cutout of the model of the form (1) is understood as a self-balanced "own" stress state in the neighborhood of an irregular border point obtained as a solution of a plane homogeneous boundary value problem.

2.3 Example of Stress State Analysis Obtained on a Flat Model with a Straight End

The stress state is analyzed in the neighborhood of the irregular point of the boundary where the gap of forced deformations exits, using the straight end model as an example. According to the theoretical model [1–4, 12, 13], the stress state in the neighborhood of an irregular point of the area boundary can be represented as:

$$\sigma_{ij} = \sigma_{ij}^o + \sigma_{ij}^l \quad (2)$$

where σ_{ij}—stresses in the vicinity of an irregular point of the area boundary; σ_{ij}^o—eigensolution of the homogeneous boundary value problem in the vicinity of an irregular boundary point; σ_{ij}^l—stress caused by the action of the specified loads or the total stress field.

Internal stresses σ_{ij}^o in the area of the end of the model with a straight end have a radial view [1–4, 12, 13]:

$$\sigma_r = \frac{\alpha ET}{r}(c_1 \cos\theta_0 + c_2 \sin\theta_0), \quad \sigma_\theta = \tau_{r\theta} = 0 \quad (3)$$

With a continuous distribution of stresses in the area Ω_2 (similarly in the area Ω_1) of the form (3) in the cross-section $r = r_0$ there will be a point at which $\sigma_r = 0$. Consider one area of the straight-end model, for example, Ω_2: E, ν, αT. Pattern of the model strips for the area Ω_2 is given in Fig. 2b. According to the experimental data in area II of the model, the point with the net shear area is observed in the cross section at an angle $\theta_0 \approx 60°$. Figure 2 shows the "pure shear" line, which has a slope angle $\theta_0 \approx 57.5°$ in the vicinity of point $O(0, 0)$. According to the experimental findings

$$\sigma_r = 0 \text{ at } \theta_0 \approx 57.5° \tag{4}$$

Taking into account the theoretical distribution (1), the radial stresses will be recorded:

$$\sigma_r = \frac{c_0}{r}(c_1 \cos\theta + c_2 \sin\theta) = 0, \tag{5}$$

or

$$c_1 = -c_2 \frac{\sin\theta_0}{\cos\theta_0} = -c_2 \frac{\sin 57.5°}{\cos 57.5°} = -1.57 c_2 \tag{6}$$

Stress (3) taking into account (6) will write over:

$$\sigma_r = \frac{c}{r}(-1.57\cos\theta + \sin\theta) \tag{7}$$

where $c = c_0\, c_2$, c is an unknown multiplier.

Determine what force is statically equivalent to the radial stresses (7), acting over a small radius cross section $r = r_0$ of the area Ω_2 of the model end [4, 12, 16, 17].

$$P_V = \sum X = \int_0^{\frac{\pi}{2}} \sigma_r \cos\theta \mathrm{d}F = \int_0^{\frac{\pi}{2}} \sigma_r \cos\theta r t \mathrm{d}\theta \tag{8}$$

where t is a model thickness. Considering (7), force P_V equals:

$$P_V = \int_0^{\frac{\pi}{2}} \frac{c}{r}[-1.57\cos\theta + \sin\theta]\ \cos\theta r\ t\mathrm{d}\theta = -\frac{ct}{r}(1.57\pi + 1) \tag{9}$$

Let's find the horizontal component of the acting forces:

$$P_H = \sum Y = \int_0^{\frac{\pi}{2}} \sigma_r \sin\theta\ \mathrm{d}F = \int_0^{\frac{\pi}{2}} \sigma_r \sin\theta r t \mathrm{d}\theta \tag{10}$$

Considering (7), force P_H equals zero:

$$P_H = \int_0^{\frac{\pi}{2}} \frac{c}{r}(-1.57\cos\theta + \sin\theta)\sin\theta r t d\theta = 0 \tag{11}$$

Then force action angle γ, statically equivalent to radial stresses of the form (7) in the area Ω_2: $tg\gamma = \frac{P_H}{P_V} = 0 \to \gamma = 0^0$.

Finally, the radial stresses of the form (7):

$$\sigma_r = \frac{c}{r}(-1.57\cos\theta + \sin\theta) \tag{12}$$

in the small radius cross section $r = r_0$ in the area Ω_2 are statically equivalent to the action of a vertical force directed along the axis OX:

$$P_V = \sum X = -\frac{ct}{4}(1.57\pi + 1) \tag{13}$$

and the horizontal thrust is zero: $P_H = \sum Y = 0$.

The resulting scheme of force action P_V, P_H in the small radius cross section $r = r_0$ of the model end coincides with the corresponding forces for the wedge with an opening angle $\theta = \frac{\pi}{2}$ under the action of a concentrated force at the top of the wedge [4, 12, 16, 17].

3 Results

The experimentally obtained position of the pure shear line in the irregular point area of the model boundary shows full coincidence with the position of the neutral axis of the radial stress state for the wedge in the area of the top $O(0, 0)$ under concentrated force action.

The coincidence of the slope angle of the pure shear line in the area of the model end and the neutral axis of the model wedge during the action of the radial stress state, the coincidence of the isocline parameter in the model isochrome tops and the calculated slope angle of the main sites confirm experimentally the existence of the self-balanced radial stress state in the neighborhood of the model end top, corresponding to the internal stress state σ_{ij}^o in the general presentation of the SS: $\sigma_{ij} = \sigma_{ij}^o + \sigma_{ij}^l$ of the model end.

4 Discussion

Experimentally, the singular area on the model is defined as the neighborhood of the top of the corner cutout of the boundary, in which the interference fringes are blurred and the isochrome pattern is not readable at any magnification of the neighborhood fragment. A contradiction arises with a continuous change of stresses in the punctured vicinity of an irregular boundary point, "sharp angles" of isochromes occur. The above analysis is applicable to the neighborhood of an irregular point of the boundary adjacent to the singular point from which the irregular point is removed. The validity of the analysis data in the selected area of the irregular boundary point is confirmed by the coincidence of the experimental and theoretical stress values.

5 Conclusions

The analysis of SS in the area of the irregular point of the boundary, in which the gap of forced deformations exits, shows the existence of a self-balanced SS σ_{ij}^{o}, corresponding to the solution of the homogeneous boundary value elasticity problem. It explains the growth of strips orders observed from the inside of the stress concentration area, and not at the very top of the cutout of the area. The inequation to zero of orders of the strips in area II of pure shear in the neighborhood of the top of the model end shows the existence of a self-balanced SS caused by the action of the given loads or the total stress field.

References

1. Cherepanov GN (2012) Fracture mechanics. Computer Research Institute, Izhevsk
2. Kuliev VD (2005) Singular boundary value problems. Nauka, Moscow
3. Feodosiev VI (2016) Strength of materials. Publ. House N. E. Bauman MSTU, Moscow
4. Parton V, Morozov E (2017) Elastic-plastic fracture mechanics: special problems of the fracture mechanics. Moscow
5. Bakushev SV (2013) Geometrically and physically nonlinear continuum mechanics. Plane problem. LIBROKOM, Moscow
6. Bakushev SV (2020) Differential equations and boundary value problems of deformable solid body mechanics. LENAND, Moscow
7. Matviyenko Yu (2006) Fracture mechanics models and criteria. Fizmatlit, Moscow
8. Albaut GN, Tabanjuhova MV, Harinova NV (2004) Determination of the first stress intensity factor in elements with angled cut-out. In: Experimental mechanics and calculation of structures (Kostinsky readings). Moscow State University Press, Moscow
9. Albaut GN, Kharinova NV, Sadovnichij VP, Semenova JI, Fedin SA (2011) Nonlinear problems of mechanics of destruction. Vestn Lobachevsky Univ Nizhni Novgorod 4:1344–1348
10. Vardanjan GS, Frishter LJu (2007) Int J Comput Civ Struct Eng 3:75–81
11. Pestrenin VM, Pestrenina IV, Landik LV (2013) Vestn TGU Math Mech 4:80–87
12. Razumovskij IA (2007) Interference-optical methods of deformable solid mechanics. Publisher MGTU named after N. Je. Bauman, Moscow, p 240

13. Frishter LYu (2017) Photoelasticity-based study of stress-strain state in the area of the plain domain boundary cut-out area vertex. In: Advances in intelligent systems and computing, EMMFT 2017, vol 692. Springer, pp 836–844
14. Makhutov NA, Moskvichev VV, Morozov EV, Goldstein RV (2017) Unification of computation and experimental methods of testing for crack resistance: development of the fracture mechanics and new goals. Ind Lab Diagn Mater 83(10):55–64
15. Kobayashi A (1990) Experimental mechanics, vol 1. Mir, Moscow, p 615; vol 2. Mir, Moscow, p 551
16. Hesin GL et al (1975) The photoelasticity method, vol 3. Stroyizdat, Moscow, p 311
17. Frocht MM (1948) Photoelasticity. In: Prigorovskiy NI (ed), vol 2. GITTL, Moscow, Leningrad
18. Durelli A, Riley W (1970) Introduction to photoelasticity. In: Prigorovskiy NI (ed). Mir, Moscow
19. Aleksandrov AJa, Ahmetzjanov MH (1974) Polarization-optical methods for deformable solid mechanics. Nauka, Moscow, p 576
20. Koshelenko AS, Poznjak GG (2004) Theoretical foundations and practice of photo mechanics in mechanical engineering. Granica, Moscow

Influence of the Particle Number on Mixture Quality

Konstantin Fudin, Vladimir Konovalov, Aleksey Chupshev, and Vyacheslav Teruyshkov

Abstract The aim of the research was to establish regression relationships establishing the interrelation between the number of measurement units in the samples and the mixture quality variability index (coefficient of variation of the control component content in the samples) for the same mixture. As a result of simulation of the obtained models for comparable parts of the particle number, we received practical coincidence of numerical values of different models in the interval of 10…30 particles in the sample, with the greatest coincidence of the models around 20…25 pc. Decreasing the number of particles significantly increases the scatter of variation coefficient values. Increasing the number of controlled components increases the variation of the obtained values. Increasing the size of the particles increases the values of the coefficient of variation. For example, for lentils at 60 (pcs./sample), the coefficient of variation is 15.5%, for barley—10.7% and for buckwheat—9.9%. That is, the differences in the two small components are insignificant and comparable to the error of the experiments. The estimated number of particles in the samples taken is at least 19 pcs. Reducing the calculated value to 14 pcs. (lentils) increases the variability of the index.

Keywords Mixture · Mixture quality · Mixture unevenness · Coefficient of variation · Number of particles in the sample

1 Introduction

Modern scientific and technological progress is not possible without improvements in the materials used. Composites and mixtures based on different types of components are developed first and foremost [1–3]. The components used can be either single-phase (dry, liquid, gaseous) or mixture ingredients in different phase states can be used [4–6].

K. Fudin · V. Konovalov (✉)
Penza State Technological University, 1A/11, pr. Baydukova/Gagarina St., Penza, Russia

A. Chupshev · V. Teruyshkov
Penza State Agrarian University, 30 Botanicheskaya St., Penza, Russia

A. Beskopylny and M. Shamtsyan (eds.), *XIV International Scientific Conference "INTERAGROMASH 2021"*, Lecture Notes in Networks and Systems 247,
https://doi.org/10.1007/978-3-030-80946-1_9

Mixture formation from different components have advantages and disadvantages. Gaseous components mix easily and require turbulence to maintain mixture homogeneity [3, 5]. Depending on the viscosity, the liquid components are less well mixed [1, 4]. In this case, if the components are not soluble, the prepared mixture may stratify. Preparation of a mixture from dry components is also energy consuming [7, 8]. However, mixture formation is not as problematic as the formation of mixtures where small quantities of liquid are added to a dry filler [9, 10].

In the study and justification of the mixer parameters, the main attention is focused on optimizing the design and kinematic parameters [11]. In this case, the task is to ensure a given productivity [12, 13], as well as to reduce energy costs [14–16]. Additional research is also required to establish the workability zone of justified mixer parameters [7]. That is, the justification of the operating intervals of technological parameters to ensure compliance with the technological requirements for the quality of the mixture [17]. If the requirements are not met, the implemented energy costs are correspondingly inappropriate, as the prepared mixture (e.g. liquid concrete) will not provide the proper quality of the subsequent composite [1].

Various techniques are used to assess the quality of the mixture. Traditionally, the coefficient of variation (irregularity) distribution of one of the mixture components is used [17]. However, this method is labour intensive. Therefore, attempts are made to determine the quality by examining the uniformity of the mixer torque [1], or, more commonly, by spectroscopy of the resulting mixture [18–20]. However, this method is far from perfect and inaccurate at this stage of its application. Therefore, determination of the coefficient of variation from the content of the control component in the samples remains the main method.

Different branches of industry and agriculture use different control components and their proportion in the prepared mixture. At the moment there is no reliable system for converting the unevenness (or its corresponding homogeneity) depending on particle size, control portion weight and proportion of control component. Therefore, it is problematic to compare the mixture quality obtained in different branches and by different researchers.

The aim of the research was to establish regression relationships establishing the relationship between the number of units in the samples and a measure of mixture quality variability (coefficient of variation of the control component content in the samples) for the same mixture.

2 Methods and Materials

The research methodology involved experimental investigation of the quality of the same mixture depending on the mass of samples taken, the number of samples, the number of control component units in the sample (or particle size), with the establishment of regression models and their analysis.

During the experiment, 50 kg of different components were loaded into a drum mixer [17], which formed a mixture during the mixing process. The mixture consisted of 5 controlled components, each constituting 1% of the mass of the final mixture: maize, beans, lentils, barley and buckwheat. The mass of a thousand grains (seeds), g: beans—353.105; maize—269.825; lentils—71.410; barley—50.205; buckwheat—23.515. After 3 min of mixing, 100 samples were taken from the total volume of the mixture formed, each weighing 0.1 kg of the mixture. Estimated number of grains in the selected samples, pcs.: beans—2.8; maize—3.7; lentils—14.0; barley—19.9; buckwheat—42.5. Using a sieve classifier, the samples were divided into the components and weighed by components. As the data were obtained (i.e. randomly) the component data were recorded in a table. To obtain data as the sample mass changed, the data from the table was used. To do this, the values of a particular controlled component were added sequentially from the available sample a corresponding number of times (1; 2; 3; 4; 5; 6 consecutive cells), forming the corresponding mass of the sample M (Table 1), kg. From the samples formed, the mass of the control components was converted to the number of grains of these components. The mean value of the grains of the control component was calculated for each component as well as the coefficient of variation (fractions or 0.01%). The calculated values are summarised in Table 1.

Since the transition from the mass of the control component to the number of grains was used, it was necessary to estimate the error in the value of the coefficient of variation associated with the random absence (or ingestion) of one grain in the sample taken. For this purpose the Excel computer program was used and tables were made where the number of cells Z (pieces) corresponded to the number of samples (in the studied case—10; 15; 20; 35; 50; 100 pieces). Stable number of grains n (5; 10; 15; 20; 25; 35; 50; 100 pcs. respectively) were written in the indicated cells [1...(Z − 1)]. The last cell was used to record the corresponding value of grains of the control component—(n − 1). The mean value, standard deviation and coefficient of variation were determined.

Using the data obtained, further statistical processing of the results was carried out with the computer program Statistica.

3 The Results of the Studies

As a result of the aforesaid experiment and the primary processing of the data, the data shown in Table 1 was obtained. Using the data from Table 1, further statistical processing was carried out. Since in [17] it was recommended to use an exponential function, in addition to the natural values of the coefficient of variation, their exponent and natural logarithm were used as separate blocks. The obtained values were processed with a degree function.

The results of statistical processing are summarised in Table 2. Regression models for the specified lists of control components, their Pearson R-correlation coefficient values and F-test data were obtained. Adequate models were used to describe the mixing process for which the correlation and F-test results should be at least 0.9 (90%).

Table 1 The results of the initial statistical processing of the experimental data

No.	Sample mass M, kg	Type of control component	Average value of the control component n, pcs.	Coefficientof variation v	No.	Sample mass M, kg	Type of control component	Average value of the control component n, pcs.	Coefficientof variation v
1	0.1	Beans	0.524	0.875	16	0.4	Beans	3.269	0.477
2	0.1	Maize	0.818	0.891	17	0.4	Maize	2.092	0.417
3	0.1	Lentils	3.571	0.527	18	0.4	Lentils	14.398	0.254
4	0.1	Barley	4.379	0.381	19	0.4	Barley	17.546	0.217
5	0.1	Buck-wheat	10.322	0.272	20	0.4	Buck-wheat	41.396	0.113
6	0.2	Beans	1.046	0.615	21	0.5	Beans	2.615	0.289
7	0.2	Maize	1.635	0.693	22	0.5	Maize	4.086	0.415
8	0.2	Lentils	7.199	0.366	23	0.5	Lentils	17.997	0.195
9	0.2	Barley	8.773	0.298	24	0.5	Barley	21.932	0.210
10	0.2	Buck-wheat	20.698	0.183	25	0.5	Buck-wheat	51.745	0.127
11	0.3	Beans	1.539	0.542	26	0.6	Beans	3.196	0.295
12	0.3	Maize	2.449	0.449	27	0.6	Maize	5.004	0.420
13	0.3	Lentils	10.691	0.276	28	0.6	Lentils	21.984	0.237
14	0.3	Barley	13.232	0.244	29	0.6	Barley	26.432	0.173
15	0.3	Buck-wheat	31.064	0.153	30	0.6	Buck-wheat	62.108	0.100

Table 2 The results of the regression analysis of the statistical data

No.	Composition of control components	Type of regression equation	Correlation coefficient, R	F-test
I	Beans, maize, lentils, barley, buckwheat	$v = -0.008 + 0.7025 \cdot$ n^(−0.404)	0.94575	0.765902
		$\ln(v) = 1.92 - 2.3327 \cdot$ n^(0.139)	0.95745	0.750761
		exp(v) = 1.06 + 0.9575 · n^(−0.6)	0.94222	0.750761
II	Maize, lentils, barley, buckwheat	$v = -0.0094 + 0.796 \cdot$ n^(−0.443)	0.97014	0.304702
		$\ln(v) = 2.05 - 2.355 \cdot$ n^(0.1464)	0.97844	0.739213
		$\exp(v) = 1.083 + 1.145$ · n^(−0.7)	0.96862	0.133935
III	Lentils, barley, buckwheat	$v = -0.025 + 0.9359 \cdot$ n^(−0.4037)	0.97417	0.915319
		$\ln(v) = 2.567 - 2.7965 \cdot$ n^(0.133)	0.97678	0.923941
		$\exp(v) = 1.02 + 1.397 \cdot$ n^(−0.65)	0.97117	0.90538
IV	Barley, buckwheat	$v = -0.287 + 0.9129 \cdot$ n^(−0.2094)	0.91783	0.97959
		$\ln(v) = 0.495 - 1 \cdot$ n^(0.2482)	0.98563	0.962565
		$\exp(v) = 276 - 274.2 \cdot$ n^(0.0005)	0.98428	0.957315
V	Lentils	$v = 0.11 + 1.149 \cdot$ n^(−0.7915)	0.98538	0.974994
		$\ln(v) = 429 - 429.27 \cdot$ n^(0.0012)	0.96006	0.931355
		$\exp(v) = 1.1 + 1.876 \cdot$ n^(−0.94)	0.98917	0.981512
VI	Barley	$v = -0.351 + 0.954 \cdot$ n^(−0.179)	0.99571	0.992708
		$\ln(v) = 0.259 - 0.8268 \cdot$ n^(0.266)	0.99248	0.987189
		$\exp(v) = 0.812 + 1 \cdot$ n^(−0.293)	0.99631	0.993718
VII	Buckwheat	$v = 113.79 - 113.31 \cdot$ n^(0.0008)	0.97537	0.95826
		$\ln(v) = 77.8 - 77.89 \cdot$ n^(0.0068)	0.97671	0.959879

(continued)

Table 2 (continued)

No.	Composition of control components	Type of regression equation	Correlation coefficient, R	F-test
		$\exp(v) = 1.04 + 1.55 \cdot$ n^(−0.738)	0.99043	0.983683

The obtained regression models for the whole list of controlled components (No. I in Table 2) show good correlation with the experimental data (R > 0.9), however, the models are not adequate because the experimental points are out of 90% confidence interval (F-test < 0.9). This is confirmed by Fig. 1(I), where the experimental points for the section up to (n < 6) are significantly removed from the line of theoretical dependence. The analysis of the points revealed a deviation of values from the function for components with a small number of grains. Therefore, an attempt was made to derive a regression model by sampling in the absence of numbers for beans (No. II in Table 2). The resulting model (Fig. 1(II)) has a closer location of the points from the theoretical function, but the pattern of deviation of the points poorly fits the normal law of distribution (see F-test).

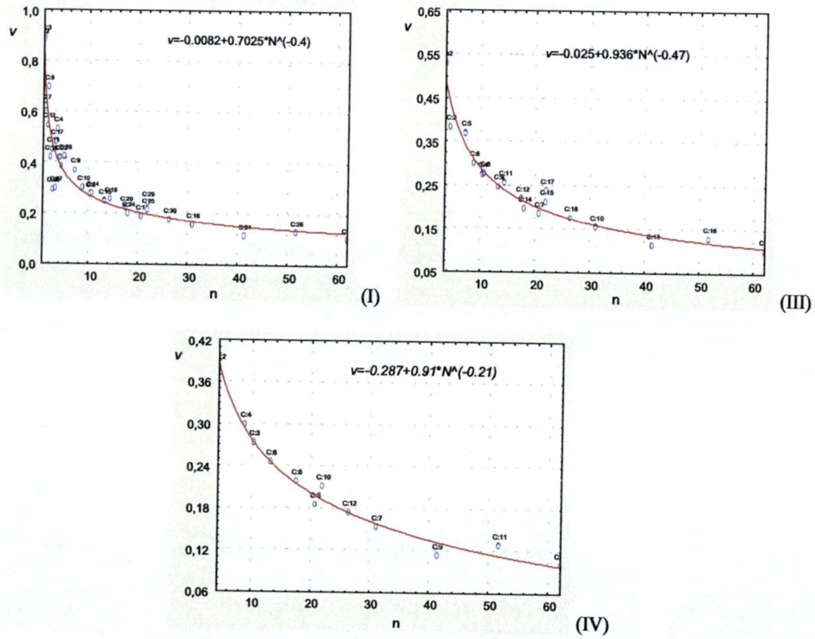

Fig. 1 The influence of the number of units of the control component n (pcs.) on the calculated value of the coefficient of variation ν (fraction): I—for a mixture of control components, comprising beans, maize, lentils, barley, buckwheat; III—for a mixture of control components, comprising lentils, barley, buckwheat; IV—for a mixture of control components, comprising barley and buckwheat

Removing the data for maize the regression model (No. III in Table 2) adequate with 90% probability (F-test > 0.9) was obtained. Correlation coefficient R and F-test information best describes a degree model by the logarithm of the coefficient of variation for the given mixture of control components. This conclusion is also valid for the mixture of control components in barley and buckwheat (No. IV in Table 2). In this case, the adequacy of the models increases to 95% probability.

The analysis of the models for individual crops confirmed high adequacy of the models obtained (over 95% probability; R > 0.96). The models derived on the basis of the exponent of variation coefficient values have the best performance. Experimental points (Fig. 2(V–VII)) fit well to the graph of the function.

The analysis of all the graphs shows that as the number of average grains in the samples (and therefore the number of units) increases, the coefficient of variation decreases. The coefficient of variation changes most intensively when the number of grains in the sample is up to 10. The values of the coefficient of variation are more stable if the number of grains in the sample is more than 20–25. In this case, the coefficient of variation corresponds to about 20%. When the number of grains is about 60, the coefficient of variation decreases to 13%. Reducing the average number of grains in the sample to 10 pieces increases the coefficient of variation to 30%.

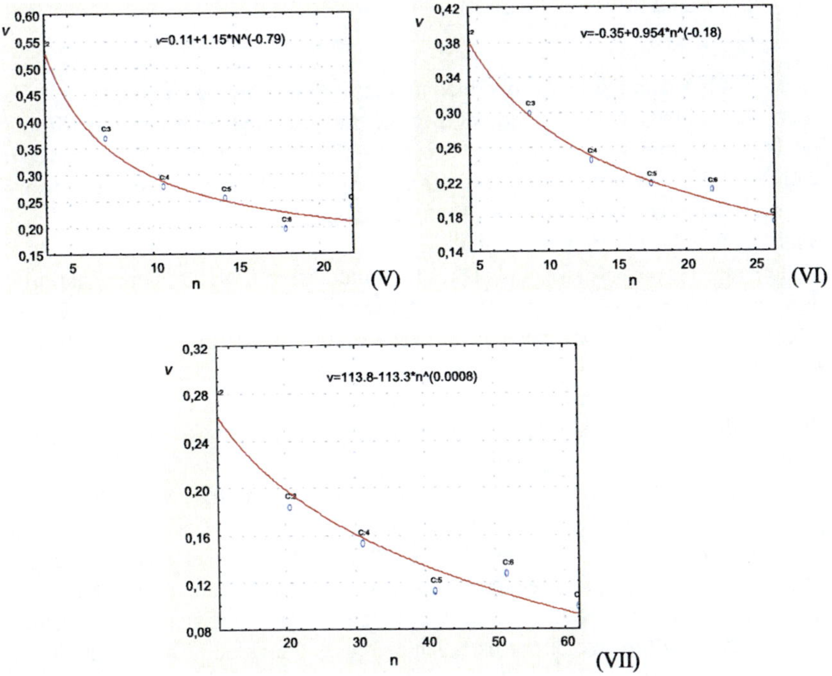

Fig. 2 The influence of the number of units of the control component n (pcs.) on the estimated value of the coefficient of variation ν (fraction): V—for the control component—lentils; VI—for the control component—barley; VII—for the control component—buckwheat

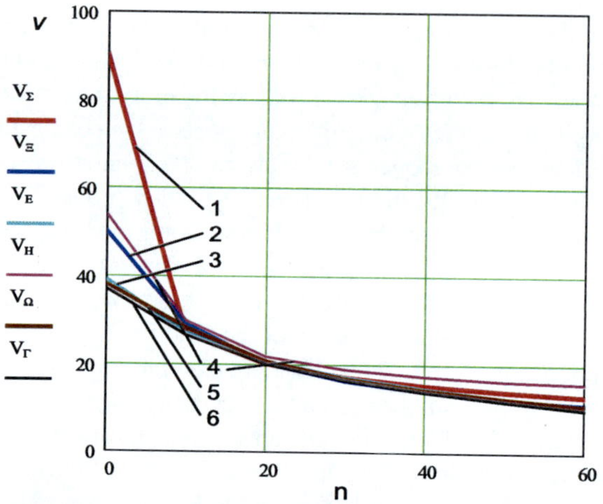

Fig. 3 The influence of the number of units of the control component n (pcs.) on the estimated value of the coefficient of variation ν (fraction): 1—$\upsilon_{\sum}$—for a mixture of control components, consisting of beans, corn, lentils, barley, buckwheat; 2—υ_{Ξ}—for a mixture of control components, consisting of lentils, barley, buckwheat; 3—υ_{E}—for a mixture of control components, consisting of barley and buckwheat; 4—υ_{H}—for a control component—lentils; 5—υ_{Ω}—for a control component—barley; 6—υ_{Γ}—for a control component—buckwheat

As a result of simulating the obtained models for comparable parts of the number of particles (Fig. 3) a practical coincidence of the numerical values of the different models in the range of 10…30 particles is seen. The best coincidence is observed in the range of 20…25 particles. Decreasing of particles significantly increases dispersion of variation coefficient values. An increase in the number of components leads to an increase in the variation of values.

An increase in particle size contributes to an increase in the coefficient of variation values. So for lentils with n = 60, the coefficient of variation corresponds to 15.5%, for barley to 10.7%, and for buckwheat to 9.9%. That is, the differences of the last two components are insignificant and comparable to the error of the experiments. At the same time the estimated number of particles in the samples taken is not less than 19 pcs. Reducing the calculated value to 14 pcs. (lentils) increases the variability of the index.

As a result of the simulation of obtained models on comparable parts of the particle number (Fig. 3) a practical coincidence of numerical values of different models in the interval of 10…30 particles can be seen. The best coincidence is near 20…25 particles. A decrease of particles significantly increases dispersion of variation coefficient values. An increase in number of components leads to increase of variation values. Increasing the size of the particles increases the value of the coefficient of variation. So for lentils with n = 60, the coefficient of variation corresponds to 15.5%, for barley—to 10.7%, for buckwheat—to 9.9%. That is, the differences of the last two components are insignificant and comparable to the error of the experiments. At

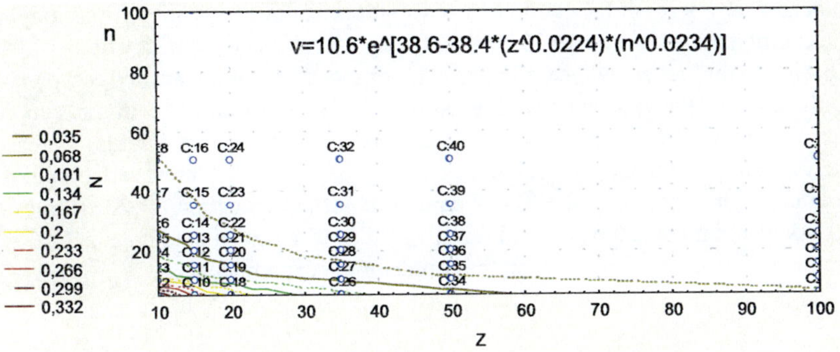

Fig. 4 The influence of the number of samples taken z (pcs.) and the estimated number of control component units n (pcs.) in the sample on the error value of the coefficient of variation (%)

the same time estimated number of particles in the samples taken is not less than 19 pcs. Reducing the calculated value to 14 pcs. (lentils) increases the variability of the index.

A regression model of the numerical modelling (simulation) to determine the error of the coefficient of variation if one grain is absent (present) in the sample (v, fraction or 0.01%; Fig. 4):

$$v = 1.06 \cdot e^{[38.6 - 38.4 \cdot z^{0.02235} \cdot n^{0.0234}]}.$$

Pearson correlation coefficient corresponds to R = 0.99977, the results of F-test = 0.978638, which indicates the adequacy of the model.

The error of the coefficient of variation of the fact of random failure of one grain for 20 samples of 20 calculation units is 0.045% and for 20 samples of 25 calculation units is 0.036%, which has no significant effect on the results of the experiments.

4 Conclusion

The laboratory studies of the mixture quality depending on the particle size of the control component, the number of particles of the control component in the sample, and the sample mass of the same mixture allowed to establish the degree character of the adequate regression dependencies. As a result of simulation of the obtained models for comparable parts of the particle number, we see practical coincidence of the numerical values of different models in the interval of 10…30 particles in the sample, with the highest coincidence of the models around 20…25 pcs. Decreasing the number of particles in the sample considerably increases the dispersion of the coefficient of variation values. Increasing the number of controlled components leads to an increase of the variation of obtained values. Increasing the size of the particles increases the values of the coefficient of variation. So for lentils at 60 (pcs./sample), the coefficient of variation is 15.5%, for barley—10.7% and for buckwheat—9.9%. That is, the differences in the two small components are insignificant and comparable to the error of the experiments. The estimated number of particles in the samples taken is at least 19 pcs. Reducing the calculated value to 14 pcs. (lentils) increases the variability of the index. The calculated error of the coefficient of variation of the coefficient of variation of the fact of random failure of one grain for 20 samples of 20 calculated units is 0.045% and for 20 samples of 25 calculated units is 0.036%, which has no significant effect on the results of the experiments.

References

1. Ravichandran M, Castorena C (2021) Critical evaluation of dynamic shear rheometer (DSR)-based methods to measure asphalt binder diffusion. J Test Eval 49. https://doi.org/10.1520/JTE20200432
2. Bahrani SA, Herbaut R, Royon L, Azzouz K, Bontemps A (2019) Experimental investigation of thermal and flow mixing enhancement induced by Rayleigh-like streaming in a milli-mixer. Therm Sci Eng Progr 14:100434. https://doi.org/10.1016/j.tsep.2019.100434
3. Orekhova TN, Prokopenko VS, Okushko VV, Kachaev AE (2019) The rotating layer mechanics of polydisperse particles in the continuous action pneumatic mixers. IOP Conf Ser Mater Sci Eng 698:022023. https://doi.org/10.1088/1757-899X/698/2/022023
4. Fuwad A, Hossain Sh, Ryu H et al (2020) Numerical and experimental study on mixing in chaotic micromixers with crossing structures. Chem Eng Technol 43(9):1866–1875. https://doi.org/10.1002/ceat.201900523
5. Wang B, Kronenburg A, Stein OT (2019) A new perspective on modelling passive scalar conditional mixing statistics in turbulent spray flames. Combust Flame 208:376–387. https://doi.org/10.1016/j.combustflame.2019.07.016
6. Emeljanova IA, Blazhko VV, Dansheva SO, Shyshko NS (2019) Simulation model for the collision of the individual components of dry mixes with the inner surface housing during their preparation in high-speed twin-rotor turbulent mixer. IOP Conf Ser Mater Sci Eng 708:012004. https://doi.org/10.1088/1757-899X/708/1/012004
7. Reznichenko IYu, Borodulin DM, Chistyakov AM, Ruban NYu, Komarov SS (2020) Determination of technological parameters for the production of enriched flour mixtures in a drum mixer. IOP Conf Ser Earth Environ Sci 613(1):012124. https://doi.org/10.1088/1755-1315/613/1/012124
8. Ulyanov V, Utolin V, Luzgin N, Borontova M et al (2019) Mixer for dry concentrated feed. IOP Conf Ser Earth Environ Sci 403:012143. https://doi.org/10.1088/1755-1315/403/1/012143
9. Omata R, Hattori Y, Sasaki T, Sakamoto T, Otsuka M (2020) Elucidation of the molecular mechanism of wet granulation for pharmaceutical standard formulations in a high-speed shear mixer using near-infrared spectroscopy. Pharmaceuticals 13(9):226 (1–16). https://doi.org/10.3390/ph13090226
10. Bratishko V, Milko D, Achkevych O, Kuzmenko V (2020) Results of experimental studies of process of preparation of feed mixtures with their moistening. In: 19th international scientific conference engineering for rural development, Jelgava, 20–22 May 2020. https://doi.org/10.22616/ERDev.2020.19.TF349
11. Cherpitskii SN, Tarshis MY, Korolev LV, Kapranova AB (2020) An investigation of the process of mixing of bulk materials in a drum-blade mixer. Chem Petrol Eng 56(1–2):3–10. https://doi.org/10.1007/s10556-020-00731-w
12. Harish VVN, Cho M, Shim J (2019) Effect of rotating cylinder on mixing performance in a cylindrical double-ribbon mixer. Appl Sci 9:5179. https://doi.org/10.3390/app9235179

13. Sirovatka V, Dorokhov A et al (2020) Study results of the on-board weight control system as exemplified by feed mixture preparation using a trailed feed mixer-and-distributor. E3S Web Conf 175:03006. https://doi.org/10.1051/e3sconf/202017503006
14. Samad PAA, Shalij PR (2020) Experimental investigation and CFD simulation of power consumption for mixing in Gyro Shaker. Int J Comput Sci Eng 21(2):188–201. https://doi.org/10.1504/IJCSE.2020.10027422
15. Savinyh P, Aleshkin A, Turbanov N, Ivanovs S (2020) Investigation of impact of technological and structural parameters upon energy indicators of work of mixer. In: 20th international scientific conference "engineering for rural development", Jelgava, 26–28 May 2021. https://doi.org/10.22616/erdev.2020.19.tf334
16. Shulaev NS, Laponov SV, Shulaeva TV, Bondar' KE (2020) Energy consumption of small-since mixers in emulsification processes. IOP Conf Ser Mater Sci Eng 862:062062. https://doi.org/10.1088/1757-899X/862/6/062062
17. Fudin K, Konovalov V, Zaitsev V, Teryushkov V (2021) A model of the influence of the degree of filling of the drum mixer on the quality of the feed mixture. IOP Conf Ser Earth Environ Sci 624:012067. https://doi.org/10.1088/1755-1315/624/1/012067
18. Barone A, Glassey J, Montague G (2019) Towards online near-infrared spectroscopy to optimise food product mixing. J Food Eng 263:227–236. https://doi.org/10.1109/ISED48680.2019.9096242
19. Dai X, Xu Y, Zheng J, Ma L, Song H (2020) Comparison of image-based methods for determining the inline mixing uniformity of pesticides in direct nozzle injection systems. Biosyst Eng 190:157–175. https://doi.org/10.1016/j.biosystemseng.2019.12.007
20. Choiri S, Sulaiman TNS, Rohman A (2020) Development and application of ATR-FTIR spectroscopy for mixture homogeneity in pharmaceutical application. Int J Pharm Res 12(4):7–16. https://doi.org/10.31838/ijpr/2020.12.04.003

Increasing the Efficiency of Calculation of Magnetic Fields of Magnetostriction Level Gauges

Edvard Karpukhin

Abstract The paper shows the results of the studies of applied magnetostrictive level gauges (AMRLG) in order to increase their efficiency. For the research proposed to use the methods of mathematical modeling and numerical methods. In particular, during the simulation of magnetic fields AMRLG iterative procedure Richardson has been involved and shown the best way to calculate the acceleration of the convergence of the parameters using the rotation method. Its efficiency by reducing the number of iterations of Richardson modified technique for solving systems of differential equations and magnetic fields AMRLG shown has been described in detail in the work. The proposed method of numerical simulation of magnetic fields can also be used to study other similar devices, which proves its versatility and effectiveness.

Keywords Mathematical modeling of magnetic fields · Numerical methods · Richardson's method · Method of rotation

1 Introduction

When they perform mathematical modeling of magnetic fields for applied magnetostrictive level gauges (AMRLG), there is the problem of solving the systems of finite-difference equations with a large number of unknowns. This problem was formulated and solved in the works of many authors, for example in [1, 2]. The determination of the sought magnetic potentials was carried out using numerical methods that require significant computational resources and time when they simulate an AMRLG with a complex geometry of the acoustic path for bypass systems.

However, the practice of solving the systems of equations, similar in structure to the finite-difference equations obtained for the AMRLG magnetic field, makes it possible to develop some ways to reduce the calculation time while maintaining and sometimes increasing the result accuracy [2]. In particular, a modification of the Richardson method with the calculation of the optimal set of parameters by the

E. Karpukhin (✉)
Penza State Technological University, Pr. Baidukova/Ul. Gagarina, 1a, 11, 440039 Penza, Russia

A. Beskopylny and M. Shamtsyan (eds.), *XIV International Scientific Conference "INTERAGROMASH 2021"*, Lecture Notes in Networks and Systems 247,
https://doi.org/10.1007/978-3-030-80946-1_10

rotation method gives a good gain in time and accuracy. Let's consider this technique in more detail.

As is known, the system of Maxwell's equations, describing, in particular, the distribution of AMRLG magnetic fields, can be approximated by a system of finite-difference equations of the form [3] via one method or another:

$$Au = b \tag{1}$$

where $A = \|a_{i,j}\|$—the matrix of the system coefficients, u—the column of unknowns (potentials), b—the column of the right parts.

Moreover, it is known about the matrix A that it is sparse and ill-conditioned, and the system of equations has a large number of unknowns, commensurate with the number of grid nodes [4].

It is known from various sources [3–6] that to solve the systems of difference equations of the form (1) in the computational domain with complex geometry, it is most expedient to use the Richardson method, the alternating triangular method, and various iterative methods of the variational type. Moreover, all of them, except for the first one, require a preliminary calculation of many parameters, while allowing solve the problem in the areas with a pronounced inhomogeneity of the media.

Within the framework of the problem being solved, the computational domain of the AMRLG type for bypass systems does not have significant heterogeneity, therefore, in this case, the use of the Richardson method will be effective. The main difficulty of this method use is the partial problem of finding the eigenvalues λ_{max} and λ_{min} of the array A. At that, when they solve the problem on a computer, it seems most expedient to use the rotation method to find the indicated eigenvalues [4].

2 Materials and Methods

The approach to an iterative Richardson scheme development is to study the behavior of the error $\delta = f(n)$. Such an analysis makes it possible to choose a parameter τ, considering the nature of the error δ^n change during the transition from the nth to (n + 1)th iteration, and

$$\delta^n = u^n - u^T \tag{2}$$

where u^T—the array of exact potential values.

Indeed, boundary conditions are specified at the computational domain boundaries, and their error is equal to zero. Therefore, inside the region, the function δ^n can be expanded in a Fourier series, which will have the following form:

$$\delta^n_{i,j} = \sum_{k,m} C_{km} \sin\frac{k\pi}{N} i \cdot \sin\frac{m\pi}{M} j \tag{3}$$

where the expansion coefficients C_{km} depend on the parameter τ the iteration number n [4]. The smaller the value of the coefficient C_{km}, the less influence the k, m harmonic makes on the total error δ^n.

Therefore, the choice of the optimal value τ_o should be carried out from the criterion of error harmonics best suppression in the middle part of the spectrum. We also take into account that the harmonic composition of the error δ^n can be changed from one iteration to another, and a new value τ_o^n should be chosen at each step for the maximum efficiency of the method [4].

The main advantage of the Richardson method is in the use of a set of optimal values τ_o^n. The slow convergence of other methods (simple iteration, Seidel, upper relaxation method) is explained by the fact that the low and high frequency harmonics of the error δ^n are suppressed at the same rate and the overall convergence of the method is determined only by the extreme boundaries of the error spectrum. The introduction of a set of optimal values τ_o^n ensures successive suppression of all error harmonics and its uniform rapid decrease during a small number of iterations.

Let's consider the ways to obtain a set of optimal values τ_o^n.

For this, assuming that τ^n depends on the iteration number, we put down the Richardson iteration scheme in the following form [4]:

$$u^{n+1} = \left(1 - \tau^n A + b\right) u^n \tag{4}$$

Due to the fact that a relation similar to (4) will be fulfilled for an array of exact potentials u^T and taking into account (2), it is possible to write the following for the error δ^n:

$$\delta^{n+1} = \left(1 - \tau^n A + b\right) \delta^n \tag{5}$$

Then, after denoting the initial error (for n = 0) via δ^0, we get an expression for the error after n_1 of iterations:

$$\delta^{n_1} = \delta^0 \prod_{n=0}^{n_1} \left(1 - \tau^n A + b\right) \tag{6}$$

Using (6), it can be shown that for the best suppression of the error with n_1 of iterations, the parameters τ^n should be selected based on the condition [4]:

$$\left\| \prod_{n=0}^{n_1} \left(1 - \tau^n A + b\right) \right\| = \min \tag{7}$$

In practice, the search for a set of parameters τ^n, minimizing the norm (7) is usually replaced by the search of $\tau^n \in \left[\lambda_{max}^{-1^{-1}_{min}}\right]$, at which the Chebyshev polynomials of the first kind of the degree n_1 take the values closest to zero. Then, as is known [4]:

$$\tau^n = 2\left(\lambda \cos \frac{(2n-1)\pi}{2n_1} \min \underset{1}{\overset{-1}{\max}} \underset{\max}{\min}\right) \tag{8}$$

3 Results

Calculated in accordance with (8), the first elements of the sequence τ^n have the order $1/\lambda_{max}$, and therefore, the error harmonics corresponding to the right side of the spectrum are most actively suppressed during the first iterations. The components of the left side of the spectrum are suppressed slowly with these iterations. However, they are actively suppressed by the higher order sequence elements τ^n, with the order $1/\lambda_{min}$, i.e., at $\tau^n \to \tau^{n_1}$. Thus, there is a significant uniform decrease of the error δ^n during n_1 of iterations [2].

When they solve practical problems on a computer by the Richardson method, they are usually set with a relatively small number $n_1 \sim 10 \ldots 50$, the parameters τ^{n_1} are calculated using the formula (7), and a series of n_1 iterations (4) with the same parameters τ^{n_1} until the accuracy criterion is met.

Richardson's method is characterized by a high convergence rate. It is known that when they use an optimal set of parameters τ_o^n the number of iterations n on a mesh with $N \times M$ of nodes depends on the specified accuracy as follows [4]:

$$n = 0.32NM \ln \frac{2}{\varepsilon} \tag{9}$$

Since the main difficulty during the implementation of this method is finding the eigenvalues λ_{max} and λ_{min} of the array A, we will further consider some approximate methods for their calculation, suitable for the arrays with the marked properties.

Currently, a large number of methods are known solving such problems. These are the methods of direct expansion, iteration, rotation, etc. [1, 4–8]. All these methods are classified into partial (allowing find only some, often arbitrary, eigenvalues of a matrix) and complete one (finding all eigenvalues). Since in the context of this problem it is necessary to find only the minimum and maximum eigenvalues, the use of partial methods is ineffective, due to the arbitrariness of the obtained eigenvalues. Therefore, we will search for all eigenvalues by the full method, and then select the maximum and minimum from them.

The most effective complete method used for large symmetric matrices is the rotation method [2]. The essence of this method is as follows.

The method is based on the transformation of the original symmetric matrix A similarity using the orthogonal matrix H. An orthogonal matrix is taken as a matrix H for the rotation method, such that $HH^T = H^T H = E$, where E is the single matrix.

Because of the orthogonality property of the similarity transformation, the original matrix A and the matrix $A^{(i)}$, obtained after the transformation retain their trace and eigenvalues λ_i, i.e. the following equality holds:

$$\mathrm{tr}A = \sum_i a_{ii} = A\sum_i \lambda_i = \mathrm{tr}A^{(i)} \tag{10}$$

The rotation method idea is in repeated application of a similarity transformation to the original matrix:

$$A^{(k+1)} = \left(H^{(k)}\right)^{-1} \cdot A^{(k)} \cdot H^{(k)} = \left(H^{(k)}\right)^{T} \cdot A^{(k)} \cdot H^{(k)}, \quad k = 0, 1, 2, \ldots \tag{11}$$

The formula (11) defines an iterative process, during which an orthogonal matrix $H^{(k)}$ is determined at the kth iteration, for an arbitrarily chosen off-diagonal element $a_{ij}^{(k)}, i \neq j$, that transforms this element and the element $a_{ji}^{(k)}$ into $a_{ij}^{(k+1)} = a_{ji}^{(k+1)} \approx 0$. In this case, at each iteration, the off-diagonal element $a_{ij}^{(k)}$ is selected among the maximum ones. At that, the matrix $H^{(k)}$ is called the Jacobi rotation matrix. It depends on the rotation angle $\phi^{(k)}$, determined from the following expression:

$$tg2\phi^{(k)} = \frac{2a_{ij}^{(k)}}{a_{ii}^{(k)} - a_{jj}^{(k)}} \tag{12}$$

At that $\left|2\phi^{(k)} < \frac{\pi}{2}\right|$, $i < j$, and it has the following form [3]:

$$H^{(k)} = \begin{pmatrix} 1\ldots 0 & 0 & 0\ldots 0 & 0 & 0\ldots 0 \\ \vdots\ddots\vdots & \vdots & \vdots\ddots\vdots & \vdots & \vdots\ddots\vdots \\ 0\ldots 1 & 0 & 0\ldots 0 & 0 & 0\ldots 0 \\ 0\ldots 0 & \cos\phi^{(k)} & 0\ldots 0 & -\sin\phi^{(k)} & 0\ldots 0 \\ 0\ldots 0 & 0 & 1\ldots 0 & 0 & 0\ldots 0 \\ \vdots\ddots\vdots & \vdots & \vdots\ddots\vdots & \vdots & \vdots\ddots\vdots \\ 0\ldots 0 & 0 & 0\ldots 1 & 0 & 0\ldots 0 \\ 0\ldots 0 & \sin\phi^{(k)} & 0\ldots 0 & \cos\phi^{(k)} & 0\ldots 0 \\ 0\ldots 0 & 0 & 0\ldots 0 & 0 & 1\ldots 0 \\ \vdots\ddots\vdots & \vdots & \vdots\ddots\vdots & \vdots & \vdots\ddots\vdots \\ 0\ldots 0 & 0 & 0\ldots 0 & 0 & 0\ldots 1 \end{pmatrix} \tag{13}$$

In the course of the iterative process (10) at $k \to \infty$ the moduli of all off-diagonal elements $a_{ij}^{(k)}$, $i \neq j$ tend to zero, and the matrix itself $A^{(k)} \to \mathrm{diag}(\lambda_1, \lambda_2, \ldots \lambda_n)$. The criterion for achieving the required accuracy of the rotation method is [4]:

$$\max_{i \neq j} \left| a_{ij}^{(k)} \right| < \varepsilon \tag{14}$$

4 Discussion

The implementation of the rotation algorithm for finding the eigenvalues of the system ratio matrix made it possible to evaluate the efficiency of the Richardson method with an optimal set of parameters τ_o^n, calculated by the formula (8). The experimental dependences of the number of iterations n on a given accuracy ε, obtained at various sets of parameters τ^n, are shown on Fig. 1.

As you can see, the number of iterations n with a set of parameters τ_o^n, calculated by the formula (8) using the rotation method (line 2) differs slightly from the ideal theoretical one (line 3) determined by the expression (9). With a random selection of a set of parameters τ^n (line 1), the number of iterations increases by 2–3 times.

Figure 2 shows the results of the magnetic field calculations of AMRLG, when YUNDK24B alloy is selected as a permanent magnet [9].

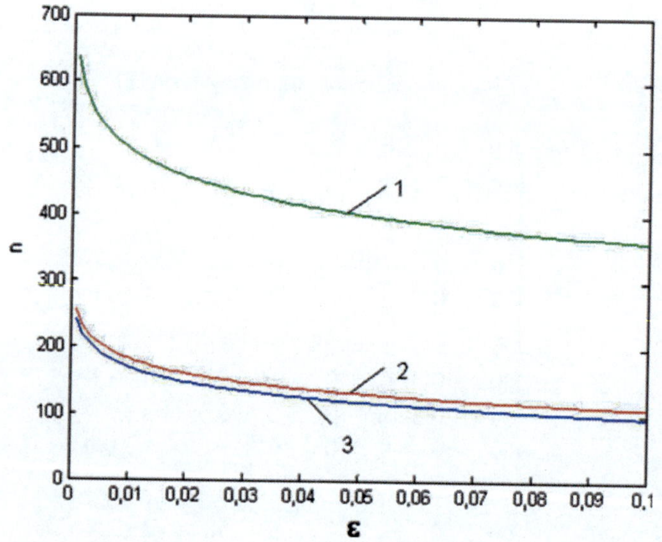

Fig. 1 Influence of a set of parameters τ^n on the number of iterations n according to the Richardson method

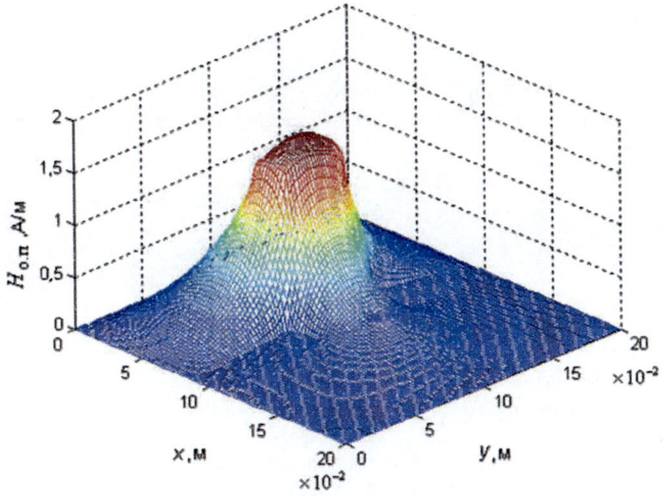

Fig. 2 The picture of the magnetic field strength of the applied AMRLG

Figure 3 shows the results of the magnetic field modeling of AMRLG using ELCUT system, when YUNDK24B alloy is selected as a permanent magnet [10–13].

So, the results of calculating the magnetic field of AMRLG by the proposed method and using the known modeling system turned out to be similar.

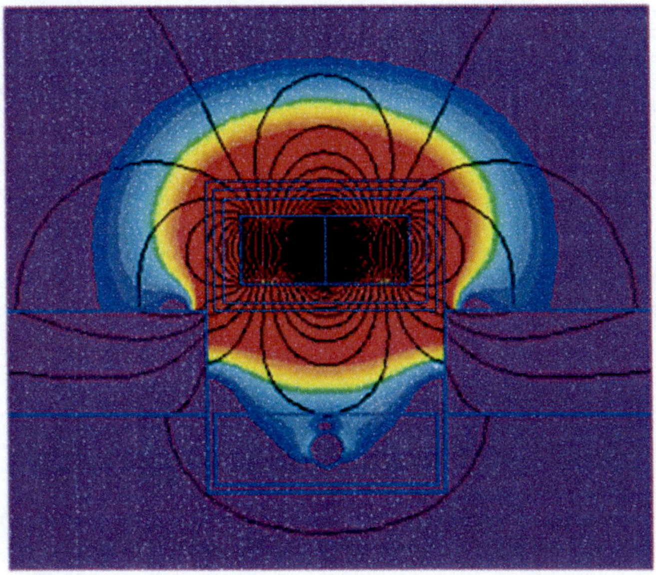

Fig. 3 The picture of the magnetic field strength of the applied AMRLG in ELCUT

5 Conclusion

Thus, the use of the developed methodology to calculate AMRLG magnetic fields for bypass systems reduces the number of iterations and the solution time, proving its effectiveness.

References

1. Demin SB (2002) Magnetostrictive systems for the automation of technological equipment: monograph. IIC PSU, Penza
2. Karpukhin EV, Mokrousov DA, Demin ES, Demin SB (2014) Research bypass measuring system with magnetostrictive converter of a level by method mathematical modeling. Mod Probl Sci Educ Surg 4
3. Karpukhin EV, Demin SB (2014) Mathematical modeling magnetic fields of applied magnetostrictive level gauges: monograph. PenzSTU, Penza
4. Demirchyan KS, Neiman LR, Korovkin NV (2009) Theoretical basis of electrical engineering. Peter, Saint Petersburg
5. Vasileva AB, Tihonov NA (2021) Integral equations. Lan', Moscow
6. Demidovich BP, Modenov VP (2021) Differential equations. Lan', Moscow
7. Neoh SS, Ismail F (2021) Time-explicit numerical methods for Maxwell's equation in second-order form. Appl Math Comput 392
8. Angelis PD, Marchis RD, Martire AL (2020) A new numerical method for a class of Volterra and Fredholm integral equations. J Comput Appl Math 379
9. Bakhvalov NS, Zhidkov NP, Kobelkov GM (2003) Numerical methods. Binom, Moscow
10. Alekseev ER (2006) Solving computational mathematics problems in packages Mathcad 12, MATLAB 7, Maple 9. NT Press, Moscow
11. Andreeva EG, Shamec SP, Kolmogorov DV (2002) Finite element analysis of stationary magnetic fields using software package ANSYS. OmSTU, Omsk
12. Dubickij S, Podnos V (2001) ELCUT—engineering modeling system for 2D physical fields. CADMaster Surg 1
13. Karpukhin EV, Mokrousov DA, Djatkov VS, Demin SB (2014) Complex of programs for calculation of parameters plated magnetostrictive converters of the level with complex geometry of the acoustic path. Mod Probl Sci Educ Surg 3

Specific Features of Surface Morphology During Plasma Electrolytic Processing

Alexandr Popov, Vitaly Novikov, Alexander Scherbakov, Oleg Kuzmin, and Alla Popova

Abstract This article presents the features of the morphology of the surface layer of stainless, tool steels, copper in different types of electrolytic-plasma treatment. It is shown that the change in the surface layer of materials depends on the type of processing, the electrode-tool used, the initial parameter of the surface layer roughness, the electrolyte flow rate and the quality of surface preparation. The study uses different electrode tools to obtain a product with high precision and low surface roughness. The effect of plasma electrolytic treatment on the parameters of sample roughness and weight loss of samples made of stainless steels, tool steels and copper grade was studied. The minimum depth of material removal is fixed when using different technological modes. The reasons for the appearance of microcraters in the surface during electrolytic-plasma treatment are analyzed. The results of the work will make it possible to use the knowledge gained for finishing and roughing plasma electrolyte treatment.

Keywords Electrolytic jet · Electrolytic bath · Finishing · Roughing · Hollow cathode · Ganged current lead · Plasma electrolyte treatment

1 Introduction

Responsible mechanical engineering products, as a rule, have significant limitations in accuracy, the Ra parameter of the surface roughness, and the quality of the surface layer. These requirements are a tight tolerance corresponding to 5–6 grade of accuracy, a surface roughness parameter of less than $0.1 \leq$ Ra, the presence of defects in the surface layer of less than 10 μm. Obtaining products with these requirements is a complex and time-consuming technological task. Achievement of the specified

A. Popov (✉) · A. Popova
Peter the Great St. Petersburg Polytechnic University, Polytehnicheskaya Street, 29, Saint-Petersburg 195251, Russia

V. Novikov · A. Scherbakov · O. Kuzmin
Saint Petersburg State University of Architecture and Civil Engineering, Vtoraya Krasnoarmeiskaya Street, 4, Saint-Petersburg 190005, Russia

A. Beskopylny and M. Shamtsyan (eds.), *XIV International Scientific Conference "INTERAGROMASH 2021"*, Lecture Notes in Networks and Systems 247,
https://doi.org/10.1007/978-3-030-80946-1_11

results is possible using the operation of finishing surface polishing using manual processing on polishing machines, using felt wheels using glued abrasive grains [1]. This method has a number of disadvantages associated with the unstable quality of grinding wheels, as well as insufficient stability of the dimensions performed using manual labor.

New surface finishing methods are replacing the traditional methods of surface finishing. These are methods of plasma electrolytic treatment by immersing the product in an electrolytic bath [2–4] or surface treatment with one or more electrolyte jets [5, 6]. The method of jet plasma electrolyte treatment includes processing the surface of a part (anode) with an electrolyte jet supplied from a nozzle connected to a negative potential [7]. The movement of the cathode is carried out using the program. It can have from two to five coordinate axes. The features of the method of jet plasma electrolyte treatment include low energy consumption, low processing temperatures, and a local processing zone [8, 9]. This method achieves high precision and low roughness of the processed surface. This method allows the polishing of complex surfaces such as dies, molds, turbine blades and other products made from various conductive materials. Undoubtedly, the method of jet plasma electrolytic treatment is applicable for accurate high-quality surface treatment. However, in the literature, the issues are insufficiently covered by issues related to the technological accuracy of the method, the minimum dimensions of the depth of the removed layer and the features of the formation of the morphology of the surface layer. Therefore, the purpose of our study is to study the effect of plasma electrolyte treatment on the accuracy, roughness and features of the formation of surface morphology. The tasks of our work are to study the parameter Ra of the surface roughness under various processing modes, to determine the amount of material loss, to establish the minimum value of the removed layer from the surface of the initial material, to study the features of the formation of surface morphology.

2 Materials and Experimental Setup

Experiments on plasma jet electrolytic treatment of samples were carried out on a pilot plant (Fig. 1). The installation consists of a system for precise positioning of the electrolyte supply system and a DC power supply unit made according to Larionov's scheme.

The voltage between the electrodes was set equal to 20–500 V. The nozzle speed was determined by the ratio of the distance traveled over the time measured by the stopwatch .The morphology of the samples was analyzed using an LV-41 metallographic microscope (Lomo, Russia), a Solver P-47 Pro scanning probe microscope (Russia), and a Zeiss Supra 55/55VP analytical field emission electron microscope (Germany). The depth of removal of the processed material was measured with a digital electronic indicator "Micron" with a graduation of 1 μm. The weight of the samples was measured on a BM213M laboratory balance with a graduation of 0.001 mg. To reduce the average measurement error, each plate was weighed at

a) b)

Fig. 1 Schematic diagrams of plasma electrolyte surface treatment: **a** 1—current lead holder, 2—hollow current lead, 3—electrolytic cathode, 4—sample, 5—electrolyte receiving bath, 6—electrolyte supply pipeline, 7—insulated supports, 8—DC power supply current; **b** ganged current lead

least 7 times before and after processing. The studies used a hollow current lead and a type-setting current lead, a magnetron spray head and an electrolytic bath. The electrolytes used were solutions of ammonium sulfate ((NH_4) $2SO_4$)—10 g and ammonium citrate ((NH_4) $3C_6H_5O_7$)—10 g in tap water, ammonium sulfate ((NH_4) $2SO_4$) and sodium sulfate (Na_2SO_4)—30 g in tap water. The electrolyte was fed in a stream through a copper tube with an inner diameter of 3 mm with a volume flow rate of 4.6–9.0 L/h. The measurement of the volumetric flow rate of the electrolyte was carried out by passing the electrolyte weighing 100 g with fixing the transit time using a stopwatch. The experiment was carried out at atmospheric pressure on marked 25 samples with dimensions 45 × 50 mm, 100 × 200 mm made of stainless steel 08Kh18N9T, AISI 304, corrosion-resistant high-temperature steel 20Kh13, alloy tool steel KhVG, chromium-silicon-manganese steel 30KhGSA and copper M1.

3 Experiment Results

3.1 Determination of Surface Roughness Parameters

To obtain the minimum values of the parameters Ra of the surface roughness of the samples in the range of operating voltages from 20 to 500 V, samples from different materials were used. In the experiment, a fixed ganged current lead was used. The minimum values of the parameter Ra of the surface roughness of the samples were obtained by polishing AISI 304 stainless steel (Fig. 2). It is shown that the value curve is not monotonic. In the voltage range from 220 to 380 V, a sharp decrease in the

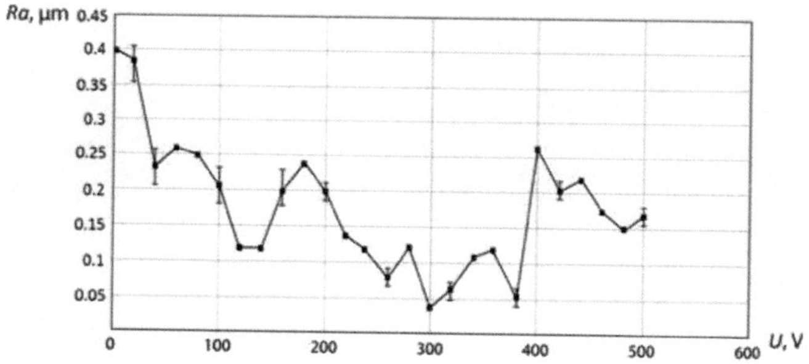

Fig. 2 Dependence of the roughness parameter on voltage during plasma electrolytic treatment of AISI 304 steel using a ganged current lead

roughness parameter to Ra 0.034 µm is observed. Obviously, this type of processing significantly affects the macro- and micro-reliefs of the surface.

With an increase in the operating voltage with a step of 20 V, a transition from a rough matte surface to a polished surface with a characteristic mirror-like shine is traced (Fig. 3). Different processing modes to varying degrees contribute to the removal of scratches and a decrease in the number and depth of depressions (holes) of the surface.

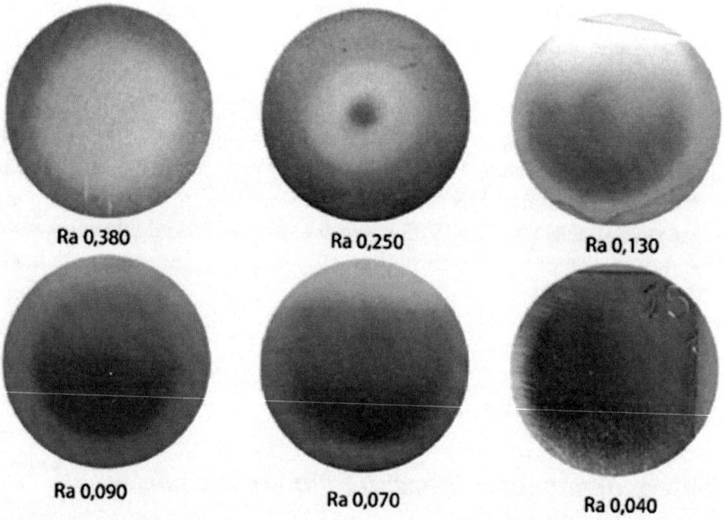

Fig. 3 Appearance of the surface of AISI 304 steel samples during plasma jet electrolytic treatment using a ganged current lead

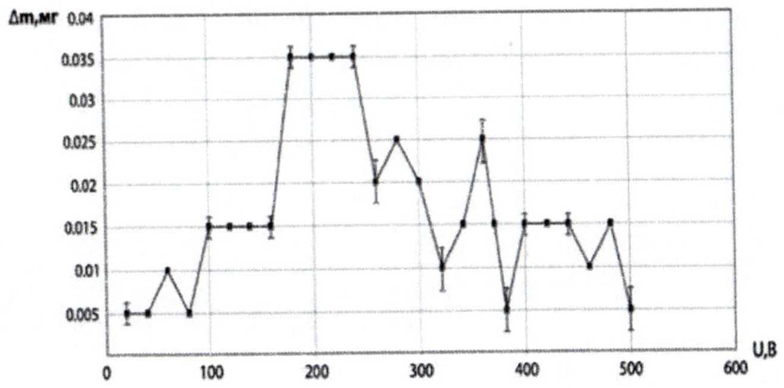

Fig. 4 Dependence of the sample weight loss on voltage during plasma electrolytic treatment of KhVG steel using a hollow current lead

3.2 *Determination of Material Loss*

The influence of jet electrolyte treatment on the value of the sample weight loss is most indicative for the KhVG steel when using a hollow current lead in the voltage range from 20 to 500 V (Fig. 4). Experiments show that the range from 160 to 300 V is characterized by increased metal removal in the electric discharge zone. This confirms the possibility of rough treatment with an electrolytic jet plasma in this voltage range. Also characteristic is the extreme value of material loss from 340 to 370 V. These extremes of material loss were also recorded for stainless steels at 340–460 V.

3.3 *Formation of a Polished Surface with a Minimum Cut-Off Depth*

However, for precision machining, a low Ra surface roughness is required. This is possible by removing the surface layer located between the maximum height of the tops and the minimum depth of the valleys of the microprofile. If this condition is met, a flat surface can be formed. Experiments show that the use of combined processing with the sequential use of ganged and hollow current leads allows you to effectively remove the tops of the micro-profile to a flat metal surface with subsequent polishing of the surface. The parameters of the electrolytic-plasma treatment of steel 20Kh13 using a hollow current lead are recorded in Table 1.

The parameters of the minimum depth of material removal using hollow and ganged current leads are fixed. The electrolyte solution in tap water contains ammonium sulfate and sodium sulfate. Research data are recorded in Table 2.

Based on the experimental results obtained, it can be concluded that the value of the minimum removal depth from the metal surface was 0.016 mm with a minimum surface roughness parameter Ra 0.083 μm.

Table 1 Depth of material removal from the sample surface using a hollow current lead

Length of the processed surface l, mm	Current I, A	Voltage U, V	Feed rate S, mm/min	Volume flow rate of electrolyte, l/h	Number of nozzle passes n, pcs	Depth of material removal from the sample surface, mm
50	1.02	225	34.1	9.0	1	0.112
50	0.71	230	17.1	9.0	1	0.130
50	0.65	230	17.1	6.1	1	0.058
50	0.40	250	17.1	4.6	1	0.146
50	1.42	250	17.1	4.6	1	0.205
58	0.33	260	17.1	4.6	1	0.144
58	0.33	260	17.1	4.6	3	0.092

Table 2 Depth of material removal from the sample surface with sequential use of ganged and hollow current leads

Processing area, cm^2	Current I, A	Voltage U, V	Volume flow rate of electrolyte, l/h	Number of passes by ganged/hollow current lead n, pcs	Depth of material removal from the sample surface, mm	Parameter Ra of surface roughness, μm
36	1.2–1.5	250–260	4–6	2/1 (3)	0.020	0.062
12	0.8–1.1	270	5–6	3/1 (4)	0.031	0.068
4	1.1	260	5	2/1 (3)	0.035	0.067
4	1.1	260	5	3/2 (5)	0.045	0.036
48	1.1	260	5	1	0.016	0.083

3.4 *Analysis of the Morphology of the Surface Layer*

The morphology of the surface of the part during plasma electrolytic treatment is of great importance. It determines the quality of the surface layer of the working surfaces of parts, has a significant effect on the service life of units and elements of any mechanisms. Under cyclic loads, special requirements are imposed on the defectiveness of the surface layers of parts. The presence of stress concentrators leads to the development of microcracks and the subsequent destruction of parts.

Only in rare cases does a defective surface relief play a positive role, such as when honing the inner surface of "cylinders". In most other cases, it leads to component failure.

We carried out research using a Solver P-47 Pro scanning probe microscope with a resolution of 5 nm. They showed that during plasma electrolytic treatment in a bath, not only the quantitative indicators of surface roughness change, but also the nature of the roughness changes. This combination defines the term "surface roughness". This means a change in the surface morphology of the treatment object. However, treatment in a bath is not always possible to completely smooth out the existing surface relief (Fig. 5). Depending on the size and shape of the part, significant surface defects can remain. They can be classified into three groups, traces of previous processing, areas of polishing, areas of selective explosive transfer of structural elements into the electrolyte.

The explosive nature of the transfer of structural elements into the electrolyte is also observed for the plasma jet electrolyte treatment.

This is due to several main factors, the initial level of surface roughness, the presence of surface defects, plastic deformation of the surface layers, and the presence of non-conductive surface contamination. The high initial roughness parameter of the surface layer Ra << 0.8 μm complicates the formation of a smooth surface with high reflectivity due to the comparable value of the plasma electrolyte layer, equal to 50–100 μm. Surface defects in the form of microcraters, deep scratches, areas of plastic deformation require long-term surface treatment or the use of additional technical solutions. Surface contamination leads to the formation of a non-conductive layer in the local area.

As a rule, these are areas in deep depressions of the surface contaminated with non-conductive oxide or grease films or products of material erosion. The presence of such areas leads to the accumulation of a surface charge of the opposite sign and the explosive development of a discharge in the form of spark discharges of higher

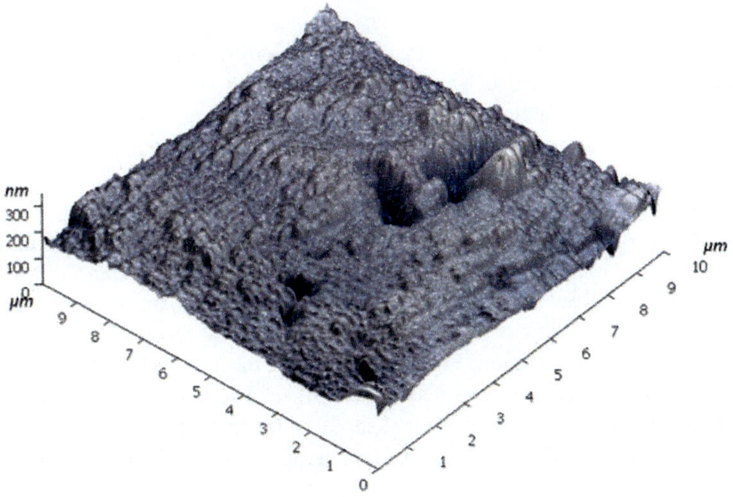

Fig. 5 Surface morphology of steel 30HGSA after plasma electrolytic treatment in a bath

power. This leads to the destruction of local surface zones mainly in the extended zone of deep grooves. In general, a change in the state of the surface layer occurs due to the intense formation of micro-discharges at the tops of microroughnesses.

In this case, when the electric field strength reaches, according to various estimates, 10^4–10^6 V/m, the discharge flows down to the top of the microroughness [10]. These microroughnesses can be formed as a group of grains, dislocations, or individual atoms. The removal of groups of atoms or individual atoms occurs, as a rule, already for an atomically aligned surface. Each type of current lead has its own peculiarity in the formation of the surface morphology. It is noted that during treatment with an electrolytic jet cathode using a hollow current lead, the formation of streamers is observed during the formation of a multichannel discharge. This may indicate the presence of uneven distribution of the electromagnetic field over the cross section of the electrolytic jet. Perhaps due to this, the localization of the maximum intensity of the electromagnetic field is present at the top of microroughnesses. The area of grooves on the surface has a lower potential, and in the presence of fatty films, zero or negative. This leads to the fact that when the charge of the opposite sign is accumulated, a discharge occurs between the non-conducting local zone and the zone with a high electric field strength. This leads to the erosion of microvolumes between the tops of microroughnesses (Fig. 6).

The ganged current lead has a higher uniformity of the field strength distribution in the area located directly under the current lead. Due to this, we have a field strength between the pointed peaks of microroughnesses and the opposite peaks of the ganged current lead. This facilitates the acceleration of the anodic processes of transfer of surface atoms into the solution and a higher rate of removal. The surface obtained with the use of a ganged current lead has a more uniform relief, the least roughness parameter and the minimum number of defects associated with the removal of microvolumes from the space between the vertices.

Mechanism of surface morphology change is well illustrated in Fig. 6. For the initial surface without processing, the tops of microroughnesses, blocks, foreign inclusions are clearly recorded. The presence of deep grooves (Fig. 6a) on the surface leads to additional formation of local depressions in them. Characteristic for the formation of surface morphology is the process of processing deep defects in the form of pits, obtained due to plastic deformation of the metal or having another origin. In the case of a shallow defect on the surface during processing, the radius also increases when passing to the main plane, with a smooth transition to the main surface and the complete disappearance of the microdefect. Another characteristic feature is the predominant influence of the chemical mechanism during the plasma electrolyte treatment. This is observed at high volumetric electrolyte flow rates $\geq$ 8 l/h and high salt concentration (Fig. 6i). In this case, the microrelief is removed with the identification of grain boundaries on the surface. The features of the magnetron spray head are as follows. This is the presence of a magnetic trap that traps electrons due to the magnetic field and makes them rotate due to the application of an electric field.

A significant number of collisions with atoms of the vapor–gas shell increases the percentage of ionization and increases the rate of processes on the surface. As

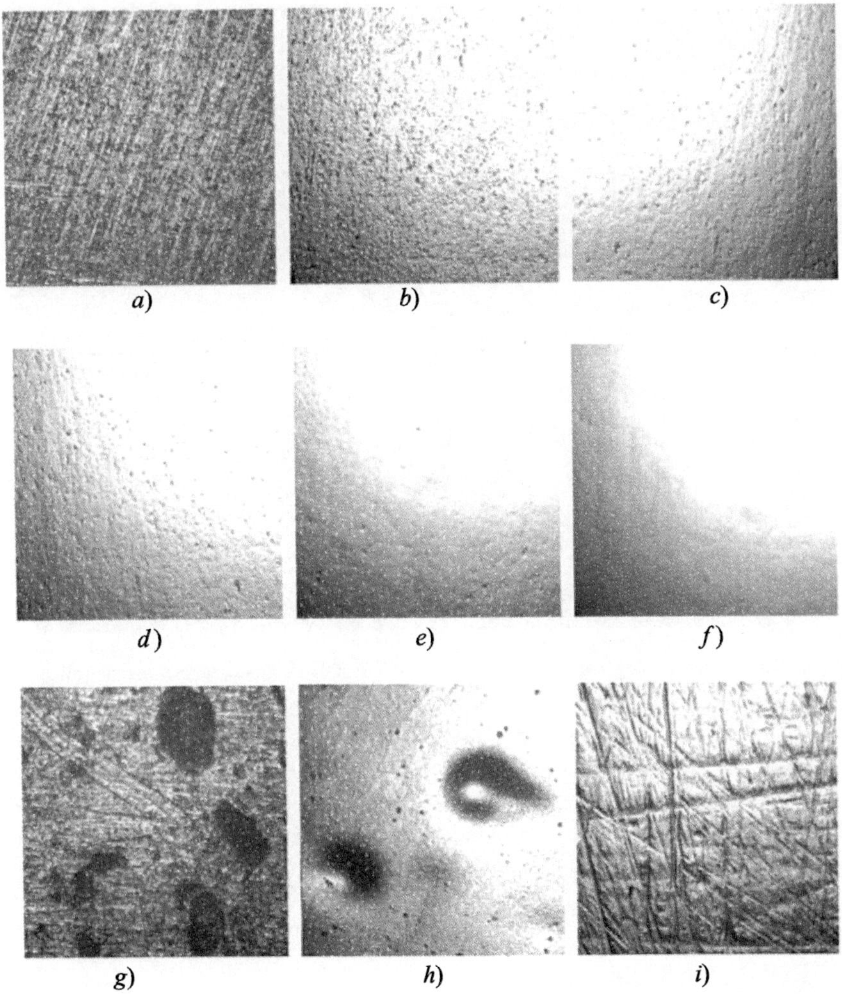

Fig. 6 Changes in the surface morphology of samples in the course of plasma electrolytic treatment when using sequentially ganged and hollow current leads: **a–f** AISI 304 steel; **g, h** steel 08Kh18N9T; **i** copper M1

a result, this type of processing has a higher reflectivity. The mechanism of surface morphology change when using a magnetron spray head when processing 20Kh13 steel after sawing with a band saw is shown (Fig. 7). Evaluation of the surface morphology using a magnetron spray head shows that in the process of removing microrelief, the tops of microroughnesses are smoothed out. The treatment is carried out evenly over the surface of the ring formed by the magnetic field of the magnetron spray head. In this case, the appearance of microdefects < 10 μm in size between the depressions of the microrelief is also recorded. Due to the fact that the initial

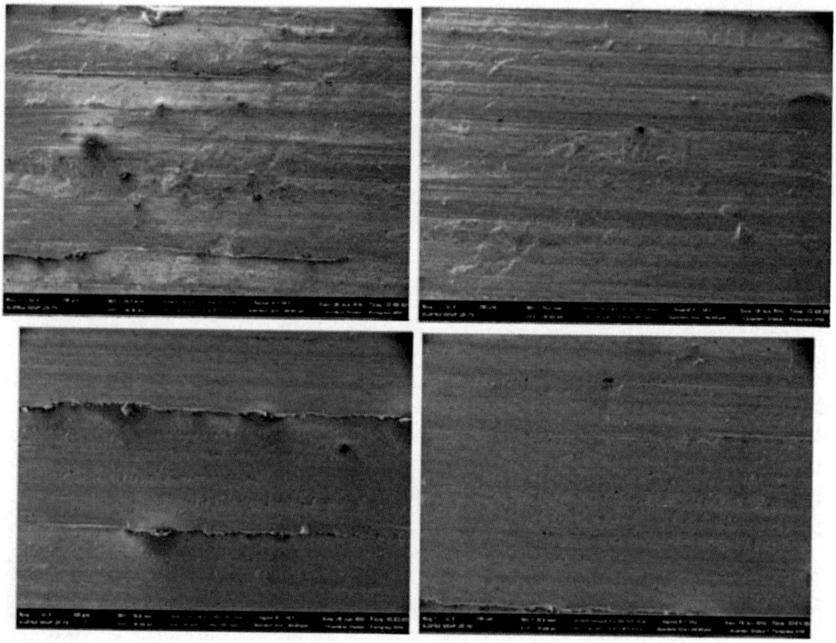

Fig. 7 Surface morphology of 20Kh13 steel specimens after treatment with a magnetron spray head

surface of the samples is very developed, a sufficient period is necessary for the final smoothing of the surface. In this case, a layer up to 0.6–0.8 mm thick is removed from the surface. The appearance of local zones of material discontinuity associated with the initial plastic flow of the material during sawing is noted.

4 Results and Discussion

As a result of numerous experiments carried out to study the formation of surface morphology, one can come to the following conclusions regarding the depth of plasma electrolyte action on the surface of samples from different materials. The minimum processing depth primarily depends on the tasks at hand. On the example of copper samples M1 it is shown that the effect of surface polishing with high reflective properties is manifested at a sufficiently developed surface roughness. This may indicate the removal of a thin layer from the surface with a thickness of less than 0.01 μm. For the depth of material removal from a developed surface with a roughness parameter Ra $\geq$ 0.8 μm, it is necessary to remove a technological allowance equal to 15–20 μm from the surface. In this case, we can obtain a surface without a defective layer with a roughness parameter Ra $\geq$ 0.1 μm. This is required in order to level the surface to the bottom of the depressions of the roughness, and then to the bottom of the microcraters formed during the initial surface treatment operations. An important point in the study of surface morphology is the physical essence of the phenomenon of removal of surface layers with a developed surface topography. It is significant that the removal of the surface layer occurs both at the tops of microroughnesses and in the depressions between two tops. Removal of metal from the surface of sharp tops of microroughnesses can be explained by a high level of electromagnetic field intensity. The processes occurring at sharp peaks are similar to physical phenomena observed in the form of spark discharges at pointed ends. The formation of microcraters between the tops of microroughnesses, shown above, is much more complex. Many authors have attempted to describe the polishing mechanism. However, there is still no understanding why there is the formation of numerous microcraters between the tops of microroughnesses and why this process stops after removing the vertices and smoothing the surface. Our hypothesis for this process is as follows. Numerous microdischarges are formed on the sample surface at the tops of microroughnesses and much less in the depressions between the tops. Smoothing of the tops of microroughnesses occurs with a smooth increase in the radii of their vertices. The formation of microcraters occurs simultaneously with the spraying of the volume of metal from the microcrater into the electrolyte. The surface of the depressions well retains various contaminants in the form of fatty films, oxide films, small dust grains, particles of the material of the tops, which are deposited during the plasma electrolyte treatment.

It is these elements of the system, being in the space between the vertices, are charged with respect to the treated surface with a negative electric charge. The growth of the potential at a local place of the surface determines the high intensity of the electric field between the non-conductive region and the main surface. The accumulation of charge in the local non-conductive region occurs by transfer of electric charge from the plasma electrolyte layer. When the threshold value of the breakdown voltage is reached, a spark discharge occurs between the main surface and the zone of opposite potential. This leads to erosion and the formation of microcraters in the space between the peaks. With a decrease in the conductivity of the electrolyte-plasma layer, an increase in the power and size of microcraters in the surface is also observed. In the general case, it can be assumed that an increase in the area and thickness of the zone contributes to the power of the spark discharge and its duration. It was revealed that after preliminary treatment with a type-setting current lead in the voltage range corresponding to the electrolysis mode, a surface is created containing a minimum number of non-conducting zones. This makes the plasma electrolytic polishing process more uniform. As a result, we have a higher quality surface with a lower Ra of the roughness and less machining time.

Acknowledgements The authors would like to thank graphic designer Diana Popova for preparing illustrations.

References

1. Nazarov NG, Rudenko NR, Sinavchian SN (2017) Modeling of the process flow of the abrasive slurry in the finishing process. Mach Build Sci Tech J 6:18–23
2. Kulikov IS, Vashenko SV, Kamenev AY (2010) Electrolytic plasma processing of materials. Belarusian Science, Minsk, p 232
3. D'yakov IG, Belkin VS, Shadrin SY et al (2014) Peculiarities of heat transfer at anodic plasma electrolytic treatment of cylindrical pieces. Surf Eng Appl Electrochem 50:346–355. https://doi.org/10.3103/S1068375514040061
4. Danilov I, Hackert-Oschätzchen M, Zinecker M, Schubert A, Meichsner G, Edelmann J (2019) Process understanding of plasma electrolytic polishing through multiphysics simulation and inline metrology. Micromachines 10(3):214
5. Fortov VE (2008) Encyclopedia of low-temperature plasma. High Temp 46:1–2. https://doi.org/10.1134/S0018151X0801001X
6. Barinov YA, Shkol'nik SM (2016) Discharge with a liquid nonmetallic cathode (tap water) in atmospheric-pressure air flow. Tech Phys 61(11):1760–1763. https://doi.org/10.21883/jtf.2016.11.43833.1833
7. Popov AI, Novikov VI, Radkevich MM (2019) Characteristics of the development of electric discharge between the jet electrolyte cathode and the metal anode at atmospheric pressure. High Temp 57(4):483–495
8. Popov AI, Radkevich MM, Teplukhin VG (2020) Thinnest finishing treatment with a focused jet of electrolytic plasma. In: Lecture notes in mechanical engineering, pp 139–149
9. Popov AI, Tyukhtyaev MI, Radkevich MM, Novikov VI (2016) The analysis of thermal phenomena occurring under jet focused electrolytic plasma processing. St. Petersburg Polytech Univ J Eng Sci Technol 4(254):141
10. Aliakseyeu YG, Korolyov AY, Parshuto AE, Niss VS (2017) Electrolyte-plasma treatment under non-stationary mode in a high-gradient electric field. Sci Tech 16:391–399. https://doi.org/10.21122/2227-1031-2017-16-5-391-399

LU-Factorization of the System of Magnetic Field Equations of Magnetostrictive Level Transducers

Edvard Karpukhin

Abstract In this paper, we consider the problem of calculating the magnetic fields of magnetostrictive transducers (MPs), in particular magnetostrictive level converters, using numerical methods using SDN-networks. Such calculations lead to the need for a numerical solution of systems of linear algebraic equations with a large number of unknowns. In this case, it is possible to significantly simplify and improve the efficiency of the solution by preliminary LU-factorization of the coefficient matrix of the system. The existing LU-factorization algorithms have a number of shortcomings and can not be effectively used to calculate the magnetic field of an MP. The article describes the application of the modified LU-factorization algorithm, designed to work in the SDN-network and shows its effectiveness. The proposed algorithm is characterized by simplicity, versatility and can be easily modified for the study of other measuring devices, where information transfer is carried out by means of magnetic field parameters.

Keywords Numerical methods · LU-factorization · Magnetic fields

1 Introduction

As you know, [1–4], when they calculate the magnetic fields of magnetostrictive devices (MD), including magnetostrictive level transducers, there is the problem of numerical solution of linear algebraic equation systems (SLAE) with a large number of unknowns.

A very significant way to improve the efficiency of all numerical methods to solve SLAE is the preconditioning of the system ratio matrix.

One of the simplest and most common preconditioning methods is matrix factorization or LU-factorization [3]. However, this algorithm is poorly suited for working with sparse matrices in the classical version, which is the ratio matrix of the system

E. Karpukhin (✉)
Penza State Technological University, pr. Baidukova/ul. Gagarina, 1a, 11, 440039 Penza, Russia

A. Beskopylny and M. Shamtsyan (eds.), *XIV International Scientific Conference “INTERAGROMASH 2021”*, Lecture Notes in Networks and Systems 247,
https://doi.org/10.1007/978-3-030-80946-1_12

of equations of the MD magnetic field. Therefore, its application requires its modification, which frees the method from this drawback. Another approach to efficiency improvement is the SDN architecture use.

The concept of software defined IT infrastructure was formulated at the end of the last decade as a logical development of virtualization technologies. The key idea is to separate the logical layer from physical devices as much as possible and automate the programming of network logic. While the traditional approach usually involves many different devices (switches, routers, firewalls, etc.), each of which needs to be configured separately, in the case of SDN, managing the network infrastructure is reduced to the centralized development of rules and policy templates that determine the processing logic and traffic transmission for the entire network [5]. The application of this concept to reorganize the existing computer network facilitates its configuration and allows a significant increase in performance, which makes it possible to implement LU-factorization algorithms based on conventional computers of average power.

In this paper, we will consider the modified LU-factorization algorithm using the rotation method implemented in the computer network of the SDN architecture and will estimate its efficiency.

2 Materials and Methods

As the studied scheme for modeling and implementing the LU-factorization algorithm, we will choose the MD, the structural diagram of which is shown on Fig. 1 and will consider its horizontal section at the height h_x [4].

Any material with magnetic permeability $\mu \approx 1$ can be used as the material of the non-magnetic wall of the tank 1, for example, the non-magnetic steel N24X2 or the alloy K40NXM, coated with a polymer to reduce the corrosive effect of the aggressive environment 3.

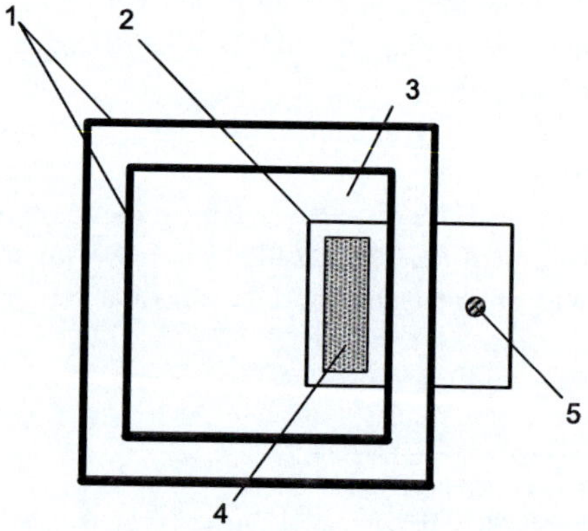

Fig. 1 Structural diagram of the MD. 1—The reservoir with a non-magnetic wall S, 2—the float with a permanent magnet 4, 3—a controlled aggressive environment, 5—magnetostrictive sound conduit

The body of the float 4 with a permanent magnet M, as a rule, is made in the form of a parallelepiped or a cylinder from the same steel grades as the tank 1. When modeling, we will consider the following dimensions of the float 4: length—0.075–0.127 m, width—0.075–0, 09 m, height—0.05–0.015 m. The permanent magnet of the float 4 can be made of the cast alloy YUNDK, since modern industry produces the magnets of various shapes and sizes from this alloy. This will allow to select a real magnet of the required size and shape with a coercive force $H_c = 40–200$ kA/m based on the simulation results.

Focusing on the parameters of known materials designed for use in the MD, we will develop a mathematical model of the permanent magnet magnetic field for the float 4 of the level gauge.

It is known [6–10] that the functioning of such MD occurs through the interaction of magnetic fields, the calculation of which is reduced to the SLAE solution by numerical methods.

Let's consider a SLAE with a basic matrix of coefficients A. In the simplest case, the matrix A is represented as the following product:

$$A = L_A U_A \tag{1}$$

where L_A and U_A are some lower and upper triangular matrices, respectively.

However, as is known [2], such a decomposition leads to the filling of the portrait; i.e., to the appearance of a significant number of nonzero elements in the matrices L_A and U_A which complicates the calculation and leads to a sharp increase of computer memory amount required for storing matrices.

Preservation of the sparsity of factors can be achieved by modifying the factorization algorithm. The existing factorization algorithms to solve the SLAE of the magnetic field of most MDs often turn out to be ineffective and require modification due to the peculiarities of the latter. Let's consider the possible ways of this approach implementation as the part of a network based on the SDN architecture.

So, if we look for the decomposition of the matrix A in the form

$$A = LU + R \tag{2}$$

Then one can use the following approach to find the matrices L and U. Suppose that the first ones $(k - 1)$ have already been found and it is necessary to find the kth. For this purpose, we write the first k rows of expansion in a block form (2) [1]:

$$\begin{pmatrix} A_{11} & A_{12} \\ a_{21}^T & a_{22}^T \end{pmatrix} = \begin{pmatrix} L_{11} & 0 \\ l_{21}^T & 1 \end{pmatrix} \begin{pmatrix} U_{11} & U_{12} \\ 0 & u_{22}^T \end{pmatrix} + \begin{pmatrix} R_{11} & R_{12} \\ r_{21}^T & r_{22}^T \end{pmatrix} \tag{3}$$

where l_{21}, u_{22}, r_{21} and r_{22} are some vectors.

Performing multiplication on the right side of (3), we get the following:

$$\begin{pmatrix} A_{11} & A_{12} \\ a_{21}^T & a_{22}^T \end{pmatrix} = \begin{pmatrix} L_{11}U_{11} + R_{11} & L_{11}U_{12} + R_{12} \\ l_{21}^T U_{11} + r_{21}^T & l_{21}^T U_{12} + u_{22}^T + r_{22}^T \end{pmatrix} \tag{4}$$

From the equality of matrices in (4) it follows that the sought vectors l_{21} and u_{22} must be determined by the following formulas:

$$l_{21}^T U_{11} + r_{21}^T = a_{21}^T \tag{5}$$

$$u_{22}^T + r_{22}^T = a_{22}^T - l_{21}^T U_{12} \tag{6}$$

Taking into account that $l_{kk} = 1$, we determine the coefficients of the kth lines of the decomposition matrices $l_{k1}, \ldots, l_{k,k-1}, u_{kk}, \ldots, u_{kn}$.

Now, assuming that $l_{k1}, \ldots, l_{k,j-1}$ have already been found, we calculate the value of l_{kj} from (5). This is possible because j < k is always met for the coefficients of the lower triangular matrix L. Since if then this means that $l_{kj} = 0$, otherwise $r_{kj} = 0$, (5) can be rewritten in the form (taking into account that at $i > j$ all the elements of the upper triangular matrix U are equal to zero):

$$\sum_{i=1}^{j} l_{ki}u_{ij} = \sum_{i=1}^{j-1} l_{ki}u_{ij} + l_{kj}u_{ii} = a_{kj} \tag{7}$$

Then l_{kj} is calculated as follows:

$$l_{kj} = \frac{1}{u_{jj}}\left(a_{kj} - \sum_{i=1}^{j-1} l_{ki}u_{ij}\right)$$

Similarly, taking into account $l_{jj} = 1$ from (6) we find u_{kj} (in this case, we assume that the elements of the lower triangular matrix L are equal to zero for $i > k$):

$$u_{kj} = a_{kj} - \sum_{i=1}^{k-1} l_{ki}u_{ij} \tag{8}$$

The equality (8) is valid for those cases when $a_{kj} \neq 0$, otherwise, it immediately implies $u_{kj} = 0$.

Based on the obtained expressions (7), (8), the following modified LU-factorization algorithm can be proposed (Fig. 2).

After the operation of this algorithm on the original matrix A of the MD system of equations coefficients, the matrix LU will be obtained that satisfies all the requirements for the preconditioner matrix. Indeed, the resulting matrix will be close to the original A, since it accurately reproduces it on the set of indices P_A, it is easily computable by the algorithm on Fig. 2, and easily reversible, due to the fact that it is the product of two triangular matrices [5].

For $k = \overline{1, n}$

  For $j = \overline{1, k-1}$

    If $(k, j) \notin P_A$

     then $l_{kj} := 0$

      else $l_{kj} := \frac{1}{u_{jj}}\left(a_{kj} - \sum_{i=1}^{j-1} l_{ki}u_{ij}\right)$

  $j{+}{+}$

  $l_{kk} := 1$

  For $j = \overline{k, n}$

    If $(k, j) \notin P_A$

     then $u_{kj} := 0$

      else $u_{kj} := a_{kj} - \sum_{i=1}^{k-1} l_{ki}u_{ij}$

  $j{+}{+}$

$k{+}{+}$

Fig. 2 Modified LU-factorization algorithm

Thus, the use of preconditioning in accordance with the algorithm on Fig. 2 together with the modified Richardson method described, for example, in [1], will significantly increase the efficiency of calculating the MD magnetic field by reducing the number of iterations and increasing the result accuracy. Let us estimate the efficiency of the preconditioning technique by counting the number of iterations required to achieve the set accuracy ε. To do this, we solve the problem on a computer using the modified Richardson method with and without preconditioning. The execution of all computational algorithms will be carried out on computers united in an SDN network, which will ensure their implementation with the least time costs.

3 Results

In the course of a computational experiment to calculate the magnetic fields of the studied MD, their main characteristics were obtained. So, the dependence of the magnetic flux in the working gap on the width of the non-magnetic wall S of the tank during various materials use of a permanent magnet is shown on Fig. 3.

Analysis of the simulation results (Figs. 3, 4, 5 and 6) shows an insignificant (less than 0.2 m Wb) change in the magnetic flux Φ_p with the width S increase of the non-magnetic tank wall by 0.1 m. At the same time, the equivalent magnetic resistance r of the MD magnetic circuit with the same increase of the width S for the nonmagnetic wall makes ~30 A/Wb.

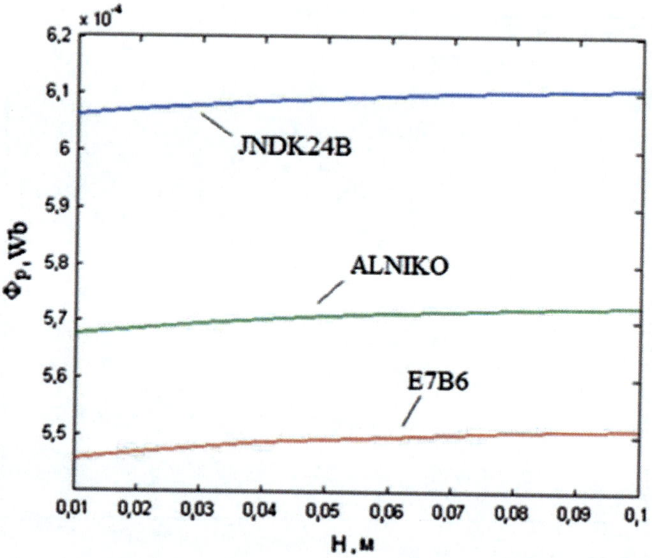

Fig. 3 Dependence of the magnetic flux on the width of the non-magnetic tank wallю

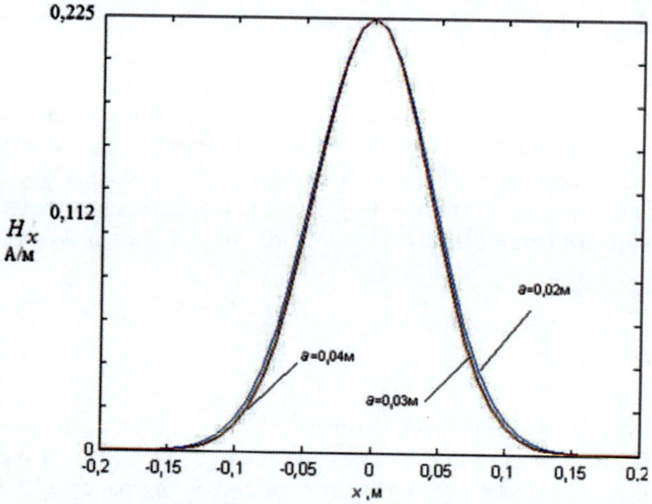

Fig. 4 Dependence of the MD magnetic field strength on the coordinate

The study of the resulting magnetic field of the MD at different widths of the working gap a, the dependences of the strength of which are shown on Fig. 4, makes it possible to evaluate the efficiency of the converter operation as a whole.

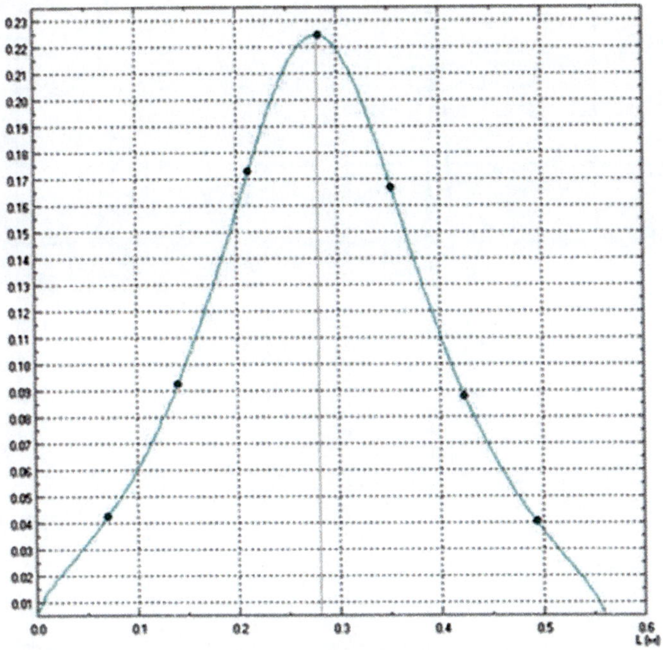

Fig. 5 Simulation of the MD magnetic field in the ELCUT system

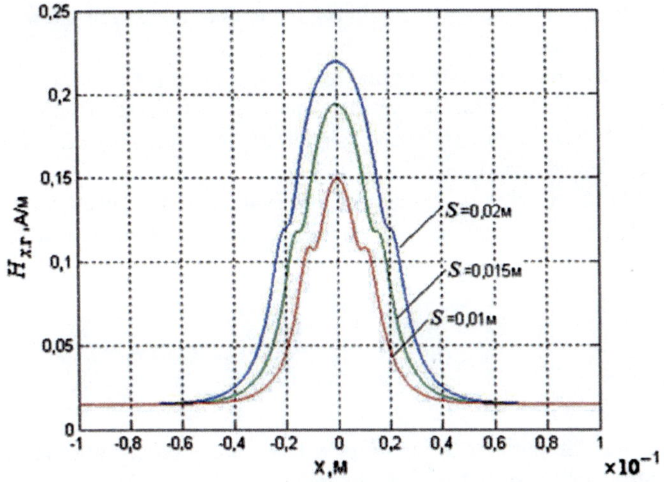

Fig. 6 Study of the non-magnetic wall S width effect on the magnetic field strength of the MD

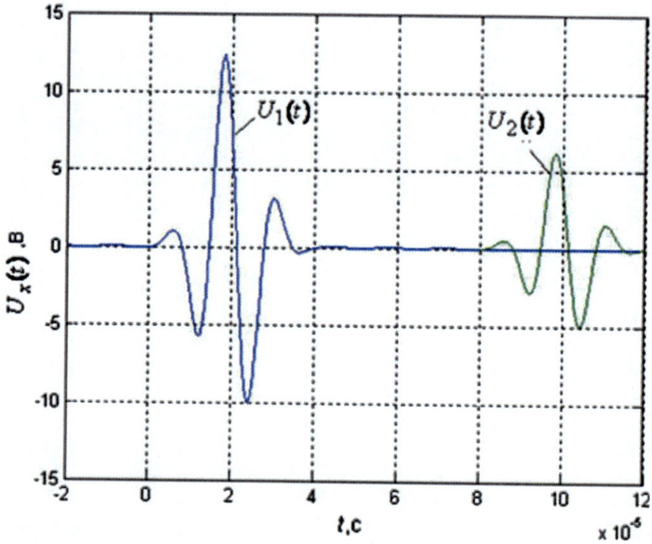

Fig. 7 MD output signals

Comparison of the obtained data with the results of modeling the MD magnetic field in the ELCUT system (Fig. 5) also allows us to conclude about the obtained result adequacy.

The results of the MD magnetic field modeling as a function of the nonmagnetic wall width are shown on Fig. 6.

The analysis of these data shows that for confident reading of information by the MD electro-acoustic transducer, it is sufficient to select a non-magnetic wall with the width of $S > 0.01$ m. In this case, the signals at the information output of the MD will have the form shown on Fig. 7.

Let's count the number of iterations n, depending on the specified accuracy ε, obtained during the computational experiment. In graphical form, these dependencies are shown on Fig. 8.

4 Discussion

As you can see, the number of iterations *n* when using preconditioning (graph 2) does not differ significantly from the theoretical one (graph 3), determined by the expression (8). When solving the problems without preconditioning (graph 1), the number of iterations increases significantly.

Thus, the use of the modified Richardson method together with the algorithm on Fig. 2 significantly reduces the number of iterations and the solution time, which increases the calculation efficiency of the MD magnetic field of the MF.

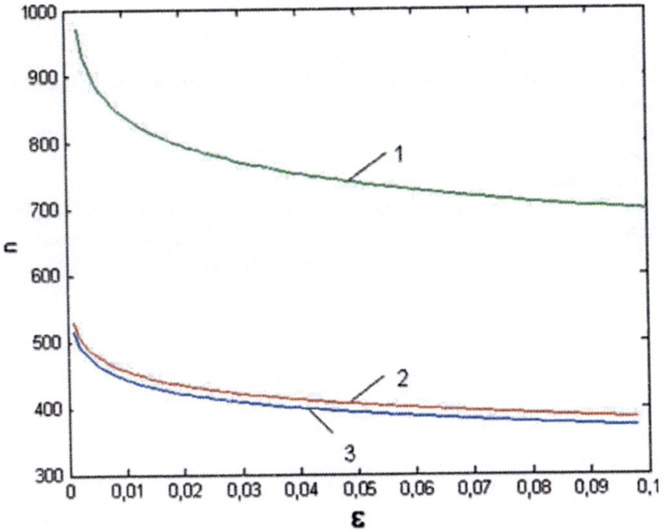

Fig. 8 Evaluation of preconditioning use effectiveness

Besides, the analysis of the literature [2, 9–12], which describes the solution of similar problems, allows us to identify additional ways to increase the efficiency of the MD magnetic field calculation.

For example, it is known that the number of iterations required to solve the SLAE of the MD magnetic field strongly depends on how close the initial values of the unknowns were to the true ones. If you set a good initial approximation, then the number of iterations during the solution can be reduced by several times. Hence, there is another way to increase the efficiency of iterative methods, which is the selection of a more accurate initial approximation. Usually, this choice is made using various systems to model magnetic fields such as ELCUT, Littlemag, ANSYS, etc. [4].

The initial values of the node potentials close to true ones can be obtained quite simply if the calculation is first performed not on the required grid, but on another one containing a smaller number of nodes. In the future, the data obtained using well-known interpolation formulas can be simply transferred to another, finer grid [1, 4].

5 Conclusion

Thus, performance of LU factorization using a modified algorithm for implementation in an SDN network will significantly simplify the MD magnetic field calculation, and at the same time, will increase the solution convergence rate and the accuracy.

References

1. Duravkin EV, Tkacheva EB, Saad I (2015) Architecture of SDN. Analysis of major development challenges. Inf Process Syst 3:92–98
2. Karpukhin EV, Mokrousov DA, Demin ES, Demin SB (2014) Research bypass measuring system with magnetostrictive converter of a level by method mathematical modeling. Mod Probl Sci Educ Surg 4
3. Karpukhin EV, Dementjeva ES, Selivanov SV (2016) Application of the rotation method to enhance the efficiency of the calculation of the magnetic fields of overhead magnetostrictive level gauges. Mod Syst Netw Econ Eng Nat Soc 2:171–177
4. Karpukhin EV, Demin SB (2014) Mathematical modeling magnetic fields of applied magnetostrictive level gauges: monograph. PenzSTU, Penza
5. Demirchyan KS, Chechurin VL (1986) Machine calculations of electromagnetic fields. High School, Moscow
6. Kuhling H (1983) Physics handbook. Mir, Moscow
7. Levshina ES, Novickii PV (1983) Electrical measurements of physical quantities. Energoatomizdat, Leningrad
8. Magirov AV, Polanin AD (1999) Methods for solving integral equations. Reference book. Factorial, Moscow
9. Karpukhin EV, Mokrousov DA, Demin ES, Demin SB (2014) To the issue of improving the modeling accuracy of magnetic fields of magnetostrictive level meters of the overlay type with complex geometry of the acoustic path. In: Aktuální vymoženosti vědy—2014: Materiály X mezinárodní vědecko—praktická conference, pp 17–21
10. Karpukhin EV, Mokrousov DA, Djatkov VS, Demin SB (2014) Complex of programs for calculation of parameters plated magnetostrictive converters of the level with complex geometry of the acoustic path. Mod Probl Sci Educ Surg 3
11. Karpukhin EV, Mokrousov DA, Demin ES, Demin SB (2014) Increase of efficiency of numerical methods of modeling magnetostrictive level gauges of overlay type with complex geometry of acoustic path. In: Aktuální vymoženosti vědy—2014: Materiály X mezinárodní vědecko—praktická conference, pp 13–16
12. Karpukhin EV, Mokrousov DA, Djatkov VS, Demin SB (2013) Application of numerical methods for calculation of magnetic fields in magnetostrictive level gauges. Bull Izhevsk State Tech Univ 3:102–104

The Distribution of Stalked Particles During the Operation of the Levelling Device

Vladimir Teterin, Mikhail Kostenko, Olga Teterina, Dmitry Blagov, and Sergey Mitrofanov

Abstract When cattle breeding and keeping, the main factor is to provide animals with the necessary amount of balanced feed. One of the main factors affecting the profitability of livestock production is the quality of roughage. When harvesting hay into bales, balers with a constant volume bale chamber are most often used. In the process of picking up a hay swath, uneven distribution of the swath along the width of the bale chamber is observed. In order to improve the uniformity of distribution of the hay swath across the width of the roll baler, the levelling device in the improved baler is made in the form of two discs with inclined fingers rotating in opposite directions. The study of the trajectories of movement of the stalked particles showed that the displacement of the stalked particles is more intense towards the periphery of the disc. Therefore, the fingers are located on the disc in two rows with a predominant location along the periphery. The analysis of the obtained trajectories makes it possible to establish the basic parameters of the levelling device: the distance between the fingers is 0.15–0.2 m, the degree of compaction when levelling is 20–25%, the angular velocity is 23–30 rad/s, the diameter of the levelling disc is 0.74 m.

Keywords Roll baler · Bale chamber · Compressed hay · Levelling device

1 Introduction

When cattle breeding and keeping, the main factor providing high milk productivity, excellent reproductive ability and prevention of various nutritional diseases is to provide animals with high quality feed, which supply the body with the necessary nutrients. This factor largely depends on the animal husbandry system [1–3].

The common methods of cattle keeping in the Russian Federation include year-round stall and stall-pasture systems. The presented livestock management systems are used in farms with a high concentration of livestock and provided with fodder

V. Teterin (✉) · M. Kostenko · D. Blagov · S. Mitrofanov
VIM Federal Scientific Agro-Engineering Center, Moscow, Russia

M. Kostenko · O. Teterina
Ryazan State Agrotechnological University Named after P.A. Kostychev, Ryazan, Russia

A. Beskopylny and M. Shamtsyan (eds.), *XIV International Scientific Conference "INTERAGROMASH 2021"*, Lecture Notes in Networks and Systems 247,
https://doi.org/10.1007/978-3-030-80946-1_13

lands. As a result, providing livestock with high quality feed becomes a paramount task, which requires advanced technologies and technical means for the preparation of rough, juicy and concentrated forage. Rough feeds such as hay, haylage, straw provide the animal not only with nutrients, but are also able to stimulate rumen motility, as a result of which cows have increased salivation and an increase in the number of chewing movements, thereby achieving a better mechanical processing of merycisin, which subsequently has positive impact on health and milk production [4–6]. Therefore, the use of new technologies and technical means for harvesting first-class hay, haylage, etc. is an urgent task today.

The generally accepted technology for harvesting stalk forage is using roll balers. This technology allows to reduce the loss of nutrients, improve preservation, reduce labor costs and transportation costs.

When harvesting hay in rolls, balers are used with a bale chamber of variable and constant volume. The process of forming a roll in them consists in continuously winding the plant mass onto the core up to a predetermined diameter [7, 8].

When harvesting hay with the use of balers with a constant volume bale chamber, the shape of the swath is an important factor. This factor can significantly affect the storage processes of the hay, since it leads to the formation of zones with insufficient density of hay in the process of roll formation, that contributes to the penetration of moisture, the development of microbiological processes [9–11].

2 Materials and Methods

During the research, the process of movement of stalk particles in the process of picking up a swath of hay and supplying it into the bale chamber of a baler with a levelling device installed on it was theoretically considered.

In addition, studies were carried out to determine the profile of hay rolls obtained on natural forage lands as a result of the operation of a GVK-6 rake in an aggregate with an MTZ-80 tractor. For this purpose, a profiler (Fig. 1) with a length of 1.6 m with measuring rods was used. In the course of the research, the profiler was installed

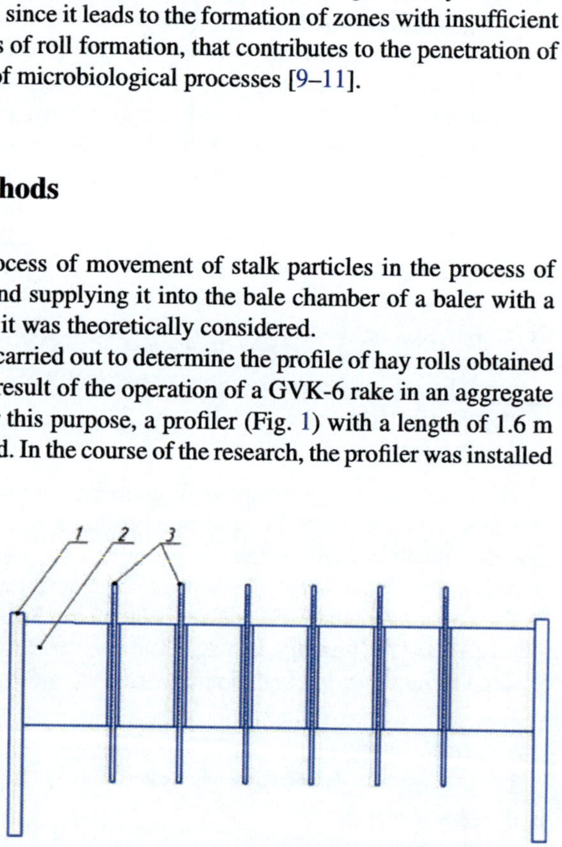

Fig. 1 Diagram of the profiler for determining the swath shape. 1—Frame; 2—ground; 3—measuring rods

across the hay roll every 0.5 m on registration plots, which were located diagonally across the field, while at least 5–7 registration plots were selected, the length of which was 25 m. The data obtained were averaged and tabulated, then their statistical analysis was carried out.

3 Results

The uneven distribution of stalk particles across the width of the pick-up is primarily due to the shape of the hay swath that enters the bale chamber. To determine its shape, a series of experiments was carried out, as a result of which a statistical model of the distribution of stem particles in the cross section of the roll was obtained (Fig. 2).

In order to improve the uniformity of the distribution of the swath of hay across the working width of the pick-up in the developed roll baler, the levelling device is made in the form of two discs with inclined fingers rotating in opposite directions (Fig. 3) [12]. The swath of hay raised by the pick-up enters the gap between the pick-up fingers and the inclined fingers of the rotating discs, while the discs above the pick-up are pushed forward to improve the distribution of the swath, forming a descending gap.

Let us write down the differential equations of particle motion

$$\begin{cases} m\frac{d^2x}{dt^2} = F_n + F_d\cos(\omega t) \\ m\frac{d^2y}{dt^2} = -F_d\sin(\omega t) \end{cases} \quad (1)$$

where Fn is the impact force of the pick-up on the hay roll, N;

Fd is the impact force of leveling disks on the hay roll, N;

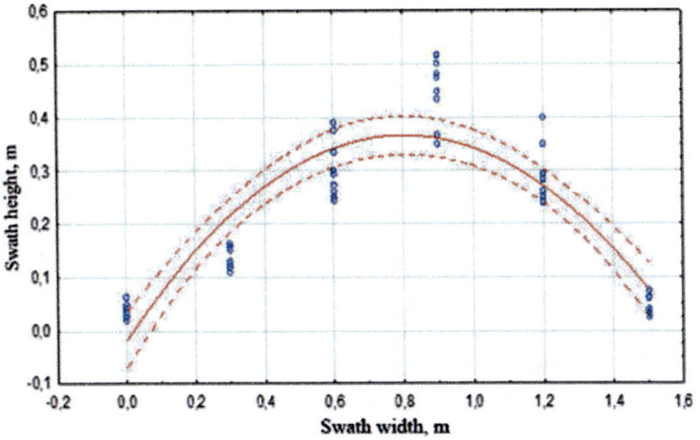

Fig. 2 The results of studies of the uniformity of a hay roll in cross-section

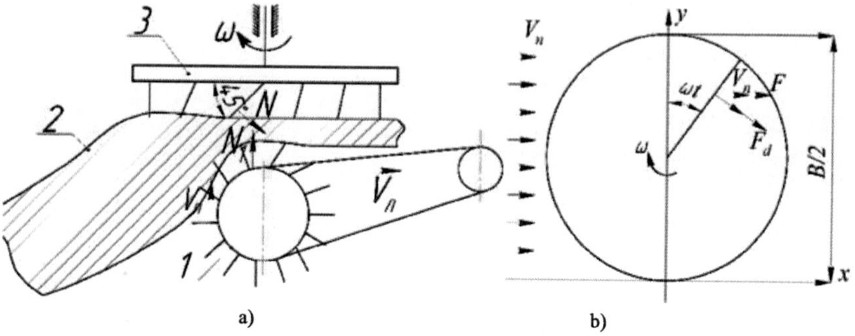

Fig. 3 The scheme for determining the law of particle motion when levelling a roll. 1—Pick-up; 2—hay roll; 3—levelling discs

ω is rpm of disks, s^{-1}.

Since the swath is unevenly distributed over the pick-up width of the baler, the degree of compaction and, accordingly, the forces will vary across the pick-up width.

As a result of studying the transverse profile of the hay swath before pressing, the regression equation is obtained:

$$H(b) = -0.016 + 0.9498b_{\mathrm{B}} - 0.5926b_{\mathrm{B}}^2 \tag{2}$$

Statistical analysis of the model shows that the correlation coefficient is R = 0.89, the determination coefficient showing the adequacy of the model is 0.79, which exceeds the limit value of 0.75, which indicates the performance of the model.

Considering that the swath is evenly distributed during the operation of the levelling discs, the average height of the swath across the working width is determined.

To do this, the cross-sectional area of the swath is determined

$$S = \int_0^b H(b)\mathrm{d}b = \int_0^b \left(-0.016 + 0.9498b - 0.5926b^2\right)\mathrm{d}b \tag{3}$$

After performing integral calculations, one gets:

$$h_{aver} = \frac{S}{b} = \frac{-0.016b + 0.9498\frac{b}{2^2\frac{b^3}{3}}}{b} \tag{4}$$

Compaction of the swath before pressing is

$$\Delta h = h_{aver} - h_0 = -0.016 + 0.9498b - 0.5926b^2 - 0.1 \tag{5}$$

where h0 is the clearance between levelling disc fingers and the pick-up (h0 = 0.1 m), m;

haver is the average height of the swath, m.

Therefore, the elastic force of the hay swath will be determined according to the following expression:

$$F_{elast} = c\Delta h = c\left(-0.116 + 0.4749b - 0.1975b^2\right) \quad (6)$$

where b is the roll baler width, m.

Since the velocity of the hay swath is constant, the interaction of the disc with hay can be investigated statically. Therefore, the impact force of the levelling discs on the hay swath will be performed according to the following expression:

$$F_{\Pi} = N_2 \cdot f_{if} = f_{if}\left(\frac{F_{elast}}{n_f} + P_{sw}\right) \quad (7)$$

where N2 is the normal pressure per unit of swath volume from the pick-up side (from below), N;

fif is the hay roll internal friction coefficient;

nf is the number of fingers, pcs;

Psw is the weight per unit of swath volume, N.

Based on the assessment of the impact of the levelling device on the hay swath and mathematical transformations of the expression (2), the law of motion of stem particles in the zone of operation of levelling discs with fingers was obtained

$$\begin{cases} x = x_0 + V_{0x} \cdot t + \frac{1}{m_s} \cdot f_{if} \cdot \left[\left(\frac{F_{elast}}{n_f} + P_{sw}\right) \cdot \frac{t^2}{2} - \frac{F_{elast}}{n_f \cdot \omega^2} \cdot \cos(\omega t) + \frac{F_{elast}}{n_n f \cdot \omega^2}\right] \\ y = y_0 + V_{0y} \cdot t + \frac{1}{m\,s} \cdot f_{if} \cdot \frac{F_{elast}}{n_f \cdot \omega} \cdot \left(\frac{1}{\omega} \sin(\omega t) - t\right) \end{cases} \quad (8)$$

where x0, y0 are coordinates of the initial position of stalked particles, m;

V0x, V0y is initial velocity of stalked particles before levelling, m/s.

Based on the obtained expression, the trajectories of movement of stem particles were studied using MathCad program. The following parameters were set as the initial ones: the degree of compaction (decrease in the height of the swath when levelling), the coefficient of friction, the angular speed of the levelling discs, the number of fingers, the speed of the hay swath. As a result of the research, a graph of the trajectories of movement of stem particles during the operation of the levelling discs is made (Fig. 4).

It is also found that the displacement of stalk particles is significantly influenced by the number of fingers in the levelling discs and the distance between them, the angular velocity of the levelling discs and the amount of compaction of the hay swath.

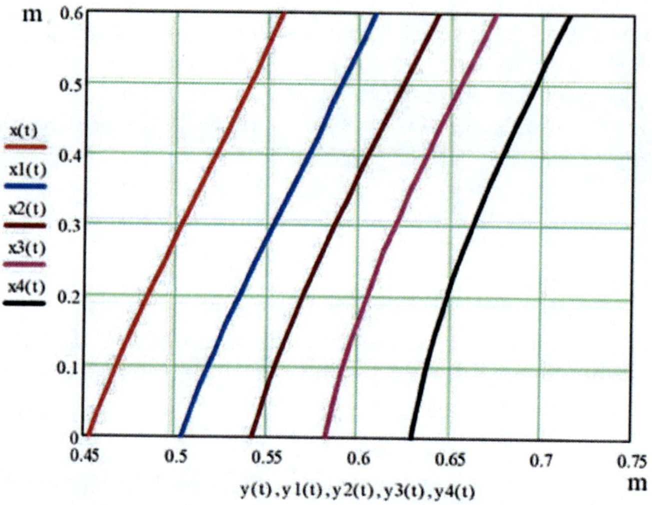

Fig. 4 Trajectories of movement of stem particles

4 Discussion

Analysis of the graph (Fig. 2) shows that the swath is extremely uneven in width, has the greatest height in the middle part, and on the sides of the swath the hay is located at an angle of natural slope (internal friction). Taking into account the complex structure of the hay swath and the multidirectional stalks, it can be concluded that the standard operations of the baler including picking, transportation and pre-pressing do not sufficiently contribute to the uniform distribution of hay across the width of the baler. When baling, the problem is aggravated, as the middle of the roll is approximately in the middle of the bale, as a result of which the middle part of the bale has a higher density, and the edges of the bale have a lower density, which leads to moisture penetration and hay spoilage.

The graph clearly shows that the displacement of stalk particles located closer to the center of the levelling disc is slightly lower, therefore the fingers are located on the disc in two rows with a predominant location along the periphery. The analysis of the obtained trajectories makes it possible to establish the basic parameters of the levelling device: the distance between the fingers is 0.15–0.2 m, the degree of compaction when levelling is 20–25%, the angular velocity is 23–30 rad/s, the diameter of the levelling disc is 0.74 m.

5 Conclusion

The conducted experimental studies have shown that the swath of hay entering the roll baler has uneven structure and shape, and therefore in the process of roll formation zones with lower density will appear, which leads to moisture penetration and hay spoilage.

Since the uneven supply of the hay swath into the bale chamber cannot be completely eliminated by stress relaxation during the formation and subsequent storage of the roll, it is relevant to use an additional operation to align the roll before pressing.

References

1. Wei X, Shangli S, Juan Q, Squires VR, Limin H et al (2013) Effect of different processing methods on the quality of Alfalfa Hay. J Anim Vet Adv 12(6):689–693
2. Uijttewaal A, Chapuis S, Crocq G, Lépée P (2016) Advances in methods for harvesting and storing forage legumes. Fourrages 227:157–166
3. Ulyanov V, Utolin Vl, Luzgin N, Borontova M, Kiryanov Al, Polunkin A (2019) Mixer for dry concentrated feed. In: XII international scientific conference on agricultural machinery industry. IOP Conf Ser Earth Environ Sci 403:1–11
4. Castagnino DS, Ying Y, Allen MS, Gervais R, Chouinard PY, Girard CL (2017) Short communication: apparent ruminal synthesis of B vitamins in lactating dairy cows fed *Saccharomyces cerevisiae* fermentation product. J Dairy Sci 100(10):8161–8164
5. Li C, Beauchemin KA, Yang W (2020) Feeding diets varying in forage proportion and particle length to lactating dairy cows: effects on ruminal pH and fermentation, microbial protein synthesis, digestibility, and milk production. J Dairy Sci 130(5):4340–4354
6. Johnston DJ, Theodoridou K, Ferris CP (2019) The impact of field bean inclusion level in dairy cow diets on cow performance and nutrient utilization. Livest Sci 220:166–172
7. Fang J, Zhang Y, Yang M, Wang A, Wang J, Liu D, Gao J, Li H (2018) Stress relaxation behavior and modeling of alfalfa during rotary compression. Nongye Gongcheng Xuebao. Trans Chin Soc Agric Eng 34(16):50–56
8. Gao D, Wang D, Li J, Liu Y, Qiao X (2015) Design and test of logarithmic spiral round baler chamber. Nongye Jixie Xuebao. Trans Chin Soc Agric Mach 46(7)
9. Byshov NV, Borychev SN, Teterin VS, Uspenskiy IA, Kostenko MYu, Rembalovich GK (2015) Investigation of the pressed hay density distribution inside the roll. Herald Ryazan State Agrotechnol Univ P.A. Kostychev 4(28):48–52
10. Terentyev V, Andreev K, Anikin N, Novikov N, Teterin V, Blagov D, Mitrofanov S (2020) The study of forming a roll of pressed hay and the influence of its density on quality parameters. E3S Web Conf 04013. In: Ecological and biological well-being of flora and fauna (EBWFF-2020)
11. Lei J, Wang D, Zhang Q, Yang X, Li L, Li C (2015) Experiment on stress relaxation characteristics of intact rice straw during rotary compression. Trans Chin Soc Agric Eng 31(8):76–83
12. Kostenko MYu, Teterin VS, Kostenko NA, Teterina OA. Pat. 157147 Russian Federation, IPC A01 A 15/07. Roll baler. Patent holder: FSBEI HE RSATU. No. 2015121102/13; declared 06/02/2015; publ. 20.11.2015, bul. No. 32

Modeling Carbon Dioxide Injection to Improve ORF in Low Permeability Reservoirs

Dmitry Klimov and Ernest Zakirov

Abstract For oil and gas fields with low-permeability formations, the most realistic way of their development is the mode of depletion of reservoir energy. Often, additional production is facilitated by the implementation of multi-stage hydraulic fracturing in production wells with their various modifications. Carbon dioxide is often used as an agent for increasing oil recovery factor, which is also considered within the framework of projects for its utilization and underground storage in depleted deposits, salt-bearing strata and shale rocks. Carbon dioxide is also actively used as an agent for the intensification of hydrocarbon production in the development of oil and gas fields. It is important to understand the ongoing physical and chemical changes occurring within the underground formation in case of carbon dioxide injection. Examples of these changes include dissolution, chemical reactions, convective mixing, advective processes, and dispersion. Computer modeling of the ongoing processes is seen as a very important task for the correct functioning of such projects. This study is a summary of the results of computer modeling of a method for developing oil and gas fields with low-permeability formations based on maintaining reservoir pressure by injection of carbon dioxide using the commercial software Navigator.

Keywords Mathematical modeling · Low-permeability reservoirs · Hydrocarbons · Hydraulic fracturing · Carbonated water · Carbon dioxide · Depletion mode · Pressure maintenance mode · Oil recovery factor

1 Introduction

Every year the share of hard-to-recover reserves is increasing, and the active part of light oil reserves is being rapidly developed. At the same time, the share of reserves, for example, high-viscosity oils, is increasing more and more. Due to the growth in the share of hard-to-recover reserves, it becomes necessary to increase the efficiency of their extraction, to improve the technologies for their production. Moreover, the oil

D. Klimov (✉) · E. Zakirov
Oil and Gas Research Institute, Russian Academy of Sciences (OGRI RAS), Gubkina st., 3, Moscow, Russian Federation 119333

A. Beskopylny and M. Shamtsyan (eds.), *XIV International Scientific Conference "INTERAGROMASH 2021"*, Lecture Notes in Networks and Systems 247,
https://doi.org/10.1007/978-3-030-80946-1_14

recovery factor (ORF) by traditional methods in fields with hard-to-recover reserves rarely exceeds 30%, and in high-viscosity oil fields it is even lower.

For oil and gas fields with low-permeability formations, the most realistic way of their development is the mode of depletion of reservoir energy. However, it is known that the depletion mode is usually characterized by the minimum values of the oil and gas recovery factors. Often, additional production is facilitated by the implementation of multi-stage hydraulic fracturing in production wells with their various modifications.

It is known that the most significant of the properties of productive formations is the permeability coefficient. Wells production rates for oil, gas, condensate and other development indicators depend on the values of the reservoir permeability coefficient.

Until recently, reservoirs with a permeability of 1 mD or more were not considered viable development targets. Today the situation has changed. Thus, in the United States, they began to successfully develop oil and gas fields with shale, low-permeability formations. In such formations, the permeability is about or markedly below 1 mD. Extraction of shale oil and shale gas begins to develop in other regions [1].

The era of 3D computer modeling that began in the 2000s forced a change in the attitude towards low-permeability formations. Thus, the need to include low-permeability formations with their own values of porosity and permeability in 3D-geological and 3D-hydrodynamic models of the reservoir or the field as a whole was justified. Thus, the authors had the following fundamental idea. The highest oil recovery factor and the least negative consequences of waterflooding will occur if oil from low-permeability reservoirs is displaced by a working agent into high-permeability reservoirs. In turn, oil from highly permeable reservoirs can be displaced to production wells by oil inflowed from low-permeability areas.

2 Materials and Methods

The situation with low-permeability reservoirs in the last decade has changed dramatically in connection with the development in the United States and other countries of oil deposits with ultra-low values of permeability [2]. It is known that this became possible due to the drilling of long horizontal wells and carrying out in them reusable, multi-stage hydraulic fracturing of the formation. This technology is widely used in the development of both conventional and unconventional fields.

As an example, Fig. 1 shows the layout of planned wells in the largest shale oil field in the United States, Barnett Shale Play. The peculiarity of the development of this and other similar fields is as follows: most of them are developed in the regime of depletion of reservoir energy. It is known that in such cases the flow rates of the wells decrease rather quickly in time. Therefore, in order to maintain a constant level of production at the field, more and more wells are drilled with the implementation of

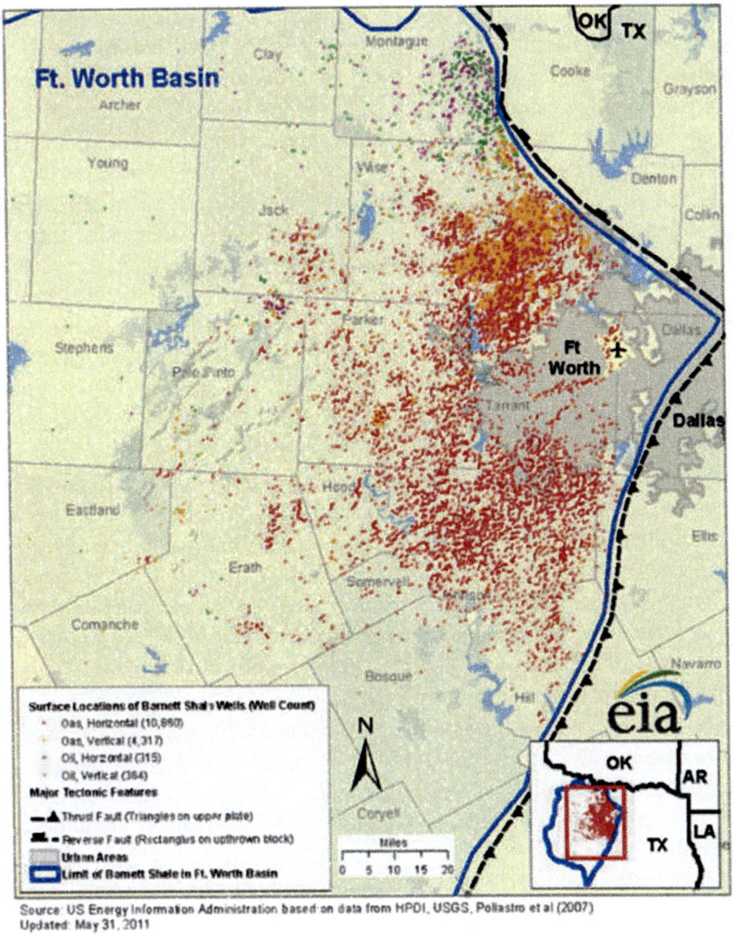

Fig. 1 Map of gas wells in the Barnett Shale, Texas. *Source* [3]

multi-stage hydraulic fracturing in the production wells. Judging by the publications [2], this development technology turns out to be cost-effective.

In some wells, the fracturing operation has to be repeated many times to maintain a stable flow rate. However, even despite the use of horizontal drilling and hydraulic fracturing technologies, the productivity of existing wells decreases much faster than in traditional fields. So, if the average "life" of gas wells in traditional US fields is 30–40 years, then about 15% of shale wells drilled in 2003 have completely exhausted their resources in 5 years.

Due to the low permeability of the reservoirs, gaseous agents deserve attention as a working agent. Among the world practice of patent applications and issued patents for the production of gas and oil from unconventional reservoirs, the experience of using hydraulic fracturing using carbon dioxide is of interest in the framework of our research [2]. This method is carried out by injecting supercritical CO_2 to treat the formation through the wellbore at a pressure higher than the fracturing pressure. The formation being treated may have a permeability of less than 1 mD. By pumping a CO_2 agent into the formation at a pressure higher than the fracture pressure, the formation integrity can be effectively compromised, which stimulates the flow of methane and other hydrocarbon gases into the well. That is, the use of CO_2 during hydraulic fracturing is promising from the point of view of increasing the total gas production due to the ability of CO_2 to displace methane.

The feasibility and efficiency of various technologies used in the development of traditional oil fields, based on the injection of carbon dioxide in various modifications, is currently beyond doubt. Moreover, there is an abundance of anthropogenic sources of carbon dioxide, and its modern emissions into the atmosphere are enormous. Therefore, the injection of carbon dioxide into oil reservoirs in various states and forms (for example, in the form of carbonated water) seems to us in the framework of the study of low-permeability reservoirs reasonable, logical, which at the same time requires specific laboratory studies and the compilation of mathematical models.

It should be noted that some experts agree that oil and gas production based on hydraulic fracturing technology can be successfully combined with technologies for utilization, sequestration and long-term storage of CO_2 in various geological formations. Geological sequestration and hydraulically fractured oil and gas share more than a location. They also face common issues such as groundwater contamination risks, water management issues, seismic risks, and general public acceptance [4]. Carbon dioxide sequestration refers to the process of capturing the excess CO_2 present within the atmosphere toward long-term storage. CO_2 storage arises from the need to mitigate the effects of global warming and serves as an avenue to reduce the rate of accumulation of greenhouse gases (GHG) arising from anthropogenic activities [5–7].

Sedimentary basins are considered to be promising targets for CO_2 storage as they have high-porosity and high-permeability layers represented by sandstones, and low-permeable and low-porosity caprocks represented by shale which limit the transfer of CO_2 toward the surface [5]. Thus, shale rocks, in addition to a new hydrocarbon source, can also represent underground storage-traps within the framework of international projects and programs for long-term conservation and utilization of carbon dioxide [6, 7].

Experimental study by Kang et al. [8], which examined the possibility of shale-rich core samples to be suitable for carbon dioxide storage, showed good results. According to the authors, despite the wide popularity of low-permeability sedimentary rocks with low porosity, organic shale has the ability to retain a significant amount of gas for a long time due to its ability to capture gas in an adsorbed state through finely dispersed organic matter (i.e., kerogen). The suitability of shale for CO_2 storage is also attractive because the spatial and thermodynamic effects of the processes occurring are similar to coal seams in methane extraction technologies.

Numerous research works [5, 9–17] have been proposed to understand the ongoing physical and chemical changes occurring within the underground formation in case of carbon dioxide injection. Examples of these changes include dissolution, chemical reactions, convective mixing, advective processes, and dispersion. There are some works devoted to modeling various processes of injection and storage of carbon dioxide in various formations, which should be mentioned.

A considerable amount of research has been conducted in simulating and modeling CO_2 sequestration in the subsurface. Calabrese et al. [9] studied the physical and chemical processes during CO_2 sequestration in a depleted gas reservoir located in the north of Italy. They concluded that to maximize the volume of CO_2 injected, an optimum rate has to be defined. At higher rates, the gas channels through high permeability streaks, and hence the storage capacity is reduced. While at lower rates, the denser CO_2 falls to the bottom of the gas zone and dissolves in the aquifer. They also concluded that molecular diffusion, dispersion, and geochemistry were not important factors for assessing the CO_2.

Seo and Mamora [10] performed experimental and simulation studies to evaluate the feasibility of sequestering supercritical CO_2 in depleted gas reservoirs. They performed experimental studies to obtain relative permeability curves; then, 3D simulation models of one-eighth of a five-spot pattern were used to evaluate the injection of CO_2.

Hesse [11] presented a compact multiscale finite volume (CMSFV) method for the numerical simulation of CO_2 storage in large-scale heterogeneous formations. The authors identified that dissolution is an important trapping process if CO_2 is present in a structural trap. They also concluded that high permeability aquifers favor dissolution trapping. The authors indicated that high permeability, gently dipping, and deep saline aquifers are the optimal targets for CO_2 sequestration.

Momeni et al. [12] presented a simulation study using ECLIPSE E300 (compositional simulation model) of a synthetic geologic model used to sequestrate CO_2. They concluded that the operating expenditure for sequestration in a depleted oil reservoir is less than in an aquifer because of lower well corrosion during injection. This is due to the higher brine concentration in an aquifer which increases the probability of corrosion.

Zhang and Agarwal [13] performed an optimization based on genetic algorithms to optimize the sequestration operation. They used the TOUGH2 solver developed by the US Department of Energy. They used both horizontal and vertical wells. Based on the results of the optimization, they concluded that the horizontal wells were much better when compared to vertical wells in aspects such as reduced migration and pressure build-up which could contribute to cap rock fracture and gas leakage.

Bao et al. [14] performed a large-scale CO_2 sequestration by coupling reservoir simulation with molecular dynamics (MD). The simulation was performed on massively parallel high-performance computing systems. They believed the coupling of molecular dynamics would provide better predictability of fluid properties under varying geological conditions. In their flow equations, they assumed the flow to be incompressible and used Darcy's equation to model the velocity in the reservoir. The advection diffusion equation was used to model the transport of CO_2 in the porous media. In the MD simulation, they solved Newton's equation of motion and the Leonard Jones and Coulomb interactions were used to represent the interaction between two atoms.

Hao et al. [15] developed a methodology to combine reservoir simulation, rock physics theory, and seismic modeling to simulate and monitor a sequestration process in an idealized geological model located in the Sleipner field. They modelled CO_2 injection using the two-phase flow model and solved the equations using the IMPES method. Then, they analyzed the effects of fluid saturation and pressure change on the elastic wave velocity based on the Gassman equation, Hertz–Mindlin theory, and effective fluid theory. Finally, seismic modeling was performed using P-wave potential equations and the symplectic stereomodeling (SSM) method on the transformed geologic model obtained from reservoir simulation.

Foroozesh et al. [16] performed a field scale simulation of an aquifer consisting of one well injecting CO_2 for ten years. The model was run for 100 years with the results showing that CO_2 solubility trapping was the main mechanism of sequestration. The simulation was performed using CMG-GEM. The results demonstrated that good vertical permeability and lower injection pressures are important factors in reducing leakage.

Mkemai and Bin [17] investigated the optimal injection strategy to enhance CO_2 storage. Their results concluded that an optimum injection pressure needs to be maintained as the pressure build-up created by injection may fracture the cap rock, which would then lead to CO_2 leakage. The authors highlighted that an optimum CO_2 sequestration does not lead to excessive migration of the injected gas.

The author's specialized laboratory experiments were also devoted to the search for effective technologies for the utilization and use of carbon dioxide as an agent for displacing hydrocarbons in oil and gas reservoirs. They revealed previously unknown physicochemical phenomena and polycondensation mechanisms of the interaction of carbon dioxide with rock models, which were expressed in the generation of hydrogen, methane and its homologues [18].

The authors' lack of the necessary set of required geological, physical and other information on a real field, as well as a software product for taking into account the experimentally identified physicochemical processes in predictive calculations, led to the solution of the synthetic model problem considered below.

3 Results

The results of the laboratory experiments [18] incite interest in injecting CO_2 into the oil bed in that or another phase conditions. However, the idea of injecting CO_2 into the oil bearing beds faces the limits determined by the permeability values. Therefore, it is for a good reason that most shale deposits are developed under of reservoir energy depletion. Unfortunately, the oil recovery factor (ORF) in such cases does not, as a rule, exceed ten percent. Therefore the computer assisted experiments have been performed for the oil carrying bed with the limit value of the effective permeability of 1 mD.

For the objectives of the study the model of multidimensional multiphase flow was used implemented within the commercial software package tNavigator developed by Rock Flow Dynamics [19]. This simulator implements an extended model of non-volatile oil, and the calculation results themselves are as close as possible to the results of the industry standard for multiphase filtration modeling—the Schlumberger Eclipse simulator.

The modeled productive bed is of low permeability (1 mD), it is not profitable by definition and belongs to "non-reservoirs" (non-pay reservoirs). Other initial data are as follows: initial reservoir pressure—23.3 MPa, oil saturation pressure—0.5 MPa, oil viscosity—1 MPa s, oil formation volume factor—1.6 m^3/m^3, formation thickness—20 m. As the bottom-hole pressure in the producing wells decreases down to 3 MPa further on it is kept unchanged. The bottom-hole pressure in the injection well is constant and it is equal to 30.3 MPa. The producing wells are shut when the oil production rate is reduced to 1 m^3/day (per a whole well). The relative permeability curves are assumed to be diagonal because of the high solubility of carbon dioxide in oil. The flow simulation model was represented by the Black Oil model. With such initial data the comparative calculations for alternative scenarios have been performed and are described below.

An element of a 5-spot development pattern is 500 × 500 m in size. It has been drilled with horizontal wells with lengths of 200 m. Around the horizontal part of the well bore the one-cell ring of the grid has been refined and the permeability values of these cells are 10 times higher than the permeability of the reservoir itself which simulates the technogenic fractures resulting from the multistage hydraulic fracturing. The schematic layout of the simulation element is shown in Fig. 2.

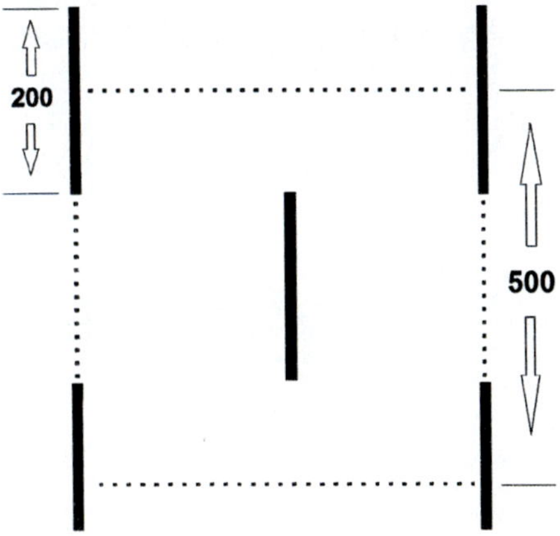

Fig. 2 Schematic layout of 5-spot development element

4 Discussion

The effect of CO_2 interphase exchange accompanied by dissolution in water and oil was not taken into account. No account was made for the volumetric properties of the phases depending on the amount of the dissolved CO_2. The ability of the water dissolved CO_2 to mix with oil was approximately accounted for by means of the diagonal relative permeabilities for the oil–water system.

The dimensions of the simulation grid were 43 × 43 × 10 grid cells. The grid is not uniform in the horizontal plane, with cell sizes being reduced down to 1 m × 1 m in the area of each well. Then the sizes of the cells were increasing exponentially preserving the preset total distance between the wells. Over the vertical plane the grid is uniform.

In «case 1» all wells in the development element are producing wells, i.e. the development is carried out under the depletion drive, the wells are producing at the bottom-hole pressure of 3 MPa.

In «case 2» one of the wells (the central one in the development element) becomes an injection well. Carbonized water of 1 MPa s viscosity is injected under the bottom-hole pressure of 30.3 MPa.

The results of the simulations for the cases under consideration are shown graphically in Figs. 3 and 4.

Figure 3 is of special interest here; it presents the comparison of the dynamics of ORF in the case with the depletion drive and in the case with the pressure maintenance. Hence, it is advisable to maintain the reservoir pressure in the low permeability reservoir under consideration. Though, here the dynamics of ORF in the case with the maintained pressure is somewhat too high, because the simulations were based on the model where the permeability of the oil reservoir was homogeneous and also some other assumptions have been made. However, there are several technological

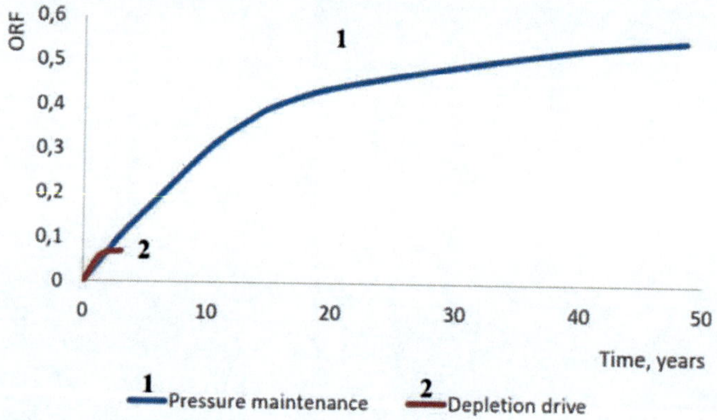

Fig. 3 Dynamics of oil recovery factor (ORF). *1*—pressure maintenance, *2*—depletion drive

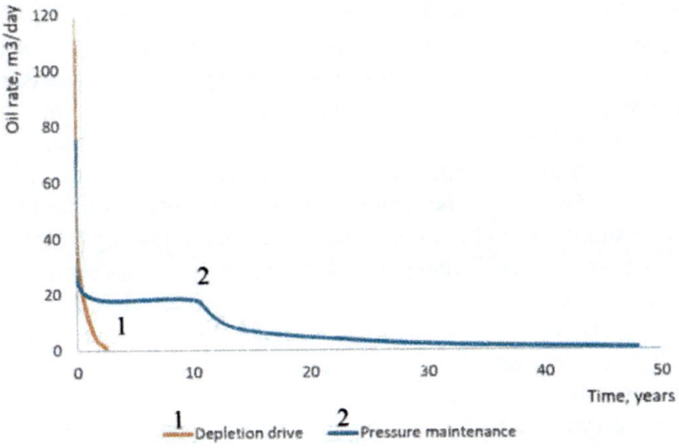

Fig. 4 Dynamics of oil production rate. *1*—depletion drive, 2—pressure maintenance

methods for taking into account the inhomogeneity of the reservoir and for alleviating its negative implications. There are still additional reserves when, for example, the lengths of the horizontal wells are assumed to be equal to 1000 or 2000 m, etc.

The results presented in Fig. 3 can be understood in more detail upon considering Fig. 4 that shows the comparison of the oil production dynamics in different modes for the same development element. Here, it has to be noted that in the depletion mode the oil production from the development element appears to be somewhat higher in the beginning as compared to the case of pressure maintenance, because in the depletion mode a larger number of producing wells are in operation.

5 Conclusion

The results of the simulations correspond to an isolated case and are by no means absolute. Obviously, in this case a large number of cases have to be investigated; however, this was not included in the scope of this study. It is also obvious that some non-apparent effects will appear, inasmuch as in real environment the layered or zonal heterogeneity of the reservoir properties often affects negatively the economics of the oil field development.

Nevertheless, the undertaken numerical experiments with the simulation model prove at the qualitative level that developing the oil deposits with low permeability reservoirs by pressure maintenance using carbonized water injection could be very efficient and can result in higher ORF values. High oil recovery factors can be achieved due to the effects of additional hydrogen generation obtained in laboratory experiments [18].

The considered approach to the development is realistic not only when applied to the oil deposits but also when applied to gas-condensate and, in some cases, to gas deposits with low permeability reservoirs. Thus, if the reservoir pressure is not maintained in the gas-condensate field then the condensate will drop-off in the reservoir becoming immovable. That is, in case of the gas-condensate deposit the pressure maintenance is meant to solve the problem of condensate recovery in the first place. In case of the depletion mode the dropped-off condensate will drastically reduce permeability to gas which will lead to lower gas recovery factor (GRF). That means that maintaining the reservoir pressure in gas-condensate deposit will help improve GRF. Maintaining the pressure in the oil deposit and in gas-condensate deposit will make it possible to postpone the period of the compressor-driven exploitation. Besides, if the pressure is maintained in these deposits the volume of the low pressure gas will decrease considerably. Thus, ideally, when the CO_2 injected agent displaces all of the natural gas, there will be no low pressure gas left in the development formation.

References

1. Alvarado V, Manrique E (2010) Enhanced oil recovery: field planning and development strategies. Gulf Professional Publishing
2. Khan R, Al-Nakhli AR (2012) An overview of emerging technologies and innovations for tight gas reservoir development. Paper SPE 155442 presented at the SPE international production and operations conference and exhibition, Doha, Qatar, 14–16 May 2012
3. US Energy Information Administration. Map of gas wells in the Barnett Shale, Texas. http://www.eia.gov/oil_gas/rpd/shaleusa1_letter.pdf
4. Nicot J-P, Duncan IJ (2012) Common attributes of hydraulically fractured oil and gas production and CO_2 geological sequestration. Greenhouse Gas Sci Technol 2:352–368
5. Kalam S, Olayiwola T, Al-Rubaii MM et al (2021) Carbon dioxide sequestration in underground formations: review of experimental, modeling, and field studies. J Petrol Explor Prod Technol 11:303–325
6. Price PN, Oldenburg CM (2009) The consequences of failure should be considered in siting geologic carbon sequestration projects. Int J Greenhouse Gas Control 3(5):658–663
7. Intergovernmental Panel on Climate Change (2018) Global warming of 1.5 °C. An IPCC Special Report on the impacts of global warming of 1.5 °C above pre-industrial levels and related global greenhouse gas emission pathways, in the context of strengthening the global response to the threat of climate change, accessed from https://www.ipcc.ch/sr15/
8. Kang SM, Fathi E, Ambrose RJ, Akkutlu IY, Sigal RF (2013) Carbon dioxide storage capacity of organic-rich shales. SPE J 16:842–855
9. Calabrese M, Masserano F, Blunt MJ (2005) Simulation of physical–chemical processes during carbon dioxide sequestration in geological structures. In: SPE annual technical conference and exhibition, Society of Petroleum Engineers
10. Seo JG, Mamora DD (2005) Experimental and simulation studies of sequestration of supercritical carbon dioxide in depleted gas reservoirs. J Energy Res Technol 127(1):1–6
11. Hesse MA (2008) Mathematical modeling and multiscale simulation of CO_2 storage in saline aquifers. Ph.D. thesis, Stanford University, Department of Energy Resources Engineering
12. Momeni A, Aghajani M, Zargar G (2012) A simulation study of carbon dioxide sequestration in a depleted oil reservoir. Pet Sci Technol 30:751–765

13. Zhang Z, Agarwal RK (2012) Numerical simulation and optimization of CO_2 sequestration in saline aquifers for vertical and horizontal well injection. Comput Geosci 16:891–899
14. Bao K, Yan M, Allen R, Salama A, Lu L, Jordan KE, Sun S, Keyes D (2016) High-performance modeling of carbon dioxide sequestration by coupling reservoir simulation and molecular dynamics. SPE J Soc Pet Eng
15. Hao Y, Yang D, Zhou Y (2016) A feasibility study of CO_2 geologic sequestration integrating reservoir simulation, rock-physics theory, and seismic modeling. Geophysics 81:M71–M82
16. Foroozesh J, Dier MA, Rezk MG (2018) A simulation study on CO_2 sequestration in saline aquifers: trapping mechanisms and risk of CO_2 leakage. MATEC Web Conf 225:03004
17. Mkemai RM, Bin G (2019) A modeling and numerical simulation study of enhanced CO_2 sequestration into deep saline formation: a strategy towards climate change mitigation. Mitigation and Adaptation Strategies for Global Change
18. Barenbaum AA, Zakirov SN, Zakirov ES, Klimov DS, Serebryakov VA (2015) Physical and chemical processes during the carbonated water flooding in the oilfields. SPE Russian Petroleum Technology Conference, 26–28 Oct 2015, Moscow, Russia. SPE 176729-MS
19. Harrison S (2015) tNavigator—breaking reservoir simulation speed limits in Europe! First SPE Norway Mag 5:22–27

Phased Passive Fluxgate Control of Structural Changes in Low-Carbon and Low-Alloy Steels of Construction Machines

Alexander Scherbakov, Anna Babanina, Elena Kuzbagarova, and Artur Kuzbagarov

Abstract The initial structure of metals and alloys determines their mechanical properties. They can differ significantly in various elements of welded metal structures, as well as in the zones and sections of welded joints. Plastic deformation can occur in hazardous areas of stress concentration of elements of operated metal structures. In this work, studies were carried out on samples of low-carbon steel 08 ps and low-alloy steel 10KhSND, which were subjected to thermal cycling in the following conditions: as delivered state; as delivered + annealing at 900 °C; as delivered + cold rolling for the degree of deformation $\varepsilon = 50\%$. The formation of a fine-grained structure during thermal cycling was monitored by a passive fluxgate method, as well as by microstructural analysis and hardness testing. The value of the stray magnetic field strength H_p was measured. As a result of the conducted studies, a relationship was established between the magnetic parameter H_p and structural changes in steels during thermal cycling.

Keywords Welded metal structures · Construction machines · Passive fluxgate control

1 Introduction

In the manufacture of welded metal structures of construction machines, low-carbon and low-alloy steels are widely used, which in the as delivered state (hot-rolled, cold-rolled, after various types of heat treatment) have a different initial microstructure

A. Scherbakov (✉) · E. Kuzbagarova · A. Kuzbagarov
Saint Petersburg State University of Architecture and Civil Engineering, Vtoraya Krasnoarmeiskaya Street, 4, Saint Petersburg, Russia 190005

A. Babanina
Peter the Great St. Petersburg Polytechnic University, Polytechnicheskaya Street, 29, Saint Petersburg, Russia 195251

A. Beskopylny and M. Shamtsyan (eds.), *XIV International Scientific Conference "INTERAGROMASH 2021"*, Lecture Notes in Networks and Systems 247,
https://doi.org/10.1007/978-3-030-80946-1_15

(coarse-grained, fine-grained, plastically deformed). In addition, structural heterogeneity is inherent in welded joints: a cast structure in the welded seam, coarse-grained in the overheating area, and fine-grained in the area of complete recrystallization. Considering that the initial structure determines the mechanical properties of metals and alloys, they can significantly differ in various elements of welded metal structures, as well as in the zones and sections of welded joints. Besides, plastic deformation can occur in hazardous areas of stress concentration of elements of operated metal structures. In this regard, it is advisable to conduct study to assess the degree of influence of the metal structure on the operational reliability of elements of metal structures of construction machines.

Numerous experimental studies carried out show that the strength of annealed metals and alloys tested at sufficiently low temperatures, when the recovery processes are difficult, increases with grain refinement. Thus, with a decrease in the grain size in the range from 10 to 1 μm, the yield stress, flow stress at various values of deformation, ultimate strength, hardness and fatigue strength of a significant number of metals and alloys increase.

The fine grain structure has a significant effect on structural strength. Iron and other metals with a body-centered lattice tend to transition from a ductile to a brittle state at a certain tensile test temperature and low strain rates. At temperatures below the brittleness threshold, the sample collapses without necking at low plasticity values. The impact toughness of steel Jn-744 greatly depends on the grain size. When it changes from 2 to 25 μm, the transition temperature from ductile to brittle fracture shifts towards higher temperatures from −130 to −45 °C. The observed difference in the properties of steel is caused solely by the difference in the dispersion of the structure, since the chemical composition of the phases of both fine-grained and coarse-grained material was the same.

Since with a decrease in the grain size at low temperatures, the yield stress, strength, hardness, fatigue strength, and impact toughness increase, obtaining a fine-grained structure is of independent importance for increasing the strength properties of metals. This phenomenon can be used when strengthening the elements of metal structures of construction machines in hazardous stress concentration zones due to the refinement of the metal structure according to the modes of thermal cycling developed in the dissertation work.

One of the conditions for ensuring a given level of quality of metal structures is the correct use of existing and development of new devices and methods of non-destructive testing. At the same time, in order to reduce production costs and reduce the cost of products, it is desirable to carry out non-destructive testing not only after final processing, but also at individual stages of their manufacture. Such possibilities are provided by the passive fluxgate control method, which was used for magnetic control of structural transformations in steels during thermal cycling.

Thermal cycling is one of the most effective ways to obtain a microstructure with a given degree of dispersion. It is based on the constant accumulation, from cycle to cycle of heating and cooling, positive changes in the structure of metals. At the same time, an important feature of the metal heating–cooling cycle is its intensity, and what is the most important is the absence or presence of short exposures at extreme temperatures, as well as the optimal range of temperature variation. During thermal cycling, additional sources of influence on the structure appear, characteristic only of the process of continuous temperature change, the main of which are phase transformations, temperature gradients, thermal (volumetric) and interfacial stresses caused by the difference in thermophysical characteristics of the phases constituting the structure [1–5].

The use of thermal cycling to obtain structures with a given degree of dispersion is important for two reasons. First, it becomes possible to obtain metal microstructures characteristic of the factory-supplied rolled metal used in the manufacture of welded metal structures and structural heterogeneity of welded joints. Secondly, to obtain a fine-grained structure, which is unattainable with conventional types of heat treatment [6–9]. This, in turn, makes it possible to use the method of thermal cycling to restore the metal structure and enhance hazardous stress concentration zones in the elements of welded metal structures and welded joints with structural heterogeneity, due to the formation of a finer-grained structure with increased strength properties in them. It should be noted that the development of thermal cycling modes in each specific case has an individual character and cannot be mechanically transferred from one material to another.

2 Materials and Methods

For research, samples were cut out of 08 ps mild steel and 10KhSND low-alloy steel, which were subjected to thermal cycling in the following conditions: as delivered condition; as delivered + annealing at 900 °C; as delivered + cold rolling for the degree of deformation $\varepsilon = 50\%$.

The formation of a fine-grained structure in the course of thermal cycling was controlled by the passive fluxgate method, as well as by microstructural analysis and hardness testing. The measurement of the stray magnetic field strength Hp was carried out by a two-channel fluxgate transducer installed perpendicular to the sample surface. On each sample, in its middle part along the axial line, the Hp values were recorded in three control zones spaced 30 mm apart from each other. The Hp values were measured both before the first thermal cycle and at the end of each heating–cooling cycle. To plot the dependences of the Hp values on the number of cycles, the averaged Hp value of the indicated control zones was taken. The number of cycles was chosen as follows: 1, 2, 3, 4, 5, 7, and 10.

3 Results and Discussion

Figure 1 shows the dependence of Hp on the number of cycles during thermal cycling of steels 08 ps and 10KhSND in different initial states.

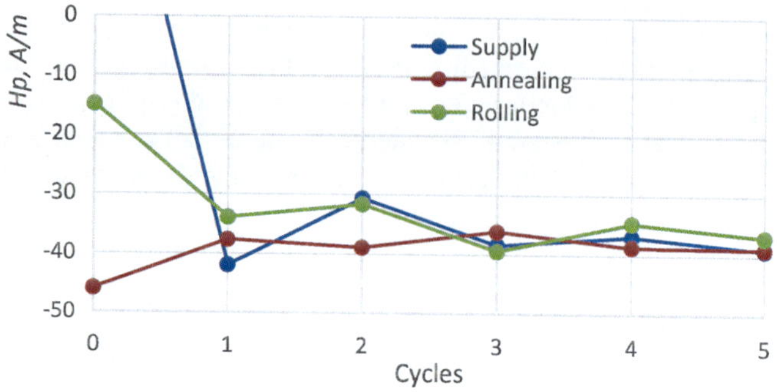

A)

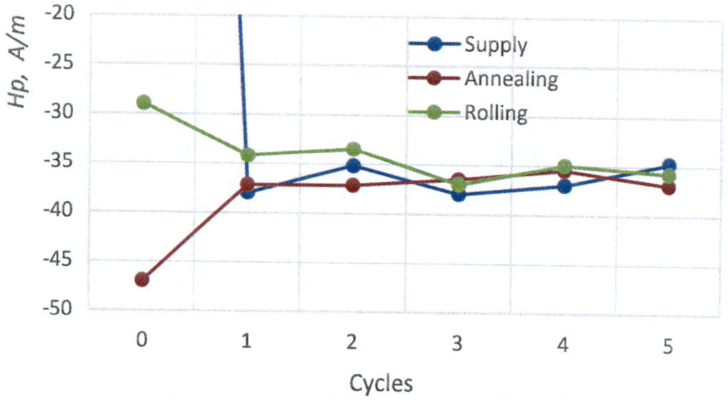

B)

Fig. 1 Dependence of the magnetic field strength H_p on the number of cycles during thermal cycling of steels 08 ps (**a**) and 10KhSND (**b**) in different initial structural states

We can see that the greatest changes in the values of Hp for all studied steels are observed during the first two or three thermal cycles. In this case, the magnetic prehistory of the samples is erased, as evidenced by the change from the positive sign to negative and the convergence of the Hp values. With an increase in the number of cycles, the magnetic field strength values stabilize, which is noticeable from a decrease in the spread in Hp values. An increase in the number of cycles from 5 to 7 and 10 has almost no effect on the change in the magnetic parameter Hp.

During the thermal cycling of the samples in the as delivered state, there is a sharp change in the value of the stray magnetic field strength after the first cycle (Fig. 2). During subsequent thermal cycles, negative values of Hp do not change their sign.

Annealing before thermal cycling shifts the initial values of Hp to the region of negative values (Fig. 3a), while no further significant changes in Hp are observed. Cold plastic deformation has a significant effect on the change in Hp in the first three cycles (Fig. 3b), then this effect decreases, which indicates a significant convergence of the structural state of the steels when the number of cycles is more than three. This

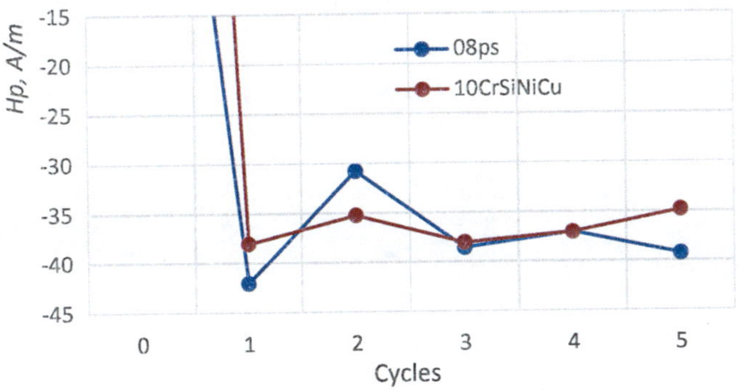

Fig. 2 Dependence of the magnetic field strength H_p on the number of cycles during thermal cycling of 08 ps and 10KhSND steels as delivered

dependence of Hp on the number of thermal cycles has a positive character, since it makes it possible to judge the completeness of the structure refinement process in hazardous zones of stress concentration in the elements of metal structures of construction machines under passive fluxgate control, where plastic deformation somehow took place during operation.

Thus, it can be seen that, regardless of the steel grade, after the fifth heating–cooling cycle, the difference between the maximum and minimum values of Hp of the steels under study (for different initial structural states) is approximately the same and amounts to 4 A/m, which indicates a relatively close final structural state of these steels after thermal cycling. An increase in the number of cycles to 10 almost does not reduce the scatter of the Hp parameter values.

The chemical composition of steels has a significant effect on the change in the stray magnetic field strength H_p. As we can see from Fig. 4, the H_p parameter undergoes the greatest changes in low-carbon steel 08 ps and the smallest in low-alloy steel 10KhSND.

The initial structure of steels also has a significant effect on the change in Hp depending on the number of thermal cycles. Thus, steels with the coarse-grained and more equilibrium structure (as delivered state + annealing at 900 °C), as compared to finer-grained steels in the as delivered state (Fig. 3a and Fig. 2, respectively), have smaller changes in Hp, starting from the 1st cycle. Steels that have undergone preliminary cold plastic deformation are characterized by more significant fluctuations in the values of Hp, especially at the 2nd and 3rd thermal cycles (Fig. 3b).

It should be noted that changes in the strength Hp depending on the chemical composition of steels, the initial microstructure, and the number of cycles are associated with structural changes in steels that occur during thermal cycling. Typical microstructures of steels after thermal cycling are shown in Figs. 5, 6, 7, 8 and 9. Metallographic studies have shown that, regardless of the initial state of the samples, a fine-grained structure is formed in all studied steels after the 3rd cycle. A further increase in the number of cycles less significantly refines the structures, which is

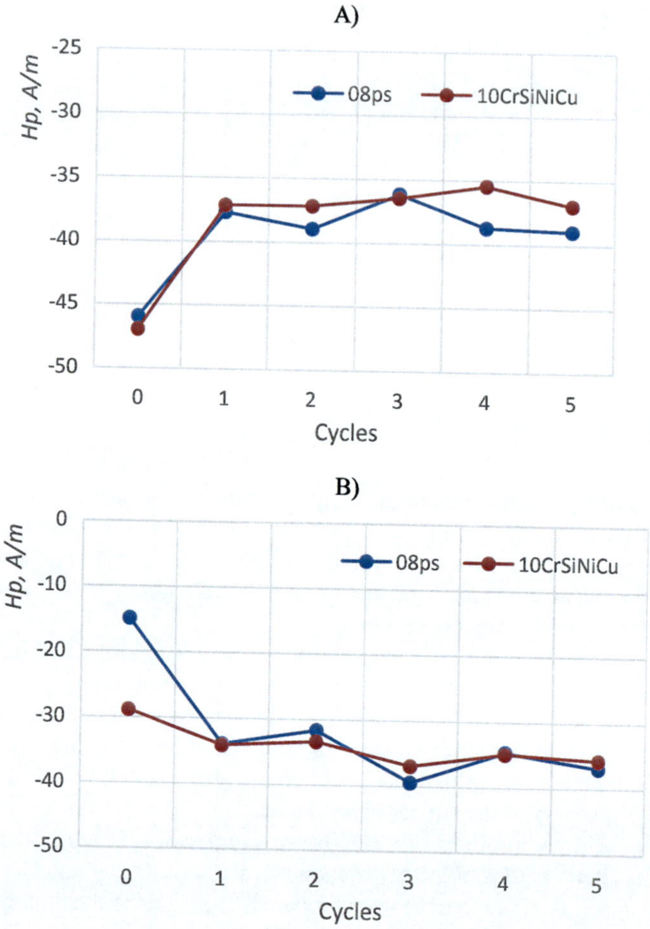

Fig. 3 Dependence of the stray magnetic field strength Hp on the number of cycles during thermal cycling of steels 08 ps and 10KhSND after preliminary annealing at 900 °C (**a**) and cold plastic deformation by $\varepsilon = 50\%$ (**b**)

consistent with the data on changes in the magnetic field strength: in subsequent cycles, a decrease in the amplitude of fluctuations in the magnetic field strength is noted.

Significant grain refinement in some areas in comparison with others after the 1st thermal cycle was also established by metallographic studies in [10]. A further increase in the number of cycles refined the structure in all areas.

It should be noted that in low-alloy steel 10KhSND, in comparison with low-carbon steel 08 ps, a finer-grained structure is formed after 5-fold thermal cycling. In addition, in alloy steel, a greater number of thermal cycles is required to obtain a fine-grained structure without noticeable grain variation than for carbon steel. Probably,

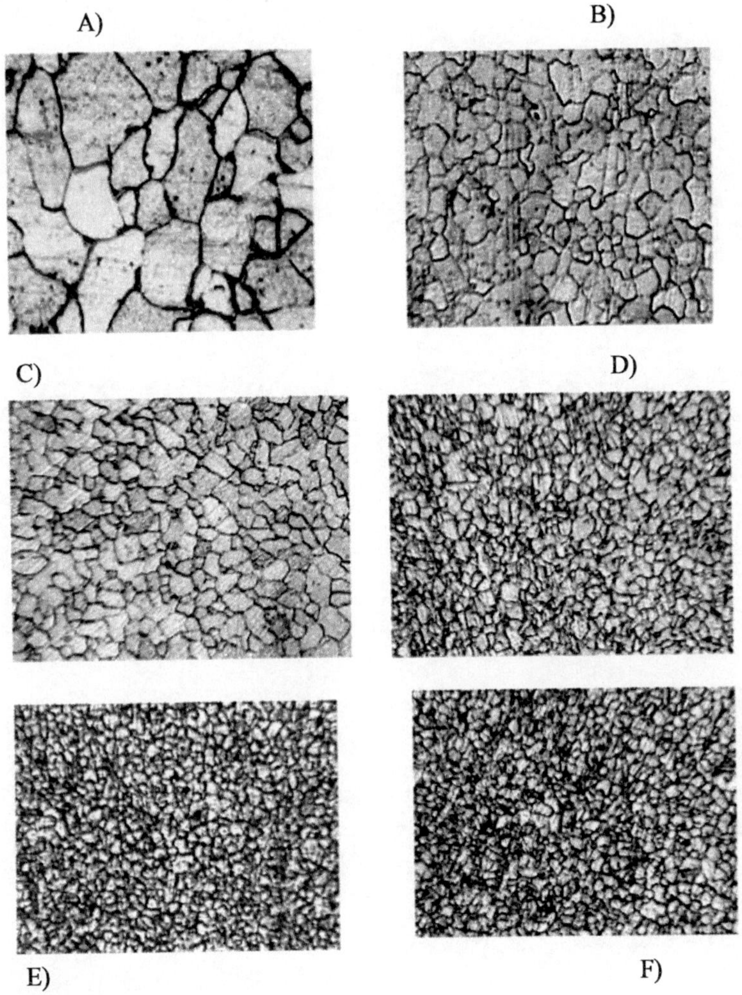

Fig. 4 Change in the structure of 08 ps steel during thermal cycling, ×650: **a** as delivered, **b–f** after 1–5 cycles, respectively

the answer must be sought in the influence of alloying elements on the formation of the final structure of steel during thermal cycling.

Alloyed steels differ in that the thermodynamic activity of carbon in them is lower than in carbon steels. Therefore, a slowdown in the diffusion processes of cementite dissolution and its release from austenite leads to a shift of the C-shaped curve of isothermal decomposition of austenite to the right, making supercooled austenite more stable. Moreover, the previously described methods of thermal cycling of carbon steels can be applied to alloyed steels of the pearlite class without significant

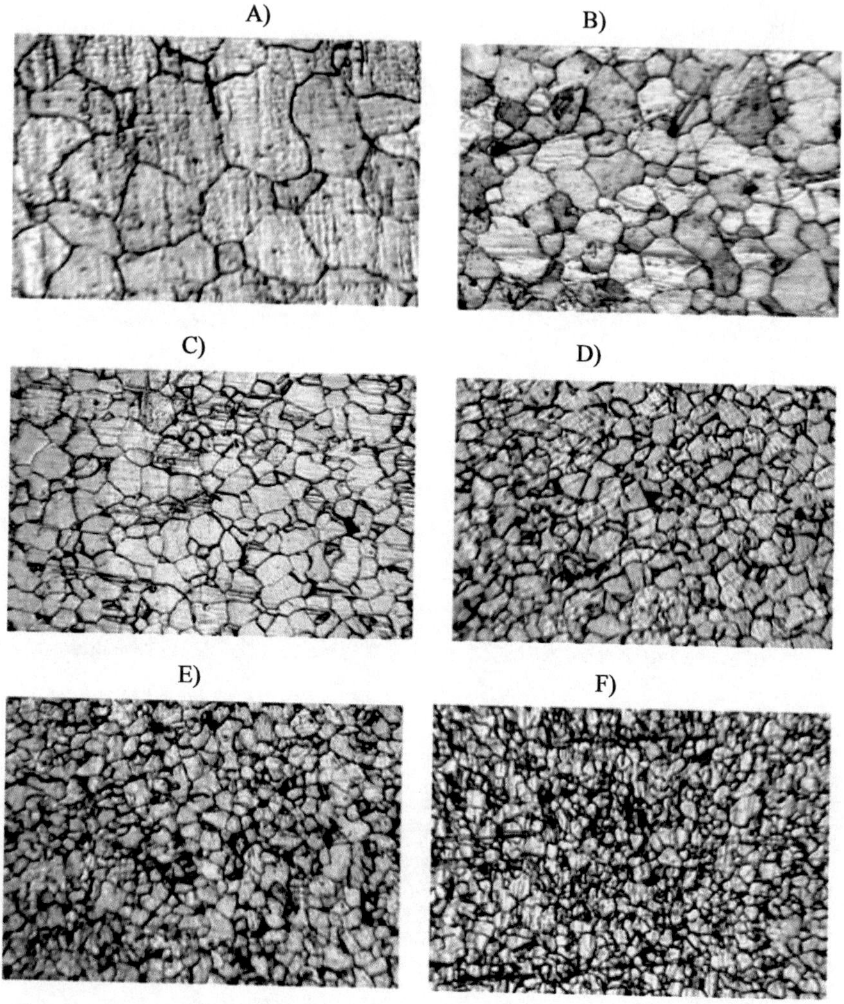

Fig. 5 Change in the structure of 08 ps steel during thermal cycling, ×650: **a** state (as delivered + annealing at 900 °C), **b–f** after 1–5 cycles, respectively

changes in technology, when the total content of alloying elements does not exceed 5%.

During thermal cycling, phase and structural transformations in steels are accompanied by the formation, displacement and annihilation of point and linear defects, as well as the redistribution of alloying elements. In the process of thermal cycling, the existing dislocations are set in motion, while the formation and multiplication of new dislocations takes place. During thermal cycling of iron, an increase in the dislocation density begins immediately after the start of treatment. With an increase in the number of cycles (thermal cycling time), the dislocation density increases.

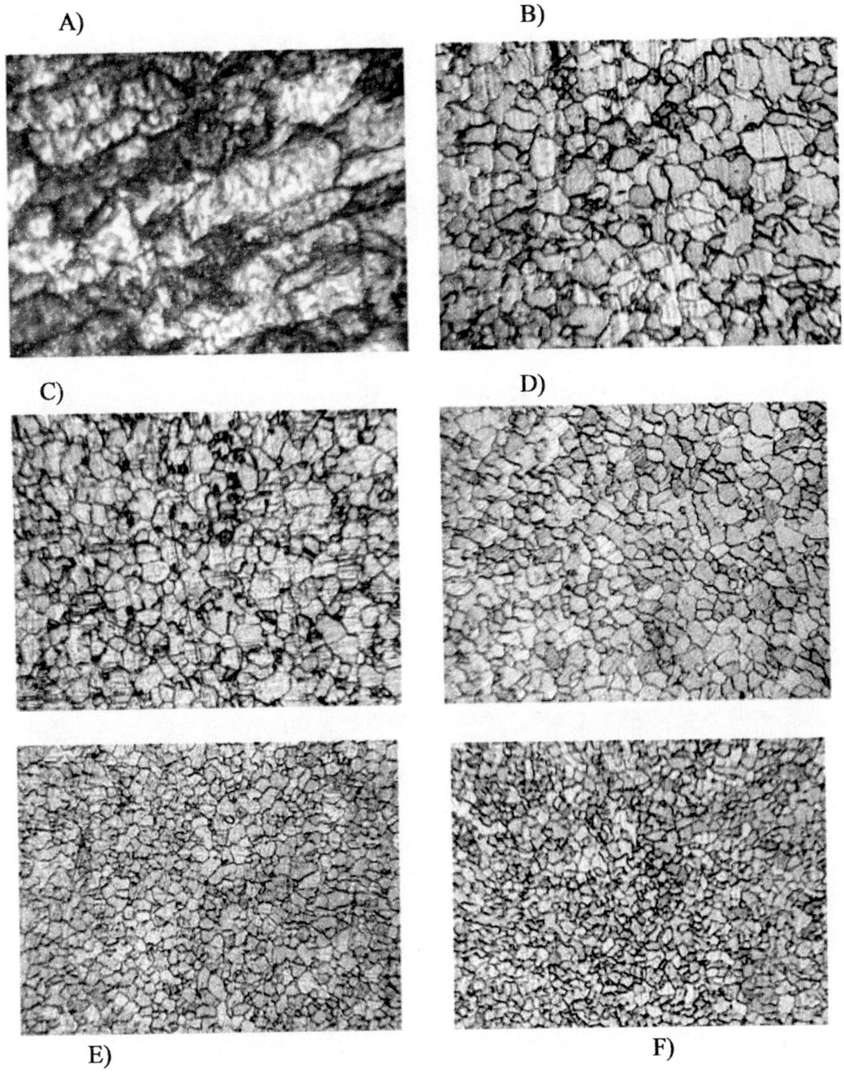

Fig. 6 Change in the structure of steel 08 ps during thermal cycling, ×650: **a** condition (as delivered + rolling by $\varepsilon = 50\%$), **b**–**f** after 1–5 cycles, respectively

Consequently, the efficiency of thermal cycling will be determined by the degree of generation of defects and, first of all, dislocations.

During thermal cycling, there is a sharp change in the microstructure, substructure, and dislocation structure. The main structural change in steel is significant grain refinement, be it pearlite, ferrite, martensite, or other structure. Moreover, the

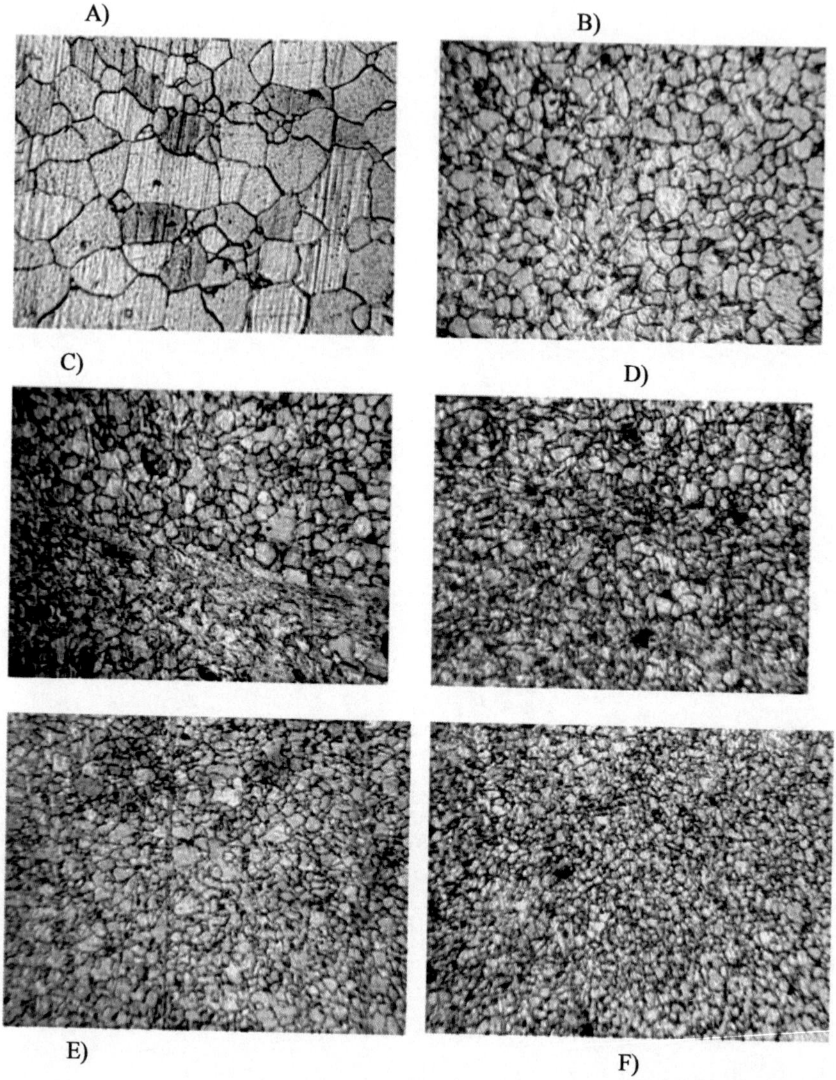

Fig. 7 Change in the structure of steel 10KhSND during thermal cycling, ×650: **a** as delivered, **b–f** after 1–5 cycles, respectively

grinding of grains, as a rule, is accompanied by a decrease in grain size and a more uniform distribution of chemical elements.

The main reason for such changes in the structure is the intensification of diffusion processes due to the intensification of the effect of thermophysical factors. In the process of thermal cycling, the existing dislocations are set in motion, while the formation and multiplication of new dislocations takes place. During thermal cycling

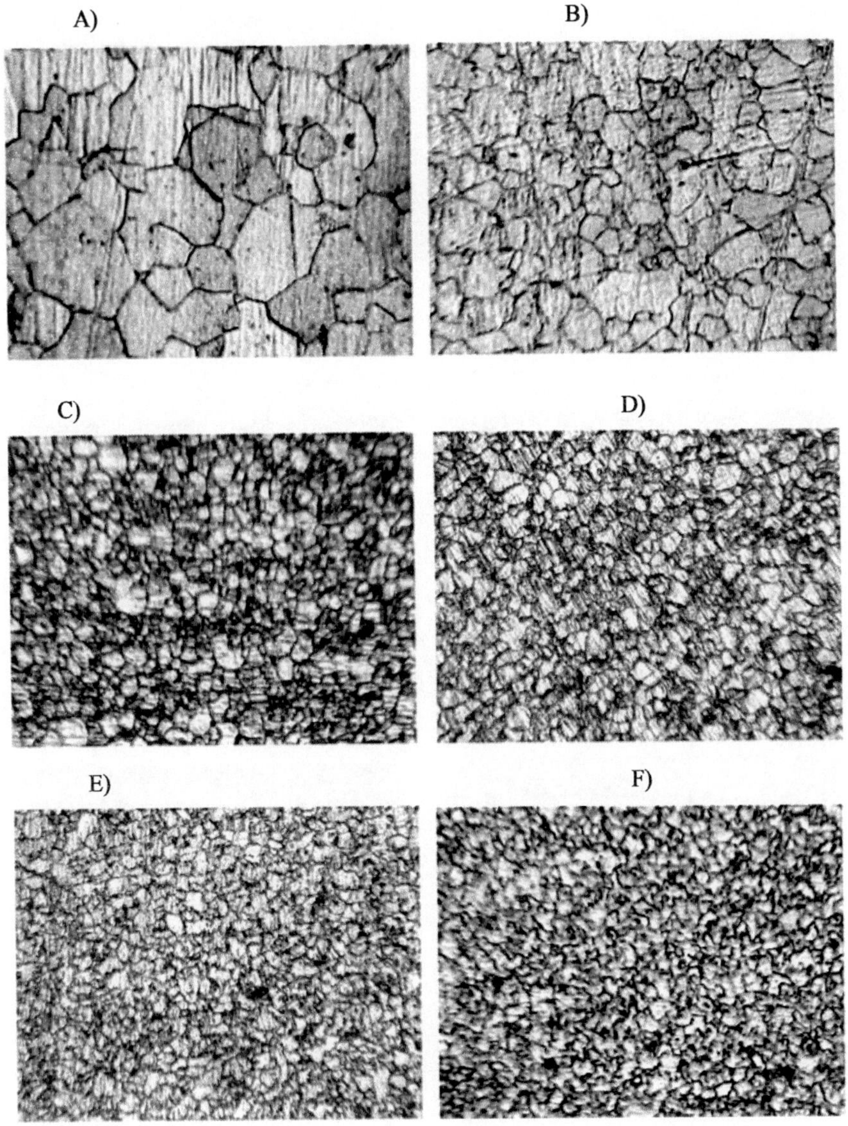

Fig. 8 Change in the structure of steel 10KhSND during thermal cycling, ×650: **a** state (as delivered + annealing at 900 °C), **b–f** after 1–5 cycles, respectively

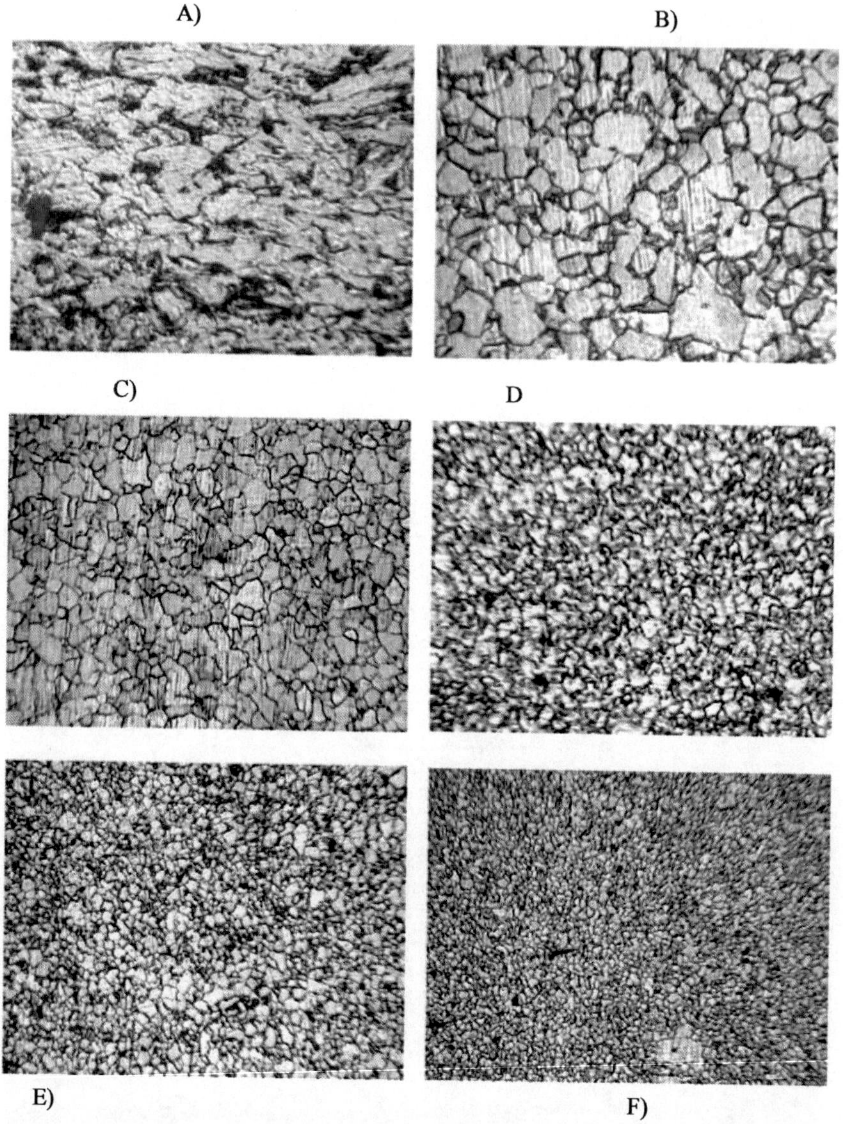

Fig. 9 Change in the structure of steel 10KhSND during thermal cycling, ×650: **a** condition (as delivered + rolling by $\varepsilon = 50\%$), **b–f** after 1–5 cycles, respectively

of iron, zirconium, tungsten, as well as low-temperature thermal cycling of molybdenum, an increase in the dislocation density begins immediately after the start of treatment.

The accumulation of dislocations and the formation of a polygonal substructure after thermal cycling of pure iron and low-carbon steel suggest that polymorphic transformations are responsible for the formation of a dislocation structure during thermal cycling of steels, leading to phase hardening mainly due to the difference in specific volumes and elastic moduli of the formed phases. Phase hardening is sometimes accompanied by recrystallization processes, which, with the accumulation of deformation, are monotonously repeated from cycle to cycle.

Recrystallization centers are formed, first of all, in those areas of the lattice that are most distorted, including at the grain boundaries and their joints, which leads to the formation of a fine-grained structure. The recrystallization process during thermal cycling can be represented as a multiple alternation of small deformations and recrystallization annealing.

In the light of the above, it becomes clear why in low-alloy steel 10KhSND, in comparison with low-carbon steel 08 ps, the process of decreasing grain size is shifted towards a larger number of cycles. This is explained by the fact that in alloyed steels, the thermodynamic activity of carbon is lower than in carbon steels, and, consequently, structural changes during thermal cycling in them occur more slowly. Therefore, the influence of alloying elements on the number of cycles during thermal cycling is such that with an increase in their percentage in steel, an increase in the number of cycles is required to obtain an equiaxed fine-grained structure.

The microstructure of steels before thermal cycling significantly affects the final grain size. The coarser-grained initial structure of steels (as delivered + annealing at 900 °C), in comparison with the as delivered structure, also corresponds to the coarser-grained structure after thermal cycling (Figs. 4, 5, 7 and 8). Figure 5 shows the structure of steel 08 ps after five-time thermal cycling. Before thermal cycling, the as delivered samples were annealed for 30 min at 900 °C, as a result of which an equilibrium structure was formed in them. In this case, as in the thermal cycling of as delivered samples, after the first cycle, a structure with different grain sizes is formed in different areas. However, the graininess in this case is somewhat less, and the grain size is larger. In subsequent cycles, further refinement of the grains occurs, and after the fifth cycle, a fairly homogeneous structure with fine grains is formed. The structure of steel 10KhSND at different stages of processing, in comparison with the previous steels, has a lower grain size difference, and after the fifth cycle, it has a homogeneous fine-grained structure. It should be noted that with an increase in the degree of alloying of steels, a finer-grained structure is formed in them, both in the state (as delivered + annealing at 900 °C) and in the as delivered state.

Consequently, during thermal cycling, when grain growth is suppressed by low austenitization temperatures and short holding times, the optimal initial structure will be such a structure that provides the maximum density of carbides at high-angle boundaries of ferrite grains. In our case, steel 10KhSND may have such an opportunity.

The presence of a deformed structure before thermal cycling makes significant adjustments for both carbon steel 08 ps and low-alloy steel 10KhSND (Figs. 7 and 8). A significant difference between the thermal cycling of pre-deformed samples is the obtaining of a finer-grained structure than that of the as delivered samples and in the state (as delivered + annealing at 900 °C), which is associated with the activation of structural changes that occur during heating and cooling.

A significant refinement of the structure during thermal cycling after cold plastic deformation is explained by the fact that cold deformation redistributes and increases the density of imperfections in the crystal structure of dislocations, vacancies, stacking faults, and, in addition, promotes the formation and development of low and high angle boundaries. Since crystal lattice defects strongly affect the formation of the structure of alloys during phase and structural transitions, plastic deformation in front of them, as well as during their passage, can be effectively used to create an optimal structure during thermal cycling of steels.

A more intense passage of structural changes during thermal cycling of cold-worked steels and the final formation of an equiaxed fine-grained structure after the third cycle is confirmed by a change in the H_p values during cyclic annealing. So, from Fig. 3c, it can be seen that the values of H_p for the studied steels after the 4th and 5th cycles are almost equal, which indicates the end of the effective refinement of the grain size and the formation of a fine-grained structure. Consequently, using the magnetic method, it becomes possible to control the formation of the finest-grained structure in low-carbon and low-alloy steels at the early stages of their thermal cycling.

4 Conclusions

As a result of the conducted studies, a relationship was established between the magnetic parameter Hp and structural changes in steels during thermal cycling. It is shown that the magnitude of the stray magnetic field strength Hp depends on the initial microstructure, the chemical composition of the steels, and the number of thermal cycles.

The greatest change in the values of the stray magnetic field strength H_p is observed during the first heating–cooling cycles. Their further increase decreases the magnetic amplitude, which is associated with a less significant refinement of the structure during subsequent cycles. This type of change in H_p values can be used to control the formation of a fine-grained structure in low-carbon and low-alloy steels during thermal cycling.

References

1. Scherbakov AP (2020) Experimental studies of heat treatment influence on the welded connections properties in working mechanisms of road building. Russ Automob Highway Ind J 17(6):664–675
2. Scherbakov A, Monastyreva D, Smirnov V (2019) Passive fluxgate control of structural transformations in structural steels during thermal cycling. E3S Web of Conferences, vol 135, 03022. https://doi.org/10.1051/e3sconf/201913503022
3. Scherbakov A, Babanina A, Graboviy K (2021) Acting stresses in structural steels during elastoplastic deformation. Adv Intell Syst Comput 1259:298–311
4. Lupachev AV, Pavlyuk SK (2019) Determination of electrode potentials on local surfaces of metals and welded joints. Factory laboratory. Diagnost Mater 8:43–46
5. Scherbakov AP, Pushkarev AE, Vinogradova TV (2021) Influtnce analysis of 09Г2С and 30MnB5 steels thermocyclic treatment on strength characteristics of road construction machines working bodies. Russ Automobile Highw Ind J 18(2):180–190. (In Russ.) https://doi.org/10.26518/2071-7296-2021-18-2-180-190
6. Scherbakov A, Babanina A, Kochetkov I, Khoroshilov P (2020) Technical condition of welded load-bearing metal structures of operated agricultural hoisting cranes. E3S Web of Conferences, vol 175, 11005. https://doi.org/10.1051/e3sconf/202017511005
7. Scherbakov AP (2020) Material and method selection for increasing the wear resistance of construction machines components. Russ Automob Highway Ind J 17(4):464–475
8. Fireushin AM, Fatkullin MR, Ulyabaev RR, Chernyatyev RR, Kadoshnikova AV (2018) Improving the technology for repairing cases of thick-walled machines and assemblies. Neftegazovoe delo 16(3):76–83
9. Plotnikov DG (2015) Methods for assessing the strength of metal structures of hoisting-and-transport machines. Nauchno-tekhnicheskie vedomosti SPbGPU 1(214):186–193
10. Scherbakov A, Babanina A, Matusevich A (2021) Passive probe-coil magnetic field test of stress-strain state for welded joints. Adv Intell Syst Comput 1259:312–323

Indicators of Road Safety as a Phenomenon of National Security of the State

Anatoli Kvitchuk, Margaryta Kvitchuk, Sergey Evtyukov, and Egor Golov

Abstract The article is devoted to the actual problem of ensuring road safety. In this article, on the basis of official statistical data, the issues of the organization of complex statistical analysis are considered. The following methods of statistical analysis were used to assess road safety: descriptive analysis (description of initial data, a priori analysis); relationship analysis (correlation and regression analysis, variance analysis); multivariate statistical analysis (linear and nonlinear analysis, cluster analysis, factor analysis, etc.); time series analysis (dynamic models and forecasting). The paper emphasizes the importance of quantitative analysis, and suggests some methods for assessing road safety. The study analyzed traffic accidents in Russia in the period from 2009 to 2019, the conclusions on the dynamics of change and also determined that, in assessing road safety as a phenomenon of national security used by each of the methods has its advantages and disadvantages, which are manifested in varying degrees, depending on purpose and depth of analysis of the research object, the technical capabilities, therefore, the most effective will be sharing discussed in the article methods.

Keywords Monitoring · Road safety · a priori analysis · Correlation and regression analysis · Time series

1 Introduction

Road traffic accidents cause colossal material, social and demographic damage to society. In Russia, according to the Main Directorate for Traffic Safety of the Ministry of Internal Affairs of Russia for 2019, the total number of road accidents was equal

A. Kvitchuk · M. Kvitchuk
Saint Petersburg University of the Ministry of the Interior of the Russian Federation, 1, Letchika Pilyutova street, Saint-Petersburg, Russia

S. Evtyukov · E. Golov (✉)
Saint Petersburg State University of Architecture and Civil Engineering, 2-Krasnoarmeyskaya st., 4, Saint Petersburg, Russia

A. Beskopylny and M. Shamtsyan (eds.), *XIV International Scientific Conference "INTERAGROMASH 2021"*, Lecture Notes in Networks and Systems 247,
https://doi.org/10.1007/978-3-030-80946-1_16

to 164,358 accidents [1]. The number of deaths due to road traffic accidents was 16,981 people.

The first stage of road safety analysis is statistical observation or monitoring—a system for observing indicators, which in turn is information and analytical support for the process of making managerial decisions in the departments of the State Traffic Inspectorate of the Ministry of Internal Affairs of the Russian Federation.

Of course, when organizing monitoring, it is necessary to use the usual principles of statistical research—representativeness and economy. This means that monitoring should be aimed at obtaining facts that are statistically significant for a given system, allowing them to be used to assess processes and phenomena of a system-wide nature and have a general focus and organization of observations [2].

During monitoring, such a problem arises as the manipulation of information reflected in quantitative indicators. In solving this problem, it is necessary to use methods of quantitative analysis to bring the observation results into an information-analytical form [3].

2 Methods

Let's list some methods of statistical analysis in assessing road safety: descriptive analysis (description of the initial data, a priori analysis); relationship analysis (correlation and regression analysis, analysis of variance); multivariate statistical analysis (linear and nonlinear analysis, cluster analysis, factor analysis, etc.); time series analysis (dynamic models and forecasting), etc.

The first group includes a priori analysis. What is the essence and significance of this method? By the methods of a priori analysis, we mean the totality of specific forms of theoretical and practical approach to the collection and analysis of information (data) on the scale of road traffic accidents, including:

- identification of economically justified and significant cause-and-effect relationships between signs and phenomena;
- assessment of the homogeneity of the studied population;
- analysis of the nature of the distribution of the population according to the studied characteristics.

An important task of research in the field of road safety at the stage of a priori analysis is to identify homogeneous groups [4]. Grouping is done to study the structure of a given population or for the relationship between the indicators that make up the population. Using this method, you can identify how individual units of the population affect the average total indicators.

When assessing the level of road safety using a priori analysis, it is possible to study the anomalies of the observed accident rate on the roads of Russia [5].

Average values are of no small importance in research, their calculation depends on the goal of the study, on the type and relationship of the studied characteristics, as well as on the nature of the initial data. Figure 1 shows the classification of average

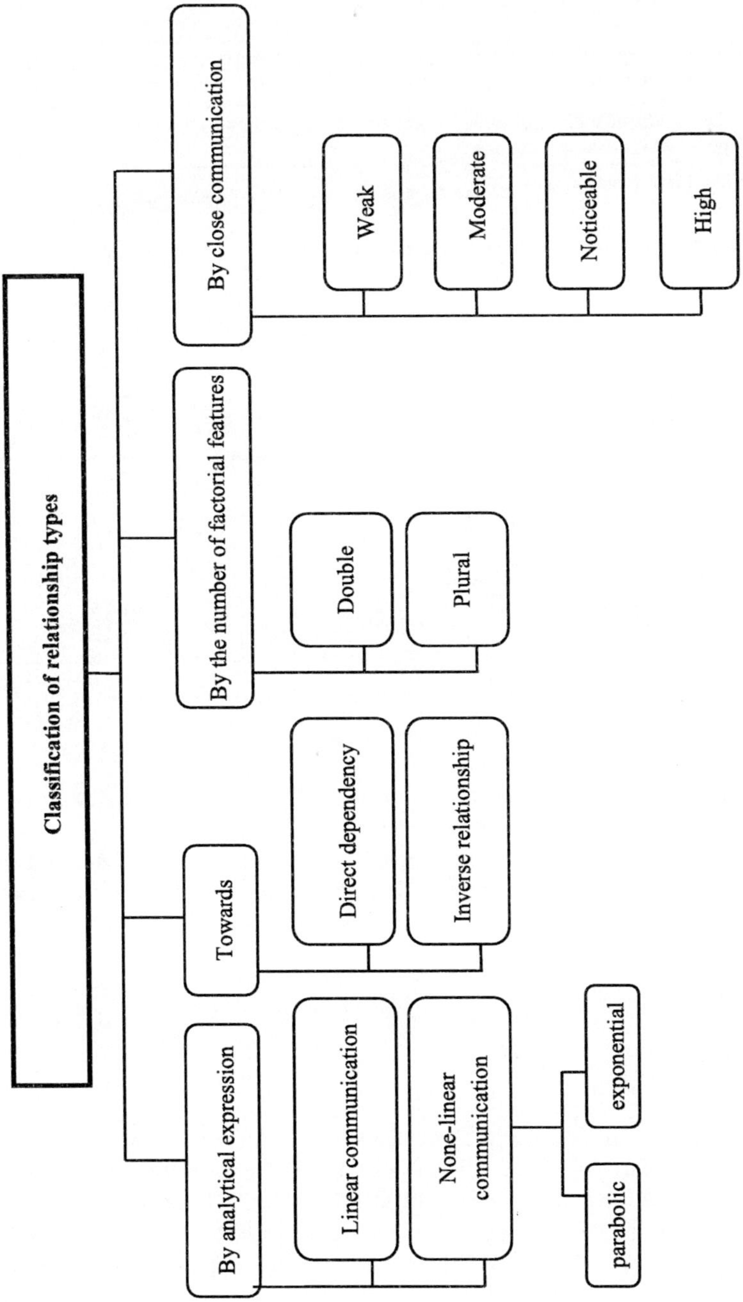

Fig. 1 Classification of relationship types

Table 1 Types of power averages and their applications

K	Name average	Average formula	
		simple	weighted
1	Arithmetic	$\overline{X}_{ar} = \frac{\sum X_i}{N}$	$\overline{X}_{ar} = \frac{\sum X_i f_i}{\sum f_i}$
It is used when replacing individual values of a characteristic; the total volume of the characteristic must be saved without changing			
– 1	Harmonic	$\overline{X}_{\mathbf{HM}} = \frac{N}{\sum \frac{1}{X_i}}$	$\overline{X}_{\mathbf{HM}} = \frac{\sum f_i}{\sum \frac{f_i}{X_i}}$
It is used when variants of a feature are known, its volumetric value, but frequencies are not known			
0	Geometric	$\overline{X}_{geom} = \sqrt[N]{\prod_{i=1}^{N} X_i}$	$\overline{X}_{geom} = \sqrt[N]{\prod_{i=1}^{N} X_i^{f_i}}$
Used for averaging chain dynamics indices			
2	Quadratic	$\overline{X}_{qd} = \sqrt{\frac{\sum X_i^2}{N}}$	$\overline{X}_{qd} = \sqrt{\frac{\sum X_i^2 f_i}{\sum f_i}}$
Used when measuring the variation of a feature			
3	Cubic	$\overline{X}_{cub} = \sqrt[3]{\frac{\sum X_i^3}{N}}$	$\overline{X}_{cub} = \sqrt[3]{\frac{\sum X_i^3 f_i}{\sum f_i}}$
Used to calculate population poverty indices			
1	Chronological	$\overline{X}_{CH} = \frac{\frac{X_1+X_N}{2} + \sum_2^{N-1} X_i}{N-1}$	$\overline{X}_{CH} = \frac{\sum (X_i+X_{i+1}) f_i}{2\sum f_i}$
Used for averaging moment statistics			

values [6].

The shape of the mean depends on the exponent k. Table 1 presents all the main averages used in socio-economic calculations.

One of the practical methods is correlation—regression analysis, which is used to identify factors affecting the number of road accidents. For example, the dependence of the number of accidents on the day of the week, month, type of incident or lighting [7]. The classification of the types of relationship is shown in Fig. 1.

3 Results

In the study, we will consider the impact of economic factors on the number of road traffic accidents in Russia for 2005–2019, using the multiple correlation method. The obtained official statistical data will be presented in Table 2. The resultant indicator (y), of course, will be the number of road accidents, as factor indicators we will take: the inflation rate (x1), the average monthly nominal wages of employees of organizations in Russia (x2) and the value living wage (average per capita; rubles per month) (x3) [8–14].

Table 2 Indicators for assessing relationships using multiple correlation

Year	Inflation rate	Average monthly nominal accrued wages of employees of organizations in Russia	Living wage (average per capita; rubles per month)	Number of road accidents in Russia
2005	10.91	8555	3018	223,342
2006	9.00	10,634	3422	229,140
2007	11.87	13,593	3847	233,809
2008	13.28	17,290	4593	218,322
2009	8.80	18,638	5153	203,603
2010	8.78	20,952	5688	199,431
2011	6.10	23,369	6369	199,868
2012	6.58	26,629	6510	203,597
2013	6.45	29,792	7306	204,068
2014	11.36	32,495	8050	199,720
2015	12.91	34,030	9701	184,000
2016	5.38	36,709	9828	173,700
2017	2.52	39,167	10,088	169,432
2018	4.27	43,724	10,287	168,099
2019	3.05	47,468	10,890	164,358

At the second stage, we build an empirical curve for each factor attribute separately (Figs. 2, 3 and 4).

According to the graphs presented, a preliminary conclusion can be made about the presence between the factor signs and the number of accidents in Russia in the first case of a direct strong relationship, and in the last two an inverse strong relationship. We will use the capabilities of Microsoft Excel and calculate the correlation coefficients (Tables 3, 4 and 5) [15–19].

Fig. 2 An empirical curve and its direction. Inflation rate and number of road accidents in Russia

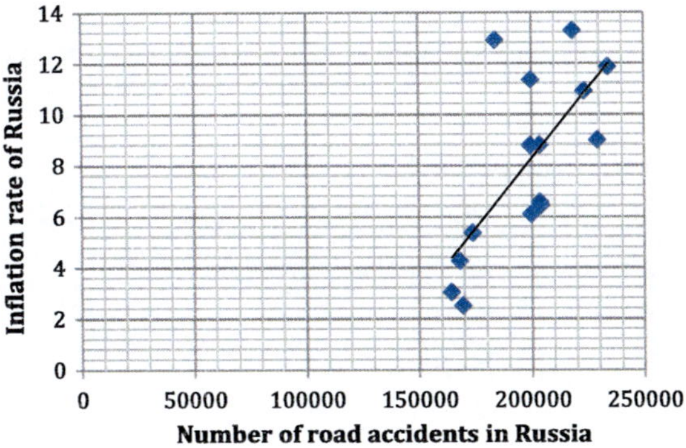

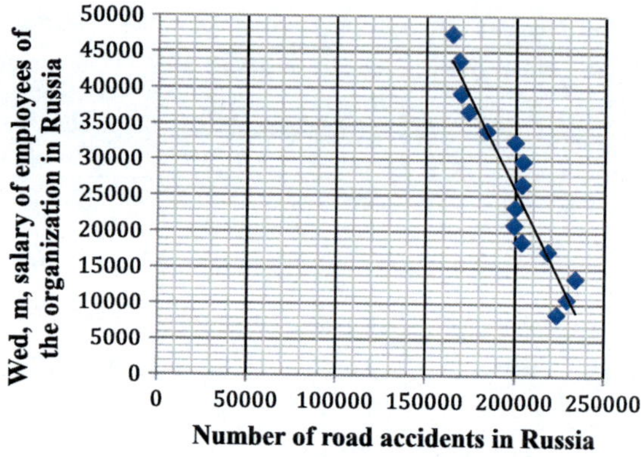

Fig. 3 An empirical curve and its direction. Average monthly nominal accrued wages of employees of organizations in Russia and the number of road accidents in Russia

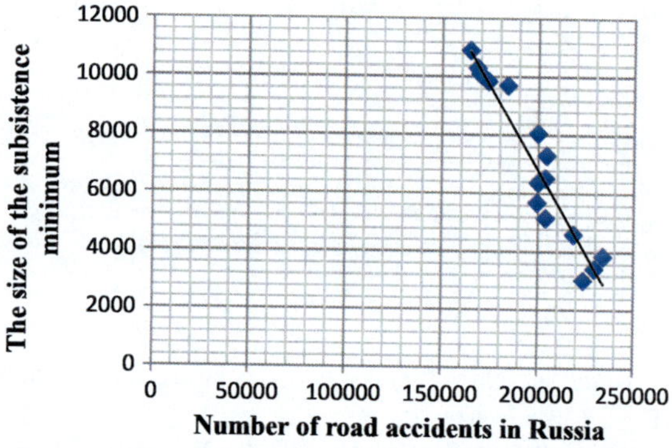

Fig. 4 An empirical curve and its direction. The value of the subsistence minimum and the number of road accidents in Russia

Table 3 Calculation table

R =	Y	X1	X2	X3
Y	1.000	0.700	−0.942	−0.956
X1	0.700	1.000	−0.650	−0.605
X2	−0.942	−0.650	1.000	0.985
X3	−0.956	−0.605	0.985	1.000

Table 4 Calculation table

R11	R22	R33	R44	R12	R13	R14	R23	R24	R34
0.015	0.002	0.040	0.057	0.003	−0.006	−0.019	−0.005	−0.006	0.043

Table 5 Calculation table

r y x1	0.700	r y x1/(x)	−0.562
r y x2	−0.942	r y x2/(x)	0.254
r y x3	−0.956	r y x3/(x)	0.634
r x1 x2	−0.650	r x1 x2/(yx)	0.461
r x1 x3	−0.605	r x1 x3/(yx)	0.533
r x2 x3	0.985	r x2 x3/(yx)	−0.893

$$\begin{aligned} ry(x) &= 0.970 \\ Ry(x) &= 0.941 \end{aligned} \tag{1}$$

Thus, the indicators of the inflation rate, the average monthly nominal accrued wages of employees of organizations in Russia, the value of the subsistence minimum (on average per capita; rubles per month) have a strong influence on the number of road accidents in Russia. To a greater extent, the number of road accidents is influenced by the subsistence minimum, in second place it is necessary to note the average monthly nominal accrued wages of employees of organizations, a moderate relationship is confirmed between the level of inflation and the number of road accidents. Multiple correlation proves that all phenomena in society are interconnected, i.e. there is a relationship between them, which is expressed in the form of causality. In general, the number of road traffic accidents in Russia is decreasing. But still, the state needs to take all measures to improve the situation on the roads, comprehensively covering all areas. This will greatly affect the national security of the state [20–24].

Let us consider the dynamics of the number of road traffic accidents in the Russian Federation as an element of the national security of the state using the time series method.

Table 6 provides information on the number of accidents on the territory of the Russian Federation for 2009–2019.

The dynamics of the number of road traffic accidents in the Russian Federation will be presented graphically by constructing a bar chart by chain rates of growth (Fig. 5).

4 Discussion

Thus, the number of road traffic accidents in 2010 compared to 2009 decreased by 2.049%; in 2011, the number of road accidents compared to 2010 increased by 0.219%; in 2012, the number again decreased by 24.878% in comparison with 2011, and in 2013, on the contrary, a large increase in road accidents (by 35.915%) is noticeable compared to 2012, these changes are clearly visible in the presented diagram. Since 2014, there has been a gradual decrease in road traffic accidents in comparison with the previous year: in 2014—by 2.131%, in 2015—by 7.871%, in

Table 6 Indicators of the dynamics of the number of road traffic accidents on the territory of the Russian Federation for 2009–2019

Year	Number of accidents on the territory of the Russian Federation (pcs)	Absolute gain		Growth rate (%)		Rate of increase (%)	
		Chain	Basic	Chain	Basic	Chain	Basic
2009	203,603	–	–	–	–	–	–
2010	199,431	−4172	−4172	97.951	97.951	−2.049	−2.049
2011	199,868	437	−3735	100.219	98.166	0.219	−1.834
2012	150,144	−49,724	−53,459	75.122	73.744	−24.878	−26.256
2013	204,068	53,924	465	135.915	100.228	35.915	0.228
2014	199,720	−4348	−3883	97.869	98.093	−2.131	−1.907
2015	184,000	−15,720	−19,603	92.129	90.372	−7.871	−9.628
2016	173,694	−10,306	−29,909	94.399	85.310	−5.601	−14.690
2017	169,432	−4262	−34,171	97.546	83.217	−2.454	−16.783
2018	168,099	−1333	−35,504	99.213	82.562	−0.787	−17.438
2019	164,358	−3741	−39,245	97.775	80.725	−2.225	−19.275

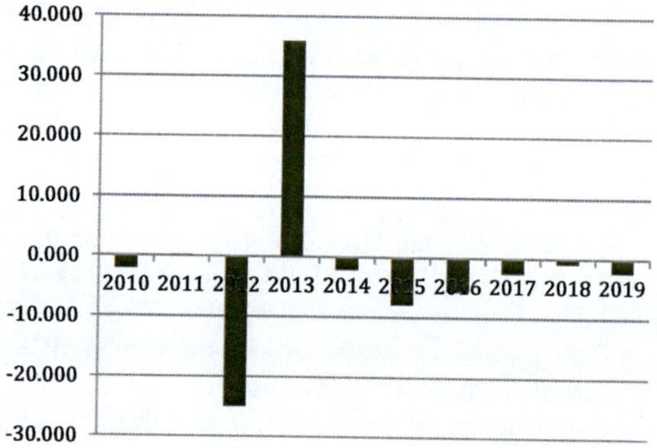

Fig. 5 Chain rates of increase in the number of accidents on the territory of the Russian Federation

2016—by 5.601%, in 2017—by 2.454%, in 2018—by 0.787%, and in 2019—by 2.225% [25–29]. Based on the basic indicators of growth rates, we conclude that the number of road accidents in the territory of the Russian Federation in 2019 decreased by 19.275% in comparison with 2009. Having calculated the average indicators of dynamics, we can formulate that over 10 years the number of road accidents on the territory of the Russian Federation has decreased by 0.807 times. On average, the number of accidents on the territory of the Russian Federation annually decreases by 3924.5 cases, which in relative terms amounted to a decrease of 2.12% per year.

5 Conclusions

Thus, when assessing road safety as a phenomenon of national security of the state, each of the listed methods has its own advantages and disadvantages, which are manifested to one degree or another depending on the purpose and depth of analysis, the object of research, and technical capabilities. Therefore, it will be most efficient to share the methods discussed in this article.

References

1. Information about the indicators of the state of road safety: site. http://stat.gibdd.ru (date of treatment 05/20/2020)
2. Kvitchuk MA (2009) Methods for monitoring the scale of the shadow economy as a threat to the economic security of the state: Diss ... Ph.D. in Economics: 08.00.05. SPb
3. Kvitchuk MA, Kvitchuk AS (2018) A priori analysis and its role in the road safety system: materials of the interdepartmental scientific and practical conference on Actual problems of management of road safety activities (state, problems, ways of improvement), St. Petersburg
4. Kvitchuk MA (2016) Research of macroeconomic indicators of the economic security of the state using statistical methods. Econ Entrep 10–2(75–2):119–121
5. Statistics. Educational visual aid (album of schemes)/comp. M.A. Kvitchuk. SPb.: Publishing house of St. Petersburg University of the Ministry of Internal Affairs of Russia, p 38 (2014)
6. Kvitchuk AS, Kvitchuk MA (2017) Statistical and legal analysis in assessing road safety. Econ Entrep 9–4(86–4):713–716
7. Evtukov S, Golov E (2017) Transp Urals 2(53):85–89. https://doi.org/10.20291/1815-9400-2017-2-85-89
8. Evtiukov S, Golov E, Ginzburg G (2018) Finite element method for reconstruction of road traffic accidents. Transp Res Procedia 36:157–165. https://doi.org/10.1016/j.trpro.2018.12.058
9. Medres EE, Golov EV, Babenko TI (2017) The factors influencing uniformity of the movement of the motor transport in the conditions of saturated transport streams. Transp Bus Russ 2(129)
10. Elena K, Sergei E, Grigory G (2020) Systemic indicators of road infrastructure at accident clusters. Arch Eng 5(1):51–58. https://doi.org/10.23968/2500-0055-2020-5-1-51-58
11. Ziakopoulos A, Theofilatos A, Papadimitriou E, Yannis G (2019) A meta-analysis of the impacts of operating in-vehicle information systems on road safety. IATSS Res 43(3):185–194
12. Monedero BD, Gil-Alana LA, Consuelo M, Martíneza V (2021) Road accidents in Spain: are they persistent? IATSS Research Available online 30 January 2021 In Press, Journal Pre-proof
13. Song Y, Thatcher D, Li Q, McHugh T, Wu P (2021) Developing sustainable road infrastructure performance indicators using a model-driven fuzzy spatial multi-criteria decision making method. Renew Sustain Energy Rev 138:110538
14. Castillo-Manzano JI, Castro-Nuño M (2012) Driving licenses based on points systems: efficient road safety strategy or latest fashion in global transport policy? A worldwide meta-analysis. Transport Policy 21:191–201
15. Kurakina EV, Riazanov SV (2020) Comprehensive analysis of accident rate and causes of complication of road traffic situation. Bull Civ Eng 4(81):189–196. https://doi.org/10.23968/1999-5571-2020-17-4-189-196
16. Kurakina EV, Sklyarova AA (2020) Road safety improvement in road traffic participant—vehicle—road—external environment system. Russ Automob Highw Ind J 4(74):488–499. https://doi.org/10.26518/2071-7296-2020-17-4-488-499
17. Kurakina EV, Lutov DA, Meike UN (2019) Assessment of the road maintenance and risk factors of automobile roads. Bull Civ Eng 1(72):177–183. https://doi.org/10.23968/1999-5571-2019-16-1-177-183

18. Rajczyk P, Knapiński M, Kurakina E (2018) The influence of surface topography on the safety of road and utility surfaces. Transp Res Procedia 36:640–648. https://doi.org/10.1016/j.trpro.2018.12.139
19. Kurakina EV (2018) On the effectiveness of studies carried out at places of road traffic accident concentration. Bull Civ Eng 2(67):231–237. https://doi.org/10.23968/1999-5571-2018-15-2-231-237
20. Elena K, Sergey E (2017) Improvement of the system for accounting of parameters during construction of motor roads. Arch Eng 2(3):34–42
21. Lazarev Y, Bondarenko A, Medres C, Raty J (2017) Method of assessment and prediction of temperature conditions of roadway surfacing as a factor of the road safety. Transp Res Procedia 20:396–400. https://doi.org/10.1016/j.trpro.2017.01.064
22. Lutov DA, Shimanova AA (2018) Perspective directions of improving the traffic safety system in the Russian Federation. Bull Civ Eng 2(67):238–245. https://doi.org/10.23968/1999-5571-2018-15-2-238-245
23. Petrov AI, Evtiukov SA, Petrova DA (2019) Statistical modelling of orderliness of regional road safety provision systems. IOP Conf Ser: Earth Environ Sci. https://doi.org/10.1088/1755-1315/224/1/012033
24. Sokolov NS, Pushkarev AE, Evtiukov SA (2019) Methods and technology of ensuring stability of landslide slope using soil anchors. Geotechnics fundamentals and applications in construction: new materials, structures, technologies and calculations. In: Proceedings of the international conference on geotechnics fundamentals and applications in construction: new materials, structures, technologies and calculations, GFAC 2019, pp 347–350. https://doi.org/10.1201/9780429058882-67
25. Khattak AJ, Wali B (2017) Analysis of volatility in driving regimes extracted from basic safety messages transmitted between connected vehicles. Transp Res Part C: Emerg Technol 84:48–73
26. Heyns T, Heyns PS, de Villiersa JP (2012) A method for real-time condition monitoring of haul roads based on Bayesian parameter estimation. J Terramech 49(2):103–113
27. Shen Y, Hermans E, Bao Q, Brijs T, Wets G (2013) Road safety development in Europe: a decade of changes (2001–2010). Accid Anal Prevent 60:85–94
28. Tselentis DI, Vlahogianni EI, Yannis G (2019) Driving safety efficiency benchmarking using smartphone data. Transp Res Part C: Emerg Technol 109:343–357
29. Hermans E, Van den Bossche F, Wets G (2009) Uncertainty assessment of the road safety index. Reliab Eng Syst Saf 94(7):1220–1228

Technical Rationale for Protective Equipment Propane Tanks in a Model Fire

Vladimir Popov, Eugene Asmankin, Philipp Sukhov, and Julia Bolandova

Abstract Railway transportation of liquefied petroleum gases in special tank wagons requires special attention to safety on the railway. A new method for analyzing the scenario of tank cars of liquefied hydrocarbon gases is proposed. It is based on the methodology for calculating the behavior of a rail tank car with liquefied gas in a fire. This method simulates processes using the «Fobot» software package. This mathematical model considers a metal tank with LPG, the outer surface of which contains a heat-insulating layer consisting of a layer of a porous non-combustible material (mineral wool) and a foaming fire-retardant paint. With the help of "Fobot", calculations of an emergency situation were made using various methods of fire protection. Certain patterns were deduced, conclusions were drawn to assess the degree of elaboration of this topic. The possibilities for improving this complex have become known, as well as the ability to design various emergency situations using all kinds of fire-fighting materials and technically improved devices.

Keywords Emergency situations on the railway · Railway tanks · Thermal risk · Simulation of emergency situations

1 Introduction

A large amount of liquefied petroleum gases (LPG) is transported by rail in special tank cars. A number of accidents with fires and explosions that occurred with railway tank cars for the transportation of LPG indicate a high fire and explosion hazard of the processes of transportation of these substances. Often accidents of LPG tank cars proceed according to the scenario when the tank with LPG is exposed to the fire source, as a result of which an explosion of the tank is possible, with consequences that can be catastrophic. One of the methods of fire protection of LPG tank cars is to choose a safety valve of such a passage section, which, on the one hand, prevents the explosion of the tank car for a given period of time under various accident scenarios, on the other hand, provides a minimum discharge of LPG into the environment.

V. Popov · E. Asmankin (✉) · P. Sukhov · J. Bolandova
Russian University of Transport (MIIT), Moscow, Russia 127994

A. Beskopylny and M. Shamtsyan (eds.), *XIV International Scientific Conference "INTERAGROMASH 2021"*, Lecture Notes in Networks and Systems 247,
https://doi.org/10.1007/978-3-030-80946-1_17

There are few studies of such emergency scenarios. Therefore, the purpose of this work is to develop a more accurate mathematical model of the behavior of LPG tank car located in the fire center and the calculated state of the LPG tank car in a model fire center during a full-scale experiment.

Often, accidents in LPG tank cars proceed according to the scenario when a tank with LPG is exposed to a fire source, as a result of which an explosion of the tank is possible, with consequences that are sometimes catastrophic. To prevent such accidents, various methods of fire protection of LPG tank cars are proposed:

- Increasing the thickness of the bottom of the tanks;
- Equipping tanks with protective screens to protect them from impacts;
- Arrangement of safety arches on tank hatches;
- Use of thermal insulation;
- Application of a fire retardant coating on the outer surface of tanks;
- Increase in flow area of safety devices.

These and other measures for fire protection of LPG tank cars can be developed on the basis of experimental and theoretical research.

In the process of calculating the critical parameters for the operation of the safety valves of tanks, it is necessary to obtain positive or negative results of the selected methods of fire protection of LPG tank cars.

2 Materials and Methods

In this mathematical model, a metal tank with LPG is considered, on the outer surface of which there can be a heat-insulating layer, consisting partly of a layer of porous non-combustible material (such as mineral wool), and the remaining part of the outer surface of the tank can be covered with foaming fire-retardant paint. It is assumed that at the moment of time $\tau = 0$ the temperature of LPG and the temperature of the layers of the two-layer tank wall are equal to the temperature of the ambient air TV and on a certain part of the outer surface of the tank F_0 (heat-insulating layer, if any), an external surface heat source with power q. If there is a fire-retardant coating on the walls of the tank, the heat flux causes its quasi-instant swelling with the formation of a heat-insulating layer with a thickness of L_{frc}, which depends on the initial thickness of the coating layer and its properties. It is assumed that the temperatures of the tank walls, as well as the temperature and pressure of the LPG, do not depend on the coordinates; the ambient temperature is considered constant. The thermophysical properties of the metal of the tank walls, the heat-insulating layer and the layer of the intumescent fire-retardant coating are assumed constant. The convective and radiative heat exchange on the outer surface of the tank with the ambient air temperature TB is taken into account. In the tank boiler, depending on the combination of LPG operating parameters, the following heat transfer modes are possible: surface, bubble or film boiling of liquid; natural convection of a liquid or gas in a large volume. When the safety valve is opened, a critical outflow of the two-phase LPG medium into the surrounding space occurs.

Two-phase region:

$$P < P_{cr}; T < T_{cr}; \tag{1}$$

Continuity equation

$$V\frac{d\left[\rho'(1-\beta)+\rho''\beta\right]}{d\tau} = -G, \tag{2}$$

where, V—tank volume; ρ', ρ''—density of the liquid and gas phases at the saturation line, respectively;

β—volumetric gas content; G—critical flow rate of a gas–liquid mixture flowing out of a safety valve

$$G/S = \left[\frac{(1-\beta)\rho'}{k} + \beta\rho''\right]\sqrt{\gamma RT_S\left(1+\frac{1-\beta}{\beta}\cdot\frac{\rho''}{\rho'}k^2\right)}; \tag{3}$$

$k = w''/w'$—slip coefficient;

$$k = 0.17x^{0.18}\left(\frac{\rho'}{\rho''}\right)^{(1/2)} \text{ – Fauske model} \tag{4}$$

$$k = \left(\frac{\rho'}{\rho''}\right)^{(1/3)} \text{ – Moody model;} \tag{5}$$

$$\gamma = c'_P/c''_P; \tag{6}$$

R—universal gas constant; S—safety valve flow area.

Energy equation

$$V\frac{d[\rho' i'(1-\beta)+\rho'' i''\beta]}{d\tau} = -Gi_k + V\frac{dP}{d\tau} + \alpha_в F_в (T_W - T_S), \tag{7}$$

where i', i''—enthalpy of liquid and gas phases on the saturation line, respectively; i_k—enthalpy of outflowing gas–liquid mixture; P—pressure in the tank boiler; $\alpha_в$—heat transfer coefficient on the inner surface of the boiler tank; $F_в$—internal surface area of the boiler tank; T_w—temperature of the inner surface of the tank boiler; T_S—LPG temperature at saturation line.

$$P = f(T_S) - \text{saturation line} \tag{8}$$

$$\begin{aligned}&\text{Single phase} - P > P_S;\, T < T_{cr};\quad \text{and}\\&\text{supercritical region} - P > P_{cr};\, T > T_{cr};\end{aligned} \tag{9}$$

Continuity equation

$$V\frac{d\rho}{d\tau} = -G, \tag{10}$$

where V—tank volume; $\rho = f(P, T)$—density of single-phase LPG; G—critical flow rate of boiling single-phase liquefied gas flowing out of the safety valve.

Energy equation:

$$V\rho\frac{di}{d\tau} = -G(i_k - i) + V\frac{dP}{d\tau} + \alpha_в F_в(T_W - T_S), \tag{11}$$

where $i = f(P, T)$—enthalpy of LPG; T_W—temperature of the inner surface of the tank boiler.

To calculate the critical pressure in the tank, I took the following parameters:

- model of 4-axle LPG tank 15-1200-02;
- safety valve SV(32–2.25);
- LPG—propane;
- design temperature 20 °C.

3 Results

To calculate (simulate) the development of the accident according to the BLEVE scenario, the «Fobot» software package was used.

The use of technical means of protecting LPG tanks in the fire site should at least prevent the development of an accident according to the BLEVE scenario within the time (4 h) required for the deployment of the fire department [1–11].

The following methods of fire protection are supposed to be applied in the fire center:

- Equipping the tank with safety valves:

  Standard valve—(SV(32–2.25)).
  Response pressure—2.25 MPa.
  Nominal diameter of the flow area $d_y = 32$ mm;

- Covering 95% of the outer surface of the tank with thermal insulation thickness $\delta_{in} = 20$ mm.

  Thermal properties of insulation (identical to mineral wool):

- density—$\rho = 150$ kg/m^3,
- heat capacity - $c = 1000$ J/(kg K),
- coefficient of thermal conductivity—$\lambda = 0.047$ W/(m K);

  – Application of a foaming fire retardant coating SGK-1 in thickness $\delta_{foa} =$ 20 mm on the remaining outer surface of the tank. Thermophysical properties of SGK-1:

- density—$\rho = 10$ kg/m^3,
- heat capacity—$c = 1000$ J/(kg K),
- coefficient of thermal conductivity—$\lambda = 0.07$ W/(m K).

Enter data into «Fobot», start modeling the situation.

Detailed results are structured, summarized and presented calculations are presented in Tables 1 and 2.

4 Conclusion and Discussion

The four-axle tank car for LPG 15–1200-02 is equipped with an SV(32–2.25) safety valve with an opening pressure of 2.25 MPa and a closing pressure of 1.95 MPa. Figure 1 shows that the pressure graph in the tank is limited to 2.25 MPa. After the pressure (Fig. 1) and temperature (Fig. 2) in the tank have dropped, the safety valve SV(32–2.25) closes [12–14].

The «Fobot» program is a complex of databases that can contain the data necessary for the calculation for any types of railway tanks, liquefied gases, safety valves, heat-insulating materials and fire-protective coatings, as well as software interpolation and other auxiliary modules, modules for calculating heat transfer characteristics and numerical solution of a system of differential equations.

Calculations using the «Fobot» software package show that the selected fire protection means:

- equipment of a propane tank (model 15-1200-02) with a safety valve (actuation pressure 2.25 MPa, nominal bore diameter $d_y = 32$ mm);
- covering 90% of the outer surface of the tank with thermal insulation $\delta_{is} = 20$ mm;
- foaming fire retardant coating thickness $\delta_{is} \approx 20$ mm on the remaining outer surface of the tank, ensure that the tank remains in the model fire source for 4 h without the development of an accident according to the BLEVE scenario [15].

Table 1 Results of calculating LPG—Propane. Tank 15-1200-02

№	View Accidents, Duration (4 h)	Remedies			Start WorkValve (min)	Condition of the tank at the end of the accident					
		Valve type	SGK-1δ_c (mm)	Thermal insulation (mm)		LPG parameters			LPG pressure change law		Outcome of the accident, boiler state
						state of LPG	P (MPa)	T (°C)	Valve adjustment	Unregulated growth	
1	3 degrees (T_a = 20 °C)	valve standard d_y = 32 mm	20	20	96.5	Gas	2.07	58.7	+	–	Not deformed

Table 2 Results of calculating LPG - Propane. Tank 15–1200-02

Event №	Event/process	Time (min)	Temperature (°C)	Pressure (MPa)
1	The beginning of the process. Two phases	0	20.15	0.8403
2	One phase-liquid	95.84	59.74	2.12
3	SV(32–2.25) opened	99.00	60.09	2.25
4	Two phases	99.00	59.63	2.11
5	SV(32–2.25) closed	99.06	55.56	1.95
6	One phase-liquid	101.93	60.67	2.16
7	SV(32–2.25) opened	104.15	60.91	2.25
8	Two phases	104.18	60.46	2.15
9	SV(32–2.25) closed	104.21	55.61	1.95
10	One phase-liquid	107.99	61.63	2.20
11	SV(32–2.25) opened	109.32	6175	2.25
12	Two phases	109.33	61.29	2.19
13	SV(32–2.25) closed	109.37	55.61	1.95
14	SV(32–2.25) opened	114.32	62.56	2.25
15	SV(32–2.25) closed	114.36	55.64	1.95
16	SV(32–2.25) opened	119.11	62.57	2.25
17	SV(32–2.25) closed	119.17	55.62	1.95
18	SV(32–2.25) opened	124.10	62.57	2.25
19	SV(32–2.25) closed	124.16	55.61	1.95
20	SV(32–2.25) opened	129.28	62.57	2.25
21	SV(32–2.25) closed	129.35	55.66	1.95
22	SV(32–2.25) opened	146.76	62.57	2.25
23	SV(32–2.25) closed	134.76	55.62	1.95
24	SV(32–2.25) opened	140.27	62.57	2.25
25	SV(32–2.25) closed	140.36	55.65	1.95
26	SV(32–2.25) opened	146.04	62.56	2.25
27	SV(32–2.25) closed	146.14	55.66	1.95
28	SV(32–2.25) opened	151.98	62.56	2.25
29	SV(32–2.25) closed	152.08	55.63	1.95
30	SV(32–2.25) opened	158.18	62.57	2.25
31	SV(32–2.25) closed	158.29	55.65	1.95
32	SV(32–2.25) opened	164.59	62.56	2.25
33	SV(32–2.25) closed	164.72	55.66	1.95
34	SV(32–2.25) opened	171.25	62.57	2.25
35	SV(32–2.25) closed	171.38	55.66	1.95

(continued)

Table 2 (continued)

Event №	Event/process	Time (min)	Temperature (°C)	Pressure (MPa)
36	SV(32–2.25) opened	178.10	62.57	2.25
37	SV(32–2.25) closed	178.25	55.64	1.95
38	SV(32–2.25) opened	185.19	62.57	2.25
39	SV(32–2.25) closed	185.35	55.66	1.95
40	SV(32–2.25) opened	192.56	62.57	2.25
41	SV(32–2.25) closed	192.73	55.66	1.95
42	SV(32–2.25) opened	200.19	62.57	2.25
43	SV(32–2.25) closed	203.76	55.67	1.95
44	SV(32–2.25) opened	208.09	62.57	2.25
45	SV(32–2.25) closed	208.29	55.67	1.95
46	SV(32–2.25) opened	216.24	62.56	2.25
47	SV(32–2.25) closed	216.45	55.66	1.95
48	SV(32–2.25) opened	224.67	62.56	2.25
49	SV(32–2.25) closed	224.92	55.68	1.95
50	SV(32–2.25) opened	233.48	62.56	2.25
51	SV(32–2.25) closed	233.72	55.67	1.95
52	Process completed	240	58.66	2.2

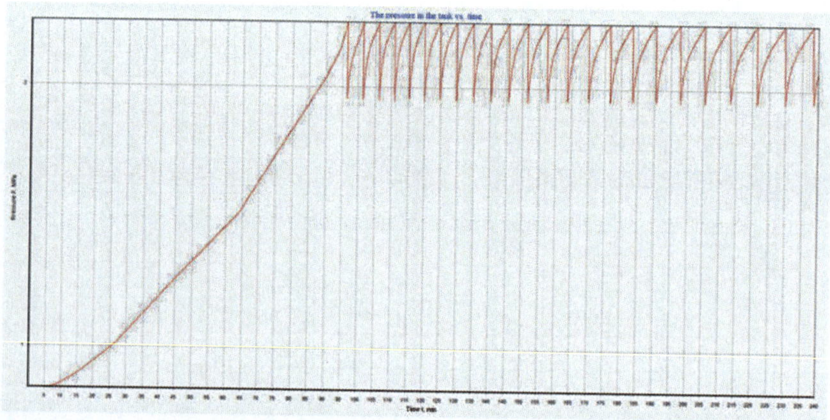

Fig. 1 The pressure in the tank vs. time

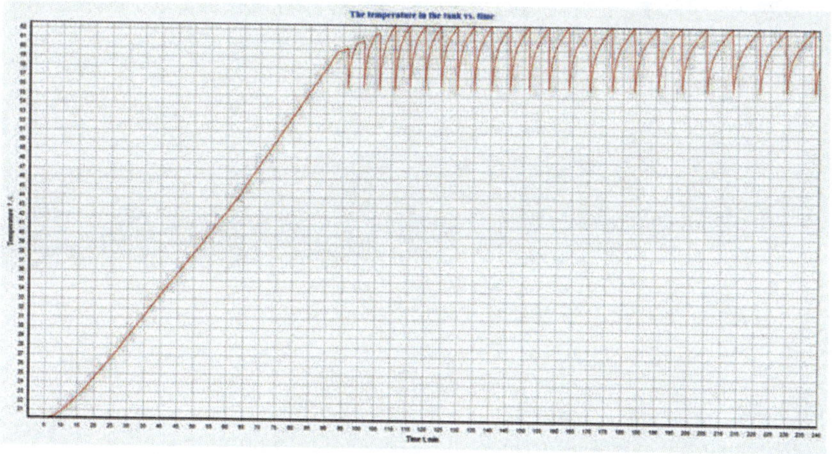

Fig. 2 The temperature in the tank vs. time

References

1. Jingde L, Hong H (2021) Process Saf Environ Prot 145:94–109. https://doi.org/10.1016/j.psep.2020.07.038
2. Yuanyuan WX, Gu LX, Yong PY, Ni SW, Wei Z (2020) J Loss Prev Process Ind 68. https://doi.org/10.1016/j.jlp.2020.104319
3. Jingde L, Hong H (2020) J Loss Prev Process Ind 65. https://doi.org/10.1016/j.jlp.2020.104107
4. Sellami I, Manescau B, Chetehouna K, de Izarra C, Nait-Said R, Zidani F (2018) J Loss Prev Process Ind 54:69–84. https://doi.org/10.1016/j.jlp.2018.02.010
5. Mingzhi L, Zhenyi L, Yi Z, Yao Z, Xuan LDZ (2018) J Hazard Mater 342:634–642. https://doi.org/10.1016/j.jhazmat.2017.09.002
6. Hemmatian B, Planas E, Casal J (2017) Process Saf Environ Prot 106:138–149. https://doi.org/10.1016/j.psep.2017.01.007
7. Bhushan KM (2016) J Loss Prev Process Ind 41:60–67. https://doi.org/10.1016/j.jlp.2016.03.003
8. Qian-xi Z, Dong L (2013) Procedia Eng 52:602–606. https://doi.org/10.1016/j.proeng.2013.02.192
9. Hemmatian B, Planas E, Casal J (2016) J Loss Prev Process Ind 40:81. https://doi.org/10.1016/j.jlp.2015.12.001
10. Zhao Y, Zhenyi L, Xiaohui S, Xinming Q, Yi Z, Deping Z, Qing L (2015) Energy Procedia 75:880–885. https://doi.org/10.1016/j.egypro.2015.07.204
11. Abo-Elkhier M, Muhammad K (2020) Eng Failure Anal 110. https://doi.org/10.1016/j.engfailanal.2020.104401
12. Li YZ, Ingason H (2018) Tunn Undergr Space Technol 81:568–589. https://doi.org/10.1016/j.jhazmat.2011.01.136
13. Brambilla S, Manca D (2010) J Hazard Mater 182:346–357. https://doi.org/10.1016/j.jhazmat.2010.06.039
14. Pasman H (2015) Risk analysis and control for industrial processes—gas, oil and chemicals, 1–31. https://doi.org/10.1016/j.matpr.2020.08.524
15. Liu L, Ni L, Jiang J-C, Zhang W-X (2018) Procedia Eng 211:479–487. https://doi.org/10.1016/j.proeng.2017.12.039

Requirements to Check Rails of Railroad Switches

Irina Shishkina

Abstract The paper investigates the requirements to check rails of railroad switches (turnouts). Railroad switches of the most commonly used structures with rigid frogs always involve check rails. Check rails not attached to running rails allow performing the adjustment of flangeways under operation, which thereby significantly increases the durability of structures involving check rails. The safety criterion is the first and key requirement, since check rails as part of turnouts ensure safe passage of a wheelset of a rolling stock through the section with a crossing frog. The second most important criterion is a dynamic-kinematic one. The third criterion pertains to the provision of robustness of check rails. All the aforementioned criteria represent the essential and adequate requirements that must serve as the basis for working out permissible rate of wear for check rails and determining the size of flangeways for them. For the purposes of examining the stress–strain state of check rails, the reasonability of a the experimental-calculation method was estimated, protector checkrails were considered and efficiency and drawbacks of their implementation were analyzed.

Keywords Check rail structure · Check rail not attached to running rail · Requirements · Criteria · Wear · Width of flangeway · Protector checkrail

1 Introduction

Railroad switches of the most commonly used structures with rigid frogs always involve check rails. They are intended to direct wheelsets and prevent the wheel that passes through the frog from taking a wrong flangeway or striking against a tongue of the frog.

Railroad switches with check rails not attached to running rails provide the opportunity to adjust flangeways under operation, which thereby significantly increases the durability of check rails. In this regard, there arises an issue on regulation of the permissible position of a check rail in the structure, its wear and sizes of its

I. Shishkina (✉)
Russian University of Transport (MIIT), Chasovaya str. 22/2, Moscow, Russia 125190
e-mail: shishkinaira@inbox.ru

A. Beskopylny and M. Shamtsyan (eds.), *XIV International Scientific Conference "INTERAGROMASH 2021"*, Lecture Notes in Networks and Systems 247,
https://doi.org/10.1007/978-3-030-80946-1_18

flangeways that proceed from this parameter [1, 2]. A worn check rail should ensure meeting the requirements imposed on railroad switches as much as a new one.

2 Research Technique and Results

Since check rails as part of turnouts ensure safe passage of a wheelset of a rolling stock through railroad sections with a crossing frog or sections of other types that involve check rails, then safety is the most important criterion and requirement [3, 4]. Parameters of the check rail and the section involving check rails should provide appropriate passage of wheelsets with excluding any impacts or striking against a tongue of the frog (item 5, Fig. 1) rolling onto a wing rail up to the throat (item 3, Fig. 1), onto the pick-up part of a wing rail or pick-up part of a check rail (item 1, Fig. 1). They also must prevent wheelsets from sprawling between a check rail and an inoperative wing rail (item 6, Fig. 1).

Another undesirable situation is when wheels roll onto an inoperative wing rail between the throat and the point of the frog tongue (item 4, Fig. 1).

The second key criterion is a dynamic-kinematic one. This criterion implies that abrupt displacements of wheelsets when passing through the frog crossing zone, rail overturning, destruction of the check rail zone and its failure after a short period of operation should be excluded. Quantitative expression of this criterion is the parameter associated with kinetic energy loss when a wheel strikes against the check rail (item 2, Fig. 1). This criterion limits bending angles of flare and pick-up parts of the frog crossing and the check rail in accordance with a design speed on the railroad switch [5, 6].

The third criterion is the provision of robustness of the check rail itself. The impact of wheels on the check rail should not cause dangerous defects or fracturing [7, 8]. This criterion is checked by comparing stresses arising in the most loaded part of the check rail with permissible stresses.

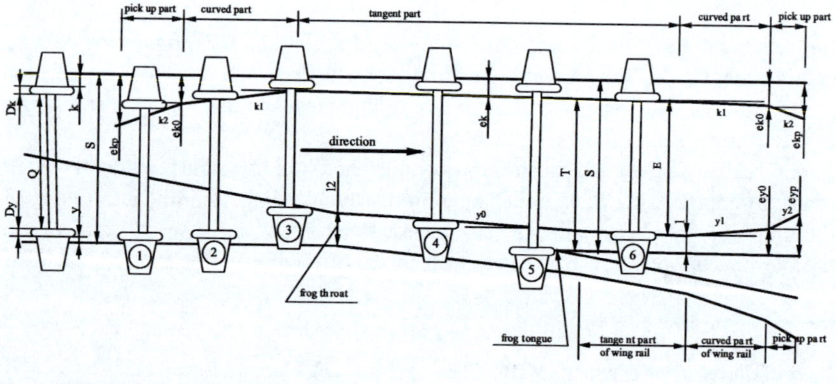

Fig. 1 Wheelset passing the frog section

All the aforementioned criteria represent the essential and adequate requirements that must serve as the basis for working out permissible rate of wear for check rails and determining the size of flangeways for them.

The levels of dynamic effects arising when trains pass through the railroad switch depend on the relative position of wheelsets and turnout elements, which is determined by a combination of their basic geometric dimensions [9, 10]. At each specific frog section, the track gage width has the value of S; dimensions of the flangeways of rails in their straight part, at the beginning of a flare and a pick-up part, respectively, are ek, eko, ekp (Fig. 1). Sizes of the flangeways of the frog in the throat, straight part of wing rails at the beginning of the flare and the pick-up part, respectively, are en, ey, eyo, eyp.

Let the wheelset entering the frog section have bottom distance between the inner edges of wheels Q, thickness of wheel flanges (with taking into account of fluting on the bottom side of a wheel) on the side of a wing rail Dy and on the side of a check rail Dk. When the wheelset approaches the frog crossing, there is a gap between the wheel flange and a gauge face of the running rail (on the side of a check rail) δk and a similar gap between the gauge face of a wing rail and the flange of the second wheel δy. Depending on particular values of all of the above dimensions, the following cases are possible when the wheelset passes through the frog crossing [11, 12]:

The bottom part of the wheel moving along the running rail strikes against the check rail at its pick-up part. The condition for this case to occur can be written as follows:

$$(\delta_k + D_k) > e_{k0} \tag{1}$$

The bottom part of the wheel moving along the running rail strikes against the check rail at its flare part. The condition for this case is as follows:

$$e_{k0} \geq (\delta_k + D_k) \geq e_k \tag{2}$$

The bottom part of the wheel moving along the frog crossing strikes against the wing rail of the crossing up to the throat. The condition for such a passage of the wheelset is as follows:

$$\left(\delta_y + D_y\right) > e_t \tag{3}$$

The bottom part of the wheel flange moving along the frog crossing strikes against the wing rail after passing the throat. This type of passing occurs under the following condition:

$$e_y \leq \left(\delta_k + D_y\right) < e_t \tag{4}$$

Impact (striking against) of the wheel moving along the frog crossing on the frog point. This can occur if:

$$Q + D_k + D_y + \delta_k \geq S - e_k = T \tag{5}$$

where T is the distance between the frog point and the gauge face of the check rail. Sprawling of the wheelset by the check rail and wing rail. This occurs on condition:

$$Q \leq S - e_k - e_y = E \tag{6}$$

where E is the distance between the gauge face of the check rail and the "inoperative" wing rail.

In addition, wheels may strike against the flare part and pick-up part of the check rail [13, 14]. Conditions entailing these cases are the same as conditions (1) and (2). It is also possible for bottom parts of wheels to strike against the pick-up and flare parts of wing rails of the frog crossing [15, 16]. The conditions for these phenomena are obtained from conditions (1) and (2) respectively by replacing index "k" with index "y":

$$(\delta_k + D_k) > e_{y0} \tag{7}$$

$$e_{y0} \geq \left(\delta_y + D_y\right) > e_y \tag{8}$$

Besides the aforementioned cases, wheels may freely pass the frog crossing zone. The condition for such a passage is opposite to conditions (1)–(8). To meet all safety requirements, dimensions of the track and flangeways, values T and E should be assigned so that conditions opposite to (1), (3), (5)–(7) are fulfilled for all possible combinations of sizes of wheelsets and frog crossings. Methods for analyzing such combinations can be different [17, 18].

Structurally, check rails consist of the main part lk0 that covers the dead space, two flares lk1 forwarding wheelsets to the frog flangeway of the needed line and two inoperative pick-up parts lk2 designed to ensure safety of train movement in cases when rules of maintaining the rail track are being violated and in other unusual situations (Fig. 2) [19, 20].

Lengths of check rail parts are determined as follows:

$$\begin{cases} \mathrm{l}_{k0} = (\mathrm{t_n} + \mathrm{w_c})\mathrm{N} + 2\mathrm{e}_k \\ l_{k1} = \frac{t_{k1} - t_{k0}}{\sin \gamma_{k1}} \\ l_{k2} \geq [l_{k2}]\min \\ l_k = l_{k0} + 2l_{k1} + 2l_{k2} \end{cases} \tag{9}$$

where N is the brand of the frog crossing;

γ_{k1} is check rail flare angle;

tn is the flangeway of the frog crossing throat;

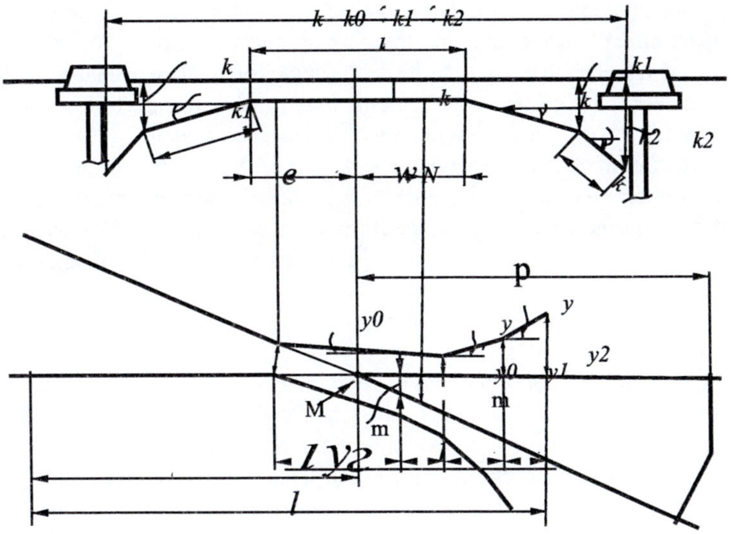

Fig. 2 Structural scheme of the check rail and frog crossing

t_{k0} is the flangeway at the main part of the frog crossing;

t_{k1} is the flangeway at the end of the flare;

w_c is the width of the tongue of the frog in the cross-section where the wheel completely rests on it;

e_k is the margin of covering the dead space;

$[l_{k2}]$ min is the minimum length of the pick-up part.

The angle of the check rail flare should not exceed the permissible angle of impact on the flare of the check rail, which is determined through the permissible value of the "impact effect" W_k according to the formula:

$$\sin \gamma_{k1} = \frac{W_k}{V_n} \tag{10}$$

where V_n is the design impact speed along the main track.

Large dynamic forces acting on the check rail entail the breakage of check rail bolts in structures that involve check rails of special RK-profile. Such a situation requires taking immediate measures, even up to interruption of train traffic (when two bolts in a row are broken) [21, 22]. In assemblies with check rails that are not connected to the running rail, the increased impact leads to fractures of check rail chairs. Wheelset displacements cause the appearance of saddles in rails of the frog during operation. This situation reduces the service life of rails and requires limiting the speed of trains [23, 24].

The level of impact of wheels on the rail is regulated by rationing values of the "impact effect" on the flare and the pick-up part of the rail. According to the standardized values of the "impact effect", the angles of the flare and the pick-up part of rails are determined. These angles, taking into account manufacturing tolerances, are included in the design documentation of the railroad switch.

The structure of turnouts involving check rails of special PK-profiles, which are structurally attached to the running rail by filler blocks implies extremely hard adjustment of flangeways, which is rarely performed under operation. If the wear rate of the check rail does not ensure safety and required dimensions of flangeways, the check rail is replaced together with the rail of the frog crossing [25, 26].

The use of structures with check rails that are not connected with running rails allows adjusting the size of flangeways during operation by installing spacers between the check rail and the supporting part of check rail chairs. This makes it possible to extend the service life of check rails, since their wear does not obstruct setting the standardized sizes of flangeways and rates of safety.

Since wheels impact check rails along their length unevenly, check rails become worn out irregularly. The middle part of the check rail is mostly subjected to wear, so, when setting adjusting shims, flare angles of the check rail increase [27, 28]. The "impact effect" determined in accordance with formula (10) increases.

The mostly manifested wear of the check rail is limited by the maximum permissible angle of wheels running onto the check rail.

The criterion is checked by comparing stresses arising in the most loaded part of the check rail with the permissible stresses: $\sigma max \leq [\sigma]$.

As the check rail wears out, dimensions of its cross section change. In this case, components of bending and torsion stresses increase, while geometric characteristics of the section (moment of inertia and moment of resistance) decrease. To resolve the issue of the permissible wear of check rail, we should proceed from the need to ensure robustness requirements not only for a new, but also for a worn out check rail. The impact of wheels on the check rail should not lead to critical defects or fracturing of a check rail. It is expensive and almost impossible to carry out tests covering all possible operating conditions for check rails of different wear rate at various combinations of sizes of a track and flangeways [29, 30]. Therefore, it is advisable to apply the experimental-calculation method to examine the stress–strain state of check rails.

The work is to be performed in two stages.

The first stage includes dynamic-strength tests with a detailed study of the parameters of the stress–strain state of check rails. The purpose of this stage is to obtain data on the most loaded sections of the check rail and ratios that allow checking the adequacy of models that are further used to calculate the robustness of the check rail and points of its attachment to sleepers.

The second stage implies carrying out calculations of the stress–strain state of check rails under various combinations of structure sizes and different rates of wear of the check rail itself. This stage is purposed to control the robustness of the check rail and points of its mounting to sleepers, as well as to determine the allowable rate of wear of check rails according proceeding from robustness conditions.

The simplest check rail model is a continuous beam of finite length resting on N supports. Such a model allows obtaining the demanded design stresses in the check rail. However, data obtained from such a calculation leads to unjustified margin of safety and unnecessary restrictions on values of permissible wear of check rails.

Models based on the finite element method give much more accurate results; therefore, when performing calculations, it is advisable to use finite element models based on using one of the standard computer software distribution packages.

Protector check rails are installed in turnouts before switch rails in the front offset of a stock rail. They are designed to improve the conditions of passing through switch panels of turnouts and to reduce the wear of switch rails and stock rails of turnouts, along which trains move to the diverging line in face direction.

It is also advisable to install protector check rails at symmetrical turnouts of hump-yards at the directions of priority movement.

The structure of protector checkrails and their attachment to bars are the same as that of check rails of frog crossings. Size of flangeways of protector checkrails, the allowable wear of check rails should be determined with applying the same methods as for check rails of frog crossings.

The most expedient length of a guiding part of the protector check rail is 10% more than the distance between the axles of wheelsets of a bogie of the most common rolling stock. In practice, length of the guiding part of the check rail is limited by options of its placement on the track. Protector check rails usually have such a length that allows them to be placed in the front offset of the stock rail of the turnout. In this case, the effect of their use is reduced, yet the expediency of using protector check rails remains.

The effectiveness of using protector check rails consists in 3–8 times increase in service life of the "stock rail—curved switch rail" set in terms of operating time until its wear rate reaches the lower limit.

The drawback of using protector check rails is a possible need to limit the speed of trains moving along the straight track of the turnout, since angles of wheels rolling on the flare of the check rail are high (restrictions on dynamic-kinematic criteria). First of all, this regards to shortened protector check rails, which have the angle of flare that can significantly exceed check rail flare angles at frog crossings.

References

1. Loktev AA, Korolev VV, Poddaeva OI, Chernikov IYU (2018) Mathematical modeling of antenna-mast structures with aerodynamic effects. IOP Conf Ser: Mater Sci Eng 463
2. Loktev AA, Korolev VV, Shishkina IV (2018) High frequency vibrations in the elements of the rolling stock on the railway bridges. IOP Conf Ser: Mater Sci Eng 463
3. Glusberg B, Korolev V, Shishkina I, Loktev A, Shukurov J, Geluh P, Loktev D (2018) Calculation of track component failure caused by the most dangerous defects on change of their design and operational conditions. MATEC Web Conf 239
4. Korolev V, Loktev A, Shishkina I, Zapolnova E, Kuskov V, Basovsky D, Aktisova O (2019) Technology of crushed stone ballast cleaning. IOP Conf Ser: Earth Environ Sci 403

5. Gluzberg B, Korolev V, Loktev A, Shishkina I, Berezovsky M (2019) Switch operation safety. E3S Web Conf 138
6. Loktev AA, Korolev VV, Gridasova EA (2019) Influence of high-frequency cyclic loading on mechanical and structural characteristics of rail steel under extreme conditions. IOP Conf Ser: Mater Sci Eng 687
7. Savin A, Kogan A, Loktev A, Korolev V (2019) Evaluation of the service life of non-ballast track based on calculation and test. Int J Innov Technol Explor Eng 8(7):2325–2328
8. Lyudagovsky A, Loktev A, Korolev V, Shishkina I, Alexandrova D, Geluh P, Loktev D (2019) Energy efficiency of temperature distribution in electromagnetic welding of rolling stock parts. E3S Web Conf 110
9. Savin AV, Korolev VV, Shishkina IV (2019) Determining service life of non-ballast track based on calculation and test. IOP Conf Ser: Mater Sci Eng 687
10. Savin A, Korolev V, Loktev A, Shishkina I (2020) Vertical sediment of a ballastless track. Adv Intell Syst Comput 1115:797–808
11. Korolev V (2020) Switching Shunters on a Slab Base. Adv Intell Syst Comput 1116:175–187
12. Savin A, Suslov O, Korolev V, Loktev A, Shishkina I (2020) Stability of the continuous welded rail on transition sections. Adv Intell Syst Comput 1115:648–654
13. Loktev A, Korolev V, Shishkina I, Illarionova L, Loktev D, Gridasova E (2020) Perspective constructions of bridge crossings on transport lines. Adv Intell Syst Comput 1116:209–218
14. Glusberg B, Savin A, Loktev A, Korolev V, Shishkina I, Alexandrova D, Loktev D (2020) New lining with cushion for energy efficient railway turnouts. Adv Intell Syst Comput 982:556–570
15. Loktev A, Korolev V, Shishkina I, Chernova L, Geluh P, Savin A, Loktev D (2020) Modeling of railway track sections on approaches to constructive works and selection of track parameters for its normal functioning. Adv Intell Syst Comput 982:325–336
16. Korolev V (2020) Check rail operation of lateral path of railroad switch. Adv Intell Syst Comput 1115:621–638
17. Glusberg B, Savin A, Loktev A, Korolev V, Shishkina I, Chernova L, Loktev D (2020) Counter-rail special profile for new generation railroad switch. Adv Intell Syst Comput 982:571–587
18. Gridasova E, Nikiforov P, Loktev A, Korolev V, Shishkina I (2020) Changes in the structure of rail steel under high-frequency loading. Adv Intell Syst Comput 1115:559–569
19. Glusberg B, Loktev A, Korolev V, Shishkina I, Alexandrova D, Koloskov D (2020) Calculation of heat distribution of electric heating systems for turnouts. Adv Intell Syst Comput 982:337–345
20. Korolev V (2020) The study of rolling stock wheels impact on rail switch frogs. E3S Web Conf 164:03033
21. Glusberg B, Korolev V, Lokteva O, Shishkina I (2020) Contact stresses in point frogs. E3S Web Conf 164:03034
22. Loktev A, Korolev V, Shishkina I, Savin A, Illarionova L, Lokteva O, Kuskov V (2020) Dynamic behavior model of the ballastless railroad track segment considering wave processes. E3S Web Conf 164:03035
23. Loktev A, Korolev V, Shishkina I, Lokteva O, Gridasova E (2020) Span operational aspects under offsetting the axis of the track panel. E3S Web Conf 164:03037
24. Korolev V (2020) Selecting a turnout curve form in railroad switches for high speeds of movement. Adv Intell Syst Comput 1258:156–172
25. Korolev V (2020) Change of geometric forms of working surfaces of turnout crosspieces in wear process. Adv Intell Syst Comput 1258:207–218
26. Glusberg B, Loktev A, Korolev V, Shishkina I, Berezovsky M, Trigubchak P (2020) Deformations and life periods of the switch chairs of the rail switches. Adv Intell Syst Comput 1258:184–196
27. Loktev AA, Korolev VV, Poddaeva OI, Stepanov KD, Chernikov IY (2018) Mathematical modeling of aerodynamic behavior of antenna-mast structures when designing communication on railway transport. Vestnik Railway Res Inst 77(2):77–83
28. Loktev AA, Korolev VV, Loktev DA, Shukyurov DR, Gelyukh PA, Shishkina IV (2018) Perspective constructions of bridge overpasses on transport main lines. Vestnik Railway Res Inst 77(6):331–336

29. Korolev VV, Loktev AA, Shishkina IV, Gridasova EA (2020) Features of the bridge span performance when shifting the axis of the rail-sleeper grid. VNIIZHT Sci J 79(3):127–138
30. Loktev AA, Korolev VV, Shishkina IV, Berezovsky M (2020) Calculation of the speed of movement along the side path for flat marks of turnouts. Transp Urals 2(65):52–56

Assessment of Stability from Overturning of Unladen Containers from Railway Platforms

Vladimir Popov, Philipp Sukhov, and Julia Bolandova

Abstract The article considers a method for assessing the stability of empty containers from tipping over on wind-dependent railway sections. The article considers the statistics of road accidents caused by the separation of containers of various types from specialized flat cars under the influence of squally winds of different speeds. The purpose of this study is to develop a methodology for determining the conditions that lead to the overturning of empty containers from specialized railway cars under the influence of wind load. An expression is obtained that allows determining the minimum wind speed that leads to the overturning of an empty container. The article presents the calculation of the wind speed that leads to the overturning of empty containers of various types on straight and curved sections of the railway track. The results of the study can contribute to the development of universal technical solutions for various regions of the world, ensuring the stability of an empty container under the influence of wind loads.

Keywords Stability · Reliability · Railway transport · Container transportation · Rolling stock · Wind load · Traffic safety violations · Traffic accidents

1 Introduction

Modern regulation acts of the Russian Federation are aimed to solving problems of STI items of rolling stock, elements of the permanent way and other problems. The modes of force action and the permissible stresses under dynamic influences on a loaded car were usually taken as standard parameters (Russian State Standard «Box-cars for 1520 mm gauge main line railways. General specifications»).

Normative acts that regulate design of containers are the documents of the register of marine shipping as in relation to modes of movement of containers on railway rolling stock is only considered normative values of vertical and lateral accelerations present in loaded containers, connection system container to the frame of the car is not specified too. The increase in the speed of movement, the imperfection of the

V. Popov · P. Sukhov · J. Bolandova (✉)
Russian University of Transport (MIIT), Obraztsova str., 9, Moscow, Russia 127994

A. Beskopylny and M. Shamtsyan (eds.), *XIV International Scientific Conference "INTERAGROMASH 2021"*, Lecture Notes in Networks and Systems 247,
https://doi.org/10.1007/978-3-030-80946-1_19

system of connection containers on the frame, changing weather factors, all of it makes the problem of stability from overturning actual.

Railway sections intended for the transportation of containers differ are different in climatic conditions, railway and operating conditions. On the Russian railway network and the European Union railway network, there are many cases of falling containers from fitting platforms (Table 1), these accidents are initiated by traffic speeds, dynamic and wind loads.

Similar incidents occur on the railways of the United States and Europe, not only with container trains, but also with a different type of wagons.

On January 3, 2018, in Switzerland, a train was derailed by a hurricane wind of a speed of more than 50 m/s. To prevent negative impact of natural disaster the Swiss Federal Railways announced cancellation of transportation on the popular tourist route Jungfraujoch [1].

On March 13, 2019 in the US state of New Mexico, near the settlement of Logan, 26 freight train wagons derailed. The situation was unusual as the wagons fell from the bridge under the influence of a strong wind [2].

Several studies [3–5] included experiments that made it possible to determine local aerodynamic load on rolling stock caused by lateral and headwind. These studies were aimed at studying the stability of traction high-speed rolling stock, while intermodal container transportation was not given due attention.

Table 1 Cases of containers falling from fitting platforms on the railway network of the Russian Federation

Date	Railway section	Number of dropped containers	Driving speed (km/h)	Climate condition
05.07.11	The Privolzhskaya railway. Krasavka-Kologrivova	6	69	Wind gusts of 21–23 m/s in the open area
21.04.14	Northern Railway. Obskaya-Harp	11	42	Squally wind up to 32–34 m/s blizzard
17.12.14	Far Eastern Railway. Chmielowski-Nahodka-B	3	43	Crosswind, about 20 m/s Wind gusts up to 32 m/s
13.08.16	Trans-Baikal Railway. Argali-Domikan	16	17	Gusty over 33 m/s in the open area. Thunderstorm
23.06.19	Trans-Baikal Railway. Urul'ga-Tarski	1	–	Gusty wind up to 31 m/s. Thunderstorm
23.09.19	South Ural Railway. Sara-Khalilovo	1	–	Wind from "storm" to "hurricane" (on the Beaufort scale), more than 22 m/s

By order of Russian Railways [2], several measures were approved to reduce the risk of traffic accidents when empty containers are transported on specialized railway platforms in conditions of forecasting strong winds, squalls or hurricanes along their route. These orders are aimed solely at the transportation process and do not take into account other factors.

The objective of the present study is associated with the necessity to develop an engineering methodology for assessing the characteristics of wind load, leading to the overturning of empty containers in freight trains. The methods of theoretical mechanics, comprising moment of force equations. And as a result, improving transport safety in particular and improving the safety culture in general.

2 Materials and Methods

In intermodal transportation, there are uniform requirements for transported containers, their overall dimensions, weight and fastening devices. These requirements are regulated by ISO standards [6]. In case of transportation by rail, the container is secured to the railway flat wagon with four lower corner fittings using universal cones. Such fastening should prevent sliding and overturning under the action of longitudinal and transverse forces, however, under the influence of additional wind load, this fastening does not justify itself.

Conditions of the container overturning caused by influence of the wind load based on the equation of the moments of force applied to the container relative to the axis of its rotation when the freight train is moving on the curved section of the railway track (see Fig. 1). Russian technical operation rules establish the maximum superelevation of the outer rail on the curved section of the railway track at 150 mm.

The main distributed wind load w (Pa) is replaced by the concentrated force of the wind pressure applied to the geometric center of the leeward surface of the container and acting normally to it. We take into account the gravity of the container FG, as well as the friction force of the surface of the container along the fittings F_{fr}, acting along the friction surface and arising from the force of wind pressure on the container and from the normal component of the gravity of the container to the friction surface.

Then the equation of the moments of forces applied to the container, relative to the axis of its rotation, will have the following form:

$$\sum_i M_i \geq 0 \quad (1)$$

where $M_i = F_i.l_i$ is product of the force modulus F_i and of the arm force.

Wind pressure force applied to the geometric center of the leeward surface of the container and acting normally to it:

$$\overrightarrow{F_w} = \vec{w} \cdot S, \mathrm{N} \quad (2)$$

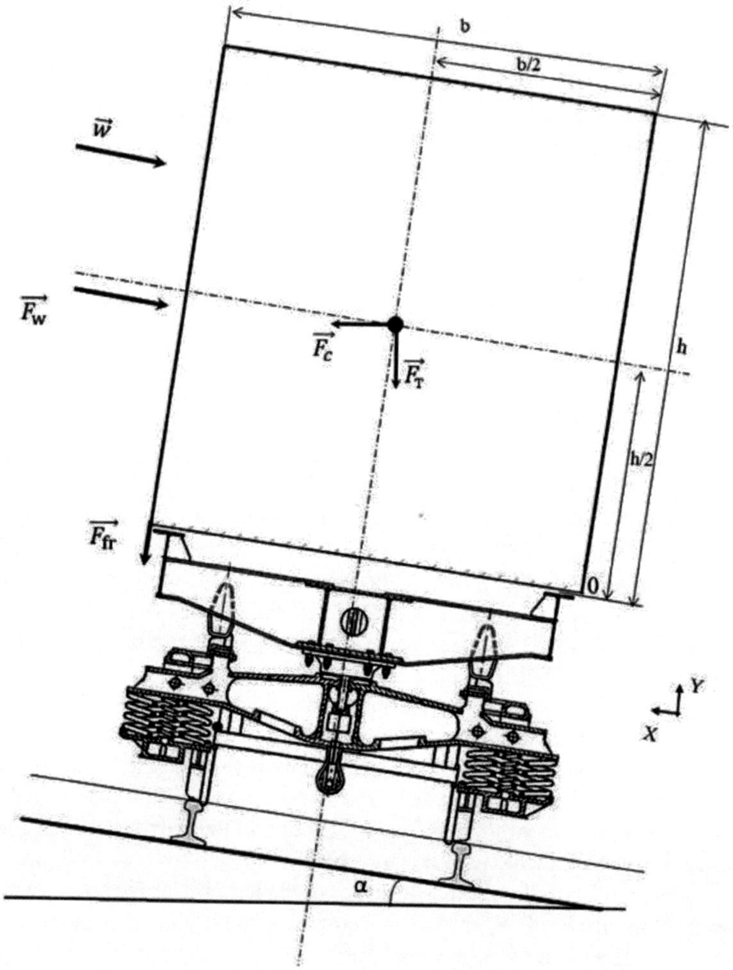

Fig. 1 Scheme of action of forces in the middle section of the container

where S = L · h is area of the leeward surface of the container, m^2 (here L, h are length and height of the container, respectively).

Gravity force of the container will act in the center of masses

$$\overrightarrow{F_G} = \vec{g} \cdot m, \text{N} \tag{3}$$

where m is weight of an empty container.

Friction force of the container on the fittings F_{fr}, acting along the friction surface:

$$\vec{F}_{fr} = k_{fr} \cdot F_w, \text{N} \tag{4}$$

where k_{fr} is coefficient of static friction.

The centripetal force will act at the center of mass of the container in a direction parallel to the earth's surface

$$\vec{F}_c = 0.077 \cdot \frac{mV_t^2}{R} \cdot \vec{\imath}, \text{N} \tag{5}$$

where V_t are the speed of the rolling stock in the curve (km/h), and R is the radius of the curve(m), $\vec{i}$ is the ort in the direction of the x-axis.

Figure 1 is a diagram of the middle section of the container, showing the main forces acting on the container when passing a curved track section.

In the middle section the following moments of forces act on the container relative to the axis of rotation:

$$M_1 = F_w \cdot \frac{h}{2}, \tag{6}$$

$$M_2 = -F_G \cos\alpha \cdot \frac{b}{2} + F_G \sin\alpha \cdot \frac{h}{2}, \tag{7}$$

$$M_3 = -F_c \cos\alpha \cdot \frac{h}{2} + F_c \sin\alpha \cdot \frac{b}{2} \tag{8}$$

$$M_4 = -F_{fr} \cdot b = -k_{fr} F_w \cdot b \tag{9}$$

From Eq. (1), taking into account Eqs. (2–7), we obtain the equation for the main wind load, leading to overturning of the container:

$$w \geq \frac{m\left[g(b \cdot \cos\alpha - h \cdot \sin\alpha) + 0.077\frac{V_t^2}{R}(h \cdot \cos\alpha + b \cdot \sin\alpha)\right]}{\left(h - 2k_{fr} \cdot b\right) \cdot L \cdot h}, \text{Pa} \tag{10}$$

In accordance with the normative value of the main wind load w is defined as the sum of the middle w_m and pulse w components:

$$w = w_m + w_p. \tag{11}$$

In turn, the normative value of the middle component of the main wind load w_m depending on the equivalent height of the container z_e above the ground is determined by the formula:

$$w_m = wk(z_e)c, \tag{12}$$

where $w_0 = 0.43\ v_2$ is normative value of wind pressure;

$k(z_e)$ is coefficient, taking into account the change in wind pressure for equivalent height z_e;

c is aerodynamic coefficient;

V is wind speed, m/s, at the level of 10 m above the surface of the ground.

Then the normative value of pulse component of the main wind load w_p at the equivalent height of the container z_e is determined by the following formula:

$$w_p = w_m \cdot \zeta(z_e) \cdot \nu, \tag{13}$$

where $\zeta(z_e)$ is coefficient of pulsation of wind pressure;

ν is coefficient of spatial correlation of pulsation of wind pressure.

Taking into account Eqs. (10–12), Eq. (9) can be represented in the following form with respect to V (m/s) which is wind speed at the level of 10 m above the surface of the ground:

$$V \geq \sqrt{\frac{m\left[g(b \cdot \cos\alpha - h \cdot \sin\alpha) + 0.077\frac{V_t^2}{R}(h \cdot \cos\alpha + b \cdot \sin\alpha)\right]}{0.43 \cdot k(z_e) \cdot c \cdot h \cdot l \cdot [1 + \zeta(z_e)\nu]\left(h - 2k_{fr}b\right)}} \tag{14}$$

3 Results

We calculate the wind speed leading to the overturning of empty containers of various types, both for straight sections of the track and for curved sections of the track of railways, taking into account the maximum superelevation of the outer rail of 150 mm.

The results of calculating the wind speed leading to the overturning of empty containers of various types are presented in Table 2.

The results obtained are confirmed by the mathematical modelling of stability of fastening of empty containers, carried out by LLC Hexa on the demand of PJSC Transcontainer.

An analysis of wind speed and the risk of its occurrence shows that the highest frequency of strong winds is observed in the coastal regions of the North and the Far East, and in the steppes in the continental part of the country. In the North of the country, wind speed varies between 28 and 35 m/s, in the Far East its range is 31–38 m/s, in the steppes of the North Caucasus if is of 28–31 m/s [7–11].

These sections are most dangerous for overturning empty containers transported by freight trains, especially on curved sections of the railway track.

The significance of the study's findings increases as the total transit capacity of Russian container terminals increases [12, 13].

Table 2 Calculation results

Container type	Curved section of railway track R ≥ 250 м, $V_t \leq 30$ km/h			Curved section of railway track R = 250 м, $V_t = 30$ km/h			Straight section of railway track		
	Wind load, w, (Pa)	Wind speed V, (m/s)	Classification of winds [2]	Wind load, w, (Pa)	Wind speed V, (m/s)	Classification of winds [2]	Wind load, w, (Pa)	Wind speed V, (m/s)	Classification of winds [2]
40 foot high container (HighCube), 40' HC	1057.39	32.67	Hurricane	1192.8	34.69	Hurricane	1165	34.29	Hurricane
40-foot standard container (Dry Van) 40' DV	1319.2	36.49	Hurricane	1466.904856	38.47	Hurricane	1435.5	38.06	Hurricane
20-foot standard container, (Dry Cube) 20' DC	1466.85	38.48	Hurricane	1631.07	40.57	Hurricane	1596.16	40.13	Hurricane
Five-ton container, UK-5	2444.6	49.67	Hurricane	2671.1	51.91	Hurricane	2620.39	51.42	Hurricane

4 Conclusion

Organizational measures do not exclusively solve the problem of neutralizing the risks of overturning empty containers. It is necessary to develop technical solutions to ensure the stability of the empty container under the influence of wind load, one of solutions might be to changing the design of the fitting stop.

The research results showed that it is necessary to make changes to the used standards in order to define the requirements for the strength and dynamic qualities of wagons.

References

1. Storm Burglind causes havoc in Switzerland, derails train. [Electronic resource]: https://www.swissinfo.ch/eng/business/wind-up_switzerland-battered-by-hurricane-speed-winds/43795876. Last accessed 17.10.2018
2. Wind pushes train off track sending 26 railcars into the ground below. [Electronic resource]: https://www.kob.com/new-mexico-news/wind-causes-train-to-derail-in-eastern-new-mexico/5277830/. Last accessed 17.10.2018
3. Baker C, Reynolds S (1992) Wind-induced accidents of road vehicles. Accid Anal Prevent 24(6):559–575. https://doi.org/10.1016/0001-4575(92)90009-8
4. Baker C, Jones J, Lopez-Calleja F, Munday J (2004) Measurements of the cross wind forces on trains. J Wind Eng Ind Aerodyn 92(7–8)547–563. https://doi.org/10.1016/j.jweia.2004.03.002
5. Sanquer S, Barre C, Dufresne M, Cleon L (2004) Effect of cross winds on high-speed trains: development of a new experimental methodology. J Wind Eng Ind Aerodyn 92(7–8):535–545. https://doi.org/10.1016/j.jweia.2004.03.004
6. ISO 668 «Cargo containers. Outside dimensions and maximum gross mass» [ISO 668«Gruzovie konteinery. Naruzhnie razmery i maksimalnaya massa brutto»]
7. Atlas of natural and technological hazards and risks of emergency situations in the Russian Federation [Atlas prirodnykh i tekhngennykh opasnostei i riskov chrezvychainykh situatsii v Rossiiskoi Federatsii]. EMERCOM of Russia; RAS; Roskartografiya - geogr. foundation; Editorial: S. K. Shoigu (head) and others; Chief ed. cards N. B. Trokhina. Moscow, DIK publ., 2005, 1 atl. (269 p)
8. Pietrzyk K (2020) Analysis of wind speed and wind pressure on the facades in frequency domain for the modelling of air change rate. E3S Web Conf 172. https://doi.org/10.1051/e3sconf/202017209005
9. Agdas D, Webster GD, Masters FJ (2012) Wind speed perception and risk. PLoS ONE 7(11):e49944. https://doi.org/10.1371/journal.pone.0049944
10. Yarnal B (1993) Synoptic climatology in environmental analysis. Belhaven Press, London, p 195
11. Popov V, Sukhov P, Bolandova J (2020) Safe train route options. In: Popovic Z, Manakov A, Breskich V (eds) VIII international scientific Siberian transport forum. TransSiberia 2019. Advances in intelligent systems and computing, vol 1116. Springer, Cham. https://doi.org/10.1007/978-3-030-37919-3_89
12. Belt and Road Transport Corridors: Barriers and Investments—Saint Petersburg: EDB Centre for Integration Studies, 2018, 44 p
13. Druzhinin AG, Dong Y (2018) One belt—one road initiative: a window of opportunity for Russia's Western Border Regions. Balt Reg 10(2). https://doi.org/10.5922/2079-8555-2018-2-3

Microsphere Effect on the Properties of Elastomer Compositions

Victor Kablov, Oxana Novopoltseva, Vladimir Kochetkov, Alexander Sychev, and Elizaveta Orlova

Abstract In order to expand the range of products made of elastomeric materials, increase their operational stability and areas of application, new components are being sought. One of the promising directions for solving this problem is the use of microspheres. Many technical problems where weight reduction is required with low thermal conductivity, high strength and volume savings, increased resistance to erosion and corrosive environments can be solved with the use of microspheres. The influence of hollow corundum microspheres grades HCM-S and HCM-L on the rheometric properties of rubbers and the physical and mechanical properties of vulcanizates based on ethylene-propylene diene rubber is considered, and also studies of vulcanizates containing corundum microspheres on the fire and heat-shielding properties based on EPDM rubber are given. -40. It is shown that the investigated microdispersed additives increase the wear resistance of the composites and decrease the specific density of the material. The addition of corundum microspheres to rubber compounds enhances the heat and fire retardant properties of rubbers based on ethylene propylene diene rubber. The degree of influence on the complex of the investigated properties of the introduced corundum microspheres depends on their fractional composition and the nature of the rubber used.

Keywords Microspheres · Fire-and-heat protection properties · Physical and mechanical properties · Elastomer material

1 Introduction

The strength properties of the compositions are greatly influenced by the characteristics of the fillers: the size and shape of the particles, their distribution in the material. The most widely used fillers in the tire and industrial rubber goods industry are carbon black and white soot. In order to expand the range of products made of elastomeric materials, increase their operational stability and areas of application,

V. Kablov · O. Novopoltseva · V. Kochetkov (✉) · A. Sychev · E. Orlova
Volzhsky Polytechnical Institute (Branch) VSTU, 42a Engelsa Street, Volzhsky, Russia 404121

A. Beskopylny and M. Shamtsyan (eds.), *XIV International Scientific Conference "INTERAGROMASH 2021"*, Lecture Notes in Networks and Systems 247,
https://doi.org/10.1007/978-3-030-80946-1_20

Table 1 Chemical composition of the corundum hollow microspheres

Substance	Content (%) (mass.)
Aluminum oxide Al_2O_3	99.7
Iron (III) oxide Fe_2O_3	0.003
Calcium oxide CaO	0.01
Silicon oxide SiO_2	0.025
Sodium potassium oxide Na_2O	0.26

Source [4]

the search for new effective technical and technological additives is underway. One of the promising directions for solving this problem is the use of microspheres.

Spherical fillers or microspheres are classified into two classes: solid and hollow.

Hollow microspheres are increasingly used in the production of plastics and elastomeric materials with reduced density [1, 2].

The uniqueness and value of solid (incomplete) glass microspheres as fillers lies mainly in their shape. The particles obtained in industry are perfect spherical formations with a smooth surface, which have almost all the qualities of an ideal filler. Continuous microspheres have a smooth surface and have a minimal effect on the viscosity and flow of the polymer matrix, provide ideal packing of filler particles, which results in the absence of uneven distribution of stresses around the particles and, as a result, improves the physicomechanical properties of the filled polymers.

In addition to glass spheres, there are a number of other solid microspheres—polymer microspheres, zinc powder, metal shot, carbon microspheres, microspheres obtained from sand or isolated from industrial smoke emissions.

One of the promising types of microspheres is corundum microspheres.

Corundum microspheres by 99.7% consist of refractory aluminum oxide, which, together with the hollow structure of the microsphere itself, makes it possible to use them to reduce the density and thermal conductivity of the final product [3]. The chemical composition of corundum microspheres is presented in Table 1.

Known application of corundum microspheres in compositions based on a combination of chloroprene and butadiene-nitrile rubbers used for the manufacture of rubber products [4].

The presence of roughness and grooves of different directions makes it possible to expect the deposition of various components and reactive viscous components on their surface (Fig. 1).

2 Materials and Methods

The effect of the dispersion of hollow corundum microspheres (HCM) on the properties of heat-shielding materials is investigated.

Investigations of HCM as a target additive for increasing the efficiency of fire-and-heat protective materials operating under extreme conditions were carried out

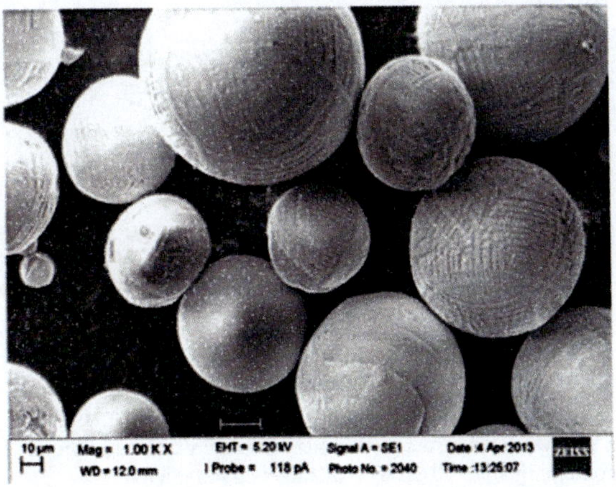

Fig. 1 Micrograph of hollow corundum microspheres. *Source* [5]

in elastomeric compositions based on ethylene-propylene diene rubber with a sulfur vulcanizing group.

To study the effect of HCM sizes on the complex of properties of heat-shielding materials, fractions with diameters of 0–40 (S), 40–70 (M) and 70–100 (L) microns were used.

The content of microspheres varied from 1 to 10 wt h.

3 Results

A study of the rheometric characteristics (MDR 3000 Professional, ASTMD2084-79) has shown that when microspheres are added to the elastomer composition, additional extrema appear on the kinetic curve of vulcanization (Fig. 2).

The appearance of the second stage of vulcanization can be associated with the processes of sorption–desorption on the surface of microspheres, leading to the appearance of additional active centers.

Physicomechanical, fire and heat protection and thermophysical properties of the investigated compositions are presented in Table 2. The physical and mechanical properties of the vulcanizates were determined using a Shimadzu AG–X plus tensile testing machine [6–10].

The fire and heat resistance of the samples was estimated from the time of heating the unheated side of the sample to 100 °C when exposed to the open flame of the plasmatron. The temperature created on the surface of the sample is 2000 °C.

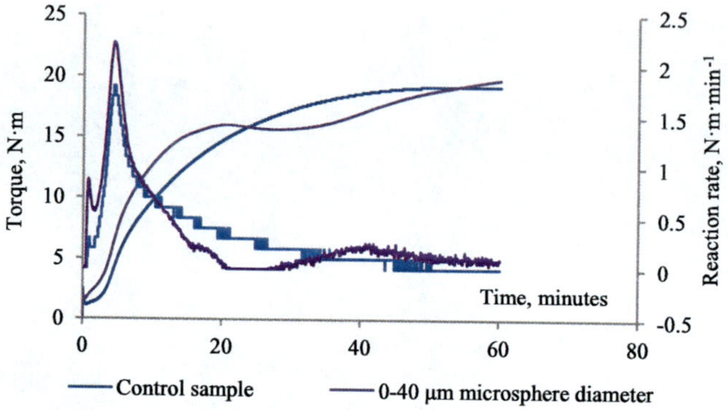

Fig. 2 Kinetic curves of vulcanization and reaction rate curves of the control mixture and a mixture containing 1 wt. HCM with diameters 0–40 microns

4 Discussion

Studies have shown that the introduction of microspheres helps to reduce abrasion by 2–12% (HCM-S) and 2–9% (HCM-L) and at the same time the heat resistance of the material increases – the heating time of the unheated side of the sample to 100 °C increases by 50–60% (Table 2). The use of microspheres with a smaller fractional diameter contributes to an increase in the heating time of the unheated sample surface to 100 °C and a decrease in the weight loss of the samples by 5–30% during testing, which may be associated with the formation in this case of a denser and more uniform coke structure. Microspheres of relatively large diameter (70–100 microns) reduce the thickness of the polymer matrix between the pores (microspheres), which leads to accelerated destruction of the material, its rapid heating and an increase in the rate of linear combustion. An increase in the content of microspheres leads to a decrease in the density of the compositions by 2–4%.

Small dosages of microspheres (1–5 parts by weight) practically do not affect the conventional tensile strength and elongation at break. Increasing the content of microspheres to 5 wt. including leads to a slight decrease in the conventional tensile strength up to 12%, which is explained by the fact that microspheres break the polymer matrix and create areas in which stresses are concentrated. Elongation at break is reduced to 18%.

The presence of microspheres practically does not affect the change in the properties of vulcanizates after thermo-oxidative aging. To evaluate the flame and heat resistance of the samples, we determined the following parameters using the developed procedure: dependence of the temperature on the unheated sample surface on the time of the action of open flame from a plasma torch, the sample weight loss, and the linear combustion velocity [8]. In high-temperature heating, the temperature created on the sample surface was of the order of 2000 °C.

Table 2 The content and size of the microspheres used and the physicomechanical, fire and heat protection and thermophysical properties of vulcanizates containing HCM[a]

Indicator	Control sample	S1	S5	S10	M1	M5	M10	L1	L5	L10
Microsphere diameter (μm)	-	0–40			40–70			70–100		
Content (wt. p)	-	1	5	10	1	5	10	1	5	10
Tensile strength (Mpa)	12,8	13,8	11,8	11,9	12,1	10,0	9,5	12,8	9,2	11,3
Elongation at break (%)	643	640	540	530	610	500	570	640	620	600
Density (g/cm^3)	1060	1017	1029	1038	1011	1019	1030	1010	1013	1025
Warm-up time of unheated sample surface to 100 °C (s)	220	330	350	370	210	240	310	190	170	160
Weight loss during heating (%)	28,78	15,08	19,34	22,10	12,59	9,12	18,63	46,73	39,91	27,39
Linear burning rate (mm/min)	43,64	41,38	40,00	30,00	64,86	48,00	53,83	53,83	60,00	70,00

[a]Vulcanization mode 165 °C × 45 min

Introduction of microspheres decelerates the speed of the sample heating in high-temperature tests (Fig. 3).

The study of the morphology of the material after high-temperature tests was carried out using a Versa 3D scanning electron microscope. The forming coke structure is more uniform, and the pore diameter in it is smaller than in the control sample. The mean pore diameter in the prepyrolysis zone decreases from 350–420 to 110–120 μm (Figs. 4 and 5).

The results of differential thermal analysis (derivatograph Q-1500 D-V1326, atmosphere—air) confirm the efficiency of using microspheres—the introduction of the studied additives leads to an increase in the temperature of maximum weight loss from 458 to 475 °C. An additional positive factor is an increase in the coke residue by 16% and an increase in the area of the endothermic peak by 61%, which indicates energy-intensive processes occurring during heating, which contribute to an increase in the fire-heat-shielding efficiency of the material.

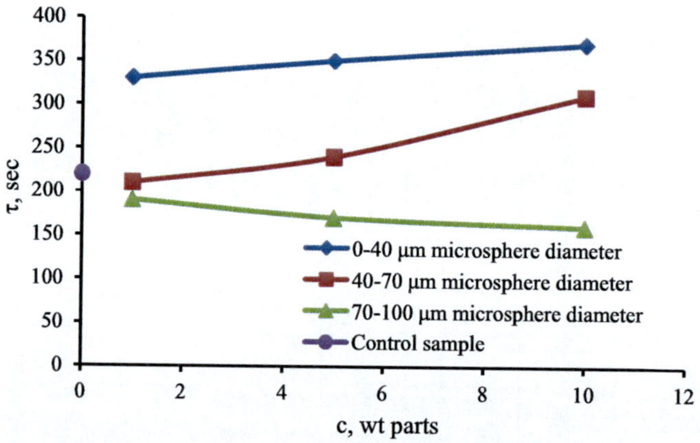

Fig. 3 Time τ in which the unheated sample surface is warmed up to 100 °C as a function of the microsphere content c

Fig. 4 Structure of the prepyrolysis layer of the samples after the test. Control sample (**a**) and sample containing 5 wt. parts of HCM (**b**)

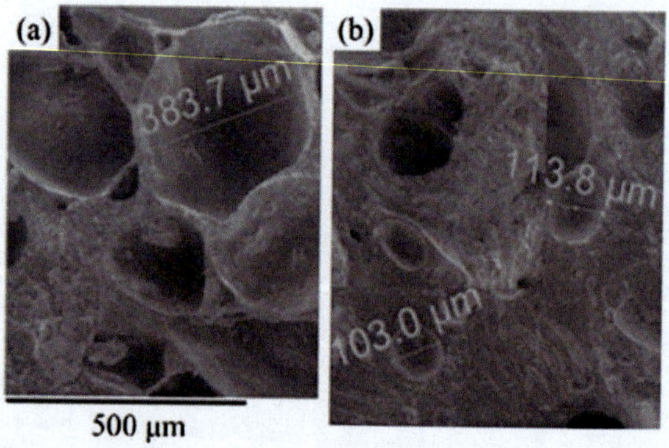

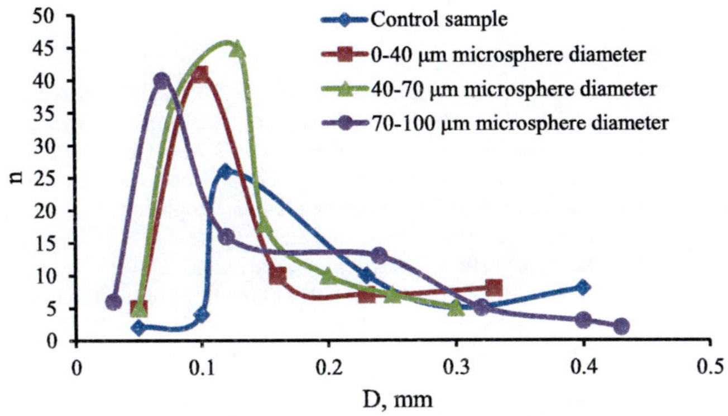

Fig. 5 Pore diameter distribution in the prepyrolysis layer of the samples. Relative amount of pores (n) and pore diameter (D). Control sample (1) and sample containing 5 wt parts of MSF (2)

The study of thermophysical characteristics has shown that the introduction of microspheres can reduce the thermal conductivity of the elastomeric composition by 8%. Improvement of the thermophysical characteristics of the material may be due to the fact that the diameter of the microspheres in the range of 20–40 microns is commensurate with the wavelength of thermal infrared radiation, which ensures the conditions for its maximum absorption by the microspheres.

5 Conclusions

The use of hollow corundum microspheres as components of elastomer fire-and-heat-protecting materials enhances the performance of these compounds owing to a decrease in the thermal conductivity and density of the item, with the optimum level of physicomechanical properties preserved. At a relatively low content of HCM, an increase in the conventional tensile strength and elongation at break occurs and a decrease in density compared to the control sample. For rubbers based on EPDM, the introduction of microspheres decreases the abrasion of the material and increases its heat resistance, which is important for products operating under severe operating conditions (in the mining and coal industries).

Acknowledgements The study was supported by a grant from the President of the Russian Federation for state support of young Russian scientists—candidates of sciences (MK-1802.2020.3).

References

1. Kablov VF, Novopol'tseva OM, Kochetkov VG et al (2017) Physicomechanical, thermal, and flame-retardant properties of elastomer compounds based on ethylene–propylene–diene rubber and filled with hollow aluminosilicate microspheres. Russ J Appl Chem 90:257–261
2. Stephen E (2015) Amos: 4—Hollow glass microspheres in rubbers and elastomers. In: Amos SE, Yalcin B (eds) Plastics design library, hollow glass microspheres for plastics, elastomers, and adhesives compounds. William Andrew Publishing, pp 107–122
3. Kit-Stroy SPb LLC Homepage, http://t-tiss.com. Last accessed 2021/03/03
4. Celyh EP, Khodakova SY, Malyutin VI et al (2017) Properties of elastomeric compositions modified with hollow corundum microspheres of aluminum oxide. Ind Prod Use Elast 1:37–40
5. Omega Minerals group of companies Homepage, www.omegaminerals.ru. Last accessed 2021/03/03
6. Amado JCQ, Ross PG, Sanches NB, Pinto JRA et al (2020) Evaluation of elastomeric heat shielding materials as insulators for solid propellant rocket motors: a short review. Open Chem 18:1452–1467
7. Kablov VF, Novopoltseva OM, Kochetkov VG et al (2017) Elastomer heat-shielding materials containing aluminosilicate microspheres. Russ Eng Res 37:1059–1061
8. Blyznyuk O, Vasilchenko A, Ruban A et al (2020) Improvement of fire resistance of polymeric materials at their filling with aluminosilicates. Mater Sci Forum 1006:55–61
9. Barabash A, Barabash D, Pertsev V, Panfilov D (2019) Polymer-composite materials for radiation protection. In: Murgul V, Pasetti M (eds) International scientific conference energy management of municipal facilities and sustainable energy technologies EMMFT 2018. EMMFT-2018 2018. Advances in intelligent systems and computing, vol 983. Springer, Cham
10. Bartholmai M, Schartel B (2004) Layered silicate polymer nanocomposites: new approach or illusion for fire retardancy? Investigations of the potentials and the tasks using a model system. Polym Adv Technol 15:355–364

Formation of Thermal Regime in a Permafrost Area Mine

Aleksandr Galkin and Vladimir Pankov

Abstract The main regularities of the formation of thermal regime in a mine designed in a permafrost area are studied. The concept of dimensionless temperature, which is a complex physical quantity and most completely characterizes the thermal processes in mine workings, was utilized to provide a generalized analysis and determination of qualitative and quantitative peculiarities of mine thermal regime formation. Multiple variant calculations of mine workings chains were carried out. These calculations permitted to establish the main regularities of thermal regime formation in the mine workings and the frozen rocks that surround them. An assessment of the impact of air expenditure, mine working cross-section and the natural (initial) temperature on the formation of thermal regime in the mine was carried out. It was demonstrated that the natural temperature influences the formation of thermal regime only during the first year of the exploitation of the mine.It was determined that when the air expenditures in mine workings are large, a gradual cooling of the active layer of rocks around the mine workings and movement of the isotherm to a significant distance along the length of the ventilation path. At that, the depth of the active layer participating in heat exchange of air and the rocks is relatively small and is about 1.7–2.5 m thick. 2D and 3D charts allowing to quickly assess the expected thermal conditions (average monthly air temperatures) in mine workings during the yearly ventilation cycle of the mine were created.

Keywords Mine · Permafrost area · Design · Thermal regime · Mine working

A. Galkin (✉)
Melnikov Permafrost Institute of the Siberian Branch of the Russian Academy of Sciences, Merzlotnaya str., 36, Yakutsk, Russia 677010

V. Pankov
North-Eastern Federal University, Belinsky str., 58, Yakutsk, Russia 677027

A. Beskopylny and M. Shamtsyan (eds.), *XIV International Scientific Conference "INTERAGROMASH 2021"*, Lecture Notes in Networks and Systems 247,
https://doi.org/10.1007/978-3-030-80946-1_21

1 Introduction

For the justification and choice of the energetically and economically efficient methods and means of air conditioning in design of underground structures of different purposes it is necessary to know the main regularities of thermal regime in mine workings during the period of their exploitation [1–8]. This question is especially relevant when designing mines in the permafrost areas given the significant energy expenses on thermal regime regulation to assure the normative parameters of microclimate in mine workings and to limit the negative impact of the thermal factor on the main and auxiliary processes of mining [9–15]. In this article, the general regularities of thermal regime formation in a mine designed in a permafrost area will be considered. A peculiarity of this mine is the large length of the mine workings, the terms of development of the under-frozen horizons and a significant range of the air expenses from 1.7 m^3/s in the field horizon to 89.1 m^3/s in the main air-supplying mine workings. The formation of thermal regime in such mine workings has its specificities. In particular, among the specificities are the gradual cooling of the active layer of rocks around the mine workings and a movement of the yearly average isotherm by a significant distance along the ventilation path. The latter is explained by the fact that the coefficient of heat transfer from the air to the rocks has a nearly linear dependency (specifically, $\alpha = f(V0.8)$) and for this reason, the increase in air expenditure in the mine workings by, for example, two times will lead to the increase of heat transfer coefficient by 1.7 times. The higher is the heat transfer coefficient, the faster will the cooling of the active layer of rocks be and the faster and further will the yearly average isotherm move.

That is, for mines with a small air expenditure the temperature along their length changes from the initial (outside) air temperature to the natural temperature of the rocks at the depth of the mine development. At high expenditures, typical for the mines in permafrost areas, given the high level of heat exchange, the natural temperature has an impact on the formation of the thermal regime only for a short period of time. And, starting from the second or third year of exploitation, an average temperature of the rocks within the active layer forms which serves as the determining temperature.

2 Methods

Models [16–18] modified for the conditions under consideration and a special computer program developed on their basis were used to forecast the thermal regime. The program contains means for forecasting the thermal regime in the mine workings in permafrost areas. An algorithm of numerical solution of thermal balance equation of a perforating mine working forms the foundation of the program. The calculation methods included in the program are applicable for solution of specific problems thermal physics in mining. In particular, problems related to the process of the formation of thermal conditions in the drifts and mines of the permafrost areas at both the natural and regulated thermal regime from the start of the mine's (or mine drift's) exploitation to the establishment of a quasi-stationary regime. The program allows to determine the sought parameters of interest (temperature at the end of the mine working section being calculated and a time at which the thermal regime reaches a quasi-stationary level) for mine workings of any form, length, different methods of fixing and thermal physical properties of the surrounding rocks depending on the network properties and relation of all mine workings among each other. The program is applicable at both the regular ventilation method and the reverse ventilation stream.

To assess the formation of the thermal regime of the rocks around the mine workings within the boundaries of the active layer a special program was developed which allows to determine the average integral value of the active layer temperature during different periods of the year and depending on the month in which the ventilation of the workings starts and duration of the ventilation.

3 Discussion of Results

On Fig. 1 are displayed the charts characterizing the change in heat transfer coefficient α (W/m^2 K) from the air to the rocks depending on the speed V (m/s) of the ventilation stream. Additionally, the heat transfer coefficient depends on the temperature and humidity of the air. Dry, cold air is always heavier during the winter period than the humid and warm air during the summer period of the year (May–September).

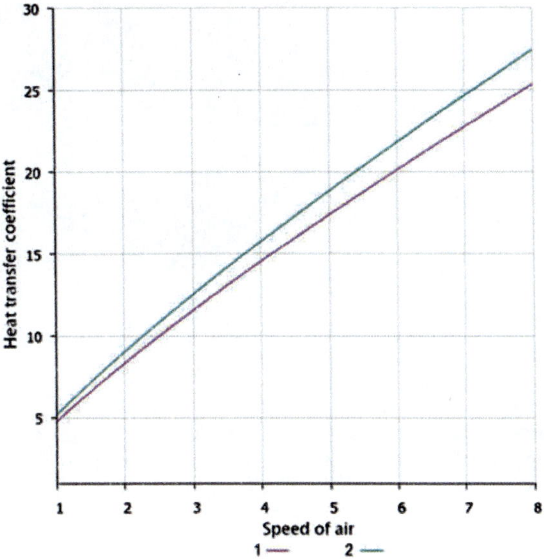

Fig. 1 Change in heat transfer coefficient (W/m^{2}K) from the air to the rocks depending on the speed of air in the mine working (m/s). *1*—during the warm period of the year; *2*—during the cold period of the year

Correspondingly, the heat transfer coefficient from the air to the rocks is always greater during the winter period than during the summer period. The higher the velocity of the air, the greater is the impact of its temperature and humidity on the heat transfer coefficient (charts 1 and 2).

This fact also facilitates the progressive cooling of the active layer of rocks. The depth of the active layer in a mine depends on the cross-section and duration of the mine working ventilation and on the physical properties of the rocks mass: thermal conductivity coefficient λ (W/mK); total heat capacity—Cp (J/kg K) and density ρ (kg/m^3). Since in the calculation model the average monthly air temperature was considered, in an analysis it is worthwhile to speak not of the entire active layer but of the radius of the thermal impact during the month, as the heat exchange of the rocks with the air is limited by this radius. Using previously obtained formulas [17], charts showing the change in the thermal impact radius depending on the typical range of change of the properties of the rocks for different mine workings were constructed. Those charts are shown on Fig. 2. The plane 1 characterizes the change in the thermal impact radius (m) and the plane 2 shows the depth of the active layer which is determined as the difference between the radius of the thermal impact and the equivalent radius of the mine working (m), equal to R0 = (S/π)0.5.

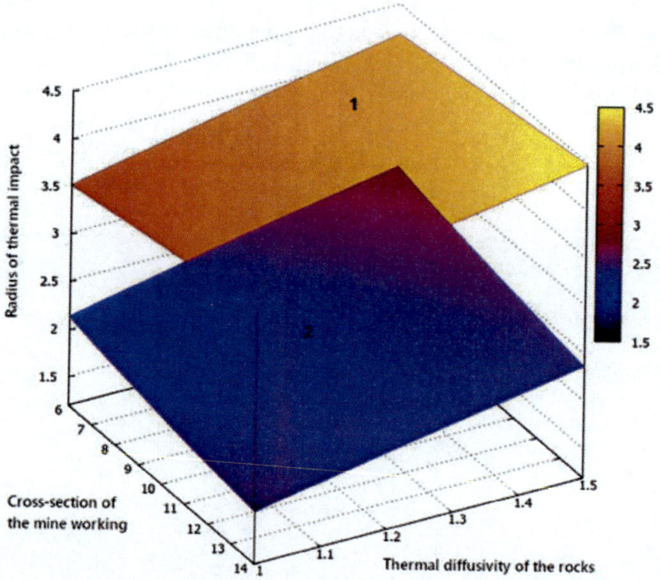

Fig. 2 Change in the radius of thermal impact (m) depending on the mine working cross-section (m^2) and thermaldiffusivityoftherocks (m^2/s). *1*—radius of thermal impact; *2*—depth of the active layer

From the figure it follows that the depth of the active layer participating in the heat exchange between the air and the rocks is relatively thin and has thickness of 1.7–2.5 m. In practice, the thermal regime of this layer determines the formation of the thermal regime in the mine workings.

The assessment calculations carried out have shown that already in the first year of the mine working's exploitation the average temperature of the mass during the cold period of the year is significantly lower than the natural temperature of the rocks. For example, the average yearly temperature of the active layer rocks after a year of ventilation is −14.3 °C which is almost three times lower than the natural temperature of the rocks ($T_e = -5$ °C).

For a generalized analysis and determination of qualitative and quantitative specificities of the formation of the thermal regime in the mine working a concept of a dimensionless temperature will be introduced. Dimensionless temperature is a complex physical quantity and is determined by the following expression: $\theta = (t_k - T)/(t_0 - T)$ where t_k—temperature at the entry to the mine working, °C; t_o—temperature at the exit from the mine working, °C; T—average integral temperature of the rocks within the boundary of the active layer, °C. For newly developed mine workings and mine workings under construction, the natural temperature of the rocks at the depth of the mine working is taken to be T. The parameter θ varies from 0 to 1. If the dimensionless temperature equals or is close to zero, it means that the air temperature at the end of the mine working, for example for a newly developed mine working or a mine working currently being excavated, is equal to the natural temperature of the rocks.

For mine workings ventilated for a long time, longer than a year, the air temperature at the end of the mine working will be equal to the average integral temperature within the boundary of the active layer. If the dimensionless temperature is equal or close to one, it signifies that there is virtually no heat transfer between the air and the rocks and the temperature at the end of the mine working is equal or close to the temperature at the entry. In practice, this is only possible for short mine workings with large air expenditures. For example, for ventilation channels or mine workings ventilated for a long time with air with constant temperature.

On the Figs. 3 and 4 the change in dimensionless temperature depending on the length of the mine working at different air temperatures is shown. Analysis of the graphs shows that the smaller is the air expenditure and the longer is the mine working, the faster the dimensionless temperature nears zero. That is, at the end of the calculated section the air temperature will be close to the temperature of the rocks. For large expenditures (40–80 m^3/s) this length is greater than 2000 m, while for small expenditures (lower than 5 m^3/s) it is almost three times smaller (Fig. 4).

On Fig. 5, the change in dimensionless temperature depending on the air expenditure in mine workings of varying lengths is shown. This supports the quantitative dependency. For shorter mine workings the impact of air expenditure on the change in dimensionless temperature is smaller than for long mine workings.

From the charts on Fig. 5 it can be seen that the gradient of dimensionless temperature when the length of the mine working is 1000 m is almost two times smaller than when the mine working is 100 m long. To clearly show, the conducted analysis, a

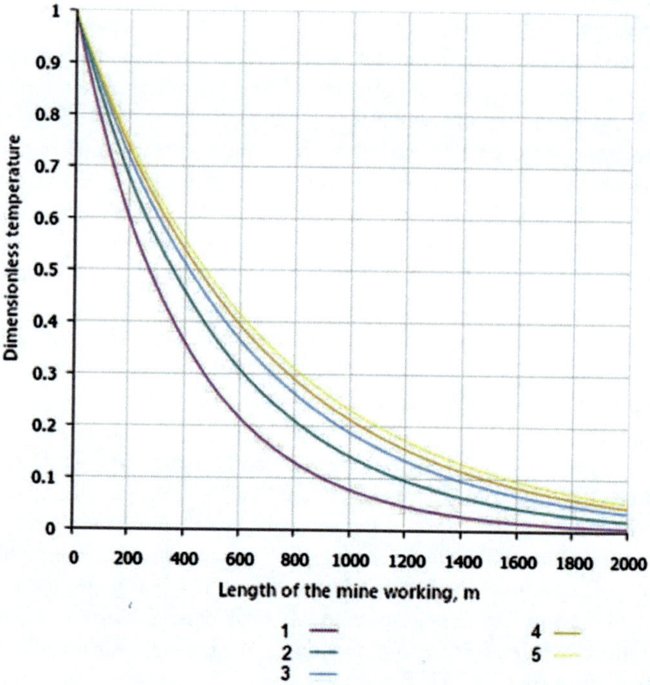

Fig. 3 Change in the dimensionless temperature along the mine working length at high air expenditures. *1*—5 m^3/s; *2*—20 m^3/s; *3*—40 m^3/s; *4*—60 m^3/s; *5*—80 m^3/s

general 3D chart on Fig. 6 shows the change in dimensionless temperature depending on air expenditures and mine workings length.

The transition from dimensionless temperature to a dimension temperature can be done using the following simple formula: $t_x = T + (t_0 - T) * \theta$, °C.

Analytical approach with the use of dimensionless temperature and a 3D graph allows to assess expected air temperatures in the mine without resorting to complicated mathematical calculations.

4 Conclusion

According to the results of the conducted analysis, the following conclusions can be made about the formation of thermal regime in a permafrost area mine. At low air expenditures (less than 20 m^3/s) the main regularities of thermal regime formation correspond to the traditional ideas about the heat exchange processes of the mine air with the rocks. At high expenses in the mine workings (higher than 40–80 m^3/s) progressive cooling of the active layer of rocks around the mine workings and a movement of yearly average isotherm a significant distance along the ventilation

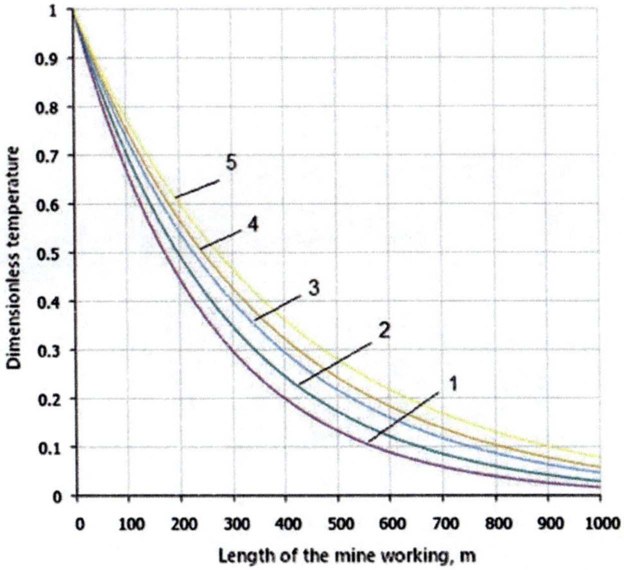

Fig. 4 Change in the dimensionless temperature along the length of the mine working at low air expenditures. *1*—0.5 m^3/s; 2—1 m^3/s; *3*—2 m^3/s; *4*—3 m^3/s; *5*—5 m^3/s

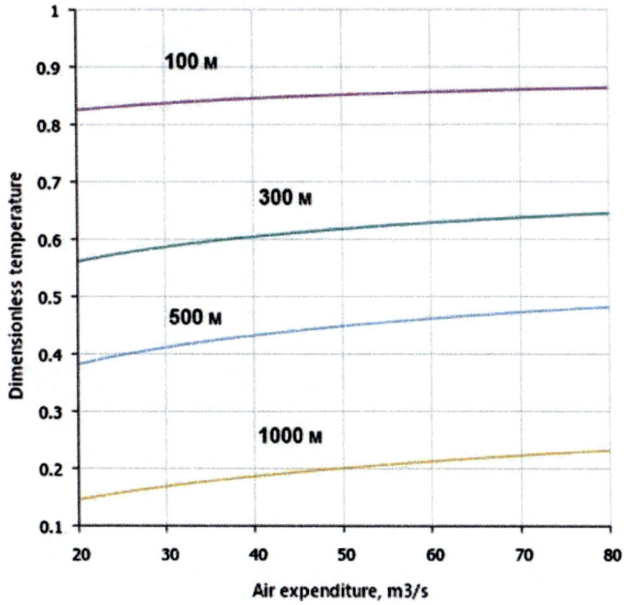

Fig. 5 Change in dimensionless temperature depending on the air expenditure at different mine working length

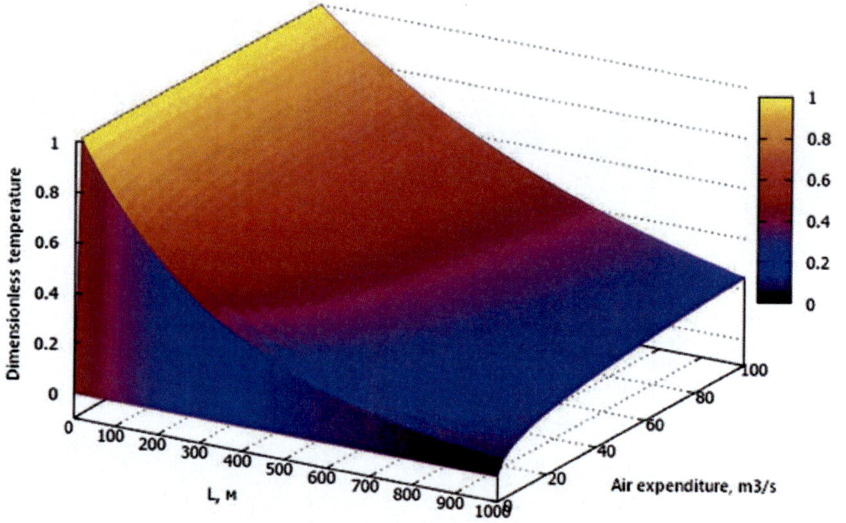

Fig. 6 General chart of the change in dimensionless temperature depending on the air expenditure and mine working length

path takes place. This can be explained by the fact that the heat transfer coefficient from the air to the mine workings has a nearly linear dependence. For this reason, the increase of air expenditure in the mine working by a factor of, for example, 2, leads to the increase of the heat transfer coefficient by a factor of 1.7. It was determined that the depth of the active layer participating in the heat exchange of air and rocks is relatively small and is 1.7–2.5 m deep.

References

1. Arens VZh, Dmitriev AP, Dyadkin YuD et al (1988) Thermal-physical aspects of minerals exploitation. Nedra, Leningrad
2. Kuvayev VA, Kuzmin GP (2018) Underground cryostorage of plant seeds in permafrost zones. Geol Geogr Glob Energy 4:150–155
3. Kuzmin GP (2002) Undeground structures in permafrost zones. Nauka, Novosibirsk
4. Shatz MM (2018) Preservation of biodiversity of cultural plants in cryostorages located in permafrost zones. Use Protect Nat Resour Russia 1:41–48
5. Greth A, Roghanchi P, Kocsis K (2017) A review of cooling system practices and their applicability to deep and hot underground US mines. In: Brune JF (ed) 16th North American Mine Ventilation Symposium, vol 11, Colorado School of Mines, Golden, CO, pp 1–9
6. Danko G (2012) Ventilation and climate control of deep mines. In: McGraw-Hill Yearbook of Science and Technology, pp 296–299
7. Kazakov BP, Zaytsev AV (2014) Research of thermal regime formation processes in deep mine workings. Bull PNIPU. Geology. Oil Gas Ind Mining 10:91–97

8. Vernigor VM, Morozov KV, Bobrovnikov VN (2013) On approaches to design of thermal regime of mines in permafrost conditions. Bull Mining Univ 205:139–140
9. Voronov ET, Bondar IA (2010) Influence of temperature factor on safety and effectiveness of underground mining in permafrost zone. Bull ChitGU 5(62):85–93
10. Naumov AA, Nikolaeva DV (2015) Sectioning of the mining regions of Russia according to the expenses on mine air conditioning. Fundam Res 8:312–317
11. Chebotarev AG, Afanasyeva RF (2012) Physiological-hygienic assessment of microclimate on workplaces in mine tunnels and open pit mines and methods of preventing its negative impacts. Mining Ind 6:34–40
12. Epstein Y, Moran DS (2006) Thermal comfort and heat stress indices. Ind Health 44(3):388–398
13. Parsons K (2006) Heat stress standard ISO 7243 and its global application. Ind Health 44(3):368–379
14. Dormidontov AV, Galkin AF (2015) Thermal regime of the Julietta mine. Mining Inf-Anal Bull (Sci Tech J) 60(2):91–96
15. Shuvalov YuV, Galkin AF (2010) Theory and practice of optimal regulation of thermal regime of underground structures of permafrost areas. Mining Inf-Anal Bull (Sci Tech J) 8:365–370
16. Galkin AF (2000) Thermal regime of the underground structures of the North. Nauka, Novosibirsk
17. Galkin AF (2015) Rational ventilation mode of mountain manufactures in cryolite zone. Metall Mining Ind 1:62–65
18. Galkin AF, Voloshina AA, Software State Registration Certificate No. 2017610789

The Stresses Distribution in a Two-Cell Ceramsite Concrete Block Cross-Section

Arturs Proskurovskis, Levon Nazinyan, Anna Tarasova, Nikita Belousov, Svetlana Belyaeva, Olga Nikonova, and Aleksandr Panfilov

Abstract One of the notable trends in modern engineering practice is the refinement of methods for calculating the building's structural loads in order to increase the efficiency of building materials' use, what subsequently will significantly reduce the final cost of the project. The purpose of this article is to study possible ways of increasing work efficiency and reducing the cost of a cellular ceramsite concrete fixed formwork block by modifying its cross section. After modeling the main stresses' distribution in different sections of the block in the Abaqus program, it was revealed that in the side walls of the modified block, the values of the maximum principal tensile stresses are much larger than those in the old block. This fact allows concluding that the load should be significantly reduced. The development of digital twins' industry may as well be useful for the use of novel construction materials in the structures subjected to significant loads.

Keywords Simulation · Abaqus · Structural analisys · Stress distribution · Stress concentration analisys

1 Introduction

One of the notable trends in modern engineering practice is the refinement of methods for calculating the building's structural loads in order to increase the efficiency of building materials' use, what subsequently will significantly reduce the final cost of the project.

Despite the significant number of recommendations and teaching aids of the design of buildings and structures, people often miss the factor of low efficiency of the building materials themselves. In order to increase this indicator, manufacturers often carry out preliminary modeling of the work of structural elements in software systems that basically use the finite element method.

A. Proskurovskis · L. Nazinyan · A. Tarasova · N. Belousov · S. Belyaeva · O. Nikonova (✉) · A. Panfilov
Peter the Great St. Petersburg Polytechnic University, Polytechnicheskaya, 29, St. Petersburg, Russian Federation 195251

A. Beskopylny and M. Shamtsyan (eds.), *XIV International Scientific Conference "INTERAGROMASH 2021"*, Lecture Notes in Networks and Systems 247,
https://doi.org/10.1007/978-3-030-80946-1_22

In previous studies, according to the authors, the most promising form of the block was designed, and its composition was selected. Subsequently, heat engineering tests of the composition were performed, and the cast block was tested for strength [1, 2]. At the moment, the optimal indicators of the block are being selected to increase its competitiveness in the building materials market.

The articles [3–8] present various methods for calculating and modeling loads in work processes using software systems that include Abaqus. The authors of the article [3] demonstrate a plugin for the Abaqus commercial FE code allowing for fatigue predictions based on the results of Abaqus FE analyses. In the study [4] a numerical investigation is conducted to simulate the total shrinkage strain of High-Strength Concrete (HSC) and Steel Fiber Reinforced High-Strength concrete (SFRHSC) by means of the transient thermal-stress analysis. The paper [5] proposes a new type of wall with built-in structural configuration. In the article [6] a correlation has been established between the formwork surface functional signatures and its adherence propensity to concrete. In the study [7] Finite element analysis (FEA) using ABAQUS software was performed to investigate the behavior of post-tensioned concrete beams. The paper [8] reports the details of numerical models used to predict the performance of two 3.6 m-high well-instrumented wrapped-face walls.

The articles [9–20] discuss various cases of crack formation and their propagation, methods for modeling crack formation using both manual calculations and software systems using the finite element method and including Abaqus. The study [9] describes method that demonstrates crack coarse model and submodel data by parameter way, including meshing, load and boundary condition. In the article [10] a new Python based numerical tool implementing a Sequentially Linear Analysis (SLA) procedure, which is able to call the FEM software Abaqus, is described and tested. The study [11] describes the behavior of the bond between a steel bar and the concrete along a lap splice region for structures subjected to cyclic loading. An experimental programme was carried out in the article [12] to investigate the mechanical behavior of masonry elements made of concrete and natural sisal fibers. The adaptability of crack detection models was considered in the paper [13]. The authors of the article test the neural network on two separate sets of image data of cracks in concrete and asphalt and offer three methods of adapting the network to increase the accuracy of detection of cracks. The study [14] presents a numerical study of steel-plate concrete (SC) composite walls using the general-purpose finite element (FE) program ABAQUS. The authors of the paper [15] propose a new approach to modeling the mechanism of crack formation based on the study of cohesion zone models (CZM). The article [16] proposes an on-line updating Gaussian process (GP) measurement model within the PF based crack prognosis framework. The study [17] offers an efficient numerical method based on the fiber beam-column model is developed for predicting the full-process nonlinear crack width development of composite beams. The paper [18] deals with an experimental and numerical study of stone block masonry beams externally reinforced using glass fiber-reinforced polymers (GFRP) strips. In the paper [19] a 3D discrete element viscoelastic and inhomogeneous microstructure-based pavement model subjected to thermal straining was introduced. In the study [20] a theoretical study of a crack by the Vickers index is carried out and two models of finite elements are developed, one of which is based on fracture mechanics, and the second on tensile stress criteria.

The authors of articles [21, 22] talk about methods for optimizing stress concentrators in sections of structural parts. In the paper [21], using the example of a cantilever beam and a structural bracket, the role of shape optimization in optimizing section stress concentrators is considered. Article [22] talks about the result of reducing weight and stress concentration without the need to repeat the generation of a numerical model by rounding the corners of a structural bracket.

In the above sources, various methods of modeling the loads in software complexes using the finite element method, the cases of crack formation and their spreading, as well as stress concentration processes are considered, however, so far, the topic of optimizing the cross section of fixed formwork blocks of this kind is not disclosed.

The purpose of this article is to study possible ways of increasing work efficiency and reducing the cost of a cellular ceramsite concrete fixed formwork block by modifying its cross section.

2 Materials and Methods

In this article we modeled the main stresses' distribution in the cross section of modified ceramsite concrete fixed formwork block when the stress of 15 MPa was applied to it. To analyze the mechanical behavior and make a comparative analysis of the ceramsite block, we used a finite elements analysis method. We used Abaqus software in order to perform such an analysis. Abaqus employs solution technology ideal for static and low-speed dynamic events where highly accurate stress solutions are critically important [23].

The following coefficients were adopted for the calculations: Poisson's ratio—0.2 and elastic modulus of ceramsite concrete—$E = 23 * 109$ Pa.

In the previous study [1] it was revealed that the main stresses' concentration in the places of a sharp cross-section change decreased due to the creating a smoother block geometry. This allowed the block to withstand bigger loads. Considering the aforementioned factor, it was proposed to round the inner corners of the block cells and remove one of the inner partitions to reduce the weight and cost of the block. The old and new section of the block is shown in Fig. 1.

3 Results and Discussions

In the light of the above, there is still one unresolved question: whether the optimized block will withstand the applied load of 15 MPa. The same load was applied to the block presented in the previous study [1].

After modeling the main stresses' distribution in Abaqus software the concentrations of dangerous stresses were analyzed and the fields of their distribution and their

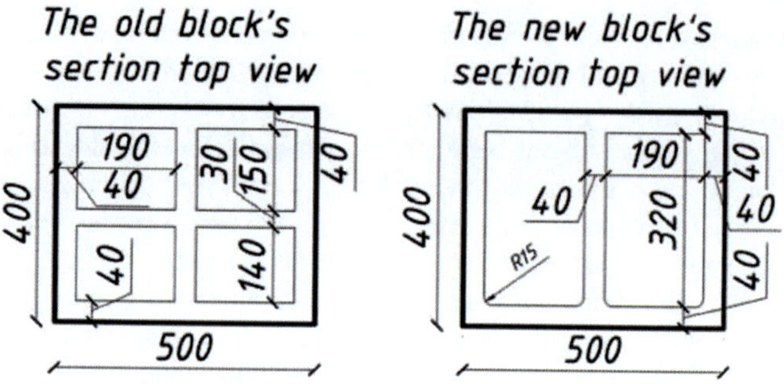

Fig. 1 The old and new block's cross section top view

values were examined. Then, based on these parameters, it was predicted whether the modified block would collapse.

For clarity of comparison, in this article we analyze the distribution of the main stresses in the old and modified sections of the block comparing the results obtained with the use of Abaqus.

In Fig. 2 the results of the distribution of the main stresses in the section of an "old block" is presented.

In Fig. 3 the distribution of the main tensile stresses in the section of the "old block" is presented.

The distribution of the main stresses in the section of the "new block" is present in Fig. 4.

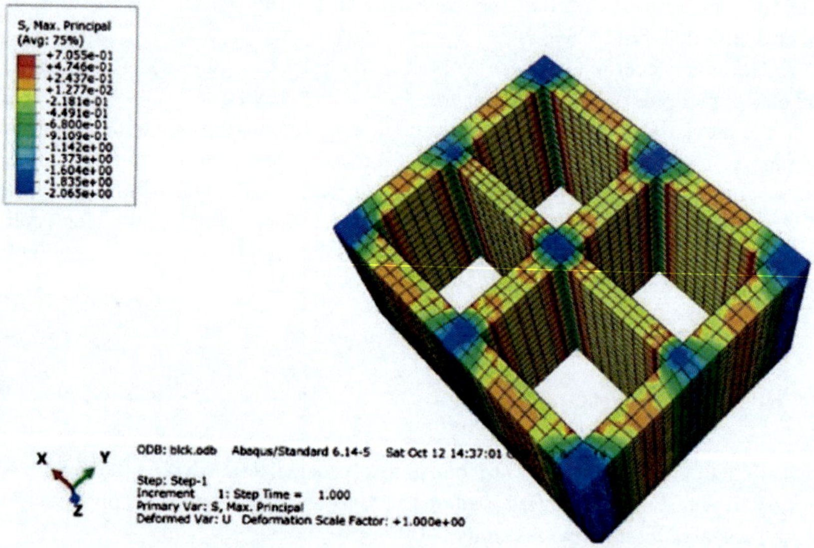

Fig. 2 The distribution of the main stresses in the section of the "old block"

Fig. 3 The distribution of the main tensile stresses in the section of the "old block"

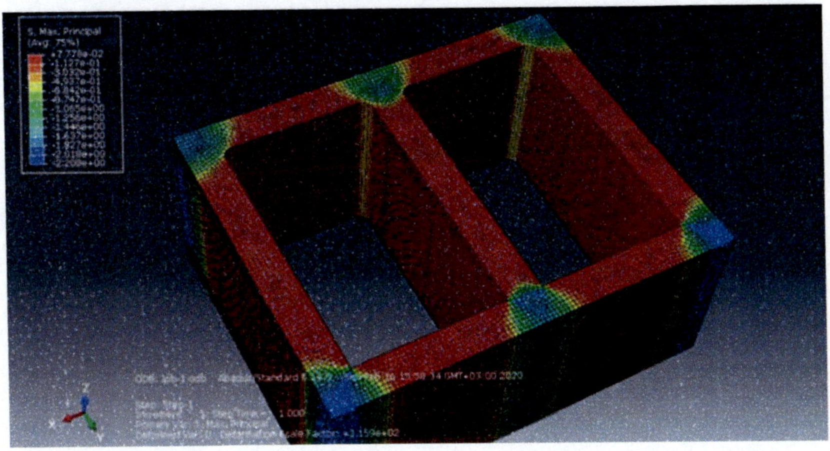

Fig. 4 The distribution of the main stresses in the section of the "new block"

The distribution of the main tensile stresses in the section of the "new block" is present in Figs. 5 and 6.

Analyzing the distribution of the main stresses in the cross sections of the old and modified blocks, several important patterns can be noted.

Firstly, it should be noted that the values of tensile stresses in the side walls of the new block increased significantly. If in the old block they reached 0.0127 MPa, then in the new block they already are 0.0778 MPa, what indicates a more than six-time increase.

Secondly, the areas of maximum main principal tensile stresses' distribution in the new block increased significantly.

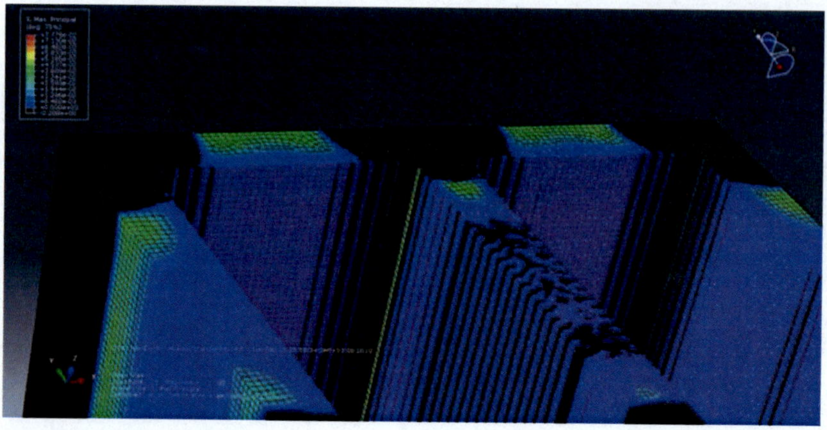

Fig. 5 The distribution of the main stresses in the section of the "new block" 1

Fig. 6 The distribution of the main stresses in the section of the "new block" 2

Thirdly, it is worthwhile to focus on the distribution of the maximum main compressive stresses on the inner edges of the block. So, in the new block, on the two inner edges of which was a rounding with a radius of 15 mm, the values of the maximum compressive stresses are less by about 10%, in comparison with the old block. At the same time on the other two edges (without rounding) maximum compressive stresses are identical to the old block.

Fourthly, it is important to point out that tensile stresses are one of the most unfavorable working conditions for ceramsite concrete and concrete in general. And since the distribution zones of the main tensile stresses have increased, we can conclude that the zone of potential cracking has increased. Moreover, it is not known whether a block with a new cross section will be able to hold even two to three times smaller loads, since expanded clay concrete is a very heterogeneous material and cracking can begin much earlier due to defects in the structure of the material.

4 Conclusion

Values of tensile stresses in the side walls of the new block increased significantly. If in the old block they reached 0.0127 MPa, then in the new block they already are 0.0778 MPa, what indicates a more than six-time increase. The areas of maximum main principal tensile stresses' distribution in the new block increased significantly.

In the new block, on the two inner edges of which was a rounding with a radius of 15 mm, the values of the maximum compressive stresses are less by about 10%, in comparison with the old block. At the same time on the other two edges (without rounding) maximum compressive stresses are identical to the old block.

It is important to point out that tensile stresses are one of the most unfavorable working conditions for ceramsite concrete and concrete in general. And since the distribution zones of the main tensile stresses have increased, we can conclude that the zone of potential cracking has increased. Moreover, it is not known whether a block with a new cross section will be able to hold even two to three times smaller loads, since expanded clay concrete is a very heterogeneous material and cracking can begin much earlier due to defects in the structure of the material.

All the above facts allow us to conclude that if the old block was destroyed due to the action of such a load, then the block with the modified section will also collapse. Moreover, the values of the main tensile stresses in the side walls of the modified block are much larger, which indicates that the applied load should be significantly reduced.

It is necessary to conduct compression tests and obtain experimental values of the compressive strength of this cellular ceramsite concrete block with a modified cross section to conduct further refinement of the cross section.

The material-efficient design is very important in nowadays construction sector that experiences the ever-increasing need of optimization of primary energy needs for materials production and energy-efficiency optimization [24]. The use of the innovative materials, recycled materials, lightweight materials and other interesting material-efficient solutions will lead to the overall decrease of the materials production. One good way highly discussed in literature [25–29] to make an efficient use of the classic construction material is to add particles that would ensure the change of material properties to its mass.

Additionally, in order to control the materials behavior to use them in the structures that may be subjected to high structural loads, it is necessary to monitor the material's behavior in real time. In order to provide a dynamical monitoring of the mechanical behavior of structures of high responsibility where the use of ceramsite concrete can be beneficial, the element's digital twin can be considered. The digital twin is a digital dynamical representation of an existing physical object in the digital environment. With the use of such a model, one can monitor all the mechanical deformations and deflections that occur in the structure in real time. The development of digital twins' industry may as well be useful for the use of novel construction materials in the structures subjected to significant loads.

References

1. Proskurovskis A, Nazinyan L, Tarasova A, Bespalov V, Gaujena B (2020) The bearing capacity of an expanded clay concrete block of permanent shuttering without reinforced concrete filling. Mater Today Proc
2. Proskurovskis A, Nazinyan L, Tarasova A, Belousov N, Belyaeva S et al (2020) The distribution of the main stresses in the section of the ceramsite concrete fixed formwork block with four cells. E3S Web Conf 08013
3. Nesládek M, Španiel M (2017) An Abaqus plugin for fatigue predictions. Adv Eng Softw 103:1–11
4. Bouziadi F, Boulekbache B, Haddi A, Djelal C, Hamrat M (2018) Numerical analysis of shrinkage of steel fiber reinforced high-strength concrete subjected to thermal loading. Constr Build Mater 181:381–393
5. Zhou X, Du J, Peng Q, Chen P (2019) Hollow block masonry wall reinforced by built-in structural configuration: Seismic behavior analysis. Soil Dyn Earthq Eng 126:105815
6. Spitz N, Coniglio N, El Mansori M, Montagne A, Mezghani S (2018) Quantitative and representative adherence assessment of coated and uncoated concrete-formwork. Surf Coatings Technol 352:247–256
7. Lee SH, Abolmaa A, Shin KJ, Lee HD (2020) ABAQUS modeling for post-tensioned concrete beams. J Build Eng 101273
8. Yu Y, Bathurst RJ, Allen TM (2017) Numerical modelling of two full-scale reinforced soil wrapped-face walls. Geotext Geomembranes 45:237–249
9. Sun Y, Zhai J, Zhang Q, Qin X (2019) Research of large scale mechanical structure crack growth method based on finite element parametric submodel. Eng Fail Anal 102:226–236
10. Távara L, Moreno L, Paloma E, Mantič V (2019) Accurate modelling of instabilities caused by multi-site interface-crack onset and propagation in composites using the sequentially linear analysis and Abaqus. Compos Struct 225:110993
11. Alyousef R, Topper T, Al-Mayah A (2018) Crack growth modeling of tension lap spliced reinforced concrete beams strengthened with fibre reinforced polymer wrapping under fatigue loading. Constr Build Mater 166:345–355
12. Soto Izquierdo O, Ramalho MA, Taliercio A (2017) Sisal fiber reinforced hollow concrete blocks for structural applications: testing and modeling. Constr Build Mater 151:98–112
13. Alipour M, Harris DK (2020) Increasing the robustness of material-specific deep learning models for crack detection across different materials. Eng Struct 206:110157
14. Nguyen NH, Whittaker AS (2017) Numerical modelling of steel-plate concrete composite shear walls. Eng Struct 150:1–11
15. Li C, Guo J (2019) Cracking simulation of asphalt concrete beam specimen using cohesive zone model. Constr Build Mater 214:49–60
16. Chen J, Yuan S, Wang H (2020) On-line updating Gaussian process measurement model for crack prognosis using the particle filter. Mech Syst Signal Process 140:106646
17. Xu LY, Nie X, Tao MX (2018) Rational modeling for cracking behavior of RC slabs in composite beams subjected to a hogging moment. Constr Build Mater 192:357–365
18. Fayala S, Limam O, Stefanou I (2016) Experimental and numerical analysis of reinforced stone block masonry beams using GFRP reinforcement. Compos Struct 152:994–1006
19. Wang H, Buttlar WG (2019) Three-dimensional micromechanical pavement model development for the study of block cracking. Constr Build Mater 206:35–45

20. Kadin Y, Mazaheri M, Zolotarevskiy V, Vieillard C, Hadfield M (2019) Finite elements based approaches for the modelling of radial crack formation upon Vickers indentation in silicon nitride ceramics. J Eur Ceram Soc 39:4011–4022
21. Groth C, Chiappa A, Biancolini ME (2018) Shape optimization using structural adjoint and RBF mesh morphing. Procedia Struct Integr 8:379–389
22. Porziani S, Groth C, Biancolini ME (2018) Automatic shape optimization of structural components with manufacturing constraints. Procedia Struct Integr 12:416–428
23. Abaqus/Standard, available from: https://www.3ds.com/products-services/simulia/products/abaqus/abaqusstandard/. Last accessed 2021.02.26
24. Belussi L, Barozzi B, Bellazzi A, Danza L, Devitofrancesco A, Fanciulli C et al (2019) A review of performance of zero energy buildings and energy efficiency solutions. J Build Eng 25:100772
25. Frolov A, Chumadova L, Cherkashin A, Akimov L (2014) Prospects of use and impact of nanoparticles on the properties of high-strength concrete. Appl Mech Mater 584:1416–1424
26. Frolov A, Cherkashin A, Akimov L, Vatin N, Koltsova T, Nasibulin A et al (2015) An impact of carbon nanostructured additives on the kinetics of cement hydration. Appl Mech Mater 725:425–430
27. Zinkevich I, Philipp M, Skvortsova O, Nikonorov A, Volkova Y, Nikonova O et al (2019) Innovative methods of concrete dams' inspection. IOP Conf Ser: Earth Environ Sci 403(1):012240
28. Akimov L, Ilenko N, Mizharev R, Cherkashin A, Vatin N, Chumadova L (2016) Composite concrete modifier CM 02–10 and its impact on the strength characteristics of concrete. MATEC Web Conf 53:01022
29. Rasheed A, San O, Kvamsdal T (2020) Digital twin: Values, challenges and enablers from a modeling perspective. IEEE Access 8:21980–22012

Method for Generating Artificial Accelerograms of Regional Earthquakes for Calculating Buildings and Structures

Vladimir Lapin, Syrymgali Yerzhanov, and Nurakhmet Makish

Abstract Methods for calculating buildings and structures based on direct numerical integration of the equations of dynamics are very relevant. For this, a new method for generating artificial accelerograms has been proposed. The method is based on the representation of the correlation function as a sum of cosine-exponential terms. The approximation parameters are determined based on the least squares method using the Curve Fitting Toolbox SCM MATLAB package. Artificial accelerograms are generated using an autoregressive (moving average) calculation formula. The generation accuracy is illustrated by the calculation of the spectral curves. The accelerogram of the 1990 Baisorun earthquake is taken as an example. The area of application of the accelerograms obtained in this way are earthquake-resistant construction and seismic microzoning, calculation of systems with dampers of viscous and dry friction, seismic isolation systems, different non-linear systems.

Keywords Artificial accelerogram · Seismic Impact · Earthquake-resistant construction · Eurocodes · Modeling

1 Introduction

The main problem of seismic-resistant construction is to reduce seismic risk and ensure seismic safety of the population of seismic-prone areas. This can be achieved either by refining the seismic action models, or refining the design models of buildings and structures, the parameters of which should be determined on the basis of experimental data. This article proposes to clarify the seismic impact models. The problems of increasing the reliability of earthquake-resistant construction are discussed.

The problem of ensuring the reliability of earthquake-resistant construction over the past years has not become easier or more studied. The main obstacles: incompleteness of information about external impact random in time and space and complex

V. Lapin (✉) · S. Yerzhanov · N. Makish
JSC Kazakh Research and Design Institute of Construction and Architecture, 21 Solodovnikova, Almaty, Kazakhstan
e-mail: lapin_1956@list.ru

A. Beskopylny and M. Shamtsyan (eds.), *XIV International Scientific Conference "INTERAGROMASH 2021"*, Lecture Notes in Networks and Systems 247,
https://doi.org/10.1007/978-3-030-80946-1_23

nonlinear work of building structures—remain unresolved. Some studies in previous years were carried out [1, 2].

Several interesting seismic events took place, instrumental records of which are of interest for simulation studies [3].

In recent decades, in the theory of earthquake-resistant construction, much attention has again begun to be paid to impact models that take into account the regional characteristics of seismic hazard and built for specific territories and sites. This is partly due to the introduction of European building codes (Eurocodes) in the CIS, in particular Eurocode 8. Since July 1, 2014, European building codes (Eurocodes) have been introduced in the Republic of Kazakhstan, which are beginning to be used on a par with old regulatory documents in construction.

The easiest way to do this is to use instrumental records of earthquakes recorded in seismogenic zones as close as possible to the city.

The task is to propose a rather general way to simulate real seismic effects by artificial accelerograms. In the city of Almaty there is a network of stations of the engineering and seismometric service [4], the instrumental records of which have up-to-date information for calculating buildings and structures.

2 Methods

The most notable seismic event of the 90s of the last century for the region of Almaty is the Baisorun earthquake of November 12, 1990.

The earthquake occurred in the Northern Tien Shan within the highly active seismically and very dangerous for the city of Almaty andKungey-Zailiyskiy zone. The tremors covered a vast territory, including the Alma-Ata, Taldy-Kurgan and Dzhambul regions of Kazakhstan, as well as the Issyk-Kul region of the Republic of Kyrgyzstan. The earthquake was most intense in the village of Kuturgan, located 20 km east of the epicenter. In Alma-Ata, an earthquake manifested itself with an intensity of 5–6 points. Earthquake magnitude: M = 6.3. Focal depth: H = 15–20 km [5].

Maximum intensity at the epicenter—8 points.

There is a two-component earthquake record recorded by the Kurmenty station at a distance of 35 km from the earthquake source, with digitization step of 0.008 s.

This record can be used to develop a seismic impact model for the Almaty region.

Table 1 shows some parameters of the most intense component (N-S) of the instrumental recording-accelerogram (Fig. 1). This entry is quite interesting—it contains two closely spaced peaks in the spectral density or reaction spectrum.

A quick analysis of Table 1 shows that the accelerogram is well centered. High-frequency instrumental recording. Dominant periods and carrier frequencies are determined from spectral density using the Signal Processing Toolbox MATLAB package. With a total duration of exposure over 8 s, the effective duration (duration of oscillations with an amplitude of more than half of the maximum) is 2.5 s, i.e.

Table 1 Accelerogram parameters

Parameters	Component (N-S)
Maximum of accelerogram (cm/s^2)	704.3
Minimum of accelerogram (cm/s^2)	−578.94
Dispersion (cm^2/s^4)	70.72
RMS value (cm/s^2)	9.21
Range (cm/s^2)	1301.06
Average value (cm/s^2)	5.004
Median value (cm/s^2)	1.599
Maximum spectral density	39.19; 37.01
Frequency (1/s)	3.59; 4.69
Period (s)	0.28; 0.09
Duration of oscillations with an amplitude of more than half of the maximum (effective duration) (s)	2.49

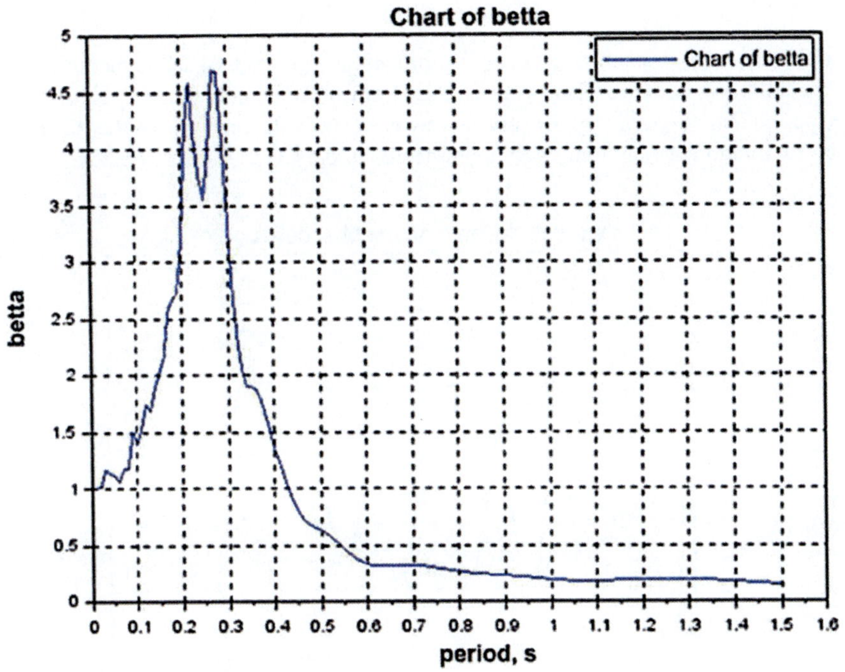

Fig. 1 Spectral curve (N-S component)

approximately 30% of the total duration. Thus, there is an impulse-type effect with two closely spaced peaks.

Figure 1 shows the spectral graph obtained with the value of the oscillation decrement 0.3. The calculations were performed using the MATLAB computer mathematics system using an integrator with automatic step selection (Runge–Kutta method).

The most universal models of seismic action are based on modeling the process of ground vibrations by stationary or non-stationary random processes [6].

Under the assumption of ergodicity, using the Signal Processing Toolbox package, an experimental correlation function of a stationary random process was constructed for a normalized accelerogram (Fig. 2).

We take the expression for the correlation function in the form of a line segment [6]:

$$K(\tau) = \sum_{i=1}^{N} A_i e^{-\alpha_i \tau} \cos(\omega_i \tau), \tag{1}$$

where A_i, α_i, ω_i—parameters to be determined, N—number of series members.

Parameters α, ω are usually determined from several characteristic points of the experimental curve in Fig. 3. Another way is to apply the methods of the theory of approximation of functions, well implemented in the Curve Fitting Toolbox.

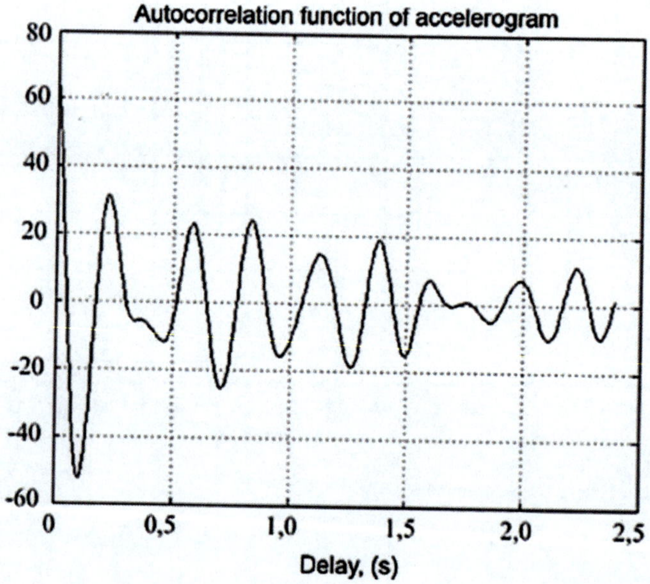

Fig. 2 Correlation function (N-S component)

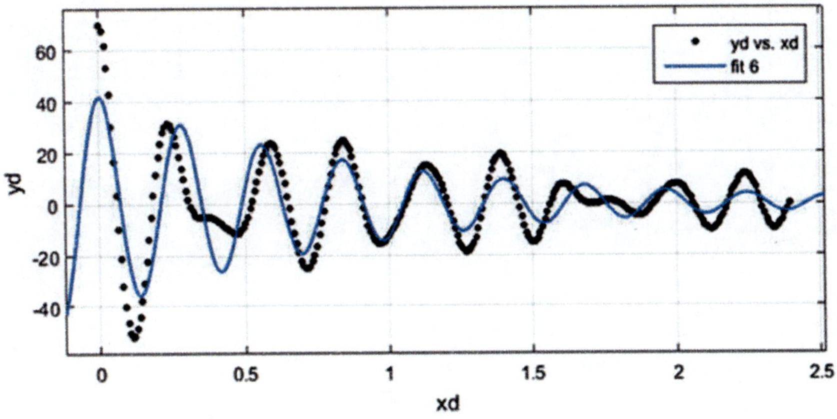

Fig. 3 Approximating (N = 1) and experimental correlation functions

3 Results

Table 2 shows the experimental values for several methods for determining the above parameters in the case of a one-term approximation. The coincidence of the experimental and approximating functions is accepted at the initial section on the first wave in the negative part of the abscissa axis (method 1). This method approximates only the initial portion of the correlation function, including the half-wave in the negative region. Other calculations were performed using the Curve Fitting Toolbox package (methods 2–4), the Levenberg-Marquardt algorithm using methods that are resistant to outliers in the original data. Below, the least squares method (LSM) in Fig. 4 is used, the method taking into account the bisquare weight (Bisquare Weights—BW)—with automatic selection of weights to reduce the effect of outliers in the data on the obtained approximation, as well as the case of minimizing the modules of residuals (least absolute residuals—LAR).

Tables 3 and 4 show the results of calculations using the Curve Fitting Toolbox package of the expansion parameters (1) for the cases of two- and three-term approximations.

Table 3 shows the results of determining the parameters by the methods of LSM (3rd column) and BW (4th column). There is no significant difference between the values of the coefficients, so Fig. 5 shows an approximating curve only by the LSM

Table 2 Parameters of the correlation function of the N-S component (N = 1)

Methods	A (cm/s^2)	α (1/s)	ω (1/s)
Method 1	69.64	6.01	40.11
Method 2 (LSM)	41.72	1.082	22.4
Method 3 (BW)	24.82	0.5518	22.47
Method 4 (LAR)	36.3	0.8400	22.50

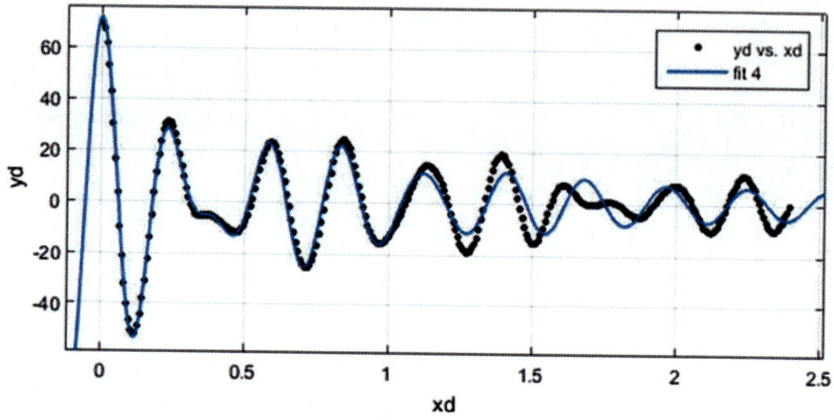

Fig. 4 Approximating (N = 2) and experimental correlation functions

Table 3 Parameters of the correlation function of the N-S component (N = 2)

№	Parameter	Parameter value by LSM	Parameter value by BW
1	A (cm/s^2)	31.26	30.54
2	A (cm/s^2)	40.76	41.47
3	α_1 (1/s)	0.716	0.6631
4	α_2 (1/s)	2.433	2.454
5	ω_1 (1/s)	22.37	22.3
6	ω_2	30.48	30.29

Table 4 Parameters of the correlation function of the N-S component (N = 3)

№	Parameter	Value
1	A (cm/s^2)	30.54
2	A (cm/s^2)	121.1
3	A_2 (cm/s^2)	−78.34
4	α_1	0.6949
5	α_2	2.700
6	α_3	2.20
7	ω_1	23.01
8	ω_2	32.05
9	ω_3	32.88

method. A more complicated method of approximation N = 3, associated with the solution of essentially 9 nonlinear equations, leads to the same result as the two-term approximation (Fig. 6).

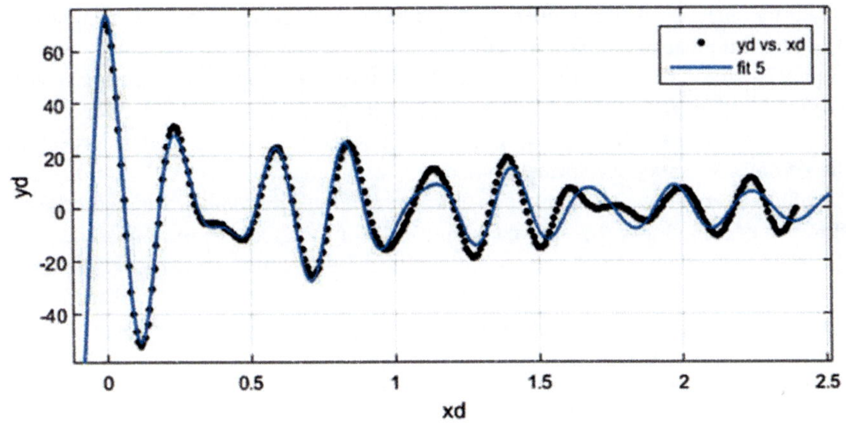

Fig. 5 Approximating (N = 3) and experimental correlation functions

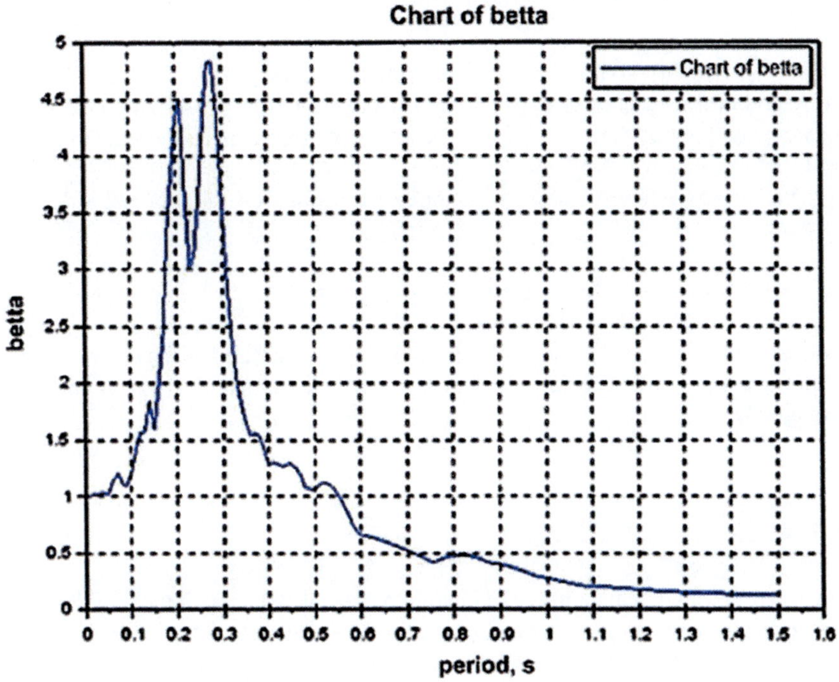

Fig. 6 Spectral curve of an artificial accelerogram

Further increase in the number of approximation terms (1) is impractical, although this can be done using the CurveFittingToolbox package. Approximation (1) is quite simple, which will make it possible to develop efficient algorithms for generating artificial accelerograms that take into account local features of earthquakes.

It is known that when several independent stationary normal random processes are summed, a stationary normal random process is formed, the correlation function of which is equal to the correlation function of the terms. Thus, if the correlation function of a random process is the sum of two or more correlation functions, then discrete realizations of this process can be formed by summing two or more independent realizations obtained using, for example, an algorithm by Prokhorov [7]. To generate independent implementations, independent sequences of independent normal random numbers are used.

If two-terms of series (1) are retained, one can write

$$\begin{aligned} \xi_1[n] &= a_0^{(1)} x_1[n] + a_1^{(1)} x_1[n-1] + b_1^{(1)} \xi_1[n-1] + b_2^{(1)} \xi_1[n-2], \\ \xi_2[n] &= a_0^{(2)} x[n] + a_1^{(2)} x[n-1] + b_1^{(2)} \xi_2[n-1] + b_2^{(2)} \xi_2[n-2], \end{aligned} \tag{2}$$

where $\xi_1[n], \xi_2[n]$—discrete realizations of a random process with single-term correlation functions; $x_1[n], x_2[n]$—independent sequences of normally distributed random numbers; $a_0^{(K)}, a_1^{(K)}, b_1^{(K)}, b_2^{(K)}$—coefficients determined according to [7]. (K = 1.2).

Then the discrete realization of a stationary random process for N = 2 is defined as the sum

$$\xi[n] = \xi_1[n] + \xi_2[n]. \tag{3}$$

One of the fastest and most convenient is the algorithm based on recursive filtering (autoregression) [8]. The main calculation formula for autoregression (moving average) is:

$$y_k = \sum_{i=0}^{m} a_i x_{k-i} - \sum_{i=1}^{n} b_i y_{k-i}, \tag{4}$$

where x_k—discrete realizations of a random process at the input of the system, y_k—discrete implementations of a random process at the output (in this case, accelerograms); a_i, b_i—moving average autoregressive model coefficients; m and n—positive numbers ($m \leq n$). This method has no methodological error, and the parameters of the modeling algorithm are expressed explicitly through the parameters of the correlation function.

A program for generating artificial accelerograms and determining spectral curves has been compiled in the MATLAB programming language. Figure 6 shows the spectral curves of the generated artificial accelerograms. The curves in Fig. 1 (real accelerogram) and Fig. 6 are generally quite close.

4 Discussion

Thus, the proposed algorithms for generating artificial accelerograms can be used when applying the new regulatory framework of the Republic of Kazakhstan to calculate the reliability of building structures of buildings and facilities, taking into account the local features of the seismic impact for the Almaty region. The criterion for the proximity of spectral curves of real and artificial accelerograms can be used in Monte Carlo schemes for digital modeling of the dynamics of linear and nonlinear systems under seismic action.

The developed method for generating artificial accelerograms can be used:

1. Determination of the seismic stiffness of the soil at the construction site.
2. Seismic zoning of territories.
3. Calculation of buildings and structures for real seismic effects according to earthquake accelerograms.
4. Calculation of buildings and structures as non-linear systems under real seismic impact, including seismic isolation systems [9, 10].
5. Determination of values of reliability (probability of failure-free operation), structural failure and seismic risk of buildings and structures.
6. Drawing up special technical specifications for the design of buildings and structures.
7. Development of regulatory documents in construction.
8. Statistical modeling of earthquakes.
9. When studying the records of the engineering and seismometric service on buildings [11] using digital instrumentation systems [12, 13].

5 Conclusion

1. Approximation (1) anticipates well the experimental correlation functions of a stationary random process. Two terms of series (1) are a sufficient approximation of the experimental correlation function. A simple analytical expression (1) will make it possible to efficiently generate realizations of a random process for the problem of numerical modeling of nonlinear systems under conditions of real seismic impact.
2. The Curve Fitting Toolbox package of the MATLAB computer mathematics system is a computational tool for processing experimental data, including correlation functions of accelerograms of strong earthquakes.
3. The algorithm of recursive filtering makes it possible to generate realizations of a narrow-band Gaussian random process (artificial accelerograms) with an accuracy sufficient for practical calculations within 10–20% in terms of the spectral characteristic. The method is very convenient for calculations using the systems of computer mathematics MATLAB and MAPLE.

Acknowledgements The studies were carried out using the funds of grant AP 05130702 of the Ministry of Education and Science of the Republic of Kazakhstan.

References

1. Alderucci T, Muscolino G, Urso S (2019) Stochastic analysis of linear structural systems under spectrum and site intensity compatible fully non-stationary artificial accelerograms. Soil Dyn Earthquake Eng 126. https://doi.org/10.1016/j.soildyn.2019.105762
2. Falamarz-Sheikhabadi M, Zerva A (2018) Two uncertainties in simulating spatially varying seismic ground motions: incoherency coefficient and apparent propagation velocity. Bull Earthquake Eng 16(10):4427–4441, 15. https://doi.org/10.1007/s10518-018-0385-x
3. Fischer E, Fischer P (1998) Quasi-resonance effects observed in the 1994 Northridge earthquake, and others. Shock Vib 5(3):153–159. https://doi.org/10.1155/1998/418528
4. Lapin V, Yerzhanov S, Aldakhov Y (2020) Statistical modeling of a seismic isolation object under random seismic exposure. J Phys: Conf Ser 1425:012006. https://doi.org/10.1088/1742-6596/1425/1/012006
5. Nurmagambetov A, Sydykov A et al (1996) Baisorun earthquake on November 12, 1990, in collection: Earthquake in the USSR in 1990, Moscow, IPE RAS, pp 36–39
6. Lapin V (1990) Modeling of seismic impact taking into account polyextremal nature. In collection: research of seismic resistance of structures and structures. Almaty 16–17(26–27):14–19
7. Prokhorov S (2001) Mathematical description and modeling of random processes. State Aerospace un-t, Samara, 209 p
8. Bykov V (1971) Digital modeling in statistical radio engineering. Sovetskoe radio, Moscow, 400 p
9. Bulat A, Dyrda V, Lysytsya M, Grebenyuk S (2018) Numerical simulation of the stress-strain state of thin-layer rubber-metal vibration absorber elements under nonlinear deformation. Strength Mater 50(3):387–395. https://doi.org/10.1007/s11223-018-9982-9
10. Dyrda V, Kobets A, Bulat I, Lapin V, Lysytsia N, Ahaltsov H, Sokol S (2019) Vibroseismic protection of heavy mining machines and structures. E3S Web Conf 109:00022. https://doi.org/10.1051/e3sconf/201910900022
11. Yerzhanov S, Lapin V, Aldakhov Y (2020) Monitoring the changes of dynamic characteristics of a high building. J Phys: Conf Ser 1425:012008. https://doi.org/10.1088/1742-6596/1425/1/012008
12. Zhang Q, Li W, GuoFeng YZ, Qiao S, Zhang Q (2019) Development of a new distributed hybrid seismic and electrical data acquisition station based on system—on-a-programmable-chip technology Geosci Instrum Methods Data Syst 8:241–249. https://doi.org/10.5194/gi-8-241-2019
13. Liu G, Liu Y, Feng X, Wang D (2019) Simulation of spatially variable seismic underground motions in saturated double-phase media with overlying water excited by SV-wave and difference from P-wave incidence. Soil Dyn Earthquake Eng (0267–7261) 123:144–161, 18. https://doi.org/10.1016/j.soildyn.2019.04.009

Green Roof as Ecological Security Design Solution During Reconstruction of Buildings

Svetlana Manzhilevskaya, Lubov Petrenko, Sofia Kravchenko, Valeria Mytsyk, and Aleksey Pnev

Abstract The presented article discusses new technological solutions for the construction of green roof structures. The developed technologies are aimed at solving the problems of improving the environmental safety of the construction industry as a whole, increasing the comfort level of the urban environment and improving the town-planning appearance. The green roofs systems act as a way to reduce the concentration of dust pollution and its spread, including from construction production by introducing this structure into the design, construction of residential neighborhoods and the reconstruction of architectural objects. In this article, the characteristics of two developed green roof structures are presented, which can be used in the design and construction of new facilities, as well as in the reconstruction of existing buildings, in addition, device technologies are described. These construction decisions let to save labour input for carrying out construction work by reducing their cost.

Keywords Green roofs · Roof · Ecological safety · Ecology in engineering · Architecture · Phytoecology · Atmosphere protection

1 Introduction

Now a days, the practice of civil and housing building and reconstruction of heritage architectural buildings gains momentum with application of green roof construction and the using of natural resources in design to ensure the maintenance of public health and a comfortable environment. The «green roof» concept acquired its social and environmental significance relatively recently, but it has been proven that the construction of such roofs is a good way to combat air pollution with fine $PM_{0.5}$-PM_{10} dust (particle matter) without increasing the sanitary protection zone in areas of close urban development.

S. Manzhilevskaya · L. Petrenko · S. Kravchenko (✉) · V. Mytsyk · A. Pnev
Don State Technical University, Sq. Gagarina, 1, Rostov-on-Don, Russia 344010
e-mail: krav4enko.sofie@yandex.com

A. Beskopylny and M. Shamtsyan (eds.), *XIV International Scientific Conference "INTERAGROMASH 2021"*, Lecture Notes in Networks and Systems 247,
https://doi.org/10.1007/978-3-030-80946-1_24

More recently, a lot of European cities decided to introduce green roofs systems like a regular practise in housing and residential building. Architects and arch contractors guide by the wishes of their customers build many buildings are successfully built using green roofs as a system for purifying atmospheric air and open space for the leisure of the population. This includes accessible, comfortable areas that will be very useful in a large city and its dense development. This decision was facilitated by the analysis of the use of territories that are subject to conventional roofing and asphalting, the calculation of the area of simultaneously built-up territories, where construction processes form environmental pollution, especially with fine dust particles that adversely affect the human health [1, 2]. The volume of construction of civil engineering objects, the building reconstruction, including objects of architectural heritage in the Russian Federation are growing rapidly, therefore, there is a growing need for the maximum exploitation of limited natural resources, urban areas and the expansion of the functionality of their use and application.

One of mane strategies for introduction of modern, cost-effective ways to combat construction dust and its harmful properties around existing residential areas and complexes, as well as improving the quality of life of the city's population, is the use of economical, functional, constructive and environmentally friendly solutions in the arrangement of exploited green roofs. In our days it is necessary to provide a sufficient, comprehensive, systematic approach to understanding the use of green roof structures in the design of new and reconstruction of existing buildings in the Russian Federation, in order to improve the environmental safety parameters of the construction industry itself and the overall composition of the renovated urban environment.

This article is aimed at solving the following tasks:

- to indicate an understanding of negative influences and consequences which have ordinary roofs and their constructions for natural environment in a big city;
- to explain necessity of implementation of green roof systems in the design, reconstruction and construction of architectural buildings for increase of ecological safety of environment and citizens health;
- to propose new innovative ideas that will change the perception, external aesthetic appearance and operation of the roof in the natural and cultural environment;
- to define ecological, aesthetical, multi-functional, social and economical benefits from redevelopment of existing structures;
- to demonstrate the detailed concept of their design and building;
- to show green roofs functional and aesthetical part in architectural sphere on the example of 3D-visualizations.

2 Materials and Methods

Traditional roof constructions (inclined and flat) are a covering or upper part of an installation structure that keeps unwanted weather precipitation outside, and also provides the most comfortable conditions for thermal and noise conditions inside the building. Modern systems of green roofs with new energy-efficient, inexpensive materials are also returned for consideration in the feasibility study of design solutions for the future building [3]. It should be note that both types of traditional constructive decisions have very big disadvantage—such roofs became such hot and right sunbeams, actually in the summer. On the flat roofs, the heat gain is stronger because the all roof is always on the sun and its direct effect on the roof surface.

Recently, the number of respiratory diseases among residents of the city of Rostov-on-Don has increased [4, 5]. Air pollution on Rostov-on-Don city territories caused actually by emissions from vehicles, heat and power facilities and construction production. Construction production for the construction of facilities and reconstruction of existing ones takes the 3rd place in air pollution in Rostov-on-Don. Over the past 2 years, the degree of air pollution at construction sites has worsened by 1.5 times. The composition of pollutants in the total amount of emissions in Rostov-on-Don is shown in Fig. 1.

During the monitoring of the air dust pollution in the urban area where repair and construction works were performed, unfavorable values were obtained [6]. The range of changes in the size of dust particles in a house with repairs were from 0.5 to 10 microns, the most common dust size was from 8 to 10 microns—up to 90%. During the monitoring of dust pollution in residential buildings under reconstruction and buildings next to them, a significant amount of fine dust was found directly near the object under construction along the entire height of the construction work. The research results showed that dust pollution was concentrated locally at the site of construction and repair work. Then it spread both in height and length in different directions, from house to house. The lower threshold level of air pollution was exceeded by about 20 times [7]. The growth in the volume of reconstruction of buildings in cramped conditions and the lack of structural and architectural planning solutions, including technologies with high economic efficiency, the use of which will help to comprehensively affect both the reduction of dust pollution of the urban environment, and harmoniously fit into the urban space, determined the use of green roof systems as a comprehensive solution to the modernization of urban development. In the context of progress in construction technology and methods of

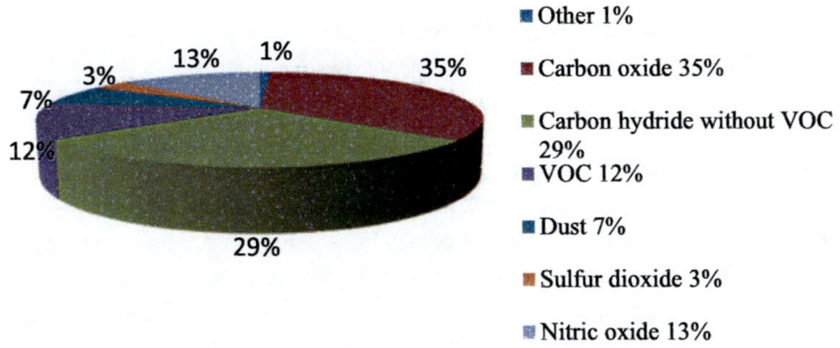

Fig. 1 Composition of pollutants in the atmosphere air of Rostov-on-Don city

reducing the environmental load, green roof structures should become a competitive and promising option when choosing design ideas and complement traditional and standard structures, since their implementation in design and construction will solve many problems at once:

- The ecological safety increase tasks.

The green roofs constructions consider as a way to reduce the concentration of dust pollution and its spread, including from construction production by introducing this technology into the reconstruction of architectural objects construction of residential neighborhoods [8].

Operated green roofs also called "breathable" because they filter out harmful dust particles, promote the processing of carbon dioxide and release oxygen, which is important and necessary for all living things, which leads to an improvement in the urban atmosphere as a whole. Such roof system let storm water pass through themselves and retain many unfavorable elements. Also green roof creates comfortable and ecological around a person, both in exteriors and interiors.

Roofs always locate under the right the influence of ultraviolet radiation from the sun and temperature changes. Roofs can «suffer» and damage because of huge warm impacts on its upper surface during the day and the year, the temperature can vary up to 100 °C. Original green roofs were created almost two centuries ago and in that period it was typical for covering bitumen with 6 cm of sand to protect it from fire. The sand also extended a life of the waterproof layer with nature vegetation. In nowadays modern green roofs demonstrated that their constructions increase the life of special waterproofing membranes, they are contained under a green roof and create a barrier that protects the waterproofing of the structure from harm. The ordinary dark roof absorbs solar radiation and can heat up to 90 °C, the surface of green roofs will never heat up so much (maximum up to 30–35 °C). Global worming leads to climate change, this situation can lead to an increase in the frequency and intensity of precipitation, then buildings must be designed to meet these new challenges, especially the introduction of methods to reduce storm runoff, this is becoming a requirement in new designs. In the summer green roofs can save 80% precipitation but in winter they save up to 35%. The difference is in the combination of winter rainfall and less plant evaporation, growth is not as vigorous during the winter period. The correct orientation of the building to the cardinal points allows the rational distribution of sunlight and heat (principle of "cooperation" with the sun).

- Social and economic tasks

The modern technology using let to build beautiful gardens, leisure time zones on the green roof, which makes it possible to improve the comfort of housing estates and detached buildings with the constraint of urban development. Here every space has its own function for reducing the cost of work while maintaining the quality of performance, the area of improvement of the territory of residential buildings increased. It is a very important factor when choosing a place of residence of the population and competitive among developers, and this increases the economic functionality of the building. This green space can be available to owners, tenants or employees. Green roofs create additional nature space without the cost for acquiring a separate land. The arrangement of the green roof can be called a response to the threat of the current crisis due to the inaccessibility of city-wide territories.

- Urban planning tasks

The use of a green roof structure in design and construction improves the architectural appearance of the city. A green roof can replenish some natural resources; its use increases the areas of greening of the city's territory and turns the places of its use into full-fledged platforms for the leisure of the population. Urban landscapes have a much higher proportion of dense, dark and impenetrable surfaces that have a fairly low albedo—reflectivity. Consequently, they absorb heat, unlike plants, which reflect it. The stored heat is emitted again at night, when the territory of the city warms up more than the surrounding area and nearby territories and areas.It can make city centres when heated up to 7 °C higher than the surroundings due to the effect of the heat island, which has adverse consequences for the area and population. City Warm Island will increase as summer temperatures rise and therefore will become an even bigger problem in the near future. In the hot summertime temperatures at night reach 8–9 °C higher than remote rural areas on a number of occasions. The green roof installation can become a good way for decision of such problem [9].

- Aesthetical architectural tasks

Green roofs systems in many cases can be used for the reconstruction of architectural structures. Such decisions can bring favorable ideas of innovation in renovation. Territories and spaces will appear that will be functional and profitable.

This use of a green roof will have a positive effect on the building's resistance to natural conditions and will be environmentally friendly, which will be a huge plus. Competently executed and compositionally verified reconstruction of existing buildings creates favorable video environment (the volume reduction for new building process). Using vertical landscaping as blinds to control room lighting and prevent overheating/hypothermia of facades (the principle of "cooperation" with the sun) [10].

The harmonic integration of the object into the environment depends on the compositional solution (the principle of respect for the place).

Aesthetical and commercial merit of green roof:

- creation of various kinds of landscapes and landscapes on the roof and a beneficial combination with the city landscape;
- additional space for people leisure time (sport space, cafe, the bureau, presentations spaces, equipped observation space, winter garden);
- an increase in the cost of the upper floors, as well as the entire architectural structure, up to 30%

3 Results and Discussions

To implement the tasks, it was necessary to develop modern, cost-effective and technologically simple design solutions that can be implemented and used already at the design stage and can be easily implemented, both in the construction of new architectural objects and in the reconstruction of existing buildings with different functional characteristics.

During the scientific research, design, and testing the two types of green roof structures were developed for installation in civic building of different functions, as well as in residential buildings and residential quarters. The use of these technological solutions in construction design is primarily aimed at reducing harmful fine dust particles $PM_{0.5}$-PM_{10} pollution of the air near the residential buildings, which often experience negative impacts from construction production.

The task of the first type of green roof structure is to increase the mobility of the roof covering, which can be disassembled in the cold season. The structure of this green roof structure includes: a reinforced concrete floor slab, roofing felt made of a strong non-rotting base, the adjustable feet "Forest Style" made of high—strength propylene with the possibility of increasing the support due to an additional filling piece of 60 mm, two layers of the asphalt roof coating Ondulin. There is a waterproofing with liquid rubber in the middle and in the central part of a decking. The drainage layer of the soil with plantings is located along the edges of green roof structure. The technical result of constructions is getting economical and social effect from its use throughout the year. The type and composition of the structure of such a green roof is shown in the Fig. 2 [11].

The structure of the second green roof structure includes: the reinforced concrete floor slab, the layer of the asphalt roof coating with glass-fiber mat which consists of a solid base, heat retention lagging "Is over 3", which creates heat protection, the adjustable feet "Forest Style" made of high—strength propylene with thermal endurance from −20 to +85 °C. The drainage layer "ENKADRAIN PREMIUM" is laid along the edges, it is resistant to many harmful substances and it also protects the waterproofing layer. The two layers of the asphalt roof coating Ondulin are placed

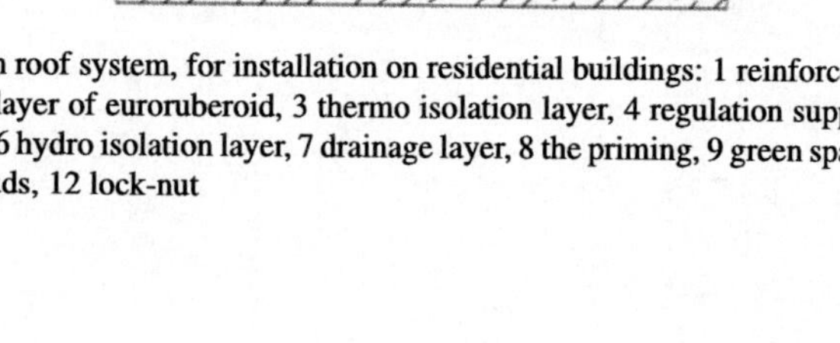

Fig. 2 The green roof system, for installation on residential buildings: 1 reinforced concrete floor slab, 2 covering layer of euroruberoid, 3 thermo isolation layer, 4 regulation supports, 5 covering layer of ondylin, 6 hydro isolation layer, 7 drainage layer, 8 the priming, 9 green spaces, 10 decking, 11 galvanized studs, 12 lock-nut

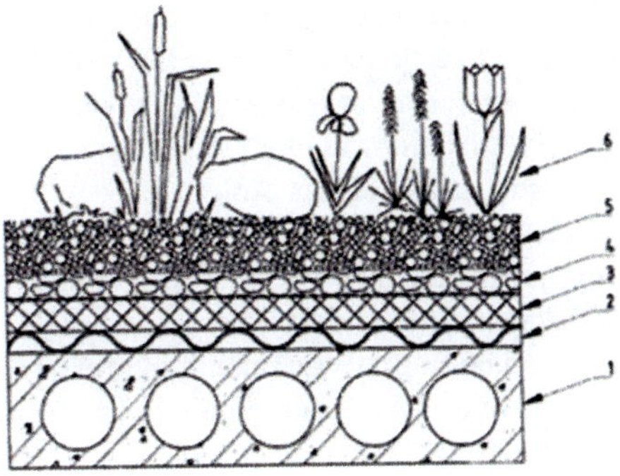

Fig. 3 The green roof construction intended for residential buildings: 1 reinforced concrete slab, 2 protective layer—ondulin, 3 heat-insulating layer of expanded polystyrene, 4 filter layer which made of peat or gravel, 5 fixing soil mixture, 6 plant layer with a substrate with plants

for the entire width of the roof. They provide strength and frost resistance for the green roof structure. The waterproofing with liquid rubber is situated in the middle of the structure. It is resistant to deformations and biological influences. At the end of installation the substrate is laid. It is an environmentally safe product. In the middle of the structure the decking Board "CM Decking nature 10" is laid, which is resistant to rot. The last step is green flooring installation. It is laid along the edges—the green spaces of the structure.Such roof's installation includes: there are two types of covering—first type: side part, it lets to use the green roof all year round, and central part for swimming pool and café.

The second type of such construction needs for installation in civil buildings and also for winters gardens and terraces. This system has a simple construction. It is cost effective and has high thermal insulation and waterproofing properties. Such type of green roof includes reinforced concrete floor slab, protective, filtering and vegetative layers, ondulin is used as a protective layer, additionally includes a layer of anchoring soil, the type and composition of the developed structure is shown in Fig. 3 [12].

The technical result is: the composition of the fixing mixture includes soil which consists of loam, waste of monomeric acids and the solution of calcium chloride. The priming is mixed with mixed liquid glass with density of 1.13 g/sm^3. The big plus of such layer in that it combines the function of several layers: soil, waterproofing layer, drainage layer and a layer that accumulates moisture. The composition of the coating includes a heat-insulating layer 2–10 cm high, which significantly reduces the heat loss of the entire building.

The technologies for arranging such green roofs were applied at the design stage of reconstruction for two buildings—architectural monuments located in the Rostov-on-Don city (Russia).

The first architecture building is architectural city heritage site which called Schepilo D. E. profitable house (1912) is located on st. Socialisticheskaya 168 in historical centre of the city. Now it is 3-storey residential building. According to the reconstruction project, it will be converted into non-residential premises with office premises. The arrangement of the green roof has become part of the reconstruction of this architectural monument, therefore, it is possible to trace the introduction of roof greening technologies into the project of the historical object. The system is arranged as a green garden on the roof of the building, it has its flower beds, paths and leisure areas. It is important to note that green roof construction accounted for 4% of all labour costs for the reconstruction of this building. Figures 4 and 5 show the visualization of this project.

The second architectural project is located on Pushkinskaya Street, house 65. It is a project of reconstruction of liftless apartment building—architectural monument «Profitable house of S.N. Mnatsakanov». The green roof system became a part of reconstruction project of a building for a modern residential complex and amounted to 4.5% of all labour costs included in the repair and construction work. On the roof of the building there is a recreation area, walking paths and numerous plantings. The project visualization is demonstrated on Figs. 6 and 7.

According to calculation economical effectiveness of developed design solutions, expenses for the green roof installations on the building on Socialisticheskaya Street 168 is over 1.3% of the total cost of all building reconstruction. In a multi-storey

Fig. 4 The green roof installation during the reconstruction of architectural residential building into an office building. 3D-visualization

Fig. 5 Plan of housing building roof with application of the green roof construction

Fig. 6 The green roof project for liftless apartment building on the Pushkinskaya street 65. 3D-visualization perspective

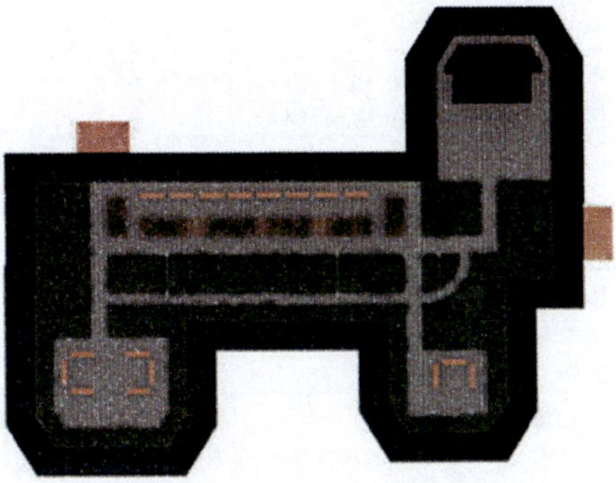

Fig. 7 Plan of an elevator-free apartment building's roof on Pushkinskaya street 65 with green roof construction

housing building this part of costs amounted to 4% of the total cost of reconstruction. It should be noted that when comparing the long-term operation period in 40 years cost of ordinary roofs (without vegetation) and green roofs are almost same.

To calculate an effectiveness of an implementation of these design solutions for the green roof, Table 1 shows the cost of installation of structures for the two options presented.

General cost of the green roof construction, which need for installation on civil buildings, in three-storey building on Socialisticheskaya Street 168 is 12359 USD. Installation of the green roof construction for housing building which need for building reconstruction on Pushkinskaya street 65—33,024 USD.

4 Conclusion

The one of tasks for green roofs using is reducing air pollution with dust particles, the solution to this problem is the composition of the plants necessary for planting green roofs in the soil in order to create a dust suppression process. This solution

Table 1 The cost of installing green roof structures per 1 m^2 according to the proposed options

Option 1. The green roof construction for installation in civil building		Option 2. The green roof construction for installation in housing buildings	
Technological process	Cost, USD	Technological process	Cost, USD
Sealing joints with cement	From 1.96	Sealing joints with cement	From 1.96
Applying bituminous coating on a concrete base and euroruberoid HMP	From 4.97	Applying bituminous coating on a concrete base and euroruberoid HMP	From 4.97
Thermal insulation works with insulation ISOVER 3	From 1.83	Installation of the ondulin flooring layer	From 2.62
Installation of the adjustable feet "Forest Style"	From 15.7	Thermal insulation works with insulation Penoplex	From 1.83
Installation of the ondulin flooring layer	From 2.62		
Applying the layer of liquid rubber waterproofing Hydra	From 9.81	Arrangement of a gravel / peat filter bed	From 1.96
Laying the drainage layer ENKADRAIN PREMIUM	From 2.62	Fixing soil mix and landscaping	From 5.23
Laying soil with landscaping	From 2.49		
Installation of decking for summer café and swimming pool	From 13.08		
In total – installation with greening – installation with landscaping and covering by decking	From 41.98/m^2 from 55.06/m^2	In total	From 18.57/m^2

provides an additional improvement in environmental safety due to additional air purification.

The cost of dust control measures will increase, but the level of environmental protection of construction site workers and the public is significantly increase. It will let to save a labour cost during the construction work, because there will be more less growth in the incidence of workers (up to 15%) and, accordingly, the cost of construction work will decrease by approximately the same 15%.

It is really important to note that green roofs add a big aesthetical value of all types of buildings in the urban and housing environment, also the cost and investment attractiveness of architectural objects increases.

Green roofs constructions open a lot of useful decisions in civil engineering. Green roofs are additional and comfortable spaces and zones for living, spending leisure time, doing sports. Also these are new possibilities for commercial using.

References

1. Wu Z, Zhang X, Wu M (2016) Mitigating construction dust pollution: state of the art and the way forward. J Cleaner Prod 112(2):1658–1666. https://doi.org/10.1016/j.jclepro.2015.01.015
2. Bespalov VI, Gurova OS, Samarskaya NS (2016) Main principles of the atmospheric air ecological monitoring organization for urban environment mobile pollution sources. Procedia Eng 150:2019–2024. https://doi.org/10.1016/j.proeng.2016.07.286
3. Köhler M, Clements AM (2012) Green roofs, ecological functions. In: Meyers RA (eds) Encyclopedia of sustainability science and technology. Springer, New York, NY. https://doi.org/10.1007/978-1-4419-0851-3_207
4. Manzhilevskaya SE, Azarov VN, Petrenko LK (2018) The pollution prevention during the civil buildings construction. In: MATEC web conference, vol. 196, pp 1–7. https://doi.org/10.1051/matecconf/201819604073
5. Ilyichev V, Emelyanov S, Kolchunov V, Bakayeva N, Kobeleva S (2015) Estimation of indicators of ecological safety in civil engineering. Procedia Eng 117:126–131. https://doi.org/10.1016/j.proeng.2015.08.133
6. Azarov VN, Petrenko LK, Manzhilevskaya SE (2019) The study of local dust pollution of atmospheric air on construction sites in urban areas. Adv Intell Syst Comput 983:430–439. https://doi.org/10.1007/978-3-030-19868-8_43
7. Zuo J, Rameezdeen R, Hagger M, Zhou Z, Ding Z (2017) Dust pollution control on construction sites: awareness and self-responsibility of managers. J Cleaner Prod 166:312–320. https://doi.org/10.1016/j.jclepro.2017.08.027
8. Ryzhova LV, Titova TS, Gendler SG (2017) Ensuring environmental safety during the construction and operation of tunnels in residential areas. Procedia Eng 189:404–410. https://doi.org/10.1016/j.proeng.2017.05.064
9. Köhler M, Michael Clements A (2013) Green roof infrastructures in urban areas. In: Loftness V, Haase D (eds) Sustainable built environments. Springer, New York, NY. https://doi.org/10.1007/978-1-4614-5828-9_209
10. Brenneisen S, Gedge D (2012) Green roof planning in urban areas. In: Meyers RA (eds) Encyclopedia of sustainability science and technology. Springer, New York, NY. https://doi.org/10.1007/978-1-4419-0851-3_208.RIS
11. Petrenko LK, Manzhilevskaya SE, Voit AV (2016) Patent RF No. 163334U1, 10/06/2016. Green roof construction, Russian Patent No. 163334U1, Bull. No. 19
12. Petrenko LK, Manzhilevskaya SE, Timoshenko EV, Semernikova AD (2019) Patent RF No. 191863U1, 26/08/2019. Green roof construction, Russian Patent No. 191863U1, Bull. No. 24

Heat and Moisture Regimes in the Facade Thermal Insulation Composite System with Polystyrene Insulation

Kirill Zubarev and Vladimir Gagarin

Abstract Facade thermal insulation composite systems with polystyrene insulation are extremely popular in modern construction sphere. Polystyrene and expanded polystyrene insulation plates have increasingly higher level of thermal resistance than classical insulation materials. It has to be stated that the vapour permeability coefficient of polystyrene is ten times less than the vapour permeability coefficient of mineral wool. Thus, the moisture regime process of enclosing structures which contain polystyrene as an insulation layer is significantly more complex than the moisture regime of enclosing structures with mineral wool. The plane of the moisture maximum was calculated using created computer program. The result showed that the moisture maximum was inside insulation layer. It is certain that in terms of providing evidence of this result, it is necessary to calculate the moisture behaviour inside researched facade thermal insulation composite system with polystyrene insulation by an unsteady-state moisture regime mathematical model. The mathematical model based on the moisture potential theory was chosen for assessing unsteady-state moisture behaviour. The second computer program, which solves the moisture transfer equation by finite-different method, was developed. It was shown that the plane of the moisture maximum has the same position according to both used methods.

Keywords Moisture regime · The moisture maximum plane · Mathematical model · Facade thermal insulation composite systems · Polystyrene

K. Zubarev (✉) · V. Gagarin
Moscow State University of Civil Engineering, 26, Yaroslavskoye Shosse, Moscow, Russia 129337

Research Institute of Building Physics of Russian Academy of Architecture and Construction Sciences, Moscow, Russia 127238

K. Zubarev
Peoples' Friendship University of Russia (RUDN University), 6, Miklukho-Maklaya Street, Moscow, Russia 117198

A. Beskopylny and M. Shamtsyan (eds.), *XIV International Scientific Conference "INTERAGROMASH 2021"*, Lecture Notes in Networks and Systems 247,
https://doi.org/10.1007/978-3-030-80946-1_25

1 Introduction

Moisture transfer is a vital problem in construction industry. There are various methods of evaluating moisture distribution. They can be based on the separate moisture transfer potentials or the unfitting potential—moisture potential [1–10].

Moisture regime in building materials affects human health [11–13] and heat losses through enclosures [14–19].

Developing the moisture transfer methods for engineers is a crucial task for modern scientists. First method for engineers was created by K. F. Fokin. This method takes into account only vapour moisture movements [20–26].

The equation which describes vapour transfer can be written as:

$$\frac{\partial}{\partial x}\left(\mu\frac{\partial e(w,t)}{\partial x}\right) = 0. \quad (1)$$

where e—water vapor partial pressure, Pa; μ—vapor permeability coefficient of a building material, kg/(m·s·Pa); x—coordinate, m.

To solve the Eq. (1) it is necessary to know: boundary conditions on enclosing structure surfaces and between enclosing structure layers for the water vapor partial pressure e:

$$\begin{cases} -\mu_1 \frac{\partial e}{\partial x}\Big|_{i=1} = \frac{1}{R_{m.ext}}(e_{ext} - e_1) \\ \mu_2 \frac{\partial e}{\partial x}\Big|_{i=N} = \frac{1}{R_{m.in}}(e_{in} - e_N) \\ -\mu_1 \frac{\partial e}{\partial x}\Big|_{i=v-0} = -\mu_2 \frac{\partial e}{\partial x}\Big|_{i=v+0} \end{cases}. \quad (2)$$

where e_N—water vapor partial pressure next to enclosing structure outer surface, Pa; $R_{m.ext}$—resistance to moisture exchange between outside air and the surface of the enclosing structure, (m^2·s·Pa)/kg; e_1—water vapor partial pressure next to enclosing structure inner surface, Pa; v—joint of materials in an enclosing structure; μ_1—vapor permeability coefficient of enclosing structure layer which is the nearest to the building outside surface, kg/(m·s·Pa); $R_{m.in}$—resistance to moisture exchange between inside air and the surface of the enclosing structure, (m^2·s·Pa)/kg; μ_2—vapor permeability coefficient of enclosing structure layer which is the nearest to the enclosing structure inner surface, kg/(m·s·Pa); e_{in}—inside water vapor partial pressure, Pa; e_{ext}—outside water vapor partial pressure, Pa.

The regulatory standards, which were in force in Russia until 2015, were based on this mathematical model.

This steady-state mathematical model has been confirmed by unsteady-state mathematical model, which was based on the same transfer potential—water vapor partial pressure.

The moisture transfer differential equation for unsteady-state moisture regime was formulated as:

$$\gamma_0 \frac{\xi_0(w)}{E_t(t)} \frac{\partial e(w,t)}{\partial \tau} = \frac{\partial}{\partial x}\left(\mu \frac{\partial e(w,t)}{\partial x}\right). \tag{3}$$

where E_t—saturated water vapor pressure, *Pa*; γ_0—dry material density, kg/m^3; ξ_0—relative vapor capacity, kg/kg; *w*—material humidity, % by weight; τ—time, *s*; *t*—temperature, °C.

With boundary conditions (2) the expression (3) gives the possibility to calculate all building enclosing structures in sorption zone of humidification.

However, modern moisture transfer theory gives an opportunity to improve accuracy of moisture regime calculations since the methods, which allow scientist to find moisture behaviour constantly, were developed [27–30].

2 Problem

To obtain the moisture maximum plane position for the facade thermal insulation composite system with polystyrene insulation by steady-state and unsteady-state calculation methods, which take into consideration sorption and ultrasorption zones of humidification.

3 Materials an Methods

3.1 Assessment of the Moisture Maximum Plane Position

The modern method of the steady-state moisture transfer is based on the moisture potential *F*:

$$F(w,t) = E_t(t) \cdot \varphi(w) + \frac{1}{\mu} \int_0^w \beta(\zeta) d\zeta. \tag{4}$$

where *F*—moisture potential, *Pa*; β—moisture conductivity coefficient, kg/(m·s·kg/kg); ζ—current material moisture value, kg/kg; φ—relative air humidity.

The steady-state moisture transfer formula can be written as:

$$\frac{\partial}{\partial x}\left(\mu \frac{\partial F(w,t)}{\partial x}\right) = 0. \tag{5}$$

The result of obtaining the maximum of the moisture transfer function can be described by following analytical expression:

$$f_i(t_{m.m}) = 5330 \frac{r_{t.v} \cdot (t_{in} - t_{ext.neg})}{R_t \cdot (e_{in} - e_{ext.neg})} \cdot \frac{\mu_i}{\lambda_i}. \tag{6}$$

where λ_i—thermal conductivity coefficient of i-th layer material, W/(m^2·°C); $t_{ext.neg}$—outdoor air average temperature in the period of monthly average temperature below zero, °C; $e_{ext.neg}$—partial pressure of outdoor air water vapor in the period of monthly average temperature below zero, *Pa*; μ_i—vapor permeability coefficient of i-th layer material, kg/(m·s·Pa); f_i—function that corresponds to the temperature of the layer i in the maximum moisture zone «maximum wetting complex», ((°C)^2/Pa); $r_{t.v}$—vapor permeability total resistance of enclosing structure, (m^2·s·Pa)/kg; t_{in}—inside air average temperature, °C; R_t—heat transfer total resistance of enclosing structure, (m^2·°C)/W; e_{in}—partial pressure of inside air water vapor, *Pa*; $t_{m.m}$—maximum wetting temperature, °C.

The modification of the moisture transfer Eq. (5) is a graphical method of determining the maximum wetting plane position [27, 28]. The computer program based on the graphical method was designed to find the exact position of the maximum wetting plane for the facade thermal insulation composite system with polystyrene insulation.

3.2 The Mathematical Model of Unsteady-State Moisture Regime

The unsteady-state moisture transfer equation was formulated [29, 30]:

$$\frac{\partial F(w,t)}{\partial \tau} = \kappa(w,t) \cdot E_t(t) \frac{\partial^2 F(w,t)}{\partial x^2}. \tag{7}$$

where κ—material heat-humidity characteristic coefficient, m^2/(s·Pa); x—coordinate, *m*; τ—time, *s*.

Formula (7) takes into account humidification in sorption and ultrasorption zones. Moreover, the moisture potential F depends on calculation time.

It is possible to use various approaches to solve Eq. (7). In current work the finite-different method was chosen for solving:

$$\begin{cases} F_1^{k+1} = F_1^k + \Delta\tau \frac{\kappa_{F1}^k}{h^2} E_{t1}^k \left(-\left(1 + \frac{h}{\mu R_{m.ext}}\right) F_1^k + F_2^k\right) + \Delta\tau \frac{\kappa_{F1}^k}{h^2} E_{t1}^k \frac{h}{\mu R_{m.ext}} F_{ext}^k \\ F_i^{k+1} = F_i^k + \Delta\tau \frac{\kappa_{Fi}^k}{h^2} E_{ti}^k \left(F_{i+1}^k - 2 \cdot F_i^k + F_{i-1}^k\right),\ i = 2, \ldots, N-1,\ k = 0, 1, \ldots \\ F_N^{k+1} = F_N^k + \Delta\tau \frac{\kappa_{FN}^k}{h^2} E_{tN}^k \left(F_{N-1}^k - \left(1 + \frac{h}{\mu R_{m.in}}\right) F_1^k\right) + \Delta\tau \frac{\kappa_{FN}^k}{h^2} E_{tN}^k \frac{h}{\mu R_{m.in}} F_{in}^k \end{cases} \tag{8}$$

where F_{ext}^k—moisture potential of outer air at the k-th time step, *Pa*; F_{in}^k—moisture potential of inner air at the k-th time step, *Pa*; F_1^k—moisture potential in the first section of the wall at the k-th time step, *Pa*; E_{t1}^k—saturated water vapor pressure in the first section of the wall at the k-th time step, *Pa*; F_2^k—moisture potential in the second section of the wall at the k-th time step, *Pa*; E_{ti}^k—saturated water vapor pressure in the *i*-th section of the wall at the *k*-th time step, *Pa*; F_1^{k+1}—moisture potential in the first section of the wall at the (*k* + *1*)-th time step, *Pa*; E_{tN}^k—saturated water vapor pressure in the *N*-th section of the wall at the *k*-th time step, *Pa*; F_{i-1}^k—moisture potential in the (*i-1*)-th section of the wall at the *k*-th time step, *Pa*; *h*—coordinate step, *m*; F_N^k—moisture potential in the *N*-th section of the wall at the *k*-th time step, *Pa*; κ_{F1}^k—material heat-humidity characteristic coefficient in the first section of the wall at the k-th time step, m²/(s·Pa); F_i^k—moisture potential in the *i-th* section of the wall at the *k*-th time step, *Pa*; κ_{Fi}^k—material heat-humidity characteristic coefficient in the *i*-th section of the wall at the *k*-th time step, m²/(s·Pa); F_i^{k+1}—moisture potential in the *i*-th section of the wall at the (*k* + *1*)-th time step, *Pa*; $\Delta\tau$—time step, *s*; F_{N-1}^k—moisture potential in the (*N-1*)-th section of the wall at the *k*-th time step, *Pa*; κ_{FN}^k—material heat-humidity characteristic coefficient in the *N*-th section of the wall at the *k*-th time step, m²/(s·Pa) F_N^{k+1}—moisture potential in the *N*-th section of the wall at the (*k* + *1*)-th time step, *Pa*.

The computer program based on the presented numerical method was created to find the maximum wetting plane position in investigated enclosing structures.

4 Results and Discussion

4.1 *The Facade Thermal Insulation Composite System with Polystyrene Insulation for Calculation*

Parameters of the facade thermal insulation composite system with polystyrene insulation were chosen for comparing the moisture maximum plane position (Table 1).

Table 1 The wall structure and parameters of its layers

Structure of the wall		Thicknesses of wall layers, m	Thermal conductivity of wall layers, W/(m·°C)	Vapor Permeability of wall layers, kg/(m·s·Pa)
1	A plaster layer of an outer wall surface	0.007	0.93	3.6×10^{-11}
2	Insulation plates of polystyrene	0.12	0.044	1.4×10^{-11}
3	Base blocks of aerated concrete	0.3	0.15	6.4×10^{-11}
4	A plaster layer of an inner wall surface	0.02	0.93	2.5×10^{-11}

Table 2 Data of microclimate parameters and a construction region

Construction region	Average outdoor temperature during the period with negative monthly average temperatures, °C	Average outdoor partial pressure of water vapor during the period with negative monthly average temperatures, *Pa*	Indoor air temperature, °C	Indoor relative humidity, %
Russia Moscow	−4.58	364	20	55

The Table 1 presents typical wall structure with polystyrene insulation. Data of microclimate parameters and a construction region are also presented in Table 2.

4.2 Comparison of the Moisture Regime Calculations

The enclosing structure was calculated by created computer program of steady-state moisture regime.

Calculation of the position of the moisture maximum plane is presented (Fig. 1).

As a result, the moisture maximum plane position is located inside the polystyrene layer.

To check the first result the parameters of the researched building wall were entered in the second computer program, which gives an opportunity to calculate unsteady-state moisture behaviour.

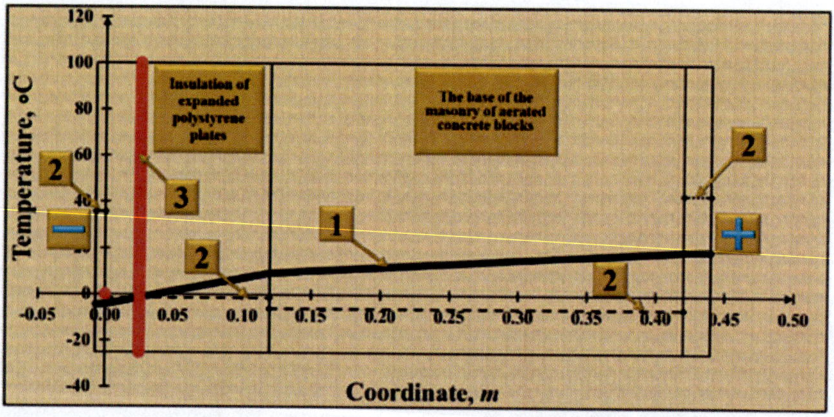

Fig. 1 Determination of the position of the moisture maximum plane in the facade thermal insulation composite system with polystyrene insulation (1 temperature distribution, 2 maximum wetting temperature distribution, 3 the position of the maximum moisture plane inside the insulation layer)

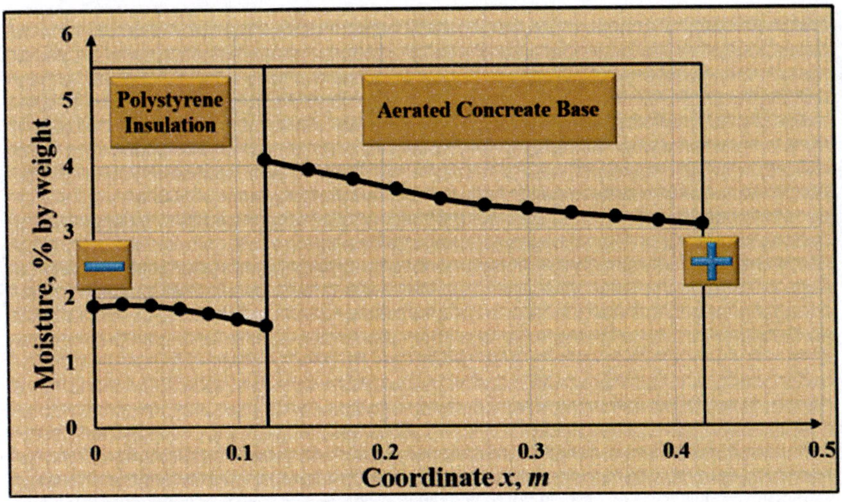

Fig. 2 Result of the moisture behaviour calculation for the facade thermal insulation composite system with polystyrene insulation

The moisture distribution inside the facade thermal insulation composite system with polystyrene insulation is presented (Fig. 2).

Thus, obtained moisture maximum coincides with the moisture maximum plane position according to steady-state method.

This coincidence allows engineers to use the graphical method of determining maximum moisture plane position for the facade thermal insulation composite system with polystyrene insulation.

This result can be implemented in real work because it was shown that engineers do not have to use numerical method like finite-difference method to find the exact position of the moisture maximum. This can be done using a steady-state one.

5 Conclusion

The position of the moisture maximum plane was obtained for the facade thermal insulation composite system with polystyrene insulation. It was demonstrated that the moisture maximum is located inside the polystyrene layer. This position of the moisture maximum plane was found by two moisture regime separate methods: steady-state and unsteady-state. Engineers are able to use the steady-state method to calculate the position of moisture maximum and they do not need to solve unsteady-state moisture transfer equation. The results of this paper can be used in practical work.

References

1. Bai H, Zhu J, Chen X, Chu J, Cui Y, Yan Y (2020) Steady-state performance evaluation and energy assessment of a complete membrane-based liquid desiccant dehumidification system. Appl Energy 258(114082)
2. Cui Y, Zhu J, Zoras S, Chen X, Bi H, Qiao Y, Soleimani Z (2019) State-of-the-art review of 3DPV technology: structures and models. Energy Conver Manage 200(112130)
3. Zubarev K, Gagarin V (2021) Determining the coefficient of mineral wool vapor permeability in vertical position. Adv Intell Syst Comput 1259:593–600
4. Zhang N, Chen X, Su Y, Zheng H, Ramadan O, Zhang X, Chen H, Riffat S (2019) Numerical investigations and performance comparisons of a novel cross-flow hollow fiber integrated liquid desiccant dehumidification system. Energy 182:1115–1131
5. Cui Y, Zhu J, Twaha S, Chu J, Bai H, Huang K, Chen X, Zoras S, Soleimani Z (2019) Techno-economic assessment of the horizontal geothermal heat pump systems: a comprehensive review. Energy Conver Manage 191:208–236
6. Petrichenko M, Rakova X, Vyatkin M, Musorina T, Kuznetsova D (2015) Architectural renovation of quarter in Mannheim, Germany. Appl Mech Mater 725–726, 1101–1106
7. Musorina T, Olshevskyi V, Ostrovaia A, Statsenko E (2016) Experimental assessment of moisture transfer in the vertical ventilated channel. In: MATEC web of conferences (02002)
8. Gamayunova O, Radaev A, Petrichenko M, Dmitrieva E (2019) The increase in energy efficiency of residential buildings of military towns. In: E3S web of conferences 110(02144)
9. Zaborova D, Vieira G, Musorina T, Butyrin A (2017) Experimental study of thermal stability of building materials. Adv Intell Syst Comput 692:482–489
10. Castro JCL, Zaborova DD, Musorina TA, Arkhipov IE (2017) Indoor environment of a building under the conditions of tropical climate. Magazine of Civil Engineering 8(76):50–57
11. Vorobyeva IV (2017) Long-term results of treatment of patients with non-proliferative diabetic retinopathy angioprotectors. Oftalmologiya 14(1):67–77
12. Vorobyeva IV, Moshetova LK, Shcherbakova EV (2017) Therapy results in patients with primary open angle glaucoma and diabetic retinopathy. Oftalmologiya 14(1):40–46
13. Vorobyeva IV (2018) Monitoring of separate pathogenetically significant biochemical markers in lacrimal fluid, ophthalmological parameters with combined pathology of diabetic retinopathy and age-related macular degeneration on the background angioprotective and antioxidant therapy. Oftalmologiya 15(2):189–199
14. Jumabekova A, Berger J, Foucquier A, Dulikravich GS (2020) Searching an optimal experiment observation sequence to estimate the thermal properties of a multilayer wall under real climate conditions. Int J Heat Mass Transfer 155(119810)
15. Zhang H, Yang J, Wu H, Fu P, Liu Y, Yang W (2020) Dynamic thermal performance of ultra-light and thermal-insulative aerogel foamed concrete for building energy efficiency. Sol Energy 204:569–676
16. Family R, Celik S, Menguc M (2020) Coupled heat transfer analysis and experiments to evaluate the radiative cooling potential of concrete and green roofs for buildings. Heat Mass Transf 56(8):2605–2617
17. Xu B, Xie X, Pei G, Chen X (2020) New view point on the effect of thermal conductivity on phase change materials based on novel concepts of relative depth of activation and time rate of activation: the case study on a top floor room. Appl Energy 266(114886)
18. Cabrera V, Lopez-Vizcaino R, Yustres A, Ruiz MA, Torrero E, Navarro V (2020) A functional structure for state functions of moisture transfer in heritage building elements. J Building Eng 29(101201)
19. Lafond C, Blanchet P (2020) Technical performance overview of bio-based insulation materials compared to expanded polystyrene. Buildings 10(85)
20. Sadlowska-Salega A, Was K (2020) Risk of moisture in diffusionally open roofs with cross-laminated timber for Northern Coastal climates. Buildings 10(10)
21. Cho HM, Wi S, Chang SJ, Kim S (2019) Hygrothermal properties analysis of cross-laminated timber wall with internal and external insulation systems. J Clean Prod 231:1353–1363

22. Busser T, Berger J, Piot A, Pailha M, Woloszyn M (2019) Comparison of model numerical predictions of heat and moisture transfer in porous media with experimental observations at material and wall scales: an analysis of recent trends. Drying Technol 37(11):1363–1395
23. Kaczorek D (2019) Moisture buffering of multilayer internal wall assemblies at the micro scale: experimental study and numerical modelling. Appl Sci-Basel 9(3438)
24. Ferrari S, Riva A (2019) Insulating a solid brick wall from inside: heat and moisture transfer analysis of different options. J Arch Eng 25(04018032)
25. Petrov A, Ivantsov A (2020) Design and calculation of the internal roof drain system structure in terms of thermal protection and moisture condensation. IOP Conf Series Mater Sci Eng 890(022141)
26. Ivantsov A, Petrov A (2020) The influence of architectural and construction parameters of residential buildings on energy efficiency in Russian Federation. IOP Conf Series Mater Sci Eng 890(022142)
27. Gagarin VG, Akhmetov VK, Zubarev KP (2020) Graphical method for determination of maximum wetting plane position in enclosing structures of buildings. IOP Conf Series Mater Sci Eng 753(022046)
28. Gagarin VG, Akhmetov VK, Zubarev KP (2020) Position control of maximum wetting plane for building walls with foam polystyrene heat insulator. IOP Conf Series Mater Sci Eng 753(022045)
29. Gagarin VG, Akhmetov VK, Zubarev KP (2018) Unsteady-state moisture behavior calculation for multilayer enclosing structure made of capillary-porous materials. IOP Conf Series Earth Environ Sci 177(012021)
30. Gagarin VG, Akhmetov VK, Zubarev KP (2018) The moisture regime calculation of single-layer enclosing structures on the basis of the discrete-continuum method application. IOP Conf Series Mater Sci Eng 456(012105)

Heat and Moisture Transfer in Building Enclosing Structures

Kirill Zubarev and Vladimir Gagarin

Abstract Nowadays, there are various mathematical models which give an opportunity to determine the moisture distribution inside building walls. It has to be stated that some models calculate steady-state moisture transportation in the capillary-porous media, while other models assess unsteady-state moisture movements. Equations of steady-state moisture regime usually can be usually solved by analytical expressions. It is certain that in terms of solving unsteady-state moisture transfer equations, scientists use numerical methods, which cannot be operated without computer. The discrete–continuous approach for evaluating the moisture behaviour of enclosing structures is proposed. Two computer programs were created. One of them calculated moisture regime by finite difference method using explicit difference scheme, whereas the second one assessed moisture behaviour by discrete continuous method using obtained analytical expression. As a result, efficiency of the discrete continuous method was demonstrated because the solution of the unsteady-state moisture transfer equation according to proposed method has similar quantity and quality distribution of moisture content with solution of the unsteady-state moisture regime by the finite difference method. It allows scientists and engineers to calculate moisture regime by analytical expression and not to use numerical calculations.

Keywords Moisture regime · Mathematical model · Discrete–continuous approach · Finite difference method · Enclosing structures

K. Zubarev (✉) · V. Gagarin
Moscow State University of Civil Engineering, 26, Yaroslavskoye Shosse, Moscow, Russia 129337

Research Institute of Building Physics of Russian Academy of Architecture and Construction Sciences, Moscow, Russia 127238

K. Zubarev
Peoples' Friendship University of Russia (RUDN University), 6, Miklukho-Maklaya Street, Moscow, Russia 117198

A. Beskopylny and M. Shamtsyan (eds.), *XIV International Scientific Conference "INTERAGROMASH 2021"*, Lecture Notes in Networks and Systems 247,
https://doi.org/10.1007/978-3-030-80946-1_26

1 Introduction

1.1 The Position of the Moisture Regime of Enclosing Structures in Construction Science

Moisture regime of building enclosing structures is a vital problem in modern construction industry. It should be noted that the importance of this topic is increasing as new building materials and new wall structures are being developed. The number of layers in modern enclosures is more than in classic walls. This process affects the difficulty of the moisture distribution calculation since the moisture transport in multilayer enclosing structures has a complicated physic base [1–9].

It has to be stated that moisture inside building materials influences durability of the buildings and heat conductivity coefficients [10–13]. Moreover, the moisture condition of enclosing structure materials affects human health [14–16].

It is obvious that moisture transportation in a capillary-porous media has various directions of researches. Firstly, scientists carry out experiments exploring moisture content in constructed enclosing structures and speed of moisture transfer in building materials [17]. These can be field studies or laboratory experiments [18, 19]. Secondly, mathematical modeling is being developed. These investigations consist of obtaining new mathematical formulas describing physic process of moisture transport, creating computer programs which give an opportunity to calculate moisture regime and finding new approaches helping to solve moisture transfer equations [20–24]. Finally, scientists have to assess the impact of the moisture regime on building life cycle and mode of heating and ventilation systems [25, 26].

1.2 Development of the Moisture Transfer Theory

Nowadays, the most common theories based on some moisture transfer potentials are: water vapour partial pressure, capillary pressure, temperature and filtration [27, 28]. One of the most interesting theory takes into consideration vapour and liquid moisture movements in sorption and ultrasorption zones.

The moisture transfer equation is formulated as [29]:

$$\frac{\partial F(w,t)}{\partial \tau} = \kappa(w,t) \cdot E_t(t) \frac{\partial^2 F(w,t)}{\partial^2}. \tag{1}$$

where F—moisture potential which was proposed by Gagarin and Kozlov, *Pa*; E_t—saturated water vapor pressure, *Pa*; x—coordinate, *m*; w—material moisture, % by weight (1 kg/kg = 100% by weight); τ—time, *s*; κ—material heat-humidity characteristic coefficient, m^2/(s·Pa); t—temperature, °C.

Although, for the first side, the Eq. (1) has only one moisture transfer potential—F, this moisture potential F depends on temperature and moisture content and takes into account two private potentials: gradient of moisture and gradient of water vapour partial pressure. Therefore, the use of formula (1) gives great advantages. On the one hand, there is a possibility to work with a complicated moisture transfer process; on the other hand, the form of the Eq. (1) is rather simple, which allows researchers to introduce solution methods [29].

1.3 Development of the Discrete–Continuous Approach for Solving the Unsteady-State Heat Transfer Equation

It has to be stated that in 2010 the new approach of solving heat transfer equation, which was called the discrete continuous approach, was obtained [30].

The heat transfer equation is presented [30]:

$$\frac{\partial t}{\partial \tau} = a\frac{\partial^2 t}{\partial^2}. \tag{2}$$

where a—thermal diffusivity coefficient, m^2/s.

The solution of the Eq. (2) was found as [30]:

$$\bar{U}(\tau) = e^{A\cdot\tau} \cdot \bar{U}_0 - A^{-1}\left(E - e^{A\cdot\tau}\right) \cdot \bar{S}. \tag{3}$$

where $\bar{U}_0$—initial temperature distribution column vector; $\bar{U}$—temperature distribution column vector; $\bar{S}$—boundary conditions column vector; A—coefficient matrix, $e^{A\cdot\tau}$—matrix exponent.

The formula (3) gives an opportunity to calculate the unsteady-state moisture transfer Eq. (2) by the final formula without using numerical methods.

2 Problem

To develop the new mathematical approach for calculating the unsteady-state moisture transfer equation based on the moisture potential F using discrete–continuous approach. To compare moisture behaviour inside a wall by the finite difference method and the discrete–continuous method to provide evidence of efficiency of proposed mathematical formula.

3 Materials and Methods

3.1 Finite Difference Method Using Explicit Difference Scheme for the Unsteady-State Moisture Transfer Equation

In formula (1) the partial derivates are replaced with finite differences and if the third-order boundary conditions are input into obtained equation, the system of equations will be formulated as:

$$\begin{cases} F_1^{k+1} = F_1^k + \Delta\tau \frac{\kappa_{F1}^k}{h^2} E_{t1}^k \left(-\left(1 + \frac{h}{\mu R_{m.ext}}\right) F_1^k + F_2^k\right) + \Delta\tau \frac{\kappa_{F1}^k}{h^2} E_{t1}^k \frac{h}{\mu R_{m.ext}} F_{ext}^k \\ F_i^{k+1} = F_i^k + \Delta\tau \frac{\kappa_{Fi}^k}{h^2} E_{t_i^k} \left(F_{i+1}^k - 2 \cdot F_i^k + F_{i-1}^k\right), \ i = 2, \ldots, N-1, \ k = 0, 1, \ldots \ .. \\ F_N^{k+1} = F_N^k + \Delta\tau \frac{\kappa_{FN}^k}{h^2} E_{tN}^k \left(F_{N-1}^k - \left(1 + \frac{h}{\mu R_{m.in}}\right) F_1^k\right) + \Delta\tau \frac{\kappa_{F_N^k}}{h^2} E_{tN}^k \frac{h}{\mu R_{m.in}} F_{in}^k \end{cases} \tag{4}$$

where κ_{Fi}^k—material heat-humidity characteristic coefficient in the *i*-th section of the wall at the *k*-th time step, m^2/(s·Pa); $\Delta\tau$—time step, *s*; F_{ext}^k—moisture potential of outer air at the *k*-th time step, *Pa*; F_2^k—moisture potential in the second section of the wall at the *k*-th time step, *Pa*; F_{in}^k—moisture potential of inner air at the *k*-th time step, *Pa*; F_1^k—moisture potential in the first section of the wall at the *k*-th time step, *Pa*; κ_{F1}^k—material heat-humidity characteristic coefficient in the first section of the wall at the k-th time step, m^2/(s·Pa); E_{ti}^k—saturated water vapor pressure in the *i*-th section of the wall at the *k*-th time step, *Pa*; E_{t1}^k—saturated water vapor pressure in the first section of the wall at the *k*-th time step, *Pa*; κ_{FN}^k—material heat-humidity characteristic coefficient in the *N*-th section of the wall at the *k*-th time step, m^2/(s·Pa); F_1^{k+1}—moisture potential in the first section of the wall at the $(k + 1)$-th time step, *Pa*; $E_{tN}k$—saturated water vapor pressure in the *N*-th section of the wall at the *k*-th time step, *Pa*; F_{i-1}^k—moisture potential in the (*i*-1)-th section of the wall at the *k*-th time step, *Pa*; *h*—coordinate step, *m*; F_N^k—moisture potential in the *N*-th section of the wall at the *k*-th time step, *Pa*; F_i^{k+1}—moisture potential in the *i*-th section of the wall at the $(k + 1)$-th time step, *Pa*; F_i^k—moisture potential in the *i-th* section of the wall at the *k*-th time step, *Pa*; F_{N-1}^k—moisture potential in the (*N-1*)-th section of the wall at the *k*-th time step, *Pa*; F_N^{k+1}—moisture potential in the *N*-th section of the wall at the $(k + 1)$-th time step, *Pa*.

The computer program was created based on the system (4).

3.2 Discrete–continuous Approaches for the Unsteady-State Moisture Transfer Problem

The discrete–continuous approach was used: derivations in X-direction are replaced with finite differences (this transformation is similar to finite difference method), but in Y-direction (time τ) the solution in a function form was found. Obtained system of equations can be presented as a matrix formula. The analytical expression which is the solution of this matrix formula can be written as [40]:

$$\begin{aligned}\bar{F} &= p \cdot ((E_t \cdot A)^{-2} \cdot e^{E_t \cdot A \cdot \tau} - \tau \cdot (E_t \cdot A)^{-1} - (E_t \cdot A)^{-2}) \cdot \bar{L} \\ &+ (E_t \cdot A)^{-1}(e^{E_t \cdot A \cdot \tau} -) \cdot \bar{B} + e^{E_t \cdot A \cdot \tau} \cdot \bar{F}_0 \qquad (5)\end{aligned}$$

where p—the coefficient of the external boundary condition for a single-layer building enclosing structure, Pa/s^2; $\overline{F}_0$—initial moisture potential distribution column vector; A—matrix of coefficients before functions of the moisture potential F for a multi-layer enclosing structure; E—unit matrix; E_t—matrix of the saturated water vapour pressure; $\bar{B}$—a column vector, the first and last elements of which describe the boundary conditions on the outer and inner surfaces of the enclosing structure, other elements are equal to 0 for a single-layer enclosing structure; $\bar{L}$—a column vector, the first element of which is equal to one, other elements are equal to 0 for a single-layer enclosing structure.

The second computer program was designed according to Eq. (5).

4 Results and Discussion

4.1 Researched Enclosing Structure

To assess efficiency of the obtained discrete–continuous formula (6) the single-layer enclosing structure with aerated concrete blocks and plaster layers was chosen (Table 1). The wall structure is presented (Table 1).

To make simulation according to developed computer programs Moscow was chosen as a city for calculation. Climate of Moscow was used as the outer boundary condition on the wall surface. Parameters of microclimate have been taken from Table 2.

Table 1 The wall structure and parameters of its layers

Structure of the wall		Thicknesses of wall layers, m	Thermal conductivity of wall layers, W/(m·°C)	Vapor permeability of wall layers, kg/(m·s·Pa)
1	A plaster layer of an outer wall surface	0.02	0.93	2.5×10^{-11}
2	Base blocks of aerated concrete	0.3	0.15	6.4×10^{-11}
3	A plaster layer of an inner wall surface	0.02	0.93	2.5×10^{-11}

Table 2 Data of microclimate parameters and a construction region

Construction region	Indoor air temperature, °C	Indoor relative humidity, %
Russia Moscow	20	55

4.2 *Moisture Behaviour Distribution Inside Researched Enclosing Structure in Winter and Summer*

Moisture behaviours inside the building wall in January (the coldest month in winter) and July (the hottest month in summer) by finite difference and discrete–continuous approaches were calculated (Figs. 1 and 2).

As we can see, the moisture distribution obtained according to the discrete–continuous approach coincides with moisture behaviour obtained according to the finite difference method. Thus, the new effective approach which simplifies operations with moisture regime calculations was found.

Figures 1 and 2 illustrate that the volume of moisture content in winter is higher than the volume of moisture content in summer.

4.3 *Moisture Behaviour Distribution Inside Researched Enclosing Structure During a year*

The result of the moisture regime calculation for the single-layer wall made of aerated concrete during a year is illustrated (Fig. 3).

As it is presented in Fig. 3, average moisture inside the enclosing structure during a year according to the discrete–continuous method coincides with average moisture inside the enclosing structure during a year according to the finite difference method. The time of maximum humidification of researched enclosing structure is the first of

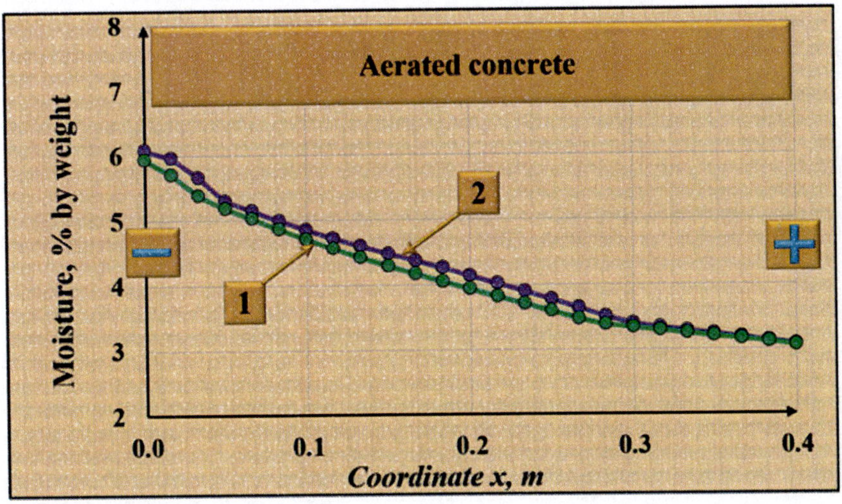

Fig. 1 The result of the moisture regime calculation for the single-layer wall made of aerated concrete (density 400 kg/m^3) in January (winter) (1 the moisture distribution inside enclosing structure according to the discrete–continuous method; 2 the moisture distribution inside enclosing structure according to the finite-difference method)

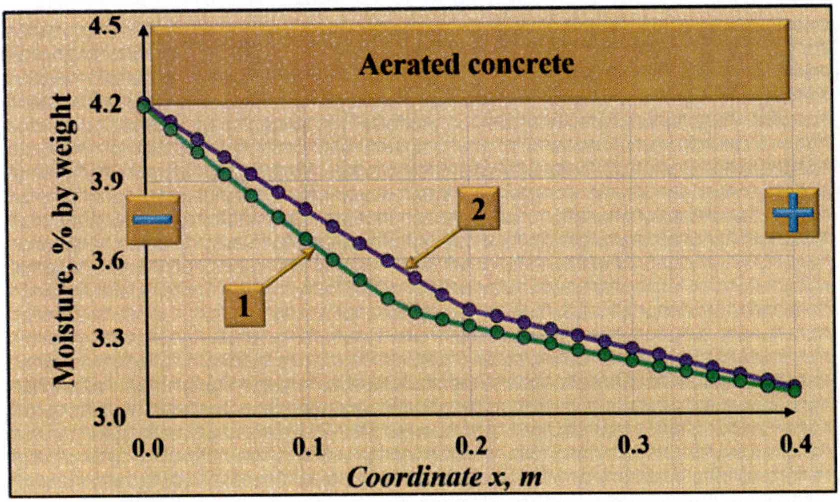

Fig. 2 The result of the moisture regime calculation for the single-layer wall made of aerated concrete (density 400 kg/m^3) in July (summer) (1 the moisture distribution inside enclosing structure according to the discrete–continuous method; 2 the moisture distribution inside enclosing structure according to the finite-difference method)

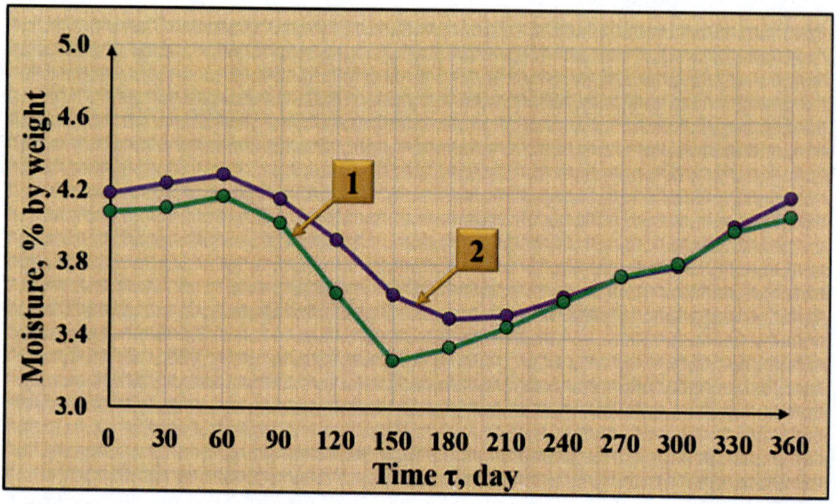

Fig. 3 The result of the moisture regime calculation for the single-layer wall made of aerated concrete (density 400 kg/m^3) during a year (1 the moisture content during a year according to the discrete–continuous method; 2 the moisture content during a year according to the finite-difference method)

March. This example demonstrates that both methods define extremely crucial result for determining the heat conductivity coefficients in building materials.

In all pictures (Figs. 1, 2 and 3) the accuracy of the finite difference method is higher than the accuracy of the discrete–continuous method. However, the moisture distribution according to the discrete–continuous method is obtained by the final formula (5).

5 Conclusion

The new mathematical approach for calculating the unsteady-state moisture transfer equation based on the moisture potential F using discrete–continuous approach was proposed. The final formula that calculated the moisture regime distribution was presented.

Moisture behaviour calculations inside the wall by the finite difference method and the discrete–continuous method were performed. It was shown that the moisture distribution inside the wall in January and in July according to the discrete–continuous method coincides with moisture behaviour according to the finite difference method. The distribution of average moisture during a year obtained by the discrete–continuous formula also coincides with the distribution of average moisture during a year obtained by the finite difference method. Moreover, both approaches have the same time of maximum humidification of the single-layer enclosing structure.

It gives a possibility to use proposed formula (5) for engineers' calculations.

References

1. Chen X, Su Y, Aydin D, Reay D, Law R, Riffat S (2016) Experimental investigations of polymer hollow fibre heat exchangers for building heat recovery application. Energy Buildings 125:99–108
2. Jarimi H, Aydin D, Yanan Z, Ozankaya G, Chen X, Riffat S (2019) Review on the recent progress of thermochemical materials and processes for solar thermal energy storage and industrial waste heat Recovery. Int J Low-Carbon Technol 14(1):44–69
3. Chen X, Su Y, Omer S, Riffat S (2016) Theoretical investigations on combined power and ejector cooling system powered by low-grade energy source. Int J Low-Carbon Technol 11(4):466–475
4. Jarimi H, Devrim A, Zhang Y, Ding Y, Rabadab O, Chen X, Dodo A, Utlu Z, Riffat S (2018) Materials characterization of innovative composite materials for solar-driven thermochemical heat storage (THS) suitable for building application. Int J Low-Carbon Technol 14(3):313–325
5. Gumerova E, Gamayunova O, Shilova L (2017) The optimal decision of insulation in cladding structures for energy efficient buildings. In: MATEC Web of conferences 106(06020)
6. Gamayunova O, Radaev A, Petrichenko M, Shushunova N (2019) energy audit and energy efficiency of modular military towns. E3S web of conferences 110(01088)
7. Petrichenko M, Ostrovaia A, Statsenko E (2015) The glass ventilated facades—research of an air gap. Appl Mech Mater 725–726:87–92
8. Zaborova D, Vieira G, Musorina T, Butyrin A (2017) Experimental study of thermal stability of building materials. Adv Intell Syst Comput 692:482–489
9. Castro JCL, Zaborova DD, Musorina TA, Arkhipov IE (2017) Indoor environment of a building under the conditions of tropical climate. Magz Civil Eng 8(76):50–57
10. Lakatos A (2020) Investigation of the thermal insulation performance of fibrous aerogel samples under various hygrothermal environment: laboratory tests completed with calculations and theory. Energy Buildings 214(UNSP 109902)
11. Reuge N, Collet F, Pretot S, Moisette S, Bart M, Style O, Shea A, Lanos C (2020) Hygrothermal effects and moisture kinetics in a bio-based multi-layered wall: experimental and numerical studies. Construct Building Mater 240(UNSP 117928)
12. Han W, Chen M, Song W, Ge C, Zhang X (2020) Construction of hexagonal boron nitride@polystyrene nanocomposite with high thermal conductivity for thermal management application. Ceramics Int 46(6):7595–7601
13. Makaveckas T, Bliudzius R, Burlingis A (2020) The influence of different facings of polyisocyanurate boards on heat transfer through the wall corners of insulated buildings. Energies 13(1991)
14. Vorobyeva IV (2018) Monitoring of separate pathogenetically significant biochemical markers in lacrimal fluid, ophthalmological parameters with combined pathology of diabetic retinopathy and age-related macular degeneration on the background angioprotective and antioxidant therapy. Oftalmologiya 15(2):189–199
15. Vorobyeva IV, Moshetova LK, Shcherbakova EV (2017) Therapy results in patients with primary open angle glaucoma and diabetic retinopathy. Oftalmologiya 14(1):40–46
16. Vorobyeva IV (2017) Long-term results of treatment of patients with non-proliferative diabetic retinopathy angioprotectors. Oftalmologiya 14(1):67–77
17. Petrov AS, Kupriyanov VN (2016) About Operational factor influence on vapor permeability of heat-insulating materials. Int J Pharmacy Technol 8(1):11248–11256
18. Zubarev K, Gagarin V (2021) Determining the coefficient of mineral wool vapor permeability in vertical position. Adv Intell Syst Comput 1259:593–600
19. Petrov AS, Kupriyanov VN (2018) Determination of humidity conditions of enclosing structures by the color indicator method. IOP Conf Series Mater Sci Eng 463:022064

20. Gagarin VG, Akhmetov VK, Zubarev KP (2020) Graphical method for determination of maximum wetting plane position in enclosing structures of buildings. IOP Conf Series Mater Sci Eng 753:022046
21. Berger J, Busser T, Colinart T, Dutykh D (2020) Critical assessment of a new mathematical model for hysteresis effects on heat and mass transfer in porous building material. Int J Thermal Sci 151:106275
22. Gagarin VG, Akhmetov VK, Zubarev KP () Position control of maximum wetting plane for building walls with foam polystyrene heat insulator. IOP Conf Series Mater Sci Eng 753:022045
23. Bastien D, Winther-Gaasvig M (2018) Influence of driving rain and vapour diffusion on the hygrothermal performance of a hygroscopic and permeable building envelope. Energy 164:288–297
24. De Mello LA, Moura LM, Mendes N (2019) A model for predicting heat, air and moisture transfer through fibrous materials. Int J Thermal Sci 145(UNSP 106036)
25. Maliki M, Laredj N, Missoum H, Bendani K, Naji H (2015) Numerical modelling of coupled heat, air and moisture transfer in building envelopes. Mech Industry 16(509)
26. Arregi B, Garay-Martinez R, Astudillo J, Garcia, M, Ramos J (2020) Experimental and numerical thermal performance assessment of a multi-layer building envelope component made of biocomposite materials. Energy Buildings 214(UNSP 109846)
27. Berger J, Dutykh D, Mendes N, Rysbaiuly B (2019) A new model for simulating heat, air and moisture transport in porous building materials. Int J Heat Mass Transf 134:1041–1060
28. Shen X, Li L, Cui W, Feng Y (2018) Coupled heat and moisture transfer in building material with freezing and thawing process. J Building Eng 20:609–615
29. Gagarin VG, Akhmetov VK, Zubarev KP (2018) Unsteady-state moisture behavior calculation for multilayer enclosing structure made of capillary-porous materials. IOP Conf Series Earth Environ Sci 177(012021)
30. Gagarin VG, Akhmetov VK, Zubarev KP (2018) The moisture regime calculation of single-layer enclosing structures on the basis of the discrete-continuum method application. IOP Conf Series Mater Sci Eng 456(012105)

Comparative Analysis of the Reactions of Low-Storey and Multi-Storey Buildings During an Earthquake

Vladimir Lapin, Syrymgali Yerzhanov, Kamadiyar Kassenov, Nurakhmet Makish, and Dauren Kassenov

Abstract During the earthquake on August 31, 2018 with a source in the Almaty region, the stations of the engineering and seismometric service registered instrumental records on rigid and flexible buildings. Accelerograms were obtained on a four-storey frame building and an 11-storey building with a steel frame. The reaction of two buildings at a given earthquake is analyzed. Spectral curves were plotted at the registration points on the roof and basement parts of buildings. The spectral curves at the basement level differ in spectral composition and spectrum irregularity. The influence of a tectonic fault on the response of an 11-storey building is revealed, and a criterion for the operation of a tectonic fault is proposed. The presence of resonance phenomena for a rigid building was established. Instrumental records are included in the accelerograms database of KazRDICA JSC. The results will be used to identify dynamic building models. This will allow you to take into account local features of seismic impact.

Keywords Accelerograms · Earthquake-resistant construction · Seismic impact · Seismic station · Tectonic fault

1 Introduction

Over 50% of the territory of the Republic of Kazakhstan is affected by earthquakes. On the territory of the Almaty region, there are sources of all catastrophic and strong earthquakes that have taken place over the past 100–150 years [1, 2]. Therefore, instrumental observations are carried out for a significant number of buildings.

Engineering and seismometric service on buildings (hereinafter—ESS) has been operating in Kazakhstan since 1968 and is currently represented by 12 stations located

V. Lapin (✉) · S. Yerzhanov · N. Makish · D. Kassenov
JSC Kazakh Research and Design Institute of Construction and Architecture, 21 Solodovnikova, 050043 Almaty, Kazakhstan
e-mail: lapin_1956@list.ru

K. Kassenov
Satbayev University, 22a Satbayev st, 050013 Almaty, Kazakhstan

A. Beskopylny and M. Shamtsyan (eds.), *XIV International Scientific Conference "INTERAGROMASH 2021"*, Lecture Notes in Networks and Systems 247,
https://doi.org/10.1007/978-3-030-80946-1_27

on buildings of various designs. 4 stations have both digital and analogue equipment. This network of stations on buildings is the only one in the Republic of Kazakhstan.

The stations are dominated by old analog devices. Currently, various measuring systems and instruments have been developed [3–7]. The stations are being modernized gradually.

Engineering seismometric stations register tangible earthquakes of various intensities in real time.

According to the operational data of the Data Center of the Institute of Geophysical Research, on August 31, 2018, at 06:21 PM Nur-Sultan time (at 12:21 PM GMT), an earthquake occurred in the Almaty region, 50 km southeast of Almaty. The coordinates of the epicenter: 43.10° north latitude, 77.42° east longitude. Magnitude mb = 4.8. Energy class K = 11.7. Depth h = 4.5 km. The earthquake was felt in Almaty with an intensity of 4 points; Talgar, Kotyr-Bulak—3–4 points; Castek—2–3 points; Karatobe, Saty—2 points.

The following tasks are set:

- Explore the vibrations of an eleven-story steel-framed building near a tectonic fault;
- To identify the possibility of resonance phenomena in such buildings in case of considerably close local earthquakes.
- To use instrumental records of the real earthquake on August 31, 2018 to solve the indicated problems.

Thus, experimental research methods (instrumental recording of accelerations) and theoretical methods based on the use of computer mathematics systems, such as MATLAB, are combined.

2 Method and Subject of Research

The entire territory of the Almaty region is prone to earthquakes, which can pose a danger to buildings and structures on the territory of Almaty [1, 2]. Since the beginning of 2018, several earthquakes have occurred both with foci in the city and around it [8].

Instrumental data were obtained by stations of the engineering and seismometric service of KazRDICA JSC. Comparison of instrumental recordings on rigid and flexible buildings is of interest. Such a comparison has already been performed in [8] during the earthquake on March 26, 2018.

2.1 *Rigid Building*

Seismic station № 1 is in the southern part of the city at the intersection of Gagarin and Dzhandosov streets. Seismic station № 1 "Institute" is mounted in a 4-storey

Fig. 1 Low rise building (seismic station №1)

L-shaped frame building. The frame is reinforced concrete. Soil is boulder type. The building houses the largest commercial bank in Kazakhstan (Fig. 1).

The seismic station was installed in 1968. This is historically the first station of the engineering and seismometric service in Almaty. The devices are located in the basement on the foundation and on the roof of the building. They were modernized in 2008. Currently, the seismic station is equipped with both analog sensors and a digital instrumentation and measuring system RSM-8.

No tectonic faults were found near the building.

For 50 years of operation of seismic station № 1, instrumental records of over 100 earthquakes were obtained.

2.2 *Flexible Building*

Station № 11 "Abay Prospect" is located on an 11-storey administrative building made with a metal frame. It is worth noting that this station is one of the very first in Almaty—it began operating in 1970. In 2010, the first modernization of the station took place—a digital instrumentation and measuring system was installed.

The second upgrade was carried out in 2016—the ADXL accelerometers were replaced by AT 1105.

The soils on the construction site of the building are boulders.

The building has dimensions 15 × 36 m, height 43.6 m.

From a constructive point of view, the building is frame-panel. Exterior wall panels are made of aluminum with glass and stemalite filling.

The frame consists of longitudinal and transverse closed frames with a transverse and longitudinal column spacing of 6 m. The height of the basement is 3.4 m, the first floor is 4.2 m and all subsequent ones are 3.6 m.

Concrete tape foundations. The basement walls are made of monolithic concrete. The columns of the frame are anchored to the basement walls with the help of metal shoes.

Floor slubs of the first and basement floors are made of monolithic reinforced concrete; over the rest of the floors, the floors are made of prefabricated reinforced concrete ribbed slabs of 3 × 6 m in size.

The partitions are made of prefabricated expanded clay concrete panels, as well as from precast gypsum concrete and slag concrete slabs.

At the top of the building there is a technical floor (12th floor), where a part of the control and measuring equipment is located—the upper registration point (mark 40.2 m). The lowest registration point is in the basement. The devices are located at the −4.4 m mark (the basement itself is at the −3.4 m mark).

Earlier, instrumental records (accelerograms) were obtained on this building during the earthquake on August 9, 2017 in China (Xinjiang Uygur Autonomous Region) [5].

The building has a unique location. It is located between two local tectonic faults. The faults are approximately 700–800 m away from the building. Therefore, the building is a very interesting object for instrumental observation.

In [10] the results of earthquakes registration by seismic station № 1 "Institute" are given. To a certain extent, the results of measurements by the instruments of this station of various seismic phenomena are reference.

For each accelerogram it is possible to construct spectral curves that will characterize the dynamic effect in the indicated direction [2].

This is the time to consider the question of spectral curves constructing. The most common way is to use the Voigt model, which describes the vibrations of a single-mass elastic system with energy dissipation overcoming viscous friction—as it was customary to say the Voigt body with a parallel connection of elastic and viscous elements. This method is the most common and numerous bibliographies are devoted to it.

Therefore, for the analysis of instrumental records of the engineering and seismometric service stations, it is most convenient to use the classical Voigt model, which explicitly depends on two parameters—the oscillation decrement and the oscillation period of the elastic system T:

$$\ddot{x} + 2\frac{\partial}{T}\dot{x} + \left(\frac{2\pi}{T}\right)x = -\ddot{x}_0 \quad (1)$$

The integration of Eq. (1) was performed using the MATLAB computer mathematics system. All calculations were performed with high accuracy. The values of the dimensionless spectral coefficients, the values of the effective duration and the periods of the spectrum maximum are calculated.

3 Results

Figures 2 and 3 show instrumental records (accelerograms) in the levels of the roof and basement during the earthquake of August 31, 2018. Spectral curves at a decrement of 0.3 are shown in Figs. 4 and 5.

Table 1 shows the results of measurements and calculations (h is the sampling step of accelerograms).

The current scale MSK-64 (K) is compiled for earthquakes with an intensity of 5–10 points. This earthquake does not fall within the specified interval in terms of the values of the registered accelerations. The indicated scale contains accelerations from 16 cm/s^2 to more than 900 cm/s^2. Here the acceleration at the basement level in the horizontal plane is 0.7–2.2 cm/s^2.

Therefore, it is fair to attribute the earthquake of 08/31/2018 to a seismic event with an intensity of no more than 2–3 points near the seismic station № 1 "Institute".

The accelerations at the roof level are approximately 4–6 times higher than the accelerations at the level of the basement of the building for all components.

The spectral characteristics of the instrumental records at the basement level characterize the frequency composition of the seismic effect at the level of the building foundation. The periods of prevailing fluctuations are 0.10–0.14 s.

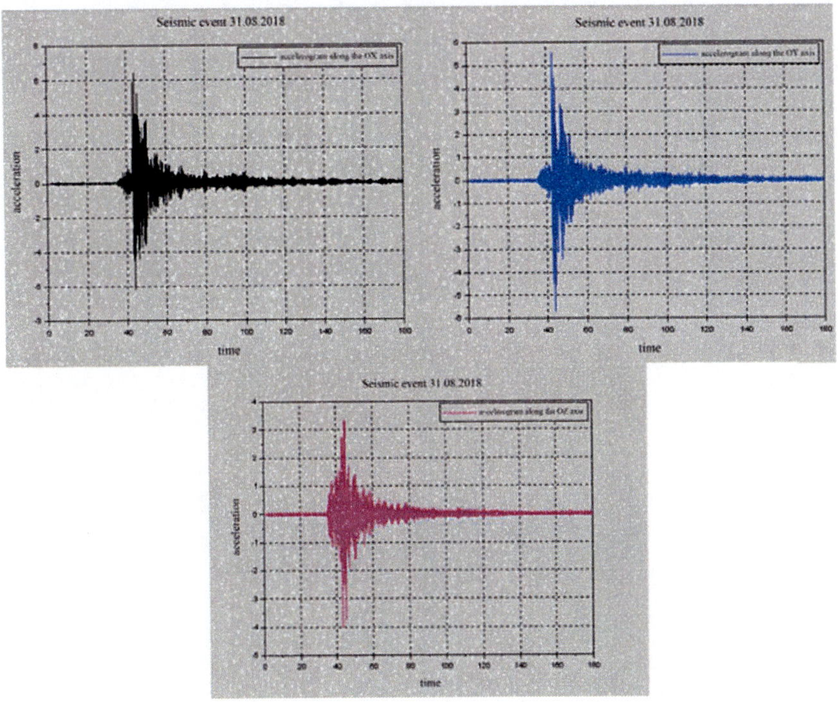

Fig. 2 Instrumental recordings (accelerograms) at the roof level (seismic station № 1)

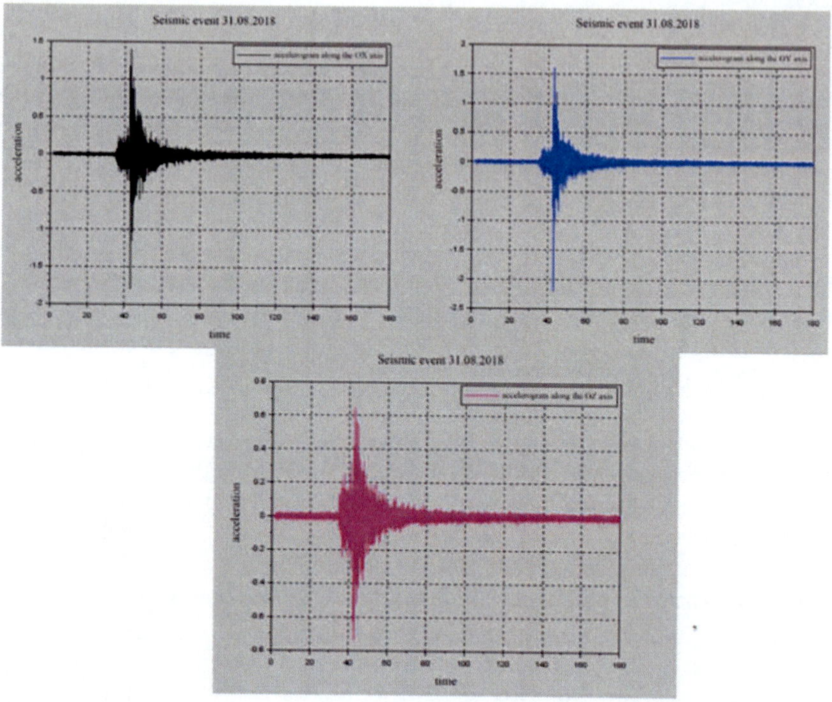

Fig. 3 Instrumental records (accelerograms) at the basement level (seismic station № 1)

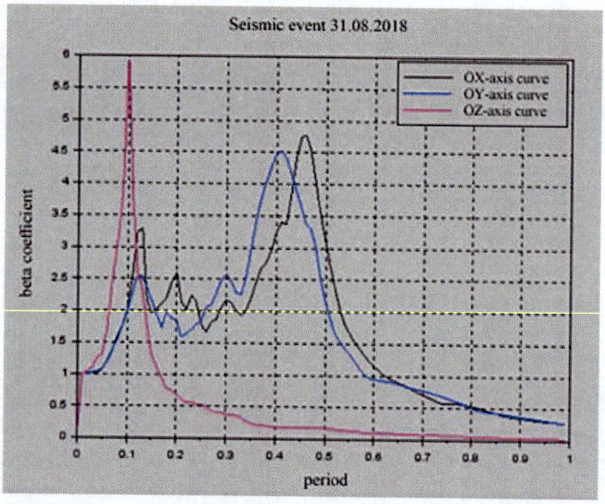

Fig. 4 Spectral curves at the roof level (seismic station № 1)

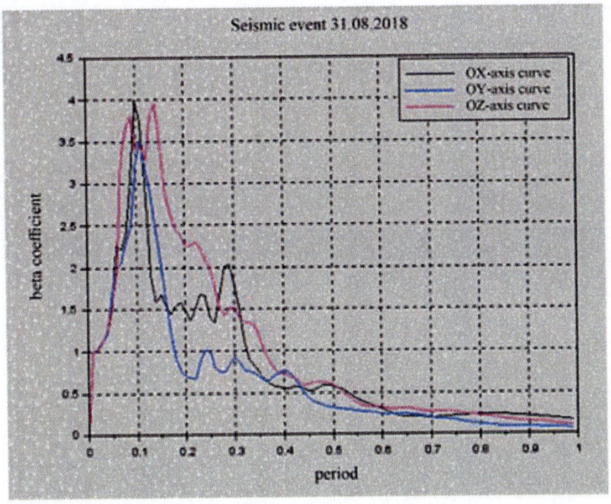

Fig. 5 Spectral curves at the basement level (seismic station № 1)

Table 1 Maximum values of accelerations and parameters of accelerograms during the earthquake of 31.08.2018 (h = 0.0064c)

Component	Acceleration, cm/s^2	Effective duration, s	Spectral coefficient	Spectrum maximum period, s
03.09.18–2-1-u1, Roof OX	6.45	7.09	4.77	0.46
03.09.18–2-1-u2, Roof OY	5.71	4.80	4.53	0.41
03.09.18–2-1-u3, Roof OZ	4.00	2.02	5.84	0.10
03.09.18–2-1-u6, Basement OX	1.80	1.54	3.13	0.10
03.09.18–2-1-u7, Basement OY	2.20	1.33	3.52	0.11
03.09.18–2-1-u8, Basement OZ	0.74	4.41	3.95	0.14

Table 2 shows the peak acceleration values and instrumental record parameters for an 11-story steel framed building (seismic stations № 11).

Figures 6 and 7 show the recorded accelerograms at the roof and basement levels. Figures 8 and 9 show the corresponding spectral curves.

Here the acceleration at the basement level in the horizontal plane is 2.33–2.65 cm/s^2.

Therefore, the intensity of the earthquake on August 31, 2018 is approximately 2–3 points in the area of seismic station № 11 “Prospect Abai”. Compared to the rigid building of the Institute seismic station, the acceleration values in the azimuthal plane at the roof level do not exceed the acceleration values in the basement of the building.

Table 2 Maximum values of accelerations and parameters of accelerograms during the earthquake of 08/31/2018 (h = 0.0064 s)

Component	Acceleration, cm/s^2	Effective duration, s	Spectral coefficient	Spectrum maximum period, s
03.09.18–11-39–38-u1, roof OX	2.88	3.18	4.90	0.37
03.09.18–11-39–38-u2, roof OY	3.38	2.62	4.86	0.40
03.09.18–11-39–38-u3, roof OZ	2.99	8.60	4.02	0.14
03.09.18–11-39–38-u6, basement OX	2.33	1.26	2.30	0.23
03.09.18–11-39–38-u7, basement OY	2.65	2.35	3.48	0.13
03.09.18–11-39–38-u8, basement OZ	1.61	1.63	3.44	0.25

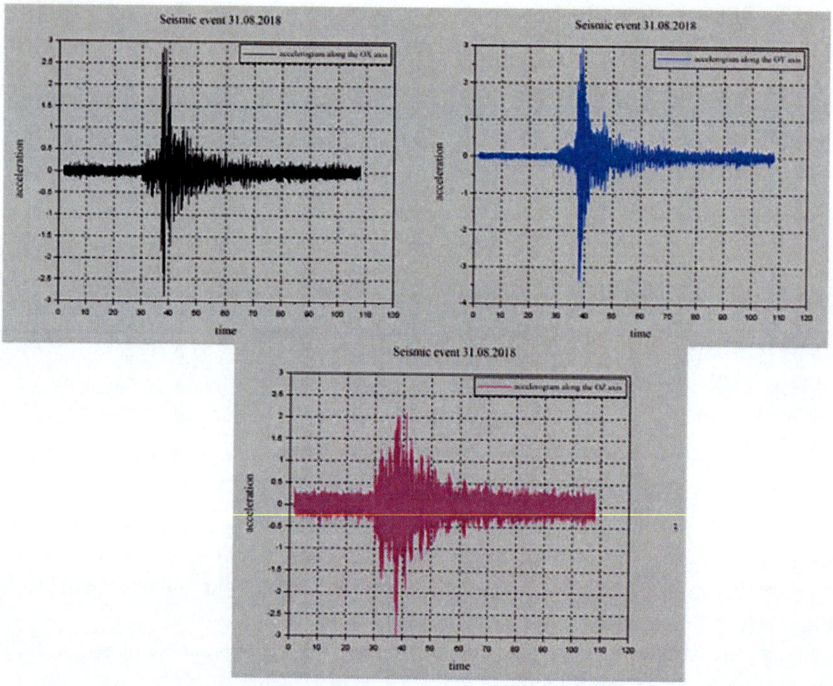

Fig. 6 Instrumental recordings (accelerograms) at the roof level (seismic station № 11)

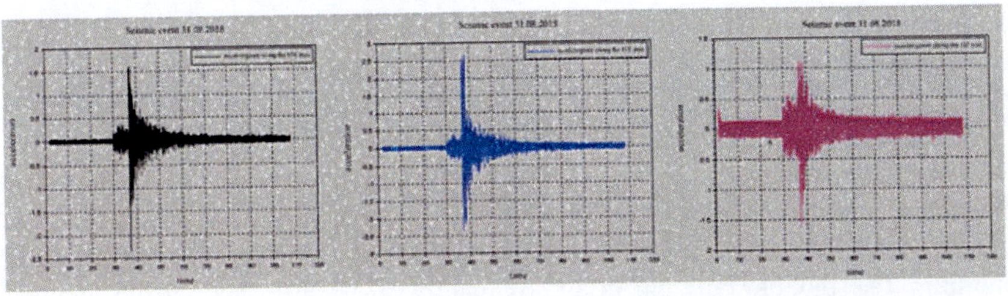

Fig. 7 Instrumental recordings (accelerograms) at the basement level (seismic station № 11)

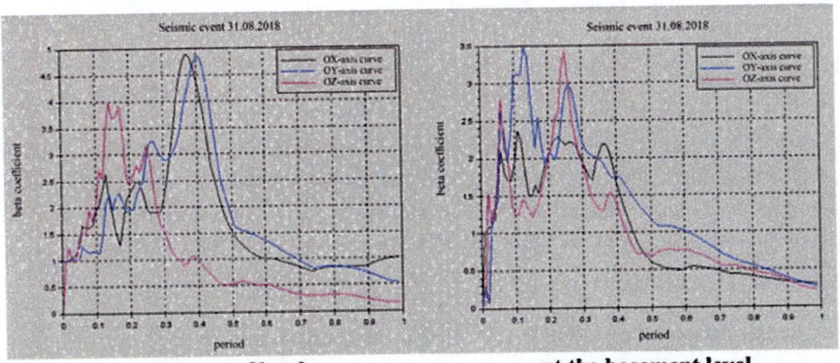

at the roof level **at the basement level**

Fig. 8 Spectral curves (seismic station № 11)

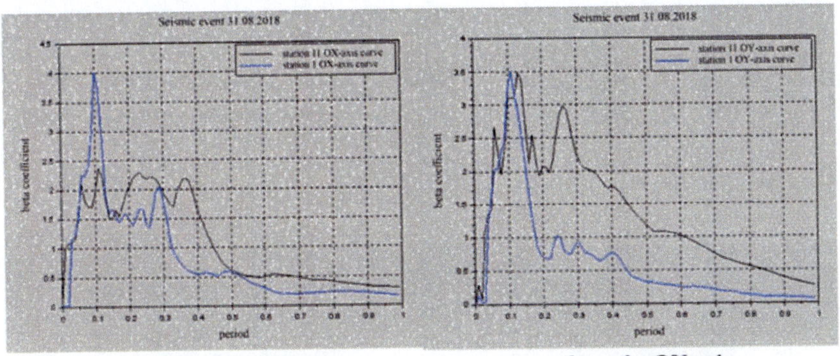

along the OX axis **along the OY axis**

Fig. 9 Spectral curves, basement

The effective durations of accelerograms vary within wide limits.

The periods of the spectrum maximum at the basement level in the horizontal plane are 0.13–0.25 s.

In Figs. 8, spectral curves are grouped along the axes at the basement level of each building. It is worth noting that the soil conditions are the same—boulders. The spectral curves characterizing the dynamic effect of seismic action differ quite strongly here. It is interesting that no tectonic fault was observed near the rigid building. The flexible building is approximately 500–800 m away from tectonic faults on both sides of the building.

4 Discussion

During the earthquake of March 26, 2018 [9], vibrations of rigid and flexible buildings were also studied. The shapes of the spectral curves at the basement level on both axes practically coincided, Fig. 10. The ground conditions are the same. The rigid building is far from the faults, and the flexible one is next to them. We can say that there is no influence of tectonic faults on the parameters of the building's reaction at the basement level.

For the earthquake of August 31, 2018, the spectral curves are very different. On the OY axis, the peaks of the spectrum practically coincided (Fig. 9), but the shapes of the spectra differ greatly. It can be concluded that in this seismic event the influence of tectonic faults is the reason for these differences.

Thus, a criterion was obtained to assess the influence of a nearby tectonic fault on the response of a building.

In contrast to a steel-framed building, resonance phenomena were observed in a low-rise building.

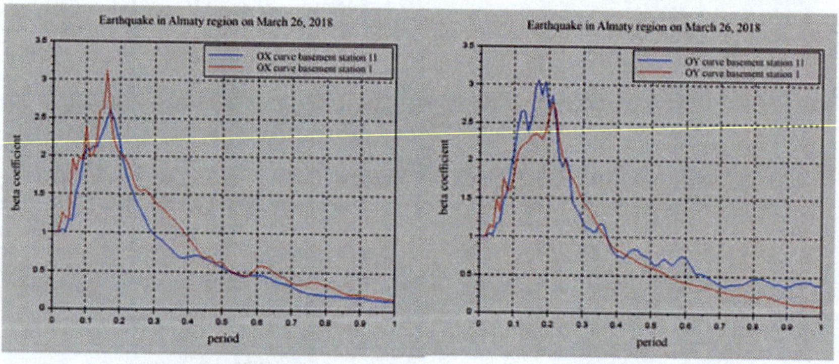

along the OX axis **along the OY axis**

Fig. 10 Spectral curves, basement during the earthquake of 03/26/2018

It should be noted that two buildings with seismic stations № 1 and № 11 form a sample, convenient for studying the spectral composition of the impact, which makes it possible to study even the influence of the presence of tectonic faults on the reactions of buildings. These are, in a sense, two classical pendulums with quite different periods of oscillation in the fundamental tone, having the same ground conditions, differing in their location relative to the tectonic faults of Almaty.

Seismic stations on buildings provide real-time experimental information on the behavior of buildings during an earthquake. Therefore, such stations should be equipped with buildings with non-typical design solutions, including buildings with seismic-insulating foundations and active seismic protection systems [11–16]. Instrumental information will contribute to the development of new methods for calculating such buildings [13].

5 Conclusion

- The periods of the spectrum maximum for instrumental records at the level of the last tier of a rigid building (seismic station № 1) characterize the period of oscillation of the building in terms of the fundamental tone, taking into account the strain capacity of the foundation.
- Oscillations of a flexible building are vibrations summing of several vibration modes. Resonance phenomena are not observed.
- The period of the spectrum maximum at the basement or first floor level mainly characterizes the spectral composition of the seismic effect rather than the frequency response of the building.
- For flexible and rigid buildings, the prevailing vibration periods of the foundation are different. Note here that the soil conditions are also the same—boulders. It was found that the spectral curves differ both in frequency composition and irregularity.
- A seismic event on August 31, 2018 is an earthquake with an intensity of no more than 3 points in the territory of Almaty. The source of the earthquake is located on the territory of the Almaty region to the south-east of the city of the city.
- The accelerograms given in the article will add to the database of accelerograms of JSC KazRDICA.
- Two aforementioned buildings form a sample of rigid and flexible buildings that is convenient for studying the spectral composition of the impact. They are, in a sense, two pendulums with quite different periods of oscillation along the fundamental tone. One of the pendulums is located near a tectonic fault.

Acknowledgements The research was carried out using the funds of the grant AP 05130702 of the Ministry of Education and Science of the Republic of Kazakhstan.

References

1. Mikhaylova NN, Sokolova IN, Velikanov AE, Sokolov AN (2015) Earthquakes in the city of Almaty. Bull Natl Nuclear Center 3:87–93
2. Galperin RM, Nersesov IL, Galperin EI (1985) Seismic regime of Almaty for 1972–1982 years. Science, Moscow
3. Deng M, Wang M, Jin S, Zhang Q, Chen K, Cheng SA (2019) A wriless monitoring system for a high-power borehole-ground electromagnetic transmitter. Geosci Instrument Methods Data Syst 8:13–19. https://doi.org/10.5194/gi-8-13-2019
4. Zhang X, Zhang Q, Wang M et al (2017) Development of a full-wave form voltage and current recording device for multichannel and current recording device for multichannel transient electromagnetic transmitters. Geosci Instrument Methods Data Syst 6:495–503. https://doi.org/10.5194/gi-6-495-2017
5. Yang L, Stehly L, Paul A (2018) Working Group High-resolutional surface wave tomography of the European crust and uppermost mantle from ambient seismic noise. Geophys J Int 214(2):1136–1150. https://doi.org/10.1093/gji/ggy188
6. Qiao S, Duan H, Zhang Q et al (2018) Development of high-precision distributed wireless microseismic acquisition stations.Geosc Instrument Methods Data Syst 7:253–263. https://doi.org/10.5194/gi-7-253-2018
7. Groukamp L, Rossow JJ, Fourie CJS, Cormbrinck L (2011) Foundation requirements for a lunar laser ranger facility at matjiesfontein South Africa. Afr J Geol 114(3/4):573–576. https://doi.org/10.2113/gssajg.114.3-4.573
8. Yerzhanov SY, Lapin VA, Daugavet VP, Devyatkh AA (2018) Analysis of instrumental recordings of the March 26, 2018 earthquake obtained on rigid and flexible buildings. Bulletin KazRDICA JSC 7(83):38–49
9. Yerzhanov SY, Lapin VA, Daugavet VP, Devyatkh AA (2017) Research of instrumental recordings obtained on an 11-story building with a steel frame, during the earthquake on August 9, 2017. Bulletin KazRDICA JSC 11(75):20–27
10. Yerzhanov SY, Lapin VA, Daugavet VP, Devyatkh AA (2017) Instrumental registration of 2 earthquakes on 21.04.2017 in Kyrgyzstan by ISS Institute station». Bulletin KazRDICA JSC 4(68):10–18
11. Calvi PM, Calvi GM (2018) Historical development of friction-based seismic isolation systems. Soil Dyn Earthquake 106(3):14–30. https://doi.org/10.1016/j.soildyn2017.12.003
12. Lapin VA, Yerzhanov SY, Aldakhov YS (2020) Statistical modeling of a seismic isolation object under random seismic exposure. J Phys Conf Ser 1425:012006. https://doi.org/10.1088/1742-6596/1425/1/012006
13. Bulat AF, Dyrda VI, Lysytsya MI, Grebenyuk SM (2018) Numerical simulation of the stress-strain state of thin-layer rubber-metal vibration absorber elements under nonlinear deformation. Strength Mater 50(3):387–395. https://doi.org/10.1007/s11223-018-9982-9
14. Ahmad S, Ghani F, Md Raghib A (2009) Seismic friction base isolation performance using demolished waste in masonry housing. Constru Building Mater 23:146–152. https://doi.org/10.1016/j.conbuildmat.2008.01.012
15. Dyrda VI, Kobets A, Bulat AF et al (2020) Present problems of vibration isolation in heavy mining machines at long-term cyclic loads. E3S Web Conf 168:00042. https://doi.org/10.1051/e3sconf/202016800042
16. Yerzhanov SY, Lapin VA, Aldakhov YS (2020) Monitoring the changes of dynamic characteristics of high-rise building. J Phys Conf Series 1425:1–9. https://doi.org/10.1088/1742-6596/1425

Laser and Photogrammetric Modeling of Roads Surface Damages

Altynbek Kuduev, Zhypargul Abdykalyk kyzy, and Boris Shumilov

Abstract The problem of diagnosing the current state of highways is considered from the standpoint of the successful functioning of a complex cyber-physical system that ensures transport accessibility of territories and settlements as an integral part of the digital economy. The prospects of using the data of laser scanning of road surfaces for interpolating surfaces and calculating the volume of repair and reconstruction of highways are outlined. In order to increase mobility and reduce the cost of data collection for diagnostics, it is proposed to use a computer vision algorithm based on a photogrammetric method for processing video recording data, which allows both to identify the objects of transport infrastructure under study and to detect their defects. Examples of determining the spatial coordinates of three-dimensional objects from photo-images obtained from different angles are given. The results of numerical experiments on the imposition of the projected road on the processed points of laser scanning are also presented. The presented algorithm can be used when planning road repairs, when analyzing road accidents, when processing requests from road users, etc.

Keywords Roads · Modeling · Damages · Laser · Photogrammetry · Calibration

1 Introduction

Timely recognition of roads surface defects reduces the economic losses inevitable due to the fact that cracks and potholes become too significant as the roads are used. Creation of automatic diagnostic systems for assessing defects and destruction of the road surface is possible using video cameras and image processing algorithms. However, the problem with systems based solely on raw video recording is their inability to filter out such shadows in images as road markings, oil spills, brake marks and shadows of passing cars, and fresh patches [1].

A. Kuduev · Z. Abdykalyk kyzy
Osh State University, 331 Lenina, Osh 723500, Kyrgyzstan

B. Shumilov (✉)
Tomsk State University of Architecture and Building, 2 Solyanaya, Tomsk, Russia 634003

A. Beskopylny and M. Shamtsyan (eds.), *XIV International Scientific Conference "INTERAGROMASH 2021"*, Lecture Notes in Networks and Systems 247,
https://doi.org/10.1007/978-3-030-80946-1_28

Fig. 1 A cloud of laser scanning points

One of the classical methods for extracting information about the depth of a three-dimensional (3D) object is the method of photogrammetric processing of stereoscopic images of a 3D object obtained from different angles [2]. The disadvantage of traditional digital photogrammetry is the significant involvement of the human operator, as well as the need to pre-scan analog images using high-precision scanners. A breakthrough in solving problems of diagnostics of highways was made using systems based on mobile laser scanning [3] (Fig. 1).

Because the results of laser measurements of scanning points include reflections from people, equipment, vegetation, etc. on the object, they need to be preliminary processed for cleaning from noise and filling in gaps to produce the surface—the section of the projected road. Moreover, the matter is complicated by the fact that this technology turned out to be very expensive and unaffordable for the municipal and regional budgets. The use of modern inexpensive photogrammetric methods [4–6] to determine damage and unevenness of the road surface and structural elements of the road gives a new impetus to the development of digital technologies for designing road repairs, increasing mobility and reducing the cost of work.

The considered above autonomous mobile road data collection and analysis devices, which operate as part of a single measuring and processing complex, are an affordable example of a cyber-physical system in transport [7, 8]. This also adds a system of stationary city security cameras for road safety and, possibly, a system of unmanned vehicles, which, having a complex software system of artificial intelligence on board, are able to analyze independently the surrounding situation, make decisions and learn from previous experience.

As a result, the totality of such devices forms a heterogeneous distributed system of artificial intelligence, enriched by some hierarchical structure, which would provide the distribution of tasks between devices, their interaction with the environment and the exchange of information in accordance with fundamental goals and settings (Fig. 2). This implies the existence of protocols for their joint work at the lower level of interaction, as well as protocols for sharing in data processing centers (DPC) at the mid-level and further for transfer to a single data center and development of management decisions at the upper level in order to form a global cyber-socio-physical urban traffic management system as part of the concept of creating "smart cities". The appearance of the word "socio" as applied to the road here is by no

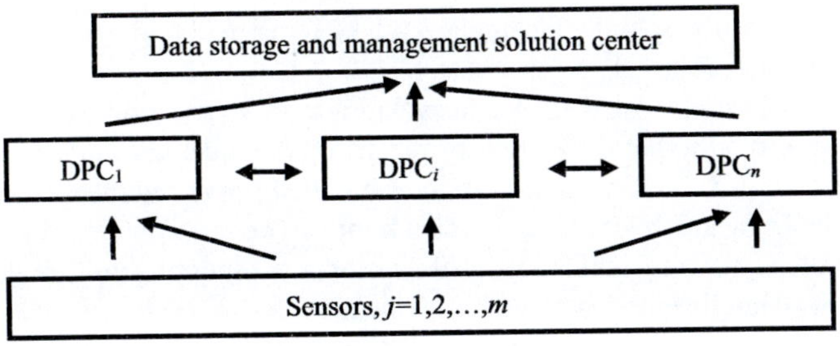

Fig. 2 Block diagram of a cyber-physical urban traffic control system [9]

means accidental, since the road is not a factory or field with the prospect of a deserted production technology, but an indispensable element of the "smart city", providing accident-free operation and maximum throughput of the transport system, with the presence in it of active road users, creating the effect of unpredictability of conscious or unconscious behavior of people, drivers, pedestrians and etc.

Based on the foregoing, the development of autonomous mobile devices capable of "on the spot" to perform the tasks of collecting and analyzing road data, including laser scanning, photographs and video, in a form suitable for designing roads repairs and reconstruction is an urgent task.

2 Methods and Results

2.1 *Stereo-Measurements Applications for Determining Coordinates of Surface's Points*

The key component of the proposed cyber-physical system for diagnosing transport infrastructure objects are the data processing centers (DPC), the principal role of which is the technical vision system, based on the method of photogrammetric processing of stereoscopic images of a three-dimensional 3D object obtained from different angles.

Highlighting the boundaries of flat images is one of the important auxiliary tasks in recognizing defects in transport infrastructure objects, in particular, damages of the road surface. They contain comprehensive information about their form for subsequent analysis. In [10, 11] there were suggested suitable algorithms for linear approach localization and identification of projections of interpolation points on images determined from reference frames of a video-sequence and forming stereo-pairs. As a preliminary conclusion, we note that the images of objects of transport infrastructure are characterized by the presence of angles formed by the inter-sections of generators (curves or straight lines). Therefore, for the analysis of such images, it is advisable to use the so-called «angular» filters, in particular, the use of the Harris detector is popular [12, 13].

Finally, [14] was devoted to the formulation and verification of a new "one-point" method of external calibration of the smartphone camera. Indeed, knowing the geometric transformation matrix for the calibrated frame and the rotation matrix of the second frame relative to the calibrated frame, it is easy to obtain a linear system of two equations with respect to the transfer vector of the shooting point of the second frame, since all nonlinearity associated with the Euler angles is analytically excluded. The resulting solution allows carrying out 3D measurements on a smartphone without much effort on the part of the road master.

2.2 *Experimental Results*

For numerical experiments, a real stereo-pair was used (Fig. 3) with an image of a typical traffic cone against the background of damage to the road surface.

The coordinates of the seven points of the corners of the cone were measured with a ruler. The coordinates of the corresponding points on the images were fixed in a graphical editor using the "mouse" manipulator. The solution was found in the MathCAD system. An attempt to restore the 3D coordinates of the points of the corners of the cone gave a good match with the original values. Moreover, in the course of the calculations, it was possible to correct the measurements in the photo and typos when taking readings. Given the measured values of a fixed point of the road surface on the left and right images, the accuracy of the photogrammetric method can be estimated. In our case, the distance from the top of the cone to the asphalt pavement, calculated according to the Pythagorean theorem, was 31.975 cm, which is 0.078% different from the passport value of the cone height of 32 cm.

Predetermining the internal parameters of the camera reduces the number of calibration points to three. Recall that these are the points that the road master is supposed

Fig. 3 Stereo-pair of two photographs of a traffic cone against the background of damage to the road surface

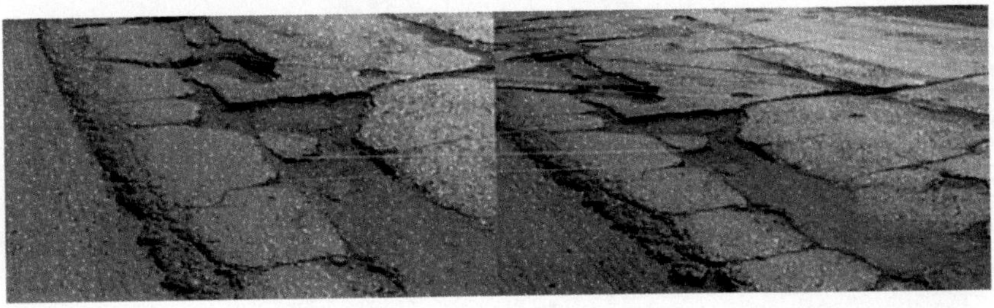

Fig. 4 A stereo-pair of two photographs of a real road surface

to find on each picture for a successful solution of the photogrammetry problem. The resulting "three-point" method gave a cone height of 21.174 cm, which provides a relative measurement error of 33.8%. Nevertheless, adding only one (fourth) point in this context gave a value of 30.756 cm, which provides a quite acceptable relative measurement error of 3.8%. The numerical experiments carried out make it possible to recommend, for video measurements on the ground, the use of such a device as a folding emergency stop sign, which has exactly four easily recognizable reference points. The more important task of calibrating the internal parameters of the camera can be successfully performed in laboratory conditions using a much larger number of marker points, for example, in the method of Tsai, there are at least 25 of them.

The last experiment was to test the ability to automatically search for graphic markers and anchor points in photographs. Remind that precisely these points must be found on each image in order to successfully solve the problem of photogrammetry. In particular, in Fig. 4 show the use of the Harris detector for detecting conjugate points on the left and right frames of a stereo-pair by depicting damage to the road surface.

2.3 *Wavelet-Based Applications for Modeling of Roads Surface*

Using the algorithms constructed, the following tasks of processing laser scanning data are solved:

- import of textual scan data to database to speed up the processing,
- dividing a set of data points into existing laser scans,
- deleting the repeated scans, if they are used to be,
- selection of points belonging to the surface of the roadway from sets of data points,
- sampling data to build a three-dimensional model of the road surface,
- presenting an image of laser scanning, tracks and road surface in selected profiles and plans with any degree of accuracy,
- presenting a visual 3-D representation of the road surface,
- –wavelet-analysis of the selected laser scan and any track of the roadway;

- wavelet analysis of the roadway surface using bicubic wavelets with obtaining statistical estimates of the error arising from the compression of information.

2.4 Numerical and Practical Results of Modeling the Road Surface

Using the developed software package for modeling the road surface, many specific numerical results obtained, among which the following can be indicated. For example, in Fig. 5 the points indicate laser measurements after preliminary processing, cleaning from noise and filling in gaps, the surface—3D-view of the section of the projected road.

When data is compressed using bicubic wavelets with a compression ratio of 3.97, the resulting root-mean-square error of the surface approximation does not exceed 5.7 mm. The mathematical expectation of the deviation is slightly shifted—by 9.4×10^{-4} m. A three-dimensional image of a part of the surface before and after bicubic wavelet compression shown in Figs. 6 and 7 respectively.

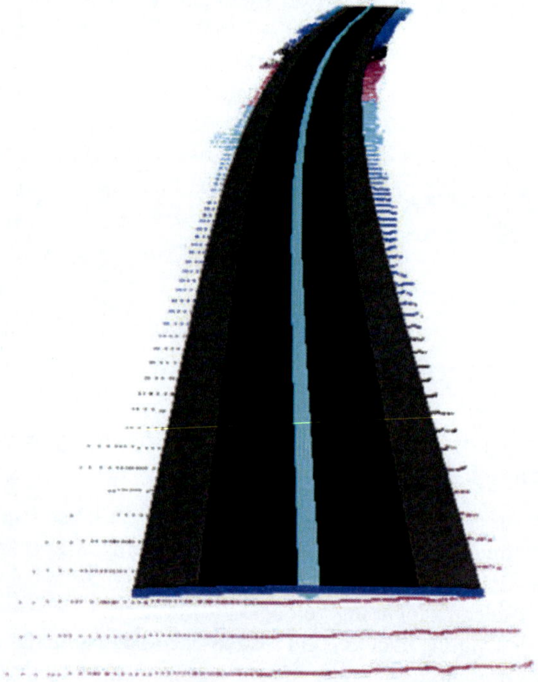

Fig. 5 Results of preliminary processing of laser measurements

Fig. 6 A piece of surface before compression by bicubic wavelets

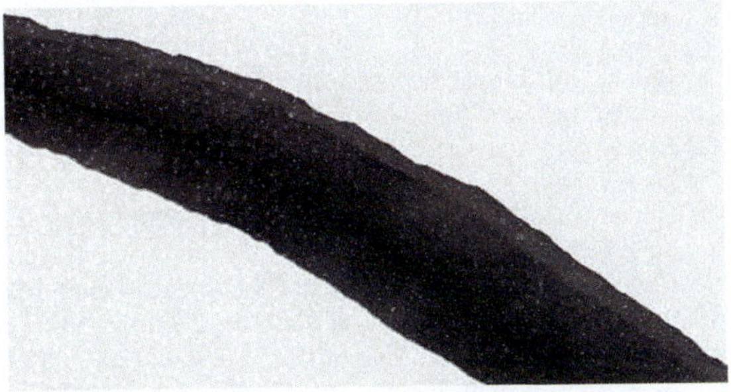

Fig. 7 A piece of surface after bicubic compression wavelets

3 Discussion

From the results obtained follows that smoothing the surface using wavelets can be used as the basis for calculating the required amount of work for cutting off protrusions and filling potholes when designing surface repairs.

The practical application of photogrammetric diagnostics methods of the transport infrastructure is quite wider, including, fixing and determining the real size of the damage motor vehicle on the measured coordinates of surface's points of the body parts [15]. Examination of the nature and list of injuries motor vehicle provides fixation of the damage to determine whether their formation and involvement in the studied event. The results of visual inspection of the damaged vehicle and photographing are issued by the relevant act. Photographs and an inspection of the damaged motor vehicle shall be attached to the expert opinion or valuation report of the repair motor vehicle. Of course, the photogrammetric method is a strong argument to determine the decision of disputable situations, and to determine actual damage of motor vehicle elements on photos and video is the actual task.

It is also worth considering that because of the peculiarities of image defects of a roadbed (the lack of clear boundaries, the presence of extraneous objects, the insignificance of some of the defects) should be pre-see the possibility of «manual» marking of characteristic point on images.

4 Conclusion

The implementation of the resulting algorithms and the automation of the searching process for graphic markers and reference points on the photos are supposed to be implemented on a smartphone running the Android OS. In conclusion, we must say that the mobile collection of data about defects of transport infrastructure, particularly road surface can be much more effective with the use described in these report methods.

References

1. Bao G (2010) Road distress analysis using 2D and 3D information. The University of Toledo, The University of Toledo Digital Repository Theses and Dissertations
2. Rogers DF, Adams JA (1990) Mathematical elements for computer graphics. McGraw-Hill, New York
3. Kuduev AZ, Esharov EA, Shumilov BM (2020) Wavelets and modeling of road surfaces. Osh State University Press, Osh, 119 p
4. Chiuso A, Favaro P, Jin H, Soatto S (2002) Structure from motion causally integrated over time. IEEE Trans Pattern Anal Mach Intell 24:523–535
5. Davison AJ, Reid ID, Molton ND, Stasse O (2007) MonoSLAM: real-time single camera SLAM. IEEE Trans Pattern Anal Machine Intell 29:1052–1067
6. Elugachev P, Shumilov B (2018) Development of the technical vision algorithm. MATEC Web Conf 216:04003
7. Kupriyanovsky V, Namiot D, Sinyagov S (2016) Cyber-physical systems as a base for digital economy. Int J Open Inf Technol 4(2)
8. Tsvetkov VY (2017) Control with the use of cyber-physical systems. Perspect Sci Educ 3(27):55–60
9. Elugachev PA, Shumilov BM, Esharov EA, Lakhodynova NV (2021) Mathematical modeling of cyber-socially physical systems in transport. Transp Res Procedia 54:136–143
10. Tsai R (1987) A versatile camera calibration technique for high-accuracy 3D machine vision metrology using off-the-shelf TV cameras and lenses. IEEE J Robot Automat 3:323–344
11. Sonka M, Hlavac V, Boyle R (2008) Image processing. In: Analysis and machine vision. Thomson, Toronto
12. Harris affine region detector Homepage, https://en.wikipedia.org/wiki/Harris_affine_region_detector. Last Accessed 11 June 2018
13. Harris C, Stephens M (1988) A combined corner and edge detector. In: Fourth Alvey vision conference. Manchester, UK, pp 147–151
14. Shumilov B, Gerasimova Y, Titov A (2018) On simplifications in the technical vision algorithm. Adv Sci Eng 10(2):39–47
15. Morales A, Sanchez-Aparicio LJ, Gonzalez-Aguilera D, Gutierrez MA, Lopez AI, Hernandez-Lopez D, Rodriguez-Gonzalvez P (2017) A new approach to energy calculation of road accidents against fixed small section elements based on close-range photogrammetry. Remote Sensing 9(1219):1–18

Controlling Temperatures in Low-Temperature District Heating: Adjustment to Meet Fluctuating Heat Load

Stanislav Chicherin, Andrey Zhuikov, Mikhail Kolosov, Lyazzat Junussova, Madina Aliyarova, and Aliya Yelemanova

Abstract One of the fundamental considerations in space heating (SH) consists in the variability of heat loads over time as result of external conditions. This paper shows the novel way to adapt a variable-flow system provide space heating (SH) and domestic hot water (DHW) to multiple buildings with varying loads. It would be ideal to provide only as much heating and DHW as the zone actually requires at any particular moment. Constant-volume systems come close to this ideal. One needs to know both indoor and outdoor design temperatures, mixed-water temperature, and the required sensible and latent heat loads. Only through analyzing the real network behavior one can gain useful insights related to the consumption patterns of the users. The investigated district heating (DH) system of Zavertyaevo District, Omsk, Russia consists of pipework within a concrete surface trench covered with earth, and two gas boilers with a rated SH supply capacity of 13.3 MW and design efficiency of 82.6%. This system is considered in a case where the SH temperature equals 70 °C and the post-SH return temperature equals 55 °C. This part is related to an idea based on combining lower supply/return temperatures and local heat pump placement, which is in line with the abovementioned LTDH concept. When maximal water flow is reached, if the zone is still too cool, SH is partially provided by a thermostatically controlled electric heater or a heat pump. This means there may be some energy wasted in the constant-volume system due to simultaneous electric heating and drops in the supply temperature. However, this energy waste is far less significant than in variable-flow systems, since the supply temperature is reduced to a minimum before the warm weather starts. These control operations can be summarized in the following chart often referred to as control logic. Advantages of constant-volume systems include low capital costs and low maintenance costs, whereby the former become lower as the system can use the current pipeline dimensions and only requires a simple

S. Chicherin (✉)
Omsk State Transport University (OSTU), 35, Marx avenue, Omsk, Russia 644046

A. Zhuikov · M. Kolosov
Siberian Federal University, Svobodny Ave. 79, Krasnoyarsk, Russia 660041

L. Junussova · M. Aliyarova · A. Yelemanova
Non-Profit JSC Almaty University of Power Engineering and Telecommunications, Almaty 050013, Kazakhstan

A. Beskopylny and M. Shamtsyan (eds.), *XIV International Scientific Conference "INTERAGROMASH 2021"*, Lecture Notes in Networks and Systems 247,
https://doi.org/10.1007/978-3-030-80946-1_29

control at the plant and at the substations to be placed while renovated; the latter costs are lower because the constant volume of water, and thus the boiler capacity and pump power closely follow the actual building-specific heat consumption.

Keywords Network · Pipework · Pipe · Return · Design · Indoor · Outdoor · Flow · Pressure

1 Introduction

One of the fundamental considerations in space heating consists in the variability of heat loads over time as result of external conditions [1]. While conventional district heating (DH) network supply temperature in an approximate range from 80 to 120 °C depending on the outdoor temperature, small-scale prosumers such as heat pumps are more efficient at lower supply temperatures. They are thus easier to install in so-called low-temperature district heating (LTDH) networks, often seen as a part of the concept referred to as the fourth generation DH. This concept typically includes cogeneration, renewable sources, storage technologies, and booster heat pumps [2]. Another option is a central heat pump unit with a high thermal capacity and that is part of a main power plant as shown in [3]. In most cases, all these systems are site-specific [4]. The normal supply temperatures discussed for LTDH range around 45–55 °C. Im et al. [5] defined LTDH as a DH system operating in the range between 50 °C and 70 °C (supply temperature) and 25–40 °C (return temperature) that ensure comfortable indoor temperature while supplying domestic hot water (DHW).

Variable-volume systems are often designed with a pump powered by a variable-speed drive (VSD) alternating current (AC) motor. Wang et al. [6] proposed a new hydraulic control method for a district heating system with distributed variable-speed pumps. Such pumps act as a throttle allowing sufficiently hot water to run to the consumer. Designing or selecting appropriate variable-speed pumps is especially challenging if a fluctuating renewable heat source is in operation [7]. Proposed below is constant-volume system that adjusts heating loads in different zones by setting the supply temperature for each zone individually. Vandermeulen et al. [8] reviewed methods applicable to controlling the supply temperature. Regardless of the variations in the heating load, a minimum flow in a DH network is always provided and care must be taken to ensure that water at plant outlets is of required temperature. In a constant-volume system, as the zone becomes cooler, the heating load increases, and the hot water flow to the zone increases as well.

Annual water volumes decrease as the thermal performance of buildings is improved [9]. Tunzi et al. [10] addressed the specific need for considering lower heat supply temperature in the UK heating sector as a prerequisite for lower heat loads. Cross-cutting solutions to decarbonize the heat market involve both demand- and supply-based measures and require overcoming the issues arising from supply–demand mismatch [11]. In contrast to the rest of the world, China is experiencing drastic increases in floorspace and industrial outputs [12]. To some extent, heat losses are related to the total volume of a building, as more connections usually requires more pipes [13]. Delangle et al. [14] developed a methodology to find the best DH network expansion strategy under a set of given constraints. Kabalina et al. [15] examined the impact of a reduction in heating, cooling and electricity demands on the performance of a polygeneration DH system. Recent works in the literature demonstrate a vivid interest for larger-scale use of cogeneration, renewable sources, and storage technologies [16]. The variability of renewable-source facilities, which is an unfavorable factor for DH systems, can be eliminated using multiple heat sources connected in parallel [17]. This has led to a demand for more renewable energy and more resource-efficient energy systems, thus heralding new solutions, e. g. so called 'prosumers'. A prosumer is a customer who both produces and consumes energy [18]. Own electricity or heat generation leads to decreased operation costs [19].

Aside from prosumers, there are consumers, typically located in the cold regions of China or Russia, where heating is always centralized. For example, a boiler room can produce hot water that is piped around the building or buildings to provide heat. According to [20], the number of fault components increases as design outdoor temperature decreases as well. Each building that needs heating has hot water pipe-supplied to it.

Several authors carried out thermodynamic and hydraulic modelling, e.g. [21] or [22]. However, some of them [23, 24] are mostly devoted to the reliability assessment of the network, considering a service disruption or failure/repair time. According to the patterns in reliability indices, there are some key elements, and proactive diagnosis of such elements helps improve the reliability of a DH network [25].

This paper is to show the novel way to adapt a variable-flow system provide space heating and domestic hot water DHW to multiple buildings with varying loads. Having read the paper, the reader should be able to:

Identify, describe, and chart the operating principle;

Understand the energy efficiency of the new supply-temperature control logic for LTDH and its ability to operate in line with fluctuating heat loads;

Explain why systems that serve many consumers and have a constant water volume, are more energy-efficient than their variable-volume counterparts.

2 Methods

Buildings that are located in continuously cold regions or specific indoor spaces in any climate require heating rather than cooling. For these situations, it would be ideal to provide only as much heating and DHW as the zone actually requires at any particular moment. Constant-volume systems come close to this ideal. To estimate the hydraulic performance of centralized heating in case of varying heat loads, Wang et al. [6] also used a steady-state hydraulic model.

Design of such a system is fairly straightforward. If the heat-emitting radiators in DH-served buildings are linked to the heat network directly, then:

$$G_{c.p.} = \frac{Q_{o.p.}}{c.(\tau_{1.p.} - \tau_{2.p.})} \quad (1)$$

where $Q_{o.p.}$ is design heat rating of a SH system;
$\tau_{1.p.}$ is design network temperature;
$\tau_{3.p.}$ is design SH supply temperature, i. e. mixed-water temperature;
$\tau_{2.p.}$ is design return temperature after SH system.

The mixed-water temperature can be calculated based on the return-water temperature, the outdoor air temperature, and the required proportion of supply/return water (refer to Eq. (2)):

$$u = \frac{\tau_{1.p.} - \tau_{3.p.}}{\tau_{3.p.} - \tau_{2.p.}}. \quad (2)$$

Thermal load factor as defined by Babiarz&Blokus-Roszkowska [1] has the same structure but with a computed indoor temperature, a real outdoor temperature, and a computed external temperature. If the indoor temperature is lower than required, as the outdoor air temperature drops, the mixed-water temperature will rise.

Minimum allowable DHW temperature is set at 50 °C. As stated by Tunzi et al. [10], in low and ultra-low DH networks with supply temperatures below 50 °C, DHW temperature would need to be increased to deliver sanitary water safely.

Referring back to the supply water temperature as shown in Eq. (3), one needs to know both the indoor and the outdoor design temperature, the mixed-water temperature, and the required sensible and latent heat loads:

$$\tau_{1.o.} = t_{6.p.} + \Delta t_{o.p.} \cdot \left(\bar{Q}_o^p\right)^2 + \left(\delta\tau_{o.p.} - \frac{\theta^p}{2}\right) \cdot \bar{Q}_o^p, ^\circ\text{C}. \quad (3)$$

Because there is no independent connection, the mixed-water temperature depends on the return- and supply-water temperatures. The dampers open, and network water is brought into the SH system, upstream of the mixed-water sensor. Conversely, if it is cold outside, as the outdoor air temperature drops, the mixed-air temperature will increase. In this situation, it is common to set the control system to provide the design indoor temperature somewhere between 18 and 22 °C. The control system can simply adjust the position of the dampers to maintain the set mixed temperature.

The current heat load depends on the difference between the baseline temperature and the outdoor temperature for outdoor temperatures lower than the baseline temperature:

$$\bar{Q}_0 = \frac{\tau_{1.o.} - t_{H.}}{t_{6.p.} - t_{H.p.o.} + \frac{\Delta t_{o.p.}}{\bar{Q}_o^{0.2}} + \frac{0.5+u}{1+u} \cdot \frac{\delta\tau_{o.p.}}{\bar{G}_{c.}}} \quad (4)$$

where $\bar{Q}_o = \frac{Q_{o.}}{Q_{o.p.}}$ with actual $Q_{o.}$ and design $Q_{o.p.}$ heat rating of the radiator system;
$\bar{G}_{c.}$ is similarly defined required waterflow;

$t_{6.p.}$ is design indoor temperature;
$t_{H.p.o}$ is design outdoor temperature;
$\Delta t_{o.p.} = \frac{\tau_{3.o.p.}+\tau_{2.o.p.}}{2} - t_{6.p.}$ is design temperature difference;
$\delta\tau_{o.p.} = \tau_{1.o.p.} - \tau_{2.o.p.}$ is the temperature difference between the supply and return lines.

Post-SH return temperature:

$$\tau_{2.o.} = \tau_{1.o.} - \frac{\bar{Q}_o}{\bar{G}_{c.}} \cdot \delta\tau_{o.p.}, {}^\circ\text{C}. \quad (5)$$

A distant gas-fired boiler has the advantage of lower operating and environmental costs, but control can be an issue. Inexpensive SH systems are "on–off" or "high-low-off" rather than fully modulating. As a result, the indoor temperature is adjustable discreetly rather than smoothly. If the radiators run continuously with the heat being turned on and off, then the supply temperature will go up and down within a single heating season, and the residents' experience might be rather unpleasant with the indoor temperature:

$$t_{6.} = t_{H.} + \bar{Q}_o \cdot (t_{6.p.} - t_{H.p.o.}), {}^\circ\text{C} \quad (6)$$

where $t_{H.}$ is an indoor operative temperature;
θ^p is radiators mean temperature difference:

$$\theta^p = \frac{\delta\tau_{o.p.}}{1+u}. \quad (7)$$

Generally, the control system needs to be informed of all the temperatures. The mixed-temperature sensor is obviously to be installed downstream of a substation. It may be located even further (relative to the substation) along the water stream to obtain an average temperature. If circulation is poor, then the average temperature will be incorrect. When a plant starts up, the return water flows through the return damper of a plant and over the return temperature sensor with a temperature:

$$\tau_{2.o.} = t_{6.p.} + \Delta t_{o.p.} \cdot \left(\bar{Q}_o^p\right)^2 - \frac{\theta^p}{2} \cdot \bar{Q}_o^p. \quad (8)$$

Operative mixed-water temperature is:

$$\tau_{3.o.} = t_{6.p.} + \Delta t_{o.p.} \cdot \left(\bar{Q}_o^p\right)^2 + \frac{\theta^p}{2} \cdot \bar{Q}_o^p, \quad (9)$$

Load profiles can now be calculated. In Ref. [24] a scenario with load profile and both centralised and local storages was applied to an actual case study. Only through analyzing the real network behavior can one gain useful insights related to the consumption patterns of the users [13].

3 Case Study

The system under study is a modification of a variable-volume system. This system enables zone control by adjusting the water flow to the value required for a particular condition.

The consumers are located close in the area and are controlled by an indoor thermostat. The bulk of this DH service consists of two products: SH and DHW, both of which are distributed by dual-pipe networks from a heat-only boiler plant to an individual substation. DHW is heated at a heat source, i. e. this is an open system. The existing DH system of Zavertyaevo District, Omsk, Russia consists of pipework within a concrete surface trench covered with earth, and two gas boilers with a rated SH supply capacity of 13.3 MW and design efficiency of 82.6%. Heat-only boiler plant can achieve a maximum heat supply capacity of 20.9 MW, including 19.9 MW of SH, ventilation, and DHW and circa 1 MW of heat losses. A typical for Russia deep-bury tunnel [26] implies low initial costs and convenient construction and maintenance, whereas its disadvantages are related to poor failure detection accuracy in case of severe fault, as well as due to the impossibility to locate insignificant faults. Like in Andric's case [4], this particular district was selected based on a conclusion that it was representative of Russian residential construction in general.

The DH network is currently operated in a variable-volume mode, hence, as noted above, there is a problem with leakage. However, leakage detection and special detection tools represent a widely-studied phenomenon on an international scale [27]. There are other issues with variable-volume systems. These include: poor water circulation in the heated space at lower flow rates; high distribution heat losses at low flow rates; and inadequate DHW supply to the area. Some issues may be addressed by closing some parallel (looping) pipelines in order to avoid the emergency of 'low-flow' sections as it is currently made in Kaunas, Lithuania [23]. Improved and/or VSD-equipped pumps have also enabled designers to avoid dumping and poor room circulation. However, the problem of inadequate temperature of return water needs additional care when renovating the system.

4 Results

We consider this system in a case where the SH temperature equals 70 °C and the post-SH return temperature equals 55 °C. This part is related to an idea based on combining lower supply/return temperatures and local heat pump placement, which is in line with the abovementioned LTDH concept [28]. When the outdoor temperature is −37 °C, 100% return water will have the required 70C. When the outdoor air temperature is above −37 °C, the required secondary temperature of 70 °C can be achieved by mixing supply and return water. As the outdoor temperature drops, the percentage required to maintain 70 °C will decrease.

Fig. 1 Supply and return temperature control logic

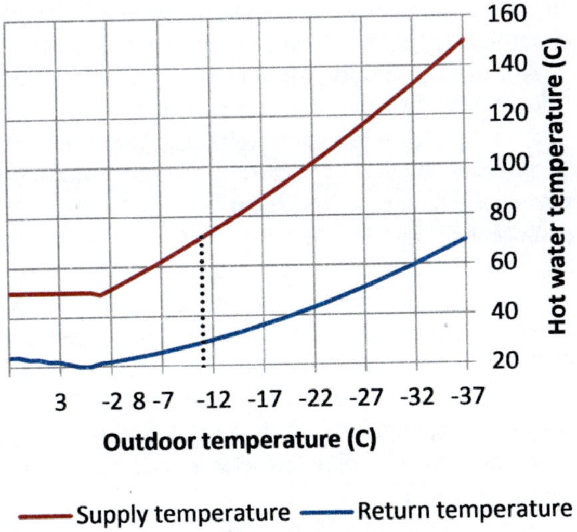

Eventually it reaches the minimum value necessary for the adequate circulation and DHW supply, see Fig. 1.

The requirement for a minimum flow in a DH network means that the controller must have a way of measuring the water volume. This can be achieved in a number of ways. We have earlier discussed water volumes without stating any specific numbers. Note that the weight (in case of an open system) of cold water entering a boiler plant must equal the weight of DHW that leaves a building.

For example, when the demand peaks:

30.7 tons of cold water (5 °C) enters a boiler plant per hour.

It is heated, piped, and leaves the buildings as 30.7 tonnes per hour at 50 °C (10 times greater temperature, same weight). However, the weight of water that is entering and leaving will usually be different, since the volume increases with greater demand for DHW.

When maximal water flow is reached, if the zone is still too cool, SH is partially provided by a thermostatically controlled electric heater or a heat pump. This means there may be some energy wasted in the constant-volume system due to simultaneous electric heating and drops in the supply temperature. However, this energy waste is far less significant than in variable-flow systems, since the supply temperature is reduced to a minimum before the warm weather starts.

To the author's best knowledge, there is no similar equation-based analysis currently available. In Ref. [9] results are obtained using different return temperatures at design outdoor temperatures (DOT) ranging between 25 and 40 °C. Sartor et al. [3] suggest decreasing supply and return temperatures at which a DH network operates and using high-temperature heat pumps connected locally at a substation to generate steam on demand. Shan et al. [20] illustrate structural and functional reliability indices of repairable DH networks depending on an outdoor temperature, rather than supply and return temperatures as such.

Any outdoor temperature above −11 °C will cause the mixed temperature to drop below 50 °C. In this situation, the mixed water will be cooler than 50 °C and will have to be heated to maintain 50 °C. Again, an electric heater will then turn on to provide enough heat to raise the mixed-water temperature (as measured by the mixed-temperature sensor) to 50 °C.

In Malmö, Sweden the DH heats the cold water from 10 to 30 °C and electricity heats the water from 30 to 60 °C [18]. In Ningbo Hi-Tech District, China, terminal equipment such as air conditioners and water heaters are used to transform energy from the supply side so that it can be used directly, e.g. for cooling, heating, and DHW supply [12]. Another option is to introduce an additional heat source such as a heat pump to meet peak heating requirements [17]. In this case, when the network supply temperature is 45 °C, the levelized cost of energy is on average 13–16% lower than with a 30 °C supply temperature and 25–27% lower than with a 15 °C supply temperature [2].

Now consider what happens when the outdoor air temperature rises above 0.3 °C. Starting from 8 °C, the temperature of the supply water will be as much as 50 °C, so it would seem best to use 100% supply water until the outdoor temperature reaches 0.3 °C. Previous results [19] showed that DH network heat losses may be reduced by 35% when the supply temperature is reduced from 70 to 50 °C. In practice, using 100% supply water until the outdoor temperature reaches 0.3 °C is not feasible, as the DHW requirements will influence the decision. In modern systems, the buckling point will be set much lower than 0.3 °C due to demand side measures (DSM). DSM is a term summarizing efforts taken to manipulate the consumption profiles through thermal insulation, load management, and the adaption of the load profiles [11]. Wang et al. [7] obtained the optimal rated capacities of all pumps as well as variable-speed hydraulic characteristic curves in desired operating modes from 100 to 40% of full heat load.

Now we shall discuss how the system is controlled, a step we refer to as the control operation. These control operations can be summarized in the following chart, often referred to as the control logic:

- When the outdoor temperature is -37 °C, 100% return water will have the required 70 °C.
- When system is in operation, if the outdoor temperature is below 0.3 °C, adjust dampers to provide SH at 70 °C and a post-SH return temperature of 55 °C.
- When the system starts, if the outdoor temperature is above 0.3 °C, modulate dampers to maintain a 50 °C mixed temperature with a minimum use of return water.

In Malmö, Sweden the baseline temperature of the control logic was set at 16 °C, since this was the specified temperature at which prosumer office buildings would begin to cool down [18]. Above the pre-determined buckling temperature (e. g. 0.3 °C), a substation may revert to the minimum heating, e.g. 20% of the design capacity. Combining with the relationships obtained in Eqs. (3), (8), and (9), an advanced control logic could be obtained to minimize the life cycle costs.

The supply temperature in a constant-volume system will vary as the substation dampers adjust the temperature for each zone. Therefore, a boiler should be able to run with a varying return temperature. Temperature must be controlled without allowing the return pressure to rise excessively or to drop below the pressure required by the SH system for its safe and proper operation. Such control is often implemented by using a temperature sensor at a plant outlet to adjust a fan-speed control unit. Similarly, the make-up pump is controlled to maintain the slightly varying DH network volume. Paper [8] gives a well-clarified presentation of the evolution of supply temperature control systems.

At a full load, all the boilers at a plant are running, and the valves in the boiler circuits are fully open. As the load decreases, the temperature sensors in front of each circuit start to close their valves, restricting the flow through the boiler. The flow sensor in the hot-water pipe coming from the boiler senses the flow reduction and restores flow by opening the bypass valve to maintain flow in the supply line. Heat consumption alters as outdoor temperatures rise or drop; it also depends on whether it is a weekday or a weekend. To estimate the actual load profile for different outside temperatures, it is imperative to have temperature sensors and mass flow meters in place [16].

When the load drops below 50%, one of the boilers can switch off, leaving one pump and one boiler to serve the load. The check valve in front of the pump that is off closes to prevent the hot water from flowing back through it. As the heating load in the district raises, the return-water temperature will fall, and the boiler will turn back on to avoid over-cooling the DHW. The output of each boiler is adjusted to maintain the set supply water temperature. Heating load has a direct impact on the performance of polygeneration DH systems, and a lesser impact on systems powered by heat-only boiler plants [15].

5 Conclusions

Advantages of the constant volume system are the low capital costs and low maintenance costs, whereby the former become lower as the system can use the current pipeline dimensions and only requires a simple control at the plant and at the substations to be placed while renovated; the latter costs are lower because the constant volume of water, and thus the boiler capacity and pump power closely follow the actual building-specific heat consumption. Simultaneous electric heating and supply temperature drops are not associated with a considerable loss in efficiency.

Then, has been shown above, in a constant-volume system where all the zones require 70 °C in a SH system, while the outdoor temperature is −37 °C, maintaining 70 °C in the return line ensures that each zone receives 100% of design load. In a constant-volume system, one cannot maintain the supply water temperature continuously. It means that, as the zone temperatures are reduced due to high thermal consumption, the supply water temperature needs to be increased. Finally, nearly constant water flow is maintained under any weather conditions. This can be achieved in a few ways, but the process does not have to be sophisticated, and less expensive control systems can be used than those needed in variable-volume systems.

Acknowledgements The reported study was funded by the Administration of Krasnoyarsk Krai and the Krasnoyarsk Krai Foundation for Support of Scientific and Technical Activity. This research was also supported by the Government of the Russian Federation under Project No. 860 (August, 8 2017 Decree). The Authors thank the Omsk District Heating Supply Company ('OmskRTS', JSC) for a cooperation and kindly provided information. The authors would like to acknowledge the valuable comments and suggestions of the reviewers, which have improved the quality of this paper as well. The author also expresses gratitude to the unnamed editor who has proof-read and edited the text.

References

1. Babiarz B, Blokus-Roszkowska A (2015) Probabilistic model of district heating operation process in changeable external conditions. Energy Build 103:159–165
2. Vivian J et al (2018) Evaluating the cost of heat for end users in ultra low temperature district heating networks with booster heat pumps. Energy. https://doi.org/10.1016/j.energy.2018.04.081
3. Sartor K, Lemort V, Dewallef P (2017) Improved district heating network operation by the integration of high-temperature heat pumps. Int J Sustain Energy 37(9):842–856. https://doi.org/10.1080/14786451.2017.1383409
4. Andrić I, Fournier J, Lacarrière B, Le Corre O, Ferrão P (2018) The impact of global warming and building renovation measures on district heating system techno-economic parameters. Energy. https://doi.org/10.1016/j.energy.2018.03.027
5. Im Y-H, Liu J (2018) Feasibility study on the low temperature district heating and cooling system with bi-lateral heat trades model. Energy 153:988–999
6. Wang N et al (2018) Hydraulic resistance identification and optimal pressure control of district heating network. Energy Build. https://doi.org/10.1016/J.ENBUILD.2018.04.003
7. Wang H, Wang H, Zhou H, Zhu T (2018) Modeling and optimization for hydraulic performance design in multi-source district heating with fluctuating renewables. Energy Convers Manag 156:113–129
8. Vandermeulen A, van der Heijde B, Helsen L (2018) Controlling district heating and cooling networks to unlock flexibility: a review. Energy. https://doi.org/10.1016/j.energy.2018.03.034
9. Averfalk H, Werner S (2018) Novel low temperature heat distribution technology. Energy 145:526–539
10. Tunzi M, Boukhanouf R, Li H, Svendsen S, Ianakiev A (2018) Improving thermal performance of an existing UK district heat network: A case for temperature optimization. Energy Build 158:1576–1585
11. Coss S, Verda V, Le-Corre O (2018) Multi-objective optimization of district heating network model and assessment of demand side measures using the load deviation index. J Clean Prod. https://doi.org/10.1016/j.jclepro.2018.02.083
12. Hou J et al (2018) Implementation of expansion planning in existing district energy system: a case study in China. Appl Energy 211:269–281
13. Noussan M, Jarre M, Poggio A (2017) Real operation data analysis on district heating load patterns. Energy 129:70–78

14. Delangle A, Lambert RSC, Shah N, Acha S, Markides CN (2017) Modelling and optimising the marginal expansion of an existing district heating network. Energy 140:209–223
15. Kabalina N, Costa M, Yang W, Martin A (2018) Impact of a reduction in heating, cooling and electricity loads on the performance of a polygeneration district heating and cooling system based on waste gasification. Energy 151:594–604
16. Guelpa E, Barbero G, Sciacovelli A, Verda V (2017) Peak-shaving in district heating systems through optimal management of the thermal request of buildings. Energy 137:706–714
17. Sayegh MA et al (2018) Heat pump placement, connection and operational modes in European district heating. Energy Build 166:122–144
18. Brange L, Englund J, Lauenburg P (2016) Prosumers in district heating networks—a Swedish case study. Appl Energy 164:492–500
19. Cai H, You S, Wang J, Bindner HW, Klyapovskiy S (2018) Technical assessment of electric heat boosters in low-temperature district heating based on combined heat and power analysis. Energy 150:938–949
20. Shan X, Wang P, Lu W (2017) The reliability and availability evaluation of repairable district heating networks under changeable external conditions. Appl Energy 203:686–695
21. Schweiger G, Larsson P-O, Magnusson F, Lauenburg P, Velut S (2017) District heating and cooling systems—framework for Modelica-based simulation and dynamic optimization. Energy 137:566–578
22. Zarin Pass R, Wetter M, Piette MAA (2018) thermodynamic analysis of a novel bidirectional district heating and cooling network. Energy 144:20–30
23. Valinčius M, Vaišnoras M, Kaliatka A (2018) Study and demonstration of pressure wave-based leak detection in a district heating network. Struct Infrastruct Eng 14:151–162
24. Badami M, Fonti A, Carpignano A, Grosso D (2018) Design of district heating networks through an integrated thermo-fluid dynamics and reliability modelling approach. Energy 144:826–838
25. Chicherin SV (2016) New approach to determination of corrosion damage degree of pipeline system elements. Bull Tomsk Polytech Univ Geo Assets Eng 327(12):110–115
26. Chicherin SV (2017) Unlocking a potential of district heating network efficient operation and maintenance by minimizing the depth of a trench system. Bull Tomsk Polytech Univ Geo Assets Eng 328(9):49–56
27. Zhou S, O'Neill Z, O'Neill C (2018) A review of leakage detection methods for district heating networks. Appl Therm Eng 137:567–574
28. Sernhed K, Lygnerud K, Werner S (2018) Synthesis of recent Swedish district heating research. Energy. https://doi.org/10.1016/j.energy.2018.03.028

Effects of Technogenic Raw Materials on the Strength of Concrete

Evgeniy Gaishun, Khungianos Yavruyan, Alexander Khalyushev, Anastasiya Filippova, and Vyacheslav Bobin

Abstract The introduction of dispersed mineral additives has a beneficial effect on many properties of concrete. This is either due to the physical effect, which means that small particles usually have a finer granulometric composition than Portland cement and present themselves as a "microfiller". On the other hand, this may be due to their pozzolanic activity, which means that silica and alumina interact with calcium oxide hydrate and form hydrosilicates and calcium hydroaluminates. In this work, the activity of dispersed mineral components was evaluated by the theoretical value of the coefficient of hydration activity, which was determined by their chemical composition. The actual coefficient of hydration activity was calculated from the maximum value of the mass gain of dispersed industrial by-products by the absorption of Ca $(OH)_2$ by these additives from a saturated solution of lime. The X-ray phase analysis was used in order to determine quantitative and material composition of crystalline compounds in industrial by-products. Experimental data showed that the actual pozzolanic activity turned out to be much lower than the theoretical value. Therefore, the by-products of the industry were arranged in the following sequence: ash-and-slag mixture → by-product of coal enrichment → granular blast furnace slag, which is confirmed by the results of X-ray phase analysis and is consistent with the values of the limit compressive strength of concrete tested in different periods of hardening.

Keywords X-ray phase analysis · Ash and slag · Granulated blast-furnace slag · Waste of coal preparation · Pozzolanic activity · Compressive strength of concrete

1 Introduction

The analysis of literary sources indicates that an effective direction for the use of mineral by-products of the industry is the use of ash and slag from thermal power plants [1]. The use of TPP ashes in building materials is an important ecological as well as scientific and technical problem that requires its own development, including

E. Gaishun (✉) · K. Yavruyan · A. Khalyushev · A. Filippova · V. Bobin
Don State Technical University, Ploshad' Gagarina 1,, Rostov-on-Don, Russia 344002

A. Beskopylny and M. Shamtsyan (eds.), *XIV International Scientific Conference "INTERAGROMASH 2021"*, Lecture Notes in Networks and Systems 247,
https://doi.org/10.1007/978-3-030-80946-1_30

the experimental and industrial development of new technologies for the use of TPP ashes and their implementation at construction industry plants [2, 3].

If such an approach to concrete technology had a certain economic justification in the 70s of the last century, then the concept of concrete science at the present stage of development of society dictates completely different needs, due, first of all, to a significant increase in the cost and shortage of energy and raw materials. According to Malinina and Batrakov [4], the general direction of new generation concretes should be the solution of the problems of further improving concrete quality indicators, as well as energy and resource conservation in its production.

The modern concept of development in concrete technology includes the creation of multicomponent cements thankfully to the increased content of mineral additives of various genesis in them, which is associated with a decrease in energy consumption and the possibility of recycling large-tonnage waste of various industries [4–7].

The structure and properties of cement concretes are formed as a result of complex chemical and physicochemical transformations in the hardening system, occurring in parallel from the moment the components of the concrete mixture are mixed with water [8]. The process of cement hardening itself is possible due to the chemical interaction between the solid phase and the mixing liquid [9].

Ash and slag waste from thermal power plants of Donbass has been well studied by specialists from the Research Institute “Donetsk PromstroyNIIproekt” [1, 10]. In the dissertation of Popov [1] it is shown that the ash removal mainly consists of spherical formations ranging in size from 2 to 110 microns, particles with a diameter of 12–80 microns predominate and consist of aluminosilicate glass of variable composition.

Ashes from electrostatic precipitators have slightly smaller spheres: the maximum diameter is 75 microns, the size of 2–20 microns prevails. Slag grains are 80–90% pure glass, mostly dense with poorly developed porosity.

According to [11], ground granulated blast-furnace slag is represented by particles in the range from 10 to 45 microns with a significant content of particles less than 10 microns in diameter. Just like fly ash, which is predominantly vitrified, although the shape of the grains differs significantly from the ash spheroids.

The addition of mineral components has a beneficial effect on many properties of concrete. This could be due to the physical effect, because small particles usually have a finer granulometric composition than Portland cement, or their pozzolanic activity during cement hardening [12], which is shown in the ability of silica and alumina of the mineral additive to interact with oxide hydrate according to the following reaction equations [13–15]:

$$2SiO_2 + 3Ca(OH)_2 \rightarrow 3CaO \cdot 2SiO_2 \cdot H_2O \tag{1}$$

$$Al_2O_3 + 3Ca(OH)_2 + 3H_2O \rightarrow 3CaO \cdot Al_2O_3 \cdot H_2O \tag{2}$$

Herewith, according to the intensity and depth of such interaction, the result of which is the formation of cementing phases, pozzolanic additives are characterized by such a parameter as the index of pozzolanic activity. According to the institute’s terminology, the ACI index of pozzolanic activity is determined by the strength of binder systems containing hydraulic cement with and without pozzolana [16].

At the same time, there are other indirect methods for studying the pozzolanic activity of mineral additives, in particular X-ray phase analysis, electron microscopy, which allows it to quantitatively determine the content of amorphous and crystalline phases in a material, which makes it possible to chemically interact with calcium hydroxide [17, 18].

There are two main fundamentally different methods of adding mineral components into concrete: a mineral part is added into concrete by volume instead of cement or sand. Pozoolanic activity appears in any way of adding the additive, and the effect of "microfiller"—only with an increase in the content of small particles in the mixture [19].

The goal of this work is to study the influence of industrial by-products on their pozzolanic activity as well as physical and mechanical properties of concrete.

2 Materials and Methods

The following mineral additives were used: unburned coal by-product, ash and slag waste from TPPs and granulated blast-furnace milled slag GGBSS 4500 (TU 0799–001-99,126,491–2013). The chemical composition of mineral additives is shown in Table 1.

The activity of mineral additives was assessed in accordance with the method described in [13]. According to the obtained experimental data, the coefficient of hydration activity was calculated by the formula [19]:

$$\Gamma_{\infty\partial} = \ln(\infty_{\mathrm{CaO}})^{1/3} \tag{3}$$

Based on the type of mineral additive (ash and slag, slag), the corresponding formula (4) or (5) was used to calculate the theoretical coefficient of hydration activity:

$$K_{\propto} = \frac{\mathrm{RO} + \mathrm{Al_2O_3}}{\mathrm{SiO_2} + \mathrm{Fe}_{gen}} \tag{4}$$

$$K_{\propto} = \frac{\mathrm{SiO_2} + \mathrm{Al_2O_3}}{\mathrm{RO} + \mathrm{R_2O}} \tag{5}$$

3 Results

The results of calculations of the chemical composition are presented in Table 2.

Table 1 Chemical composition of mineral additives

Additive name	Oxide content, %										
	SiO_2	Al_2O_3	Fe_{gen}	TiO_2	CaO	MgO	K_2O	Na_2O	SO_3	MnO	P.P.P
Unburned coal by-product	50.1	20.3	5.44	0.94	0.30	1.5	4.01	1.01	0.8	0.05	15.3
Ash and slag waste from TPP	55.7	22.4	15.0	0.75	2.1	1.6	2.26	0.78	0.01	-	0.02
Granulated blast-furnace milled slag	34.9	13.4	0.51	2.25	34.2	12.3	0.45	0.56	0.35	0.54	2.7

Table 2 Values of the coefficient of hydration activity of industrial by-products

№	Name of the material	The value of the coefficient of hydration activity		
		calculated		factual
		(f.2)	(f.3)	
1	Ash and slag mixture	0.369	1.58	1.12
2	Coal by-product	0.398	1.24	1.05
3	Granulated blast-furnace milled slag GGBSS 4500	1.7	–	0.92

According to the maximum values of the mass gain of dispersed industrial by-products, the coefficient of hydration activity was calculated and compared with theoretical values (Fig. 1).

All actual values of dispersed industrial by-products showed lower values compared to theoretical ones. This is probably due to the fact that the ambient conditions during the experiment differed from those recommended in [13].

X-ray phase analysis of industrial by-products showed the following results. The X-ray diffraction pattern of the TPP ash-and-slag mixture shown in Fig. 2a is characterized by the presence of a significant number of peaks corresponding to various compounds in the crystallized state: quartz (d = 0.228; 0.334; 0.426 nm), hematite (d = 0.160; 0.251; 0.268; 0.369 nm), mullite (d = 0.206; 0.207; 0.232 nm), feldspar (d = 0.294; 0.303 nm).

The X-ray diffraction pattern of a by-product of coal enrichment (Fig. 2b) contains a large number of compounds in a crystallized state: hydromica (d = 9.99 nm), kaolinite (d = 7.07 nm), aluminum oxides (d = 3.34; 3.19; 2.56; 2.45; 2.12; 1.81 nm).

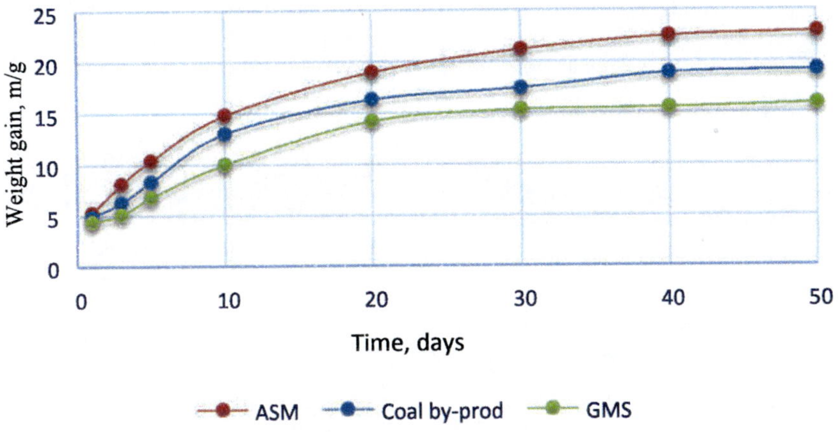

Fig. 1 Kinetics of absorption by industrial by-products of Ca $(OH)_2$ from a saturated lime solution: 1 with the addition of TPP ash and slag; 2 with the addition of coal preparation; 3 with the addition of GGBSS 4500 granulated blast furnace slag

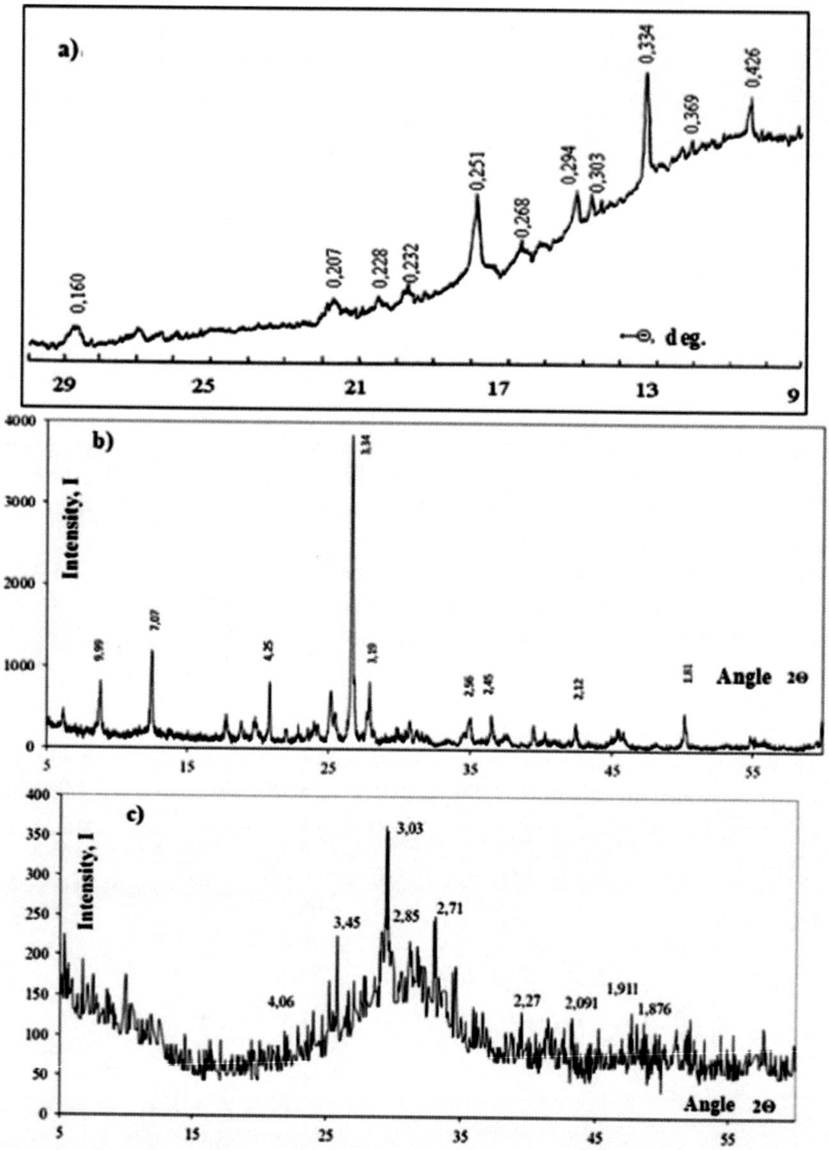

Fig. 2 X-ray diffraction patterns of samples: **a** ash-and-slag mixture of TPP; **b** unburned coal waste; **c** granulated blast-furnace milled slag GGBSS 4500

Crystalline compounds in the form of minerals wolastonite (d = 0.406 nm), melilite (d = 0.285 nm), merwinite (d = 0.187 nm) and other compounds were revealed in the composition of granulated blast-furnace slag (Fig. 2c). At the same time, it is known that the content of crystalline compounds of the glass phase accelerates the grinding of the slag, but slows down the processes of structure formation, especially in the early stages of hardening.

It is known from concrete technology that the compressive strength of concrete is significantly influenced by high-quality fillers. Therefore, fine-grained sand increases the consumption of cement in concrete because it has an increased water absorbtion. The use of coarse sand in concrete leads to an overconsumption of cement due to the high intergranular poreness. That is why the optimal ratio between different sand and rubble was pre-established for the preparation of concrete with a dense structure, which was determined by the maximum average density.

To study the kinetics of the growth of the ultimate strength of concrete during compression at the age of 3, 7, 14, and 28 days of normal hardening, a concrete mixture of the BSTV35P5 grade was prepared. The nominal composition of concrete by the consumption of components was as follows, kg/m^3: cement CEM I (42.5) produced by OAO "Novoroscement"—410; natural sand "Maiskaya Zarya" (Mk = 2.1)—200; natural sand "Latnenskoe" deposit (Mk = 1.7)—560; rubble of fraction 5–10 mm produced by AO "Pavlovsk Nerud"—390; rubble of fraction 10–20 mm produced by AO "Pavlovsk Nerud"—615; filler GPM powder, produced by ZAO "NP TsMID"—40 kg; water 150 leters. The superplasticizer GPMzh Ultra was added into the concrete mixture in the amount of 4.8 kg of the cement mass. The mobility of the concrete mixture was determined according to the standard procedure given in GOST. After the slump test, standard cubes sized 10 × 10 cm were molded, which were kept in a normal hardening chamber for 3, 7, 14, and 28 days. The developed concrete compositions are presented in Table 3.

Table 3 Compositions of concretes based on industrial by-waste

№	Material consumption per kg/m^3							
	Aggregates				Cement CEM I	Filler	Chem. additive	By-prod
	Sand		Rubble					
	$M_K = 1.7$	$M_K = 2.1$	Fr. 5–10	Fr. 10–20				
K1	560	200	390	615	410	40	19.7	-
C2 ASM	560	200	390	615	348	40	19.7	61.5
C3 Coal by-prod								
C4 GMS								

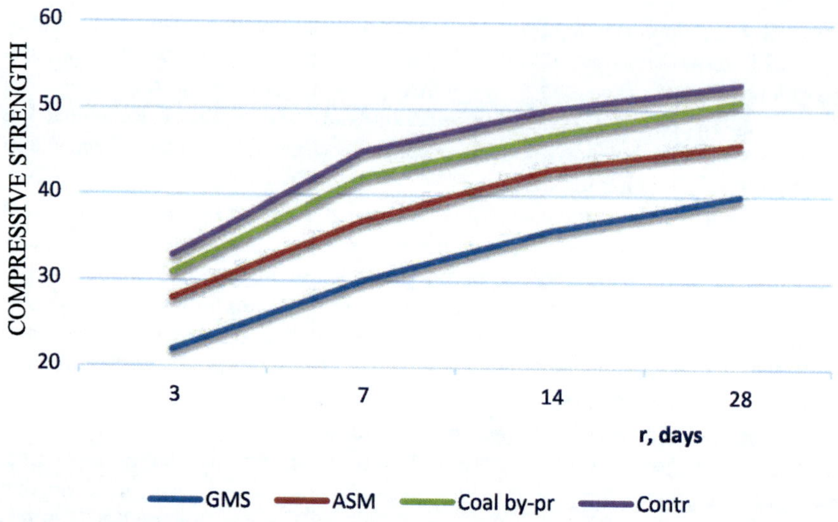

Fig. 3 Change in the ultimate strength of concrete in compression: 1 with the addition of granulated blast furnace slag; 2 with the addition of TPP ash and slag; 3 with added coal by-product; 4 control without additive

4 Discussion

The increase in the strength of samples with the addition of TPP ash and slag at the age of three days is 7% lower than the control composition; by the age of 28 days, this increase has decreased and amounted to 5% (Fig. 3).

For similar compositions with the addition of coal by-products, the decrease in strength is also insignificant and amounts to 15% each at the age of three and 28 days. Most likely, it is due to the higher pozzolanic activity and dispersion of ash and slag particles from TPPs in comparison with waste from coal by-products. The smallest increase in strength was shown by blast-furnace granular slag in comparison with the control composition.

5 Conclusions

Based on the results of the studies of assessing the quality of technogenic raw materials from industry by-products in heavy concrete, the following conclusion can be made: from the analysis of the chemical composition, the greatest theoretical hydration activity has the granulated blast furnace slag ground GGBSS 4500, then the ash-and-slag mixture and the by-product of coal enrichment. At the same time, due to the large amount of substances in the crystallized state, the actual pozzolanic activity of granulated blast furnace slag turned out to be much lower than the theoretical value and therefore can be put in the following order: ash and slag mixture → by-product of coal enrichment → blast-furnace granular slag, which is confirmed by the results of X-ray phase analysis and is consistent with the values of the ultimate strength of concrete in compression tested in different periods of hardening.

References

1. Popov SV (2003) Fine-grained dense concrete with special properties based on aggregates made of coal and anthracite ash and slag materials from TPP of Donbass: dis ... cand tech Sciences: 05.23.05/Popov Sergey Vladimirovich. Makeevka
2. Belyakova ZhS, Velichko EG, Komar AG (2001) Environmental, material-science and technological aspects of the use of TPP ash in concrete. Building Mater 3:46–48
3. Reznichenko PT, Chekhov AP (1979) Environmental protection and use of industrial waste. Directory. Dnepropetrovsk: Promin
4. Malinina LA, Batrakov VG (2003) Concrete science: present and future. Concr Reinforced Concr 1:2–6
5. Ferraris CF, Obla KH, Hill R (2001) The influence of mineral admixtures on the rheology of cement paste and concrete. Cem Concr Res 31(2):245–255
6. Moranville-Regourd M (1999) Portland cement—based binders—cements for the next millennium. In: Proceedings of the international conference creating with concrete. Dundee
7. Naik TR, Kraus RN (1999) Use of industrial by-products in cement-based materials. In: Proceedings of the international conference creating with concrete. Dundee
8. Mchedlov-Petrosyan OP (1977) Chemistry of inorganic building materials. M.: Stroyizdat
9. Volzhensky AV (1986) Mineral binders: textbook. for universities. 4th edn. Rev. and add. M.: Stroyizdat
10. Braginsky VG (1977) Economical types of concrete with the use of fuel waste from the state district power station. Modern problems of construction: annual. scientific and technical Sat. Donetsk, Donetsk PromstroyNIIproekt
11. Popov SV (2007) Rational use of industrial mineral waste in construction. Modern problems of construction: annual. Sci Tech Sat 5(10):123–127
12. Ramachandran VS, Feldman RF, Collepardi M et al (1988) Additives to concrete: [ref. manual] M.: Stroyizdat
13. Li G (2004) Properties of high-volume fly ash concrete incorporating nano-SiO_2. Cem Concr Res 34(6):1043–1049
14. Daderko G Specifying blended cements for performance and strength. The construction specifier. http//:www.lafarge-na.com
15. Sabir BB, Wild S, Bai J (2001) Metakaolin and calcined clays as pozzolans for concrete: a review. Cem Concr Comp 23(6):441–454
16. ACI Concrete Terminology, http: //www.concrete.org/Technical/CCT/FlashHelp/ACI_Concrete_Terminology.pdf
17. Testing methods for pozzolanas (Practical Action Brief), http://www.appropedia.org/Testing_methods_for_pozzolanas
18. Shvarzman A, Kovler K, Schamban I et al (2001) Influence of chemical and phase composition of mineral admixtures on their pozzolanic activity. Adv Cem Res 13(1):1–7
19. Shvitaridze RE (2005) Chemistry of cement and methods of accelerated forecasting of compressive strength (activity) of cements with mineral additives and concretes. Concr Reinforced Concr 3:6–11

The Impact of Nuclear Power Plants on the Cooling Pond

Irma Martyn, Yaroslav Petrov, Sergey Stepanov, and Artem Sidorenko

Abstract The impact of nuclear power plants on the cooling reservoir is considered. In the course of the work, the main impact was identified. Potential impacts are considered in connection with the identified thermal pollution of the reservoir, and an assessment of the ecological state and its development with further pollution is given. The importance of monitoring the cooling reservoir during the operation of nuclear power plants was noted. The results obtained can be used to improve the environmental monitoring of nuclear power facilities.

Keywords Ecology · Environmental monitoring · Nuclear power

1 Introduction

Nuclear power plants are large industrial facilities with a high degree of environmental stress. One of the important aspects is the need to use a large amount of water to cool the turbine condensers. The greatest stress is placed on hydrological characteristics as a consequence of the ecosystem of cooling reservoirs. In addition to thermal pollution, cooling reservoirs are subject to radioactive and chemical contamination [1]. The main factors of potential technogenic impact on water bioresources are the intake of natural water together with organisms and the discharge of technical waters involved in the technogenic cycle of a nuclear power plant. Humanity needs more and more energy, and this will increase the impact on the environment. Studying the nature of this impact on the environment is an urgent task of our time [2].

Currently, the issues of distribution of thermal emissions and discharges and their impact on the environment are not sufficiently developed. The influence of thermal emissions on the thermal regime and hydrological regime of rivers and reservoirs has been studied more successfully. The impact of cooling systems on the environment is currently a subject of attention due to the fact that the thermal emissions of nuclear power plants are becoming commensurate with the energy of some atmospheric

I. Martyn · Y. Petrov (✉) · S. Stepanov · A. Sidorenko
Department of Applied Computer Science, Russian Federation, Russian State Hydrometeorological University, 79 Voronezhskaya Street, St. Petersburg, Russia 192007

A. Beskopylny and M. Shamtsyan (eds.), *XIV International Scientific Conference "INTERAGROMASH 2021"*, Lecture Notes in Networks and Systems 247,
https://doi.org/10.1007/978-3-030-80946-1_31

processes. This effect was considered by some researchers, but mainly concerned natural heat sources with low power [3, 4].

In conclusion, conclusions are drawn about the impact of a nuclear power plant on the hydrological state of the cooling reservoir under study and the possible impact on other systems of the reservoir.

2 Materials and Methods

This work is based on the analysis of full-scale hydrological data describing the hydrological state, as well as affecting the hydrobiological parameters of the cooling reservoir in the area of the Leningrad nuclear power plant in the Koporskaya Bay of the Gulf of Finland. Meteorological data were taken using a portable weather station, and five cross-section measurements were made using GPS to determine the location when entering the study area. The depth of the place was determined by an echo sounder, after which the speed and direction of the current were measured using the Vector hydrological device, and the temperature was determined by the SBE-19 probe. The obtained results of the instruments were presented in the form of a spatial distribution in depth over the studied section performed in the Surfer program.

3 Results

On the day of research, July 13, 2016, a table of meteorological data was compiled (Table 1) and based on the data obtained, the wind rose is plotted for this day (Fig. 1).

Sections are constructed based on the temperature obtained from the SBE-19 probe, along the section from the coast (station 1) to the open part of the Koporskaya Bay (station 5) (Fig. 2). The depth profile of the flow velocity obtained with the help of the "Vector" device is constructed from the same section (Fig. 3).

4 Discussions

Water temperature is the most important factor affecting the flow of physical, chemical and bioogical processes in the reservoir, which determines the oxygen regime and self-purification processes. The change in the temperature background of the Gulf of Finland serves as a daily and annual change for the temperature of the atmosphere of the surrounding area, which is especially noticeable for the summer and winter months. According to the study, the influence of an increase in the temperature background of the water of a narrow strip of the coast, which quickly weakens when moving away from the coast, is noted (Fig. 2). In the area of the leningrd Nuclear power plant, water is discharged at high speeds, which leads to intensive mixing of

Table 1 Weather data for the city of 13.07.2016 in the area of the Leningrad nuclear power plant (Koporskaya Bay)

Time	11:35	16:35	9:45	16:55	10:46	17:30
Latitude	59°53′37″	59°53′37′	59°53′37″	59°53′37″	59°53′37″	59°53′37″
Longitude	29°3′2″	29°3′2″	29°3′2″	29°3′2″	29°3′2″	29°3′2″
Air temperature					18.9	19.8
Dry countdown	19	19.8	18	19	17.4	17.9
Wetted countdown	16.1	17.1	16	17	18.9	19.3
Humidity						
Elasticity e, mb	16.2	17.6	16.6	16.7	18.9	19.3
Rel. humidity f, %	72	75	81	83	85	82
Dew point, ° C	13.9	15.2	14.6	14.6	16.4	16.7
Atmosphere pressure						
Countdown	100.85 kPa	100.9 kPa	101.3 kPa	101.6 kPa	101.1 kPa	106 kPa
Wind						
Direction	NW	NW	NE	NW	NE	SE
Start					1075	2505
End countdown					1424	3145
Difference in 100 s					349	637
Speed, m/s	4.9	3.2	4.1	6.2	3.49	6.37
Cloudiness						
Total amount	9	8	8	3	10	8
Number of lower tier	9	8	6	4	10	8
Top tier shapes				Ci		
Lower tier shapes	Sc diur	Cu conf	Ac		Sc trans	Ci int
Middle tier shapes			Sc,Cu	Cs		
Bottom height			0.6–1.5	0.8–1.5	6–7	9
Weather on time	Mainly cloudy	light rain	Mostly cloudy	clear	Mainly cloudy	Mainly cloudy

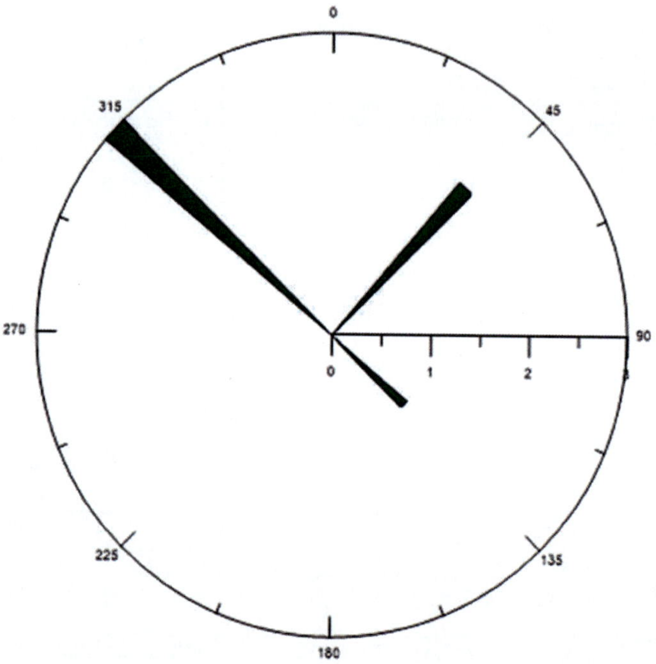

Fig. 1 Wind rose on 13.07.2016 in the area of the Leningrad nuclear power plant (Koporskaya Bay)

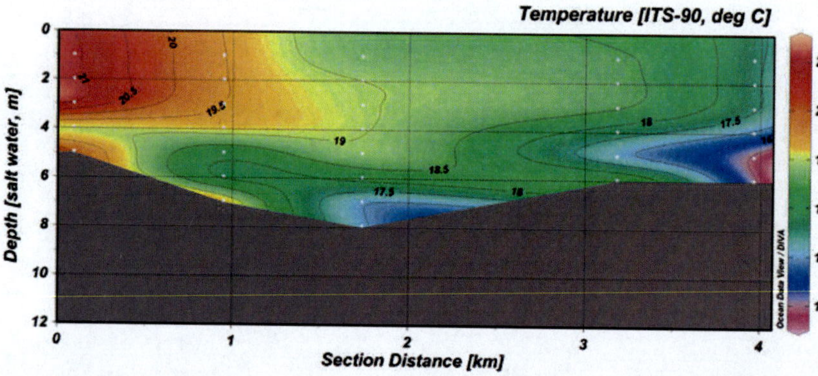

Fig. 2 Water temperature section for 13.07.2016 in the area of the Leningrad nuclear power plant (Koporskaya Bay)

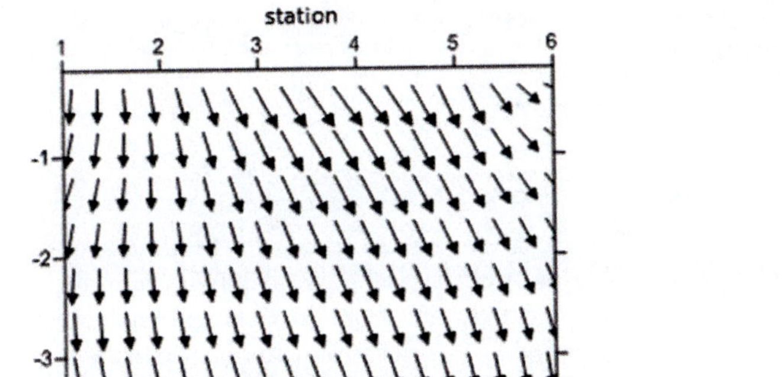

Fig. 3 Depth section for the speed and direction of the current on 13.07.2016 in the area of the Leningrad nuclear power plant (Koporskaya Bay).

the discharged water with the mass of water in the reservoir. Thus, in the coastal zone, an elevated temperature background is formed to the bottom in the coastal zone a little more than 1 km, since downwelling was observed in the study area at this time (Fig. 3), which promotes mixing of waters, as well as prevents mixing of heated coastal water with relatively cold flows of bottom water penetrating from the open part of the Bay.

On the surface of the water area, the maximum speed is 27 cm/s, and the minimum speed is 6 cm/s at the bottom (Fig. 3). The flow rate decreases exponentially with depth. This can lead to the spread of surface heated waste water and, as a result, to an increase in the area of elevated sea surface temperature.

An increase in water temperature accelerates the circulation of substances in the ecosystem, in particular, primary production, which causes eutrophication of reservoirs, which currently affects the coast of the Gulf of Finland during the warm season. Along with this, changes in the flora and fauna of the reservoir occur, causing shifts in the structure and function of the original ecosystems in the unfavorable direction [5–7].

5 Conclusions

As a result, it is noted that most of the water area, both in space and depth, is not affected by the discharge of heated water from the Leningrad nuclear power plant, this effect is mainly experienced by a very narrow coastal strip in the discharge area. There is a significant temperature difference between the temperature of intake water and discharge water of the order of 3 °C. The heat distribution zone strongly depends on the currents and wind conditions of the area. In comparison with previous years, the environmental situation has improved in the summer of 2016, which indicates that environmental safety measures have been taken during the operation of the nuclear power plant.

References

1. Maksimov AA (2006) long-term changes in macrozoobenthos as an indicator of eutrophication of the Eastern part of the Gulf of Finland. Collection of scientific papers Gosniorh Issue 331, vol. 2, pp 77–91
2. Final Technical Report Under the Agreement 1/LEN2/1521 (2008) Production of engineering and geological surveys and research of the Leningrad NPP-2 construction site. Regime hydrogeological observations. Neftegazgeodeziya LLC, Saint Petersburg
3. Golubkov SM, Maksimov AA, Shilin MB (2004) Research of eutrophication characteristics of the Eastern part of the Gulf of Finland. Compl Res Process Charact Res Russian Seas North European Basin Apatity Publish House Kola Sci Center Russian Acad Sci 1:250
4. Shuisky VF (1995) Assessment of the level of local thermal pollution. In: The water reservoirs-coolers, Influence of thermal and organic pollution on the biota of cooling reservoirs: SB. nauch Tr Gosniorh, Issue 314, pp 82–86
5. Martyn I, Petrov Y, Stepanov S, Sidorenko A (2020) Spatial-temporal variability of ice cover of the Bering sea. IOP Conf Series Earth Environ Sci 539:012198. https://doi.org/10.1088/1755-1315/539/1/012198
6. Istomin E, Petrov Y, Stepanov S, Sidorenko A, Martyn I About the methodology of geo-risk management in forestry. IOP Conf Series Earth Environ Sci 507:012006. https://doi.org/10.1088/1755-1315/507/1/012006
7. Sidorenko A, Stepanov S, Petrov Y, Martyn I Application of a remote sensing data processing method for assessment ice cohesion in the Arctic navigation. IOP Conf Series Earth Environ Sci 539:012128. https://doi.org/10.1088/1755-1315/539/1/012128

Experimental Determination of the Load of Mounted Working Bodies, Aggregated with Tractors

Aleksey Rogachev and Anatoliy Korsakov

Abstract The characteristics of a tractor unit are determined by the drag indicators of the towed agricultural machine, which require experimental determination. Experimental determination of the working bodies loading requires the use of a system of sensors of forces acting on the working bodies in different planes. Common disadvantages of known devices that provide measurements during volumetric dynamometry of working bodies of tillage machines are the complexity of the design and high measurement error. Force sensors are manufactured mainly using strain gauges that are glued directly to the loaded parts. To determine the forces in the traction of the hydraulic suspension system of tractors, strain gauge fingers are used. Calibration of force sensors is usually carried out individually on special stands. The best calibration results will be achieved when all strain gauges are installed in their proper positions and all sensors are calibrated simultaneously. The design and technological scheme of the device is proposed, the kinematic scheme of which allows using a single measuring device. This technical solution simplifies the technology of measuring the force on the working body and increases its accuracy. Analytical expressions for calculating the main design parameters are given. The results of experimental calibration of the proposed device when changing the thickness of the adjustment washers are presented. The proposed device reduces the error when measuring the total horizontal force on the tractor, acting from the side of the mounted tool or trailer.

1 Introduction

Determination of energy and fuel-economic indicators of transport and technological units based on wheeled and tracked tractors is an urgent task, the solution of which is based on experimental research [1]. In the process of operation of tractor trailers and

A. Rogachev (✉)
All-Russian Research Institute of Irrigated Agriculture, Volgograd, Russian Federation

A. Korsakov
Volgograd State Agrarian University, Universitetskiy Ave, 26, Volgograd, Russian Federation
400002

A. Beskopylny and M. Shamtsyan (eds.), *XIV International Scientific Conference "INTERAGROMASH 2021"*, Lecture Notes in Networks and Systems 247,
https://doi.org/10.1007/978-3-030-80946-1_32

working bodies of mounted agricultural machinery and implements are subjected to dynamic loads. The efficiency of working tractor units, mainly for transport, technological and agricultural purposes [2–4] is determined by the traction resistance of the towed machine, which has a stochastic nature.

When moving a tractor agricultural unit, the total resistance to movement from the soil increases with increasing loosening depth, which negatively affects their resource, requiring hardening of the working surfaces [5–7]. When experimentally measuring by spatial dynamometry of working bodies of tillage machines and tools, a significant measurement error and design complexity are a technical problem (patents US 4,354,390, SU 1,522,054 and 1,644,730, RU 2,589,217, etc.). The review of patents in the field of dynamometry of loading of working bodies of tractor units has shown that improving the accuracy of measurements requires improvement of methods and measurement tools that reduce the error. A possible method for reducing the error is to separate the measured forces in the vertical and horizontal plane (patents RU 2,684,441, 2,703,423, 2,703,910 RU, etc.). This determines the relevance of theoretical and applied research in this area.

2 Materials and Methods

A classic approach to the experimental determination of the loading capacity of tractor units is the method of control dynamometry, which allows determining the dependence of the traction resistance on the speed of the unit at various operations (transport, main tillage, intra-soil fertilization, and a number of others [4, 8].

Experimental determination of the efficiency of the MTA requires the use of force sensors that affect the working bodies in several planes. In General, force sensors are manufactured using strain gauges [7, 9–11], which are often glued directly to parts with different cross-sections.

To determine the forces in the traction of the hydraulic suspension system of agricultural vehicles. tractors use strain gauge fingers, which are cylinders with strain gages glued to them. The pull force P of the hydraulic suspension system can be decomposed into three mutually perpendicular directions-vertical PV, axial PO, and horizontal Rh (Fig. 1).

For experimental determination of the horizontal forces typically use strain gauge fingers 1 of circular cross-section mouth-nalivaete rod hydraulic hinged system 2 of the tractor and prose-HN adapter plate 3. The strain gages secured in a diametrical horizontal plane in the points 1, 2, 3, 4 (Fig. 2) and is connected via the bridge circuit (Fig. 3).

If the resistances of all four strain gauges are equal, the voltage between points C and D (Fig. 2) will be zero and the signal of the recording device will also be zero.

If the strain gauge is loaded with the force RG indicated in Fig. 2, the stresses at points 1 and 4 (and hence the strain gages) will be tensile, and at points 2 and 3-compressive.

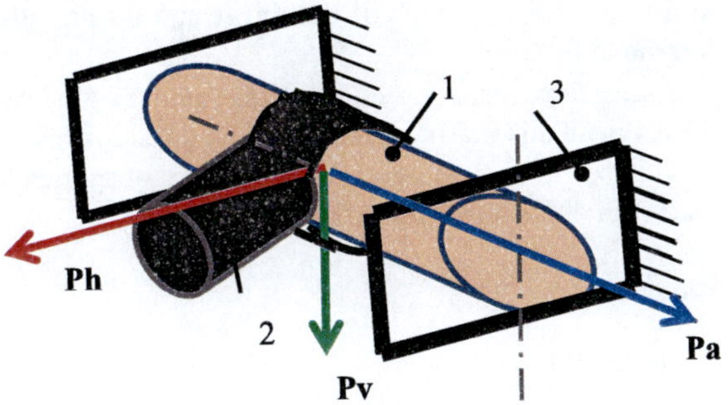

Fig. 1 Diagram of the application of forces from the tractor hydraulic suspension system rods

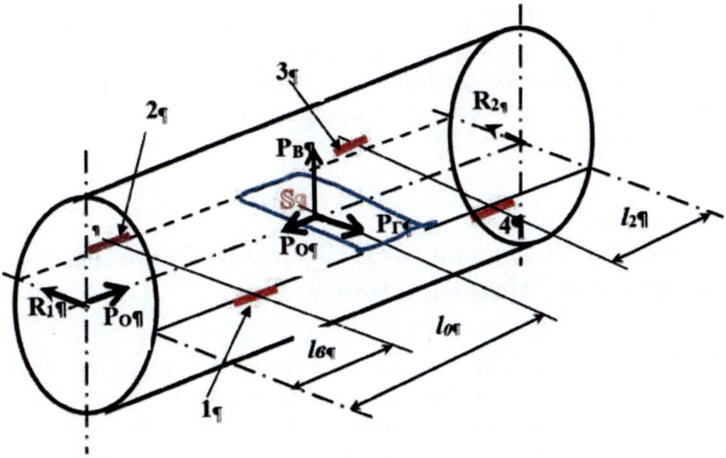

Fig. 2 Arrangement of strain resistors for measuring the horizontal load

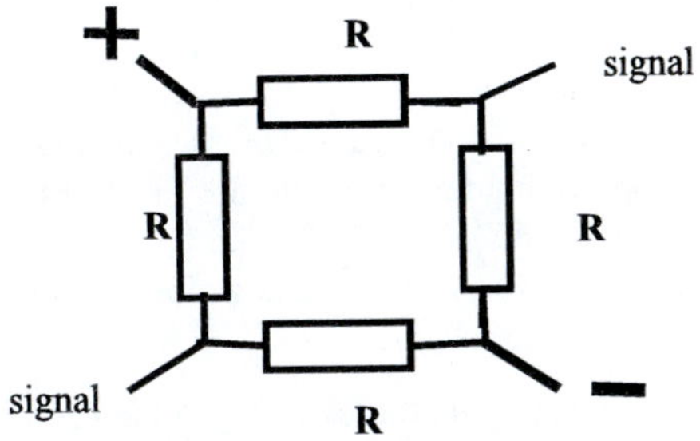

Fig. 3 Bridge connection diagram for strain gages

The appearance of strain gauge deformations causes changes in their resistance and changes in the display of the recording device. The stresses at these points from the RF force will be equal to zero, the Stresses from the axial force of the PO will be the same at all points of the finger and do not affect the determination of the horizontal force.

In this case, there will be no change in the signal value in the measuring bridge. However, all this will be observed only if the strain resistors have the same resistance, are glued at the same distance from the point of application of the force, and are glued exactly in diameter at points 1–4.

3 Results

Note that it is almost impossible to fix the strain resistors at the points indicated above, and in this case they will also react to the vertical load. In addition, the transfer of effort from one part to another occurs not in points, but on sites.

It is not known where the resultant force is applied at these sites. At the same time, these points are displaced during operation, which reduces the accuracy of determining the horizontal force.

3.1 Theoretical Justification of the Design and Technological Scheme of the Measuring Device

To improve the accuracy of horizontal force detection, it is necessary to eliminate the impact of vertical and axial forces on the horizontal force sensor. This can be achieved by passing all three forces through a mechanical force filter, which is a two-arm lever with one degree of freedom, mounted on a vertical axis on the support. If one arm of the lever at a distance of l_1 applies a force along the longitudinal axis of the lever PO, vertical RV and horizontal RG, then to keep the lever in balance, it is sufficient to apply one force on the second arm of the lever-a horizontal RD at a distance of l_2 (Fig. 4).

The axial and vertical forces are transmitted along the lever only up to the bearing support and are perceived by the axis of rotation.

Acting on the axis with the forces RO, M and RB (Fig. 4), only the horizontal force P, which will be defined below, will be transmitted to the second side of the lever. Internal force factors acting along the length of the lever are represented by the following diagrams:

- Qv transverse forces acting in the vertical plane;
- Qg transverse forces acting in the horizontal plane;
- N longitudinal force;
- Mg bending moment acting around the horizontal axis passing through the center of the hinge;
- Mo bending moment around the longitudinal axis passing through the center of the hinge.

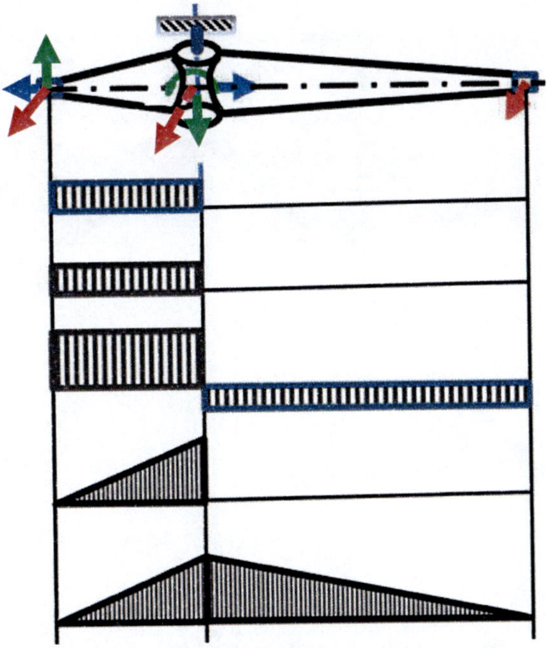

Fig. 4 The scheme of application of force and bending moment diagrams for duplicem lever

The equilibrium condition of the lever:

$$\sum m_A = P_h * l_1 - \mathbf{P_D} * l_2 = 0 \quad (1)$$

where we get it from

$$l_2 = l_1 * \mathrm{P}_h / P_D \quad (2)$$

Usually, the mounted tool is attached to the tractor using three rods. By transmitting the forces from the rods using three double-arm levers and summing them, you can get the total horizontal force from the arm to the tractor.

A design solution that implements this approach is shown in Fig. 5. On the transition plate 1, the axes 2 are vertically installed, on which two double-arm levers are fixed with the help of angular contact bearings. At the outer ends of these levers, the lower rods of the hydraulic suspension system are attached, and links 3 are attached to the inner ends, transmitting forces to the summing axis 4.

In addition, a third double-arm lever is vertically mounted on the plate 1 using bearings on the horizontal axis. The upper rod 5 of the tractor's hydraulic attachment system is attached to its outer end. The inner end is connected to the summing axis 4 by means of a fork 6.

Thus, the forces from all the tractor hitching rods are transmitted to one summing axis 4, and from it to the horizontal force sensor 7.

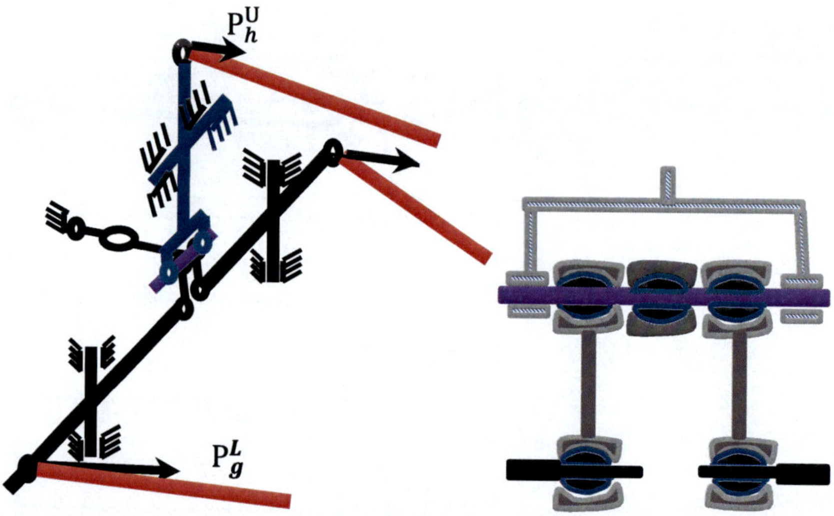

Fig. 5 The constructive solution, summarizing the efforts of three rods with duplicate levers

The forces from the hitch pull can generally be represented as Ph-horizontal force, Pv-vertical force, Po-from the axial force (Fig. 1).

Only the horizontal Ph force that can be determined will be transmitted to the second side of the lever:

$$P_D = P_h \frac{l_1}{l_2} \tag{3}$$

This force is then transmitted to the horizontal force sensor. For all double-arm levers, the ratio of the lengths of the outer arms to the inner arms should be the same, which can be achieved by changing the lengths of the outer arms of the lower rod levers by installing the corresponding washers between the lower rod hinge and the end of the double-arm lever.

3.2 Experimental Study of the Developed Measuring Device

The experimental determination of the load characteristics of the developed measuring device on the load stand is shown in Fig. 6.

Experimental calibration graphs obtained for the upper rod and for one of the lower rods for the variants without a washer, with a washer thickness of 1 mm and with a washer thickness of 15 mm are shown in Fig. 7. The diagram in Fig. 7 shows that the graphs for the upper thrust and for the lower thrust are almost identical.

Fig. 6 Experimental determination of the characteristics of the proposed device on a load stand

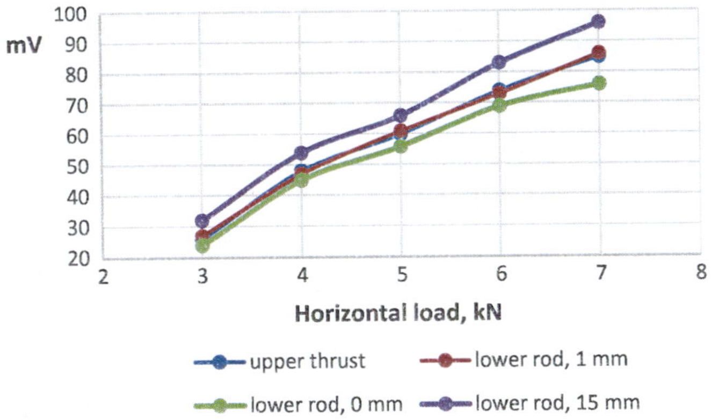

Fig. 7 Calibration graphs for the upper rod and for one of the lower rods when changing the thickness of the adjusting washer

Thus, the proposed device reduces the error when measuring the total horizontal force on the tractor, acting from the side of the mounted tool or trailer.

4 Conclusions

The study of the design and technological schemes of known devices for experimental measurement of forces on a tractor, delivered from the side of a mounted tool or trailer, showed that the proposed device provides an increase in measurement accuracy. This is achieved due to the proposed design and technological scheme, which reduces the number of intermediate links.

Technical solutions for patents RU 2,682,085, 2,684,441, 2,703,910 and 2,703,423, developed by the authors, allow for direct determination of the horizontal component of the traction force transmitted to the working body during the operation of the tractor unit from the soil. The proposed device provides the use of a single measuring device—a force sensor, to determine the total horizontal load on the working body of the tractor unit.

References

1. Arzhenovsky A (2017) Methods for determining energy and fuel and economic indicators of machine-tractor units Agricultural machines and technologies. J Phys Conf Series 6:36–40
2. Rogachev A (2012) Modeling and automation of the synthesis of loading and transport units based on structural graphs Izvestiya Nizhnevolzhskogo agrouniversitetskogo complex: science and higher professional education. 2 (26):165–170
3. Rogachev A, Karsakov A, Kosulnikov R, Gapich D (2020) Devices for experimental determination of the loading of MTU working bodies by strain measurement method. Sci life 15(7–107):980–990
4. Tseplyaev V, Matasov A, Tseplyaev A, Rogachev A (2017) Justification of the use of a rotary unit for weed control by the method of heating Izvestiya nizhnevolzhskogo agrouniversitetskogo complex: science and higher professional education. 2(46):256–263
5. Kupryashkin V, Ulanov A, Naumkin N (2018) Substantiation of the design of a dynamometer module for the study of a ploughshare and dump plow of a motor-block and its practical approbation using reverse engineering technologies. Bull Mordovian Univ 28(3):400–415
6. Novikov A, Motorin V, Lamskova M, Filimonov M (2018) Composition and tribological properties of hardened cutting blades of tillage machines under abrasive. Deteriorat J Friction Wear 39(2):158–163
7. Panneerselvam S, Martis C, Putatunda S, Boileau J (2015) An investigation on the stability of austenite in Austempered ductile cast iron (ADI). (Article) Mater Sci Eng 626:237–246
8. Berezhnov N, Averichev L (2018) Justification of the energy saturation of a forest tractor according to the data of control dynamometry of the agribusiness. Achiev Sci Technol 32(12):76–81
9. Korsakov A, Kosulnikov R, Fomin S, Nazarov E (2018) Features working caterpillar tracks. The proceedings of lower Volga agro-University complex: science and higher professional education. 3(51):355–363
10. Gavryushin S, Skvortsov P (2017) Evaluation of output signal nonlinearity for semiconductor strain gage with ANSYS software. Solid State Phenomena 269:60–70

11. Wang H, Zhu X, Fang Y, Zhang H, Zhang S (2018) Study on the detection arm accuracy of a leaf spring caliper for detecting internal convex defects in gas pipelines. Measure Sci Technol 29(8):085901
12. Berezhnov N, Sherbakov A (2017) Obtaining baseline information for the assessment of operational indicators of combined seeding machine based on the data of the control dynamometrical. Herald ISAA 81(1):56–62
13. Kosulnikov R, Karsakov A, Fomin S, Privalov V (2019) forming the hook loading of a tractor as part of a machine-tractor unit Izvestiya nizhnevolzhskogo agrouniversitetskogo complex: science and higher professional education. 1(53):371–377
14. Parkhomenko S, Parkhomenko G (2018) Mathematical modeling of the functioning of the tillage machine-tractor AG-regatta. Tract Agricul Mach 4:55–62
15. Parkhomenko S, Parkhomenko G (2016) Measurement of the traction force on the hook of a tractor in an aggregate with a mounted agricultural machine. Tract Agricul Mach 4:15–19

Application Layer Services in the RW.Ring Platform

Sergey Medvedev, Vitaly Terleev, and Olga Shevchenko

Abstract Production process models are the intellectual core of decision support systems in agriculture. The tasks faced by the users of these systems are quite varied and may require the operation of the same model of the production process in different environments. At the same time, for the practical application of models in solving applied problems, a convenient user interface is required in any environment. The.NET Framework standard library includes the System. Component Model namespace, which describes the abstract user interface entities that are not tied to a specific technology to ensure that the user interface works efficiently in different environments. The RW.Ring platform extends this library with services that provide an interface for working with background tasks, notifying the user with information messages, and managing an application instance. It should be noted that the scope of this development is limited to the client sphere: it is obvious that the architecture of the Web server, which generates html in response to client requests, does not allow creating dialogs from server logic. However, even in distributed applications, as a rule, there is a specialized client where this architecture is quite applicable.

Keywords RW.Wing platform · Decision support systems · Production process models · Agriculture

S. Medvedev · V. Terleev (✉)
Agrophysical Research Institute, 14, Grazhdansky pr., St. Petersburg, Russian Federation 195220

V. Terleev
Peter the Great St. Petersburg Polytechnic University, Polytechnicheskaya, 29, St. Petersburg, Russian Federation 195251

St. Petersburg State Agrarian University, Peterburgskoe shosse, 2, St. Petersburg-Pushkin, Russian Federation 196601

O. Shevchenko
Kuban State Agrarian University Named After I. T. Trubilin, Street Kalinina, 13, Krasnodar, Russian Federation 350044

A. Beskopylny and M. Shamtsyan (eds.), *XIV International Scientific Conference "INTERAGROMASH 2021"*, Lecture Notes in Networks and Systems 247,
https://doi.org/10.1007/978-3-030-80946-1_33

1 Introduction

Dynamic crop simulation models are the intellectual core in decision support systems [1]. This circumstance presupposes the reuse of the same models of the production process in different systems designed for different applied problems [2–4].

For example, the dynamic model AGROTOOL [5] is used both as a standalone Windows application, which allows performing single calculations and analyzing the results in detail at each step of the model up to the distribution profiles of moisture and mineral nutrients in the soil, and as part of the APEX polyvariant calculation system [6].

As a rule, studies using the AGROTOOL model are carried out in the APEX system, but in addition, there is the CropMap project [7], which is a Web portal for farms, in which model calculations are also carried out using the AGROTOOL model. Thus, the same dynamic model of the production process, which is a complex computational algorithm programmed as an executable module, can be used in different software products with completely different user interfaces.

The opposite challenge is to provide a unified user interface for different dynamic models of the production process. One of the first products that attempted to solve this problem was GUICS [8]. This program supports a special script language with which you can describe the format of the input and output files of models, as well as describe how this model can be run. These files are edited through the generic tabular interface and the models can be run. However, GUICS requires that the model be written in a specific way in order to be able to connect to it. In addition, the functionality itself for the design and processing of the results of computational computer experiments in GUICS is rather poor and does not allow to effectively automate the solution of complex problems that face decision support systems in agriculture.

The APEX polyvariant calculation system is a much more flexible and powerful solution [6]. It imposes on the model, which can be connected to it, the minimum restrictions, which are subdivided into technical (the ability to run the polyvariant calculation system and the model in the same environment), functional (the ability of the model to work without user participation) and semantic (the model is precisely the dynamic model of the production process cultivated plants). In addition, the APEX polyvariant calculation system contains rich functionality that will allow to carry out a wide range of computational computer experiments with various models of the production process [9]. However, this system does not allow solving problems of ensemble calculations and does not support work in the global network [10]. To develop a new version of the APEX polyvariant calculation system, in which it is planned to overcome these shortcomings, the RW.Ring platform was developed, which provides typical functions for the operation of such applications [11].

One of the features of a universal framework for dynamic models of a production process is to provide a generic user interface for a large number of different,

Table 1 Tasks solved by application-level services

Application functionality	Service providing this functionality
User authentication when working in the global Internet	Displaying the login and password input dialog
Launching a dynamic model of the production process for mass (polyvariant) calculation	Displaying a dialog with the progress of the operation
Opening multiple documents in one application	Managing an Application Instance and Its Restarts
Handling errors and warnings during bulk calculations	Display of information messages of varying degrees of difficulty

but similar models. In addition, dealing with a large number of possible environments, bearing in mind the distributed nature of the new version of the APEX polyvariant computing system, may require that the user interface functions be technology neutral. The C# standard library includes the System. Component Model namespace, which provides such technology-neutral classes for describing the user interface. In particular, it is possible, using the abstractions described in this namespace, to describe a dialog intended for editing several parameters grouped by categories; in this case, by default, the parameters are edited through a text box, which, if necessary, can be replaced by a drop-down list. Also, for each parameter, both a display name and a detailed description can be specified, which is displayed to the user as a hint. It also supports the notification of the user interface about the programmatically change of certain properties edited through the dialog. This standard library is a very powerful tool for developing user interfaces in a technology-neutral manner, but it does not solve all the problems that you might encounter when developing general-purpose frameworks for dynamic production models. Therefore, the RW.Ring platform libraries supplement it with application-level services (Table 1).

2 Materials and Methods

The main architectural solution that makes it possible to implement a system of application-level services is the Service Locator pattern [12]. According to this pattern, a service is an abstract type, an instance of which is stored in a static field. The main application logic accesses this field every time it needs the functionality of this service, but this functionality can be implemented in different ways. When the application starts, a static field that stores an instance of a service is initialized with a real object, the type of which is usually not available in an assembly that declares an abstract service type and a static field. This is what allows to make a service a technologically neutral abstraction, the concrete implementation of which becomes available at the time of application initialization.

In the RW.Ring platform, the Service Locator pattern is implemented in the static class Notung.Services.AppManager, whose properties are access points to static fields that store service instances. Services are interfaces implemented in various RW.Ring classes. In the current version of the platform, there are five of these services:

- Non-visual services, as a rule, do not require substitution of implementation:
  - Configurator—provides access to application settings, which can be changed during its operation by user actions;
  - Assembly classifier—a service that provides initialization of loaded assemblies and support for plugins. Both of these services have already been analyzed in another work [11].
- Visual services. Such services consist of logic that is persistent and a view that can be spoofed. Views always implement the special interface ISynchronizeProvider, which provides access to the synchronization object, to synchronize with the user interface.
  - Service for working with the progress of long-term operations;
  - Service for displaying information messages.
- A semi-visual service that manages an instance of the current application. Like visual services, it has a placeable view, however, unlike visual services, a view is not a collection of graphical user interface elements.

Thus, all application-level services that are the subject of this work are arranged according to the following scheme (Fig. 1).

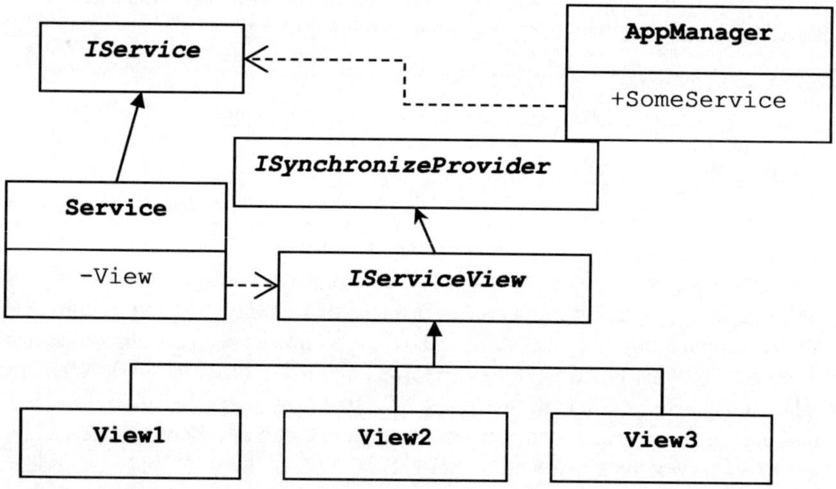

Fig. 1 Schematic diagram of an application-level service device

The decision support system algorithm accesses the required service through the corresponding static property of the AppManager class. The service does its job by accessing the view, which can be different depending on how the application was initialized. At the same time, the properties of the AppManager class are implemented in such a way as to never return null: if the field in which the property value is stored has not yet been initialized, it is initialized with an instance of the class that implements the service interface with a default view. In Notung.dll, all default views have a simple console line interface by default. The initialization of these fields on the first call is thread-safe. In addition, if the services need to be replaced during the initialization of the application, this will happen without creating unnecessary service instances. The Notung.Helm.dll library contains a generic view for all application-level services implemented in the MainFormView class, which implements these interfaces using Windows Forms dialogs. The decision support algorithm calls the required service through the corresponding static property of the AppManager class. The service does its job by accessing the view, which can be different depending on how the application was initialized. At the same time, the properties of the AppManager class are implemented in such a way as to never return null: if the field in which the property value is stored has not yet been initialized, it is initialized with an instance of the class that implements the service interface with a default view. In Notung.dll, all views have a simple command line interface by default. The initialization of these fields on the first call is thread safe. In addition, if the services need to be replaced during the initialization of the application, this will happen without creating unnecessary service instances. The Notung.Helm.dll library contains a generic view for all application-level services implemented in the MainFormView class that implements these interfaces using Windows Forms dialogs.

Services with views, which are the subject of this work, solve the following problems. The service for managing an application instance provides service information about the application: console line arguments, a link to the current process and the path to the file that started this process, an event that occurs when the application exit, and methods that allow only one copy of the application to be launched and restarted. The service for displaying information messages must analyze the message for display, choose the most suitable presentation method for it and display it on its presentation in the selected way. A service for working with long-term calculations should block user actions for the duration of the calculation, but so that the user can see that the application is not frozen and displays the progress of the operation. Microsoft recommends using either a marquee or a progress bar for this, and the service provides access to all these controls through a single call [13].

3 Results and Discussions

A non-visual service for accessing application settings is represented by the static AppManager.Configurator property of the IConfigurator data type. A non-visual service for classifying assemblies and loading plugins is represented by the AppManager.AssemblyClassifier property of the IAssemblyClassifier data type. These services provide access to such functions of the RW.Ring platform as the configurator and classifier of assemblies, which were considered in another work [11].

A semi-visual service for managing an application instance is represented by the static AppManager.Instance property of the IAppInstance data type. It is an interface that has the following members:

- CommandLineArgs property - arguments passed to the current instance of the application via the command line at startup. The property type is a read-only string collection;
- The ApartmentWrapper property is a decorator of operations for performing them in a single thread apartment. Property type—IOperationWrapper interface. Used to run production process models that require execution in a single threaded apartment, such as AGROTOOL;
- CurrentProcess property—the current process instance as a special component from the System.ComponentModel standard library. Used to avoid creating this component every time it is needed;
- The Restarting property is a boolean value indicating that the application is currently being restarted. This value depends on the call to the Restart method, which will be described below;
- The Terminating property is a boolean value indicating that the application is currently terminating. This value depends on the calls to the Restart and AllowOnlyOneInstance methods, which will be described below;
- The StartupPath property is the path to the executable file that launched the application. The data type is string;
- Exit event—occurs when the application ends;
- Method AllowOnlyOneInstance—Checks if the application is already running, and if it is running, terminates the current instance of the application by passing command line arguments to the previous instance. Has no parameters and no return value;
- Restart method restarts the application. Has no parameters and no return value. The base implementation of this interface is the AppInstance class, to which you must pass a view, an instance of the IAppInstanceView interface, when you create it. This interface inherits from the ISynchronizeProvider interface, which provides access to the synchronization object, and adds the following members to it;
- The Restart method restarts the application using the same technology that is used to develop the graphical interface. For example, Windows Forms contains a complete implementation of reloading and makes it trivial to implement this method, but without using Windows Forms, a more complex implementation is required. The method has two parameters: the path to the current application and a list of command line arguments;
- The Boolean property SupportSendingArgs returns a value indicating whether the application supports sending command line arguments from one copy to another.

This is used to open multiple files associated with an extension with a given application in instance copy of the application;
- The SendArgsToProcess method sends command line arguments to the previous copy of the application. The method has two parameters, a process instance with a previous copy of the application and command line arguments, and returns a boolean value indicating whether the command line arguments were successfully sent;
- The ReliableThreading Boolean property returns a value indicating whether the view can provide reliable information about the state of the application's main thread.

The last of the described properties of the application instance view is related to the fact that when the AppInstance class instance is initialized, the logger is configured [11]. This class itself implements the IMainThreadInfo interface and provides the logger with information about the state of the main application thread, which, in turn, it receives from the view. If this information is reliable, then the logger switches to a faster asynchronous mode.

For maximum efficiency when working with an instance of the current process, the AppInstance class uses the Global helper class. This class contains several static read-only fields that are initialized the first time the class is accessed. It contains a link to an instance of the current process and the path received from it to the executable file that launched the application. In addition, this file contains a commonly used empty array of objects and references to the most commonly used assemblies: the assembly that launched the application and the assembly Notung.dll.

When the AllowOnlyOneInstance method is called, a mutex is created in the background thread, the name of which is formed from the full path to the executable file that started the current process. If the mutex has already been created, it checks to see if command line arguments have been passed to the current instance of the application. If they are transferred, the application tries to transfer them to the previous copy and writes to the log about the success of this operation. Regardless of the command line arguments passed, the application exits. Otherwise, the thread continues to run until the application ends, keeping the mutex open.

The visual service for displaying informational messages is represented by the static property AppManager.Notificator, which has the INotificator data type. This interface has three methods, each with three overloads:

- The Show method displays an informational message without requiring the user to make any decisions. The method does not return a value;
- The Confirm method shows the user an informational message that requires him to make a decision with two choices (yes / no). The method returns a boolean value;
- The ConfirmOrCancel method shows the user an information message that requires him to make a decision with three answer options (yes/no/cancel). The method returns an nullable boolean value.

The three overloaded versions of each of these methods have the following parameter sets:

- One parameter of type Info. This type is a class with the following properties:
  - Message—message text;
  - Level—message level (Debug, Information, Warning, Error, Fatal);
  - InnerMessages—attached messages to this Info instance, also having the Info data type;
  - Details—additional data associated with the message.
- Two parameters—a line with the message text and a message level;
- Two parameters—a list of Info messages and a summary description of the list of messages, which is optional.

The main logic of the Notificator class, which is the basic implementation of the INotificator interface, is to select the most appropriate way of displaying them based on the structure of the transmitted messages. The display methods are described in the view interface for this INotificatorView service, an instance of which must be passed to an object of the Notificator class upon creation. This interface inherits from ISynchronizeProvider for accessing the synchronization object and declares two versions of the Alert method, the first of which allows you to display the message text and its level (for example, using the standard Windows message box, in which the text is written on the window, and the level is displayed with an icon), and the second is designed to display a complex message. In addition to the displayed data, these methods have a ConfirmationRegime parameter, which indicates from which service method the view method was called in order to display the required number of buttons on the dialog, by clicking on which the user makes his choice.

If the message that wanted to be displayed is one and does not contain messages attached to it, then simple display is selected. The same simple display will be selected to display the message list if it contains only one simple message. In addition, if there are several messages, and a summary description of the list of messages is not specified, then the generalized description of messages will be selected automatically at the highest level among all messages that need to be displayed at a time. The service also checks whether its method is called on the user interface thread and, if not, redirects the call to the view to the user interface thread using the synchronization object supplied by the view.

The visual service for starting long-running operations is represented by the static property AppManager.OperationLauncher, which has the data type IOperationLauncher. It is an interface in which the following members are declared:

- SyncWaitingTime property—the time span during which the service waits for the completed operation to complete before showing the user a dialog with a progress bar or marquee;
- The SynchronizationContext property is a synchronization context that can be accessed from any thread. In the standard library System.ComponentModel, the

synchronization context must be passed to the component at the time of creation through the current thread. This property allows you to remove this restriction;
- Run method—starts the task for execution. As a parameter, the method accepts an instance of the IRunBase interface—the task to be performed, and, optionally, additional settings for launching the LaunchParameters task. The method returns TaskStatus, an enumeration from the standard library showing the result of the task.

The IRunBase interface itself contains only two members: the Run parameterless method and return value and the ProgressChanged event, similar to the event of the same name in the BackgroundWorker standard library component. This provides the ability to start a task and track the progress of its execution. Progress boundary values are declared as constants in the ProgressPercentage class: Started, Completed, and Unknown. The first two values represent the two boundary values of the progress bar in the graphical user interface, and the third is used when a scrolling line is required instead of a progress bar. In addition to numerically expressing the progress of a task, the ProgressChanged event argument contains an arbitrary State object, which can be either a string or a special LaunchParametersChange enumeration that is responsible for changing the task display settings. If this line, then it will be displayed next to the progress bar or creeping line. In addition, it is assumed that the dialog with these controls has a title and an image that visualizes the task in progress. The enumerated type LaunchParametersChange, which contains two values: Caption and Image, is intended to change these two properties during task execution. To get the window title, the ToString method of the task or the value of the standard DisplayName attribute obtained from the task is used; to get the image, the task must implement the IServiceProvider interface, declared in the standard library System.ComponentModel.

Some tasks support cancellation. For this, the Notung.dll library contains the ICancelableRunBase interface, which inherits from the IRunBase interface and adds the following members to it:

- The CancellationToken property allows to set the task cancellation token. This property is of type System.Threading.CancellationToken from the standard library and is accessed in the usual way;
- Boolean property CanCancel returns the ability to cancel the task at the current moment. Even if the task supports cancellation, during its execution there may be critical moments when it cannot be canceled;
- The CanCancelChanged event occurs when the value of the CanCancel property changes.

It is assumed that if the task supports cancellation, a cancel button appears on the dialog displaying the status of the running task, which becomes inactive when the CanCancel property returns false. Checking the status of the cancellation token must be implemented in the logic of the task itself.

Both task interfaces have a base implementation in the abstract classes RunBase and CancelableRunBase. If it does not need to inherit specific tasks from specific classes, it is recommended to use these base implementations.

In addition to displaying a dialogue with a progress indicator, the service supports two special modes of operation. The first is related to the use of the LaunchParameters task launch settings. In this case, the instance of the IRunBase task is wrapped with another instance of this interface, which partially changes its behavior in the part in which it concerns the initial display settings—the task title and the image associated with the task. In addition, this class has the CloseOnFinish property, which allows, after the completion of the task, not to close the dialog displaying the status of the task before the user clicks a special button. This mode allows displaying the image and title of the task without writing them in the class of the task itself.

The second special mode of operation of the service is associated with the fact that tasks can be performed in other domains into which plugins for the main application are loaded. To do this, all AppManager services, except AssemblyClassifier, support sharing between domains. When creating an application domain for a loaded plug-in, a link to an instance of each of the services from the current domain is passed to this domain, so that all application models work with the same application-level service instances that are instantiated only in the main application domain. However, this is not enough for task management, as tasks notify progress of execution, and more complex infrastructure is required to support bidirectional communication. A link to a service is not just transferred to another domain, but wrapped in a special implementation of OperationLauncherProxy, which, in turn, wraps all tasks transferred to it with special wrappers that ensure the operation of the task in another domain. If the task implements the ICancelableRunBase interface, then this wrapper is the CancelableRunBaseCallerWrapper class, otherwise it is RunBaseCallerWrapper. These wrapped tasks are passed to the real service, which, upon learning that the task belongs to one of these classes, for its part also wraps them with the CancelableRunBaseProxyWrapper and RunBaseProxyWrapper classes, respectively. All four of these classes ensure that all task event arguments are correctly passed across application domain boundaries. However, for the application code, all these things happen completely transparently. If the plugin is running in a default application domain, it can run the task as if it were running in the main domain, and the entire GUI will also run in the main application domain, while the task logic will run in the plugin domain (Fig. 2).

In addition to two special modes of launching tasks, the operation of the service is closely related to such a feature of the RW.Ring operation as support for switching the user interface language during the program's operation. The fact is that in.NET, the user interface language is a property of the thread, and the service creates new threads to perform long-term operations, in which text messages are generated that are displayed in the user interface. To switch between the languages of the user interface, a special LanguageSwitcher component is used, which can simply put on the form in the designer and subscribe to its LanguageChanged event. In the handler for this event, you need to take data from localized resources using the standard.NET mechanism. To switch between languages, the LanguageSwitcher.Switch static method is

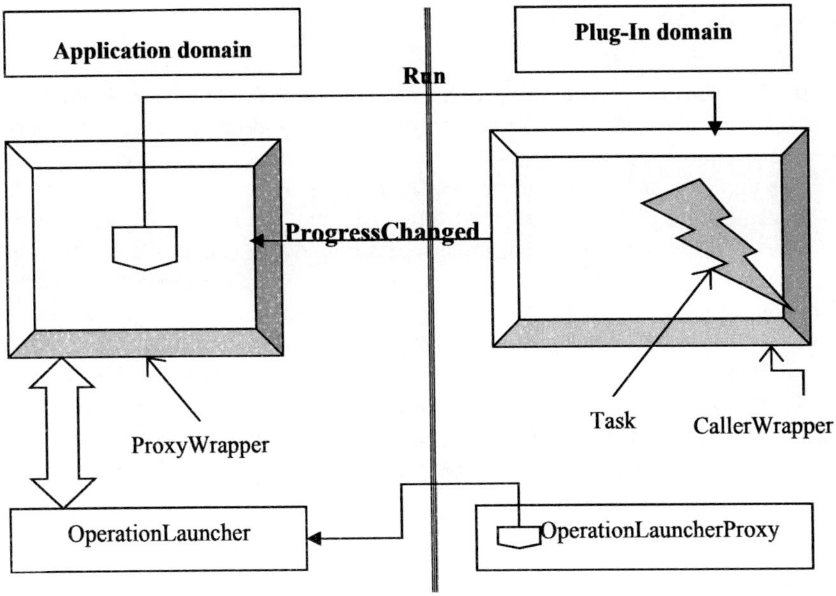

Fig. 2 How the task works in different domains

intended, which takes as parameters information about the user interface language to switch to, in the form of a string or an object of the standard CultureInfo class. This component uses the ThreadTracker class to get a list of threads currently running in the application to change their language. At the same time, when a task starts in a background thread, the task management service registers this thread in the ThreadTracker, and when the task ends, it removes it from the list of registered ones.

The view for the IOperationLauncherView task management service contains the ShowProgressDialog method, which takes a special LengthyOperation object as a parameter, which contains full information about the status of the task running in the background thread and its display settings, and a Boolean parameter indicating whether this dialog should be closed after the task will end. In addition, this interface contains the ShowError and ShowMessages methods that display an error message or a set of messages generated during the execution of a task. These messages from the task, as well as the image displayed on the dialog, are obtained through the IServiceProvider interface, declared in the System.ComponentModel standard library.

4 Conclusion

A set of services has been developed that solve the problem of creating a standard user interface that can be used repeatedly. Among the developed services, there is currently no one that is responsible for entering a login and password, but it is obvious that its architecture will be similar to that shown in Fig. 1, and there are no fundamental difficulties in this. At the same time, already developed services are successfully used in software products such as Schicksal. The experience of using such solutions shows that they can significantly simplify representations when developing presenters in a graphical interface [12], since the main application logic can access services to solve typical tasks [14–21]. It should be noted that the scope of this development is limited to the client sphere: it is obvious that the architecture of the Web server, which generates html in response to client requests, does not allow creating dialogs from server logic. However, even in distributed applications, as a rule, there is a specialized client where this architecture is quite applicable.

References

1. Badenko V, Terleev V, Topaj A (2014) AGROTOOL software as an intellectual core of decision support systems in computer aided agriculture. Appl Mech Mater 635–637:1688–1691
2. Poluektov RA, Oparina IV, Terleev VV (2003) Three methods for calculating soil water dynamics. Russ Meteorol Hydrol 11:61–67
3. Dunaieva I, Barbotkina E, Vecherkov V, Popovych V, Pashtetsky V, Terleev V, Nikonorov A, Akimov L (2021) Spatial and temporal databases for decision making and forecasting. Adv Intell Syst Comput 1259:198–205
4. Dunaieva I, Mirschel W, Popovych V, Pashtetsky V, Golovastova E, Vecherkov V, Melnichuk A, Terleev V, Nikonorov A, Ginevsky R, Lazarev V, Topaj A (2019) GIS services for agriculture monitoring and forecasting: development concept. Adv Intell Syst Comput 983:236–246
5. Poluektov RA, Fintushal SM, Oparina IV, Shatskikh DV, Terleev VV, Zakharova ET (2002) Agrotool-A system for crop simulation. Arch Agronomy Soil Sci 48(6):609–635
6. Medvedev S, Topaj A (2011) Crop simulation model registrator and polyvariant analysis. IFIP Adv Inf Commun Technol 359:295–301
7. Assessment of the state of the field, forecast of the yield and the dates of the onset of the phases of the growing season. https://cropmap.ru/. Last Accessed 10 Apr 2021
8. Acock B, Pachepsky YA, Mironenko EV, Whisler FD, Reddy VR (1999) GUICS: A generic user interface for on-farm crop simulations. Agron J 91(4):657–665
9. Medvedev S, Topaj A, Badenko V, Terleev V (2015) Medium-term analysis of agroecosystem sustainability under different land use practices by means of dynamic crop simulation. IFIP Adv Inf Commun Technol 448:252–261
10. Medvedev SA, Cheryaev AS (2020) Prospects for creating a universal service for remote ensemble calculations of dynamic models of the production process of cultural plants. Agrophysics 3:45–52. (rus)
11. Medvedev S, Terleev V, Vasilyeva, O.: Non-visual platform components for a system of polyvariant calculation of dynamic models of the production process. E3S Web Conf 244:09008
12. Fowler M (2002) Patterns of enterprise application architecture. Addison-Wesley Longman Publishing Co., Inc, Boston
13. Progress Bars, https://docs.microsoft.com/en-us/windows/win32/uxguide/progress-bars. Last Accessed 10 Apr 2021

14. Terleev V, Nikonorov A, Togo I, Volkova Yu, Garmanov V, Shishov D, Pavlova V, Semenova N, Mirschel W (2016) Modelling the hysteretic water retention capacity of soil for reclamation research as a part of underground development. Procedia Eng 165:1776–1783
15. Terleev V, Ginevsky R, Lazarev V, Nikonorov A, Togo I, Topaj A, Moiseev K, Abakumov E, Melnichuk A, Dunaieva I (2017) Predicting the scanning branches of hysteretic soil water-retention capacity with use of the method of mathematical modeling. IOP Conf Series Earth Environ Sci 90:012105
16. Badenko V, Badenko G, Topaj A, Medvedev S, Zakharova E, Terleev V (2017) Comparative simulation of various agricultural land use practices for analysis of impacts on environments. Environments 4(4):1–15
17. Terleev V, Nikonorov A, Togo I, Volkova Yu, Ginevsky R, Lazarev V, Khamzin E, Garmanov V, Mirschel W, Akimov L (2017) Hysteretic water-retention capacity of sandy soil. Magz Civil Eng 2:84–92
18. Terleev V, Nikonorov A, Ginevsky R, Lazarev V, Togo I, Topaj A, Moiseev K, Pavlova V, Layshev K, Arkhipov M, Melnichuk A, Dunaieva I, Mirschel W (2018) Hysteresis of the soil water-retention capacity: estimating the scanning branches. Magz Civil Eng 1:141–148
19. Terleev V, Mirschel W, Nikonorov A, Ginevsky R, Lazarev V, Topaj A, Moiseev K, Layshev K, Arkhipov M, Melnichuk A, Dunaieva I, Popovych V (2018) Five models of hysteretic water-retention capacity and their comparison for sandy soil. MATEC Web Conf 193:02036
20. Terleev V, Mirschel W, Nikonorov A, Lazarev V, Ginevsky R, Topaj A, Moiseev K, Pashtetsky V, Dunaieva I, Popovych V, Melnichuk A , Arkhipov M (2019) Models of hysteresis water retention capacity and their comparative analysis on the example of sandy soil. Adv Intell Syst Comput 983:462–471
21. Terleev V, Mirschel W, Nikonorov A, Ginevsky R, Lazarev V, Pavlova V, Topaj A, Pashtetsky V, Dunaieva I, Popovych V, Melnichuk A, Layshev K (2019) Improved Hydrophysical functions of the soil and their comparison with analogues by the Williams-Kloot test. Adv Intell Syst Comput 983:449–461

Training of Specialists in the Field of Agriculture

Learning Path of Distance Education in Regional Universities: Challenges and Opportunities

Sergey Kokhan, Sviatlana Vlasava, Asylbek Eshiev, Kasiet Musabayeva, and Gulnara Anarbaeva

Abstract The global online transition has become a real challenge to the traditional form of education which have led universities to overhaul the delivery of education services. A sociological study was conducted, based on the study of the relationship of academic teaching staff, their assessment of the effectiveness of the measures taken by the universities in organizing the transition to DL format of education. The study was attended by lecturers from regional universities in Russia, Kyrgyzstan and Mongolia. The results of the study revealed the most organized and optimistic category of lecturers from Mongolian universities and focused on the main problems in the implementation of the DL of all universities at the ATS level. The normalization of the epidemiological situation in the future will make it possible for universities to define their approaches to the use of DLT and their role in the traditional face-to-face education system in accordance with the needs of the students and the characteristics of social rehabilitation of the economic level of development of each country.

Keywords Lecturers · University · Pandemic · Region · Higher education · Coronavirus

S. Kokhan (✉)
Trans-Baikal State University, Chita 672039, Russia

S. Vlasava
Belarusian State Medical University, Minsk 220116, Belarus

A. Eshiev
Osh State University, Osh 720900, Kyrgyzstan

K. Musabayeva
Kyrgyz National Agrarian University named after K. I. Skryabin, Bishkek 720016, Kyrgyzstan

G. Anarbaeva
Jalal-Abad State University, Jalal-Abad 720900, Kyrgyzstan

A. Beskopylny and M. Shamtsyan (eds.), *XIV International Scientific Conference "INTERAGROMASH 2021"*, Lecture Notes in Networks and Systems 247,
https://doi.org/10.1007/978-3-030-80946-1_34

1 Introduction

The global online transition has become a real challenge to the traditional form of education [1]. The shift in the focus to education using distance-learning technologies (DLT) continues to cause alarm and concern among higher education lecturers because of the risk of losing the fundamental nature of the educational process [2].

With the exception of some developed countries, whose universities had been providing online education for a long time, others universities had faced various difficulties during the pandemic [3].

The COVID-19 pandemic has had significant socio-economic consequences in all countries. The protective measures used in the world are extremely polar, ranging from strict quarantine in some European countries to inaction, as in Sweden, where only collective immunity was relied upon.

The challenges of the times have led universities to overhaul the delivery of education services. In order to facilitate the transition to online learning during the coronavirus pandemic, each university was based on its own capabilities, depending on the level of technical equipment and available methodological support. The efficiency of DLT implementation is determined by many factors, chief among which are the readiness of direct participants in the educational process to use information technologies, competence, psychological and material preparedness [4].

Before the pandemic, DLT was not in demand in Russian universities, which was supported by many experts in the field of distance learning (DL), who skeptically and without enthusiasm discussed the advantages of DL on the pages of scientific editions. In their opinion, at the start of the COVID-19 pandemic, the role and place of DL in higher education on Post-Soviet states territory was not defined [5].

This position was controversial, as the use of information and communication technologies (ICT) in training has been effective in foreign countries for a long time [6].

In addition, there has been a transformation of educational activities in response to the intensive development of mobile technologies and the integration of the latter into the educational process [7].

Withal, the existing experience in creating online courses and digital learning materials has not been sufficient to organize DL in universities. The emergency transition during the COVID-19 pandemic to the DL has revealed a number of problems related to different views on the functioning, interaction, development and content of educational activities [8].

At the same time, there was a different view in the scientific pedagogical community about the introduction of DLT in Russian universities. Studies carried out by Menter [9], Kolesnikova [10] have identified trends in higher education development in the digital age, where the role and place of ICT in the educational activities of universities have been clearly defined. However, the theory and practice, in reality, was far from each other.

DL involves interaction between lecturer and learner at a distance and incorporates in the learning process almost all the components: methods, objectives, organizational forms, content, including sometimes the means of learning [11]. The implementation of DL is carried out by means of specific telecommunication technologies, which provide for interactivity of the educational process, depending largely on the attitude of the university administration [12].

The pedagogy of the Coronavirus pandemic has made us realize and rethink the direction of further development of pedagogical education as a corporate system of lecturer training; the need to change the ways and principles of interaction between educators and student youth, as well as to change attitudes towards oneself in digital learning and work [13].

According to UNICEF, Kyrgyzstan is among the countries that have made significant progress in accessing DL. The Government and the Ministry of Education of Kyrgyzstan provided all students in the country with free access to remote learning through online platforms, national television channels and mobile network applications. However, the distribution of the coronavirus served as a characteristic stress test to determine the depth of the problem of «digital difference». The lack of modern information devices and Internet communication among poor families makes it difficult to DL. Many lecturers were not fully prepared to go DL online. Regional universities have problems with technology and the availability of quality Internet. The average age of lecturers in the country is 50–60 years and many of them do not know information technology. DL has slowed considerably due to frequent disruptions in the internal information systems of educational institutions [14].

Mongolia is a low-density country. On January 27, 2020, the Cabinet of Ministers of Mongolia adopted a resolution on the closure of all educational institutions and the implementation of other restrictive measures. The measures recommended by the Government and the universities to counter the pandemic were urgently communicated to the population and quickly implemented. The strategy of containment and serious prevention among the population has yielded results—no deaths have been recorded.

Radical changes and reforms in Mongolia's higher education system in recent years were aimed at guaranteeing the quality of higher education in accordance with world standards. Mongolia is adopting and localizing the American model in education. The accreditation system was the main form of ensuring quality control of education in universities. Higher education in Mongolia is in transition, which includes the creation of a more stringent environment for the development of higher education, as well as the improvement of the accreditation system, standards of international and regional quality networks [15]. Of course, the focus on information computer technology (ICT) development in central Mongolian universities is a factor in limiting the widespread adoption of ICT in the regions of the country due to insufficient funding of the programme.

The research carried out by Bolortuya et al. [16] on the introduction of e-learning in the regional universities of Darkhan revealed a number of significant problems. Therefore, lack of technical equipment, computer literacy, time-consuming preparation of online lecture and practical materials, feedback and assessment of knowledge students hinders the development of online technologies in Mongolia's higher education system. The implementation of e-learning in regional universities during the pandemic has shown satisfactory results. Statistically significant results have been found in the training of lecturers for the DL and in the availability of accessible, in electronic form, educational literature by directions of study.

Meanwhile, most researchers focused on the evaluation of DL by learners. There are virtually no publications on the results of studies on changing the reference system and values of lecturers in remote regional universities as participants in DL during the COVID-19 pandemic. This updated our research and defined its design in order to improve the system of higher education in a distance format.

2 Materials and Methods

The study was attended by lecturers from regional universities in Russia (TBSU, Chita—98 lecturers), Kyrgyzstan (OSU, Osh—91) and Mongolia (Darkhan—92). The total number of participants was 281.

From September to October 2020, a sociological study was conducted, based on the study of the relationship of academic teaching staff (ATS), their assessment of the effectiveness of the measures taken by the universities in organizing the transition to DL format of education. The author's questionnaire (19 questions) was made and translated into Mongolian and Kyrgyz.

The questionnaire was analyzed and approved for using by independent experts from among the professors of three universities with at least 25 years' experience.

Survey results were provided remotely through Google forms.

The research did not violate the rights of respondents to the questionnaire. ATS were informed of the objectives of the study and provide with electronic informed consent before participating in the sociological study. The Research Ethics Committee of TBSU (Chita, Russia) has considered, approved and agreed with partner universities on the proposed research protocol.

The study was a pilot. The study database was formed on the data obtained and was used for statistical analysis.

Qualitative and quantitative responses were recorded in the survey. They were represented by the absolute number of answers for each question, the relative (%) and average values (M $\pm$ SD).

The null hypothesis (that there was no difference between the observed feature distribution and the theoretically expected normal distribution) was tested using the W-Shapiro-Wilk criterion due to its highest power compared to other criteria. With $P > 0.05$, the analyzed distribution did not differ from the normal distribution, so a parametric variant of variance analysis was applied. The Student's t-test and dispersion analysis (one-way ANOVA) were used. Post-hoc pairwise comparison was performed. The data analysis also used Pearson's chi-squared test or the Fisher's exact test. 95% of the error-free prognosis was considered to be the minimum sufficient ($p < 0.05$). Statistical analysis methods were used to determine the lowest value of the expected event. The resulting value of the criterion χ^2 was compared to the critical

Table 1 Distribution of participants in the pilot study by age and pedagogical length of service, Chita-Osh-Darkhan, 2020

	Group 1	Group 2	Group 3
Answer options	$M_1 \pm SD$	$M_2 \pm SD$	$M_3 \pm SD$
Average age of participants, years	42.2 ± 7.28	42.8 ± 8.25	42.9 ± 7.64
Average pedagogical length of service (full years), years	18.2 ± 6.38	15.7 ± 7.65	15.3 ± 7.08

value for degrees of freedom at a minimum value $p < 0.05$ (according the table of Critical values of the Chi-squared Distribution).

The correlation between the sings was estimated by computing the Spearman's rank correlation coefficient (in the case of a non-parametric type of distribution) taking into account the statistical significance of the probability of error for the null hypothesis of no correlation between the attributes ($p < 0.05$).

For the processing and analysis of empirical data the statistical methods of the licensed version of programs (STATISTICA 10.0, MS Exsel 2010) were used.

For the convenience of gathering, Grouping and presentation of the results of the study, all participants were identified in Groups according to the affiliation to the studied universities: Group 1—ATS TSBU, Group 2—ATS OSU, Group 3—ATS of the Association of Higher Education Institutions of Dharkan.

Recruitment was limited by age (25–60 years) and length of service at the university (2–30 years). Participants in the sociological study were comparable in age and sex.

More than 70% of lecturers in Groups 2 and 3 had no science degree or academic rank, while in Groups 1 the corresponding indicators were 26.5%.

Position held at the place of work of ATS is presented in Table 2.

A significant number of respondents taught humanitarian disciplines. Data on the profile of the disciplines taught are presented in Table 3.

Table 2 Position held at the place of work of academic teaching staff, Chita-Osh-Darkhan, 2020 ($\chi^2 = 10.6, p > 0.05$)

	Group 1		Group 2		Group 3	
Answer options	Abs	%	Abs	%	Abs	%
Teaching staff by the main occupation	76	77.6	65	71.4	74	80.4
Teaching staff (internal part-time)	6	6.1	5	5.5	5	5.4
Teaching staff (external part-time)	2	2	10	11	2	2.2
Leader (Head of Department, Dean, etc.)	14	14.3	11	12.1	11	12

Table 3 Profile of the disciplines taught, Chita-Osh-Darkhan, 2020

	Group 1		Group 2		Group 3	
Answer options	Abs	%	Abs	%	Abs	%
Humanitarian	64	65.3	65	71.4	22	23.9
Technical	13	13.3	3	3.3	21	22.8
Natural science	11	11.2	9	9.9	24	26.1
Culture, art, sport	9	9.2	9	9.9	1	1
Other (specify)[a]	1	1	5	5.5	24	26.1

[a]Other: Economics, international relations, legal affairs, nursing, clinical practice, agriculture, health, pharmacology and biomedicine, medical sciences

3 Results

3.1 Academic Teaching Staff Awareness of Coronavirus Infection

According to the results of the study, a significant number of Group 3 ATS monitored continuous coronavirus information (75%).

The lowest number of people interested in COVID-19 was in Group 1 (44.9%), while Group 2 had the highest number of ATS indifferent to the spread of the pandemic in their country.

ATS from 2 and 3 Groups considered as «high» COVID-19 infection risk. Group 1 respondents in 55.1% of cases rated the chances of this virus infection as «high». This indicates an underestimation of the COVID-19 risk by ATS from the Russian Federation, and consequently a disregard for precautions and social distance. The opposite result was recorded for the "low" assessment, ranging from 6.5% in Group 3 to 19.4% in Group 1. The large number (25.5 and 16.5%, respectively) of respondents from Russian and Kyrgyz universities who was difficult to answer about the virulence of COVID-19 compared to the indicators of respondents from the Mongolian Group confirms the low awareness of participants.

In order to support citizens and organizations during the pandemic, the State authorities developed and implemented various measures that were actively promoted through the media. Mongolian lecturers from Group 3 (82.6%) were well informed, and to a lesser extent ATS from Groups 1 and 2 (50 and 38.5%, respectively). However, between 11.2 and 12.1% of ATS reported a low level of awareness, despite information in the press and on television about the measures taken by the Government. This may indicate a general lack of interest in the issue (Table 4).

The user-friendliness of the information perception greatly influence the formation of a sustainable interest and should be taken into account in the shaping of information content for the respective focus groups (Table 5). At present, the Internet has been the primary resource for obtaining information on coronavirus support for all three

Table 4 Results of the evaluation of respondents' knowledge on public authority support measures, Chita-Osh-Darkhan, 2020 ($\chi^2 = 44.8, p < 0.01$)

	Group 1		Group 2		Group 3	
Answer options	Abs	%	Abs	%	Abs	%
Well informed	37	37.8	43	47.2	76	82.6
Information at the middle level	49	50	35	38.5	14	15.2
Ill-informed	11	11.2	11	12.1	1	1.1
Hard to answer	1	1	2	2.2	1	1.1

Table 5 Respondents' evaluation of the source relevance study on Government Support Measures, Chita-Osh-Darhan, 2020 ($\chi^2 = 14.9, p > 0.05$)

	Group 1		Group 2		Group 3	
Answer options	Abs	%	Abs	%	Abs	%
Television, radio, print media	32	32.7	16	17.6	38	41.3
Internet: news aggregators, social networks, forums etc	55	56.1	61	67	46	50
From friends/acquaintances, colleagues	5	5.1	6	6.6	1	1.1
Information materials and activities of State bodies	5	5.1	7	7.7	6	6.5
All of the above	1	1	1	1.1	1	1.1

Groups as the study has shown. Television, radio and print media ranked second. Lower levels of trust were shown by other sources of information.

An analysis of the position of the ATS Group 3 regarding the diversity of proposed approaches in taking measures to prevent the propagation of coronavirus in a country revealed statistically significant differences ($\chi^2 = 20.7, p < 0.01$) (Table 6). 29.7% of respondents from Kyrgyzstan (Group 2) believe that radical action is not necessary because of the pandemic. The smallest number of ATS with the same opinion and loyalty to restrictive measures were from Mongolian universities (Group 3–10.9%).

It is very important to note that all Groups were dominated by the desire of ATS to do their utmost and to take all necessary urgent measures to prevent the spread and eliminate coronary infections. The other options were not critical.

3.2 *Effectiveness of Preventive Institutional Measures at the Regional Level*

The evaluation of the effectiveness of the measures taken by the country's leadership and the Ministry of Education revealed significant differences in the perception and analysis of what is taking place among the ATS surveyed. The lowest average score (3.3 ± 1.06 points) is obtained in the Russian university. Average performance indicators were found in the Kyrgyz higher education institution (4.0 ± 1.58 points)

Table 6 Distribution of preferred pandemic response approaches for respondents, Chita-Osh-Darkhan, 2020

	Group 1		Group 2		Group 3	
Answer options	Abs	%	Abs	%	Abs	%
1. Not to take radical measures: many people will inevitably get coronavirus, some may die, but collective immunity will develop faster	21	21.4	27	29.7	10	10.9
2. Do everything possible to avoid the spread of infection before a vaccine or effective treatment is found	74	75.5	61	67	79	85.8
3. Take no drastic measures, but observe security and disinfection measures	1	1	1	1.1	1	1.1
4.Take measures to keep infected with less negative impact on the economy	1	1	1	1.1	1	1.1
5. There will be no collective immunity, but people get infected twice. But prevention is needed	1	1	1	1.1	1	1.1

and the highest confidence in the work carried out by the country's governing bodies and higher education institutions was noted by Mongolian colleagues (5.2 ± 1.47 points). Respondents' assessment of the Government's response to the pandemic differed statistically from one study Group to another (Table 7).

Table 7 Respondents' evaluation of the study on the effectiveness and adequacy of educational measures taken by the Government in the context of a pandemic in accordance with the proposed scale (1 point—not effective, 7 points—effective), Chita-Osh-Darkhan, 2020 ($\chi^2 = 112.1, p < 0.01$)

	Group 1		Group 2		Group 3	
Answers in points	Abs	%	Abs	%	Abs	%
1	7	7.1	11	12	3	3.3
2	11	11.2	2	2.2	1	1
3	38	38.8	19	20.9	3	3.3
4	36	36.7	24	26.4	24	26.1
5	4	4.1	25	27.5	21	22.8
6	1	1	3	3.3	17	18.5
7	1	1	7	7.7	23	25

Table 8 Results of the survey respondents' evaluation of combat measures taken in the region with the spread of coronavirus in accordance with the severity of the pandemic, Chita-Osh- Darkhan, 2020 ($\chi^2 = 63.7, p < 0.01$)

	Group 1		Group 2		Group 3	
Answer options	Abs	%	Abs	%	Abs	%
Corresponding to the hazard	26	26.5	27	29.7	69	75
Rather superfluous	10	10.2	5	5.4	8	8.7
Rather insufficient	51	52	48	52.8	12	13
Hard to answer	8	8.2	10	11	2	2.2
Other	3	3	1	1.1	1	1.1

Lecturers rated the pandemic prevention work of the regional authorities low from Groups 1 (3.2 ± 1.07 points) and 2 (4.7 ± 1.7 points) compared to Group 3 (5.2 ± 1.6 points). The results of the evaluation of the measures taken by the administration of the region differed statistically from one Group to another (Table 8).

Among the main judgments and comments that respondents to the study pointed out in the option «Other» were indicated «measures are largely absurd, illogical and contradictory»; «measures are harmful rather than really helpful»; «insufficiently justified»; «people behave too irresponsibly», «not enough control over organizers of mass events», «absence of fines not only to organizers, but also to participants of the event».

The results of the studies on trust and evaluation of the performance of the institutions of higher education during the pandemic did not reveal significant differences between the Groups of studies. The study found the highest rates in Group 3 (5–7 points), the average level (3–7 points) was recorded in Group 2 and the ATS Group 1 level was very low (2–4 points). The majority of lecturers from the TBSU consider that the work carried out by the administration of the higher educational establishments is inefficient and insufficient.

Contradictory results obtained regarding the effectiveness of the recommended coronavirus prevention measures at the universities during the pandemic. Whereas in OSHUs and Darkhan higher educational establishments the indicators of compliance assessment of extent of fulfilment in the full in universities of «all recommended measures» amounted to 42.8 and 38.0% respectively, in TBSU—no more than 8%. Between 33.0 and 42.9% of respondents indicated that the «most preventive measures» had been implemented. As «partial measures», the results of the university measures were evaluated by a large number of the ATS Group 1 (in twice) compared to the rest of the participants of the Groups 2 and 3 (Table 9).

An important element in the organization of activities at the university is the existence of normative-regulatory documents defining the specific conditions for coordination of ATS activities in the transition to a DL at university. According to Groups 3 respondents (88%), the university leadership approved and explained the application of all the necessary documents to ensure that the university operates in the face of a pandemic. A smaller number of ATS have been identified in the remaining

Table 9 The results of the survey respondents' evaluation of the organizational measures taken by the university leadership "in full" in DL transition under pandemic conditions, Chita-Osh-Darkhan, 2020 ($\chi^2 = 43.4, p < 0.01$)

	Group 1		Group 2		Group 3	
Answer options	Abs	%	Abs	%	Abs	%
Yes, all measures were followed in the organization	8	8.2	39	42.8	35	38
Yes, most of the measures were implemented in the organization	42	42.9	30	33	36	39.1
Yes, but only a few measures have been implemented in the organization	46	46.9	17	18.7	20	21.7
No, no measures were followed in the organization	2	2	5	5.5	1	1.1

Groups as indicating that the adopted regulations in universities are sufficient. The negative response rate to this question was almost the same in all three Groups (from 7.1 to 11%). It was difficult to answer this question for a rather high percentage of ATS from TBSU and OSU. In Darkhan universities, the figure was only 2.2%.

3.3 Prospects for Post-pandemic Higher Education

The prospects for the development of post-pandemic distance-learning higher education are inextricably linked to its quality and effectiveness.

The benefits of the DL are obvious, but there are also disadvantages to its implementation.

The results of the study show a serious impact on the quality of education of a remote format (Table 10). For example, a significant number of respondents, ranging from 52.5% of ATS in Group 3 to 74.5% of ATS in Group 1, tend to recognize the negative side of online learning. At the same time, 35.9% of Mongolian colleagues rated the influence as insignificant.

The polarity of the ATS opinions once again proves the relevance of studying the issue of implementing a standard higher education.

In assessing the impact of post-pandemic changes in higher education, participants in Group 1 of the study suggested that the requirements for ATS would change significantly in the future. Almost all respondents recognize the likelihood of changes in methodological approaches to the formation of educational content, ranging from

Table 10 Respondents' evaluation of a study on the impact of DL in a pandemic on the quality of university education, Chita-Osh-Darkhan, 2020 ($\chi^2 = 23.5, p < 0.01$)

	Group 1		Group 2		Group 3	
Answer options	Abs	%	Abs	%	Abs	%
Has a strong influence—quality has decreased	73	74.5	64	70.3	48	52.2
Has a strong influence—quality has grown	1	1	8	8.8	4	4.3
Had little impact	18	18.4	11	12.1	33	35.9
Had no influence at all	1	1	2	2.2	2	2.2
Hard to answer	5	5.1	6	6.6	5	5.4

52% in Russian higher education to 42.4% in Mongolian higher education institutions. A sufficient number of TBSU ATS (20.4%), who was it difficult to identify future changes, may still not have been able to clearly assess all the pros and cons in the perspective of DL implementation (Table 11).

The normalization of the epidemiological situation in the future will make it possible to define this approach in accordance with the needs of the learners and the specific features of the restoration of the socio-economic level of development of each country.

Table 11 Respondents' evaluation of a study on the possibility of changing the country's higher education system following the lifting of restrictive measures related to the pandemic, Chita-Osh-Darkhan, 2020

	Group 1		Group 2		Group 3	
Answer options	Abs	%	Abs	%	Abs	%
Quality of education, competencies and skills of graduates will be changed	38	38.8	41	45	49	53.3
Requirements for professors and professors will change	55	56.1	37	40.7	21	22.8
Possible organizational changes (fewer universities, greater DL)	36	36.7	22	24.2	17	18.5
There will be changes in the methodology and organization of work with students (there will be more use of digital technologies in teaching and interaction between lecturers and students)	51	52	40	44	39	42.4
There will be no special change	11	11.2	15	16.5	14	15.2
Hard to answer	20	20.4	8	8.8	10	10.9

4 Discussion

The negative effect of the coronavirus pandemic COVID-19 is felt worldwide. Whether they live in mega-cities or remote regions of the world, no one is safe [17, 18].

The coronavirus crisis has had a significant impact on the state and development of the world economic and social situation, including the university system. The Russian Federation, Kyrgyzstan and Mongolia were no exception.

The spread of the pandemic has contributed to significant changes in the social interaction of people in all spheres of life, including higher education. Understanding the importance of ATS awareness and public health education for preventive measures while moving to DL to reduce the risks of infection, adhere to social distancing measures is a key factor in decision-making at different levels [19].

The results presented in this study open up prospects for improving cooperation and coordination of the efforts of the society and representatives of the university leadership and remote regions of Russia, Kyrgyzstan and Mongolia.

Basic social demands and assessments regarding awareness, effectiveness of measures taken at various levels of government and university administration, as well as possible changes in the development and improvement of higher education after the end of the COVID-19 pandemic can influence the direction of education development in these countries and require further monitoring. Regardless of the existing infrastructure and learning management system at the university, there are obvious gaps in the detailed planning of activities to ensure full effective online learning.

The results of the study indicate a serious attitude towards the pandemic in Mongolia. The extent and level of COVID-19 contamination in this country was posed a serious threat, both economically and medically and socially at the beginning. The Government's efforts to contain the pandemic were supported by all Mongolians. As a result of the prompt and high-quality work of all branches of government, high responsibility and awareness of citizens, there have been isolated cases of coronavirus morbidity without fatalities.

The public's perception of the prognosis of coronaviruses infections varies widely from country to country, as do the measures they take. While some countries try to comply fully with isolation measures, others, on the contrary, deny the very existence of the virus and, as a result, refuse to comply with all restrictive measures [20].

The continued risk of a coronavirus disease for all citizens without exception, including lecturers in educational institutions, places increased demands on the coordination of measures and efforts at all levels of government and society [21].

However, half of Russia's population does not consider a dangerous coronary infection, and 48% of respondents consider the danger of COVID-19 to be an exaggeration in the media and the Internet. Kutumova et al. [22] studies confirm low levels of awareness of COVID-19 problems of Russian educators.

When considering the results of the study, regional differences in the implementation of online education should be understood. Thus, in the central regions of Russia, despite the considerable financial, economic, scientific and technical capabilities of universities, the developed infrastructure, cultural potential and experience in the application of DL, certain difficulties have arisen in the transition to online education. Remote regions lacking such resources and practices in DLT and digital education and student communication management have had to adapt to the challenges of the times in a more difficult environment.

According to our research, a large number of members of the university community, more to the extent of colleagues of Mongolian universities, are well informed about the prevention of coronavirus infection. This gives a healthy sense of optimism about the replication of knowledge and the possibility of raising public awareness of COVID-19, and hence of widespread support and compliance with recommended preventive measures.

Russian lecturers are more inclined to believe that changes in PPP attitudes are inevitable in the post-pandemic period of higher education of restructuring, which is consistent with the studies of foreign scientists Lassoued et al. [23] and De [24].

5 Conclusion

The personal readiness of ATS to perceive, accept and adapt to the challenges of time associated with the negative consequences of coronavirus proliferation, an adequate assessment of the prospects for the development of higher education in regional universities is due in large part to the peculiarities of socio-economic development and traditions of higher education in Russia, Kyrgyzstan and Mongolia. The Mongolian colleagues (82.6%) were the most responsible for their own health and COVID-19 situation monitoring, compared to the TBSU ATS and Kyrgyzstan ATS (50 and 38.5% respectively) ($\chi^2 = 44.8, p < 0.01$).

The ineffective and insufficiently productive work of the administration of the university was noted by representatives of TBSU (8%), who were not fully satisfied with the implementation of the proposed preventive measures. In OSU and Darkhan universities, compliance assessment rates in the full extent of «all recommended measures» were 42.8 and 38.0% respectively ($\chi^2 = 43.4, p < 0.01$).

The ATS Group 1 (3.2 ± 1.07 points) and Group 2 (4.7 ± 1.7 points) compared to Group 3 (5.2 ± 1.6 points) were less satisfied by regional authorities' efforts to prevent the pandemic.

Representatives of Russian (74.5%) and Kyrgyz (70.3%) universities emphasized that DL had a strong influence on the quality of education, compared to a similar opinion of Mongolian ($\chi^2 = 23.5, p < 0.01$).

Attitudes towards prospective changes in higher education since the removal of restrictive measures have varied considerably from university to university. In assessing possible post-pandemic changes in higher education in each country, participants from the TBSU Group (56.1%) estimated that the requirements for ATS would change significantly in the future; a 20.4% of respondents in this Group could not clearly assess all the pros and cons in the perspective of the realization of the DL.

The majority of respondents to all Groups identified the need for changes in methodological approaches to the formation of educational content, ranging from 52% in Russian universities to 42.4% in Mongolian universities.

Thus, the results of the study revealed the most organized and optimistic category of lecturers from Mongolian universities and focused on the main problems in the implementation of the DL of all universities at the ATS level.

The normalization of the epidemiological situation in the future will make it possible for universities to define their approaches to the use of DLT and their role in the traditional face-to-face education system in accordance with the needs of the students and the characteristics of social rehabilitation of the economic level of development of each country.

However, it is already clear that there is a need for continuous monitoring of feedback from the university community implementing online technologies; financial, technological and methodological support from Governments and investors for the adaptation and continuation of educational services in a new format; development and accessibility of Internet communications in various regions, together with technical support for universities; lecturers professional training in Internet technologies to continue to improve the efficiency of any university in the face of contemporary challenges.

References

1. Joshi A, Vinay M, Bhaskar P (2020) Impact of coronavirus pandemic on the Indian education sector: perspectives of teachers on online teaching and assessments. Interact Technol Smart Educ https://doi.org/10.1108/ITSE-06-2020-0087
2. Alea LA, Fabrea MF, Roldan RDA, Farooqi AZ (2020) Teachers' Covid-19 Awareness, Distance Learning Education Experiences and Perceptions towards Institutional Readiness and Challenges. Int J Learn Teach Educ Res 19(6):127–144. https://doi.org/10.26803/ijlter.19.6.8
3. Mercader C, Gairín J (2020) University teachers' perception of barriers to the use of digital technologies: the importance of the academic discipline. Int J Educ Technol High Educ 17(4):1–14. https://doi.org/10.1186/s41239-020-0182-x
4. Sipilä K (2014) Educational use of information and communications technology: teachers' perspective. Technol Pedagogy Educ 23(2). https://doi.org/10.1080/1475939X.2013.813407
5. Pekker PL (2015) Distance learning: experience of Moscow universities. Hum Educ 2(43):66–71
6. Chekalina TA, Tumandeeva TV, Maksimenko NV (2018) The main directions and prospects of online learning. Prof Educ Russia Abroad 3(31):44–52
7. Gafurov IR, Ibragimov HI, Kalimullin AM, Alishev TB (2020) Transformation of higher education during the pandemic: pain points. High Educ Russia 29(10):101–112. https://doi.org/10.31992/0869-3617-2020-29-10-101-112
8. Skidan AV, Melnichenko AM (2020) The attitude of teachers of universities of the rostov region to the system of distance learning in the conditions of emergency transition. State and municipal administration. Sci Notes 3:9–54. https://doi.org/10.22394/2079-1690-2020-1-3-49-54
9. Menter I, Valeeva R, Kalimullin A (2017) A tale of two countries—forty years on: politics and teacher education in Russia and England. Eur J Teach Educ 40(5):616–629. https://doi.org/10.1080/02619768.2017.1385060

10. Kolesnikova IA (2019) Post-pedagogical syndrome of the digimodernism age. High Educ Russia 28(8–9):67–82. https://doi.org/10.31992/0869-3617-2019-28-8-9-67-82
11. Listera K, Pearson VK, Collins TD, Davies GJ (2020) Evaluating inclusion in distance learning: a survey of university staff attitudes, practices and training needs. Innov Eur J Soc Sci Res. https://doi.org/10.1080/13511610.2020.1828048
12. Kim J (2020) Learning and teaching online during Covid-19: experiences of student teachers in an early childhood education practicum. Int J Early Childhood 52:145–158. https://doi.org/10.1007/s13158-020-00272-6
13. Frantsuzskaya EO (2020) Distance education: university instructor and student perspective. Scientific and Pedagogical Review. Pedagogical Rev 6(34):82–90. https://doi.org/10.23951/2307-6127-2020-6-82-90
14. Kanatova A (2020) Attitudes of kyrgyz and russian university students to switching to distance studying due to the covid-19 pandemic: a pilot study. Sci J «Vestnik of Pedagogy and Psychology of Ûžnaja Siberia», p 3. ISSN 2303-9744
15. Tserendagva B, Jamts T (2017) The rise of quality assurance in Asian higher education. Elsevier Ltd., pp 143–159
16. Bolortuya B, Khongorzul B, Bolormaa Ts, Munkhchimeg E, Battamir U, Khandaa B, et al (2020) From the results of the research on the implementation of the electronic form of learning. The era of higher education. J Darkhan 12(04):93–95
17. Azizaha YN, Rijalb MK, Rumainurc Rohmahd UN, Pranajayae SA, Ngiuf Z, Mufidg A, Purwantoh A, Maui DH (2020) Transformational or transactional leadership style: which affects work satisfaction and performance of Islamic University Lecturers During COVID-19 Pandemic? Multifaceted Rev J Field Pharm 11(7):577–588
18. Suleiman Y, Ishola MA (2020) Assessment of the impact of coronavirus (COVID-19) pandemic on educational system in Nigeria: implication for stakeholders. UMT Educ Rev 3(2):111–126
19. Paakkari L, Okan O (2020) COVID-19: health literacy is an underestimated problem. Lancet Public Health 5(5):e249–e250. https://doi.org/10.1016/S2468-2667(20)30086-4
20. Chotchhaev AN (2020) Changes in the structure of higher education in the context of the COVID-19 pandemic. Modern Sci 7(2):236–241. ISSN: 2414–9918
21. Alqabbani S, Almuwais A, Benajiba N, Almoayad F (2020) Readiness towards emergency shifting to remote learning during COVID-19 pandemic among university instructors. E-Learning and Digital Media, pp 1–20. https://doi.org/10.1177/2042753020981651
22. Kutumova OYu, Babenko AI, Pukhova E, Babenko EA (2020) Public awareness of prevention of coronavirus infection. Sci J "Current problems of health care and medical statistics", p 3. https://doi.org/10.24411/2312-2935-2020-00062
23. Lassoued Z, Alhendawi M, Bashitialshaaer R (2020) An exploratory study of the obstacles for achieving quality in distance learning during the COVID-19 pandemic. MDPI. Educ Sci 10:232. https://doi.org/10.3390/educsci10090232
24. De S (2020) Impacts of the covid-19 pandemic on global education. Nova Southeastern University, NSU Works, pp 84–94. https://doi.org/10.26524/royal.37.6

The Problem of Determining the Video Conferencing Platform Criteria for Online Learning

Mikhail Smolenskiy and Nikolay Levshin

Abstract The purpose of this study is to determine the criteria that must be taken into account when choosing a platform for organizing educational and scientific videoconferences. The relevance of achieving the goal is due to the fact that, in the context of the implementation of measures extended to the fight against viral infections, many unprepared people regularly face the need to solve the problem of choosing a suitable video platform. In order to accumulate the experience gathered by the respondents in the field of organizing videoconferences, the authors use the methods of polling, questioning, experiment and observation. Methods of statistical and logical analysis were used to process the collected data. The authors gathered and analyzed data from higher education workers from Russia, Ukraine, Kazakhstan, the United States and a number of other countries. Along with this, the authors studied a large volume of domestic issues and foreign studies devoted to the organization of video conferencing. As a result of the study, a number of key and optional criteria were identified that can be checked when choosing a platform for organizing online learning. The key criteria were highlighted: availability at the location of the conference, the size of the restriction on the maximum number of conference participants, broadcast quality, the organizer's ability to forcibly mute the conference participant, the organizer's ability to forcibly exclude the conference participant, operating costs. The optional results included values that do not have a critical impact on the organization of videoconferencing, but acquire significance in some situations. At the same time, the authors draw the attention to the fact that the assessment of key criteria requires a large amount of preparatory work, and any optional criterion can, under certain conditions, become key.

Keywords Online learning · Video conferencing · Virtual meeting

M. Smolenskiy
Don State Technical University, Gagarin Sq., 1, Rostov-on-Don 344003, Russia

M. Smolenskiy · N. Levshin (✉)
Rostov State Transport University, Rostovskogo Strelkovogo Polka Narodnogo Opolcheniya sq., 2, Rostov-on-Don 344038, Russia
e-mail: info.law.expert@yandex.com

A. Beskopylny and M. Shamtsyan (eds.), *XIV International Scientific Conference "INTERAGROMASH 2021"*, Lecture Notes in Networks and Systems 247,
https://doi.org/10.1007/978-3-030-80946-1_35

1 Introduction

Global implementation of restrictive measures due to the need to combat the threat of the spread of viral infections, has led to the fact that in 2020 teachers massively confronted with the necessity of choosing a convenient and reliable tool for video conferencing. And, since new participants regularly join the educational process, this problem of choice does not lose its relevance to this day.

In order to choose the right tool, you need to understand what criteria and requirements should be imposed on its use. Even a person who is well versed in information technology will not immediately be able to determine the criteria that are important to consider when choosing a tool for organizing video conferencing, if this person does not have practical experience in using such tools.

Many of those who first encountered the problem of choosing a platform for organizing video conferencing have low qualification in the application of information systems and technologies and do not have the opportunity to promptly consult with qualified specialists. These people can be helped by informing them what criteria should be followed when choosing a platform for organizing video conferencing.

The authors of this study aim to summarize the experience accumulated by people involved in organizing educational and scientific video conferencing, and to prepare universal recommendations that will help other people quickly and confidently make the best choice of tool for organizing video conferencing.

2 Materials and Methods

The authors of the study analyze the experience of using video platforms accumulated by videoconferencing organizers living in Russia, Ukraine, Kazakhstan, the United States and other countries. At the same time, methods of polling, questioning, experiment and observation are used to collect information. Along with the use of the listed methods, a review of the results of related studies presented in Russian and English databases. Methods of statistical and logical analysis are used to analyze the accumulated data.

3 Literature Review

In scientific periodicals, a lot of studies are presented on the use of videoconferences, the assessment of their advantages (as shown in the study by Foramitti et al. [1]) and the specifics of the conduct (as in the work of Standaert [2]). For obvious reasons, a significant part of modern research concerning the problems of organizing video conferencing is associated with the use of video conferencing technologies in the field of healthcare [3–12]. Nevertheless, despite their specificity, these studies can also provide interesting experience that will be useful in the preparation of educational and scientific videoconferences.

As noted by the authors of the research devoted to the study of the influence of the human factor on online meetings and video conferencing [3], good video communication requires significant bandwidth on each of the Internet connection nodes. If the bandwidth is too low on at least one of the nodes, it will lead to connection problems, blurry images and audio problems. All this can lead to loss of performance or complete disconnection of the conference participant. This allows to talk about the high importance of such a criterion as the quality of the broadcast.

S. Dhawan draws our attention to such a criterion for choosing a tool for organizing online training as the cost of operation [13]. In particular, he says that due to the high cost of digital devices and tariffs for data transmission on the Internet, representatives of low-income segments of the population may be at a disadvantage. At the same time, A. Roy and his colleagues separately note the value of the cost of tariffs for the use of video conferencing services [4].

The authors of a number of studies [5, 6] draw attention to the need for trainees to have the skills to use information and communication technologies. In turn, this indicates a certain importance of such a criterion for evaluating a video conferencing service as the presence of an intuitive interface that a person without special training can easily understand. However, this criterion cannot be considered a key one, since, as noted in a review prepared by A. Banbury and his colleagues [7], people quickly learn to use video conferencing systems.

L. Billingsley, in his study of mobile applications for video conferencing [8], talks about how users of the popular Zoom application faced a massive social phenomenon called "zoombombing". This is when an uninvited person joins the meeting and interferes with the work of the videoconference participants. Such incidents can jeopardize the conduct of any conference. This allows to mention the importance of such a criterion as the ability of the conference organizer to forcibly exclude a conference participant.

The authors of a number of studies [3, 9, 10] note the unintended negative influence of external stimuli. For example, such as email notifications, messenger alerts or phone calls. This spotlights the importance of such a criterion as the ability of the conference organizer to forcibly turn off the sound of the conference participant.

A study by Tiosano and his colleagues [11] draws our attention to the potential significance of such an option as the ability to share your screen with other videoconference participants. In their opinion, it was this option that made it possible to provide a full-fledged professional development of ophthalmologists in a pandemic. Kim [12] also noted the importance of the "screen sharing" function, saying that it allows learners to demonstrate and discuss the achieved results.

Kemery D.C. and Goldschmidt K. note that the words of a person speaking against a simple, uncluttered background are best perceived [14]. In this regard, such a criterion as the possibility of using a virtual background acquires significance. This feature has already been implemented in some modern video conferencing systems.

A number of studies (for example, in the work of Mahr and his colleagues [15]) pay attention to such important issues as ensuring the security, privacy and confidentiality

Table 1 Key criteria for choosing a platform for video conferencing

Criterion	Average score (1–10, rounded)
Max number of participants	9
Broadcast quality	9
Availability at the location of the participant	7
Ability to forcibly mute the conference participant	7
Ability to forcibly exclude a conference participant	6
Availability of screen sharing	6
Operating cost	6

of users of the video conferencing system. But we see no reason to single them out as a key criterion, since in practice the assessment of such a criterion can be carried out only by highly qualified specialists of a narrow profile.

4 Results

As a result of the study, criteria were identified that must be taken into account when choosing a platform for video conferencing. According to the degree of their impact on the possibility of holding a conference, the selected criteria were divided into key and optional.

Key criteria are critical when choosing an online video conferencing platform, as failure to comply with a key criterion may result in the inability to host a conference (Table 1). Average scores for the significance of key criteria among study participants range from 6 to 9.

Along with the key criteria, a number of optional criteria were identified that do not have a critical impact on the conduct of a videoconference, but are important in some situations (Table 2). Average scores for the significance of optional criteria among study participants range from 3 to 5.

Despite the fact that the listed optional criteria do not have a critical impact on the process of organizing a video conference, they were often mentioned by respondents as criteria for choosing a video platform for organizing online learning.

5 Discussions

Discussing the presented results, it should be borne in mind that the estimates of some of the considered criteria may differ depending on the geographic location of an individual user and the quality of communication services provided to him by the

Table 2 Optional criteria for choosing a platform for video conferencing

Criterion	Average score (1–10, rounded)
Ability to text chat	5
Ability to record video	5
User-friendly interface	5
Ability to "raise your hand"	4
Pass checks to gain access	4
Lifespan of the link to connect to the conference	4
Requirement to create an account for each participant	4
Ability to transfer files	4
Native applications for various systems	4
Noise-canceling filters	4
Ability to use a virtual background	3

Internet provider. For example, if the conference organizer does not have problems (delays and interruptions) with the quality of the video and audio signal, this does not guarantee that all other participants do not have such problems. This allows us to say that some criteria require evaluation by all conference participants.

Along with the quality of communication, the availability of using the video conferencing system at the point of the conference participant connection is one of the criteria that requires evaluation by all conference participants. The most common problem is the blocking of entertainment services in the local networks of educational organizations. So, if the use of a social network is blocked at the university, then the conference participant will not be able to use the video conferencing system, which is part of this social network. Even temporary lifting of such administrative restrictions may not be easy to achieve.

A slightly different kind of the described problem is the blocking of the service on the territory of the entire settlement. For example, as a result of the sanctions taken in response to the conflict over the territorial ownership of Crimea, a number of American services were forced to stop working in the "disputed" territories (as discussed in the study by Nitsevich and his colleagues [16]). This led to the fact that people living in this territory were deprived of the opportunity to normally use American services, including educational purposes.

Obviously, the considered restrictions associated with blocking the video conferencing service at a specific connection point will not be presented in the description of the information system. Such limitations can only be detected experimentally. Therefore, it is very important to conduct preliminary testing of the platform with the connection of all prospective participants in the video conference from the places from which they plan to connect at the time of the conference.

Another reason for conducting preliminary testing of the video conferencing system may be the need to check the influence of the environment of the conference participant. The influence of the immediate environment is considered in the previously mentioned work by M. Gross and his colleagues [10]. They note that the unusual "conference atmosphere" in the workplace or at home can limit the ability of a conference participant.

The degree of negative influence of ambient stimuli largely depends on whether the conference participant is a listener or a speaker. For example, if a neighbor of one of the conference participants begins repair work, then the listener can just turn off his microphone, while the lecturer will have to be distracted by looking for a compromise with the neighbor and thus wasting the time of all conference participants. Thus, at the stage of preparation for the conference, it is necessary to make sure that during the conference there will not be any extraneous sounds.

Many respondents and researchers (for example, Oeppen, Gross and their colleagues [3, 10]) note the importance of adhering to the culture of holding conferences. Of course, the entire responsibility for maintaining a culture of communication rests with the person, and not with the system used to organize the video conference. However, in this system, auxiliary tools can be implemented to facilitate the observance of the communication culture. First of all, this is the opportunity to "raise your hand", ie. demonstrate to other conference participants that you have a comment and are waiting for you to be given the opportunity to voice it.

If the "raise your hand" option is not implemented in the system, then as an alternative, a technique can be used when everyone except the active speaker turns off their cameras and turns them on only to let the moderator know that they have a comment on the speaker's speech. This technique is described in a study by Oeppen and colleagues [3]. It also notes the obvious advantage of this technique, which is that turning off the cameras by all conference participants, except for the active speaker, can significantly reduce the bandwidth load of the communication channel and, due to this, increase the broadcast quality.

However, one of the disadvantages of the proposed approach is that the organizer cannot control whether all conference participants are closely following the speech of an active speaker. This can be important, for example, when the teacher needs to make sure that all students pay due attention to the activity and are not distracted from the educational process. Therefore, having a built-in "raise hand" function in the system is preferable.

Along with the ability to "raise your hand", among the capabilities of systems that contribute to the development of a culture of conferencing, one can single out such a tool as a text chat. This tool allows participants to express their opinions or agree on the order of subsequent speeches, without interrupting the presentation of an active speaker.

Along with the above, one should discuss such a criterion as the ability to connect to the conference by reference, without creating an account in the service database. This can be very convenient as it saves participants time and does not require complex actions from those who are poorly versed in information technology. However, the use of this function opens up opportunities for sabotage and other abuse. Therefore, it is desirable that the system has the ability, if necessary, to restrict the connection to the conference with a password.

The video conferencing system must support all the variety of devices that can be used to access the conference. Therefore, the system should have applications for all common platforms, and, even better, if there is a universal web interface that works in any web browser.

Another challenge is the need to meet the criteria set by the employing organization. So at the Rostov State Transport University, the teacher is given the task of preloading links to the conference connection to the schedule of his classes for a week. Obviously, the solution to this problem is possible only when using a service with a long lifespan of links to connect to the conference. Thus, for teachers of a specific university, such an optional criterion as the life expectancy of a link becomes key.

Also, among the criteria set by the employing organization, one can include the requirement to create and store video recordings of all videoconferences held. This requirement can be met by using additional programs for screen recording, but it is much more convenient to use video conferencing systems that have a built-in recording option that can work in automatic mode.

6 Conclusion

Based on the results and discussion, we can make a general conclusion that the definition and understanding of the criteria for choosing a tool for videoconferencing minimizes the likelihood of problems arising during the conference, but does not provide a complete guarantee of their absence. Therefore, the process of organizing a conference must, without fail, include a preparatory stage. At this preparatory stage, it is necessary to exclude the likelihood of a negative impact from the external environment, to make sure that the communication channel allows each potential participant to take full participation in the videoconference, in accordance with the role assigned to him, and also to agree with all participants backup communication channel in case of unforeseen problems with the main channel.

References

1. Foramitti J, Drews S, Klein F, Konc T (2021) The virtues of virtual conferences. J Clean Prod 294
2. Standaert W, Muylle S, Basu A (2021) Business meetings in a post-pandemic world: when and how to meet virtually? Business Horizons
3. Oeppen RS, Shaw G, Brennan PA (2020) Human factors recognition at virtual meetings and video conferencing: how to get the best performance from yourself and others. Br J Oral Maxillofac Surg 58(6):643–646

4. Roy A, Andruska N, Orlowski HLP, Pipkorn P, Daly MD (2020) The novel use of a commercially available video-conference platform to facilitate multidisciplinary target volume review and delineation for skull-base radiation therapy during the coronavirus disease 2019 pandemic. Adv Rad Oncol 6(2):100598
5. Hai L, Nguyen X (2020) Impact of female students' perceptions on behavioral intention to use video conferencing tools in COVID-19: data of Vietnam. Data in Brief, p 32
6. Wotherspoon R, Mannion C, Harlow RE (2020) Maintaining medical team communication using video conferencing during the COVID-19 lockdown. Br J Oral Maxillofac Surg 58(10):326–327
7. Banbury A, Nancarrow S, Dart J, Gray L, Parkinson L (2018) Telehealth interventions delivering home-based support group videoconferencing: systematic review. J Med Internet Res 20(2):e25
8. Billingsley L (2020) Using video conferencing applications to share the death experience during the COVID-19 pandemic. J Radiol Nurs 39(4):275–277
9. Kerawala C, Riva F, Paleri V (2020) Videoconferencing for multidisciplinary team meetings in the coronavirus disease era—human factors awareness and recognition. J Laryngol Otol 134(12):1118–1119
10. Gross MW, Läubli H, Cordier D (2021) Multidisciplinary tumor boards as videoconferences—a new challenge in the COVID-19 era. Ann Oncol 32(4):572–573
11. Tiosano A, Hindi I, Bahar I, Nahum Y (2020) Continuity of ophthalmology education by combining video conferencing application with a slit-lamp camera, in times of a pandemic. Canadian J Ophthalmol 56(2):e44–e46
12. Kim J (2020) Learning and teaching online during Covid-19: experiences of student teachers in an early childhood education practicum. Int J Early Child 52:145–158
13. Dhawan S (2020) Online learning: a panacea in the time of COVID-19 crisis. J Educ Technol Syst 48:1–18
14. Kemery DC, Goldschmidt K (2020) Can you see me? Can you hear me? Best practices for videoconference-enhanced telemedicine visits for children. J Pediatr Nurs 55:261–263
15. Mahr A, Cichon M, Mateo S, Grajeda C, Baggili I (2020) Zooming into the pandemic! A forensic analysis of the zoom application. Forensic Sci Int: Digital Inv 36
16. Nitsevich VF, Moiseev VV, Glagole SN, Sudorgin OA (2019) Western sanctions and their consequences for Russia. Smart technologies and innovations in design for control of technological processes and objects: economy and production. FarEastCon 2018. Smart Innovation, Systems and Technologies, vol 138. Springer, Cham, pp 65–79

Inclusiveness of Online Education: Problems and Prospects

Larisa Guterman

Abstract This study attempts to assess the problems and prospects of online education related to its compliance with the requirements of inclusivity. The authors argue that the inclusiveness of online education is poorly understood and requires a comprehensive scientific understanding. The research is based on a cognitive approach that defines a person as a system that knows the world and is influenced by communication that occurs in the learning environment. There are two indicators of the inclusiveness of online education: socio-cultural and technical adaptability. The first is related to the willingness of students and teachers to accept people with disabilities as full-fledged subjects of the educational process, and the second determines the compliance of the technical tools of online education with the requirements of inclusivity. The empirical basis is a sociological study conducted by the authors, during which students and teachers of the Southern Federal University were interviewed. Based on the data obtained from the survey, it was found that students and teachers are tolerant of people with disabilities and are ready to cooperate with them on an equal basis. The most serious problem is considered to be the compliance of online education with the technical requirements of inclusivity. It is concluded that the software used for online education does not meet the requirements of inclusivity and should be replaced with an analog intended for educational purposes.

Keywords Inclusive education · Online education · Online learning · Distance learning · Tolerance

1 Introduction

One of the priority directions of the organization of the educational process in the Russian Federation is the formation of an inclusive tolerant environment that is conducive to the education of persons with disabilities (hereinafter referred to as

L. Guterman (✉)
Southern Federal University, Rostov-on-Don, Russia
e-mail: laguterman@sfedu.ru

A. Beskopylny and M. Shamtsyan (eds.), *XIV International Scientific Conference "INTERAGROMASH 2021"*, Lecture Notes in Networks and Systems 247,
https://doi.org/10.1007/978-3-030-80946-1_36

HIA). This environment, in turn, consists of two elements: socio-cultural and organizational. The first implies that students, teachers and the university administration treat people with disabilities with tolerance, do not hinder their participation in the educational process, and also provide them with support, while the second is built around the compliance of the online learning infrastructure with the needs of people with disabilities. Before the coronavirus pandemic, Russian higher education gradually increased its inclusiveness both technically and at the university policy level, but the sudden transition of the absolute majority of higher education institutions to an online format revealed significant problems related to the inclusiveness of the educational space. At first glance, it may seem that online education is a much more inclusive environment than classical education, but in practice, online education, implemented in Russia, does not have the proper characteristics to call it truly inclusive. The scientific study of inclusivity in higher education is popular both in Russia and abroad, where many researchers conduct research on the development of concepts for building an inclusive environment, but attempts to scientifically understand inclusivity are rarely made, since the online learning format itself is not so popular in higher education, especially in the Russian Federation.

Let us focus on the issue of scientific development of the topic of inclusive online education. Stein describes inclusivity as an important direction in the development of information technologies as such [1], Kormos and Nijakowska reveal that online education has a positive impact on the work of teachers with people with disabilities [2]. Armstrong found that the use of electronic textbooks developed on inclusive principles helps to improve the academic performance of students studying online [3], Nightingale and his colleagues experimentally proved that the use of inclusive additional lecture material has a positive impact on the academic performance of people with disabilities [4], Cidral focused on studying the satisfaction of students studying online with an educational product and identified, that students ‘ satisfaction with the tools of online education directly affects their academic performance [5]. Russian researchers are also interested in the problems of online education. Grechushkina writes about the prospects of using online courses in inclusive education [6], Shutova deduces criteria for assessing the availability of educational Internet resources for people with disabilities [7], and Yurchenko wrote about the psychological and pedagogical support of participants in inclusive education in distance learning [8]. It is worth noting that there are quite a lot of works devoted to online education, but studies that consider online education in the context of inclusion are not enough to form a holistic view of the problems and prospects of this phenomenon. We can say that the scientific community tends to study online education and inclusivity separately, which, in our opinion, does not meet the requirements of modernity. Online education, according to its idea, should be inclusive and designed to work with people with disabilities, but in practice this does not happen due to the fact that teachers and students are still poorly adapted to online learning and experience serious difficulties with it. At the moment, it is not possible to make online education truly inclusive due to many factors that will be considered in this paper. The aim of the work is to determine the socio-cultural and technical factors that affect the effectiveness of inclusive online education.

2 Materials and Methods

The research is based on the cognitive approach, which considers a person as a system that knows the world and is influenced by the communication that occurs in the learning environment. In other words, the research is based on the understanding of the educational process as communication, where its qualitative characteristics determine the effectiveness of learning [9]. This approach is especially relevant in the context of inclusivity, since communication with people with OVS can become a factor that determines a student's academic performance and involvement in the educational process [10].

The empirical basis of the study is the data obtained from a sociological study conducted by us, the purpose of which was to analyze the degree of formation of the socio-cultural environment at the Southern Federal University, which allows us to assess the degree of readiness of an educational organization to implement an inclusive approach in teaching people with disabilities. The above research serves as a foundation for identifying the problems and prospects of building inclusive online learning in the context of accessible communication with students with disabilities. In the course of the study, 414 students, 255 teachers and 108 representatives of the administrative and managerial staff of SFU were interviewed. 30% of the respondents were boys and 70% were girls. 80% of the respondents are undergraduate students. Research objectives:

(1) find out the degree of awareness of actors about persons with disabilities;
(2) identify attitudes and stereotypes towards people with disabilities;
(3) determine the attitude to people with disabilities.

All the above tasks are aimed at achieving the goal of determining the formation of an inclusive socio-cultural environment at the Southern Federal University, which contributes to the high involvement of people with disabilities in the educational process within the framework of online learning, as well as identifying the compliance of the technical tools of online education with the requirements of inclusiveness.

3 Results

Let us start by considering the issue of students 'and teachers' awareness of people with disabilities, which will determine their attitude to them. Awareness is directly related to stereotypes, which are a defensive reaction to something unknown, which often leads to misunderstandings and resentments [11]. People who have real contacts with people with disabilities are less prone to stereotypes and condescending behavior towards them, while the information vacuum will always be filled with the simplest and most understandable information—stereotypes [12] (Table 1).

According to empirical data, 70% of teachers interact with people with disabilities, while only 39% of students deal with them. The next step is to determine how students

Table 1 Are there people with disabilities among your family, friends, or acquaintances?

Statements	Students (%)	Teachers (%)
Yes, there are such people	39.1	69.1
No, there are no such people	60.9	30.9

and teachers position individuals with disabilities. Here there are two ways: to consider them as people with certain mental and physical characteristics/deviations, or to perceive them as the same as everyone else.

The survey shows that 70% of teachers and 52% of students consider people with disabilities a special category of people, while only 12% of teachers and 8% of students considered them "like everyone else". As a rule, students and teachers get all the information about people with disabilities from the Internet (32% of students, 20% of teachers) and from television (25% of students, 20% of teachers), and much less often from scientific and popular science literature (6% of students, 4% of teachers) (Table 2).

Let us turn to the consideration of attitudes and stereotypes associated with persons with disabilities. Online education focuses on asynchronous communication between the teacher and the student [13], and the effectiveness of interaction avoids many problems associated with misunderstanding of the material, lagging behind classmates and other problems typical of online learning. It should be noted that in full-time education, the communication process is much easier than in online mode, so the issues of the socio-cultural environment of the educational institution have a significant impact on how students will communicate with people with disabilities. In our study, we formulated stereotypical statements about people with disabilities, which can be indicators of the inclusiveness of the online educational environment (Table 3):

(1) people with disabilities look and behave differently from everyone else and find it difficult to fit into society;
(2) people with disabilities require significant and unjustified expenses on the part of the state to ensure their lives;
(3) people with disabilities are no different from others and can do the same as others;
(4) people with disabilities are more similar to ordinary people than it seems at first glance.

Table 2 Who do you think people with disabilities are?

Statements	Students (%)	Teachers (%)
People who have certain physical or mental characteristics	52.7	70.9
People with physical or mental deviations	27.8	7.3
People just like everyone else	8.2	12.7
Sick people with physical or mental disabilities	10.1	9.1
Other	1.2	0

Table 3 Sources of information about the problems of people with disabilities (ranked list)

Rank	% of the total number of students	% of the total number of teachers
1	The Internet—32.2%	The Internet—20.5%
2	A television—25.8%	A television—19.9%
3	Acquaintances, friends—12.5%	Personal experience of communicating with people with disabilities—16.3%
4	Personal experience of communicating with people with disabilities—12%	Acquaintances, friends—12.7%
5	Popular science and scientific literature—6%	Teachers at the university—10.8%
6	Teachers at the university—4.5%	Popular science and scientific literature—9.6%
7	Teachers at the university—4.1%	Educational institutions—8.4%
8	Not familiar with their problems—2.6%	Not familiar with their problems—0.6%

As can be seen from the above, the first statement is categorical and is characterized by a negative attitude towards people with disabilities. It indicates the unwillingness to accept people with disabilities into their social circle. The second statement is even more categorical and criticizes people with disabilities for being objects of state support. The third statement is as tolerant as possible, and the fourth is moderately tolerant.

According to the results of an empirical study, the first statement was chosen the least (8% of students and 2% of teachers), the second statement also did not find popularity among respondents, since 62% of students and 76% of teachers did not agree with it. It should be noted that in these matters, teachers show themselves to be more tolerant than students. Almost 90% of respondents agreed with the third and fourth statements, which allows us to state that even in the absence of regular contacts with people with disabilities, the educational environment is generally favorable to them (Table 4).

To the question of what, in the opinion of respondents, is not enough for people with disabilities, the most popular answer was chosen: "The opportunity to feel like full members of society", and in second place was the option associated with the material support of the state.

Summing up the intermediate results of the empirical analysis of the readiness of the educational environment of the Southern Federal University to work with people with disabilities in the framework of online education, we can say that students have a lack of real contacts with people with disabilities, but they are friendly and tolerant to them. This allows us to conclude that the transition of persons with HIA to the online format will not be associated with significant difficulties.

Next, we will proceed to substantiate the technical factors that determine the degree of inclusiveness of the digital environment. It was found that the subjects of the educational process are tolerant, but the question of whether the tools of online education meet the requirements of inclusivity remains in question. Very few

Table 4 Indicators of the inclusiveness of the online educational environment, formulated on stereotypical statements about people with disabilities

Answer options	% of the total number of students	% of the total number of teachers
People with disabilities look and behave differently than other citizens, therefore they do not fit into society	8	1.8
Ensuring the life of people with disabilities does not require significant unjustified expenses from the state	62.3	76.4
People with disabilities can work as productively as people without health restrictions	89.4	94.5
There are much more similarities than differences between ordinary people and people with disabilities	60.1	70.9

scientific papers have been written on this topic, and the written works are narrowly focused on testing the online interfaces of specific educational institutions. We will try to fill this gap and formulate indicators of the inclusiveness of online education in technical terms. To begin with, the software used for educational purposes should be designed to meet the needs of people with disabilities. The simplest and most obvious solution is to use fonts and colors for people with color blindness. A large number of software developers are introducing Colorblind functionality, which allows them to boast about the inclusiveness of their product, but in our opinion, the inclusiveness of the software depends on many other features. Among these are the following:

1. subtitle generator;
2. information visualizer;
3. text-to-speech utilities;
4. font size management;
5. the ability to use alternative mouse and keyboard manipulators and input devices;
6. gesture and eye control.

The subtitle generator is an invaluable assistant for people suffering from hearing disorders. In classical full-time education, it is necessary to use gadgets, while online education is implemented using software, therefore, every word of the teacher can be broadcast in the format of subtitles. However, at the moment, the Microsoft Teams program, which is the main platform for implementing distance and online learning in SFU, does not have such functionality. To be precise, it does, but only in English. The information visualizer is a popular solution for people suffering from cognitive disorders and experiencing difficulties in recognizing text information. Text-to-speech and Speech-to-text utilities are the most important functionality that allows people with speech or hearing disabilities to communicate with teachers and students [14], but this functionality is rarely used in video communication programs, which leaves the above category of people without the ability to communicate. Font size control is often used, but not always. For example, in Microsoft Teams, selected by the SFU, there is no such functionality. The ability to use something other than a mouse and keyboard to enter data on a computer is also an important element for people suffering from motor-related illnesses. Gesture and eye control is derived from and related to the previous problem, but is aimed at a different audience of people with disabilities.

It follows from the above that the main indicator of the inclusiveness of the educational environment from the technical side is the adaptability of the software and web tools used to meet the needs of people with disabilities. All the functions described above will not be necessary for students who do not experience health problems, but for the disabled, they may be able to participate in the educational process and work on an equal basis with all students.

4 Discussion

Thus, we have identified two indicators of the inclusiveness of the online educational environment. These are: socio-cultural and technical adaptability. The first is directly related to the willingness of students and teachers to accept persons with disabilities as full-fledged subjects of the educational process and interact with them on an equal footing, and the second is determined by the compliance of the technical tools of the programs through which online learning is carried out with the requirements of inclusivity. By inclusive, we mean an environment in which people with disabilities can fully implement educational and communication practices.

Let us consider the correspondence of the online educational environment to the first indicator of inclusiveness on the example of the Southern Federal University.

The adaptability of the socio-cultural environment of SFU can be assessed as satisfactory, which is explained by the tolerant attitude of the teaching staff and students to persons with disabilities. This allows us to conclude that at the level of federal universities in Russia, the attitude towards people with disabilities and persons with special needs is generally positive or neutral.

The technical adaptability of the means of implementing online education in SFU does not meet the criteria of inclusiveness. We can say that for people with disabilities, the transition from full-time education to online education will be associated with many difficulties and obstacles in communication. The main problem here is the choice of the SFU administration of an unsuitable platform for working online. Microsoft Teams is a platform for business meetings, not for education, which is also noted by foreign researchers [15]. The non-compliance of the technical tools of this program with the requirements of the field of education imposes serious restrictions on the communication of persons with disabilities with teachers and students. The most appropriate solution is to switch to another platform or create your own, focused not only on the needs of the educational sphere, but also on the need to include people with disabilities in the online education process.

The experience students gain during online education is determined by their communication. In our opinion, the educational process, in its essence, is communication. This means that it is subject to all the rules of information transmission, such as the presence of noise and communication failures. In face-to-face communication, noise can be eliminated by repeatedly broadcasting the same message in different formulations, and communication failure can be eliminated with the help of improvised tools, smartphones, writing materials and other devices. As part of online education, a student with HIA will not be able to connect auxiliary applications for communication, counting, reading and other operations to their computer and transmit this information to their classmates and the teacher. This means that online education for people with disabilities is associated with many challenges that can be eliminated only with the help of the organizational intervention of the university administration.

Thus, the greatest problems are observed in the technical adaptation of online education to the needs of students with disabilities. The socio-cultural environment of an educational institution is assessed as satisfactory, but technology and software, unlike people, can become tolerant and inclusive only after the intervention of developers and organizers of the educational process. Online education in SFU cannot be considered inclusive until another platform is chosen for the implementation of online education, which has the necessary functionality, since what is currently used does not meet the requirements of inclusiveness.

5 Conclusions

Thus, online learning, despite the obvious advantages, has many factors that prevent the full inclusion of students with disabilities in the educational process.

We have identified two indicators of compliance of the online educational environment with the requirements of inclusivity: socio-cultural and technological adaptability. The first one is rated as satisfactory, and the second one can be considered a source of a large number of problems. It was found that the software used by students of SFU does not meet the requirements of inclusivity and "turns off" persons with disabilities from the educational process. The main recommendation is to create your own online educational platform with tools that meet the needs of students and teachers and at the same time have all the necessary utilities and add-ons for people with disabilities. It should be noted that the solution to technical problems can also be the choice of another platform, in particular, the rental of popular platforms for conducting mass open online courses, which are considered more inclusive than what is currently used.

Thus, the inclusiveness of the online education space has an impact on the experience of students with disabilities and will determine the degree of their involvement in the educational process. Online education creates new requirements for inclusivity that remain largely unexplored and require scientific reflection. We see further research prospects in conducting additional empirical studies related to the technical support of online education and identifying the tools that are needed by Russian students who are experiencing difficulties in using digital tools of online education.

References

1. Ziegler M, Sloan D (2017) Accessibility and online learning. University of Pennsylvania Press, Philadelphia, pp 158–181
2. Kormos J, Nijakowska J (2017) Inclusive practices in teaching students with dyslexia: second language teachers' concerns, attitudes and self-efficacy beliefs on a massive open online learning course. Teach Teach Educ 68:40–41. https://doi.org/10.1016/j.tate.2017.08.005
3. Armstrong C (2019) Adoption of an electronic textbook with inclusive access and adaptive learning reduces DFWI rate in an online introductory nutrition course. J Nutr Educ Behav 51(7):81–82. https://doi.org/10.1016/j.jneb.2019.05.486
4. Nightingale PK, Anderson V, Onens V, Onens S, Fazil Q, Davies H (2019) Developing the inclusive curriculum: is supplementary lecture recording an effective approach in supporting students with Specific Learning Difficulties (SpLDs)? Comput Educ 130:13–25. https://doi.org/10.1016/j.compedu.2018.11.006
5. Cidral WA, Oliveira T, Di Felice M, Aparicio M (2018) E-learning success determinants: Brazilian empirical study. Comput Educ 122:273–290. https://doi.org/10.1016/j.compedu.2017.12.001
6. Grechushkina NV (2019) Prospects for online courses in inclusive education. Bull VGPU, pp 66–72
7. Shutova AS (2018) Open education for people with disabilities: design challenges. Acad Bull UralNIIProject RAASN 1(36):85–90
8. Yurchenko JV (2020) Psychological and pedagogical support of inclusive education participants in distance learning conditions. Innov Aspects Sci Technol Dev 2:169–173
9. Sergeev SF (2016) Cognitive pedagogy: peculiarities of adult learning and education. Pedagogy Psychol Philos 3(03):30–35
10. Allan J (2010) The sociology of disability and the struggle for inclusive education. Br J Sociol Educ 31(5):603–619. https://doi.org/10.1080/01425692.2010.500093
11. Clément-Guillotin C et al (2018) Implicit and explicit stereotype content associated with people with physical disability: does sport change anything? Psychol Sport Exerc 38:192–201. https://doi.org/10.1016/j.psychsport.2018.06.014
12. Rubinstein RS, Jussim L, Stevens ST (2018) Reliance on individuating information and stereotypes in implicit and explicit person perception. J Exp Soc Psychol 75:54–70. https://doi.org/10.1016/j.jesp.2017.11.009
13. Csikosova A, Teplicka K, Senova A (2012) Communication and humanization of University Education through e-learning. Procedia Soc Behav Sci 46:2978–2982. https://doi.org/10.1016/j.sbspro.2012.05.600
14. Hamiti M, Dika A (2010) Learning opportunities through generating speech from written texts. Procedia Soc Behav Sci 2(2):4319–4324. https://doi.org/10.1016/j.sbspro.2010.03.686
15. Pal D, Vanijja V (2020) Perceived usability evaluation of Microsoft Teams as an online learning platform during COVID-19 using system usability scale and technology acceptance model in India. Children Youth Serv Rev 119:105535. https://doi.org/10.1016/j.childyouth.2020.105535

Competency Models of the Professional Institute of Education in the Conditions of Digitalization of Public Life

Yaroslava Zinchenko and Eleonora Morgunova

Abstract The purpose of the article is to study the current areas of professional development of subjects of learning, educational and administrative processes of the transforming institute of education in the context of the requirements of digital society. Education, based on an institutional approach, is considered as a professional institution. Classical scientific approaches to the sociology of professions are given in the research. A comparative analysis of state educational and professional standards, standard job descriptions, as well as curricula for the training of the professional corps of the Institute of Education made it possible to draw conclusions about the inconsistency of the competencies formulated in them with the request of the employer, society, state and the requirements of the new social reality. The data of the All-Russian study of the competencies and qualifications of teachers, current amendments to the legislation on education that integrate new subjects into its social structure are given and interpreted. The work offers specialist models supplemented with current competencies, demanded by modern realities. Resources and scope of application of online training for formation of actual competencies of specialists of professional institute of education are considered. The project of the online simulator "Electronic Teacher Mentor" is presented.

Keywords Professional institute of education · Competence · Qualification

1 Introduction

Education in scientific discourse is most often considered as a value, process, outcome or as a system of social organizations providing educational services, as one of the main institutions of the socialization of the individual. However, the reduction of the institute of education to the performance of the function of social reproduction and the organization of the provision of significant social services does not give a holistic idea of this social phenomenon. Education is also a professional institute, uniting one of

Y. Zinchenko (✉) · E. Morgunova
South-Russia Institute of Management, Branch of Russian Presidential Academy of National Economy and Public Administration, Rostov-on-Don, Russia
e-mail: yazinchenko@uriu.ranepa.ru

A. Beskopylny and M. Shamtsyan (eds.), *XIV International Scientific Conference "INTERAGROMASH 2021"*, Lecture Notes in Networks and Systems 247,
https://doi.org/10.1007/978-3-030-80946-1_37

the largest and extremely heterogeneous in its composition professional corporation. This is a set of norms for the implementation of learning, educational, mediatory and administrative activities that determine the behavioral models of specialists in the field of pedagogy, psychology, management and their relationship with students, their families, regulatory and supervisory state bodies.

The term "professional institute" appeared in sociological science thanks to Herbert Spencer, who used it in relation to "liberal" professions, including teachers. Spencer believed that the institute of professions adapts society to changing conditions and that it preserves and expands national life. Linking science and education, the scientist noted: "Various applications to the practice of scientific discoveries, the various mental interests that a person of science may excite, as well as the universal education that such a person promotes—all this serves to expand life" [1].

The social division of labor and its filling with functional content led to the emergence of a profession. In the chapter "Division of Labor and Production of Capital", Karl Marx wrote that the process of division of labor begins with the separation of various professions [2].

Emil Durkheim believed that a professional institute performs specialized functions assigned to specific social groups, thereby integrating professionals to protect their interests. The scientist designates these social groups as professional corporations—social institutions that create and support positive social patterns of behavior and contribute to the formation of organic solidarity. In the work "On the Division of Social Labor," Durkheim emphasizes the importance of professions for society as follows: "In a professional group, we see, first of all, a moral force that can restrain individual egoism, support a sense of general solidarity" [3].

The understanding of professional morality was continued by Max Weber in research on professional ethics in two of his famous works connecting divine vocation and profession: "Politics as a vocation and profession" and "Science as a vocation and profession." The second work was presented to scientists in a report on November 7, 1917 at the University of Munich in front of students whom Weber turned to as future scientists and teachers to explain the essence of their future vocation. The scientist outlined the teacher's task as follows: "And I consider it irresponsible to use the fact that students should attend teachers' lectures for the sake of their future and that there is no one there who could criticize him; to use their knowledge and scientific experience not in order to benefit students—what the teacher's task is in particular—but rather, in order to instill in them their personal political views "[4].

Pitirim Sorokin emphasized that the significant social properties of the individual are formed due to the mastery of the profession: "It mechanically, in addition to the will and desire of the individual, reconstructs him, creates him in its image and semblance, determines his interests, beliefs, tastes, aspirations and desires; in short, it determines his whole nature" [5].

In such a way, the vocational direction of institutional development of education is due to:

- the public need for specialized activities to educate and foster the younger generation of society;

- the existence of distinctive features of the teaching profession;
- the formalized status-role structure of the Institute of Education: teachers, psychologists, administrators;
- standardization of qualification requirements reflected in State educational and professional standards and standard job descriptions;
- the competencies necessary for the performance of functionality in a dynamically changing society;
- the need for professionalization through competitive selection, certification and continuous professional development of personnel through the development of relevant knowledge, skills and expertise;
- the formation of a professional pedagogical culture;
- specific mechanisms of professional socialization;
- a sense of professional identity of a professional corporation;
- the distinctive motivation for the professional activities of teachers;
- a public assessment of the prestige of teaching.

The challenge to the life of any social institution is due to public need. Institutional education is designed to meet the need for social reproduction of new generations of individuals of a particular society, to connect them to its culture. These functions are carried out in the system of continuing education and are professionally carried out by teachers, as well as accompanied by professionals from other areas of practical activity—psychologists, administrators, IT specialists, lawyers, etc.

The main distinctive features of the pedagogical profession are its creative, humanistic and adaptive character.

The status and role structure of institutional education is supplemented in accordance with the requirements of the time by new entities: tutors, educational advisers and digital development managers.

The regulatory function of a professional educational institution manifests itself by the desire of members of a professional corporation to increase its professional potential, motivates teachers to develop professionalism and competence. The forced rapid transition to a remote format of interaction with students and to a variety of online learning practices, determines the need to change regulatory requirements for the level of professional training and the quality of professional competencies of specialists who implement and accompany the educational process.

Like other socially significant professions, the profession of a teacher is regulated as much as possible by the State by forming a state order for the training of qualified personnel in accordance with educational, and now professional standards accredited by universities. In the future, in carrying out professional activities, the teacher is guided by administrative and official regulations, a contract, regulations of federal, subjective legislation, departmental and local regulatory acts.

The competence of the teacher and pedagogical competence are variables of the teacher's professional status. Competence usually refers to the content and scope of an employee's activities. Competence means the quality of the teacher's activities. Competency includes skills such as performing a task or procedure, as well as the ability to solve problems and interact critically [6]. Competence is the end result, indicating the ability of a specialist to achieve the goal and consists of such elements as: subject and operational knowledge; skills, expertise; ability and readiness to use them in activities; responsibility for the results of these activities [7].

The priority areas of professionalization of pedagogical personnel include: (1) vocational training, retraining, advanced training, mentoring; (2) career growth on a competitive basis; (3) certification; (4) rating; (5) conclusion of the contract.

It is the responsibility of the Russian legislator to carry out his/her activities at a high professional level. The legislator defines another important duty of the teacher—to systematically increase its professional level and the right to additional professional education in the profile of pedagogical activity at least once every three years. Federal legislation provides for the improvement of the professional skills of pedagogical workers.

Professional culture refers to the way in which teachers develop their activities in educational institutions. Thanks to this, the work done becomes important, and young teachers learn to solve their problems, gradually integrating into the professional community. The professional culture of teachers includes beliefs, values, habits and supposed ways of performing things that are separated within a certain group or in the pedagogical community of which they are a part, and which can be observed in their relationships [8].

Professional identity is defined as the individual perception of oneself as a professional in the professional community, which includes a set of ideas about one's profession and is based on attributes, beliefs, values, motives and experience. Professional identity creates a personal framework that determines perception, interpretation and actions in work-related situations [9].

Studies of modern foreign scientists show that teachers are motivated more by internal than external factors, that is, related to the content of work: the nature of work, the duties assigned or assumed, promotion, recognition of contribution and efforts, personal growth and career development [10].

Traditionally, the teaching profession received high public marks and respect. However, this trend has been changing in recent years. According to the Foundation for Public Opinion, published in September 2020, 46% of Russians are convinced that the profession of a school teacher is respected, 45% are sure of the opposite. It is obvious that the feminization of the profession and low wages contribute to the formation of these assessments.[1]

2 Materials and Methods

For the research tasks of the article, statistical and sociological data from the information bases of Federal State Statistics Service, Russian Public Opinion Research Center, Public Opinion Foundation and Yandex were used. The study was based on an institutional approach and a comparative analysis method.

[1] https://fom.ru/Nauka-i-obrazovanie/14464.

3 Results

The professional corporation of teachers is one of the largest and heterogeneous in its composition. According to Federal State Statistics Service, 2,179,027 people are employed in the field of general education, including: high-level personnel—147,020, pedagogical workers—1,347,854, of which teachers-psychologists—27,965, educators—114,539, tutors—5479. The largest age cohort are persons of 45–49 years of age—321,831; young professionals who most recently completed the vocational training cycle—under 25 years of age—97,953 people. The vast majority of them are women.[2]

The Ministry of Labor of the Russian Federation acted as the customer of a study identifying gaps between official ministerial qualifications—professional standards and real demand in the labor market among teachers, educators, teachers of preschool institutions and university professors. The executors of the study—the All-Russian Center for the Study of Public Opinion and the National Agency for the Development of Qualifications have tested an automated system for collecting data on available and promising qualifications in the economy of Russia. The study involved 4,000 employers embodied in frontline managers. 502 descriptions of what professionals actually do at work (teachers, educators, teachers of schools and preschool institutions, university professors) were compared with descriptions of professional standards of the Ministry of Labor. Only 15% of the descriptions provided by employees did not coincide at all with the descriptions of the registers of the Ministry of Labor. In the remaining 85%, the descriptions of workers either completely coincided (18%) or meaningfully repeated the requirements of the registries, but were stated by respondents "in their own words" (67%).[3]

The above data was submitted in January of this year. It is worth noting that the new social reality, due to the limitations of the Covid-19 pandemic has actualized the need to review the competent models of specialists implementing and accompanying the educational process.

The Russian federal state educational standard for higher education in the direction of training "Pedagogical Education" contains 29 competencies; 99 labor actions contain a professional standard. In comparison with this number, foreign researchers significantly optimize the competent models of teachers by the number of incoming elements. For example, a group of Slovak scientists from the University of Žilina consider the optimal competent model, consisting of eight positions: moral and ethical, role, technical (expert), competence of a mature person, scientific competence; critical thinking competence; copyright, pedagogical, communication, motivational competence [11]. Romanian scientists distinguish scientific and professional competence by the specialty, psychological and pedagogical competence, as well as psychosocial competence and managerial competence [12].

---

[2] https://docs.edu.gov.ru/document/ed3ca74f26a1dc055a313991f66d2fa3.

[3] https://wciom.ru/analytical-reviews/analiticheskii-obzor/kvalifikacija-uchitelei-v-rossii-normy-i-realnost.

First of all, the information and analytical competencies of teachers, tutors, psychologists and administrators need updating. Digital competence is a reflection of digital skills and knowledge, which are politically considered necessary for full participation in our increasingly technological society [13]. The professional standard of the teacher already in 2013 stipulates the need to possess the following information and communication competencies: general user, general pedagogical, subject-pedagogical and professional.[4] Given competencies, that appeared in two years after the professional standard, were not reflected in the updated educational standard of the higher pedagogical education. The curriculum for the training of a teacher of history and social studies of one of the best pedagogical universities in the country and the developer of the professional standard for 2019—Moscow State Pedagogical University, contains only two educational disciplines that develop these competencies during a total of 60 academic hours of contact work with a teacher. In 2020, in the conditions of abrupt and full transition of educational process and its participants to a remote format, teachers and managers quickly increased the digital competence-based arsenal through development of the principles and mechanisms of work of the software allowing to conduct educational process remotely; learned to inform personnel, parents and students on organizational issues; learned to estimate results of study and work. However, the information educational environment provides much more opportunities to the professional to influence own activity and activity of students, creation of the professional image and digital reputation. First of all, it concerns opportunities of digital analytics of big data on the basis of digital traces of students which allows to draw conclusions about efficiency of a technique of training and estimation of results of students. Competently building a digital image of a professional on the network and managing a digital reputation will help position yourself as a specialist demanded by the educational services market and capitalize your professional competencies. Of course, the skills of searching and verifying information, optimizing information flows, and effective communication in a virtual environment will help not only the teacher, but it will also help adapt his student to life in a digital society. Strong digital competence of teachers is a key component of the development of skills that allow students to make the most of new technologies [14].

The most effective way to develop the digital competence of teachers is to provide them with experience in integrating digital technologies into their professional training [15].

Last fall, Yandex, together with experts from Moscow State Pedagogical University, evaluated digital skills from 63,365 Russian teachers. Teachers who underwent testing showed a good level of digital literacy—61% scored 50–75 points for the corresponding tasks, 26%—75–100 points. However, it turned out that the higher the experience of the teacher, the harder he or she is given tasks that test digital literacy. Therefore, it is important to specifically help senior teachers, who make up the numerical majority of a professional corporation with the mastery of new technologies and tools. Young teachers, in turn, also need to develop their competencies,

[4] http://www.consultant.ru/document/cons_doc_LAW_155553/.

mastering a training course at a higher school in a situation where the requirements of the State, profession, client and educational program are broken. The virtual educational product "Digital Teacher Mentor" is intended to help both.

3.1 Online Simulator Project "Digital Teacher Mentor"

The standard job description of a teacher of a general education school is based on a competent approach. The structure of the job description has a model consisting of seven main blocks: general provisions, main components of competence, functions, duties, rights, responsibility, relationships and relations by position. The main components of the teacher's competence include: professional competence, communicative competence, information competence and legal competence.

The simulator "Digital Teacher Mentor" is a project for self-education and the development of teacher competencies, a digital tool for working out managerial and pedagogical decisions.

The simulator models a virtual school (educational space) with real situations of pedagogical activity.

The teacher himself or with the help of an invited tutor will be able to:

- to work out the legal and regulatory framework in the field of school student's education;
- to learn to analyze the indicators of the educational process;
- on the basis of data analysis, to learn to develop strategies for the development of the quality of education and learning;
- to formulate objectives;
- to distribute tasks and implement them in a virtual environment;
- to deal professionally with pedagogical situations;

The following categories of employees can use this digital tool:

- young teachers;
- teachers who have encountered difficulties in their teaching activities;
- candidates of competitive selection for a position "teacher", "educator", "additional education teacher"—students of pedagogical higher education institutions;
- heads of educational institutions for personnel solutions.

The basis of the simulator is the portfolio or profile of management competencies of a modern teacher, which includes:

- resource management;
- results management;
- information management;
- lesson Management
- class management;
- management of the parent community;

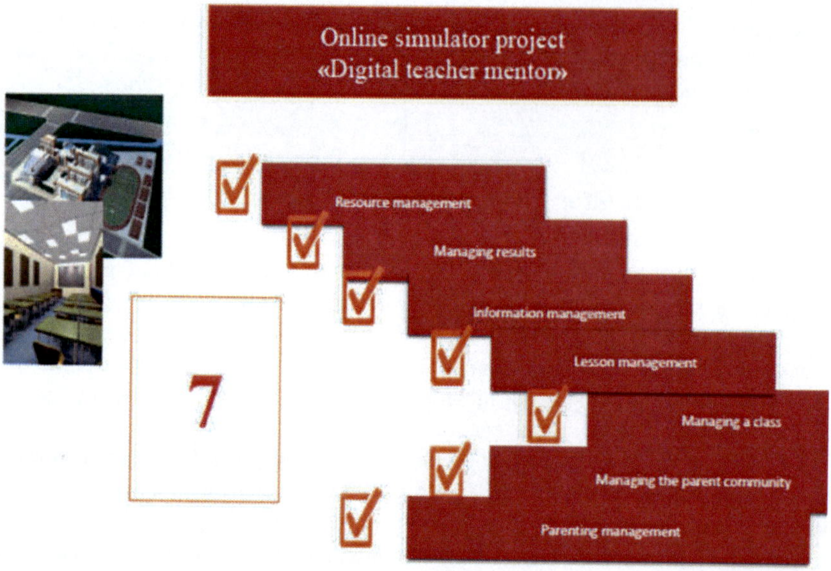

Fig. 1 Online simulator project «Digital teacher mentor»

- education management (security, socialization, general personality culture, healthy lifestyle, development).

The student is given an opportunity to form 7 managerial competencies through the "Digital Teacher Mentor" simulator; 7 levels of passing training tasks in various areas of competence are provided as well (Fig. 1).

4 Discussion

In January of last year, the Ministry of Education of the Russian Federation launched the national project "Education" with a solid total amount of funding—784.5 billion rubles. Basically, the project events affect the field of general education, but there are those that concern the higher education system as well. Therefore, 30 universities located in different regions of the country, having passed the competitive selection, will receive serious state support for updating development programs taking into account the national goals of the Russian Federation. 80 universities representing 40 regions of the country will also receive state support and will be included in the list of educational institutions that train personnel for the basic sectors of the economy and social sphere. In order to occupy the tenth place in the world ranking of the top 500 universities in 2024, 60 universities are tasked with moving to the implementation of 5 educational programs that have passed international accreditation. Moreover, higher educational institutions will be actively involved in the activities of volunteer public associations. The state will also support continuing education projects—more than 14 million people will undergo additional vocational education courses. Updating the content of education and creating modern infrastructure in terms of the university system is associated with distance learning technologies. 20% of students will master educational disciplines in the format of online courses, universities will provide this educational process, ensuring that the quality of training of students meets the world level.[5]

Over the past year, educational institutions and participants in the educational process have faced unprecedented challenges of a new social reality. In a short timeframe, teachers and students entered new formats of interaction and communication, and felt a variety of effects of digitalization in everyday life themselves.

How do students who, in accordance with the goals of the national project, have to master the educational products of the country's leading universities evaluate distance learning?

The study of the opinions of students primarily indicates their consensus on a number of problems, which emphasizes the commonality of the social attitudes of participants in the educational process and characterizes them as representatives of a truly new digital generation, to the needs of which adequate pedagogical approaches have yet to be found that will fit into the technological capabilities of educational platforms.[6]

Regardless of the situation of transition to a new learning format, modern students call the development of potential, competencies (38%), achievement of high results (35%) and pleasure from educational process (34%) as the main motivations for involvement in the educational process.

What event are of interest to students? First of all, cultural, mass and sports events (57%), followed by educational events in the format of open lectures and master classes (51%), public (27%) and volunteer (26%) activity.

What happened after the transition to the regime of quarantine and after the cancellation of events of involving collective activities? Half of the respondents admit that in the self-isolation mode they did not take part in any form of educational, creative or scientific activity. Only a small part of students (15%) participated in open educational and scientific events.

The degree of satisfaction with the organizational conditions of the educational process, both before and after the transition to remote format conditions, is highly appreciated, and the assessment decreased only by one indicator—information about extracurricular and scientific events (42% of students were fully satisfied before the transition to the distance learning, after—only 26%), which is natural, since many planned events were postponed indefinitely.

---

[5] National Education Project//[Electronic Resource] https://edu.gov.ru/national-project/.

[6] A pilot sociological study on the impact of the COVID crisis on the problems of the educational process was conducted in September this year. The survey was attended by 127 s-year full-time students studying in the direction of training "State and Municipal Administration".

Has it become easier to learn in a new format? Most respondents are sure of this. Answer options: "much easier" and "rather easier" have been chosen by 19% and 39%, respectively. Only 2 people indicated the answer "much harder."

Students positively assess their performance as well. The situation did not worsen for 43% of students, slightly increased for 30% and significantly increased for 17% of students who answered this question. Such a positive trend is reasonable in order to analyze the number of positive and negative assessments obtained during the control events during the semester with those put forward by teachers during the interim certification. And here there are opportunities for digital humanities—analytical work with large data—statistical and sociological, the totality of which can become the basis for making decisions regarding the quality of the educational process and the provided educational service.

But what do students think about the quality of knowledge? The answers of most also record a positive assessment, but it is already more restrained. The fact that the quality remained at the same level was stated by 40% of people; only 17% see a slight increase; the quality of knowledge increased significantly for only 3% of respondents.

What difficulties did students encounter when moving to a distance learning format? It is characteristic that the communication problems of the student community did not arise, nor did they arise with the preparation and attendance of online classes and intermediate certification events. The problems of distance learning are related only to the technical side—low quality of Internet connection and power outages, and these rates are not so high (15 and 11%) for the educational process to stagnate.

How involved are the students with online classes in the learning process? 48% claim that they have never left the virtual audience in order to engage in extraneous affairs, and 46% admit that in about a quarter of cases they allowed themselves such freedom.

The absolute majority of respondents (over 90%) claims that there were no cases of inappropriate, conflicting behavior on the part of both students and teachers, and this is also a good trend—generational features did not prevent the building of a constructive and civilized dialogue.

Only a quarter of respondents sought advisory support from administrative structures, but more than 80% turned to teachers for help and received that help and were satisfied. It is necessary to pay tribute to the teaching staff, as communications with students have intensified significantly—40% of students sought advice at least once a week.

More than half of respondents do not want to abandon remote formats of various training events in the future. Most of all, students enjoyed attending online lectures (84%) and taking exams/credits online (78%). However, it should be noted that 30% of them are wishing to refuse to study in open courses of educational platforms, who obviously believe that such educational products should not replace communication with teachers of their native university who read author's courses corresponding to the direction of training.

Students are also unanimous in assessing the readiness of the teaching staff to conduct classes in a distance format (85%) and adapt educational disciplines to online formats (79%).

93% of respondents say that they adapted to distant learning and were able to improve their digital skills: working with relevant software, independent training, time distribution and electronic communications.

Quarantine restrictions also contributed to the transition to adulthood in 20% of young people, they first found employment.

About a third of respondents noted a deterioration in sleep and rest, deterioration in nutrition and physical and psychological well-being. However, the vast majority claims that drinking alcohol, energy drinks, psychoactive substances, smoking and game addiction during the period of self-isolation did not affect them in any way.

Respondents admit that at this difficult time they felt anxiety (40%), a decline in strength (51%), irritation (39%), however, despite this, more than half of the students tried to maintain the usual level of activity, and 36% took advantage of the free time to learn something new and implement plans for which there was not enough time before.

Therefore, it is possible to draw conclusions about the readiness of teachers and students to build activities in a virtual space using all the technical achievements of our time. This readiness allows us to be optimistic in the near future (despite the COVID crisis)—the future associated with state initiatives: ensuring the global competitiveness of Russian education and the education of a harmoniously developed and socially responsible person.

5 Conclusions

Competencies form the basis of any skillful working behavior, and their maturity is crucial for successful work in the relevant profession. From this point of view, the competence of teachers is of exceptional importance, mainly because teachers form the basis for the creation of new knowledge and new values that are useful for students and their subsequent path of continuous education, and it is useful for a society being socially reproduced by high-quality generation.

The updating of regulations governing the process of professional training of such a professional corporation, as teachers, which is strategically important for any society, should be based on sociological analysis methods and begin with the social design of a competent model of a specialist.

In order to maintain the level of qualification and growth of professionalism, teachers of any age and experience need mentoring at the moment of a problem situation. This task can be solved by modern digital online educational products that have the architecture of official regulations and contain content that ensures the mastery of each functional model of the teacher.

References

1. Spencer G (1898) Social professions. Foundations of Sociology, SPb, vol 2, pp 484–485
2. Marks K, Engels F, Gospolitizdat M (1955) (eds) Capital, vol 1, K. Marx, F. Works, 2nd edn, vol 23, p 363
3. Durkheim E (1996) On the division of social labor. Canon, Moscow, p 25
4. Weber M (1990) Selected works. Progress, Moscow, p 712
5. Sorokin PA (1993) The system of sociology: in 2 volumes. Social analytics. The doctrine of the structure of complex social aggregates. Moscow, vol 2, p 183
6. Olofsson AD, Fransson G, Lindberg JO (2019) A study of the use of digital technology and its conditions with a view to understanding what "adequate digital competence" may mean in a national policy initiative. Educ Stud 5–17. https://doi.org/10.1080/03055698.2019.1651694
7. Zinchenko YG (2011) Professionalization of the state civil service. Rostov-on-Don, p 98
8. Chipaco E, Branco ML, Professional culture and teacher professionality in higher education. In: Conference: 10th international conference on education and new learning technologies. https://www.researchgate.net/publication/326716984_PROFESSIONAL_CULTURE_AND_TEACHER_PROFESSIONALITY_IN_HIGHER_EDUCATION
9. Richter E, Brunner M, Richter D (2021) Teacher educators' task perception and its relationship to professional identity and teaching practice. Teach Teach Educ 101:103303. https://doi.org/10.1016/j.tate.2021.103303
10. Benteaa C-C, Anghelachea V (2012) Teachers' motivation and satisfaction for professional activity. Procedia—Soc Behav Sci 33:563–567. https://doi.org/10.1016/j.sbspro.2012.01.184
11. Blašková M, Blaško R, Kucharčíková A (2014) Competences and competence model of University Teachers. Procedia—Soc Behav Sci 159:457–467
12. Lăzărescu Mihaela P (2015) Dimensions of teaching staff professional competences. In: The 6th International Conference Edu World 2014 "Education Facing Contemporary World Issues", 7th–9th November 2014. Procedia—Social and Behavioral Sciences, vol 180, pp 924–929
13. Ilomaki L, Paavola S, Lakkala M, Kantosalo A (2014) Digital competence—an emergent boundary concept for policy and educational research. Educ Inform Technol 21(3):655–679. https://doi.org/10.1007/s10639-014-9346-4
14. Howard SK, Tondeur J, Ma J, Yang J (2021) What to teach? Strategies for developing digital competency in preservice teacher training. Comput Educ 165:104149. https://www.researchgate.net/publication/348826715_What_to_teach_Strategies_for_developing_digital_competency_in_preservice_teacher_training/link/601ba63aa6fdcc37a8ff84e6/download
15. Farjon D, Smits A, Voogt J (2019) Technology integration of pre-service teachers explained by attitudes and beliefs, competency, access, and experience. Comput Educ 130:81–93. https://doi.org/10.1016/j.compedu.2018.11.010

Online Education as an Indicator of Effective Public Administration of Higher Education

Tatiana Anistratenko, Tembot Bersyrov, and Alexander Dyatlov

Abstract The relevance of this article is determined by two social processes related to the institutionalization of online education in the higher education system of the Russian Federation. The first is that the total number of students studying in electronic format is steadily increasing. The second one is manifested in the state's desire to use online education as an indicator that evaluates the effectiveness of public administration in the field of educational policy. The author hypothesizes that online education cannot be an indicator of the effectiveness of public administration, since the ill-considered implementation of this form of educational process will inevitably lead to a decrease in the quality of student education and a systemic deterioration in the training of specialists. To substantiate this hypothesis, an empirical study was conducted using the group interview method in student groups. The analysis of the obtained data fully confirmed this hypothesis.

Keywords Online-learning · E-learning · Distance education · Higher education

1 Introduction

The relevance of the research topic is determined by some regulatory acts that have been adopted and are valid as formal rules governing the social institution of higher education in Russia. One of the fundamental documents that play an important regulatory role is the State program of the Russian Federation "Development of education" for 2018–2025, adopted on December 26, 2017. It contains two indicators, the achievement of which is defined as a condition for completing the specified document. Thus, according to the criterion of "quality of education", it is required to ensure by 2025. Employment in the developed educational areas is up to 59%, at least 10 Russian universities have entered the top 100 world rankings for at least

T. Anistratenko · A. Dyatlov (✉)
Southern Federal University, Bol'shaya Sadovaya st. 105/42, Rostov-on-Don, Russia

T. Bersyrov
Maikop State Technological University, Pervomayskaya st., 385000 Maikop, Adygea Republic, Russia

A. Beskopylny and M. Shamtsyan (eds.), *XIV International Scientific Conference "INTERAGROMASH 2021"*, Lecture Notes in Networks and Systems 247,
https://doi.org/10.1007/978-3-030-80946-1_38

two consecutive years, and according to the "online education" criterion, 5 million students have completed online courses. Our focus is on only one indicator—online education. It will be the subject of research in the proposed article.

The development programs were based on the pre-defined premise that online education is a certain progressive phenomenon, the spread of which in Russian higher education should be fully and maximally encouraged. The growth of students' involvement in online learning is estimated, therefore, as a result of improving the efficiency of higher education management. In our opinion, such a decision contains certain groups of risks that can potentially destroy higher education. Digitalization of the educational process should be carried out in a meaningful way. It should be implemented only where it will be for the benefit, where there will be no deterioration in the quality of training.

From this point of view, the purpose of the work can be defined as an analysis of the possibility of using online education as an indicator of effective management of higher education.

2 Materials and Methods

We will present this part of the article in two sections. The first section will describe the research methodology, and the second section will describe specific empirical methods that have been used in empirical practices.

First, let us define the methodology and a few basic concepts that underlie our concept.

Neo-institutional theory has been chosen as a methodological platform. First of all, we will decide whether it is acceptable to use it in the practice of working as a sociologist. Including when analyzing the social institution of higher education.

All scientists working within the framework of neo-institutionalism proclaim the principle of methodological universalism. In their opinion, the conceptual approaches developed when considering economic processes are quite acceptable to apply to the study of social reality. If in the studies of classical economists of the XVIII and XIX centuries, as well as the first half of the XX century, the actual economic problems were analyzed using methods developed in sociology, political science, philosophy, law and other Sciences, then neo-institutionalists began to transfer the approaches constructed in the study of Economics to the theoretical and methodological Arsenal of socio-humanitarian Sciences, including sociology [1]. Following this direction further, we note that neoinstitutionalism is characterized by a complex understanding of economic and social processes, because of which a deductive type of thinking is declared, based on the assumption that it is impossible to understand individual segments of social reality without taking into account all factors that somehow affect the behavior of social actors [2]. At the same time, the economy plays a leading role. That is why it becomes the intellectual platform on which the most productive, according to neo-institutionalists, doctrines addressed to the socio-humanitarian Sciences are created [3].

The choice of a neo-institutional doctrine, in addition to the above, is due to a methodological revolution that is significant for the subject of our research. This refers to the transfer of cognitive interest associated with it from studying collectives and social groups to understanding the individual behavior of social actors. The problem of the social and the individual has always been a stumbling block for representatives of the social Sciences. On the one hand, the holistic principle that the whole is something more than the sum of its constituent particles, which was most consistently implemented in the methodology of Durkheim [4] and Parsons [5], put pressure on social activists. For the proponents of this approach, social reality was presented in the form of a frozen socio-cultural matrix, in which actors accepted already established, non-alternative models of social behavior. The key to understanding activities and behaviors is the social context. The opposite is true for studies of the incentive spectrum of activity, where the point of reference is to consider the individual, mainly, needs and desires of social actors. This trend stems from the sociology of M. Weber, who first set out to understand social reality through the value-motivational sphere of an individual [6]. Neoinstitutionalists, in particular J. Coleman, proposed their own way to resolve the differences between the two opposing doctrines. It is based on the fact that the social activity of actors is basically rational, based on a free choice of target directions, but, at the same time, is subject to formal and informal restrictions with which they are forced to coordinate their social behavior [7]. Rational choice of constraints is accepted as an essential condition for achieving individual goals.

This methodology contains a great potential for studying the Institute of higher education, which is inherently characterized by the dialectic of institutional, reflecting strict regulation of activities by many formal and informal rules, and individual, which is inherent in it due to the highly intellectual nature of work, based on free choice, without external compulsion.

The application of this methodology allows us to define the basic concepts in the perspective we are interested in.

Public administration *is a separate type of social administration. Its main feature is that management activities are carried out by state authorities specially authorized for their implementation.*

In each state, these bodies have their own structure, name, range of functions and scope of powers. All these characteristics may differ based on what understanding of the essence of the state dominates, what methods of management are mainly used, and what socio-cultural features distinguish the society to which the management influence is directed. First of all, the scope of power exercised by state authorities, and ultimately, the dialectic of organization and self-organization will depend on all this [8]. In the tradition of Russian statehood, it is necessary to concentrate the maximum possible set of power resources, since the state acts as a leading actor in social management. This, as a rule, does not cause protests from the Russian society, which has such properties as inertia, dependency, and dependence [9]. These distinctive features are not the only ones and in no case can they be addressed to the whole society, but they occupy a prominent place in the system of established social values.

There are several approaches to determining efficiency. We believe that it will be difficult to assess the effectiveness of higher education from the point of view of comparing the amount of resources and time relative to the result obtained. First of all, because of its subject specifics. It is rather difficult to measure the final product created by science and educational activities in terms of time and material characteristics. And this is due to the fact that it cannot be defined as a standardized product by its characteristics. The first two approaches are more applicable for conveyor production or any kind of typical operations, where the invariance of production conditions allows you to calculate all types of costs incurred for them with detailed accuracy. This makes it possible to consider and measure all the existing factors that affect efficiency in both negative and positive terms. The institutional sphere of higher education deals with a product of special quality. The value it creates—"knowledge"—manifests itself in the form of scientific innovations or relevant educational information. In contrast to the activities of a commercial organization, where material resources and production time are primarily considered, in higher education there is a complex synergy of invested efforts and obtained results, the effectiveness of which most often does not have a direct correlation with material resources and labor duration. Yes, and it can be evaluated after an indefinite time and by indefinite subjects (employers, graduates, patients, clients, colleagues, etc.).

So, we assume that *efficiency is the optimal use of various types of resources within the established stable ideas, which contributes to the achievement of goals that are relevant for social actors, organizations, society, and the whole.*

In this understanding, the main problem of achieving efficiency is the harmonization of the ideas indicated in the above definition regarding its qualitative content between the three parties—the social actor, the organization and society. A potentially probable solution lies in the theoretical approach we have chosen. The value of the declared neo-institutional methodology is that it provides an opportunity to justify the procedural side of reaching the expected performance indicators and, at the same time, shows the prospects for solving problems of non-achievement of efficiency that are universal for all institutional spheres.

The analysis carried out is sufficient to formulate our ideas about what we understand by *the effectiveness of public administration.* First of all, it is a question of purpose. The goal should be clear to social actors and achievable from the point of view of the resources allocated to them by the state authorities. Simplifying the above definition *of efficiency, we will fix it in relation to the activities of the state as the achievement of socially significant goals with optimal expenditure of public resources.* This understanding stems from the assessment of the state as a socio-political institution that implements public power and is aimed at maintaining a fair order in society, harmony between social groups, a high level of economic well-being, basic values, geopolitical interests, and strengthening its own sovereignty [10].

It should also be determined with our understanding of higher education. A social institution of higher education can be defined as a stable set of formal and informal norms that guide its social actors in their professional and educational activities, being focused on achieving their individually defined goals, the vectors of which do not contradict the General institutional goal.

Finally, understanding online education is of particular importance. In real educational practices, several models are used that have different learning technologies and functional purposes. However, all of them are united by the common name "online education". In our opinion, this becomes possible only because of the lack of attention to this problem on the part of science. To clarify the use of terms, you can use some conceptual revision. First of all, we should abandon the outdated concept of "distance learning". It is advisable to replace it with a more successful legitimate definition of "distance learning technologies". And the concept of "e-learning", introduced by the law on education, seems more appropriate than the concept of "distance learning", since it more accurately reflects the specifics of the educational process in the information society [11]. The specificity of online learning, which differs from e-learning, consists in the organic use of distance technologies and contact interaction, which together creates a new quality that requires a different conceptual design. The following working definition is proposed: "*Online education is a formalization of the educational process implemented in a remote mode of simultaneous interaction between the teacher and the student, provided through the use of information and telecommunications networks*" [12].

When studying the subject of research, the author of the article used the empirical results obtained by V. V. Kovalev, A. V. Dyatlov and K. V. Vodenko through the use of the group interview method [13]. In total, the work was carried out in individual and student groups of the southern Federal University. Six student groups, 143 first-and second-year students participated in the interview. Of the six groups, two are represented by the areas of "Sociology", "Foreign regional studies" and "Conflictology". The average interview time is 3.5 h. At the end of the survey, students were asked to write an essay on the topic "do I see the possibility of using online education in the educational process, and if so, in what part of it?" The total array of essays was 137 papers.

3 Results

The main idea was to move away from evaluating the convenience of learning and its accessibility, focusing on the positional advantages of classroom-contact and online education in ensuring the quality of educational services. At the same time, quality was understood not as the criteria defined by the Ministry of education and science, but as the expectations of students related to the acquisition of universal and professional competencies. The final goal of the study is to identify students' expectations of obtaining a high-quality education in the course of educational activities within the framework of classroom-contact work and training on the Teams platform, and if there is personal experience—in the e-learning format.

The main question of the group interview is formulated as follows: "What are the advantages and disadvantages of classroom-contact and online learning in achieving high-quality results of your educational activities?". All related concepts were previously operationalized and included in the guide in the form of target definitions and variables that reflect the logical scope of the concepts. Variables were derived from empirical indicators. The interview participants were offered two main indicators: learning and intellectual development. Achieving the goal of high-quality training was measured with such variables as: (1) mastering knowledge, skills and abilities, (2) gaining practical experience, (3) developing abilities, (4) forming students' motivation. Achievement of intellectual development was evaluated using: (1) intelligence, (2) sensitivity, (3) curiosity, (4) verbalization, (5) erudition. The content of indicators was explained to students in the required sequence. At the same time, the concept of online education itself was discussed in advance, and three of its discursive models were formulated and characterized by students with the help of a moderator. When making their judgments, respondents were asked to specify which form of online (or e-learning) training they should use.) they evaluate. The remaining variables were derived from the direct discussion process, since they were not directly related to its purpose, and were aimed at establishing the individual characteristics of individual groups.

The following coding categories were used to interpret the data obtained: education, training, intellectual development, online education, indicators of the quality of education and their variables, type of education, form of education, hybrid education, problems of classroom-contact education, solving problems of classroom-contact education. These phrases and terms are most relevant to the goal and provide an opportunity to group data based on the most General terms to variables of empirical indicators, as well as to identify potential ways to solve educational problems related to improving the quality of educational services.

Education. In most cases, students focused on the fact that education is a process of mastering professional knowledge. However, in this perspective, their ideas were expressed exactly until the moment when the idea was expressed during the discussion that this is also "personal development, which is also of great importance." This statement rarely raised objections, and almost all groups quickly adopted a consolidated view that education should not be reduced to vocational training alone.

Training. This concept took the shortest time to clarify the ideas of the respondents, since almost all statements were reduced to the acquisition of professional skills. To the moderator's question about the availability of alternative judgments, all the answers were typical: "we agree with the speakers."

Intellectual development. As a rule, the answers to this question turned into the actualization of personal growth, motivation, and upbringing. For most respondents, this concept was poorly reflected in clearly formulated ideas. This shows that students do not have their own algorithm and reliable criteria for reflection. They rely entirely on teachers to define mechanisms, guidelines, and control measures for testing intellectual development.

Education quality indicators and their variables. The most intense debate was caused by the very first indicator—"mastering knowledge, skills and abilities". Overall, there were 1–2 online education advocates in each group who were convinced that online education has the potential to be a reliable tool for acquiring professionally relevant knowledge. The main argument is: "anyone who wants to study will study in any format." Opponents of online education on average in groups were 5–6 people. There are three main arguments against:" there is no necessary environment where everyone is used to learning"," the presence of a teacher for learning is necessary, because he acts as an expert in mastering the material", "learning with live communication is more interesting and productive". Most of the participants in the groups behaved passively, but in front-line blitz surveys, positive statements about the ability of online education to provide high-quality educational services were extremely rare. The remaining eight variables were discussed for a shorter period of time, as the level of consolidation within groups around the rejection of online education sharply increased and, accordingly, the subject for discussion disappeared.

Type of education and form of education. The discussion of these concepts was conducted in order to solve the following problem: to determine the place of online education in the Russian educational system. Most agreed that it has both specific and form-forming properties. Moocs are a separate subspecies of additional education, and work in contact online is its form for higher professional education. With the exception of 1–2 respondents, students were convinced that moocs will not be able to replace classical education, and work on Teams does not provide the necessary quality.

Hybrid education, problems of classroom-contact education, solving problems of classroom-contact education. Almost all respondents supported hybrid education, which includes elements of e-learning and "live work with a teacher in contact". Lecture classes received the most critical comments. Many of the students, even online opponents, offered to record them in video mode. There were no interesting opinions about solving the problems of classroom-contact teaching, which obviously shows the lack of awareness of students on this issue. Or, alternatively, it reveals a limited resource on this aspect of moderation of a group interview by a teacher (Table 1).

The discussion revealed that most students expressed a negative opinion about the opportunities to ensure the proper quality of education in all the online formats discussed above. Many people periodically tried to translate the dialogue into a conversation about the convenience and accessibility of learning using electronic resources, but after turning the vector towards quality assessment within the framework of the proposed indicators, almost all concluded that classroom-contact education has a greater potential. However, we do not consider these results to be so significant in terms of students' ability to reconsider their point of view.

Table 1 Distribution of supporters and opponents of online education

Statements	Results
I support online education	14
I am against online education	33
I support hybrid education	106

4 Discussion

The problem of the quality of education causes numerous disputes among specialists of different subject areas. Next, we will present several approaches to this problem, and then, in the section "Conclusions", we will relate all the issues considered to the concept of the quality of online education and the permissibility of determining the effectiveness of public administration in the field of higher education through such an indicator as the ubiquity of higher education.

V. M. Polonsky by quality means compliance with a certain level of education, which is fixed according to the traditional aspects of personality development in Russian pedagogy. These facets include competence (knowledge, skills), intellectual (development of cognitive abilities that form memory, attention, and imagination), physical (improvement of the individual's bodily resources), and moral (readiness to accept the spiritual values of society as their own) components [14]. This understanding can be called typical for Russian teachers, including those working in the higher education system. V. A. Kachalov and B. A. Prudkovskii quality associated with the successful functioning of the educational system, is able to achieve the students a level of proficiency that corresponds to certain normative standards [15]. A. A. Andreev and V. I. Soldatkin writing about the quality of education in the context of its utility in certain types of socially significant tasks. This is not just about the standard or its levels of achievement, but about the practical applicability of this standard [16].

K. V. Pavlenko investigated the multivariate assessment of the quality of higher education depending on the target orientations of the subjects of the evaluation [17].

From the perspective of the subject-oriented approach, sociologists were able to identify and characterize the main disagreements that have arisen on quality issues among different social actors of the Institute of higher education. However, such studies are usually conducted in isolation of practice from theory, which largely deprives them of scientific value. Their significance is localized only by those universities where opinions were measured. The practical efforts of sociologists clearly lack theoretical and methodological foundations. Meanwhile, scientists periodically attempt to develop them in order to develop a universal theory of quality.

B. G. Ushakov proceeds from the fact that the higher education system in its relations with society through the concept of "quality of education" has a multi-level character. It identifies the levels of state, regional, University and the subject of training itself. The quality of education is understood by him in the dichotomy of learning goals (derived from values) and achievement of the level of learning (measured quantitatively). In his opinion, it is impossible to come to absolute harmony regarding the quality of education at different levels, since the standard of quality of education cannot be described in a normative way. Hence, the author emphasizes the importance of dialogical forms of interaction between all subjects of the educational space as a condition for overcoming obstacles that prevent them from receiving high-quality education [18].

An attempt to develop a concept of the quality of education was made by E. N. Zaborova and N. V. Shchipacheva. As evaluation indicators, the degree of satisfaction of the main actors of the educational process, i.e. students and teachers, was determined. By the quality itself, they mean "…characteristics of the institution of education, the process of education and the result of education, their measure, compliance with the needs and expectations of society, the state, and various social groups" [19]. Based on this definition, the authors suggest evaluating three basic variables: characteristics of educational institutions, properties of the educational process, and the level of education of graduates. As a result of the formalization of variables, subjective assessments of actors and objective achievements of universities serve as the basis for measuring the quality of education.

5 Conclusions

The conducted analysis of group interviews confirmed our hypothesis that in the conditions of online education there are great risks of reducing the quality of educational process results. If education is not presented in its pure form as the transmission of knowledge, then its goals cannot be reduced solely to the technological process of high-quality packaging of educational content in some form that can be separated from the source of knowledge creation itself. At this stage, education cannot be subject-free or quasi-subject. It must necessarily be organized in such a way that the one who creates the educational environment is, at the same time, the main inducer for the perception of socially useful knowledge. This conclusion means that the indicator adopted in the development Program for 2018–2025 in the form of online education, according to which 5 million students should attend online courses, is erroneous. This indicator cannot and should not be an indicator by which the effectiveness of public administration is assessed. The very implementation of online courses in the curricula of higher education institutions in the space of quantitative assessments is nothing more than an extremely unsuccessful attempt by officials from the Ministry of science and higher education to enter the mainstream of the fashionable phenomenon of modern pedagogy and higher education, the desire to be progressive without a clear and clear understanding of the goals for which new educational technologies are introduced. Without any doubt, it is the goal that should be the basic criterion for evaluating the relevance of a particular educational technology. It would be a mistake to introduce online education just because the world's leading universities have become participants in this process. Moreover, they use this tool primarily for the sake of extracting commercial profit by expanding the number of students studying on a commercial basis. In our universities, online education is introduced for full-time students who receive education on a budget basis. It is obvious that the presence of highly qualified teaching staff actualizes the question of the expediency of e-learning, which obviously has a lower pedagogical potential in comparison with contact classroom work. One of the main goals of higher education is to train qualified personnel. So, is it worth calling it into question for the sake of imitation indicators of imaginary managerial efficiency?

References

1. Tyukavkin NM (2015) Systematization of the theoretical foundations of neo-institutionalism. Bull Samara State Univ Ser "Econ Manage" 9/1(131):209–216
2. Moskovskiy AV (2005) Limits of institutionalism. Economist 6:74–81
3. Demsetz H (1988) Efficiency, competition, and policy. Oxford, pp 164–184
4. Durkheim E (1991) On the division of social labor. Science, Moscow, p 595
5. Parsons T (2002) On social systems. Akadem-Project, Moscow, p 534
6. Weber M (1990) On some categories of understanding sociology. Selected works. Progress Publ, Moscow, pp 494–546
7. Coleman JS (1993) The role of rights in a theory of social action. J Inst Theor Econ 149:213–232
8. Miroshnikov SN (2018) The Essence and content of the concepts of "public administration" and "development management" in modern conditions. Prospect Sci 9:40–44
9. Gnatyuk MA (2019) Methodological aspect of inertia research. Bull Modern Res 1.6(28):68–74
10. Grachev NI (2017) The Essence of the state. Bull Saratov State Law Acad 4:11–24
11. Vaskov MA, Kovalev VV, Gafiatulina NKh (2020) Online education in the higher school of Russia: the main actors of institutionalization and social consequences 9(3):45–57
12. Kovalev VV, Kasyanov VV, Manucharyan AK (2020) Online education in higher education in Russia: a factor of destruction or a source of development? Hum South Russia 9(3):2–91
13. Kovalev VV, Dyatlov AV, Vodenko KV (2020) Assessment of the quality of online education (based on focused interviews in student groups). Bull South Russian State Univ (Novocherkassk Polytechnic Institute) 13(3):6–14
14. Polonsky VM (1995) Dictionary of concepts and terms on the legislation of the Russian Federation on education. Moscow Institute of Educational Systems Development, pp 231–232
15. Kachalov V, Prudkovsky B (1999) About the assessment of the quality of higher education. High Educ Russia 2:54–58
16. Andreev AA, Soldatkin VI (2002) Applied philosophy of open quality: a pedagogical aspect. Alfa Publisher, Moscow, p 234
17. Pavlenko KV (2010) Neoinstitutional approach to assessing the quality of education in higher education. J Sociol Soc Anthropol 13(1):90–100
18. Ushakov BG (2008) The quality of education and/or education quality (problems of methodology, practice and evaluation). Manage Consult 3:149–161
19. Zaborova EN, Shchipacheva (2004) Quality of education in the higher school system: the experience of sociological analysis. News of Ural State University of Economy 9:78
20. Karasik LV (2019) The concept and principles of implementing public administration functions. Acad J 10:126–129

A New View to the Concept "Virtual Digital Rights" in the Online-Learning Process

Tatyana Skvortsova, Gennady Pratsko, and Tatyana Epifanova

Abstract The purpose of the work is to study approaches to some legal categories, such as "virtual-digital rights" in the context of digital transformation of educational activities. One of the most important steps in the legal form of relations with the use of virtual-digital techniques (smart contracts, blockchain and cryptocurrencies) to include virtual-digital rights in the Civil code of the Russian Federation. In the process of teaching in the era of digital transformations in the fields of economy and law, there are problems of adequate perception of new concepts in these fields by students. To achieve the research goals, the authors analyze various views on the issue of determining the legal nature of digital rights, and examine the concept of virtual-digital rights written in the Civil code of the Russian Federation. The authors used the system and structural–functional methods of scientific knowledge, the method of the interpretation of civil law norms, as well as the logical method. The study showed to us that virtual-digital rights are classify as property rights by the law. The legal definition of virtual-digital rights is overloaded and has disadvantages. The authors proposed to present article 141.1 of the Civil code of the Russian Federation, containing the concept of digital law in the amended version. The results of the study can be used in teaching legal disciplines to students studying "Jurisprudence", "Economics", "Management" and others.

Keywords Teaching law · Digital transformation of educational activities · Digital rights · Civil rights object · Property rights · Digital techniques · Smart contract · Blockchain · Cryptocurrency

T. Skvortsova (✉)
Rostov State Transport University, Rostovskogo Strelkovogo Polka Narodnogo Opolchenya Square, 2, Rostov-on-Don 344038, Russia

G. Pratsko
Don State Technical University, Gagarin Square, 1, Rostov-on-Don 344003, Russia

T. Epifanova
Rostov State University of Economics, B. Sadovaya str., 69, Rostov-on-Don 344012, Russia

A. Beskopylny and M. Shamtsyan (eds.), *XIV International Scientific Conference "INTERAGROMASH 2021"*, Lecture Notes in Networks and Systems 247,
https://doi.org/10.1007/978-3-030-80946-1_39

1 Introduction

In the era of digital transformations in the economy and the legal sphere, there is a need of highlighting at the proper professional level the basic categories of legal science, which include, among other things, the category of "object of law".

Modern economic realities [1–5] are characterized by various forms of expressing relations that legally enshrined in the norms of current legislation.

Thus, nowadays, contractual relations with the use of digital techniques, in particular, smart contracts based on blockchain and cryptocurrencies, are more and more spread [6]. Digital techniques are used in various spheres of public life, including corporate relations [7], intellectual property relations [8], notary [9].

The digitalization of public relations has a significant impact on human and civil rights [10, 11].

The most important field of application of digital technologies is their use in the educational process. And with the spread of the new coronavirus infection COVID-19 this need become particularly acute, because of the introduction of restrictive measures. At the same time, scientists note some difficulties in digitalizing of legal business education [12–15], as well as problematic aspects of the use of intellectual property objects in the learning process [16].

It seems that the problematic aspect is not only the introduction of various digital technologies in the educational process, the use of copyright objects in the learning process. It is also difficult to understand new legal terms in the process of learning law. One such of that terms which requires deep understanding is the term "virtual-digital rights".

As a result of the progressive movement in the scientific and technological sphere there are arising of new law objects, also legal relations of such objects in the civil sphere are transforming. In this regard, it is logical to recognize virtual-digital rights as an object of civil rights from October 1, 2019. The updated version of the article 128 of the Civil Code of the Russian Federation establishes that virtual-digital rights are a kind of property in the civil-legal sense [17, 18].

The article 141.1 of the Civil Code of the Russian Federation by the redaction of Federal Law No. 34-FZ of 18.03.2019 "On Amendments to Parts One, Two and Article 1124 of Part Three of the Civil Code of the Russian Federation", provides that virtual-digital rights are obligations and other rights, which contains and conditions for the implementation of which are determining in accordance with the rules of the information law system.

We intentionally use the concept of "virtual-digital" rights in this article, since we consider this name to be the most characteristic of the essence of the phenomenon under consideration.

The Civil Code sets out the general conditions for the circulation of virtual-digital rights.

Special normative legal acts may establish other rules that provide for the transfer of this objects. Currently, Federal Law No. 259-FZ of 02.08.2019 "On Attracting investments using Investment Platforms and on Amendments to certain Legislative

Acts of the Russian Federation" (regulates relations on attracting investments by commercial organizations or individual entrepreneurs using information technologies, as well as defines the legal basis for the activities of operators of investment platforms for the organization of retail financing (crowdfunding)) adopted; also as Federal Law of 31.07.2020N 259-FZ "On Digital Financial assets, Digital Currency and on Amendments to certain Legislative Acts of the Russian Federation" (aimed at regulating relations that arise when virtual-digital rights and digital financial assets are created, registered and traded).

Thus, the norms of the Civil Code of the Russian Federation create the basis for the legal regulation of virtual-digital rights relations and are a basis for outlining the general model of legal regulation of public relations, the object of which is virtual-digital rights [19]. The turnover of virtual-digital rights is currently regulated by special legislation, which, in our opinion, is quite justified.

The analysis of the above normative legal acts and individual norms of the Kazakh legislation makes it possible to assert that investment activities in the field of virtual-digital rights and the circulation of digital financial assets within the framework of which virtual-digital rights are objectified are regulated by mandatory norms of legal acts. Previously, before the adoption of the above-mentioned legislative acts, such activities were not subject to regulatory regulation. Therefore, only those subjects of civil turnover who have acquired rights in relation to the objects under consideration within the framework of information systems operating on the basis of recently adopted legal regulation will be able to protect their rights in court.

2 Methodology

The article uses traditional and proven methods of scientific research of civil law processes. The analysis of virtual-digital rights as objects of civil turnover is based on the use of systematic and structural–functional methods. This analysis is based on the logical method of cognition, as well as on the method of interpretation of the norms of law, which are contributing to the consideration of various aspects of the subject of research.

3 Results/Discussion

Russian legislator has defined virtual-digital rights as a type of property. Place of virtual-digital rights in the system of objects of civil rights (see Fig. 1).

The exercise, disposal, including transfer, pledge, encumbrance of virtual-digital rights by other means or restriction of the disposal of virtual-digital rights is possible only in the information system without a third side. The owner of a virtual-digital right is a person who, in accordance with the rules of the information system, has the opportunity to dispose of this right, if the law is not provide other position. The

Fig. 1 Place of virtual-digital rights in the system of objects of civil rights

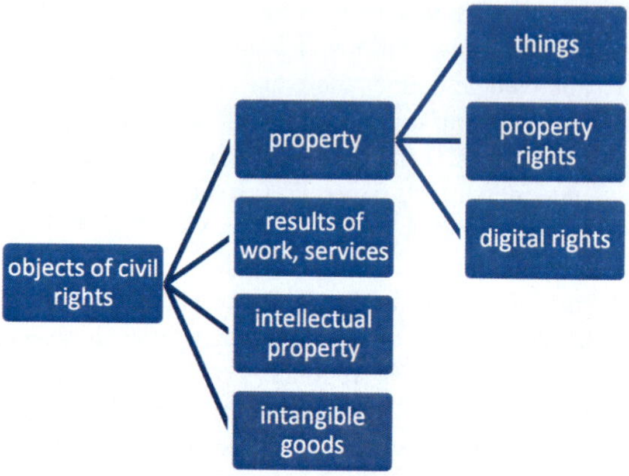

transfer of a virtual-digital right on the basis of a transaction does not require the consent of a person who is considered obligated in the legal seems.

Property relations with the use of virtual-digital rights as an object are becoming more widespread. In particular, payments using cryptocurrencies are becoming more common. So, the bitcoin rate has grown over the past time by millions of times (Fig. 2).

The introduction of the category of virtual-digital rights as new objects of civil law relations has caused an expected discussion among leading civil law scholars and practitioners.

Thus, according to S. Sarbash, these novels "are a fundamental legislative mistake and they make no sense in the theory of civil law" [20], since, in his opinion, virtual-digital rights" are not a new object of civil law also as some new right, that is fresh in the civil law seems. This is the main legislative problem.

Fig. 2 Bitcoin exchange rate from 2009 to 2021. *Source* https://bytwork.com/articles/btc-chart-history

According to the author, "Digital law is really just a form, i.e. the formalization of any known right: law of obligations, real, corporate, exclusive, personal right. Digitalization itself and the digitalization of the economy is only a method of registration, but not the content of property relations" [20].

In general the other scientists' evaluation of these legislative innovations is positive. As Shestakova and Zhirkova [21] point out, digital techniques and electronic transactions are developing dynamically. Many countries take measures to accelerate the digitalization pace.

32 of 36 countries member of the OECD, as well as 6 partner-countries of this organization have national virtual-digital strategies. United States, China and the European Union also have huge programs for the development of the virtual-digital economy. Therefore, it is obvious that changes will take place in Russia as well.

L. Y. Vasilevskaya is quite right when she believes that regulations in the digital sphere are becoming today very important tasks of civil law as a separate science and as a branch of the law [22].

This approach should be reflected in the teaching of law to economists, lawyers, managers, etc. At the same time, the use of active online technologies in teaching gifted students is promising [23]. Optimization of library services in the digital environment is also required [24].

Please note that most teachers of higher education institutions do not support the online learning format (Fig. 3) and believe that it is not convenient for students and will worsen the quality of education.

We believe that the distance learning format in law can exist along with full-time training. Especially when it comes to highlighting the legal regime of digital rights, digital learning methods can and should be applied.

The very idea of the need to create a legal framework for the development of blockchain platforms and business systems in civil turnover, should be undoubtedly supported.

However, the emergence of virtual-digital rights in the list of right objects indicates a new trend in the development of civil turnover, i.e. its digitalization. The manifestation of this trend is possible by a new approach (to constructing the objects of rights), implemented in the application of information techniques in the modeling of

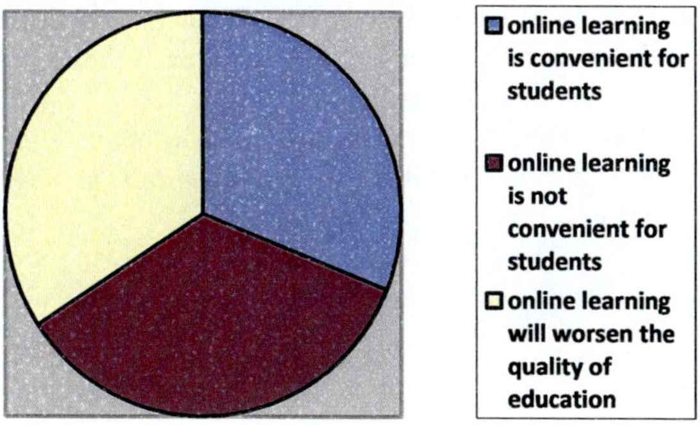

Fig. 3 Results of a survey of teachers of Russian universities on the impact of online learning on the quality of education

digital tools (tokens, mining, cryptocurrency, etc.) that ensure the turnover of digital rights. Practitioners give a positive evaluation of innovations in “digital rights".

So, A. Tolkachev believes that “in general the proposed model corresponds to world practice and it is considered to be successful if a special regulation is formed that specifies the possibility of applying virtual-digital rights and it is based on an understanding of the capabilities and properties of new techniques” [20].

At the same time, he draws attention to a number of shortcomings of the introduced legislative regulation of digital rights: “The concept of virtual-digital money was not included in the law, as in the previous version of the draft, but this concept could help to determining the legal status of cryptocurrencies. A number of offers in sphere of smart contracts and virtual-digital transactions were rejected by the lawmaker, in fact, enshrined the practice of recent years. In other words, the legislator didn’t make a breakthrough in the competent legal structuring of relations". A number of researchers point out some shortcomings of writing the concept of “digital law” into legislation. Thus, according to A.M. Erdelevsky the legislator could do without the following words “law of obligations and other” in paragraph 1 of article 141.1 of the Civil code of the Russian Federation since they do not contribute in any way to the comprehension of the concept of digital law [25].

The representative of the presidential Commissioner For the protection of entrepreneurs’ rights in the field of intellectual property, A. Semenov, also draws attention to this problem. He believes that “the ambiguous wording “law of obligations and other” makes:

- either assume that the word “law of obligations” is mentioned accidentally, since “other” rights include all the possible property rights, such as absolute rights, obligations from delicts and condition, etc.;
- or conclude that the legislator mean “law of obligations and other similar rights”, i.e. the legislator is talking only about relative rights, and not, for example, absolute rights;
- or in a more severe interpretation, proceed from the fact that the legislator means rights from transactions, and not from delicts and condition [20].

A. M. Erdelevsky think that the virtual-digital right owner is a person who has the ability to dispose of this right, but only by the rules of the information law system [25]. Also should pay attention to the redundancy of the introductory phrase “unless otherwise provided by law”, since the sentence, which states “in cases and on the grounds provided by law, another person may be recognized as the digital rightsholder”, completely eliminates its need.

In our opinion, the inclusion of the concept of “digital rights” as an object of civil rights is justified, but it is necessary to eliminate these inaccuracies in the wording.

4 Conclusion/Recommendations

In the process of teaching law in the modern stage of digital transformation of the economic and legal system of society, there is a need to develop the right approaches to understanding new categories of legal science. This article analyzes one of these concepts—"virtual-digital rights".

The legal literature widely discusses the validity of the inclusion of a new object of civil rights, i.e. "virtual digital right".

In our view, it is necessary to introduce the concept of "virtual-digital rights" as an object of civil rights. But the legal definition of "virtual-digital law" needs to change.

We believe that article 141.1 of the Civil code of the Russian Federation should be set out in the following wording (Table 1).

Table 1 Proposed amendments to Article 141.1 of the civil code of the Russian Federation

Paragraph of Article 141.1	Proposed revision	Current version
1	Virtual-digital rights are those rights named in such a way in the law, the content and conditions of which are determined in accordance with the rules of the information system that meets the criteria set in the law. The implementation, disposal, including transfer, pledge, encumbrance of virtual-digital rights by other means or restriction of virtual-digital rights disposal is possible only in the information system without re-course to a third party	Digital rights are recognized as binding and other rights named in this capacity in the law, the content and conditions for the implementation of which are determined in accordance with the rules of the information system that meets the criteria established by law. The exercise, disposal, including transfer, pledge, encumbrance of digital rights by other means or restriction of the disposal of digital rights is possible only in the information system without contacting a third party
2	The virtual-digital right owner is a person who has the ability to dispose of this right, but only in accordance with the rules of the information law system. The law pro-vides cases and the grounds, when another person is recognizing to be new digital right owner	Unless otherwise provided by law, the holder of a digital right is a person who, in accordance with the rules of the information system, has the opportunity to dispose of this right. In the cases and on the grounds provided for by law, another person is recognized as the owner of the digital right
3	The transfer of a virtual-digital right on the basis of a transaction does not require the con-sent of a person who is considered obligated in the legal seems	The transfer of a digital right on the basis of a transaction does not require the consent of the person obligated under such a digital right

The proposed changes will allow us to develop a clearer approach to understanding the essence of the regulated social phenomenon and, accordingly, will contribute to a better assimilation of the relevant legal category by students when teaching law.

References

1. Hanafizadeh P, Kim S (2020) Digital business: a new forum for discussion and debate on digital business model and digital transformation. Digit Bus 1(1):100006. https://doi.org/10.1016/j.digbus.2021.100006
2. Kohtamäki M, Parida V, Oghazi P, Gebauer H, Baines T (2019) Digital servitization business models in ecosystems: a theory of the firm. J Bus Res 104:380–392
3. Parka H, Anderson TR, Seo W (2021) Regional innovation capability from a technology-oriented perspective: an analysis at industry level. Comput Ind. https://doi.org/10.1016/j.compind.2021.103441
4. Shi Y, Chen M, Qu T, Liu W, Caic Y (2020) Digital connectivity in an innovative joint distribution system with real-time demand update. Comput Indus. Available online 2. https://doi.org/10.1016/j.compind.2020.103275
5. Epifanova TV, Shatkovskaya TV, Romanenko NG, Mosienko TA, Tkachenko M (2017) Legal provision of clustering in Russia as environment for development of Innovations. Int J Trade Glob Market 2–3(10):217–225
6. Skvortsova TA, Mosienko TA, Ulezko AYu, Nikolaev AV, Arzumanyan AA (2020) Development of digitization in contractual relations. In: Popkova EG, Sergi BS (eds) Digital economy: complexity and variety vs. rationality: Lecture Notes in Networks and Systems. Springer, Berlin, vol 87. pp 1025–1032. https://doi.org/10.1007/978-3-030-29586-8_117
7. Skvortsova TA, Skorev MM, Kulikova TV, Nesterova NV, Merkulov MM (2020) Prospects for the further digitization of corporate relations. In: Popkova EG, Sergi BS (eds) Digital economy: complexity and variety vs. rationality: Lecture Notes in Networks and Systems. Springer, Springer, vol 87, pp 1017–1024. https://doi.org/10.1007/978-3-030-29586-8_116
8. Shatkovskaya TV, Shumilina AB, Nebratenko GG, Isakova JI, Sapozhnikova EY (2018) Impact of technological blockchain paradigm on the movement of intellectual property in the digital space. Eur Res Stud J S1(21):397–406
9. Sukhovenko AV (2020) Electronic notarial system as a new social institution in a digital economy: quality, availability, security. In: Popkova EG, Sergi BS (eds) Digital economy: complexity and variety vs. rationality: Lecture Notes in Networks and Systems. Springer, vol 87, Springer, Heidelberg, pp 1033–1039. https://doi.org/10.1007/978-3-030-29586-8_118
10. Perry S, Roda C (2016) Human rights and digital technology: digital tightrope. In: Human rights and digital technology: digital tightrope, pp 1–203. https://doi.org/10.1057/978-1-137-58805-0
11. Shakhrai SM (2018) Digital constitution: fundamental rights and freedoms of an individual in a totally informational society. Her Russ Acad Sci 88(6):441–447
12. Azarko E, Abakumova I, Kupriyanov I (2020) Assessment of resource and self-regulation of behavior of students with different sense-forming strategies during the session in remote format behavior. In: Innovative technologies in science and education (ITSE-2020). E3S web of conferences, vol 210, p 18004. https://doi.org/10.1051/e3sconf/202021018004
13. Abakumova I, Zvezdina G, Grishina A, Zvezdina E, Dyakova E (2020) University students' attitude to distance learning in situation of uncertainty. In: Innovative technologies in science and education (ITSE-2020). E3S Web of Conferences, vol 210, p 18017. https://doi.org/10.1051/e3sconf/202021018017
14. Enkova EE (2019) Difficulties of digitalization of legal business education. Bus Law 3:69–74
15. McAvinia C (2016) Online learning and its users, 1st edn. Chandos Publishing, Oxford

16. Sushkova OV (2020) Features of using the results of intellectual activity as part of audiovisual works for effective digitalization of the educational process in a law school. Actual Probl Russian Law 5:144–150
17. Kirillova EA, Bogdan VV, Surkov AN, Baumshteyn AB, Novichkov VE (2019) Virtual-digital rightsand digital objects: features of control and regulation. J Adv Res Dyn Control Syst 11(S10):567–573
18. Fedorenko NV, Hejgetova SE (2020) Virtual-digital rights in civil legislation of Russia development of digitization in contractual relations. In: Popkova EG, Sergi BS (eds) Digital economy: complexity and variety vs. rationality: Lecture Notes in Networks and Systems. Springer, Berlin, vol 87, pp 1011–1016. https://doi.org/10.1007/978-3-030-29586-8_115
19. Bychkov A (2019) E-wallet: simple, convenient, secure. New Account 6:134–144
20. Novoselova L, Gabov A, Savelev A et al (2019) Virtual-digital rightsas a new object of civil law. Law 5:31–54
21. Shestakova E, Zhirkova A (2019) Virtual-digital rightsand electronic transactions. What is the future of the digitalization of legal support for business? Financial Newspaper 28:10–11
22. Vasilevskaya LYu (2019) Token as a new object of civil rights: problems of the digital law legal qualification. Actual Probl Russian Law 5:111–119
23. Irina A, Irina B, Anastasia G, Elena D (2019) Active learning technologies in distance education of gifted students. Int J Cognit Res Sci Eng Educ 7(1):85–94
24. Ryan B (2019) Optimizing academic library services in the digital milieu, 1st edn. Chandos Publishing, Oxford
25. Erdelevsky AM (2019) On digital rights. Reference legal system “Consultant Plus”

Students' Manipulative Techniques for Passing Attestation in Distance Learning

Olga Efremova, **Larisa Kobysheva**, **and Svetlana Shalova**

Abstract The article is devoted to the use of manipulative techniques by students for the current and interim attestation in distance learning which is considered as a consequence of "learned helplessness" (M. Seligman) and "double bind" (G. Bateson). The problem of psychological manipulation in the "student–teacher" system is first studied on the basis of the analysis of students' class activities (written works, spoken replies). Students' imitation of academic progress is carried out by using in the MOODLE system such techniques as the presentation of plagiarism, collectively tasks performed, copying of tasks done by other students, answers to didactic tests based on random selection, and many others. "Double bind" of students is resulting from the contradictory requirements to high quality and at the same time to a sizeable portion of tasks in a limited period of time. In the situation of "double bind" is also the teacher, who is guided by contradictory attitudes towards the implementation of the criterion (high quality) and statistical approach (comparative analysis without taking into account the criteria of the required quality) in assessing academic work and students' answers. Based on the analysis of the products of students' class activities, the frequency of their use of manipulative techniques for imitating educational competencies when passing credits and exams is characterized. Correlations were found between the frequency of students' use of manipulative techniques and the level of "learned helplessness" felt by them, as well as the intensity of the "double bind" among students. The necessity of dosing the teaching load of students when teaching in a distance learning format and special control of authorship of students' academic works presented in the electronic educational milieu is emphasized.

Keywords Psychological manipulation · Distance learning · Learned helplessness · Double bind · Pedagogical control · Pedagogical assessment

O. Efremova (✉) · L. Kobysheva · S. Shalova
Taganrog Institute of A. P. Chekhov (Branch), Rostov State University of Economics, Taganrog 347936, Russian Federation

A. Beskopylny and M. Shamtsyan (eds.), *XIV International Scientific Conference "INTERAGROMASH 2021"*, Lecture Notes in Networks and Systems 247,
https://doi.org/10.1007/978-3-030-80946-1_40

1 Introduction

In the case of numerous unsuccessful attempts by the subject to influence unsatisfactory external circumstances, he may develop a state that Seligman [1] designated as "learned helplessness". It is characterized by tactics of avoiding further failure, passivity, loss of a sense of freedom and the ability to control. This state can manifest itself in a student who finds himself in a situation of constant choice between the high quality of the products of academic work and their sizeable portion, which must be provided in a short time. This situation has become very common in connection with the transition of universities to distance learning during the coronavirus pandemic, since the volume of written tasks that a student must complete and submit in the electronic information and educational milieu has significantly increased. In fact, the students found themselves in the conditions of the "double bind" described by Bateson [2], when they receive mutually exclusive requirements that must be fulfilled, but cannot be disputed. The teacher is also sink into the situation of "double clamping", forced to evaluate the work of students in accordance not with criteria (requirements for the quality of work), but with a statistical approach (orientation to the level demonstrated by the majority).

The urgency of the problem of overcoming "double bind" in higher education is substantiated in the work of Mayer [3], in one of our studies [4]. V. P. Dolmatov explodes the myth about the advantages of distance learning at the present stage [5, p. 14] and notes that in practice it does not yet provide methodological advantages, but it imposes strict requirements on the skills of students' self-organization, which for many turn out to be insufficiently formed. As a result of "double bind", a student in a state of "enforced helplessness" either does not fulfill the teacher's tasks, or save them "for later" (academic procrastination), or imitates their fulfillment, manipulating the teacher's trust and his desire to provide a satisfactory academic progress.

Delay in completing class assignments (academic procrastination) can be carried out in a passive and active form. In case of passive procrastination, the subject postpones the fulfillment of assignments due to his unreadiness to make prompt decisions; in the case of active (intentional) procrastination, the subject is driven by the desire to focus on solving more important tasks and to ensure a higher result [6–10]. Although the authors note a higher level of academic progress among students who are active procrastinators, we suppose that their grades may depend not only on rational self-organization, but also the use of manipulative techniques for imitating academic achievement.

The analysis of the problems of manipulation in the "student (pupil)–teacher" system goes back to the works of Shostrom [11], Bern [12], and mainly the methods of manipulation in communication were studied. In particular, E. Shostrom describes methods of manipulation by students (flattery, comparing teachers with each other, etc.), which can indirectly affect the teacher's assessments. In this tradition, some Russian authors (for example, Tarelkin [13], Sheinov [14]) analyze manipulation by students as a latent influence on the psychologically vulnerable aspects of the teacher's personality to control his opinion and future assessments. Among these manipulation techniques—the student's emphasized demonstration of interest in the subject, a request to the teacher to tell in more detail about the phenomenon being studied (the target of the influence is the teacher's love for the subject). Manipulation in the process of interaction between a student and a teacher during assessment can also be considered in a personally detached aspect. On the part of the teacher, such manipulations of assessment are possible, as avoiding effective assessment criteria and replacing them with procedural ones; presentation of individual successful answers as an integral characteristic of the student's knowledge; creation of conditions for write-off; skipping errors when checking works, etc. [15]. On the part of the student, such techniques are used to manipulate upcoming assessments, as plagiarism, presentation of borrowed texts as creative works [16], performing tasks with the search for answers using a smartphone [4], substitution of test items and many others. At the same time, students perceive fraud as a normal phenomenon and consider it acceptable and justified. In one study, it was empirically shown that in students machiavellianism is positively correlated with unethical behavior [17]. To denote the described manipulations, we propose the term "academic machiavellianism", that reflects the distorted perception of the norms of educational ethics by students.

The adequacy of student attestation in distance learning has become one of the pressing issues during the coronavirus pandemic. Therefore, it is necessary to study the manipulative deformations of the educational activity of students in distance learning, focused on the falsification of academic progress.

2 Materials and Methods

The main material for the article was the concepts of "double bind" by Gregory Bateson and "learned helplessness" by Martin Seligman, as well as studies on the problems of psychological manipulation in the "student–teacher" system carried out by Russian and foreign researchers. Methods used: phenomenological and problem analysis; observation; inquest (a point assessment by students of the degree of "forced helplessness" and "double bind" felt by them on a scale from 0 to 10); analysis of students class assignment presented in the MOODLE educational milieu, analysis of students' spoken replies in the classroom in the disciplines of the psychological and pedagogical cycle in the distance learning system on the ZOOM platform, Spearman correlation analysis, assessment of the statistical significance of differences between the series of indicators using the chi-square test (χ^2).

Typical manipulative techniques of "performing" class assignment by students were highlighted: replacing note-taking of primary sources and performing creative work with texts from the Internet, replicating work done by groupmates, presenting the same works in different academic disciplines (similar in content), laying out assignments, not relevant to the topic, with the hope that the teacher will not be able to view a large amount of information, answers to didactic test tasks randomly, faked references to literature when citing, completing only part of the tasks to get credit, and others. Some specific manipulation techniques are used by students and in classes conducted in the form of videoconferences. For example, taking advantage of the limited time of the ZOOM conference, students are more active, demonstrating syllable answers to simple questions, in the hope of future automatic grades on the exam; in case of difficulties, they quietly take a peek at the answers, placing prompts out of the teacher's field of view; in case of serious difficulties, they turn off, referring to the instability of the Internet connection and computer freeze. We supose that manipulative techniques may differ for different forms of control: a credit or an exam. The analysis of 1800 written works of students (reports, abstracts, essays, creative projects, answers to questions at the seminars, practical work) was carried out in 29 academic disciplines (17—with final control in the form of a credit, 12—in the form of an exam).

We also tested the hypothesis about the presence of a correlation between the assessments of "learned helplessness" and the intensity of the "double bind", on the one hand, and the frequency of falsification of the performance of class assignment by students, on the other. To test this hypothesis, 80 students of the Faculty of Psychology and Social Pedagogy were interviewed and their work, presented in the electronic educational milieu MOODLE, was studied. Students were asked to rate their state of health in points from 0 to 10 according to two parameters: (1) "How often do you acutely feel the impossibility of fulfilling the teacher's requirements for a subject in distance learning?", (2) "How intensely do you feel a contradiction between the pursuance of high quality of work presented in MOODLE, and the difficulty of task performance due to the sizeable portion of them that need to be completed in a short time? " For the convenience of recording students' work in various academic subjects, the frequency of their use of manipulative techniques was assessed by teachers on the following scale: 0 points—never uses, 1 point—sometimes, 2—often, 3—in most cases, 4—always.

3 Results

The analysis of students' written works presented in the electronic educational milieu, in general, showed that students more often use manipulative techniques in order to falsify the picture of academic progress in studying disciplines in which the form of control is credit. At the same time, the quality of written work is lower than in disciplines in which the form of control is exam. More often, students use such manipulations as downloading works from the Internet, partial performance of the assignment (incomplete answers to questions, ignoring difficult tasks). The statistical estimation of the data obtained using the nonparametric chi-square test showed the presence of statistically significant differences at $P < 0.05$ in connection with the preparation of students for passing a credit or exam: the criterion value was 12.25 with the number of degrees of freedom $k = 4$. Table 1 shows the percentage indicators of the frequency of students' appeal to various techniques-manipulations. In total,

Table 1 Frequency of students' use of manipulative techniques for passing certification in different forms of control in distance learning (results of analysis of students' written works; number of works—percentage data)

Manipulative techniques	Control form	
	Credit	Exam
1. Download from the Internet	36.3	25.5
2. Replication of works	9.0	6.5
3. Skipping tasks	10.6	7.2
4. Inconsistency with the topic	11.1	7.1
5. Partial task performance	12.4	11.8
Proportion of manipulative answers in the total number of works	79.4	58.0

830 works were studied in disciplines with a credit form of control and 970 papers in disciplines with an examination form of control.

Knowingly we have not carried out a quantitative analysis of the manipulation techniques used by students in verbal communication with the teacher in distance classes via videoconferencing. Let us note some manipulative techniques used by students in seminars and practical classes in the form of ZOOM conferences: refusal to answer under the pretext of instability of the Internet connection, mute audio and video, replacing speeches on the lesson with monosyllabic answers during a conversation with a teacher—demonstration of activity in order to avoide extensive knowledge testing, unmotivated exit from the videoconference. There was a higher activity of students in the classroom in disciplines that involve an examination form of control.

Comparison of students' assessments of the degree of their internal tension in the distance learning process showed a high degree of correlation (R = 0.792; $P < 0.01$) between the parameters of experiencing the impossibility of completing tasks ("learned helplessness") and the subjective acuteness of the contradiction between the desire for high quality work and understanding its unreality for a given volume in a limited time ("double bind"). For both parameters, significant correlations were noted with the use by students of such manipulative techniques that falsify the faithful fulfillment of class assignment, such as downloading ready-made information products from the Internet and missing some tasks, the failure of which, according to the student, may go unnoticed. According to the parameter "double bind", a significant at $P < 0.05$ correlation with the "partial performance of individual task" technique was also noted, manipulation consists in demonstrating activity and expecting that the teacher will not cavil about minor points, because something (if not in full) was made and presented at MOODLE.

Some of the data obtained allow us to note the presence of individually-specific techniques for imitating the performance of tasks by students. Thus, a highly significant ($P < 0.01$) negative correlation between downloading from the Internet and replicating other people's work (R = −0, 0.348) indicates that some students find ready-made works on the Internet, while others claim as own of their classmates. The

Table 2 Matrix of correlations (according to Spearman, two-sided criterion) of subjective indicators of the intensity of the educational process and the use of falsification techniques by students

Scales	1	2	3	4	5	6	7	8
1. «Learned helplessness»	1000	0.792**	0.497**	– 0.109	0.303**	– 0.240*	0.192	0.159
2. «Double bind»	0. 792**	1000	0.551**	– 0.104	0.279*	– 0.307**	0.225*	– 0.171
3. Internet downloads	0.497**	0.551**	1000	– 0.348**	0.100	– 0.139	– 0.137	0.176
4. Replication of other people's works	– 0.109	– 0.104	– 0.348**	1000	0.112	0.136	– 0.101	– 0.283*
5. Skipping tasks	0.303**	0.279*	0.100	0.112	1000	– 0.063	0.171	0.045
6. Inconsistency with the topic	– 0.240*	– 0.307**	– 0.139	0.136	– 0.063	1000	– 0.097	– 0.191
7. Partial performance	0.192	0.225	0.137	– 0.101	0.171	– 0.097	1000	0.195
8. Random choice of answers when testing	0.159	0.171	0.176	– 0.283*	0.045	– 0.191	0.195	1000

negative correlation (R = –0, 0.283; $P < 0.05$) between the use of someone else's work and passing tests by random selection reflects the presence of individual styles of deliberate use of someone else's knowledge or blind hope of luck, at random.

Table 2 shows the data of this study. Sign "**" denotes correlations that are significant at $P < 0.01$, and the sign "*"—at $P < 0.05$.

4 Discussion

In general, the conducted observations show that in the case of distance learning, when the teacher's ability to resist student's manipulations is limited, academic machiavellianism manifests itself in students quite often. In particular, a teacher can very limitedly use of such methods of countering manipulations described by Sheinov [14], as ethics, unpredictable behavior, firm refusal, practiced mainly through direct contact. In distance learning, the teacher evaluates the numerous written works of students presented by them in the electronic educational milieu, spends time downloading each work and viewing it on the monitor (which is less convenient than viewing the paper version), has difficulty comparing the works of different students that are needed open one by one to make sure they are not replicated.

When checking the written works of students from the standpoint of a criterial approach (requirements for the quality of work), the teacher states numerous deviations from the requirements, the use of manipulative techniques by students to manipulate the result. When checked from the standpoint of a statistical approach, the teacher is guided by the level demonstrated by the majority of students, and sometimes he himself is forced to become a manipulator in order to ensure a satisfactory picture of progress. He, as it were, "does not notice" the tricks of the students (plagiarism, duplication of works in a group, partial completion of tasks, etc.). The rejection of the statistical approach is highly relevant, since in more than half of the cases, manipulative distortions were found when students performed class assignment (79.4% of cases when preparing for a test, 58.0% when preparing for an exam).

An interesting fact discovered as a result of the study is the more frequent use of manipulative methods of demonstrating educational activity in disciplines by students, suggesting a form of control in the form of credit. Since the exam provides for the differentiation of grades, students perform work in the relevant disciplines with higher quality. This indicates that students can provide a higher quality of educational work and strive to do so in the presence of external motivation.

Analysis of the frequency of students' use of manipulative techniques when performing written works presented in the MOODLE system allows us to highlight some aspects that are subject to special control: compliance of student works with the criteria of originality (no plagiarism), independence of implementation, completeness, and accuracy of reflecting the topic. The allocation of student manipulation techniques in the classroom conducted in the videoconference mode allows formulating some requirements for students in the process of distance learning communication: do not turn off the sound and video, work out classes in which the connection to the videoconference failed, pre-inform in the chat about difficulties and unresolved questions when completing assignments. It is necessary to intensify the work of students in distance learning, involving a credit form of control.

On the basis of the research carried out, at least two types of student manipulators can be distinguished: the first—using ready-made information products from the Internet, the second—who prefer to replicate the work of their group mates. The first type of manipulators, who independently search for ready-made information, often show an orientation toward guessing the answers during testing. The second type of manipulators is characterized by a focus on the use of other people's knowledge, they tend to pass off the work of other students as their own, to use hints during testing. The teacher needs to record and prevent these tendencies, to specifically control the independence of students in performing class assignment.

5 Conclusions

1. In the course of distance learning at a university, students often use such methods of falsifying the results of educational work as the use of ready-made information products from the Internet, replicating the work of fellow students, skipping part of the tasks in the hope of difficulties in control from the teacher, inconsistency of work with the set goals, incompleteness disclosure of individual issues.
2. At seminars and practical classes held in the form of videoconferences, students use the following methods of psychological manipulation: refusal to answer under the pretext of instability of the Internet connection, mute the sound and video, frequent demonstration of episodic activity in order to avoid a detailed test of knowledge, unmotivated exit from the videoconference.
3. Manipulative techniques designed to imitate the fulfillment of the teacher's requirements are much more often used by students when performing assignments in disciplines that provide for a credit form of control. Greater independence and a less pronounced share of formalism in the work in the disciplines for which the exam is taken, testifies to the fundamental possibility of improving the quality of students' work in a distance format.
4. The disadvantage of the teacher's control and evaluation activity in the distance learning mode is the assessment of students' work in accordance with the statistical (orientation towards the level demonstrated by the majority), and not the criterial approach (orientation towards the quality of work). The volume of assignments assigned by students in the electronic educational milieu and subject to control should be determined in relation to the teacher's ability to check them informally.
5. Carrying out an individual approach to monitoring and assessing students' knowledge, the teacher should pay attention to two fundamentally different types of manipulators: uncertain students who prefer to use ready-made information products from the Internet and focus on randomly choosing the right answers during testing, and students who present the work of fellow students as their own and hint-oriented when testing.
6. The elimination of the appearance of states of "forced helplessness" and "double bind" in distance learning among students implies a clear designation of the criteria for evaluating written works, a clear dosage of the teaching load on the part of the teacher, an increase in the share of creative works to avoid plagiarism, a special formulation of the requirements for the originality of works and control over the fulfillment of this requirement.

References

1. Seligman MEP, Reivich K, Jaycox L, Gillham J (1995) The optimistic child: a proven program to safeguard children against depression and build lifelong resilience. Houghton, Mifflin and Company, Boston, MA, US
2. Bateson G (1972) Steps to an ecology of mind. Collected essays in anthropology, evolution and epistemology. Chandler Publishing Company, San Francisco
3. Majer BO (2012) About the "double bind" pattern in modern society and education. Bull Novosibirsk State Pedagogical University 3(7):44–55. https://doi.org/10.15293/2658-6762
4. Efremova O, Kobysheva L, Shalova S (2020) "Double bind" in educational process as a factor of activization of students' clip way of thinking. In: E3S Web of Conferences 210, 18086.ITSE-2020. https://doi.org/10.1051/e3sconf/202021018086
5. Dolmatov VP (2017) The mythological genesis of pedagogy the problem of introducing new FGOS. Bull Samara Humanit Acad. A Ser Psychol 1(21):133–146
6. Barabanshchikova VV, Marusanova GI (2015) Perspectives for research of the procrastination phenomenon in professional work. Natl Psychol J 4(20):130–140. https://doi.org/10.11621/npj.2015.0413
7. Chu AHC, Choi JN (2005) Rethinking procrastination: positive effects of «active» procrastination behavior on attitudes and performance. J Soc Psychol 145(3):245–264. https://doi.org/10.3200/SOCP.145.3.245-264
8. Demeter DV, Davis SE (2015) Procrastination as a tool: exploring unconventional components of academic success. Creat Educ 4(7):144–149. https://doi.org/10.4236/ce.2013.47A2018
9. Kim S, Fernandez S, Terrier L (2016) Procrastination, personality traits, and academic performance: when active and passive procrastination tell a different story. Pers Individ Differ 108:154–157. https://doi.org/10.1016/j.paid.2016.12.021
10. Lee D (2013) Differences in passive/active procrastination, happiness, and depression by the clusters of perfectionism in college students. Korea J Couns 14(2):857–875. https://doi.org/10.15703/kjc.14.2.201304.857
11. Shostrom EL (1967) Man, the manipulator. The inner journey from manipulation to actualization. Abingdon Nashville, TN
12. Berne E (2012) Games people play: the psychology of human relationships. Penguin Books Ltd., UK
13. Tarelkin AI (2009) Manipulation in pedagogic communication in a higher school. Psychol Educ 5:56–79
14. Sheinov VP (2014) Manipulation and protection from manipulation. Peter, SPb
15. Efremova O (2016) Mythologization of the educational evaluation as a means of psychological manipulation in the teacher's activity. Psychol Educ 12:80–95
16. Shalova S, Efremova O (2019) The opinion poll of students on the use of internet activity in education. In: Trends in the development of psycho-pedagogical education in the conditions of transitional society (ICTDPP-2019), 08036, EDP Sciences, LesUlis. https://doi.org/10.1051/shsconf/20197008036
17. Tang LP, Chen YJ (2008) Intelligence vs. wisdom: the love of money, Machiavellianism, and unethical behavior across college major and gender. J Bus Ethic 82(1):1–26. https://doi.org/10.1007/s10551-007-9559-1

The Usage of Communication Systems and Networks as a Means of Modern Teacher's Work in a Higher Educational Institution

Anna Balykina, Andrey Ageev, Valentina Pashchenko, Yuri Pikulin, Aleksandr Sergeev, and Svetlana Smolina

Abstract The purpose of this work is to theoretically substantiate and empirically prove the need to introduce a digital educational environment into the educational process. The theoretical and methodological basis of the study is given, on the basis of which the research sample is built. Taking into account all the interrelated main blocks of activity the general methodology is considered for a higher educational institution. To obtain the empirical base of the study, we introduce samples of respondents who are offered a research methodology in the form of selected questionnaires, the validity of which is empirically proven. Taking into account the purpose of the study all questionnaires are selected to identify the usefulness and necessity of introducing a digital educational environment into the educational process. Using the methods of mathematical statistics, empirical results are processed and a new term "the phenomenon of problematic Internet use" is introduced. Correlation analysis allows you to identify differences between samples and formulate patterns of their dependencies. Based on the conducted research and subsequent analysis, the authors prove the need for a more in-depth study of the introduced phenomenon by developing a multi-factor questionnaire in order to create a complex of psychological and pedagogical support designed to increase the level of efficiency of the educational process in higher education.

Keywords Communication systems · Teachers · Higher education institution · Mathematical statistics · Students · Samples · Questionnaires

1 Introduction

This paper sets the goal, which is currently the most relevant for the modern educational process: to identify psychological factors that contribute to the design and effective use of computer systems and communication networks included in the management of complex automated university's systems by teachers in their professional activities.

A. Balykina (✉) · A. Ageev · V. Pashchenko · Y. Pikulin · A. Sergeev · S. Smolina
Russian New University (RosNOU), Moscow 105005, Russian Federation

A. Beskopylny and M. Shamtsyan (eds.), *XIV International Scientific Conference "INTERAGROMASH 2021"*, Lecture Notes in Networks and Systems 247,
https://doi.org/10.1007/978-3-030-80946-1_41

As you know, communication (from the Latin word "make common, connect") is the process of transferring information from a source to a recipient in order to change their knowledge, attitudes, or explicit behavior. Effective communication is important for success in management, since the solution of many management tasks is based on the direct interaction of people (chief with subordinate, subordinates with each other) within the framework of different events. Communication is the best way to discuss and resolve issues that are characterized by uncertainty.

The effectiveness of communication is influenced by such factors as:

- communication skills,
- settings and experience.

the mental abilities of the communication's subjects, which form the specifics of their message's perception and the emotional background.

In the interpersonal communication process, there are six main variables: sender/encoder; message; channel; receiver/ decoder; perception; feedback [1].

The sender is responsible for the wording of the message that accurately conveys the message to the recipient. The process of thought's translating into a message is called encoding.

The sender's task is to search and use the communication symbols and skills that will lead to the correct reflection of the message in the recipient's mind.

The message consists of verbal and nonverbal symbols that represent the information we want to convey. Every message we send is an attempt to convey a thought to the recipient.

The process of translating a message into a thought is called decoding, and this is the task of the recipient. How well the recipient perceives the information depends on the following factors:

the recipient's knowledge of the conversation's topic;
the probability of perceiving the sender's message appropriately;
experience of communication between the sender and the recipient.

The recipient is described by two aspects of behavior:

ability to listen,
ability to provide feedback to the sender.

Feedback can be verbal or nonverbal; written or printed, or electronic. Feedback provides guidelines for the next message we send to the recipient. We can use feedback to evaluate the effectiveness of our communication. Therefore, it is very important to master the skill of accurately interpreting feedback.

In the presence of feedback, the sender and recipient change their communication roles. The original recipient becomes the sender and goes through all the stages of the information exchange process to transmit its response to the original sender. Feedback can significantly improve the efficiency of management information exchange. Although the two-way exchange of information is slower, it is more accurate and increases confidence in the correct interpretation of messages. Feedback increases the chances of an effective information exchange, allowing both parties to eliminate interference.

In order to achieve the most effective teaching process in the digital educational environment, namely, achieving the most effective interaction between the participants of the educational process—teachers and students, it is necessary to identify the main factors affecting the success of the educational process in the digital educational environment, as well as to develop a complex of psychological and pedagogical support to increase the level of effectiveness of those factors in which there is a low level of efficiency.

2 Materials and Methods

The digital educational environment of a higher educational institution includes the following elements: a set of technological means of information and communication technologies (ICT): computers and other ICT equipment, communication channels, a system of modern pedagogical technologies, as well as a set of information educational resources, including digital educational resources. Their complex application in the modern digital educational environment of an educational institution allows to significantly increase the effectiveness of the educational process.

When organizing the learning process in the modern digital educational environment, the training function is subject to the greatest changes (less-educational). Let's consider the main features of such a learning process in comparison with the traditional one. Traditional training includes leadership, information, and control functions. Learning in a digital educational environment includes organizing, orienting, presenting, and pedagogical support functions.

Currently, the key to ensuring an effective educational process is open and accessible interaction between its participants (students, teachers, heads of structural divisions, system administrators), and the role of such interaction continues to grow. Changes in the legislation of the Russian Federation (Decree of the Government of the Russian Federation No. 2040 of December 7, 2020 "On conducting an experiment on the introduction of a digital educational environment") that have occurred in recent years have significantly influenced many educational processes. After the creation of educational complexes (university-college), in order to increase the effectiveness of the educational process, there was a need to revise the existing structure of interaction between its participants.

In this interaction, three separate blocks can be distinguished, united by a common goal—to increase the efficiency and accessibility of education. The first block ensures the interaction of the complex administration at all levels of the educational process. The second block provides interaction at the level of "head-teacher". The third block is at the levels "head—system administrator—teacher—student—methodologist". Let's take a closer look at each of them.

The first block is the interaction of the complex administration at all levels of the educational process and its characteristic features. In practice, the buildings of one educational complex are often located at a significant distance from each other, and its administration is usually located in the "head" building. This situation negatively affects the effectiveness of the implementation of various management administration's decisions, as the distance makes it difficult to exchange information between remote elements, if it uses only standard tools. This problem can be effectively solved by the widespread use of remote computer technologies such as online meetings, internal corporate mail and internal chat. At the same time, it is desirable to create a working email address for teachers and administrative employees of an educational organization on a separate domain of the organization. In the case of using a separate domain, the administration can create an internal site. The use of the internal site significantly expands the opportunities for interaction between the participants of the educational process at all levels and increases the efficiency of the tasks facing the organization. Each teacher (employee of the administration) has the opportunity to access the internal information of the university under their username and password, which is necessary to organize an effective educational process, regarding them.

Administrative interaction can also be carried out using internal video conference calls instead of regular face-to-face meetings. In this case, there is no need for the arrival of administrative staff of the organization in the "head" building, which saves their working time and increases the efficiency of information dissemination. The implementation of this interaction's type requires the administration to purchase the necessary equipment and lay (set up) a separate internal network.

The second block is the interaction "administration–teacher". At this level, a significant problem is the impossibility to personally attend important organizational meetings of the administrative and pedagogical staff for various valid reasons (participation in classes, illness, etc.). This necessitates timely communication of important information considered during the meeting and decisions taken to the absent staff. The best way to quickly communicate any important information (office meetings, organizational and methodological documents) necessary for the organization's employees to effectively perform educational and administrative functions is to send it electronically through specially created resources for this purpose. Using the official website of the organization here is irrational, since it does not provide the necessary speed and volume of information exchange between employees. Significantly greater opportunities open up when creating a local network of the organization. Placing all the information required by the employee on such a network significantly increases the speed of information exchange in the organization. This, in turn, improves manageability and ultimately leads to an improvement in the quality of educational services. Of course, the implementation of such network capabilities requires additional financial costs and is an additional burden on network administrators. However, such a network provides a significant expansion of its information exchange. This suggests that these costs are absolutely justified and effective.

All of the above should be applied systematically. This requires specialists who can organize the work of collecting, systematizing and placing the materials necessary for the organization's employees in its local network. Depending on the network topology used, this can be either a professional network administrator or a person with basic knowledge of network administration. As a result of their work, a situation should be provided when the teacher turns on the computer at the beginning of the working day and sees all the operational information he needs at least for the current working day (week, month, semester and academic year). It provides a view of the office meetings materials, with the necessary office documents, as well as access to the working email address, the internal university chat and online meetings.

The third block is the interaction "teacher-student-methodologist". Currently, teachers can use various training management systems to organize distance learning. Their correct application makes it possible to implement flexible and effective training aimed both at meeting the needs of an individual student and at achieving the requirements of the Federal State Educational Standard. In universities, such resources are most often used to solve this problem. The most extensive opportunities in the organization of distance learning are provided by the Moodle Remote Learning System. It is currently one of the most popular e-learning platforms. This is due to its features such as: functionality, reliability in operation, availability of content development and knowledge verification tools, usability. It allows the teacher to create and manage a course, add students, monitor and analyze their progress, and differentiate access to educational materials and course blocks.

As ready-made solutions suitable for implementing a sustainable remote education system, it is also possible to use integrated services and applications included in a modern software product from Microsoft Office 365. Among its applications and services, the following have the greatest capabilities for this task: the Microsoft Teams Communication Center, Outlook Mail, and the Skype for Business corporate communications server.

The Microsoft Teams Communication Center is a convenient online communication platform that provides a remote format for teachers to communicate with students. Includes audio–video calls, content demonstrations, chat, file sharing and remote storage. Teams allows you to create a unique educational environment for teacher interaction with students, communicate with them, create and post assignments and training materials, develop and check control tasks. Skype for Business has similar but more limited features. It was created as a finished product for use in commercial structures and is less suitable for educational purposes. The program provides users with the ability to exchange instant messages, a system of audio, video and web conferences, as well as telephone communication between users.

Microsoft Outlook Mail is a convenient personal information manager with the functions of an email client. Provides the teacher with additional opportunities for organizing information exchange in the format of distance learning. Among its features are the creation of shared mailboxes, task folders, calendars, conferences, and document negotiation. It is especially convenient to share the Outlook and Teams applications, which organically complement their functionality.

An important electronic document that ensures interaction between students, teachers and methodologists, and controls students' attendance at classes, has become an electronic "journal-diary". Its use allows students to get acquainted with the received grades, comments of the teacher and exchange messages with him. This electronic resource is designed to monitor the performance of tasks of independent training, attendance and academic performance of students. The teacher's subject course has advanced features and allows him to post the necessary educational information in the form of documents, presentations, links to external sites, e-learning resources (electronic educational resources). It is possible to interact with students through tests and interactive tasks. It provides the opportunity to communicate on the subject in forums and personal messages.

The main tasks of the information and educational environment of an educational institution:

1. ensuring remote interaction of educational process participants;
2. planning the educational process;
3. planning the resource support of the educational process;
4. providing procedures for creating, searching, collecting, analyzing, processing, storing and presenting information;
5. information and methodological support of the educational process;
6. monitoring and recording the progress and results of the educational process;
7. remote interaction of educational institutions with social organizations (institutions of culture, sports, leisure, health).

The creation of an information educational environment requires the investment of financial resources, as well as solving some organizational and technical tasks by the leadership of the educational organization. Among them, we can highlight such as the need to have ICT specialists in the organization's staff who are capable of technical implementation of the digital educational environment and maintaining its continuous functioning, as well as increasing the level of ICT competence of the entire organizations' teaching staff to a level that allows full work in this environment. Another problem that needs to be solved is the inertia and unwillingness of some teachers, and in some cases, the administration, to change the previously established approaches and principles of administrative interaction. On the contrary, the understanding by the management of an educational organization of the prospects that open up in the case of the formation of a digital educational environment in it is the key to the successful solution of this most important task.

Informatization of the educational environment is a rather complex and long process that has been actively conducted in recent years. In the coming decades, we should expect an active continuation of this process, which will require updating the technological base, developing and mass mastering new pedagogical technologies, updating the content, methods and forms of organizational work of mass higher education. In the short term, it is a purposeful process and is controlled to some extent by the relevant local programs of informatization (development) of the school. In the long term, it develops abruptly and is difficult to predict.

For the success of the digital educational environment process, it is essential to understand the importance of its creation and maintenance on the part of all participants in educational relations. Effective interaction of all participants in the educational process of an educational organization is possible if an environment is created in which each element functions in accordance with its tasks with a common systematic approach.

Let's analyze the most important communication process in the "teacher-student" system, for which we will determine its essence. One of the definitions of the concept of "system" is usually considered as "a set of elements that are in relations and connections with each other, forming a certain integrity, unity". It should be noted that the concept of "system" is constantly evolving and currently has no clear boundaries and limits of its understanding in science. In addition, each area of concept application complements it with its own elements. Consider some important additions to this concept put forward by its researchers. So Agoshkova [2] that "the definition of a system should include not only the collection, the composition of elements and relations, but also the integral property of the object itself, relative to which the system is built". Bezrukova [3] draws attention to the importance of the environmental component: "the system is characterized by the presence of components, connections (structure) between them, the presence of a leading link, an inseparable unity with the environment, in which the system expresses its integrity". In addition, she identifies the features of pedagogical systems, which consist of subsystems that include people, processes, objects, actions, and writes: "in these systems, the behavior of each element affects the behavior of everyone and the system as a whole."

Based on the above, it can be assumed that the teacher—student system should include the following elements: teacher, student (and/or a group of students), as well as the educational environment. This allows us to formulate the definition of the teacher-student system as a set of relations between the teacher, the student (group of students) and the educational environment, forming the unity of teaching and training. In order to study the communicative processes occurring in this system, we will consider its structure. The first element is the participants of the educational process: the teacher, the student (and/or a group of students).The second element is the relationship between the teacher and the student (group of students).The third element is communication between all participants of the educational process. The fourth element is the joint activity of a teacher and a student (a group of students); the digital educational environment.

Let's consider in more detail the third element of the "teacher–student" system, for this purpose we will use the definition of I. N. Demina: "Communication is a semantic and ideally meaningful aspect of social interaction. Communicative actions are actions that are deliberately focused on their semantic perception. The main function of communication is to achieve social community while maintaining the individuality of each of its elements" [4]. The authors present the structure of the simplest communication in the following form:

1. "participants-communicants";
2. the situation (or situations) that they seek to comprehend and understand;
3. texts expressing the meaning of the situation in the language;
4. the motives and goals that make the texts directed, i.e., what motivates the subjects to address each other;
5. the process of material transmission of texts.

Let us focus on the historical perspective of the study of the communication phenomenon by various sciences. In philosophy, it began to study at the turn of the

XIX–XX century, and for a long time these researches were limited. They limited themselves to studying only certain aspects: the origin of social norms, morality, law and the state (the theory of the social contract), as well as the available means of organizing philosophical communication itself (the problem of dialogue). At the beginning of the XXI century, a new direction in philosophy appeared and was actively developed—media philosophy. It studies the place and role of communication and communication technologies in public spheres, the change of which is associated with the intensive development of modern communication tools.

The social sciences were also interested in the phenomenon of communication at the beginning of the XX century, and they advanced much further in their research. Philosophers studied communication in a general sense, and sociological sciences studied social communication, turning it into a new scientific direction. Currently, social communication is considered as an interdiscipline that studies the processes of dissemination and circulation of information in society.

I. N. Demina identifies five levels of communication in her work [4]:

1. intrapersonal communication as "conversations with oneself", information processing by an individual (between parts of the same person);
2. interpersonal communication as interpersonal interaction;
3. group communications as studies of group dynamics, they also include studies of (organizational) communication;
4. mass communications as messages sent to a mass audience through mass media, often for political or commercial purposes;
5. communications that study the exchange of messages with non-human objects, primarily with machines, computers (computer games, using ATMs, etc.).

Communication technologies emerged since the advent of writing. However, they gained a revolutionary development at the turn of the XX–XXI century due to the emergence of the first global Internet network, and then social networks, which became a means of social communication of modern society with extremely wide opportunities. Note that they ensure the implementation of all five levels of communication. The concept of a social network was introduced into scientific use by Barnes [5].

In work "Classes and Meetings in a Norwegian Island Parish", he said: "Each person has a certain circle of friends, and these friends, in turn, have their own friends. Some of one person's friends know each other, others don't. I found it convenient to talk about such social fields as networks. Under this, I see a system of points, some of which are interconnected. The points of this system are people, and the connection lines of these points indicate what people interact with each other" [5]. The researchers of this issue noted a rapid further development of social networks.

Following Bitkov [6] in our analysis, we proceed from the acceptance of the thesis that "a social network (on the Internet) is an automated, interactive, multi-user service created for the interaction of people in a group or groups, which is based on the "friends" system and the "community" system; the content of the service is downloaded by the network participants themselves."

The importance of social networks as a means of social communication for their users and modern society as a whole is high and continues to grow rapidly. Many researchers indicated this in their works. According to Sergodeev, "social networks contribute, firstly, to the organization of communication between people, and secondly, to the realization of their basic needs" [7]. In network communication, they distinguish a number of such specific features as virtuality, interactivity, hypertextuality, globality, creativity, anonymity and mosaicity. For the purpose of their further study we will give their brief author's formulation: "virtuality is a characteristic of symbolic reality, opposed to the objective world and replacing it with images, simulators and analogues. Interactivity is associated with a focus on individual and selective use of information. Hypertextuality is based on the idea of the text, which allows the user to model and organize the text space in a convenient way. The globality of communication is determined by the ultimate expansion of the space in which various types of communication occur. Creativity is due to the fact that the virtual space provides a person with maximum opportunities for constructive activity, allowing them to invent many self-presentations. The anonymity of social network communication contributes to the creation of a relaxed atmosphere for self-presentation. The mosaic of communication is based on the absence of an ordering center and an ordered periphery, due to which the socio-cultural space inevitably forms in some way chaotic (mosaic) structures."

The analysis of the above allows us to conclude that social network communications provide a significant amount of social and information functions for their users. In particular, they perform and increase information exchange, help to orient in large amount of information, as well as in self-improvement and integration into the selected community. These functions of social network communications allow them to apply a humanistic approach in learning. Need to note that the nature of such communication is computer-mediated, that is, the exchange of written messages, communication and the establishment of television-visual connections occurs exclusively through personal computers and data communication networks.

Let's have a look at the most important elements of social networks. Shipitsin [8] identifies two such elements: an Internet site (service), as well as associations, clubs and communities of people. This guides our research to the analysis of Internet sites and communities of people, especially teachers and students. It is necessary to consider the implementation features and the significance of computer-mediated communication for each element of the teacher-student system.

To implement computer-mediated communication, the teacher needs constant, constructive and effective interaction with the participants of the educational process. According to Rosina [9]. "A modern teacher needs to have skills and abilities, a willingness to build effective communication with other people directly involved in the pedagogical process (colleagues, students), and supporting, administering the learning process in an electronic environment (administrators, technical staff), as well as managing the educational institution, the education system (administration of the educational institution, governing authority)."

In order to ensure its effectiveness the student, as an object of the educational process, is also obliged to create the necessary level of communication with other participants –teachers, administration, technical staff and other students. According to Rozina, " the acquisition and development of these skills is possible only during of activity, including various forms of computer-mediated communication (interpersonal, group, intercultural)" [9].

A careful study of the interaction of computer-mediated communication and the educational environment, as the next element of the teacher-student system, shows a significant decrease in the influence on the educational process of the space–time component. These changes were made thanks to the technical capabilities of modern personal computers. Among them are such opportunities as distance learning, saving messages for the learner at a convenient moment of accessing them, the use of asynchronous technologies (chat, e-mail, messages on the network, virtual whiteboard, website, blog, etc.). When using them, pedagogical communication is implemented in an environment with other spatial and temporal characteristics formed by the corresponding information and communication technologies (text, sound, television, computer), different from traditional educational environments.

Rozina [9] divides computer-mediated communications into synchronous, asynchronous, offline communication technologies (a message is sent and received at different times: e-mail, teleconferences) and online technologies (real-time correspondence, chat, social network messages).

It is also necessary to consider such an important difference for the modern educational environment and the educational process as computer-mediated technologies that do not require the mandatory presence of all participants in the learning process in the same place (virtual or real). Such a distance learning format significantly expands the teacher's possibilities within the educational process. According to Rozina [9], who describes the combination of space–time parameters of Internet technologies, all of them differ in terms of conducting a computer-mediated communication process. Let's list the tools of the educational process, through which different combinations of spatial and temporal parameters are implemented in the social network:

1. one time/one location (virtual, cyberspace)—chat-synchronous communication;
2. one time/different location—instant messages-synchronous communication;
3. different times/one location (virtual, cyberspace)—messages and comments on the "wall" of a personal page or group of the studied discipline—asynchronous communication;
4. different time/different location—messages in your personal account—asynchronous communication.

3 Results

To study the effectiveness of the introducing a digital educational environment into the educational process, we turn to the main reason for the use of the Internet by different layers of students. Everyone knows that some students have Internet addiction and, in this case, we can no longer talk about the useful properties of the introducing Internet resources into the educational process. The cognitive-behavioral approach is used to differentiate and use Internet resources in the digital educational environment. The theory of this approach is based on the formulation of Davis [10], who is the founder of the theory that Internet addiction is based on cognitive symptoms which are the cause of affective or behavioral symptoms. The model reviewed has a main focus on the dysfunctional beliefs associated with Internet addiction.

Two student samples were taken for the empirical base of the study. The first sample (in a further we will call it sample No. 1–47 people) was made up of students of natural sciences who study using traditional methods, and filled out blank versions of questionnaires in paper form. The second sample (sample No. 2–51 people) consisted mainly of active Internet users who filled out questionnaires online, and in the training of which computer and Internet technologies are widely used (students of natural science training areas).

The study was divided into blocks, each block of methods included different information: socio-demographic questions (age of the respondent, gender, education, marital status, work experience as an Internet user, the average amount of time spent on the Internet). In addition to the basic information about the respondent, each student was asked to take 6 psychological questionnaires (Internet preferences questionnaire, V. V. Stolin self-attitude test questionnaire, depression scale, K. Yang Internet dependence questionnaire, V. G. Romek confidence test, subjective control level questionnaire).

The results of each questionnaire were recorded in separate tables, which were further processed by mathematical statistics using the SPSS package. The order of mathematical data processing is as follows:

1. Description of general statistical indicators;
2. Checking the differences in the results of the study depending on the gender and marital status of the subjects;
3. Checking the reliability of the questionnaires, calculating the Cronbach's alpha for each scale for all samples;
4. Validation of questionnaires:
    a. intercorrelation of scales;
    b. factor validity;
    c. construct validity;
    d. convergent validity;
    e. correlation analysis.

The results of the empirical data after the correlation and factor analysis are presented in Tables 1 and 2.

Extraction method: Analysis of main components; rotation method: Varimax with Kaiser normalization.

Extraction method: Analysis of main components; rotation method: Varimax with Kaiser normalization.

Table 1 Factor matrix after rotation (sample No. 1)

Questions	Factors								
	1	2	3	4	5	6	7	8	9
Online is the best place for me	0.07	0.40	0.26	0.46	0.42	0.09	0.23	−0.18	0.02
Only a few people like me besides my online acquaintances	0.56	0.34	0.48	0.07	−0.04	−0.19	0.20	−0.12	0.36
I feel most secure on the Internet	0.54	0.01	0.29	0.55	0.14	0.02	0.26	−0.06	−0.10
I often keep thinking about something I've experienced on the Internet for a long time after leaving it	0.00	0.03	0.83	0.14	0.20	0.08	0.10	−0.01	−0.01
When I'm online, I often experience a kind of excitement or emotional uplift	0.43	0.01	0.65	0.21	0.21	0.19	0.07	0.05	−0.15
It is easier to know a person intimately through the Internet than through direct communication	0.24	0.26	0.21	0.27	0.62	−0.20	0.02	−0.23	0.13
I often feel comfort on the Web	0.27	0.18	0.07	0.22	0.65	0.00	0.06	0.42	−0.18
I can be myself online	0.08	0.06	0.34	0.02	0.11	−0.17	0.15	0.82	0.10
The network treats me with more respect than in real life	0.72	0.05	0.16	−0.13	0.30	0.12	0.21	0.02	0.22
I spend more time on the internet than I should	0.76	0.12	0.11	0.11	0.11	0.24	0.12	−0.13	0.07
People around me complain that I use the Internet too much	0.28	0.24	0.10	0.41	0.26	0.50	0.28	−0.11	0.12
I never go online longer than I plan	0.18	0.13	0.43	−0.09	0.04	0.33	0.27	−0.11	−0.05
The network accepts me for who I am	−0.21	−0.09	−0.28	0.13	−0.01	0.02	−0.04	0.74	0.09
Online relationships with people are more profound than in real life	0.11	0.11	0.26	0.00	0.73	0.24	−0.05	0.20	0.35
I often think about the Internet being outside	0.35	−0.04	0.71	0.31	0.13	0.13	0.25	0.03	−0.07
I reveal myself from the best side in the network	−0.02	0.21	0.33	0.66	0.25	0.21	0.10	0.15	−0.01
The world out of the network fascinates me less than being on the Internet	0.22	0.25	0.19	0.43	0.32	0.21	0.58	−0.12	0.14
I would like my friends and family to know how much I am respected in the network	0.07	0.34	0.07	0.73	0.10	0.06	0.09	0.09	0.16

(continued)

Table 1 (continued)

Questions	Factors								
	1	2	3	4	5	6	7	8	9
The Internet is "more real" than real life	0.05	0.39	0.29	0.29	0.50	−0.18	0.32	−0.11	0.08
I don't think about my responsibilities when I'm online	0.19	0.16	0.12	0.71	0.04	0.42	−0.01	0.30	0.10
I can't stop thinking about the internet	0.12	0.15	0.50	0.18	0.48	0.48	0.13	0.10	−0.16
I'm not so lonely when I'm online	0.29	0.21	0.59	0.17	0.04	0.17	−0.01	0.44	0.23
I can't imagine that I could ever do without the Internet for more or less a long time	0.31	0.19	0.13	0.13	0.07	0.22	0.71	0.00	0.20
The Internet is an important part of my life	0.23	0.20	0.28	0.06	−0.05	0.05	0.73	0.27	−0.09
Without internet access I feel myself helpless	0.49	0.33	−0.04	0.36	0.33	0.19	0.39	−0.07	−0.02
On the Internet, I say and do things that I would never be able to say or do outside of the Internet	0.63	0.14	0.43	0.12	0.15	−0.01	0.20	0.08	−0.12
I connect to the internet when I have nothing to do	0.07	−0.02	−0.13	0.14	0.12	0.04	0.09	0.16	0.86
In my opinion, I often connect to the network when I should be doing something else	0.06	0.23	0.16	0.19	−0.10	0.78	0.29	−0.03	−0.01
When I'm online, I don't have to think about problems outside of the network	0.58	0.36	0.15	0.36	−0.08	0.00	0.17	0.32	0.02
Sometimes I use the Internet to delay more necessary cases	0.42	0.35	−0.06	0.16	0.23	0.37	0.23	−0.03	−0.15
I can leave all my worries behind when I'm online	0.16	0.78	0.04	0.30	0.20	0.07	0.18	0.24	0.01
I often use the Internet as an excuse to get away from unpleasant activities	0.20	0.81	−0.01	0.08	0.24	0.19	0.25	−0.05	−0.03
The Internet is a way to forget about what needs to be done, but what you don't want to do	0.02	0.83	0.16	0.32	0.06	0.21	0.14	0.03	−0.13
Sometimes I want to start using the Internet less, but I can't do it	0.51	0.52	0.26	0.06	−0.13	0.39	−0.09	−0.02	0.32
It bothers me that I can't stop using the internet so much	0.29	0.67	0.00	0.07	0.22	0.41	0.00	−0.12	0.31

(continued)

Table 1 (continued)

Questions	Factors								
	1	2	3	4	5	6	7	8	9
It seems that sometimes I am not able to control the time of my stay on the Internet	0.20	0.32	0.35	0.29	0.05	0.62	−0.10	−0.07	0.16

Table 2 Factor matrix after rotation (sample No. 2)

Questions	Factors							
	1	2	3	4	5	6	7	8
Online is the best place for me	0.33	0.57	0.23	0.07	0.42	0.15	0.06	0.05
Only a few people like me besides my online acquaintances	0.11	0.18	−0.13	0.55	0.05	0.49	−0.03	0.25
I feel most secure on the Internet	0.06	0.34	0.44	0.54	0.14	0.00	0.14	−0.20
I often keep thinking about something I've experienced on the Internet for a long time after leaving it	0.45	0.50	0.05	0.42	0.17	−0.05	−0.03	−0.03
When I'm online, I often experience a kind of excitement or emotional uplift	0.41	0.43	0.18	0.43	0.21	0.08	0.02	−0.16
It is easier to know a person intimately through the Internet than through direct communication	0.07	0.75	0.14	0.04	0.20	0.28	0.07	−0.04
I often feel comfort on the Web	−0.03	0.37	0.45	0.27	0.09	0.22	0.42	−0.08
I can be myself online	0.10	0.13	0.04	0.32	0.12	0.01	0.72	−0.09
The network treats me with more respect than in real life	0.19	0.15	0.19	0.73	0.06	0.10	0.12	0.11
I spend more time on the internet than I should	0.45	0.19	0.18	0.28	0.48	0.25	0.00	0.01
People around me complain that I use the Internet too much	0.55	0.36	0.21	0.13	0.30	0.10	0.12	−0.15
I never go online longer than I plan	0.40	−0.01	0.05	0.26	0.49	0.05	−0.08	−0.29
The network accepts me for who I am	−0.05	0.03	0.14	−0.14	0.01	−0.02	0.75	0.03
Online relationships with people are more profound than in real life	0.22	0.66	0.07	0.26	0.13	0.06	0.19	0.24

(continued)

Table 2 (continued)

Questions	Factors							
	1	2	3	4	5	6	7	8
I often think about the Internet being outside	0.40	0.48	0.24	0.44	0.23	0.04	0.09	−0.22
I reveal myself from the best side in the network	0.39	0.52	0.37	0.25	−0.09	−0.09	0.20	−0.06
The world out of the network fascinates me less than being on the Internet	0.28	0.60	0.39	0.13	0.14	0.03	−0.11	−0.04
I would like my friends and family to know how much I am respected in the network	0.32	0.37	0.29	0.16	−0.33	0.16	0.33	0.12
The Internet is "more real" than real life	−0.02	0.64	0.24	0.17	0.16	0.37	0.18	−0.22
I don't think about my responsibilities when I'm online	0.40	0.16	0.69	−0.04	0.03	−0.06	0.27	−0.07
I can't stop thinking about the internet	0.45	0.53	0.06	0.33	0.27	−0.06	0.29	−0.09
I'm not so lonely when I'm online	0.42	0.33	0.08	0.40	0.18	0.07	0.41	0.21
I can't imagine that I could ever do without the Internet for more or less a long time	0.28	0.27	0.22	−0.02	0.70	0.11	0.13	0.02
The Internet is an important part of my life	0.22	0.30	0.02	0.27	0.62	0.21	0.14	−0.09
Without internet access I feel myself helpless	0.27	0.33	0.40	0.08	0.56	0.28	0.08	0.09
On the Internet, I say and do things that I would never be able to say or do outside of the Internet	0.10	0.18	0.57	0.47	0.27	0.02	−0.05	0.11
I connect to the internet when I have nothing to do	−0.01	−0.06	0.03	0.08	−0.05	−0.11	−0.02	0.85

(continued)

Table 2 (continued)

Questions	Factors							
	1	2	3	4	5	6	7	8
In my opinion, I often connect to the network when I should be doing something else	0.73	0.01	0.25	0.14	0.16	0.20	−0.04	−0.14
When I'm online, I don't have to think about problems outside of the network	0.18	0.09	0.76	0.25	0.12	0.20	0.08	0.01
Sometimes I use the Internet to delay more necessary cases	0.36	0.03	0.15	0.17	0.18	0.71	0.07	−0.18
I can leave all my worries behind when I'm online	0.12	0.26	0.75	−0.01	0.15	0.32	0.11	0.06
I often use the Internet as an excuse to get away from unpleasant activities	0.29	0.24	0.32	0.03	0.29	0.66	−0.11	−0.02
The Internet is a way to forget about what needs to be done, but what you don't want to do	0.44	0.23	0.39	−0.05	0.06	0.59	0.03	−0.23
Sometimes I want to start using the Internet less, but I can't do it	0.74	0.20	0.07	0.18	0.12	0.31	0.11	0.13
It bothers me that I can't stop using the internet so much	0.73	0.31	0.04	0.02	0.20	0.22	0.00	0.26
It seems that sometimes I am not able to control the time of my stay on the Internet	0.69	0.18	0.26	0.14	0.29	0.16	0.00	−0.06

4 Findings

The study proved that the phenomenon of problematic use of the Internet is a multi-dimensional characteristic, including the main factors necessary to achieve the effectiveness of the educational process, such as a sense of social comfort on the Internet, loneliness in real life, the level of self-control, the use of the Internet as a means of avoiding more important tasks. Also, according to the results of the study, it can be concluded, although with a certain degree of caution, that dependence on the Internet is directly proportional to the level of individual's depression, and absolutely the same dependence is observed between the level of self-attitude and the level of self-confidence. After conducting a correlation analysis, it can be concluded that the scales of the questionnaires are characterized by sufficient reliability and validity and can be used to assess the level of respondents ' dependence on the Internet, while the factor validity of the scales was not confirmed. For further and more in-depth research, it was decided to create a multi-factor questionnaire that can reflect the full structure of the phenomenon of pathological Internet use.

Conclusion. Empirical research has produced and processed statistics that demonstrate that there is a phenomenon of problematic Internet use in the form of a characteristic that includes many different factors. In addition, using the methods of mathematical statistics, the high validity and legitimacy of the chosen methodology was proved, however, there is a need to create a multi-factor questionnaire for a more in-depth study of the formulated phenomenon.

We think that the goal set in this work has been achieved. The analysis of the communication process in the "teacher-student" system, which is implemented by means of social networks, was carried out. The novelty of our approach is the refinement of the concept of "teacher-student system" on the basis of the studied scientific literature. We consider that the results of the analysis and clarification of the key concept will help in further work on the pedagogical design of the innovative (virtual) educational environment in the "teacher-student" system, including the development and testing of the appropriate structural and functional model.

References

1. Soldatova IYu, Chernysheva MA (eds) (2006) Fundamentals of management: textbook. Dashkov and K, Moscow
2. Agoshkova EB (2009) Category "system" in modern thinking. Questions Philos 4:57–71
3. Bezrukova VS (2013) All about the modern lesson at school (question–response training). September, Moscow
4. Demina IN (2012) Economic communication: the rationale for the term. News Irkutsk State Econ Acad 3:175–179
5. Barnes JA (1975) Class and committees in Norvegian Island Parish. Human Relations. Hafner Press, New York. https://doi.org/10.1177/001872675400700102
6. Bitkov LA (2010) Social networks: between mass communication and interpersonal communication. Bull Chelyabinsk State Univ Philology. Art Hist 70(28):36–38
7. Sergodeev VA (2013) Networked Internet communities: essence and socio-cultural characteristics. Bull Adygeya State Univ. Ser: Reg Stud Philos Hist Sociol Law Polit Sci Cult Stud 1:122–127
8. Shipitsin AI (2011) The phenomenon of social networks in modern culture. News Volgograd State Pedagogical Univ 3(57):36–40
9. Rozina IN (2004) Computer-mediated communication in the practice of education and business. In: Theory of communication & applied communication. Bulletin of the Russian Communication Association, pp 217–226. IUBiP, Rostov-on-Don
10. Davis RA (2001) Cognitive-behavioral model of pathological Internet use. Comput Human Behav 17:187–195. https://doi.org/10.1016/S0747-5632(00)00041-8

Students' Perception of Distance Learning During COVID-19 Pandemic

Tatiana Kiseleva, Andrey Simanovsky, Inna Melnikova, and Svetlana Elifantieva

Abstract The article investigates university students' perception of the distance learning during the COVID-19 pandemic. The authors surveyed 211 undergraduate students (ages 18–22) in different years of study at the Schools of Special Education of Yaroslavl State Pedagogical University named after K. D. Ushinsky to find out their level of satisfaction with distance learning. Participants were asked to evaluate various aspects of distance learning related to students' motivation for taking courses, the subjective difficulty of the courses and their suitability for online mode, and instructors' professional qualification and readiness for remote teaching. The survey data were used to establish significant statistical differences (Student's t-test) and correlations (Spearman's ρ-coefficient) between various parameters influencing students' perception of the online courses. Based on the correlation analysis of various factors, the authors suggest emphasizing the following aspects while creating and teaching an online course: novelty of the course content, its significance for professional development and career, clear structure and coherent presentation of the course content, carefully crafted sequence of assignments. It is highly recommended to consider creating an out-of-class space (online platform) where the students can communicate with the instructors and each other.

Keywords Educational psychology · Distance learning · Satisfaction level

1 Introduction

Within the last decade distance learning has made a long way from an experimental approach to a regular practice implemented by many educational institutions. The results of such implementation have been carefully collected and analyzed to define the possible barriers to distance learning, as well as to share with the teaching community the most effective ways of organizing the course content, navigating various educational platforms, balancing synchronous and asynchronous modes, and building community online [1–5].

T. Kiseleva (✉) · A. Simanovsky · I. Melnikova · S. Elifantieva
Yaroslavl State Pedagogical University named after K. D. Ushinsky, Yaroslavl, Russia

A. Beskopylny and M. Shamtsyan (eds.), *XIV International Scientific Conference "INTERAGROMASH 2021"*, Lecture Notes in Networks and Systems 247,
https://doi.org/10.1007/978-3-030-80946-1_42

The natural process of exploring and slowly integrating distance learning into traditional teaching models was abruptly interrupted by the COVID-19 outbreak when schools and universities in Russia and other countries had to move all their courses fully online with little or no warning. Such fast-tracking brought up a wide range of issues [6–9], which we allocate to three broader groups: those of technical nature, those related to course administration, and those related to pedagogics and psychology of education [10].

The technical issues are mostly attributed to the lack or imperfection of devices and gadgets, accessibility of the software, getting accustomed to online educational platforms and platforms for videoconferencing [7, 11, 12].

Revealed issues with organization of the learning process and the course administration are associated with the new format of communication between students, faculty, stuff, and university administration, changes to the scheduling process, and the lack of ways to control the quality of online courses and students' attendance and participation [13].

Pedagogical issues are mostly caused by the lack of communication between students and instructors [14, 15], declining quality of education [16], and the lack of self-motivation among students [17]. Recent research has also shown the importance of the emotional component, and the impact it has on the effectiveness of the distance learning [18]. Most issues in the last group are directly related to students' subjective perception of the distance learning, which, in turn, affects their motivation and the level of satisfaction with the learning outcome. Hence, the purpose of this research is to find out which internal and external factors make an impact on students' perception of distance learning, and to outline the ways to overcome the existing barriers and negative tendencies.

2 Materials and Methods

The authors conducted an online survey among 211 undergraduate students at the School of Special Education of Yaroslavl State Pedagogical University named after K. D. Ushinsky. The survey was distributed to students in all years of education (first through fourth), ages 18–22, and aimed to define the level of satisfaction with the distance learning during COVID-19 pandemic in 2020, as well as the problems the students encountered. Students were asked to rank the courses by the level of difficulty and evaluate the following parameters: (a) student's motivation level for studying each course; (b) communication with the instructors (c) compatibility of the course material with online learning mode; (d) success rate for moving courses online.

All the collected data were analyzed statistically, using Student's t-distribution for assessing the statistical significance of the differences and Spearman's rank correlation coefficient (Spearman's ρ) for determining correlation.

3 Results

The first stage of our research included analysis and comparison of students' responses on the difficulty of online courses, individual motivation for taking them, number of skipped classes for each course, subjective success rate for adapting courses to online mode. The authors hypothesized that students' attitude towards distance learning would correlate with the following parameters for each course taken online:

Practicality: usefulness of the material for the future job and career in education;
Professional development: usefulness of the material for the future career, not necessarily in education;
Novelty of the content;
Difficulty level of the course;
Attendance: how often students chose to skip classes without a valid reason;

Subjective evaluation of the suitability and preparedness of courses for online mode.

Students were asked to rate each parameter based on their importance. The results of the analysis are presented in Table 1.

The analysis showed that the students in the first and the fourth years of study gave all the courses they took higher points for practicality and importance in the future teaching career. Their results are statistically different from those of the students in

Table 1 Analysis of the basic parameters influencing students' satisfaction with the online courses

Attitude toward a course		1–2 years	1–3 years	1–4 years	2–3 years	2–4 years	3–4 years
Practicality	t	**2.606**	0.744	**– 2.968**	– 0.419	– 0.246	0.566
	p	0.028	0.476	0.013	0.685	0.811	0.366
Professional development	t	**4.192**	**– 4.57**	0.557	**– 5.71**	0.716	3.566
	p	0.007	0.001	0.324	0.004	0.236	0.041
Novelty	t	**2.516**	– 0.32	– 0.26	0.316	0.572	– 0.47
	p	0.031	0.592	0.782	0.641	0.355	0.437
Difficulty	t	**2.566**	**2.572**	**2.419**	– 0.26	**– 2.72**	– 1.77
	p	0.032	0.036	0.023	0.737	0.024	0.101
Attendance	t	– 0.342	– 0.28	0.66	0.568	– 0.56	0.581
	p	0.492	0.681	0.455	0.352	0.311	0.266
Preparedness for online mode	t	– 0.47	**2.458**	**2.565**	**1.780**	**2.357**	0.557
	p	0.618	0.021	0.041	0.113	0.043	0.351

The table shows Student's t-distribution (t) and Spearman's rank correlation coefficient (ρ). *Statistically* significant differences are shown **in bold** second and third years. The students in the first three years also marked the courses they took as important for professional development.

The novelty of the course content played a role only for the first-year students, who also assigned a lot of importance to the difficulty of the material. The last parameter has proven to be less significant for students in their second and third years of study, but regained importance in the answers of the fourth-year students.

No statistically significant difference was found in relation to the attendance for students in different years, although the attended classes amounted to 80% of all classwork.

Finally, junior and senior students positively evaluated the preparedness of all the courses they took for distance learning, while their counterparts in the first and second years were much more critical.

The second stage of this research was dedicated to analyzing student-instructor communication during the pandemic and assessed students' satisfaction level of the instructors' availability and preparedness to teach remotely. This part of the survey was focused on the following parameters:

instructor's qualification
clarity of requirements
clarity and coherence of instructor's presentation of the course content
contact with the audience
instructor's availability outside class
instructor's preparedness to teach remotely.

The results of the analysis are shown in Table 2.

The qualification of instructors has proven to be of particular importance for junior and senior students. Their responses were statistically different from those of the freshmen. Sophomore students have not taken a definitive stance responding to this prompt.

Against our expectations, the clarity of requirements during the distance learning has not been marked as an important parameter. Significant difference was discovered only between the sophomore and junior students.

Coherence, consistency, and clarity of presentation of the course material was considered highly important by all the students. By contrast, all students also deemed instructor's ability to establish and maintain contact with the audience as very insignificant.

Instructor's availability outside the class time was highly valued and requested by the students in second, third and fourth years, while the first-year students considered this parameter not important in distance learning.

Responses regarding instructors' preparedness to work in a fully online mode varied, with freshmen students assigning their instructors the highest points and sophomore students—the lowest. Junior and senior students showed confidence that the faculty was able and ready to teach remotely, with senior students giving higher points.

Finally, in the third stage of our research, all the parameters revealed in previous stages undergone correlation analysis. Intercorrelation matrix is shown in Table 3.

Table 2 Analysis of student-instructor communication in distance learning

Attitude towards instructor		1–2 years	1–3 years	1–4 years	2–3 years	2–4 years	3–4 years
Instructor's qualification	t	– 0.537	**– 1.909**	**– 2.611**	– 0.161	– 1.023	– 1.206
	p	0.60	0.08	0.03	0.88	0.33	0.26
Requirements' clarity	t	– 1.548	– 1.114	– 1.481	**2.023**	1.274	0.563
	p	0.16	0.26	0.17	0.07	0.24	0.59
Clear and coherent presentation	t	**– 1.87**	**– 2.71**	**4.02**	– 1.01	0.566	– 1.63
	p	0.09	0.02	0.01	0.26	0.51	0.14
Contact with audience	t	– 0.171	– 1.177	– 0.537	– 0.161	– 1.023	– 1.348
	p	0.868	0.269	0.604	0.876	0.333	0.370
Instructor's availability	t	– 1.548	**– 2.611**	**– 1.874**	**– 2.400**	**1.941**	– 0.372
	p	0.156	0.028	0.094	0.040	0.084	0.541
Instructor's preparedness to teach remotely	t	**4.745**	**1.909**	**1.881**	**– 1.976**	**– 2.497**	– 1.206
	p	0.021	0.089	0.094	0.085	0.039	0.259

Statistically significant differences are shown **in bold**

The results turned out to be as follows: the difficulty of the course content did not matter for the students oriented towards making career in education. They attended online classes regularly, understood instructor's requirements and particularly valued the ability to communicate with the instructor out of class.

Students, who were oriented towards professional development or towards acquiring knowledge in general, valued the most the novelty of information presented to them during online classes. Similar to the first group, these students also attended classes with regularity and appreciated instructors' availability out of class.

For all students, the difficulty level of the course correlated with the high number of skipped classes, lower points for suitability of the course for online mode and for instructor's qualification. These students tended to evaluate course requirements as unclear, and the presentation of the content—as confusing and not coherent.

Students who evaluated their instructors as highly qualified, also gave them high points for establishing and maintaining contact with the audience.

The clarity of course requirements directly correlated with the coherent presentation of the material, good contact with the audience and high evaluation of instructor's preparedness to teach remotely.

Correlation has also been found between the subjectively high evaluations of instructor's preparedness to teach remotely, and, firstly, the clarity and coherence in presenting material; secondly, good contact with the online audience; and thirdly, the possibility to reach out to the instructor out of class.

Table 3 Correlation analysis of parameters indicating the level of satisfaction from the course and interactions with faculty

	Attendance	Course suitability for online mode	Qualification	Clear requirements	Clear and coherent presentation	Contact with audience	Availability outside of class	Instructor's preparedness to teach remotely
Difficulty level	*− 0.38*	**− 0.57**	*− 0.43*	**− 0.50**	**− 0.53**	− 0.19	− 0.09	− 0.217
Attendance	X	0.034	0.218	*0.309*	*0.387*	0.021	0.147	0.117
Course suitability for online mode		X	0.038	0.194	0.269	0.019	0.094	0.597
Qualification			X	0.298	0.289	*0.457*	0.297	0.427
Clear requirements				X	**0.498**	*0.409*	0.078	***00.72****
Clear and coherent presentation					X	0.011	0.183	***00.69****
Contact with audience						X	**0.588**	**0.601**

Statistically significant differences are shown *in cursive* for ρ = 0.1; **in bold** for ρ = 0.05; ***in bold cursive with an asterisk*** for ρ = 0.01.

4 Discussion

The authors have used the data analysis presented above to create recommendations for improving distance learning on college level.

Two groups of students—freshmen and seniors—have demonstrated that professional orientation is an important factor influencing their evaluation and satisfaction of the distance learning. In both cases, the students are likely going through the social transition (from school to college and from college to independent professional practice/life). Hence, it would seem beneficial to support this orientation by inviting to class successful professionals and alumni, who can share their first-hand experience with the students. Apart from webinars, demo lessons and roundtables, distance learning allows for discovering and working with professional websites and groups in various social networks, where said professionals present their materials and share experience. Faculty can navigate the process and assist the students with choosing and analyzing methods and pedagogical technologies used by the professionals. Such tasks and forms of work are also a good networking opportunity for the students.

Only first-year students marked the novelty of information related to the chosen professional field as one of the most important parameters in their course evaluations. This trend is likely explained by the fact that these students have just started their college education and value new professional knowledge the most. This position shows some informational indiscrimination: the students do not yet have a solid theoretical foundation in the chosen field and tend to absorb all the information that seems even distantly related. This feature becomes even more prominent in distance learning when students need to work with the material independently. In this regard, we suggest incorporating more tasks related to information search and filtering while working with the first-year students.

Our analysis has revealed that professional development, taken broadly, and cognitive engagement through studying is statistically significant for the third-year students as well. However, we suspect that their interest towards the informational content is of different nature. While general and professional curiosity of the freshmen students is oriented more towards quantity of the material (the more information the better), the junior students demonstrate a quantum leap: from the current of all the available information, which sharply increases in the distance learning, they want to receive only the quality content which responds to their professional interests and can be used in teaching practice. This difference suggests that the distance learning for the third-year students shall be organized much differently: the system shall include masterclasses, workshops, demo- and open classes with schoolchildren, led by either faculty or leading specialists in the field. One of the successful techniques used by the authors is inviting student to attend professional competitions (The Teacher of the Year, The Special Education Educator of the Year, The Therapist of the Year) held online.

The difficulty of the material is marked as important for evaluating courses by first- and fourth-year students. This parameter likely becomes crucial for the freshmen students since they must navigate the online mode and study completely new subjects while getting used to the college education model which differs significantly from what they were used to in school. As their information literacy develops more, the importance of this parameter is projected to go down. The following assignments can be beneficial for sharpening students' information skills: precis-writing, abstracts, oral presentations, conference- and research papers. Significant differences in evaluations of the course difficulty by the senior students has been unexpected. It is likely that in this case the power of habit played against them: during the first three years they have developed the skills necessary to succeed at college education but lacked flexibility and mobility to quickly adapt to the completely new learning mode.

The evaluation of course suitability for online mode concludes the first part of our analysis. Only freshmen and sophomore students evaluated most of their courses as ready for the distance learning. This can be explained by the lower level of instructor's control and the ability to use technological issues as an excuse: students can easily hide behind the black screen during the discussion and not participate. As long as first-year students tend to demonstrate external, rather than internal, motivation for studying, are oriented towards the grade rather than knowledge and skills, and often desire to achieve results with minimum efforts, their assessment of distance learning efficiency shall be taken with a grain of healthy skepticism. Consequently, we recommend faculty working with the first-year students use methods and techniques aiming to help students develop subjective position and internal motivation for studying.

Accepting the immensely important role faculty plays in any education system, in this research we paid special attention to students' evaluation of the professional and personal qualities of their instructors.

As the students get older and acquire more profession-related skills, they tend to assign more importance to the instructor's professional qualification and experience. We connect this change to the practical experience students get during their teaching internship in school and kindergarten, when they have a chance to see for themselves the effectiveness of instructors' recommendations, as well as the necessity of professional knowledge. Underclassmen are lacking this experience, hence from their point of view qualification and professional competence of the instructor does not affect the quality of education. Therefore, we suggest moving part of the online classes for first- and second-year students to the internship locations, where students can see their instructors working with school-age children and try their own hand in teaching. It would also be beneficial for students at School of Special Education to start helping people with disabilities as early on as possible, so we propose adding a separate training practice to the first-year curriculum.

The change of federal standards of high education fell on the third-year students and became one of the possible reasons why this cohort especially values the "clarity of instructor's requirements". Another parameter— "clarity and coherence of instructor's presentation of the course content"—has been deemed important by all underclassmen, since the traditional in-person learning mode in Russia provides a much lighter the intellectual workload on the students compared to online mode. In a traditional classroom instructor guides the learning process, organizes, controls, and explains the sequence of actions. Distance learning requires the students perform all these activities by themselves. Open informational content, on the one hand, allows for more freedom and flexibility in learning new material, but on the other—exacerbates students' feeling of being drowned in the sea of new information. Consequently, distance learning requires providing students with an extremely clear educational "roadmap", which positions the material in the most efficient way.

The survey has revealed an interesting trend related to students' need in instructor's support out-of-class. Only the first-year students claimed they did not need any extra time with faculty, possibly, taking it as excessive control and coddling, but also because this cohort does not yet have enough field-related questions to discuss with the faculty members. Students in all other years of study, on the contrary, emphasized the importance of this parameter, making it the most significant for their evaluation of instructor's preparedness to teach remotely. Students' desire for the instructors to be available for consultation at any given time suggests the necessity of creating online forums and chats, where students can communicate with the instructors out-of-class and get additional support without putting excessive pressure on the faculty.

First-year students prove to be the most optimistic and generous in evaluating both instructors' preparedness to teach remotely and courses suitability and preparedness for distance learning. We consider this to be another example of this cohort lacking a definitive opinion about the online mode. On the contrary, the sophomore students have shown skepticism towards instructors' technical competencies. This can be attributed to second-year students being more used to the college format and their tendency to overrate their professional knowledge and skills, combined with the lack of practical experience in teaching. Students of the third- and fourth years, who have already had teaching internships, are much more prone to recognize the difficulties encountered during the move from in-person to remote teaching, and, consequently, are much more forgiving in rating instructors' readiness to teach online.

Correlation analysis has been performed to better understand the interaction and mutual influence of various factors and parameters making an impact of students' perception of distance learning. The analysis has shown that students' evaluation of instructors' preparedness to teach remotely is closely connected by two parameters—clarity of course requirements and cohesiveness and clarity of instructor's presentation of the course content. This connection can be explained by the abrupt shift from in-person to remote teaching in the Spring 2020, which required faculty to completely rebuild the courses they were teaching on the go. Naturally, this enormous task was performed with varying degrees of success, which likely prompted students' increased attention to the two parameters mentioned above. We expect their importance to decline as more and more courses are being created specifically for distance learning.

The next important parameter, which needs to be considered while planning and holding online courses, is students' strong preference for having more one-on-one meetings with instructors out-of-class and instructors' general openness to constructive and productive communication.

Our research has proven once again that attendance directly correlates with successfully reaching the course learning goals, understanding course requirements, and developing a coherent picture of the subject field.

The last parameter that has significant impact on students' level of satisfaction with distance learning is the subjective difficulty level of the course. While revealing the mechanisms behind assigning a course as more or less difficult is beyond of the scope of this research, the data show that if students perceive a course as hard, they are reluctant to take it remotely and demonstrate strong preference for studying it in-person.

The research has shown various parameters students in different years of study take into consideration as they evaluate their satisfaction with distance learning during COVID-19 pandemic. These parameters, factors and tendencies can and shall be used to increase the quality of distance and hybrid learning, since the benefits of both modes cannot be ignored.

5 Conclusion

The following outcomes can be drawn from the conducted research:

- College students' informational literacy which would allow them adapting to the new learning mode quickly is not as high as was previously thought. Therefore, the sudden move to distance learning presented a significant difficulty for them.
- Students with external motivation for studying who have not fully developed their subject position, evaluate online courses higher, since for them distance learning means less control from the faculty and more freedom for the students.
- Students' evaluations of course's suitability for online mode and instructor's preparedness for remote teaching is poorly differentiated: they tend to either over- or underrate both parameters.
- Attendance affects not only the comprehension and mastering of the course content, but also the subjective rating of the course suitability for online mode and the level of satisfaction from taking this course.
- Subjective perception of a course as difficult leads to the students expressing preferences for studying it in-person, reluctance and unreadiness to take it online.
- In a distance learning, all students expect the instructors to help them navigate in the floods of information related to their chosen field, in order to broaden their professional horizon and see the ways to apply the learned material in practice.
- Professional qualification, practical experience and technical skills of the instructor play immense role in moving courses online and ensuring their preparedness and positive response from the students.
- While developing a course, it is important to clearly articulate the requirements and expectations for the course in general, and to all the offered assignments, and ensure that each element, section, or theme are positioned logically and cohesively.
- Students' perception of distance learning and hybrid mode will be more positive, and if the instructor is able to demonstrate that the course is directly related and important for students' professional career, and to enables communication with working professionals in the field and networking.

- Quality online-course shall necessarily offer an out-of-class space for the students to communicate with each other and the instructors, which will work towards both, maintaining the sense of community, and getting much needed support.

References

1. Hale S (2007) Being online. Academe 93(6):28–32. http://www.jstor.org/stable/40253659
2. Bergstrand K, Savage S (2013) The chalkboard versus the avatar: comparing the effectiveness of online and in class courses. Teach Sociol 41(3):294–306. http://www.jstor.org/stable/43186514
3. Isman A, Gazi Z, Aksal F (2010) Students' perceptions of online learning. Educ Technol 50(3):53–54. http://www.jstor.org/stable/44429807
4. Ni A (2013) Comparing the effectiveness of classroom and online learning: teaching research methods. J Public Aff Educ 19(2):199–215. http://www.jstor.org/stable/23608947
5. Baum S, McPherson M (2019) The human factor: the promise & limits of online education. Daedalus 148(4):235–254. https://doi.org/10.2307/48563401
6. Malkus Nat (2020) Reopening in the Shadow of COVID-19: beginning the first full coronavirus school year. American Enterprise Institute. www.jstor.org/stable/resrep26789
7. Abuhammad S (2020) Barriers to distance learning during the COVID-19 outbreak: a qualitative review from parents' perspective. Heliyon 6(11):1–5
8. Aguilera-Hermida PA (2020) College students' use and acceptance of emergency online learning due to COVID-19. Int J Educ Res Open. https://doi.org/10.1016/j.ijedro.2020.100011
9. Herodotou C, Rienties B, Hlosta M et al (2020) The scalable implementation of predictive learning analytics at a distance learning university: insights from a longitudinal case study. Internet High Educ 45. https://doi.org/10.1016/j.iheduc.2020.100725
10. Wagner N, Hassanein K, Head M (2008) Who is responsible for e-learning success in higher education? A stakeholders' analysis. J Educ Technol Soc 11(3):26–36. http://www.jstor.org/stable/jeductechsoci.11.3.26
11. Yoo M, Jin S (2020) Development and evaluation of learning analytics dashboards to support online discussion activities. Educ Technol Soc 23(2):1–18. https://doi.org/10.2307/26921130
12. Arulogun OT, Akande ON, Akindele AT, Badmus TA (2020) Survey dataset on open and distance learning students' intention to use social media and emerging technologies for online facilitation. Data Brief 31:1–8. https://doi.org/10.1016/j.dib.2020.105929
13. Wargadinata W, Maimunah I, Dewi E, Rofiq Z (2020) Student's responses on learning in the early Covid-19 pandemic. Jurnal Keguruandan Ilmu Tarbiyah 5(1):141–153. https://doi.org/10.24042/tadris.v5i1.6153
14. Gill S (2003) Myths and reality of e-learning. Educ Technol 43(1):20–24. http://www.jstor.org/stable/44428804
15. Costa RD, Souza GF, Valentim RAM, Castro TB (2020) The theory of learning styles applied to distance learning. Cogn Syst Res 64:134–145. https://doi.org/10.1016/j.cogsys.2020.08.004
16. Chu H-C (2014) Potential negative effects of mobile learning on students' learning achievement and cognitive load—a format assessment perspective. J Educ Technol Soc 17(1):332–344. http://www.jstor.org/stable/jeductechsoci.17.1.332
17. Kokoç M, Akçapınar G, Hasnine M (2021) Unfolding students' online assignment submission behavioral patterns using temporal learning analytics. Educ Technol Soc 24(1):223–235. https://doi.org/10.2307/26977869
18. Shen L, Wang M, Shen R (2009) Affective e-learning: using "emotional" data to improve learning in pervasive learning environment. J Educ Technol Soc 12(2):176–189. http://www.jstor.org/stable/jeductechsoci.12.2.176

Communication and Motivation for Success in Distance-Learning

Zinaida Ryabikina, Lyudmila Ozhigova, Natalia Barysheva, and Natalia Lupenko

Abstract The article discusses differences in the characteristics of communication and motivation for success among young people studying on a distance and face-to-face basis. The adolescent's personality is formed through interaction with peers and adults, in sociocultural interaction with other people. The educational space of the school allows the adolescent to gain experience in various ways of communicating and implementing the motivations of the self-realization. Distance learning makes it possible to optimize the temporary and emotional aspects of an individual's personality, but it reduces the real interaction of the adolescent. It has been found that young people who attend school full-time have a normal level of communication and sociability, but tend to escape from failure. Young people who study remotely are more willing to socialize, but are less forthcoming and more closed-minded, cautious in their contacts and more motivated to succeed. It has been shown that the higher the level of distance learning among adolescents, the more motivated they are to achieve success, the greater their motivation to learn, and the higher their educational motivation, the lower their need for communication. In other words, young people who are learning from a distance tend to have more cognitive motives than social motives.

Keywords Communication · Adolescent personality · Motivation for success · Motivation to avoid failure · Distance learning

1 Introduction

Adolescence is considered one of the most difficult time. The reason for this is the physiological restructuring of the body, and the main feature of this period is the transition from childhood to adulthood. The teenager tries to understand his identity, to understand the meaning of life [1].

Complex processes in the adolescent's emotional intelligence, family processes and values are specifically linked to aggressive behavior frequently [2].

Z. Ryabikina · L. Ozhigova (✉) · N. Barysheva · N. Lupenko
Kuban State University, 149 Stavropol'skaia St., Krasnodar 350040, Russia

A. Beskopylny and M. Shamtsyan (eds.), *XIV International Scientific Conference "INTERAGROMASH 2021"*, Lecture Notes in Networks and Systems 247,
https://doi.org/10.1007/978-3-030-80946-1_43

The fragile inner world of the adolescent is constantly in touch with the changing social reality of the modern world, which are a current, uncertain, two realities—objective and digital [3].

Especially during 2020–2021. At the time of the COVID 19 pandemic, all of the adults, children and adolescents alike had to show flexibility, resilience and the ability to move quickly to new ways of interacting in a virtuous environment.

The relevance of the study therefore lies in the importance of assessing the possibilities and the impact on the adolescent's personality of the consequences of changes in the education system, namely, the digitization of education and the use of a distance learning platform.

The purpose of this article is to present the results of a study aimed at identifying communication links and motivations for success among young people on distance learning.

Literature analysis shows that the impact of information and communication technologies on society is now at the peak of popularity in almost all the humanities and natural sciences. Various issues relating to the emotional content of cyberspace are being investigated, and questionnaires are being developed to monitor the online history of adolescents and youth [4].

In this context, the study of the impact of information and communication technologies on psychological and pedagogical technologies in education and training is becoming a separate area. Distance learning is becoming a fairly common phenomenon in the world and in Russia, which implies the interaction of real subjects of education in virtual space.

Distance learning is not a new level of education, but a form of distance interaction between teacher and student, reflecting all the components of the learning process (objectives, content, methods, organizational forms, means of learning) and enacted by specific Internet technologies or other interactive media [5].

Distance learning is chosen parents and adults, not only as a consequence of a pandemic, but also as a convenient form of education, which is crucial to many of the socio-economic problems of the family. For example, major issues such as distance from the place of education or the student's physical disabilities were of concern.

The distance also reduces the psychological stress of students. The stress associated with the collections, the road to the place the educational institution, the fear of being called to the board and ridicule—all these issues are resolved by the switch to the remote educational platform. Most teenagers learn from home, from a place where they are comfortable and safe, which makes them more comfortable.

At the same time, distance learning complicates the transmission of sociocultural experiences, impedes group and personal identification and depersonalizes students.

Full communication in this type of education is impossible due to the absence of direct emotional and suggestive contact. This leads to sensory deprivation, which is a disabling factor for adolescents at the stage of development of their cognitive and emotionally empathic sphere [6, 7].

In general, communication encourages young people to be proactive and affects the range of emotional experiences. This creates favorable conditions for personal development, self-consciousness and optimism. It also develops the ability to see in a communication partner a self-valuable person, to understand the thoughts of the partner, his experiences, and allows the teenager to define himself.

The problem of communication in adolescents, and how they are generally affected by a lack of precision in knowledge and experience in this area, exists in itself and without distance learning. The advent of the Internet and distant forms of interaction had added new dimensions to the theme of adolescent communication.

As the result of the Internet influence, adolescents are developing new values and virtual communication is becoming an integral part of their lives. Many people equate virtual communication with real communication. So they do not see or notice the fundamental difference between the two types of contacts. However, that virtual communication cannot really be called full human interaction, as it does not convey the position of the subject in space, gestures, odors, facial expressions, emotions, etc.

Hence, virtual communication and the absence of a teenager in a real educational space provokes a sense of loneliness. Absence from school leads to reduced social interaction and may result in some children losing all contact with their friends. In a quantitative study of 4526 adolescents aged 13–19 years, researchers from the Norwegian Institute of Public Health examined factors that positively and negatively affect the psychological health of young people. From the factors they chose to study, they saw that "social support from friends" and "spending free time with friends" were the strongest protective factors against mental disorders among adolescents [8].

Being socially excluded is not synonymous with loneliness. However, there is a correlation between social exclusion and loneliness. People with few contacts were more likely to be alone than people with more contacts [9].

In addition to the fact that some children and young people feel alone, research has also found a link between loneliness and mental health problems. Many empirical studies of adults and teenagers show the relationship between loneliness and depression [10].

Interestingly, studies have shown that loneliness associated with friendship is more likely to explain depressive symptoms among teenagers than loneliness associated with parents. One possible explanation is that friends are the preferred source of social support in adolescence. Thus, communication is a leading personal need of the teenager, during realization of which the overall development of the teenager, formation of his self-concept and personal identity.

Teenagers give a special meaning to communication, from accepting them and searching for something in common within each other to searching for common themes and interests. This shapes their employment, distributes time, and strengthens self-esteem through a sense of belonging to a group. This is an important component of personal development and prevention of deviant and suicide behavior [11].

Virtual communication helped to make personal communication skills redundant, so that modern adolescents were less social, more likely to communicate indirectly. The teenager chooses more individualized communication strategies and develops his or her criteria for assessing his or her success. Motivation for self-assertion, self-expression, success is one of the most relevant for adolescence. It is related to the fact that the young person seeks to increase or maintain self-esteem and respect for others [12].

The development of motive, the balance of desire for success and the desire to avoid failure may differ: some adolescents are more eager to succeed, they are more willing to take on complex problems, others are trying to preserve what they have already achieved, not to take risks.

Many of the motivations for achievement relate to the adolescent's subjective understanding of the objectives of the activities and the difficulties in achieving them, of himself and his abilities, of the level of results and of the possible reasons for them. The motivation of students is also related to health preservation, good academic performance, self-regulatory skills, sustainability, critical thinking, successful career, psychological and physical well-being [13].

The social value of motivating the achievement of adolescents underscores the importance of studying the patterns of their manifestation and development and of finding ways and means of shaping them in a targeted manner [14].

So the question that emerges is: can distance learning be a complete substitute for face-to-face communication, obscure the need to belong to a group and help motivate individuals to succeed? What is the motivation of a distance-learning teenager?

2 Methods

On-line In order to diagnose the motivations for success and communication in teenagers, the following methods were used: a method allowing to determine the main tendencies of the motivational sphere of personality in the continuum «striving for success—avoiding failure»; a method for diagnosing the level of sociability, which makes it possible to determine what percentage of adolescents are closed, careful in their communication, what percentage is outgoing and what percentage of the subjects are outgoing within the norm.

There were 60 participants, 30 were teenagers and students from a distance and 30 were full-time students in a general secondary school, age 12–15.

Young people who studied remotely had previously been learning face-to-face, but for various personal reasons had switched to distance learning. The students selected for this group were those who had moved to distance education at least three years ago. There are a number of Internet forums where parents of such pupils communicate and the selection was made directly through them. There was no gender separation among adolescents. On the basis of the results of the survey, statistical methods of data processing were applied: calculation of average trends, validity of differences, correlation analysis.

3 Results

The results of the diagnostic of social characteristics show that 60% of young people in the full-time education group have a normal level of communication and 24% of sociability, which has a significant difference relative to the group of young people who study remotely. Distance students predominate a higher percentage of isolation and caution in communication. Only 37% of adolescents have normal sociability.

Caution in communication is found in 30% of young people on distance learning and in only 10% of students in full-time. So, young people in distance learning are willing to communicate, but cautious about new connections and acquaintances.

Table 1 shows the assessment of adolescent sociability on distant learning and full-time education.

The Mann–Whitney U-criterion was used to assess the validity of the differences. The combined rankings for the group of teenagers studying at the station are 736.5 and for the group of teenagers studying full-time are 1093.5. The empirical value of the Mann–Whitney U-criterion is determined by the formula half-tea Uemp = 271.5. U is critical of the table for 1% error is 292. The existence of a significant difference between the level of the topic in the samples considered is recognized, provided that Uemp < U is critical. Since 271.5 < 292, respectively, differences in the expression of social indicators in groups are statistically significant.

As a result, it was found that the isolation of students on distant learning was expressed and distinguishes them from full-time students. The closed-mindedness is here interpreted as a lack of dialogue and a preference for solitude in the society of others. Full-time students are more forthcoming than students on distant learning. Sociability is expressed in curiosity, conversation-intensity and impulsiveness. Sociable teenagers prefer society, new acquaintances, and being the center of attention, but by learning from a distance, it's rather problematic.

The diagnosis of the level of motivation for success had shown that young people who were educating remotely were much more motivated to succeed than those who were studying full-time. These adolescents have higher success rates. At the same time, teenagers of the full-time study group have a higher percentage on the «motivation to avoid failure» scale and are much more inclined to it.

Table 1 Assessment of adolescent sociability

Scales	Distance education (%)	Full-time education (%)
Closure	27	6
Discretion in communication	30	10
Proper interpersonal skill	37	60
Sociability	6	24

Table 2 Results of motivation for success and avoiding failure on distant learning and full-time education

Scales	Distance education (%)	Full-time education (%)
Motivation to avoid failure	20	37
The Tendency to avoid failure	7	30
The Tendency towards motivation for success	33	20
Motivation for success	40	13

Assessing motivation for success and avoiding failure

Table 2 shows the results of motivation for success and avoiding failure on distant learning and full-time education.

The data were analyzed statistically to assess the differences between the two independent samples (Mann–Whitney U-criterion).

The combined rankings for the group of teenagers studying at the station are 743 and for the group of teenagers studying full-time are 1087. Define the empirical value of the Mann–Whitney U-criterion by formula, giving Uemp = 278. U critical for 1% error is 292. The existence of a significant difference between the level of the topic in the samples considered is recognized, provided that Uemp < Ucritical. Since 278 < 292, respectively, differences are considered valid, assuming that there are significant differences in motivation between students from a distance and students from a teenager.

Thus, young people in distance learning have a higher motivation for success than their peers in full-time education. It has been statistically proven that line up teens are less motivated to succeed and more inclined to avoid failure. It is interesting that the majority of teenagers in both groups did not gain high scores on the scales «motivation to success» or «avoiding failure» but were distributed on the scales «tendency to success» or «tendency to avoid failure». This can be explained by the uncertainty surrounding adolescent self-identification.

That age was characterized by a fluid attitude and principles and a lack of maturity with regard to self-determination. Opinions and attitudes in adolescence are only emerging, and can change significantly under the influence of the environment and important adults. Teenagers who show a tendency are still in the process of determining their own position, and this once again indicates the importance of what is happening at that age.

The presence of differences between students, both in person and remotely, in relation to motivation and level of sociability, made it possible to move on to the analysis of correlational relationships by means of the Spearmen rank correlation coefficient. A comparison of the responses from both methods in the remote-format group has so far shown that the maximum scores under the methodology of diagnostic motivation speak of a high motivation for success, and the maximum scores under the methodology of consent—a tendency towards closed communication. The estimation of the rank correlation coefficient of Spearman revealed a significant rank correlation ($p = 0.9895$).

The calculation results in a direct link. That meant that the more closed-minded young people in a remote group, the more motivated they were to succeed, and the more motivated they were to learn. The higher the educational motivation of the adolescent, the lower the need for communication. In this situation, cognitive motives prevail over social motives.

4 Discussion and Conclusions

Distance learning uses interactive technologies and communication systems to improve learning. It has the potential to change the way that people learn and learn in all directions. This can raise standards and increase participation in lifelong learning. It cannot replace teachers and teachers, but, along with existing methods, it can improve the quality and coverage of teaching, as well as reduce the time spent on administration. That could enable everyone to realize their potential and help to create an educational workforce capable of change.

However, the social environment plays a significant role in the behavior and development of adolescents. Students are influenced by their circumstances and respond to this environment in a certain way. Environmental factors play a significant role in the development of the personality of adolescents. In line with the results of our research, the surrounding environment of a young student at a distance limits his or her development in the social and cultural sphere, and therefore the active development of the cognitive sector begins. Isolation appears where there is no environment for expression of emotion and creates a sense of alienation. A growing person that needs to be implemented develops what it can. A commitment to success could be seen as a desire for hierarchical self-assertion. So, the psyche seeks to acquire a sense of belonging to society in a different way, by asserting its own knowledge, its own success.

Society is obligatory for every living person, and it is very important that the teenager learn to interact with it. As our research shows, offline studying teenagers feel well and open to new ways of life, which is not the case for the group in the distance. Communication, recesses at school, dining in the dining room, moving from room to room are a living dynamic, a process in which the adolescent learns skills, and this is a huge part of the educational process for the individual. The example that a teenager takes when observing a person with the involvement of the senses, sensing their physical presence, is priceless for the psyche.

Distance learning thus focuses on the content of the subject, that is, on cognitive motives and reduces social orientation, that is, social motives. A student sitting in front of a computer cannot fully appreciate what other people are doing, so he is more focused on himself and what is required of him—learning and cognition. However, in distance learning students have fewer ways of acquiring experience and knowledge through peers, hence the personal and social identification of teenagers is going through a more complicated way, connected with managing risks of «blurred» false self-identification, supported by virtual space [15].

Our study did not answer the question of whether the isolation of the young person was the reason for the distance learning, or whether it was acquired in connection with such education, but it has been found that there is no need to establish contacts within the school community, no need to spend time on the horn and no stress from fear of being summoned to the board or being ridiculed in public, increases the learner's desire for success.

In the light of the information received, it is advisable for distance learning students to attend sports sections and face-to-face groups in order to communicate live with the presence of physical contact of peers and important adults; to harness the potential of success and to develop a diverse cognitive and emotional field, taking into account the area of near development. Adults (parents and teachers) need to maintain adolescents' self-reliance and discipline, helping them to organize their education in homework environments.

References

1. Negru-Subtirica O, Pop EI, Luyckx K, Dezutter J, Steger MF (2016) The meaningful identity: a look at the interplay between identity and meaning in life in adolescence. Dev Psychol 52(11):1926–1936
2. Pérez-Fuentes MDC, Molero Jurado MDM, Barragán Martín AB, Gázquez Linares JJ (2019) Family functioning, emotional intelligence, and values: analysis of the relationship with aggressive behavior in adolescents. Int J Environ Res Public Health 16(3):478. https://doi.org/10.3390/ijerph16030478. Last accessed 06 Feb 2019
3. Mossberger K, Tolbert CJ, McNeal RS (2008) Digital citizenship: the internet, society, and participation. MIT Press, Cambridge
4. Zych I, Ortega-Ruiz R, Marin-Lopez I (2017) Emotional content in cyberspace: development and validation of e-motions questionnaire in adolescents and young people. Psicothema 29(4):563–569. https://doi.org/10.7334/psicothema2016.340
5. Moore MG, Kearsley G (eds) (2015) Distance education: a systems view. Belmont, CA, Wadsworth, vol 2/4, pp 107–118
6. Lopes PN, Mestre JM, Guil R, Kremenitzer JP, Salovey P (2012) The role of knowledge and skills for managing emotions in adaptation to school: Social behavior and misconduct in the classroom. Am Educ Res J 49(4):710–742
7. Meyer DK, Turner JC (2006) Re-conceptualizing emotion and motivation to learn in classroom contexts. Educ Psychol Rev 18:377–390
8. Myklestad R (2012) Tambs. Risk and protective factors for psychological distress among adolescents: a family study in the Nord-Trøndelag Health Study. Soc Psychiatry Psychiatric Epidemiol 47(5):77–91
9. Meeuwesen L, Hortulanus R, Machielse A (2017) Social isolation in modern society. Nguyen, vol 5, no 2, pp 159–171
10. Qualter B, Munn R (2018) Examining psychometric properties of the kutcher adolescent depression scale using multidimensional item response theory. Eur Child Adolesc Psychiatry 19:493–501
11. Calvete E, Orue I (2010) Cognitive schemas and aggressive behavior in adolescents: the mediating role of social information processing. Spanish J Psychol 13(1):190–201

12. Ståhlberga J, HetaTuominena A-T, Niemivirtad M (2019) Maintaining the self? Exploring the connections between students' perfectionistic profiles, self-worth contingency, and achievement goal orientations. Personality Individ Differ 151(1):109495. https://doi.org/10.1016/j.paid.2019.07.005
13. Zaccoletti S, Camacho A, Correia N, Aguiar C, Mason L, Alves RA, Daniel JR (2020) Parents' perceptions of student academic motivation during the COVID-19 lockdown: a cross-country comparison. Front Psychol 11:592670. https://doi.org/10.3389/fpsyg.2020.592670
14. Kim J, de Long SPA, Gorelik W, Penwell K, Donovan C, Chung H (2020) Family orientation and achievement goal orientations among the children of immigrant and non-immigrant families. Int J Educ Psychol 9(2):132–160. https://doi.org/10.17583/ijep.2020.4422
15. Ryabikin I, Bogomolova E, Lyudmila Ozhigova N (2020) Personality identity in the terms of virtualization of being. In: Innovative technologies in science and education (ITSE-2020). E3S Web of conferences 210:20016. https://doi.org/10.1051/e3sconf/202021020016

Information Technologies in Training of the Extreme Profile Athletes

Anatoly Nepomnyashchiy and Anastasiya Bersova

Abstract The problem of safe and effective use of modern digital information and communication technologies in the course of training of athletes of an extreme profile is considered. It is shown that rates of information technology development advance rates of development of consciousness of their users in this connection in activity of the person his attention is generally concentrated on digital simulations of reality, but not on her. At the same time in the extreme activity the athlete has to be most concentrated on reality in general—on the inner world and the outside world that is necessary for information support of decision-making in critical and limit situations. It defines a key part of the problem of a research and the task following from it—search of a way of safe application of all types of natural and technical information support of athletes training. The solution of this task by correlation of stages of development of scientific knowledge with stages of development of human consciousness on whom affordable conscious management of the athlete limit opportunities is proposed. Restrictions, ways and methods of use of digital information technologies in practice of extreme profile athletes training are proved. Justification is executed in the context of the main methodology of post-nonclassical science—integral vision and a method of the focusing generalizations and also with application of metasystem and deductive approaches to the solution of scientific problems.

Keywords Sport · Extreme · Training · Information · Technologies

1 Introduction

The training of athletes of higher qualifications is always associated with the study of the ultimate capabilities of a person and with the study of those restrictions that

A. Nepomnyashchiy
Southern Federal University, Bolshaya Sadovaya St., 105/42, Rostov-on-Don, Rostov Region 344006, Russia

A. Bersova (✉)
Russian State University of Physical Education, Sport, Youth and Tourism, 4, Sireneviy b-r, Moscow 105122, Russia

A. Beskopylny and M. Shamtsyan (eds.), *XIV International Scientific Conference "INTERAGROMASH 2021"*, Lecture Notes in Networks and Systems 247,
https://doi.org/10.1007/978-3-030-80946-1_44

prevent a person from detecting these limits and trying to overcome. In this regard, any high-level sport can be called extreme. However, there are sports in which these extreme opportunities (extremes) can manifest themselves at the very beginning of athletes trying. Such species include: mountaineering, freediving, mountain skiing and a number of others. Naturally, extreme activities exist not only in sports, but also in a number of professional fields. But in preparation for professional activity there is its own specificity, which is not considered in the present work. Here, the object of the study is precisely the training of extreme athletes, and the possibility of using modern digital information and communication technologies (ICTs) for this training.

The need to consider ICTs from this perspective is dictated, above all, by the fact that globalization processes penetrate ICT into all aspects of modern human life and sports, including. The ever-increasing intensity of data exchange creates the illusion that the whole modern world exists and develops only by intensifying this exchange. At the same time, not everyone recalls that "data" is the simplest semantic aspect of the concept of "information," except for which there are many others. Among this set there is also the asymptotically most complex definition given in the Bible (the Gospel of John). "In the beginning was the Word, and the Word was with God, and the Word was God. He was with God in the beginning. Through him all things were made; without him nothing was made that has been made." Here "Word" (information) is obviously positioned as the substance that controls the creation and existence of everything in the universe.

In this regard, the problem arises of defining the safe use of the simplified definition of "information" (as a collection of data) and the corresponding formalized ICTs in the training of athletes who comprehend extreme (limit) possibilities in an environment of multidimensional reality, and not in a medium of two-dimensional computer simulations.

The purpose of the study is to show and substantiate the methodology for the safe use of modern digital information and communication technologies in the training of extreme athletes.

2 Materials and Methods

This problem can be represented in the form of a combination of its individual questions, the main ones of which, in our opinion, are the following: the genesis of the problem; the current state of scientific knowledge in the field of research of publicly available and extreme activities; methodological approaches and specific solutions.

Research methods: deductions, integral vision, orienting generalizations, digital processing of gas discharge visualizations.

3 Results of the Study

The shortest way to a vision of the genesis of any problem is the use of the method of orienting generalizations proposed by the American philosopher and specialist in the field of transpersonal psychology [1]. The essence of the method is the integration of the metasystem approach and the principle of deduction in cognition. The first, according to K. Gödel's theorem "On insufficiency," makes it possible to see the true causes and consequences, the second—to rise to the level of the basic laws of the universe to see why the detected causes give rise to precisely such consequences. In other words, looking at the general laws of the world order, we can navigate and see the necessary way to solve a specific problem.

Following this method and looking at the world of the Earth from the position of a meta-observer, it is easy to find that everything living on the planet, except modern man, does not need artificially created means of providing information exchanges either in everyday life or in extreme situations. Therefore, every living creature on the planet has natural means built into its structure of obtaining, analyzing and transmitting the information necessary for its life support and ensuring collective, interconnected, safe life.

Looking at rationally not developed tribal aboriginal communities, we understand that such opportunities are embedded in the structure of each person, but a modern "civilized" person, falling under the magic of technical comfort, simply forgot about this or does not want to remember, as R. Kegan and L. L. Lahey say, the existence in a person of "immunity to change" [2]. This immunity also exists because the human community, like everything in this world, is organized according to the principle of hierarchy (governance from above). And, if "from above" they said that ICT is very good, and behind them the future, most, "without wanting cunning," thoughtlessly repeats this as a spell, so as not to fall out of the structure of established and familiar life support and comfort.

In an ordinary condition of consciousness, the ordinary person accepts the statement "lowered" from above as true, following simple logic of a reasoning: if it is "from above", so it is correct because from above is more visible what should be done and as. Not to be in a conflict with the system of life support, the person looks for and quickly finds what validates his conclusion. What doesn't confirm this correctness isn't taken into account or just isn't noticed because it doesn't pass through the filter of the realized perception existing in consciousness of each person who didn't reach as speak in the east, conditions of an enlightenment. In the prevailing majority of cases all this looks as an ontologization—giving by the person the status of reality to some set of ideas of the world or the studied object. At the same time at the extra-mental levels of the consciousness such compromiser can be also in a metasystem, but doesn't allow to bring new knowledge to conscious level the ontologization myth and fear to lose a habitual support.

Such condition of blind belief to tops and disbelief to itself is fixed owing to existence of the second law of thermodynamics according to which and the electron in atom seeks to occupy the steadiest, low-energy state. And in this plan of people differs in nothing from an electron. Even the great politician U. Churchill said that he never stood if it was possible to sit down, and never sat if it was possible to lay down. At the same time, according to the same law, what isn't used passes into a rudimentary state or in general atrophies. Going down a course of life, it is easy for person to forget and lose the opportunities, and here anew to update them very difficult and not everyone manages, even very wishing.

What are the possibilities for receiving, processing and transmitting information in a person initially, even if he is not aware of them at the level of personal perception? In our opinion, these possibilities are most fully described in the works of Wilber [1, 3, 4], which refers to four human "eyes" of knowledge: "eye of the body"—sensory perception of the world; "eye of mind"—a vision of meaning in conceptual devices and semantic systems; "soul eye"—sensual and super-sensitive perception of the energy information state of space; "eye of the spirit" is an intuitive vision through time and space by visualizing what is learned.

Since it follows from the works of Nobel laureate I. Prigogine «life»—are purposeful exchanges of substance, energy and information, universal means of exchanging information (an eye of the soul and an eye of the spirit) should initially be built into all forms of living, which proves a lot of experimental studies. There must also exist a universal language of communication of all living things, which is the language of energy information exchanges—the language of thought. This is the very language that all the inhabitants of the earth consciously spoke until, according to the Bible, they began to build a "Babylonian tower." This is the very language in which the so-called "rainbow children" still communicate perfectly with representatives of all worlds of the Earth—minerals, plants, animals, birds, reptiles, creating many new troubles for those who consciously cannot use the language of thought, and therefore use the tongue to communicate with themselves.

Now, to fully draw the genesis of the problem under consideration, it is necessary to show how the representative of modern human civilization forgot natural abilities and eventually came to the need to use digital ICTs.

The beginning of this process is well stated in the Bible, in the first book of Moses "Life", the Old Testament. According to the 9th section, having landed on the land when waters of a flood subsided, "Noah, a man of the soil, proceeded to plant a vineyard. When he drank some of its wine, he became drunk and lay uncovered inside his tent. Ham, the father of Canaan, saw his father naked and told his two brothers outside" … When Noah awoke from his wine and found out what his youngest son had done to him, he said, "Cursed be Canaan! The lowest of slaves will he be to his brothers."

For what did such cruel punishment follow? Noah cursed not Ham, but his son Canaan that the first and his brothers well remembered that it is impossible to give big secrets even to the closest. And the essence of a secret is that alcohol immediately closes an eye of soul and an eye of spirit and the person becomes "nude", i.e. unprotected energetically and informationally. Any person who is interested in the meaning of the dreams knows that to dream itself nude meant by approach of diseases and other troubles. This nakedness was seen by Ham who told about result of alcohol use to the brothers.

And meanwhile, the point of schooling of the person to alcohol was in that he, having lost sensual and intuitive perception, began to move from an archaic condition of mind and reason to development of rational mind. But on the way to rationality there is one more step—mythical consciousness [5] which can't also be passed. In a condition of mythical consciousness of people, lost ability consciously to use the eye of soul and a spirit eye, gets automatically as the lack of information for decision-making results him in need a lot of things to accept on the basis of trust to information arriving from an eye of a body and external sources. In the beginning such person gets to the power of breeding and confessional myths, then gets under influence of myths of classical and nonclassical science. But, gradually developing the rationality to high level, he begins to understand need of the further development and restoration lost. For this purpose, he also addresses science post-nonclassical which in the basis puts synthesis of knowledge gained by culture, science of the previous stages, philosophy and mysticism.

Fundamentals of methodology of post-nonclassical science is the integral vision [3–5] reached by conscious updating of all four systems of perception by the person of the external and internal worlds mentioned above. At the same time objectively registered forms of existence of any individual of this world it is studied by an objective method, and the consciousness of these forms which doesn't have objectively registered borders [6] is studied in the subjective way by contemplate the practician [7].

Thus, the post-nonclassical science using a triad formulation of conservation law of substance, energy and information [8] as the paradigm is included into full compliance with provision from "eternal" philosophy on triad structure of the person—"body–soul–spirit" [9]. But on it the history doesn't come to an end as there is a lot of adherents of the post-nonclassic in natural sciences, especially among developers of information weapon, and in humanitarian field there are a few of them.

The humanities are still locked in a cage of classical science that a people at large didn't guess the divine origin and the free will granted to them by the Creator. And as in a paradigm of classical science, there is neither energy, nor information as creative force, from her position in the person there can be neither eye of soul, nor a spirit eye, "because it can never be". Such is reaction of classical science to everything, not holding in her paradigm, known since the time of G. Galilei.

But, at exclusively objective, classical perception of the world, arises theoretically, almost ineradicable problem of a lack of information at decision-making as sensor systems of the person (body eye) have very narrow bandwidth, and a mind eye in general works only with semantic systems which possibilities in information transfer are strictly limited to the final alphabet and syntax of the language used.

To somehow reassure those who are concerned about the complexity of decision-making in the face of uncertainty, they are offered as a panacea the digitization and chipping of everything and everyone. At the same time, it is claimed that the information capacity of the Internet is growing unabated, and in the near future it will be possible to find and download any amount of information there. At the same time, they forget to say that only data is accumulated on the Internet, that is, not integral images of reality, which are transmitted only from person to person in the language of thoughts, but the simplest version of objectified information is data that will never help anyone solve the problem of uncertainty.

This is particularly the case in extreme activities, in which it is not possible to use external means to inform the security decision-making process. In extreme activity, the score does not go for minutes and seconds, but for milli and nanoseconds, and technical means of decision-making only interfere. Imagine a situation in which a climber is at risk of disruption. Here, according to the theory of "classics" and «digitalists», it's time to look into the Internet to find out further actions. Understanding the absurdity of such an idea, adherents of the classical view of man try to convince people that this problem can be solved by connecting the brain online using chipping. At the same time, ignorance or intent, no one explains that the eyes of the soul and spirit will be blocked by this chip for conscious perception for the rest of the life of this body and brain. That is, a person can easily turn into a biological robot controlled online.

Since the problem of exploring extreme possibilities is far from new and has always worried many for a variety of reasons (from the desire to know yourself, to the desire to rule over others), there is a need to briefly dwell on already tested general approaches and technologies. To do this, we turn to large-scale projects implemented in recent history and existing today.

Such large-scale projects certainly include the creation and functioning in the USA of the "Institute for Integral Human Development," called Esalen [10]. The institute was founded in 1962 on the west coast of California by major financiers with the involvement of two American psychologists as methodologists—Michael Murphy and Richard Price. The first conducted research on human reserve capabilities, the second studied and practiced gestalt therapy. Both studied the traditions of Yoga and other eastern systems of human spiritual development.

In 1992, based on the results of 30 years of research, Michael Murphy published the fundamental work The Future of the Body. A Study of Further Human Evolution [11], in which he summarized the data and results of more than 10,000 studies conducted at that time in Esalen, indicating the possibility of a global transformation of man in terms of expanding the marginal possibilities in his activities. According to experts, this is the most thorough work ever written in the world of Western science on the relationship between consciousness and body. At the same time, this is the most significant Western experiment not only in the field of studying the ultimate capabilities of a person, but also in the field of integral education, built on the philosophy of unity of existence.

The concept and activities of the Esalen Institute were based on the absence of any technical means to support the training and development of its staff. This is still the most important aspect of the institute's activities and its success, which confirms the conclusion made above that it is unacceptable to use a kind of technical "crutches" in the practice of studying and expanding the maximum capabilities of a person in general and in the training of athletes of extreme profile, in particular.

Naturally, Esalen is not the only example of such large-scale experiments on Earth. So, the idea of creating the institute came to Murphy and Price after they visited the adept community of the integral yoga, founded by Sri Aurobindo Ghosh. This Sri

Aurobindo Ashram has existed in India since 1926 to this day. In 1951, under the leadership of Sri Aurobindo student and collaborator Mirra Alfasa (Mother), whom back in 1926 Sri Aurobindo entrusted the spiritual and material leadership of the ashram, the Sri Aurobindo International Educational Center was created. In 1968, Mother founded the city of Auroville, which was built in the state of Tamil Nadu (India), near the city of Puducherry under the auspices of UNESCO [12]. It was in Auroville where its founders created a unique institution of integral education, based on the philosophy of unity of all being, to which humankind has been always called by eternal philosophy [9]. According to the founders, the city and its educational institution are called upon to become a place where men and women from any country could live in peace and in developing harmony, following world universals, and being independent of nationality, politics and religious preferences.

In Auroville, even during its construction, the use of technical means was minimized. Sri Aurobindo and Mother presented humanity with more than 20 volumes of description of spiritual experience and the foundations of integral yoga. And all these grandiose results in the field of human limit research were achieved only using oneself as an instrument of cognition.

These examples of research and educational centers naturally do not exhaust the list of institutions of the world in this direction. Not only are institutions of narrow vocational training developing and multiplying in the world, where only objective methods, methods and means of training, including ICT, are used, but also educational institutions of an integral type, where a person comes to the need for self-knowledge, knowledge and expansion of opportunities.

Bhagavan Sathya Sai Baba, who devoted 35 years to the creation of the Integral University in India, made a special contribution to the development of integral education based on the philosophy of holism. The main goal of the university is to promote the education of an integral personality, in which "a harmonious synthesis of spiritually mature intelligence, a compassionate heart and golden hands is manifested; it is a wonderful combination of nobility and skill. Nobility without skill is useless to society, and skill without nobility is dangerous to it. Spiritual education is the basis of integral education. This is what Bhagavan calls «educare» (education of universal human values). In the end, all education should come down to this." [13].

Of great interest in terms of the problem under consideration is the individual experience of extreme athletes, which is multiplying day by day and is constantly enriched with new achievements. Analysis of climbing dynamics to peaks with a height of more than 8000 m. shows that the mass nature of climbs is determined by two factors—the growth of human internal capabilities and the improvement of technical support for athletes, especially in terms of climbing equipment. Of course, better clothing (light and warm), shoes, safety equipment and specially prepared food allow to climb heights and those people who are not so much concerned with knowing and testing themselves as with image nurturing. These people are not taken into account in our analysis, since they do not make a significant contribution to solving the task of exploring limit possibilities.

Of interest are people such as Reinhold Messner—the first person to climb all 14 peaks, more than 8000 m high, mainly alone and without the use of compressed oxygen cylinders. In book Crystal Horizon [14], he describes in some detail the system of preparation, the key aspect of which was the development of the ability to feel the world and, in particular, the mountains as full-fledged living partners in experiments on the study of altitude extremity. And this is not a metaphor, since it was this ability that allowed him to stay alive, while many trying to repeat his achievements either could not do this, left alive, or died.

Messner himself, answering the question of how he managed to do this and survive, answers in this book that he never tried to climb to the top if, when asked to the mountain: "Can I climb You?" The mountain answered him: "No." In all such cases, Messner decided to return. And it never depended on the money already spent on the expedition organizing.

This is his key to success, which cannot be understood by a person with an exceptionally rational way of thinking, for whom everything that does not run, does not crawl, flies and does not speak the human language is dead. That is, it is necessary to go beyond rational thinking, which is achieved only with the expansion of the area of conscious perception of the world by means of integral preparation. It is in this way that the activity of the eye of the soul and the eye of the spirit is brought to the level of the conscious and it becomes possible to communicate with the world in the language of thought-like. At the same time, we are talking not only about the world of mountains, but about the world in general. This was confirmed by Messner, communicating in extreme journeys not only with the "pole of altitude," but also with the geographical poles of the planet and with its deserts.

Naturally, the question arises: "At what stage of development should the human consciousness be in order to ensure such contact with the extreme environment of interaction, and, accordingly, the safety of his extreme activities?".

At the first stages (archaic, magical, mythical and rational), this is unattainable, since a person sees the world "lopsided" on them—either only subjectively (sensually intuitive), or only objectively (sensory-logical). The further, superrational stages of the human consciousness development are as follows: pluralistic (a person understands that all individuals of this world have the right to live, and this is a consequence of the conscious manifestation of the activity of the eye of the soul); holistic (a person realizes and understands that everything in the world is totally interconnected, and this is a further "discovery" of the eye of the soul with the simultaneous beginning of actualization of the eye of the spirit); causal consciousness (allows you to see by all four eyes the true causes of what is happening); not dual—consciousness of the cognizant (in Eastern-Buddhas) [1, 5].

With the achievement of each new, superrational stage of development, a person moves towards an integral, holistic perception of the world—sensationally, logically, ethically and intuitively, while reducing situational uncertainty. That is why everyone who strives for the development, realization and expansion of limits always begins with a general integral education in order to update and harmonize all systems of perception. On the contrary, narrowly specialized education, especially with the widespread use of ICTs, figuratively speaking, "locks" a person in a cell of objective, excluding the development of ethical and intuitive perception of the world. With this formation, uncertainty in decision-making increases immeasurably. In this case, to ensure the safety of life, extreme activities must be excluded.

Since post-nonclassical science accommodates all types of knowledge (ordinary, scientific of all previous stages, philosophical and obtained in the process of religious contemplative practices), this conclusion can also be substantiated using the seven-level human structural model proposed at the beginning of the last century by the representative of Rosicrucianism Heindel [15].

According to this model, a person is structurally a hierarchy of seven nested bodies that have substances different from each other. The form most manifested for sensory systems has the lower level of this hierarchy—the biological body. This is the so-called protein-nucleic form of life, in which objectively available information for fixation is propagated electro-physically and electrochemically, i.e., with a large delay. At the same time, the trillions of cells that make up this body have a built-in mechanism of "conscience"—connecting the lead—that allows cells at the level of their consciousness to "hear" each other uninterrupted, that is, without delay of the signal, which allows the body to exist both normally and in extreme. This effect was discovered in non-classical science during the development of quantum physics and found reflection in Bell's theorem on the non-locality of quantum effects, which if occur, then everywhere at the same time.

The second, "vital" human body is some kind of source of power for a biological body, and is invisible to a usual human eye. It can register by method of gas-discharge visualization in the field of currents of high frequency and has huge value for extreme the practician, the demanding essential power expenses.

The third body, even more distinguished and inaccessible for touch visual perception, plays a motivator role—the stimulus to action therefore M. Heindel calls it "a body of desires". The Substance structure of the body of desires is still inaccessible to orthodox science and is investigated only by adherents of subjective experience. The body of desires plays the important role in motivation of the person to self-knowledge through extreme situations.

The fourth human body—the thought body consisting of a body of a concrete thought, a body of mind and a body of an abstract thought. At even more distinguished substance structure of the fourth body there is an electromagnetic component that allows now scientists and engineers to work on mentally operated computers creation.

Three hierarchically top levels of the structural organization of the person (in this model 5, 6 and the 7th bodies) are triune Spirit, structurally including human spirit (5th body), vital spirit (6th body) and God Spirit (7th body) that in Christianity is called as the Holy Trinity (God Son, God Father and God Holy Spirit). It is what is called as man consciousness, his hierarchically highest operating system, having in it all information of previous experience of existence at incalculable quantity of the most various forms. The man consciousness is immortal.

The first four bodies in the set form what is called as the personality. It is the temporary tool of consciousness which exists one century, but the experience accumulated by it accumulates by consciousness and is used as required in the course of it infinite development.

For the solution of the task set in this research, the offered level of consideration of M. Heindel model is quite enough as from him directly and inevitably the conclusion follows.

In any hierarchy strategic control is exercised from top levels as from it is most information provided that reduces degree of uncertainty of the lived situation. Naturally, extreme situations, especially, can't cope the lower levels of hierarchy. Therefore, any means of assistance to decision-making by the person which are on the complexity at the level lower, than a biological human body, have to be surely excluded as extremely dangerous, distracting the personality from contact with the own operating system.

However, if we consider the stages of general physical and psychological training of future extreme athletes, on which the basis of extreme achievements is laid, then here we can see the possibility of widespread use of modern information technologies and digital methods for exploring human capabilities.

So, in our studies of the level of readiness of athletes for extreme activity, we used the method of gas-discharge visualization of the psychophysical state of a person in combination with the method of computer processing of GDV-grams developed at the Leningrad Institute of Precision Mechanics and Optics [16].

Our experiment involved 168 athletes of different levels of readiness for extreme activity, which, according to the results of the study, were divided into 4 groups. Figure 1 shows illustrative examples of GDV-grams that illustrate well the differences in the training levels of athletes.

From this figure, it is quite obvious that representatives of the first two groups for security reasons cannot be allowed to extreme activities due to significant violations in their structural, energy organization (life body in the model of M. Heindel). These

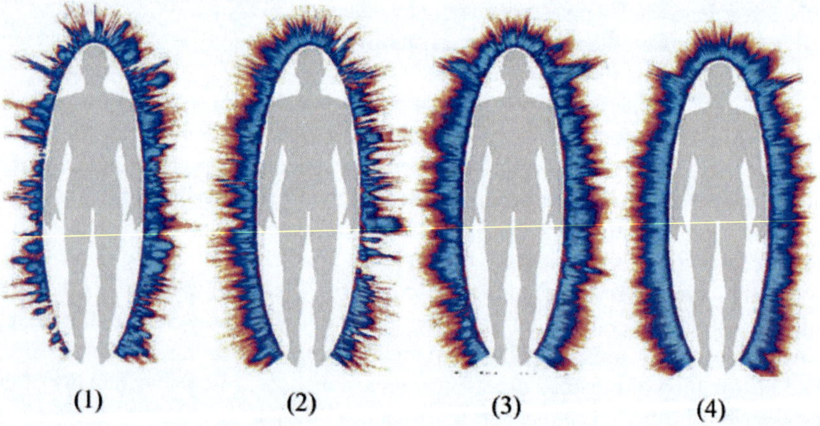

Fig. 1 Examples of the GDV Energy Fields for four groups of subjects: (1) low level of readiness for extreme activities, admission only to general physical training; (2) average level, admission to propaedeutics training; (3) sufficient level, admission to initial extreme training; (4) readiness for training and competition in extreme conditions

Table 1 Digital parameters of sportsperson readiness level assessment

Readiness levels and number of subjects (in parentheses)	Normalized value of energy field glow area	Symmetry of energy field (%)
High level (42)	22 and over	91 and over
Sufficient level (41)	20–21	86–90
Average level (42)	18–19	76–85
Low level (43)	10–17	60–75

deficiencies in preparation can also be displayed as numerical parameters as shown in Table 1.

Here, the normalized value of the luminous area of the energy field means the luminous area, as it is already seen as a result of the computer two-dimensional reconstruction of the human energy (life) body (Fig. 1). The primary data for such processing is a photo of a gas discharge around the fingers of a person arising in the field of high frequency currents (Kirlian Effect). The symmetry of the energy field is determined by the same method—computer analysis of the areas of the left and right halves of the GDV-gram.

4 Discussion

In order to avoid discrepancies, it is necessary to emphasize that in itself digital and, in general, information and communicative technologies are only modern instruments of human activity and consequently do not carry in itself any good or evil. With the help of a scalpel, you can save a person, or you can kill. A question in what hands this tool is used and in what purposes.

So, in principle, there is nothing bad that in the world the data exchange by means of digital technologies is intensified. Thanks to opportunities the Internet we, for example, can learn about practice of extreme activity "the ice person" Wim Hof [17], get acquainted with his method and even to register in pupils. We can, without leaving the house, to visit the Esalen website and to receive the invitation to visit institute for the purpose of cooperation. We can listen to a speech of the employee of Sberbank A. Kurpatov, during a business breakfast in Davos [18] and learn that hobby for gadgets and digital technologies harmful influences a brain as his centers intended for processing of integrated images of reality at the same time begin to pass into a rudimentary state. And it for us is very important information proving justice and validity of our conclusions.

We can by digital means create and use biological feedback systems and get, as shown above, very important results for the effective training of athletes at the initial levels of its organization.

At last, we can thumb through continuously in the days and at the nights the uncountable number of the pages the Internet containing numerous descriptions and results of experience of athletes and just researchers of an extreme profile. And it will be the end of our own achievements as our body, a brain and the systems of perception will regress, rolling not only in "useful" data, but also in the postmodern simulacra and computer simulations [19] born in attempts of authors of text information to explain all unclear with means of classical science.

5 Conclusions

The analysis of the genesis of the indicated problem and the experience of extreme achievements in sports (including author's) shows that in the process of extreme activity itself, even at the training stages, the technical means of information support for the decision-making process cannot be used, since it is possible to learn extreme activity only in its process, and not distracted from it. Therefore, we can only talk about the stage of propaedeutics—the stage of preliminary, introductory acquisition of the knowledge necessary to begin the preparatory process itself—the process of knowing yourself and opportunities.

One of the most important aspects of such propaedeutics is the above-mentioned integral education, the need for which is dictated by the problem of ensuring the safety of extreme experience, which can be solved only by the preliminary actualization and harmonious development of all four human systems of knowledge of the world discussed above—sensory, logical, sensual and intuitive. This approach allows you to significantly reduce the relevance of the problem of uncertainty in the decision-making process, since an integral vision allows you to know not only the manifested (consequences), but also the essential causes of the occurrence of extreme states and situations, as well as ways to overcome them.

At the same time, information technologies can be widely used at those levels of general training of athletes, when their initial training in self-organization, self-control and self-regulation is carried out. At this stage of preparation means of objectification of psychophysical state of athlete are very important and widely used. For this purpose, in particular, a wide range of technical and information means of biological feedback has been created, which allows to significantly accelerate the process of preliminary training of athletes.

References

1. Wilber K (1997) The eye of spirit. An integral vision for a world gone slightly mad. Shambhala Publications, Boston and London
2. Kegan R, Lahey LL (2009) Immunity to change. How to overcome it and unlock the potential in yourself and your organization (Leadership for the common good). Harvard Business Press, Boston
3. Wilber K (2000) Eye to eye. The quest for the new paradigm. Shambhala Publications, Boulder

4. Wilber K (2011) A theory of everything: an integral vision for business, politics, science & spirituality. Shambhala Publications, Boulder
5. Wilber K (2017) A brief history of everything, 20th anniversary edn. Shambhala Publications, Boulder
6. Wilber K (1981) No boundary. Eastern and western approaches to personal growth. Shambhala Publications, Boulder-London
7. Wilber K (2011) Integral spirituality: a startling new role for religion in the modern and postmodern world. Integral Books, Boston and London
8. Kuhn TS (1970) The structure of scientific revolutions, 2nd edn. University of Chicago Press Ltd., Chicago, London
9. Huxley A (2012) The perennial philosophy. Harper Perennial, New York
10. Esalen Center for Theory & Research. Retrieved from https://www.esalen.org/about. Last accessed 16 Mar 2021
11. Murphy M (1992) The future of the body. Explorations into the further evolution of human nature. Jeremy P. Tarcher, Putnam Books, New York
12. Auroville, the city of dawn. Retrieved from https://auroville.org/. Last accessed 15 Mar 2021
13. Sahni S (2021) Guide light of integral education. Retrieved from http://www.saiorg.ru/?id=2420. Last accessed 15 Mar 2021
14. Messner R (2000) Everest solo. "Der gläserne Horizont". S. Fischer, Frankfurt am Main
15. Heindel M (2015) The Rosicrucian Cosmo-conception or mystic Christianity. An elementary treatise upon man's past evolution, present constitution and future development. Scholar's Choice, London
16. Korotkov K, Williams B, Wisneski L (2004) Biophysical energy transfer mechanisms in living systems: the basis of life processes. J Altern Complement Med 10(1):49–57
17. (2021) Become strong, happy & healthy. Retrieved from the official website of the Wim Hof method. https://www.wimhofmethod.com/. Last accessed 15 Mar 2021
18. Kurpatov A (2020) Performance in Davos at the business breakfast of Sberbank on 23 Jan 2020. https://www.youtube.com/watch?v=WkbMLCH1tlo&ab_channel=TASS. Last accessed 15 Mar 2021
19. Baudrillard J (1994) Simulacra and Simulation. University of Michigan Press, Ann Arbor

Symmetrical and Asymmetrical Approaches to Communication in Education in Distance Learning

Liubov Ryumshina, Alla Belousova, Yuliya Berdyanskaya, and Ireedui Altan-Avdar

Abstract The goal of the empirical research is a qualitative and quantitative analysis of publications on the topic of "communication in education". The study used content analysis, statistical methods. 198 publications and 231,870 words were analyzed. 3 groups of publications were identified: "communication in education" as activity, as communication, as communication and activity. The largest number of publications belongs to the group "communication in education" as activity, the smallest—"communication in education" as communication, where it is considered as the basis of humanistically oriented education. The number of publications devoted to it has been decreasing over the years, but there is an assimilation of humanistic ideas with national traditions. When approaching "communication in education" as activity, the educator turns out to be the main subject of the educational process, which imposes a special responsibility on him and when this activity becomes more complex, as in the case of distance learning, it leads to stress and overstress of educators. At the same time, the greatest difficulties they face are caused by the inability to fully control the learning process and the violation of traditional status-role positions. When treating "communication in education" as communication, the educator and students share the responsibility for the learning process. It is not of fundamental importance by what means it is mediated. As for "communication in education" as communication and activity, it all depends on how much the teacher is focused on communication and how much on the management of the educational process.

Keywords Communication in education · Digitalization of education · Distance learning

L. Ryumshina (✉) · Y. Berdyanskaya · I. Altan-Avdar
Southern Federal University, St. Bolshaya Sadovaya, 105/42, 344006 Rostov-on-Don, Russia
e-mail: ryumshina@sfedu.ru

A. Belousova
Don State Technical University, Gagarin Square, 1, 344000 Rostov-on-Don, Russia

A. Beskopylny and M. Shamtsyan (eds.), *XIV International Scientific Conference "INTERAGROMASH 2021"*, Lecture Notes in Networks and Systems 247,
https://doi.org/10.1007/978-3-030-80946-1_45

1 Introduction

Today, the term "digital technology", which appeared at the end of the twentieth century, has become widespread in a wide variety of fields of science and practice. As for education, the use of digital technologies [1–4] is widely discussed abroad, both as additional tools [5, 6] and the basis of education [7]. This topic is also becoming relevant for Russian science. The pandemic has added to the popularity of digitalization of education, the whole world has temporarily switched to distance learning. However, its popularity was forced, so education system was not fully prepared for this. This is especially true of Russian educational institutions, where traditionally a crucial role was assigned to the educator, to be more accurate, to communication in education, contributing to the performance of the educator's professional functions. Therefore, in order to understand how the digitalization of education is proceeding, what hinders and contributes to it, one must first understand the meaning of the concept of "communication in education".

It should be noted that communication is one of the most important categories of Russian psychology, which began to be intensively developed in the 1920s and 1930s. In this regard, it is no coincidence that the term "communication in education" has also been used for a long time, although, as is commonly known, its understanding as an important psychological and pedagogical category was started by A. A. Leontiev and V. A. Kan-Kalik only in the mid-1970s [8–11]. In its most general form, communication in education is the professional communication of a teacher with students in an integral pedagogical process, but the role of communication in it can be interpreted in different ways [12–14].

In the second half of the twentieth century, in Soviet psychology, the study of activity becomes relevant, which could not but be reflected in the concept of "communication". "Communication" and "activity" are becoming one of the main categories of Soviet psychology. A discussion arose, in which, in its most general form, the case touched on whether communication is an activity, a separate type of activity (a side of an activity, its element), or an independent phenomenon (a separate side of social being). Let's see how this was reflected in the idea of communication in education, as communication between teachers and students, but, first of all, let us consider the essence of these two concepts, fundamental for Soviet psychology.

So, the essence of activity is focus and efforts to change external conditions, which ultimately leads to a change in the person himself. One can agree with a number of scientists [8], that activity is always an asymmetric process, since an active subject, based on one's goals, motives, skills and abilities, affects a relatively passive object. As for communication, it is more symmetrical, associated with the position of equality, since it comes from universal human characteristics (the need for communication that any person has, the patterns of communication, regardless of the status and age of subjects, etc.), is characterized by the efforts of both subjects, even if outwardly they do not demonstrate them. Such communication is aimed at the development of the personality, its formation, thus, contributing to learning. In fact, this equality is characteristic of humanistically oriented education [15]. In the concept of Carl Rogers, for example, such communication provided by educators-facilitators is the main condition for the formation of a personality in the process of education [16, 17]. It should be noted that the humanization of education at the end of the twentieth century was in the center of attention of Russian scientists [18], although it was not so widespread as abroad. Of course, the degree of efforts of the subjects can be different, but if one of the subjects is passive, equality will be lost, communication actually develops into an asymmetric process, i.e. activity [8]. Therefore, it is important that the learner is an active and equal co-participant in the educational process.

A lot of time has passed since that discussion, but Russian traditions are quite strong in national science in the study of various psychological phenomena. We assumed that in modern conditions and in the analysis of communication in education, there will also be a place for echoes of that discussion.

In order to confirm this assumption, we turned to the publications presented in the scientific electronic library eLIBRARY.RU—the largest Russian information and analytical portal in the field of science, technology, medicine and education. This library contains abstracts and full texts of over 35 million scientific publications and patents, including electronic versions of about 8000 Russian scientific journals since 1990. We limited ourselves to the analysis of publications characterizing the communication of educators with students, published over the past 15 years.

2 Methods

So, the **aim** of the study is a qualitative and quantitative analysis of publications characterizing the communication of educators with students.

The **object of the study** was publications on communication in education from 2006 to 2021, posted in eLIBRARY.

The study used a combination of theoretical and empirical **methods**.

Theoretical: analysis of theoretical sources on the issue under study. Content analysis was chosen as the main empirical method. Since this method now has many modifications, we used one of them, highlighting the fixed unit of analysis, and the units of context that reveal its content [19].

The category "communication in education" was taken as a fixed unit, which was mentioned in the title of the publication, abstracts and key words. It was assumed that by such a frequent fixation of attention on this term, the author (s) emphasized its significance.

The units of context that reveal the content of "communication in education" were complete statements that differed from others in meaning, allowing them to be attributed to activity or communication. Based on the main difference between them, described earlier, the statements were supposed to characterize the symmetry or asymmetry of communication in education.

Statistical methods that used in the study: Kruskal-Wallis H test, Mann-Whitney U test, Pearson's correlation coefficient.

The study was conducted from December 2020 to February 2021. The procedure was carried out as follows. First, the search included the category "communication in education" in all publications (books, articles in journals, conference materials, deposited manuscripts, dissertations). Then, they were selected for analysis of their publication content, where the desired category was also mentioned in the abstracts to publications and keywords.

3 Results

The eLIBRARY search engine has produced 584 publications in response to a request for publications on "communication in education" over the past 15 years. However, among them there were those in the name of which other categories were indicated, and not the one we had chosen. For example, "psychological communication in education", "effective communication", "social communication in education", "professional communication in education" and others. Sometimes such terminology was absent in the title, but there were indications of the analyzed topic ("interaction of students and teachers in foreign language classes", "development of a culture of communication between students and teachers", etc.) However, in accordance with the aim of the study, these publications were not analyzed.

There were 198 publications in which the category "communication in education" was mentioned in the title, abstract and keywords. The abstract to them was subject to further semantic analysis.

If necessary, when the semantic context was not clear from the abstract, the authors addressed the full text of the publication.

In total, 231,870 words were analyzed that reveal the content of "communication in education" as an activity or communication.

The group of "communication in education" as an activity included publications in which it was characterized as influence, management, etc.

An example of such a publication is an article on communication in education as a way of influencing the target audience. The abstract notes that the article "is devoted to the issues of psychological and pedagogical influence on the consumer of information in the process of communication", "the forms of speech influence are considered", "the conditions that increase the effectiveness of persuasive influence are analyzed." Another example of an article where communication is considered by analogy with the management activity of a leader, in connection with which "its characteristic features are studied: focus on solving educational problems", "creating pedagogically expedient influences in management communication." "The conditions for the use of communication in education in the management activity of the leader are identified." An extreme expression of this position can be the consideration of communication in education by analogy with the operation of technical devices, where "educator acts as a computer operator" who changes his "influences on the system (student group) in accordance with the feedback." There were 118 publications in total in this group.

As for "communication in education" as communication, there were 11 such publications. These included the ones where communication was considered as the basis for the development of the personality and society, the role of the communication process in the development of personality and self-development, mutual understanding, the ability to listen and understand, etc.

An example would be an article that talks about the "importance of "lively", emotionally colored communication", "emphasizes the importance of listening to the student", "give an opportunity to express one's point of view."

Or publications where communication is analyzed as "a condition for personal and professional growth", as "a process that is generated by the needs for collaborative activity of people."

Interestingly, such articles often consider not only communication with pupils or students, but also with the administration, communication with colleagues, etc.

However, it was difficult to assign some of the publications to one of these groups. They represent an attempt to combine "communication" and "activity", which is more consistent with the previously described discussion. Often, when referring to the full text of the article, it turns out that symmetry, which we have taken as the most important criterion for dividing into groups, is not observed everywhere. So, for example, recognizing the importance of the dialogical relations of the educator with the students and indicating their equal responsibility for the learning process, the educator, nevertheless, is assigned a leading role according to certain indicators. He should "demonstrate a desire to communicate, … encouraging students to social interaction", "in resolving conflicts, the main role is assigned to the educator," thereby leveling the role of students in creating and resolving conflict situations. Most often, such asymmetry is observed when describing the "educational functions of an educator", where it is assumed that "communication in education serves as a tool for influencing the personality of the student." Since such publications, in which "communication in education" appears both as communication and activity, were difficult to analyze from the point of view of asymmetry/symmetry, we conditionally designated "communication in education" as communication and activity. The group included 40 publications.

Finally, 19 publications could not be attributed to any of the selected groups. Most often, they considered the communication of educators with students without indicating their own point of view. In other words, publications of other authors were cited without their corresponding analysis, the types of communication, its means, communication styles were described, but it was not clear how the author (authors) proposes to take this into account in real communication in education.

The ratio of publications assigned to a particular group can be represented in the form of a figure (see Fig. 1).

The use of the Kruskal-Wallis *H* test showed significant differences between the groups at a significance level of $p < 0.001$. Therefore, we applied the Mann-Whitney *U* test. The value of the Mann-Whitney *U* test turned out to be significantly higher for the group "communication in education" as communication and activity than for the

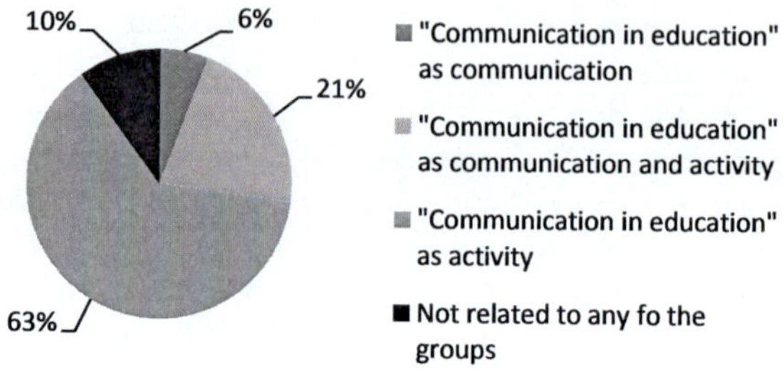

Fig. 1 Representation of publications by groups

group "communication in education" as communication ($p < 0.01$). It is also higher for the group "communication in education" as activity than for the group "communication in education" as communication and activity ($p < 0.05$). And it is higher for the group of "communication in education" as activity than for "communication in education" as communication ($p < 0.001$). In other words, most publications are devoted to "communication in education" as activity, second place is taken by publications where "communication in education" appears as communication and activity, and publications from the group "communication in education" as communication are extremely rare.

Thus, the largest number of publications is presented in the group "communication in education" as activity where the educator is the main subject of the educational process (sometimes even the only one, in the sense that students are rarely mentioned and only to emphasize the importance of the educator). There, the educator performs a variety of functions, but above all, controlling ones and is responsible for the entire educational process.

Least of all publications are in the group "communication in education" as communication. However, it should be noted that this category of publications included those that in fact represent humanistic education, where the role and responsibility for learning is assigned to both the educator and the students. Such education is quite rare, including, as noted earlier, in Russian education. Without dwelling on the reasons for this, you can familiarize yourself with them, for example, in [20], we will pay attention to the following. If we combine this group of publications with more common publications, where "communication in education" appears as a combination of communication and activity, then their sum is 27%, almost a third of the total number of analyzed publications. Such a combination is also justified from the standpoint of considering the identified groups by years (see Fig. 2).

As can be seen from Fig. 2, the number of publications where "communication in education" appears as communication decreases over the years. At first glance, it seems that the interest in the humanization of education, aroused at the end of the twentieth century largely under the influence of foreign scholars of a humanistic orientation, primarily Carl Rogers, is fading away. However, most likely, it becomes meaningful and is assimilated with Russian psychological traditions, which leads to an increase in the number of publications that we referred to the group of "communication in education" as communication and activity. In them, as well as

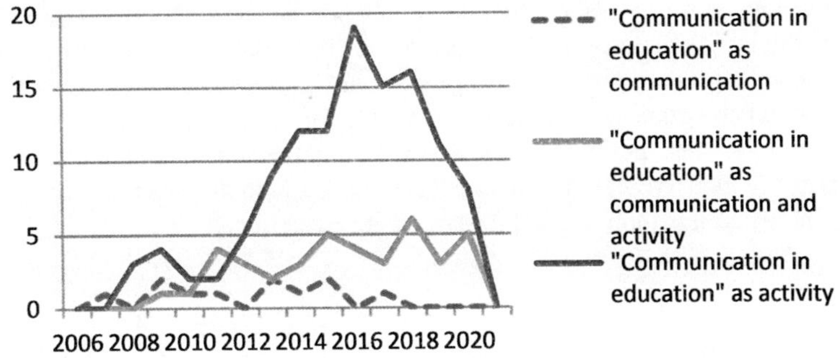

Fig. 2 Distribution of groups of publications by years

in the publications of the group "communication in education" as communication, humanistically oriented terminology (empathy, active listening, self-actualization, dialogical communication, etc.) is widely used.

The influence of national psychological traditions is also evidenced by the fact that at this time (2015–2018) the number of publications that we referred to the group of "communication in education" as activity and the use of methods of mathematical statistics also increases.

Thus, a strong correlation was found between the publications "communication in education" as activity and "communication in education" as communication and activity (Pearson's correlation coefficient r = 0.999 at $p = 0.03$). This serves as a mathematical confirmation of our assessment of these groups of publications as quite like-minded. In both cases, communication in education is considered as the influence of the educator on the students. The difference is that in the first case, this influence appears to be authoritarian. In the second case, it is softer and more humane. However, it is still not truly humanistic.

The same strong correlations were found between publications on different levels of education (at $p = 0.05$). Publications on school education correlate both with publications on higher education (r = 1) and with the general group (r = 0.998). Figure 3 shows the number of publications distributed by educational level—school

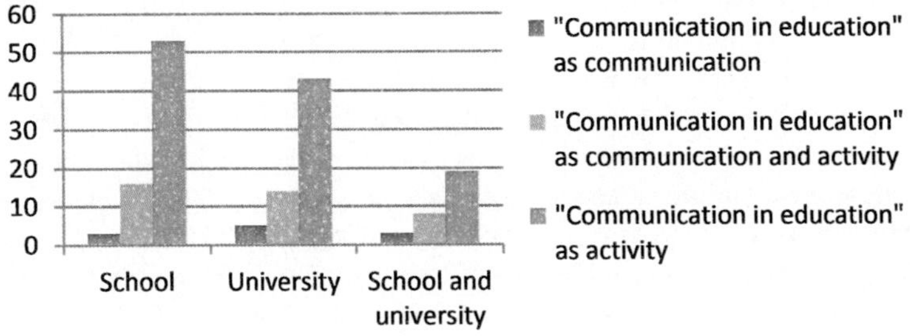

Fig. 3 The correlation of groups of publications dedicated to different levels of education

and university, as well as the number of publications whose authors did not take into account the level of education. In addition, we found three publications related to kindergarten and two related to management, not educational topics. They are not shown on the chart due to their scarcity (see Fig. 3).

As one can notice, publications in which "communication in education" appears as activity are more typical for school education, and "communication in education" as communication is more typical for university education.

4 Discussion

Thus, returning to the topic of digitalization of education, let us see what can hinder and what can contribute to this process with different approaches to communication in education.

The analysis of publications shows that the priority in Russian psychology is given to the consideration of the communication of an educator with students as activity. It consists of various functions, where the importance of the impact of the teacher's communication on students is often made absolute. The asymmetry of this approach leads to the fact that the educator is the main subject of the educational process. Everything depends on him: skills, abilities, knowledge of students, their moral traits, personal development, well-being, general prosperity, etc. In other words, the educator is responsible for the entire learning process and student development.

In this case, the transition to distance learning creates many problems for him, because the educator cannot fully control the effectiveness of his influences. There are problems with maintaining the interest of students in the class. Sometimes the educator cannot even be completely sure how many students are present at the lesson, to track what they are doing at this time, i.e. distracted or not, how often and what they are distracted by. In our opinion, this is one of the main reasons hindering the effectiveness of distance learning from these positions. However, there are other reasons as well.

It should be noted that there were very few publications where communication in education would be analyzed from the standpoint of digital technologies. In addition to the above article, which we referred to the group "communication in education" as activity, where this topic is raised in relation to computer education to one degree or another, we were able to find only two more publications. One belongs to the student, his experience of mastering distance learning and it does not carry special information about communication in education. In another article, also from the group "communication in education" as activity, the experience of distance learning is also described, but this is already the teachers' view of it. The article makes the same attempt, as we do, to analyze the problems of distance education, albeit from a different position [21]. The authors—university professors, rather rely on their own experience of distance learning, which is also of some interest for understanding the issue. Agreeing with them that distance learning forced by the pandemic causes discomfort among educators due to problems with technical equipment, including insufficient skills of working with them, we note other problems described by the authors.

The new form of communication in education, according to the authors, violates the status and role positions that are customary for educators, the "teacher-student" subordination traditional for the educational process and is associated with the loss of personal space and intimacy due to students' access to the personal profile of teachers on Facebook, VK etc. In this regard, students get access to the elements of the private life of teachers, the range of their interests, lists of friends and subscriptions, photos and videos. Another problem these educators see in the reduction of communication distance related to the home environment, possible interference caused by family members or pets. Moreover, sometimes educators are forced to seek help from students who are better at computer technology, which, according to the authors, also undermines their authority. In other words, the teacher ceases to be the main subject of the educational process, the one that knows everything, thus becoming one of such subjects.

"Communication in education" as communication, at first glance, also suffers from digitalization, for example, its non-verbal component is lost. However, non-verbal means are replaceable to a certain extent (blind, speech-impaired people can also communicate), their compensation is possible (in the absence of eye contact, the load increases on intonation, etc.). Besides, technical means now, albeit in part, allow the non-verbal expression of emotions, for example, using various symbols.

The symmetry, characteristic of this view on communication in education, brings it closer to dialogical interaction, the necessary attributes of which are trust, mutual understanding, and mutual respect. Shifting attention to them, of course, facilitates the process of digitalization, since in a humanistically oriented education, the educator, students and even their parents are equal subjects of the educational process and are equally responsible for the knowledge, skills, etc. acquired by students.

At the same time, the educator does not teach, he helps to master the material. As you know, the motto of humanistic pedagogy is: "The goal of the teacher is to become redundant for the student." Since this is a specially organized communication, it does not really matter what digital technologies it is mediated by.

This is also true for the concept of "communication in education" as communication and activity, if there remains equality and responsibility of all subjects of the educational process for the acquisition of knowledge by students. Although in this case, with distance learning, problems may arise in the implementation of certain functions and educational influences, the educators may experience some discomfort if it is impossible to control them.

5 Conclusions

The authors did not set out to review publications on digital education. Our goal was to try to consider the existence of problems in such education, based on the idea of communication in education as the basis of any educational process. With an asymmetric approach to communication in education as activity, the educator turns out to be the main subject of the educational process, which imposes a special responsibility on him, and when this activity becomes more complex, as in the case of distance learning, it leads to stress and overstress of educators.

When treating communication in education as communication (symmetric approach), the educator and students share the responsibility for the learning process. It is not of fundamental importance by what means it is mediated.

As for communication in education as communication and activity, it all depends on how symmetrical/asymmetrical it is. In other words, to what extent the educator is focused on communication and on the activity and management of the educational process.

The study allows a more meaningful approach to the digitalization process, developing the most convenient strategy for the educator to interact with students.

References

1. Vélez AP, Zuazua II (2017) Digital literacy and cyberconvivencia in primary education. Procedia Soc Behav Sci 237:110–117
2. Pitkänen K, Iwata M, Laru J (2020) Exploring technology-oriented fab lab facilitators' role as educators in K-12 education: focus on scaffolding novice students' learning in digital fabrication activities. Int J Child-Comput Interac 26:100207
3. Lacka E, Wong TC, Haddoud MY (2021) Can digital technologies improve students' efficiency? Exploring the role of virtual learning environment and social media use in higher education. Comput Educ 163:104099
4. Cvetković BN, Stošić L, Belousova A (2018) Media and information literacy—the basis for applying digital technologies in teaching from the discourse of educational needs of teachers. Croatian J Educ 20(4):1089–1114
5. Yildirim T, Yavuz AO (2012) Comparison of traditional and digital visualization technologies in architectural design education. Procedia Soc Behav Sci 51:69–73
6. Blackwell CK, Lauricella AR, Wartella E (2014) Factors influencing digital technology use in early childhood education. Comput Educ 77:82–90
7. Verner I, Merksamer A (2015) Digital design and 3D printing in technology teacher education. Procedia CIRP 36:182–186
8. Smirnov SD (2011) Pedagogical activity and communication in education. Vestnik Moskovskogo Universiteta. Seriya 14. Psikhologiya. Moscow University Psychology Bulletin 3:5–18
9. Aliev SN, Rakhmatova S (2018) Communication in education as a form of educational cooperation of the teacher and students in the conditions of the higher education. Vestnik Pedagogicheskogo Universiteta 5–2(77):34–37
10. Akhmerova AF (2016) The role of communication in education in the formation of the student's personality. Vysshee Obrazovanie Segodnya 4:33–36
11. Sigacheva NA, Baranova AR (2018) Communication in education in educational process. Azimut Nauchnyh Issledovanij: Pedagogika i Psihologiya 1(7):205–207
12. Belousova A, Kozhukhar G (2008) Emotional characteristics and styles of pedagogical communication of educators. In: INTED 2018 Proceedings. 12th International technology, education and development conference, 5–7 Mar 2018, Valencia, Spain. https://doi.org/10.21125/inted.2018 (Web of Science), 4/2018. https://doi.org/10.15516/cje.v20i4.3001
13. Christensen BM, Nilsson S, Stensson M (2020) Developing communication support for interaction with children during acute radiographic procedures. Radiography 26(2):96–101

14. Belousova A, Kozhukhar G, Ryumshina L (2015) Educational environment as a factor of pre-school children development. In: ICERI2015 Proceedings, Palma de Mallorca, Spain
15. Santos H, Batista J, Marques RP (2019) Digital transformation in higher education: the use of communication technologies by students. Procedia Comput Sci 164:123–130
16. Servant-Miklos VFC (2019) From Flexner to Rogers: an inquiry into the intellectual origins of problem-based learning at McMaster University Medical School. Health Prof Educ 5(2):93–102
17. Bryan VD, Lindo J, Weaver S (2015) Using Carl Rogers' person-centered model to explain interpersonal relationships at a school of nursing. J Prof Nurs 31(2):141–148
18. Rjumshina LI (2000) Empirical study of pedagogical activity styles. Voprosy Psikhologii 1:148–149
19. Titscher S, Meyer M, Wodak R, Vetter E (2009) Methods of text and discourse analysis. Humanitarian Centre, Kharkiv, Ukraine. ISBN 978-966-8324-61-1
20. Ryumshina L (2014) Humanization of education: from traditional to humanistic oriented school. In: SGEM International muitidisciplinary scientific conferences on social sciences and arts, vol 3, pp 369–374. https://doi.org/10.5593/sgemsocial12014B11
21. Baranov AYU, Malkova TV (2020) Problems of communication in education in the context of distance learning. Voprosy Pedagogiki 6–2:38–40

Features of Emotional Intelligence and Self-acceptance in Students of Various Training Profiles

Anna Kukulyar, Sergey Shaginyan, and Elena Eresko

Abstract The modern social environment in general, and the educational environment in particular, impose a number of requirements on the personality of the student as a future representative of the professional community. This manifests itself in the need to acquire professional knowledge and skills in the face of constantly changing requirements, plan your activities for a short-term and/or long-term period, as well as multiple behavioral flexibility. This flexibility, in our opinion, can be realized through the formation of emotional intelligence. This article presents the results of a study that involved 200 students from various universities in Rostov-on-Don. Research hypotheses: (1) The features of emotional intelligence and self-acceptance among students of different training profiles (legal and economic) will differ; (2) The features of emotional intelligence and self-acceptance among students of different training profiles (boys and girls) there may be differences. In the course of the empirical study, we used the following methods: (1) Self-attitude questionnaire of V. V. Stolin and S. R. Panteleev (OSO). (2) Hall's "Emotional Intelligence" test. The results obtained can be used in consulting and correctional work with students. The results of the study can be used in the process of training psychologists (bachelors, masters) in the areas of training 37.00.00 "Psychology", 38.00.00 "Economics" and 40.00.00 "Jurisprudence", etc., as well as to create a program of advanced training and retraining of psychologists, economists, lawyers, social workers and teachers working in the field of education.

Keywords Emotional intelligence · Self-attitude · Psychological characteristics of students' personality · Self-acceptance · Students

A. Kukulyar (✉)
Faculty "Psychology, Pedagogy and Defectology", Rostov Don State Technical University, Rostov-on-Don 344000, Russia

S. Shaginyan · E. Eresko
Department of Economics, Rostov State Transport University, Rostov-on-Don 344000, Russia

A. Beskopylny and M. Shamtsyan (eds.), *XIV International Scientific Conference "INTERAGROMASH 2021"*, Lecture Notes in Networks and Systems 247,
https://doi.org/10.1007/978-3-030-80946-1_46

1 Introduction

The term "self-acceptance" is a complex multi-faceted phenomenon that has been studied in sufficient detail in humanistic and existential psychology, in particular by A. Maslow, K. Rogers, A. B. Orlov, and others. The degree of self-acceptance of a person depends on the relationship between the real Self and the ideal self, the greater the gap between them, the more difficult it is for a person to accept himself (Z. Freud, K. Horney, A. Bandura, K. Rogers) [1, 2].

The concept of self-acceptance is closely related to such concepts as identity, authenticity, congruence, and personification. Self-acceptance of personality is achieved through openness to experiences, understanding the value of one's own individuality, as well as the rejection of the desire to meet the expectations of others (V. Frankl, J. Bujenthal, K. Rogers). Self-perception is a necessary condition for the psychological health of the individual and an integral quality of the self-actualizing person (V. Frankl, J. Bujenthal, K. Rogers, A. Maslow) [1, 2].

Thus, in the context of our research, we will understand self-acceptance as a nuclear formation of the student's personality structure, which is manifested in a positive emotional and value attitude to oneself, adequate self-esteem, self-understanding, reflection of one's inner world and one's actions, self-respect and acceptance of other people, awareness of the value of oneself, one's inner world. On this basis, we consider it interesting to identify the specifics of the relationship between the various components of self-acceptance and the specifics of the manifestation of emotional intelligence at the stage of training at the university [3–8].

To date, there are two alternative models on which the measurement of the level of emotional intelligence is based. Mixed models include cognitive, personality, and motivational traits, and ability models define emotional intelligence as a set of abilities measured by tests consisting of tasks that have correct and erroneous answers [9, 10].

The most well-known models of emotional intelligence are J. Mayer and P. Salovey's model of abilities, R. Barton's non-cognitive model of emotional intelligence, D. Goleman's theory of emotional competence, and D.V. Lucin's two-component model [11].

Emotional intelligence, as a psychological education, is formed in early childhood. Interpersonal communication with adults and peers, the development of empathy, and the idea of emotions as an important source of information have a positive effect on the development of emotional intelligence. The prerequisites for the development of emotional intelligence can be both biological and social [12, 13].

There are several points of view about the development of emotional intelligence: emotional intelligence cannot be developed; it is possible and necessary to develop; development is possible under special conditions. The specifics of the manifestation of emotional intelligence in the student age will be revealed by us in the framework of an empirical study [3–8].

2 Discussion

The concept of "emotional intelligence" appeared in 1964 in the work of Michael Beldoch (Michael Beldoch) Sensitivity to expression of emotional meaning in three modes of communication, and in 1966 in the work of B. Leuner Emotional intelligence and emancipation [14].

In 1995, Daniel Goleman published the book "Emotional Intelligence" ("Emotional Intelligence"). He defined emotional intelligence as "the way, method, form of a person's treatment of themselves and others" [15]. Since that time, an active study of the nature and regularities of the formation of this phenomenon and its application in practice began.

D. Goleman proposed a model of emotional intelligence, where he represents that it consists of five components [15]:

- Empathy. It is manifested in the ability to empathize with others, in the ability to take into account the feelings of others when making decisions.
- Self-knowledge. The ability to recognize your emotions, identify personal motivations, follow the logic of problem solving, recognize and accept your weaknesses and strengths, as well as identify life goals and values.
- Motivation. The ability to set a goal and strive to achieve it.
- Social skills. The ability to build relationships with others. This category includes the ability to manipulate people, push them to make decisions that are beneficial to the individual.
- Self-regulation. The ability to control and restrain your emotions.

J. Mayer, P. Salovey, and D. Caruso distinguish four components of emotional intelligence [9]:

The first component of emotional intelligence, the authors call the perception of emotions. This category includes the ability to recognize other people's emotions, by verbal and nonverbal signs, and the ability to identify your own emotions.

Understanding emotions is the next component of emotional intelligence. It is based on the ability to determine the cause of the manifestation of emotion, the ability to recognize the relationship between thoughts and the emotions caused by them. This understanding over time contributes to the development of the ability to predict different emotions, makes it possible to interpret them in relationships, to understand complex (ambivalent, ambiguous) feelings [9].

Next, J. Mayer, P. Salovey, and D. Caruso highlight the ability to use emotions to stimulate thinking. This ability is usually not realized by a person, it awakens the process of creative thinking, while emotions act as a motivating factor [9].

The ability to awaken and direct one's own emotions and those of others is identified by the authors as the fourth component of emotional intelligence. This component allows a person to achieve their goals, when building logical chains and solving logical problems, as well as when choosing a certain behavior model [10, 12, 13].

R. Bar-On defined emotional intelligence as "the non-cognitive ability, knowledge, and competence that enable a person to successfully cope with various life situations" [9].

The author deduced five spheres of emotional intelligence: intrapersonal, adaptive, interpersonal relations, stress regulation, and mood sphere. R. Bar-on created the so-called EQ-i questionnaire to measure the above-mentioned spheres in order to determine the emotional coefficient (analogous to the intelligence coefficient). What is fundamentally new is that, unlike those methods that mainly studied emotional intelligence in adults, this questionnaire has a real opportunity to study the child population (from 6 to 18 years old) [10, 12, 13].

According to D. V. Lyusin, the concept of "emotional intelligence" should include a cognitive component, and if there is no such component, then it makes no sense to use the term "intelligence" [14].

In Russian science, the theory of the unity of intellect and affect is reflected in the works of L. S. Vygotsky, A. N. Leontiev, S. L. Rubinstein, and others.

A similar opinion was held by A. N. Leontiev, who believed that thinking has an emotional regulation. If a person has a certain knowledge about the objects and phenomena of the surrounding reality, then the attitude to the latter will have a changeable character [11, 14].

S. L. Rubinstein, noting the relationship between emotions and intelligence, concludes that emotions "represent the unity of the emotional and intellectual, as well as cognitive processes". Defining emotionality as one side of cognitive processes, he came to the conclusion that emotional and cognitive processes are not comparable.

Emotional intelligence is defined as a person's ability to recognize emotions, understand the intentions, motivations and desires of other people and their own, as well as the ability to manage their own emotions and the emotions of other people in order to solve practical problems [11].

The foundation of emotional intelligence is the mature values of self-development and responsibility of a person in the relations of himself, society and the whole world around him. In a life that is built on such values, there is no place for manipulation. The main components of emotional intelligence, as noted by S. I. Semenaka, are self-awareness, self-control, motivation, sensitivity, communication [14]. "Self-awareness is defined as the ability to recognize and understand one's own emotions, moods, inner motivations, etc. Self-control is expressed in the ability to exercise control over their emotions, to be responsible for their own behavior, actions, etc. Motivation is characterized by a passionate desire to achieve a goal and is manifested in optimism, creativity, perseverance and perseverance. Sensitivity is expressed in the ability to feel the emotions and moods of other people, to understand their point of view and to take an active interest in their concerns. Communication allows a person to feel and inspire trust in others, find mutual understanding, accurately express their ideas, thoughts and feelings, and in the same way understand the messages of others, resolve conflicts, eliminate tension and disagreements, and cooperate with others.

In our opinion, such a component of the personal structure as self-acceptance is an integral component in the formation of emotional intelligence and its manifestations.

Self-acceptance—a sense of the value of your own personality, the strength of your "I", a sense of self-esteem; trust in your feelings, faith in yourself and your abilities, openness of character, understanding your own weaknesses, defending the right to be as it is, and have your own point of view. Weakly expressed fear, anxiety, fear of failure and negative evaluation, guilt, desire for dominance, orientation to external forms.

The concept of self-acceptance is an important psychological problem. Many scientists consider self-acceptance a necessary component of mental health of the individual. M. Yagoda included self-acceptance as a high self-esteem and a strong sense of identity in the criteria of mental health [14].

L. S. Vygotsky came to the conclusion that emotions are mediated by intelligence and there is a close connection between them, and the level of development of one determines the development of the other. He believed that the one who "tore thinking from the very beginning from the affect, he forever closed the way to the explanation of the causes of thinking itself" and "made it impossible to study the reverse influence of thinking on the affective, volitional side of mental life" [14].

Self-acceptance is a nuclear formation of the personality structure and manifests itself in a positive emotional and value attitude to oneself, in an adequate self-assessment, in self-understanding, reflection on one's inner world and one's actions, self-respect and acceptance of other people, in awareness of the value of oneself, one's inner world. Self-acceptance depends on the relationship to others and is adequate when these relationships become a value. Self-acceptance is based on moral values.

Self-acceptance, as defined by S. L. Bratchenko and M. R. Mironova, means self-recognition and unconditional love for myself as I am, an attitude to myself as a person worthy of respect, capable of independent choice, faith in myself and my abilities, trust in my own nature, the body [14].

According to D. A. Leontiev, self-acceptance is part of a broader concept-self-relation. The most superficial manifestation of self-attitude is self-esteem—a general positive or negative attitude towards yourself. However, just one sign can not describe the self-relationship. First, we should distinguish between self-esteem—an attitude towards myself as if from the outside, conditioned by some of my real advantages or disadvantages—and self-acceptance-an immediate emotional attitude towards myself, regardless of whether there are any traits in me that explain this attitude. Often there is high self-acceptance with relatively low self-esteem, or vice versa. Secondly, no less important characteristics of self-attitude than its evaluation mark are its degree of integrity, integration, as well as autonomy, independence from external assessments [14].

A different interpretation of the structure of self-acceptance is proposed by V. F. Safin. Based on the concept of a holistic self-assessment (and not a ratio) of the subject, he identifies a number of aspects, among which he particularly highlights self-esteem. This type of self-assessment is the result of correlating one's attitude, assessment with the attitude of others to the subject, with the assessment of his "significant others". But self-respect is a deeper, more global education than individual self-esteem. If "I-images" are more cognitive formations, then self-esteem is a holistic emotional-value relationship. Although V. F. Safin somewhat contradicts himself, defining relatively independent types of self-esteem as an emotional-cognitive-value attitude to oneself as an active subject of activity. The first of the aspects can be called a self-assessment of the sphere of intellectual capabilities, the second—a self-assessment of the sphere of motivational and need forces, the third—a self-assessment of relatively stable, fixed existing properties (physical, psychophysiological, and psychological) [14].

Awareness of your importance to your immediate environment may be included as an element of self-esteem, but it may not coincide with it. Therefore, it can be assumed that self-esteem is an attitude to one's importance to others, based on one's ideal, which ensures self-affirmation and satisfaction of the subject with himself. It turns out that self-satisfaction is the result of a global, holistic self-attitude that includes "I-images"—all forms of self-esteem that generalize self-esteem.

It is worth noting that a different mechanism of self-esteem is proposed by V. M. Raeva, according to which self-esteem is an emotional-value attitude to oneself, reflecting the level of general positive or negative self-esteem, based on the coincidence of the level of claims and the level of achievements [14].

3 Research Methods and Techniques

In preparation for the empirical study, we selected 210 students studying in the direction of 38.03.01 "Economics" (105 boys and girls) and 40.03.01 "Jurisprudence" (105 boys and girls) 1–2 courses in universities of Rostov-on-Don. Research hypotheses: (1) The features of emotional intelligence and self-acceptance in students of different training profiles (legal and economic) may differ; (2) The features of emotional intelligence and self-acceptance in students of different training profiles (boys and girls) there may be differences. In the course of the empirical study, we used the following methods: (1) Self-attitude questionnaire of V. V. Stolin and S. R. Panteleev (OSO). (2) Hall's "Emotional Intelligence" test, as well as methods of mathematical statistics.

4 Research Results

According to the OSO V. V. Stolin questionnaire, our study was conducted according to gender criteria and the specifics of the training profile. The results are shown in Table 1.

The most pronounced signs of self-interest among all students (84.35 and 81.56), which indicates the willingness to accept new emotional experiences when entering the university, as a new round in life, meeting with a new team, a new level of specificity of the educational organization and the teaching staff.

Table 1 Average indicators on the scales of the V. V. Stolin OSO methodology among the entire sample of respondents

	Young men	Young woman	Economics students	Law students
S scale (integral)	*45.52*	*38.67*	*41.1*	*44.8*
Self-esteem scale (I)	***57.67***	***46.87***	***52.87***	***59.43***
Autosympathy scale (P)	***57.86***	***53.45***	***56.23***	***51.43***
Scale of expected attitude from others (W)	*45.51*	*32.12*	*43.15*	*48.34*
Self-interest scale (IV)	***79.61***	***72.24***	***80.16***	***68.74***
Self-confidence scale (1)	*47.65*	*34.32*	*49.52*	*52.45*
Scale of attitude from others (2)	*36.12*	*28.43*	*43.23*	*44.26*
Self-acceptance scale (3)	***69.21***	***70.34***	***66.78***	***67.26***
Self-consistency scale (4)	***60.09***	***56.78***	***56.83***	***64.25***
Self-incrimination scale (5)	***59.44***	***67.23***	***62.45***	***54.32***
Self-interest scale (6)	***77.22***	***76.13***	***84.35***	***81.56***
Self-understanding scale (7)	***57.41***	***60.98***	***56.28***	***59.43***

In the course of the study, we identified the following fact that among the total number of respondents, trends were expressed on the following scales: self-esteem, autosympathy, self-interest, self-acceptance, self-consistency, self-blame, self-interest and self-understanding.

The predominance of these scales can also be noted at the level of comparison by gender of students.

For clarity, we present these values in the form of a graphical profile relative to the training profile and gender. The results obtained are shown in Figs. 1 and 2.

However, if we consider the data obtained in comparison, we can see that the integral scale is weakly expressed in girls, and the scales of expected attitude from others and the scale of self-confidence, attitudes from others are not expressed, this indicates a low level of self-esteem, lack of confidence, and distrust in girls at the beginning of training.

Next, we will describe the results of Hall's emotional intelligence method, which are shown in Fig. 3. Within the framework of this methodology, the average indicators are the distribution of values from 8 to 13. The figure shows that in both groups 1 and 2 of students, the values on all scales are within the norm.

This may indicate an adequate awareness and understanding of their emotions, and this requires constant replenishment of their own vocabulary of emotions; emotional responsiveness, emotional flexibility, etc., In other words, respondents of both groups can arbitrarily manage their emotions, their behavior by managing emotions, as well as understand the emotions of other people, empathize with the current emotional state of another person, and are ready to provide support if necessary. They are able

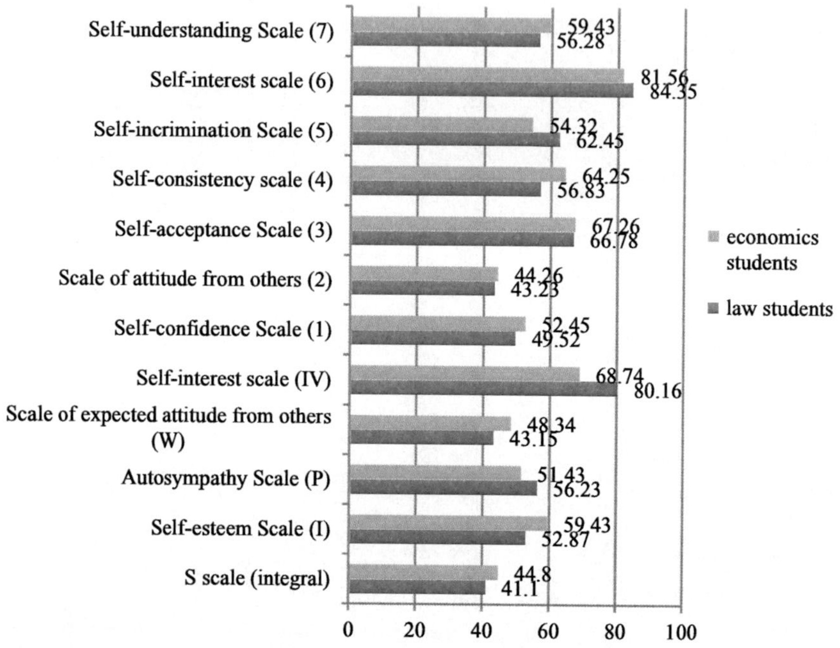

Fig. 1 Results of V. V. Stolin's OSO among the entire sample of students (in points)

to understand the state of a person by facial expressions, gestures, shades of speech, posture, but also know how to influence the emotional state of other people.

Next, we will go to the description of the results of the Hall's method of emotional intelligence for young men and women students, which are presented in Fig. 4.

The figure shows that both in the group of boys and in the group of girls, the values on all scales are within the norm. This may indicate an adequate awareness and understanding of their emotions, and this requires constant replenishment of their own vocabulary of emotions; emotional withdrawal, emotional flexibility, etc., In other words, respondents of both groups can arbitrarily manage their emotions, their behavior, by managing emotions, also understand the emotions of other people, are able to empathize with the current emotional state of another person, and are also ready to provide support. They are able to understand the state of a person by facial expressions, gestures, shades of speech, posture, but also know how to influence the emotional state of other people. However, girls have more developed self-motivation, and boys have more developed recognition of other people's emotions.

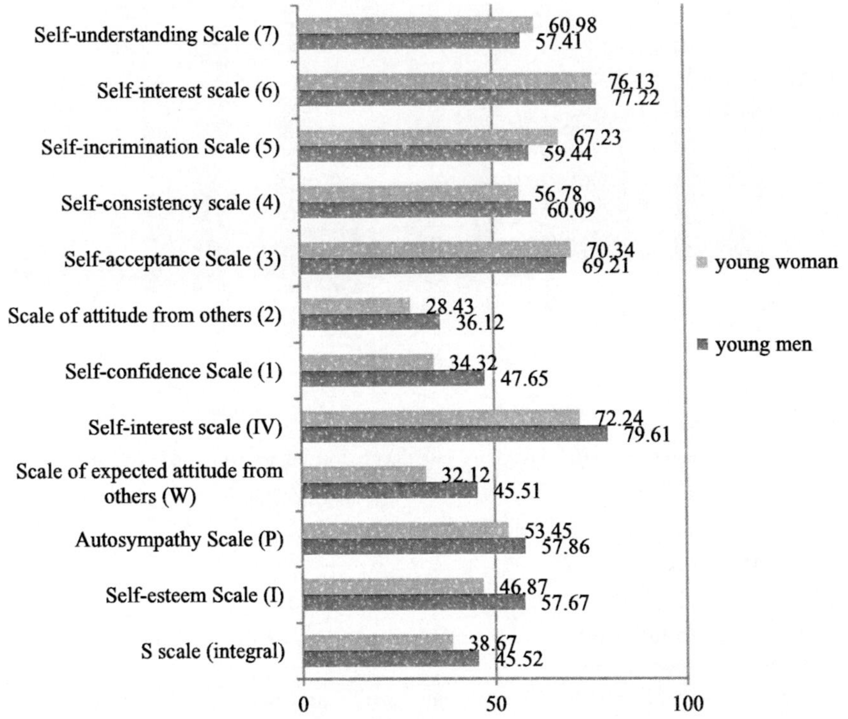

Fig. 2 Results of V. V. Stolin's OSO among young men and women students (in points)

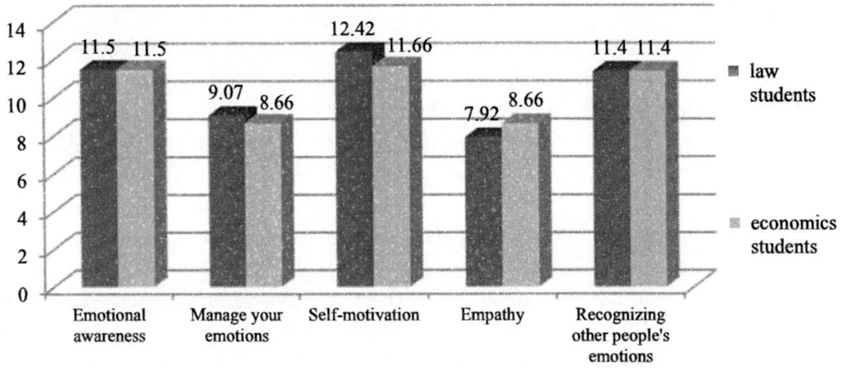

Fig. 3 Average scores on the Hall Emotional Intelligence test among the entire group of respondents (in points)

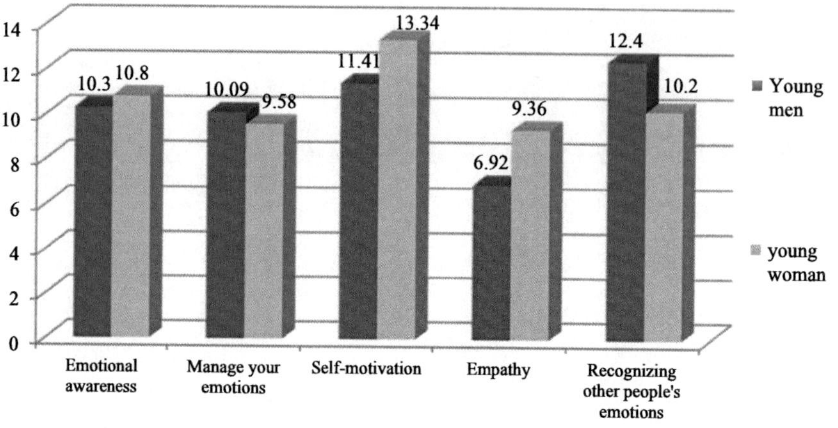

Fig. 4 Average scores on the Hall Emotional Intelligence test among male and female students (in points)

5 Conclusion

Next, we turn to the results of the statistical analysis of the data obtained by the Mann-Whitney U-criterion.

Regarding hypothesis 1, that the features of emotional intelligence and self-acceptance among students of different training profiles (legal and economic) will differ. We carried out a statistical analysis using the method of V. V. Stolin's OSO and Hall's method of emotional intelligence in relation to two groups of respondents-law students and economics students. The obtained statistical data allow us to state that there are significant differences among students of the legal and economic profile with respect to the following scales of V. V. Stolin's OSO: "Self-esteem scale (I)": U = 239.5, *p*-value = 0.016; "Autosympathy scale (P)": U = 475.3, *p*-value = 0.018; "Self-interest scale (IV)": U = 589.8, *p*-value = 0.021; "Self-acceptance scale (3)": U = 452.4, *p*-value = 0.042; "Self-consistency scale (4)": U = 263.2, *p*-value = 0.023; "Self-blame scale (5)": U = 471.6, *p*-value = 0.036; "Self-interest scale (6)": U = 968, 7, *p*-value = 0.019; "Self-understanding scale (7)": U = 247, *p*-value = 0.022.

The data obtained allow us to state that there are significant differences among students of the legal and economic profile in relation to the following scales of the Hall's emotional intelligence methodology: "Emotional awareness": U = 522, *p*-value = 0.014; "Self-motivation": U = 263.5, *p*-value = 0.041; "Recognition of emotions of other people": U = 256.5, *p*-value = 0.034.

Summarizing the obtained statistical data, we can conclude that there are statistically significant differences among students of different training programs regarding their level of self-acceptance and the features of the manifestation of social intelligence. As a result, hypothesis 1 that the features of emotional intelligence and self-acceptance in students of different training profiles (legal and economic) will differ, was fully confirmed.

Further, regarding hypothesis 2, that in the features of emotional intelligence and self-acceptance in students of various training profiles (boys and girls) there may be differences, a statistical analysis was carried out using the OSO method of V. V. Stolin and the emotional intelligence method of Hall in relation to two groups of respondents, male and female students. The obtained statistical data allow us to state that there are significant differences among students of boys and girls with respect to the following scales of the OSO methodology of V. V. Stolin: "Self-esteem scale (I)": U = 164.5, *p*-value = 0.027; "Autosympathy scale (P)": U = 228, *p*-value = 0.187; "Self-interest scale (IV)": U = 163.5, *p*-value = 0.026; "Self-acceptance scale (3)": U = 341, *p*-value = 0.006; "Self-consistency scale (4)": U = 136, *p*-value = 0.005; "Self-blame scale (5)": U = 238, *p*-value = 0.022; "Self-interest scale (6)": U = 376, *p*-value = 0.028; "Self-understanding scale (7)": U = 172.5, *p*-value = 0.036.

The obtained statistical data suggest that there are significant differences among male and female students with respect to the following scales of the Hall's emotional intelligence methodology: "Emotional awareness": U = 341, *p*-value = 0.012; "Managing your emotions": U = 238, *p*-value = 0.008; "Self-motivation": U = 475.5, *p*-value = 0.037; "Recognizing the emotions of other people": U = 539, *p*-value = 0.026.

Summarizing the obtained statistical data, we can conclude that there are statistically significant differences among young men and women students of different training profiles regarding their level of self-acceptance and features of emotional intelligence. As a result, hypothesis 2, that in the features of emotional intelligence and self-acceptance in students of various training profiles (boys and girls) there may be differences, fully confirmed.

References

1. Bernard M, Vernon A, Terjesen M, Kurasaki R (2013) Self-acceptance in the education and counseling of young people. In: Strength self-acceptance, vol 1, pp 155–192
2. Korableva SV, Biryukova AA (2017) Students' self-relationship development during university preparation. Interact Sci 5(15):71–74
3. Abakumova I, Brizhak Z, Kukulyar A, Kolenova A (2020) Features of self-presentation of modern young people in social networks. In: E3S Web of conferences, vol 210, 19003. https://doi.org/10.1051/e3sconf/202021019003. Last accessed 2021/03/10
4. Abakumova I, Zvezdina G, Grishina A, Zvezdina E, Dyakova E (2020) Models of information behavior: changes in psychological boundaries of internet users. In: E3S Web of conferences, vol 210, 20015 (2020). https://doi.org/10.1051/e3sconf/202021020015. Last accessed 2021/03/10
5. Abakumova I, Zvezdina G, Grishina A, Zvezdina E, Dyakova E (2020) University students' attitude to distance learning in situation of uncertainty. In: E3S Web of conferences, vol 210, 18017. https://doi.org/10.1051/e3sconf/202021018017. Last accessed 2021/03/10
6. Brizhak ZI, Kolenova AS, Kukulyar AM (2021) Research of psychological features of the personality students with varying degrees of Internet addiction. Int J Med Psychol 4:78–84

7. Grishina AV, Achina AV, Kukulyar A, Chinarova ON (2019) Perception of social advertising on a value-semantic level by students of humanitarian orientations. In: Trends in the development of psycho-pedagogical education in the conditions of transitional society (ICTDPP-2019). SHS Web of conferences, vol 70, 6. URL: https://www.shs-conferences.org/articles/shsconf/pdf/2019/11/shsconf_ictdpp2018_06004.pdf. Last accessed 2021/03/10
8. Kolenova AS, Kukulyar AM, Bessonova NN, Gudz NA (2021) Research of psychological features of the personality of students showing addictive behavior. Int J Med Psychol 4:38–44
9. Dhani P, Sharma T (2016) Emotional intelligence; History, models and measures. Int J Sci Technol Manage 5:189–201
10. Dolev N, Leshem S (2017) Developing emotional intelligence competence among teachers. Teach Dev 21(1):21–39
11. Kosonogov VV, Vorobyeva EV, Kovsh EM, Ermakov PN (2019) A review of neurophysiological and genetic correlates of emotional intelligence. Int J Cogn Res Sci Eng Educ 7(1):137–142. https://doi.org/10.5937/IJCRSEE1901137K. ISSN 2334-847X. Last accessed 2021/03/10
12. Maguire R, Egan A, Hyland P, Maguire P (2017) Engaging students emotionally: the role of emotional intelligence in predicting cognitive and affective engagement in higher education. High Educ Res Dev 36(2):343–357
13. Zysberg L, Orenshtein C, Gimmon E, Robinson R (2017) Emotional intelligence, personality, stress, and burnout among educators. Int J Stress Manage 24(S1):1–16. Last accessed 2021/03/10
14. Murensky SL (2000) The relationship between emotional intelligence, personality, critical thinking, and the effectiveness of organizational leadership at the highest levels of management. Theses of International Dissertations: Section B: Science and Technology 61(2-B):1121
15. Goleman D (2011) Brain and emotional intelligence: new ideas. More Than Sound LLC

Digitalization of Trading Zones: Education and Psychology in a Post-modern Society

Tatiana Pavlova, Irina Antipova, and Natalya Lomova

Abstract Digitalization leads not only to an increase in the speed and volume of information processing, but to the transformation of a person and society. Changes in science are described and evaluated in different ways, including using the concept of trading zones (R. Galison). Psychology implements the role of social technologies (I. T. Kasavin). Digitalization increases the importance of the role of interaction mediators. In trading zones, philosophers point to alienation (A. M. Dorozhkin), therefore it is necessary to study possible alienations generated by intermediaries. The goal is to study the change in the mediative role of the education system and psychology as a result of digitalization and to consider the danger of alienation. The study is carried out in line with the program of social epistemology (I. T. Kasavin) and is based on the difference between Galisonian trading zones and Humboldtian trading zones (A. M. Dorozh-kin). Conclusions—(1) as a result of digitalization, knowledge, implying reflection and considerable time to master it, is alienated, (2) education system and psychology as a social technology are implemented most effectively in Galisonian trading zones (TZ), (3) digitalization of education and psychology as a social technology implies, that the most effectively formed trading zones (TZ) are between a society that expects specialists with instrumental knowledge, and the sciences that pro-vide quickly and easily applicable knowledge.

Keywords Digital technology · Digitalization · Psychology of education · Trading zones · P. Galison. Galisonian trading zones · Humboldtian trading zones · Non-Humboldtian trading zones · Partial alienation · Social technologies

T. Pavlova (✉) · N. Lomova
Don State Technical University, Gagarin Sq., 1, Rostov-on-Don 344003, Russia
e-mail: zabrodinka@bk.ru

I. Antipova
Southern Federal University, Bolshaya Sadovaya Str., 105/42, Rostov-on-Don 344006, Russia

A. Beskopylny and M. Shamtsyan (eds.), *XIV International Scientific Conference "INTERAGROMASH 2021"*, Lecture Notes in Networks and Systems 247,
https://doi.org/10.1007/978-3-030-80946-1_47

1 Introduction

Digitalization leads to a change not only in the means of transmitting, receiving and accumulating the information, but also in individuals and society. The interaction between various systems and subsystems of society, between socio-cultural communities, between science and post-modern society is intensified. This society is distinguished by large volumes and speed of information transfer, openness to changes. Science evokes significant criticism from non-classical positions, but continues to be the most important subsystem of culture. Changes in science are described and evaluated in different ways, including the use of the concept of trading zones (R. Galison). This concept, taken from anthropology, is used to study the issues of interdisciplinarity in science, and is beginning to be used to comprehend other types of interaction. It is possible to apply the concept of trading zones in the study of interaction between science and society. Various actors are at play here as mediators, but the educational system and psychology as a social technology carry out the greatest load in this zone. Digitalization changes the role of the interaction mediator, and there is a transformation of interacting agents (science and society). Reflection, assessment of changes and possibilities of rational decisions turns out to be hindered because the assessment methodology is also changing. But the problem of assessing the role of interaction mediators cannot but arouse interest, because the value of the transmitted information depends on the actions of intermediaries (education system and psychology as a social technology). In trading zones, philosophers see alienation (I. T. Kasavin, A. M. Dorozhkin), therefore, it seems relevant to study possible alienations generated by intermediaries. The goal is to study the change in the mediation role of the education system and psychology as a result of digitalization and to consider the danger of alienation.

2 Literature Review

Digitalization opens up tremendous opportunities and leads to risks that are difficult to predict. The warnings of the philosophers of the last century are becoming acutely relevant in connection with the spread of digital technology. M. Heidegger wrote about the closed nature of technology, about the fact that technology is not just a neutral tool, but requires and forms a human proportionate to the essence of technology [1]. The origins of the philosophy of technology can be seen in antiquity. Plato began to thematize technology as an imitation of nature. K. Mitcham calls this type of philosophy of technology "humanities philosophy of technology" (Stanford Encyclopedia of Philosophy, electronic resource).

At present, the philosophy of technology is changing, acquiring acute relevance, as is shown by the analysis of V. G. Gorokhov, "at the end of the twentieth century, the philosophy of technology was considered as a peripheral discipline of modern philosophical science" [2, p. 181], but the philosophy of technology changes significantly due to changes in technology. According to A. A. Voronin, "The identity of technology and culture has changed, many philosophers interpret it as an attack of technology on culture, as a threat to man" [3, p. 106]. And this threat is discussed when strengthening the capabilities of a person in the construction of oneself and the surrounding objects. Digital technology is changing the way traditional problems are articulated. "Information—the basis of today's culture, production and thinking—does not allow rigor, localization, certainty and determinism. There is still no single definition of information. The exchange of information between man and nature and between people is already fundamentally different. The secrets of nature are revealed not by fixation, experiment and interpretation, but by calculation, modeling, construction" [3, p. 105]. Digitalization opens up opportunities to design not only various objects, but also criteria for recognizing their objectivity and reality. But this possibility of constructive activity "now finds itself in the "hands" of the computer. The world has become a product of a human genius, but with the permission of … a computer" [3, p. 107]. With digital technology comes pluralism in the criteria for assessing the realism of structures, but philosophers find no reason to completely abandon truth and reality. In short, digital technology opens up wide possibilities for constructing reality and can lead to skepticism and doubts in the classical picture of reality, but it does not give an exact answer and grounds for avoiding reality. This leads, as can be considered, to the observed spread of eclecticism and naive economic and social positivism. It is not freedom from classical reality, but the choice of simplified representations and ready-made methods for constructing reality are a consequence of the wide possibilities of technology. Consequently, A. A. Voronin warns about the possibility of an irrational solution with all the rationality of the use of technology. "Constructive … drives a person forward, leaving behind rationally set goals, rationally calculated risks, rationally arranged institutions, rationality as such" [3, p. 106]. Digital design is changing reality, and it seems to be canceling science as a rational study. This problem is not new to philosophy. "We … meet here with Kantian constructivism … we learn what we ourselves have created" [4, p. 49], but, O. E. Stolyarova explains this problem of science, material culture and techno-nature as a referent of science "imposes ontological frameworks" [ibid., p. 49], and therefore relativism was sown from digitalization, but did not give a final answer as a fulcrum. Rather, it limits the ability to think and question the limitations of our traditional views. A. J. Antonovsky points out the limitations of reflection in connection with the spread of technology, "in the most general sense, technology can be conceptualized as a program for the implementation of socially relevant tasks … by specifically optimizing activities and communication in time and space: as a resource for saving time for comprehension and reflection" [5, p. 9]. Technology provides the economy of space, but not of time for the development and use of technology. This is compensated by the limitation of critical thought and reflection. Digital capabilities turn out to be twofold both for assessing the effectiveness of the introduction of technology and for a person's sense of reality. The philosophy of technology conducts research into the problems of introducing technical innovations and V. G. Gorokhov presents this introduction as a social technology [6, p. 123]. I. T. Kasavin understands social technologies more broadly as the management of social processes and the formation of a social world view [7]. Digital technology not only increases the speed and volume of information processing, but transforms society, "technology acquires new social functions" [3, p. 96]. Researching the technology, A. A. Voronin comes to the conclusion that "In social terms, we are witnessing no less significant shifts than those at the turn of the first two eras. A new form of social connection is emerging (following the Internet)" [3, p. 106]. Therefore, digital technology, digitalization can be considered as a change in sociality, one can observe how new opportunities appear for interaction between various subsystems of society, between socio-cultural communities. These changes are diverse and are studied from different perspectives. The concept of trading zones, proposed by P. Galison is promising [8]. This concept (TZ) was originally used in anthropology to describe trade exchange between Aboriginal tribes. The interaction led to cultural innovation in the form of jargon (fragmented lexical units), pidgin (complex structures), and creole (holistic system). P. Galison applied this concept for the philosophical study of science, showing TZ as a space of interdisciplinary interaction of scientists. P. Tagart, discussing the problems of interdisciplinarity in cognitive science, describes "trading zones" [9]. As I. T. Kasavin highlights, the concept of TZ is used for the tasks of "conceptualizing the communicative content of science … and is today in the focus of social research in science" [10, p. 10]. O. E. Stolyarova uses the concept of TZ in the study of philosophical problems of techno-science. G. Collins "focuses on the construction of trading zones in natural science and technology and the intermediary role played in this process by a scientist–humanist" [10, p. 13]. V. N. Porus connects the problem of mediation in trading zones with metaphilosophy, which is ready to become a space where scientific disciplines and philosophy meet. A. M. Dorozhkin sees the problem in the fact that "almost all authors touch upon the processes of functioning of trading zones, but do not consider … the emergence of such zones" [11, p. 21] and connects mediation with the work of a philosopher.

S. V. Shibarshina, N. N. Voronina, A. N. Tkachev, L. V. Shipovalova address the mediative actions of philosophy in scientific trading zones. The philosopher addresses the issue of the possibility of justifying this communication. The conventionality of reality is questioned and opened for comprehension. The social trend in the philosophy of science (A. J. Antonovsky) is correlative to the digital one and this is discussed in the collective monograph (Social and Digital Science Studies, 2019). The problems of network collaborations in science (S. V. Shibarshina,), digitalization of trading zones, challenges of information technologies in science are addressed. According to S. V. Shibarshina, "the scientific community … acquires the character of a set of networks" and P. Virilio's pessimism in assessing the applicability of technology turns out to be less and less relevant, as well as the "alertness and rejection" of scientists [12, p. 77]. The trend in the recognition of the digitalization of science is discussed by S. V. Shibarshina who shows that with an increase in the speed of information transfer, the latter replaces knowledge. David W. Orr regarded the twentieth century as a "century of rapid knowledge" and formulated the propositions that "(1) only that which can be measured can be considered true knowledge, (2) the more knowledge we accumulate, the better, (3) there is no significant difference between knowledge and information, (4) miscalculations and mistakes on the path of progress can be corrected through new knowledge" [12, p. 77]. The education system has always performed the work of an intermediary between science and society, and psychology, as a social technology, implements mediative work. This mediation can be thought of as work at TZ.

The concept proposed by P. Galison began to be used to describe the interaction of various systems. TZ can be viewed as social interaction at the level of both personal and public relations. A. M. Dorozhkin reasonably speaks about TZ, being both intra-scientific and extrascientific [11]. The education system, which includes psychology as a social technology, changes significantly in its mediative work due to changes in science, society, and technical means of interaction between actors. Science becomes a space for exchange, receives a social interpretation and perspective, while society is presented in a non-classical way as an emergent object. Non-classical criticism, flexible rationality, digitalization provide ample opportunities and lead to significant problems. The application of the TZ concept to psychology and the education system as social technologies opens up an opportunity to discuss the positive and negative aspects of digitalization and the social trend of science. On the positive side, the openness of trading zones presupposes the elimination of bureaucratization and formalism. Pluralism is widely discussed in the education system, and the interaction of non-normativity and non-normativity can be represented as a zone of exchange in education, the education system is changing "in the direction of reducing and compensating for various forms of inequality" [13, p. 65]. This is correlative to the rejection of simplified understanding in assessing normativity, from the essentialist approach to assessing non-normality. D. V. Vorontsov, shows that queer studies "at the present stage of development of social knowledge open up opportunities not only for a new look at the phenomenon of sexuality. The questions that are raised here lead us to the general principles of the functioning of the psyche, which has the quality of sociality" [13, p. 161]. Digitalization is presented as a dehumanization of education, but this criticism does not take into account significant humanistic trends in the formation of the educational environment (see Psychological aspects of the formation of the educational environment—Belousova, Kozhuchar). A. A. Kostrigin speaks positively about digitalization in psychology. "Currently, there is an increased interest in digital methods in science, including in historical works. This area is called Digital Humanities (DH)—digital social science, digital humanitaristics. The prospects and advantages, as well as the difficulties and disadvantages of using DH methodology in the history of psychology are discussed" [14, p. 125].

There are also negative aspects in modern trends in the interaction of science and society, in the implementation of the mediative work of the education system and psychology as social technologies. I. T. Kasavin, in his article on the social trend and trading zones, formulates a relevant issue of the education system: "In the modern knowledge society, there is a high need for highly qualified scientists and engineers. At the same time, the conditions of the consumer society reduce the prestige of intellectual activity, which is becoming one of many goods. There is also a sharp contradiction between the growing specialization and differentiation of sciences, on the one hand, and everyday consciousness, on the other, which does not have time to master the achievements of science" [15, p. 8]. This problem is qualified by I. T. Kasavin as a gap between science and society. The proposed solution is associated with the development of "new tools (social and humanitarian technologies) of a special type—trading zones" [15, p. 8], including psychology, with the integration of science "into a cultural society" [ibid., p. 8].

The analysis shows that the education system and psychology as social technologies represent intermediaries in the trading zones between society and science. Changes in science and society in connection with digitalization make it necessary to clarify the positive and negative aspects in the actions of the trading zones mediators.

3 Methodology

The goal is to study the change in the mediative role of the education system and psychology in the interaction zone (TZ) of science and society as a result of digitalization and to consider the danger of alienation.

The study continues the development of the issue of the functioning of the education system and psychology in trading zones, started in the article [15].

The study is carried out in line with the program of social epistemology (I. T. Kasavin), on the difference between Galisonian trading zones and Humboldtian trading zones [11], on A. M. Dorozhkin's understanding of partial alienation in education (Figs. 1 and 2).

The methodological provision for the study was the representation of the education system and the humanities as social technologies for the formation of a social picture of reality [11, 15]. I. T. Kasavin, distinguishes his interpretation of social technologies and manipulation techniques as "technocratic imbalance." I. T. Kasavin

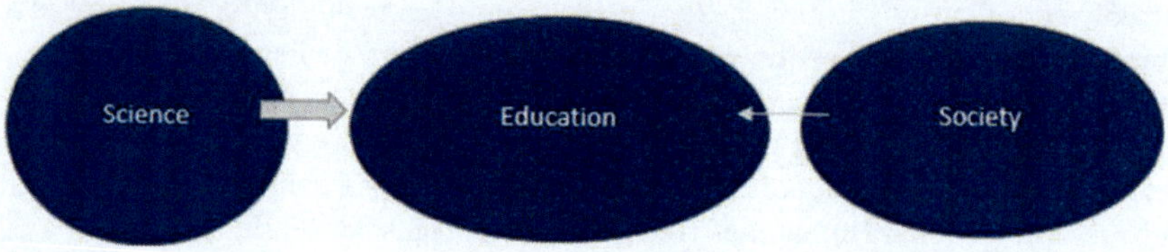

Fig. 1 Galisonian trading zone

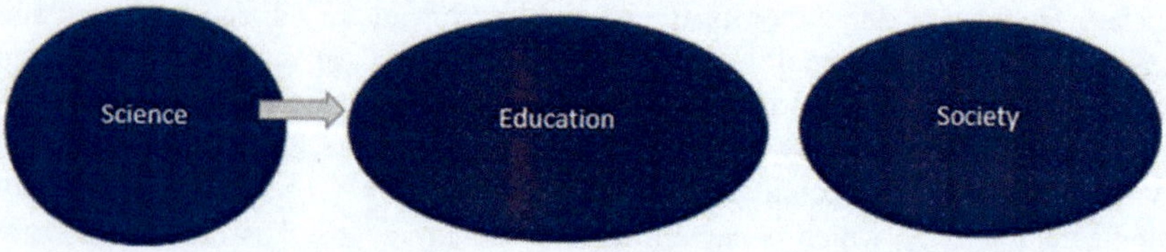

Fig. 2 Humboldtian trading zone

emphasizes that the modern view of social technologies (ST) from the standpoint of social constructivism leads to a fundamental revision of their technocratic interpretation to the extent that ST are identified with "conversational technologies" [16, p. 9]. The primitivism of understanding and the technocratic emphasis is a consequence of the fact that "ST are understood as a certain guaranteed algorithm for managing social subjects and processes that does not take into account a number of fundamental differences between society and nature" [16, p. 10].

4 Results

The problems of the education system and psychology as social technologies, changes in their mediating roles in the context of digitalization can be studied only in connection with changes in science (social trend, non-classical criticism of fundamentalism, pluralism) and society (intensification of interaction between various subsystems of society and cultural communities).

Let us analyze the digitalization trend in the education system and psychology as a social technology from both positive and negative sides. As a positive side, one can indicate an increase in the speed and volume of processing and transmission of information, as negative aspects, we will present only the main ones, an increase in the level of mental tension and a limitation of reflection and humanitarian knowledge in the preparation of students. Traditionally, the education system is implemented as an intermediary between society and science, and the selection of the content transmitted to students is guided by (1) didacticism (2) pro-social values (3) possibility that students will be applying the acquired knowledge.

Digitalization opens up the possibility of broadcasting content that is available for "digitization" and is consistent with the mercantile orientation in the application of technology. This is expressed in a change in the selection of educational content, the most accessible material for transmission to students is material that can be quantified, dosed quantitatively, the most preferable is educational material that is easily and effectively used by students in practical work. Digitalization provides a significant opportunity to save space, digital textbooks do not take up any space, students located at a considerable distance from educational institutions can participate in educational activities on an equal basis with those who can come to classrooms. These space savings exceed the time spent on maintenance and use of technology and on bureaucratic procedures that inevitably increase in importance due to the fact of being equipped with technology. Lack of time, bureaucratization and an increase in the mercantile orientation in the activity of education lead to the limitations of reflection. One can regrettably state that there is a shift towards an increase in the importance of indicators without direct connection with the actual results. The initially wary perception of digitalization in science is replaced by acceptance, even the humanities and philosophers are accepting the digitization of their activities. Partial alienation analyzed in the education system by A. M. Dorozhkin, affects

Table 1 Branches of knowledge and possibilities for alienation in the context of digitalization

Branches of science	Possible alienation
Natural, engineering and exact sciences	Methodological knowledge
Humanitarian sciences	Philosophical existential knowledge

both complex systems of knowledge and philosophical and humanitarian knowledge, assuming a significant reflective activity of students during its acquisition. Psychology as a social technology cannot be presented as an integral, unambiguously assessed system of the formation of human consciousness, but the general points in the change in social technologies can be presented. Digitalization with an increase in the speed and processing of information and the implementation of social contacts cannot but affect the direction of social technologies. The latter are forced to deal with the correction of mental stress, formation of a positive assessment of digitalization and changes in social systems, and the limitation of reflection. Existentialist questioning does not fit into a digitized existence and is evaluated as mental stress. It has become bad form to discuss the tragedy of being, and philosophical reflection limits the acquisition of educational material. As an objection, one can cite facts indicating that there are manifestations of reflection and interest in the humanities in education. But since it is not a pattern that is discussed here, but the observed trends in the education system as a result of digitalization, these facts do not call into question the general conclusions (Table 1).

5 Discussion

The analysis of the described trends in the education system and psychology as social technologies as trading zones can be carried out using the difference between Galisonian trading zones and Humboldtian trading zones [11]. A. M. Dorozhkin distinguishes between the Galisonian zones of mutual striving for interaction and the Humboldtian zones, characterized by the activity of one side. This distinction is important for understanding what kind of partial alienation can develop in education, what partial alienation is inspired by the education system, aimed at transmitting practically applicable educational material with limited reflection. Let us present and apply an additional trivial difference of directly practically applicable knowledge and knowledge that presupposes philosophical reflection and does not include direct instructions for practice. It can be stated that given the mercantile orientation of teachers, students and society, trading zones will be built with a mutual striving for interaction. Whereas the humanities and philosophy, that inevitably presuppose reflection, will experience a greater alienation. The education system and psychology as a social technology are realized as intermediaries in the trading zones between science and society. The possibility of alienation in the context of digitalization is associated with limited reflection, alienation in the education system is explained by the limitation in the transmission of philosophical problems of studies of man and technology.

6 Findings

1. as a result of digitalization, knowledge, implying reflection and considerable time to master it, is alienated.
2. education system and psychology as a social technology are implemented most effectively in the Galisonian trading zones (TZ).
3. digitalization of education and psychology as a social technology suggests that most effectively formed TZ are between a society that expects specialists with instrumental knowledge, and the sciences that provide quickly and easily applicable knowledge.

References

1. Heidegger M (1993) The question of technology. Heidegger M. Time and being. Articles and speeches. Translation from German. Respublika, Moscow, pp 221–237
2. Gorokhov VG (2011) The place and role of the philosophy of technology in modern philosophy and its organic connection with the philosophy of science. In: Arshinov VI, Gorokhov VG (eds) Philosophy of science. Issue 16: Philosophy of science and technology. Russian Academy of Sciences, Institute of Philosophy. IP RAS, Moscow, pp 181–199
3. Voronin AA (2020) Technology 1–2–3. Philos Sci Technol 25(1):96–109
4. Stolyarova OE (2011) The historical context of science: material culture and ontology. Epistemol Philos Sci XXX(4):32–50
5. Antonovsky AJ (2011) Towards a systemic and communicative interpretation of language technologies. In: Kasavin IT (ed-in-chief) Science and social technologies. Russian Academy of Sciences, Institute of Philosophy. IP RAS, Moscow, pp 9–32
6. Gorokhov VG (2011) The concept of "technology" in the philosophy of technology and the characteristics of social and humanitarian technologies. Epistemol Philos Sci XXU111(2):110–123
7. Kasavin IT (2011) Introduction. In: Kasavin IT (ed) Science and social technologies. Russian Academy of Sciences, Institute of Philosophy. IP RAS, Moscow, pp 3–8
8. Galison P (2004) Trading zone: coordinating action and belief. Questions of the history of natural science and technology (1). URL: http://vivovoco.astronet.ru/VV/JOURNAL/VIET/GALISON.HTM
9. Tagart P (2014) Interdisciplinarity: trading zones in cognitive science. Logos 97(1):36–60
10. Kasavin IT (2017) Trading zones as a subject of social philosophy of science. Epistemol Philos Sci 51(1):8–17
11. Dorozhkin AM (2017) Problems of construction and typology of trading zones. Epistemol Philos Sci 54(4):20–29
12. Shibarshina SV (2019) Scientific communications and collaborations on the web as possible trading zones. Sociol Sci Technol 10(2):75–92

13. Vorontsov DV (2015) The inclusion of sexuality in the educational process: socio-psychological aspects of normativity in educational activity. Educ Psychol 9:65
14. Kostrigin AA (2019) Digital history of psychology: characteristics, methods and directions. Social and digital science studies: collective monograph. In: Argamakova AA, Maslanova EV, Slyusareva VV, Khusyainova TM (eds) Publishing house "Russian Society of History and Philosophy of Sciences", Moscow, pp 125–135
15. Antipova I, Krishchenko E, Pavlova T, Lomova N (2020) Psychology of education as a social technology: designing exchange zones. In: E3S Web of conferences, vol 210, 20014. ITSE-2020
16. Kasavin IT (2010) Social technologies and scientific knowledge. Epistemol Philos Sci XXY1(4):5–15

The Relationship Between Tolerance for Uncertainty and Creativity Among High School Students with a Critical Style of Thinking

Alla Belousova, Alena Samarskaya, Ekaterina Belousova, and Ekaterina Kryazhkova

Abstract The article presents the results of a study of the relation of tolerance to uncertainty and creativity among senior schoolchildren with a critical style of thinking. The authors use valid diagnostic methods: A. Belousova's thinking style questionnaire, S. Badner's uncertainty tolerance questionnaire, and P. Torrance's test. The sample consists of 90 senior schoolchildren, both boys and girls. As a result of the study, the groups of schoolchildren with a dominant critical and practical style of thinking have been identified. The authors show that there are significant differences in the development of relationships between uncertainty and creativity tolerance scales, between uncertainty tolerance and critical thinking style, creativity and critical thinking style. The differences have been obtained in the development of tolerance for uncertainty, creativity in relation to the practical style of thinking.

Keywords Uncertainty tolerance · Creativity · Non-verbal creativity · Style of thinking · Critical style of thinking · Senior school age

1 Introduction

Tolerance to uncertainty belongs to the categories whose importance is growing, since there is a real need for practical recommendations and technologies for providing psychological assistance to people who are in a situation of variability, transitivity of socio-economic changes. Suroedova and Tushnova [1] consider the problem of adaptation of young people, their socio-psychological settings that contribute to the development of constructive behavior. Dautov et al. [2] raise questions about the development of clip thinking among modern youth who live and study in conditions of uncertainty and the development of digital technologies. The attitude of young people to digital technologies as a component of distance learning in the context of growing uncertainty raises many questions for modern teachers [3]. Dynamically developing socio-economic transformations are increasingly creating situations of uncertainty, forming possible economic and psychological risks in the social behavior

A. Belousova · A. Samarskaya (✉) · E. Belousova · E. Kryazhkova
Don State Technical University, Gagarin Sq., 1, Rostov-on-Don 344003, Russian Federation

A. Beskopylny and M. Shamtsyan (eds.), *XIV International Scientific Conference "INTERAGROMASH 2021"*, Lecture Notes in Networks and Systems 247,
https://doi.org/10.1007/978-3-030-80946-1_48

of young people. This is especially true for senior students who are susceptible to the challenges of our time and the contradictions that they reveal in social sphere.

Older students are an age group characterized by: ease of perception, acceptance and assimilation of new ideas; emotional sensitivity, anxiety; the search for personal and life choices, the desire to rethink moral and value truths—which are combined with categorical, straightforward, imperative decisions [4]. Senior schoolchildren, due to their age characteristics, are focused on the new, different from the ordinary or standard and stereotyped. In such conditions, it is important for high school students to develop skills to critically evaluate information and situations, the ability to accept the possibility of the existence of uncertainties, ambivalence in problem solving, skills to quickly assess and make decisions, and the ability to creatively approach design and problem solving.

We believe that the uniqueness of the psychological characteristics that senior schoolchildren's personalities have, together with the instability of the formed value-semantic structures of consciousness, are the psychological basis through which new information is filtered, values are perceived, interests and motives are formed, and the adaptation to socio-cultural conditions takes place. One of the ways that make it possible to form and develop constructive behavioral strategies in terms of behavior, adaptability, an active life position is the development of the sovereignization of the individual, his independence. Human thinking is one of the ways to develop the sovereignty of the individual: the ability to perceive, comprehend the information received, reflect on it, analyze, separate the essential and the nonessential, highlight reliable and inaccurate information, critically process the information received—these are all operations that relate to thinking and mental activity of a person.

In this regard, one of the most important tasks is the formation of a thinking person who has his own life position, is capable of setting and developing life goals, arguing his position and possessing the skills of supporting his position. In accordance with the ideas existing in the literature, these requirements are met by critical thinking, which implies the ability to see problems, readiness to solve problems, reflection of one's intellectual activity [5].

The problem of the development of critical thinking has been studied by many foreign and domestic psychologists (E. N. Volkov, L. Ginny, D. Johnson, I. O. Zagashev, E. S. Zair-Beck, D. Kluster, S. Kurfis, G. Lindsay, K. Meredith, S. Matthews, R. Paul, R. Paul, R. Sternberg, D. Steele, C. Temple, D. Halpern), who study the characteristics of critical thinking in senior schoolchildren.

D. F. Halpern believes that critical thinking is balanced, logical and focused. It is characterized by the use of cognitive skills and strategies that increase the probability of getting the desired result. Critical thinking is characterized by controllability, validity and purposefulness [5].

At the same time, there are not many works devoted to the study of the characteristics of the critical thinking style and such abilities of students as uncertainty tolerance and creativity.

The study of the problem of tolerance to uncertainty originates from the works of E. Frenkel-Brunswik, who emphasized the concept of ambivalence. Ambivalence was defined as the possibility of existence and reflection of the positive and negative qualities of an object by a person. In this construct, uncertainty tolerance was considered as a cognitive component. In turn, intolerance to uncertainty was defined as a binary perception of situations, in other words, the black-and-white type of perception [6].

In Russian psychology, a stable tradition of analyzing theoretical and empirical studies of tolerance to uncertainty has developed, which is presented in the review articles by A. N. Guseva, I. N. Leonova, A. G. Matushanskaya, B. S. Alisheva, E. S. Fominykh, M. N. Yurtaeva, N. S. Glukhanyuk. The presented analytical articles allow us to say that, perhaps, starting from the moment E. Frenkel-Brunswik used this term, the problem of understanding the essence of the phenomenon of tolerance to uncertainty arose, which resulted in the differentiation of approaches: on the one hand, understanding of tolerance to uncertainty as a personality trait (S. Bakalis and T. Joiner, D. T. Kenny and R. Ginsberg, B. T. Hazen, J. L. Herman, J. Litman, A. P. MacDonald, B. D. Naemi, R. W. Norton, N. G. Rotter and A. N. O'Connell, J. Sidanius, M. Trottier, F Zenasni) [7], on the other hand, uncertainty tolerance was considered as a cognitive phenomenon (D. T. Kenny, R. Ginsberg, D. L. MacLane, E. Frenkel-Brunswik) [8].

Hallman [9] considered tolerance to uncertainty as the ability to accept the conflict and tension that arise in a situation of duality, to resist the incoherence and inconsistency of information, to accept the unknown, not to feel uncomfortable facing uncertainty.

Jach and Smillie [10] in the framework of the development of the theory of psychological entropy Hirsh et al. [11], DeYoung [12] investigated the tolerance to uncertainty and personality traits of the Big Five. The authors show that openness to experience is associated with tolerance to uncertainty through intellectual curiosity, which is included as one of the parameters in the structure of the "Openness to experience" factor (intellectual curiosity, aesthetic sensitivity and creative imagination). They believe that participants with higher intelligence are more tolerant of uncertainty because behind this is the need to reflect on complex problems. Openness to experience and extraversion are positively correlated.

Lauriola et al. [13] found that uncertainty tolerance correlated with openness to experience, need for complexity and novelty. Zenasni et al. [14] found a positive correlation between creativity and uncertainty tolerance among adolescents. According to scientists, the more adolescents are tolerant of uncertainty, the higher their creativity indicators. The research by Mahmoud N. E., Kamel S. M., Hamza T. S. shows that there is a significant correlation between student creativity and uncertainty tolerance as measured by the Torrance test. The authors conclude that students' ability for divergent thinking is related to their ability to accept uncertainty [15]. In the K. Stoycheva's study, it was found that tolerance to uncertainty is a factor that has a positive effect on the adolescents' creative abilities: a high level of tolerance to uncertainty among adolescents contributes to the generation of more original and unusual ideas [16].

A. V. Karpov considers uncertainty tolerance as an integral ability and understands it as "resistance to the action of the uncertainty factor of the external and internal environment, which is professionally one of the main important qualities of a leader. It is determined by a combination of cognitive abilities towards removing uncertainty and an emotional tendency, towards perceiving uncertain situations as difficult, but not traumatic" [17, p. 562]. I. N. Leonov considers it possible to talk about tolerance to uncertainty as a personality trait that changes "under the influence of new experience or purposeful activity of the subject himself", "takes an active part in dynamic processes associated with overcoming or generating uncertainty (primarily at the cognitive level) by changing existing ideas and concepts, as well as by creating new ones" [7]. Thus, I. N. Leonov tries to integrate two approaches, combining the understanding of tolerance to uncertainty both as a personality trait and as a cognitive characteristic. According to the author, the changing of "existing ideas and concepts" and the creation of new ones are the characteristics of human mental activity [18]. A similar position is taken by T. V. Kornilov, proposing the concept of intellectual and personal potential, in which the ideas of dynamic regulatory systems and multiple cognitive choices made by a person in a decision-making situation are assimilated [19, p. 71]. E. G. Lukovitskaya considers uncertainty tolerance as a social attitude that has a traditional three-component composition with its three-component structure: cognitive, emotional and behavioral. "Since it contains the assessment of uncertainty, and emotional response and a certain behavioral response" [7]. Kholodnaya [20, p. 71] considers tolerance to uncertainty as a cognitive style, calling it "tolerance to unrealistic experience" which is characterized by the ability to accept impressions that do not correspond to a person's personal experience.

Thus, in domestic studies, tolerance was presented as an integral personality characteristic, studied as psychological stability, a system of personal attitudes and a set of multilevel individual properties.

With all the diversity of scientists' ideas regarding the nature of tolerance to uncertainty, generalizing the ideas of Leonov [7] and Kholodnaya [20], we believe that this phenomenon is based on a person's ability to accept conflicting information.

There are not many studies devoted to the study of tolerance to uncertainty, individual psychological characteristics among adolescents with different styles of thinking.

Creativity is one of the most significant parameters that influence the style of thinking.

According to E. P. Ilyin, there are different approaches to the problem of creativity: as the ability to bring something new to experience (F. Barron), generate original ideas for solving problems (M. Wallach), formulate hypotheses about new components of the situation (E. Barron). Torrence, to form new ways of thinking (D. Gilford) [21].

In our study [22], conducted on students of the conservatory, it was shown that students with a high level of nonverbal creativity have a pronounced desire to find new, unusual solutions.

Abolghasem et al. [23] analyzes the relationship between thinking styles and students' creativity. As a basis the authors take the classification of thinking styles by R. Sternberg and a questionnaire on Abedi's work. We have found positive correlations with the creativity of the following thinking styles: legislative, global, liberal, hierarchical, anarchic, internal, external; at the same time, local and conservative styles of thinking are negatively associated with creativity. Examining the features of motivation and the severity of the young managers' thinking styles, Belousova A., Mochalova U., have revealed a significant positive relation between the critical thinking style and achievement motivation [24].

Hamid [25] tries to determine the relationship between students' critical and creative thinking with their cognitive abilities. In the study the author uses questions related to critical and creative thinking and the creative style test. The study shows that almost 47% of students have critical thinking, 33% of them have the highest level of critical thinking, and 20% of the students have average values of the creative style of thinking. The analysis of the data allowed the author to conclude that the styles of students' critical and creative thinking are interrelated with cognitive abilities.

Malekian [26] has conducted a study among Iranian university students to determine the relationship between thinking styles, creativity, innovation orientation and academic motivation. The author base his statements on the classification of thinking styles by R. Sternberg. The results shows that there are significant relations among executive, external, judicial, legislative, individualistic, secondary, conservative, and creativity thinking styles.

Zhua and Zhang [27] have conducted a study on a large sample of students from several universities in China to explore the relationship between thinking styles and perceptions of creativity in students. Significant correlations were identified between thinking styles and ideas about creativity. Using regression analysis, the authors found the influence of thinking styles on students' perceptions of creativity.

Thus, our review of the research on the critical style of thinking has shown that a variety of patterns of relationships with personal and cognitive characteristics of an individual are possible.

We proceed from the functional nature understanding of the style of thinking [24, 28, 29]: the style of thinking is defined as a peculiar set of functions: generation, selection, meaning transmission and implementation actualized by a person in different situations of problem solving. The direction of the thinking style is determined by the type of the dominant function that determines the nature of the thinking style: the dominance of the generation function is manifested in the development of the proactive style of thinking, the dominance of the selection function—in the development of the critical style, the dominance of the meaning transfer function—in the development of the managerial style, the dominance of the implementation function—in the development of the practical style of thinking. In accordance with these ideas, the critical style of thinking develops with the dominance of the selection function aimed at selection, screening of information, which in the behavioral terms appears as the dominance of criticism, critical statements and actions. In connection with this understanding, it can be assumed that the critical style of thinking may have a negative relationship with the tolerance of uncertainty, which is understood as a person's ability to accept conflicting information, because adolescents with a critical style of thinking are likely to have a negative attitude towards information that is contrary to their ideas. As a second hypothesis, we assume that there may be positive relationships between non-verbal creativity and the critical style of thinking, since the critical style of thinking involves the generation of hypotheses, ideas aimed at achieving a goal. This corresponds to the indicators of creativity.

2 Materials and Methods

The aim of the study was to investigate the relationship between uncertainty tolerance and creativity in senior schoolchildren with a critical style of thinking. The sample consisted of 90 senior schoolchildren aged 15–18 years. The study involved 37 are boys and 53 are girls. The following methods were used in the study: to identify the style of thinking—the questionnaire "Methodology for measuring the style of thinking" by Belousova and Pishchik [29]; to identify the level of non-verbal creativity—P. Torrens test "Completion of pictures", adaptation of Voronin [30]; to identify the level of tolerance to uncertainty, we use the questionnaire "Methodology for determining tolerance to uncertainty" by S. Badner, adapted by Soldatova and Shaigerova [31]. Methods of mathematical statistics were used: descriptive statistics, Pearson's linear correlation coefficient and Student's t-test. The analysis of the results was carried out using a computer program for statistical data processing "SPSS 23.0 for Windows".

3 Results

At the first stages of our research, the diagnostics of the thinking style of senior schoolchildren was carried out, the results of which are presented in Table 1. The respondents were singled out who had a pronounced certain style of thinking. As part of our research, we selected respondents with a high level of expression of the critical thinking style. The percentage is shown in Table 1.

The study identified differences in the severity of thinking styles among senior schoolchildren; 35 respondents (39%) have a pronounced Critical thinking style; 31 subjects (34%) have a pronounced Practical thinking style; Managerial thinking style prevails in 17 (19%) senior schoolchildren, Initiative thinking style prevails in 7 (8%) senior schoolchildren. Based on the results obtained, we can say that Critical and Practical thinking styles dominate in high school students, while Managerial and Initiative thinking styles have a low level of representation.

Table 1 The ratio of respondents with a pronounced style of thinking

Thinking style	Number of respondents	Percentage (%)
Initiative	7	8
Critical	35	39
Managerial	17	19
Practical	31	34

Table 2 Significant correlations between uncertainty tolerance, nonverbal creativity, and thinking styles

	Originality	Uniqueness	Critical	Practical
Complexity	−0.291* при $p = 0.015$	−0.318** при $p = 0.007$	−0.331** при $p = 0.005$	0.354** при $p = 0.003$
Undecidability	−0.317** при $p = 0.008$	−0.256* при $p = 0.032$	−0.324** при $p = 0.006$	0.342** при $p = 0.004$
General indicator of TN	−0.345 при $p = 0.003$	−0.348** при $p = 0.003$	−0.341** при $p = 0.004$	0.421** при $p = 0.000$
Originality			0.259* при $p = 0.030$	−0.444** при $p = 0.000$
Uniqueness			0.261* при $p = 0.029$	−0.469** при $p = 0.000$

The results obtained allowed us to distinguish two groups of high school students: those with a high degree of Critical Thinking style and those with a high degree of Practical thinking style.

To identify the relationships between the level of tolerance to uncertainty, creativity and thinking styles, we used the Pearson rank correlation coefficient. This proposal indicates that the result is presented in Table 2.

Analysis of tabular data reveals significant connections between the style of thinking, as well as tolerance for uncertainty and creativity in older students.

The critical style of thinking has significant inversely proportional connections with the uncertainty tolerance scales: "Complexity" (r = −331**, at $p = 0.005$), "Undecidability" (r = −324**, at $p = 0.006$), "General indicator TH" (r = −341**, at $p = 0.004$). And at the same time, direct positive connections were found with the scales of creativity: Originality (r = 259*, at $p = 0.030$), Uniqueness (r = 261*, at $p = 0.029$).

Tolerance to uncertainty (TN) has significant inversely proportional relationships with the scales of Creativity (K): the scale of Complexity (TN) has inverse relationships with the scale of Originality (K) (r = −291*, at $p = 0.015$), the scale of Uniqueness (K) (r = −318**, at $p = 0.007$); scale Undecidability (TN) has negative relationships with scales Originality (K) (r = −317**, at $p = 0.008$), Uniqueness (K) (r = −256*, at $p = 0.032$); scale General indicator of tolerance to uncertainty has a negative relationship with the scales of Creativity: Originality (r = −345**, at $p = 0.003$), Uniqueness (r = −348**, at $p = 0.003$).

The results of the correlation analysis are presented in the form of a correlation constellation (Fig. 1).

In relation to the Practical Style of Thinking, results were obtained that show different relationships.

The practical style of thinking has significant direct links with the uncertainty tolerance scales: "Complexity" (r = 354**, at $p = 0.003$), "Undecidability" (r = 342**, at $p = 0.004$), "General indicator of TH" (r = 421**, at $p = 0.000$). And at

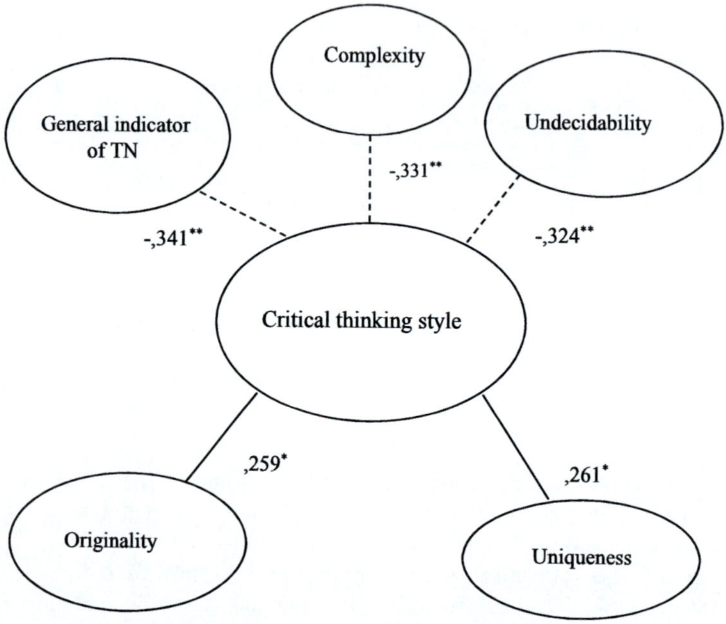

Fig. 1 Correlation galaxy "The relationship between the critical thinking style and the components of tolerance for uncertainty and creativity"

the same time, reverse negative relationships were found with the creativity scales: Originality (r = −444**, at $p = 0.000$), Uniqueness (r = −469**, at $p = 0.000$).

The results of the correlation analysis are presented in the form of a correlation pleiad (Fig. 2).

4 Discussion

The results obtained show that in the sample of senior schoolchildren, boys and girls with a pronounced critical and practical style of thinking dominate. This allows us to say that for senior schoolchildren, the dominant tendency is the desire, first of all, to evaluate everything: plans, hypotheses, goals, knowledge, in general—the personality and activities of others; the second tendency is the desire to achieve, implement plans, ideas in practice, which is consistent with the general characteristics of this age group [4].

Analysis of the results of the relationship of the Critical Style of Thinking with creativity and tolerance to uncertainty shows that the lower the level of the components of tolerance to uncertainty ("Complexity", "Insolubility", "General indicator of TH"), the higher the development of creativity (Originality, Uniqueness) and Critical style thinking.

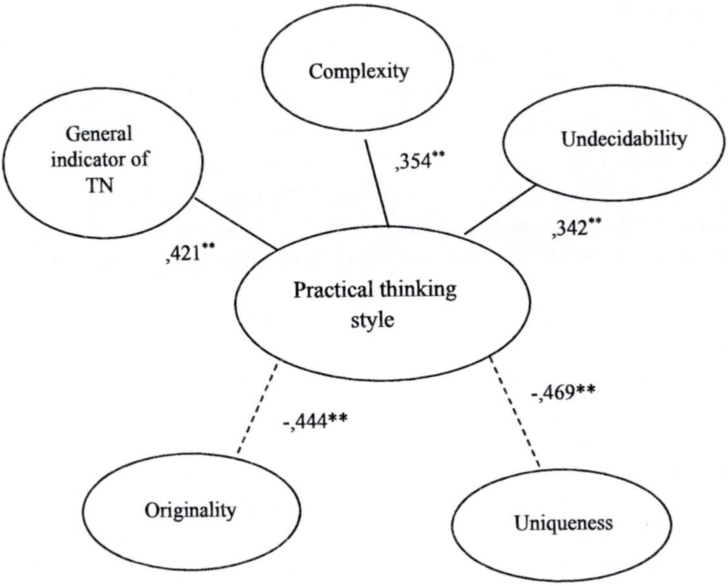

Fig. 2 Correlation galaxy "Interrelation of the practical style of thinking and the components of tolerance to uncertainty and creativity"

Comparing the obtained results with the studies available in the literature, we find that similar results were obtained in studies [23, 26], which showed a positive relationship between different styles of thinking (in accordance with the classification of R. Sternberg) and creativity. Our results are also consistent with the results of [14–16], which emphasize the positive relationship between uncertainty tolerance and creativity in adolescents.

In our sample, the emphasis was placed on senior students with a developed critical style of thinking, which gives a unique picture. In our opinion, in the sample under study, negative relationships between tolerance and the critical style of thinking arise precisely because the critical style of thinking prevents students from accepting contradictions, reflecting the complexity and insolubility of the situation due to the dominance of evaluative activity, which stops and stops productive activity. At the same time, the fact of the presence of direct positive links with creativity is emphasized. We believe that the results obtained underline the complex dual nature of the critical thinking style, in which two contradictory tendencies collapse [28, 29]: the ability to detect contradictions and the ability to evaluate, criticize everything that does not fit into the existing a person has knowledge and experience. In our opinion, this explains the negative links with uncertainty tolerance and positive links with creativity. We can continue the line of reasoning of Abakumova et al. [32], which speaks about the transformation of the individual. In this connection, we emphasize that a person opens up the possibility of changing himself through the perception of uncertainty as a source of activity that transforms the situation and his life world.

5 Conclusions

As a result of an empirical study of uncertainty tolerance and creativity in high school students with a critical style of thinking, we made the following conclusions:

1. The critical style of thinking has significant inversely proportional connections with the scales of tolerance to uncertainty: "Complexity", "Insolubility", "General indicator of tolerance to uncertainty", i.e. the lower the level of uncertainty tolerance components, the higher the development of the Critical Style of Thinking.
2. The critical style of thinking has significant direct positive connections with the scales of creativity: "Originality", "Uniqueness", i.e. the higher the level of development of the scales of creativity "Originality", "Uniqueness", the higher the level of development of the Critical Style of Thinking.
3. Tolerance to uncertainty (TH) has significant inversely proportional connections with the scales of Creativity (K): the scale of Complexity (TH) has inverse relationships with the scale of Originality (K), the scale of Uniqueness; the scale of Undecidability (TN) has a negative relationship with the scales Originality (K), Uniqueness (K); scale General indicator of tolerance to uncertainty has a negative relationship with the scales of Creativity: Originality, Uniqueness.
4. The practical style of thinking has significant direct links with the uncertainty tolerance scales: "Complexity", "Undecidability", "General indicator of TH".
5. The practical style of thinking has significant negative inverse relationships with the scales of creativity: Originality, Uniqueness.

References

1. Suroedova E, Tushnova Y (2020) Subjective assessment of social status and socio-psychological attitudes of student youth. In: E3S Web of conferences, vol 210, 19004. https://doi.org/10.1051/e3sconf/202021019004
2. Dautov D, Korochentseva A, Al Hussini MKM (2019) Features of clip thinking and attention among representatives of generations X and generations Z. In: SHS Web of conferences, vol 70, 06001, Jan 2019. https://doi.org/10.1051/shsconf/20197006001
3. Abakumova I, Zvezdina G, Grishina A, Zvezdina E, Dyakova E (2020) University students' attitude to distance learning in situation of uncertainty. In: E3S Web of conferences, vol 210, 18017. https://doi.org/10.1051/e3sconf/202021018017
4. Craig G, Bokum D (2019) Developmental psychology. Peter, St. Petersburg
5. Halpern DF (2003) Thought and knowledge: an introduction to critical thinking, 4th edn. Lawrence Erlbaum Associates, Inc. Publishers, Mahwah
6. Matushanskaya AG, Alishev BS (2011) Tolerance to uncertainty and related psychological constructs of cognitive psychology. Scientific Notes of Kazan University 153(5)
7. Leonov IN (2014) Tolerance to uncertainty as a psychological phenomenon: the history of the formation of a construct. Bulletin of the Udmurt University Philosophy, Psychology, Pedagogy
8. Rokitskaya YA, Manucharyan NSh (2019) Formation of tolerance to uncertainty in the process of professional self-determination of high school students: monograph. ZAO «A. Miller Library», Chelyabinsk

9. Hallman RJ (1967) The necessary and sufficient conditions of creativity. In: Gowan JC et al (eds) Creativity: its educational implications. Wiley, New York
10. Jach HK, Smillie LD (2019) To fear or fly to the unknown: tolerance for ambiguity and big five personality traits. J Res Pers 79:67–78. ISSN 0092-6566. https://doi.org/10.1016/j.jrp.2019.02.003
11. Hirsh JB, Mar RA, Peterson JB (2012) Psychological entropy: a framework for understanding uncertainty-related anxiety. Psychol Rev 119(2):304. https://doi.org/10.1037/a0026767
12. DeYoung CG (2013) The neuromodulator of exploration: a unifying theory of the role of dopamine in personality. Front Hum Neurosci 7:762. https://doi.org/10.3389/fnhum.2013.00762
13. Lauriola M, Foschi R, Mosca O, Weller J (2016) Attitude toward ambiguity: empirically robust factors in self-report personality scales 23(3):353–373. https://doi.org/10.1177/1073191115577188
14. Zenasni F, Besançon M, Lubart T (2008) Creativity and tolerance of ambiguity: an empirical study. J Creative Behav 42(1):61–73
15. Mahmoud NE, Kamel SM, Hamza TS (2020) The relationship between tolerance of ambiguity and creativity in architectural design studio. Creativity Stud 13(1):179–198. https://doi.org/10.3846/cs.2020.9628
16. Stoycheva K (2010) Tolerance for ambiguity, creativity, and personality. Bul J Psychol 1–4:178–189
17. Karpov AV (2005) Management psychology: a study guide. Gardariki, Moscow
18. Tikhomirov OK (2002) Psychology of thinking. Moscow Academy
19. Kornilova TV (2013) Psychology of uncertainty: the unity of intellectual and personal regulation of decisions and elections. Psychol J 34(3):89–100
20. Kholodnaya MA (2004) Cognitive styles. On the nature of the individual mind. Peter, SPB
21. Ilyin EP (2012) Psychology of creativity, creativity, giftedness. Peter, SPB
22. Belousova E, Kryazhkova E, Stošić L (2020) Verbal and non-verbal creativity of the students of the conservatory. In: E3S Web of conferences. Innovative technologies in science and education (ITSE-2020), vol 210. https://doi.org/10.1051/e3sconf/202021018081
23. Abolghasem P, Dehghankar L, Jafarisani M, Badiee SE, Tatari M, Khalafi A (2016) Studying the association between thinking styles and creativity among students. Nova J Humanit Soc Sci 5(2):1–7. https://doi.org/10.20286/nova-jhss-050202
24. Belousova A, Mochalova U (2020) The relationship of thinking style and motivation features of sales and advertising managers. Behav Sci 10(3):68
25. Hamid A (2017) Analysis of students' critical and creative thinking style and cognitive ability on chemistry. In: 5th South East Asia development research (SEA-DR) international conference. Advances in social science, education and humanities research, vol 100, pp 302–304
26. Malekian F (2016) Investigating the relationship between thinking styles and creativity, tendency to Innovation and academic motivation in students. Iran J Educ Sociol 1(1):95–103
27. Zhua C, Zhang L-F (2011) Thinking styles and conceptions of creativity among university students. Educ Psychol 31(3):361–375
28. Belousova AK, Pishchik VI (2011) Thinking style. Rostov n/D Publisher SFEDU
29. Belousova A, Pishchik V (2015) Technique of thinking style evaluating. Int J Cogn Res Sci Eng Educ (IJCRSEE) 3(2):1–8
30. Druzhinin VN (2019) Psychology of general abilities. Yurayt Publishing House, Moscow
31. Soldatova GU, Shaigerova LA (eds) (2008) Psychodiagnostics of personality tolerance. Meaning, Moscow
32. Abakumova IV, Ryadinskaya EN, Golubov M (2017) Systemic and structural analysis of theories of psychological transformations of personality. Russ Psychol J 14(1):10–24

Matrix Thinking in the Fractal Digitalization of Education

Elena Pinevich, Karina Mekhantseva, Aleksandr Volokhov, and Svetlana Gargatsova

Abstract This article is devoted to improving the motivation of students to study and, in general, the quality of education in the context of global digitalization. Along with numerous existing methods, the internal component of thinking and perception as an integral part of it plays a significant role. Research has shown that the mathematical logic used in the creation of most algorithms is not identical to the logic of thinking. The combined fuzzy and genetic logic that guides the student only partially includes clear logical structures. Moreover, this inclusion is manifested in different students with different degrees of belonging. If the teacher understands and uses the degree of such inclusion, then students will be able to show results of education that seem to exceed their own capabilities. Examples of the implementation of matrix thinking with students of military accounting, technical and creative specialties, as one of the ways to digitalize knowledge, are given. An example of the implementation of digitalization of training with students of applied programmers is considered on the example of creating software based on the material passed in a related IT discipline.

Keywords Digitalization · Data processing · Matrixes · Data fields · Training · Quality of education · Teaching methods · Innovative approaches · Non-standard pedagogy · Increasing student interest · Learning theory · Innovative pedagogy · Interest in knowledge · Matrix thinking

E. Pinevich (✉)
Don State Technical University (DSTU), 1 Gagarin Square, Rostov-on-Don 344000, Russia

K. Mekhantseva
Rostov State University of Economics, 69, B. Sadovaya St., Rostov-on-Don 344006, Russia

A. Volokhov · S. Gargatsova
Rostov State Transport University (RSTU), 2 Rostovskogo Strelkovogo Polka Narodnogo Opolcheniya Sq., Rostov-on-Don 344038, Russia

A. Beskopylny and M. Shamtsyan (eds.), *XIV International Scientific Conference "INTERAGROMASH 2021"*, Lecture Notes in Networks and Systems 247,
https://doi.org/10.1007/978-3-030-80946-1_49

1 Introduction

Information consists of various classes of nondeterministic sources. Therefore, the nondeterministic logic that a person uses when making decisions and obtaining information often contradicts the mathematical logic of numbers, formulas and schemes. A person needs a digit with an image, and objects need a digital prototype. Mechanical and electronic devices are controlled by algorithms of clear mathematical logic. And the human brain controls actions, guided by a combination of fuzzy and genetic logic. As a result, a multi-component field of knowledge, thoughts and opportunities (including learning) is formed. Such fields have only some properties of mathematical fields, namely commutativity and multiplicativity. The results of the activity of aggregate logic can be written in matrices, over which addition and multiplication operations can be performed, but not multiplication by a number. Various matrix multiplications: maximin, minimax, maximultiplicate, minimultiplicate, midimultiplicate, as well as products with block and variable multiplicativity are processed by the human brain [1–4].

This is probably why the topic of "Matrices and actions on them" so often evokes positive emotions during training. Arrays of data written on the board in the form of matrices are primitive analogues of the aggregate fields of information in the head. How can you not get lost in the numbers while learning? How do I keep my interest in abstract digital prototypes alive?

2 Materials and Methods

This article presents the results of an analytical study of the possibility of using system-based educational blocks in the technology of teaching in higher education. The essence of the method is to conduct disciplines at the intersection of numbers and creativity, numbers and the history of the Fatherland, numbers and physical culture, numbers and patriotic education of a citizen of society, numbers and narrow-profile technical specialization, and so on. The quintessence of such a conglomeration should be a digit with a software implementation [5–9].

3 Main Part

Matrix thinking can be implemented in different ways in different groups of students.

1. Military accounting specialties

With students studying in military accounting specialties, in the field of matrices, you can study material with a military-historical and patriotic orientation.

Let's consider an example of studying military-historical dates in the process of studying matrices:

$$\begin{pmatrix} 1\,2\,4\,0 \\ 1\,2\,4\,2 \\ 1\,7\,7\,0 \\ 1\,8\,1\,2 \end{pmatrix}.$$

The first line of this matrix contains the year of the Battle of the Neva, the second line-the year of the Ice Battle, the third line—the Battle of Chesma, and the fourth line-the Battle of Borodino.

Thus, it is possible to create a system-educational block of historical-military-mathematical orientation. This type of training is aimed at increasing the level of attention, efficiency and evokes positive emotions in students.

2. Technical specialties

Let's consider a system-educational block that implements and consolidates knowledge in the aggregate in the disciplines "Algebra", "History of the Fatherland", "Organization of operation of means of mechanization and automation of lifting and transport, construction and road works".

Example 1 Buses PAZ 320435-04 Vector NEXT are delivered in RTC 1, RTC 2, RTC 3, RTC 4 in the amount of 21, 09, 13, 80 units, respectively, where the numbers written in this sequence are the date of victory in the Battle of Kulikovo of the Russian troops under the command of Prince D. Donskoy over the Mongol-Tatar regiments.

LiAZ 529265 buses are delivered to RTC 1-4 in the amount of 24, 12, 17, 90 units, respectively, where the numbers written in this sequence are the date of the capture of the Turkish fortress of Izmail by the Russian troops under the command of A. V. Suvorov.

NEFAZ 5299-30-52 buses are delivered to RTC 1-4 in the amount of 02, 02, 19, 43 units, respectively, where the numbers written in this sequence are the date of the victory of the Soviet troops in the Battle of Stalingrad.

If we remove the names of the rows and columns from the table, and enclose the numeric values in round or square brackets, we get an example of what is called a matrix in mathematics:

$$A = \begin{pmatrix} 21 & 09 & 13 & 80 \\ 24 & 12 & 17 & 90 \\ 02 & 02 & 19 & 43 \end{pmatrix}.$$

For example, as a result of insufficient funding, the equipment of the new equipment of RTC 1 and RTC 2 is postponed to the next year, and the provision of ground vehicles for RTC 3 and RTC 4 remains unchanged.

This will cause the first and second columns to disappear. We will get another example of the matrix, where the above days of military glory will be associated with the years of these events:

Table 1 Control of the distribution of vehicle supplies

Name of the vehicle	Distribution by road transport companies (units)			
	RTC 1	RTC 2	RTC 3	RTC 4
PAZ 320435-04 Vector NEXT	21	09	13	80
LiAZ 529265	24	12	17	90
NEFAZ 5299-30-52	02	02	19	43

$$B = \begin{pmatrix} 13\ 80 \\ 17\ 90 \\ 19\ 43 \end{pmatrix}.$$

Example 2 Let Table 1 reflect the distribution of new special equipment at municipal enterprises this year:

$$\mathrm{A} = \begin{pmatrix} 21 & 09 & 13 & 80 \\ 24 & 12 & 17 & 90 \\ 02 & 02 & 19 & 43 \end{pmatrix}.$$

The rows of this matrix also display the days of military glory of Russia.

Let the distribution of the same types of new vehicles for the same utilities in accordance with the plan, next year is given by the matrix:

$$\mathrm{B} = \begin{pmatrix} 21 & 09 & 13 & 80 \\ 09 & 05 & 19 & 45 \\ 07 & 07 & 17 & 70 \end{pmatrix}.$$

In order to prepare documents for the purchase of new vehicles, it is necessary to determine the total number of new equipment by position, in this and next years:

$$\mathrm{A} + B = \begin{pmatrix} 21 & 09 & 13 & 80 \\ 24 & 12 & 17 & 90 \\ 02 & 02 & 19 & 43 \end{pmatrix} + \begin{pmatrix} 21 & 09 & 13 & 80 \\ 09 & 05 & 19 & 45 \\ 07 & 07 & 17 & 70 \end{pmatrix} = \begin{pmatrix} 30\ 17\ 30\ 94 \\ 11\ 07\ 38\ 88 \\ 08\ 19\ 35\ 123 \end{pmatrix}.$$

Example 3 Let $A = \begin{pmatrix} 05 & 12 & 19 & 41 \end{pmatrix}$-vector—a string that defines the number of FD 18 forklifts purchased at four wholesale-retail bases. The elements of the matrix-row display the day of the beginning of the Soviet counteroffensive in the battle of Moscow.

Let $B = \begin{pmatrix} 0.3 \\ 1 \\ 1.5 \\ 3 \end{pmatrix}$-vector—a column that determines the cost of special equipment, million rubles. It is required to determine the total cost of special equipment.

Solution

$$A \cdot B = \begin{pmatrix} 5 & 12 & 19 & 41 \end{pmatrix} \cdot \begin{pmatrix} 0.3 \\ 1 \\ 1.5 \\ 3 \end{pmatrix} = (5 \cdot 0.3 + 12 \cdot 1 + 19 \cdot 1.5 + 41 \cdot 3) = (165).$$

4 Research Results

The results of the implementation of this method are given on the examples of specialties of creative orientation and software-applied orientation.

1. Creative specialties

With students who have a creative orientation, you can use matrices to study the dates of the lives of great people of art, consider various statistics on the life of interesting people, and study the main milestones and results of their work.

During the implementation of project activities at the Don State Technical University, the following results were obtained. The works of the poets of the Silver Age were taken for consideration and their work on rhyming and theme was analyzed. The works of young writers of the twenty-first century were also considered.

The results of the analysis of works of poets of different time periods are presented according to the creativity of each in the form of diagrams and then in the aggregate in the form of a matrix.

For the study, 50 poems of the authoritative literary figure of the first third of the nineteenth century, A. S. Pushkin, whose genius has not yet been surpassed, were taken. As a result, it was concluded that A. S. Pushkin used a cross rhyme in 56% of cases, a ring rhyme in 24% of cases, and a pair rhyme in 20% of cases.

The results of the analysis of 50 poetic works by A. S. Pushkin are presented in Fig. 1.

Next, 35 works of the authoritative poet and playwright of the Silver Age, Yu. M. Lermontov, were taken for consideration. As a result, it was found that M. Yu. Lermontov mainly used a cross rhyme—60%, a pair rhyme—34%, a ring rhyme—6%.

The results are shown in the diagram (see Fig. 2):

Further, 31 works of the young Russian poet of the early twenty-first century, D. V. Impersky, were considered, as a result it was found that the poet places the greatest emphasis on cross-rhyme—90%. He uses the ring rhyme in 7% of cases, the pair rhyme—3%. The results of the research are shown in the diagram (see Fig. 3).

40 works of the young Russian poet of the beginning of the twenty-first century O. I. Boronenko were also considered. As a result, the following conclusions were obtained: cross-rhyme is involved in 75% of cases, pair-in 15%, ring-in 10%. The results of the research shown in the diagram (see Fig. 4).

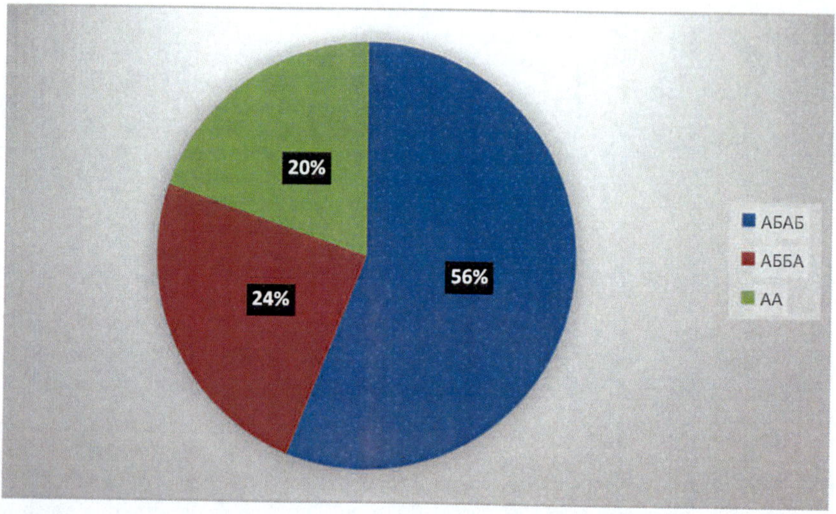

Fig. 1 Percent ratio of rhyming types in A. S. Pushkin's poems

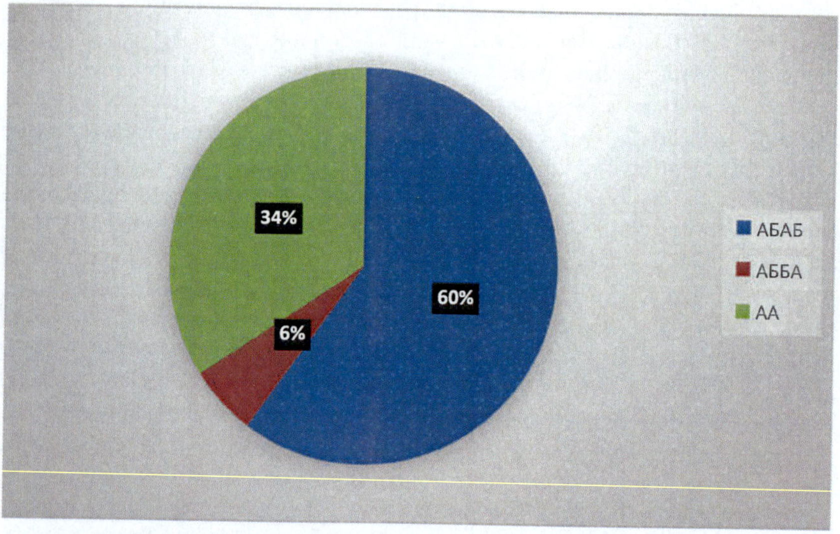

Fig. 2 Percentage ratio of rhyming types in the poems of Yu. M. Lermontov

It is convenient to write all the conclusions for further analysis in the form of a matrix:

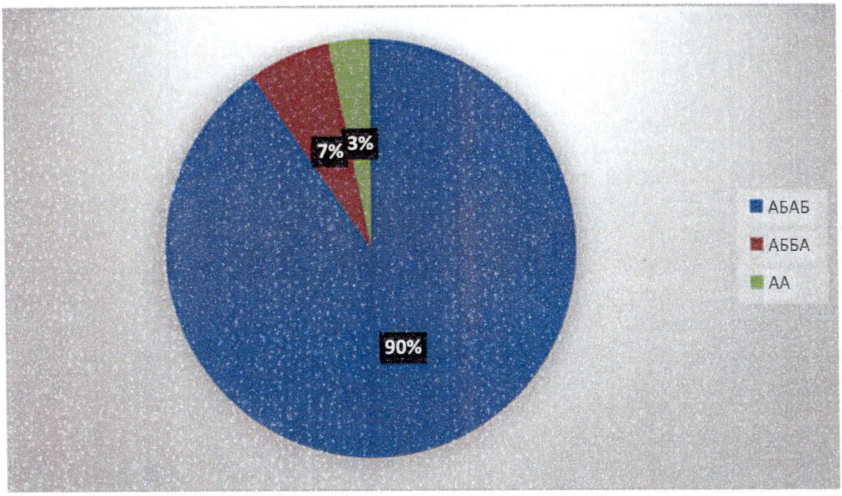

Fig. 3 Percentage ratio of rhyming types in D. V. Impersky's poems

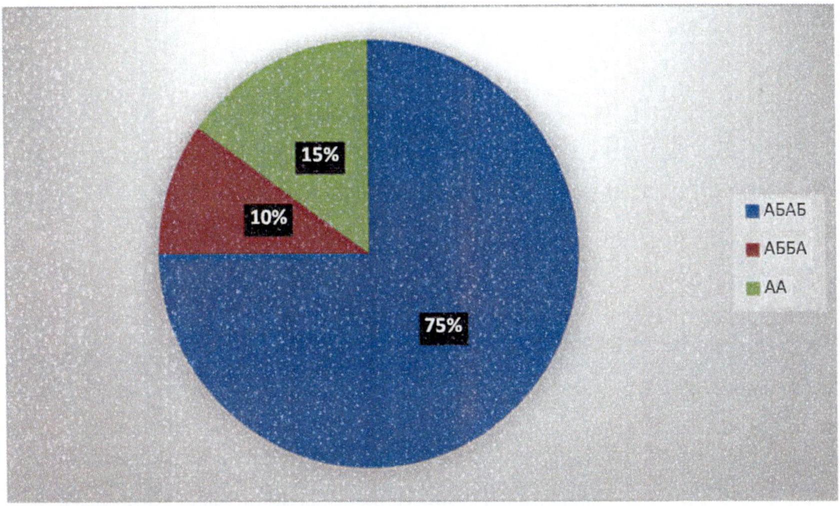

Fig. 4 Percentage ratio of rhyming types in O. I. Boronenko's poems

$$A = \begin{pmatrix} 56 & 60 & 90 & 75 \\ 20 & 34 & 3 & 15 \\ 24 & 6 & 7 & 10 \end{pmatrix},$$

where the rows display the types of rhyming: cross, pair, ring, respectively; the columns correspond to the names of great poets: A. S. Pushkin, Yu. M. Lermontov, D. V. Impersky, O. I. Boronenko.

2. Domain-specific programming

You can directly digitize knowledge matrices and capabilities with students of applied programmers. The set-theoretic operations studied in the course of the course can be visualized, gamified, giving the study of the topic a certain intrigue and understatement, and then provide an opportunity to encode the studied algorithms. Students of applied programming were asked to come up with various tasks for the geometric representation of sets. As a result, the students developed a training manual on set-theoretic operations. Here are some examples from this collection.

Example 1 A puzzle problem. After performing operations on sets, you need to answer the question: "Where does our programmer live?"

$$\begin{aligned}
A &= \{(x, y)| -8 \le x \le 8,\ \ 0 \le y \le 8\},\\
B &= \{(x, y)|y \ge 8\},\\
C &= \{(x, y)|y \le -|0.8 * x| + 16\},\\
D &= \left\{(x, y)|x^2 + (y-12)^2 \le 3\right\},\\
E &= \{(x, y)| -2 \le x \le 2,\ \ 3 \le y \le 7\},\\
F &= \{(x, y)| -6 \le x \le -4,\ \ 8 \le y \le 14\},\\
G &= \{(x, y)|4 \le x \le 7,\ \ 0 \le y \le 5\},\\
H &= \{(x, y)|0 \le y \le 14,\ \ -0.3 \le x \le 0.3\},\\
I &= \{(x, y)| -2 \le x \le 2,\ \ 4.7 \le y \le 5.3\},\\
J &= \{(x, y)| -2 \le x \le 2,\ \ 11.7 \le y \le 12.3\},\\
K &= \{(x, y)|4.1 \le x \le 6.9,\ \ 0 \le y \le 4.9\},\\
Q &= (A \backslash E)\mathrm{G} \cup B \cup C \backslash D \cup F \cup H \cup I \cup J \cup K.
\end{aligned}$$

Example 2 The task—a joke.

After performing operations on the sets, you need to finish the phrase: "Nothing so invigorates a cybersecurity specialist in the morning as…"

$$\begin{aligned}
&\mathrm{A} = \left\{(x, y, z)|\frac{x^2}{4} + \frac{y^2}{4} = 2z\right\},\\
&B = \{(x, y, z)|0.5 \le z \le 3\},\\
&C = \left\{(x, y, z)|x^2 + (y-3)^2 + (z-1.25)^2 + 1 - 0.75^2 - 4\left(z^2 + (y-3)^2\right) = 0\right\},\\
&Q = (A \cap B) \cup (C \backslash A).
\end{aligned}$$

When students created methodological guidelines on set theory, difficulties arose in displaying the drawings given by the formulas of analytical geometry. As a result, a software tool was developed that displays the shaded areas as a set of mappings of various inequalities of analytical geometry by applying Boolean operations on logical expressions to them.

The main application window is divided into two zones, in the first one the user can enter analytical inequalities describing the areas bounded by second-order curves. The program displays sets defined by inequalities. Inequalities can be displayed as a list and then the intersection, union, and difference operations can be applied to them. Since each operation on sets corresponds to a certain logical connection, then applying Boolean algebra, we get the corresponding mapping. By combining different functions, you can create absolutely any drawing (see Fig. 5).

The data processing system allows you to define display zones. The algorithm remembers the inequalities and creates a separate set in memory for further combinations (see Fig. 6).

The function outputs an array of points on the plane. The created algorithm finds points that are repeated in several objects and, depending on the specified logical operations, performs actions on these formulas (see Fig. 7).

The program is designed in such a way that the user has the opportunity to save the lists of functions entered by him with the possibility of further copying and editing them. In the input field, you can adjust the size of the drawing. The image display field contains a grid with a Cartesian coordinate system, the OX and OY axes which

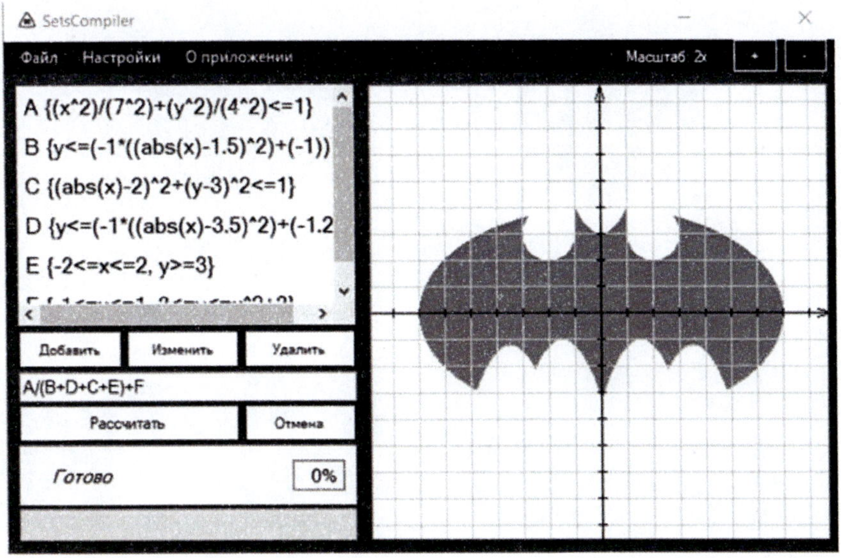

Fig. 5 Input and output windows

Fig. 6 Buffer of sets

A {-4<=x<=4, -4<=y}

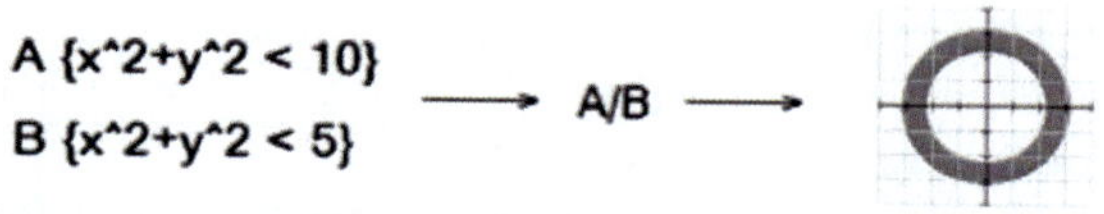

Fig. 7 Visualization of actions on formulas

is convenient for determining the coordinates of points and selecting the coefficients of inequalities. This system allows you to program drawings in any operating system. In the settings, there is a function for controlling the image quality of the drawing, which allows you to study the result in detail for the changes of interest (see Fig. 8).

The program allows you to save the drawing in various modes. The user can control the display of the axes and the grid (Fig. 9). The image can be saved in jpg format with further translation to other formats (see Fig. 9).

Computational operations are performed asynchronously with the main program, which allows the user to collapse the application or see the calculation process with a significant number of operations on sets (see Fig. 10).

The program developed by students to teach other students is universal, practical and visual. This software application can be used in the course of training in the course of set theory, as well as used in solving various problems of visualization of operations on sets.

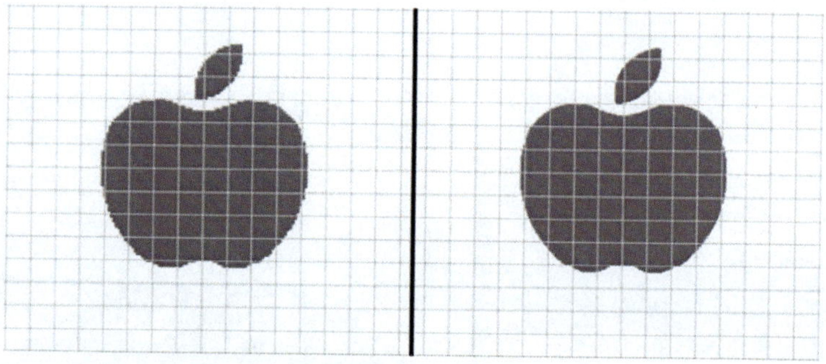

Fig. 8 Setting the image quality

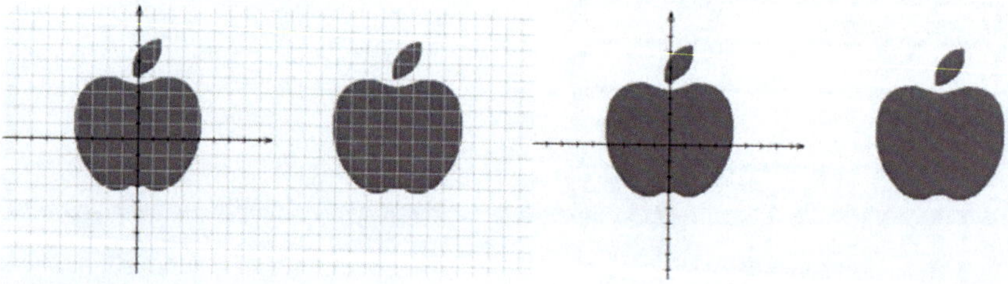

Fig. 9 Image modes

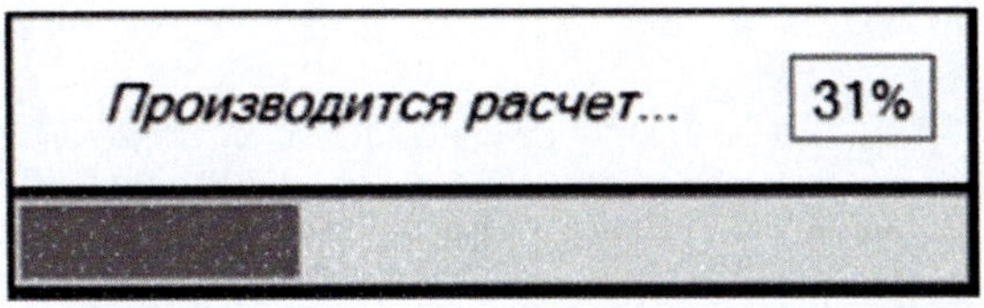

Fig. 10 Calculation field

5 Conclusions

Education in the era of digitalization should be of a multi-component research nature, integrating several disciplines at the same time. A versatile approach to the study of matrix algebra makes it possible for students to understand that matrices are used not only in the development and use of databases, but also are an integral part of the perception of data, as well as software. Almost all information in the computer is processed and stored in matrix form. Human thinking also uses matrix algebra as an element of aggregate logic [10]. The block-based nature of learning, combined with matrix thinking and an analytical approach, provides the basis for building mathematical models and writing program codes. And clear logical schemes, intertwined with the soul and creativity, with an interest in the specialty, can increase the effectiveness of training and digital literacy, as its integral fractal component.

6 Discussion

The matrix as an object of data visualization makes it possible not only to briefly and uniformly present them, but also, having a well-developed theory, allows you to work with large amounts of information rolled up in it. It implements the basic principles of visualization as a quality tool—clarity of data, reduction of visual noise, the use of different sign systems, the possibility of instant pre-reading. This is how it is possible to ensure the fulfillment of the key requirement of modern students to the educational process—reducing the time of perception and working with new data in disciplines representing different areas of knowledge. The quality of modern education is reflected in two aspects—as a process and as a result. The use of the matrix as an abstract visualization tool in the educational process will provide both an increase in its quality in the perception of new information, and an increase in the quality of the result—the acquisition of competencies in the three levels of "know-know-own" [11–16].

The matrix as an object that visualizes structured information on several objects at once at a minimum of space and has the ability to perform operations with it and the like, is one of the preferred forms of knowledge transfer in the educational process, providing an increase in the quality of education as a whole.

References

1. Blaisdell AP (2017) Cognitive dimension of operant learning/learning and memory: a comprehensive reference, 2nd edn, pp 85–110
2. Tsuchiya N, Andrillon T, Haun A (2020) A reply to "the unfolding argument": beyond functionalism behaviorism and towards a science of causal structure theories of consciousness. Conscious Cogn 79:102877. https://doi.org/10.1016/j.concog.2020.102877
3. Manolescu M (2013) School competence between behaviourism and cognitivism or the cognitive approach to schooling. Procedia Soc Behav Sci 76:912–916
4. Wölfling KJ, Müller KW, Beutel ME (2012) AS32-02—treating internet addiction: first results on efficacy of a standardized cognitive-behavioral therapeutic approach. Eur Psychiatry 27:1
5. Kesim M, Altınpulluk H (2015) A theoretical analysis of MOOCs types from a perspective of learning theories. Procedia Soc Behav Sci 186:15–19
6. Wu W-H, Chiou W-B, Kao H-Y et al (2012) Re-exploring game-assisted learning research: the perspective of learning theoretical bases. Comput Educ 59(4):1153–1161
7. Fatemi Aqda M, Hamidi F, Ghorbandordinejad F (2011) The impact of constructivist and cognitive distance instructional design on the learner's creativity. Procedia Comput Sci 3:260–326. https://doi.org/10.1016/j.procs.2010.12.044
8. Iqbal HM (2015) Cognitive development, educational theories of international encyclopedia of the social & behavioral sciences, 2nd edn, pp 51–57
9. Kurt S (2011) Use of constructivist approach in architectural education. Procedia Soc Behav Sci 15:3980–3988. https://doi.org/10.1016/j.sbspro.2011.04.402
10. Pinevich E, Safaryan O (2020) Methods for improving the educational potential of students. In: E3S Web of conferences, vol 210, 18034
11. (2020) Learning of students? Internet High Educ 45:165–194. https://doi.org/10.1016/j.sbspro.2011.04.402
12. Duane BT, Satre ME (2014) Utilizing constructivism learning theory in collaborative testing as a creative strategy to promote essential nursing skills. Nurse Educ Today 34(1):31–34
13. Loureiro A, Bettencourt T (2014) The use of virtual environments as an extended classroom—a case study with adult learners in tertiary education. Procedia Technol 13:97–106. https://doi.org/10.1016/j.protcy.2014.02.013
14. Gallagher JR (2020) The ethics of writing for algorithmic audiences. Future Gener Comput Syst
15. Klement M, Chráska M, Klementová S (2015) Explanation of instruments and procedures used by the recipients of education in order to build their own learning network, based on the generic model cognitive process. Procedia Soc Behav Sci 12(174):1614–1622
16. Gueya C-C, Chengb Y-Y, Shibatac S (2010) A triarchal instruction model: integration of principles from behaviorism, cognitivism, and humanism. Procedia Soc Behav Sci 9:105–118. https://doi.org/10.1016/j.sbspro.2010.12.122

Formation of Visual Thinking of Students in the Information Environment

Lyubov Savenkova

Abstract The text reveals the research directions of development problems of visual thinking and visual literacy of students in the process of mastering the information space through visual activity, conducted directly by the author of this article. An attempt was made to select the pedagogical conditions of the organization of the educational process and the basic competencies of students, allowing the teacher to find effective educational forms and directions of work adequate to the mental age characteristics of children in the process of mastering different types and directions of artistic and creative activity within the framework of polyart education. Their implementation in practice will allow students to get the necessary level of knowledge, ideas and skills, corresponding to the education of a competent and thinking viewer who can read and understand the graphic "text" and visual images created by artists, directors, screenwriters, etc. The author highlights the important components of visual literacy, depending on the age and interests of students; clarifies the concepts of "visual thinking" and "visual literacy" from the perspective of a variety of artistic activities in the learning process; the author defines and formulates competencies aimed at the formation of visual literacy through the development of the visual art alphabet in the context of integrated learning; identifies the pedagogical conditions for the introduction of the presented model into the training system; reveals the dynamics of the visual literacy development based on the development directions of students based on the mental characteristics of the children's age.

Keywords Visual thinking · Visual literacy · Polyart education · Integrated learning · Artistic creativity · Competencies · Development directions

1 Introduction

In the modern information world, the formation problem of the visual thinking and visual literacy is becoming more and more relevant and significant. Although history

L. Savenkova (✉)
Institute of Art Education and Cultural Studies of the Russian Academy of Education, Pogodinskaj, 8/1, 119121 Moscow, Russia

A. Beskopylny and M. Shamtsyan (eds.), *XIV International Scientific Conference "INTERAGROMASH 2021"*, Lecture Notes in Networks and Systems 247,
https://doi.org/10.1007/978-3-030-80946-1_50

shows that the process of researching this direction is not new, but rather a long forgotten old one, which is experiencing a new birth today, like much in science. An example of this is the study of George Nelson (1908–1986), set out in his book "How to See: Visual Adventures in a World God Never Made", where the author points out the following: "education has long been recognized as the basis of civil society, and yet most people remain visually illiterate, unable to decipher information provided by non-verbal means and do not feel the slightest concern about this [1]".

It is difficult to disagree with this, since the problem has not only been insufficiently studied yet, but has now become more in demand, necessary and, in our opinion, can be solved with the help of the "round dance of arts" [2]—polyart education and integrated education. The current situation in the world requires education to quickly solve a number of problems that dictate both the need for schools to turn around 180° and offer other forms of education, as well as other goals and objectives. But we have already passed this in history not so long ago (just a hundred years ago) under the slogan "destroy everything to the ground, but ...", but then, unfortunately, it is very difficult to build without a foundation. Therefore, the basic values of education, such as morality, spirituality, aesthetic and artistic, universal and special, should not slip away, but rather, on the contrary, be revealed with an unprecedented breadth, since only in alliance with innovative technologies they will help create not only new and original, but, most importantly, it is consonant and necessary for a modern person. And in this case, the following opinion should be given about preschool education by Mazunova L. K., Gubaidullina M. I. which should be considered "not as a preparatory, but as a starting stage that determines the effectiveness of all subsequent ones. This is due to the fact that every physically and mentally healthy child is born with a great potential for performing and creative abilities, but only those of them are developed for which effective external conditions are created. And the timing of the start of development and the stimulating conditions are crucial here" [3].

We also agree with Kudryavtseva E. L., referring to world studies, claims that 65% of children who went to first grade today will work in completely new positions that do not yet exist today. She connects this with the direction that is increasingly evident in society and education, it is about career growth, which is formed in children when mastering Soft skills [4–6]. But in our opinion, in this paradigm, the most important thing is missing: a person who cannot remain in this situation outside the emotional and sensory sphere, so as not to turn into a robot. Only art in its close connection with technology will allow you to form a harmonious person. As Nelson G. writes, referring to Joshua Taylor, who stated that "To see is to think. To think means to put together and put in order random bits of individual experience. Vision is not a unique divine gift, but an academic discipline. This suggests that it can be learned [1]". According to him, in this sense, art is very close to science, because it is also looking for ways to understand the world. The difference between the two lies not so much in ends as in means. The scientific method is based on precise measurements and searches for laws that unite the phenomena of the material world. Art relies more on an intuitive-visual approach. But even these differences are not manifested in all cases: many outstanding natural scientists and mathematicians widely used intuition in their work, and a convincing intellectual basis is visible in the works of artists [1].

Today, school teachers try more and more to use various technical means, information and communication technologies, including video presentations, documentaries and less and less fiction, fine arts, music, excerpts from feature films, and this is what allows the most intelligibly and quickly, relying on emotions and intuition, on a person's ability to simultaneously see, hear and act to achieve the desired result. Based on the foregoing, the study made an attempt to isolate pedagogical conditions and basic competencies, as well as didactic techniques and technologies for organizing the process of perception and creative thinking, allowing a teacher at school, a teacher in a preschool organization to find age-appropriate educational forms and directions of activity with children in conditions additional education, as well as to achieve the necessary level of education in the development of a competent thinking viewer who is able to read and understand not only the text as such, but also the visual images created by artists, writers, film and theater directors. This is where we see the relevance of the proposed research.

2 Methods

To achieve this goal, we needed to: (1) identify the components of visual literacy depending on age; (2) to clarify the concepts of "visual thinking" and "visual literacy" from the standpoint of a polyart approach; (3) to define and formulate basic competencies that make it possible to form visual literacy through mastering the alphabet of fine art and polyart education; (4) highlight the pedagogical conditions for the introduction of this model into the training system; (5) to build the dynamics of the development of visual literacy and visual thinking based on the development line of students, the basis of which is the age-related mental characteristics of students.

The designated areas of research were lined up in the process of introducing into the training system a complex integrated polyart education based on the triad of the perception process of art by Yusov B. P.; three positions of the theory of creativity of Berdyaev N.; psychology of mastering any information of Razhnikov V. G.; theory of developmental education by Davydov V. V.; subject-spatial approach of Savenkova L. G; three spheres of human activity by Solovyov V. S.. The dynamics of training was built taking into account the features of communication and real artistic and creative activities (individual and group), which were based on integrated technologies (author Savenkova L. G.), such as: "classes united by one cross-cutting theme", "lessons-travel", "integrated warm-up"; "I am in changed circumstances", "creative project", "creative tasks"; "work on a hint", etc.

In the practice of mastering the diversity of the manifestation of visual language, there is a model of active interaction between the teacher and students, as well as students among themselves, based on the understanding and creation of real creative products through well-organized design and artistic activities. The latter is especially important, since obtaining pure knowledge and theoretical information about how to do something outside of real creative activity is not enough. In this we agree with the studies of Shaklein et al. [7], Gershunsky [8], which rely on communication-oriented learning, where they highlight the form and structural relationships of open communication between students based on the proposed content; pay attention to the timeliness and relevance of the material being studied; to work in small groups; reliance on the interests of everyone; individualization of the learning process and personal experience. Moreover, Pasov E. I. clarifies that the key concepts of education aimed at self-realization of a maturing person "like homo moralis" are the following: "values and culture"; "spirituality and morality", "creativity and development", "freedom and responsibility", "communication and self-determination". Summarizing the above, relying on Pasov E. I., we can say that "education is the creation of an image: themselves, the world, their actions in the world", that is (here Pasov quotes Zinchenko V. P.)—"the whole space of images". And concludes that from this "all education is ultimately self-education [7]".

3 Research Novelty

An attempt has been made to create a basic practice-oriented model of mastering the basics of visual literacy and visual thinking by students in conditions of immersion in a variety of types, directions and forms of polyart education and integrated learning, the core of which is the formation of a complex of relevant and interdependent competencies (motivational, ideological, social, informational, natural-scientific, creative, communicative, professional, technical), based on the development directions of students, built on the basis of the age-related mental characteristics of children and their innate ability to see, hear, feel, think and act simultaneously.

Here it should be recalled that the main task of any educational organization is to develop the abilities given to them by nature, equally. Without different types of art, it is impossible to fully develop these abilities, including thinking, since nature has made it possible for a person to think not only logically, but also figuratively. For some reason, the latter is almost ignored in the education system, most often.

Another problem is a huge variety of types and directions of artistic activity with children of different ages, but which are most often aimed at mastering the figurative and expressive language of any one type of artistic activity, which, in our opinion, does not correspond to the nature of child development, since the child's consciousness is syncretic (undivided), and therefore the selection of a single art (especially at a young age) inhibits its development. On the other hand, the limitation of information, the reliance in training on the development of knowledge and specific skills and abilities, for example: a picture on a sheet, playing an instrument, actions on stage, movement in a dance, performed out of connection with the environment and in isolation from communication, collective discussions and reflections, does not form motivations for action in the student, does not develop ideas, how and how knowledge and skills are applicable in society, does not create preconditions for research in creativity and self-realization and self-study.

The above is addressed to the concept of personal culture development in the broadest sense of the word, since only with the breadth and diversity of the learning process itself, a creative personality is formed. At present, there are also a huge number of cultural concepts, each of them implies a certain way of thinking, an original turn of thinking. This only means that culture is comprehensive and present everywhere: in communication, in art, in production, in sports, in the organization of lifestyle, in science and education, and so on, but in all of them, according to Kudryavtseva E. L., the component of the concept of culture "is a person who perceives (the recipient) of the past and creates the future, as an individual and at the same time as a component of society, people, social group, etc. [3]". In this case, it is necessary to give the definition of Vereshchagin E. M. and Kostomarov V. G.: "Culture is the product of social activity of human collectives. It has a historical genesis and plays a decisive role in the formation of a separate human personality [9]"—this is directly related to education and upbringing. To this understanding of culture, we should add the definition of Pasov E. I.: "Culture is an individually (personally) mastered spiritual values. And in order for each person to be able and willing to master them individually, it is necessary to humanize the content of education and its technologies. After all, any knowledge has a value, cultural content, when it is appropriated by a person in an activity. Therefore, any education should pursue the goal of developing the individual as a subject of culture [7]". And as a generalization of what has been said, we should cite the definition of Yusov B. P.: "culture is the organization of the elements of being [2]", which allows us to understand the importance of art in life, in the education of a competent consumer and a creator of culture at the same time. It should be concluded that the attitude to fine art should also be understood broadly in this context, since the depth of its development will allow students to form those elements of visual literacy and visual thinking that are outside of painting, graphics, sculpture, architecture, design, applied and folk art—it is impossible to fully disclose this problem. Since each of the listed areas can be represented in the widest and most diverse forms and types of work. Nelson's reasoning pushed us to the decision of the indicated, about "how is it that children who absorb visual information in huge quantities grow up visually illiterate? I would say that this early-born ability in them is successfully knocked out in the learning process. … It is necessary to consider what kind of education children receive. …. A technological society needs people who can read and write, add and subtract [1]". In order to highlight the basic components of the process of gradual and sequential formation of visual literacy and visual thinking, it is necessary to highlight the formulations of these concepts from the standpoint of education in conditions of immersion of students in the diversity of the artistic expressive language of visual creativity, a component of which is the content. The importance of this is justified by the fact, as scientists prove, that almost (90% of the perceived information a person receives with the help of sight). Therefore, consider what we mean by the concepts of "visual literacy" and "visual thinking".

- "Visual literacy" is not only a person's ability to "interpret, communicate and derive meaning from information presented in the form of an image", as indicated in a number of its definitions, but also the ability to understand the language of artistic expression, comprehension of the language of forms, and its correlation with time creation and the skill of interpreting it with their own ideas and experiences.
- "Visual thinking" is a concept that belongs to Arnheim R., as the translation of information "from the language of an image into the language of understanding [10]"—that is, it is one of the types of human thinking based on visual images, the harmonious formation of which is possible only under conditions competently organized real practice-oriented various artistic and creative activities, the elements of which, according to scientists, are: "point, line, tone, color, structure, size, scale, direction, movement".

Continuing what has been said, to confirm the importance of fine art, it is necessary to quote the statement of Spengler [11] that a person knows numena (the essence of things apprehended by the mind), and the impressions he interprets the word, where the word he interprets as sound, as a concept, "an element of scientific prose" and "world of forms of science is fully consistent with the corresponding world of forms, math, religion and fine arts [11]". Hence, one natural conclusion should be drawn that a growing person must be prepared for a competent perception of the surrounding visual world, where elements as parts of a complete visual image, recognizable by the perceived, are the basis for reading visual information, as "look and see". Imagination plays a huge role in this, and as scientists (Vygotsky L. S., Roem D., Yusov B. P., etc.) prove, it is necessary to start forming visual thinking from an early age. So Roem D., the author of one of these methods, believes that visual thinking is a natural ability of a person—"to see mentally", which, in his opinion, has a serious impact on cognitive activity. To do this, it is necessary to depict the "exciting question": "Who/what?", "Where/when?" and "Why/why?" which is a strategy [12]. This is also a learning process, namely, the child begins to learn to select the necessary information and, using his imagination, supplement, change and explain it.

Today, researchers often relate visual literacy to audiovisual and media culture as the basis for comprehending the fundamentals of analyzing the products of the film industry, which is a synthesis of different types of art. But the authors defending such a point of view turn to the elements of the language of cinema art, among which they distinguish: montage, foreshortening, plan composition, frame, etc., the basis of which is the language of the visual arts.

4 Results

To match the aspect of the proposed model of education to the selected task, it is necessary that art is not dosed in the learning system, but from the first years of life it surrounds the child daily and naturally enters his life. And as already mentioned, it is important to develop the viewer's mind, visual memory, emotions and feelings not in isolation of one art from another, but in the interaction of a set of various types of artistic activity, relying primarily on the mental characteristics of the age, interests and preferences of the one who consumes it, and not at the discretion of usefulness and significance from the standpoint of an adult. Only in this case, the child's desire to do everything himself will be natural, and his need to express his thoughts through an artistic form will be justified and understood by his peers, since the graphic language used by the child will be adequate to his thinking. What does this mean? The point is that when choosing the direction of work with a child, first of all, you need to pay attention to what the child sees and hears himself, and not what he should (according to the adult). Understanding the nature of a child's thinking allows you to create a set of stimuli and motivations to stimulate his desire to learn more, to learn more broadly, the need to explore himself. The above is equally important not only for any type of work with children, but also in the field of art knowledge, this is what most accurately reflects the meaning. In this regard it is necessary to isolate the leading lines of the child's development in the dynamics of the most important components of visual thinking: form, color, space, composition and general developmental directions based on the interaction of different types of art, namely polyart education.

The directions (lines) of development of children in the model are allocated for each age for the development of general artistic and expressive means of the fine arts: form, color, space, composition. Below are the most important of them, which have passed many years of experimental testing in a number of regions of Russia directly under the guidance of the author of this article Savenkova L., according to the program "Fine Arts and Environment" for grades 1–11 and fine art textbooks from grades 1–8 and are fully presented in the book "Education in the space of peace and culture: integration in art pedagogy" [13, 14].

- General developmental: Development of ideas about the surrounding world of nature and awareness of oneself in harmony with this world. Interest and responsiveness to the beauty of nature and works of art. A variety of arts in human life, their development based on ideas: form and content, subject and environment, artistic word in a work of literature, sound and music, form and color, mood and color, color-form-space, rhythm and mood, nature–human environment. The artistic and expressive language of various arts. The desire to express oneself in any kind of creativity. Integrity and interdependence of the life history of the people and their art. Figurative and semantic relationships in folk art. Era and art: style-character-time-man. The connection of times in art: art of the present, past, future, cosmic. The movement of life: the accumulation of quality, cyclicality, spiral, symmetry in the development of life, thought, art on Earth. Harmonization of relations between a person and his environment. Dynamics of form, color, space, composition in the history of art. Thought-idea-embodiment.
- Space: Development of sensations, perceptions, impressions and ideas about the immediate and distant environment. Human habitation of the space of the earth.

The idea of the space of the native land, of the dependence of the customs and traditions of the people on the characteristics of the surrounding landscape. Changes in the perception of space in nature at different times of the year. Architecture and surrounding landscape. A sense of the nearest living space. Nature-space-man-environment. The space of the cosmos, its infinity. Cosmic motives in different types of art. Light and color on Earth and in Space. Biostructure, bioarchitecture. "Eternal ideas" and "eternal forms" in space and environment. Era-subject-environment-architecture. Ecology of space. Geometry of space. Artistic image of space.

- Form: The singular around us. Development of the ability of ambiguous perception of surrounding objects. The creation of form in nature, in the history of the Earth. The idea of folk art: beauty, durability, the use of household items. Creating a form based on the works of folk crafts. Architectural forms and natural forms. The human measure of folk and temple architecture. The subject environment as a special space of each person. Form in the decorative arts. The sign and its significance in the history of different peoples. The beauty and expediency of external and internal forms in nature. Time-color-shape-composition in real space and on a plane. Time and form (style-character-space-person).
- Color: Interest in the variety of colors, light, shapes, sounds, objects, gestures, movements, smells, intonations in nature and life. Mood in nature and art. Color and light in the space of the world. Space and color in the painting. Man is the center of space. Color in the architectural space. Color and shape. Color and symbol. Silhouette-shape-color-tone-chiaroscuro-material-space. Color and artistic image. Color-shape-space-composition.
- Composition: Development of the ability to master the limited plane of a sheet of paper of different sizes. Composition on a plane, in volume, in space. The connection of the elements of the composition (musical, subject, decorative, etc.). Composition in art (sounds, words, colors, form, space). Composition in the history of housing architecture (Surroundings). The semantic and stylistic unity of the elements of the composition in any art, in history. Composition as an expression of the consciousness of the creative artist: rhythm, proportions, symmetry, dynamics, balance. Composition in the expression of a specific image in space (Fig. 1).

The learning process is based on a subject-spatial approach (author Savenkova L.) [15], its implementation involves the formation of a child's ideas about: (1) the subject and architectural space in which a person studies, works, lives, rests, (2) the surrounding nature and their role in life. What this space will be for everyone depends only on us. The task is to familiarize schoolchildren with understanding and independent conscious activity to create an artistically organized space, the development of the desire to express oneself in independent creative work, the fostering of the desire for constant self-improvement. The goal is to prepare schoolchildren for independent thinking and creating an environment in complete harmony with their character and interests. Areas of work make it possible to involve in the classroom topics that are

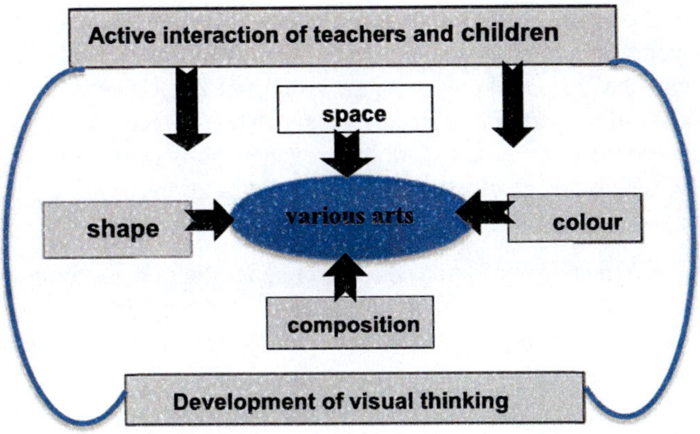

Fig. 1 Development of visual thinking

studied in parallel at the lessons of history, literature, geography, natural history, and also rely on the diversity of the surrounding visually rich information environment.

To solve these problems, the content of the model identified seven main areas of student development, varying in importance at different age levels: visual-external observations and the development of differentiated vision; transfer of the observed into an artistic form (drawing, construction, music, literary presentation); own creativity is an independent product of a creative product; social aspect, social significance of the studied material; visual and external differences of concepts: natural environment, object and artificial environment; "Child and art" (the ability to create "your own" artistic image of an object through your own "I", relying on the variety of activities and materials mastered by children); material, technique, tools.

5 Discussion

Putting this into practice is possible only in the process of an integrated approach to learning, the pedagogical conditions of which are: going beyond school education and the subject under study; reliance in education on the age-related interests and preferences of children; shifting the emphasis on the independence and own creativity of each child; collective cooperation and co-creation of students; interaction of teachers in the team; game form of organizing the educational process; individual and collective design and research activities based on the personal motivation of students, self-organization and self-assessment of their activities; variety: integrated areas of work, technologies, materials and tools.

The foregoing cannot be fully realized without highlighting those competencies that will help the teacher to navigate himself in the material mastered with children and independently edit his activities, introducing non-standard forms and methods of communication with children and joint creativity. To do this, you should not close yourself in your shell and it is not enough to rely only on special skills and abilities associated with the process of mastering the visual arts, but it is necessary to be based on the diversity of modern life. This also means that it is necessary to highlight the competencies, skills and areas of work that are characteristic of the coming twenty-first century, which are much broader than reality and are based on a variety of external and internal relationships, both in art and in life.

- Motivational competencies—the desire to organize their own activities; find arguments for gaining interest from the work begun, the desire to finish on their own, to complete the work begun; to form a willingness to offer different options for work on the same topic.
- Worldview competencies—to understand the importance and significance of the work performed for oneself; to be aware of and highlight important areas of work and topics of creative activity; to understand why and why it is necessary to master artistic and creative activity.
- Information competence—to know and be able to work with different texts (literary, visual); to be able to find the needed information and work with it on their own; to be able to tailor the information you need to work direction and to transmit it to another; to have an idea about the specifics of work in this area, including the phasing and sequence work (creative design).
- Natural science competencies—to have an idea of research activities and their importance in life; to know and be able to put into practice the knowledge gained; the need to develop the ability to conduct research activities: to observe and record the learned, open; the desire to generalize and present the results of the research to the audience;
- Creative competence—the desire and ability to create independently creative works in different types of visual activity; to be able to convey the content of one art in the language of another; to create compositions on the plane, in volume, in the subject space to convey the concept (content, mood, state). To understand that the creative works of authorship must not repeat the well-known earlier works of other authors (artists, sculptors, architects) and to be original.
- Social competencies—to know and apply in practice the ability to interact with different people (students and adults); to feel the mood of another person; to find the opportunity to work together in a variety of circumstances, regardless of personal likes and interests;
- Communication skills and leadership skills—to be able to work in a group and in a couple; the ability to perform collective work; to distribute responsibilities in a group taking into account the interests of each; to be able to take the initiative; to help a friend when working on a joint project; to be able to listen and hear the other.
- Professional competence—know and be able to work in various types and techniques of fine art; know the types of artistic and expressive means of different skills (general and special); to own techniques work for different art materials

(drawing, painting, sculpture, decorative and applied arts; architecture); understand and be able to work with form, space, color, composition on a plane and in space.

- Self-improvement competencies—independently find ways of personal growth (intellectual, physical, moral); form a set of cultural skills (listening to music, perceiving pictures, watching movies and theater performances) necessary for the profession; strive to learn more about culture; understand what knowledge, skills and abilities are necessary for self-development.

The highlighted competencies are formed in students gradually, which requires patience and a benevolent attitude of the elders to the younger ones. As basic incentives for this work with students to instill a desire collectively (in a small group) to reason, reflect, make their judgments and the ability to substantiate their position, their views on the work performed, since this forms the main thing in a person—visual literacy and observation, the ability to "look and see", which allows in the future to develop and improve visual thinking, the ability to capture nuances and improve intuition are the most important mental abilities of a person, without which it is impossible to create and achieve success in art. The language of images in the visual arts is based on form, color and light, on line and rhythms, on contrast and nuance; on movement and statics; on texture and structure; design and space, as Roehm D. points out that the most "exciting questions" in this are: «Who/What?», «Where/When?» and «Wherefore/Why?» which is understood as a strategy.

Acknowledgements The article was prepared within the framework of the project of fundamental scientific research "The formation of cultural and cultural-creative competencies among students and teachers as the basis for the formation of a culture of social relations".

References

1. Nelson G (2020) How to see: visual adventures in a world god never made. Mann, Ivanov and Ferber, pp 18–20
2. Yusov BP (2004) The relationship of cultural factors in the formation of modern artistic thinking of the teacher of the educational field "Art". Sputnik, Moscow, pp 16–20
3. Mazunova LK, Gubaidullin MI (2018) Ontology of children's giftedness: monograph. RIC Bash GU, Ufa, p 27
4. Kudryavtseva EL, Mazunova LK, Pulyaevskaya AM (2018) Game and gaming technologies as a mandatory tool and content of lifelong education. Rep Bashkir State Univ 3(10):152–156
5. Kudryavtseva EL (2014) Road card bilingual: a modern tool for targeted psychological and medical-pedagogical support of bilinguals. In: Zakharov GP (ed) Bilingual (polylingual) education and intercultural communication in the XXI century: a collection of scientific articles. Chuvash State Pedagogical University. Chuvash State Pedagogical, Cheboksary, pp 150–153
6. Walther E, Preckel F, Mecklenbräuker S (2010) Befragung von Kindern und Jugendlichen: Grundlagen, Methoden und Anwendungsfehler. Hogrefe Verlag, Göttingen, p 236
7. Shaklein VM, Pasov EI, Kuzovleva NE (2011) Fundamentals of communicative technology of foreign language education. Bulletin of the Peoples' Friendship University of Russia. Series: Russian and Foreign Languages and Methods of Teaching Them 2:108–109

8. Gershunsky BS (2002) Philosophy of education for the 21st century. Pedagogical Society of Russia, Moscow
9. Vereshchagin EM, Kostomarov VG (2005) Language and culture. Indrik, Moscow. Kostomarov VG (2015) Pamphlets about the language. Native, acquired and Russian in Eurasia. Flinta, Moscow
10. Arnheim R (2012) Art and visual perception. Architecture-S, Moscow
11. Spengler O (2017) Sunset of the Western World. Complete collection in one volume. ALPHA-KNIGA, Moscow, pp 328–428
12. Roehm D (2011) Visual thinking. How to "sell" your ideas with drawings
13. Savenkova LG (2014) Education of a person in the space of peace and culture: integration in art pedagogy. MAGMU-RANEPA, Moscow, pp 57–70
14. Savenkova LG (2019) International conference "Scientific research of the SCO countries: synergy and integration". Part 4: Participants' reports in English, 11–12 Feb 2019, Beijing, PRC, pp 92–101. ISBN 978-5-905695-93-3
15. Fomina N, Savenkova L (2017) Object-spatial approach to studying fine art: development of three-dimensional and spatial perception among students. Rupkatha J Interdisc Stud Humanit IX(2):282–292. https://doi.org/10.21659/rupkatha.v9n2.29. Full Text: http://rupkatha.com/V9/n2/v9n229.pdf

The Features of the Personal Value-Semantic Sphere of Internet Users with a Deviant Type of Internet Socialization

Angelika Luchinkina, Liliya Zhykhareva, Tatiana Yudeeva, Ruslan Zekeriaev, and Evelina Rasina

Abstract Virtual space leads to the formation of completely new values and meanings, which a person embeds in his personal-semantic sphere by selecting functional (tested in practice) semantic constructs. Recognizing the importance of many modern studies, it should be noted that in them the value-semantic sphere was not a complex object of research, but its individual components were touched upon. Despite the otherness of the Internet in relation to the real environment, the features of the system of values and meanings of the individual, in the context of its functioning in the Internet space, were practically not considered. The purpose of the study: to identify the psychological features of the value-semantic sphere of the personality of Internet users. At the stage of the ascertaining experiment, we used: focused interviews with Internet users; a questionnaire by A. I. Luchinkina aimed at identifying the level of Internet activity of the individual; the research questionnaire "Value-semantic sphere of the individual in virtual space" (R. I. Zekeriaev); the questionnaire "Personality in virtual space" (A. I. Luchinkina). As a result of the study, it was found that the value-semantic sphere of the Internet user in the virtual space is represented by specific values-beliefs and values-actions. Internet users with deviant (asocial) behavior in the Internet space are characterized by the following values and beliefs. Also, differences in the ways of implementing values-beliefs were found in this category of users, not only from the standard respondents, but also among themselves.

Keywords Virtual reality · Internet space · The personal value-semantic sphere · The personal virtual image · Values-beliefs · Values-actions

A. Luchinkina (✉) · L. Zhykhareva · T. Yudeeva · R. Zekeriaev · E. Rasina
State Budget Educational Institution of Higher Education of the Republic of Crimea «Crimean Engineering and Pedagogical University the named of Fevzi Yakubov», Uchebnyy Ln., 8, 295015 Simferopol, Republic of Crimea
e-mail: aluch@ya.ru

A. Beskopylny and M. Shamtsyan (eds.), *XIV International Scientific Conference "INTERAGROMASH 2021"*, Lecture Notes in Networks and Systems 247,
https://doi.org/10.1007/978-3-030-80946-1_51

1 Introduction

At the present stage of social development, an increasing part of human activity is transferred to the Internet space. Virtual reality leads to a change in interpersonal interaction, the formation of new ways of communication and behavior, and contributes to the active construction of new personal meanings and values. At the same time, network communities are often focused on the preferences of Internet users and produce content aimed at promoting values, norms and attitudes, which may differ from generally accepted and be destructive in its nature.

The virtual space makes it easier to realize deviant aspirations without the danger of being held accountable (including criminal responsibility). The study of the factors that determine the manifestation of deviant forms of behavior on the Internet environment is significant. The dynamics of the social interaction nature development, the spread of new communication forms social networks; all kinds' online services require a new understanding of the behavior norms and understanding of deviations from the norm.

The important role in the integral image formation and the perception of other subjects' images is played by the personal value-semantic sphere. During the perception of a virtual image, there is a process of objectification the psychological qualities of a person, an internal (psychological) connection with a sign-symbolic one [1]. The important place in this process are occupied by the individual values and meanings, which participate in the social categorization of the surrounding world. Leontiev D. A. defines the classification and assessment tools by the subject of objects significant characteristics and reality phenomena as sense-forming sources for the personal value-semantic environment. These tools determine the process of perception and construction of the personal virtual image [2].

The formation of meanings in such processes occurs with the help of parameters by identifying categories that determine its importance value and its semantic characteristics. These parameters are the sense-forming units of the personal value-semantic sphere. At the same time, not only the meaning of the virtual image is assessed, but also the main features, which make it possible to divide the perceived images into categories and assign a certain meaning to them, are identified. Of great importance in the processes of perceiving the personal virtual image of the interaction partner is the fact that each individual subject of interaction forms different sets of attributes (parameters) for the semantic categorization of images, which have a decisive influence on the perception of the same image by different people [2].

The connection between the personal meaning of objects and their images, which are endowed with specific meanings and the assessment of these objects according to individual, differentiated categories, received the definition of a semantic construct. The semantic construct of a personality is a set of categories or a categorical scale, which is expressed in the distribution of characteristics according to their significance for the subject. At the same time, with the help of semantic constructs, person not only categorizes and organizes the images around him, but also evaluates them by attributing a certain meaning. The semantic construct allows to dividing and combining virtual images into classes based on personal and semantic characteristics. In that way, the affective generalization takes place, in which the test subjects' emotional attitude (biased) is important. In addition, semantic constructs are highly generalized [2].

At the same time, values and meanings affect not only the virtual image perception, they have a significant impact on the process itself and the result of the image construction, being, in fact, a reflection of the personality-semantic sphere. The formation of a virtual image includes factors such as motives, attitudes, views (ideas), personal characteristics, as well as values and meanings. It is often quite difficult to separate its meaning from the image. The personality-constructed image in the virtual space is the carrier of information about the user. Usually only its meaning-forming function is recognized. The idea formation of a particular person occurs through social perception. The above factors affect of the virtual image construction and can become decisive for the disclosure of its meaning.

The virtual space, due to its specific properties, leads to the formation completely new values and meanings, which a person builds into his personal-semantic sphere by selecting functional (tested in practice) semantic constructs.

The emergence of new values and meanings on the Internet is related with the characteristics of the space and the virtual personality properties, which construct the image, based on available means, tools and guided by its motives, needs, attitudes, etc. Among the properties of the virtual personality, there is the possession of priorities that is inaccessible to a real person. It goes about the desire for replication (living several lives) [3].

Motivational sphere changes of interpersonal interaction in the virtual space also affect motives and values formation and their variation, which are not characteristic of the real world or their variations. A. I. Luchinkina identified the following motives of users: the motive of the designated presence, the motive of the contribution, the motive of personal space, the motive of incarnation into a role, the motive of replication, which largely affect the value-semantic sphere of the personality and semantic constructs involved in the construction of the personal virtual image [4].

The motivational component of the virtual personality image can also be related with the users' own expectations, and illusions. In the book devoted to interaction of people and modern technology, Sherry Turkle highlights vulnerability and fear of intimacy as one of the reasons for using social networks. According to the scientist, the virtual space creates the impression of a firm confidence in control over the communication situation. In addition, a professor at Massachusetts Institute of Technology, MIT in her work identifies a number of illusions that are formed among users of Social networks in the course of building interaction with them: the illusion of control, the illusion of care and the illusion of loneliness lack [5].

These illusions create the need for presence in the information space and the need for self-presentation in it, which occurs with the help of an image created by the user and affect the consolidation of semantic constructs and the personal meaning of the subject, which is formed by the presentation of objects, phenomena and people in their images. Thus, personal meaning is a form of the attitude and assessment reflection of the surrounding world in images, the basis of which is the personal values and meanings. One of the most important functions of the personal value-semantic sphere is the semantic regulation of life activity. This personal meanings function is realized through two different psychological mechanisms—the mechanism of emotional indication and the mechanism of image transformation. Both of these mechanisms find their application during the construction of interpersonal interaction in the virtual space and the personal virtual images perception [6].

Being one of the most important parts of the personality, its basis the value-semantic sphere determines the motives, intentions and behavior of the subject in social interaction. The values and meanings form the system of personality's personal meanings [7].

V. Frankl singled out the concept of "the meaning of life" as the central part of the value-semantic sphere, which he characterized as the actual meaning of life for a separately considered situation of interaction and defining it as the individual meaning of a person. The individual meaning is vividly manifested in practical activities, particular in creativity [8].

K. Rogers considered values and meanings from the point of the problem of constructing the self-concept view and paid attention to the importance of experiencing one's Self for the formation of the personality value core. The scientist believed that the personal value-semantic sphere is formed during interaction with the social environment and forms the basis of the self-concept and is a system of self-perception. A related structural concept, according to K. Rogers, is the ideal "I", which implies the desired image of oneself, which one wants to correspond to Alicke et al. [9].

E. Spranger defines the value-semantic sphere of personality as a tool for cognition and reflection of the surrounding world, building the framework of the world picture. This tool assumes the division of people and objects into types and categories, based on ratings of them [10].

M. Rokeach systematized personal values and divided them into two large classes. The first class contains the values of a terminal nature, which manifest themselves in the form of self-sufficient beliefs and goals that can be pursued in their own right. These include the desire for love, the desire for recognition, the need for freedom, the achievement of harmony, social equality, etc. The second class includes instrumental values, which are not valuable in themselves and serve to achieve other goals. Such values include polite communication, a high level of creativity, broadening one's horizons, a responsible attitude, etc.

Within the psychological direction in the sphere of our interests are the works, which determine the motivations of human activity on the Internet space (O. N. Arestova, V. E. Voiskunsky), investigate the motives for entering the Internet space (A. I. Luchinkina) and study the mechanisms of personality socialization on the Internet environment (E. P. Belinskaya, G. Yu. Soldatova, V. A. Pleshakov, A. I. Luchinkina, etc.). Researchers have revealed that a person being in the hyperreality system performs actions, which are based on the desire for self-expression, which is not only present in the virtual space, but also becomes the most relevant and leading in it. According to scientists, the virtual personality formation occurs as a result of replication in new constructs and the desire to live several lives.

Despite the great importance of these studies, it should be noted that, the value-semantic sphere was not a complex object of research, but its individual components were touched upon. Notwithstanding the differentness of the Internet and the real world, the peculiarities of the personal values and meanings system were practically not considered in the context of its functioning on the Internet space.

Thus, there is a need to identify the features of the value-semantic sphere of the Internet users' personality, which affect the choice of behavioral strategies, including deviant ones. Taking into account the specifics of the Internet users' value-semantic sphere will make it possible to create a system of psychological support of a person in the virtual space, which maximally ensures his psychological safety of a person.

Purpose of the study: to reveal the psychological characteristics of the Internet users' personal value-semantic sphere.

2 Materials and Methods

The research was carried out during 2017–2020 in real and virtual spaces. The study involved 400 respondents, including: respondents aged 14–16 years—180 people, 17–22 years old—220 people.

At the stage of the ascertaining experiment, the following methods were applied: to determine the subjective criteria of normative behavior on the Internet space—focused interviews with Internet users; to identify the level of personal Internet activity—the A. I. Luchinkina's questionnaire; to identify the Internet users personal values and meanings in the virtual space—the research questionnaire "The personal value-semantic sphere in the virtual space" (R. I. Zekeriaev); to determine the direction of personality behavior in the virtual space—the questionnaire "The personality in the virtual space" (A. I. Luchinkina).

The research of the personal value-semantic sphere on the Internet space has identified the problem of the absence of questionnaires and methods that could be used to assess the degree of significance of the properties, processes and phenomena inherent in the process of life in the hyperreality system. The closest diagnostic tool for determining the features of these phenomena in virtual space is the questionnaire "The personality in virtual space" by A. I. Luchinkina. However, it should be noted that the researcher singled out the motives that are inherent in the personality in the Internet socialization process, which can determine the activity of the Internet user, but did not consider the possibilities and resources inherent exclusively in the virtual space from the standpoint of subjective value for the individual. This fact led to the need for a three-stage creation of the corresponding author's research questionnaire "The personal value-semantic sphere in the virtual space". At the first stage, the respondents had to compose a short story on the topic "The Internet is ..." and "On the Internet, I like ...". The resulting narratives were subjected to content analysis, during which two classes of values inherent in the values virtual space were identified: values-actions (value of another life, value of easy success, value of geographic tolerance, value of information accessibility, value of entertainment) and value-belief ("The Internet is a safe/dangerous space", "You can/cannot show your creativity on the Internet", "There is/is not responsibility on the Internet for committed acts", "There are/are not rules and norms for communication and behavior on the Internet", "There is a lot/little useful information on the Internet").

At the second stage, the questionnaire "The personal value-semantic sphere in the virtual space" was developed, which includes three blocks of tasks aimed at studying values-attitudes, values-actions and direction of personality behavior on the Internet space. At the third stage, the reliability of the questionnaires was statistically substantiated: stability (retest method), constancy (method of replacing the experimenter), and instrumentality (Guttmann's half splitting). Their validity was also substantiated: meaningful (method of expert assessments) and constructive (methods of convergent and discriminant validation). All indicators exceed the theoretical value of the confidence level of 0.05 and the confidence level of 0.01, which testifies to the reliability of the created measuring tool.

Besides, we used the Kolmogorov–Smirnov test and the normality of the distribution of the study sample was proved on all scales. With the help of descriptive statistics, the boundaries of high, medium and low levels of severity of the studied properties were determined. The factor analysis made it possible to substantiate statistically the decrease in the dimensionality of the questionnaire "Intention of personality behavior on the Internet space" from the initial generating scales to a general scale characterizing the personality's tendency to antisocial behavior on the Internet. Statistical analysis of the experimental data was performed using the SPSS 22.0 application software (Kruskal–Wallis test, Wilcoxon test, and descriptive statistics).

3 Research Results

To identify the features of normative behavior in real and virtual space, a series of focused interviews was conducted among young people of adolescence and teenagers. The respondents were asked to tell about their vision of the content of the following categories:

1. a normal person in real space (what he is, what he does/does not do …);
2. abnormal (deviant) person in real space (what he is, what he does/does not do …);
3. a normal person in the virtual space (what he is, what he does/does not do …);
4. abnormal (deviant) person in the virtual space (what he is, what he does/does not do …).

As a result of the data analysis, it was found that young people refer to the normative behavior in the virtual space as the same forms of behavior as in the real space (respects the feelings of others, treats without judgment, uses normative vocabulary, does not violate boundaries). The following manifestations were attributed to the forms of deviant behavior in the virtual space by the respondents: use information taken out of context, throw in untrue and unverified information, live on the network under a false name, violate confidentiality, implement ridiculous hype, use profanity, negatively comment, violate personal borders, compensate for life's problems, ask for sex photos, hack pages, steal content, send spam, distribute dangerous life hacks, economic scammers, threats and extortion.

At the next stage of the study, the sample was divided according to the levels of Internet activity. Four groups have been identified: passive Internet users (27 people), who spend no more than 2 h a week on the Internet as needed; situational users (36 people), who spend 1–2 h daily on the Internet space, if such a need arises for work or study; active users (207 people)—respondents, who are on the Internet every day from 3 to 7 h, have a wide circle of contacts and are well-versed on the Internet space; overly active Internet users (130 people), who actively spending more than 7 h on the Internet.

The study of respondents' values characteristic with different Internet activity in the virtual space allowed us to draw conclusions about the presence of significant differences in the manifestation of the significance of the personal values-actions on the Internet space between groups of respondents with different levels of Internet activity, which were reflected in the following coefficients of asymptotic significance: value of another life (χ^2emp = 159.091, $\rho < 0.005$), value of easy success (χ^2emp = 139.938, $\rho < 0.005$), value of geographic tolerance (χ^2emp = 157.572, $\rho < 0.005$), value of information accessibility (χ^2emp = 148.264, $\rho < 0.005$) (see Fig. 1).

As can be seen in Fig. 1, the values-actions of respondents with low Internet activity are not significantly expressed. However, the largest number of Internet users in the groups of passive and situational users, in whom these values are expressed, belong to adolescence.

The study of values-beliefs allowed us to conclude that active and over-active Internet users are characterized by high levels of indicators on the scales "Internet is a safe space", "You can show your creativity on the Internet", "There are rules and norms of communication and behavior on the Internet", "There is a lot of useful

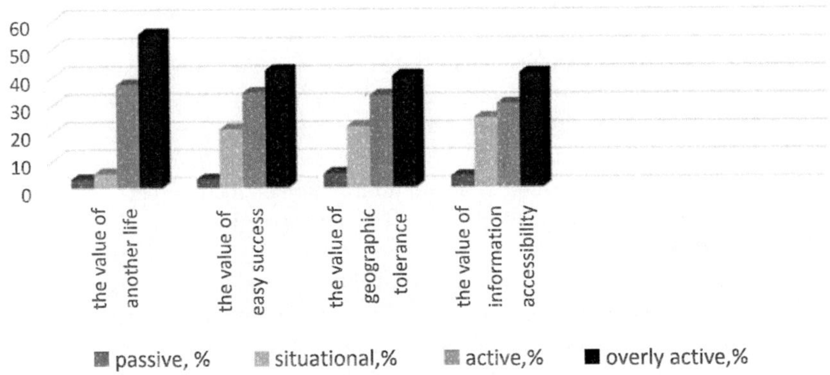

Fig. 1 Values-actions of respondents with different levels of internet activity

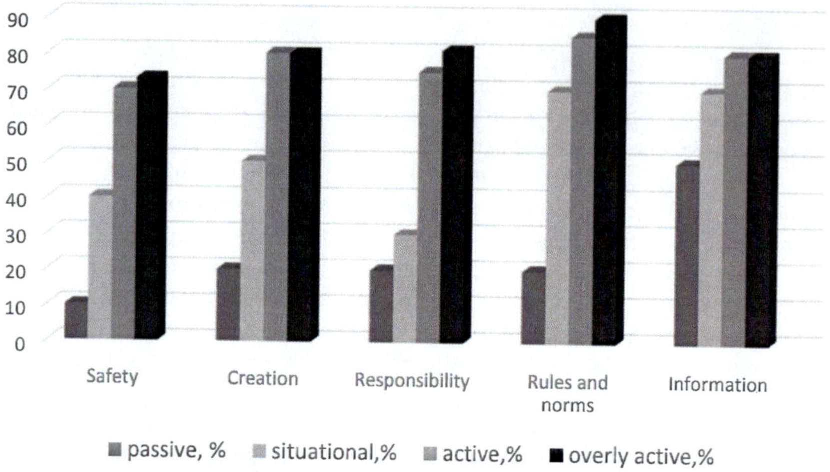

Fig. 2 The values-beliefs of respondents with different levels of internet activity

information on the Internet". These results may be due to the fact that these users tend to spend a lot of time on the Internet, this allows them to expand their social experience, looking for new information, creating new content, observing the rules of netiquette, and also making sure that the network Internet is relatively safe (see Fig. 2).

At the same time, the respondents' existence with a low level on the scale "There is responsibility for actions on the Internet" can be noted among the respondents of the passive and situational Internet users' groups. This fact can be explained by the communities functioning peculiarities on the Internet do not always allow identifying the committing asocial acts facts in order to apply appropriate measures to the offender [11].

During the study of the values-beliefs in groups with different Internet activity, significant differences were revealed in the manifestation of the significance of the personal values-beliefs in the Internet space between the groups of respondents with different levels of Internet activity, which were reflected in the following coefficients of asymptotic significance: "Internet safe space" (χ^2emp = 151.208, $\rho < 0.005$), "You can show your creativity on the Internet" (χ^2emp = 30.612, $\rho < 0.005$), "There are rules and norms of communication and behavior on the Internet" (χ^2emp = 24.440, $\rho < 0.005$), "There is a lot of useful information on the Internet" (χ^2emp = 270.333, $\rho < 0.005$), "There is responsibility for actions on the Internet" (χ^2emp = 31.215, $\rho < 0.005$).

The research of the personality behavior intention in the Internet space made it possible not only to identify a group of respondents with antisocial intentions, but also to analyze the severity of their individual values of the virtual space [12, 13].

Thus, the sample included 86 respondents with antisocial intentions. These participants were inclined to violate the rules and regulations of the virtual world, gravitated towards bullying, trolling. According to the results of the research, two groups were identified among the respondents with an antisocial orientation:

Respondents with high and excessively high Internet activity, who have antisocial intentions and values-beliefs such as there is no responsibility on the Internet for committed actions; there are no rules and regulations for communication and behavior on the Internet. Among the values-actions, the most pronounced were the values of another life, information accessibility, entertainment. They are characterized by a plurality of fictitious accounts from which they can conduct destructive virtual communication, create situations of trolling, cyberbullying and, in general, violation of the rules of netiquette.

Respondents with excessively high Internet activity, having antisocial intentions and values-beliefs such as there is no responsibility on the Internet for committed actions; there are no rules and regulations for communication and behavior on the Internet. Among the values-actions, the most pronounced were the values of easy success. The preferred activity of the respondents in this group is visiting sites with unlicensed content, creating and distributing such information, phishing attempts and hacking other people's pages on the Internet, as well as creating malicious software.

4 Discussion

Why was it not enough for the study to use the theory of systematization of values by M. Rokeach? Moreover, how is the proposed classification related to the main research in the field of value-semantic sphere?

The M. Rokich's systematization of the values, known in psychological science, refers to the life of an individual in real space and leaves the specifics of the virtual environment outside the scope of consideration. Terminal and instrumental values, which in real space only approximately describe the personal value-semantic sphere, are insufficient in the Internet environment. Meanwhile, the analysis of the A. I. Luchinkina's works, showed that Internet-active respondents have not only stable beliefs about the role of the Internet in human life, but also other ways of realizing these achievements than in real space. According to D. A. Leontiev, personal values are represented by personal constructs and personal dispositions.

At the same time, we can attribute the following to personal constructs:

"The Internet is a safe/dangerous space",
"You can/cannot show your creativity on the Internet",
"There is/is not responsibility on the Internet for committed acts",
"There are/are not rules and norms of communication and behavior on the Internet",
"There is a lot/little useful information on the Internet".

Personal beliefs—the value of a different life, the value of easy success, the value of geographic tolerance, the value of information accessibility, the value of entertainment [14, 15].

These considerations lead us to the introduction of new classes of values: values-beliefs and values-actions. In this case, the semantic load is carried by the values-beliefs, but the ways of realizing these values are reflected in the values-actions and can be different. For example, an Internet user with the belief that "there is a lot of useful information on the Internet" with a high level of value "another life" can engage in hacking, trolling, and with a low level of expression of the same values-actions, use the Internet as a reference [16]. With the same values-beliefs, a high level of expression of the value "easy success" can characterize a player.

5 Conclusions

1. The value-semantic sphere of the Internet user in the virtual space is represented by values-beliefs: "The Internet is a safe/dangerous space", "You can/cannot show your creativity on the Internet", "There is/is not responsibility on the Internet for committed acts", "There are/are not rules and norms of communication and behavior on the Internet", "There is a lot/little useful information on the Internet" and values-actions: the value of another life, the value of easy success, the value of geographic tolerance, the value of information accessibility, the value of entertainment.
2. Internet users with deviant (asocial) behavior in the Internet space are characterized by the following values—beliefs: "There is no responsibility for committed actions on the Internet", "There are no rules and norms of communication and behavior on the Internet", "There is a lot of useful information on the Internet".
3. Internet users with deviant (asocial) behavior in the Internet space differ in the ways of realizing values-beliefs not only from normative respondents, but also among themselves.

References

1. Nilsson A (2014) Personality psychology as the integrative study of traits and worldviews 18–32
2. Gendina N (2017)Information culture and information literacy as a scientific direction and a field of educational activities in Russia. In: Pathways into information literacy and communities of practice, teaching approaches and case studies, pp 167–202
3. Luchinkina A (2014) Specifika motivacii internet-pol'zovatelej. Perspektivy Nauki i Obrazovaniya 105–109
4. Luchinkina A (2018) Psihologicheskie zakonomernosti socializacii lichnosti v virtual'nom prostranstve. IP Hoteeva L. V., Simferopol
5. Turkle Sh (2011) Alone together. Why we expect more from technology and less from each other. Basic books. A member of the Perseus books group, New York
6. Adolescents' and teachers' outlook on leisure activities: personal values as a unifying framework. https://doi.org/10.1016/j.paid.2016.04.095. Last accessed 2021/03/03

7. Cooper C (2021) Pitfalls of personality theory. Pers Individ Differ. https://doi.org/10.1016/j.paid.2019.109551. Last accessed 2021/03/03
8. Altmaier EM (2020) Chapter 1—Meaning in life amidst life transitions. In: Navigating life transitions for meaning, pp 3–12
9. Alicke MD, Zell E, Guenther KL (2013) Chapter 4—Social self-analysis: constructing, protecting, and enhancing the self. In: Advances in experimental social psychology, pp 173–234
10. Ding Q, Li D, Zhou Y, Dong H, Luo J (2017) Perceived parental monitoring and adolescent internet addiction: a moderated mediation model. Addict Behav 74:48–54
11. Luchinkina A, Yudeeva T (2020) Deviant forms of socialization of personality in the internet space. ARPHA Proc 3:1511–1518
12. Luchinkina A, Zhikhareva L, Yudeeva T, Luchinkina I (2020) Cross-border and digital socialization of personality. In: E3S Web conference, vol 210, 8
13. Grigoriev Yu, Grigoriev O (2013) Cellular and health. Electromagnetic environment, radiobiological and hygienic problems. Economics, Moscow
14. Voiskunsky A, Soldatova G (2019) Counseling psychology and psychotherapy 27(3):22–43
15. Peer victimization, deviant peer affiliation and impulsivity: predicting adolescent problem behaviors. https://doi.org/10.1016/j.chiabu.2016.06.008. Last accessed 2021/03/03
16. Pervushin N (2015) Internet as a new space for life: phenomenological view. Procedia Soc Behav Sci 227–233

Non-verbal Communication in Meanings Transmission

Irina Abakumova, Natalya Mironenkova, and Evgeny Pronenko

Abstract The article presents various aspects of non-verbal communication in the educational process from the standpoint of a sense-centered approach. In the materials and methods, the authors present the works of Russian and foreign authors in order to determine: the ratio of verbal and non-verbal means of communication; types of non-verbal communication tools used in the educational process. We analyzed the influence of non-verbal communication on the quality of teaching and interaction with students. It was revealed that non-verbal communication has a culturological peculiarity, that's why the use of the same gesture will have a different semantic meaning for each culture. We determined the functional purpose of non-verbal communication: informational, compensatory, emotional, evaluative. As a result, we identified the types of non-verbal communication, with the help of which it is possible to start the process of meanings transaction or to give a personal-semantic orientation to education. These include: kinesics; vocal; smells, the ambiance, the teacher's external image (associated with the sensory perception of the environment); network inclusions-symbols (smiles, stickers, emoticons). Non-verbal communication is a means of a student's self-expression, his ability to reflect emotions and feelings, to express personal meaning. The authors emphasize the importance of using non-verbal communication to create a motivating, comfortable educational environment. They substantiate the importance of the teacher's complex use of verbal and non-verbal communication to ensure the educational process on a semantic basis.

Keywords Non-verbal communication · Semantic communication · Kinesics · Gestures · Vocal · Artifacts · Symbols · Emoji · Emoticons · Emotions · Feelings

I. Abakumova
Don State Technical University, 1, Gagarina Sq., Rostov-on-Don 344000, Russian Federation

N. Mironenkova (✉)
Platov South-Russian State Polytechnic University, Novocherkassk, Russian Federation
e-mail: vim-mir@bk.ru

E. Pronenko
Southern Federal University, Rostov-on-Don 344006, Russian Federation

A. Beskopylny and M. Shamtsyan (eds.), *XIV International Scientific Conference "INTERAGROMASH 2021"*, Lecture Notes in Networks and Systems 247,
https://doi.org/10.1007/978-3-030-80946-1_52

1 Introduction

Communication is one of the most important components of the various social institutions functioning and management, a necessary element of different individuals' interaction, during which the transfer and mutual transfer of information, feelings, assessments, meanings, senses is carried out.

The importance of using various types of communication and its influence in the educational process is indicated by a number of works [1–7].

In the context of the sense-oriented approach, communication plays an important role in the meaning formation process [8], for example, dialogue is viewed as a didactic initiation of the students' meaning formation in the educational process practice [9]. In semantic didactics, semantic communication is an essential component of pedagogical communication. Semantic communication is understood as an intention—a teacher's actual want to enter into communication with a student, as a value interaction with the aim of reflexive-semantic going beyond the limits of existing knowledge into a new context [10, p. 137]. However, there are practically no works devoted to the importance of using non-verbal communication in the meanings generating process. Therefore, the purpose of this study is to highlight the possibilities of the meaning formation process between the subjects of the educational process by means of not only verbal, but also non-verbal communication.

2 Materials and Methods

Verbal communication involves the use of language (speech), and non-verbal communication is based mainly on the use of gestures, actions, intonation, facial expressions, postures, laughter, tears. It is obvious that verbal communication plays a much more important role in the transfer of knowledge than non-verbal communication. Nevertheless, in the process of complete transmission (broadcast) or understanding of the educational content, it is important to combine both forms, as indicated by a number of studies [6, 11, 12].

Haneef et al. [13] affirm that non-verbal communication more often occurs unconsciously. Whereby we need to be aware that non-verbal messages also convey a powerful message. Body language, eye contact, appearance, and voice tone provide the audience with meaningful information. Non-verbal communication is considered to be more reliable and effective than verbal, for example, if a mother asks her child: "Did you take the food?" and the child answers "Yes", but shakes his head and gives a message "No", in that case, the mother accepts the child's non-verbal message and ignores the verbal one [13, p. 513].

Non-verbal signals are of particular value because they are spontaneous and unconscious and, unlike words, are always sincere [14]. It is known that many elements of non-verbal behavior do not have an adequate system of notes, therefore non-verbal behavior from the point of view of practical tasks of fixation remains elusive, probabilistic [15].

Non-verbal communication plays an important role in education, it is a means of transferring meaning. A theoretical approach to non-verbal communication, non-verbal behavior as a direct expression in communication of the semantic sphere, the semantic position of the individual is developed in the study of Ekintsev [2]. A gesture is viewed as a condition for an internal dialogue in the mechanism for solving a problem for meaning. According to the authors, the assessment of the semantic position severity is expressed precisely in gestures. In the situation of solving heuristic, visual tasks, a gesture, as a means of expressing a semantic position, is addressed by a person to himself, as a subject of activity, acting as a reflexive means that consolidates meaning at the stage of its transition from an irrational form to a rational (verbal) form. In other words, non-verbalized meanings-signs (gestures) pass into the form of verbalization.

Surkamp [16] endows non-verbal behavior with a compensatory function: non-verbal communication can help in learning, for example, a foreign language, since the lack of vocabulary or the inability to reproduce speech can be compensated by decoding speech signals or transmitting part of the communicative intention of the gesture modality. In other words, non-verbal communication can understand the sense and meaning of information, but at the same time help the student to express his thoughts and feelings.

Thus, the emotional function of non-verbal behavior provides valuable information about the emotions and intentions of their interlocutor in linguistic emergencies. Note that, first of all, this is information about the personality of the communicator (his emotional state at the moment of communication, communicative competence, and in some cases we can even get his idea of self-esteem).

In addition, the use of the phatic function of non-verbal communication can be a strategy for speakers to use their own facial expressions and/or gestures to give their interlocutor feedback on their reaction to the perceived communication process, or, conversely, to interpret whether the communication was successful or not, by deciphering non-verbal behavior of the interlocutor [16, p. 16].

The quantity and quality of non-verbal signals depends on a person's age, sex, temperament, nationality, social status, health status, and the level of his culture [17].

Strunina [17, p. 14] defines the gender peculiarity of non-verbal communication, endowing women with the ability to read the partner's body language better, conditioned by the fact that they had experience of raising children at a pre-verbal level. For example, in the Western business world, women are more often sent to negotiate.

Indeed, knowledge of cultural differences is especially important in business negotiations. For example, experienced communicators-negotiators have the ability to correctly read information through eye contact [18]. Often, the Middle East representatives wear very dark glasses in business negotiations, so as not to betray their true emotions and reactions by signals of non-verbal information. Therefore, non-verbal communication is also determined by the person's culture conditionality, his belonging to the people (nation).

Ignatieva [19] distinguishes the following types of non-verbal means of communication: "kinetic (facial expressions, postures, gestures); proxemic means (spatial relationships aspects between students); paralinguistic means (special aspects of pronunciation, speech timbre, volume, tempo); extralinguistic means; artifacts. We are interested in artifacts. These are non-verbal signals associated with the various objects use in the communication process (pens, pencils, pointer, glasses, cigarette) and with the use of clothing style, its color, various smells, and so on, transmitting information about a person, his social status, psychological state and so on" [19, p. 51].

There is a study [7] on the non-verbal communication use in English classes, in which a foreign teacher used three types of non-verbal communication with students: (1) kinesics (includes gestures: emblems and illustrators, head movements and posture, eye contact and facial expressions); (2) vocals; and (3) proxemics. At the same time, tactile sensations, touches were completely reduced to naught, since in most countries of the world this is not accepted and can be considered as a violation of the student's personal space. At the study's final stage, the results of the students' responses analysis were clearly in favor of using non-verbal communication as well. It was revealed that 89.6% of students reacted positively to the non-verbal communication of a foreign English teacher in the educational process, and 10.4%—negatively [7]. The students described the given by the following: we became happier; full of enthusiasm; obtained a desire to learn more.

Most teachers understand that the use of non-verbal communication in the educational process is a good way to support students' understanding, to motivate them for further learning activities, and, most importantly, to grasp non-verbalized meaning, so unstable, but true, based on the subconscious.

Another study also identified the importance of the non-verbal communication use in education [20]. The students agreed that the use of non-verbal communication in the education process actually works. The students were able to correctly understand the whole expected meaning of non-verbal communication. All of the teacher's non-verbal communication that the researcher focused on, namely gestures, eye contact, and tone of voice, worked well. Most students agreed that the teacher's non-verbal communication motivated them to learn, focused their attention and made them speak with enthusiasm, and sometimes the students were even proud of this [20].

Due to the prevalence of network communication, the emotional component transfer in Internet communication can occur through emoticons, emoticons, stickers. Smagina [21] endows emoticons with the ability to evoke positive or negative emotions in the interlocutor, to make this or that impression, consciously or unconsciously. However, communicators need to remember that emoticons are paralinguistic symbols that only supplement speech, they cannot replace words [22]. In the context of the sense-centered approach [23], a symbol is "an internal semantic core, a storehouse of meanings. The symbol's influence is enormous for mastering the content side of certain content. Concept-symbols are universal, significant, connected with the assessment level of events as a public consciousness component. Using concept-symbols in stories and descriptions, we have a tendency to scope, globalize; the emphasis is placed on a wide range of the comprehension subject. Therefore, education with the use of symbols in various situations has a sufficient effect in the meanings translation."

3 Results

Due to the fact that non-verbal signs and signals were developed earlier as a means of communication (children's pre-verbal age) than verbal language, they turned out to be stable in their original functions and are often used unconsciously, which determines a close connection with personal meanings. At the same time, they are perceived more directly and often carry a more reliable picture of human behavior [13], so they can often have a stronger effect on perception, decision-making, choice in various situations, despite their short duration. Non-verbal communication can convey the subtlest nuances of an attitude, an emotional and sensual assessment of what is happening. This means that it can contribute to the student's self-expression at the level of feelings and emotions, which is very important in the transaction of meanings process. Due to this, non-verbal communication between the educational process subjects can serve as an additional attribute of personality self-expression.

Note that gestures are the most popular means of non-verbal communication in the educational process. Sign language facilitates the possibility of self-expression, increases the effectiveness of establishing and maintaining contact [4].

As a result, we will single out the types of non-verbal communication, with the help of which it is possible to start the meanings transaction process or to give a personal-semantic orientation to education.

The first one is kinesics, which includes gestures (gestures-symbols, gestures-illustrators, gestures-adapters), head movements and posture, eye contact and facial expressions (mimics). The next type is vocal (paralinguistics)—vocal effects, namely: tone, speech speed, voice strength, presence or absence of pauses, sound intensity. Intonation, vibration, timbre, cry, breathing make sense and sound inseparable, and often sounding (intonation) can replace or substitute for the meaning of what is said. The third group: smells, the use of various things (objects), the temporal aspects of interaction, the environment, the background, the external image of the teacher. Note that this group is closely related to the individual's emerging associations, and the associations are closely related to the human sensory system. Revealing the unique nature of the student's perception, we create a comfortable environment for the information transfer.

We should add that due to the development of distance learning forms, elements of non-verbal communication have appeared that can complement network interaction in education. In other words, let us single out the non-verbal means of communication in the network Internet communication. These include: emoji (smileys), emoticons, stickers. From the semiotic point of view, they contain a coded message, which recipients are able to interpret as a description of the virtual interlocutor's personality. However, their semantic potential is not as great as, for example, the use of facial expressions or eye contact. This is due to the presence of some ambiguity in these network symbols of communication, i.e. the same emoticon in a certain situation can be interpreted in different ways, and sometimes the meaning is not revealed to the maximum and the interlocutor has to supplement it, speculate, basing on his subjective experience, imagination.

In addition, due to the peculiarity of the network "ready-made" elements, which consists in the brightness and clarity of representations, there is an opportunity for a stronger elimination of ideological and semantic evaluation and expressive transmission of the message (text, content). This phenomenon is capable of disrupting or even interrupting the intergenerational cultural and semantic dialogue, which means the impossibility of education as a pedagogical process of transmitting cultural traditions.

There is a study in which Tandyonomanu and Tsuroyya [24] recommend to mobile messaging providers: (1) to create natural emoticons and communication settings in order to increase user comfort; (2) to reduce ambiguity and increase the role of emoticons in online communication. In other words, get closer to real non-verbal communication.

It should be noted that various sciences are currently engaged in the study of the non-verbal communication phenomena: psychology, sociology, cultural studies, anthropology, linguistics, semiotics, etc., each of which considers it from their own positions. Due to the variety of researchers' conceptual attitudes, there is no single definition of this concept [14]. Non-verbal communication has different meanings for different people and cultures.

Therefore, the non-verbal communication use must be taken into account in accordance with the student's cultural affiliation. Today it is relevant due to the development of the popularity of international academic student exchange programs, as well as replenishment of foreign students in schools. Considering this, it is possible to learn on a value-semantic basis.

Very interesting results are presented in the study by Jihan Ananda, Dadang Solihat, Yayan Suryana, they are presented in Table 1 [7].

4 Discussion

Mimics, gestures, pantomime enhance the impression of the teacher's speech, save time for the lesson, add semantic nuances, and allow to highlight the main thing [25].

With the help of non-verbal communication, the meaning is presented to the partner, the struggle of value-semantic positions is carried out [2]. In the course of joint activity (primarily mental), the participants exchange not only verbal, but also non-verbal information. Semantic positions are an essential feature of the dialogue. In an internal dialogue, semantic positions are developed by the speaker himself.

Non-verbal means can convey information that is difficult or for some reason inconvenient to express in words. Symbols, gestures, facial expressions, background, color, voice, eye contact, etc.—all these elements, within the appropriate framework of the situation and under certain conditions, outline or complement the communication general picture in education!

Table 1 The student's questionnaire result [7]

No.	Statement	Answer	
		Yes	No
1	I always feel excited learning English when the teacher is coming to the class	30	1
2	I feel interested to learn English	28	3
3	Learning English with Mr. Miguel is fun	31	0
4	I like learning English with Mr. Miguel	31	0
5	I actively involved in learning English with Mr. Miguel	29	2
6	I understand when Mr. Miguel puts his index finger on his lips while making "sshh" sound is to keep students quiet and I feel motivated to be silent	30	1
7	I understand when Mr. Miguel pointing up his index and middle fingers while saying "two" is because number two. I feel more understood and motivated	31	0
8	When Mr. Miguel touches his ear while saying "listen", he asks students to listen. I feel motivated to listen	31	0
9	I understand the movements such as making a circle in the air with both hands while saying "everyone" means "everyone." I feel motivated to listen more	26	5
10	I feel motivated when Mr. Miguel nods his head when my answer is correct	30	1
11	When Mr. Miguel stands up straight while explaining the material, I feel motivated and more ready to study	31	0
12	I feel motivated and interaction is active when Mr. Miguel makes eyes contact with me	25	6
13	I feel embarrassed when Mr. Miguel makes eyes contact with me	15	16
14	I feel motivated because communication signals become strong when Mr. Miguel smiles at me	27	4
15	I feel motivated listening to Mr. Miguel's voice that is clear (such as the pitch, volume, rate, and verbal fillers)	25	6
16	English words pronounced by Mr. Miguel are poorly understood	12	19
17	Communication runs smoothly when Mr. Miguel is next to me	26	5
18	I feel motivated and love to communicate directly with Mr. Miguel	27	4

(continued)

Table 1 (continued)

No.	Statement	Answer	
		Yes	No
19	I often feel sleepy when studying English with Mr. Miguel	2	29
20	I often daydream when studying English with Mr. Miguel	3	28
21	I often concentrate when studying English with Mr. Miguel	30	1
22	Studying English with Mr. Miguel increases my motivation to learn	31	0
	Total	540	142
		682	
	Positive	611	89.6%
	Negative	71	10.4%

Based on the characteristics of non-verbal behavior, its interpretation in the communication process turns into a creative process aimed at reconstructing not always obvious psychological and socio-psychological senses, meanings of non-verbal behavior, at establishing connections between it and the psychological, socio-psychological characteristics of the personality and communication. Partners turn to various "situational clues" at the moment of interpreting and understanding non-verbal behavior, not only because its psychological essence is difficult to verbalize, but also because communication and non-verbal behavior are interconnected, that the observed and interpreted non-verbal behavior is the result of interaction between the person and the situation communication, which leads to the new meanings formation and dynamics of the non-verbal communication semantic space, to its situationality and "fluidity" [26].

It follows that conceptual analysis in intercultural communication should be based not only on prototypical, but also on instance categorization, include the entire complex of the linguistic community's knowledge and extra-linguistic experience, a set of linguistic and extralinguistic information, which differ significantly in various linguistic cultures [18].

5 Conclusions

Non-verbal communication reflects the student's emotions and feelings, and, therefore, can be a means of his self-expression. It is directly related to the subconscious, associations, sincerity, meanings, although it is often short-lived. The role of non-verbal communication is not to reflect the situation, but to express the meaning, values, value-semantic position of the individual (assessment of the situation, its details, the individual himself and his willingness to act in relation to the situation, to its details, to oneself).

References

1. Pogue L, Ahyun K (2006) The effect of teacher nonverbal immediacy and credibility on student motivation and affective learning. J Commun Educ 3(55):331–344
2. Ekintsev VI (2009) Non-verbal communication and its role in the self-organization of mental activity. Siberian Psychol J 34:93–97
3. Lipatova TV (2015) Verbal and non-verbal means of communication and their place in the language system (system-typological approach). Bulletin of the Peoples' Friendship University of Russia. Series: Philosophy (4):115–123
4. Pyastolova NB, Tretyakova EA (2016) Nonverbal communication at physical education classes. Physical Culture, Sport, Tourism, Motor Recreation 2(1):112–114
5. Wahyuni A (2018) The power of verbal and nonverbal communication in learning. In: 1st International conference on intellectuals' global responsibility (ICIGR 2017). Atlantis Press, pp 80–83. https://doi.org/10.2991/icigr-17.2018.19
6. Sutiyatno S (2018) The effect of teacher's verbal communication and non-verbal communication on students' English achievement. J Lang Teach Res 2(9):430–437. https://doi.org/10.17507/jltr.0902.28
7. Ananda J, Solihat D, Suryana Y (2020) Nonverbal communication performed by foreign English teacher. Indonesian EFL J 2(6):175–188. https://doi.org/10.25134/ieflj.v6i2.3424
8. Abakumova IV, Kagermazova LTs, Ermakov PN (2016) Technologies of directed translation of meanings in the educational process practice. Credo, Moscow
9. Abakumova IV, Kagermazova LTs, Savin VA (2013) Dialogue of cultures as a sense technology for the attitudes formation of tolerant consciousness and behavior of university students. Russ Psychol J 1(10):46–58
10. Kagermazova LTs (2014) Semantic communications as a type of interaction between teacher and students in "Our New School". Russ Psychol J 3(11):49–57. https://doi.org/10.21702/rpj.2014.3.5
11. Ali SAM (2011) The use of non-verbal communication in the classroom. In: 1st International conference on foreign language teaching and applied linguistics, 5–7 May, Sarajevo, pp 1096–1099
12. Kuleshova NM (2015) Verbal, non-verbal and paraverbal in intercultural communication and their consideration in the process of foreign language teaching and translation from a foreign language into Russian and from Russian into a foreign one. In: Russian language and culture in the mirror of translation, vol 1, pp 342–351
13. Haneef M, Faisal MA, Alvi AK, Zulfiqar M (2014) The role of nonverbal communication in teaching practice. Sci Int (Lahore) 26(1):513–517
14. Martynova EM (2014) Non-verbal communication: theories and opinions. Bulletin of the South Ural State Humanitarian Pedagogical University 8:227–233
15. Labunskaya VA (2014) On the semantic space of non-verbal communication. Russ Psychol J 3(11):93–109. https://doi.org/10.21702/rpj.2014.3.9
16. Surkamp C (2014) Non-verbal communication: why we need it in foreign language teaching and how we can foster it with drama activities. Scenario 2:12–27. https://doi.org/10.33178/scenario.8.2.3
17. Strunina EN (2013) On the question of the plastic communication semantics. Bulletin of NSTU named after R.E. Alekseev. Series "Management in Social Systems. Communication Technologies" 3:13–18
18. Molchanova GG (2014) Cognitive non-verbal as a polycode means of intercultural communication: kinesics. Moscow University Bulletin. Series 19: Linguistics and Intercultural Communication 2:13–30

19. Ignatieva NYu (2016) Non-verbal means of communication when teaching students foreign language speech in the modern communicative space. Baltic Humanit J 4(5):50–53
20. Marianti M, Sale F, Tasrim I (2020) Students' perception on the teacher's non-verbal communication in teaching English at SMAN 14 Bombana. J Teach Engl 3(5):219–229. https://doi.org/10.36709/jte.v5i3.13859
21. Smagina ES (2017) Non-verbal methods of impact in internet communication. Meteor-City 2, 58–62
22. Maksimov AA, Golubeva NM (2015) Emoticons as a way of expressing emotions. Problems and Prospects for the Education Development in Russia 33:126–130
23. Abakumova IV, Ermakov PN, Koltunova EA (2016) Symbol in the context of psychoanalytic and semantic interpretation. Russ Psychol J 2(13):225–233
24. Tandyonomanu D, Tsuroyya T (2018) Emoji: representations of nonverbal symbols in communication technology. In: IOP Conference series: materials science and engineering, vol 288. https://doi.org/10.1088/1757-899X/288/1/012052
25. Chirkova EI (2017) The functions of non-verbal communication in pedagogical communication. Mil Eng 2(4):55–63
26. Abakumova IV, Ermakov PN (2007) The search for the personal in the personality: the chronology of the sense-centrism formation in Russian psychology. Psychology of personality. Eksmo, Moscow

Teachers and Students in the Digital Age: Attitude to Online Learning, Analysis of Aspects of Communication and Meaning Transfer

Evgeny Pronenko, Kamilla Tsakhilova, Darya Popova, and Margarita Belikova

Abstract This research examines the problem of the attitude of students and teachers to the online learning format, to which many universities in the world were forced to switch due to the COVID-19 pandemic occurred in 2020. This research aims to study how teachers and students relate to the online learning format, as well as to analyze aspects of communication and meaning transfer in online and classroom lessons. The object of the study was 104 students of higher educational institutions, as well as 38 teachers of higher educational institutions. Students answered questions from February 14 to March 3, 2021, and teachers from February 16 to March 12, 2021. The responses were collected using Google Forms. To study different aspects of the perception of online learning and aspects of transferring understanding of the content of academic disciplines among students of higher educational institutions, 2 original questionnaires were developed: a questionnaire for students, which included 17 questions, and a questionnaire for teachers, which included 21 questions. The questions concerned the attitude toward the online learning format, aspects of communication, and meaning transfer during class, as well as aspects of transferring the understanding of academic subjects. The results obtained indicate that teachers and students perceive the online learning format in different ways. Teachers have some difficulties in working in an online format, while students do not see significant difficulties in this format in terms of communication and understanding of academic subjects.

Keywords Teachers · Students · Online-learning format · Communication · Understanding of academic disciplines · Meaning transfer

---

E. Pronenko (✉) · K. Tsakhilova · D. Popova · M. Belikova
Southern Federal University, Rostov-on-Don 344006, Russian Federation
e-mail: pronenko@sfedu.ru

A. Beskopylny and M. Shamtsyan (eds.), *XIV International Scientific Conference "INTERAGROMASH 2021"*, Lecture Notes in Networks and Systems 247,
https://doi.org/10.1007/978-3-030-80946-1_53

1 Introduction

1.1 Determination of Online Learning

Online learning is the process of transferring knowledge, forming skills and abilities during an interaction both between a teacher and a student as well as between them and an interactive source of an information resource reflecting all the components inherent in the educational process carried out in the context of the introduction of computer technologies. The concept of online leaning applies to the form of teaching in which the teacher and students are separated by distance, which introduces specific forms of interaction into the educational process. "Online learning is a new form of education that already exists alongside full-time, part-time, and part-time education. And it should be viewed precisely as an independent learning system" [1, p. 1].

1.2 The Relevance of Researching the Problems of Online Learning

As S. Bezoluk et al. notes, "in the context of the modern education transformation, the transition to the official online education, the form of interaction between the teacher and the student is changing significantly". The "real" communication is being replaced by the "virtual", namely by means of the Internet communications. The lack of engagement into the dialogue, the decrease of the empathy and the loss of emotional connection are some of the negative consequences of "online education" [2, pp. 1–2].

Online learning undoubtedly allows you to get a high-quality professional education due to the possibility of access to educational and additional materials, a quick way of transferring information in the process of interaction between a student and a teacher. However, as noted by L. I. Tikhomirova, L. P. Dolgova, "for a classical university, the forced transition to full online learning due to the threat of the spread of the new coronavirus infection COVID-19 requires careful analysis related to both the assessment of learning efficiency and the formation of student motivation." [1, p. 4]. The need for such an analysis determines the relevance of this study.

1.3 Modern Problematics of Online Learning

The Internet has made it possible to use new forms of online learning with virtual lessons. In the literature based on various studies conducted during 2020, the main platforms used for learning were Zoom, Microsoft Teams, and Google Classroom [3]. Gonçalves et al. [3] note that the classes have become too intense for students, with too many tasks, there are big problems associated with conducting laboratory classes.

Transfer of learning process to online platforms can have a variety of features that, according to Hunter [4], can help the learner practice while online learning. Among the important features of online learning are: freedom of space and environment in which the student is located, simplified access to the lesson, and free access to educational materials [5].

Kartensi's review [6] found that open online courses have the same benefits traditionally attributed to online learning, such as increased availability, asynchronous access, and self-directed, and therefore student-initiated, learning. However, the danger lies in the overly superficial use of available tools and resources by teachers and students.

According to Almarashdeh [7], the quality of online learning remains tied to the quality of the technologies used, the individual competence of the teacher, technological access to the platform, and issues including the problems of expectation, motivation, and time management. In turn, subsequent evidence to support the value of open online courses and online learning remains inconclusive [5].

The digitalization of learning brings new activities and behaviors to the teaching and learning landscape. Moreover, it challenges the traditional ways and situations of interaction between teachers and students. Some students find space in the virtual classroom that they previously lacked, while others "disappear behind the screen" [8]. However, it is impossible to say that the digitalization of society and education only supports the learning process or, otherwise, hinders it. It changes how we learn and how we come to interpret teaching and learning [8]. Of important aspect is also how well the transfer of meaning occurs in the context of distance learning [9].

Based on this, we set a goal: to study how teachers and students relate to the online learning format, and to analyze, in this regard, aspects of communication and meaning transfer in online and classroom lessons.

2 Materials and Methods

The object of the study was 104 students of higher educational institutions (Southern Federal University, Don State Technical University) (of which 92 (88.5%) women and 12 (11.5%) men), the age range was 18–23 years and 38 teachers of higher educational institutions (Southern Federal University, Don State Technical University) of which 30 (78.9%) women and 8 (21.1%) men aged 23–78 years (median—48.5).

Students answered questions from February 14 to March 3, 2021, and teachers from February 16 to March 12, 2021; collecting of responses was carried out using Google Forms. The respondents took part in the survey voluntarily.

To study various aspects of the perception of online learning and the aspects of transferring understanding of the content of academic subjects among students of higher educational institutions, 2 author's questionnaires were developed: a questionnaire for students included 17 questions, a questionnaire for teachers—21 questions.

According to the structure, the questionnaires consisted of 3 types of questions: open-ended questions, multiple-choice questions, and questions where the respondents were asked to choose an answer on a scale from 1 to 10. The questions related to attitudes towards online learning, aspects of communication in the classroom, and aspects of transmission of understanding of academic subjects.

To process the data obtained, the following methods of mathematical statistics were used: Mann–Whitney U-test, The Sign test.

3 Results

To analyze the answers of teachers to questions with scales, the medians were determined for the answers of the respondents. The questions, the extreme values, and the medians are shown in Table 1.

In question 7, half of the respondents choose options 1, 2, and 3. These results indicate that students, according to teachers, rarely ask them again about the correctness of the meaning of the words spoken during a online lesson. In question 9, half of the teachers more often found answers 7, 8, and 9, indicating serious difficulties in interpreting students' nonverbal patterns in online learning. Question 11 shows that the most frequent answers for half of the respondents are 1 and 2. Therefore, it is very easy for teachers to establish contact with students in the first class in the classroom. Question 13 showed that teachers do not have any difficulties and the transfer is carried out completely freely—this is indicated by the fact that half of the respondents choose answers 9, 10. A similar situation is observed in a similar question 14. The only difference is in what happens in an online environment. The common answers of half of the respondents are 8, 9, and 10. As for other questions, there is a wide range of teachers' opinions regarding the perception of differences in the transmission of meaning during classroom and online learning.

When analyzing the data on questions 2 and 3 using the Sign test, it was found that the median between the answers to these questions is zero (at a significance level of 0.05). Consequently, from the point of view of teachers, there are no differences in the degree of students' perception of the meaning of the topic under consideration when it is transmitted in classroom and online classes.

The median of differences between the answers in questions 4 and 5 is equal to zero (at a significance level of 0.05). Thus, the frequency of questions from teachers to students about whether they understood them correctly does not differ in classroom and online learning.

The median of differences between the answers in questions 6 and 7 is equal to zero (with a significance level of 0.05). It follows from this that the frequency of students' questions to teachers about the correctness of the meaning of the words they said does not differ in classroom and online learning.

When analyzing questions 11 and 12, statistically significant differences were found between the answers (at a significance level of 0.05). Consequently, it is more

Table 1 Formulation of questionnaire questions for teachers with scales and values of medians

№	The question	The extreme values	The median
1	To what extent do online learning technologies help you convey the meaning of the topic discussed in the lesson?	1—greatly facilitate 10—makes it very difficult	5
2	Evaluate, on average, the degree of perception of the meaning of the topic under consideration in the lesson when it is transmitted to students during the classroom lesson?	1—the meaning is fully preserved 10—the meaning is completely lost	5
3	Rate on average the degree of perception of the meaning of the topic under consideration when it is transmitted during a online lesson?	1—the meaning is fully preserved 10—the meaning is completely lost	5
4	Do you often have to ask the students again if you understood them correctly during the classroom?	1—never 10—always	4
5	Do you often have to ask the students again whether you understood them correctly during a online lesson?	1—never 10—always	5
6	How often do students ask you about the correctness of the meaning of the words you said during the classroom?	1—never 10—always	4
7	Do students often ask you about the correctness of the meaning of the words you said during an online lesson?	1—never 10—always	3
8	Do you often interpret the meaning of a student's words with the help of his non-verbal patterns (facial expressions, facial expressions, intonation, etc.)?	1—never 10—always	6.5
9	How difficult is it for you to interpret the non-verbal patterns of students in a online class?	1—very easy 10—very difficult	7
11	How easy is it for you to connect with students in the first class of the semester in class?	1—very easy 10—very difficult	2
12	How easy is it for you to connect with students in the first lessons of the semester in online learning?	1—very easy 10—very difficult	4
13	Rate how freely you could convey your understanding of the subject to students during a lecture or answering a question in a classroom lesson?	1—such a transfer was practically impossible 10—I could do it completely free	9

(continued)

Table 1 (continued)

№	The question	The extreme values	The median
14	Rate how easily you could transfer your understanding of the subject to students during a lecture or answering a question in an online lesson?	1—such a transfer was practically impossible 10—I could do it completely free	8
15	Do you perceive the long silence of the students after you ask the audience a question during the classroom session as a serious difficulty in understanding the subject?	1—I never perceive in this way 10—I always perceive this way	5
16	Do you perceive the long silence of students after asking a question to the audience during an online class as a serious difficulty in understanding the subject?	1—I never perceive in this way 10—I always perceive this way	5

difficult for teachers to establish a contact in the first lessons of a semester in a online format compared to face-to-face training.

When analyzing questions 13 and 14, statistically significant differences were found between the answers (at a significance level of 0.05). Thus, teachers have greater difficulty in conveying their understanding of the subject during a lecture or answering a question in online learning than in classroom lessons, where the transmission is completely free.

The median of the differences between the answers in questions 15 and 16 is zero (at a significance level of 0.05). The data indicate that there is no difference in the perception of long silence by students as a serious difficulty in understanding the subject during classroom and online learning.

One of the questions was aimed at clarifying the attitude of teachers to the situation when students do not turn on video cameras in a remote lesson. A quantitative analysis of the data showed that such a situation seems uncomfortable for 42.1% of respondents, rather uncomfortable—32.6%, rather comfortable—18.4%, comfortable—7.9% (Fig. 1). Thus, the inability to see the rest of the participants during online learning causes discomfort for 73.7% of teachers. This suggests that for most teachers, this aspect of teaching is very important.

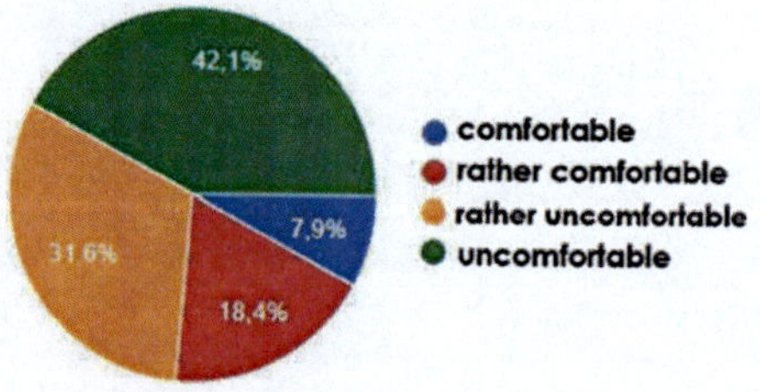

Fig. 1 Respondents' answer to the question "How comfortable is it for you to be in the lesson remotely, without the opportunity to see the rest of the participants (when students do not turn on their video cameras)?"

In the questionnaire, there is a question about the means of interpreting the meaning of students' words, which are used by teachers in an online lesson in a situation where there is no video connection (Table 2). The respondents could choose several answers.

The sample of teachers was divided into three groups depending on the answer to the question "Is it possible, in your opinion, to equate full-time and online education from the point of view of the successful transmission of the meanings and meanings of the subject to students? If not, what do you think are the differences?". As a result, the first group included respondents who categorically deny the possibility of identifying full-time and online learning—23 teachers (60.5%), the second group accepted the possibility of identification, and it included 11 respondents (29%), the third group included teachers whose answers cannot be clearly attributed to either the first or the second position—4 respondents (10.5%).

The first and second groups of respondents were selected for further analysis.

Using the Mann–Whitney U-test, two groups of teachers (1 and 2) were analyzed by age. Thus, no statistically significant differences in the age of the respondents were found—the bilateral asymptotic significance was 0.685. Accordingly, the age of teachers is not a factor that determines the choice of position regarding the identification of online and full-time learning formats.

Next, an analysis of the attitude to online learning was carried out in connection with the comfort of being in a lesson without the opportunity to see the participants, its results are presented in Table 3.

Thus, being able to see other participants is a key aspect for most educators who do not equate to face-to-face and online learning.

To analyze students 'answers to questions with scales, the medians were determined for the respondents' answers. The wording of the questions, the wording of the extreme values, and the meanings of the medians are shown in Table 4.

In Question № 2, half of the students choose options 1 and 2. This indicates that when students do not see each other directly during an online lesson, most students almost never have difficulties in communicating with the group. Also, according to the students, there are no difficulties in the distribution of the sequence of answers in the group in the process of online learning. This is indicated by the answers of half of the respondents to question № 4, where half of the students chose options 1 and

Table 2 Answers of teachers to the question about the interpretation of the words of students during the online class

Means of interpreting students' words	Percentage (%) of teachers who chose this means of interpreting students' words
Intonation	76.3
Laughter	26.3
Pauses	53.2
Volume	28.9
Text add-ons in chat	78.9

Table 3 Quantitative analysis of the attitude of teachers to online learning and the comfort of being in class without the ability to see the participants

Attitude towards the online learning format	Is it comfortable to be in a online lesson without the ability to see other participants (without video communication)?	
	Comfortable/rather comfortable	Uncomfortable/rather uncomfortable
Identify face-to-face and online learning	6	5
Do not equate face-to-face and online learning	4	19

Table 4 Formulation of questionnaire questions for students with scales and values of medians

№	Question	Poles	Median
1	How comfortable is it for you to perceive the meaning of what has been said if the teacher does not turn on video communication during a online lesson?	1—very comfortable 10—very uncomfortable	5.5
2	Do you have difficulty communicating with the group if you cannot see them directly during the online class?	1—never occurs 10—always occurs	2
3	How much do you feel you are fully present in the meeting if it is held remotely?	1—always occurs 10—never occurs	4
4	Do you find it difficult to prioritize responses in a group during online learning?	1—very simple 10—very difficult	2
5	How easy is it for you to conduct a dialogue with a teacher if you do not see him directly during a online lesson?	1—very simple 10—very difficult	4
7	Rate the general level of understanding of studied subjects before the transfer to online learning	1—totally clear 10—totally unclear	3
8	Rate the general level of understanding of studied subjects after the transfer to online learning	1—totally clear 10—totally unclear	4
9	Do you easily interpret the meaning of the teacher's words with the help of his non-verbal patterns (facial expressions, intonation, etc.) during a online lesson?	1—very simple 10—very difficult	4
10	Has education lost its meaning for you, having made the transition to online learning?	1—absolutely lost 10—didn't lost	3
11	Rate, how freely you could transfer your understanding of the subject to other participants of the lesson during a discussion or when answering the question?	1—almost impossible 10—easily possible	8

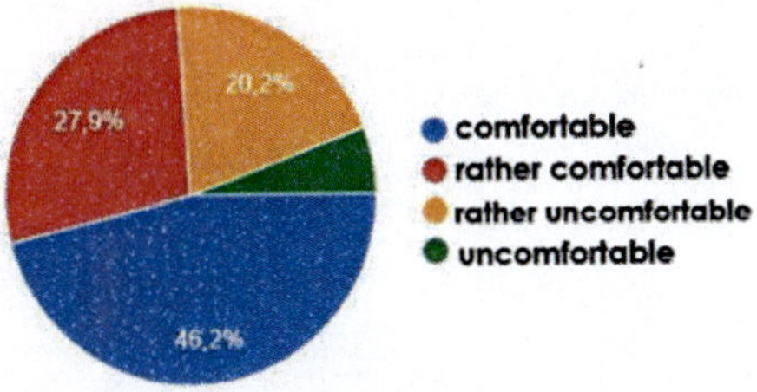

Fig. 2 Respondents' answers "How comfortable is it for you to be in the lesson remotely, without the opportunity to see the rest of the participants?"

2. Question № 7 shows that half of the respondents choose answers 1, 2, 3, which indicates a high level of understanding of academic subjects before the transition to a online learning format, from the point of view of students. In question № 10, half of the respondents choose answer options 1, 2, and 3, indicating that for many students, education has not lost its meaning, having made the transition to online learning. At the same time, half of the respondents chose answer options 8, 9 and 10 when answering Question № 11 about the degree of freedom in transferring their understanding of the subject to participants in online learning when answering a question or in a discussion, which indicates a completely free transfer of understanding. In the answers to the remaining questions, there is a wide range of students' opinions regarding aspects of online learning.

When analyzing the data on Questions № 7 and № 8 using the G-test of Signs, it was found that the median between the answers in these questions is zero (at a significance level of 0.05). Consequently, the transition to online learning does not affect how students perceive their level of understanding of academic subjects. The questionnaire contained a question aimed at clarifying the attitude of students to a situation when participants do not turn on video cameras during a online lesson. A quantitative analysis of the data showed that this situation seems comfortable for 46.2% of respondents, rather comfortable—27.9%, rather uncomfortable—20.2%, uncomfortable—5.8% (Fig. 2). Thus, the inability to see the rest of the participants during online learning gives discomfort to only 26% of students. It may be assumed that for most students this aspect of conducting classes is not essential.

The questionnaire asked about the impact of interference and communication problems on the quality of online learning. The analysis showed that for 54.8% of students these aspects insignificantly affect the educational process, 1.9% of respondents believe that they do not influence at all, and for 43.3% of students, interference greatly affects the process of conducting classes (Fig. 3). It can be concluded that for many students this aspect is not a significant factor that can influence the quality of the learning process.

Noteworthy are the answers of students to the question of how the higher level of mastering the subject is achieved in the format of online learning. As a result of the quantitative analysis, the following data were obtained: 61.5% of the respondents choose classes in real time with audio and video; for 18.3% of students, real-time lessons with only audio will be sufficient for a high level of assimilation; 17.3% of respondents prefer video recording as the most effective type of activity, and 2.9%

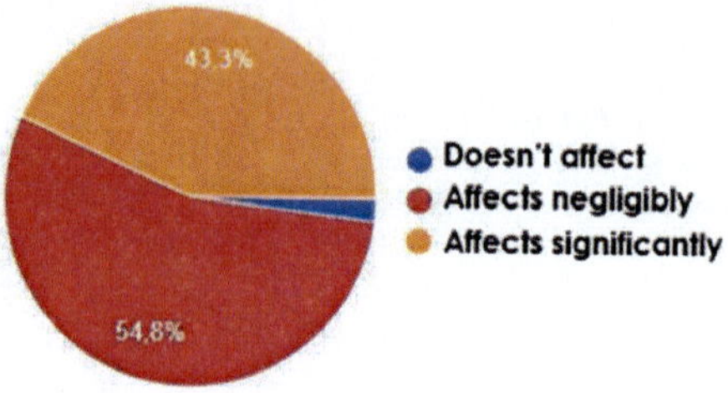

Fig. 3 Respondents' answers to the question "Do you think interferences and communication problems affect the process of conducting classes?"

prefer audio recording (Fig. 4). Therefore, we can conclude that for most students, real-time audio and video communication play an important role in the level of mastering subjects in the classroom in the online space.

The sample of students was divided into three groups depending on the answer to the question "Is it possible, in your opinion, to equate full-time and online education? If not, what do you think are the differences?". The first group included respondents who categorically deny the possibility of identifying full-time and online learning—56 students (53.8%), the second group agreed with the possibility of identification, and it included 40 respondents (38.5%), the third group included students whose answers cannot be clearly attributed to either the first or the second position—8 respondents (7.7%).

The first and second groups of respondents were selected for further analysis.

We conducted a quantitative analysis of students' attitudes towards online learning and comfort in a lesson without being able to see the participants. The results are presented in Table 5.

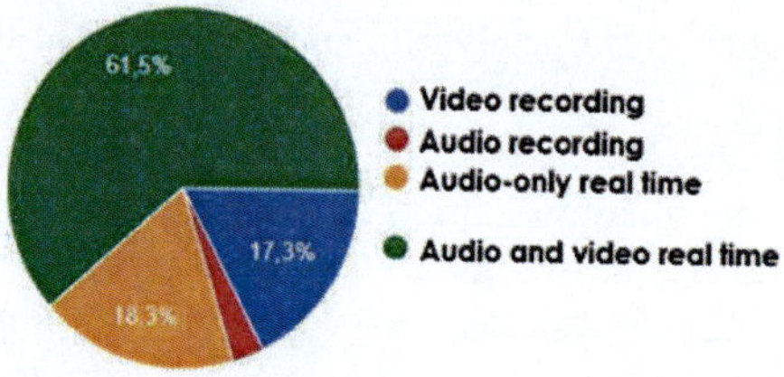

Fig. 4 Respondents' answers to the question "In what form is a higher level of mastering the subject in the format of online learning achieved?"

Table 5 Quantitative analysis of students' attitudes towards online learning and comfort in a lesson without being able to see the participants

Attitude towards the online learning format	Is it comfortable to be in a online lesson without the ability to see other participants (without video communication)?	
	Comfortable/rather comfortable	Rather uncomfortable/uncomfortable
Identify full-time and online learning	35	5
Do not equate full-time and online learning	36	20

Thus, it can be concluded that for many students who equate full-time and online learning, the ability to see other participants in online learning is not a key aspect of conducting classes.

4 Discussion

4.1 Perception of Teachers of Aspects of Online Learning Format

The results obtained during this study are in a wide field of discussion regarding online learning. One of the important aspects of this discussion is teachers' perception of the online learning form. In our study, we obtained the following results on this topic:

1. Most of the teachers we interviewed (60.5%) strongly disagree with the fact that full-time and online education can be equated in terms of the success of transferring the meanings of the subject to students.
2. At the same time, we believe that the fact that this attitude towards online education turned out to be independent of the age of teachers is quite significant, since there is a widespread opinion that the attitude towards the online learning format depends on age and this format is more negatively perceived by older teachers.

In the works devoted to this problem, several significant problems of the online learning format are highlighted, which cause concern for teachers.

According to N. Milman, even though online learning is becoming more attractive for many students, higher education institutions need to carefully assess whether their courses are suitable for transferring to an online format [10]. While, universities must ensure that teacher satisfaction contributes to the overall online learning experience [5].

One of the factors influencing the attitude of teachers to the online learning format is previous experience. Thus, A. Youde, examining teachers' perceptions of blended learning, found that they perceive online learning as negative, relying on their first experience of working in a new environment. This experience led them to feel isolated from the outside world, disorientated: "tutors outlined limited engagement with peers in online environments, with reasons given including time, superficial discussions, contributions a "tick-box exercise", and a lack of trust that would have developed in face-to-face meetings. Further, four tutors found the online elements impersonal and lacking human contact." [11, p. 6].

In addition, the author notes that some teachers considered these difficulties to have an impact on students' perception of the quality of teaching in an online environment and, accordingly, their academic performance: "Daisy was confident in face-to-face communication, but her limited self-efficacy online-context and with a mixed approach to teaching in general could negatively affect students' perception of quality during the module" [11, p. 10].

It corresponds our findings: the teachers interviewed felt that they can easily establish contact in the classroom and have no problems communicating understanding of the subject in the classroom. At the same time, they have more difficulties with the same aspects in online learning.

One of these difficulties is the insufficient level of non-verbal communication. The study by S. Willermark indicates that teachers indicate a decrease in interaction due to the lack of feedback in the broad sense of the word, due to the lack of facial expressions and small gestures, as well as due to questions and reflections of students in the learning situation. This raises questions about student participation in the learning situation [8].

Our research shows that video communication in class is very important for teachers. 73.7% of respondents-teachers feel discomfort if they do not see other participants in the online lesson. Without seeing students, teachers most often rely on intonation and text messages in chat. We believe that seeing students means for the teacher to see their involvement in the lesson, in the general psychological situation. According to V. Abou-Khalil, involvement is an indicator of the success of an activity for a teacher, which encourages him to continue or improve it [12]. On the other hand, if the teacher does not see the guidelines for student engagement, then this gives him discomfort.

4.2 Perception of Students of Aspects of Online Learning Format

Analysis of the work on this topic shows that students perceive the online learning format differently than teachers. They also have a wide range of opinions.

On the one hand, the common transition of universities to online learning has had a serious impact on the life and educational activities of students. Ferraro et al. [13], analyzing the situation in Italian higher education, note that the whole situation had a significant negative impact on students, increasing the number of cases of mental disorder, with surges of depression, anxiety, and stress.

This situation is also influenced by the negative sides of the online learning format itself. When students recall what they experienced while familiarizing themselves with this form of education, the most common responses become feelings of insecurity, disorientation, and dissatisfaction with the organization of classes and communication with the teacher. They also experienced tension, fatigue, and stress "due to insufficient time management." [14, p. 179]. New complicated barriers appear for learners in the online learning process: "low level of Internet connectivity, limited access to technology, low resources, and lack of financial support" [12, p. 1]. They make it difficult for students and teachers to interact and create significant gaps in the transmission of meanings.

On the other hand, there is a sufficient amount of data that students perceive online learning as a more comfortable and simplified form of learning [1].

As part of our research, results were obtained that also speak in favor of the fact that students do not experience serious difficulties in online learning in the field of communication and the transfer of understanding.

1. Students easily interact with their study group, easily distribute the sequence of answers.
2. Students can easily convey their understanding to all participants in the online lesson.

According to students, their level of understanding of academic subjects has not significantly decreased, and education for the majority has not lost its meaning.

F. V. Ferraro et al. in their study also note that relationships with equal in age and teachers tend to be perceived as identical to what they were before isolation (that is, during classroom activities). According to the authors, the new generation of "Digital Natives" is defined as a hyper-cognitive generation that can interact with equal in age through social media platforms (such as TikTok, Facebook, Twitter, and Snapchat) more effectively than previous generations [13]. We agree with this opinion and assume that since young people talked a lot on social networks before online learning, the online format in terms of communication did not cause them any difficulties.

Another confirmation that the online format is convenient for students is that only 5.8% of students note that they are uncomfortable in a online lesson without the opportunity to see other participants (versus 73.7% of teachers).

As F. V. Ferraro et al. points out, online learning is a teaching and learning method that was already available before the pandemic. Learning can be synchronous (when the teacher and students are in the same room, both in real and virtual) and asynchronous (using e-mail or online platforms [13]. Asynchronous learning is most often considered in the framework of Massive Open Online Courses (MOOCs), which are now extremely actively researched [15–17].

However, according to our research, 79.8% of the surveyed students prefer the synchronous format. In our opinion, this may be due to both the insufficient development and prevalence of MOOC in Russia, and the fact that many students do not want to reduce the level of contact with teachers and classmates.

5 Conclusion

This study showed that teachers and students approach online learning from different perspectives. These differences are especially noticeable regarding the comfort of being in an online class without the ability to see other participants, which significantly reduces the ability for teachers to evaluate students' non-verbal patterns. This aspect makes it difficult for teachers to communicate and establish contact with students. At the same time, it was found that the online format does not have a serious impact on the level of understanding of academic subjects by students, does not significantly complicate the transfer of understanding of the subject between participants in online learning. However, most students prefer direct contact with teachers in real time. For teachers, there is a significant difference in conducting online and classroom lessons, but for students the level of communication and understanding of subjects remains the same.

References

1. Tikhomirova LI, Dolgova LP (2020) The experience of distance learning for students of Altai State University in the conditions of COVID-19. ASEDU-2020: Advances in science, engineering and digital education, 012149, Krasnoyarsk. J Phys Conf Ser 1691. https://doi.org/10.1088/1742-6596/1691/1/012149
2. Bezoluk S, Azarko E, Sannikova V, Derezko O (2020) The initiation of the students' meaning formation via the meaning essay. In: Innovative technologies in science and education (ITSE-2020), 18059. E3S Web of conferences, vol 210. https://doi.org/10.1051/e3sconf/202021018059
3. Gonçalves SP, Sousa MJ, Pereira FS (2020) Distance learning perceptions from higher education students—the case of Portugal. Educ Sci 10(12):374. https://doi.org/10.3390/educsci10120374
4. Hunter P (2015) The virtual university. Sci Soc 16(2):146–148. https://doi.org/10.15252/embr.201440016
5. Corsby CLT, Bryant A (2020) "I felt like I was missing out on something": an evaluation of using remote technology in the classroom. Educ Inf Technol 25:4897–4914. https://doi.org/10.1007/s10639-020-10207-2
6. Kartensi T (2013) The MOOC. What the research says. Int J Technol High Educ 10:23–37
7. Almarashdeh I (2016) Sharing instructors experience of learning management system: a technology perspective of user satisfaction in distance learning course. Comput Hum Behav 63:249–255
8. Willermark S (2021) Who's there? Characterizing interaction in virtual classrooms. J Educ Comput Res. https://doi.org/10.1177/0735633120988530
9. Suroedova E, Tushnova Y, Belousova E (2020) Agriculture faculty students' meanings translation in conditions of the development of communicative competence. In: XIII International scientific and practical conference "State and prospects for the development of agribusiness—INTERAGROMASH 2020", 15028. E3S Web of conferences, vol 175. https://doi.org/10.1051/e3sconf/202017515028
10. Milman N (2014) Is online learning for all learners? Distance Learn 11(4):13–15
11. Youde A (2020) "Face-to-face trumps everything": an exploration of tutor perceptions, beliefs and practice within blended learning environments. Educ Sci 10(5):147. https://doi.org/10.3390/educsci10050147
12. Abou-Khalil V, Helou S, Khalifé E, Chen MA, Majumdar R, Ogata H (2021) Emergency online learning in low-resource settings: effective student engagement strategies. Educ Sci 11(1):24. https://doi.org/10.3390/educsci11010024
13. Ferraro FV, Ambra FI, Aruta L, Iavarone ML (2020) Distance learning in the COVID-19 era: perceptions in Southern Italy. Educ Sci 10(12):355. https://doi.org/10.3390/educsci10120355
14. Oliveira T, Morgado L (2020) IImpacto da dinâmica emocional na aprendizagem em cursos a distância no ensino superior: O papel da presença emocional e das microlideranças. Revista Portuguesa de Educação 33(2):177–199. https://doi.org/10.21814/rpe.14331
15. Galikyan I, Admiraal W, Kester L (2021) MOOC discussion forums: the interplay of the cognitive and the social. Comput Educ 165:104133. https://doi.org/10.1016/j.compedu.2021.104133

16. Kim D, Jung E, Yoon M, Chang Y, Park S, Kim D, Demir F (2021) Exploring the structural relationships between course design factors, learner commitment, self-directed learning, and intentions for further learning in a self-paced MOOC. Comput Educ 166:104171. https://doi.org/10.1016/j.compedu.2021.104171
17. Deng R, Benckendorff P (2021) What are the key themes associated with the positive learning experience in MOOCs? An empirical investigation of learners' ratings and reviews. Int J Educ Technol High Educ 18:9. https://doi.org/10.1186/s41239-021-00244-3

Career Orientations, Personal Values and Types of Socio-cultural Adaptation of International Students

Elena Yashchenko, Ekaterina Shchelokova, and Olga Lazorak

Abstract The article presents an overview of modern studies of career orientations, personal values, and social adaptation, which showed the appropriateness of studying these phenomena in their interrelations and differences depending on the type of adaptation. The aim of the study is to determine the interrelations of career orientations and personal values with different types of socio-cultural adaptation of international students studying in the UK, London. 29 foreign students aged 18–20 were examined. The research techniques are "The Career Anchors" questionnaire by Schein, the personality adaptation questionnaire to the new socio-cultural environment by Yankovsky, and the diagnostic technique of the axiological orientation of the personality by Kaptsov. Statistical methods of processing empirical data include descriptive statistics by groups, Pearson correlation analysis. Software for processing the results of an empirical study was the SPSS 22.0, for data visualization—Excel 2007, Corel Draw 2018. In general, international students prefer such career orientations as "Lifestyle" and "Pure challenge", and personal values "Values of profession" and "Values of education". Career orientations are not associated with adaptive and interactive types of socio-cultural adaptation; there are no interrelations of career orientations and personal values with any type of socio-cultural adaptation. Polyvariability was found in the choice of types of socio-cultural adaptation to a new environment.

Keywords Acculturation · Internationalization · Intercultural adaptation · Strategies of socio-psychological adaptation · International students · Career · Life values

E. Yashchenko
Emperor Alexander I St. Petersburg State Transport University (PGUPS), Saint Petersburg 190031, Russia

E. Shchelokova · O. Lazorak (✉)
Federal State Autonomous Educational Institution of Higher Education "South Ural State University (National Research University)", Chelyabinsk 454080, Russia

A. Beskopylny and M. Shamtsyan (eds.), *XIV International Scientific Conference "INTERAGROMASH 2021"*, Lecture Notes in Networks and Systems 247,
https://doi.org/10.1007/978-3-030-80946-1_54

1 Introduction

The conditions for the development of modern society and higher education are largely unpredictable and uncertain that poses serious risks for students and forces them to acquire new competencies that would contribute to successful and adaptive behavior, socio-cultural mobility, the ability to work with huge information arrays, to allocate relevant for building a career in accordance with life values and meanings. Career orientations, including professional self-esteem, motives, and own values, play a significant role in the process of building a career [1]. The problem of a career is interdisciplinary; it is multilaterally considered both in international and Russian science: in psychology, sociology, pedagogy, cultural studies, conflict science, and others. In connection with the policy of university internationalization, studies of the acculturation and management practices of universities focused on international markets are now relevant [2–5]. In this regard, several approaches can be distinguished.

One of them is the search for interrelations of career orientations with various dependent and independent variables. One of them is the search for interrelations of career orientations with various dependent and independent variables. Fouad and Byars-Winston investigated the relationship of a career with ethnicity, culture with career choices, concluding that racial/ethnic differences do not affect career aspirations, but there are differences in perception of career-related opportunities and obstacles [6]. Yu et al. identified the impact of social nets on the acculturation process of Chinese students studying at universities in the UK, both in psychological and behavioral aspects, whether acculturation of international students predetermines academic achievement, value, and loyalty of the university. In particular, the results showed that the psychological acculturation of Chinese international students in the host culture has no significant connection with their academic performance [7]. Gregor et al. investigated the relationship between agency characteristics (initiative and persistence of personal growth), barriers (perception of academic and educational barriers and overcoming obstacles), and career adaptability in college students; their proposed model explains 55% of deviations in career adaptation [8]. Furness shows which psychological phenomena predetermine career adaptability [9]. Sari investigated the role of the sense of life and life satisfaction in the self-effectiveness of career decisions of Turkish university students, determined that life satisfaction and sense of life are positively related to the self-effectiveness of career decisions and they are their important predictors [10]. Kaspi-Baruch studied the interrelations between personality and versatile and limitless career orientation, as well as the mediating role of motivational orientation [11].

The second approach involves works on the creation of psychodiagnostic techniques, methods, and models for career orientations' study. For example, Peng and Wu created a scale (structural model) for measuring the intercultural adaptation of international students in eastern countries and studied the main ways to promote their intercultural adaptation: communication competence and social communication of the host party, basic intercultural transformation, advanced intercultural transformation [12]. Wiernik and Kostal conducted a meta-analysis of studies of versatile limitless career orientations (VLCO); a general factor was identified—proactive career orientation, predicting behavior in the field of career self-management and career satisfaction, but to a lesser extent related to non-career-oriented relationships, objective success and behavior in the field of physical mobility; an integrative model of how VLCO affect career behavior is proposed [13].

The third approach to the study of career orientations is related to the problem of employment. Cortellazzo et al. discussed a basic career orientation, its behavioral predecessors, and employment outcomes [14]. Xin et al. identified career success criteria as predictors of employment outcomes [15]. Rodrigues et al. conducted one of the first longitudinal studies identifying and empirically studying the role of predecessors of versatile and unlimited career orientations—basic self-esteem, estimated employment opportunity, and social capital of university students [16].

The problems of socialization and adaptation are also given a lot of attention in modern psychological research. Quinton identified predictors of socialization of international students with host citizens [17]; Sadewo et al. considered an intercultural adaptation of international students in terms of social selection and social influence [18]; Toth-Bos et al. defined the interactive effect of achieving the goal and the importance of the goal for acculturation and well-being [19].

At the same time, the interrelations of career orientations and personal values of students, depending on the type of socio-cultural adaptation, have not yet been studied enough.

The aim of this study is to determine the interrelations of career orientations and personal values with different types of socio-cultural adaptation of international students (using the example of students studying in the UK, London). The empirical base of the study was 29 international students at the age of 18–20.

The hypothesis of the study: there are differences in the interrelations of career orientations and personal values with different types of socio-cultural adaptation—adaptive, interactive, conformal, estranged, depressed, and nostalgic—in international students studying in the UK.

2 Methods

The techniques of the study are the questionnaire "Career Anchors" by Schein, the personality adaptation questionnaire to the new socio-cultural environment by Yankovsky, adapted by Konstantinov; the diagnostic technique of the axiological orientation of the personality by Kaptsov.

Statistical methods of processing empirical data include descriptive statistics by groups, Pearson correlation analysis. Software for processing the results of an empirical study was the SPSS 22.0, for data visualization—Excel 2007, Corel Draw 2018.

3 Results

The general characteristic of career orientations showed that international students had the highest average score of the indicator of such a career anchor as "Lifestyle", and the lowest score—of "General management" (Table 1).

The general characteristic of the personal values of international students is presented in Table 2.

Using the correlation analysis, the following interrelations of career orientations and personal values with different types of socio-cultural adaptation of international students were obtained (Figs. 1, 2 and 3). In this research, adaptation is presented in the positive, negative, and conformal strategies which we described in the previous publication [21]. With the help of empirical research, the structure of socio-psychological adaptation of international students, which includes the following strategies: positive (types of adaptability, interactivity), conformal (conformal and estranged types), and negative (depressive and nostalgic types of adaptation) was determined.

However, 14 significant interrelations of personal values with the types of "adaptability" and "interactivity" were identified, forming together a positive adaptation strategy: in international students, adaptive-type indicators have a close positive interrelationship with values of self-development (r = 0.785; $p \leq 0.01$), spiritual satisfaction (r = 0.511; $p \leq 0.05$), social contacts (r = 0.469; $p \leq 0.05$), achievements (r = 0.564; $p \leq 0.05$), financial position (r = 0.512; $p \leq 0.05$), values of profession (r = 0.655; $p \leq 0.01$), values of education (r = 0.512; $p \leq 0.05$), values of the family (r = 0.541; $p \leq 0.05$) and of public life (r = 0.519; $p \leq 0.05$), as well as with the integral indicator "humanistic orientation of the personality" (r = 0.631; $p \leq 0.01$); interactive-type indicators have a close positive interrelationship with values of self-development (r = 0.576; $p \leq 0.05$), financial situation (r = 0.602; $p \leq 0.05$), values of profession (r = 0.647; $p \leq 0.01$), values of education (r = 0.455; $p \leq 0.05$), as well as with the integral indicator "pragmatic orientation of personality" (r = 0.553; $p \leq 0.05$).

Table 1 Descriptive statistics of the scales of Schein's "Career Anchors" questionnaire

Scale	CA-1	CA-2	CA-3	CA-4	CA-5	CA-6	CA-7	CA-8
M	14.97	14.28	15.52	13.66	11.86	13.31	14.10	14.28
σ	2.69	2.33	2.68	2.86	3.28	3.72	2.81	2.79
min	9.00	10.00	10.00	8.00	7.00	6.00	9.00	6.00
max	20.00	19.00	20.00	19.00	19.00	20.00	19.00	18.00

Note CA—scales of Schein's "Career Anchors" questionnaire: 1—Autonomy and Independence, 2—Pure Challenge, 3—Lifestyle, 4—Security and Stability, 5—General Management, 6—Entrepreneurial Creativity, 7—Service and Dedication to a Cause, 8—Technical and Functional Competence

Table 2 Descriptive statistics of the diagnostic technique of the axiological orientation of the personality by Kaptsov

Scale	val-1	val-2	val-3	val-4	val-5	val-6	val-7	val-8	val-9	val-10	val-11	val-12	val-13	val-14	val-15
M	7.15	8.74	4.93	5.44	0.22	5.81	4.37	3.70	12.63	12.11	8.44	3.63	3.56	26.26	14.11
σ	5.95	5.52	6.24	5.24	5.77	4.56	5.15	5.44	7.96	8.49	8.06	7.51	8.49	19.78	16.60

Note It should be noted that at the level of significance, values have the property of bipolarity: the test provides the detection of the positive and negative significance of personal values [20]. val—scales of the diagnostic technique of the axiological orientation of the personality by Kaptsov: 1—self-development, 2—spiritual satisfaction, 3—creativity, 4—social contacts, 5—prestige, 6—achievements, 7—financial situation, 8—preservation of individuality, 9—values of profession, 10—values of education, 11—values of the family, 12—values of public life, 13—values of leisure, 14—humanistic orientation of the personality (a total indicator of 1–4 scales), 15—pragmatic orientation of personality (a total indicator of 5–8 scales)

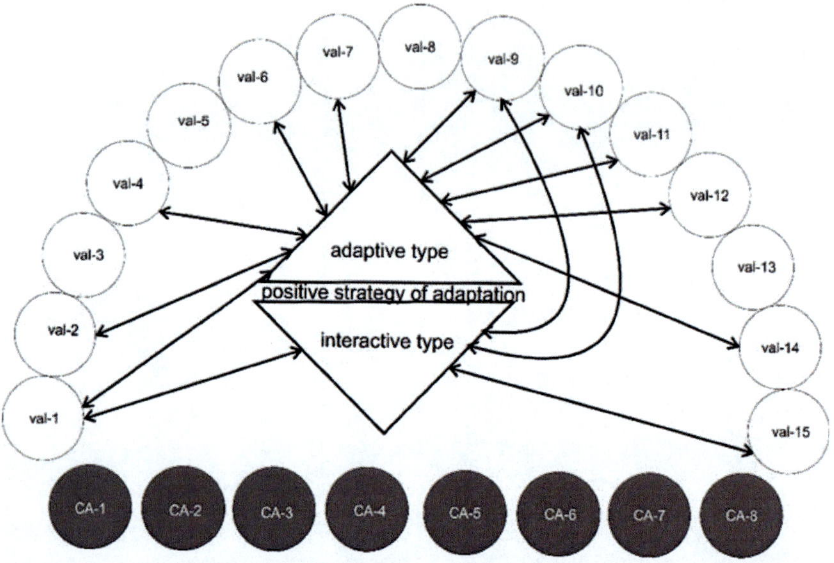

Fig. 1 Interrelations of career orientations and personal values of international students with types of adaptability and interactivity, which together form a positive strategy of socio-psychological adaptation. *Note* val—scales of the diagnostic technique of the axiological orientation of the personality by Kaptsov: 1—self-development, 2—spiritual satisfaction, 3—creativity, 4—social contacts, 5—prestige, 6—achievements, 7—financial situation, 8—preservation of individuality, 9—values of profession, 10—values of education, 11—values of the family, 12—values of public life, 13—values of leisure, 14—humanistic orientation of the personality (a total indicator of 1–4 scales), 15—pragmatic orientation of personality (a total indicator of 5–8 scales); CA—scales of Schein's "Career Anchors" questionnaire: 1—Autonomy and Independence, 2—Pure Challenge, 3—Lifestyle, 4—Security and Stability, 5—General Management, 6—Entrepreneurial Creativity, 7—Service and Dedication to a Cause, 8—Technical and Functional Competence; the scales of the personality adaptation questionnaire to the new socio-cultural environment by Yankovsky: adaptive type, interactive type (positive strategy of adaptation)

Figure 2 presents significant interrelations of career orientations and personal values with conformal and estranged types, forming together a conformal strategy of socio-psychological adaptation, which can also be considered neutral or compromise.

One direct interrelation of career orientation "Autonomy and Independence" with an estranged type of socio-cultural adaptation (r = 0.745; $p \le 0.01$) and 2 direct interrelations of career orientations "Pure Challenge" (r = 0.506; $p \le 0.05$) and "Service and Dedication to a Cause" (r = 0.473; $p \le 0.05$) with a conformal type of socio-cultural adaptation were found.

Figure 3 shows the interrelations of career orientations and personal values with types of nostalgia and depressiveness, which together form a negative strategy of socio-psychological adaptation.

2 interrelations of personal values with the nostalgic type were found in international students with depressive and nostalgic types, forming together a negative

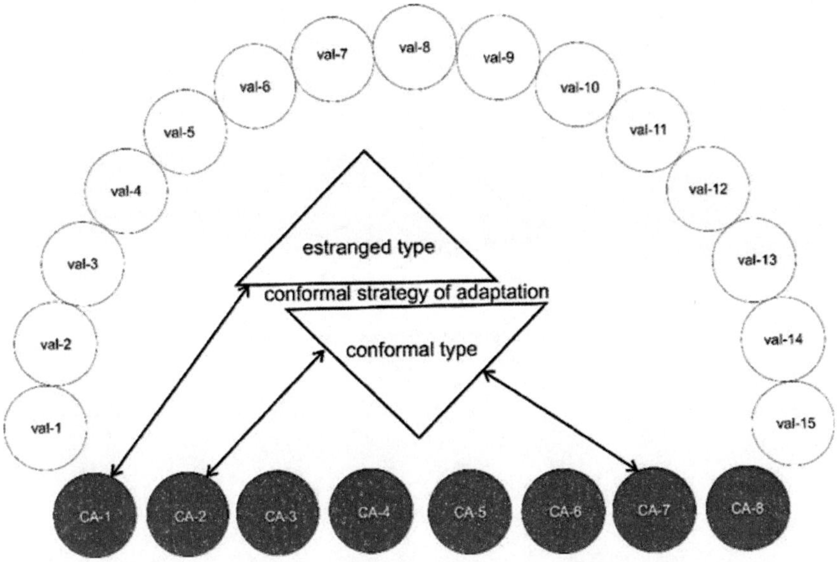

Fig. 2 Interrelations of career orientations and personal values of international students with conformal and estranged types, forming together a conformal strategy of socio-psychological adaptation. *Note* val—scales of the diagnostic technique of the axiological orientation of the personality by Kaptsov: 1—self-development, 2—spiritual satisfaction, 3—creativity, 4—social contacts, 5—prestige, 6—achievements, 7—financial situation, 8—preservation of individuality, 9—values of profession, 10—values of education, 11—values of the family, 12—values of public life, 13—values of leisure, 14—humanistic orientation of the personality (a total indicator of 1–4 scales), 15—pragmatic orientation of personality (a total indicator of 5–8 scales); CA—scales of Schein's "Career Anchors" questionnaire: 1—Autonomy and Independence, 2—Pure Challenge, 3—Lifestyle, 4—Security and Stability, 5—General Management, 6—Entrepreneurial Creativity, 7—Service and Dedication to a Cause, 8—Technical and Functional Competence; the scales of the personality adaptation questionnaire to the new socio-cultural environment by Yankovsky: estranged type, conformal type (conformal strategy of adaptation)

strategy of socio-psychological adaptation: one is direct with the value "preservation of individuality" (r = 0.457; $p \leq 0.05$) and one is inverse with the value of "prestige" (r = −0.515; $p \leq 0.05$); one inverse interrelation of the personal value of the profession (r = −0.490; $p \leq 0.05$) with a depressive type. 2 inverse interrelations of career orientations—"Lifestyle" with the depressive type (r = −0.458; $p \leq 0.05$), and "Security and Stability" with a nostalgic type of socio-cultural adaptation (r = −0.469; $p \leq 0.05$) were also found.

Meanwhile, there are separate interrelations between career orientations with values. 1 significant correlation of the "Pure Challenge" indicator with values of preservation of individuality (r = 0.407; $p \leq 0.05$) was revealed. 2 significant correlations of the indicator "Lifestyle" with values of spiritual satisfaction (r = 0.508; $p \leq 0.01$) and education (r = 0.390; $p \leq 0.05$) were found. 5 significant interrelations of the "Technical and Functional Competence" scale with the values of spiritual

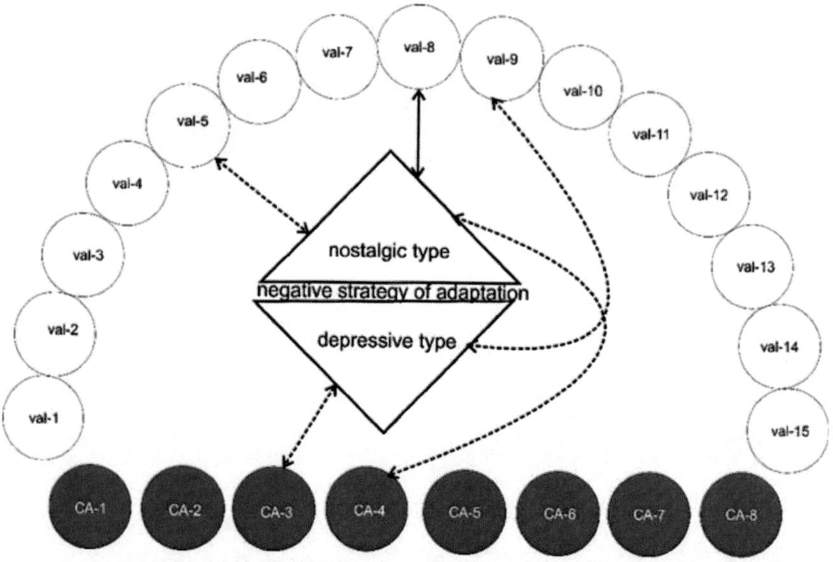

Fig. 3 Interrelations of career orientations and personal values of international students with nostalgic and depressive types, forming together a negative strategy of socio-psychological adaptation. *Note* val—scales of the diagnostic technique of the axiological orientation of the personality by Kaptsov: 1—self-development, 2—spiritual satisfaction, 3—creativity, 4—social contacts, 5—prestige, 6—achievements, 7—financial situation, 8—preservation of individuality, 9—values of the profession, 10—values of education, 11—values of the family, 12—values of public life, 13—values of leisure, 14—humanistic orientation of the personality (a total indicator of 1–4 scales), 15—pragmatic orientation of personality (a total indicator of 5–8 scales); CA—scales of Schein's "Career Anchors" questionnaire: 1—Autonomy and Independence, 2—Pure Challenge, 3—Lifestyle, 4—Security and Stability, 5—General Management, 6—Entrepreneurial Creativity, 7—Service and Dedication to a Cause, 8—Technical and Functional Competence; the scales of the personality adaptation questionnaire to the new socio-cultural environment by Yankovsky: nostalgic type, depressive type (negative strategy of adaptation)

satisfaction (r = 0.383; $p \leq 0.05$), social contacts (r = 0.502; $p \leq 0.01$), preservation of the individuality (r = 0.381; $p \leq 0.05$), profession (r = 0.472; $p \leq 0.05$), and education (r = 0.444; $p \leq 0.05$) were revealed.

4 Discussion

For people with a high orientation towards integration, career, according to Schein, is associated with a general lifestyle that contributes to the balance of personal needs and organizational requirements. For international students, a management orientation is less important, that is, the integration of the actions of other people to solve organizational problems. Perhaps they understand the boundaries of their opportunities at the stage of vocational training and do not build illusions by studying abroad. It is important to note that the largest variation of values was obtained by the indicator of "Entrepreneurial Creativity", and the smallest—by the indicator of "Pure challenge". Based on this, we can assume that international students are a relatively homogeneous group of young people in the sense that in general, they are focused on accepting the "challenge". Perhaps their decision to study in another country is due to this quality. The heterogeneity of this group is that international students choose "the targets" for a challenge in different ways. For some students, the challenge may be studying a more complex educational program, for others—competition and interpersonal relations, for the third—orientation towards entrepreneurship, building their own business (that is the reason for the high spread of values in this career orientation, see Table 1).

Based on the data of descriptive statistics (see Table 2), it can be noted that international students obtained the highest average score according to the indicator of such personal value as "Value of profession" and "Value of education" (which once again reflects the specificity of our sample and draws its socio-psychological portrait), and the lowest score—according to the value of prestige. Moreover, the average score is not just lower than the rest in the hierarchy of values but tends to negative values. This means that the social approval of important people is not just less noticeable, but also they do not see differences in the approval of their actions by people with different social status. It can also be noted that in general, the humanistic orientation of the personality prevails over the pragmatic one in international students.

As shown in Fig. 1, career orientations do not have significant interrelations with types of adaptability and interactivity, forming together a positive adaptation strategy, based on which we can assume that successful adaptation to new socio-cultural conditions, can be achieved with any type of career orientation. This is consistent with the results of previous studies of the problem of self-actualization on other samples: Yashchenko shows that self-actualization can be achieved in any type of personal realization and the period of early maturity (from 17 to 25 years old) [22]; Shchelokova that achieving a high level of comprehension of life and self-actualization is possible with any type of career orientation, which has qualitatively different self-actualization content [23].

However, there may be another explanation—the development of career orientations most likely can be accompanied by maladjustment and tension. However, the revealed 14 significant interrelations of personal values with the types of "adaptability" and "interactivity", forming together a positive strategy of socio-psychological adaptation can mean that for international students with an interactive type of adaptation, the development financially and socially (realization of pragmatic values) is of great value, and for students with an adaptive type—value and meaning development (realization of spiritual values). This is consistent with the interpretation of the scales of the questionnaire by Yankovsky [24]: the scale of adaptability characterizes emigrants, who have a sense of belonging to society, strive for self-realization, plan their future based on their capabilities, values and past experience, while the scale of interactivity characterizes emigrants, who are determined to enter the environment actively, expand their social bonds; they are distinguished by their readiness to change their behavior taking into account the social attitudes of a new society, self-realization through gaining financial independence, as well as their focus on a given goal and submission to this goal.

The findings let us suggest that career orientation towards autonomy may be connected with the complexities of socio-cultural adaptation, first of all, communicatively, and the interrelation of career orientations on challenge and service with conformity can serve as an indicator of problems in socio-cultural adaptation, first of all, academically, if the reference group, to which an international student with a conformal type of adaptation will be oriented, will not support the values of education. There are no significant interrelations of personal values with conformal and estranged types of socio-cultural adaptation. This can be taken into account in the process of psychological support of international students: students with difficulties in adaptation can be recommended to build their academic and personal life based on the principles of stability and integration.

Found separate interrelations between career orientations with values mean that the most closely related career orientation to the value components of a personality is "Technical and Functional Competence"; it means that, although it is not a systemically forming element of career orientation, it is most firmly included in the value-sense core of a personality, compared to other career orientations. It is also important to note that international students with high scores of the "Technical and Functional Competence" and "Lifestyle" scales, most likely, will not have problems with learning, since the values of education are important for them.

Summarizing the results presented in Figs. 1, 2 and 3, it is important to note that correlation analysis on the determination of interrelations between personal value scales and types of socio-psychological adaptation showed that types of adaptability and interactivity have the largest number of direct interrelations with personal values. Perhaps the presence of these interrelations forms a socio-psychological mechanism that contributes to a positive strategy of international students' adaptation. It was revealed that the types of adaptation that form the conformal strategy of adaptation do not form significant interrelations with personal values. This can be interpreted as a mechanism of socio-psychological adaptation under the type of estrangement and conformity: abandoning values, the personality adapts to a new socio-cultural environment. It was also revealed that the types of adaptation that form a negative strategy of adaptation have a small number of significant interrelations with personal values, most of them are negative. This can be explained not only as a rejection of values but also as a lack of contact with them (access to them). At the same time, international students with a type of depressiveness fall into the academic risk group, since the expressiveness of this type has a significant negative interrelation with the value of the profession. International students with a nostalgic type of adaptation have the value of preservation of their individuality expressed, but at the same time, they reject the value of prestige, thereby becoming "cultural recluses" in a new socio-cultural environment.

5 Conclusion

Thus, a psychological portrait of international students studying in the UK is compiled. Such career orientations as "Lifestyle" and "Pure challenge" are generally significant for international students, which characterizes them as ready for challenges and overcoming insurmountable obstacles, managing unsolvable problems, when a career is associated with a common lifestyle, balancing the needs of a person, family, and career; as for personal values, professional life and the field of training are significant, values of prestige are not. Career orientations are not associated with adaptive and interactive types of socio-cultural adaptation, as they imply rather maladjustment and tension to realize career aspirations. At the same time, no interrelations of career orientations with personal values were also found—in international students, career orientations do not fit into the structure of life values, they are not significant in value for the personality. Perhaps the choice of career development does not have value for the person. For international students with an interactive type of adaptation, the realization of pragmatic values is of great value, and for students with an adaptive type, it is the realization of spiritual values. The interrelation of career orientation "Autonomy and Independence" with the estranged type of adaptation and the interrelations of career orientations "Pure Challenge" and "Service and Dedication to a Cause" with the conformal type of socio-cultural adaptation emphasize the specifics of career development in another country when it is necessary to maintain strength, act autonomously, realize service and overcome difficulties with minimum obstacles. Significant relationships of personal values with conformal and estranged types of socio-cultural adaptation are not revealed, most likely, such styles of behavior are not close to international students. International students in the host country are nostalgic for maintaining their individuality, and the more significant the nostalgic type of adaptation to them, the smaller the role of prestige and social assessment; the higher the value of the profession and career orientation "Lifestyle", the less depressive style of behavior is; the more significant the value "Security and Stability", the less relevant is the nostalgic type of socio-cultural adaptation. Polyvariability in the choice of types of socio-cultural adaptation to the new environment was noted. The hypothesis of the study was confirmed in the results. There are differences in interrelations of career orientations and values with different types of socio-cultural adaptation in international students. The study can be continued in terms of determining the interrelations of academic success, the level of subjective well-being of international students with different career orientations, and the types of socio-cultural adaptation they choose.

References

1. Schein E (2006) Career anchors: participant workbook. Pfeiffer Essential Resources for Training and HR Professionals

2. Berry JW (2015) Global psychology: implications for cross-cultural research and management. Cross Cult Manag 22(3):342–355. https://doi.org/10.1108/CCM-03-2015-0031
3. Bierwiaczonek K, Waldzus S (2016) Socio-cultural factors as antecedents of cross-cultural adaptation in expatriates, international students, and migrants: a review. J Cross Cult Psychol 47(6):767–817. https://doi.org/10.1177/0022022116644526
4. Pekerti AA, Van De Vijver FJR, Moeller M, Okimoto TG (2020) Intercultural contacts and acculturation resources among international students in Australia: a mixed-methods study. Int J Intercult Relat 75:56–81. https://doi.org/10.1016/j.ijintrel.2019.12.004
5. Gladkova AA (2017) Psychological and socio-cultural adaptation of international journalism students in Russia: the role of communication skills in the adaptation process. Psychol Russ State Art 10(4):45–59. https://doi.org/10.11621/pir.2017.0405
6. Fouad N, Byars-Winston A (2005) Cultural context of career choice: meta-analysis of race/ethnicity differences. Career Dev Q 53(3):223–233. https://doi.org/10.1002/j.2161-0045.2005.tb00992.x
7. Yu Q, Foroudi P, Gupta S (2019) Far apart yet close by: social media and acculturation among international students in the UK. Technol Forecast Soc Change 145:493–502. https://doi.org/10.1016/j.techfore.2018.09.026
8. Gregor MA, Weigold IK, Wolfe G, Campbell-Halfaker D, Martin-Fernandez J, Ganginis Del Pino HV (2021) Positive predictors of career adaptability among diverse community college students. J Career Assess 29(1):115–128. https://doi.org/10.1177/1069072720932537
9. Furness M (2020) What predicts career adaptability? An application of achievement goal theory and adult attachment theory. J Career Dev 47(6):671–685. https://doi.org/10.1177/0894845318815610
10. Sari SV (2019) Attaining career decision self-efficacy in life: roles of the meaning in life and the life satisfaction. Curr Psychol 38(5):1245–1252. https://doi.org/10.1007/s12144-017-9672-y
11. Kaspi-Baruch O (2015) Motivational orientation as a mediator in the relationship between personality and protean and boundaryless careers. Eur Manag J 34(2):182–192. https://doi.org/10.1016/j.emj.2015.10.004
12. Peng R-Z, Wu W-P (2019) Measuring communication patterns and intercultural transformation of international students in cross-cultural adaptation. Int J Intercult Relat 70:78–88. https://doi.org/10.1016/j.ijintrel.2019.03.004
13. Wiernik B, Kostal J (2019) Protean and boundaryless career orientations: a critical review and meta-analysis. J Couns Psychol 66(3):280–307. https://doi.org/10.1037/cou0000324
14. Cortellazzo L, Bonesso S, Gerli F, Batista-Foguet JM (2020) Protean career orientation: behavioral antecedents and employability outcomes. J Vocat Behav 116:103343. https://doi.org/10.1016/j.jvb.2019.103343
15. Xin L, Zhou W, Li M, Tang F (2020) Career success criteria clarity as a predictor of employment outcomes. Front Psychol 11:540. https://doi.org/10.3389/fpsyg.2020.00540
16. Rodrigues R, Butler CL, Guest D (2019) Antecedents of protean and boundaryless career orientations: the role of core self-evaluations, perceived employability and social capital. J Vocat Behav 110:1–11. https://doi.org/10.1016/j.jvb.2018.11.003
17. Quinton WJ (2020) So close and yet so far? Predictors of international students' socialization with host nationals. Int J Intercult Relat 74:7–16. https://doi.org/10.1016/j.ijintrel.2019.10.003
18. Sadewo GRP, Kashima ES, Gallagher C, Kashima Y, Koskinen J (2020) International students' cross-cultural adjustment: social selection or social influence? J Cross Cult Psychol 51(6):490–510. https://doi.org/10.1177/0022022120930092
19. Toth-Bos A, Wisse B, Farago K (2020) The interactive effect of goal attainment and goal importance on acculturation and well-being. Front Psychol 11:704. https://doi.org/10.3389/fpsyg.2020.00704
20. Kaptsov AV (2016) Contradictions in the personality's consciousness value basis. World Psychol 2(86):233–242
21. Yashchenko E, Shchelokova E, Lazorak O (2018) Strategies and types of social and psychological adaptation of international students studying in London. SHS Web Conf EDP Sci 55(03006):1–5. https://doi.org/10.1051/shsconf/20185503006

22. Yashchenko EF (2016) Theoretical-methodological and empirical aspects of the study of self-actualization in the context of a metasystem approach: psychological concept. In: Psychology of self-actualization and the formation of a professional in modern education: monograph. Part 1. ELVI-Print, Saint Petersburg, pp 9–39
23. Shchelokova EG (2013) Valuable-sense priorities of future professionals with different career orientations. Fundam Res 11(4):788–794
24. Konstantinov VV (2018) Socio-psychological adaptation of migrants: theory and empirical research: monograph. Pero, Moscow

Game Simulation of Organizational Conflicts with Due Regard for the Type of Temperament

Amulanga Shakirova, Mikhail Rozin, Valery Svechkarev, and Anton Ivanov

Abstract The article shows the possibility and practicality of building game simulations of organizational conflicts with due regard for the type of temperament. In this case, information about the types of temperament prevailing in the team is included in the leader training procedures in the process of the simulation game. The adaptable simulation model used in the process of a business game should be functionally expanded by means of elements and connections that ensure the causality of the psychological attitude of the team and the leadership style chosen by the leader. To analyse this causality, a cognitive model of game simulation is suggested that is implemented in the course of further research in the form of a simulation model within AnyLogic simulation software. The problem of studying the influence of temperament types on the emergence and escalation of organizational conflicts is analysed on the basis of characteristics of four types of temperament (sanguineous, melancholic, choleric and phlegmatic types). The article shows the possibility of game training of a leader proceeding from psychological attitudes when choosing a management style.

Keywords Game simulation · Simulations model · Organizational conflicts · Behaviour strategies · Temperament · Leadership style

1 Introduction

It is known that modern educational game technologies allow achieving the maximum activity of students despite the increased intellectual and psychological stress during the game. Moreover, a participant in the game learning process needs to demonstrate his/her knowledge in an integrated form in the conditional practice environment, which is the most important advantage, for example, for higher professional education or professional retraining of specialists [1]. This is why the tasks of both the organizational and methodological formalisation of the game learning process and

A. Shakirova · M. Rozin (✉) · V. Svechkarev · A. Ivanov
Southern Federal University, Pushkinskaya St., 140, Rostov-on-Don 344006, Russia
e-mail: idpr@bk.ru

A. Beskopylny and M. Shamtsyan (eds.), *XIV International Scientific Conference "INTERAGROMASH 2021"*, Lecture Notes in Networks and Systems 247,
https://doi.org/10.1007/978-3-030-80946-1_55

the construction of a variety of simulation environments adjusted to the educational tasks to be solved are so relevant [2].

Nowadays, educational game technology in the field of organizational systems—this is a class of the so-called business games—is becoming more widespread. Using computer simulators and adaptable simulation models of production processes in the toolkit for building business games makes this learning technology especially attractive. In the scientific and educational environment, there are examples of successful implementation of business games (see, for example, [3, 4]). There is a steady growth in the audience of users. There is observed the ambition of developers and teachers to expand functionality [5], focus on methodological innovations [6, 7], and finally approbate traditional educational technologies in the new computer environment [8, 9].

Examples of the most complex combination of such ambitions include the development of the game simulation environment of organizational conflicts based on computer simulators. In the process of game simulation, a student should be able to make a forecast of the conflict escalation and make a management decision based on the assessment of the current situation, for example, relying on the choice of a leadership style corresponding to the current parameters in a specific conflict situation [10]. However, the disadvantages of the mentioned example and similar implementations of computer simulators should be noted. First of all, this is a too schematic, algorithmically unambiguous procedure for making a decision on the choice of a leadership style, avoiding behavioural uncertainties and socio-psychological personal or collective attitudes. In this paper, we propose to consider a variant of building the game simulation environment of organizational conflicts based on simulation models with due regard for the type of temperament of employees of a subordinate team.

2 Materials and Methods

It is suggested to use convergently integrated methods of the theory of management of organizational systems [11] and practical psychodiagnostics [12] as a theoretical and methodological basis for this article. The first one attracts by its strategy of describing management functions within the framework of decision-making models. At the same time, it substantiates the fundamental need to take into account the purposefulness of behaviour of participants in organizational systems [13]. The methods and tools for such accounting are given as part of the second component of integration, namely, practical psychodiagnostics. Evaluation of a stable personality characteristic, a type of temperament, and possible strategies for behaviour of a particular person or a team as a whole in a conflict in an organization forms an image of a conflict situation and should be taken into account when making a management decision. Therefore, the mutual influence of methods, their convergence are the natural foundations for building a game simulation of managing organizational conflicts.

In order to describe the obtained model solutions, it is suggested to use the methods of cognitive analysis that represent game simulation models of organizational systems in the form of a visual topological image. Examples of the successful use of analytical tools of cognitive models for the study of educational problems have already been recognized and developed within the framework of professional education (see, for example, [14, 15]). Cognitive models provide access to the analysis of the structure of the game at the level of basic concepts, their target relationship and causal interaction, i.e. the dynamics of functioning; they also make it possible to reveal the semantic attitudes characteristic of a real conflict situation. Formally, cognitive models are already a prototype of simulation models, which predetermines the rationality of joint consideration of cognitive and simulation models. In this work, we propose to use AnyLogic simulation software to build a simulation model, which allows creating dynamic models of organizational management systems and provides ample opportunities for building a virtual educational environment with elements of game simulation and artificial intelligence [5].

3 Results

The structure of an organizational conflict includes the following basic concepts: Pre-conflict situation, Conflict, Escalation and De-escalation [10]. The cycle of initialization and escalation of a conflict is described in cognitive analysis by a single-loop model of interaction of these concepts, namely, Pre-conflict situation → Conflict → Escalation → De-escalation → Pre-conflict situation. The game environment is constructed by supplementing the cognitive model of the organizational conflict with elements of formation of a pre-conflict situation and management of the launch and escalation of the conflict. Such supplementing is intended to create the level of a conflict situation required from the point of view of training. The student's reaction to the conflict escalation is assumed in the form of a choice and assignment of a leadership style in accordance with a specific situation. The purpose of the business game within the framework of the theory of management of organizational systems [11] is to develop the leader's ability to adapt his/her leadership style for a specific conflict situation, achieving its effective overcoming [4]. For implementing the function of choosing a style in the game, the provision is made for the player to construct a leadership style that will de-escalate the conflict. Figure 1 presents the cognitive model of a business game for constructing a leadership style in a conflict situation.

First of all, attention should be paid to the initial structural and dynamic orientation of the game to counteract the conflict escalation. The game loop forms a closed loop with negative feedback, i.e. the player's choice of the leadership style should enhance the tendencies to de-escalate the conflict and defuse the pre-conflict situation. The level and rate of the conflict escalation in the cognitive model are set through the concepts of Data on conflict risk factors, Data on the level of conflict and Data on escalation rates. The combination of each of these concepts with the corresponding concept of the game cycle of the conflict (Pre-conflict situation, Conflict, Escalation) forms the nodes of integration of the causes of the conflict situation and, as a result, generates a "spin-up"—intensification of the conflict. In turn, the effective

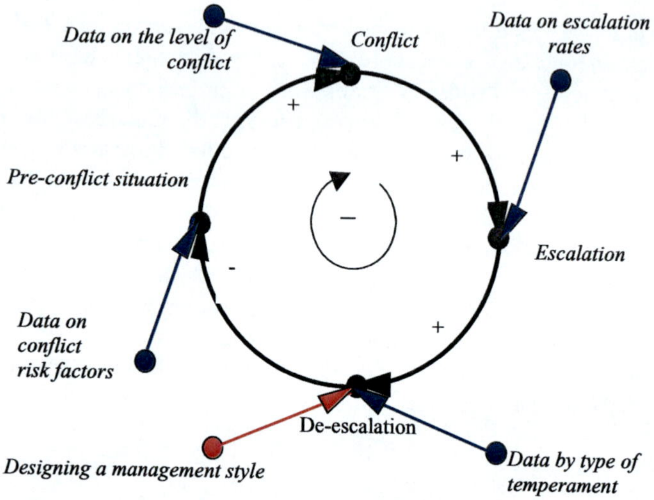

Fig. 1 The cognitive model of a business game for constructing a leadership style in a conflict situation

construction of the leadership style through its connection with the concept of De-escalation forms a resistance node where multidirectional tendencies of escalation and de-escalation of the organizational conflict are integrated.

However, this construction of the model [10] does not take into account the factors influencing implementation of the structure and dynamics of the process of de-escalation of the conflict, in particular, behavioural uncertainties and socio-psychological personal or collective attitudes. The need to build a type of business game in accordance with the specifics of a particular business is indicated, for example, in [16]. In our case, this means the need to take into account the influence of the types of employees' temperaments prevailing in the team on specific procedures for the conflict de-escalation. The De-escalation node is a more complex structure where the consequences of three directional influences are integrated, namely, the multidirectional tendencies of escalation and de-escalation of the organizational conflict already mentioned above and data on the type of temperament that can both intensify and weaken the de-escalation procedures. In this case, temperament is understood as a set of individual mental properties of a person manifested in his/her attitude to the surrounding reality, in particular, in possible behaviour in a conflict situation. Information about the prevailing types of temperaments should enable the player to design the leadership style that will most effectively counteract the organizational conflict.

To move from a cognitive model to a simulation model when forming the game simulation environment, it is necessary to expand on the structural implementation of procedures for de-escalation of organizational conflicts. In [10], it is suggested to structure the de-escalation procedures in accordance with the classical typology that distinguishes authoritarian, democratic and liberal styles. Each of the styles corresponds to one or more de-escalation procedures that effectively promote the appropriate conflict resolution strategy, for example, from the series: cooperation, compromise, rivalry, avoidance and evasion. The specified conflict resolution procedures are based on specific strategies of behaviour. However, these strategies of behaviour, in turn, stem from the type of temperament, namely, sanguineous, melancholic, choleric and phlegmatic temperaments [17], characteristic of employees. Let us consider how a game simulation of an organizational conflict can be built taking into account two origins of setting a behaviour strategy. As in [10], we will use for construction AnyLogic simulation software which implements the cognitive model of game simulation described above in terms of system dynamics using flow diagrams and accumulators, variables and other tools [5]. The simulation model of countering the organizational conflict with due regard for the type of employees' temperament is shown in Fig. 2.

Basically, the model corresponds to the description given in [10] and includes as accumulators (analogues of concepts of the cognitive model) Pre-conflict situation, Conflict, Escalation, De-escalation. Moreover, to implement the de-escalation procedure, the model provides for the following accumulators (corresponding to one of the five recommended strategies of behaviour): Cooperation, Compromise, Rivalry, Avoidance, Evasion. The game leader sets data on the factors of conflict risk in a particular situation using the interval (0–100) setters Resource constraints, Management shortcomings, Psychological climate, Authority of the Manager and

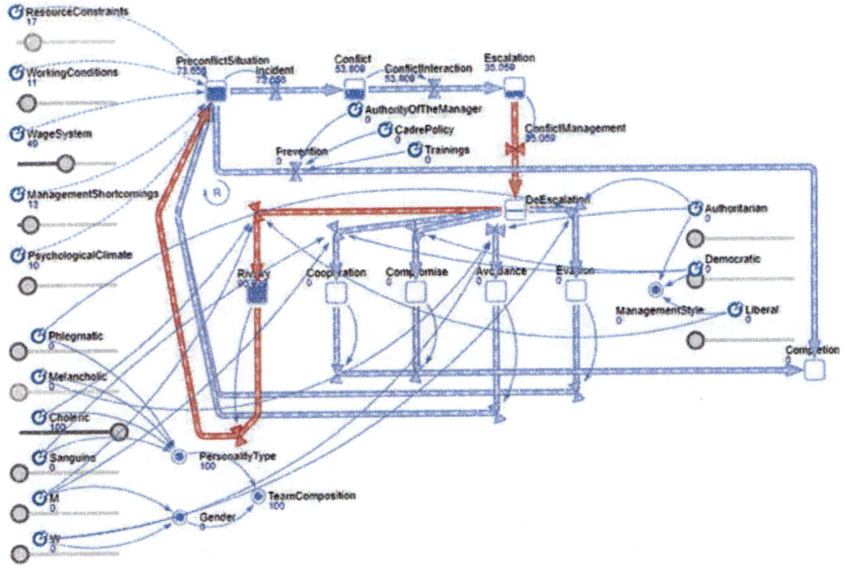

Fig. 2 The simulation model of countering the organizational conflict with due regard for the type of employees' temperament

Cadre policy. The data on the level of conflict and data on escalation rates, respectively, are determined by the flow characteristics of Incident and Conflict interaction. In contrast to the model described in [10], this one suggests implementation of the mechanism for taking into account the type of temperament. For this purpose, the game leader has the opportunity to form the characteristics of prevalence of choleric, melancholic, phlegmatic and sanguineous temperaments in the team of employees with the help of interval (0–100) setters of the Personality type family (in the future, there is a possibility to include the prevailing gender characteristics using the Team composition, Gender parameters during the game).

For a player trained using such a simulation model, the task of constructing a leadership style in a specific conflict situation becomes much more complicated. He/she needs to solve the traditional problem of constructing a style based on the entire set of initial data, i.e. not only the characteristics of the organizational conflict cycle, but the restrictions imposed on a possible solution by the preferences of employees in terms of strategies of behaviour in the conflict. For example, let us assume that choleric people predominate in the team, respectively; the Choleric slider is set to 100 by the game leader. As far as is known [17], choleric people are characterized as hot-tempered, strong, fast, expressing their feelings and emotions vividly, due to such a temperament they prefer the behaviour strategy of Rivalry in the conflict. In Fig. 2, the cycle of the organizational conflict with the behaviour strategy of Rivalry is highlighted in red. But such a strategy corresponds only to the liberal leadership style [10] (but not completely). In this case, the player must set the level of the style type of Liberal to 100 and Authoritarian, Democratic to 0 (see Fig. 2) using the interval setter. However, the choice of the liberal style can aggravate the conflict situation that is emerging in accordance with the data on the conflict risk and the escalation rate. The player finds himself/herself already in a state of an internal conflict when it is necessary to construct more complex combinations of leadership styles taking into account the revealed contradictions of the initial data.

4 Discussion

A similar situation is observed with other types of temperament. Melancholic people, in contrast to choleric people, are prone to strong and rather long emotional stress; therefore, in a conflict they prefer the strategy of Avoidance. Another type of temperament is phlegmatic. They are characterized as slow and balanced, calm and non-conflict, i.e. avoiding conflicts, hence the strategy of their behaviour is Evasion. And the last fourth type of temperament is sanguineous. Sanguineous people are similar to choleric people; they are quick and agile, with a vivid expression of feelings and emotions but less strong and quickly fading away. This temperament is characterized by the Cooperation strategy. None of the types of temperament is strictly unambiguously consistent through the strategy of behaviour with any of the leadership styles. We remind that the liberal style is based on the strategies of Rivalry and Evasion, the authoritarian style is based on Avoidance and Evasion, the democratic style is based on Cooperation and Compromise.

It is known that it is quite rare to meet bright representatives of one or another type of temperament in real life, basically people combine traits of different types of temperament, among which one predominates and the others complement. So, in reality, there is no pure temperament type of a conflict. The situation with the predominance of some type of temperament in a team of employees looks even more complicated. A team is invariably a conglomerate, an interaction and a mutual influence of temperaments. But using a pure type of leadership style is also hard to imagine. Therefore, the game leader can offer the player an almost infinite number of options for tasks based on the variety of actually observed organizational conflicts, and the player is faced with an interesting and difficult task of constructing the most effective leadership style in the context of emerging preferences of the strategies of behaviour of employees in accordance with the type of temperament. The training process can be as close as possible to the real situation through the preliminary collection of data about the organization. Plunging into a specific conflict situation, the player is forced to leave the usual leadership style and, overcoming significant intellectual and psychological stress, construct a variable leadership style that is most consistent with reality. The leader's game experience worked out in critical conflict situations will allow avoiding a sketchy character of organizational decisions in the future and developing aт ambition for a variable leadership style in the context of organizational conflicts.

The player's reaction to the initial and current situation already removes him/her from the perception of the game environment and allows him/her to evaluate his/her decisions at the level of forecasting real conflict management in the work team. In this form, the model can be used not only as a game training one but also as an advisory or forecasting one in managing and resolving workplace conflicts.

5 Conclusion

We have suggested a cognitive model of a business game for constructing a leadership style in a conflict situation. The model is a contour-nodal structure where nodal inclusions in the conflict cycle allow the game leader and the player to build game strategies taking into account the influence of the prevailing types of temperament of employees in the team on specific conflict de-escalation procedures.

A simulation model is implemented in the paradigm of system dynamics and in the notation of AnyLogic simulation software for the game-based training of a leader for counteracting the organizational conflict in the context of variability of authoritarian, democratic and liberal leadership styles. The proposed simulation model represents an organizational conflict not in the form of separate procedures with explicit causal dependencies but in the form of a dynamic system where the initial data of the conflict risk, organizational structural and psychological attitudes form a complex conflict situation with implicit integrative characteristics. This is how a new, higher level of intelligence in the game is achieved, i.e. a fundamental withdrawal from "coaching" for primitive leadership schemes.

When simulating the game, taking into account the prevailing types of temperament in the team that influence specific procedures for de-escalating the conflict fundamentally changes the psychological environment. The player finds himself/herself in a state of internal conflict when it is necessary to construct more complex combinations of leadership styles with due regard for the revealed contradictions in the data on the conflict risk and on the preferences in the strategies of behaviour of employees in accordance with the type of temperament.

The presented simulation model is focused on the game-based training of a leader in countering organizational conflicts; however, the range of tasks solved on its basis can be significantly expanded.

Acknowledgements The study has been carried out with the financial support of Southern Federal University as part of the scientific project No. ВнГр-07/2020-05-ФП.

References

1. Vasilenko VG (2014) Game methods of conducting training sessions in higher school. Vestn RMAT 1(10):84–93. URL https://cyberleninka.ru/article/n/igrovye-metody-provedeniya-uchebnyh-zanyatiy-v-vysshey-shkole/viewer
2. Zizovic MR, Plecicb KG (2017) Multicriteria assessment of possible use of computer games in teaching. Vojnoteh glas 3(65):641–650. https://doi.org/10.5937/vojtehg65-13000
3. Trusevich NE (2012) Imitational modeling of organizational conflicts personality—group. In: Proceedings of BSTU: publishing and printing, vol 9, pp 64–71. URL https://cyberleninka.ru/article/n/imitatsionnoe-modelirovanie-organizatsionnyh-konfliktov-lichnost-gruppa/viewer
4. Komarov VF, Aloyan GN (2017) A study of Kurt Lewin's styles of leadership by business games method. World Econ Manag 2(17):110–120. https://doi.org/10.25205/2542-0429-2017-17-2-110-120
5. Borshchev A, Mahdavi A (2019) Simulation models as virtual environments to train and test artificial intelligence for business applications. In: Proceedings of the conference "simulation modeling. Theory and practice (IMMOD-2019)", Yekaterinburg, pp 20–29. URL https://www.elibrary.ru/item.asp?id=41462098
6. Averchenko LK, Doronina IV, Ivanova LN (2013) Imitation business game as a method of professional competence development. High Educ Today 10:35–39. URL https://cyberleninka.ru/article/n/imitatsionnaya-delovaya-igra-kak-metod-razvitiya-professionalnyh-kompetentsiy
7. Mustata IC, Alexe CG, Alexe CM (2017) Developing competencies with the general management II business simulation game. Int J Simul Model 16(3):412–421. https://doi.org/10.2507/IJSIMM16(3)4.383
8. Vaganova O, Egortseva N, Yavorskiy Y, Zotova I, Bulaeva M (2020) Situational and game technologies of training in students' professional competence development. Rev Amazon Investig 9(28):83–91. https://doi.org/10.34069/AI/2020.28.04.10
9. Oliveira PHP (2018) RPG method improving business games. Independ J Manag Prod 9(1):111–124. https://doi.org/10.14807/ijmp.v9i1.669
10. Shakirova AE, Rozin MD, Svechkarev VP, Ozerov AA, Ivanov AV (2020) Designing the management style of organizational conflicts on simulation models. Eng Bull Don 10. URL ivdon.ru/ru/magazine/archive/n10y2020/6587

11. Novikov DA (2005) Theory of management of organizational systems. MPSI, Moscow, p 584. URL http://www.methodolog.ru/books/tyos2005.pdf
12. Raigorodsky DYa (2011) Practical psychodiagnostics. Methods and tests. Bayrakh-M, Moscow, p 672. URL 1raygorodskiy_d_ya_prakticheskaya_psikhodiagnostika_metodiki.pdf
13. Novikov DA, Rusyaeva EYu (2011) Cultural foundations of the theory of management of organizational systems. Quest Cult Stud 7:63–68. URL https://www.elibrary.ru/download/elibrary_16857491_36449258.pdf
14. Rozin M, Svechkarev V, Nesmeyanov E, Ryabtseva I, Yusov S (2020) Model of integration of competencies in a professionally-oriented education. E3S Web Conf 210:7. https://doi.org/10.1051/e3sconf/202021022038
15. Tarasenko LV, Rosin MD, Svechkarev VP, Avanesyan KA (2018) Cognitive analysis in the context of media education: an oriented graph modelling approach. Media Educ (Mediaobrazovanie) 58(4):130–137. https://doi.org/10.13187/me.2018.4.130
16. Magradze T (2020) Application of game theory simulation in enterprise management. Colloq J 23(60):136–140. https://doi.org/10.24411/2520-6990-2020-11559
17. Borodkin FM, Koryak NM (2009) Attention: conflict. Nauka. Sib. otd-nie, Novosibirsk, p 154

Professional Representations of Students in Context of the Development of Digital Technologies

Svetlana Zholudeva, Irina Ulybysheva, and Lyudmila Ivanova

Abstract In this paper, the authors reveal the peculiarities of the ideas of students of pedagogical specializations about the profession in the context of the development of digital technologies. The purpose of the research is to study the peculiarities of the ideas of students of pedagogical specializations about the profession in the context of the development of digital technologies. The paper analyzes the main theoretical approaches to the study of the peculiarities of the formation of ideas about the profession of teaching personnel. The specific features of professional pedagogical activity are considered. The possibilities of vocational education in the context of the development of digital technologies are analyzed. The study involved 122 respondents—students of pedagogical specializations of the Southern Federal University. The research methods were psychodiagnostic techniques and methods of mathematical statistics. The data obtained as a result of the study will allow identifying areas in the context of the development of digital technologies, which will determine effective methods of personal self-realization in the context of the profession being acquired, optimize the motivational component, and allow adjusting the programs, content and forms of training for future professionals.

Keywords Representation · Students · Profession · Teachers · Digitalization

1 Introduction

The relevance of the study, the peculiarities of the ideas of students of pedagogical specializations about the profession in the context of the development of digital technologies, is determined by the importance of recognizing and studying the image of the future profession among students of pedagogical specializations, as well as

S. Zholudeva · I. Ulybysheva (✉)
Academy of Psychology and Pedagogy, Southern Federal University, Rostov-on-Don, Russia
e-mail: Iulybysheva@sfedu.ru

L. Ivanova
South-Russia Institute of Management—Branch of Russian Presidential Academy of National Economy and Public Administration, Rostov-on-Don, Russia

A. Beskopylny and M. Shamtsyan (eds.), *XIV International Scientific Conference "INTERAGROMASH 2021"*, Lecture Notes in Networks and Systems 247,
https://doi.org/10.1007/978-3-030-80946-1_56

determining the specifics of the formation of ideas in the learning process using digital technologies. The increasing digitalization of the educational and professional component of the students' life sets special trajectories of development in the profession, which has a significant impact on the formation of the image of the future profession. The purpose of this study is to identify the peculiarities of the ideas of students of pedagogical specializations about the profession in the context of the development of digital technologies, which will make it possible to determine effective methods of personal self-realization in the context of the acquired profession, optimize the motivational component and adjust the programs, content and forms of training of future professionals in the context of the development of digital technologies.

The theoretical basis of this research is represented by such fundamental works as: Bayerl et al. [1], Horton, Jacobs, Klimov, Kondratyev, Petrovskaya, in which the problems of the image of the future profession, its formation and development are considered. The stages of professionalization and professional self-determination disclosed in the studies of Fernet, Alexander [2], Morin, Austina, Gagnec, Litaliend, Lavoie-Tremblaye, Forest et al. Professionally important personality traits were revealed in the research by Donaldson [3] and others. Study of the image of the profession from the standpoint of motivation and personality orientation: Rogova, Pankratova, Zholudeva, Sheveleva, Naumenko, Skrynnik, Ulybysheva [4, 5] and others. Peculiarities of pedagogical professions: Gargallo et al. [6], Yessimgaliyeva et al. [7], Alexander and others. Specificity of development of digital technologies in education: Hamalainen et al. [8] and others.

The study of ideas about the profession is the basis for the formation of a personality focused on professional development. The image of the profession lays down the main directions for the development of a specialist in the professional sphere. Ideas, as a mental reflection of reality, are not formed by themselves, but exclusively as a result of purposeful human activity. Professional training, as a specific type of human activity, especially in the context of global digitalization, should have a fundamental goal—to form adequate, effective and high-quality professional knowledge, skills and abilities for effective professional development and formation, as well as to provide students with conscious mechanisms for shaping the image of the future profession. Markauskaite and Goodyear [9] study the image of the profession as an intrapersonal education, which reflects the process of raising professional awareness by comparing the subjective, personal development of a person and the external, specific content of a specific professional sphere. The authors show that in the process of learning, students, mastering certain disciplines, build their own, individual concepts of the image of the profession and their place in the professional sphere. Then, in the process of professionalization, the image of the profession is detailed and becomes more specific, but the basis for emotional and promising development in the profession is laid at the initial stages—during the period of professional education.

Park and Schallert [10] analyzed various theories and concluded that the formation of the image of the profession occurs through the process of professional identity, mediated by mutual interaction with four subprocesses: assessment of congruence between past and current identity, development of disciplinary knowledge and skills, ideas about the possible future "I", and representations of oneself in the community of professional practice.

Features of education at a university with digital learning were studied by Lohr et al. [11], which showed that in the process of development of digital technologies applied in professional education, there are three levels at which teachers initiate digital learning activities: low level (powerpointers), moderate level (clickers), and high level (digital pros). The authors revealed the relationship between the level of proficiency of digital technology teachers and the level of student involvement. Thus, in the context of the development of digital learning, it is important to use an integrated approach to teaching, combining the traditional aspects of learning and the digital component, which will contribute to the successful teaching and learning in higher education institutions.

Merono et al. [12] conducted a large-scale study devoted to the study of: technological pedagogical knowledge content (TPACK) model and co-education; the relationship between the perception of TPACK by teachers of pedagogical activity and their academic achievements. The main results showed that the experimental groups improved their perception of the TPACK model, and their academic achievement increased. The use of these teachers can influence the development of digital competence of future teachers and contribute to the formation of a positive outlook on pedagogical activity in general. Increasing the digital competence of future teachers is an essential aspect, given the current socio-pedagogical scenario.

Cetin [13] conducted a research aimed at studying the process of digital learning development. The research results show that the development of the digital component in education brings a positive contribution to the development of professional education. But such development is possible only under the condition of high competencies of university teachers in the field of digital technologies, which involve the widespread use of the latest technologies in organizing virtual professional tests and immersing students in initial professionalization.

In their study on the students' personal qualities, Vitulic and Zupancic [14] prove the existence of a direct dependence of the success of training and future professionalization on personal characteristics. The authors show the hierarchy of personality traits, which constitute the core of professional ideas and have a systemic nature, which ensures satisfaction with the process and the result of professional activity.

Ahn [15] conducted a study based on the theory of self-determination (SDT), according to which the teacher's motivation indirectly affects the motivation of students by supporting and meeting the basic psychological needs of students. The authors tested the complete model: teacher motivation (autonomous, controlled and amotivation); perceived methods of supporting needs (support for autonomy, structure and involvement); the student needs satisfaction (autonomy, competence and kinship); motivation of students (autonomous, controlled and amotivation); student achievement. The multilevel model of structural equations showed evidence supporting: (1) a positive relationship between the autonomous motivation of teacher and student and (2) its mediation by students perceived as needy, and the autonomy and competence of students, which emphasizes the importance of learning about the profession during the studying phase.

In their work, Stenberg and Maaranen [16] examined the trajectory of the teacher's personality development starting from the stage of professional training and came to the conclusion that the struggle at the beginning of professional training is rooted in fundamental issues related to the student's personality traits, and not only in questions about how to teach effectively. On the basis of the obtained results, the role of pedagogical education in the training of teacher-students and the importance of this stage in the further development of a professional and his attitude to professional pedagogical activity are discussed.

Among students, the formation of an idea about the profession occurs through the filling of objective properties and actions with personal meaning. The completeness and versatility of the image of the profession depends on the characteristics of training. Considering the digitalization of education as a given of obtaining professional education, it is important to take into account the relevance and significance of this stage of professionalization for future professionals. The formation of an adequate image of the profession contributes to the development of the student's readiness to perform certain professional functions and occupy a certain stage in the social world. An adequately formed image of the profession carries a motivational function that contributes to interiorization. It is especially important to study the ideas of the profession among students of pedagogical specializations, since the style of individual pedagogical activity and the development of internal motivation of the future pedagogical worker will depend on how motivationally and emotionally the image of the future profession is colored.

2 Materials and Methods

To solve the set tasks, the methods of theoretical analysis and generalization of scientific data were used. In order to identify the meaningful characteristics of the career orientations of the respondents, the methodology "Career anchors" by Schein and the components of a motivational personality profile "Motivation Management" by Ritchie and Martin were used. To process the results obtained, the Spearman correlation coefficient was used. The study involved students of pedagogical specializations of the Academy of Psychology and Pedagogy of the Southern Federal University in Rostov-on-Don. A total of 122 students, aged 18–20, took part in the study.

3 Results

As a result of empirical research, we obtained data indicating the desire of students of pedagogical specializations for the reliable and stable workplace. This result was shown by 43% of respondents.

Career orientation toward service holds the second place in terms of occurrence (31% of respondents). This orientation implies a focus on interacting with people and serving society as a whole. A high percentage of respondents with the dominance of this motivation characterizes the area of their chosen professional activity, namely pedagogical, which is focused on the upbringing and training of the younger generation, the transfer of knowledge.

The next most common meaningful characteristic of the profession of students of pedagogical specializations is career orientation towards the desire to integrate lifestyles—22% of respondents. Integration of different spheres of life, their harmonious combination and balance determines the way of life of a person, his desire to achieve balance between family, work and career. The frequency of occurrence of this career orientation among respondents in pedagogical specializations is also explained by the fact that pedagogy is based on harmonious and all-round development of the personality, which is reflected in the preferences of the respondents themselves.

The least pronounced characteristics of the content component of the image of the profession of students of pedagogical specializations are: career orientation towards professional competence (2%), autonomy (1%), challenge (0.6%), and stability of residence (0.4%). A graphic representation of the percentage of career orientations of students of pedagogical specializations is shown in Fig. 1.

In our opinion, the fact of a low orientation towards professional competence is of interest. However, we assume that since the respondents were primary students, professional self-awareness and the position of a "professional" have not yet been formed, and a more expanded focus than a focus on professional skill dominates at the optant stage.

The study of the motivational profile of the personality of students of pedagogical specializations showed a high level of development of motivational factors, which are the driving force of personality development in the chosen profession. These factors include the need for self-improvement (41%), the need for interesting and socially useful activities (40%). Dominant needs characterize a conscious professional choice of pedagogical professions that meets the request of the respondents themselves.

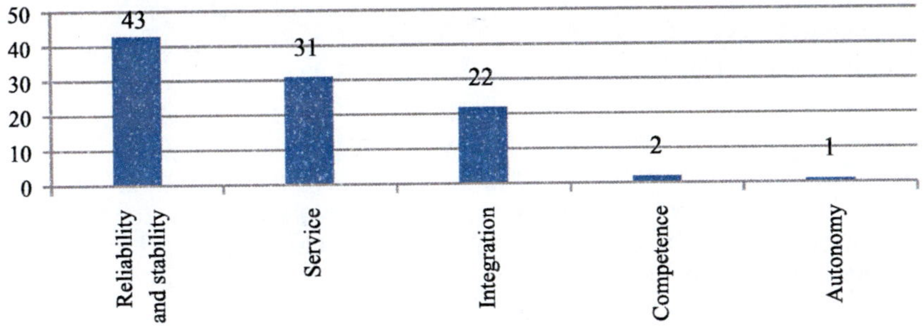

Fig. 1 Distribution of career orientations in the research sample

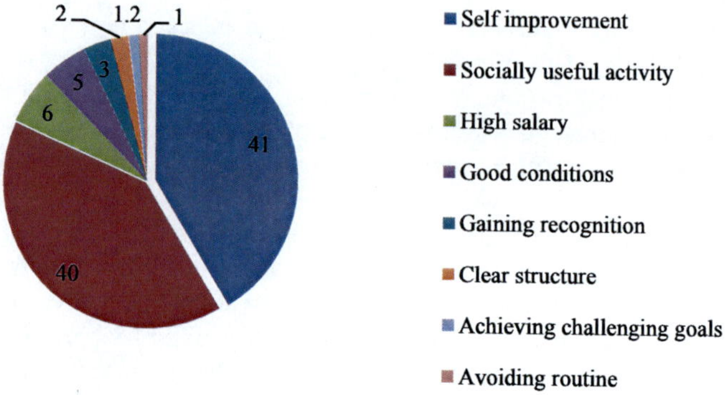

Fig. 2 Motivational personality profile of respondents

The motivational factors of the profession's image, which are not expressed, are: the need for high salary (6%); the need for good working conditions (5%); the need to gain recognition (3%); the need for clear structuring of work (2%); the need to set difficult goals for oneself and achieve them (1.2%); tendency to avoid routine (1%); the need for change (0.5%), and the need to be creative (0.3%). Figure 2 illustrates the distribution of the motivational components of the profession image of the respondents of this study.

The hierarchy of the needs of students of pedagogical specializations in importance is compared with the very purpose of pedagogical activity and reflects not only the respondents' internal need for self-improvement in professional activity, but also an orientation towards high pedagogical tasks.

To identify the relationship between the motivational profile of the profession and the image of the future profession among students of pedagogical specializations, the Spearman correlation coefficient was used. At the level of statistical significance of 0.05 and 0.01, it was found that the orientation of students towards high salary corresponds to the orientation towards the integration of the efforts of other people, the completeness of responsibility for the final result and the combination of various functions of the organization (management ($r = 0.371$, $p = 0.013$)), as well as the integration of their life ($r = 0.422$, $p = 0.004$). The revealed relationships showed that there were no significant relationships with the dominant motivation and orientation of the respondents participating in the survey. But the correlation of the scales allows concluding that the need for a high salary (which is not dominant among our respondents), even with low indicators, implies cooperation in a team of colleagues, the acceptance of personal responsibility for the implementation of the set goal and interchangeability in the performance of various job responsibilities. The indicator of integration of all spheres of life recorded a significant relationship, but the need for high salary was not identified by our respondents as a dominant need.

Requirements for good working conditions correlate with the instability and unreliability of the work itself ($r = -0.375$, $p = 0.012$). The identified indicators reflect the instability in the respondents' ideas about the place of work, conditions and comfort, which entail an orientation towards fulfilling the set goals, and the stability of employment, depending on the consistency and reliability of the work (employment) itself for a long period of time.

Stable relationships are interconnected with ideas about the stability of the place of residence ($r = 0.425, p = 0.004$). Ideas about the embodiment of their ideals and goals, full realization in work ($r = -0.381, p = 0.011$), as well as the stability of the place of residence ($r = -0.340, p = 0.024$) are interconnected with the complexity of the goals and the ability to achieve them. The more difficult the goal, the lower self-realization and stability of living. However, the presence of influence and power contributes to the release from organizational rules, regulations, restrictions and gives a certain degree of freedom and autonomy ($r = 0.396, p = 0.008$).

4 Discussion

The results obtained in the course of empirical measurements and their further statistical processing allow us to draw the following conclusions:

- students of pedagogical specializations, when forming the image of the profession, place the semantic emphasis on the reliability and social significance of the profession. In educational work, it is important to include in the disciplines an orientation towards the ethical and moral aspects of the professional formation of the self-consciousness of the I-professional;
- orientation to service, in which constant interpersonal interaction in the process of performing professional functions, providing feasible help and support to other people, striving to improve the professional activity and the surrounding reality of students of pedagogical professions are characteristic, reflects a number of professional competencies of graduates of a pedagogical specialization;
- the desire to harmonize all spheres of life reflects the need to gain knowledge and master planning;
- the integration of all spheres of activity, on the one hand, is associated with professional development, on the other, with well-being;
- respondents associate the material side of professional activity with organizational and integration skills. With regard to the analysis of career orientations identified as a result of the study, personnel management should be highlighted. However, in the dominant career orientations and motivational needs, these indicators are not dominant.

The main dominant orientation contributes to the development of professional activity at the level of vocation, when a person considers it his duty to do his job with high quality, to receive approval and acceptance from society. The vector of professional development of such respondents is aimed at personal and professional growth, which is characterized by a pronounced need for constant professional development, improvement of professional skills. We can also talk about the aim of students of pedagogical specializations, who are respondents of this study, to interesting and socially significant professional activities. This orientation is the most desirable for pedagogical professions.

The analysis of professional orientations of the profession image of students of pedagogical specializations testifies to the predominance of stability and reliability of the workplace as the main characteristics, as well as the recognition of professional activity as socially desirable and accepted, and the desire to have a certain status approved by society.

In perceptions that are characterized by good working conditions, there is a tendency to associate this with the instability and unreliability of the workplace, which can be explained by an unstable social environment. On the contrary, stable professional relationships have a clear connection with the constancy of the place of residence and the absence of frequent business trips and relocations. Students associate development in the professional sphere with the increasing complexity of professional functions and actions.

Analysis and interpretation of the data obtained indicates that in the process of forming the image of the profession, students of pedagogical specializations have a desire for self-realization, which is caused by personal development, professional self-improvement, and the development of a trajectory of successful professional activity. It is important to note that the personal growth of students in their ideas about the profession is reflected through the desire for autonomy and freedom of professional activity. Also, an important component of the image of the pedagogical profession is the desire of students to be socially accepted and positively assessed by the society.

5 Conclusions

Thus, it can be concluded that the formation of the image of the profession of students of pedagogical specializations is accompanied by a constant comparison of the individual capabilities of a person with the requirements of the profession, as well as with the changing conditions of interaction. In the process of professionalization, the personal qualities of the subject are compared with such components of professional activity as tasks, working conditions, controversial situations, as a result of which contradictions arise between the requirements of professional activity and the level of personality development. Such contradictions act as a driving force for the development of a professional and the formation of professional identity.

In order to increase the adequacy and relevance of the image of the profession of students of pedagogical specializations, the necessary adjustments should be made to the construction of a model of professional education and educational work in the context of the development of digital technologies. The development of digital technologies has enough opportunities to create the basis and adjust the image of the profession corresponding to the realities of the modern world and society. Students of pedagogical specializations, in order to compare their subjective image of the profession with the real one, must identify possible inaccuracies for themselves as early as possible and correct them for further development in the profession as a successful specialist. Pedagogical activity has its own specific characteristics, which are clearly manifested in the image of the profession of the respondents of this study. This is both the focus on service and the desire to engage in socially desirable and demanded activities. The digitalization of education makes it possible to make the process of forming the image of the profession of students more detailed and holistic, which will contribute to the concretization of professional ideas and the development of internal motivation to develop in the pedagogical sphere.

References

1. Bayerl PS, Hortona KE, Jacobsa G (2018) Investigating collective identity configurations across professions. J Vocat Behav 107:168–181
2. Alexander PA (2008) Charting the course for the teaching profession: the energizing and sustaining role of motivational forces. Learn Instr 18(5):483–491
3. Donaldson SI (2001) Overcoming our negative reputation: evaluation becomes known as a helping profession. Am J Eval 22(3):355–361
4. Rogova EE, Pankratova IA, Zholudeva SV, Sheveleva AM, Naumenko MV, Skrynnik NE (2016) Studying of interrelation of professional motivation and professional concepts of specialists with different professional orientation as a basis of career building. Int J Environ Sci Educ 11(17):10137–10153
5. Zholudeva SV, Ulybysheva IN (2020) Innovative method of developing professional ideas of high school students. Int J Psychosoc Rehabil 24(3):2951–2957
6. Gargallo B, Morera I, García E (2015) Innovative methodology at the university. Its effects on learning processes of university students. An Psicol 31(3):901–915
7. Yessimgaliyeva T, Ageleuova A, Beisenbayeva A, Zhansugurova K, Zhorabekova A (2020) Forming diagnostic competence of psychology teachers in a university setting. Think Skills Creat 38:100708
8. Hamalainen R, Nissinen K, Mannonen J, Lamsa J, Leino K, Taajamo M (2021) Understanding teaching professionals' digital competence: what do PIAAC and TALIS reveal about technology-related skills, attitudes, and knowledge? Comput Hum Behav 117:106672
9. Markauskaite L, Goodyear P (2020) Learning for knowledgeable action: the construction of actionable conceptualisations as a unit of analysis in researching professional learning. Learn Cult Soc Interact 100382
10. Park JJ, Schallert DL (2020) Reciprocity between doctoral students' emerging professional identity and their envisionment of a possible future self in real and imagined communities of practice. Learn Cult Soc Interact 26:100434
11. Lohr A, Stadler M, Schultz-Pernice F, Chernikova O, Sailer M, Fischer F, Sailer M (2021) On powerpointers, clickerers, and digital pros: investigating the initiation of digital learning activities by teachers in higher education. Comput Hum Behav 119:106715
12. Merono L, Calderon A, Arias-Estero JL (2021) Digital pedagogy and cooperative learning: effect on the technological pedagogical content knowledge and academic achievement of pre-service teachers. Digital pedagogy and cooperative learning: effect on technological and pedagogical knowledge of content and academic performance in initial teacher training. J Psychodydiact 26(1):53–61
13. Cetin E (2021) Digital storytelling in teacher education and its effect on the digital literacy of pre-service teachers. Think Skills Creat 39:100760
14. Vitulic HS, Zupancic M (2014) Robust and specific personality traits as predictors of adolescents' final grades and GPA at the end of compulsory schooling. Eur J Psychol Educ 28(4):1181–1199

15. Ahn I, Chiu MM, Patrick H (2021) Connecting teacher and student motivation: student-perceived teacher need-supportive practices and student need satisfaction. Contemp Educ Psychol 64:101950
16. Stenberg K, Maaranen K (2021) A novice teachers teacher identity construction during the first year of teaching: a case study from a dialogical self-perspective. Learn Cult Soc Interact 28:100479

Technologies for Creating a Positive Image of Educational Organizations (On the Example of the Federal Penitentiary Service of Russia)

Alexander Parshkov, Alexander Machkasov, Svetlana Chudakova, Tatyana Simakova, and Dmitry Gorban

Abstract Modern realities of educational activities in most countries are characterized by the expansion of cross-cultural communications, increasing the role of intersubjective interaction and voluntary decision-making by subjects in all spheres of life, including education. In the context of competition in the educational services market, such parameters as brand, image and reputation have become intangible assets of educational organizations that help them simultaneously achieve their goals and optimize their development. In connection with the above, there is no doubt about the relevance of activities to create the image of educational organizations as a phenomenon of interaction with civil society institutions and individual citizens, as well as targeted interaction with the media. In order to conduct an empirical study of technologies for the formation of a positive image of educational organizations in the example of the Federal penitentiary service of Russia, the authors used the method of questioning cadets of 1st, 3rd and 5th courses, the teaching staff and employees of other subdivisions, including employees of press services of educational organizations of the Federal penitentiary service of Russia, concerning their attitude to the image of their educational institution, as well as their personal contribution to its formation.

Keywords Image · Educational organizations · Management · Federal penitentiary service of Russia · Formation · Mass media

1 Introduction

In recent years, much attention has been paid to the concept of "image" in the scientific literature. Scientists have developed a whole philosophy of image, which

A. Parshkov (✉) · A. Machkasov · S. Chudakova · T. Simakova
Academy of Federal Penitentiary Service of Russia, Ryazan 390000, Russia

D. Gorban
Saint Petersburg University of the Federal Penitentiary Service of Russia, Saint Petersburg, Pushkin 196602, Russia

A. Beskopylny and M. Shamtsyan (eds.), *XIV International Scientific Conference "INTERAGROMASH 2021"*, Lecture Notes in Networks and Systems 247,
https://doi.org/10.1007/978-3-030-80946-1_57

consistently proves the influence of an external, purposefully formed image of an object on its future fate [1].

Image is a universal category and can be applied to any object of study of social and humanitarian disciplines [2]. The image has always existed, but in our time its importance has increased significantly. This may be due to the fact that the image is a natural product of processing a large flow of information [3].

Russian researches Semyonov and Maslova characterize image as a purposeful or spontaneous formation of an object reflection in the minds of people, and the object reflected in the image is a person, a group of people, an organization. This definition characterizes the combination of purposeful activity and the spontaneous image formation, not excluding the influence of unconscious factors, that, in its turn, is the evidence of the complexity of the image formation process [4].

It is appropriate to note that the presence of an image is necessary not only for the person, but also for the organization. This image is called "corporate". Corporate image—the image of an organization formed in the public consciousness, that is, the positioning technique is used not only to create the image of individuals, but also of the organization (institution, firm, etc.) as a whole. At present, there is a significant amount of research by both foreign and domestic experts, who have studied the problem of image in our society [5].

Among the works of foreign authors devoted to the image of organizations and corporate image, we should point out the works of A. Deyan, B. Karlof, F. Kotlar, H. McKay, F. Rogers, Lee Iacocc. In the early 20s, the foundation of foreign image-making was laid by such works as "Public opinion" by W. Lippman, "Herd instincts" by W. Trotter, "Crowd behavior" by E. Martin, "Crowd Psychology" by G. Lebon.

Corporate image is also the subject of research by some modern foreign scientists [2, 6]. According to public relations specialist Schoenfeld, the corporate image should contain the following four components: product image; managerial and financial image; public image; corporate image as an employer [7].

Stokes, president of the management firm "Stokes, Wood and Associates" identified three grounds for the image of a corporation: first, it must be presented as a "person"; second, the corporation must have its "reputation"; and third, the corporation must show its "character" [8].

Levinson justified the opinion that the perception of any organization is a mixture of what it does and what people think it should be. If these two things coincide—the organization's image is harmonious [9].

The process of image formation is also typical for educational organizations of various levels. According to Guchanova, the main directions of optimizing the image of an educational organization are the following parameters:

- a system for providing targeted information to consumers about their potential, success, and offered educational services;
- bright, catchy, up-to-date information materials for external representation[10].

In the context of analyzing the problem of an educational organization's image, the University brand model developed by Selyukov and Shalygina, based on the values and associations of consumers of educational services, deserves attention [11]. Thus, all the above gives reason to believe that the image is a representation that is under control, it is created by people consciously and thoughtfully as an image aimed at forming an attitude to themselves, as well as to something or someone.

At the present stage of development of Russian society, one of the most pressing problems is the formation of a positive image of law enforcement agencies, including the Federal penitentiary service (hereinafter—the FPS of Russia). In particular, educational organizations of the FPS of Russia come on the foreground in this aspect.

Educational organizations of the FPS of Russia occupy a special place in the formation of the overall image of the Penal system of the Russian Federation.

For educational institutions of the FPS of Russia's image has important functional and status value as an indicator of the level of confidence of the population and the criterion of a society of management efficiency and, by and large, the government's reforms. Thus, the image largely determines the behavior of citizens in relation to the FPS of Russia and, in particular, their employees. The formation of a positive image of educational organizations of the Federal penitentiary service of Russia with the help of mass media is the most important condition for the effective functioning and development of this social institution.

2 Materials and Methods

To conduct an empirical study of technologies for creating a positive image of educational organizations on the example of the FPS of Russia, the questionnaire method was used. Based on the structure of the corporate image of educational organizations of the FPS of Russia, the authors developed a special survey method for collecting empirical material (questionnaire). An independent survey method (questionnaire) was formed for each survey group, taking into account the specifics and professional characteristics of the respondents.

In order to obtain the most complete and reliable results of the survey a group of 150 respondents was interviewed in 8 educational organizations of the FPS of Russia:

Vologda Institute of Law and Economics (VILE, Vologda);
Samara Law Institute (SLI, Samara);
Academy of the FPS of Russia (Ryazan);
Pskov branch of the Academy of the FPS of Russia (Pskov);
Voronezh Institute (Voronezh);
Perm Institute (Perm);
Vladimir Law Institute (Vladimir),
Kuzbass Institute (Novokuznetsk).

The total number of respondents was 1200.

The General group of respondents to the questionnaire was divided into four subgroups:

Junior cadets (1st and 2nd years);
3rd year cadets;
graduate students (4 and 5);
teaching staff (both military and civilian personnel), including employees of departments and services of educational organizations of the FPS of Russia (educational department, training units, etc.).

The majority of respondents had less than five years of service in the Penal system, due to the fact that the majority of them are cadets of educational organizations.

To determine professional opinion and expert assessments of technologies of positive image formation among educational organizations, for example, the FPS of Russia in a specially designed questionnaire were interviewed 15 employees and workers of press services of 8 the above mentioned educational institutions of the FPS of Russia.

3 Results

When analyzing technologies for creating a positive image of educational organizations on the example of the FPS of Russia, the concept of "corporate image of an educational organization of the FPS of Russia" takes an important place.

Based on approaches to the corporate image of educational organizations available in the scientific literature, four concepts of corporate image were identified [12].

The first question for respondents was formulated as follows: "What is the definition of the corporate image You seem to be the most accurate?"

The analysis of answers revealed that the most accurate practically in all universities of the Federal penitentiary service of Russia (except the Voronezh Institute and Vladimir Law Institute of the FPS of Russia) believe the following:

corporate image is a holistic perception of various community groups that are formed based on information stored in their memory information about the various aspects of its activities.

The percentage of respondents' answers to this concept is shown in Table 1.

Forming a positive image of an educational organization of the FPS of Russia, it is necessary, first of all, to determine the purpose of this process. In foreign scientific literature, the goal of forming a positive image of an educational organization is to increase its competitiveness [13, 14]. However, an educational organization of the FPS of Russia can determine other goals of this activity forming a positive image.

In this regard, we asked the respondents: "Why does an educational organization of the FPS of Russia need a properly created image?"

Most of the respondents expressed their opinion in order to raise the prestige of the profession of the Penal system's employee.

The percentage of respondents' answers to this concept is shown in Table 2.

Table 1 Results of the survey methodology on the definition of corporate image

Educational organization	VILE, Vologda	SLI, Samara	Academy	Pskov branch
Results of the survey (%)	42.8	38	57.3	39.8
Educational organization	Voronezh Institute	Perm Institute	Vladimir Law Institute	Kuzbass Institute
Results of the survey (%)	28.5	35.8	26.7	55.7

Note VILE, Vologda = Vologda Institute of Law and Economics; SLI, Samara = Samara Law Institute

Table 2 Results of the survey methodology on the definition of corporate image

Educational organization	VILE, Vologda	SLI, Samara	Academy	Pskov branch
Results of the survey (%)	40.1	38	58.7	37.4
Educational organization	Voronezh Institute	Perm Institute	Vladimir Law Institute	Kuzbass Institute
Results of the survey (%)	26.3	27.7	35.3	61

Note VILE, Vologda = Vologda Institute of Law and Economics; SLI, Samara = Samara Law Institute

The process of image forming of educational organizations of the FPS of Russia is quite complex in its structure and content and in addition to the goal includes a number of elements: subjects, means, factors, etc. A special place in the structure of the process of image forming of an educational organization of the FPS of Russia is occupied by the means with which this process is carried out. In this regard, we asked the following question to the respondents: "What means form the image of a modern educational organization of the FPS of Russia?"

The majority of respondents indicated that the image formation of an educational organization of the FPS of Russia depends on good specialists of this organization.

The percentage of respondents' answers to this concept is shown in Table 3.

Among the factors that influence the formation of the image of educational organizations of the FPS of Russia, the respondents indicated:

social guarantees for employees of an educational organization—VILE, Vologda (40.8%), SLI, Samara (42%), Perm Institute of the FPS of Russia (35.8%);
availability of modern educational and material base of the educational organization—Academy of the FPS of Russia (30%), Pskov branch of the Academy of the FPS of Russia (33.1%);
personal contribution of each employee to the image formation of an educational organization—Voronezh Institute of the FPS of Russia (29.6%), Vladimir Law

Table 3 Results of the survey methodology on the issue of means forming the image of an educational organization

Educational organization	VILE, Vologda	SLI, Samara	Academy	Pskov branch
Results of the survey (%)	42.2	38.7	42	40.5
Educational organization	Voronezh Institute	Perm Institute	Vladimir Law Institute	Kuzbass Institute
Results of the survey (%)	30.7	27	31.3	53

Note VILE, Vologda = Vologda Institute of Law and Economics; SLI, Samara = Samara Law Institute

Institute of the FPS of Russia (34%); participation of an educational organization in citywide holidays and celebrations—Kuzbass Institute of the FPS of Russia (40.3%).

Among other factors contributing to the formation of the image of the educational institutions of the FPS of Russia effective communication, which is performed by the press officers of these universities, is important. In turn, this factor consists of a number of elements. An important place in the process of forming a positive image of educational organizations of the FPS of Russia is the symbolism of the University. This symbolism is an integral part of the corporate identity of the educational organization of the FPS of Russia. In turn, corporate identity contributes to the formation of a positive image of the educational institution. Corporate identity is a set of permanent visual and textual elements that identify belonging to a particular educational institution and distinguish the educational institution from competitors, forming a unique image of the institution [15].

In order to find out the influence of corporate identity and types of symbols on the formation of the image of educational organizations of the FPS of Russia we asked the following question to the respondents: "Do you think that the symbolism of an educational organization of the FPS of Russia is important for the development of its image?"

The majority of respondents answered that the symbolism of an educational organization is of great importance for the development of its image. The percentage of respondents' answers to this concept is shown in Table 4.

Among the most significant types of symbols that form the image of an educational organization of the FPS of Russia are the following:

badge of a graduate—VILE, VOLOGDA (76.6%); banner—SLI, Samara (76.6%), Academy of the FPS of Russia (75.3%), Perm Institute of the FPS of Russia (76.6%), Vladimir Law Institute of the FPS of Russia (80%), Kuzbass Institute of the FPS of Russia (76.6%);
institute logo—Pskov branch of the Academy of the FPS of Russia (84.6%);
emblem—Voronezh Institute of the FPS of Russia (74%).

Table 4 Results of the survey methodology on the importance of symbols of educational organizations

Educational organization	VILE, Vologda	SLI, Samara	Academy	Pskov branch
Results of the survey (%)	73.8	67	67.6	79.3
Educational organization	Voronezh Institute	Perm Institute	Vladimir Law Institute	Kuzbass Institute
Results of the survey (%)	64.8	67	68.1	67.4

Note VILE, Vologda = Vologda Institute of Law and Economics; SLI, Samara = Samara Law Institute

The formed positive image of the educational organization of the FPS of Russia affects the level of its competition in the market of educational services, as well as the number of applicants who want to enter this educational institution. Competitiveness, as we mentioned above, is one of the main goals of the process of forming the image of an educational organization of the FPS of Russia.

Developed a positive image of the educational organization of the FPS of Russia may become a kind of measure of the degree of development of all companies, evaluating prospects of their undertakings, maturity and professionalism of the entire team, the timeliness methodological product and creativity of methodical work in educational organizations. In this regard, we asked the following question to respondents: "Did the image of your educational organization influence its choice as your place of study or service (work)?"

The majority of respondents expressed the opinion that the image of the educational organization of the FPS of Russia to some extent influenced its choice as a further place of study, although this was not a determining factor in choosing an educational institution. The percentage of respondents' answers to this concept is shown in Table 5.

Table 5 Results of the survey methodology on the impact of the image of an educational organization on its choice for admission

Educational organization	VILE, Vologda	SLI, Samara	Academy	Pskov branch
Results of the survey (%)	60.5	52	43.3	57
Educational organization	Voronezh Institute	Perm Institute	Vladimir Law Institute	Kuzbass Institute
Results of the survey (%)	20.7	28.4	36.7	32.2

Note VILE, Vologda = Vologda Institute of Law and Economics; SLI, Samara = Samara Law Institute

Table 6 Results of the survey methodology on the impact of the image of an educational organization on its choice for admission

Educational organization	VILE, Vologda	SLI, Samara	Academy	Pskov branch
Results of the survey (%)	49.6	42.7	44	43.6
Educational organization	Voronezh Institute	Perm Institute	Vladimir Law Institute	Kuzbass Institute
Results of the survey (%)	29	39.9	40	61

Note VILE, Vologda = Vologda Institute of Law and Economics; SLI, Samara = Samara Law Institute

Today, the mass media play an important and significant role in the process of forming the image of educational organizations of the FPS of Russia. We can say that the media is one of the most important factors that influence and participate in this process. The life of modern society is unthinkable without mass media, which are the largest element of mass communication in society and serve to collect, distribute information, form public opinion, and maintain stability in society.

In this regard, we asked the following question to respondents: "What are the media that have a greater influence on the image formation of an educational organization of the FPS of Russia?"

The majority of respondents indicated departmental periodicals (Table 6).

Analysis of the periodicity of information placement in internal media shows their rather high intensity. However, this cannot be said about the placement of information about the educational organization of the FPS of Russia at the level of Regional and Federal media. For example, only 30% of respondents publish information in print Federal media no more than 2 times a year. Only 40% of respondents post information on Central television 1–2 times a year. 40% of respondents do not publish information on the radio in the Federal media. In this regard, we consider it necessary to strengthen this area of activity of the press services of educational organizations of the FPS of Russia and more frequent interaction and publication of information about the activities of educational organizations at the level of Regional and Federal media.

As we can see from the answers received from respondents and employees of the press services, the official website of the University occupies a very important place in the process of forming the image of educational organizations of the FPS of Russia. We do not consider these responses to be random. Recently, the role of official websites of educational organizations of the FPS of Russia has been increasing.

The official website of an educational organization of the fps of Russia should have a special style and content, create a positive image, effectively use the capabilities of the global computer network, perform representative functions for various categories of potential visitors, and play the role of a link between the organization, its employees, students and their parents, as well as public organizations and media in the region and country. In other words, the official website of an educational organization of the FPS of Russia is a tool for spreading information, promoting interaction between all participants in the educational process and reflecting the brand of the educational organization. Placing information on the official website of the educational organization of the FPS of Russia and responsibility for this area of activity is assigned to the employees of the press services.

In addition to the official website of the educational organization of the FPS of Russia, the image is formed using social networks and messengers. Employees of the press services are also responsible for this area of activity.

4 Conclusion

The conducted empirical research of the role of mass media in the process of image formation of educational organizations of the FPS of Russia allowed us to formulate a number of main recommendations for the development of their image:

- the purpose of forming a positive image of an educational organization of the FPS of Russia is to raise the prestige of the profession of the Penal system's employee.
- in educational institutions of the FPS of Russia (as in other departments of the FPS of Russia) media liaison is handled by a special department—the press service.
- currently, it is necessary to adopt a regulatory document regulating the activities of the media in relation to such a state structure as the FPS of Russia.
- it is necessary to pay special attention to the symbols of the University: the banner of the educational institution, the badge of the graduate, the logo, the emblem. Appropriate specialists should be involved in the development of badges, logos and emblems of educational organizations of the FPS of Russia.
- important attention should be paid to maintaining the official website of the educational organization of the FPS of Russia and official accounts in social networks.
- among the most effective methods of countering negative information published in the media about an educational organization, press service employees identify the following: operational review at the scene or events management or a representative of the press service of the educational institution; organization of press-conferences, briefings with the leadership or spokesman of an educational organization; the location of the refutation in the news feed on the website of the educational organization or social networks; appeal to the court.

References

1. Mostafa RB, Lages CR, Shabbir HA, Thwaites D (2015) Corporate image: a service recovery perspective. J Serv Res 18(4):468–483. https://doi.org/10.1177/1094670515584146

2. Kehinde OA (2012) Organizational culture and its corporate image. A model juxtaposition. Bus Manag Res 1(1):121–132. https://doi.org/10.5430/bmr.v1n1p121
3. Latif WB, Islam MdA, Noor IMd, Mohamad M, Kongsompong K (2016) Imagination of brand image for tourism industry. Probl Perspect Manag 14(2–1). https://doi.org/10.21511/ppm.14(2-1).2016.02
4. Semyonov AK, Maslova EL (2010) Psychology and ethics of management and business, 2nd edn. Moscow
5. Pérez JP, Torres EM (2017) Evaluation of the organizational image of a university in a higher education institution. Contad Adm 1(62):123–140. https://doi.org/10.1016/j.cya.2016.01.007
6. Gouji RK, Taghvaei R, Soleimani H (2016) The effect of corporate image on the formation of customer attraction. Manag Sci Lett 6:655–670. https://doi.org/10.5267/j.msl.2016.8.003
7. Schoenfeld C (1963) Publicity media and methods. New York, p 371
8. Stokes PM (1968) A total system approach to management control. New York, p 160
9. Levinson H (1966) How to undermine an organization. Public Relat J 10:82
10. Guchanova AS (2016) Managing the creation of an educational institution's image. Young Sci 24:520
11. Selyukov MV, Shalygina NP (2012) Brand technologies in the management system of higher education institutions. Mod Probl Sci Educ 5:260
12. Polinkevych O, Kamiński R (2018) Corporate image in behavioral marketing of business entities. Innov Mark 14:33–40. https://doi.org/10.21511/im.14(1).2018.04
13. Arabaci IB (2010) Academic and administration personnel's perceptions of organizational climate (sample of Educational Faculty of Firat University). Procedia Soc Behav Sci 2(2):4445–4450. https://doi.org/10.1016/j.sbspro.2010.03.709
14. Sokol A, Gozdek A, Figurska I, Blaskova M (2015) Organizational climate of higher education institutions and its implications for the development of creativity. Procedia Soc Behav Sci 182:279–288. https://doi.org/10.1016/j.sbspro.2015.04.767
15. Marchenko MN, Lapchenko AK (2016) Corporate identity as a means of forming the image of an educational institution. Young Sci 11:1610–1612

Features of Professional Wellbeing of Teachers

Yulya Tushnova, Maria Vyshkvyrkina, and Natal'ia Dunaitseva

Abstract The phenomenon of professional wellbeing is included in the context of research on psychological wellbeing. The researchers point out the importance of the professional and labor sphere of human activity for ensuring the general state of subjective wellbeing. At the same time, the ability to maintain a state of subjective professional wellbeing is the main condition for the successful work of a teacher. The purpose of this study was to study the characteristics of the experience of professional wellbeing of teachers working in educational organizations of different levels. The study involved 65 teachers engaged in professional activities in preschool educational institutions and secondary schools (M = 31.3; SD = 1.2; 28% men). Methods were used: the psychological wellbeing of K. Riff in the adaptation of N. N. Lepeshinsky, the professional demand for the personality of E. V. Kharitonova, B. A. Yasko, as well as statistical methods (descriptive statistics, Mann Whitney U test, Spearman rank correlation coefficient). The differences between the experience of psychological wellbeing and the professional demand for teachers engaged in professional activities in educational institutions of different levels were established. The interrelationships of psychological well-being and professional demand of the individual in the group of teachers were also found. The prospects for further development of this topic are the expansion of the level of educational institutions, the study of factors that affect the level of psychological wellbeing of teachers and determine the level of professional demand for an individual, the study of the factor structure of professional wellbeing of teachers.

Keywords Psychology · Teachers · Psychological wellbeing · Professional wellbeing · Professional demand of the individual

---

Y. Tushnova (✉)
Don State Technical University, Gagarin Sq. 1, Rostov-on-Don 344003, Russia

M. Vyshkvyrkina
Southern Federal University, Bolshaya Sadovaya Str., 105/42, Rostov-on-Don 344006, Russia

N. Dunaitseva
South Russian Humanitarian Institute, Krasnoarmeyskaya Str., 108, Rostov-on-Don 344082, Russia

A. Beskopylny and M. Shamtsyan (eds.), *XIV International Scientific Conference "INTERAGROMASH 2021"*, Lecture Notes in Networks and Systems 247,
https://doi.org/10.1007/978-3-030-80946-1_58

1 Introduction

The process of any professional activity involves a number of problems and difficulties that are necessarily faced not only by young professionals, but also by experienced professionals.

A special aspect of pedagogical activity is its high stressfulness (the breadth of the teacher's functions, high responsibility for the duties performed, increased intellectual load, the need for constant improvement). Research in this area shows that many teachers set a low indicator of the degree of social adaptation, do not have ways to adequately respond emotionally to difficulties and unforeseen difficulties. These studies update the study of the problem of professional wellbeing of teachers. For a teacher, psychological stability and the ability to maintain a state of subjective professional wellbeing is the main condition for successful work throughout the entire period of active professional activity.

The study of professional wellbeing in modern psychology is carried out within the framework of positive psychology (Boniwell, Wong, Robins, Seligman, Friedman, Chiksentmihayi) [1, 2]. For the first time, the emphasis on the study of a healthy personality and the maintenance of its harmonious functioning was made within the framework of humanistic psychology (Rogers, Maslow, Allport) [3–5]. Thus, according to the views of Rogers, a fully functioning personality is a person in the process, open to new experience and congruent in this process [4]. Maslow the main need of the individual is the need for self-actualization, that is, the functioning of the individual is aimed at the maximum realization of its potential capabilities [3].

Ermolaeva and Lubovsky point out that the term "psychological wellbeing" has many definitions. Thus, a number of authors point out that psychological wellbeing is "a positive assessment of oneself and one's own life, a subjective state of satisfaction with one's work, one's life, and a dominant mental state. Psychological wellbeing is also considered as a set of necessary personal resources that ensure the success of the individual in the "subject—environment" system; the orientation of the individual towards positive functioning and the result of this orientation, which is expressed in the experience of happiness and satisfaction with one's own life" [6, p. 92].

Zausenko considers the concept of psychological wellbeing as an intrapsychic phenomenon, through the prism of a system of relations, highlighting its corresponding components: attitude to oneself, attitude to the surrounding reality and attitude to one's functioning [7].

The phenomenon of professional wellbeing is included in the context of research on psychological wellbeing [8–15]. Argyle [16], Kulikov [17] in their studies describe that the overall level of satisfaction is higher in working people than in the unemployed. Skripkina believes that psychological wellbeing is an indicator of personal and professional wellbeing, i.e. without satisfaction with professional activities, without optimal professional wellbeing, it is impossible to feel general psychological wellbeing [18]. A similar point of view is presented in the works of Druzhilov. According to the author, professional wellbeing is "a necessary condition for a person to feel the general wellbeing of their own life as a whole". At the same time, professional wellbeing is not just satisfaction with one's professional activity. It includes both adaptation to the activity and the professional group, and professional identity, and awareness and acceptance of professional values [19].

Berezovskaya believes that professional wellbeing is an independent phenomenon and the main problem of modern research is the definition of this concept. The author's analysis of the research allowed her to speak about the complexity of the problem of psychological wellbeing in the context of professional activity [20].

The lack of a generally accepted definition of the term "professional wellbeing" also leads to a variety of ideas about the structure of professional wellbeing. Thus, Warr considering wellbeing in the context of professional activity, believed that it is determined by the conditions and content of the professional activity itself. Developing the model of professional wellbeing, the author identifies four of its basic components: emotional wellbeing, autonomy, competence, and the desire for development and growth [20].

Ruth and Augustova adapting the model of professional wellbeing of Riff K. also identified four basic components: professional self-acceptance, professional development, autonomy in professional activities and positive relationships in the team. At the same time, each component includes components such as satisfaction with competence, professional growth, satisfaction with achievements, etc. [21].

Pahol considering the model of Ruth and Augustova insufficiently proven, developed his own model, identifying six predictors of professional wellbeing: "flow in professional activity, satisfaction with the organization and working conditions, professional motivation, satisfaction with professional choice, satisfaction with professional life, meaningfulness of professional activity, assessment of professional implementation, skill and professional identity" [22, pp. 14–15].

The analysis of the definitions of professional wellbeing allows us to generalize its components, namely, emotional (professional happiness, satisfaction with professional activities, attitude to oneself as a professional), cognitive (conscious attitude to the profession as a means of self-realization, attitude to professional aspects of life, attitude to the results of activity), behavioral (motivational-need sphere of professional activity, social subjective activity, interest in work, search for ways of self-realization in the professional sphere).

Considering the question of personal determinants of professional wellbeing of teachers, Zausenko identifies personal and motivational determinants. The author refers to the personal determinants of the qualities that determine the teacher's desire for self-development. Motivational determinants include the qualities that determine the attitude to oneself: self-worth, low self-esteem, self-guidance, psychoenergetic potential, openness, trust in one's intuition [23, 24].

Litvinova and co-authors, studying the personal and professional wellbeing of teachers of educational complexes at different stages of the innovation process, proved that the lowest level of both psychological and professional wellbeing is observed in teachers of educational organizations who have been working in the conditions of innovation for a long time. At the same time, teachers of educational institutions who are just beginning to introduce various innovative technologies feel emotionally more comfortable. The authors also proved the existence of an inverse relationship between the level of psychological wellbeing and the level of emotional burnout [25]. Panova revealed that the stages of professional socialization diverge from the process of professional wellbeing, which, in turn, depends on the experience and attitude to what is happening, individual style of activity, professional experience and taking into account the content of the leading structural components [26].

Interesting results were obtained by Laurmann and Koenig. Studying the professional competence and professional wellbeing of teachers, the authors come to the conclusion that knowledge about the peculiarities of the educational process and the characteristics of students not only allows the teacher to cope with professional tasks, but also positively affects the level of their professional wellbeing [27].

Various studies of the professional wellbeing of educators are presented [28–32]. Peele and Wolf [30] studied the mental health of educators in its relationship with professional wellbeing at the end and at the beginning of the school year. Symptoms of anxiety and depression have been shown to affect professional motivation, job satisfaction, and burnout, as well as affect the frequency of absenteeism in the school year. McMullen et al. [31] and co-authors identified the relationship between professional wellbeing of caregivers and staff turnover, and were able to identify three factors of professional wellbeing: supportive structures, including the environment and climate; relationships with colleagues; professional beliefs and values, such as-autonomy, choice, participation in decision-making.

Kwon [32] and co-authors investigated the features of professional wellbeing and working conditions of educators. The first thing that the authors pay attention to is the high proportion of educators who point to physical and psychological problems and working conditions that do not contribute to professional wellbeing. The researchers identified differences in the wellbeing of caregivers and their working conditions in different age groups of children with whom they work, as well as in groups of caregivers with different levels of education. For example, caregivers with a high level of education had more professional resources and provided a higher quality of care for children, but they had a lower level of physical and professional wellbeing than caregivers with a low level of education.

Thus, despite the high relevance of the problem of professional wellbeing of teachers, there are quite a few studies on this issue. In this regard, our research, devoted to the study of the peculiarities of the experience of professional wellbeing of teachers working in educational organizations of different levels, will expand and enrich the understanding of the professional wellbeing of teachers.

2 Materials and Methods

The object of the empirical study was 65 teachers engaged in professional activities in preschool educational institutions (26.2%) and secondary schools (73.8%), aged 22–40 years (M = 31.3; SD = 1.2). The following characteristics of the sample are presented: male-28%; 79% of respondents have higher professional education;

38.35% carry out professional activities in educational institutions of the city; 21.4% of respondents have experience up to 5 years, 5–10 years-40.4%, 38.2% above 10 years.

Methodological tools were used: the questionnaire "Psychological wellbeing" by K. Riff (adaptation by Lepeshinsky) [33]; the methodology "Professional demand for personality" by Kharitonova and Yasko [34]. The statistical methods used were descriptive statistics, Mann Whitney U test, Spearman rank correlation coefficient.

The objectives of the study included: analysis of the characteristics of the experience of professional wellbeing of teachers who carry out professional activities in educational organizations of different levels; establishing the relationship between professional wellbeing and professional demand for the personality of teachers.

The following assumptions were tested: (1) the features of the experience of professional wellbeing and professional demand of the personality of teachers who carry out professional activities in educational organizations of different levels (preschool, secondary school) may have substantive differences; (2) there may be a relationship between the components of psychological wellbeing and the professional demand for the personality of teachers.

3 Results

To test assumption (1), the nonparametric Mann Whitney U test was chosen due to the nonequivalent frequency distribution of teachers engaged in professional activities in preschool educational institutions (Group 1) and secondary schools (Group 2).

It was found that teachers who carry out professional activities in educational organizations of different levels have different components of psychological wellbeing: positive relationships with others, autonomy, environmental management at the level of 0.01 significance (Table 1). There are also trends in the manifestation

Table 1 Indicators of professional wellbeing of teachers

	Integral indicator of psychological wellbeing	Positive relation-ships with others	Autonomy	Managing the environment	Personal growth	Goal in life	Self-acceptance
	(M (SD))	(M (SD))	(M (SD))	(M (SD))	(M (SD))	(M (SD))	(M (SD))
Group 1	223.7 (43.6)	66.8 (16)	31.6 (13.3)	16.5 (9.5)	30.9 (15.8)	29.8 (14.1)	48.2 (25.8)
Group 2	265.9 (68.3)	41.9 (22.9)	59.3 (23.7)	52.7 (25.6)	33.1 (20.1)	42.2 (22.8)	36.5 (18.6)
U	262	155	154	88.5	399	271.5	311
P	0.029	0.000	0.000	0.000	0.892	0.041	0.146

of differences in the "goal in life" scale and the integral indicator of psychological wellbeing (Table 1).

Positive relationships with others are more pronounced among teachers of preschool educational institutions. Autonomy, as an indicator of psychological well-being, and environmental management have the highest values among teachers of secondary schools. Normative values indicate that teachers of pre-school institutions have a high level of desire to build positive relationships with others, a low level of autonomy and management of the environment, while teachers of secondary schools have a low level of desire to build positive relationships with others, a high level of autonomy and an average level of management of the environment.

The revealed trend of differences in the goal in life among teachers who carry out professional activities in educational organizations of different levels shows that in both groups, according to the normative indicators of the methodology, the goal in life is at a low level. Qualitative analysis indicates that secondary school teachers have higher scores on this scale than teachers of the preschool education system.

The trend in the differences in the integral indicator of psychological wellbeing shows an identical picture, the level of psychological wellbeing in both groups is low, according to the normative indicators of the methodology. A qualitative analysis indicates that secondary school teachers have higher scores for psychological wellbeing than preschool teachers.

The analysis of differences in the components of professional demand showed that teachers who carry out professional activities in educational organizations of different levels have different scales: "satisfaction with the realization of professional potential", "belonging to the professional community", "experience of professional demand", "attitude of others", "general level of professional demand of the individual" (Table 2).

Correlation of the obtained indicators of professional demand of the teachers' personality with the standards of the methodology shows that the teachers of the preschool educational institution have an average level of satisfaction with the realization of professional potential, belonging to the professional community, experiences of professional demand, a low level of evaluation of the attitude of others and a high level of professional demand of the individual. Teachers of secondary schools have a low level of satisfaction with the realization of professional potential, belonging to the professional community, experience of professional demand, evaluation of the attitude of others, and an average level of professional demand of the individual.

According to the scales of professional competence, professional authority, evaluation of the results of professional activity, self-attitude, there were no significant differences between teachers of pre-school and secondary school. The normative indicators of the methodology indicate that the data on the professional demand of the individual are expressed at low and medium levels in both groups of teachers.

To test assumption (2), the Spearman rank correlation coefficient was selected. It was found that the integral indicator of psychological well-being and its component—the goal in life-are positively correlated with professional authority (Table 3).

The component of psychological wellbeing-positive relationships with others-is positively associated with belonging to the professional community and the overall

Table 2 Indicators of professional demand for teachers' personalities

	A	B	C	D	E	F	G	H	J
	(M (SD))	(M (SD))	(M (SD))	(M (SD))	(M (SD))	(M (SD))	(M (SD))	(M (SD))	(M (SD))
Group 1	23 (6.7)	25 (4.2)	26.9 (5)	19.2 (7.6)	22.3 (7.8)	19.6 (7.9)	9.4 (3.3)	13.1 (6.1)	158.2 (4.9)
Group 2	16.9 (7.3)	19.5 (3.9)	17.1 (6.1)	21.1 (7.8)	19.5 (5.9)	17.1 (5.9)	11.4 (2.8)	13.1 (4)	135.7 (14.7)
U	224.5	145.5	93.5	356	277	336	228.5	368.5	115
P	0.006	0.000	0.000	0.435	0.050	0.279	0.006	0.548	0.000

Note A—satisfaction with the realization of professional potential, B—membership in the professional community, C—experiencing professional demand, D—professional competence, E—professional authority, F—evaluation of professional performance, G—relation to other, H—self-relation, J—the general level of professional demand of the individual

Table 3 The relationship between the components of psychological wellbeing and the professional demand for the personality of teachers

Psychological wellbeing/professional relevance of the individual	B	C	D	E	G	J
Integral indicator of psychological wellbeing	–	–	–	r = 0.257 $p = 0.039$	–	–
Positive relationships with others	r = 0.271 $p = 0.029$	–	–	–	–	r = 0.358 $p = 0.003$
Autonomy	r = – 0.308 $p = 0.012$	r = – 0.275 $p = 0.026$	–	–	r = 0.355 $p = 0.004$	–
Managing the environment	r = – 0.420 $p = 0.001$	r = – 0.320 $p = 0.009$	–	–	–	–
Purpose in life	–	–	–	r = 0.279 $p = 0.024$	–	–
Self-acceptance	–	–	r = 0.277 $p = 0.026$	–	r = – 0.280 $p = 0.024$	–

Note B—membership in the professional community, C—experiencing professional demand, D—professional competence, E—professional authority, G—relation to other, J—the general level of professional demand of the individual

level of professional demand of the individual (Table 3). The component of psychological well-being-autonomy-is negatively associated with belonging to the professional community, the experience of professional demand, and positively with the attitude of others (Table 3). The component of psychological wellbeing-environment management-is negatively associated with belonging to the professional community and experiencing professional demand (Table 3). The component of psychological wellbeing-self-acceptance-is positively associated with professional competence, and negatively—with the attitude of others (Table 3).

4 Discussion

The conducted research of differences in the characteristics of the experience of professional wellbeing of teachers who carry out professional activities in educational organizations of different levels allows us to make meaningful profiles of the studied phenomena according to the selected groups.

Teachers of pre-school educational institutions have a low level of psychological wellbeing. The specific features of teachers of pre-school education are satisfactory, trusting relationships with others. concern for the wellbeing of others, the ability to empathize, build attachments and close relationships. At the same time, pre-school teachers depend on the opinions and assessments of others, rely on the opinions of others in making important decisions, and succumb to the attempts of society to force them to think and act in a certain way. Also, this group was diagnosed with a lack of a sense of improvement or self-realization. An interesting fact is that with a low level of professional wellbeing, teachers of pre-school institutions demonstrated a high level of satisfaction with the professional demand of the individual. That is, this group is characterized by the experience of a positive attitude towards themselves as a person who has realized their professional potential. Pre-school teachers are sufficiently satisfied with the realization of their professional potential, demonstrate a differentiated view of themselves as a representative of a certain professional community and, in this regard, a positive emotional attitude to themselves and to values, demonstrate sufficient satisfaction with professional demand. At the same time, pre-school teachers are not satisfied with the attitude of others to them due to the low assessment of the results of their work by colleagues and management. The results obtained are consistent with the results of world studies [35–38], which indicate a low professional wellbeing of pre-school teachers and the relationship of wellbeing with such factors as self-efficacy, overall life satisfaction, financial stability, emotional and physical wellbeing, and autonomy.

The teachers of the secondary school also revealed a low level of psychological well-being. The features of psychological wellbeing of this group are the presence of frustration and dissatisfaction in building interpersonal relationships, a high level of self-regulation of their own behavior and autonomy, demonstration of power and competence in managing the environment when organizing activities, as well as meeting their own needs. Teachers of the secondary school showed an average level of professional demand for the individual. The experience of professional demand among teachers of secondary schools has the following features. This group is characterized by dissatisfaction with their professional status and the degree of demand for professional experience, an undifferentiated view of themselves as representatives of a certain professional community and the lack of a positive emotional attitude to themselves and to values, a painful experience of the inability to realize themselves as much as possible in professional activities, dissatisfaction with the attitude of others due to the low assessment of the results of their work by colleagues and management. Similar results were obtained in other studies. In particular, a number of authors [39–47] conclude that in addition to gender characteristics and work experience, the level of professional wellbeing depends on the assessment and feedback of colleagues and management, an adequate and correct assessment of the results of professional activity, professional cooperation with colleagues, and the needs and desires of students and the ability to meet them play an important role.

Common to teachers of pre-school and secondary school are a lack of self-development, a low experience of a sense of improvement or self-realization, as well as frustration with the events of their past, anxiety about certain personal qualities, a desire to be not who he or she is. In addition, doubts about their professional competence, the experience of their insignificance, non-reference for others, the devaluation of both the professional activity itself and its results against the background of a negative perception of the attitude to themselves as a professional from other people.

The obtained results indicate significant links between the components of psychological wellbeing and the professional demand for teachers. the higher the teachers' overall satisfaction with life and sense of direction in life, the higher the level of awareness of themselves as significant to others as a "source of information", "reference person", and authority within the framework of their professional activities.

The developed ability of teachers to build numerous interpersonal relationships with others will be accompanied by a differentiated view of themselves as representatives of a certain professional community and a positive emotional attitude towards themselves and the values produced in this field of work, as well as an attitude towards themselves as a person who has realized their professional potential. The higher the independence and independence of teachers is formed, the less differentiated the attitude towards themselves as a representative of a certain professional community will be, and there may be experiences of the inability to realize themselves as much as possible in professional activities. But the higher the teachers' satisfaction with the professional attitude of other people (colleagues, management) will be.

The higher the power and competence of teachers in managing the environment, the less differentiated the attitude towards themselves as a representative of a certain professional community will be, and there may be experiences of the inability to realize themselves as much as possible in professional activities. A positive attitude towards teachers will be accompanied by confidence in their professional competence, but dissatisfaction with the attitude of others to them due to the low assessment of the results of their work by colleagues and management.

5 Conclusions

The results of the study of the experience of professional wellbeing of teachers allow us to formulate a number of conclusions:

1. positive relationships with others are more pronounced among teachers of preschool educational institutions, and autonomy and environmental management have the highest values among teachers of secondary schools;
2. the level of psychological wellbeing in both groups is low, but secondary school teachers have higher ratings of psychological wellbeing than pre-school teachers;
3. pre-school teachers have an average level of satisfaction with the realization of professional potential, belonging to the professional community, experiences of professional demand, a low level of assessment of the attitude of others and a high level of professional demand of the individual;

4. teachers of secondary schools have a low level of satisfaction with the realization of professional potential, belonging to the professional community, experience of professional demand, evaluation of the attitude of others, and an average level of professional demand of the individual;
5. multiple interrelations of psychological wellbeing and professional demand of teachers' personality are found, which can be used as reference points for priority areas of work in the psychological support of teachers' professional activities.

The prospects for further development of this topic are the expansion of the level of educational institutions, the study of factors that affect the level of psychological well-being of teachers and determine the level of professional demand of the individual, the study of the factor structure of professional wellbeing of teachers.

References

1. Shamionov RM (2008) Subjective wellbeing of the individual in the professional environment. Probl Soc Psychol Individ 14(1):80–86. http://psyjournals.ru/sgu_socialpsy/issue/30341_full.shtml
2. Shevelenkova TD, Fesenko TP (2005) Psychological wellbeing of the individual. Psychol Diagn 3:95–129
3. Maslow A (2011) New frontiers of human nature. "Meaning", Moscow
4. Rogers KR (2016) A look at psychotherapy. Becoming a person. "Institute of General Humanitarian Research", Moscow
5. Schultz M (2008) Occupational wellbeing: the development of a theory and a measure. http://krex.k-state.edu/dspace/handle/2097/746
6. Ermolaeva MV, Lubovsky DV (2017) Psychological wellbeing of a teacher in the context of his psychological culture. TvSU Bull Pedag Psychol 3:92–97
7. Zausenko IV (2012) Psychological wellbeing of a teacher. In: Pedagogical education in Russia 2. https://cyberleninka.ru/article/n/psihologicheskoe-blagopoluchie-pedagoga
8. Borodkina EV (2012) On the issue of studying subjective wellbeing in professional and pedagogical activity. Psychol Pedag Methods Probl Pract Appl 28:31–34
9. Boyarkin MYu, Dolgopolova OA (2007) Psychological and professional wellbeing of civil servants. "Publishing House of FGOU VPO VAGS", Volgograd
10. Vodiakha SA (2012) Modern concepts of psychological wellbeing of the individual. Discussion 2:132–138
11. Zausenko IV (2011) Features of self-attitude of teachers with different indicators of psychological wellbeing. Siber Pedag J 10:191–197
12. Pishchik VI (2020) Value-semantic bases of ideas about the profession and satisfaction with the profession of higher school teachers. Int J Cogn Res Sci Eng Educ 8(2):69–81. https://doi.org/10.5937/IJCRSEE2002069P
13. Minyurova SA, Zausenko IV (2013) Personal determinants of psychological wellbeing of a teacher. Pedag Educ Russ 1:94–101
14. Osmuk LA, Safronova MV, Zahir YuS (2013) Psychosocial wellbeing and professional activity of Russian teachers. Ideas Ideals 1(3):91–104
15. Cvetković BN, Stošić L, Belousova A (2018) Media and information literacy—the basis for applying digital technologies in teaching from the discourse of educational needs of teachers. Croat J Educ 20(4):1089–1114. https://doi.org/10.15516/cje.v20i4.3001
16. Argyle M (2003) Psychology of happiness. "Peter", St. Petersburg
17. Kulikov LV (2008) Determinants of life satisfaction. Soc Polit 476–510

18. Skripkina NV (2018) Professional wellbeing and satisfaction with pedagogical activity. In: Professional wellbeing of a modern person, pp 252–258. https://core.ac.uk/download/pdf/128641657.pdf
19. Druzhilov SA (2016) Professional wellbeing of a person and psychological aspects of professional adaptation and professional destructions. Mod Sci Res Innov 12. http://web.snauka.ru/issues/2016/12/76347
20. Berezovskaya RA (2016) Professional wellbeing: problems and prospects of psychological research. Psychol Res 9(45). http://psystudy.ru/index.php/num/2016v9n45/1232-berezovskaya45.html
21. Ruth EI, Augustova LI (2017) Professional wellbeing of employees of commercial organizations: criteria and methods of evaluation. Sci Res Grad Fac Psychol St. Petersb State Univ 5:72–78
22. Pahol BE (2018) The structure of professional wellbeing: definition of the phenomenon, factor structure and predictors. Ukr Psychol J 3(9). https://www.academia.edu/38723174
23. Zausenko IV (2012) Personal determinants of psychological wellbeing of the teacher. Ural State Pedagogical University, Yekaterinburg
24. Minyurova SA, Zausenko IV (2013) Personal determinants of psychological wellbeing of a teacher. Pedag Educ Russ 1. https://cyberleninka.ru/article/n/lichnostnye-determinanty-psihologicheskogo-blagopoluchiya-pedagoga
25. Litvinova EYu, Kharchenko MA, Pogodina AV (2018) Psychological wellbeing and emotional burnout of teachers of educational complexes at different stages of the innovation process. Inst Psychol Russ Acad Sci Organ Psychol Labor Psychol 2(3):96–119. http://work-org-psychology.ru/engine/documents/document363.pdf
26. Panova NV (2011) The role of personal development of a teacher in professional wellbeing. World Sci Cult Educ 2(27):118–121. https://cyberleninka.ru/article/n/rol-lichnostnogo-razvitiya-pedagoga-v-professionalnom-blagopoluchii
27. Lauermann F, König J (2016) Teachers' professional competence and wellbeing: understanding the links between general pedagogical knowledge, self-efficacy and burnout. Learn Instr 45:9–19. https://doi.org/10.1016/j.learninstruc.2016.06.006
28. Shchekotikhina IV, Burenkova NV, Maksimova TV (2019) Professional wellbeing of teachers of preschool educational organizations. Psychological wellbeing of modern man, pp 542–546
29. Klusmann U, Kunter M, Trautwein U, Lüdtke O, Baumert J (2008) Teachers' occupational wellbeing and quality of instruction: the important role of self-regulatory patterns. J Educ Psychol 100(3):702–715. https://doi.org/10.1037/0022-0663.100.3.702
30. Peele M, Wolf S (2021) Depressive and anxiety symptoms in early childhood education teachers: relations to professional wellbeing and absenteeism. Early Child Res Q 55:275–283. https://doi.org/10.1016/j.ecresq.2020.11.008
31. McMullen MB, Lee MSC, McCormick KI, Choi J (2020) Early childhood professional wellbeing as a predictor of the risk of turnover in child care: a matter of quality. J Res Child Educ 34(3):331–345. https://doi.org/10.1080/02568543.2019.1705446
32. Kwon KA, Ford TG, Salvatore AL, Randall K, Jeon L, Malek-Lasater A, Ellis N, Kile MS, Horm DM, Kim SG, Han M (2020) Neglected elements of a high-quality early childhood workforce: whole teacher wellbeing and working conditions. Early Child Educ J. https://doi.org/10.1007/s10643-020-01124-7
33. Lepeshinsky HH (2007) Adaptation of the questionnaire "scales of psychological wellbeing" by K. Riff. Psychol J 3(15):24–27. http://elibrary.miu.by/journals!/item.pj/issue.15/article.3.html
34. Yasko BA, Kharitonova EV (2009) Experience in the development of psychodiagnostic methods "professional demand for personality". Bull Adygeya State Univ Pedag Psychol 4:282–288. https://cyberleninka.ru/article/n/opyt-razrabotki-psihodiagnosticheskoy-metodiki-professionalnaya-vostrebovannost-lichnosti
35. Hall-Kenyon K, Bullough R, MacKay K, Marshall E (2014) Preschool teacher wellbeing: a review of the literature. Early Child Educ J 42(3). https://doi.org/10.1007/s10643-013-0595-4

36. Belousova AK, Barsukova OV, Vyshkvyrkina MA, Kryschenko EP, Tushnova JA, Mochalova JA, Mozgovaya NN, Pavlova TV (2015) Personality traits of preschoolers as subjects of collaborative thinking activity. Mediterr J Soc Sci 6(5(S1)):79–83. https://doi.org/10.5901/mjss.2015.v6n5s1p
37. Day C, Gu Q (2009) Teacher emotions: well being and effectiveness, pp 15–31. https://doi.org/10.1007/978-1-4419-0564-2_2
38. McGrath B, Huntington A (2007) The health and wellbeing of adults working in early childhood education. Aust J Early Child 32(3). https://doi.org/10.1177/183693910703200306
39. Yildirim K (2014) Main factors of teachers professional wellbeing. Educ Res Rev 9(6):153–163. https://doi.org/10.5897/ERR2013.1691
40. Suroedova E, Tushnova Yu (2020) Subjective assessment of social status and socio-psychological attitudes of student youth. E3S Web Conf 210:19004. https://doi.org/10.1051/e3sconf/202021019004
41. Briones E, Tabernero C, Arenas A (2010) Job satisfaction of secondary school teachers: effect of demographic and psycho-social factors. Rev Psicol Trab Organ 26(2):115–122. https://doi.org/10.5093/tr2010v26n2a3
42. Chong WH, Klassen R, Huan V, Wong I, Kates A (2010) The relationships among school types, teacher efficacy beliefs, and academic climate: perspective from Asian middle schools. J Educ Res 103(3):183–190. https://doi.org/10.1080/00220670903382954
43. Van Horn J, Taris T, Schaufeli W, Schreurs P (2004) The structure of occupational well-being: a study among Dutch teachers. J Occup Organ Psychol 77:365–375. https://doi.org/10.1348/0963179041752718
44. Butt R, Retallick J (2021) Professional wellbeing and learning: a study of administrator-teacher workplace relationships. https://www.researchgate.net/publication/251685621_Professional_wellbeing_and_learning_a_study_of_administrator-teacher_workplace_relationships
45. Rasku A, Kinnunen U (2003) Job conditions and wellness among Finnish upper secondary school teachers. Psychol Health 18(4):441–456. https://doi.org/10.1080/0887044031000147184
46. Sann U (2003) Job conditions and wellness of German secondary school teachers. Psychol Health 18(4):489–500. https://doi.org/10.1080/0887044031000147184
47. Aelterman A, Engels N, Petegem K, Verhaeghe J (2007) The wellbeing of teachers in Flanders: the importance of a supportive school culture. Educ Stud 33(3):285–297. https://doi.org/10.1080/03055690701423085

Cooperation and Confrontation in Intergroup Interaction Based on the Prisoner's Dilemma Model

Denis Dautov

Abstract Intergroup interaction differs from interpersonal interaction in a number of specific features. Confrontation and cooperation in inter-group interaction have different rates according to the conditions in which the interaction takes place. This article presents the study results of the features of cooperation and confrontation in intergroup interaction in the framework of the game theory model—the prisoner's dilemma. With the help of a business game created on the basis of the prisoner's dilemma, the features of interaction between groups and individuals are studied. On the basis of the method of mathematical statistics of the Fisher's φ-criterion, the features of intergroup and interpersonal interaction in different conditions were compared. Significant differences were found in the ratio of confrontational and cooperative actions between the subjects of intergroup interaction under different game conditions. The differences between the subjects of inter-group and interpersonal interaction were also established. Conclusions are drawn about the qualitative differences between intergroup and interpersonal interaction.

Keywords Social dilemma · Prisoner's dilemma · Confrontation · Cooperation · Intergroup interaction · Interpersonal interaction

1 Introduction

In psychological science, the idea of two interrelated components of the activity of a small group: social and target has spread relatively long ago. The target component is represented by the organizational and activity component of the group and is aimed at implementing the achievement of the final result of its activities by this group. The social component is represented by interpersonal relations and is aimed at preserving its internal stability, cooperation between its members, or, on the contrary, at destabilizing its interpersonal relations [1–3].

As a rule, any social association of people seeks to establish cooperation both within themselves and with other groups, which is a common goal for such an

D. Dautov (✉)
Don State Technical University, Rostov-on-Don, Russia

A. Beskopylny and M. Shamtsyan (eds.), *XIV International Scientific Conference "INTERAGROMASH 2021"*, Lecture Notes in Networks and Systems 247,
https://doi.org/10.1007/978-3-030-80946-1_59

association [4, 5]. At the same time, the main obstacle to achieving this goal is often conflicts between individual and collective interests or among inter-group interests [6, 7]. Such conflicts are often realized in the form of social dilemmas, manifested in the pursuit of personal or narrow-group interests that clearly prevail over group or intergroup interests. This tendency inevitably provokes a decrease in the effectiveness of group activity and can cause the disintegration of the group itself [8–10].

The phenomenon of the "social dilemma" as a situation where the majority of group members or groups involved in interaction act solely on the basis of their personal interests has had a great impact on the understanding of collective actions and cooperative behavior [9, 11, 12].

In particular, numerous studies have demonstrated a paradoxical picture of rational and irrational behavior in interpersonal interaction [13, 14]. It turned out that the rational actions of individual group members can lead to irrational results of the group as a whole. This is confirmed by the results of the study, concluding impossibly for individual rational strategies and effective collective activities to coexist [15, 16].

The paradox of this conclusion is that a selfish, rational and extremely effective individual action, or the sum of such actions, from the point of view of the overall group result, is extremely ineffective, and often destructive for the group. On the contrary, irrational behavior from the standpoint of individual interests turns out to be extremely rational at the level of the interests of the entire group [11, 15].

Despite the fact that the group is a kind of system consisting of separate individuals, it includes a mandatory contradiction between collective and individual rationality [16, 17].

Collective rationality dictates cooperation to all participants in the interaction. If the members of the group cooperate, take into account each other's interests, then each of them wins. Selfish rationality, on the other hand, is aimed at destroying cooperation. If one of the participants in the interaction pursues personal interests, then at the individual level he\she receives the maximum benefit, against the background of the damage caused by his\her actions to the other participants. However, the effect of pursuing individual interests does not end there. As demonstrated by a number of researchers, the effect continues over time, causing harm to all participants in the interaction [11, 18]. This harm far exceeds the short-term benefit to the individual. In the long run, the irrational actions of separate individuals inevitably cause pronounced negative effects that cannot be compensated for by individual rapid successes [18]. Therefore, the social dilemma has a temporal component, according to which the selfish actions of the individual are beneficial for a short period of time, and actions in the interests of the group are beneficial for relatively long time intervals.

Thus, social dilemmas contain two main contradictions: between the selfish interests of the individual and the collective interests of the group, as well as between the short-term and long-term interests of a person and a group [17, 18].

It should be noted that scientific research on social dilemmas is conducted within the framework of game theory. Accordingly, the subject of study, first of all, are abstract models, which are conventionally called "games".

In game theory, many models have been created: "the stowaway problem", "the prisoner's dilemma", "ultimatum", "trust", "dictator", "community field" and a number of others.

One of the most developed models is the "prisoner's dilemma" model.

This model has been extensively studied by economists, psychologists, sociologists, and mathematicians. The foundations of scientific research on the prisoner's dilemma were laid in the works of Drescher, Nash, Tucker, Flood, Hofstadter [19]. This dilemma was further developed in the studies of Axelrod, Rapoport and Camerer [20, 21].

Some of the most recent works in the field of psychology, neurophysiology, and evolutionary biology are the articles by Heuer and Orland, Bruno and Guidetti, Realpe-Gómez and Montoya, Li and Han [22–25].

The Prisoner's Dilemma game is an activity model with mixed motives, offering players a choice between selfish action and cooperation. In this game, there is a clear formal structure of possible interpersonal relationships, the catalyst of which is to obtain a certain benefit. At the same time, the result of the decision of one subject of interaction inevitably affects the decisions of another subject [19–21].

In the modern version of the prisoner's dilemma, as a rule, it is proposed to play a whole series of games. This allows participants to take into account the decisions made in the previous course. In addition, it contributes to the strengthening of the influence of socio-psychological factors on decision-making between individual study participants and between groups of such participants [20–23].

In the classic form, the game provides participants with only two possible choices—egoistic and cooperative. Egoistic choice is initially more beneficial to the individual, except for the option of simultaneous choice by all parties to cooperation. Such an act from the point of view of the interests of all parties to the interaction will be the most rational and effective [17, 26].

However, numerous experiments have shown that their participants do not always have the necessary abilities or motivation to find an effective way to act in such conditions. Instead, they tend to change the strategy quite arbitrarily in each case [21–25].

In particular, it was found that the participants of the study, depending on their inclinations, assume the presence of a similar course of action in their opponents. Those who are inclined to cooperate expect the rest of the participants to cooperate, and those who prefer individualism expect the rest to act in their own interests. In other words, the participants in the game pay attention not to objective (rational) factors of decision-making, but to subjective (irrational) ones [27].

Moreover, due to the uncertainty of the players in the actions of the other participants of the interaction, their decisions can be either mixed rational-irrational, or exclusively rational or irrational, without trying to choose between them. In this case, they can demonstrate collective rational behavior, with irrational individual actions [22].

In game theory, collective rationality is understood as the receipt of individual benefits by the participants of interaction, when they implement cooperative behavior [22, 28]. It should be noted that this behavior cannot be considered altruistic, since individual participants in the interaction achieve their own success through helping others. It can only be about the optimal strategy to benefit from your decisions.

As noted by a number of researchers, the mere presence of common goals among participants in joint activities can not serve as a guarantee of the voluntary desire of group members to achieve the goal. This is true even if the realization of a common goal contributes to objective well-being at the level of an individual [29].

According to Olson, all members of the group will benefit from achieving a common goal, regardless of their individual desire to achieve it. This means that the overall benefit itself is not a sufficient justification for joint action [29].

It seems that at the forefront here are the factors of social interaction, such as the assessment of the fairness of behavior, each other by the participants of the interaction, communication between the players, the degree of their acquaintance with each other [29].

In this case, social interaction acts as a system of mutual actions, when the actions of one person are the cause and effect of the response actions of other people [29, 30]. It turns out that when choosing between cooperation and a selfish decision, people simultaneously choose between personal gain and social interaction.

Despite the numerous studies devoted to this interaction, they are mainly aimed at studying the interaction in dyads. When each side of the interaction is represented by a single individual. In the few studies where the subject of decision-making is a group, we are usually talking about groups specially created for the experiment, or about network structures that are temporary and dynamically changing in accordance with the nature of the study [31].

Currently, the situation where the subject of decision-making is natural groups that make decisions in their own interests is practically not studied.

2 Methods

The study involved 143 subjects aged 20–23 years. Subjects—bachelors of 3 courses and masters of 1 course (humanities and technical specialties) of Don State Technical University.

All the subjects were divided into four large groups. The first and second groups consisted of ten subgroups, each of which included from five to seven people. The third and fourth groups also consisted of ten subgroups, each of which included two people, forming dyads.

The formed groups were offered a game based on the principles of the prisoner's dilemma. The game was called "The Meeting of two Civilizations". The essence of the game is to decide who will own the newly discovered planet—earthlings or aliens. The game was chosen for two reasons. First, it provides a relatively neutral cultural background that has no direct connection to events in the modern world, which could distort the data obtained. Secondly, it involves the work of teams, whereas in the classic prisoner's dilemma, decisions on the conditions of the game are made by individuals. A different choice of options for team interaction leads to different

Table 1 Quantitative values of interaction options

Earthlings	Aliens	Earthlings	Aliens
Open fire	Open fire	−3	−3
Open fire	Negotiations	+5	−5
Negotiations	Open fire	−5	+5
Negotiations	Negotiations	+3	+3

results and, accordingly, brings a different number of points, the score of which is shown in Table 1.

The table describes four possible outcomes. In the first case, both teams decide to open fire by choosing the path of conflict, which ends with a loss of 3 points for each of them. In the second and third cases, if one of the teams decides to start negotiations, and the other team opens fire, the dead lose 5 points, the attackers, on the contrary, gain 5 points. In the fourth case, the teams decide to start negotiations by choosing the path of cooperation, which brings them 3 points each.

In each of the four groups, the subgroups were divided into pairs of teams that played each other. Each team was given two cards—"Open fire" and "Negotiate". After making a decision, each team passed one of the cards to the host, who opened them at the same time, announcing the decisions recorded on them and awarding points for the result. The game consisted of 12 moves, during which the winner was revealed. The multi-input option was chosen so that players can make decisions based on past experience of interacting with opponents, thereby enhancing the socio-psychological effect of their actions.

The conditions of the game were different in different groups of subjects. In the first and third groups, the team that scored more points than the other team won. In the second and fourth teams, the teams with more than 20 points won.

Thus, half of the groups (the first and third) were forced to enter into an antagonistic relationship according to the terms of the game, when a player wins at the expense of another player. Such a game is called a zero-sum game. According to the proposed situation, the only possible option of winning for one side means an imminent defeat for the other [22]. The second and fourth groups are spared from the forced antagonism. Thanks to this, they can play games with a non-zero sum. When some people win, it doesn't necessarily lead to others losing.

In addition, the first and second groups include subgroups that are able to communicate with each other before making a certain decision. Therefore, they can analyze more information about the situation of the game itself and about the behavior of the other team. The very exchange of information within a group enhances cohesion and group identity, thereby contributing to cooperative decision-making trends.

Each of the groups in the study had its own set of parameters and game conditions. The first group consisted of teams of 5–7 people each and could only win at the expense of the other team. The second, also consisted of small teams, but its victory depended only on a certain number of points scored, and not on superiority over the other team. The third group consisted of dyads, each participant of which had to fight with another individual of this dyad, for the victory. The fourth group consisted of dyads, in which individuals, when interacting with each other, could score a certain number of points to win.

To find differences in the ratio of the share of cooperation and confrontation between the groups, we used the Fisher's φ-criterion, which allows us to assess the reliability of differences between the percentages of samples.

3 Results

The results of the game participants' cooperation and confrontation choices were reduced to percentages for each of the four groups.

The results obtained are divided into four main blocks. The first block provides data on the distribution of preferences for choosing cooperation and confrontation in groups. The second block provides data on the differences between the percentages of preferences for choosing confrontation and cooperation in all four groups. The third block provides data on the distribution of mutual choices of cooperation and confrontation in groups. The fourth block provides data on the differences between the percentages of mutual choices of cooperation and confrontation in all four groups.

Figure 1 shows the results of a study of the preferences for choosing cooperation and confrontation in groups.

According to the data presented, in the first group of subjects, where decisions were made in small subgroups, in which, according to the conditions of the game, a team could win only at the expense of another team, the level of desire for confrontation was the highest among all groups (71.7%). The level of desire for cooperation, on the contrary, is the lowest in the groups (28.3%). The teams in the first group chose attacking their teammates from among the other teams as their preferred action. In less than one third of the cases, they chose to negotiate. In the second group, where decisions were also made in small subgroups, but the victory depended only on the points scored, the results were different. The level of desire for cooperation was the

	Group 1	Group 2	Group 3	Group 4
Cooperation	28.30%	49.20%	34.20%	35.80%
Confrontation	71.70%	50.80%	65.80%	64.20%

Fig. 1 Distribution of preferences for choosing cooperation and confrontation in groups

highest of all groups (49.2%), and the level of desire for confrontation was the lowest (50.8%). The propensity to attack and negotiate in the teams of this group was almost equal. However, it is noteworthy that even in these teams, the tendency to cooperate is not predominant. Even if it is insignificant, but the tendency to confrontation is higher. In the third and fourth groups, the results are similar. Both of these groups are made up of dyads, where one person plays for one side of the opposition, and the second, for the other. In them, as in the first group, the tendency to confrontation prevails, although it is somewhat less pronounced. Despite the fact that in the fourth group, the subjects do not have to defeat the opponent, but only need to score a certain number of points, the level of confrontation remains high (65.8%), not much different from the indicators of the fourth group (64.2%).

In general, in all groups of subjects, confrontation is the predominant mode of action, and cooperation is not dominant in any group, even when this mode of action is certainly capable of winning the game.

The results of the study of four groups of subjects allow us to compare the preferences for choosing cooperation and confrontation, to find differences between these groups. The calculation of the Fisher's φ-test showed the following statistically significant differences between them: between the indicators of the first and second groups ($\varphi = 3.354$, $p = 0.001$); between the indicators of the second and third groups ($\varphi = 2.897, p = 0.001$); between the indicators of the second and fourth groups ($\varphi = 2.355$, $p = 0.001$). No differences were found between the indicators of the first and third groups, the first and fourth, as well as the third and fourth.

The results obtained allow us to describe the situation with the tendency of groups to confrontation and cooperation in a general way. Equally important is the data on cases of mutual preferences for cooperation and confrontation between teams, when one team chooses to attack or negotiate simultaneously with the other team.

Figure 2 shows the results of a study of mutual choices of cooperation and confrontation in all four groups.

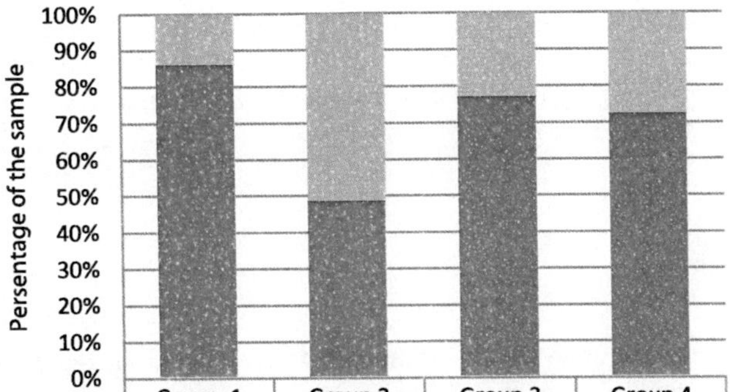

	Group 1	Group 2	Group 3	Group 4
Cooperation	13.90%	51.40%	22.90%	27.50%
Confrontation	86.10%	48.60%	77.10%	72.50%

Fig. 2 Distribution of mutual choices of cooperation and confrontation in groups

The results obtained show that in the first group, the number of mutual confrontational choices is the highest among all groups (86.1%). Teams in this group, mutually tend to choose an attack, expecting the same from their opponent. The number of mutual cooperation choices is minimal (13.9%). There are significantly fewer such coincidences than the single cooperation choices shown in Fig. 1. The teams of the first group demonstrated a pronounced mutual unwillingness to cooperate, in which the desire to attack one side was obvious and predictable for the other side. The teams of the second group showed a slight, but the predominance of the number of mutual cooperation choices (51.4), over confrontation (48.6). In this group, the desire for negotiation prevailed over the desire for attack, however, mutual attacks were only slightly less. The teams constantly alternated between confrontational actions and cooperation. The similarity of the percentages of single choices in this group with the percentages of mutual elections is noteworthy. The indicators of mutual choices of the third and fourth groups are similar both in terms of confrontation and cooperation. In both groups, mutual attacks prevailed over mutual negotiations. However, the propensity for mutual cooperation in these groups is higher than in the first, and the propensity for confrontation, on the contrary, is slightly lower. The number of single elections of cooperation exceeds the number of mutual choices, which indicates the inconsistency of actions with opponents in such cases. The percentages of confrontation in single choices, on the contrary, are higher than in mutual choices, which indicates that each side understands the desire to attack.

In the situation of mutual choices, the teams of all groups, except the third, showed a predominant tendency to attack. The coordination of confrontational actions was predominant in relation to the coordination of actions aimed at cooperation. The exception was the teams of the second group, in which these actions almost balance each other.

The results of the study of four groups of subjects allow us to compare mutual choices of cooperation and confrontation, to find differences between these groups. The calculation of the Fisher's φ-test showed the following statistically significant differences between them: between the indicators of the first and second groups ($\varphi = 2.514, p = 0.001$); between the second and third groups ($\varphi = 3.518, p = 0.001$). The indicators of differences between the second and fourth groups are in the zone of uncertainty ($\varphi = 2.139, p = 0.05$). In other cases, no differences were found between the groups.

4 Discussion

It can be assumed that the teams of the first and second groups were more sensitive to the initial conditions of the game than the teams of the third and fourth groups. The teams of both the first and second groups showed significant differences in their actions. In the first group, whose playing conditions forced the teams to antagonize, the desire for confrontation really prevailed over the desire for cooperation. In the second group, in which the conditions of the game did not require mandatory rivalry between the teams, the tendency to cooperate was the highest among the teams of all other groups. At the same time, the groups in which the subject of the game was a dyad, and the inter-group interaction was replaced by interpersonal interaction, the sensitivity to the conditions of the game itself was minimal. Despite the fact that in the fourth group, as in the second, it was not necessary to win at the expense of their opponent, the teams still preferred confrontational actions. A possible explanation for this is the very nature of the joint activity [32]. In the first and second groups, team members could communicate with their like-minded people, introducing the social dimension into the context of interaction, with its ideas of justice, the desire for mutual understanding, but at the same time, the grouping of thinking, the division according to the principle of "friend-foe". The presence of competitive conditions in the game reinforced this trend in the teams of the first group. In the absence of the need to compete within the conditions of the game, this led to increased cooperation in the interaction that the teams of the second group demonstrated [25, 30]. In the third and fourth groups, players were introduced to themselves and could only interact with their opponent, who could easily be imagined as a rival or enemy, which makes it easier to use confrontational actions against them. However, the tendency to confrontation in the teams of the first group was higher than in the third and fourth. Apparently, this is due to the greater stability of the goal setting of the group, compared to the individual. The group, as a subject of interaction, is able to adhere more consistently and persistently to a certain course of action, and is less sensitive to obstacles to the realization of the goal.

The differences in mutual elections between the groups are of particular interest, due to the sharp difference in the indicators of the second group from the others. Only the teams of the second group demonstrated a relative balance of mutual choices of confrontation and cooperation. The teams tried to coordinate their actions with their opponents, both in the attack and in the negotiations, whereas in all other groups, the teams were much better at coordinating the attack, and they were able to coordinate the negotiations in exceptional cases. This shows a higher level of trust in inter-group interaction among the teams of the second group and a lower dependence on emotions, in cases where opponents chose to attack instead of negotiations. In the fourth group, although the conditions of the game were identical to the third, there was a lack of perseverance, consistency in their actions, which prevented cooperation with opponents. Despite the fact that the level of coordinated cooperation in these teams exceeded the level of the teams of the first and third groups, the favorable conditions of the game itself were not enough for cooperation. Even though, as the researchers note, the ability to punish an opponent on the next move by choosing confrontation, if he refused to cooperate on the previous move, generally contributes to the rejection of antagonism. This is true only when the participants in the interaction plan the sequence of their actions in advance and try to anticipate the actions of their opponents. A relatively small number of moves, prevented a long confrontation and the development of appropriate experience. In particular, Realpe-Gomez and a number of other researchers have found that finding the optimal course of action in a game often occurs in each situation separately, based on repeated interactions [21, 24, 33]. This hinders the consideration of the overall picture of events and does not allow its participants to develop a rational way of action for the entire game situation as a whole, which explains the unsuccessful attempts to cooperate.

5 Conclusions

The results obtained in the study provide an opportunity to see direct inter-group and interpersonal interaction as a combination of confrontation and cooperation.

Based on the data obtained, the following conclusions can be drawn:

1. Intergroup interaction is more sensitive than interpersonal interaction to the initial conditions in which it takes place. Subjects of inter-group interaction show a greater tendency to take into account the consequences of compliance or non-compliance with established rules.
2. Intergroup interaction is more stable and rational than interpersonal interaction. Subjects of inter-group interaction are more likely to adhere to the chosen course of action and are less likely to change it than subjects of interpersonal interaction.
3. Subjects of intergroup interaction have a clearer idea of the overall picture of events and prefer to plan their actions for several moves ahead, while the subject of intergroup communication is prone to situational, tactical actions of the "here and now" type.
4. Subjects of inter-group interaction better represent the actions of their opponents and are more effectively able to coordinate actions with them. The subjects of interpersonal interaction experience difficulties in coordinating each other's actions, even in the case of direct benefits for all parties to the interaction.

References

1. Galustyan OV et al (2018) Development of group subjectivity of pupils within collaborative activities. TEM J 7(4):854–858
2. Zhuravlev AL (2009) Collective subject: main features, levels and psychological types. Psychol Mag 30(5):72–80
3. Sidorenkov AV (2011) Microgroup theory. Soc Psychol Soc 2(1):17–30
4. Koutsombogera M, Voge C (2019) Observing collaboration in small-group interaction. Multimodal Technol Interact 3(3):45
5. Driskell JE, Salas E, Driskell T (2018) Foundations of teamwork and collaboration. Am Psychol 73(4):334
6. McCaghy C et al (2016) Deviant behavior: crime, conflict, and interest groups. Routledge, New York
7. Suroedova E, Tushnova Y (2020) Subjective assessment of social status and socio-psychological attitudes of student youth. E3S Web Conf 210:19004
8. Muzdybaev K (2007) Social dilemmas and ways of their solutions. J Sociol Soc Anthropol 10(2):99–122
9. Szolnoki A, Perc M (2016) Collective influence in evolutionary social dilemmas. EPL (Europhys Lett) 113(5):58004

10. Belousova A (2020) Functions of participants in the collaborative solution of thinking problems. Int J Cogn Res Sci Eng Educ 8:29–36
11. Fujii S (2017) Strategies for solving social dilemmas: structural strategy and psychological strategy. In: Prescription for social dilemmas. Springer, Tokyo
12. Korochentseva A, Krasnova E, Moysova O (2020) Simulation role play as a contact work with students and its role in their communicative skills development. In: Innovative technologies in science and education (ITSE-2020). E3S Web Conf 210:18087
13. Bland AR et al (2017) Cooperative behavior in the ultimatum game and prisoner's dilemma depends on players' contributions. Front Psychol 8:1–11
14. Li N, Kirkman BL, Porter C (2014) Toward a model of work team altruism. Acad Manag Rev 39(4):541–565
15. Pletzer JL et al (2018) Social value orientation, expectations, and cooperation in social dilemmas: a meta-analysis. Eur J Pers 32(1):62–83
16. Cornelissen G, Dewitte S, Warlop L (2011) Are social value orientations expressed automatically? Decision making in the dictator game. Pers Soc Psychol Bull 37(8):1080–1090
17. Klein SA, Hilbig BE, Heck DW (2017) Which is the greater good? A social dilemma paradigm disentangling environmentalism and cooperation. J Environ Psychol 53:40–49
18. Ispano A, Schwardmann P (2017) Cooperating over losses and competing over gains: a social dilemma experiment. Games Econ Behav 105:329–348
19. Khadjavi M, Lange A (2013) Prisoners and their dilemma. J Econ Behav Organ 92:163–175
20. Rapoport A, Seale DA, Colman AM (2015) Is tit-for-tat the answer? On the conclusions drawn from Axelrod's tournaments. PLoS One 10(7):e0134128
21. Jones G (2008) Are smarter groups more cooperative? Evidence from prisoner's dilemma experiments, 1959–2003. J Econ Behav Organ 68(3–4):489–497
22. Heuer L, Orland A (2019) Cooperation in the prisoner's dilemma: an experimental comparison between pure and mixed strategies. R Soc Open Sci 6(7):182142
23. Dell'Aversana G, Guidetti G (2018) Developing organizational competences for conflict management: the use of the prisoner's dilemma in higher education. Front Psychol 9:376
24. Realpe-Gómez J, Vilone D, Andrighetto G, Nardin LG, Montoya JA (2018) Learning dynamics and norm psychology supports human cooperation in a large-scale prisoner's dilemma on networks. Games 9(4):90
25. Li D, Du J, Han D (2019) The evolutionary prisoner's dilemma game considering behavior interaction in a two-layered network. EPL (Europhys Lett) 126(3):30002
26. Liang F et al (2021) Selfish or altruistic? The influence of thinking styles and stereotypes on moral decision-making. Pers Individ Differ 171:110465
27. Bogaert S, Boone C, Declerck C (2008) Social value orientation and cooperation in social dilemmas: a review and conceptual model. Br J Soc Psychol 47(3):453–480
28. Colman AM, Pulford BD, Rose J (2008) Collective rationality in interactive decisions: evidence for team reasoning. Acta Physiol 128(2):387–397
29. Callan MJ et al (2010) The effects of priming legal concepts on perceived trust and competitiveness, self-interested attitudes, and competitive behavior. J Exp Soc Psychol 46(2):325–335
30. Perc MJ., Rand DG, Wang Z, Boccaletti S, Szolnoki A (2017) Statistical physics of human cooperation. Phys Rep687:1–51
31. Cason TN, Lau SHP, Mui VL (2019) Prior interaction, identity, and cooperation in the intergroup prisoner's dilemma. J Econ Behav Organ 166:613–629
32. Dautov DF, Nikolenko OF, Rashchupkina YV, Lomova NV, Tushnova YA (2019) Cognitive maps of participants in joint mental activity. In: 2nd international conference on education science and social development (ESSD 2019), pp 210–214
33. Pishchik VI, Molokhina GA, Petrenko EA, Milova YV (2019) Features of mental activity of students—eSport players. Int J Cogn Res Sci Eng Educ (IJCRSEE) 7(2):67–76

The Personality of a Psychologist Who Prefers Remote Counseling

Elena Suroedova, Julia Kreneva, Olga Korzhova, and Tatiana Ozerova

Abstract Changes in society connected with the COVID-19 pandemic have influenced the choice of the format of family psychologists work with clients. Today, we can observe that groups of specialists have formed that demonstrate their willingness to work with the client through ICT technologies, Internet platforms and social networks, and specialists who prefer the face-to-face format of consulting. The study sample consisted of 60 practicing family psychologists aged 23–67 years (the average age in the sample was 36.7 years) with an average experience of consulting work of 2.3 years, using both full-time and remote consulting formats in their work with clients. The following methods were used: a demographic questionnaire that allows us to identify the specifics of the respondents' attitude to the formats of counseling—face-to-face and distance, the method "Personal semantic differential" by Ch. Osgood in the adaptation of employees of the V. M. Bekhterev Psychoneurological Institute, the 16-factor test by R. Kettell (form C), the method "Test of professional self-attitude" by Karpinsky. Mathematical data processing included standard methods of mathematical statistics and was carried out using the following statistical methods: descriptive statistics; correlation coefficient r-Spearman; nonparametric criterion U-Mann–Whitney and H-Kruskal–Wallace; φ-Fisher. In the course of the study, the individual and personal characteristics of psychologists who prefer the remote format of counseling were established. The prospects of the research are aimed at further studying the factors and predictors of the formation of readiness for the remote format of the work of psychologists.

Keywords Psychologist · Family psychotherapy · Counseling · Personality · Remote and online counseling format · ICT technologies

E. Suroedova (✉) · O. Korzhova
Don State Technical University, 344000 Rostov-on-Don, Russia

J. Kreneva · T. Ozerova
Academy of Psychology and Pedagogy, Southern Federal University, 344006 Rostov-on-Don, Russia

A. Beskopylny and M. Shamtsyan (eds.), *XIV International Scientific Conference "INTERAGROMASH 2021"*, Lecture Notes in Networks and Systems 247,
https://doi.org/10.1007/978-3-030-80946-1_60

1 Introduction

The events of 2020 related to the COVID-19 pandemic have significantly affected all aspects of human life, professional activity, health and well-being. Forced self-isolation caused a part of the population to experience increased anxiety, depression, frustration, feelings of helplessness and loneliness. In some families, there was an increase in conflict, quarrels, resentments, and misunderstandings between spouses, parents, and children due to the need to spend a lot of time together, the inability to establish optimal emotional contact, and the psychological stress associated with frustration and fear. Difficulties and feelings of hopelessness were experienced by elderly people who were closed from society, deprived from visiting medical institutions, shops, and social institutions. Even greater difficulties, during the period of restrictions, fell to parents with children with disabilities or specially challenged. According to research, about one in five parents of a child with autism has severe symptoms of post-traumatic stress disorder (PTSD). First of all, they are connected with severe problem behavior in a child [1].

The quarantine and restrictions imposed were a stressful factor for a large number of people. This, in turn, created a demand for psychological assistance and support to the population, but counseling at a direct meeting in the office of a psychotherapist or psychologist was not possible due to the restrictions of quarantine measures. The need to provide psychological assistance to the population has led to the fact that psychologists have begun to master various information and communication technologies (ICT), social networks and Internet platforms. Specialists began to offer customers the format of working online via Skype, Zoom, WhatsApp, Google Hangouts, Meetings, etc. And if telemedicine has already been tested in working with patients [2], then the remote format of working with clients was quite new not only for psychologists and psychotherapists, but also for managers, office workers, teachers, salesmen, and consultants [3].

The use of ICT technologies by psychologists made it possible to conduct counseling during which clients could receive information, support and psychological assistance, together with a specialist to develop a plan to overcome the current difficult situation, resolve family conflicts, reduce emotional stress, improve psychological and emotional well-being. For example, Sood research confirms that online consultations and teleconsultations are a useful resource for reducing psychological problems such as anxiety and stress [4]. Scientists Lau et al. found that the cognitive behavioral therapist through online resources significantly reduces anxiety, depression and stress in women in the postpartum period [5]. Dores et al. concluded that ICT counseling is a very useful strategy for reducing anxiety, depression, and stress in a difficult situation, and psychologists play a significant role in psychological support for quarantined people as well as ordinary people [6].

Today, in Russia and other countries, we can observe the weakening of quarantine measures and the returning of specialists into their workplace. But many counseling psychologists and family psychotherapists continue to offer psychological help and counseling to clients through online platforms. This transition is linked with some positive aspects of working in an online format and a revision of the attitude of specialists to this type of consulting. So in the studies of Feijt et al. the advantages of using ICT are described: the ability to compensate for the limitations of traditional interventions (for example, travel requirements for clients or therapists), use as additional tools, easy access to Internet sites, high adaptability, flexibility and convenience, evolution according to the pace of the client, ease of compliance with the treatment regime and control of treatment, confidentiality and the possibility of anonymity, cultural adaptation, low cost and high potential for dissemination [7]. However, there are also some problems with the use of ICTs: ethical problems (for example, security, privacy, confidentiality, and lack of deontological orientation), client illiteracy in the field of ICTs, and negative attitudes towards Internet interventions [8], lack of access to technological and digital tools by some users, technological problems in their use, and changes in the environment and therapeutic relationships [5].

Despite the risks of using ICT technologies in the process of family psychological counseling in the professional environment of psychologists, there is a certain division between the preference for the format of work-online counseling and in-office counseling. But, despite the preferred forms of work, it is the personality of the psychologist that is the most important factor affecting the "recovery" of the client.

What qualities and characteristics should a modern specialist in the field of psychology and psychotherapy be endowed with?

A modern psychologist must first of all be a competent specialist, possess the necessary psychological knowledge, have professional skills and abilities [9]. In addition, the family and an individual psychologist must have a high level of development of adaptive qualities and adaptive potential. Such qualities include compliance with the law, normativity of behavior, honesty, civic courage, and integrity. In addition, the specialist has high requirements for neuropsychic and emotional stability, intellectual abilities, communication, organizational and motivational qualities [10]. Complex professional activities, increased contact with clients and care for them often leads to professional burnout of a specialist, which reduces the effectiveness of the psychologist. Sandoval confirmed the role of personality factors that predict burnout of psychologists and identified a set of personal characteristics that contribute to reducing the risk of burnout. He found that people who are balanced, free from neurotic tendencies, mature, optimistic and have a wide range of interests are less likely to burn out [11]. Efremova found that resilience and endurance reduce the risk of burnout of a consulting psychologist [12]. Karapetyan revealed that a successful psychologist is a carrier of the following qualities: vitality, psychological well-being, adaptability, resistance to professional burnout, optimism, extroversion, dominance and self-sufficiency [13]. We can say that balance, optimism, and resilience are important professional qualities of a modern family psychologist [14].

So Shevkieva together with colleagues revealed that a psychologist should show the ability to establish effective psychological contact with any client [15]. The historical analysis of the psychological qualities of the specialist allowed Shevkieva and Baitumanova to reveal that a modern psychologist should be more erudite, more skilled and experienced in the implementation of various types of professional activities. Moreover, the authors found that professional and personal changes extend to the sphere of cognitive processes, communication and organizational abilities. The need to be more creative in choosing and building psychological practices has increased significantly. Researchers explain this fact by the new freedom of choice, greater responsibility for the activities performed [16]. Rudakova studying the structure of personal qualities of a psychologist, proved that a modern specialist should have leadership qualities and reflexive self-organization [17].

Litten et al. investigating the predictors of career choice, found that cognitive empathy and high social desirability are the primary factors influencing the choice of a psychologist's profession [18]. Studying the personal factors that influence job satisfaction and the desire to pursue a career as a psychologist among young professionals, they revealed that self-efficacy and psychological flexibility are the determining properties of the need for young people to stay and develop in their profession [19–22].

Thus, the personality of a psychologist, including a family psychotherapist, is a carrier of various properties and qualities. A modern family psychologist should possess organizational, communicative and leadership qualities, high psychological stability, endurance, balance, emotional well-being, optimism, resilience, etc.

In the course of the theoretical analysis, we found that there are no studies on the personal characteristics of psychologists who prefer to conduct consulting work using ICT and remote technologies, although this format of working with clients imposes different requirements on a psychologist-consultant. Working online requires a professional to be more attentive to various signals from the client, to have the skills to conduct a dialogue correctly and tactfully, to clarify the client's emotional and physical state more often, to provide support and reinforce positive behavioral reactions, to teach self-help and self-support techniques, to tolerate some disadvantages of working with the help of ICT and remote technologies. We believe that this format of work contributes to the formation and actualization of other properties and qualities of a specialist than a psychologist who adheres to the traditional form of face-to-face meetings. This fact determined the novelty of the study, which consists in studying the personal characteristics of psychologists who prefer the online format in working with clients. We have put forward the following hypotheses: (1) the differences in personal qualities of specialists who prefer different formats of working with the client (online format, face-to-face meetings) will be revealed; (2) the relationship between the preference of the work format and the personal characteristics of psychologists will be revealed.

2 Materials and Methods

The sample in the study consisted of 60 practicing family psychologists aged 23–67 years (the average age in the sample was 36.7 years) with an average experience of consulting work of 2.3 years, using both full-time and remote consulting formats in their work with clients.

The respondents were asked to answer the questions of a demographic questionnaire, which also allows them to identify the specifics of their attitude to the formats of counseling—face-to-face and distance, and to identify the attitude to these formats and to themselves working in these formats, the method "Personal semantic differential" by Charles Osgood was used in the adaptation of employees of the V. M. Bekhterev Psychoneurological Institute [23]. The personal characteristics of the respondents were diagnosed using the 16-factor test of R. Kettell (form C) [24]. The specifics of the respondents' professional self-attitude were revealed using the method "Test of professional self-attitude" by Karpinsky [25].

Mathematical data processing included standard methods of mathematical statistics and was carried out using the following statistical methods: descriptive statistics; the r-Spearman correlation coefficient for establishing statistically reliable relationships between variables; the nonparametric U-Mann–Whitney test for comparing two independent samples; the nonparametric H-Kruskal–Wallace test for comparing more than two independent samples; the φ-Fisher angular distribution for comparing two independent samples on a nominal scale.

The survey showed that only 8 out of 60 respondents prefer the remote format of counseling, 26 of the surveyed psychologists-practitioners prefer to work with clients face-to-face, and the same number of respondents (26 people) without giving a clear preference for evaluation to any of the proposed work formats, they successfully apply both options in their practical activities.

3 Results

The study of the relationship between the preferences of family psychologists-practitioners of counseling formats with their personal characteristics revealed the following patterns, presented in Table 1.

Thus, 12 statistically accurate correlations (10 direct and 2 inverse) were revealed between the preference for the full-time format of consulting work with clients and the personal characteristics of family psychologists-consultants, 5 of which were at a high level of statistical reliability and 7-on average. In addition, one inverse statistically accurate correlation was found at a high level of statistical accuracy between the preference for the remote format of counseling and the personal characteristics of specialists.

In particular, direct correlations were found at a high level of statistical accuracy between the preference for the full-time format of counseling and the indicators on the following scales of the psychodiagnostic methods: factor F "restraint–expressiveness" of the 16-factor test of R. Kettell ($p = 0.002$), factors "Assessment" ($p = 0.006$) and "Strength" ($p = 0.009$) of the category "I am a practicing psychologist" and "Assessment" ($p = 0.004$) and "Strength" ($p = 0.001$) of the category "I am in full-time work with clients" methods "Personal semantic differential".

Table 1 Relationship of personal characteristics of family psychologists with the specifics of their attitude to the format of counseling (r-Spearman correlation coefficient)

Indicators	Correlation coefficient	Meaning (*p*)
Attitude towards the face-to-face format		
A (R. Cattell) "Is closed (−)–communicative (+)"	0.319	0.013*
B (R. Cattell) "Low intelligence (−)–high intelligence (+)"	−0.262	0.043*
F (R. Cattell) "Restraint (−)–expressivity (+)"	0.395	0.002**
G (R. Cattell) "Is low (−)–high (+) normative behavior"	0.276	0.033*
Internal conflict of professional self-attitude (Karpinsky)	−0.287	0.026*
Assessment "I am a practicing psychologist" ("Personal semantic differential")	0.354	0.006**
The power of "I am a practicing psychologist" ("Personal semantic differential")	0.335	0.009**
Activity "I am a practicing psychologist" ("Personal semantic differential")	0.314	0.015*
Power of "Digital and online technologies" ("Personal semantic differential")	0.278	0.031*
Assessment "I am in face-to-face work with a client" ("Personal semantic differential")	0.370	0.004**
The power of "I am in face-to-face work with a client" ("Personal semantic differential")	0.403	0.001**
Activity "I am in face-to-face work with a client" ("Personal semantic differential")	0.326	0.011*
Attitude towards the distance format		
L (R. Cattell) "Trusting (−)–suspiciousness (+)"	−0.402	0.001**

**The relationship is significant at the level of 0.01; *the relationship is significant at the level of 0.05

At the average level of statistical reliability, correlations were found between the preference for full-time consulting work with clients and the indicators on the following scales of the tests performed: direct correlations with the indicators for factors A "closeness–sociability" ($p = 0.013$) and G "social normativity of behavior" ($p = 0.033$), as well as an inverse correlation with the indicator for factor B "intelligence" ($p = 0.043$) of the R. Kettell test. In addition, direct correlations were found with the indicators for the "Activity" factor in relation to the categories "I am a practicing psychologist" ($p = 0.015$) and "I am in full-time work with clients" ($p = 0.011$), as well as with the indicator for the "Strength" factor in relation to the category "Digital and Online Technologies" ($p = 0.031$) of the "Personal Semantic Differential" methodology. An inverse correlation was found with the indicator on the scale "Internal conflict of professional self-attitude" of the method of Karpinsky ($p = 0.026$).

Only one inverse correlation was found between the propensity of specialists to work with clients remotely in a digital environment and their personal characteristics at a high level of statistical significance—namely, with the indicator for the factor L "trustfulness–suspicion" of the R. Kettell test ($p = 0.001$).

The following statistically accurate differences were found among the personal characteristics of the groups of respondents who prefer the face-to-face format of counseling, the remote format of counseling and do not prefer any of these formats, presented in Tables 2 and 3.

The data presented in Table 2 reflect statistically accurate differences in the personal characteristics of representatives of all three selected groups of respondents, identified using the nonparametric N-Kruskal–Wallace test.

Thus, at a high level of statistical accuracy, the experience of consulting practice differs among respondents included in different groups according to their preference for consulting formats ($p = 0.009$): respondents who prefer the remote format of consulting have, on average, longer work experience than representatives of the other two groups, and the shortest experience of consulting is noted in the group of respondents who prefer the face-to-face format.

Table 2 Differences in the personal characteristics of family psychologists who prefer different formats of counseling (H-Kruskal–Wallace criterion)

Indicators	Preferences	Average values	Meaning (p)
Experience in consulting work	Face-to-face format	0.7	0.009**
	Distance format	7.5	
	Both formats	2.4	
A (R. Cattell) "Is closed (–)–communicative (+)"	Face-to-face format	8.69	0.006**
	Distance format	4.88	
	Both formats	7.15	
F (R. Cattell) "Restraint (–)–expressivity (+)"	Face-to-face format	5.85	0.027*
	Distance format	3.13	
	Both formats	5.23	
L (R. Cattell) "Trusting (–)–suspiciousness (+)"	Face-to-face format	5.38	0.011*
	Distance format	3.88	
	Both formats	4.31	
Q_3 (R. Cattell) "Low self (–)–high self-control (+)"	Face-to-face format	6.65	0.043*
	Distance format	5.75	
	Both formats	7.35	
Activity "I am a practicing psychologist" ("Personal semantic differential")	Face-to-face format	8.77	0.017*
	Distance format	3.25	
	Both formats	5.42	

**Differences are significant at the level of 0.01; *differences are significant at the level of 0.05

Table 3 Differences in the personal characteristics of family psychologists who prefer different formats of counseling (U-Mann–Whitney criterion)

Indicators	Preferences	Average values	Meaning (p)
Those who prefer the distance format and those who prefer the face-to-face format			
Age	Face-to-face format	38	0.047*
	Distance format	32	
Experience in consulting work	Face-to-face format	0.7	0.022*
	Distance format	7.5	
A (R. Cattell) "Is closed (−)–communicative (+)"	Face-to-face format	8.69	0.003**
	Distance format	4.88	
F (R. Cattell) "Restraint (−)–expressivity (+)"	Face-to-face format	5.85	0.015*
	Distance format	3.13	
L (R. Cattell) "Trusting (−)–suspiciousness (+)"	Face-to-face format	5.38	0.035*
	Distance format	3.88	
Activity "I am a practicing psychologist" ("Personal semantic differential")	Face-to-face format	8.77	0.047*
	Distance format	3.25	
Those who prefer the distance format and those who prefer both formats			
F (R. Cattell) "Restraint (−)–expressivity (+)"	Distance format	3.13	0.043*
	Both formats	5.23	
Q_3 (R. Cattell) "Low self (−)–high self-control (+)"	Distance format	5.75	0.008**
	Both formats	7.35	
The power of "I am in distance work with clients" ("Personal semantic differential")	Distance format	4.25	0.039*
	Both formats	8.15	
Those who prefer the face-to-face format and those who prefer both formats			
Experience in consulting work	Face-to-face format	0.7	0.034*
	Both formats	2.4	
A (R. Cattell) "Is closed (−)–communicative (+)"	Face-to-face format	8.69	0.038*
	Both formats	7.15	
L (R. Cattell) "Trusting (−)–suspiciousness (+)"	Face-to-face format	5.38	0.007**
	Both formats	4.31	
N (R. Cattell) "Straightforwardness (−)–diplomacy (+)"	Face-to-face format	4.92	0.027*
	Both formats	7.50	
Overconfidence in the profession (Karpinsky)	Face-to-face format	13.19	0.038*
	Both formats	14.46	
Activity "I am a practicing psychologist" ("Personal semantic differential")	Face-to-face format	8.77	0.010**
	Both formats	5.42	

**Differences are significant at the level of 0.01; *differences are significant at the level of 0.05

At a high level of statistical accuracy, differences were found in factor A "closeness–sociability" of the 16-factor R. Kettell test ($p = 0.006$), whose indicator is higher than the rest in the representatives of the group of specialists who prefer the face-to-face format of counseling, and the lowest indicator is noted in the group of specialists who prefer the remote format.

At the average level of statistical accuracy, the results of all three selected groups differ in the factors F "restraint–expressiveness" ($p = 0.027$), L "trustfulness–suspicion" ($p = 0.011$) and Q3 "strong-willed self-control" ($p = 0.043$) of the 16-factor R. Kettell test, and, in the factors F "restraint–expressiveness" and L "trustfulness–suspicion", the indicators are higher in the group of respondents who prefer the face-to-face format of counseling, and the lowest indicators in the group of respondents, preferring the remote format. According to the Q3 factor "strong-willed self-control", respondents who prefer the remote format of counseling also have the lowest indicator, and the highest indicator for this factor is noted in the group of respondents who do not see a difference between the formats of counseling.

Also, at the average level of statistical accuracy, there were differences in the indicators of the "Activity" factor for the category "I am a practicing psychologist" of the "Personal Semantic Differential" method ($p = 0.017$), where the highest values were noted in respondents who prefer the face-to-face format of counseling, and the lowest—in respondents who prefer the remote format of work.

The results of a pairwise comparison of the three selected groups of respondents using the nonparametric U-Mann–Whitney test are presented in Table 3.

Pairwise comparison of the results obtained for the age of the groups of specialists under consideration revealed differences in the average level of statistical significance between the age of respondents who prefer the face-to-face format and respondents who prefer the remote format of working with clients ($p = 0.047$), and the respondents who prefer the face-to-face format are on average older than the respondents who prefer remote work.

The experience of consulting work among the respondents of the considered groups also statistically accurately varies. Thus, statistically accurate differences were found in the work experience of psychologists-consultants who prefer the face-to-face format of counseling, and respondents who prefer the remote format of counseling ($p = 0.022$), where the work experience is greater for those who prefer the remote format. There are also statistically accurate differences in the length of service of respondents who choose the full-time format of counseling, and those specialists who are suitable for both formats ($p = 0.034$), and the length of service is greater for respondents who do not see a difference in the effectiveness of work in different formats.

Differences in the indicators for factor A "closeness–sociability" of the 16-factor R. Kettell test were found at a high level of statistical accuracy in respondents who prefer the face-to-face format of counseling, and respondents who prefer the remote format of work ($p = 0.003$), and at an average level of statistical significance—in respondents who prefer face-to-face counseling, and respondents who do not note the difference between the formats ($p = 0.038$), and in both cases, respondents who prefer the face-to-face format of counseling, the indicator is higher.

Differences in the indicators for factor L "trustfulness–suspicion" of the 16-factor R. Kettell test were found at a high level of statistical accuracy in respondents who prefer the face-to-face format of counseling, and respondents who do not note the difference between the formats ($p = 0.007$), as well as at an average level of statistical significance in respondents who prefer face-to-face counseling, and respondents who prefer the remote format of work ($p = 0.035$), and in both cases, in respondents who prefer the face-to-face format of counseling, the indicator is higher.

At the average level of statistical accuracy, there are differences on the F scale "restraint–expressiveness" of the 16-factor R. Kettell test for groups of respondents who prefer the remote format of counseling, and respondents who prefer the face-to-face format ($p = 0.015$), and for groups of respondents who prefer the remote format of counseling, and respondents who do not note the difference between the formats of work ($p = 0.043$), and in both cases, respondents who prefer the remote format of counseling, the indicator is lower.

The differences in the indicators for the Q3 "self-control" factor of the 16-factor test of R. Kettell were revealed at a high level of statistical accuracy for the groups of respondents who prefer remote counseling, and respondents who do not note the difference between face-to-face and remote formats ($p = 0.008$). Differences at the average level of statistical accuracy were found in the indicators for the factor N "straightness–diplomacy" of the 16-factor *P* test. Kettell for groups of respondents who prefer the face-to-face format of counseling, and respondents who do not note the difference between the face-to-face and distance formats ($p = 0.027$). At the same time, the indicators for these two factors are higher for specialists who are comfortable working in both formats.

Differences at the average level of statistical accuracy were also found between the indicators on the scale of "Self-confidence in the profession" of the Karpinsky method for groups of specialists who prefer face-to-face counseling, and specialists for whom both formats of working with clients are equally comfortable ($p = 0.038$), the indicator here is also higher for respondents who do not prefer one or another format of work.

The differences in the "Activity" factor for the "I am a practicing psychologist" category of the "Personal Semantic Differential" methodology were revealed at a high level of statistical accuracy for the groups of respondents who prefer face-to-face counseling, and respondents who do not note the difference between face-to-face and remote formats ($p = 0.010$), and at an average level of statistical accuracy—for groups of respondents who prefer face-to-face counseling and respondents who prefer remote work ($p = 0.047$), and in both cases the indicator is higher for those who prefer face-to-face counseling.

At the average level of statistical accuracy, differences were found in the indicators of the "Strength" factor for the "I am in remote work with clients" category of the "Personal Semantic Differential" method for groups of respondents who prefer remote work format and respondents who do not make a choice in favor of one of the formats ($p = 0.039$), where the indicator is higher for those who do not prefer full-time or remote work format.

Table 4 Differences in preferences and experience of remote software consulting in groups of respondents with different work experience and different ages (criterion φ-Fisher's angular distribution)

Indicators	Group	Average values	φ	Meaning (*p*)
Experience in remote format				
Age	$n_2 = 31$, age 31–40	35.4	2.51	0.01**
	$n_3 = 18$, age 41–67	49.9		
Preference for remote format				
Age	$n_1 = 16$, age 23–30	27.1	1.65	0.05*
	$n_3 = 18$, age 41–67	49.9		
Experience in consulting work	$n_4 = 14$, exp. ≥1 year	2.0	2.50	0.01**
	$n_5 = 46$, exp. ≥1 year	0.4		
Preference for full-time format				
Experience in consulting work	$n_4 = 14$, exp. ≥1 year	2.0	2.74	0.01**
	$n_5 = 46$, exp. ≥1 year	0.4		

**Differences are significant at the level of 0.01; *differences are significant at the level of 0.05

The study of differences in the preferences of the work format (remote, full-time, full-time and remote) and the practice of working in an online format in groups distributed by work experience in psychological counseling, gender and age characteristics, showed some features. The differences in the φ-angular Fischer distribution are shown in Table 4.

Differences in the experience of working in a remote format and the preference for working in a remote format between groups of respondents of different ages were revealed. Differences in the preferences of remote and full-time work in groups with different experience of psychological activity are established. There were no differences in the preferences and experience of remote counseling between the groups of men and women.

4 Discussion

The results of the conducted empirical research allowed us to identify the following personal characteristics of family psychologists-practitioners, depending on their preference for different formats of working with clients.

From the identified correlations described above, it follows that specialists who prefer the full-time format of consulting work are characterized by such personal characteristics as expressiveness, cheerfulness, the desire to expand and deepen interpersonal contacts, activity, extroversion, sociability, openness, social normativity of behavior, the average level of development of verbal intelligence, their professional self-identification as a full-time practicing psychologist has an appeal for them. At the same time, respondents have confidence in their own professional capabilities, including when it comes to remote consulting in a digital environment that they do not prefer, where they probably also feel quite confident, although not quite comfortable. Specialists who prefer to work with clients remotely in a digital environment have such personal characteristics as trustfulness, openness, livability, tolerance, flexibility and compliance.

The revealed statistically accurate differences indicate such personal characteristics that distinguish specialists who prefer the remote format of consulting work.

Respondents who prefer the remote format of consulting clients have a longer experience of consulting work than their colleagues who prefer face-to-face consulting, and those who work in two formats with the same degree of comfort, while they are younger in age than their colleagues who prefer face-to-face consulting. Specialists who prefer the remote format of counseling are less sociable, more reserved, and more trusting than their colleagues who prefer to consult in person.

Compared to psychologists who prefer to consult clients in person, and psychologists who do not choose for themselves any one of the considered work formats, specialists who prefer to work in a remote format have a lower level of emotional and volitional self-control. Specialists who are equally comfortable working in two formats are more diplomatic and confident in their own professionalism than their colleagues who prefer to work only in full-time format. Specialists who prefer the remote format of work are less likely to present themselves as practicing psychologists working with clients than specialists who prefer to work full-time, while they feel less confident when working, including when working remotely, than their colleagues who actively use both formats of counseling.

The application of the Fisher's φ-angular distribution allowed us to establish significant differences in the proportion of people with remote work experience between the groups of respondents from 31 to 40 years old and 41–67 years old. In the group of older psychologists, there are more specialists who have experience working in a remote format. In the middle age group, the proportion of people with experience working on Internet sites is significantly lower than in other groups. Perhaps people in the older age group have more experience in counseling and therefore they do not have any difficulties to switch to a new format of work. There were no differences in the proportion of people who prefer the remote format of consultation between the respondents of group 1 and group 3.

According to the indicator "preferred remote format of work", it was found that the respondents of the age group from 27 to 30 years prefer the remote format, in contrast to the respondents of the older age group (41–67 years). The data obtained indicate that young people see the positive aspects of this format of work and are willing to use innovative technologies in their practice.

Differences in the preference for the format of work in connection with the experience of psychological activity are revealed. Thus, in the group of specialists with more than a year of work experience, the proportion of people who prefer the remote format is much higher than in the group of psychologists with little work experience. The full-time format of work, as shown by the statistical analysis of the work, is mostly preferred by specialists with little work experience. The result can be explained by the lack of experience in direct interaction with clients, insufficiently developed consulting skills, and the difficulty of interpreting nonverbal signals in indirect contact. It can also be assumed that there are personal predictors of the choice of face-to-face format of work with the client.

The results obtained are consistent with the studies of Portuguese colleagues Dores, Geraldo, Carvalho, Barbosa who found that psychologists with average and long-term professional experience positively assess the use of ICT in counseling. Remote consultations fully meet the needs of the client and show high results [5].

5 Conclusion

Thus, the forced transition to the remote format of individual and family psychological counseling connected with the COVID-19 pandemic, of course, allowed us to constructively expand the possibilities of providing psychological assistance to clients by minimizing organizational issues. On the other hand, as the results of the empirical study showed, practicing psychologists, especially those who provide family counseling, are probably not yet ready to switch to the remote format of working with clients.

From the data obtained, it can also be concluded that professionalism and confidence are acquired by specialists in direct proportion to their experience in consulting activities, while even psychologists who feel confident when working at a distance are not ready to completely switch to this format, which is due to the personal characteristics of specialists who choose this socionomical profession as a professional activity.

References

1. Stewart M, Schnabel A, Hallford DJ, McGillivray JA, Forbes D, Foster M, Shandley K, Gardam M, Austin DW (2020) Challenging child behaviours positively predict symptoms of posttraumatic stress disorder in parents of children with autism spectrum disorder and rare diseases. Res Autism Spectr Disord 69:101467
2. Goldschmidt L, Langa M, Masilela B, Ndhlovu LM, Mncina B, Maubane B, Bujela K (2021) Telepsychology and the COVID-19 pandemic: the experiences of psychologists in South Africa. S Afr J Psychol. https://doi.org/10.1177/0081246321993281
3. Cvetković BN, Stošić L, Belousova A (2018) Media and information literacy—the basis for applying digital technologies in teaching from the discourse of educational needs of teachers. Croat J Educ 20(4):1089–1114. https://doi.org/10.15516/cje.v20i4.3001
4. Sood S (2020) Psychological effects of the coronavirus disease-2019 pandemic. Res Humanit Med Educ 7. https://doi.org/10.1016/j.ajp.2020.102128
5. Lau Y et al (2017) Therapist-supported internet-based cognitive behavior therapy for stress, anxiety, and depressive symptoms among postpartum women: a systematic review and meta-analysis. J Med Internet Res 19(4):e138, 23–26. https://doi.org/10.2196/jmir.6712

6. Dores AR, Geraldo A, Carvalho IP, Barbosa F (2020) The use of new digital information and communication technologies in psychological counseling during the COVID-19 pandemic. Int J Environ Res Public Health 17:7663. https://doi.org/10.3390/ijerph17207663. www.mdpi.com/journal/ijerph
7. Feijt MA, de Kort YA, Bongers IM, IJsselsteijn WA (2018) Perceived drivers and barriers to the adoption of eMental health by psychologists: the construction of the levels of adoption of eMental health model. J Med Internet Res 20:e153. https://doi.org/10.2196/jmir.9485
8. Santos C, Weiderpass E, Santana R, Andersson G (2020) Portuguese psychologists' attitudes towards internet interventions: an exploratory cross-sectional study. JMIR Ment Health 7:e16817. https://doi.org/10.2196/preprints.16817
9. Zheldochenko L, Nikolenko O (2020) The problem of professional identity in the professional development of the individual. E3S Web Conf 210:22001. https://doi.org/10.1051/e3sconf/202021022001
10. Korochentseva A, Khachaturyan N (2020) Socio-psychological characteristics and nonverbal behavior of the psychologist, as components of its professional image in student representations. In: Innovative technologies in science and education (ITSE-2020). E3S Web Conf 210. https://doi.org/10.1051/e3sconf/202021018076
11. George-Levi S, Schmidt-Barad T, Natan I, Margalit M (2020) Sense of coherence and burnout among school psychologists: the moderating role of loneliness. Curr Psychol Early Access. https://doi.org/10.1007/s12144-020-00766-5
12. Efremova PR (2020) Organizational and personal factors of burnout in psychologist-consultant. Psikhol Zh 41(2):39–53. https://doi.org/10.31857/S020595920008535-4
13. Shevkieva NB, Bembeyeva NA, Bovaeva AV, Uvarova GN (2018) The profession of psychologist in the coordinates of a modern social demand. In: Social and cultural transformations in the context of modern globalism (SCTCGM 2018), vol 58, pp 812–820. https://doi.org/10.15405/epsbs.2019.03.02.93
14. Pishchik V, Vorontsov D, Breus E (2018) Socio-psychological characteristics of modern generations. In: 5th international multidisciplinary scientific conference on social sciences and arts. SGEM 2018: conference proceedings, vol 5 (3.2), pp 409–416
15. Shevkieva NB, Baytumanova ZN (2017) Dynamics of requirements to the personality of psychologist in Russia. Priority directions of science and education development. "Science and Education" (IP Gulyaev G.Yu.), Penza. https://www.elibrary.ru/item.asp?id=31234543
16. Karapetyan L (2019) Social and psychological predictors of successfulness: a comparative study among representatives of professional groups. Organ Psikol 9(3):20–34
17. Rudakova OA (2019) Acmeological development of the personality of a future psychologist as a basis of professional competence. Probl Mod Pedag Educ 63(1):469–472
18. Litten V, Roberts LD, Ladyshewsky RK, Castell E, Kane R (2021) Empathy and psychopathic traits as predictors of selection into business or psychology disciplines. Aust J Psychol 72(1):93–105. https://doi.org/10.1111/ajpy.12263
19. Schele I, Olby M, Wallin H, Holmquist S (2021) Self-efficacy, psychological flexibility, and basic needs satisfaction make a difference: recently graduated psychologists at increased or decreased risk for future health issues. Front Psychol 11:569605. https://doi.org/10.3389/fpsyg.2020.569605
20. Young EL, Moulton SE, Julian A, Smith A, Butler R (2021) Retention and job satisfaction of school psychologists. Psychol Sch 58(3):585–600. https://doi.org/10.1002/pits.22465
21. Kreneva J, Suroedova E, Korzhova O, Ozerova T (2020) Personal characteristics of wives of law enforcement officers as predictors of the formation of the psychological climate in the family system. E3S Web Conf 210:19015. https://doi.org/10.1051/e3sconf/202021019015
22. Korochentseva A, Suroedova E, Khachaturyan N, Nikolenko O (2019) Pupil's representation about the socio-psychological qualities of teachers in the process of effective meaning transmission. SHS Web Conf 70:08020. https://doi.org/10.1051/shsconf/20197008020
23. Smirnova OY, Serkin VP (2020) The development and approbation of the semantic differential "a strong-willed person". Psychol J High Sch Econ 17(2):210–222. https://doi.org/10.17323/1813-8918-2020-2-210-222

24. R. Cattell's 16-factor test (form C) (2019) Imaton, Moscow. https://imaton.com/content/files/catalog_2020_web.pdf
25. Karpinsky KV. Professional self-attitude of the individual and the methodology of his psychological diagnostics. Grodno State University, Grodno

The Knowledge Development of State and Municipal Employees Using Electronic Educational Technologies

Yulya Lebedeva, Yulya Belozerova, Olga Petrina, Michael Stadolin, and Elena Lyapuntsova

Abstract This article describes the state of the educational system in the field of developing the competencies of state and municipal employees and the features of using electronic educational technologies to improve the professionalism and competence of employees. The objectives of the study include: to define the concepts of "professional competencies", "professional competencies of a civil servant" and "competence"; to determine the features of training with the use of electronic educational technologies and to justify the effectiveness of it; to determine the degree of introduction of electronic and distance educational technologies in the state and municipal sector; identify and describe opportunities and prospects for further development and research on this topic. The methods used during the research include: the study, processing and analysis of scientific sources on the research problem; the method of content analysis; comparative analysis of data on the implementation of educational programs and projects; methods based on a systematic approach with a description of the effectiveness of solutions used in domestic practice, etc. The results of the study: it is necessary to maximize the introduction of the latest innovative technologies, equipment and new methods of education; the advantage of electronic educational technologies is the ability to attract high-class specialists without taking into account the territorial component; universal introduction of broadband Internet in all government bodies to improve the efficiency of services provided and training of employees; personalization of additional education programs; monitoring the needs of state and municipal employees to develop the necessary competencies for professional activity.

Y. Lebedeva · O. Petrina · M. Stadolin
State University of Management, Ryazanskiy Prospect, Moscow 109572, Russia

Y. Belozerova (✉)
The Institute of Cinema and TV (GITR) (ex named The Humanities Institute of TV and Radio Broadcasting), 32A Khoroshevskoye Road, Moscow 123007, Russia

E. Lyapuntsova
Bauman Moscow State Technical University, St. 2-Ya Baumanskaya, 5/1, Moscow 105005, Russia
e-mail: lev77@me.com

A. Beskopylny and M. Shamtsyan (eds.), *XIV International Scientific Conference "INTERAGROMASH 2021"*, Lecture Notes in Networks and Systems 247,
https://doi.org/10.1007/978-3-030-80946-1_61

Keywords Professional competencies · Electronic educational technologies · State and municipal government bodies · Personnel training

1 Introduction

Studies of this issue have shown that today the Russian system of state and municipal administration is undergoing a stage of modernization. However, given that at the federal level for many years there has been a fixed legislative system that regulates important aspects and features of the civil service, nevertheless, the latest innovations in the legislative framework that improve the activities of the state civil and municipal service are aimed at reforming and updating the professional characteristics (competencies) of employees at the federal, regional and municipal level.

This article describes the state of the educational system in the field of developing the competencies of state and municipal employees and the features of using electronic educational technologies to improve the professionalism and competence of employees.

The objectives of the study include: to define the concepts of "professional competencies", "professional competencies of a civil servant" and "competence"; to determine the features of training with the use of electronic educational technologies and to justify the effectiveness of the use of these methods for the training and professional development of public and civil servants; to determine the degree of introduction of electronic and distance educational technologies in the state and municipal sector, as well as the directions (competencies) of priority development; identify and describe opportunities and prospects for further development and research on this topic.

2 Materials and Methods

The methods used in the research include: the study, processing and analysis of scientific sources on the research problem; the method of content analysis, which allowed us to compare the used materials and legal acts in the field of EET (electronic educational technologies) and the educational system in the field of state and municipal administration; comparative analysis of data on the implementation of educational programs and projects of state and municipal authorities; methods based on a systematic approach with a description of the effectiveness of solutions used in domestic practice, etc.

It is possible to distinguish several groups of interest in the analysis of scientific works on the topic of this study. The first group includes scientific works devoted to the problems of terminology, the analysis of the concepts of "professional competence" and "competence". The authors of such works are: Winterton J., Delamare Le Deist F., Stringfellow E., Martynova S. and Maslennikova O., Fakhrutdinova E., Yuryeva Oksana, Burganova L., Yuryeva Olga, etc.

The second group of interest identified research in the field of electronic educational technologies, their impact on the development of competencies of employees of state and municipal government bodies, the authors include Vasilyeva E. V., Pulyaeva V. N., Yudina V. A., Belova S., Travkina N., Podkhalimov M., Kretova V.

It should also be noted that the study used static data provided by the Federal State Statistics Service (Rosstat) on additional professional education of personnel of the state civil and municipal service from 2008 to 2019. In addition, the materials presented in the database of the results of the All-Russian competition "Best Personnel Practices and Initiatives in the System of State and Municipal Administration" from 2015 to 2019, conducted by the Ministry of Labor and Social Protection of the Russian Federation, were analyzed.

3 Results

3.1 Professional Competence

The theoretical relevance of the issue under study is due to its interdisciplinary nature, the low degree of development of the problem, and the lack of a clear theoretical and methodological basis.

The Federal Law of the Russian Federation "On State Civil Service"[1] states the principle of professionalism and competence of civil servants. There are two approaches to the definition of "competence" and "competency"—personal and functional. The first approach focuses on the personality itself and its own resources, while the competency approach implies a concentration on the level of quality and functional areas of activity [1].

Studies of domestic and foreign scientists show that at the moment there is no single definition of "competence", this paper considers specifically "professional competence", i.e., manifested in the professional activity of the subject, namely in the field of state and municipal administration.

We shall consider some features of the concept of "professional competence", which various scientists have noted in their works. Thus, the definition of "competence" requires a level approach and should have a psychological and pedagogical nature [2].

Some authors [3, 4] defined competence as the ability of an employee or candidate to perform tasks assigned to him\her, taking into account their personal professional qualities, skills and experience, but the development of competencies is possible only in an emotionally safe and favorable learning environment according to the student's personality.

---

[1] Federal Law No. 79-FZ of 27.07.2004 (as amended on 08.12.2020) "On the State Civil Service of the Russian Federation".

Competency in the field of public service should be considered separately, according to Fakhrutdinova et al. [5, 6] as professional competences are indicators that characterize the professional knowledge, awareness and ability of a civil servant to effectively implement them in their official activities.

The analysis of the data of the public procurement portal showed that from the standpoint of the corporate order for professional development of state and municipal employees, the most popular topics of competencies are:

- security issues (work with personal data, security and emergency actions, protection of information and state secrets),
- anti-corruption policy,
- knowledge of information technologies (computer literacy),
- communication, psychological aspects of working with the population, emotional intelligence.

It is in these areas that the mass demand for professional development of employees is formed. There is also a demand for general management competencies (strategic planning, budgeting, motivation and personnel management, procurement management) and knowledge in the field of national policy in various industries.

Teaching the basics of creating a positive image of a civil servant through publications in the media and social networks is gaining popularity. Civil servants should be able to find a reasonable balance between information openness and corporate ethics [7, 8].

3.2 *The Importance of Electronic Educational Technologies in the Development of the Competencies of State and Municipal Employees*

In the system of state and municipal administration, there is a constant process of self-improvement, employees of institutions and public sector bodies are replaced, changes in the legislative and regulatory framework are taking place, as well as the needs of the population and regions are changing [3, 9]. As the basic abilities of the post-industrial era, a number of abilities were identified that make it possible, on the one hand, to learn and retrain, and on the other—to self-educate [4]. In this regard, the compliance of professional competencies with the needs of this field of activity is possible only with the continuous process of training (additional professional education and professional training) [10–12]. Since even the knowledge that was mastered during training in an educational institution does not give a general picture for the full performance of their duties in a certain position.

For the full implementation of the learning process and professional development, it is necessary to use the most effective forms and methods of training.

Therefore, it is necessary to consider the principles of forming programs for the training of state and municipal employees

- practice-oriented (solving practical cases and tasks);
- creation of programs in priority areas of development (sectors/areas of activity);
- differentiation of educational programs by target audience (positions and management levels);
- program duration optimization;
- specialized content of educational programs for the needs of state, regional and municipal government bodies;
- application of modern teaching methods and innovative technologies.

In this regard, we will consider the most promising direction for today in the field of additional professional education—electronic educational technologies.

E-learning technologies are opening up more and more opportunities to improve the quality of education for everyone, including state and municipal employees. Every year there are new resources, equipment and methods for training and knowledge control, which should be implemented in the system of professional development of representatives of the state and municipal sector, since they are one of the main systems for ensuring a high quality of life for the population [13–16].

Moreover, in accordance with the Decree of the President of the Russian Federation "On the main directions of development of the State Civil Service of the Russian Federation for 2019–2021",[2] the main directions of development of the state civil service include:

- introduction of new forms of professional development of civil servants, including the use of information and communication technologies;
- accelerated implementation of information and communication technologies in government agencies in order to improve the quality of personnel work.

Based on the analysis of leading practices from the database "Best personnel practices in the system of state and municipal administration" (2015–2019), it was also found that the most promising direction is e-learning and other types of distance and blended learning.

However, in practice, when performing a number of government contracts, the authors of this article encountered the following restrictions:

- large customers form applications for mass educational programs (several thousand students at a time), while the training period is quite short, this makes it difficult to individually approach students, does not allow them to form different educational trajectories in full, often limits the time for consulting when performing tasks, dictates the need for simple and universal knowledge control tools (testing);
- the poor quality of the information infrastructure in remote regions restricts the use of Internet-based training, it is often necessary to prepare a backup local version of electronic training materials on a physical medium (disk or flash card);
- increased requirements for the friendliness of the software product of electronic educational technologies (no requirements for the presence of any non-standard

[2] Decree of the President of the Russian Federation of 24.06.2019 N 288 "On the main directions of development of the State Civil Service of the Russian Federation for 2019–2021".

computer or software parameters, communication speed, additional equipment, all material must be provided with detailed step-by-step instructions);
- due to the large extent of the territory of Russia, time difference, the complexity of combining education with work processes, the advantage goes to offline classes in distance format (video recording lectures prepared material, which is available at a convenient time without reference to the online class);
- from the point of view of the educational tools preparation it is necessary to provide the maximum details of all educational materials like conceptual and technical side, personal consultations of a large number of listeners being organizationally difficult.

Based on the above principles and data, it can be concluded that in the current environment, the use of electronic educational technologies is necessary for many reasons, including, for example:

- lower financial costs;
- ability to personalize the training program;
- free access for students;
- flexibility of training (the ability to choose individually the duration and sequence of studying materials);
- improving the electronic literacy of employees;
- time saving, etc.

However, the use of advanced, non-traditional educational technologies, such as virtual and augmented reality, is hindered due to the requirements for equipment and the quality of Internet communication.

3.3 Application of Electronic and Distance Learning Technologies in Practice

According to the results of the analysis, it can be established that a significant part of those trained in additional professional education programs belong to the category of "specialists", while the number of trained managers remains low, which is partly due to the smaller number of this category of officials.

We shall consider the dynamics of the use of distance learning technologies and the directions of the development of competencies for additional education programs.

Figures 1 and 2 show the dynamics of the number of state and municipal employees, respectively, who received additional professional education, including the number of those trained using distance education technologies, as well as data on the format (on-the-job/pre-service/partially-on-the-job). It should be noted that Rosstat has not recorded information about partial separation from work since 2014 due to insignificant indicators for this format.

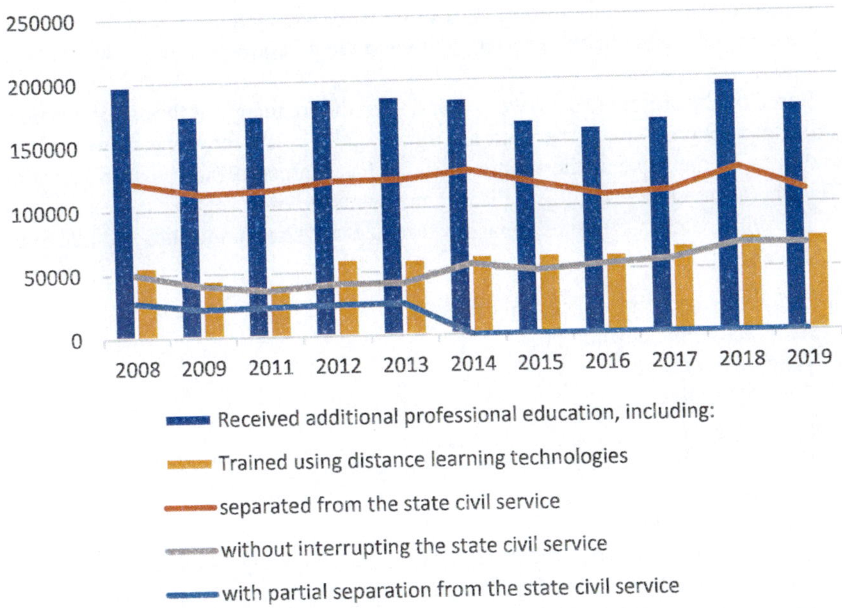

Fig. 1 Dynamics of the number of state civil servants who received additional professional education, including data on the training format from 2008 to 2019, people. *Source* Rosstat, https://rosstat.gov.ru/

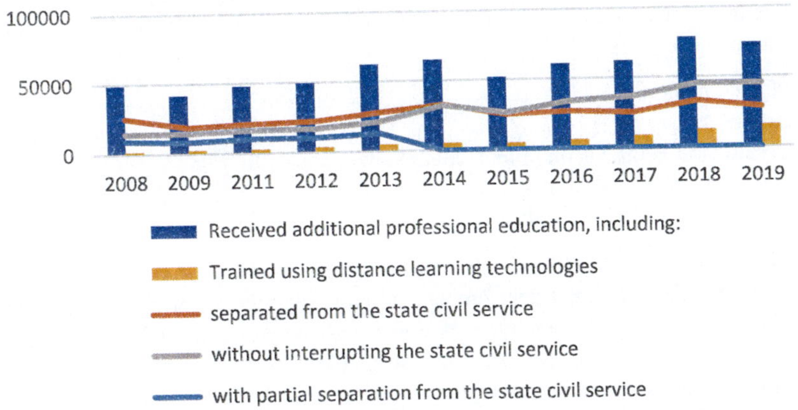

Fig. 2 Dynamics of the number of municipal employees who received additional professional education, including data on the training format from 2008 to 2019, people. *Source* Rosstat, https://rosstat.gov.ru/

It should also be noted that the statistics for 2010 were not taken into account due to the changed data collection system and there are no corresponding indicators for this year.

Based on the presented data, we conclude that the frequency of the use of distance education technologies at the state and municipal levels is growing, however, the dynamics of their use at the municipal level has increased more than 8 times in 10 years (at the state level, this indicator has increased only 1.3 times).

The priority areas of professional development for state authorities from 2015 to 2021 include:

- Combating (preventing) corruption;
- Socio-economic development;
- National Security;
- Financial Policy;
- Budget policy;
- Foreign policy;
- Competition Policy;
- Demographic policy;
- Ensuring information openness and the introduction of digital technologies in the work of government and self-government bodies;
- Improving the efficiency of the provision of services by state and municipal authorities, etc.

According to official data, no more than 58% of civil service employees received professional training in these areas of additional education (Fig. 3).

For the municipal level, Rosstat took into account the following areas of additional educational programs:

- Management;
- Legal;
- Organizational and economic development;
- Planning and financial planning;
- Information and analytical;
- Others.

It can be concluded that over the past 4 years, the number of trained personnel in priority areas of professional development has increased in state structures, but due to the lack of statistical data, it is impossible to say which specific programs and areas of activity were specialized in training, and which competencies were developed. However, the monitoring of the directions of educational programs at the municipal level has been carried out without changes since 2008. Analysis of data from 2015 shows despite the fact that about 30–35% of the programs of additional education, are classified as "others", without further specification, the rest 70–65% are key competencies required in municipal service (Fig. 3).

It should be also noted that information-analytical area of professional development for municipal employees is the last place not exceeding 5.4%, while also demonstrating low growth.

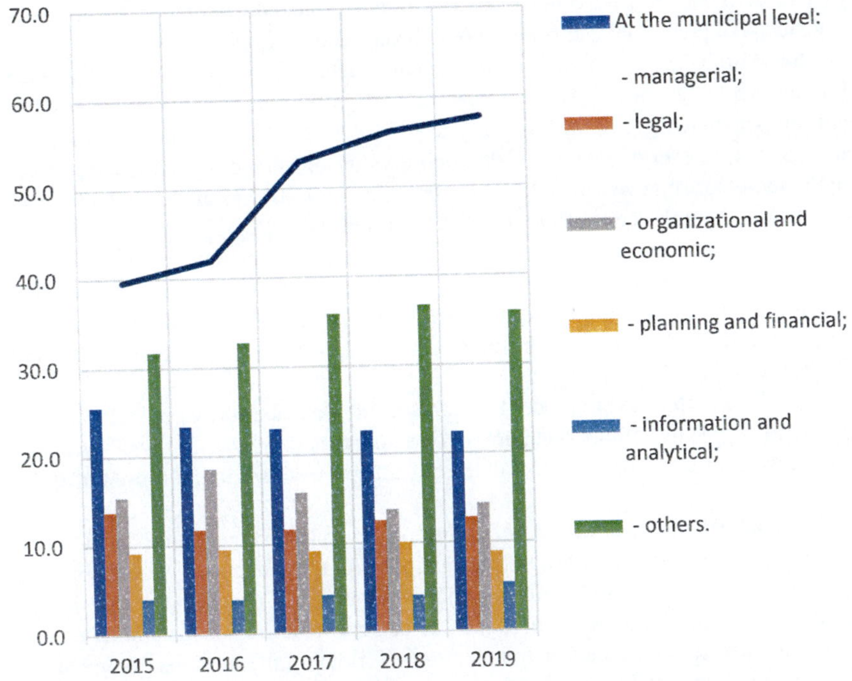

Fig. 3 Dynamics of the percentage ratio in the areas of additional professional education at the municipal and state levels from 2015 to 2019, %. *Source* Rosstat, https://rosstat.gov.ru/

It should also be noted that the share of organizations using broadband Internet in state authorities in 2018 was 90.3%, in local governments-85.9%. In general, one tenth of the authorities in Russia do not have access to the Internet, and only a quarter of all authorities use cloud services in their work [17].

4 Discussion

Thus based on the results obtained and the analysis of the literature [18, 19], it can be concluded that the competence of a state and municipal employee depends on the knowledge and skills acquired, which are respectively accumulated through experience, practice, and primarily professional training (including practice-oriented). Educational programs allow you to develop and form professional and personal qualities in state and municipal employees, which will help in making effective management decisions, especially in critical situations.

Electronic educational technologies are already an integral part of the system of professional development of state and municipal employees, as shown by research data and the work of other scientists [20]. This leads to the conclusion that the positive dynamics of the use and results of the use of electronic educational technologies in practice give favorable results and should be applied everywhere.

However, based on the analysis of statistical data, it can be established that even the significant dynamics of some indicators (such as the increase in the use of distance education technologies at the municipal level by 8 times) does not mean that the technology is used everywhere, since the total number of trained municipal employees in the remote format was only 15,924 people (4%) for 2019 (while the total number of municipal employees in Russia is 395 thousand people).

At the state level, the indicators differ significantly, since the number of trained employees in general is about one third of all trained employees, but the number of on-the-job trained employees exceeds the number of non-on-the-job trained employees by 2–3 times, which negatively affects professional achievements and the performance of official duties.

In addition, 75,275 trained civil servants in the distance format in 2019 for 855 thousand people out of the total number of employees (8.8%, respectively) is also a fairly low figure, taking into account the advantages of this training format over the rest.

The analysis of the directions of additional education showed that the study of information—analytical materials, informational, digital and innovative technologies and other related materials is not of primary importance. Consequently, this significantly reduces the efficiency of both the work of state and municipal authorities, and their ability to study the material in electronic format on the job. Electronic educational technologies and digitalization of state and municipal management are directly interrelated, since they improve e-literacy and increase employee productivity.

5 Conclusions

Therefore, from all of the above, it is possible to determine a number of the most promising ways of development:

- the maximum possible introduction of the latest innovative technologies, equipment and new techniques to provide the system of training of state and municipal employees with everything necessary (given the low cost of electronic educational technologies in comparison with full-time programs, financial costs will be much less when replacing full-time classes with remote ones using electronic educational technologies);
- increasing the use of remote and electronic educational technologies will allow municipalities and regions far from the central one to attract leading industry specialists for the practice-oriented and high-quality provision of educational services;
- widespread introduction of broadband Internet in all authorities across Russia, without access to the Internet, the capabilities of not only local authorities, but also the population who use the services of these authorities are significantly

reduced, which leads to negative consequences for the entire region and leads to a lack of ability to respond quickly to the needs of the population due to the lack of data monitoring in the territory. For such territories including, e-learning is the only possible solution for the development of professional competencies of employees and the acquisition of new practice-oriented knowledge and skills;

- personalization of additional education programs. In addition to the distance/electronic learning format, it is necessary to select the optimal programs for the development of the necessary competencies for the employee. One of the options is to select the necessary educational program from a wide range of educational services offered on the market, but this option is fraught with problems with the quality of teaching staff, funding problems or non-compliance of the declared content with the necessary requirements;
- another way, the most optimal, is to monitor the target audience. When analyzing the needs, active or absent competencies, knowledge and skills for a specific situation, etc. (in the form of a questionnaire, survey or other monitoring operations), the necessary disciplines/areas of knowledge/questions can be identified for discussion and study in additional professional educational programs. It is also necessary to constantly monitor the satisfaction, if possible, not only of employees, but also of service consumers, in order to identify the progress of training and to understand further promising areas of development.

References

1. Andersson UB, Löfgren H, Gustafson S (2019) Forward-looking assessments that support students' learning: a comparative analysis of two approaches. Stud Educ Eval 60:109–116. https://doi.org/10.1016/j.stueduc.2018.12.003
2. Belova S, Travkina N, Podchalimov M, Kretova V (2015) Creation of conditions for professional retraining and professional development of employees of the city construction organizations in Kursk Assembly Technical School. Procedia Eng 117:80–84. https://doi.org/10.1016/j.proeng.2015.08.127
3. Martynova SE, Maslennikova OG (2015) The "service" model of the competences of the municipal employee as the basis of vocational training: Russian experience. Procedia Soc Behav Sci 174:2716–2720. https://doi.org/10.1016/j.sbspro.2015.01.957
4. Mozhaeva G (2014) Network interaction in distance education: analysis of Russian experience. Procedia Soc Behav Sci 152:1124–1127. https://doi.org/10.1016/j.sbspro.2014.09.286
5. Fakhrutdinova E, Yurieva O, Burganova L, Yurieva O (2015) Role of competence-based approach in an increase of efficiency of public administration. Procedia Econ Finance 23:1064–1067. https://doi.org/10.1016/S2212-5671(15)00511-0
6. Filippova T (2015) Priority fields of E-learning development in Russia. Procedia Soc Behav Sci 206:348–353. https://doi.org/10.1016/j.sbspro.2015.10.063
7. Glebova I, Yasnitskaya Y, Maklakova N (2015) The evaluation of the innovative activity of modern Russia's higher educational establishments and ways of its fostering. Procedia Soc Behav Sci 191:2374–2381. https://doi.org/10.1016/j.sbspro.2015.04.378
8. Golitsyna I (2017) Educational process in electronic information-educational environment. Procedia Soc Behav Sci 237:939–944. https://doi.org/10.1016/j.sbspro.2017.02.132

9. Kohanovskaya II, Fatyhova AL, Golovneva EV, Golovneva NA (2020) Training of civil servants in the conditions of development of digital economy. European Publisher, pp 890–899. https://doi.org/10.15405/epsbs.2020.04.114
10. Popkova EG, Ekimova KV, Sergi BS (2020) Data set of balance of Russia's regional economy in 2005–2024 based on the methodology of calculation of "underdevelopment whirlpools". Data Brief 31. https://doi.org/10.1016/j.dib.2020.105821
11. Vasilieva EV, Pulyaeva VN, Yudina VA (2018) Digital competence development of state civil servants in the Russian Federation. Bus Inform 4(46):28–42. https://doi.org/10.17323/1998-0663.2018.4.28.42
12. Vorontsov A, Vorontsova E (2015) Innovative education in Russia: the basic tendencies analysis. Procedia Soc Behav Sci 214:1147–1155. https://doi.org/10.1016/j.sbspro.2015.11.731
13. Winterton J, Delamare Le Deist F, Stringfellow E (2006) Typology of knowledge, skills and competences: clarification of the concept and prototype. In: Cedefop reference series 64, Luxembourg
14. Skills for a high performing civil service. www.oecd.org/gov/pem. Accessed 15 Mar 2021
15. Misbakhova CA, Shinkevich AI, Belozerova YM, Yusupova GF, Stakhova LV (2016) Innovation infrastructure of engineering and small innovative business in development of national innovation system. J Adv Res Law Econ 7(2):323–331
16. Lyapuntsova E, Belozerova I, Drozdova I, Berlinova M (2018) Investment risks in the field of social entrepreneurship. MATEC Web Conf 251
17. Lyapuntsova E, Vdovichenko M, Belozerova Y, Gorbatov A (2019) Application of modern modeling methods: virtual technologies in the era of digitalization and their role in modern companies. IOP Conf Ser J Phys Conf Ser 1425:1–7. https://doi.org/10.1088/1742-6596/1425/1/012165
18. Lyapuntsova E, Belozerova Y, Drozdova I, Afanas'Ev G, Okunkova E (2018) Entrepreneurial risks in the realities of the digital economy. MATEC Web Conf 251:1–6. https://doi.org/10.1051/matecconf/201825106032
19. Lebedeva Y, Belozerova Y, Zakharova M, Volkova M (2020) Methodological efficiency of using virtual and augmented reality in the educational process. In: Innovative technologies in science and education (ITSE-2020). E3S Web Conf 210:1–9
20. Vdovichenko M, Lyapuntsova E, Belozerova Y, Drozdova I (2020) Analysis of the methods for assessing the quality of entrepreneurial education. E3S Web Conf 210:1–7
21. Dobos A (2015) The role of learning environments in civil service professional development training. Procedia Soc Behav Sci 197:565–569. https://doi.org/10.1016/j.sbspro.2015.07.194

Perceptions of Themselves and an Employee Being a Difficult Communication Partner of Managers with Different Managerial Styles

Alla Belousova, Elena Breus, and Galina Kozhukhar

Abstract The article examines the managers' ideas about the subordinate and about themselves as a difficult communication partner. We study the differences in these perceptions among managers with different management styles: directive, collegial, and permissive. The results obtained confirm that the idea of a subordinate as a difficult communication partner among managers with different management styles differs in formal and substantive parameters. It is revealed that managers with directive and permissive management styles do not consider themselves as difficult communication partners for subordinates. At the same time, subordinates, in the views of managers with directive and permissive management styles, are difficult partners. For managers with these management styles, all the studied groups of communication characteristics of subordinates (expressive-speech, social-perceptual, relations-addresses and interaction skills) cause a greater degree of difficulties in communication than for managers with a collegial management style.

Keywords Difficult partner · Ideas about a difficult communication partner · Management styles · Characteristics of difficult communication

1 Introduction

The digital revolution involves changes in all spheres of human life. From the point of view of modern management specialists, digitalization changes some of the basic principles and rules of doing business, competition, forms of obtaining information,

A. Belousova (✉)
Don State Technical University, Gagarin Sq. 1, Rostov-on-Don 344003, Russian Federation

E. Breus
Southern Federal University, 105/42 Bolshaya Sadovaya Str., Rostov-on-Don 344006, Russian Federation

G. Kozhukhar
Moscow State University of Psychology and Education, Sretenka St. 29, Moscow 127051, Russian Federation
e-mail: kozhuhargs@mgppu.ru

A. Beskopylny and M. Shamtsyan (eds.), *XIV International Scientific Conference "INTERAGROMASH 2021"*, Lecture Notes in Networks and Systems 247,
https://doi.org/10.1007/978-3-030-80946-1_62

and much more. The basic principles of office work are changing. Some employees no longer need to be tied to their place of work and the format of remote access to work appears. The implementation of various applications makes it possible to conveniently and quickly receive information and share it with subordinates, colleagues, and managers. Thus, the format of communication and interaction of business entities is changing. In fact, the ongoing transformation concerns more the organizational and content side of the activity, which affects the existing system of relations between employees with each other and with management.

One of the promising directions for understanding the changes in communication and perception of employees of each other that occur in the context of the use of digital technologies is the socio-psychological approach to the phenomenon of difficult communication by Labunskaya [1]. Labunskaya considers the idea of the other as a difficult communication partner being one of the important determinants of difficult communication [1]. If the ideas about the partner (features and characteristics of his\her behavior, communication skills) do not meet expectations, then there are difficulties in communication. The degree of these difficulties depends on many factors and affects the system of relations between partners. In management psychology, it has long been known that the system of relations among employees, and in particular the system of relations "manager-subordinate", is an important element of corporate culture and one of the most important determinants of the effectiveness/inefficiency of the organization's business processes. Belousova et al. [2] showed that the nature of the attitudes underlying the perception of subordinates depends on personal factors, among which the authors point out the orientation of the individual and self-esteem. The general orientation of the personality and self-esteem affect the overall style of the manager's activity [2]. The influence of motivational features on the style behavior of managers is reflected in the study Belousova and Mochalova [3]. The work Korochentseva et al. [4] shows the features of the ideas about professionally important qualities as cognitive components of the formed attitudes of students of the food industry to the images of professionals. Most of the studies are devoted to the study of the relationship between the effectiveness of employees' activities and their managers' style [5–7]. Thus, a group of Finnish researchers Suhonen et al. [5] studied the perception of management styles of nursing managers in the wards of medical centers in Finland. The authors have shown that the ability of nursing managers, if necessary, to change their leadership style affects the success of the unit and employee satisfaction with the work.

The study by Castro [6] revealed significant differences between the assessments of school managers using forced and affiliated management styles, especially in relation to the communication skills of school leaders. In the studies of Russian scientists Meshcheryakova [7], it is also shown that the effectiveness of the organization's employees, regardless of the field of activity, largely depends on the management style. Recently, more and more attention of scientists has been drawn to the problems of generational continuity in relation to various principles, ways of activity and interaction [8], as well as in relation to the continuity of leadership in family firms that ensure business success [9].

Brazilian researchers Ramos et al. [10], studying the management styles most often used by managers, raise the problem of developing managerial abilities. Also, in the work of the Russian Scientist Bgashev [11], attention is focused on the study of not only positive, but also negative qualities of a manager, which, in his opinion, affect the formation and development of specific management styles, which the author considers as temporary formations.

A number of studies have shown that the most effective is the combination of different styles. Thus, the study by Namiq [12] shows that many managers use a combination of all styles to achieve progress. Nezorenko [13] also notes that it is impossible to determine unequivocally which style is the best: authoritarian, liberal or democratic, since its choice depends not only on the qualities and skills of the manager himself, but also on the specific situation, the characteristics of employees and a number of other factors.

Therefore, it is advisable, from the author's point of view, to use a flexible style that combines different elements of other styles. Also, a number of studies have revealed the influence of management style on the formation of a system of relations between managers and subordinates, which, in turn, contributes to the effectiveness of employees' activities. Thus, in the work of Taucean et al. [14], when studying various styles of leadership power, it was revealed that the perception of one's organizational power was very important for achieving success or failure in work. Studying different management styles Russian scientists Kovalchuk and Tskhai [15] revealed that the main difference between the styles lies in the manifestation of the attitude of the manager to the performer. In addition, the work of Panina [16] shows that an important factor influencing the formation of individual organizational relationships between managers and subordinates are different management styles and psychological types of managers.

However, despite the fact that many factors in the formation of the system of relations between employees are already known, it is possible and necessary to fill in the gaps in the field of studying this phenomenon from the point of view of ideas about a difficult communication partner. In works devoted to the study of ideas about a difficult communication partner, socio-psychological and personal characteristics have been identified that affect the degree of communication difficulties that arise [1, 17]. However, the specifics of managers' perceptions of their employees as difficult communication partners, in connection with the management style of the manager, have not yet been studied.

The following assumptions were used as hypotheses of the study: (1) The perception of an employee as a difficult communication partner may differ among managers with different management styles. (2) Managers' perceptions of themselves and their subordinates as difficult communication partners may differ in formal and substantive parameters.

2 Methods

The following methods were used in the work: the method "Diagnostics of management style" by Zhuravlev [18] used to identify three main management styles: directive, collegial and permissive. The management style, according to Zhuravlev, is an individual-typical features of an integral, relatively stable system of methods, ways, and techniques, with which the manager implements his functions. The directive style is characterized by a rigid unity of command, as well as a weak interest in the employee as a person. In the case of collective management, the manager strives to develop collective solutions, while demonstrating an interest in the informal, human aspect of relations. Permissive (or liberal) means the complete removal of the leader from the affairs of the team. The method "Socio-psychological characteristics of the subject of communication" by Labunskaya et al. [19] is used to study the perceptions of the other as a difficult communication partner. The method presents 5 groups of characteristics: expressive-speech characteristics (ES) such as: quiet or loud speech, motionless face, long pauses in speech, sluggish gestures, unwillingness to maintain eye contact, etc.; social-perceptual characteristics (SP) such as: inability to put yourself in the place of another person, errors in assessing the feelings and moods of another person, the habit of judging a person by his appearance, etc.; characteristics of the attitude-treatment (AT) such as: an overbearing or arrogant attitude towards other people, an indifferent or suspicious attitude towards other people, etc.; skills and abilities of organizing interaction (OI) such as: the habit of interrupting a conversation, the inability to argue their comments and suggestions, the desire to impose their point of view, etc. A separate group includes such conditions of communication as: the intensity of communication, the presence of witnesses of communication, gender, age and status of communication partners, etc. In the first part of the methodology, respondents are asked to assess how much a particular characteristic of the partner's behavior makes it difficult to communicate with him (from 5 points—very difficult, to 1 point—not difficult). The second part of the methodology asks the respondent to assess how much one or another of his\her own characteristics (if this characteristic, as he believes, is inherent in him) makes it difficult for his partner to communicate (the same assessment system is used, only 0 point is added—there is no quality). Thus, the method allows us to identify the formal and meaningful parameters of representations, both about the other and about ourselves as difficult communication partners. Formal parameters are represented by quantitative indicators and describe the degree of difficulties in communicating with a partner for each group of characteristics. The content parameters are represented by the description of group portraits of a difficult partner. The method allows you to explore the idea of a difficult partner in any area of interaction and in relation to any status-role position (a spouse, as a marriage partner; a parent and child in a parent relationship; a manager and subordinate in a business relationship; a seller and customer in the service sector and in many other areas of activity). The study was conducted in 2 stages. At the first stage, the empirical object of the study was 60 middle-level managers of commercial organizations with at least 3 years of experience in a managerial position (male, aged 29–50 years). At the initial stage of the study, the managers' style was evaluated by their employees. Each manager was evaluated by 5 employees. At the same time, managers conducted a self-assessment of their management style. For these purposes, the method of Zhuravlev was used. In the second stage of the study, out of 60 managers, only 45 managers participated, whose results of the self-assessment of the management style coincided with the external assessment of their management style by their employees. The managers who passed the 2nd stage of the study were divided into 3 groups, according to their identified management styles: the first group included 16 managers with a directive management style; the second group included 17 managers with a collegial management style; the third group included 12 managers with a permissive management style. At the 2nd stage of the study, in groups of managers with different management styles, ideas about themselves and their subordinates as difficult communication partners were revealed. For these purposes, the method of Labunskaya was used. Descriptive statistics and the Mann–Whitney U-test were used to process the obtained data.

3 Results

3.1 *Analysis of the Perception of the Employee as a Difficult Partner for Managers with Different Management Styles*

To prove the first hypothesis, the Mann–Whitney U-test was applied to the obtained data (Table 1).

The analysis of the data in Table 1 allows us to conclude that there are differences in the average values of the degree of difficulty in the communication characteristics of a subordinate among managers with different management styles. To prove the first hypothesis, the Mann–Whitney U-test was applied to the obtained data. Significant differences in the degree of difficulty in managers with directive and permissive management styles (ES) were revealed in the expressive-speech characteristics of subordinates ($Z = -3.631$; at $p = 0.000$); (IS) in the characteristics of the interaction skills of subordinates ($Z = -1.857$; at $p = 0.068$) and (CC) the conditions in which communication occurs ($Z = -2.162$; at $p = 0.031$).

Based on the results obtained, it can be concluded that managers with a directive management style consider their subordinates more difficult partners in three groups of characteristics (ES, IS, CC) than managers with a permissive management style. There were also significant differences in the degree of difficulty in managers with directive and collegial management styles in relation to (ES) expressive-speech characteristics ($Z = -2.643$; at $p = 0.008$) and (CC) communication conditions ($Z = -2.439$; at $p = 0.015$). This suggests that managers with a directive style, compared to a collegial style, also view their subordinates as more difficult partners. Thus, in the views of managers with a directive management style, the subordinate will make communication difficult with his expression and interaction skills. In addition, it is

Table 1 Indicators of the degree of difficulty with the characteristics of the subordinate's communication in the views of managers with different management styles

Management style	Average values of the degree of difficulty of the manager with the characteristics of communication of the subordinate				
	Expressive-speech	Social-perceptual	Relations-addresses	Interaction skills	Conditions in which communication takes place
Directive	8.72	7.94	8.53	8.67	9.24
Collegiate	7.39	7.31	6.80	8.03	7.1
Permissive	5.72	7.25	7.69	6.33	6.27

the representatives of the directive style that cause great difficulties in a wide variety of communication conditions.

3.2 *Analysis of Self-image and Perception of an Employee as a Difficult Partner for Managers with Different Management Styles*

The Mann–Whitney U-test was used to prove the hypothesis that there are differences in the perception of oneself and the subordinate as a difficult communication partner. The average values of the characteristics of the subordinate that make it difficult for the manager to communicate, and the ideas about their own characteristics that make it difficult for the subordinate to communicate, are presented in Tables 2, 3 and 4.

Analysis of the data in Table 2 allows to conclude that there are significant differences in the views of managers with a directive management style about themselves and about the subordinate as a difficult communication partner. According to the Mann–Whitney criterion, significant differences were obtained in relation to all groups of characteristics: in relation to (ES) expressive-speech characteristics ($Z = -3.618$; $p = 0.000$); in relation to (SP) social-perceptual characteristics ($Z = -3.122$; $p = 0.002$); in relation to (AT) attitude-treatment characteristics ($Z = -2.850$; $p = 0.004$); in relation to (IS) interaction skills ($Z = -3.819$; $p = 0.000$). The data obtained suggest that managers with a directive management style view all the characteristics of a subordinate as hindering communication. Their own characteristics, in the view of managers with a directive management style, are not considered at all as hindering the communication of subordinates.

Table 2 The perception of themselves and of a subordinate as a difficult communication partner for managers with a directive management style

Directive management style	Average values of the degree of difficulty with the characteristics of the subject of communication			
	Expressive-speech	Social-perceptual	Relations-addresses	Interaction skills
The manager about the subordinate as a difficult communication partner	8.72	7.94	8.53	8.67
The manager about himself as a difficult communication partner	4.57	5.11	5.63	4.89

Table 3 The perception of themselves and of a subordinate as a difficult communication partner for managers with a collegial management style

Collegial management style	Average values of the degree of difficulty with the characteristics of the subject of communication			
	Expressive-speech	Social-perceptual	Relations-addresses	Interaction skills
The manager about the subordinate as a difficult communication partner	7.39	7.31	6.80	8.03
The manager about himself as a difficult communication partner	7.39	7.30	6.33	7.94

Table 4 The perception of themselves and of a subordinate as a difficult communication partner for managers with a permissive management style

Permissive management style	Average values of the degree of difficulty with the characteristics of the subject of communication			
	Expressive-speech	Social-perceptual	Relations-addresses	Interaction skills
The manager about the subordinate as a difficult communication partner	7.39	7.31	6.80	8.03
The manager about himself as a difficult communication partner	7.39	7.30	6.33	7.94

From the data in Table 3, it follows that the ideas about their own communication characteristics and the characteristics of the partner that make communication difficult are almost the same. The analysis of the presented data according to the Mann–Whitney U-test revealed no significant differences in any group of characteristics. This means that managers with a collegial management style do not view themselves and their subordinates as difficult communication partners.

The analysis of the data in Table 2 allows to conclude that there are significant differences in the views of managers with a permissive management style about themselves and about the subordinate as a difficult communication partner. According to the Mann–Whitney criterion, significant differences were obtained in relation to all groups of characteristics: in relation to (ES) expressive-speech characteristics ($Z = -2.721$; $p = 0.007$); in relation to (SP) social-perceptual characteristics ($Z = -2.350$; $p = 0.019$); in relation to (AT) attitude-treatment characteristics ($Z = -2.839$; $p = 0.005$); in relation to (IS) interaction skills ($Z = -2.597$; $p = 0.009$). The data obtained suggest that managers with a permissive management style view all the characteristics of a subordinate as hindering communication. Their own characteristics, in the view of managers with a permissive management style, are not considered as hindering the communication of subordinates.

3.3 *Analysis of Group Portraits of Employees as Difficult Communication Partners in the Views of Managers with Different Management Styles*

In the group portraits, the characteristics of subordinates were selected, which received an estimated score of 4 and 5 more than 50% of the surveyed managers in each group with a different management style.

Group portrait of a difficult partner-employee for managers with a directive management style: this is an employee with a frozen pose and a fixed face, who does not want to maintain eye contact, whose facial expressions do not match his words. He is characterized by a fast pace of speech, frequent touching of the partner, patting on the shoulder, putting his hand, etc.; a suspicious and hostile attitude towards the partner. He talks more than he listens, tends to take a leading position in communication, to impose his point of view. He does not know how to get out of communication, stop it in time, does not know how to argue his comments, suggestions. Focuses on his\her own feelings and thoughts.

Group portrait of a difficult partner-employee for managers with a collegial management style: this is an employee with a frozen pose and a motionless face, who systematically moves during communication, often touches the partner (puts his hand, pats on the shoulder, etc.); seeks to evaluate people based on the ideas that have developed in his environment. This is a person with an indifferent or hostile attitude towards other people, with a habit of interrupting a conversation.

Group portrait of a difficult partner-employee for managers with a permissive management style: this is an employee who does not know how to put himself in the place of another and with a suspicious attitude towards other people.

4 Discussion

The study found differences in the formal and substantive parameters of the ideas about the subordinate as a difficult partner among managers with different management styles. Managers with a rigid display of unity of command and a weak interest in the employee's personality, on the one hand, are more hampered by external, expressive features of the subordinate's behavior (speed of speech, dynamic or static posture and face, tactile touch), on the other hand, these managers are very sensitive to characteristics that violate personal equality (the desire to impose their point of view, suspicious and hostile attitude, the desire to take a leading position in communication, concentration on their own feelings and thoughts). It should be noted that the data obtained in this work confirm the type of authoritarian personality described by Fromm [20] and Maslow [21].

The obtained data indicate that for representatives of directive and permissive management styles, the subordinate, not the manager, causes communication difficulty in formal parameters. In the views of managers with directive and permissive management styles, their own characteristics cannot act as hindering the communication of subordinates, while all groups of communication characteristics of the subordinate are considered as hindering the communication process between the manager and the subordinate.

A harmonious position in relation to the ideas of themselves and the subordinate as a difficult communication partner among representatives of the collegial management style was revealed, which was manifested in the awareness of the equally possible determination of difficulties in communication, both by the characteristics of the partner and by their own communication characteristics.

The authors also found out the peculiarities of the perception of themselves and the subordinate as completely non-difficult subjects of communication among managers with a permissive (liberal) management style. This was shown in the formal indicators (the minimum values of the indicators of the degree of difficulty with their own characteristics), and in the content of the group portraits of a difficult subordinate, only 2 characteristics that hinder the manager are presented.

In general, the conducted research reflects the importance of the management style in updating the ideas about employees as difficult communication partners, which we consider as the basis for forming a system of relationships and choosing forms of interaction that motivate or demotivate the activities of employees.

References

1. Labunskaya VA (2003) Psychological portrait of a subject of hampered communication. Psikhol Zh 24(5):14–22
2. Belousova A, Kozhukhar G, Vyshkvyrkina M, Ermak V (2020) The role of perceptual attitudes of managers in the cognition of employees of the organization. In: Innovative technologies in science and education (ITSE-2020). E3S Web Conf 210. https://doi.org/10.1051/e3sconf/202021020009
3. Belousova A, Mochalova U (2020) The relationship of thinking style and motivation features of sales and advertising managers. Behav Sci 10(3):68. https://doi.org/10.3390/bs10030068
4. Korochentseva A, Dautov D, Khachaturyan N (2020) Features of ideas about professionally important qualities in food industry workers with differences in the severity of properties of the nervous system. In: "State and prospects for the development of agribusiness—INTERAGROMASH 2020". E3S Web Conf 175. https://doi.org/10.1051/e3sconf/202017515018

5. Suhonen M, Isola A, Paasivaara L, Laukkala H (2013) Nurse managers' perceptions related to their leadership styles, knowledge, and skills in these areas—a viewpoint: case of health centre wards in Finland. Int Sch Res Notices. https://doi.org/10.1155/2013/951456
6. Castro RG (2016) Management styles and organizational productivity skills: an analysis. Eur J Res Reflect Educ Sci 4(3):66. www.idpublications.org
7. Meshcheryakova EV (2019) Management styles and their impact on employee performance. HR Offic 6:95–103. https://www.elibrary.ru/item.asp?id=38196840
8. Pishchik V, Belousova A, Ryumshina L (2018) The relationship of generations representatives in the educational environment. In: EDULEARN18 proceedings 10th international conference on education and new learning technologies, Palma, Spain, pp 1782–1786. https://doi.org/10.21125/edulearn.2018.0524
9. Fries A, Kammerlander N, Leitterstorf M (2020) Leadership styles and leadership behaviors in family firms: a systematic literature review. J Fam Bus Strat. https://doi.org/10.1016/j.jfbs.2020.100374
10. Ramos P, Mota C, Corrêa L (2016) Exploring the management style of Brazilians project managers. Int J Project Manage 34(6):902–913. https://doi.org/10.1016/j.ijproman.2016.03.002
11. Bgashev M (2018) Personality and style analysis of the head of a small enterprise based on the application of the model of formation of management styles. Bull Sci Pract 4(12):368–375. https://doi.org/10.5281/zenodo.2271955
12. Namiq F (2018) Most effective management style for modern workplace. Int J Eng Manag Sci 3(3):402–411. https://doi.org/10.21791/IJEMS.2018.3.33
13. Nezorenko TK (2010) Leadership styles and their influence on the formation of the moral and psychological climate in the team. Laboratory of Books, Moscow
14. Taucean IM, Tamasila M, Negru-Strauti G (2016) Study on management styles and managerial power types for a large organization. Procedia Soc Behav Sci 221(7):66–75
15. Kovalchuk AI, Tskhai VA (2020) Modern management style: features, prospects and risks. Karelian Sci J 9(1):65–67
16. Panina OI (2016) The problem of styles in the management. Soc Humanit Technol 2(2):93
17. Breus ED (2015) "Communicative core"—socio-psychological features of a difficult communication partner in the ideas about it. In: Ryumshina LI, Pishchik VI, Vorontsov DV, Belousova AK, Vyshkvyrkina MA, Serikov GV, Shkurko TA, Gabdulina LI, Breus ED, Alperovich VD (eds) Applied psychology of communication and interpersonal cognition: a collective monograph. Credo, Moscow. https://www.elibrary.ru/item.asp?id=25135402&selid=25135407
18. Stolyarenko LD (2006) Fundamentals of psychology. Phoenix, Rostov on Don
19. Labunskaya VA, Mendzheritskaya YA, Breus ED (2001) Psychology of difficult communication: theory, methods, diagnostics, correction. Publishing Center «Academy», Moscow
20. Fromm E (2017) Beyond the chains of illusion. Bloomsbury Publishing PLC, New York
21. Maslow A (1987) Motivation and personality, 3rd edn. Addison-Wesley Educational Publishers Inc., New York

Digital Didactical Resources as Innovative Multimedia Methods in the Work of a Modern and Reflective Teacher

Jelena Maksimović, Lazar Stošić, and Łukasz Tomczyk

Abstract The demand for innovation in school and teaching is a reaction to what is perceived as the traditional school. Innovation in teaching mostly depends on the teachers themselves and how much they are ready to change and improve their own work, to be a reflective practitioner. The subject of our research is the role of innovative methods in improving students' school achievement. Innovative multimedia methods can influence student motivation, and consequently their achievements can be even more pronounced. With this in mind, the following questions were asked in the research: Do teachers use innovative teaching methods; and which innovative methods do they use the most? For what purposes do teachers use a computer in teaching? Do students create PowerPoint presentations and use the Internet to generate content? The research considered the attitudes of teachers regarding the use of educational software in teaching, as well as the role of innovative methods on students' school achievements. The following variables were considered in the research: how many years the teachers had served (up to 10 years; from 11 to 20 years; and over 20 years of work experience) and the type of school (primary-secondary school). For the purposes of this research, a questionnaire with a Likert-type scale instrument consisting of 31 items was used. The research included 105 teachers of different subjects who attended seminars on Innovative Methods through ICT. The research shows that multimedia has not been fully implemented in primary and secondary schools.

J. Maksimović
Department of Pedagogy, Faculty of Philosophy, University of Niš, Ćirila i Metodija 2, 18000 Niš, Serbia
e-mail: jelena.maksimovic@filfak.ni.ac.rs

L. Stošić (✉)
Don State Technical University, Gagarin Square 1, 344000, Rostov on Don, Russia
e-mail: lstoshih@donstu.ru

Ł. Tomczyk
Faculty of Pedagogy and Psychology, Pedagogical University of Cracow, Ingardena 4, 30-060, Krakow, Poland
e-mail: Lukasz.tomczyk@up.krakow.pl

A. Beskopylny and M. Shamtsyan (eds.), *XIV International Scientific Conference "INTERAGROMASH 2021"*, Lecture Notes in Networks and Systems 247,
https://doi.org/10.1007/978-3-030-80946-1_63

Keywords Innovative methods · Teachers · Innovative practitioners · ICT · Digital methodology · Multimedia models

1 Introduction

We define the term innovation as renewal, enrichment and improvement in order to increase quality. Innovation in teaching means introducing novel methods into the teaching process, with the aim of improving, developing and modernizing teaching. Innovation in teaching implies the introduction of innovations in relation to the existing situation. For that reason, innovations in teaching imply the introduction of pedagogical, social, technological and organizational measures aimed at raising the level and quality of education and child development [1, 2]. Every innovation used in an appropriate way represents a reform, i.e. a change for the better [3]. The constant changes to the circumstances in which we live require the constant adjustment of teaching to the needs of modern society. One of the ways in which the circumstances can be seen to change is the increasing presence of multimedia in education and teaching [4, 5]. The application of a multimedia approach provides the teacher with greater chances for more successful and high-quality teaching and project implementation, while enabling students to fully experience and form clear ideas and concepts from which objective facts, judgments and conclusions are derived [6–9]. The most significant change brought about by the new information society is the constant requirement of the individual to refresh their old knowledge and acquire new knowledge to replace that which has become obsolete due to the rapid development of the conditions that surround us [10]. Computers, the Internet and multimedia are an integral part of the educational process. In order for changes in the field of science and technology to be successful, work must be done on the modernization of the educational system. The development and application of new technologies is growing, but the question becomes whether or not teachers are trained sufficiently to keep pace with such developments [11].

1.1 The Importance of Multimedia Methods in Teaching

The knowledge and appropriate use of information technologies in the modern world is one of the basic elements of human literacy and culture. Many authors [12–14] list the press, television, and radio as basic media. The book was the first medium. What a book can do still cannot be achieved by any other medium, and it is no wonder that these authors believe that it is still the foundation and precondition for media literacy and the understanding of media-constructed reality.

Television is an audiovisual medium of very complex technology. Television offers programs that disseminate interesting ideas from different areas, as well as more specifically scientific research programs whose content refers to the problems of those different sciences [15, 16]. Radio is a medium of mass communication that in a sense can contribute to research work, because current topics discussed on the radio can stimulate the discussion of different research issues. Additionally, radio conversations with experts from the field of education can fill gaps in knowledge and resolve the doubts of many teachers and pedagogues.

Computers enable a completely new organization of the work of teaching, one that is appropriate to the individual abilities and interests of students, and which enables the faster and more efficient acquisition of knowledge. Students progress faster with the help of computers and the knowledge acquired is more permanent. Teaching and learning with the help of computers offer better outcomes in terms of acquired knowledge, the mental mobility of students, and student motivation and independence in learning. The use of computers in teaching contributes to the formation of students' IT competence, is interesting and attractive to students, and allows them to acquire extended knowledge in the area that interests them [17, 18].

The use of the Internet in all types of education (formal, non-formal and informal) has led to changes in communication that have affected education:

1. Communication is not limited by distance, i.e. the internet denies space as a dimension;
2. Communication takes place in virtual reality or cyberspace;
3. Through its inherent interactivity, the Internet enables users to engage in a two-way flow of communication [19].

The use of the Internet changes communication among users, thus changing the approach adopted in education. The Internet is becoming a medium that mediates between students and teachers. Using the Internet in educating students can develop their self-initiative for the development of their creative abilities. Teachers are trained to use innovations in teaching, but they are not sufficiently trained to implement the Internet in the classroom [20]. The use of the Internet changes educational practice and directs students to ways of learning in which they must be active participants [21]. A school organized in such a way changes its organization and directs students to joint activities, the exchange of knowledge and increased mutual communication. The education system must be able to withstand the rapid growth of world knowledge, so those working in educational institutions must themselves be participants in everyday and lifelong learning [22]. According to Novković Cvetković et al., computers and the Internet are the most widely-used digital technologies with 61.22% and 56.12% of teachers respectively [23].

1.2 *Definition of Media Literacy*

The question that drives this research is: "Do innovative teaching methods affect student success?" From this there arises the subject of the research which refers to the role of innovative methods in improving the academic success of students. The cognitive goal set in the research refers to the examination of teachers' attitudes towards the application of innovative models in teaching. Teachers' attitudes to the use of innovations in teaching and which innovative methods they use, whether students create PowerPoint presentations for the needs of classes and whether they use the Internet as a supplement to educational content were examined. Teachers' attitudes towards the role of innovative methods in the academic success of students were also examined. The following independent variables were set in the research: the length of service of the teachers (up to 10 years; from 11 to 20 years; and over 20 years of work experience) and the type of school (primary/secondary school). For the purposes of this research, a scaling technique with the Likert-type scale instrument entitled "Teachers' Attitudes of Innovative Methods (IM)" was used, which consists of 31 items with response modalities from 1 to 5. The research included 105 teachers in the city of Nis, Serbia, during 2018.

2 Methodological Approach to the Problem

The question that drives this research is: "Do innovative teaching methods affect student success?" From this there arises the subject of the research which refers to the role of innovative methods in improving the academic success of students. The cognitive goal set in the research refers to the examination of teachers' attitudes towards the application of innovative models in teaching. Teachers' attitudes to the use of innovations in teaching and which innovative methods they use, whether students create PowerPoint presentations for the needs of classes and whether they use the Internet as a supplement to educational content were examined. Teachers' attitudes towards the role of innovative methods in the academic success of students were also examined. The following independent variables were set in the research: the length of service of the teachers (up to 10 years; from 11 to 20 years; and over 20 years of work experience) and the type of school (primary/secondary school). For the purposes of this research, a scaling technique with the Likert-type scale instrument entitled "Teachers' Attitudes of Innovative Methods (IM)" was used, which consists of 31 items with response modalities from 1 to 5. The research included 105 teachers in the city of Nis, Serbia, during 2018.

3 Analysis and Interpretation of the Research Results

Innovative methods are very important for the overall work of the educational system, but above all for teaching activities and students' academic success. Teachers' attitudes about the use of innovative teaching methods were examined.

Research has shown that teachers believe that they "use" innovative models of multimedia in teaching. However, in an examination of their claims the lowest percentage is seen in programmed and e-teaching. Teachers most often apply interactive and integrative teaching in their work. Next, we examined whether the answers of the respondents differed in relation to the variable type of school. The answers of the respondents did not differ, as on all of the examined subscales (p) the value was greater than 0.05, which indicates that there is no statistically significant difference. The respondents' responses did not differ with respect to the school type variable. Homogeneity dominates in the responses of primary and secondary school teachers, $p > 0.05$ (Tables 1 and 2).

We examined whether the answers of the respondents differ in relation to the length of service of the teachers, and here again the respondents answers did not show a statistically significant difference: all of the examined factors yielded a value (p) higher than 0.05. Homogeneity dominates in the teachers' responses regardless of their length of service, $p > 0.05$ (Table 3).

In addition to the fact that we have previously determined that the least represented aspect is programmed teaching, e-teaching and similar teaching, the research has shown that teachers and students must use a computer in their work, as well as the Internet as a supplement to teaching content. The results show that the use of computers is present in primary and secondary schools, with a single item of data relevant to this research or further research on this issue. The use of computers and the Internet is more dominant in primary schools (M = 3.50) compared to secondary schools (M = 3.10), with a statistically significant difference in the responses of primary and secondary school teachers, $p < 0.05$ (Table 4).

We examined whether the answers of the respondents differ in relation to the length of service of the teachers. No statistically significant effect could be seen here, suggesting that homogeneity dominates in the teachers' answers regardless of their length of service, $p > 0.05$. The use of computers is present regardless of the length of service of the teachers (Table 5).

The research showed the unanimous attitude of teachers that educational software is not used in their school and that software cannot replace "face to face" interactions

Table 1 Application of innovative methods in teaching in relation to the type of school

Subscale	Type of school	M	SD	T-test	P
Application of innovative methods	Primary school	2.7	0.7	0.70	0.48
	Secondary school	2.8	0.8		
Individualized teaching	Primary school	3.5	0.5	−0.75	0.45
	Secondary school	3.7	0.7		
Programmed teaching	Primary school	4.0	1.0	−0.17	0.80
	Secondary school	3.9	0.9		
Interactive teaching	Primary school	2.9	0.9	1.92	0.05
	Secondary school	2.6	0.6		
E-teaching	Primary school	3.9	0.9	−0.40	0.68
	Secondary school	4.3	1.2		
Integrative teaching	Primary school	3.9	0.9	0.88	0.37
	Secondary school	3.7	0.7		

Table 2 Application of innovative methods in teaching in relation to the length of service of the teachers

Subscale	Years of service	M	SD	F-test	*P*
Application of innovative methods	25–35	2.8	0.8	0.30	0.73
	36–45	2.7	0.7		
	More than 46	2.7	0.7		
	Total	2.7	0.7		
Individualized teaching	25–35	4.0	1.0	0.23	0.73
	36–45	3.9	0.9		
	More than 46	4.1	1.1		
	Total	4.0	1.0		
Programmed teaching	25–35	3.0	0.3	1.58	0.21
	36–45	3.4	0.4		
	More than 46	3.3	0.3		
	Total	3.2	0.2		
Interactive teaching	25–35	2.6	0.7	1.61	0.20
	36–45	2.8	0.8		
	More than 46	2.6	0.6		
	Total	2.6	0.6		
E-teaching	25–35	3.8	0.8	1.75	0.17
	36–45	3.7	0.7		
	More than 46	3.5	0.5		
	Total	3.7	0.7		
Integrative teaching	25–35	3.6	0.6	2.99	0.06
	36–45	3.9	0.9		
	More than 46	3.7	0.7		
	Total	3.7	0.7		

Table 3 Application of computers in teaching in relation to the type of school

	Type of school	M	SD	T-test	*P*
Computer application	Primary school	3.5	0.5	0.113	0.04
	Secondary school	3.1	0.4		

Table 4 Application of computers in teaching in relation to the work experience of teachers

	Years of service	M	SD	F-test	*P*
Computer application	25–35	2.9	0.9	0.451	0.63
	36–45	2.8	0.8		
	More than 46	2.7	0.7		
	Total	2.8	0.8		

Table 5 Use of educational software in teaching in relation to the type of school

	Type of school	M	SD	T-test	*P*
Educational software	Primary school	3.6	0.6	−0.834	0.40
	Secondary school	3.3	0.3		

with teachers. No statistically significant effect was seen here: the answers of the respondents do not differ in relation to the variable type of school. Homogeneity dominates in the responses of primary and secondary school teachers, $p > 0.05$ (Table 6).

We examined whether the answers of the respondents differ in relation to their length of service years of work experience of teachers. No statistically significant difference can be seen in the answers of the respondents in relation to the variable of the length of service. Homogeneity dominates in the teachers' answers regardless of their length of service, $p > 0.05$. Educational software is not used among teachers of different lengths of service (Table 7).

The basic purpose of using innovative methods is to encourage the success and development of students through different ways of working in school. Teachers are reflective practitioners who implement innovative methods in class and that is why their opinion on the effectiveness of these methods is very important. In terms of percentage, the teachers mostly support the view that innovative teaching models can have a motivating effect on students. The data do not show a statistically significant difference with respect to the school type variable. Homogeneity dominates in the responses of primary and secondary school teachers, $p > 0.05$ (Table 8).

We examined whether the answers of the respondents differ in relation to the length of service of the teachers. No statistically significant difference was found and the answers of the respondents do not differ in relation to the variable of the years of service. Homogeneity dominates in the teachers' answers regardless of their length of service, $p > 0.05$. Student motivation is visibily greater in lessons that make

Table 6 Use of educational software in teaching in relation to length of service

	Years of service	M	SD	F-test	P
Educational software	25–35	3.2	0.2	2.51	0.08
	36–45	3.1	0.1		
	More than 46	3.3	0.3		
	Total	3.2	0.2		

Table 7 Innovative methods and student success in relation to the type of school

Claims	Type of school	M	SD	T-test	Df	*P*
Innovative methods and student motivation	Primary school	2.7	0.7	0.60	100	0.54
	Secondary school	2.6	0.6			

Table 8 Innovative methods and student success in relation to years work experience

	Years of service	M	SD	F-test	P
Innovative methods and student motivation	25–35	3.1	0.8	8.080	0.50
	36–45	3.0	0.9		
	More than 46	2.7	0.7		
	Total	2.8	0.7		

use of innovative teaching approaches. This is confirmed by teachers of both primary and secondary schools and regardless of the length of service of the teacher.

4 Concluding Remarks

As the first task of the research, the attitudes of teachers about the use of innovative methods in teaching were examined. Teachers' attitudes about innovative methods and whether they are applied in their pedagogical practice are very important. The respondents mostly pointed out that they use innovative methods and that interactive and integrative teaching is a great way to combine knowledge in classes, with the least application of programmed teaching, e-teaching etc. The teachers' answers do not differ in relation to the type of school and years of service to any level of statistical significance. The first hypothesis was rejected.

As the second task of the research, the attitudes of teachers about the use of computers in teaching were examined. Computers have become a modern means of communication, and at school they represent modern media in the transfer of knowledge to students. The respondents mostly pointed out that their computer is also used for personal needs, writting in Word, and not just in class. The research has shown that the use of computers and the Internet in teaching is more dominant in primary schools than in secondary schools. These data partially confirm the second research hypothesis.

As the third task of the research, the attitudes of teachers about the use of educational software in teaching were examined. Educational software is an innovation available to the modern school and enables the gradual discovery of solutions and the possibility of quickly accessing the necessary information. The respondents most often pointed out that the demands of creating educational software require a lot of investment, as well as that the computer cannot replace "the living word" of the teacher and a real textbook. The teachers' attitudes are the same regardless of the type of school they work in and regardless of their length of service. The hypothesis of the application of educational software in teaching was rejected.

As the fourth task of the research, the attitudes of teachers about the relationship between innovative methods and students' motivation to learn were examined. Innovative methods should by their very nature provide more interesting ways of working for students in school and influence the motivation of students to become more active and achieve better results in school. The respondents most often pointed out that innovative methods motivate students to work. Under the umbrella term of 'innovative methods,' the respondents focused the least on multimedia, and the most on interactive and individualized teaching. The respondents' responses did not differ at levels of statistical significance. Given that they positively evaluate innovative models in teaching, we can nonetheless suggest that the hypothesis has been partially refuted.

The research shows that multimedia has not been fully applied in primary and secondary schools, and this research should open new perspectives for future research on this problem, which can facilitate the educational process and make it much more interesting for students.

Acknowledgements The article is the result of work on the project "Pedagogical pluralism as the basis of education strategy" number 179036 (2011–2021), whose implementation is funded by the Ministry of Education, Science and Technological Development of the Republic of Serbia.

References

1. Osmanagić I (2011) Vrednovanje primjene nastavnih inovacija [Evaluation of the application of teaching innovations]. Post Scriptum 2:2–9
2. Vlajkovac S (2012) Inovativni modeli nastave i njihova primena u nastavi učenika mlađih razreda osnovne škole [Innovative teaching models and their application in the teaching of primary school students]. In: Zbornik radova sa Međunarodnog naučnog skupa Obrazovne inovacije u informacionom društvu, Užice
3. Vilotijević M, Vilotijević N (2008) Inovacije u nastavi [Innovations in teaching]. Učiteljski fakultet, Vranje
4. Tomczyk Ł, Martins VF, Eliseo MA, Silveira IF, de la Higuera Amato C, Stošić L (2020) ICT and education in Brazil—NGO, local government administration, business and higher education expert perspective. World J Educ Technol Curr Issues 12(4):401–424. https://doi.org/10.18844/wjet.v12i4.5198. Accessed 2021/04/02
5. Tomczyk Ł, Oyelere SS (2019) ICT for learning and inclusion in Latin America and Europe. Pedagogical University of Cracow, Cracow
6. Gazivoda P (2006) Obrazovne vibracije [Educational vibrations]. Zavod za udžbenike i nastavna sredstva, Podgorica
7. Gazivoda P (2004) Društvene promjene i doživotno obrazovanje [Social change and lifelong learning]. Zavod za udžbenike i nastavna sredstva
8. Iahad N, Dafoulas GA (2004) The role of feedback in interactive learning systems: a comparative analysis of computer-aided assessment for theoretical and practical courses. In: IEEE international conference on advanced learning technologies, 2004. Proceedings, Aug 2004. IEEE, pp 535–539. https://doi.org/10.1109/ICALT.2004.1357472. Accessed 2021/04/02
9. Tomczyk Ł, Szotkowski R, Fabiś A, Wąsiński A, Chudý Š, Neumeister P (2015) Selected aspects of conditions in the use of new media as an important part of the training of teachers in the Czech Republic and Poland—differences, risks and threats. Educ Inf Technol 22(3):747–767. https://doi.org/10.1007/s10639-015-9455-8. Accessed 2021/04/02
10. Soleša DD (2001) Računarski praktikum za informatiku i obrazovnu tehnologiju [Computer practicum for informatics and educational technology]. Tehnički fakultet 'Mihajlo Pupin', Zrenjanin

11. Stošić L (2015) The importance of educational technology in teaching. Int J Cogn Res Sci Eng Educ 3(1):111–114. https://doi.org/10.23947/2334-8496-2015-3-1-111-114. Accessed 2021/04/02
12. Danilović M (2003) Tehnologija i informatika kao proizvod ljudskog uma i njegove kreativnosti [Technology and informatics as a product of the human mind and its creativity]. Tehnologija, informatika, obrazovanje, 2. Institut za pedagoška istraživanja, Beograd; Centar za razvoj i primenu nauke, tehnologije i informatike, Novi Sad
13. Hutinski Ž, Aurer B (2009) Informacijska i komunikacijska tehnologija u obrazovanju: stanje i perspective [Information and communication technology in education: state and perspectives]. Informatologia 42(4):265–272. https://hrcak.srce.hr/42347. Accessed 2021/04/02
14. Vuksanović D (2007) Filozofija medija: Ontologija, estetika i kritika [Philosophy of the media: ontology, aesthetics and criticism]. Fakultet dramskih umetnosti, Univerzitet u Beogradu, Beograd
15. Radivojša B (2012) Kultura većeg broja ekrana [Multi-screen culture]. Politika, 16 Apr
16. Williams R (1974/2004) Television: technology and cultural form. Routledge, London and New York
17. Mandić DP (2001) Informaciona tehnologija u obrazovanju [Information technology in education]. Viša škola za obrazovanje vaspitača, Beograd
18. Melmed A (1995) The costs and effectiveness of educational technology: proceedings of a workshop. https://files.eric.ed.gov/fulltext/ED392409.pdf. Accessed 2021/04/02
19. Trifunović V (2010) Kultura mira i obrazovanje za medije [Culture of peace and education for the media]. Institut za sociologiju, Niš
20. Stošić L, Stošić I (2015) Perception of teachers regarding the implementation of the internet in education. Comput Hum Behav 53:461–468. https://doi.org/10.1016/j.chb.2015.07.027. Accessed 2021/04/02
21. Stanisavljević-Petrović Z (2010) Openness of the school as a human social community and multicultural development. Facta Univ Ser Philos Sociol Psychol Hist 9(1):125–138
22. Stanišić T, Leković M, Stošić L (2019) Relationship between the quality of higher education and Balkan countries' competitiveness. Int J Cogn Res Sci Eng Educ 7(3):49–59. https://doi.org/10.5937/IJCRSEE1903049S. Accessed 2021/04/02
23. Novković Cvetković B, Stošić L, Belousova A (2018) Medijska i informacijska pismenost—osnova za primjenu digitalnih tehnologija u nastavi iz diskursa obra [Media and information literacy—the basic for application of digital technologies in teaching from the discourses of educational needs of teachers]. Croat J Educ—Hrvatski časopis Za Odgoj i Obrazovanje 20(4). https://doi.org/10.15516/cje.v20i4.3001. Accessed 2021/04/02

Psychological Predictors and Determinants Forming Internet Dependence Among Students

Igor Groshevihin, Valery Rybchinskiy, Stanislav Shengelaya, Elena Gaylomazova, and Alexandr Kovtunenko

Abstract Currently, the Internet has become a new type of everyday reality and a sphere of life. The results of various sociological surveys and psychological studies show that it is the people of adolescent age who are, predominantly, a more lively group of Internet users. We set out to investigate the predictors and determinants of the formation of Internet-addicted behavior. In the study, we used the following methods: CIAS test ("Chen Internet Addiction Scale"—Chen's Internet Addiction Scale); test of affiliation motives diagnostics (A. Mehrabian); methodology for the study of volitional self-regulation, A. V. Zverkova, E. V. Eydman; Method "Express diagnostics of the level of social isolation of the individual" D. Russell and M. Fergusson.

The results of the study allow us to make the following conclusions: we found that students whose indicators on the scales testified that they had a pronounced and stable pattern of Internet-dependent behavior, were determined by a combination of motivational tendencies "low desire for people—low fear of rejection"; it was found that students who have a pronounced and stable pattern of Internet-dependent behavior, the level of social isolation is significantly higher than among students who have a minimal risk of Internet-dependent behavior; it was revealed that students who have a minimal risk of developing Internet-dependent behavior, the level of volitional self-regulation is significantly higher than students who have a pronounced and stable pattern of Internet-dependent behavior.

Keywords Internet addiction · Predictors · Determinants · Motivational tendencies

I. Groshevihin (✉)
Don State Technical University, 1, Gagarin Square, Rostov-on-Don, Russia 344000

V. Rybchinskiy · A. Kovtunenko
Southern Federal University, 13, Nagibin Avenue, Rostov-on-Don, Russia 344038

S. Shengelaya
Rostov State Transport University, 1, Ashgabat Lane, Rostov-on-Don, Russia 344038

E. Gaylomazova
Southern Federal State Budget Educational Institution of Higher Education "K.G. Razumovsky Moscow State University of Technologies and Management, (the First Cossack University", 73, Earth Vale Street, Moscow, Russia 109004

A. Beskopylny and M. Shamtsyan (eds.), *XIV International Scientific Conference "INTERAGROMASH 2021"*, Lecture Notes in Networks and Systems 247,
https://doi.org/10.1007/978-3-030-80946-1_64

1 Introduction

Relevance In modern society, the Internet has become a new kind of everyday reality and a sphere of life. In our time, during the age of information technology, being in a society without even the most basic skills in using the networks is already considered something unusual and abnormal. The vastness of the Internet provides great opportunities, using which you can open new horizons and immerse yourself in a new world. Using the networks, everyone can find the necessary resources for themselves: entertainment, communication, self-development and sources of income. This whole palette of possibilities allows a person to optimize his time as much as possible and minimize efforts to implement a particular activity in life.

Every second Russian Internet user visits social networks [1, 2]. The possibility of being constantly on the mobile network, the Internet blurs the traditional boundaries between physical and virtual spaces. The use of computer-mediated social networks leads to significant structural and functional changes in the cognitive, communicative and personal spheres [3].

The results of these sociological surveys and psychological studies, namely, persons of adolescence, which are students, are mainly a more lively group of Internet users. The development of excessive use of the Internet can affect all areas of a student's life as a representative of the youth age group. The development of Internet-dependent behavior leads to learning difficulties, communication disorders, increased individual conflicts, changes in emotional and social intelligence, and similar difficulties.

Understanding one's own and others' emotions and feelings, the ability to regulate one's emotional sphere in accordance with the emotional reactions of partners is one of the key parameters of effective interpersonal communication [4, etc.]. However, the main features of computer-mediated communication, which can affect the functioning of emotional intelligence (first, the transformation of non-verbal communication signs into language symbols—emoticons, as well as into animated pictures) can change various parameters of the functioning of emotional intelligence in active users of virtual networks [5, 6, etc.]. Despite the extensive study of the phenomenon in the literature, the research data of the authors, as well as the available conclusions, often contradict each other. Also, as a rule, Internet-dependent behavior is studied among adolescent children, without touching on in detail the development of this type of addiction among students, as representatives of youth.

In modern psychology, there are many works aimed at studying Internet addiction [7–10, etc.], including its manifestations among students [3, 6, 10–14].

Having studied the materials of these authors, one can come to the conclusion that the problem of Internet addiction is widespread in modern society and has a need for research and identification of individuals with this addiction.

Some modern experts in the field of Internet addiction deny a new type of addiction that has appeared today, while others, on the contrary, are actively studying this topic, determine the causes of Internet addiction, provide the scientific community with classifications and types of phenomena, develop methods prevention, and most importantly—they are already actively working.

The authors note that Internet addicts are characterized by uncontrolled use, which translates into many other problems associated with stress disorders, social adaptation, financial problems and educational difficulties. Consequently, the predictors of the formation of Internet addiction can be weak motivational-volitional and self-regulatory spheres of a person, impulsivity, social isolation, etc.

Thus, the study of these personal characteristics among students will help in solving the problem of Internet addiction, which, in turn, is of particular relevance.

The research problem is to search for psychological predictors and determinants of the formation of Internet addiction in student youth.

The aim of the research is to study the psychological predictors and determinants of the formation of Internet addiction among students.

The subject of the research is psychological characteristics in the formation of Internet addiction among students.

Research hypothesis—psychological predictors and determinants of Internet addiction among students include violations of the communicative, emotional-volitional, motivational spheres, which are manifested in a low level of self-confidence, aggressiveness, and social isolation.

2 Research Methods and Procedure

The study used the following methods:

1. Theoretical (analysis of psychological literature on the studied problem);
2. Psychodiagnostic (observation, testing);
3. Statistical—the mean values were calculated and comparative analysis of the results obtained using the Mann–Whitney U-test to assess the differences between two independent samples, as well as correlation analysis of variables by Spearman's test, aimed at identifying the connection between the studied phenomenon of Internet addiction and indicators of motivational-volitional qualities, methods of descriptive statistics.

Methodological research tools, in order to confirm or refute the hypothesis, we have selected the following diagnostic tools:

1. Test CIAS ("Chen Internet Addiction Scale"—Chen's Internet Addiction Scale). The technique makes it possible to simply and easily assess how seriously the problem of Internet addiction has developed in a person. When answering the questions of the test, students took into account the work and pastime both at the computer and with a tablet, smartphone, etc.
2. Test diagnostics of affiliation motives (A. Mehrabian). This test is designed to diagnose two personality motives: the desire to be accepted by the people around you (SP) and the fear of being rejected by other people (SO).

3. Methodology for the study of volitional self-regulation, A. V. Zverkova, E. V. Eidman. The technique allows diagnosing such parameters of personality traits as self-control, perseverance and the general scale of self-regulation.
4. Methodology "Express diagnostics of the level of social isolation of the individual" D. Russell and M. Fergusson. The technique allows you to identify an indicator of the social isolation of a person.

The object of the research—the students of the Southern Federal University and the Rostov Institute of Railway Engineers (Russia) were the objects of research. In total, 242 students were surveyed, aged from 18 to 23 years, of which 138 girls and 104 boys. All of the subjects lead approximately the same active image of social Internet life. And the level of social activity in real life is different, there are students who are socially inactive, and there are highly socialized personalities, as well as most of the students with the usual level of social adaptation. The entire sample is in approximately the same material position. Some of the students came to study from other cities, and the rest are local residents. The sample also includes both employed students and unemployed. All students have different interests and hobbies; there is no narrow focus in the selection of the sample.

The study did not take into account gender differences. The study was conducted in September–December 2020.

3 Results

Analyzes of test results using the CIAS methodology "Chen Internet Addiction Scale" are shown in Table 1.

The analysis of the test results using the Chen Internet Addiction Scale methodology revealed that the average statistic for all scales in the general sample indicates that the surveyed students have a tendency to develop Internet addictive behavior. We also found that girls are more prone to Internet addiction.

Comparative analysis of the results obtained using the Mann–Whitney U-test to assess the differences between two independent samples (male and female students), revealed that the empirical value of Uemp (24) at $p \leq 0.05$ is in the zone of insignificance.

Further, we selected from the general sample, test participants, those students whose indicators on the scales testified that they have:

- minimal risk of Internet addictive behavior (46 boys and 53 girls);
- a tendency to the emergence of Internet-dependent behavior (34 young men and 47 girls);
- an expressed and stable pattern of Internet-dependent behavior (24 men and 38 women).

In the experimental group, we selected 62 people, students who showed a marked and stable pattern of Internet-dependent behavior. In the control group, we also

Table 1 Values of indicators on the main scales ($\bar{x}$) according to the methodology "Internet Chen Dependency Scale" in the total sample

Chen internet addiction scale (**CIAS** «Chen Internet Addiction Scale»)	Boys ($\bar{x}$) $n = 104$	Girls ($\bar{x}$) $n = 138$
Com (compulsive symptoms)	10.11	11.22
Wit (withdrawal symptoms)	13.54	14.35
Tol (symptoms of tolerance)	9.43	12.02
IH (intrapersonal and health problems)	12.77	14.26
TM (time management problems)	11.82	13.13
KSIA—The key symptoms of Internet addiction is a state of distress (Com + Wit + Tol)	33.08	37.59
PIA—Problems associated with Internet addiction behavior (criterion of negative consequences of Internet addiction), disorders in social, professional or other significant areas of life (IH + TM)	24.59	27.39
The general indicator of Internet addiction is an integral indicator indicating the presence of Internet addicted behavior—the sum of all scales	57.67	64.98

included 62 students whose students were found to have a minimal risk of developing Internet addictive behavior.

Comparative analysis of the test results according to the CIAS method between the subjects of the experimental group and the subjects of the control group are shown in Table 2.

Comparative analysis of the results obtained using the Mann–Whitney U-test to assess the differences between two independent samples (experimental and control groups), revealed that the empirical value of Uemp (16) at $p \leq 0.05$ is in the zone of significance.

Further, we tested the students of the experimental and control groups according to the method (test) of A. Mehrabian modified by M. Sh. Magomed-Eminov (analysis of motivational tendencies). The test consists, respectively, of two scales: "Desire to accept" (SP) and "Fear of rejection" (CO) (Table 3).

Then we tested the students of the experimental and control groups according to the methodology "Research of volitional self-regulation" (A. V. Zverkov and E. V. Eydman) (Table 4).

To determine the level of social isolation of a person, the method "Express-diagnostics of the level of social isolation of a person" by D. Russell and M. Fergusson was taken.

The results of determining the level of social isolation of Internet dependent and independent students are presented in Table 5.

Table 2 Values of indicators on the main scales ($\bar{x}$) according to the methodology "Scale of Internet Chen's dependence" in the experimental and control groups

Chen's Internet addiction scales internet (**CIAS** «Chen Internet Addiction Scale»)	Experimental group ($\bar{x}$) n = 62	Control group ($\bar{x}$) n = 62
Com (compulsive symptoms)	10.14	7.91
Wit (withdrawal symptoms)	12.72	7.63
Tol (symptoms of tolerance)	12.34	6.82
IH (intrapersonal and health problems)	17.27	9.13
TM (time management problems)	15.92	7.83
KSIA—The key symptoms of Internet addiction is a state of distress (Com + Wit + Tol)	35.20	22.36
PAIA—Problems associated with Internet addicted behavior (criterion of negative consequences of Internet addiction), disorders in social, professional or other significant areas of life. (IH + TM)	33.19	16.96
The general indicator of Internet addiction is an integral indicator indicating the presence of Internet addicted behavior—the sum of all scales. CIAS score = (Com + Wit + Tol + IH + TM)	68.39	39.32

Table 3 Indicators of motivation for affiliation (A. Mehrabian in modifications by M. Sh. Magomed-Eminov)

	Level by methodology	Experimental group (n = 62)	Control group (n = 62)
Motivational trends	High (SP)–low (CO) (CO)	9	46
	High (SP)–high (CO) (CO)	20	6
	Low (SP)–high (CO) (CO)	12	6
	Low (SP)–low (CO)	21	4

4 Discussion

Comparative analysis of the test results by the method of diagnosing the motives of affiliation between the subjects of the experimental group and the subjects of the control group are shown in Table 3.

Table 4 Values of indicators ($\bar{x}$) of the level of development of volitional self-regulation according to the methodology "Research of volitional self-regulation"

Scale name	Experimental group ($\bar{x}$) n = 62	Control group ($\bar{x}$) n = 62
General scale (level of volitional self-regulation development) (B)	9.36	17.61
Subscale "persistence" (P)	6.45	12.12
Self-control subscale (S-C)	4.81	9.47

Table 5 Values of indicators of the level of social isolation ($\bar{x}$) according to the method "Express diagnostics of the level of social isolation of the individual" (D. Russell and M. Fergusson), in the experimental and control groups

Scale name	Experimental group ($\bar{x}$) n = 62	Control group ($\bar{x}$) n = 62
Social isolation	48.93	15.17

The respondents in the experimental group differ from the respondents in the control group by the combination of motivational tendencies "high desire for acceptance—high fear of rejection". This indicates a strong internal conflict between the desire for people and their avoidance, which arises every time you have to meet with strangers.

The majority of students from the control group (46 people), in contrast to students from the experimental group, expressed a desire for affiliation, which indicates that they are actively looking for contacts and communication with people, experiencing mostly only positive emotions from this.

Also, a significant part of the subjects (21 people) from the experimental group was determined by the combination of motivational tendencies "low desire for people—low fear of rejection." As the creators of this test believe, the combination of these motivational tendencies characterizes a person who, living among people, communicating with them, does not experience either positive or negative emotions from this and feels good, both among people and without them. 12 people from the experimental group were determined by a combination of motivational tendencies "low desire for people—high fear of rejection." An individual with such a combination of both motives, on the contrary, actively avoids contact with people, seeks loneliness.

Among the students of the control group, only four people showed a combination of motivational tendencies "low desire for people—low fear of rejection."

When comparing the test results of students from the experimental group and students from the control group, it turned out that students from the control group had a more marked desire for affiliation (the desire to accept, establish, maintain and strengthen good relations with people) than students from the experimental group (all differences are statistically significant according to the Mann–Whitney test, Uemp [41 at $p \leq 0.01$), UKp (56 at $p \leq 0.01$ and 72 at $p \leq 0.05$)].

Comparative analysis of the test results according to the method of assessing volitional self-regulation between the subjects of the experimental group and the subjects of the control group are shown in Table 4.

Correlation analysis (Spearman's rank correlation) revealed significant negative relationships between students' Internet addiction and their volitional self-regulation. (r = − 0.75 $p \leq 0.05$).

The analysis of the results showed that the respondents of the experimental group differ from the respondents of the control group by a low score for the level of volitional self-regulation. This indicates the emotional instability of a person experiencing self-doubt. They are characterized by impulsiveness, inability to control themselves.

The control group students showed a high score for the level of volitional self-regulation. They are distinguished by self-confidence, stability of intentions, realistic views, and a developed sense of their own duty.

Thus, the level of development of volitional self-regulation and self-control among students without Internet addiction is higher than among students with this phenomenon. Comparative analysis of the results obtained using the Mann–Whitney U-test to assess the differences between two independent samples (experimental and control groups), revealed that the empirical value of Uemp (24) at $p \leq 0.05$ is in the zone of significance.

It can be noted that addicted students who developed a marked and stable pattern of Internet-dependent behavior demonstrated a high level of social isolation (48.9). This fact can be explained as follows: the constant presence in the virtual space deprives the student of real friends, therefore, despite the extensive virtual acquaintances, the addicted student in real life is very lonely. Such students believe that they may not be interesting in real communication, that people cannot understand them, that it is extremely difficult to find someone who would share their interests, that they have increased anxiety from communicating with people, which develops a sense of isolation from the world.

Comparative analysis of the results obtained using the Mann–Whitney U-test to assess the differences between two independent samples (experimental and control groups), revealed that the empirical value Uemp (0) at $p \leq 0.05$ is in the zone of significance.

Correlation analysis (Spearman's rank correlation) revealed significant interrelationships between students' Internet addiction and their social isolation (r = 0.8 $p \leq$ 0.05).

5 Conclusions

As a result of empirical research, a group of students was identified who developed a marked and stable pattern of Internet-dependent behavior. The predictors and determinants of the formation of Internet addiction have been established. Identified personal characteristics of students addicted to the Internet.

Internet addiction, in contrast to traditional forms of addictive behavior, can be characterized by rapid development rates, as well as significant changes in the psychological characteristics of the personality of dependent students.

The results of the study lead to the following conclusions:

1. It was revealed that students who have a minimal risk of developing Internet-dependent behavior, the level of volitional self-regulation is significantly higher than students who have marked and stable pattern of Internet-dependent behavior. A negative relationship was also found between the level of Internet addiction and the level of volitional self-regulation, which suggests that the weaker the level of willpower a student has, the more marked Internet addiction is. As a result of the correlation analysis, it was found that both the indicator of volitional self-regulation and the indicator of self-control are negatively inter-connected with the level of Internet addiction, which means that the weaker these self-regulatory qualities are, the more marked Internet addiction is.
2. It was found that students who have a marked and stable pattern of Internet-dependent behavior, the level of social isolation is significantly higher than students who have a minimal risk of Internet-dependent behavior. A significant relationship was also found between the level of Internet addiction and the level of social isolation, which suggests that the stronger the student's social isolation, the more pronounced the Internet addiction. We determined that students who are dependent on the Internet are characterized by low initiative in social contacts, which can cause a lack of interest in expanding real contacts, and makes it difficult to seek help in a situation of a real problem.
3. It was found that students, whose indicators on the scales, testified that they had a marked and stable pattern of Internet-dependent behavior, were determined by a combination of motivational tendencies "low desire for people—low fear of rejection."
4. Thus, experimentally, we have established that the predictors of the formation of Internet addiction are:
5. the inability of the Internet addict to realize, understand and analyze his condition;
6. social isolation, situations that lead to alienation from the real world and people;
7. the ability to correctly perceive time (past, present, future), to manage it, to be able to plan the events of one's life has been lost.
8. compulsiveness, low level of volitional self-regulation;

   And also an analysis of studies in literary sources shows that predictors of the formation of Internet addiction can be:
9. disadaptation in real life;
10. feelings of an inferiority complex;
11. a state of distress, depression.
12. It has been experimentally established that the determinants of the formation of Internet addiction are:

13. large intervals of time spent on the Internet;
14. lack of control of loved ones;
15. irresponsibility in communication.

Thus, we have confirmed the hypothesis put forward that psychological predictors and determinants of Internet addiction among students include violations of the communicative, emotional-volitional, motivational spheres, which are manifested in a low level of self-confidence, aggressiveness, and social isolation.

6 Practical Recommendations

In the treatment of addictions (including those from the Internet), a wide range of methods are used, and many specialists are dealing with this problem. The choice of a method suitable for a particular person depends on many factors—age, availability of certain therapeutic procedures, individual psychological characteristics of the addicted person. There is no single and appropriate treatment for Internet addiction.

General practitioners rarely have the necessary qualifications and experience in treating Internet addictions. This is due to the fact that Internet addiction has not yet been recognized by the World Health Organization (WHO) as a mental illness, and therefore it does not have an approved treatment regimen. In this regard, when choosing a doctor to help a person with Internet addiction, it is better to choose a psychotherapist or psychiatrist who has experience in treating other (chemical) addictions. The main way to free a person from Internet addiction at the moment is a therapeutic conversation.

Internet addiction is a relatively new phenomenon that most clinical psychologists may not be familiar with. However, psychologists with experience of successful work with addicts can apply the same methods to treat patients with Internet addiction.

In the treatment of any addiction, the clinical psychologist should help the patient learn to recognize and manage situations that cause relapse and develop a lifestyle that will bring him enough positive emotions while abandoning the source of the addiction.

A therapy conversation is much more than just a conversation between a patient and a doctor. The therapeutic conversation uses a number of time-tested methods of helping the patient to become aware of their problems and stresses and to find effective methods to deal with them.

An experienced professional will be able to determine what exactly is causing the addiction to a particular form of Internet addiction in a patient, and then he will help to take this habit under control. This is not about developing willpower, which is by no means an effective solution, but about acquiring skills that can help you manage stress in healthy ways.

Counseling is another way to correct Internet addiction.

Counseling can take the form of a conversation with a religious leader, friend, social worker, or psychologist. For some people, this is enough. Counselors can provide advice and guidance for solving day-to-day problems and introduce some element of accountability that will help the addict to give up their addictions. Counselors should receive specific training in counseling, in particular group counseling, and its specific techniques.

For those who have suffered from addiction for a long time, and for those whose lives have become completely dependent on the addiction, the best option may be a course of treatment in a medical institution. As in the case of doctors and clinical psychologists discussed above, it is highly desirable to choose an institution that has some experience in the treatment of Internet addiction or other types of addiction. The patient should be happy with his choice and be sure that he is admitted to treatment in an institution that can provide real help.

It is very important to begin treatment with reasonable expectations. The fact that a patient has spent a certain period in the clinic does not automatically mean that the patient leaves it completely healthy. Ideally, there he should overcome the most difficult part of the recovery process and acquire the skills necessary to complete it successfully.

Another way to overcome internet addiction is to prevent reform and development.

References

1. Jun S, Choi E (2015) Academic stress and Internet addiction from general strain theory framework. Comput Hum Behav 49:282–287. https://doi.org/10.1016/j.chb.2015.03.001
2. Borisova O, Ignatieva A (2014) Internet addiction among students of Russia and Germany. Int J Experi Educ 6(1):76
3. Bogomolova E (2015) The relationship of personal characteristics with the peculiarities of the activity of users of social networks of the Internet: dis. … Cand. psych. sciences: 19.00.01, Krasnodar
4. Bae S, Jang J, Kim J (2013) Good samaritans on social network services: effects of shared context information on social supports for strangers. Int J Hum Comput Stud 71(9):900–918. https://doi.org/10.1016/j.ijhcs.2013.04.004
5. Belasheva K (2015) Internet addiction as a predictor of the formation of personality disorders with various types of interhemispheric asymmetries. Knowl Understand Skill 1:224–235
6. Zakharova I (2014) The specificity of the values of young men and women, prone to "Internet addiction", in the aspect of self-regulation of activity. Bullet Samara Sci Center Russian Acad Sci 2(2):333–338
7. Junco R (2014) Student class standing, Facebook use, and academic performance. J Appl Dev Psychol 36:18–29. https://doi.org/10.1016/j.appdev.2014.11.001
8. Lee S (2014) Analyzing negative SNS behaviors of elementary and middle school students in Korea. Comput Hum Behav 43:15–27. https://doi.org/10.1016/j.chb.2014.10.014
9. Shershneva T (2017) Dependence on virtual communication as a medical and psychological problem. In the collection: topical issues of health psychology and psychosomatics. In: Mutalimova A (ed.) Conference 2017. Moscow, Russia, pp 434–440
10. Park S, Na EY, Kim E (2014) The relationship between online activities, netiquette and cyberbullying. Child Youth Serv Rev 42:74–81. https://doi.org/10.1016/j.childyouth.2014.04.002

11. Larionova S, Degterev A (2013) Internet addiction among students: issues of prevention. J Pedagogical Educ Russia 2:232. http://elar.uspu.ru/handle/uspu/1516
12. Salehan M, Negahban A (2013) Social networking on smartphones: when mobile phones become addictive. Comput Human Behav 29(6):2632—2639. https://doi.org/10.1016/j.chb.2013.07.003
13. Rogova E, Suvorova E (2015) Internet addiction is a problem of modern youth. Sci Time 212(24):641–646
14. Lee WB, Teng CW, Hsueh NL, Li YH (2013) Using Facebook to better engage college students in learning. In: Chang RS, Jain L, Peng SL (eds) Advances in intelligent systems and applications—volume 1. Smart innovation, systems and technologies, vol 20. Springer, Berlin, Heidelberg. https://doi.org/10.1007/978-3-642-35452-6_41

Social Environment as a Predictor of Destructive Behavior in Cyberspace

Sergey Lenkov and Nadezhda Rubtsova

Abstract In recent years, psychological studies have identified many risk factors for destructive behavior in cyberspace (such as Internet addiction, cyber aggression, etc.), including personal and individual typological properties, characteristics of the family, educational environment and others. The factors associated with the broader social environment have been less studied. The purpose of the study was to identify the factors that characterize social interaction and determine various types of destructive behavior in cyberspace. The sample included 151 university students aged 18 to 30 years. We tested the influence of social factors on four types of destructive behavior: cyber aggression, cyber victimization, Internet addiction, and risky behavior on the Internet. The measurements were carried out using the author's self-report scales, including from 6 to 9 points. It has been established that the social environment as a whole has a significant impact on destructive behavior in cyberspace. At the same time, many environmental factors have a rather strong influence on several types of behavior: for example, the factor «Unfriendly environment» significantly affects cyber aggression, cyber victimization, and Internet addiction. Besides, some factors are specific to certain types of behavior: for example, the factor «Gossips» has a strong effect only on cyber victimization. The theoretical significance of the study lies in the concretization and systematization of social risk factors for destructive behavior in cyberspace. The practical significance is to substantiate the need to expand the risk factors considered in the development of programs for the prevention of destructive behavior of young people in cyberspace.

Keywords Behavior in cyberspace · Destructive behavior · Cyber aggression · Cyber victimization · Internet addiction · Risky behavior · Social environment · Risk factors

---

S. Lenkov
Russian Academy of Education, Pogodinskaya str. 8, 105005 Moscow, Russia

N. Rubtsova (✉)
Russian New University, Radio str. 22, 119121 Moscow, Russia

A. Beskopylny and M. Shamtsyan (eds.), *XIV International Scientific Conference "INTERAGROMASH 2021"*, Lecture Notes in Networks and Systems 247,
https://doi.org/10.1007/978-3-030-80946-1_65

1 Introduction

1.1 Literature Review

In modern research, the study of information behavior is usually carried out not from narrowly cognitive, but from broader methodological positions, including, for example, the concept of metacognition (e.g., [1]), and more often—within the framework of interdisciplinary research (e.g., [2]). This is facilitated by the fact that the use of cyber technology has become a widespread social practice. For example, in January 2021, out of all current humanity in the amount of 7.83 billion people: 66.6% used mobile phones; 53.6%—social networks; 59.5%—the Internet. Among users aged 16–64, the average daily use of social networks in 2020 was 2 h 25 min, and the Internet—6 h and 54 min [3].

Nowadays, cyber technologies have become an integral part of the daily life of the vast majority of people. In the methodology of science, a reflection of this fact has been the recent widespread use of a number of fundamental and interdisciplinary-universal constructs, such as informatization, digitalization, cyber socialization, etc. These include the term «cyberspace», which is widely used, including in psychological studies of the human behavior (see, e.g., [4]).

Behavior in cyberspace (as well as informational behavior included in its structure) can differ according to the principle of usefulness for the individual and society as a whole. This behavior can be constructive: the use of cyberspace for socially normative learning, personal or professional development, constructive communication, adequate rest, etc. On the other hand, behavior in cyberspace can be destructive. Among the various types of such behavior, three key ones can be distinguished: cyber addiction, cyber aggression, and cyber victimization.

Cyber addiction includes such well-studied forms as Internet addiction and video game addiction; as well as dependence on social networks, on a cell phone, etc. (see, e.g., [5]). Cyber aggression includes such forms as cyberbullying, as well as cyber or online flaming, trolling, harassment, stalking, grooming, hating, etc. (see, e.g., [6]). Cyber victimization is a kind of «pair» to cyber aggression, and therefore has many similar forms.

In addition to these three main areas, many other types of destructive behavior in cyberspace were studied, associated, for example, with involvement in destructive network communities (suicidal, asocial, criminal, etc.), the use of dangerous content, cybercrime, information overload, etc. (see, e.g., [7]).

Thus, the current situation is characterized by the total penetration of cyber technologies into the life of the population strata with a wide representation of destructive forms of behavior in cyberspace, which have a significant negative impact on the processes of education, personal development, socialization, etc., as well as on such key characteristics of person and society as a psychological well-being, satisfaction with life, social tolerance.

An additional and significant contribution to expanding the social base and increasing the intensity of the use of cyber technologies was made by the conditions of the COVID-19 pandemic: for example, in 2020, many studies recorded a significant increase in various types of digital activity of users [8]. At the same time, in many countries, the conditions of the pandemic have led to a sharp increase in manifestations of various types of destructive behavior—both in real life and in cyberspace [9].

Thus, destructive behavior in cyberspace is still expanding, and the risk associated with the COVID-19 pandemic, with its «disruptive power on the social, economic, and other aspects of human life» [10, p. 617], only contributed to this.

Risk factors for destructive behavior in cyberspace can be divided into personal and environmental ones. Personal risk factors have been extensively studied for various types of destructive behavior. In addition to gender, age, health indicators, the following factors were identified among similar factors: for various forms of cyber aggression and/or cyber victimization—impulsivity, delinquency, depression, etc. (see, e.g., [11]); for various forms of cyber addiction—neuroticism, poor academic performance, low self-control, etc. (see, e.g., [12]).

Environmental factors in general have also been extensively studied, with the main focus on the factors of the educational and family environment.

Among the risk factors of the educational environment, the following were identified:

- for cyber aggression or cyber victimization—school insecurity, negative school climate, lack of care of teachers, etc. (see, e.g., [13]);
- for cyber addiction—for example, Xin et al. [12] identified such risk factors for Internet addiction as stress from studies and exams, the presence of Internet addictive classmates, and poor relationships with teachers [12].

Among the risk factors of the family environment, the following were identified: for cyber aggression or cyber victimization—an incomplete family, low parental faith in the child's ability, etc. (see, e.g., [14]); for cyber addiction—low socio-economic status, overprotection, abuse, parental divorce and others (see, e.g., [12, 15, 16]).

Risk factors for the broader social environment as a whole have been less studied. The following studies can be cited as examples. Olenik-Shemesh et al. [17] found loneliness to be a risk factor for cyberbullying and cyber victimization [17]. Przybylski [18] found that ethnic minorities are among the predictors of cyber victimization of adolescents in mobile games [18]. Burnell and Kuther [19] found that low online social support is a risk factor for cell phone and social media addiction [19].

We can also note three important current trends in the development of research on risk factors for destructive behavior in cyberspace.

First, heterogeneous (including complex, mediating) interrelationships of various types of destructive behavior, both among themselves and with the corresponding risk factors, are increasingly revealed. For example, Burnell and Kuther [19] found that impulsivity increases cell phone and social media addiction. At the same time, the latter is also influenced by social comparisons and social online support, and dependence on social networks acts as a mediator of dependence on a cell phone

[19]. Gámez-Guadix et al. [20] found that there are different connections between problematic Internet use, cyberbullying, and meeting with strangers on the Internet [20].

Secondly, the interrelationships of heterogeneous risk factors are increasingly revealed, and their joint influence on certain types of destructive behavior is determined. For example, Ding et al. [21] found that both personal risk factors (male gender, younger age, depressive symptoms) and environmental factors, including factors of the educational environment (negative school climate, connection with deviant peers) and the family environment (conflicts between parents, low parental warmth and acceptance), are significant for cyberbullying and cyber victimization among adolescents [21].

Thirdly, more and more often, risk factors are identified not for one type of destructive behavior, but for several at once. For example, Arpaci et al. [22] found that the factor of vertical individualism significantly increases the manifestations of both cyberbullying and Internet addiction [22].

Thus, an analysis of the literature has shown that to date, a multitude of heterogeneous risk factors have been identified that contribute to certain types of destructive behavior in cyberspace. At the same time, along with undoubted significant achievements, some serious and still unresolved problems can be noted here.

First, the rapid development, diversification and rapid introduction of new cyber technologies can significantly change the conditions for the manifestation of identified risk factors. For example, if in 2014, only 35% of the world's population used the Internet, and social networks—26% [23]. And in January 2021, 59.5% of the population used the Internet, and social networks—53.6% [3].

Second, the conditions of the COVID-19 pandemic have led to a radical surge in both digital activity and its destructive manifestations [24]. At the same time, many living conditions of people associated with employment, education, social interaction, etc. have fundamentally and qualitatively changed. For example, as UNICEF states, the conditions of the pandemic have sharply exacerbated the problem of digital inequality between various groups of children and youth, creating a deep inequality of opportunities in meeting educational, social, professional and economic needs [25], which, obviously, leads to an increase in socio-psychological tension and a decrease in psychological well-being. Such profound changes may well lead to a significant transformation of risk factors for destructive behavior in cyberspace, and this assumption also needs to be tested urgently.

Third, the variety of types and forms of destructive behavior in cyberspace, the insufficient theoretical certainty of many of them and, as a consequence, the absence of their generally accepted classification also significantly complicates the comparative analysis of risk factors identified in the studies of various authors. Because of this, it is advisable to consider the risk factors characteristic of a number of broad, generalized types of destructive behavior.

Thus, the relevance of our research is provided by the following circumstances:

- the rapid development of cyber technologies and areas of their application; the corresponding rapid expansion of cyberspace and the forms of manifestation of human activity in it;
- the actual increase in the general level of manifestations of destructive behavior in cyberspace (including in the current conditions of the COVID-19 pandemic), a wide diversification of the types and forms of such behavior;
- the need to clarify and deepen the understanding of risk factors for such behavior, including those related to the wider social environment.

1.2 *Purpose and Research Issues*

The purpose of this study is to identify the social environment factors influencing destructive behavior in cyberspace using the example of a contingent of university students. We consider the following four types: cyber aggression, cyber victimization, Internet addiction, and risky behavior on the Internet. Achieving this goal involves obtaining answers to the following research questions:

(1) what factors of the social environment are advisable to choose as potential primary risk factors for destructive behavior in cyberspace and at the same time, in their totality, characterizing the overall negative potential of the social environment as a generalized risk factor?
(2) what influence do the selected primary and generalized environmental factors have on various types of destructive behavior in cyberspace?
(3) which of the identified environmental risk factors are common, and which are specific to various types of destructive behavior?

2 Materials and Methods

2.1 *Conceptual Framework*

Due to the lack of a generally accepted classification of types of destructive behavior in cyberspace, we used the following approach.

To describe aggressive behavior, we will use the term «cyber aggression», since, as rightly noted by Bauman and Baldasare [6], it is more general and expresses deliberate harmful actions performed using cyber technologies [6]. Accordingly, by cyber victimization we mean victimization under the influence of cyber aggression.

We understand Internet addiction as a general, non-specific dependence on the Internet, not related to specific functions of use. Similarly, we also understand risky behavior on the Internet as general, non-specific behavior that potentially provokes any dangers and risks.

2.2 Sampling and Procedure

The sample ($N = 151$) consisted of students of Russian universities aged 18–30 ($M = 22.93$, $SD = 3.777$), including: 64 males and 87 females; 116 full-time students and 35 part-time students; 48 undergraduate students, 19 specialist's program students, 84 graduate students; 85 students enrolled in humanitarian specialties (psychology, sociology) and 66 in technical specialties (information technology, construction, electrical engineering, etc.).

The study was conducted in September–November 2020 when COVID-19 restrictions for students were blended: some of the classes were conducted online, and some were in classrooms. Similar partial restrictions were in effect for other conditions of life in general. One part of the subjects filled out the proposed questionnaires online, and the other—in the paper form.

2.3 Measurements and Data Processing

Despite the numerous studies of risk factors for destructive behavior in cyberspace, the problem of measuring instruments for such studies remains acute: in fact, there are no generally accepted methods for measuring many types of destructive behavior (see, e.g., [26]), including those that we have chosen, defining the design of our study. A similar situation is observed in relation to the measurement of risk factors for destructive behavior in cyberspace.

Taking into account these circumstances, all measurements of psychological variables in our study were performed using the author's scales, including statements, the agreement with which was assessed on a dichotomous scale «yes–no». Such a reduced, actually initially qualitative approach is compensated by the subsequent provision of a sufficiently high reliability and factorial validity of the scales used.

Risk factors. Risk factors of the social environment were measured using the 9-point *RE* scale «Risk (negative potential) of the social environment» (short name «Total risk») (Table 1). The scale includes two subscales: the *RE1* subscale «Unfriendly social environment» (short name «Unfriendliness») includes factors *F*1–*F*4; the *RE*2 subscale «Danger in the social environment» (short name «Danger») includes factors *F*5–*F*9. The coefficient α (Cronbach's alpha) in the study sample ($N = 151$) was: 0.626 for the *RE1* subscale, 0.609 for the *RE*2 subscale, 0.706 for the *RE* scale.

Destructive information behavior. The study considered 4 types of this behavior. Cyber aggression *AG* (short name «Aggression») was measured using an 8-point scale ($\alpha = 0.618$). Examples of items: «I have had to insult or ridicule someone on the Internet», «Sometimes I impersonate another person on the Internet and post comments on his/her behalf».

Cyber victimization *VIC* (short name «Victimization») was measured using a 7-point scale ($\alpha = 0.837$), which included options for answering the question «Have

Table 1 Table captions should be placed above the tables

Label	Short name (n_{yes})	Statement
F1	Loneliness (22)	In everyday life, I often feel a sense of loneliness, my uselessness
F2	Teasing (9)	I have weak points that others tease me for
F3	Gossips (12)	I had to be very worried about the fact that rumors, gossip, fakes were spread about me
F4	Bullied (6)	My peers often bullied me
F5	No support (19)	I do not feel support from other people who are ready to help me solve any problems
F6	Discomfort [of insecurity] (6)	I often feel insecure in everyday life
F7	Violence (15)	I grew up in an environment in which people are often aggressive and violent
F8	Fear [for safety] (28)	I often fear for my safety
F9	University insecurity (13)	I do not feel safe in the university

Note n_{yes} number of «yes» answers in the sample ($N = 151$)

you had to worry a lot lately about the fact that someone…». Examples of items: «made a rude or offensive comment on your message or material», «removed you from a general chat or from a group on social networks, on a forum, in an online game».

Internet Addiction *IAD* (short name «Addiction») was measured using a 6-point scale ($\alpha = 0.646$). Examples of items: «I am very annoyed when the Internet is disconnected or unavailable», «If there was no Internet, my life would be absolutely boring and joyless».

Risky behavior on the Internet *RIB* (short name «Risky behavior») was measured using a 7-point scale ($\alpha = 0.728$). Examples of items: «I get to know people on the Internet, even if I feel that they can be dangerous», «When I make Internet contacts, I sometimes deliberately tell a lie».

Statistical analysis was performed using IBM SPSS Statistics for Windows, Version 23.0 (IBM Corporation, Armonk, N. Y., USA). To identify the influence of environmental factors, we used the nonparametric Mann–Whitney test, and for the influence of the environment as a whole—one-way analysis of variance (one-way ANOVA). To assess the strength of the influence, the «eta squared» coefficient was used, showing what part of the variance is explained by the influence of the factor.

Table 2 Means and Spearman's correlations ($N = 151$)

Variable	AG	VIC	IAD	RIB	RE
AG	0.84 (1.206)	0.172*	0.210**	0.516**	0.235**
VIC		0.38 (1.124)	0.283**	0.171*	0.276**
IAD			1.59 (1.520)	0.309**	0.386**
RIB				1.34 (1.681)	0.306**
RE					0.86 (1.433)

Note The main diagonal shows *M* (*SD*)

3 Results

3.1 Descriptive Statistics and Correlations

For generalized risk factors of the social environment, the following values were obtained: for *RE1* (Unfriendliness) $M = 0.32$, $SD = 0.744$; for *RE2* (Danger) $M = 0.54$, $SD = 0.958$; for *RE* (Total risk) $M = 0.86$, $SD = 1.433$. Spearman's correlations were (at $p < 0.01$): between *RE1* and *RE2*, rho = 0.409; between *RE1* and *RE*, rho = 0.716; between *RE2* and *RE*, rho = 0.893. The results for indicators of destructive behavior are shown in Table 2.

Thus, the key research variables are linked by correlations that are statistically significant, but only small in strength.

3.2 Influence of Environmental Factors

Checking the influence of risk factors is presented: for primary factors—in Table 3, for generalized factors—in Table 4.

A statistically significant effect on certain types of destructive behavior was found for all primary risk factors, except for factor *F6* (Discomfort), as well as for many generalized risk factors.

4 Discussion

The overall impact of the negative social environment can be further tested using one-way analysis of variance. The results obtained earlier using nonparametric methods (Table 4) were mainly confirmed by ANOVA (Table 5).

The only discrepancy is related to the effect on cyber victimization: when checking using ANOVA, here, with an overall significance $p < 0.01$, multiple comparisons were slightly inconsistent, where the best result (when comparing groups 1 and 3) was

Table 3 Checking the impact of primary risk factors

Risk factor	Destructive behavior	M_0/M_1	Z	p	η^2
F1 loneliness	*AG* aggression	0.77/1.27	– 8.295	0.000**	0.456
	IAD addiction	1.36/2.95	– 4.512	0.000**	0.135
	RIB risky behavior	1.26/1.82	– 1.670	0.095	0.018
	VIC victimization	0.32/0.73	– 1.509	0.131	0.015
F2 teasing	*AG* aggression	0.77/2.00	– 2.588	0.010*	0.044
	VIC victimization	0.32/1.22	– 2.455	0.014*	0.040
	IAD addiction	1.51/2.78	– 2.016	0.044*	0.027
	RIB risky behavior	1.29/2.22	– 1.706	0.088	0.019
F3 gossips	*VIC* victimization	0.14/3.17	– 8.458	0.000**	0.474
	AG aggression	0.75/1.92	– 2.560	0.010*	0.043
	RIB risky behavior	1.25/2.42	– 1.808	0.071	0.022
	IAD addiction	1.50/2.58	– 1.662	0.096	0.018
F4 bullied	*VIC* victimization	0.25/3.50	– 4.898	0.000**	0.159
	IAD addiction	1.52/3.33	– 2.194	0.028*	0.032
	AG aggression	0.79/2.00	– 2.161	0.031*	0.031
	RIB risky behavior	1.32/2.00	– 0.783	0.434	0.004
F5 no support	*IAD* addiction	1.39/2.95	– 3.508	0.000**	0.081
	RIB risky behavior	1.18/2.47	– 3.076	0.002**	0.063
	VIC victimization	0.28/1.05	– 2.061	0.039*	0.028
	AG aggression	0.74/1.53	– 1.526	0.127	0.015
F6 discomfort	*IAD* addiction	1.57/2.00	– 0.872	0.383	0.005
	RIB risky behavior	1.34/1.50	– 0.361	0.718	0.001
	AG aggression	0.84/0.83	– 0.262	0.793	0.000
	VIC victimization	0.39/0.17	– 0.074	0.941	0.000
F7 violence	*RIB* risky behavior	1.25/2.20	– 2.458	0.014*	0.040
	AG aggression	0.77/1.47	– 2.199	0.028*	0.032
	IAD addiction	1.49/2.53	– 1.964	0.050	0.026
	VIC victimization	0.32/0.87	– 0.601	0.548	0.002
F8 fear	*IAD* addiction	1.33/2.75	– 3.842	0.000**	0.098
	RIB risky behavior	1.23/1.86	– 2.015	0.044*	0.027
	AG aggression	0.76/1.18	– 1.771	0.077	0.021
	VIC victimization	0.26/0.89	– 0.537	0.591	0.002
F9 Univ. unsafe	*AG* aggression	0.77/1.62	– 2.264	0.024*	0.034
	IAD addiction	1.51/2.46	– 2.040	0.041*	0.028
	RIB risky behavior	1.27/2.15	– 1.876	0.061	0.023

(continued)

Table 3 (continued)

Risk factor	Destructive behavior	M_0/M_1	Z	p	η^2
	VIC victimization	0.35/0.69	– 1.446	0.148	0.014

Notes M_0, M_1 mean values in the groups that chose the answer «no» or «yes», respectively, for a statement related to the factor in question. Z, p z-statistic and two-tailed asymptotic significance for the Mann–Whitney Test. η^2 indicator of the power of influence of «eta-squared»
$^*p < 0.05$, $^{**}p < 0.01$

Table 4 Checking the influence of generalized risk factors

Scale	Destructive behavior	M_0/M_1	Z	p	η^2
RE1 unfriend	*VIC* victimization	0.15/1.22	– 4.330	0.000**	0.124
	IAD addiction	1.30/2.66	– 4.136	0.000**	0.113
	AG aggression	0.63/1.62	– 3.665	0.000**	0.089
	RIB risky behavior	1.18/1.97	– 2.660	0.008**	0.047
RE2 danger	*IAD* addiction	1.22/2.40	– 3.994	0.000**	0.106
	RIB risky behavior	1.10/1.89	– 3.351	0.001**	0.074
	AG aggression	0.68/1.19	– 2.102	0.036*	0.029
	VIC victimization	0.25/0.66	– 1.203	0.229	0.010
RE total risk	*IAD* addiction	1.14/2.31	– 4.302	0.000**	0.123
	RIB risky behavior	1.03/1.84	– 3.694	0.000**	0.090
	VIC victimization	0.15/0.74	– 3.008	0.003**	0.060
	AG aggression	0.62/1.19	– 2.409	0.016*	0.038
	VIC victimization	0.35/0.69	– 1.446	0.148	0.014

Notes as in Table 3

only $p = 0.053$. However, this circumstance is explained, in our opinion, only by significant deviations from the normal distribution, which influenced the results of multiple comparisons. It is precisely because of these limitations that we generally preferred nonparametric methods.

Summarizing the results obtained, it is possible to single out risk factors that have a significant ($p < 0.05$) and at the same time the strongest ($\eta^2 > 0.05$) effect on destructive behavior. This influence was revealed:

- for cyber aggression—from the side of the primary factor «Loneliness» and the generalized factor «Unfriendly social environment»;
- for cyber victimization—on the part of the primary factors «Gossips» and «Bulled», generalized factors «Unfriendly environment» and «Total risk of the environment»;
- for Internet addiction—on the part of the primary factors «Loneliness», «Fear» and «No support (lack of support)», generalized factors «Unfriendly environment» and «Dangerous environment» and «Total risk of the environment»;

Table 5 The general impact of a negative social environment (ANOVA)

Destructive behaviour	Group	Mean [95% CI]	ANOVA		Post hoc test	
			F	*p*	Groups	*p*
IAD addiction	1	1.14 [0.89, 1.39]	14.839	0.000^{**}	1–2	0.154
	2	1.83 [1.13, 2.52]			1–3	0.000^{**}
	3	2.63 [2.04, 3.21]			2–3	0.176
RIB risky behavior	1	1.03 [0.71, 1.36]	4.636	0.011^{*}	1–2	0.236
	2	1.65 [0.95, 2.35]			1–3	0.025^{*}
	3	1.97 [1.35, 2.60]			2–3	0.765
AG aggression	1	0.62 [0.43, 0.82]	7.132	0.001^{**}	1–2	0.863
	2	0.74 [0.32, 1.16]			1–3	0.019^{*}
	3	1.49 [0.90, 2.07]			2–3	0.094
VIC victimization	1	0.15 [0.03, 0.27]	6.850	$0.001^{**,\dagger}$	1–2	0.365
	2	0.43 [0.03, 0.84]			1–3	0.053
	3	0.94 [0.29, 1.60]			2–3	0.378

Notes Group—group corresponding to the level of severity of the general environmental factor *RE* (Total risk): 1—low ($n = 93$), 2—medium ($n = 23$), 3—high ($n = 35$). *F* —Fisher statistics for one-way ANOVA. Post Hoc Test—results of pairwise comparisons of groups (1–2, 1–3 and 2–3) using the Games-Howell test
$^{*}p < 0.05$, $^{**}p < 0.01$
†the overall significance of the effect according to one-way ANOVA was not confirmed by the results of multiple comparisons according to the Games-Howell test

- for risky behavior on the Internet—on the part of the primary factor «No support», generalized factors «Dangerous environment» and «Total risk of the environment».

Note that a number of our conclusions are somehow consistent with the data obtained by other authors. Thus, the revealed influence of factor *F1* «Loneliness» on cyber aggression is qualitatively consistent with the results of Olenik-Shemesh et al. [17], and the revealed influence of factor *F5* «No support» on Internet dependence is essentially coherent with the results of Burnell and Kuther [19], who found that low online social support is a risk factor for cell phone addiction and social media addiction [19].

5 Conclusions

The study made it possible to identify many significant risk factors for various types of destructive behavior in cyberspace.

It was found that some factors of the social environment are common to several types of destructive behavior in cyberspace:

- the primary factor «Loneliness» increases not only cyber aggression, but also Internet addiction;
- the primary factor «No support» increases Internet addiction and risky behavior on the Internet;
- the generalized factor «Unfriendly environment» increases cyber aggression, cyber victimization and Internet addiction;
- the generalized factor «Dangerous environment» increases Internet addiction and risky behavior on the Internet;
- the generalized factor «Total risk of the environment» increases cyber victimization, Internet addiction and risky behavior on the Internet;

At the same time, some factors of the social environment are specific to a certain type of destructive behavior in cyberspace. Thus, the primary factors «Gossips» and «Bulled» increase only cyber victimization, and the primary factor «Fear» – only Internet addiction.

The obtained results provide some evidence that the range of risk factors traditionally taken into account in the development of programs for the prevention of destructive behavior of young people in cyberspace and including, most often, personal factors, factors of the family environment, and factors of the educational environment should be supplemented by factors of the social environment, characterizing a more general, broader interaction between the individual and society. Such factors have a significant, sometimes very strong influence on various types of destructive behavior, and the lack of taking them into account can reduce the adequacy of the used theoretical models and the effectiveness of the used practical methods and prevention programs.

Thus, the theoretical significance of the study consists in concretizing the risk factors of the social environment that enhance the severity of destructive behavior in cyberspace, as well as in their systematization by highlighting general and specific factors and by ranking according to the strength of influence. The practical significance of the study is to substantiate the need to expand the risk factors considered in the development of programs for the prevention of destructive behavior of young people in cyberspace by including factors in their composition, which characterize the social environment.

The presented study has a number of limitations:

- the study was conducted on a sample of university students aged 18 to 30. Accordingly, its conclusions may turn out to be incorrect for other target groups that differ in age and/or educational qualifications;
- the small size of the sample caused the inapplicability of many methods of parametric statistics for data analysis (including regression models estimated using structural equation modeling), which are often more powerful than the nonparametric methods used in the presented study.

Prospects for the continuation of the study are associated with the expansion of the spectrum of the considered risk factors of the social environment and types of destructive behavior in cyberspace, the improvement of methods for their measurement, as well as an increase in the sample size, which will make it possible to move to a qualitatively new, higher level of statistical data analysis.

References

1. Wokke ME, Achoui D, Cleeremans A (2020) Action information contributes to metacognitive decision-making. Sci Rep 1(10):3632. https://doi.org/10.1038/s41598-020-60382-y
2. Palanca-Castana N, Tajadurab BS, Cofréc R (2021) Towards an interdisciplinary framework about intelligence. Heliyon 2(7):e06268. https://doi.org/10.1016/j.heliyon.2021.e06268
3. Kemp S (2021) Digital 2021: global overview report (27 January 2021). https://datareportal.com/reports/digital-2021-global-overview-report. Last accessed 2021/02/24
4. Cuadrado-Gordillo I, Fernández-Antelo I (2014) Cyberspace as a generator of changes in the aggressive-victim role. Comput Hum Behav 36:225–233. https://doi.org/10.1016/j.chb.2014.03.070
5. Andrie EK, Tzavara CK, Tzavela E, Richardson C, Greydanus D, Tsolia M, Tsitsika AK (2019) Gambling involvement and problem gambling correlates among European adolescents: results from the European network for addictive behavior study. Soc Psychiatry Psychiatr Epidemiol 11(54):1429–1441. https://doi.org/10.1007/s00127-019-01706-w
6. Bauman S, Baldasare A (2015) Cyber aggression among college students: demographic differences, predictors of distress, and the role of the university. J Coll Stud Dev 4(56):317–330. https://doi.org/10.1353/csd.2015.0039
7. Quaglio G, Millar S (2020) Potentially negative effects of internet use. European Union, European Parliamentary Research Service, Brussels (2020). https://www.europarl.europa.eu/stoa/en/document/EPRS_IDA(2020)641540. Last accessed 2021/02/24
8. Vargo D, Zhu L, Benwell B, Yan Z (2021) Digital technology use during COVID-19 pandemic: a rapid review. Human Behav Emerg Technol 1(3):13–24. https://doi.org/10.1002/hbe2.242
9. WHO Regional Office for Europe (2020) The rise and rise of interpersonal violence—an unintended impact of the COVID-19 response on families (3 June 2020). https://www.euro.who.int/en/. Last accessed 2021/02/24.
10. Huang Y, Yang C (2020) A metacognitive approach to reconsidering risk perceptions and uncertainty: understand information seeking during COVID-19. Sci Commun 5(42):616–642. https://doi.org/10.1177/1075547020959818
11. Ronis ST, Slaunwhite A (2019) Gender and geographic predictors of cyberbullying victimization, perpetration, and coping modalities among youth. Can J Sch Psychol 1(34):3–21. https://doi.org/10.1177/0829573517734029
12. Xin M, Xing J, Pengfei W, Houru L, Mengcheng W, Hong Z (2018) Online activities, prevalence of Internet addiction and risk factors related to family and school among adolescents in China. Addictive Behav Rep 7:14–18. https://doi.org/10.1016/j.abrep.2017.10.003
13. Su W, Han X, Yu H, Wu Y, Potenza M (2020) Do men become addicted to internet gaming and women to social media? A meta-analysis examining gender-related differences in specific internet addiction. Comput Hum Behav 113:106480. https://doi.org/10.1016/j.chb.2020.106480
14. Davis K, Koepke L (2015) Risk and protective factors associated with cyberbullying: are relationships or rules more protective? Learn Media Technol 4(41):1–25. https://doi.org/10.1080/17439884.2014.994219
15. Wichstrom L, Stenseng F, Belsky J, Von Soest T, Hygen BW (2019) Symptoms of Internet gaming disorder in youth: predictors and comorbidity. J Abnorm Child Psychol 1(47):71–83. https://doi.org/10.1007/s10802-018-0422-x
16. Bussone S, Trentini C, Tambelli R, Carola V (2020) Early-life interpersonal and affective risk factors for pathological gaming. Front Psych 11:423. https://doi.org/10.3389/fpsyt.2020.00423

17. Olenik-Shemesh D, Heiman T, Eden S (2012) Cyberbullying victimisation in adolescence: relationships with loneliness and depressive mood. Emot Behav Diffic 3–4(17):361–374. https://doi.org/10.1080/13632752.2012.704227
18. Przybylski AK (2019) Exploring adolescent cyber victimization in mobile games: preliminary evidence from a British cohort. Cyberpsychol Behav Soc Netw 3(22):227–231. https://doi.org/10.1089/cyber.2018.0318
19. Burnell K, Kuther TL (2016) Predictors of mobile phone and social networking site dependency in adulthood. Cyberpsychol Behav Soc Netw 10(19):621–627. https://doi.org/10.1089/cyber.2016.0209
20. Gámez-Guadix M, Borrajo E, Almendros C (2016) Risky online behaviors among adolescents: longitudinal relations among problematic Internet use, cyberbullying perpetration, and meeting strangers online. J Behav Addict 1(5):100–107. https://doi.org/10.1556/2006.5.2016.013
21. Ding Y, Li D, Li X, Xiao J, Zhang H, Wang Y (2020) Profiles of adolescent traditional and cyber bullying and victimization: the role of demographic, individual, family, school, and peer factors. Comput Hum Behav 111:106439. https://doi.org/10.1016/j.chb.2020.106439
22. Arpaci I, Abdeljawad T, Baloğlu M, Kesici Ş, Mahariq I (2020) Mediating effect of internet addiction on the relationship between individualism and cyberbullying: cross-sectional questionnaire study. J Med Internet Res 5(22):e16210. https://doi.org/10.2196/16210
23. Kemp S (2014) Digital 2014: global digital overview (8 January 2014). https://datareportal.com/reports/digital-2014-global-digital-overview. Last accessed 2021/02/24
24. Ko CH, Yen JY (2020) Impact of COVID-19 on gaming disorder: monitoring and prevention. J Behav Addict 2(9):187–189. https://doi.org/10.1556/2006.2020.00040
25. United Nations Children's Fund and International Telecommunication Union (2020) How many children and young people have internet access at home? Estimating digital connectivity during the COVID-19 pandemic. UNICEF, New York (2020). https://data.unicef.org/. Last accessed 2021/02/24
26. Chun J-S, Lee J, Kim J, Lee S (2020) An international systematic review of cyberbullying measurements. Comput Hum Behav 113:106485. https://doi.org/10.1016/j.chb.2020.106485

Features of Coping Behavior of Students in Connection with Models of Informational Behavior

Irina Abakumova, Galina Zvezdina, and Anastasia Grishina

Abstract The article is devoted to the study of the relationship of students' informational behavior with coping strategies used to overcome life obstacles and difficulties. The authors indicated the relevance of the study, analyzed theoretical approaches to understanding coping behavior, disclosed studies related to the problem of Internet behavior, virtual communications in the context of coping strategies. The article describes the procedure of empirical research on the study of the relationship of informational behavior of students, presents the results on the study of the degree of expression of normative and risky strategies of informational behavior. It was found that the orientation of students to search the information and use the Internet as a motivating force are the leading strategies of normative informational behavior. It was revealed that the leading strategy of risky informational behavior is the use of the Internet to spy on the life of others. The study of the relationship between coping strategies and models of informational behavior showed that adaptive coping strategies have inverse correlations with the normative strategies of informational behavior ("Internet as a motivating force", "Internet as a means of self-presentation"). It was found that non-adaptive coping behavior strategies (impulsive actions, manipulative, aggressive, asocial actions) have direct connections with risky strategies of informational behavior. Based on the results obtained, conclusions are drawn and the prospects for further research are determined.

Keywords Informational behavior · Coping strategies · Adaptive and non-adaptive coping strategies · Models of informational behavior · Normative (pro-social) and risky models of informational behavior · Interconnection of informational behavior and coping behavior

---

I. Abakumova · G. Zvezdina · A. Grishina (✉)
Don State Technical University, Gagarina sq, 1, Rostov-on-Don, Russia

A. Beskopylny and M. Shamtsyan (eds.), *XIV International Scientific Conference "INTERAGROMASH 2021"*, Lecture Notes in Networks and Systems 247,
https://doi.org/10.1007/978-3-030-80946-1_66

1 Introduction

1.1 Relevance

The modern digital space presents great opportunities to meet the diverse needs of a modern person. New concepts are actively entering scientific use—digital behavior [1], Internet behavior [2], informational behavior [2]. In this study, we use the concept of informational behavior, understanding informational behavior as manifested in the ability to navigate the information space, in the skills of information technology, in the ability to adequately assess and productively use the information received. The ability to create, distribute and be responsible for the dissemination of new information products also lies within the competence of informational behavior. Although the technical side of informational behavior is significant, this phenomenon cannot be considered one-sided, psychological aspects are an important component [3].

The interest of researchers in informational behavior is due to both the demands of practice and the need for a theoretical understanding of the features of modern Internet users informational behavior oand the strategies of overcoming behavior they use. Informational behavior, like real behavior, is associated with the appearance of difficulties, stresses, crises and the need to overcome them.

Coping (overcoming) behavior is actively studied by modern foreign and domestic researchers. Based on the author's theoretical position, coping behavior is understood as a response to a stressful situation and adaptation to it [4], avoidance of destruction [5], stress management [6], adaptation to influences [7].

There are different approaches to the study of coping behavior: the dispositional approach (Z. Freud), which considers whether there are certain personal qualities that contribute to better coping with difficulties, and the styles of coping with stress. The situational approach (R. Lazarus, S. Folkman, H. Faifkel, S. Strack, K. Carver) focuses on the process of coping with stress and the study of specific strategies that change in accordance with the change in specific situations. An integrative approach (R. Moos, J. Schaeffer), whose representatives consider personal and situational factors as predictors that influence the choice of coping efforts. Coping in this case is one of the aspects of a person's abilities, which, together with resources, serves to transform the situation, eliminate the threat (R. Moos, A. Billing, G. McCubbin) [8, 9].

Researchers are raising the question of the activity and passivity of coping strategies. Active strategies: active-cognitive coping (situation assessment), active-behavioral coping (intervention in a situation, avoidance) (A. Billings, R.H. Moos) [9].

With the advent of virtual communications and the Internet space, it became necessary to study the features of informational behavior and coping behavior strategies used by Internet users. The problem of coping behavior efficiency in the context of virtual communications is in the focus of modern research. In the study on Internet communication in the context of studying the coping strategies of modern youth by A. S. Agolakova, it was revealed that for active Internet users virtual communication is a means of seeking emotional support. Inactive Internet users are less likely to seek emotional help and support from other people in Internet communication [10].

Researchers I. R. Sushkov and N. S. Kozlova examined the relationship between human coping strategies and their integration into the Internet space. As a result of the study, the authors came to the conclusion that being included in the Internet space is not independent coping, but at the same time it makes it possible to show characteristic personality tendencies in terms of choosing behavior strategies [11].

Considering the motives of users' turning to the Internet space and its resources, the researchers note that "adolescents who have experience of self-damaging behavior use social networks not only as a means of communication and receiving social support, but also as a source of negative content that promotes and demonstrates self-harm [12, 13]. Teenagers directly indicate that they feel a desire to publish negative content on social networks ("stress posting"), such publications become for them a means of coping with negative emotions and "cry for help". Thus they manifest their need for support and assistance from other users [14]. There is evidence that the frequency of adolescent access to Internet content associated with self-injurious behavior directly correlates with the frequency of such behavior in him [12–15].

Taking into account the relevance and insufficient development of the problem of the relationship between informational behavior and strategies for coping behavior, we studied this problem on the contingent of student youth.

1.2 Purpose and Research Questions

The aim of this research is to study the relationship between models of informational behavior and strategies of coping behavior among students of the humanitarian profile.

The objects of the research were students of 1–2 courses of the Humanities Faculty of the university.

Based on the purpose of the study, the following hypotheses were put forward:

1. There can be a connection between the models of informational behavior of students and strategies of overcoming behavior;
2. Risky models of students' informational behavior can be associated with non-adaptive coping strategies—aggressive, impulsive, manipulative and asocial actions.

Choosing the object of research, we proceeded from the fact that students, for the most part, are active users of Internet content, and on their example, it is possible to trace various models of informational behavior. The modern Internet environment provides the user with a large number of pro-social resources, but this environment is also risky, carrying unverified, sometimes offensive and provocative information, calls for illegal and destructive activities. Considering this situation, in the author's methodology aimed at studying the models of informational behavior, we took into account the vectors of behavior based on normativity (prosociality) and deviance (riskiness).

2 Materials and Methods

The empirical research was carried out on the basis of the Don State Technical University with students of the humanity faculties, full-time and part-time departments. The study involved 121 students, aged from 18 to 29 years old, the average age was 20 years.

The following research methods were used:

- "Strategies of informational behavior" SIP (Abakumova I. V., Romek V. G., Kolenova A. S., Grishina A. V., Zvezdina G. P.).
- Methodology "Indicator of coping strategies" by D. Amirkhan (adaptation by N. A. Sirota and V. M. Yaltonsky).
- Personality questionnaire "SACS" (S. Hobfall).

Spearman's correlation coefficient was used as methods of mathematical statistics. Statistical analysis was performed using IBM SPSS Statistics for Windows, Version 23.0.

2.1 *Measurements and Data Processing*

At the first stage of empirical research, we studied the strategies of informational behavior of students using the author's methodology "SIP". The methodology is presented in two parts: normative behavior (50 statements) and risky behavior (25 statements), including five scales each. The technique allows to determine the severity of a particular strategy of informational behavior, based on motivational orientation. The questionnaire presents 10 strategies of informational behavior. Regulatory strategies include the following scales: Internet as a means of self-presentation, Internet for shopping, Internet for information search, Internet for "killing time", Internet as a motivating force (examples of others). Risky behaviors are represented by the following scales: Internet as access to alternative information, Internet to participate in communities, Internet for spying on others on social networks, Internet for realizing sexual needs, Internet for expressing religious ideas.

In order to study coping behavior strategies, two methods were used: "Coping strategies indicator" by D. Amirkhan (adaptation by N. A. Sirota and V. M. Yaltonsky) and the personality questionnaire "SACS" by L. Hobfall.

The Coping Strategy Indicator methodology is a short self-assessing questionnaire consisting of 33 statements that determines the basic coping strategies: a problem-solving strategy, a strategy for seeking social support, a strategy of avoidance, and their expression in the structure of coping stressful behavior.

The personal questionnaire of L. Hobfall is intended to study strategies and models of coping behavior (stress-overcoming behavior), as types of a person's reactions to overcome stressful situations. The questionnaire includes 54 statements, each of them must be assessed by the subject on a 5-point scale. In accordance with the key, the sum of points is calculated for each line, which reflects the degree of preference for a particular model of behavior in a stressful situation. The questionnaire contains nine behavioral models (coping strategies): assertive actions, entering into social contact, seeking social support, cautious actions, impulsive actions, avoidance, manipulative (indirect) actions, antisocial actions, aggressive actions.

3 Results

According to the results obtained according to the methodology "Strategies of informational behavior", this sample of subjects demonstrated the following models of informational behavior. The most represented model in the repertoire of informational behavior is the use of the Internet to search for information, to prepare for classes. The values of this strategy received high values, which may indicate the dominance of this strategy among humanities students (Fig. 1).

At the border of high values is the strategy of behavior—"Internet as a motivating force." For the sample of students, the example of others that is reflected in Internet is important. It provides a certain incentive for their activity, and possibly serves as an example to follow and guide in modern reality.

The strategy of informational behavior, aimed at becoming involved in the Internet outside of business and communication activities, has average values and suggests

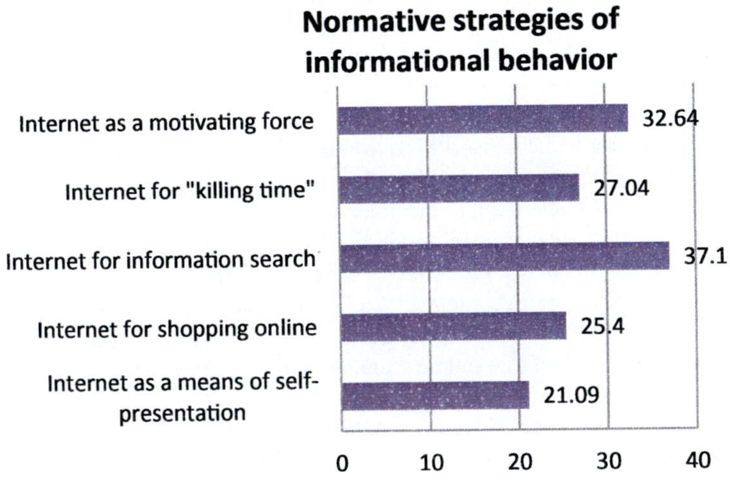

Fig. 1 Normative strategies of informational behavior among students

that surfing on the Internet is enjoyable for students and is quite common in the student environment.

Pragmatic use of the Internet in the form of shopping in online stores is also quite common among this sample of respondents.

The least important is an active strategy of informational behavior, aimed at presenting oneself in the Internet space, demonstrating one's achievements to Internet users.

The obtained results indicate the severity of the normative (pro-social) strategies of informational behavior in the repertoire of informational behavior strategies of humanities students. The use of the Internet as the main source of information is the leading strategy for student behavior.

The strategy "Internet as a motivating force", which has received high values, may indicate that students are looking for role models for their own self-confirmation in the Internet environment. Appeal to the images of the Internet, on the one hand, helps to identify, and on the other hand, sets the stereotypes of behavior for the young person.

According to E. P. Belinskaya's research on the behavior of Internet users, it is noted that "first of all, changes occur in the processes of perception and categorization of social information: after all, the natural dominance of the audiovisual channel and the experience of "clip thinking" during active Internet use leads to a decrease in reflexivity. As a result, the image becomes the leading component of a person's ideas about the social world" [16].

Analysis of the repertoire of informational strategies of behavior, non-standard from the point of view of social norms, showed that the least represented is a model of behavior that reflects the participation of users in online communities of risky content. Perhaps this reflects a need for personal safety and an unwillingness to appear in a socially disapproving light.

The strategies "Internet for realizing sexual needs" and "Internet for expressing religious ideas" received average values, which suggests that these behavioral strategies are in the arsenal of students' informational behavior and they periodically resort to them.

The strategies "Internet as an access to alternative information" and "Internet for spying on others on social networks" have become more widespread among Internet users. Virtual participation in the life of another makes it possible to satisfy the individual's need for belonging to the other, for participation in his life (Fig. 2).

The identified strategies of informational behavior of the normative and risky orientation of students were considered in connection with the used coping strategies.

We studied basic coping strategies. According to the obtained results, the scale of "Avoidance of problems" and the scale of "Search for social support" are within the values at the border of low and medium, the scale of "Problem-solving" has very low indicators (Fig. 3).

According to the obtained results, students, when overcoming difficulties that arise, more often prefer a passive strategy of avoidance, postponing the solution of the problem for "later" and seeking social support.

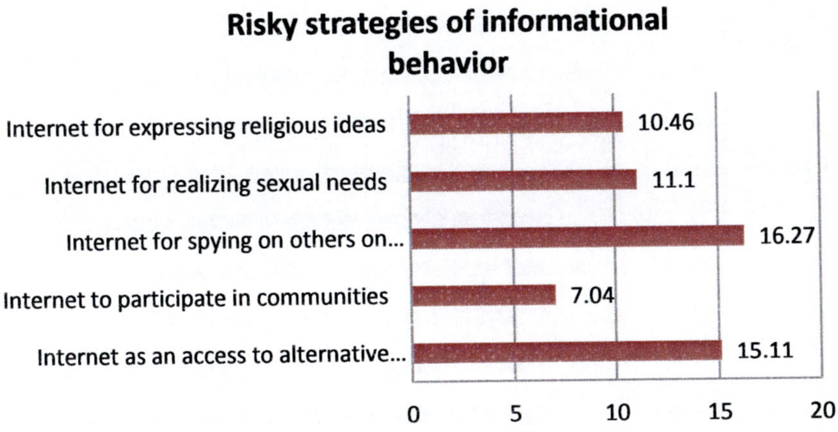

Fig. 2 The degree of expression of risky strategies of informational behavior among students

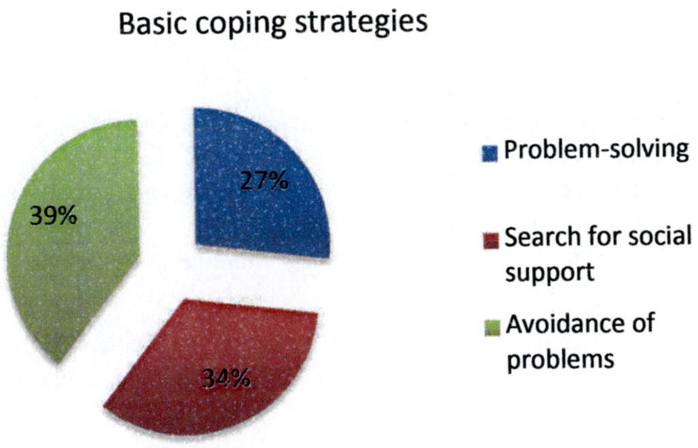

Fig. 3 The severity of basic coping strategies among students

Analysis of the coping behavior repertoire using the SACS method showed that students actively use different coping strategies, both active and passive, both direct and indirect.

However, young people resort to aggressive actions most often, which can indicate either the lack of opportunities to get out of the situation of difficulty in a more constructive way, or because of the inability to do this (Fig. 4).

The study of the relationship between the strategies of informational behavior and the used coping strategies showed that direct and inverse relationships were revealed between these indicators on the basis of Spearman's correlation analysis.

Direct weakly expressed relationships were established using Spearman's coefficient between:

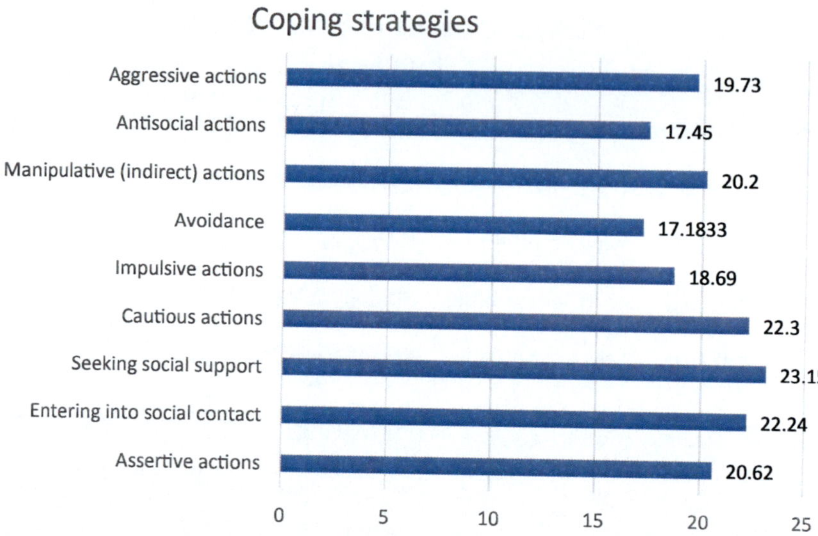

Fig. 4 Repertoire of coping behavior strategies in the studied sample of students

- the "Problem avoidance" scale and the "Internet for information search" strategy (r = 0.23, at $p < 0.03$);
- scale "Assertive actions" and the strategy of informational behavior "Internet as a motivating force" (r = 0.206, at $p < 0.05$);
- the scale "Aggressive actions" and the strategy "Internet as an access to alternative information" (r = 0.23, at $p < 0.03$).

The results obtained may indicate that students, using the Internet as a source of information, avoid solving the problems that they face, or in this way avoid solving the problems they face.

The presence of a connection between assertive actions and the motivating power of the Internet demonstrates that the positive experience of other people, their achievements and success contribute to the constructive overcoming of life's difficulties, increase confidence in oneself and one's own capabilities. It can be assumed that this strategy of informational behavior teaches a person constructive behavior, based on the demonstrated images of success.

The relationship between aggressive actions and the use of the Internet as a source of alternative, often oppositional information shows that a feeling of dissatisfaction with the social situation, rejection of the one-sidedness of information coverage of facts provokes aggressive actions as a strategy for coping with a difficult situation.

Inverse correlations were found between the following coping strategies and models of informational behavior:

- "Search for social support" and "Internet as a means of self-presentation" (r = -0.25, with $p < 0.01$);

- "Search for social support" and the strategy "Motivating power of the Internet" (r = -0.29, at $p < 0.01$);
- "Seeking social support" and the behavior strategy "Internet for spying on others on social networks" (r = -0.38, with $p < 0.01$).

The presented results indicate that with an increase in the importance of the coping strategy "Seeking social support" in the studied sample of students, the importance of using the Internet as a platform for self-presentation decreases, and the desire to spy on the lives of others, that is, students, becomes less significant. those oriented towards social real relationships, relying on real help and support of others, are less dependent on the Internet space, the Internet is not a dominant motivating factor for them.

Feedbacks have been established between the "Problem solving" scale and strategies—"Internet as a motivating force" (r = − 0.27, with $p < 0.01$) and " Internet as a means of self-presentation" (r = − 0.25, for $p < 0.01$).

According to the results obtained, we can conclude that the less the strategy of solving problems at the expense of personal resources and self-reliance is expressed, the more urgent becomes the desire to go into virtual space and present oneself, one's achievements, in isolation from life reality.

The use of the coping strategy "Problem solving" reduces the motivating role of the Internet environment, focus on the example of others from the virtual space and reduces the importance of the Internet as a means of self-presentation.

An analysis of the links between non-adaptive coping strategies and information models showed the following picture.

Impulsive coping strategy is associated with normative and risky strategies of informational behavior. Impulsive behavior in psychological science is understood as unplanned, non-purposeful. Coping actions with such a strategy lose their purposefulness and become mainly the result of the emotional tension release. According to the obtained results, Internet users with an impulsive coping strategy use the Internet platform both to demonstrate themselves and their lives, not always thinking about the consequences of violation of privacy (r = 0.20, at $p < 0.05$), and as a motivating one started on the basis of imitation of others (r = 0.22, at $p < 0.05$).

Impulsive actions are associated with the strategy of informational behavior focused on the use of alternative information (r = 0.2, at $p < 0.05$), with a strategy that allows one to express one's religious opinions and study other people's opinions (r = 0.24, at $p < 0.01$), with the motivation of users to participate in Internet communities of questionable orientation (r = 0.34, with $p < 0.01$).

Risky strategies of informational behavior of Internet users in situations of difficulty are accompanied by impulsive actions, which are manifested in reactivity, in unfocused behavioral activity, which is accompanied by a reduced criticality of assessing the situation and consumed information content.

Also, in the course of the study, direct positive links were revealed between manipulative and asocial coping strategies and models of informational behavior.

Positive links were revealed between the manipulative actions of respondents and the normative strategies of Internet behavior, such as "Internet as a motivating force" (r = 0.27, at $p < 0.01$), and “Internet for shopping online” (r = 0.22, at $p < 0.03$). Manipulative actions, being indirect actions, help respondents to solve the problem situationally, but do not provide an opportunity to solve it productively.

Manipulative actions are associated with risky Internet behavior patterns such as “Internet to participate in social communities” (r = 0.29, at $p < 0.01$) and “Internet as access to alternative information” (r = 0.32, at $p < 0.01$).

A fairly large number of positive connections were found between the coping strategy “Asocial actions” and models of informational behavior: normative—“Internet as a means of self-presentation” (r = 0.23, at $p < 0.02$), “Internet as a motivating force” (r = 0.22, at $p < 0.03$) and risky—“Internet as access to alternative information”, (r = 0.28, at $p < 0.01$) “Internet for participation in communities” (r = 0.38, at $p < 0.01$), “Internet for realizing sexual needs” (r = 0.27, at $p < 0.01$), “Internet for expressing religious ideas” (r = 0.22, with $p < 0.05$).

Thus, based on the results obtained, we can conclude that non-adaptive coping strategies are more associated with risky strategies of informational behavior.

4 Discussion

In the course of the research, we found a direct connection between the scale of “Assertive actions” and the strategy of informational behavior “Internet as a motivating force”, which may indicate that confident behavior, self-reliance among modern students depends on self-construction on the basis of virtual images, which E. P. Belinskaya “expanded opportunities for ro-left experimentation and an increase in the number of maximally manageable self-presentations (today—mainly through the creation of personal profiles and pages on social networks) can give the user both new reasons for self-categorization and new reasons for self-change” [17].

According to the results of the empirical study, risky models of informational behavior are associated with aggressive and asocial strategies aimed at violating accepted norms and rules of behavior, which is logical in principle. As evidenced by the studies of foreign authors, “that in network communication, people pay less attention to the observance of social norms, which makes such communication less formal, more trusting” [18]. Our study provides additional support for this idea.

Our experiment is also in line with works of scientists from Moscow state university (G. U. Soldatova, E. I. Rasskazova), showing that for the modern teenager, the Internet is a complex psychological “tool,” which he appropriates by probing its capabilities and opportunities for self-regulation, including coping with difficult life situations [19].

The problem of informational behavior as a coping behavior is also widely studied abroad, gaining special importance during the Covid-19 pandemic. Zhao N., Zhou G. the relationship between COVID-19 stress and addictive social media use by examining the mediating role of active use and social media flow (i.e., an intensive, enjoyable experience generated by social media use that perpetuates media use behaviors). Their findings suggest that individuals who experience more COVID-19 stress are at increased risk of addictive social media use that may be fostered by active use and flow experience [20]. This confirms our previous findings that suggest that increased informational consumption can be considered as a coping strategy of overcoming the pandemic social isolation for respondents with low hardiness and tolerance to ambiguity: searching different types of information (in particular the latest news about the COVID-19 pandemic) helps such people to overcome the ambiguity of the situation, makes their life more understandable and predictable, thereby giving confidence in the future and making today's interesting [21].

This results have a number of similarieties with Eden, A. L., Johnson, B. K., et al., findings: stress was linked to more hedonic and less eudaimonic media use, as well as more avoidant and escapist media-based coping. Anxiety, on the other hand, was linked to more media use in general, specifically more eudaimonic media use and a full range of media-based coping strategies. In turn, escapist media was linked to negative affect, while reframing media and eudaimonic media were linked to positive affect. Avoidant coping was tied to poorer mental health, and humor coping was tied to better mental health. Hedonic and need-satisfying media use were linked to more flourishing. Hope, optimism, and resilience were all predictive of media use, with the latter two traits moderating responses to stress and anxiety [22].

Wu et al. [23] investigated the associations between psychological resilience, students' characteristics (gender, major and grade) and coping styles among undergraduate students. The research revealed that females and medical students are more likely than males and non-medical students to adopt positive coping styles. Higher psychological resilience is associated with a better positive coping style [23].

Our results have s satisfactory agreement with the study of Cauberghe, V., Van Wesenbeeck, I., et al., that examines if social media are beneficial for adolescents to cope with feelings of anxiety and loneliness during the quarantine. anxious participants indicated to use social media more often to actively seek for a manner to adapt to the current situation, and to a lesser extent as a way to keep in touch with friends and family. The indirect effect of anxiety on happiness through active coping was significantly positive. Participants who were feeling lonely were more inclined to use social media to cope with lacking social contact. However, this coping strategy was not significantly related to their happiness feelings. Humorous coping was positively related with feelings of happiness, but not influenced by loneliness or anxiety. To conclude, social media can be used as a constructive coping strategy for adolescents to deal with anxious feelings during the COVID-19 quarantine [24].

Our research consists of several stages, we are faced with the task of further researching the indicated problem and conducting a comparative analysis with similar studies of domestic and foreign authors. The findings of other scientists help us to correct the futher stages of our empirical research.

5 Conclusions

Based on the obtained results, the following conclusions can be drawn:

- it was found that there are direct and feedback links between informational behavior strategies and coping strategies:
- between the adaptive strategies of coping behavior ("Seeking social support", "Solving problems") and the normative strategies of informational behavior ("Internet as a means of self-presentation", "Internet as a motivating force"), inverse correlations were revealed;
- the use by students of active and prosocial strategies for coping with difficult life situations reduces the importance of informational behavior focused on self-presentation of oneself, the need to look for virtual landmarks, the desire to be an outside observer of someone else's life, from which it can be concluded that adaptive strategies of behavior more focused on self-realization in real life, with the help of real means and methods;
- it was found that non-adaptive coping strategies of humanitarian students have direct links with risky models of informational behavior:
- students demonstrating risky informational behavior strategies to overcome stress and difficult life situations use impulsive, aggressive, manipulative, asocial actions, which more likely leads to increased risks and makes them more vulnerable to life challenges and threats.

The results obtained may indicate that the virtual (digital) space has a powerful both constructive and risk-generating resource, as evidenced by the research results. Among students with a non-adaptive style of coping behavior, risky strategies of informational behavior are most actualized, which makes it possible to outline a vector of psychological activity to prevent risky behavior of students in the Internet environment.

A prospect for further research can be the study of the features of informational behavior and other personal predictors that influence this behavior and the development of programs for psycho-preventive measures to reduce the riskiness of informational behavior.

Acknowledgements The research was supported by the Russian Foundation for Fundamental Research (Project No. 18-29-22004) (2018) «Psychological and genetic studies of users' behavior predictors that determine the perception of Internet content of various informational orientation».

References

1. Pogozhina IN, Podolsky AI, Idobaeva OA, Podolskaya TA (2020) Digital behavior and features of the motivational sphere of Internet users: logical-categorical analysis. Educ Issues 3:60–94. https://doi.org/10.17323/1814-9545-2020-3-60-94
2. Fablinova ON (2015) Behavior on the Internet as an object of study of social sciences. Sociol Almanac 6:543–549
3. Abakumova IV, Zvezdina GP, Grishina AV, Zvezdina EU, Dyakova EV (2020) Models of information behavior: changes in psychological boundaries of internet users. E3S Web Conf 210:20015. https://doi.org/10.1051/e3sconf/202021020015

4. Parker I, Endler N (1993) If it changes, it might be unstable: examine the factor structure of the ways of coping questionnaire. Psychol Assess 5(3):361–368. https://doi.org/10.1037/1040-3590.5.3.361
5. Pearlin LI, Schooler C (1978) The structure of coping. J Health Soc Behav 19:2–21
6. Lazarus RS (1966) Psyhological stress and the coping process. McGraw-Hill, N.Y.
7. Paulhan I (1992) Le concept de coping. Annee Psychol:315–319
8. Lazarus RS, Folkman S (1984) Stress, appraisal, and coping. Springer, New York
9. Kryukova TL (2005) Psychology of coping behavior in different periods of life: dis. ... Dr. Psycho. Sciences: 19.00.13. KSU. Kostroma, 476 p
10. Agalakova AS (2010) Internet communications in the context of studying the coping strategies of modern youth. In: Yu O, Shchelkova SPB (eds) Psychology of the XXI century" under scientific. Publishing House of St. Petersburg, p 513
11. Sushkov IR, Kozlova NS (2013) The relationship between the coping strategies of a person and their involvement in the Internet space. In: Psychology of stress and co-dominant behavior: proceedings of the international. Scientific—practical conference. KSU IM on the Nekrasov, Kostroma, pp. 212–214. http://www.ipras.ru/engine/documents/document5438.pdf
12. Brown R, Fischer T, Goldwich A, Keller F, Young R, Plener PL (2018) Non-suicidal self-injury (NSSI) on Instagram. Psychol Med 48(2):337–346. https://doi.org/10.1017/S0033291717001751
13. Memon AM, Sharma SG, Mohite SS, Jain S (2018) The role of online social networking on deliberate self-harm and suicidality in adolescents: a systematic review of literature. Ind J Psychiatry 60(4):384–392. https://doi.org/10.4103/psychiatry.IndianJPsychiatry_414_17
14. Radovic A, Gmelin T, Stein BD, Miller E (2017) Depressed adolescents' positive and negative use of social media. J Adolesc 55:5–15. https://doi.org/10.1016/j.adolescence.2016.12.002
15. Zhu L, Westers NJ, Horton SE et al (2016) Frequency of exposure to and en-gagement in nonsuicidal self-injury among inpatient adolescents. Archiv Suicide Res 20(4):580–590. https://doi.org/10.1080/13811118.2016.116224
16. Belinskaya EP (2013) Informational socialization of adolescents: experience of using social networks and psychological well-being. Psychol Res 6(30):5. http://psystudy.ru/index.php/eng/2013v6n30e/871-belinskaya30e.html
17. Belinskaya EP (2014) Coping with challenges in the era of new information technologies: opportunities and limitations. Psychol Res 7(38):2. http://psystudy.ru/index.php/eng/2014v7n38e/1073-belinskaya38e.html
18. Boyd D, Ellison N (2007) Social network sites: definition, history, and scholarship. J Comput-Mediat Commun 13(1):210–230. https://doi.org/10.1111/j.1083-6101.2007.00393.x
19. Soldatova GU, Rasskazova EI (2016) Adolescent safety on the Internet. Russ Educ Soc 58(2):133–162. https://doi.org/10.1080/10609393.2016.1214492
20. Zhao N, Zhou G (2021) COVID-19 stress and addictive social media use (SMU): mediating role of active use and social media flow. Front Psych 12:635546. https://doi.org/10.3389/fpsyt.2021.635546
21. Grishina AV, Abakumova IV (2020) Informational behavior in the COVID-19 pandemic: psychological predictors. Int J Cogn Res Sci Eng Educ (IJCRSEE) Curr Res Trends Cogn Sci 8:59–67. https://doi.org/10.23947/2334-8496-2020-8-SI-59-67
22. Eden AL, Johnson BK, Reinecke L, Grady SM (2020) Media for coping during COVID-19 social distancing: stress, anxiety, and psychological well-being. Front Psychol 11:577639. https://doi.org/10.3389/fpsyg.2020.577639
23. Wu Y, Yu W, Wu X et al (2020) Psychological resilience and positive coping styles among Chinese undergraduate students: a cross-sectional study. BMC Psychol 8:79. https://doi.org/10.1186/s40359-020-00444-y
24. Cauberghe V, Van Wesenbeeck I, De Jans S, Hudders L, Ponnet K (2020) How adolescents use social media to cope with feelings of loneliness and anxiety during COVID-19 lockdown. Cyberpsychology, behavior and social networking. https://doi.org/10.1089/cyber.2020.0478

Education Systems Management in Critical Situations: Potential Risks of Digitalization

Ksenia Skobeltsina, Sergey Beshenkov, and Andrei Kuznetsov

Abstract The authors address the currently vital issue of the cyber-risks arising on the grounds of such challenges as 'COVID-19 phenomenon', general digitalization of the national and international communities, change in the social and educational trends, etc. The digitalization of education proves to perform as a risk factor for the management of education systems at different levels, and the problems of the e-safety and the digital skills advancement of the children subjected to the Internet need to be solved to fight cyberbulling, cyber-victimization, and cyber-slavery that bases on the hypertrophic gaming and excessively digitalized social net-working. Within the research methodology, the key research methods were the context analysis, the content analysis of the quality Russian and international academic publications and regulatory acts. The social poll covered school students of the 7–11 grades (N = 5682), and the results were processed with SPSS and STATISTICA packages. The most crucial research findings associate with the confirmation of the fact that the absolute majority of the Russian school students are digital natives; Internet and social networks proved to be the key source of information for the schoolers (even those who recognize that some of this information is authentically and potentially faulty); the social motive is also key for the digital devotion; over 30% of the respondents faced with cyberbulling and personal interventions from strangers. The research results are presented with the background idea to facilitate the development of the managerial mechanisms in education that would ease the critical situations rooting in the digitalization risks.

Keywords Education management · Crisis management · E-safety · Cyberbulling · Children in Internet · Digitalization · Digital skills · Cyber risks · Digital natives · Critical situations · Social poll

K. Skobeltsina (✉) · S. Beshenkov · A. Kuznetsov
Institute of Education Management of the Russian Academy of Education, Moscow, Russia
e-mail: skobeltsina@iuorao.ru

A. Beskopylny and M. Shamtsyan (eds.), *XIV International Scientific Conference "INTERAGROMASH 2021"*, Lecture Notes in Networks and Systems 247,
https://doi.org/10.1007/978-3-030-80946-1_67

1 Introduction

Management of educational systems at different levels (federal, regional, municipal and institutional) implies the development and application of the management mechanisms that contribute to the implementation of the state and the regional educational policies [1]. In this regard, one of the urgent challenges is the design of management decisions in the field of the organization of the work of the regional and the municipal educational systems in the context of the need in overcoming the emergency and the critical situations. The very term, i.e. 'critical situations', is an umbrella one not only for the consequences of the natural and man-made disasters, but also the internal risks. The latter may include the sudden bursts of tension associated with the objective activation of the digitalization of the society and its education system. Such risks can be considered as an aggravating factor for the violation of the social well-being of students as well as teachers, which directly affects the degree of the controllability of the education system, which may significantly modify the organization of the educational process.

In this context too, the recent 'COVID-19 phenomenon' is discussed in the professional community. "The problem of cyber-addiction in primary school children has taken on new dimensions in the last few months in the context of the state of emergency" [2]. Not only adults, children as well are reported to spend considerably more of the screen-time for various reasons (i.e. study, the surplus of free time, information retrieval, etc.). This associates immediately with the likelihood of developing cyber-addiction in children. "It turned out that the threat of the ability of digital technologies and the Internet to flood the daily lives of children, in fact proved to be a salvation from the crisis and did not allow to cancel the school year" [2]. According to Mitchell [3], "risk assessment is considered where there is a potential danger involved in undertaking social media-related projects", including educational projects. The authors [3] point out the vulnerabilities of the project participants, and the types of threats that they may be exposed to.

The Joint Research Centre, the European Commission's science and knowledge service, report has highlighted that "today's children by the time they leave education, will have used more technology and of a far greater variety than any within the current generations of working adults" [4]. Thus, school children need to be adequately equipped for their cyber-being.

Recently, qualitative changes have also taken place in the Russian information sphere. The range of information threats—potentially possible events that can lead to the damage to information safety—has significantly expanded, and it has changed qualitatively. Today, not only teachers, but also heads of education systems within their work must take into account such cyber threats as deviation of the users from 'the actual reality', phishing, trolling, cyberbullying, cyber slavery, hypertrophied development of multimedia social networks (including the activation of sects and extremist, criminal communities being among those), expanding the availability and attractiveness of online games, etc. [5, 6]. At the same time, it is obvious that information safety is a prerequisite for the implementation and functioning of the digital educational environment. The latter is treated as "the state of protection of children, in which there is no risk associated with information causing harm to their health and (or) physical, mental, spiritual, moral development". (This definition is in accordance with Article 2 of the Russian Federal Law "On the Protection of Children from Information Harmful to Their Health and Development" of December 29, 2010 #436-FZ (version of July 31, 2020)).

'E-safety', i.e. "a person's skill to effectively respond to the challenges and opportunities offered by the internet is an important skill that needs to be taught to children from a young age in formal schooling" [7]. According to Legate [8], little is known "about how parents may protect against cyberbullying". "Parents who used more autonomy-supportive strategies —understanding the adolescent's perspective, offering choice, and giving rationales for prohibitions—had adolescents who reported engaging in less cyberbullying" [8].

The issues associating with the school cyberbullying were tackled by a number of education experts in different regions [7, 9–12]. As it was revealed, there develop certain 'cyber human values' and 'cyber-traits', and in electronic forms of bullying, both victims and bullies feature the anger, and among the key parameters here is the measure of cyberbullying and cybervictimization that is characteristic for children and adolescents. Mathur et al. [13] studied the expressions of suicidal intent on social media through leveraging of the contextual psychological connotative cues from online user communities on Twitter. Also, studies [10, 11] touched upon the effects of the cyber-interventions. As "most cyberbullying victims are public high school students, < ... > over 60% of the victims had not seeked help but dealt with the attack on their own. < ... > victims manifested sleep or eating disorders, physical/psychological symptoms or changes in their social life as a consequence of the cyber-attack" [10].

International authors [14, 15] report their positive experiences with the launching of a special programs—some of those being organized as interactive web-based learning environment—aiming at ensuring the citizens' well-being in digital space, and even those at the primary school level. "However, this program was found to be inadequate in articulating the mechanisms a community can undertake to influence the development of students' digital skills" [14]. The need is emphasized in "the students' understanding of the good practices to foster appropriate social behavior in the digital space through responsible and respectful actions" [14].

In this regard, conducting a sociological study to identify students' possession of knowledge and skills related to the recognition of information threats, as well as allowing to counteract these threats, seems to be rather relevant and timely.

By the present time, there have been accumulated a number of sociological studies related to assessing the impact of the information environment on a person's personality, and, in particular, the personality of a student. In particular, those were the large-scale studies conducted by the Institute and, later, the Center for the Sociology of Education of the Russian Academy of Education under the leadership of Academician Prof. Vladimir S. Sobkin. The researchers analyzed the importance of the new information technologies in the structure of leisure and the information space of a contemporary child [16–19]. Within the framework of these studies, the content features of students' interest in the world of computers, the role and the place of the new information technologies in the educational process were considered, alongside with the impact of the regular use of computers on the academic success of students. Moreover, researchers from the Institute of Education Management of the Russian Academy of Education previously the conducted sociological research on the availability of digitalization tools in the schools of the Russian regions [20, 21].

This article presents the results of a sociological study aimed at identifying the level of cyber threats to which modern Russian schoolchildren are exposed (based on a survey of students in grades 7–11 of secondary schools in the regions of the Russian Federation).

2 Materials and Methods

As a part of the fulfillment of the state commission of the Ministry of Education of the Russian Federation, the experts of the Institute of Education Management of the Russian Academy of Education developed a program and tools for the sociological research, i.e. a questionnaire for students in grades 7–11 of secondary schools.

To conduct a sociological study, a questionnaire was developed to identify the students' possession of knowledge and skills related to recognizing information threats and allowing them to counter these threats. The structure of the questionnaire assumes aimed at studying issues related to information safety of students in the digital environment.

The questionnaire used in the study included three content blocks: (1) features of the use of the information and communication technologies (ICT), the Internet and social networks by students; (2) awareness and knowledge of students in the realm of the information technology threats to the individual; (3) possession of methods and means of protecting information and countering information threats to the individual.

The sampling plan for the regions—which was reconciled by the Ministry of Education of the Russian Federation—was based on the cluster principle. When preparing the study, the principle of clustering of the general population was implemented, and seven regions of the Russian Federation were used as clusters (Kaliningrad Region, Leningrad Region, Moscow Region, Primorsky Kray, Republic of Crimea, Republic of Sakha (Yakutia), Tambov Region).

In the sociological study, a questionnaire survey method was applied, which makes it possible to identify the quantitative and qualitative assessments of the state of the cyber risks to which students of general education schools are exposed, as well as their readiness to resist information threats. The questionnaire included the closed, open, multiple-choice, alternative and rank questions, which provided for characterizing students' ideas about the information threats and the ways of overcoming those. The data obtained were processed by the methods of mathematical statistics using a statistical software package SPSS and STATISTICA.

The sociological survey was conducted in November, 2020. The data collection was organized in the study regions on the official website of the Institute of Education Management of the Russian Academy of Education (www.iuorao.ru), and an electronic survey service was used.

In total, during the study, 5682 students of grades 7–11 of secondary schools from different regions of Russia were interviewed. The research sample comprised 2586 males (45.5%) and 3096 females (54.5%).

3 Results

The study identified various features of the use by students of the information and communication technologies (ICT), the Internet and the social networks.

The results showed that the vast majority, i.e. 99.2% of the respondents, of modern schoolchildren regularly use the Internet. Moreover, most often they access the Internet using smartphones—76.7%, and home computers/laptops—19.7%; and only 0.6% of the surveyed students named a school computer.

During the study, adolescents were asked to indicate the most significant sources of information. When answering this question, such sources of information as the Internet and computer networks is clearly in the lead, i.e. 76.9% of the respondents. It is important to note that the opinions of the adults and peers who are in the immediate circle of the respondents are also highly significant: parents—44.7%, friends—37.4%, teachers—29.1%.

It is important to note that the use of the social networks is also extremely common among adolescents, i.e. more than 88% of the respondents are registered in one or another social network (see Table 1). The following sites turned out to be the most popular among schoolchildren: VKontakte—88.8% are registered in this network, YouTube—86.6%, Instagram—79.0%, Facebook—38.6%, Pinterest—36.8%, Google + 1—34.5%, Twitter—26.1%, Odnoklassniki—22.5%.

During the study, students were asked to rate the role of the Internet in their lives. They were offered a multiple-choice question, where three choices were expected of the answer options from the list, as well as a blank line reserved for their own answer (if relevant). The results are shown in Table 2.

As can be seen from the table, the most frequent motives for using the Internet for schoolchildren are social (helps me to communicate with friends/relatives; helps to communicate with interesting people and to follow their lives), informational (helps the students to keep abreast of the latest events) and educational (helps to study at school and receive distant education).

Since it is the social motive of using the Internet that is the most significant, it is important to address the question—with whom do teenagers usually communicate on the Internet. The study showed that the majority of schoolchildren (84.8%) usually communicate with people they know in their ‘actual reality’. However, every tenth respondent answered that (s)he communicates with people whom (s)he has never

Table 1 Use of social networks by Russian school students

Social network	Social network awareness (%)	Social network registration (%)	Social network unawareness (%)
VKontakte	9.6	88.8	1.5
Facebook	57.3	38.6	4.2
Instagram	19.3	79.0	1.7
Odnoklassniki	71.6	22.5	6.0
Twitter	66.5	26.1	7.4
LiveOnce	16.5	1.0	82.5
MySpace	21.2	1.3	77.5
LinkedIn	16.2	1.3	82.5
YouTube	11.7	86.6	1.7
Tumblr	35.2	6.1	58.7
Pinterest	22.3	36.8	40.9
LiveInternet	12.9	0.8	86.4
Habrahabr	10.0	1.1	88.9
Google + 1	36.8	34.5	28.7
LiveJournal	15.2	1.3	83.5
Vkrugudruzei	11.8	1.0	87.2
MoyMir	26.5	8.7	64.8
Last	9.3	0.8	89.9

Table 2 The significance of Internet for Russian school students

Answers	Percent (%)
Helps in communication with friends and relatives	77.1
Helps in the social discourse	49.5
Helps in school study	47.6
Helps in socializing with social celebrative	32.2
Helps in distant learning	24.2
Helps in my cultural advancement	19.8
Helps in personal development	18.0
Helps in job hunting	3.2
Of no personal significance	3.0
Another significance	3.0
Advances me in my peer community	2.5

met (11.0%). Thus, there is a danger of facing various cyber threats to the individual, since every tenth teenager communicates on the Internet with unknown people.

Only 3.6% of the respondents indicated that they do not communicate on the Internet.

A separate question concerned the topic of adolescents' communication in the social networks. The results show that most often schoolchildren say that they "just chat" on social networks (71.5%), and discuss their interests and hobbies (57.6%).

Computer games are an important component of the modern schoolchild's leisure time. In this regard, in the course of the study, it was revealed whether schoolchildren play computer games. The results showed that every second teenager (50.3%) plays network games with real partners (other people); 23.6% of the respondents indicated that they play online games with virtual partners, and 44.5%—games installed on their computer/gadget. In turn, 26.4% of the students reported that they do not play computer games. It is important to note here that the majority of the respondents (41.4%) believe that computer games cannot pose a threat to their personality, while 37.9% believe that such a risk exists.

The study paid special attention to identifying the degree of awareness of students in general education schools in the field of information and cyber threats to the individual. The knowledge of adolescents about possible threats in computer networks was revealed, as well as their experience in interacting with persons or information that could harm a person. Thus, in the course of the survey, the majority of the respondents (65.2%) answered that information can pose a threat to a personality.

In addition, the students were asked to indicate what benefits and harms computer networks can bring. The distribution of answers is shown in Table 3.

As can be seen from the data in the table, the surveyed schoolchildren are more likely to note the positive aspects of using computer networks than the negative

Table 3 Pros and cons of computer networks—according to Russian school students

Answers	Percent (%)
Give access to information diversity	69.9
Allow socializing with celebrities who are no access outside the Net	57.4
Consume lots of time	41.7
Allow freedom of creativity	34.7
Allow access to entertainment diversity	33.4
Contain authentically and potentially faulty information	30.8
Deviate users from 'actual reality'	29.1
Allow freedom of speech	21.8
May lead users to illegal activities	14.5
Corrupt users (physically, psychologically, morally)	8.5
No answer	6.3
Intervene with users' private lives	5.7
Others	1.0

ones. The most frequently indicated ones are the opportunity to receive a variety of information—some 70%, the ability to communicate with people inaccessible in real life—a little under 60%, the ability to freely engage in creativity—about 35%, the opportunity to have a variety of entertainment—33.4%. Among the negative influences of social networks, the most frequently cited are the waste of time—over 40%, inaccuracy of information—over 30%, and 'disconnection from reality'—almost 30%.

A separate question is devoted to the experience of adolescents facing with the threats on the Internet and the social networks. According to the results of the survey, almost every third schoolchild (30.0%) faced with the threats towards them on the social networks, and 12.0% of those surveyed indicated that such threats came to their friends. In addition, 29.2% of students said that they tried to mystificate any important data on the Internet, and 7.8% indicated that their friends faced similar situations. Thus, the problem of cyberbullying and fraud on social networks is relevant for modern adolescents, since one in three faced such threats or witnessed attempts to do so.

4 Discussion

The conducted sociological research in the field of cyber risks, which the students of the general education schools are exposed to, made it possible to draw a number of meaningful conclusions.

First of all, it is important to note that the vast majority of modern schoolchildren in Russia regularly use the Internet and the social networks. Most of the interviewed teenagers are registered in several social networks at the same period.

An important point when analyzing the potential cyber threats for a modern teenager is that it is the Internet and the social networks that turn out to be the most significant source of information for this category of citizens, even though about a third of the surveyed schoolchildren pointed to the risk of encountering inaccurate information on the network.

Among the most frequent motives for using the Internet for schoolchildren are the social (helps to communicate with friends/relatives; helps to communicate with interesting people, follow their life), the informational (helps me to keep abreast of the latest events) and the educational (helps me to study at school, helps receive distance education) ones. At the same time, the respondents themselves note that a significant amount of time for the Internet communication is occupied by communication on various topics ("just chatting").

Since it is the social motive of using social networks that is the most significant for the present-day adolescents, there is an obvious danger of facing various cyber threats to the individual, since every tenth teenager communicates on the Internet with people unknown to him. The urgency of the problem of cyberbullying and fraud in the social networks for the modern adolescents is also confirmed by the fact that every third respondent faced with similar threats or witnessed attempts of such influence.

5 Conclusion

Thus, the conducted sociological study showed the relevance and importance of a comprehensive and interdisciplinary study of the issues related to the cyber risks faced by modern adolescents in the context of educational systems management.

The development of the social networks and their progressing influence on all spheres of life of children and young people is a significant challenge to the society of the day. In this regard, systemic management decisions are in need so that to prevent the afore characterized cyber threats and the negative impacts of the Internet networks on Russian children and adolescents. The latter is associated with the need to ensure not only the psychological well-being of students, but also the need to ensure the feasibility of management mechanisms at the level of the regional and the municipal educational systems, which are a significant factor in preserving the social well-being of the population.

In this regard, the Russian education system—as any of the breed—bears a double responsibility: (a) direct, i.e. to conduct educational activities to counter cyber threats, and (b) indirect, i.e. to ensure social stability in the region by countering the cyber risks. It is necessary to conduct the further comprehensive psychological, sociological, pedagogical and medical scientific research to ensure the adoption of the scientifically grounded management decisions in various industries in order to upgrade the knowledge and skills of the country's students in countering the modern cyber threats.

The study also made it possible to identify some problem points in this issue, which requires further study and discussion in the academic community with the subsequent design of effective mechanisms for managing educational systems in the situations associated with the cyber risks and the cyber crises.

Acknowledgements This paper was developed as a part of the state commission of the Federal State Budgetary Scientific Institution 'Institute of Education Management of the Russian Academy of Education' with financial support from the Ministry of Education of the Russian Federation.

References

1. Kuznetsov AN, Neustroev SS, Serdyukova NA, Serdyukov VI (2021) Management decisions on optimizing regional educational infrastructure and ensuring its transport accessibility. In: Systemic challenges and changes in education: perspectives from Russia. Springer Nature, Switzerland, pp 41–52

2. Dishkova M, Papancheva R (2020) Estimation of parents' opinion how the time spent in front of a screen for learning affects the likelihood of developing cyber-addiction in children in primary school. In: ICERI 2020 proceedings. IATED Academy, pp 8925–8933
3. Mitchell I, Cockerton T, Hara S, Evans C (2018) SMERF: social media, ethics and risk framework. In: Jahankhani H (eds) Cyber criminology. Advanced sciences and technologies for security applications. Springer, Cham. https://doi.org/10.1007/978-3-319-97181-0_10
4. Brittan T, Jahankhani H, McCarthy J (2018) An examination into the effect of early education on cyber security awareness within the U.K. In: Jahankhani H (eds) Cyber criminology. Advanced sciences and technologies for security applications. Springer, Cham. https://doi.org/10.1007/978-3-319-97181-0_14
5. Shutikova M, Beshenkov S (2020) Digital educational environment and media education-platforms for transforming education system. Media Educ (Mediaobrazovanie) 4:729–737. https://doi.org/10.13187/me.2020.4.736
6. Mikhaylovsky MN, Lopatkova IV, Komarova NM, Rueva EO, Tereschuk KS, Emelyanenkova AV (2019) Cyberbulling as a new form of a threat: a physiological, psychological and medicinal aspects. Electronic Journal of General Medicine 16(6):161. https://doi.org/10.29333/ejgm/114268
7. Nicolaidou I, Venizelou A (2016) Be smart when online!: kids learn how to protect personal data, stop cyber-bullying and avoid hackers. In: ICERI2016 proceedings. IATED Academy, pp 3374–3383. https://doi.org/10.21125/iceri.2016.1789
8. Legate N, Weinstein N, Przybylski AK (2019) Parenting strategies and adolescents' cyberbullying behaviors: evidence from a preregistered study of parent-child dyads. J Youth Adolescence 48:399–409. https://doi.org/10.1007/s10964-018-0962-y
9. Lonigro A et al (2015) Is cyberbullying related to trait or state anger? Child Psychiatry Hum Dev 46:445–454. https://doi.org/10.1007/s10578-014-0484-0
10. Gkiomisi A, Gkrizioti M, Gkiomisi A et al (2017) Cyberbullying among Greek high school adolescents. Indian J Pediatr 84:364–368. https://doi.org/10.1007/s12098-016-2256-2
11. Kniffin KM, Palacio D (2018) Trash-talking and trolling. Hum Nat 29:353–369. https://doi.org/10.1007/s12110-018-9317-3
12. Yildiz Durak H (2019) Cyber human values displayed by university students in online social networking sites: the relationship of cyber human values to cyberbullying and cyber victimization behaviors displayed. In: INTED2019 proceedings. IATED Academy, pp 10035–10038. https://doi.org/10.21125/inted.2019.2531
13. Mathur P, Sawhney R, Chopra S, Leekha M, Ratn Shah R (2020) Utilizing temporal psycholinguistic cues for suicidal intent estimation. In: Jose J et al (eds) Advances in information retrieval. ECIR 2020. Lecture notes in computer science, vol 12036. Springer, Cham. https://doi.org/10.1007/978-3-030-45442-5_33
14. Lim WY, Tan CM, Nizam M, Zhou W, Tan SM (2016) Toward digital citizenship in primary schools: leveraging on our enhanced cyberwellness framework. In: Chai C, Lim C, Tan C (eds) Future learning in primary schools. Springer, Singapore. https://doi.org/10.1007/978-981-287-579-2_7
15. Ron M, Rivera O, Fuertes W, Toulkeridis T, Díaz J (2019) Cybersecurity baseline, an exploration, which permits to delineate national cybersecurity strategy in ecuador. In: Rocha Á, Ferrás C, Paredes M (eds) Information technology and systems. ICITS 2019. Advances in intelligent systems and computing, vol 918. Springer, Cham. https://doi.org/10.1007/978-3-030-11890-7_79
16. Sobkin VS, Evstigneeva YM (2001) Podrostok: virtual'naya i sotsial'naya real'nost' [Teenager: virtual and social reality]. Works Sociol Education, vol VI, Issue X. Center for Sociology of Education, Russian Academy of Education, Moscow
17. Sobkin VS, Adamchuk DV (2006) Otnosheniye uchastnikov obrazovatel'nogo protsessa k informatsionno-kommunikatsionnym tekhnologiyam [The attitude of participants in the educational process to information and communication technologies]. Center for Sociology of Education, Russian Academy of Education, Moscow

18. Sobkin VS, Adamchuk DV (2008) Monitoring sotsial'nykh posledstviy informatizatsii: chto izmenilos' v shkole za tri goda? [Monitoring the social consequences of informatization: what has changed at school within three years?]. Institute of Sociology of Education, Russian Academy of Education, Moscow
19. Sobkin VS, Skobeltsina KN (2012) Predstavleniya roditeley ob osobennostyakh obshcheniya ikh rebenka s komp'yuterom [Parents' ideas about the peculiarities of their child's communication with a computer]. Modern Preschool Educ 3:30–34
20. Skobeltsina KN, Kuznetsov AN (2019) Research on public satisfaction with educational infrastructure of advanced development territories. In: Proceedings of the 9th international conference for science educators and teachers (ICSET). Moscow, Russia, pp 36–41. https://doi.org/10.2991/icdee-19.2019.7
21. Skobeltsina KN, Kuznetsov AN (2021) Parents' satisfaction with their access to preschool education in the Russian advanced special economic zones. In: Systemic challenges and changes in education: perspectives from Russia. Springer Nature, Switzerland, pp 62–74

Plans for the Future and the Ability to Predict as Cognitive Navigators of a Person's Social Success

Julia Makarevskaya and Zinaida Ryabikina

Abstract Social success is an ambiguous category that has both external and internal criteria. Despite the difficult comprehension of the criteria and the ambiguity of their interpretation in different cultures and social strata, the desire for success in the form of various behavioral tendencies and manifestations is considered inherent in most people. The research results presented in the article were obtained on samples of different age groups: adolescence, young, adult age periods (from 17 to 45 years). The criteria of social success are considered through the prism of the leading activity of the age and are used as the basis for differentiating the samples into socially successful and unsuccessful respondents. The study is comparative in nature and shows that the cognitive markers of socially successful respondents differ from the mental trajectories of socially unsuccessful respondents. In adolescence, social success is accompanied by clear formulated goals and conscious plans for the future; in young and adult ages, while maintaining awareness of plans, socially successful respondents also have a high level of predictive abilities, expressed in the validity of the thinking process, flexibility of thinking, plasticity of ideas and the prospect of causal relationships. These cognitive characteristics allow a person to build mental processes in such a way that they accompany the person's social success.

Keywords Social success · Plans for the future · Ability to predict

1 Introduction

The category of success is discussed in various scientific fields, such as philosophy, sociology, political science, economics, in various management theories.

J. Makarevskaya (✉)
Federal State-Funded Educational Institution of Higher Education, Sochi State University, Plastunskaya Street 94, Sochi, Russia 354003

Z. Ryabikina
Federal State-Funded Educational Institution of Higher Education, Kuban State University, Stavropolskaya Street 149, Krasnodar, Russia 350000

A. Beskopylny and M. Shamtsyan (eds.), *XIV International Scientific Conference "INTERAGROMASH 2021"*, Lecture Notes in Networks and Systems 247,
https://doi.org/10.1007/978-3-030-80946-1_68

In the West, there are whole schools of development of success and success is cultivated as a certain style of behavior and life [1–6, etc.], in Russia the attitude to this topic is more complicated and ambiguous.

Currently, the features of the social situation of personality development (including economic, social, political, and cultural processes) form an example of a successful person (in personal, professional, social terms) in the public consciousness as an idea, and success, life achievements are one of the basic values of the modern Russian at the present time [7–13, etc.].

Studies by K. Levin and his followers in this area show that experiences of success and failure arise only within the zone of the subject's capabilities and in activities that are personally significant for the individual and, in addition, achievement in activity is assessed as success only if it is attributed to the individual's own efforts (K. Levin).

Despite the importance of the value of success both for a particular individual and for society, in Russian psychology, until recently, only some of the applied aspects of this problem were considered—mainly in the context of pedagogical activity (in matters of academic performance and success in learning), engineering psychology (in the area of achievement of labor productivity) [14, 15]. Recently, the terms "achievement" and "success" have begun to attract the attention of researchers in the field of the psychology of professional activity [16, 17].

Society estimates the success or failure of an individual based on the presence or absence of an objective result (goal achievement) and the significance of the activities performed in accordance with the current system of social values.

The subjective assessment of the result of the performed activity is emotionally experienced by the person as success (pleasure, joy, etc.) or failure (disappointment, resentment, anger, etc.), in contrast to the "result", the receipt of which may not cause emotional feelings if the activity turns out to be of little importance for the individual.

However, in any case, in order to be understand oneself as having achieved success, a person should at least promptly carry out certain types of activities, no worse and more original than others performing the same activities in a similar age period (that is, on time).

Thus, a personal problem in the timely achievement of social success is fixed, under visible similar objective conditions and, as a consequence, the possibility of dividing people into successful and unsuccessful in social terms.

In addition, we see a question about the difference in the criteria of social success for people of different age groups as an interesting. That is why in this study, questions about cognitive navigators of a person's social success were posed and solved by different methods. At the same time, by navigators, we mean certain mental paths (roads), directions of thoughts that accompany the social achievements of an individual.

Out of here, we formulate the assumption that the social success of an individual is accompanied by specific cognitive trajectories and abilities (opportunities).

2 Materials and Methods

Based on the content of the concept of social success, it can be assumed that the methods of its fixation may be objective criteria for the achievement of an individual in certain types of activity. At the same time, the activity should be personally significant for the subject and be essential for the formation of the social future of the individual.

This type of activity in psychology is the leading activity, the purpose of which is to ensure the completeness of mental development and the effectiveness of the social functioning of the individual.

It was according to this parameter that the types of activity of the subjects, included in the sample, were selected.

It is known that the leading types of activity in different age periods differ.

Therefore, for the subjects of adolescence, the criteria of their success in educational and professional activity were assessed, and for the subjects of adult and mature ages the criteria of their success only in professional activity were evaluated.

Thus, one of the research methods was the analysis of the selection of objective criteria for the success of the training-professional and professional activities of the respondents.

Based on these criteria, samples were formed:

- young people and girls aged 17–21. N = 100 people, average age—20.21 years, 19% men and 81% women.
- heads of various management levels. N = 30 people. First-line management (10 people), middle management (10 people) and top management (10 people). The average age in the sample in general is 31.60, in the sample of first-line management—22.5, middle management—31.3, top management—41.

Both samples underwent additional differentiation according to the level of success in the leading type of activity of the corresponding age periods.

In the sample of adolescents, the average academic performance score ($Mx =$ 4.29) and standard deviation ($\sigma = 0.61$) for this sample were calculated.

According to the bell curve, respondents with values in the range of $Mx \pm \sigma$, that is, from 3.68 to 4.9 points, were charged to the average level of social success. High values (level 3) in this sample were considered indicators of academic performance from 4.9 points—up to 5 points exactly, and respondents with values ranging from 3.68 to 2.87 points—were included in the sample of socially unsuccessful ones.

In numerical terms, the differentiation of the sample by the levels of academic performance is as follows: low-performing people—22 people (22%), average-performing people—59 people (59%), high-performing people—19 people (19%).

The sample of mature age was differentiated into levels of social success according to the criteria of the management level (position held): grass-roots, middle, highest.

That is, in a comparative study, 6 groups of respondents were identified: 3 of which are in the range of the adolescent age period, 3—in the range of adult age.

Diagnostic techniques:

- for a sample of adolescents: test "Life-meaning orientations" D. A. Leontiev (adapted version of the Purpose-in-Life Test (PIL) by James Crumbaugh and Leonard Maholic) [18];
- for managers: the "Predictive task" test [19].

The results of the LMO test reflect the general meaningfulness of life and the orientation of the individual towards conscious goal-setting, achievement of the result, concentration on the life process and/or on oneself.

An indicator of the ability to predict are the levels of success in establishing causal relationships, planning, transforming ideas, proposing and developing hypotheses by a person [18–21].

The results of the "Predictive task" test can be assessed by 19 parameters [19].

3 Results

The use of the "LMO" methodology made it possible to diagnose the general indicator of the meaningfulness of the subjects' life ($Mx = 106.91$), goals in life ($Mx = 33.1$), life saturation ($Mx = 32.51$), satisfaction with self-realization ($Mx = 27.15$), locus of I-control ($Mx = 21.66$) and locus of life-control ($Mx = 32.41$). The medians of the statistical sets of values according to the methodology scales do not significantly differ from the average, therefore, the statistical series are free from outliers. The data obtained correspond to the standards suggested by the author of the methodology for evaluating the results.

Subsequent processing of the results consisted of comparing the average values in the samples of low and highly successful subjects (1 and 3, respectively).

Student's t criterion showed that:

- the severity of signs "meaningfulness of life" ($p = 0.007$) and "purpose of life" ($p = 0.006$) in samples 1 and 3 differs at a high level of statistical significance ($p \leq 0.01$);
- the severity of signs "emotional saturation of life—process" ($p = 0.038$), "satisfaction with self-realization—result" ($p = 0.019$), "locus of I-control" ($p = 0.018$), "locus of life-control" ($p = 0.02$) in samples 1 and 3 is significantly different ($p \leq 0.05$).
- The meaningfulness of the present, past and future by a person being for a long period in conditions of social success is higher than that of a low-successful person, which can probably be expressed in the personality's determination, if not only of the meaning of life in general, but also of some of its aspects (for example, prospects for the future ($p = 0.006$), the fullness of the current moment of life ($p = 0.038$), or comprehension of the past ($p = 0.019$); as well as ideas about the strength of one's personality ($p = 0.018$) and the ability to control one's life ($p = 0.02$).

The recorded differences suggest the existence of a concatenation between social success and various aspects of the meaningfulness of the subjects' lives. To test this hypothesis, the calculation of the Pearson correlation coefficient was applied.

The results of the correlation analysis recorded the concatenation between the success of the subjects in the leading activity and the general indicator of the meaningfulness of their life ($p = 0.026$), the awareness of life goals ($p = 0.009$), the fullness of the meaning of the current activity ($p = 0.011$), satisfaction with the achieved results ($p = 0.02$), as well as ideas about the strength of their I ($p = 0.01$) and the conviction in control over their own life ($p = 0.02$).

The direct ratio of meanings and the closeness of the identified concatenation suggests that the higher the success of the individual in the leading and significant for him/her activity, the more he/she is inherent in comprehension of his/her own life, in awareness of the focus and time perspectives of the personal I, interest in current real life events, satisfaction with his/her past, consciousness of himself/herself as a strong personality with sufficient capabilities to build and control his/her life in accordance with his/her goals.

To determine the direction of the concatenation of influence between success and the components of the meaningfulness of a person's life, one-way analysis of variance (ANOVA) was used.

From the results of the analysis, it can be concluded that the level of a person's achievements in activities that are significant for his/her future affects the meaningfulness of life in general ($p = 0.054$), the level of goal-setting of the individual ($p = 0.029$), the level of development of the personality's worldview beliefs in the ability to control one's own life ($p = 0.029$) and ideas about the capability to exercise this control ($p = 0.047$).

In general, the adolescent personality under conditions of social success can more express:

- meaningfulness of life in general (r = 0.223*);
- awareness of the time perspective and direction of one's own life (r = 0.260**); (**—the correlation is significant at the 0.01 level, *—the correlation is significant at the 0.05 level).

The social success of an individual in the leading activity that is significant for his/her future efforts acts as a factor that determines the meaningfulness of life as a whole ($p = 0.054$), the level of goal-setting of the individual ($p = 0.029$), the level of development of the personality's worldview beliefs in the ability to control his own life ($p = 0.029$) and ideas about the capability to exercise this control ($p = 0.047$).

However, the tasks of social achievement are not limited by the adolescence, but rather become even more relevant, therefore stage 2–1 of the study was devoted to testing the hypothesis on a sample of young and adult ages (Table 1).

Table 1 Empirical indicators of the test "predictive task" average values for 3 groups

	Low level	Middle level	High level
Average score	25.9	31.3	37.8

Table 2 Empirical values of the student's t-test and the level of significance of the revealed differences in the groups of subjects

Predictive ability	Student's empirical t-test	Sig. (2-tailed)
Low-middle level	– 3.918	0.001
Low–high level	– 12.742	0.000...
Middle-high level	– 4.364	0.001

From the data presented in the table, it can be seen that the average values of the severity of the ability to predict differ in the samples of the low, middle and high management levels.

Pair-wise comparison of the means in 3 groups of subjects was carried out using the Student's t-test (Tables 2 and 3).

As can be seen from the obtained values, the greatest differences in the ability to predict were found between the leaders of the low and top management ($p \leq$ 0.000...).

After the primary processing of the empirical data, it can be seen that the average values for all 19 parameters of the predictive ability of the subjects increase in the direction from the low to the middle.

The measures of the central trend show that the spread of values from the low level to the high level decreases, which indicates a greater stability of the empirical values in the sample of the top management level (Table 4).

Pair-wise comparison of differences (Student's t-test) revealed significant differences between all levels of management in all parameters of the "Predictive task" test. By clarifications, it can be considered that the differences in the level of predictive ability are somewhat greater between the middle and high level management (temp. = – 4.364) than between the low and middle level management (temp. = – 3.918).

In general, according to this method, it was determined that the level of development of the ability to predict is expressed:

- at the low level: low in 30% of the subjects, medium—20%, high—50%;
- in the middle level: low in 10% of the subjects, high—30%, very high—60%;
- at the high level—very high—100%.

The data obtained indicate that the representation of a very high level of predictive ability among top managers is expressed by 100%.

4 Discussion

The results of the study, obtained on a sample of adolescents, were repeatedly checked by us and found confirmation.

Due to the fact that the criteria of social success are still controversial, it can be said that in some cases respondents who were unsuccessful in their leading activities could

Table 3 Quantitative indicators of 19 parameters of predictive ability

Tasks and parameters of predictive ability	Management levels		
	Low level (total points)	Middle level (total points)	High level (total points)
Task 1			
1. The level of verbal generalization of consequences	20	40	55
2. The validity of the highlighted consequences	20	25	50
3. Perspective of the consequences	21	27	55
4. The logic of constructing consequences	10	25	40
Task 2			
7. The level of verbal generalization of reasons	23	60	60
8. Completeness of cause-and-effect relations	25	60	60
9. The relevance of cause-and-effect relations	22	50	54
Task 3			
10. Awareness of the purpose of the plan	29	50	59
11. Completeness of planning operations	20	45	45
12. The prospect of cause-and-effect relations	10	32	40
Task 4			
13. The latitude of the association area	50	53	55
14. The variability of the association area	22	45	49
15. Plasticity of representations	15	23	38
Task 5			

(continued)

Table 3 (continued)

Tasks and parameters of predictive ability	Management levels		
	Low level (total points)	Middle level (total points)	High level (total points)
16. The breadth of the search for hypothesizing	21	33	40
17. Accounting the requirements of the conditions when hypothesizing	53	54	56
18. Flexibility of hypotheses	13	24	60
19. Validity of hypotheses	12	25	44

Table 4 Measures of the central trend in 3 groups

	Low level	Middle level	High level
Average score	21.4	38.2	49.5
Minimum	10	23	36
Maximum	53	60	60
σ	12.056	13.294	8.269

successfully fulfil themselves in other spheres of their lives and thereby neutralize the lack of achievements in activities that are significant for a given age.

That is, in interpreting the results obtained, we cannot unequivocally say that planning for the future builds social success, but we can assert that clear plans for the future accompany the recognized social achievements of young men and women in the leading activity for their age—educational and professional.

So, more socially successful young men and women statistically significantly differ from unsuccessful ones in that they have clear plans for the future in the structure of their life-meaning orientations.

Their plans are formulated as goals for the future, "which give life meaning, direction and time perspective."

General personality characteristics: purposefulness.

An additional survey of respondents (N = 178) made it possible to describe the pithy characteristics of this cognitive (mental, attributive) parallel of social success in adolescence: "I strive to achieve my goal", "I strive for knowledge", "I have a desire to learn", "I am purposeful", "I want to achieve success", "there are goals in life", "there is a goal for which I study".

That is, we want to emphasize once again that in adolescence, the achievements recognized by society in activities that are significant for the future of individual are perceived as social success and are accompanied by clearly understood plans for the future.

Do these results guarantee continued social success when leading activities change? This is the main question for the second stage of our study.

It is the samples of young and mature age, differentiated by the level of career achievements, that allow us to move further in the study of the cognitive components of a person's social success.

The results obtained on a sample of managers of different levels show how significant predictive ability is in a successful managerial career.

In this case, we cannot say that it is predictive ability that underlies the successful career of managers. Perhaps the performance of managerial functions stimulates the development of predictive mindset. However, the severity of these abilities in our study increases statistically significant with an improvement in the social status of respondents from the low to the high management level.

At the same time, it is interesting that the differences in the predictive ability of respondents between the low and middle management levels are statistically less expressed than between the middle and high management levels. That is, it is easier to cognitively rebuild and take a career step from the low management level to the middle one than from the middle to the high one.

Based on the results of a quantitative analysis of empirical data, it can be concluded that the predictive ability of top managers is much more developed in all structural parameters. This means that high level managers have more developed logical, predictive functions of thinking, as well as mental operations of generalization, analysis of information, and such qualities of the mind as plasticity, flexibility, breadth of thinking, the possibility of distraction and abstraction.

So, the quantitative and qualitative analysis of empirical data made it possible to conclude that predicting is most clearly manifested among top managers in comparison with the middle and low managers.

Thus, we understand that career aspirations are not the only criteria for social success, however, as a sign of material well-being and employment, the level of the position held is a category in cognitive schemes that describes social well-being.

5 Conclusion

On the basis of the research carried out, it can be concluded that social success, recognition of the achievements of an individual in activities that are significant for him/her cause clarity in the consciousness of the time perspective, an increase in the feeling of the strength of oneself and faith in one's own ability to manage his/her own life.

If in adolescence it is quite enough to decide on plans for the future, to have clear cognitive ideas about life goals, then in young and mature ages high social success (career achievements) is most accompanied by such mental abilities of the individual as the validity and prospectivity of predicting various consequences, prospectivity of cause-and-effect relations of events and circumstances, plasticity of ideas, flexibility and validity of hypotheses.

That is, we see that awareness of thoughts from adolescence does not leave a socially successful person, and predictive thinking abilities are acquired (or developed), which allow the person to literally see the future.

We see prospects for research in the study of the emotional and volitional components of the individual, accompanying the achievement of social success.

References

1. James W (2005) Great men and their environment. Kessinger Publishing LLC, Whitefish
2. Ryan RM, Deci EL (2000) Self-determination theory and the facilitation of intrinsic motivation, social development, and well-being. Am Psychol 55:68–78. https://doi.org/10.1037/0003-066X.55.1.68
3. Lewin K (2000) Field theory in social science: transl. from English. Sensor, St. Petersburg
4. Thomas M (2006) A new attitude: achieving personal and professional success by keeping a positive mental outlook: transl. from English by Paul, A. AST Publishing Group, Moscow, 398
5. Covey S (2017) The seven habits of highly effective people: restoring the character ethic. Alpina Publisher, Moscow, 384
6. Hill N (2006) Think and grow rich! Popuri, Minsk
7. Guseltseva MS, Asmolov AG (2020) Hidden leadership and diversity of success models in modern society. Volga Pedagogical Search 2(32):14–23. https://doi.org/10.33065/2307-1052-2020-2-32-14-23 (in Russian)
8. Soldatova GU, Chigarkova SV (2018) Social capital online: intergenerational analysis. European Proc Soc Behav Sci Future Acad 49:636–643. https://doi.org/10.15405/epsbs.2018.11.02.73 (in Russian)
9. Ryabikina ZI, Makarevskaya IG, Makarevskaya YE (2018) Specificity of the structure of the intellect of socially successful people. EpSBS ICPE 42:392–401. https://doi.org/10.15405/epsbs.2018.11.02.42
10. Leontiev DA (2020) Happiness and well-being: toward the construction of the conceptual field. Monit Publ Opin Econ Soc Changes J (Publ Opin Monit) 1. https://doi.org/10.14515/monitoring.2020.1.02 (in Russian)
11. Melia M (2012) Success is a personal matter: how not to lose yourself in the modern world. Prosveshenie publishing house, Moscow
12. Makhmudov O (2017) Psychology of success. Molodoy Uchenyy 14.1 (148.1):19–21. https://moluch.ru/archive/148/42020/. Last accessed 2021/04/02
13. Klein KG, Leontiev DA, Kostenko VY, Osin EN, Taranenko OA, Kosheleva NV (2019) Experiences in different activities: temporal dynamics and construct validity. Psikhologicheskaya nauka i obrazovanie Psychol Sci Educ 24(5):47–57. https://doi.org/10.17759/pse.2019240505 (in Russian, abstr. in Engl.)
14. Leontiev DA, Klein KG (2018) Quality of motivation and quality of experiences as characteristics of learning activity. Moscow Univ Bullet Series 14 Psychol 4:106–119. https://doi.org/10.11621/vsp.2018.04.106 (in Russian)
15. Leontiev DA, Osin EN, Dosumova SS, Rzaeva FR, Bobrov VV (2018) Related experiences and their association with psychological well-being. Psikhologicheskaya nauka i obrazovanie Psychol Sci Educ 23(6):55–66. https://doi.org/10.17759/pse.2018230605 (in Russian, abstr. in Engl.)
16. Kuznetsova AS, Titova MA, Zlokazova TA (2019) Psychological self-regulation of the functional state and professional success. J Moscow Univ Bullet Episode 14 Psychol 1:51–68. https://doi.org/10.11621/vsp.2019.01.51 (in Russian)
17. Smirnova A (2018) Psychology of success and career growth. Man and society in the digital economy: experience, problems and directions of development: materials of the all-russian

scientific and practical conference dedicated to the 100th anniversary of the Federal State-Funded Educational Institution of Higher Education "Financial University under the Government of the Russian Federation". Publishing House, South, Krasnodar. https://www.elibrary.ru/item.asp?id=35373543&selid=35373569. Last accessed 2020/04/02
18. Leontiev D (2004) Life-meaning orientation test (LMO). Psychodiagnostic series. Publishing House "Smysl"
19. Regush L (2003) Psychology of predicting: advances in understanding the future. Publishing House "Rech", St. Petersburg
20. Leontiev DA (2016) Self-regulation, resources and personality potential. Siberian Psychol J 62. https://doi.org/10.17223/17267080/62/3
21. Pishchik VI (2020) Internal professional mobility of lyceum teachers belonging to different past generations. Human Soc Sci Rev 8(4):917–930. https://doi.org/10.18510/hssr.2020.8489

Comparative Analysis of the Ratio of the Maturation Level of Inhibitory Control and Parameters of Working Memory in Ontogenesis

Elena Nikolaeva, Svetlana Burkova, Elvira Dunajevskaja, Irina Shirokova, and Nina Soboleva

Abstract Working memory and inhibitory control have the same neural basis—the prefrontal-parietal system. As children grow older, the improvement of their working memory and of inhibitory processes do not occur synchronously. At the same time, there are no reliable ideas about how specifically the processes in the working memory relate to the parameters of inhibitory control at the different stages of ontogenesis. The aim of this work was a comparative analysis of the relationship between inhibitory control and the parameters of the working memory in ontogenesis. To assess the inhibitory control, a computer version of the author's program of complex reflexometry was used. When studying working memory, the mechanisms of retrieval-induced forgetting and retrieval-based learning were assessed. The study involved 257 children: 93 children 5–6 years old; 53 children 7–8 years old, 50 children 10–11 years old; 61 children 12–17 years old. The data obtained indicate that inhibitory control, interference inhibition, and sensorimotor integration are associated with different mechanisms in the brain and have different time parameters of maturation. If the processes of the working memory corresponds with the 5 years old kids to those that are teenagers, then inhibitory control and sensorimotor integration are formed much longer into ontogenesis.

Keywords Inhibitory control · Working memory · Retrieval-induced forgetting · Retrieval-based learning · Children

1 Introduction

1.1 An Inhibitory Control

Inhibitory control is the ability of a person to control attention, behavior, thoughts and/or emotions, including suppressing strong internal urges or external temptations. This provides a mechanism for purposeful behavior driven by internal or external decisions [1].

E. Nikolaeva (✉) · S. Burkova · E. Dunajevskaja · I. Shirokova · N. Soboleva
Herzen State Pedagogical University, nab reki Moiki, 48, 191186 Saint-Petersburg, Russia

A. Beskopylny and M. Shamtsyan (eds.), *XIV International Scientific Conference "INTERAGROMASH 2021"*, Lecture Notes in Networks and Systems 247,
https://doi.org/10.1007/978-3-030-80946-1_69

A person without inhibitory control is a weather vane that turns towards the strongest external or internal stimulus, the activity of which corresponds to conditioned reflexes [2]. This is how a person becomes after a complete transection of the prefrontal areas, such as an infant that responds to any external or internal signal of a certain intensity [3]. Inhibitory control provides an opportunity for change, growth, learning, choice of action, including the most difficult one available, without resorting to the most activated habit at the moment [4].

Humans tend to develop habits that minimize effort and make the world appear predictable. Inhibitory control turns a person into a conscious being, capable of setting tasks and purposefully moving towards their solution, without being distracted by a multitude of accompanying stimuli [5].

The go/go, go/no-go and stop-signal paradigms are used to evaluate braking control. In this case, a certain reaction is generated, which must then be stopped in accordance with the instructions [6].

1.2 Working Memory

Working memory—memory that allows you to remember all the intermediate stages of a certain task [7].

The most discussed mechanisms operating in working memory are retrieval-induced forgetting (RIF) and retrieval-based learning (RBL) [8]. RIF is a deterioration in the reproduction of subsequent stimuli as a consequence of reproduction of previous stimuli that are similar in some parameter.

There is no uniform understanding of the RIF mechanism. One of the mechanisms for the deterioration of reproduction is considered to be proactive interference, that is, the imposition of memories of previous stimuli on the memorization of new information and weakening the formation of a new memory trace [9]. Theories explaining the effect of interference [10] suggest that the search signal activates several memories simultaneously, and the competition of memory traces prevents accurate memory. That is, forgetting seems to be a consequence of the associative structure of memory and the proximity of the reproduced information.

Anderson and Bjork [11] suggested that forgetting is not a consequence of competition between memory traces, but is due to a special active inhibitory process acting during the search for an answer. To facilitate memorization of the target stimulus, the non-target one is suppressed at the moment, and it is this mechanism that impairs the possibility of subsequent reproduction.

Norman et al. [12] proposed a theoretical approach that combines these two mechanisms and explains forgetting by the action of both interference and inhibition. Moreover, in this paradigm, inhibition occurs when competing units are activated. However, both inhibition and interference are of particular importance in relation to the theories that have come together in this new view.

RBL is the opposite of the RIF process, which means that each subsequent reproduction in working memory leads to better retention of information with repeated testing. It is known that RBL accompanies the memorization of stimulus material of different quality and volume [13, 14].

The result of reproduction is determined by the interaction of both processes [15].

Formulation of the problem

Working memory and inhibitory control work together. These phenomena occur simultaneously when it is necessary to calculate the subsequent actions and not to perform irrelevant, but already lost their significance, actions. To connect ideas or facts into one whole, you need to focus only on one thing and exclude everything else, while maintaining the original direction of thought.

At the same time, there are different points of view on the relationship between these two components of executive functions. Some researchers believe that working memory is primary, while inhibitory control is secondary [16]. Those researchers who create computer models [17] support this point of view. According to them, the primary goal, which is controlled by working memory and which will determine what exactly needs to be subjected to inhibitory control.

There is also an idea that these are two independent processes [15]. This follows from the results of studies in which older people were asked to pay attention to some stimuli and ignore others. Older adults were more responsive to stimuli to be attentive, but could not ignore peripheral stimuli. According to the authors of this study [18], increasing attention and ignoring are independent processes.

Working memory and inhibitory control have the same neural basis—the prefrontal-parietal system [19]. As children grow older, the improvement of working memory and the improvement of inhibitory processes do not occur synchronously [20]. At the same time, there is no reliable idea of how exactly the processes in working memory correlate with the parameters of inhibitory control at different stages of ontogenesis.

The aim of this work was a comparative analysis of the relationship between inhibitory control and parameters of working memory in ontogenesis.

2 Materials and Methods

To assess the inhibitory control, a computer version of the author's program for complex reflexometry ReBOS was used [21]. A feature of this technique is the ability to assess the child's ability to navigate in the flow of sensory signals and to identify the quality of inhibitory processes. The test consists of three series. The training series is aimed at revealing the child's understanding of the sequence of actions in the test. In the next series, created in the go/go paradigm, the child was taught to perform a certain set of actions. In the last series, based on the go/no-go paradigm, the child was required not to perform an already learned action.

The following indicators were assessed: reaction time in each series, missing stimuli, errors—reactions to a prohibited stimulus.

An original computerized technique was used [22], designed to memorize visual objects presented on a computer screen. The technique included three series, during which the same set of simple objects (30 units) was presented, but the order of presentation varied from series to series. The objects were objects familiar to children of all ages (leaves, butterflies, twigs, bugs, etc.)

The subject initially saw three objects, from which he had to select any and click on it with the mouse cursor. Subsequently, new objects were gradually added, and the subject was asked to mark with the mouse cursor the object that had not been marked earlier. If the subject made a mistake and pressed the object that he had already chosen earlier, then the series ended, and a new presentation of the same stimuli began with a different order of their appearance. Another mistake led to the end of this series and the beginning of a new one. The number of correctly reproduced elements in each series was counted and the ratio of the number of reproduced elements in different series was estimated.

The study involved 257 children: 93 children 5–6 years old; 53 children 7–8 years old, 50 children 10–11 years old; 61 children 12–17 years old.

The time of object selection was not regulated. For statistical data processing, the IBM SPSS Ver.22 software package was used.

3 Results and Discussion

At first, the indicators of working memory were assessed in all studied groups.

The amount of working memory in the groups did not change (Table 1), as shown earlier [23]. But the RBL has increased with age. We have previously shown that children under 5 years of age have a significantly low volume of working memory at the first presentation in relation to other groups. For all other groups, the spread of indicators is so great that there are no significant differences. However, in children less than 5 years of age, the RBL mechanism is practically absent [24].

A different situation is observed with the development of a simple (Table 2) and complex sensorimotor reaction (Table 3): to press a stimulus when any signal appears. The reactions become shorter and there are fewer gaps in stimuli.

Table 1 Representation of RIF and RBL in different groups

Групп (years) Groops (years)	RIF	RBL
5–6	88.1	54.2
7–8	86.8	52.8
10–12	85.1	61.7
14–15	85.2	72.7*

Note *difference with a significance level of 0.05, ** – 0.01 between the age group 5–6 years old and 14–15 years old (Fisher's exact test)

Table 2 Parameters of the first series in the go/go paradigm (mean and standard. deviation)

Groups (years)	Response time (ms)	Number of passes
5–6	401.4 ± 62.6	4.9 ± 3.4
7–8	361.8 ± 64.4	3.5 ± 2.3
10–12	315.5 ± 41.1	2.1 ± 1.9
14–15	299.2 ± 39.8***	2.0 ± 2.1*

Note designations as in Table 1

Table 3 Parameters of the second series in the go/go-no go paradigm (mean and standard deviation)

Groups (years)	Number of passes	Number of errors
5–6	9.2 ± 6.3	9.2 ± 4.5
7–8	9.7 ± 5.5	8.6 ± 3.3
10–12	2.6 ± 2.4	7.9 ± 3.2
14–15	2.4 ± 2.7*	6.4 ± 2.8

Note designations as in Table 1

Factor analysis gave a 4-factor solution with the adequacy of the Kaiser–Meyer–Olkin sample 0.663, the percentage of explained variance was 74% (Table 4).

The first factor (27.9%) included indicators of the go/no go series (response time to a stimulus and the number of missing responses), as well as age with the opposing sign. Indeed, according to the data obtained, the responses become faster with age

Table 4 Rotated component matrix

	Component			
	1	2	3	4
Mean reaction time in the first part of the go/no go	0.899	– 0.073	0.018	– 0.072
Mean reaction time in the second part of the go/no go	0.865	– 0.123	0.040	0.066
Passes in the first part of the go/no go	0.835	0.146	– 0.066	– 0.058
Passes in the second part of the go/no go series	0.813	0.116	– 0.046	– 0.041
Age	– 0.756	– 0.010	– 0.022	0.039
Errors in the first part of the go/no go series	– 0.129	0.882	0.013	– 0.083
Errors in the second part of the go/no go series	0.182	0.876	– 0.009	– 0.006
Memory capacity for third playback	– 0.026	– 0.028	0.934	0.196
RBL	0.060	0.031	0.674	– 0.399
Memory capacity at first playback	– 0.041	– 0.077	0.017	0.950
RIF	0.087	– 0.013	– 0.520	0.543

Factor extraction method: principal component analysis
Rotation method: Varimax with Kaiser normalization
[a]Rotation converged in 5 iterations

and the number of missing stimuli decreases. This reflects the gradual maturation of sensorimotor integration processes.

The second factor (16.5% of the explained variance) included errors of the go/no go series, the number of which reflects the effectiveness of inhibitory control.

The third factor (16.1%) included the volume of stimulus reproduction in the third attempt and RBL. This is explained by the fact that after the deterioration in the second attempt as a result of interference inhibition, there is an improvement in reproduction as a result of learning. Finally, the fourth factor included the volume of playback in the first attempt and RIF.

4 Discussion

The aim of the study was to compare the maturation level of inhibitory control and the parameters of working memory RIF and RBL. To assess the inhibitory control, the go/go go/no go paradigms were used: in the first series, the subject was trained to respond to all stimuli presented on the computer monitor, in the second series, it was forbidden to respond to red circles. Thus, at first, a conditioned reaction was developed, and then the effectiveness of its inhibition was analyzed.

To assess the parameters of working memory, the same set of stimuli, which were presented in a different sequence, was presented three times. This created conditions for interference inhibition. The probability of deterioration of reproduction or, on the contrary, its improvement in the last two reproductions in relation to the first was assessed.

Our data have shown that the effectiveness of inhibitory control largely depends on age and increases linearly with growing up from 5 to 14 years old. Sensorimotor integration, which can be assessed by response time to stimuli, also becomes more effective with age.

In contrast, interference control remains virtually unchanged from 5 to 14 years old. At the same time, RBL gradually grows with age. Consequently, despite the fact that, in general, the prefrontal areas are significant for all the studied processes, their maturation to varying degrees is determined by the maturity of a given brain area.

Our data are more consistent with the assumption that working memory is primary in relation to inhibitory control [16, 17].

5 Conclusion

Thus, our data indicate that inhibitory control, interference inhibition, and sensorimotor integration are associated with different mechanisms in the brain and have different temporal parameters of maturation. If the processes of working memory correspond at 5 years old to those in adolescence, then inhibitory control and sensorimotor integration are formed much longer into ontogenesis.

References

1. McKenna R, Rushe T, Woodcock KA (2017) Informing the structure of executive function in children: a meta-analysis of functional neuroimaging data. Front Hum Neurosci 11:154. https://doi.org/10.3389/fnhum.2017.00154
2. Diamond A (2013) Executive functions. Annu Rev Psychol 64:35–68. https://doi.org/10.1146/annurev-psych-113011-143750
3. Luna B, Marek S, Larsen B, Tervo-Clemmens B, Chahal R (2015) An integrative model of the maturation of cognitive control. Annu Rev Neurosci 38:151–170. https://doi.org/10.1146/annurev-neuro-071714-034054
4. Fiske A, Holmbo R (2019) Neural substrates of early executive function development. Dev Rev 52:42–62. https://doi.org/10.1016/j.dr.2019.100866
5. Razumnikova OM, Nikolaeva EI (2001) Correlation of assessments of attention and learning success. Questions Psychol 1:123–129
6. Merenkova VS, Elnikova OE, Nikolaeva EIA (2020) Comparative analysis of the interrelation between the level of executive functions, which are viewed as cognitive control parameters, and the level of the inner picture of health in students of different ages. Sci Educ Today 10(6):27–39. https://www.elibrary.ru/item.asp?id=44478880. https://doi.org/10.15293/2658-6762.2006.02
7. Alloway RG, Alloway TP (2015) Understanding working memory, (2nd edn.). Sage, Los Angeles. https://doi.org/10.1017/edp.2015.14
8. Cinel C, Mack CC, Ward G (2018) Towards augmented human memory: retrieval-induced forgetting and retrieval practice in an interactive, end-of-day review. J Exp Psychol Gen 147(5):632–661. https://doi.org/10.1037/xge0000441
9. Rowland CA, Bates LE, DeLosh EL (2014) On the reliability of retrieval-induced forgetting. Front Psychol 5:1343. https://doi.org/10.3389/fpsyg.2014.01343
10. Postman L (1976) Interference theory revisited. In: Brown J (ed) Recall and recognition. Wiley, New York, NY, pp 157–182
11. Anderson MC, Levy BJ (2010) On the relationship between interference and inhibition in cognition. In: Bjork RA, Benjamin AS (eds) Successful remembering and successful forgetting: a festschrift in honor. Psychol Press Publ 2010:107–132
12. Norman KA, Newman EL, Detre G (2007) A neural network model of retrieval-induced forgetting. Psychol Rev 114:887–953. https://doi.org/10.1037/0033-295X.114.4.887
13. Pastötter B, Schicker S, Niedernhuber J, Bäuml K-HT (2011) Retrieval during learning facilitates subsequent memory encodingю. J Exp Psychol Learn Mem Cogn 37:287–297. https://doi.org/10.1037/a0021801
14. Roediger HLIII, Karpicke JD (2006) The power of testing memory: basic research and implications for educational practice. Perspect Psychol Sci 1:181–210.https://doi.org/10.1111/j.1745-6916.2006.00012.x
15. Davidson MC, Amso D, Anderson LC, Diamond A (2006) Development of cognitive control and executive functions from 4–13 years: evidence from manipulations of memory, inhibition, and task switching. Neuropsychologia 44:2037–2078
16. Hanania R, Smith LB (2010) Selective attention and attention switching: towards a unified developmental approach. Dev Sci 13:622–635. https://doi.org/10.1111/j.1467-7687.2009.00921.x
17. Friedman NP, Miyake A (2017) Unity and diversity of executive functions: Individual differences as a window on cognitive structure. Cortex 86:186–204. https://doi.org/10.1016/j.cortex.2016.04.023
18. Zanto TP, Rubens MT, Thangavel A, Gazzaley A (2011) Causal role of the prefrontal cortex in top-down modulation of visual processing and working memory. Nat Neurosci 14:656–661
19. Gazzaley A, Nobre AC (2012) Top-down modulation: bridging selective attention and working memory. Trends Cogn Sci 16:129–135
20. Stedron JM, Sahni SD, Munakata Y (2005) Common mechanisms for working memory and attention: the case of perseveration with visible solutions. J Cogn Neurosci 17:623–631

21. Vergunov EG, Nikolaeva EI, Balioz NV, Krivoschekov SG (2018) Lateral preferences as the possible phenotypic predictors of the reserves of the cardiovascular system and the features of sensorimotor integration in climbers. Hum Physiol 44(3):320–329. https://doi.org/10.1134/S0362119718030143
22. Rasumnikova OM, Nikolaeva EI (2019) Age characteristics of inhibition control in the model of proactive interference. Questions Psychol 2:124–132. https://doi.org/10.11621/pir.2018.0206
23. Efimova VL, Nikolaeva EI (2019) The relationship of language and intelligence development to the maturity of the subcortical structures in children with specific language disorders. Psychol Russia State Art 12(1):79–88. https://doi.org/10.11621/pir.2019.0106
24. Razumnikova OM, Nikolaeva EI (2020) Age and sex differences in the formation of inhibitory control in the process of learning memorization. Izvestia Herzen State Pedagogical Univ 195:89–95 (Russian)

Association of Different HTR2A, DRD4 Gene Genotypes with the Peculiarities of the Emotional and Personal Sphere of Russian Boys and Girls

Elena Vorobyeva, Ekaterina Kovsh, Daria Alekseeva, Vitaly Babenko, and Pavel Ermakov

Abstract The article is devoted to the study of the association of different genotypes of genes of serotonin receptor 5-HT2A (HTR2A) and dopamine receptor subtype D4 (DRD4) with the peculiarities of the emotional and personal sphere of Russian boys and girls. The work was attended by Russian boys and girls, psychology students, in the amount of 100 people. The study aimed to study the association of different HTR2A and DRD4 gene genotypes with emotional and personal characteristics of Russian boys and girls living in the South of Russia. The following methods were used: Test of emotional intelligence (D.V. Lyusin); Emotional Empathy Questionnaire (A. Mehrabian, N. Epstein); «Big five» test; Toronto Alexithymia Scale. For statistical processing of the results obtained, multivariate analysis of variance ANOVA with Tukey's post hoc analysis for non-equilibrium sample sizes was applied. According to the results of the study, based on the analysis of variance, no data were obtained on the association of different genotypes of the HTR2A and DRD4 genes in terms of the characteristics of the emotional and personal spheres. However, based on the use of Tukey's a posteriori analysis for non-equilibrium samples, the differences between Russian boys and girls, carriers of different genotypes of the HTR2A and DRD4 genes, are described in terms of the characteristics of the emotional and personal spheres, such as empathy, alexithymia, and personality traits that are included in the Big Five personality traits factors.

Keywords HTR2A gene · DRD4 gene · Emotional and personal sphere · Boys · Girls

---

E. Vorobyeva (✉) · V. Babenko
Don State Technical University, Gagarin Square, 1, 344003 Rostov-on-Don, Russian Federation

E. Vorobyeva · E. Kovsh · D. Alekseeva · V. Babenko · P. Ermakov
Southern Federal University, st. Bolshaya Sadovaya, 105/42, 344006 Rostov-on-Don, Russian Federation

A. Beskopylny and M. Shamtsyan (eds.), *XIV International Scientific Conference "INTERAGROMASH 2021"*, Lecture Notes in Networks and Systems 247,
https://doi.org/10.1007/978-3-030-80946-1_70

1 Introduction

Personal characteristics manifested in behavior, interaction with other people, the ability to understand and express emotions are influenced by both hereditary and environmental factors. Survey studies note that the Big Five is the most commonly used tool for assessing the severity of personality traits such as neuroticism, extraversion, openness to experience, agreeableness, and conscientiousness [1]. The association of personality traits of the Big Five and genes of dopamine and serotonin receptors has been shown [2].

In our early works, it was shown that the HTR2A genotype affects the level of emotional intelligence according to Mayer-Salovey-Caruso, and the carriers of the minor genotype A/A had a higher level of emotional intelligence than the carriers of the major genotype G/G [3].

The level of empathy is essential for correct recognition of emotional facial expressions [4]. The genotype for the HTR2A gene affects the parameters of evoked potentials in the recognition of emotionally colored and neutral scenes [5].

In studies on samples of schizophrenic patients, it was shown that patients with schizophrenia are statistically significantly more likely to have the major C/C genotype of the HTR2A gene (it corresponds to G/G, since the rs6313 C > T and rs6311 G > A loci are in linkage disequilibrium), while such patients have a low density of type II serotonin receptors in the brain [6, 7].

In a study performed on healthy subjects, no statistically significant differences were found in the density of serotonin 2A receptors in carriers of different genotypes of the HTR2A gene, however, the trend (without statistical significance) was the same as in the patient samples [8, 9].

In a study on college students using analysis of variance, it was found that the indicators of the big five personality parameters significantly differ in carriers of different genotypes of the 5-HT2A gene. Carriers of the CC genotype had lower indicators of emotional stability compared to those with the TC genotype, and carriers of the CC genotype had a greater openness to experience than people with the TT genotype. The indicator of openness to experience also significantly differed among carriers of different genotypes of the DRD4 gene [10].

This work aimed to study the association of the genotypes of the HTR2A and DRD4 genes with the emotional and personal characteristics of young men and women, representatives of the Russian ethnos, living in the South of Russia.

2 Materials and Methods

The sample consisted of 100 psychology students studying at the universities of Rostov-on-Don (68%—girls), average age—19.3 years, it was right-handed people without pronounced health problems. In our work, we carried out genotyping of study participants for the HTR2A and DRD4 genes.

On the basis of the laboratory «Biological Solutions and Technologies» (Russia, Moscow), DNA analysis, PCR for polymorphic loci: T102C (rs6313, Tr3) of the HTR2A gene was carried out. Frequency distribution of CC genotypes—34%; CT—50%; TT—16%. C-521 T of the DRD4 gene (rs1800955). TT genotypes—39%; CT—50%; CC—11%.

The paper hypothesized that the genotypes for the HTR2A and DRD4 genes can be associated with the level of emotional intelligence and personality traits of young people, such as empathy, alexithymia, and the Big five personality traits.

Research methods. Test (self-report questionnaire) of emotional intelligence [11]. EETS Emotional Empathy Questionnaire by A. Mehrabian, N. Epstein as adapted by Orlov and Emelyanov [12]. «Big five» personality test 5PFQ, adapted by Khromov [13]. Toronto Alexithymia Scale, TAS-26 by G. Taylor et al. in adaptation by Eresko et al. [14].

For statistical processing of the results obtained, multivariate analysis of variance ANOVA (Statistica 13.0; factors «HTR2A gene», «DRD4 gene», «sex») was applied with Tukey's post hoc analysis for non-equilibrium sample sizes.

3 Results

Study of the level of empathy in carriers of different genotypes of the HTR2A and DRD4 genes. There were no significant influence of the factors «HTR2A gene» (F = 2.8, MS = 53.4, SS = 106.8, $p = 0.07$), «DRD4 gene» (F = 0.05, MS = 1, 0, SS = 2.1, $p = 0.9$) and combinations «HTR2A gene * sex» (F = 1.2, MS = 23.6, SS = 47.2, $p = 0.3$), «Gene DRD4 * sex» (F = 1.8, MS = 8.3, SS = 16.5, $p = 0.7$) per empathy level.

According to the results of a posteriori analysis, women generally have a higher level of empathy. The heterozygous CT genotype of the HTR2A gene (M = 22.3, $p = 0.01$) is associated with a higher level of empathy in men, while carriers of the homozygous CC (M = 17.5, $p = 0.000$) and TT (M = 18.8, $p = 0.004$) genotypes have significantly lower values (Fig. 1).

According to the results of Tukey's post-hoc analysis, it was revealed that the highest level of empathy in the study sample is in women with the CT genotype of the DRD4 gene (M = 26.2, $p = 0.04$), and the lowest is in men with the same genotype (M = 20.4, $p = 0.04$) (Fig. 2).

Study of the level of emotional intelligence in carriers of different genotypes of the HTR2A and DRD4 genes. There were no significant differences in the severity of the general level of emotional intelligence among carriers of different genotypes of HTR2A (F = 0.2, MS = 15.0, SS = 30.0, $p = 0.8$), DRD4 (F = 0.9, MS = 75.3, SS = 150.5, $p = 0.4$).

Study of the level of alexithymia in carriers of different genotypes of HTR2A and DRD4 genes. There were no significant influence of factors «HTR2A gene» (F = 0.2, MS = 22.5, SS = 45.0, $p = 0.9$), «DRD4 gene» (F = 1.0, MS = 127, 6, SS = 255.2, $p = 0.4$) and combinations «HTR2A gene * sex» (F = 0.6, MS = 78.8, SS

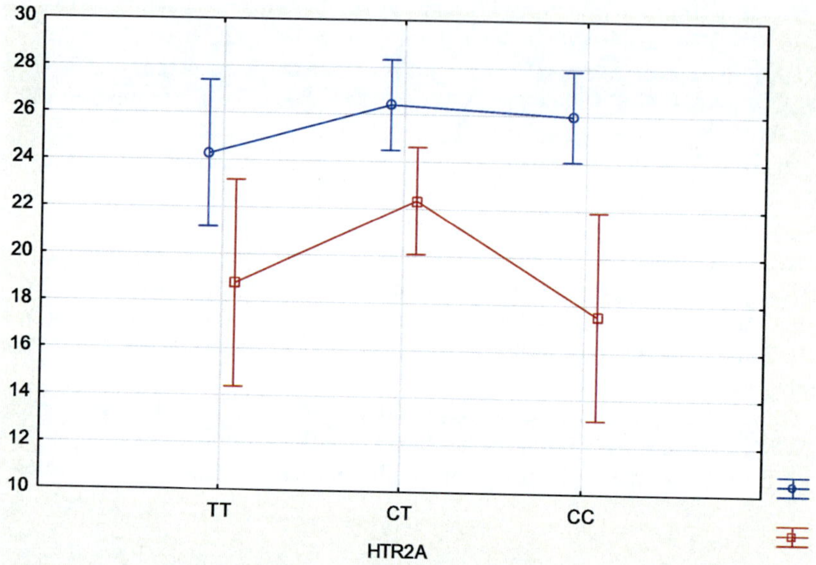

Fig. 1 Results of analysis of variance of the level of empathy (along the vertical axis) in carriers of different genotypes of the HTR2A gene, men (red line) and women (blue line) (ANOVA, $p \leq 0.05$) Designations: TT—genotype TT of the HTR2A gene; CC—genotype CC of the HTR2A gene; CT—heterozygous genotype CT of the HTR2A gene

= 157.5, $p = 0.6$), «Gene DRD4 * sex» (F = 1.8, MS = 229.4, SS = 458.9, $p = 0.2$) per level of alexithymia. According to the results of a posteriori analysis, it was revealed that women with the CC genotype of the DRD4 gene have a significantly lower level of alexithymia (M = 49.0, $p = 0.03$), compared with women with the TT genotype (M = 64.1, $p = 0.03$) (Fig. 3).

Study of personal characteristics of carriers of different genotypes of HTR2A and DRD4 genes. There were no significant influence of the factors «HTR2A gene» (F = 1.4, MS = 76.0, SS = 58.0, $p = 0.09$), «DRD4 gene» (F = 1.0, MS = 58, SS = 76, $p = 0.5$) and combinations «HTR2A gene * sex» (F = 1.0, MS = 76.0, SS = 58.0, $p = 0.5$), «DRD4 gene * sex» (F = 0.7, MS = 58, SS = 76, $p = 0.9$) for personality traits. According to the results of a posteriori analysis, the poles «Dominance» (M(CC) = 11.4, M(CT) = 9.9, $p = 0.04$), as well as «Respect for others» (M(CC) = 12.7, M(CT) = 11.4, $p = 0.04$) are more pronounced in carriers of the CC genotype, in comparison with carriers of the CT genotype of the HTR2A gene (Fig. 4).

The «Persistence» factor has significantly higher values in carriers of the TT genotype, in comparison with carriers of the CT genotype of the HTR2A gene (M(TT) = 20.3 M(CT) = 11.2, $p = 0.03$) (Fig. 4)

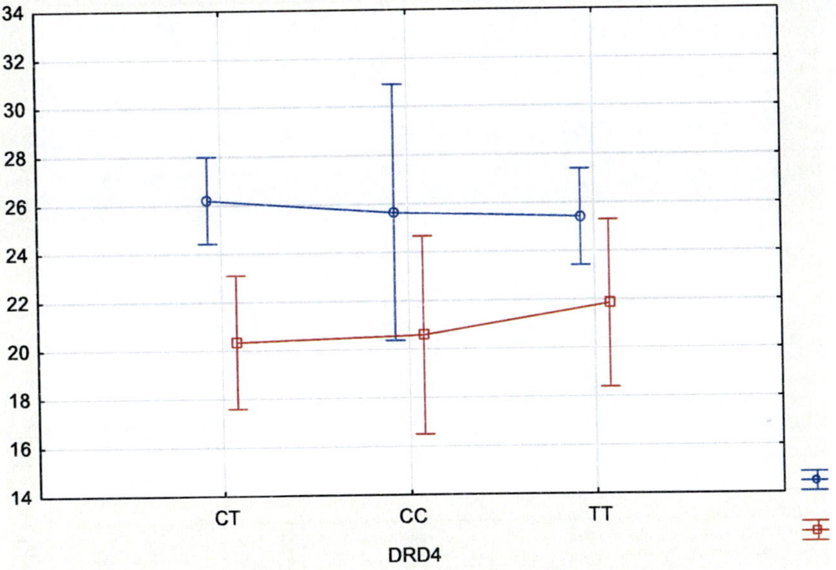

Fig. 2 Results of analysis of variance of the level of empathy (along the vertical axis) in carriers of different genotypes of the DRD4 gene, men (red line) and women (blue line) (ANOVA, $p \leq 0.05$) Designations: TT—genotype TT of the DRD4 gene; CC—genotype CC of the DRD4 gene; CT—heterozygous genotype CT of the DRD4 gene

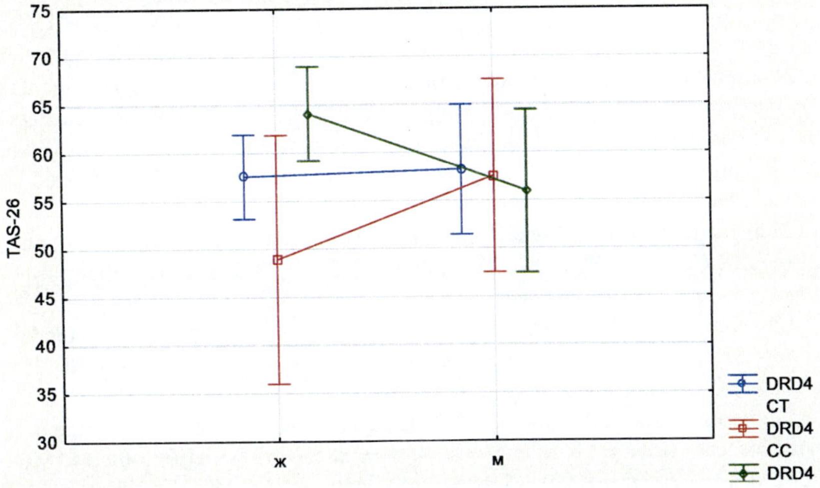

Fig. 3 Results of analysis of variance of alexithymia (along the vertical axis, TAS-26) in carriers of different genotypes of the DRD4 gene, men (m) and women (ж) (ANOVA, $p \leq 0.05$). Designations: blue line—CT, heterozygous genotype of the DRD4 gene; red line—CC, homozygous genotype of the DRD4 gene; green line—TT, homozygous genotype of the DRD4 gene

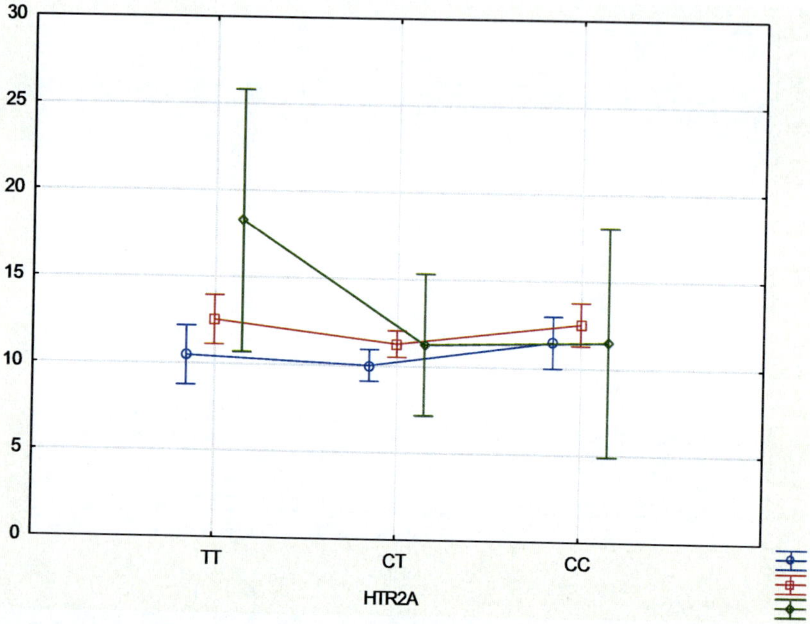

Fig. 4 Results of analysis of variance of the level of severity of various personality traits in carriers of different genotypes of the HTR2A gene (ANOVA, $p \leq 0.05$). Designations: blue line—pole «Domination»; red line—pole «Respect for others»; green line—pole «Perseverance». TT—genotype TT of the HTR2A gene; CC—genotype CC of the HTR2A gene; CT—heterozygous genotype CT of the HTR2A gene

Women carrying the TT genotype of the HTR2A gene have the highest scores on the Persistence scale in the study sample (M = 25.7, $p = 0.007$). Men, carriers of the CC genotype of the HTR2A gene, have the lowest sensitivity values (M = 7.5, $p = 0.02$) and the highest practicality indicators (M = 48.5, $p = 0.03$), compared with women, having more pronounced indicators of expressiveness.

Men in general have higher indicators of emotional stability, in particular—carriers of CC (M = 38.3, $p = 0.03$) and TT (M = 37.8, $p = 0.02$) homozygous genotypes of the HTR2A gene (Fig. 5).

Carriers of the CT genotype of the DRD4 gene have a more pronounced tendency to depression compared to carriers of the CC genotype (M(CC) = 8.0, M(CT) = 10.7, $p = 0.04$) (Fig. 6).

Women with the CT genotype of DRD4 gene have a higher score on the scales «Attachment» (Mw = 58.1, Mm = 49.4, $p = 0.049$), «Respect for others» (Mw = 12.4, Mm = 10.5, $p = 0.02$), «Self-control» (Mw = 57.7, Mm = 50.2, $p = 0.049$), «Sensitivity» (Mw = 12.6, Mm = 10.5, $p = 0.03$), compared to males carrying this genotype.

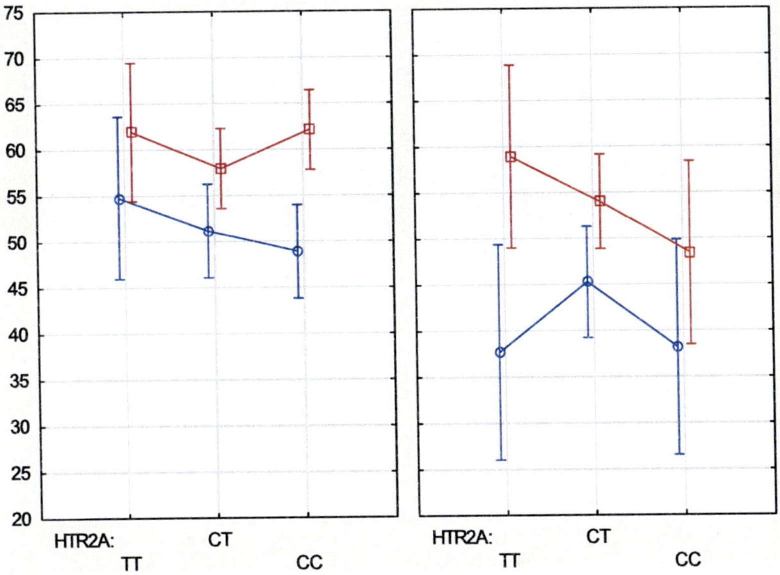

Fig. 5 Results of analysis of variance of the level of severity of various personality traits in carriers of different genotypes of the HTR2A gene, men (picture on the right) and women (picture on the left) (ANOVA, $p \leq 0.05$)
Designations: blue line—pole «Emotional stability», red line—pole «Expressiveness»
TT—genotype TT of the HTR2A gene; CC—genotype CC of the HTR2A gene; CT—heterozygous genotype CT of the HTR2A gene

Men with the TT genotype (M = 40.7, $p = 0.03$) have a significantly high level of emotional stability in the studied sample, and women with the CT genotype (M = 51.9, $p = 0.03$) of the DRD4 gene have emotional instability.

The highest level of expressiveness in the studied sample is observed in women with the CT genotype (M = 64.5, $p = 0.03$), the lowest—in men with the TT genotype (M = 52.0, $p = 0.03$) of the DRD4 gene.

According to the results obtained, women with CT (M = 11.2, $p = 0.003$) and TT genotypes (M = 9.9, $p = 0.04$) have a high level of anxiety in the sample under study, compared with men with the TT genotype (M = 7.0, $p = 0.003$) of the DRD4 gene.

The highest level of tension in the studied sample is observed in women with TT (M = 9.9, $p = 0.008$) and CT (M = 8.9, $p = 0.04$) genotypes, the lowest—in men with CC genotype (M = 5, 6, $p = 0.008$) of the DRD4 gene.

Women with the TT genotype (M = 11.8, $p = 0.04$) have a significantly higher level of emotional lability in the studied sample, the lowest—have men with the TT genotype (M = 8.7, $p = 0.04$) of the DRD4 gene.

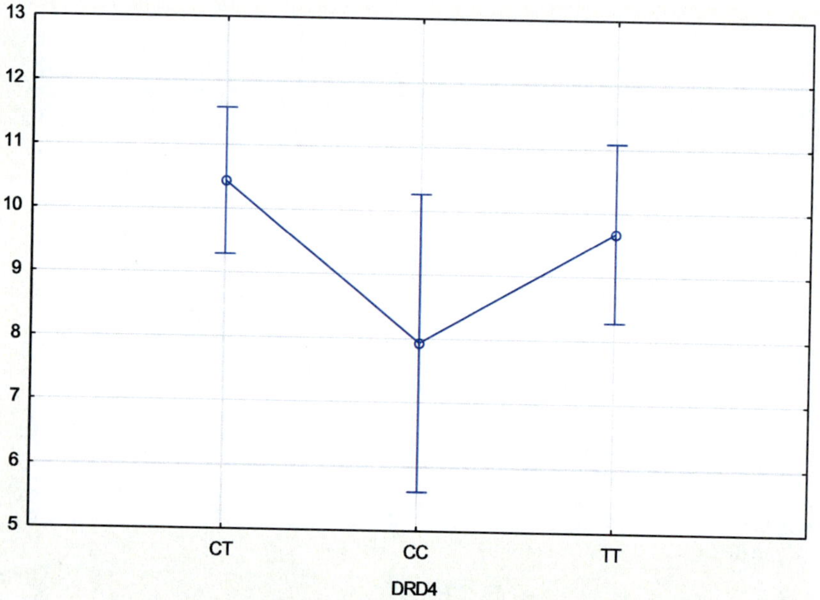

Fig. 6 Results of analysis of variance of the severity of depression (along the vertical axis) in carriers of different genotypes of the DRD4 gene, men (m) and women (ж) (ANOVA, $p \leq 0.05$) Designations: CT—heterozygous genotype of the DRD4 gene; CC—homozygous genotype of the DRD4 gene; TT—homozygous genotype of the DRD4 gene

4 Discussion

Based on the results of analysis of variance, no reliable relationship was found between the genotypes of the HTR2A and DRD4 genes with the general level of emotional intelligence, the level of alexithymia, empathy, and personality traits of Russian boys and girls. However, thanks to the use of Tukey's post-hoc analysis for non-equilibrium samples, we can describe the following features of carriers of different genotypes of the HTR2A and DRD4 genes.

HTR2A gene. Carriers of the CC genotype of the HTR2A gene (low density of type II serotonin receptors on the postsynaptic membrane of the neuron) showed a high level of propensity to dominate, and the pole of respect for other people is more pronounced.

Young men, carriers of the CC genotype of the HTR2A gene, have a low level of empathy and sensitivity, along with high emotional stability and practicality.

Carriers of the CT genotype of the HTR2A gene were found to have a low level of propensity to dominate, a low level of persistence, and a more pronounced pole of self-esteem.

Young men, carriers of the CT genotype of the HTR2A gene, have a higher level of empathy. Carriers of the TT genotype of the HTR2A gene (high density of type

II serotonin receptors on the postsynaptic membrane of the neuron) showed a high level of persistence, especially in women. Young men, carriers of the TT genotype of the HTR2A gene, have a low level of empathy along with high emotional stability.

In the works of other authors, it was previously shown that carriers of a larger number of C alleles of the HTR2A gene are more likely to be anxious when observing the pain of other people [15]. Data on the association of the HTR2A gene with the Toronto Alexithymia Scale (TAS-20) parameters were obtained [16].

DRD4 gene. Carriers of the CC genotype have a lower level of depression. Girls, carriers of this genotype, have a low level of alexithymia. Thus, a high level of expression of the D4 receptor is associated with the normative characteristics of the emotional sphere in young people [17, 18]. The low level of expression of the D4 receptor has different associations with the characteristics of the emotional and personal sphere in boys and girls.

Girls carrying one or two T alleles of the DRD4 gene have high affective tension (carriers of the TT genotype have a high level of alexithymia, emotional lability, anxiety and tension; carriers of the CT genotype have a high level of empathy, attachment, respect for others, self-control, sensitivity, emotional instability, expressiveness, anxiety, tension). Young men carrying one or two T alleles (CT, TT genotypes) have opposite characteristics. Carriers of the CT genotype, both boys and girls, have a higher level of depression.

The literature contains data confirming the association of the genotypes of the serotonin transporter gene (5-HTTLPR) and the severity of indicators according to the five-factor model of human personality, in particular, the indicator of openness to experience [19].

5 Conclusions

Based on the results of the study, the differences between Russian boys and girls, carriers of different genotypes of HTR2A and DRD4 genes, in empathy, alexithymia, and personality traits that are included in the Big Five factors of personality traits, are described.

This work has been supported by the grants of the Russian Science Foundation, RSF 20-64-47057.

References

1. Abakumova IV, Zorina E (2017) Sense-making techniques in educational process and their impact on the personal characteristics of students. Int J Cogn Res Sci Eng Educ 5(2):41–46. https://doi.org/10.5937/ijcrsee1702041A
2. Arias B, Aguilera M, Moya J, Sáiz PA, Villa H, Ibáñez MI, García-Portillo MP, Bobes J, Ortet G, Fañanás L (2012) The role of genetic variability in the SLC6A4, BDNF and GABRA6

genes in anxiety-related traits. Acta Psychiatr Scand 125(3):194–202. https://doi.org/10.1111/j.1600-0447.2011.01764.x
3. Kosonogov VV, Vorobyeva EV, Kovsh EM, Ermakov PN (2019) A review of neurophysiological and genetic correlates of emotional intelligence. (IJCRSEE) Int J Cogn Res Sci Eng Educ 7(1):137–142. https://doi.org/10.5937/ijcrsee1901137K
4. Kosonogov V, Titova A, Vorobyeva E (2015) Empathy, but not mimicry restriction, influences the recognition of change in emotional facial expressions. Q J Exper Psychol 68(10):2106–2115. https://doi.org/10.1080/17470218.2015.1009476
5. Vorobyeva EV, Ermakov PN, Borokhovski EF et al (2020) Association between categorization of emotionally-charged and neutral visual scenes and parameters of event-related potentials in carriers of different COMT, HTR2A, BDNF gene genotypes [version 2; peer review: 2 approved]. F1000Research 9:446. https://doi.org/10.12688/f1000research.22503.2
6. Parsons MJ, D'Souza UM, Arranz MJ, Kerwin RW, Makoff AJ (2004) The 1438A/G polymorphism in the 5-hydroxytryptamine type 2A receptor gene affects promoter activity. Biol Psychiatry 56:406–410
7. Polesskaya OO, Sokolov BP (2002) Differential expression of the 'C' and 'T' alleles of the 5-HT2A receptor gene in the temporal cortex of normal individuals and schizophrenics. J Neurosci Res 67:812–822
8. Spies M, Nasser A, Ozenne B et al (2020) Common HTR2A variants and 5-HTTLPR are not associated with human in vivo serotonin 2A receptor levels. Hum Brain Mapp 2020:1–11. https://doi.org/10.1002/hbm.25138
9. Walitza S, Bové DS, Romanos M et al (2012) Pilot study on HTR2A promoter polymorphism, – 1438G/A (rs6311) and a nearby copy number variation showed association with onset and severity in early onset obsessive-compulsive disorder. J Neural Transm (Vienna) 119(4):507—515. https://doi.org/10.1016/j.pnpbp.2007.12.005
10. Cam FS, Colakoglu M, Tok S, Tok I, Kutlu N, Berdeli A (2010) Personality traits and DRD4, DAT1, 5-HT2A gene polymorphisms in risky and non risky sports participation. Turkiye klinikleri tip bilimleri dergisi 30(5):1459–1464. https://doi.org/10.5336/medsci.2009-14839
11. Lyusin DV (2006) Novaya metodika dlya izmereniya emocional'nogo intellekta: oprosnik EmIn. Psihologicheskaya diagnostika 4:3–22
12. Soldatovoj GU, SHajgerovoj LA (2008) MPsihodiagnostika tolerantnosti lichnosti/Pod red. Smysl
13. Hromov AB (2000) Pyatifaktornyj oprosnik lichnosti. KGU, Kurgan
14. Eres'ko DB, Isurina GL, Kajdanovskaya EV, Karvasarskij BD, Karpova EB, Korepanova TG, Krylova GS, Tarhan AU, Chekhlatyj EI, Shifrin VB (2005) Aleksitimiya i metody eyo opredeleniya pri pogranichnyh psihosomaticheskih rasstrojstvah. SPb.: NIPNI im. Bekhtereva
15. Gong P, Liu J, Blue PR, Li S, Zhou X (2015) Serotonin receptor gene (HTR2A) T102C polymorphism modulates individuals' perspective taking ability and autistic-like traits. Front Human Neurosci 9(23):575. https://doi.org/10.3389/fnhum.2015.00575
16. Li X, He L, Liu J, Guo WX, Wang QH, Fang PP, Yang X, Zhang MF, Wang CL, Gong PY (2020) The rs6311 of serotonin receptor 2A (5-HT2A) gene is associated with alexithymia and mental health. J Affect Disord 272:277–282. https://doi.org/10.1016/j.jad.2020.03.153
17. Okuyama Y, Ishiguro H, Toru M, Arinami T (1999) A genetic polymorphism in the promoter region of DRD4 associated with expression and schizophrenia. Biochem Biophys Res Commun 258:292–295
18. Ronai Z, Barta C, Guttman A, Lakatos K, Gervai J, Staub M et al (2001) Genotyping the 521C/T functional polymorphism in the promoter region of dopamine D4 receptor (DRD4) gene. Electrophoresis 22:1102–1105
19. Rahman MS, Guban P, Wang M, Melas PA, Forsell Y, Lavebratt C (2017) The serotonin transporter promoter variant (5-HTTLPR) and childhood adversity are associated with the personality trait openness to experience. Psychiatry Res 257:322–326. https://doi.org/10.1016/j.psychres.2017.07.071

Psychological and Pedagogical Aspects of Distance Learning

Svetlana Fedorova, Olga Petukhova, and Tatyana Petrova

Abstract The article gives the psychological and pedagogical characteristics of distance learning, reveals its advantages and disadvantages from the point of view of the educational process subjects. The research was carried out at Mari State university with 2–4th year students of intramural and extramural forms of study. In total, 100 students of the psychological and pedagogical direction of training took part in the study. There were noted such problems of distance learning that worried respondents, such as insufficient technological preparation of teachers for the use of distance learning tools, insufficient control of the level of knowledge assimilation and development of skills, incomplete correspondence of distance learning programs to the educational needs of students, and low quality of educational material. From the psychological point of view, students are worried about the situation of uncertainty in the knowledge assessment system, the decrease in contacts with the teacher and fellow students, the unpreparedness of a number of teachers to use digital technologies in the educational process. It was emphasized that distance learning requires the formation of students' responsibility, dedication, independence, attention, memory, thinking, etc. It is concluded that distance learning is somewhat inferior in its psychological and pedagogical characteristics to blended learning, which is preferred by students. However, a serious analysis of a variety of experimental data on distance learning will increase its effectiveness and minimize negative risks.

Keywords Distance learning · Blended learning · University · Student · Information and educational environment · Digitalization of education

1 Introduction

The boom in distance learning, which occurred as a result of the COVID-19 epidemic spread, has also intensified scientific research in this direction in different countries. Of course, it cannot be claimed that until 2020 this problem has not been investigated in any way. National and foreign authors proved the main methodological issues

S. Fedorova (✉) · O. Petukhova · T. Petrova
Mari State University, Lenin sq., 1, Yoshkar-Ola 424001, Russia

A. Beskopylny and M. Shamtsyan (eds.), *XIV International Scientific Conference "INTERAGROMASH 2021"*, Lecture Notes in Networks and Systems 247,
https://doi.org/10.1007/978-3-030-80946-1_71

related to the organization of distance learning back in the nineteenth century since the time of "correspondent" and "remote" learning. Modern authors have analyzed the specifics and dysfunctions of distance learning [1], the possibilities of virtual reality technologies [2, 3], online learning in social networks, the use of chatbots, e-platforms [4], networks and mobile technologies in the educational process [5], the significance of digital technologies in vocational training [6], etc. The purpose of this study is to analyze the psychological and pedagogical aspects of distance learning, to identify its advantages and disadvantages from the point of view of the educational process subjects.

2 Materials and Methods

The aim of our research was to explore the positive experiences and difficulties of distance learning. The objectives of the study were to reveal the attitude of students to distance learning, to determine its advantages and disadvantages, and to formulate conclusions. The study used methods such as individual and group conversations, brainstorming, interviewing, mathematical processing and analysis of research results. The research was carried out at Mari State university with 2–4-year students of intramural and extramural forms of study. In total, 100 students of the psychological and pedagogical direction of training took part in the study.

A survey was conducted to assess the effectiveness of distance learning. The questionnaire included the following questions and tasks: What form of the educational process do you consider the most acceptable? Which of the distance learning methods did the teachers in your training use? Which of the individual forms of communication with teachers did you use? What are the most relevant criteria for distance learning? Rate how successful the distance learning process was for you, how easy it was to learn with this form. Evaluate the quality of the teaching materials (whether the teaching material is conveniently organized, meets the needs of students, etc.). Are you satisfied with the quality of the distance learning courses? Point out the main problems in the organization of distance learning. Point out the advantages and disadvantages of distance learning, etc.

3 Results

Let's analyze the data we received on some issues. So, when choosing the forms of education (see Fig. 1), students give preference to blended learning. In their opinion, it is the form that allows them to study at an individual pace, without isolating themselves from communication with fellow students and teachers. The students did not choose the distance form, although they noted that it is still attractive. It increases the level of independence in the study of various materials. It is more relaxed. During the distance examination period, there are no special experiences in contrast to the

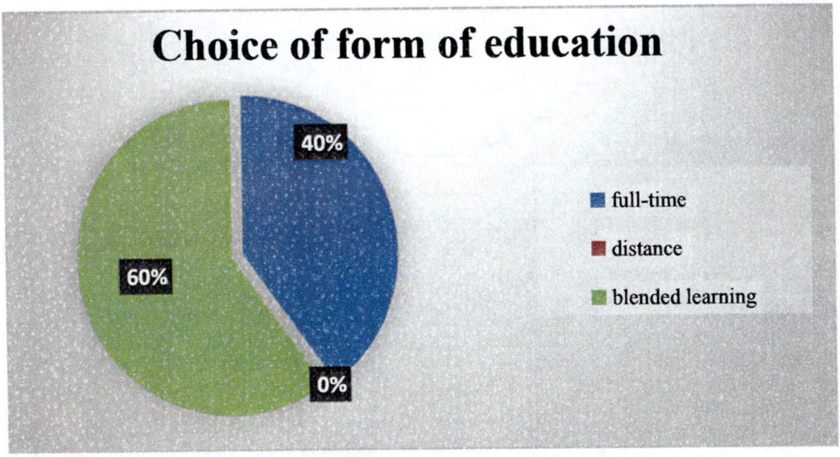

Fig. 1 The most acceptable form of the educational process for students

passing of exams/tests in full-time learning. There is minimal risk of virus infection. It is possible to refer to additional information at any time. Students who chose the full-time form of study justified their choice by the fact that it is better to study in the classroom, when you see and hear the teacher personally, the material is much better remembered and the results are higher.

Students noted that teachers use various means and methods of distance learning: electronic textbooks, computer presentations, distance learning resources, information retrieval systems, testing programs and interactive games, but in different percentages (see Fig. 2).

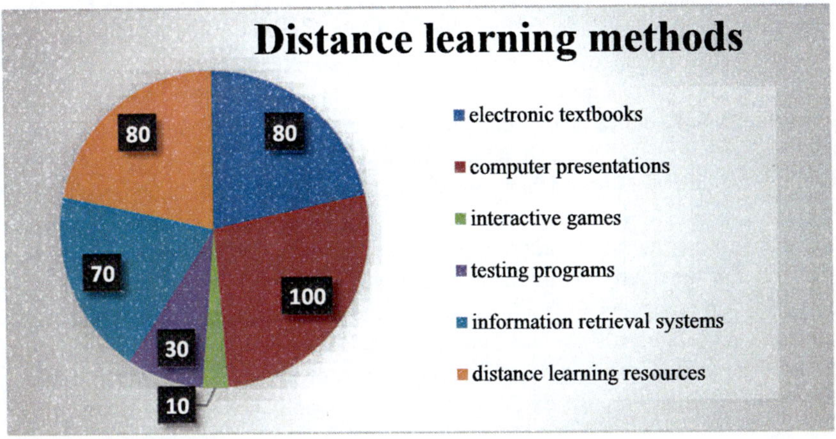

Fig. 2 Distance learning methods used by teachers in the learning process

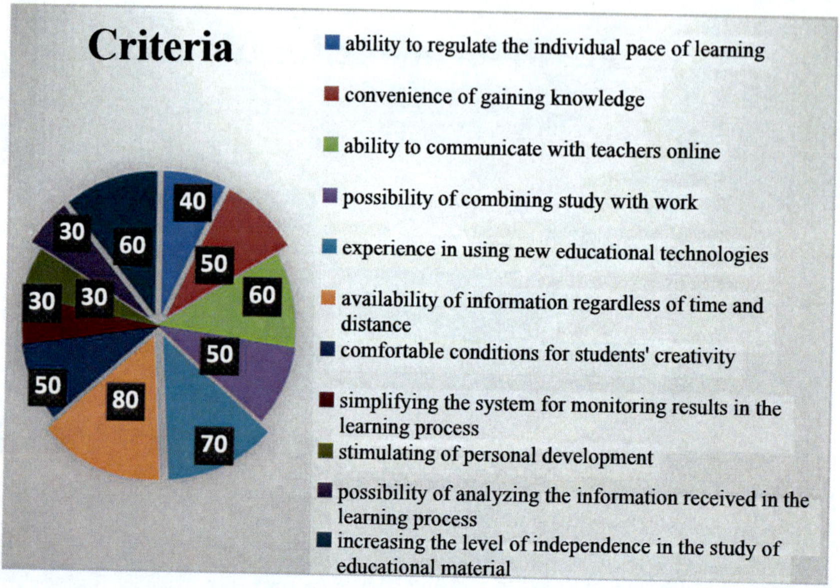

Fig. 3 Most relevant criteria for distance learning

Describing the most relevant criteria for distance learning, the respondents, first of all, note the availability of information regardless of time and distance, experience in using new educational technologies, the convenience of gaining knowledge, the ability to regulate the individual pace of learning, and the possibility of combining study with work. The possibility of analyzing the information received in the learning process, stimulating of personal development, simplifying the system for monitoring results in the learning process was also indicated. A number of students noted that it was much safer for them to study remotely knowing that the learning outcome depends only on them, and there is no social pressure (see Fig. 3).

As the most popular form of communication with teachers during distance learning, students called chat as the most efficient form of communication. Communication with the teacher on social networks and by e-mail was also actively used. Communication with the teacher via the group e-mail was practically not used (see Fig. 4).

Since distance communication is a fairly new form of interaction between teachers and students, the questionnaire included a question about the use of communication norms in communicative interaction. Some of the respondents noted that the teachers adhered to absolutely all communication norms. However, there were also comments on the clarity of the wording, the lack of punctuality of some teachers, and the literacy of speech. There were also cases of ill will and impoliteness among teachers (see Fig. 5).

In general, it must be said that the students are satisfied with the way the distance learning went. The advantages of distance learning were:

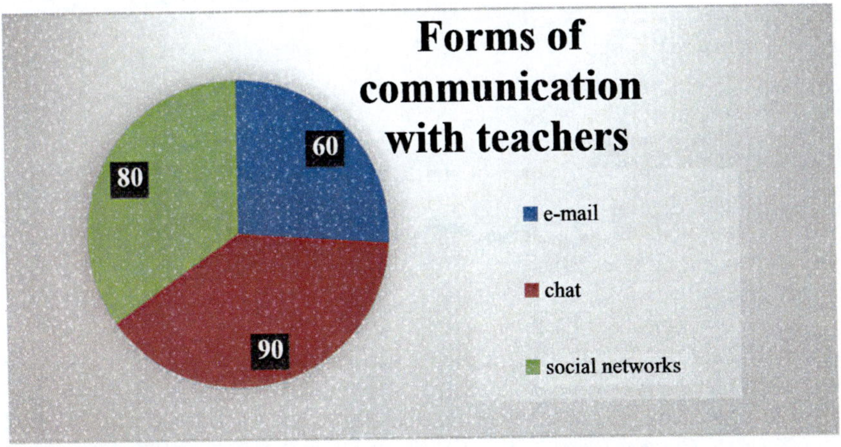

Fig. 4 Individual forms of communication with teachers used during distance learning

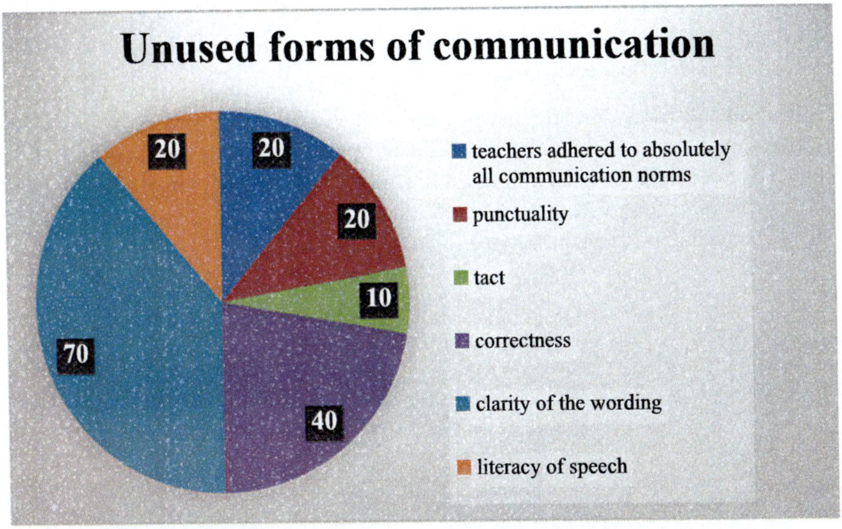

Fig. 5 Communication norms that were not used in the distance learning process

- the ability to master new technologies,
- convenience of time planning,
- an increase in the level of independence in the study of the material,
- the opportunity to study something else, get a second higher education, take online courses,
- comfortable conditions ("everything is close at hand: food, school supplies, Internet"),

- saving time on the road,
- saving material costs,
- the opportunity to study anywhere,
- no need to sit in cold classrooms.
- Although they noted some of its disadvantages:
- insufficient control of the level of knowledge acquisition,
- lack of live communication with teachers and fellow students,
- difficulties in using computer technology,
- untimely check of tasks,
- bad connection,
- the negative impact of the computer on vision,
- physical fatigue from daily sitting at the computer,
- lack of practical activity,
- misunderstanding of some information that requires clarification from the teachers ("the task was not always clear, we had to ask the teachers once again, and sometimes we could not get in touch"),
- incorrect wording of tasks,
- insufficient explanation of how to complete the task ("sometimes they just gave the task and did not explain anything, the students themselves had to figure out how to answer the questions posed").

These disadvantages made distance learning less effective, since it was difficult for them to get used to this form of educational activity. The level of control of the knowledge received by students did not satisfy them. At the same time, students noted that the educational material used during distance learning met all the needs and was conveniently organized (see Fig. 6).

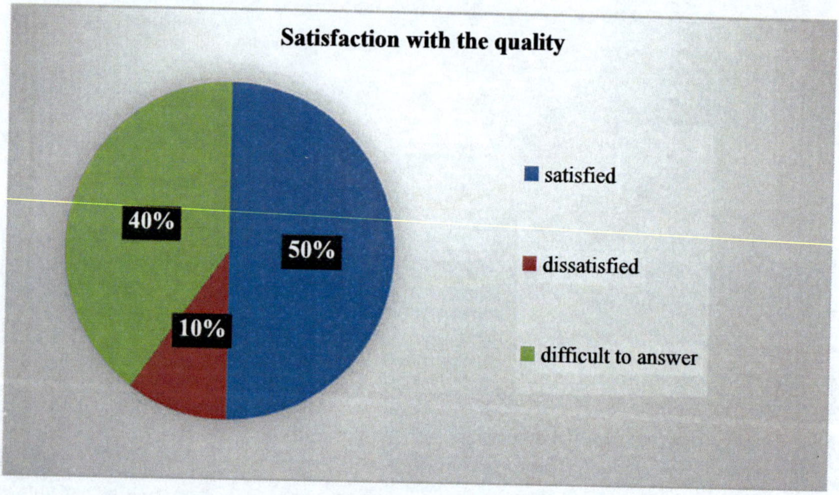

Fig. 6 Satisfaction with the quality of distance learning

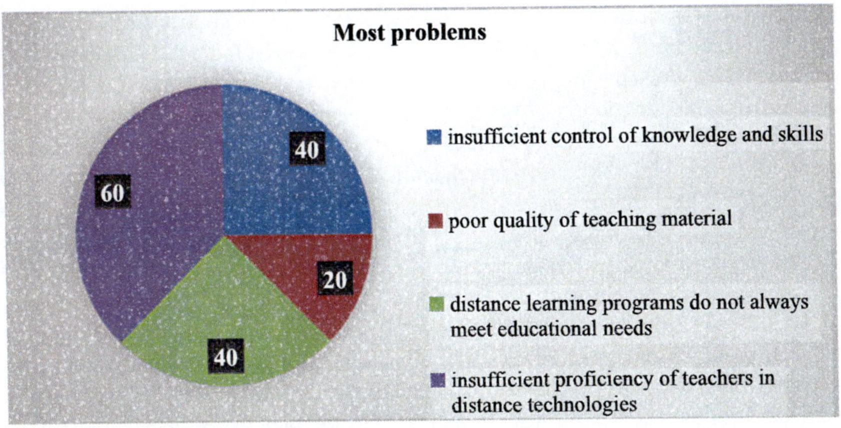

Fig. 7 The main problems in the organization of distance learning

Analyzing the answers of students about the main problems of distance learning, it can be noted that the majority of respondents associate them with insufficient technological preparation of teachers to use distance-learning tools. In the second place, there are the problems associated with insufficient control of the level of knowledge acquisition and development of skills, incomplete correspondence of distance learning programs to the educational needs of the respondents. A small number of respondents associate the difficulties of distance learning with the low quality of educational material. There were also answers related to excessive control and even a mark of daily presence from the dormitory administrations, which greatly bothered the students (see Fig. 7).

4 Discussion

The data we have obtained make us think about mechanisms for improving the quality of distance learning. Let us turn, first of all, to the pedagogical aspects. Let us start with the teachers, who have a lot of work to do in the distance learning format. It is one thing when a teacher works in a classroom with students in person, he has the opportunity to switch them to another form of activity at any time, conduct an oral survey in a short time, organize a discussion, "brainstorm", use a joke to improve the psychological atmosphere, etc. And it is completely another matter when he has to master a new pedagogical space—an information and educational environment characterized by intensive digitalization. In this environment, it is necessary to design a teaching technology, select and more often develop new didactic material for it, and not only theoretical and practical, but also entertaining (interactive games, quizzes, web quests) to keep students' attention and interest, organize productive educational and communicative interaction, use the best methods and techniques for managing the cognitive activity of students; to develop and test the forms of control over knowledge acquisition and the development of skills (with the latter it is more difficult in distance learning). This is just the procedure for preparing and conducting the lesson. And after it: checking the completion of tasks by each student (the "uploading" of attached answers takes a lot of time), comments, assessment, re-checking the materials of some careless students who may attach empty files, pictures or materials that do not correspond to the task (hoping that the teacher will not check but put a mark against his name for completing the task), etc.

The psychological aspect of distance learning also requires attention. Remote communication with students does not give the teacher the opportunity to implement fully the principle of individualization, which makes it difficult to choose the optimal style of pedagogical communication. Students are worried about the situation of uncertainty in the knowledge assessment system [1], the decrease in contacts with the teacher and fellow students [7], the lack of readiness of a number of teachers to use digital technologies in the educational process [8]. All this cannot but affect the specifics of the individual virtual educational environment, which is built by each student independently during distance learning and becomes a factor in his socialization, the formation of a new socio-cultural experience [9]. Distance learning requires students to form responsibility, dedication, independence, attention, memory, thinking, etc. Thus, distance learning is somewhat inferior in its psychological and pedagogical characteristics to blended learning, which is preferred by students. The main advantages of blended learning have been analyzed in the works of many authors [10–14]. An equally serious analysis of a variety of experimental data on distance learning will increase its effectiveness and minimize negative risks.

5 Conclusions

Most of the students preferred a blended form of education, no one chose exclusively distance learning. The main methods of distance learning that teachers used were presentations, electronic textbooks, information retrieval systems. Chat has become the most popular form of distance communication between students and teachers. The most important criteria for distance learning, students consider the availability of information regardless of time and distance, experience in using new educational technologies, the convenience of gaining knowledge, the ability to regu-late the individual pace of learning, and the possibility of combining study with work. Half of the students believe that the distance learning process was successful for them, less than half found it difficult to answer. The most difficult problem of distance learning, students believe that teachers do not adequately master distance learning technologies.

References

1. Moawad RA (2020) Online learning during the COVID-19 pandemic and academic stress in university students. Revista romaneasca pentru educatie multidimensionala 12(1):100–107
2. McGovern E, Moreira G, Luna-Nevarez C (2019) An application of virtual reality in education: Can this technology enhance the quality of students' learning experience? J Educ Bus 95(7):490–496
3. Nguyen T-L (2019) Enhancing students' learning motivation for better academic performance: an empirical case in Dong Nai Province, Vietnam. Adv Appl Sci 6(3):17–22
4. Veretekhina SV, Novikova VA (2019) Transformation of education in the digital economy. Contemp Probl Soc Work 5(2–18):30–37
5. Fedorova SN, Toktarova VI, Semenova DA (2019) Network and mobile technologies in the educational process of the Hel. In: Proceedings of INTCESS 2019—6th international conference on education and social sciences, pp 432–436
6. Eltemerov AA, Fedorova SN (2020) The use of digital technologies in the professional training of cadets. J Phys Conf Ser 1691:1–5
7. Strekalova NB (2019) Risks of introducing digital technologies into education. Bull Samara Univ History Pedagogy Philol 25(2):84–88
8. Frolova EV, Rogach OV, Ryabova TM (2020) Digitalization of education in modern scientific discourse: new trends and risks analysis. European J Contemp Educ 9(2):331–336
9. Nosov NA (2000) Virtual psychology. Agraf, Moscow
10. Twigg CA (2003) Improving learning and reducing costs: new models for online learning. EDUCAUSE Rev 38(5):29–38
11. Horn MB, Staker H (2015) Blended: using disruptive innovation to improve schools. Jossey-Bass, San Francisco, CA
12. Alammary A, Sheard J, Caebone A (2014) Blended learning in higher education: three different design approaches. Australas J Educ Technol 30(4):440–454
13. Krylova EA (2020) Blended learning technology in higher education. TSPU Bull 1(206):86–93
14. Yakovleva AG (2017) Organization of blended learning in teaching pedagogical disciplines. TSPU Bull 4(181):96–99

Information and Communication Technologies for Future Speech Therapy Teachers in the Professional Activities

Nikita Sadovoi, Tatiana Gordienko, and Nikolay Dontsov

Abstract The publication studies the relevance and importance of information and communication technologies studying for future speech therapy teachers in the professional activities, as well as the future prospects for this work area development. The article considers and analyzes a study that was conducted in the brands of conversations using Google forms with students of the first (36.7%), second (8.2%) and third (55.1%) courses of study of the State Budget Educational Institution of Higher Education of the Republic of Crimea Crimean Engineering and Pedagogical University the name of Fevzi Yakubov studying in the specialty Speech Therapy. The topic of the conversation is "The Prospects for Studying Information and Communication Technologies for the Future Speech Therapy Teachers". A quantitative and qualitative conversation analysis was carried out, on the basis of which the prospects for studying information and communication technologies for the future speech therapy teachers in their professional activities were determined. The study results showed that the future speech therapy teacher majority considers this area of work promising. As well as many students believe that for effective speech therapy activities it is necessary to introduce new innovative information and communication technologies, some students claim that they will take part in the research work in this direction. Accordingly, the conversation analysis showed an increased demand among future speech therapy teachers to the study and use of information and communication technologies in their professional activities, which suggests further research in this direction, and also determines trends in the field of education and speech therapy.

Keywords Speech therapist teacher · Information and communication technologies · Professional activity · Questionnaire survey · Conversation

N. Sadovoi · T. Gordienko
Crimean Engineering and Pedagogical University Named After Fevzi Yakubov, 8 Uchebny'j Lane, Simpheropol 295015, Russian Federation

N. Dontsov (✉)
Don State Technical University, Gagarina sq., 1, 344003 Rostov on Don, Russia

A. Beskopylny and M. Shamtsyan (eds.), *XIV International Scientific Conference "INTERAGROMASH 2021"*, Lecture Notes in Networks and Systems 247,
https://doi.org/10.1007/978-3-030-80946-1_72

1 Introduction

This study examines the future speech therapy teachers professional training features in the Borisova [1], Zodbayeva [2], Shklyar [3], Grigas [4], Jennifer [5], Johnson [6] works.

E. A. Borisova, B. D. Zodbaeva noted that the future teachers-speech therapists' professional training carried out taking into account their certain function development. The educational function reflects the provision by the speech therapist of purposeful interaction with the children with speech disorders and their immediate environment (family, relatives). The diagnostic function determines the proficiency degree in the diagnostic tools. The correctional function involves determining the main directions of correctional and developmental speech therapy, planning goals, tasks for the speech therapy activity implementation. The organizational function is focused on the creating the necessary comfortable conditions for the harmonious psychological and students' speech development [1, 2].

N. V. Shklyar states that the main condition for the professional competence of future speech therapy teachers' formation is their participation in educational and professional activities. Extracurricular activities are an additional opportunity to increase the future specialists' practical professional training effectiveness.

Thus, the author argues that the educational process together with extracurricular educational work should complement and strengthen each other, solving a common complex task—a highly qualified specialist training of [3].

B. Grigas notes that specialized training without teacher motivation is not effective, in the study, motivated behavior reflects the motivating factor result. These motivational factors can be divided into three main components: expectation, value, and affective components. The expectation component embodies the teacher's desire for success and the belief that he or she will be successful. This is also called teacher self-efficacy. A high level of self-efficacy leads to a higher level of planning and organization, a willingness to collaborate and experiment with new ideas, and greater persistence and resilience when problems arise. The value component is aimed at the goal perception by the teacher. If the goal is of a great importance or interest, the value increases. The affective component examines the emotional state of the teacher in relation to the goal, the joint team, and the organization as a whole.

Thus, in the training future specialists' process, it is necessary to form a motivational and value component and a belief in the professional abilities [4].

E.W. Jennifer writes in the research that the future teacher internal motivation is much more effective than the external one. It was found that students performed poorly in the work for which they were paid, highlighting the fact that paying for the work does not necessarily lead to a high quality product. It is noted that creativity is inhibited when a reward is offered for work, when external motivation is a priority in the school environment; the least creative ways of its implementation are chosen.

Thus, external rewards are not only ineffective, but can also negatively affect the work quality performed. In addition, external rewards tend to limit the focus of training on achieving a specific desired product. On the contrary, internal motivation is characterized by immersion in the task, getting pleasure from this task and active participation. When a student completes a task for internal reasons, such as self-interest, they usually complete the task, become more persistent, and put in more effort [5].

D. E. Johnson notes that speech therapy intervention should be comprehensive, so it is important for future speech therapy teachers to learn to work closely with all participants in the pedagogical process. Speech therapy teachers can assist general education teachers and other educators in changing the overall curriculum and curriculum in order to improve student performance in the classroom [6].

The information and communication technologies (ICT) essence in the teachers' efforts was studied in the works of Chernova [7], Gordienko [8], Shmakova [9].

E. V. Chernova in the work writes that information and communication technology is a set of techniques and software tools that are integrated with the aim of collecting, processing, storaging, distribution, display and use of information for its users with the respect to this research, for the therapy purpose [7].

T. P. Gordienko, A. P. Shmakova note that ICTs are a set of combined methods, software and hardware tools for working with information in order to reduce the complexity of the processes of using an information resource, as well as to increase their reliability and efficiency [8, 9].

Thus, it is necessary to determine the prospects and importance of studying information and communication technologies for the future speech therapy teachers in their professional activities.

2 Materials and Methods

The purpose of the work is to explore the prospects of studying information and communication technologies for the future speech therapy teachers in their professional activities.

Research objectives

1. To analyze the special psychological and pedagogical literature on the problem of research.
2. To conduct a conversation with the future speech therapy teachers on the topic "Prospects for studying information and communication technologies for the future speech therapy teachers in the professional activities", on the pre-planned issues.
3. To perform a quantitative and qualitative analysis of the study.

The research hypothesis. The study of the information and communication technologies for the future speech therapy teachers in the professional activities is a promising and important direction, due to the trends in the education development, which requires more detailed research.

Research methods. Various methods of study were used including both theoretical methods (analysis, comparison and systematization of psychological and pedagogical, methodological and special literature on the research problem) and empirical methods (conversation, qualitative and quantitative analysis of the results obtained).

We analyzed the works of various scientists, N. N. Bal, M. N. Gaidai, J. L. Colandrea, A. Velez, S. B. Ewell, V. J. Ward, in which we considered the role and importance of information and communication technologies in the speech therapist professional activity [10–15].

N. N. Bal notes that it is necessary to prepare the future speech therapy teachers for the informatization possibility of the almost any speech therapy work direction with the help of familiar and accessible modern technologies, as well as to form students' knowledge, skills and abilities in the use of special (speech therapy) software and hardware complexes [10].

M. N. Gaidai writes in the work that the traditional training speech therapy teachers system, which was characterized by determinism, uniformity, and a minimum number of innovations, today is unsatisfactory to the modern education needs. Future speech therapy teachers should master advanced innovations, including in the ICT use [11].

J. L. Colandrea reports in the study that teachers who worked in high-tech schools tended to report more extensive computer technology use in the lesson planning process, and also noted better computer technology compatibility with traditional teaching methods. The positive impact effectiveness on the students' technology computer use in such teachers was significantly higher than in other teachers. This discovery showed that school resources related to the technology level can also influence the computer technology use effectiveness in the education [12].

A. Velez notes that information and communication technologies and innovations will continue to evolve and become more complex. According to the forecasts, the world knowledge amount will increase rapidly. This "knowledge body" refers primarily to the ICTs integrated development. Therefore, it is extremely important that students master the skills that are necessary in the modern world for effective study adaptation and continuation, as well as for success in mastering higher education and further work [13].

S. B. Ewell notes that for the future teachers, practical work is extremely important, within the framework of an educational course, with the help of which they could consolidate their knowledge in practice. This study also points out that the teacher education is the teachers' professional training beginning and is not an end in itself; teacher education should promote genuine continuous professional training. It is necessary that teacher education programs do not ignore this professional training aspect, as teachers continue to grow and improve themselves throughout the careers, which implies constant knowledge replenishment and the innovations study [14].

V. J. Ward notes the communication skills importance for the teachers. The training quality is enhanced by the effective communication. The survey results in the V. J. Ward study showed that the first-year teachers demonstrate effective verbal and nonverbal communication skills.

The principals of those who participated in the study believe that the first-year teachers use technology and mass communication effectively. The appropriate implementation of the technology is an effective communication tool in education [15].

Thus, on the basis of the studied psychological, pedagogical and special literature, conversation questions were compiled, which determine the prospects for studying information and communication technologies for the future speech therapy teachers in the professional activities. Four questions were presented in the conversation (the fourth question is ranked). The questions were aimed at studying the students' attitude and opinion regarding the using and studying information and communication technologies prospects in the future professional activities. In the conversation, the future speech therapy teachers evaluated the most promising areas of studying the ICT use in the speech therapy field, and also answered questions related to the planning the research work in the field of innovative information application and implementation and communication technologies in the speech therapy field.

Further, answers' qualitative and quantitative analysis is given by the future speech therapy teachers during the conversation, and the results that determine the prospects for studying ICT for the future speech therapy teachers in their professional activities were summed up.

3 Results

The future speech therapy teachers training is carried out taking into account the educational, diagnostic, correctional and organizational functions development.

Based on the analyzed psychological, pedagogical and special literature, we note that the future speech therapy teachers training is carried out taking into account the educational, diagnostic, correctional and organizational function development. The future teacher-speech therapist readiness for the professional activity is the need to help children with the problems in speech development, the personal quality complex presence realizes the effective goal achievement of the correctional and pedagogical, speech therapy activities, the theory and methodology knowledge of correctional and developmental work with student speech disorders, modern technologies for correcting and compensating for dysontogenetic development, including the information and communication technologies using [2, 10].

Exploring the study features and information and communication technologies use by the future speech therapy teachers in the professional activities, it was noted that the education and the rapidly developing labor market informatization necessitate the specialist's ability to navigate freely in the information space. A future speech therapist should be able to use innovative technologies to access various information, recognize information, organize, process, evaluate, develop and distribute the information that is necessary to perform effectively their speech therapy duties in the information society.

The speech therapist teacher information and communication competence is the desire to master the information culture, the willingness and ability to independently use modern information and communication technologies in speech therapy to solve a wide range of speech therapy, correctional, developmental and educational tasks and to design the ways to improve skills in this field [1, 11].

Based on the analyzed literature sources, questions were compiled for the conversation, which were answered by the students, the future teachers-speech therapists, and then the study results were quantitatively and qualitatively analyzed. Accordingly, we will consider this analysis in more detail.

The first questionnaire question in the conversation was formulated as follows: "Do you think that the information and communication technology importance in the future speech therapy teachers training, as well as in the future professional activities, will increase?", there were three possible answers: "it will definitely increase", "it will remain in the same place", "it will decrease". The answers results to the first question among future speech therapy teachers of the 1–3 courses are shown in Fig. 1 (see Fig. 1).

From the survey results, it follows that the vast majority of 48 future speech therapy teachers (98%) believes that the ICT importance in the future speech therapy teachers training, as well as in the further professional activities, will definitely increase, this trend is also highlighted by some authors, pointing out that today there is an educational informatization. The future teacher-speech therapist must fully possess the ICT using knowledge and skills in the professional activities for the effective and qualified work [8, 11].

The second questionnaire question in the conversation was presented as follows: "Do you think that the new innovative information and communication technology

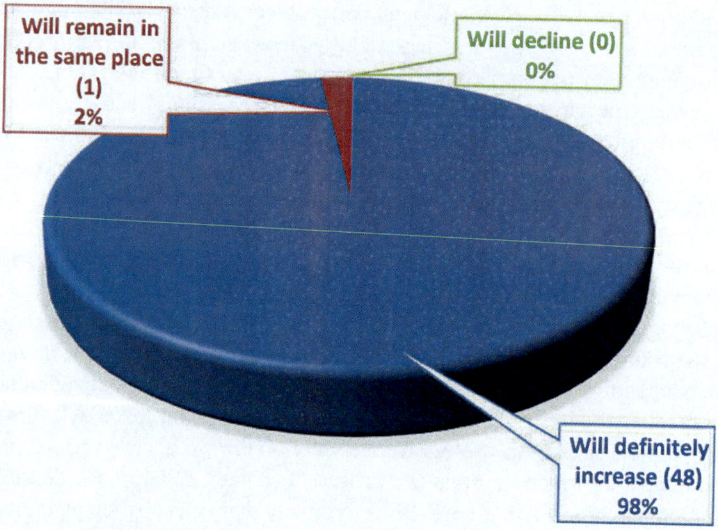

Fig. 1 Quantitative analysis of the first question' answers of the 1st–3rd year students

introduction is necessary for your future professional activity?" The answers and their quantitative indicator among the 1st–3rd year students are shown in Fig. 2 (see Fig. 2).

The answer results to the question indicate that the overwhelming majority of 43 future speech therapy teachers (87.8%) believe that there is a need to introduce innovative ICT in the speech therapy teachers' professional activities.

The technologies are being implemented in various areas of education, improving and optimizing the work. Working with the children with speech disorders requires additional visual, informational, game and interactive support, which is effectively provided by ICTs [11].

The future speech therapy teachers' responses result to the third question: "Do you plan to participate in the research work on the application and implementation of new innovative information and communication technologies in the speech therapy field?" are presented in Fig. 3 (see Fig. 3).

Thus, from Fig. 3, it follows that 24 future speech therapy teachers plan to participate sometimes in the research work on the ICT application and implementation in the speech therapy field, and 11 students said that they will definitely participate in the research work in this area.

The answers to the ranked (fourth) question are shown in Fig. 4. The participants had to evaluate the studying most promising areas of information and communication technologies in the speech therapy field. The following prospective direction levels were selected: a sufficiently promising direction—2 points, a satisfactorily promising direction—1 point, a non-promising direction—0 points (see Fig. 4).

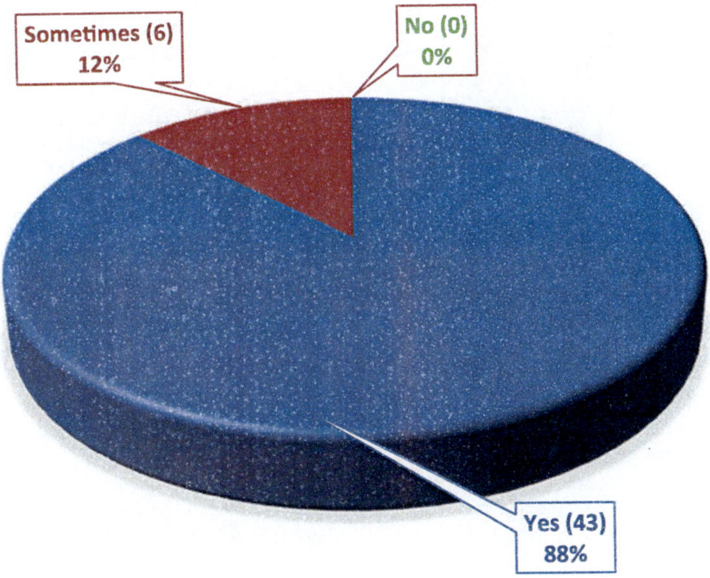

Fig. 2 The second question' answers analysis of the 1st–3rd year students

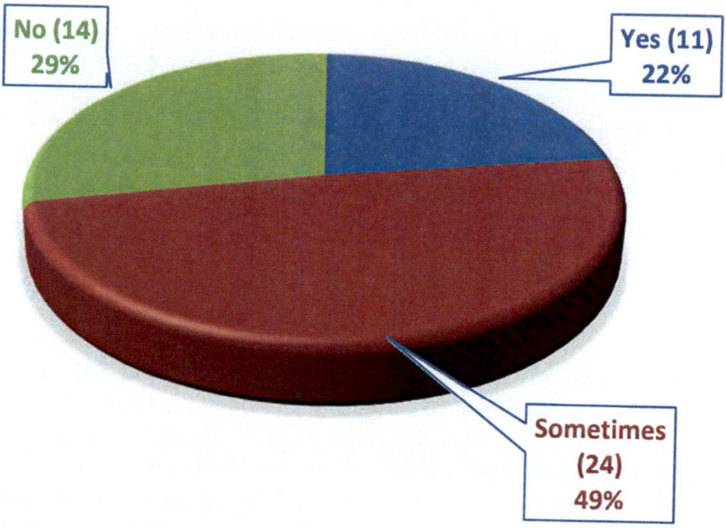

Fig. 3 The third question' answers analysis of the 1st–3rd year students

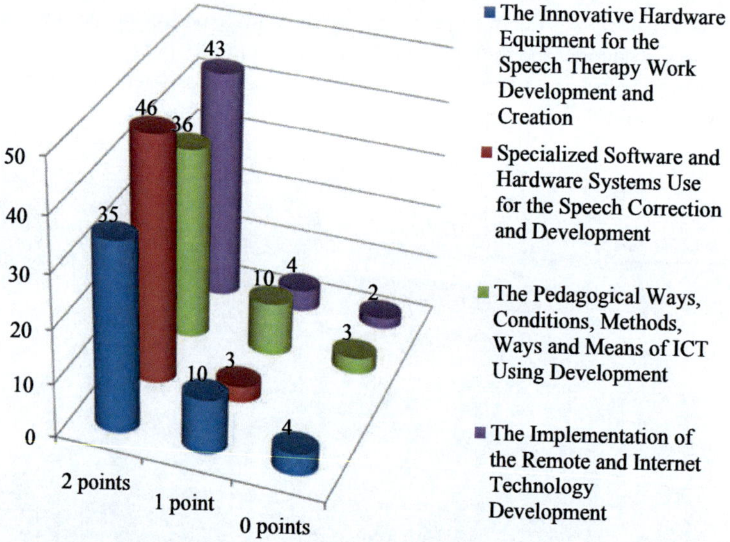

Fig. 4 The fourth question' answers analysis of the 1st–3rd year students

The data presented in Fig. 4 shows that future speech therapy teachers consider all the work areas presented in the questionnaire to be promising. However, they believe that the most promising direction is the software and hardware system use for the speech correction and development.

4 Discussion

The study results provided considering theoretical and experimental component prospects of the information and communication technology studying for the future speech therapy teachers in the professional activities.

It was revealed that the future speech therapy teachers training is represented by diverse work, which is aimed at the students' knowledge comprehensive formation, skills and abilities that contribute to the most effective and optimal performance of the official functions and duties in the future, using the necessary technologies, methods, tools and techniques of work that will increase the work effectiveness and optimize it.

One of the technology priority types used in the speech therapy is information and communication technologies, which are actively introduced into the educational system and modern civilization integral social and cultural aspects [9, 10].

A conversation conducted with students of the 1–3 courses of the specialty "Speech Therapy" showed that the students' majority consider the study and use of information and communication technologies as promising areas in speech therapy work, and also predict an unambiguous increase in educational and professional activities. In addition, the majority of future speech therapy teachers believe that the innovative information and communication technologies introduction will certainly have a positive impact on the speech therapy effectiveness. Some students are even going to carry out research work in this direction.

Also, according to the conversation results, future speech therapy teachers believe that all the directions presented in the conversation are important and necessary.

Accordingly, the opinions of the future speech therapy teachers and the analyzed theoretical calculations agree that the study and information and communication technologies use in the speech therapy field is a promising and important direction that will develop rapidly.

5 Conclusions

The conducted research confirms its relevance and importance in the direction of the information and communication technologies study and application in the speech therapist professional activity.

The various researchers' opinion unity and the students of the 1–3 courses, future teachers-speech therapists is that information and communication technologies, undoubtedly, a promising direction in speech therapy determines the direction of further research, characterized by relevance today. The largest number of students (46 people) believes that the most promising direction of ICT using in the speech therapy is the specialized software and hardware system use for the speech correction and development, which effectively complement and optimize the speech therapist work.

Considering the information and communication technology prospects studying for the future speech therapy teachers, taking into account the researchers' forecasts and the students' needs, we note that it is necessary to conduct research in the finding ways field, conditions, and further future speech therapy teacher method training to the ICT use in the professional activities, since the specialist qualification is significantly correlated with the future speech therapy teacher information and communication competence level. The trends show that this dependence will increase. Although at the present stage, the ICT development is a part of the mandatory training program for the future speech therapy teachers, improving and increasing the technology consumption in all human life require spheres of the further active research activities in this direction.

Considering the information and communication technology prospect studying for the future speech therapy teachers, taking into account the researchers and the students' forecast needs, we note that it is necessary to conduct the research in the finding way field, conditions, and methods for the further future speech therapy teacher training to ICT use in the professional activities, since the specialist qualification is significantly correlated with the future speech therapy teacher information and communication competence level. The trends show that this dependence will increase. Although at the present stage, the ICT development is the part of the mandatory training program for the future speech therapy teachers, improving and increasing the technology consumption in all human life spheres requires further active research activities in this direction.

References

1. Borisova EA (2013) Preparation of future speech therapy teachers for independent professional activity. Special Educ 4:80–86
2. Zodbaeva BD (2013) Formation of students-defectologists of practical readiness for interaction with preschool children with intellectual disabilities. School Speech Therapist 2(5):80–86
3. Shklyar NV (2013) Formation of professional competence of students-defectologists in the process of educational work. Pedagogy Psychol 1(12):289–292
4. Grigas B (2015) Collaboration begins in the classroom: future teachers discuss their knowledge, preparedness, and perceptions regarding speech-language concepts and collaboration with speech-language pathologists. Abstract of Ph.D. dissertation, California State University, Long Beach
5. Jennifer EW (2015) An investigation of struggling learners' motivation to read, abstract of Ph.D. dissertation, University of Louisiana at Lafayette, Louisiana

6. Johnson DE (2017) The perceived roles of speech-language pathologists, reading specialists, and general education teachers in the intervention of reading comprehension and decoding for elementary school students. Abstract of Ph.D. dissertation, Richard Stockton College of New Jersey, New Jersey
7. Chernova EV, Chusavitina GN (2019) Preparation of future teachers for the prevention of deviant behavior of schoolchildren in the field of information and communication technologies: monograph. Flint, Moscow
8. Gordienko TP (2019) The use of information and telecommunications technologies in the educational process of higher education: monograph. Humanitarian and Pedagogical Academy (branch) of the Crimean Federal University named after V. I. Vernadsky, Yalta
9. Shmakova AP (2019) Formation of the future teacher's readiness for pedagogical creativity by means of information technologies: monograph, 2nd edn. Ster. Flint, Moscow
10. Bal NN (2017) Improving the information literacy of speech therapy students in the process of studying the academic discipline "Information technologies as a means of speech therapy work". Special education: experience and prospects of development materials of the II Republican scientific and practical conference, pp 7–11
11. Gaidai MN (2012) Preparation of students-future speech therapists for the use of information technologies in correctional pedagogical work. Bull Univ Pedagogy Psychol Socio History Polit Sci 7:212–215
12. Colandrea JL (2012) The diffusion of computer-based technology in k–12 schools: teachers' perspectives. Abstract of Ph.D. dissertation, Graduate school of education of Fordham University, New York
13. Velez A (2012) Preparing students for the future—21st century skills. Abstract of Ph.D. dissertation, The University of Southern California, Los Angeles
14. Ewell SB (2009) Urban teacher preparation and retention: an exploration of experiences. Abstract of Ph.D. dissertation, University of North Carolina at Chapel Hill, Simply Carolina
15. Ward VJ (2015) A study of the perceptions of first-year teachers as prepared classroom teachers. Abstract of Ph.D. dissertation, Lindenwood University, St. Charles, Missouri

Peculiarities of Expressing Empathy Among Psychology Students and Law Students

Dina Volodenko, Yuri Bulygin, Liudmila Zaytseva, and Karina Tovstorebrova

Abstract Researchers approach of the concept of empathy the consideration of this concept from the point of view of the cognitive component, behavioral, activity, and affective. The ability to empathize is quite an important addition to the professional activities of many specialists. The authors conducted a detailed theoretical analysis of modern approaches to the description of empathy and identified three main approaches to its understanding. In this study, empathy is understood as the ability to consciously understand the emotional state and inner world of another person, during which the subject of empathy has a need to help the object of empathy, the desire to overcome his negative state. Using the test method "Emotional Response Scale", developed by A. Mehrabyan and adapted by N. Epstein, the level of empathy expression in students was determined. Based on the criteria of mathematical statistics, the Friedman chi-square and the Mann–Whitney U-test, we determined the leading content of empathy and the leading way of expressing empathy in groups of psychology and law students. This article examines the relationship between the chosen profession, personal characteristics and the content and method, as well as the subject of the expression of empathy. Different levels of expression of empathy towards different population groups were revealed. The results of the study can be used in the practice of improving the education system by relying on individual psychological characteristics, the development of which will contribute to increasing the empathic potential of the individual.

Keywords Empathy · Psychologists · Lawyers · Individual psychological characteristics · Multivariate questionnaire

D. Volodenko (✉) · Y. Bulygin
Don State Technical University, Gagarina sq., 1, 344003 Rostov-on-Don, Russia

L. Zaytseva · K. Tovstorebrova
Southern Federal University, B. Sadovaya str., 105/42, 344007 Rostov-on-Don, Russia

A. Beskopylny and M. Shamtsyan (eds.), *XIV International Scientific Conference "INTERAGROMASH 2021"*, Lecture Notes in Networks and Systems 247,
https://doi.org/10.1007/978-3-030-80946-1_73

1 Introduction

The success of a particular person in modern society increasingly depends on the level of development of his social and emotional qualities. Empathy is an important component of emotional intelligence. The ability to understand the essence of a person's experience, to feel his sorrows and joys, to penetrate into his inner world—all these abilities in psychology can be combined with the concept of "empathy". Good empathy is an important quality of an employee, whose activity requires understanding the mental state of another, and feeling into his inner world: this is required by professional activity in the field of psychology, art, medicine, law, journalism. However, the question remains whether empathy changes in the course of a person's professional life or whether people choose professions based on their empathic characteristics.

In modern psychological literature, the concept of "empathy" is ambiguous. This is due to the fact that a large number of clarifications are used for this concept: empathic influence, empathic reactions, empathic abilities, empathic relationships. And also, other concepts are used with similar meanings: empathy, sympathy, altruism, compassion, prosocial behavior, social sensitivity.

1.1 Theoretical Overview

The study of the empathy of specialists in socionomic professions is a relatively young trend in modern psychological research. It is already known that the ability to empathize is quite an important addition to professional activity. It is believed that psychoanalysts through empathy can get a more accurate idea of the mental and emotional state of the patient; empathy allows psychologists to correctly understand the problem states of the interlocutor; with the help of empathy, forensic scientists are able to find out some peculiarities of thoughts and feelings of victims and criminals; doctors, thanks to empathy, can understand and assess how ready the patient is for medical manipulations, how effectively the treatment is being carried out and, on this basis, make better interaction with the patient. In pedagogical activity, empathy helps to understand the child and accept him as he is; helps to sympathize and empathize with the child; promotes its recognition as an equal partner in educational interaction [1].

Thus, the recognition of the involvement of empathy in the work process of professionals makes it an important professional quality of psychologists, lawyers, doctors, and teachers. To study the issue of the initial presence and specifics of this quality in the personality of a future professional, it seems important to us to empirically study the qualitative features of empathy, to whom it is directed, and the way in which a person realizes his empathic potential in students of psychological and legal educational profiles [2].

In modern psychology, there are three main areas of consideration of empathy.

In the first approach, empathy is viewed as a mental process aimed at the perception of various open variables of the object of empathy, their understanding and construction of a plan to assist the object. The dynamic, procedural and phase nature of empathy is emphasized here. Representatives of this understanding of empathy G. Barrett-Lennard, V. Aykes in their studies distinguish three successive phases inherent in empathy as a mental process. The first phase is the perception and resonance of the listener, the stage of empathic understanding, during which the subject makes specific conclusions about the thoughts and feelings of the perceived person. The second phase is an expressive message about the onset of such a state, during which the subject expresses his judgments about the experiences of the other. The third phase is the stage of empathic communication, during which empathic understanding is tested and developed in a dialogical way of mutual understanding [3].

E. A. Ichalovskaya believes that the mechanism of the empathic process is the following successive sequences: (1) the perception of "open variables" of the object of empathy; (2) decoding of external signals; (3) attaching to the object of empathy and reading "hidden variables»; (4) decentration of the subject of empathy; (5) modeling of assistance to the object of empathy; (6) expression of assistance [4].

Domestic psychologists A. P. Sopikov and T. P. Gavrilov identified only two phases of the empathic process. During the first phase, various open variables of the object of empathy are perceived, information is obtained about the quality, sign and content of its experiences; and during the second—building in the internal plan a model of open and hidden activity of the object of empathy and correlating it with one's own values and needs [5].

In the second approach, empathy is viewed as a mental, empathic response to a stimulus. Empathy in this case is the emotional ability to respond to signals that convey the emotional experience of another person; or it is a behavioral ability that manifests itself in help, assistance, altruistic behavior in response to the experiences of another [6]. All the set of empathic reactions, scientists conditionally divided into two groups: empathic reactions in response to group behavior and empathic reactions to a certain person [4].

Empathy, according to E. P. Ilyin, this is an emotional characteristic of a personality that plays a significant role in relations between people, in their acceptance of each other, in the formation of internal unity between them [7].

In the view of T. P. Gavrilova, empathy is presented in the form of sympathy, which entails the actualization of the relationship of friendliness, honesty, guardianship, sincerity, help to the object of empathy [8].

A. A. Bodalev and T. R. Kashtanova consider empathy in connection with the mental and moral health of people, and believe that empathy is a truly positive attitude towards another person, which does not contain violence and includes love, trust, respect, can act as a moral quality of a person [9].

L. P. Vygovskaya in her research deepens the interpretation of empathy as a positive attitude towards another individual, the ability to reflect the moral level of a person. The specified researcher identifies five levels of manifestation of empathic attitude from indifferent to stable positive attitude towards another, including real help to him in problem situations [7].

So, an important point is the discovery of the addressee of empathy—a specific person or group of people: children, women, the elderly. In the context of this approach, it is important to study the content of empathy. I. M. Yusupov proposes to classify empathy based on the object to which it is directed—empathy with parents, animals, old people, children, heroes of works of art, strangers or unfamiliar people.

In the third approach, empathy is considered as a person's ability or property that has a complex affective-cognitive-behavioral nature. The study of empathy began with the study of the affective component. N. Eisenberg includes in empathy and experience, mediated by the open expression of emotions by another person, and the process of mental attempts to put oneself in the place of another [10]. The inclusion of a cognitive component involves accepting the role of another, interpreting behavior, and determining the emotional state. N. Feshbach calls the ability to recognize the emotions of another person and accept the point of view of another as a necessary but insufficient condition for empathy. At the cognitive level, empathy depends on the ability to recognize the emotional states of others and the ability to assume the role of another, that is, to see a situation in the same way as another person who is directly in this situation sees it. The third compulsory component is emotional response—experiencing the same emotions as the other person. All three components must be present in the empathic response [11].

M. Davis presented empathy as a construct that includes affective and cognitive components. The cognitive component is the ability to realize what perspectives exist for another individual and identification with him, while the emotional component is a predisposition to experience feelings of anxiety and sympathy for another individual [12].

V. V. Boyko considers empathy as a structural phenomenon and identifies 6 parameters: rational channel of empathy (RK), emotional channel of empathy (EC), intuitive channel of empathy (IC), attitudes that promote or hinder empathy (C), penetrating ability into empathy (PS), and the ability to identify (I). We propose to consider these channels of empathy as ways of expressing it [13].

In general, we consider empathy as the ability to consciously understand the emotional state and inner world of another person, during which the subject of empathy has a need to help the object of empathy, the desire to overcome his negative state [14]. We propose to conduct the study of empathy along the way of studying its content (to whom empathy is experienced) and methods (how exactly it is experienced) among students of psychological and legal profiles [15].

The aim of the study was to study the peculiarities of experiencing empathy by students with a psychological and legal profile of training in connection with their individual psychological characteristics. The subject of the research was the content and way of showing empathy among students.

The hypotheses of the part of the research presented in this review were the assumptions that: (1) the content and way of expressing empathy may differ among law students and psychology students; (2) the leading content and way of expressing empathy among psychology students are associated with their individual psychological characteristics; (3) the leading content and way of expressing empathy among law students are associated with their individual psychological characteristics.

2 Methodology and Methods

The purpose of study is expanding the understanding of the phenomenon of empathy. Research of empathy abilities of studying at different students of various humanities professions. The relationship of empathy with personality characteristics.

An empirical object of research. The study involved 58 people aged 19–29 years: of them 27 people—3–4 year undergraduate and graduate students of the Faculty of Psychology of the Southern Federal University of Rostov-on-Don; and 31 people—2–3 year undergraduate and graduate students of the Faculty of Law of the Rostov State Economic University (RINH) in Rostov-on-Don.

Research methods—testing, methods of mathematical statistics (correlation analysis, Friedman chi-square, Mann–Whitney U-test). The following were selected as specific methods: (1) test of the empathic potential of the personality of I. M. Yusupov; (2) the method of diagnosing the level of empathic abilities of V.V. Boyko; (3) a multivariate questionnaire by R.B. Cattell (16f, Form C); (4) questionnaire "The level of subjective control".

3 Results and Discussion

The leading content of empathy in the group of psychologists at the level of tendencies is empathy for children; at the same time, psychology students are not inclined to empathize with their parents and unfamiliar people. In the group of law students, the leading content of empathy (at the level of reliability ($P < 0.016$) is empathy for the elderly and parents. The data are presented in Table 1.

The leading way of expressing empathy among psychology students is the penetrating ability for empathy and attitudes, at the level of reliability ($P < 0.012$), for law students only attitudes ($P < 0.012$). The data are presented in Table 2.

The discovery of the connection between the leading content and the method of empathy with individual psychological characteristics of the personality became the next stage of the study, serving the confirmation or refutation of hypotheses 2 and 3.

Correlation analysis of the leading content of empathy was carried out according to the test "Empathic potential of personality" by I. M. Yusupov with the factors of the multi-factor questionnaire of R. B. Cattell's and the level of subjective control in the group of law students. As a result, significant positive correlations of empathy with parents were recorded with the scales of internality of failures "In" and internality of family life "Is" (with reliability $p \leq 0.05$, Table 3); empathy with elderly people

Table 1 Distribution of values of categories of empathy content

Group	Distributed categories	Sum Of Ranks	*P*—significance level (Friedman)	Conclusion on significance	Compared indicators	*P*—significance level (wilcoxon)	Conclusion on significance
Psychologists (n = 27)	With children	15.59	$P < 0.076$	Not significant	½	$P < 0.009$	Significant
	With kids	14.00			2/3	$P < 0.896$	Not significant
	With heroes	10.07			¾	$P < 0.061$	Not significant
	With the elderly	9.56			4/5	$P < 0.001$	Significant
	With parents	8.00			5/6	$P < 0.460$	Not significant
	Interpersonal	7.63					
Lawyers (n = 31)	With the elderly	21.87	$P < 0.016$	Significant	½	$P < 0.001$	Significant
	With parents	13.95			2/3	$P < 0.002$	Significant
	With heroes	13.89			¾	$P < 0.498$	Not significant
	With animals	13.52			4/5	$P < 0.006$	Significant
	With children	10.42			5/6	$P < 0.398$	Not significant
	Interpersonal	6.97					

Table 2 Distribution of values of categories of the way of expressing empathy

Group	Distributed categories	Sum of ranks	*P*—significance level (Friedman)	Conclusion on significance	Compared indicators	*P*—significance level (Wilcoxon)	Conclusion on significance
Psychologists (n = 27)	Penetrating ability	12.82	$P < 0.012$	Significant	½	$P < 0.839$	Not significant
	Settings	10.96			2/3	$P < 0.034$	Significant
	Intuitive	9.04			¾	$P < 0.619$	Not significant
	Rational	5.04			4/5	$P < 0.344$	Not significant
	Identification	3.19			5/6	$P < 0.922$	Not significant
	Emotional	2.82					
Lawyers (n = 31)	*Settings*	14.32	$P < 0.016$	Significant	½	$P < 0.017$	Significant
	Penetrating ability	13.03			2/3	$P < 0.858$	Not significant
	Intuitive	11.23			¾	$P < 0.634$	Not significant
	Emotional	11.23			4/5	$P < 0.783$	Not significant
	Identification	9.07			5/6	$P < 0.065$	Not significant
	Rational	7.52					

Table 3 Correlations between the leading content of empathy, factor "C" of the multivariate questionnaire RB Kettell and the level of subjective control in law students

		Level of subjective control		R. Cattell's multivariate questionnaire
		In	Is	C
Empathy content	With parents	0.359	0.356	
	With the elderly			0.463

Table 4 Correlations between the leading channel of empathy (VV Boyko), factor "Q1" of the Multivariate questionnaire RB Cattell, and the level of subjective control

		R. Cattell's multivariate questionnaire	Level of subjective control
		Q1	Ip
Empathic abilities	Y	0.374	0.395

with factor "C" of the multifactorial questionnaire of R. B. Cattell's (with reliability $p \leq 0.01$, sey Table 3).

Law students who have high levels of empathy with their parents tend to consider themselves responsible for the events that occur in their family life, they also prefer to blame themselves for various troubles and suffering. Low levels of empathy with parents are characteristic of law students who tend to consider their partners to be the cause of significant situations in their family. This category of law students is also distinguished by the fact that they prefer to attribute responsibility for unpleasant and painful events to other people, or they may consider such situations to be the result of bad luck.

Further, we carried out a correlation analysis of the level of expression of the leading method of empathy (according to the method of V.V. Boyko) with the factors of the multivariate questionnaire of R. B. Kettell and indicators of subjective control among law students. The data are presented in Table 4.

As a result, a significant positive relationship was found between the Attitude "Y" of empathy with the factor "Q1" and with Internality in the production sphere "Ip" (with reliability with $p \leq 0.05$).

Law students who have many attitudes that will promote or hinder empathy believe that they and their activities are an important factor in the organization of their own production activities, in their advancement in the career ladder. They also take responsibility for the emerging relationships in the team. In contrast, law students who lack attitudes that promote or hinder empathy will attribute important external circumstances in industrial relations, be it leadership, colleagues, or luck versus bad luck.

To study the connection between the content of empathy and the individual psychological characteristics of psychology students, we carried out a correlation analysis of the indicators of N. Mehrabyan's "Scale of emotional response" with 16

Table 5 Correlations between empathy content and level of subjective control

		Level of subjective control	
		General	In
Empathy content	With children	– 0.387	– 0.409

Table 6 Correlations between the level of empathic abilities and factors "E" of the multivariate questionnaire R. B Cattells

		R. Cattell's multivariate questionnaire
		E
Empathic abilities	PS	0.523

factors identified using a multifactorial questionnaire by R. B. Kettell and indicators of the level of subjective control among students—psychologists. As a result, significant negative relationships ($p \leq 0.05$) were recorded, indicating the presence of a stable relationship between the leading content of empathy for children and the Internality of the general "Io" and in relation to the failures of "In". The data obtained are presented in Table 5.

The data obtained in the course of correlation analysis can be interpreted as follows: psychology students with high levels of empathy with children showed a low level of subjective control. They consider important events in their lives, as well as sorrows and failures that occur on their way as the result of an accident or the actions of other people. They also believe that they are not able to control various events in their lives (Table 5).

We also carried out a correlation analysis of the level of empathic abilities (V.V. Boyko) and 16 factors identified using a multivariate questionnaire by R. B. Cattell's among students—psychologists. As a result, a significant positive relationship was recorded between the penetrating ability (PS)—with the "E" factor with a reliability of $p \leq 0.01$.

The penetrating ability of empathy will be better developed in disruptive, self-confident, independent psychology students who seek power and want to be admired. The penetrating ability in empathy will be less developed in soft, compliant, humble, dependent psychology students who are shy, respectful and willing to take the blame.

Thus, psychology students empathize with children, mainly using the method of penetrating empathy or empathic attitude, while law students empathize mainly with the elderly and parents, while having the leading way of expressing empathy is attitude. At the same time, the leading content of empathy for law students—empathy for parents correlates with internality in the field of failures and internality in the family sphere, the leading content of empathy for the elderly—with factor C "Emotional instability-stability»; while the leading way of expressing empathy "Attitude" is associated with the factor Q1 "Conservatism-radicalism" and the internality of professional life. The leading (at the level of tendency) content of empathy of psychology students "Empathy for children" negatively correlates with the general internality and internality in the sphere of failures; the leading way of expressing empathy "Penetrating ability" is associated with factor E "Subordination-dominance".

4 Conclusion

A comparative qualitative analysis of the statistical data obtained made it possible to find that the important features of their empathy: if the leading content of psychologists is associated with the trait of externality, then for lawyers it is internality. A reliable pattern is found in relation to the subjective control of failures.

As a result of the study, we have fully confirmed hypothesis 1: we found differences in the leading content and ways of expressing empathy by psychology students and law students; and hypothesis 2: significant correlations were found between the leading content and the ways of expressing empathy among law students. Hypothesis 3 was partially confirmed: among students of psychology, the content of empathy was not associated with the factors of the multivariate questionnaire of R. B. Cattell's.

References

1. Fang J, Wang X, Yuan K-H, Wen Z, Yu X, Zhang G (2020) Callous-unemotional traits and cyberbullying perpetration: the mediating role of moral disengagement and the moderating role of empathy. Personality Individ Differ 157:109829. https://doi.org/10.1016/j.paid.2020.109829
2. Kovalenko RK., Zvonareva NA (2020) A study on the relationship between the level of emotional intelligence and a socionic type. World Sci Pedagogy Psychol 3(8) (in Russian)
3. Basova AG (2013) Formation of empathy. Young Sci 5:631–633
4. Brazhnikova AN (2012) Empathy as a moral quality of a future professional. Psychol Sci Educ 4(4) (in Russian)
5. Buravtsova NV (2009) The relationship of the psychological space of the individual and the empathy of students-teachers and psychologists. Kazan Pedagogical J 9–10:89–97
6. Rozhkov OA (2012) Interrelation of professional deformation and motivation of the personality of employees of the penal system. Prevent Professional Deform 6:61–72
7. Gordeeva MA (2014) Emotional burnout of civil servants. Theor Pract Soc Dev 9:43–45. https://doi.org/10.17759/psylaw.2019090203
8. Zaitseva LA (2018) Features of career choice by high school students, university graduates and subjects receiving second higher education. Bull Univ 9:164–170
9. Kosonogov V, Titova A, Vorobyeva E (2015) Empathy, but not mimicry restriction, influences the recognition of change in emotional facial expressions. Q J Experi Psychol 68(10):2106–2115. https://doi.org/10.1080/17470218.2015.1009476
10. Pulido F, Herrera F (2016) La inteligencia emocional como predictora del rendimiento académico infantil: el contexto pluricultural de Ceuta. Revista Complutense de Educación 28:1251–1265
11. Franco MG, Candeias AA, Beja MJ (2011) Social and emotional intelligence in children with special educational need institutionalized. Int J Dev Educ Psychol 4(1):277–282
12. Billings C, Downey LA, Lomas JE, Lloyd J, Stough C (2014) Emotional intelligence and scholastic achievement in pre-adolescent children. Personality Individ Differ 65:14–18

13. Azpiazu L, Esnaola I, Sarasa M (2015) Predictive capacity of social support on emotional intelligence in adolescence. European J Educ Psychol 8(1):23–29. https://doi.org/10.14482/psdc.37.1.155.71
14. Ruslyakova EE, Timofeeva EA, Shestakova YV, Chepaykin DA (2020) Study of the level of empathy in adolescents who are keen on computer technology. World Sci Pedagogy Psychol 6(8)
15. Gutiérrez M, Expósito J (2015) Autoconcepto, dificultades interpersonales, habilidades sociales y conductas asertivas en adolescents. Revista Española de Orientación y Psicopedagogía 26(2):42–58. https://doi.org/10.5944/reop.vol.26.num.2.2015.15215

Formation of Inclusive Competence of the Medicine Institute Graduate

Natalia Tokareva

Abstract This paper analyzes the role of the developed adaptation discipline "Psychology of personality and professional self-determination" in the formation of inclusive competence in the educational process of a graduate of the Medical Institute. The goal of the discipline Personality of psychology and professional self-determination: mastering of psychological knowledge about the main directions of development psychology, general, individual and age peculiarities of the human psyche, associated with the principles of the organization of processes of training, education, self-education, and personal development needed in future professional activity of the doctor. The results of the study of the discipline: forming an integrated understanding of the mechanisms of functioning human psyche, individual psychological characteristics of personality, social and pedagogical aspects of interpersonal interaction; developing students professionally significant abilities and personal qualities; forming a unit of knowledge students about the inner world and human behavior, teaching students to use this knowledge in professional practice "for the benefit of the patient"; teaching students the techniques and methods of perfecting their own personal and cognitive spheres, the motivation for personal and professional growth; developing skills to work with scientific literature; forming skills to apply the obtained knowledge in clinical practice of the doctor.

Keywords Professional self-determination · Adaptation discipline · Inclusive competence

1 Introduction

The profession of a doctor makes high demands on the personality of a specialist, his professional level. In this connection, the formation of universal competencies of a graduate of the Medical Institute is of particular relevance. A special role is given to inclusive competence. Professional readiness and training of pedagogical staff for

N. Tokareva (✉)
Medicine Institute, National Research Ogarev Mordovia State University, Saransk, Russian Federation

A. Beskopylny and M. Shamtsyan (eds.), *XIV International Scientific Conference "INTERAGROMASH 2021"*, Lecture Notes in Networks and Systems 247,
https://doi.org/10.1007/978-3-030-80946-1_74

work in the conditions of inclusive education was considered in works of Alekhina [1]. Professional competence of teachers is disclosed in works of Tammets and Pata [2], Caena [3], Liakopoulou [4], works of Russian scientists Arpentyeva [5], Ibragimov and Kamasheva [6]. Educational process stimulates flexibility and operativeness of clinic thinking of doctors, enriches the principles of complex psychotherapy, allowing to solve the most actual issues of intensive treatment and prophylaxis in the form of individual or collective psychotherapy.

Matters of inclusive education are taken up in works of Cronis and Ellis [7], Loreman et al. [8], Sharma et al. [9]. Formation of inclusive competence is considered in works of Brandon and Charlton [10], Akhmetzyanova [11], Borodina [12].

The patients, people with disabilities for whom significant social interests are important, actively interact with a doctor in the process of treatment. When they are absent, a doctor faces pedagogical and psychological problems along with medical ones. According to Knapp et al. [13], Tokareva [14, 15], the most important prerequisite of the successful treatment is working out the conscious motivation to active overcoming the disease.

The basis of good resistance to the disease is not peculiarities of the organism itself, the nervous system, but their concrete embodiment in the requirements inherent to the given personality, needs, strivings, and in the activity inspired by them. Active life position and naturally, intensive counteraction to the disease, as a rule, are based on the consideration of the society's interests.

2 Material and Methods

On the basis of the Medical Institute of National Research Ogarev Mordovia State University (Saransk, Russian Federation) developed a program of adaptation discipline "Psychology of personality and professional self-determination" for students of specialties "General Medicine", "Pediatrics", "Dentistry", "Pharmacy" in accordance with the Federal State Educational Standard (dated 12.08.2020).

The discipline "Psychology of personality and professional self-determination" refers to the variational part of the educational program. Discipline "Psychology of personality and professional self-determination" is studied at the Medical Institute of National Research Ogarev Mordovia State University in the C semesters. Study of this discipline is aimed at developing of the following universal (UC) competencies in students. Universal competences: The competence of the UC-9: Ability to use basic defectological knowledge in social and professional spheres. Indicator of achievement of universal competence: Knows the concept of inclusive competence, its components and structure; features of the application of basic defectological knowledge in the social and professional spheres (UC-9.1); Knows how to plan and carry out professional activities with people with disabilities (UC-9.2); Powers the skills of interaction in social and professional spheres with persons with disabilities and people with disabilities (UC-9.3).

3 Results

The discipline "Psychology of personality and professional self-determination" refers to the variational part of the educational program. The discipline is interrelated with such disciplines as: philosophy (the fundamental question of philosophy; the laws of dialectics; the basic philosophical categories; the theory of reflection; the theory of knowledge; the role of the sensual-subject activity (labor) in the origin of man and his consciousness; social consciousness; practice as criterion of truth; social nature of personality; methods of scientific knowledge; the problem of man in philosophy, etc.), history (principle of historicism; historical nature of social phenomena; forms of consciousness in different historical epochs).

The antecedent for studying the disciplines is: public health and health care; psychiatry, medical psychology; neurology, medical genetics, neurosurgery, hospital therapy. The planned results of mastering the discipline "Psychology of personality and professional self-determination" are presented in Table 1.

Educational Technologies Used in the Study of the "Psychology of personality and professional self-determination" discipline: the training process uses multimedia lectures and seminars, case studies, psychological testing of mental processes and states, psychological discussions.

The students with special needs of Medical Institute of the National Research Ogarev Mordovia State University are provided with the following opportunities for their access to the higher education:

Table 1 The student's competencies formed as a result of mastering the "Psychology of personality and professional self-determination" discipline

Competence code	Competence	Components of the competence ("to Know—to Be Able to—to Master" requirements for the students)
UC-9	Ability to use basic defectological knowledge in social and professional spheres	To know: • The concept of inclusive competence, its components and structure; features of the application of basic defectological knowledge in social and professional spheres (UC-9)—I To be able to: • To plan and carry out professional activities with persons with disabilities and people with disabilities (UC-9)—I To master: • Skills of interaction in social and professional spheres with persons with disabilities and people with disabilities (UC-9)—I

Table 2 The topics lectures of the "Psychology of personality and professional self-determination" discipline (in hours)

№	The theme of the lecture	Semesters
		C
1.	Individual, individuality, personality	C
2.	Character	C
3.	Personality theory	C
4.	Total hours during the academic year	6 h

- The students with special needs and impaired vision have an excess to the version of the official website of the Institute for the visually impaired individuals; a free access is given to the support information on the curriculum presented in the adapted form (data on the support board is printed with large relief-contrast font (on white or yellow background) and duplicated in Braille); presence of an assistant who provides the student the needed support; Issue of alternative formats of printed materials (in large font or in audio files); Using computer hardware and software aiding users with the impaired vision, which converts computer information accessible to blind and visually impaired individuals (sound reproduction, Braille- or enlarged text), and enables working independently using a standard personal computer with a general-purpose software; access of the visually impaired students with a guide dog to the facilities of the Institute;
- Students with special needs in hearing: providing with adequate sound reproducing devices;
- For students with special needs, who have disorders of the musculoskeletal system, access to the lecture and classrooms, dining rooms, restrooms, as well as staying them in these areas (ramps, handrails, expanded doorways, locally lowered barriers-racks) are provided.

The topics lectures of the "Psychology of personality and professional self-determination" discipline is presented in Table 2.

The topics Practical classes of the "Psychology of personality and professional self-determination" discipline is presented in Table 3.

The content of the "Psychology of personality and professional self-determination" discipline is presented in Table 4.

4 Discussion

The study of the adaptation discipline the "Psychology of personality and professional self-determination" form universal competencies aimed at professional activities with persons with disabilities.

The doctor as a pedagogue when communicating with a patient realizes the principle "recovery through consciousness" more fully realizes and the known to many clinists conception, that a patient should be not only treated of but also educated

Table 3 Practical classes

The number of the section	The theme of the practical class (seminar)	Workload (h)
1	Individual, individuality, personality, the subject of activity	2
1	Methodological problems of personality psychology	2
1	Personality development	2
1	Needs-motivational sphere of the individual	2
1	Emotionally strong-willed personality sphere	2
1	Temperament: the physiological bases and psychological characteristics	2
1	Character and its formation	2
1	Psychology of abilities	2
1	The essence of professional self identity	4
2	Approaches to understanding personality	7
2	Personality theory	8

and brought up. To teach means to add knowledge's to a man which deliver him from great suffering high level of anxieties and a sense of fear. Upbringing envisages repeated, patient, friendly repeating a number of conceptions with the help of which patients get rid of weaknesses mistakes, harmful habits prejustices, tendentionsness and superstitions.

People's culture is formed and achievers great heights only when the following 4 aspects are united: upbringing, education, knowledge and constant personality's perfection. Upbringing relies on the succession of human experience which is developing and perfecting from generation to generation. A doctor dealing with medical pedagogics, possessing psychological knowledge continuously uses the collective experience both in his profession and in the sphere of a adjacent science.

5 Conclusions

1. The development of professionally significant abilities and personal qualities of medical students plays an important role in mastering the competencies of the doctor's profession.
2. Teaching students techniques and methods to improving their own personal sphere, motivation at personal and professional growth are the key to successful professional activity.
3. Mastering the Adaptation discipline "Psychology of personality and professional self-determination" contributes to the formation of a holistic view of the

Table 4 Content of the "Psychology of personality and professional self-determination" discipline

No.	Section name of the discipline	Section content
1.	Individual, individuality, personality, the subject of activity	The notion of the individual. The individual as a representative of the species and the product of biological evolution. Development of phenotypic features of individual in ontogenesis. Individual as a product of historical-evolutionary and ontogenetic development. Self-development of personality. The concept of individuality. Methods and methodology for personality psychology. Personality as a developing system. Principles of the study of personality. Methods of studying personality: observation, interviewing, standardized tests, experiment, projective methods Age and psychological features of the development of the individual. Maturation and development. The problem of periodization of personality development. Scheme of periodization of mental personality development in childhood (D. Jel'konin). The problem of periodization of personality development in adulthood. Crises of development and their role in the situation of the individual formation. Personality as a representative of a certain system of governmental public relations. General characteristics of the notions "social role", "social group", "social status". Provision on public functions, roles, and their place in the structure of personality. Socialization of personality. Requirement and motivational sphere of the individual. The concept on needs, motives and motivation. The concept on emotions and feelings. The functions of emotions. Theory of emotions. Types of emotions. Manifestation of emotions. Higher feelings: moral, intellectual, aesthetic, and practical. Social conditionality of feelings. The concept of free will. Will function. The theory of will. Deliberate and volitional actions. Structure of volitional actions Volitional qualities of a personality. Temperament. The concept of temperament. Teachings on temperament in psychology. Physiological basis of temperament. Types of temperaments. Type of higher nervous activity and temperament. Modern studies of the properties of the nervous system. The structure of temperament. Temperament and individual style of activity. Methods of diagnosis of temperament. Nature. The concept of nature. The structure of nature. Character traits. Character and temperament. Natural and social preconditions in nature. Typology of characters. Body composition and character. Accent nation of character traits. The formation of character. Diagnostic methods of character traits. Ability. Problems of psychology of abilities. History of the research of problematics abilities. The concept of abilities. Nature's abilities. Signs of abilities. The inelinations, their role in the development of abilities. The influence of social factors on development of abilities. Classification of abilities. Skills and activities. Levels of development of abilities. The notion of "professional self-determination of personality". Professional self-determination of personality factors. The essence and the result of professional self determination of personality. Psychological "space" for self-determination of the individual. The main line of the subjects professional and personal self-determination

(continued)

Table 4 (continued)

No.	Section name of the discipline	Section content
2.	Personality theory	Psycho-anthropometric views of L. Vygotsky. Personality theory of S. Rubinstein. Personality theory of A. Leontiev. Personality theory of K. Platonov. Z. Freud: Classic psychoanalysis; A. Adler: basic concepts of individual psychology; K. Jung: basic concepts of analytical psychology; K. Horni, E. Fromm: basic concepts of humanistic psychoanalysis; G. Sullivan: basic concepts of Interpersonal psychotherapy; G. Allport: basic concepts of psychology of individuality; R. Cattell: factor theory of features; G. Eysenck: factor type theory; K. Rogers, Maslow: humanistic theory of personality; D. Kelly: basic concepts of the theory of personal constructors

individual psychological characteristics of personality, psychological characteristics and conditions of mental processes of a person presented in various theoretical concepts and schools.

References

1. Alekhina CB (2013) Preparation of pedagogical staff for inclusive education. Pedagogical Mag 44(1):26–32
2. Tammets K, Pata K (2014) The model for implementing learning and knowledge building in the extended professional community: a case study of teachers' accreditation. Syst Res Behav Sci 31(1):127–143. https://doi.org/10.1002/sres.2138
3. Caena F (2014) Teacher competence frameworks in Europe: policy-as-discourse and policy-as-practice. Eur J Educ 49(3):311–331. https://doi.org/10.1111/ejed.12088
4. Liakopoulou M (2011) Teachers' pedagogical competence as a prerequisite for entering the profession. Eur J Educ 46(4):474–488. https://doi.org/10.2307/41343395
5. Arpentyeva MR (2018) Inclusive culture and pedagogy of mutual aid. Scientific Notes of the St. Petersburg State Institute of Psychology and Social Work 29(1):97–10
6. Ibragimov GI, Kamasheva YL (2010) An assessment of quality of an educational methodological support of the main educational programs in higher education institution. Practice-oriented monograph, Kazan
7. Cronis T, Ellis D (2000) Issues facing special educators in the new millennium. Education 120(4):639–648
8. Loreman T et al (2010) Inclusive education: supporting diversity in the classroom, 2nd edn. London and New York
9. Sharma U, Shaukat S, Furlonger B (2015) Attitudes and self-efficacy of pre-service teachers towards inclusion in Pakistan. J Res Spec Educ Needs 15(2):97–105
10. Brandon T, Charlton J (2011) The lessons learned from developing an inclusive learning and teaching community of practice. Int J Incl Educ 15(1):165–178
11. Akhmetzyanova AI (2015) Anticipation and prediction interrelation neuropsychological mechanisms at youthful age. Soc Sci 10:399–401
12. Borodina OS (2014) Formation of inclusive competence of the future teacher health bases. Professional Education in Russia and Abroad 1(13):75–79
13. Knapp S, Gottlieb M, Handelsman M (2015) Ethical dilemmas in psychotherapy: positive approaches to decision making. American Psychological Association, Washington, DC

14. Tokareva NG (2001) Methodology for studying the characteristics of medical and social work with the population. Integr Educ 3:92–94
15. Tokareva NG (2018) The role of psychological knowledge in the formation of a professional portrait of a doctor. Humanities 4(44):39–44

Problems and Challenges of Educational and Methodological Activities in Higher Education in the Context of Digitalization of Education

Olga Vitchenko and Sergey Shcherbakov

Abstract The article deals with the problems of educational and methodological activity in higher education institutions in the light of the digital transformation of education, analyzes the principles, technologies and tools that can be effectively applied to improve it and bring it into line with modern requirements of society. The existing challenges are explored: the challenge of big data; the challenge of project work; the challenge of distance and blended learning; the challenge of technological change; call for individualization. To respond to these challenges, based on a comparative analysis of educational and methodological activities at the university with related fields, the requirements for the system of educational and methodological support are formulated, including: automation of the management of educational and methodological documentation, integration of the documentation management system with the LMS, synchronization of online and offline education forms, the use of educational analytics, etc. Possible solutions to the identified problems are considered using the experience gained in the framework of mass online courses, as well as taking into account modern principles and trends in business process management. At the same time, such ideas as: patterns and antipatterns, lightweight (Agile) development methodologies adapted for educational and methodological activities, and management of educational and methodological documentation based on automated systems can be successfully involved.

Keywords University · Educational and methodological activities · Agile · Management · Efficiency

O. Vitchenko (✉)
Don State Technical University, Gagarin Square, 1, 344003 Rostov-on-Don, Russia

S. Shcherbakov
Rostov State University of Economics, B. Sadovaya Str., 69, 344002 Rostov-on-Don, Russia

A. Beskopylny and M. Shamtsyan (eds.), *XIV International Scientific Conference "INTERAGROMASH 2021"*, Lecture Notes in Networks and Systems 247,
https://doi.org/10.1007/978-3-030-80946-1_75

1 Introduction

Educational and methodological activities have always been an integral part and a necessary component of higher education. Any university has sufficient experience in the methodological design and implementation of a system of deep and systematic education, the development of competencies. All of the above, along with immersion in the academic environment, determined the high potential of the university graduate as a specialist.

But it should be recognized that at present, in the context of the digital transformation of education, it is the methodological support of the university that ceases to be its advantage. The reasons can be identified as follows:

- general inertia of university educational programs;
- bureaucratization of educational and methodological work to the detriment of its content aspect;
- the development of new technologies and tools that are not yet included in university education or are poorly adapted to its features.

Next, we will describe the problems and challenges identified during the analysis of educational and methodological activities that determine the trends of its development.

2 Materials and Methods

As part of the study of the educational and methodological activities of the university, we will analyze it by comparing it with similar or alternative types of activities: mass online courses, corporate information technology schools, project and educational intensive courses. At the same time, we will record the emerging challenges for the higher education system in the field of educational and methodological activities.

Mass open online courses (MOOCs) have recently become one of the most popular forms of education [1]. We compared the standard university discipline and the mass open online course (MOOC), and found that MOOCs have a more effectively organized educational and methodological cycle, much wider opportunities for receiving feedback, and the ability to quickly analyze the results of training to adjust the educational and methodological and organizational support [2].

It should also be noted that MOOCs, like other forms of non-university education, actively use IT platforms. For example, the Yandex system used in Yandex corporate programs. The contest allows you to automatically check tasks, accumulate and study data, the digital footprint of students, which allows you to analyze the effectiveness of training in a timely manner, find weaknesses and strengths in its educational and methodological support. Another effective tool is GitHub, a software version control service that has now become a powerful tool for sharing experiences, for interaction, and for storing data [3, 4].

The next challenge relates to the scope of project activities. Indeed, the effective formation of a specialist (for example, in the field of information technology) is largely determined by the projects completed by him [5], the experience of overcoming problems in the course of their development. The importance of project work is recognized and encouraged not only by the university, but also by the employer partners. One of the indicators of the student's success is his active participation in various events and competitions from the university to the federal level. One can, for example, single out a series of hackathons "Digital Breakthrough", a successful performance at which is the subject of legitimate pride of students and their universities [6].

At the same time, project activities are poorly integrated into the educational and methodological support of the university. Here it exists mainly in the form of local projects within the same discipline. Team, long-term projects, which represent the greatest educational value, are taken out of the framework of the educational process of the university.

The next serious problem of the university is distance and mixed education, which "broke" into practice in the wake of the COVID-19 epidemic. Updating the existing problems, the resulting "shake-up" has benefited many universities, determining the need to update the arsenal of technologies, resources and methods of education, forcing teachers and managers to acquire new competencies, as well as get rid of a number of outdated provisions and requirements.

One of the identified problems was the lack of integration between the educational and methodological documentation and the electronic information and educational environment of the university, including, among other things, LMS (for example, Moodle).Their synchronization falls on the shoulders of the teacher, which is even more complicated in the conditions of mixed education, when often in an arbitrary order (for example, dictated by the epidemiological situation), it is necessary to combine online and offline forms of work with students.

The next important challenge is individualization, designing the individual educational trajectory of the student. Thanks to modern technologies, it is possible to adapt the educational process to the real needs of each student, as required by the Federal State Educational Standards.

At the same time, the students' choice of disciplines (modules) remains largely limited. Meaningfully, each discipline is linked to the competencies of its educational program (educational standard). And, if a student wants to study something else (not in another country, but in a neighboring classroom), he can not do it because of organizational and methodological restrictions. Moreover, providing an individual educational trajectory in the realities of a modern university will lead to an increase in the cost of education.

Thus, in order to meet the current level of training, the university needs to improve the system of educational and methodological support.

3 Results

The identified problems in the higher education system with regard to educational and methodological activities can be successfully overcome if the system of educational and methodological support is able to adapt the opportunities available in other forms of education.

We will formulate the main requirements that the system of educational and methodological support in the university should meet in the light of the problems and challenges we have identified (Table 1). The requirements correspond to the identified challenges and are the result of comparing the educational and methodological activities of the university with one or more alternative areas of education.

In fact, when creating a new system of educational and methodological support, it is necessary to find a certain balance between simplicity and flexibility, between control and academic freedoms, between innovation and the preservation of traditions.

Solutions to the described problems can be found in the case of applying experience from related fields: massive open online courses, corporate training, software development industry.

Figure 1 shows one of the possible life cycles of educational and methodological support, taking into account the requirements defined above (based on the MOOC life cycle [8]) (see Fig. 1).

It should be noted that such training and methodological support is more detailed, which makes it possible to apply digital analytics and make changes quickly [9]. For example, you can determine which task is causing difficulties for students and how you can change the parameters of the relevant educational content so that the material is absorbed better Hypotheses can also be tested by providing different groups of students with alternative versions of educational content [10].

The next source of possible tools for improving educational and methodological support is the IT software development industry, which faces many of the same problems as education: the need to maintain the relevance of documentation, find a balance between its flexibility and accuracy, and the need to accelerate the life cycle.

To solve these problems, the software industry has put forward a number of ideas, and first of all, the concept of lightweight development methodologies (Agile) [11].

Let's consider the basic principles of the Agile methodology and analyze the possibilities of their application in the educational and methodological activities of the university [12] (see Fig. 2).

In the classical formulation, the ideas of Agile can be formulated as follows (Manifesto for Agile Software Development, http://agilemanifesto.org):

- people and their interaction are more important than processes and tools;
- a working software product is more important than comprehensive documentation;
- cooperation with the customer is more important than agreeing on the terms of the contract;
- willingness to change is more important than following the original plan.

Table 1 Requirements for the system of educational and methodological support

Requirement	Description	Source
Flexibility	The adjustment of the educational and methodological support should be carried out in an operational mode. Instead of paper documents, modern software should be used, for example, based on a wiki engine, which will allow you to quickly change the educational and methodological content if necessary. The right to make such changes to a certain part of the educational and methodological support should be delegated	MOOCs
Automation	The formation, storage, modification, publication and analysis of educational and methodological support are effective only in the presence of an automated educational management environment	MOOCs, software industry
Integration with LMS	The integration of the LMS and the training software should occur automatically (the schemes of such integration may be different) [1]	MOOCs, corporate training
Synchronization of online and offline education forms	Educational and methodological support education is unified for various forms of conducting specific classes	Corporate training
Using modern IT communication tools	Educational and methodological support should be focused on the implementation of its modern IT, electronic resources, digital technologies [7]	Software industry, corporate training
Accounting for project work	The student's participation in project activities should be reflected in some way in the educational and methodological support to ensure the motivation of students and managers	Design and educational intensive courses

(continued)

Table 1 (continued)

Requirement	Description	Source
Analytics	The system of educational and methodological support should make it possible to collect and analyze data within the educational process. The results of the analysis will allow for adjustments to the educational and methodological support	MOOCs
Flexibility	The adjustment of the educational and methodological support should be carried out in an operational mode. Instead of paper documents, modern software should be used, for example, based on a wiki engine, which will allow you to quickly change the educational and methodological content if necessary. The right to make such changes to a certain part of the educational and methodological support should be delegated	MOOCs
Automation	The formation, storage, modification, publication and analysis of educational and methodological support are effective only in the presence of an automated educational management environment	MOOCs, software industry
Integration with LMS	The integration of the LMS and the training software should occur automatically (the schemes of such integration may be different) [1]	MOOCs, corporate training
Synchronization of online and offline education forms	Educational and methodological support education is unified for various forms of conducting specific classes	Corporate training
Using modern IT communication tools	Educational and methodological support should be focused on the implementation of its modern IT, electronic resources, digital technologies [7]	Software industry, corporate training

(continued)

Table 1 (continued)

Requirement	Description	Source
Accounting for project work	The student's participation in project activities should be reflected in some way in the educational and methodological support to ensure the motivation of students and managers	Design and educational intensive courses
Analytics	The system of educational and methodological support should make it possible to collect and analyze data within the educational process. The results of the analysis will allow for adjustments to the educational and methodological support	MOOCs

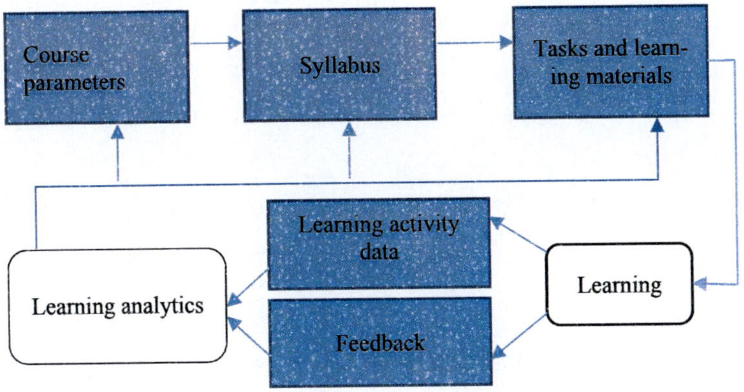

Fig. 1 Stage of the life cycle of educational and methodological support

These principles are quite relevant for the educational and methodological support of higher education—it becomes possible to design the educational process that allows you to effectively take into account changes, manage the work of the team, and coordinate the interests of various parties [13].

Another idea that can be borrowed is the idea of patterns and antipatterns (the author of the idea is Christopher Wolfgang Alexander, designer and architect, creator of more than 200 architectural projects) [14].

A pattern is a typical proven solution to a common problem that fits into the context of the problem. Accordingly, the antipattern is a typical unsuccessful solution. At the same time, you should distinguish the antipattern from the error. In case of an error, the system does not work, while in case of a solution based on an antipattern, it works, but it is inflexible and fragile [15].

An example of an antipattern shown on Fig. 3.

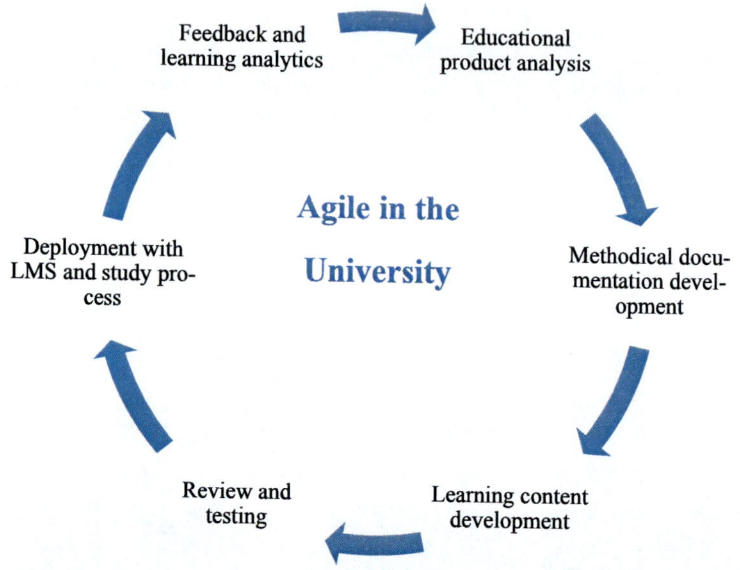

Fig. 2 The cycle of development and use of educational and methodological support in the Agile environment

Name Potential unemployed		***Level*** Educational program
Description New educational programs that are not called upon to specific professions. A «catch phrase » in the educational program title with an incomprehensible place of further work.		
Causes The desire to attract applicants and / or public funding		***Connected with*** Bite off more than you can chew
Troubles		
For student Lack of competence and vague prospects in the labor market	***For teacher*** Problems with generating learning content	***For university*** Low ratings

Fig. 3 Potential unemployed antipattern card

Finally, about solving the problem of labor costs for educational and methodological documentation, for efforts to synchronize it with the real educational process. It suggests the use of special systems, such as wiki, which will allow the implementation of the relationship between the LMS and the management system of educational and methodological documentation will reduce the labor costs of the teacher and eliminate the likely gap between the educational process and its educational and methodological support.

4 Conclusions

As a result of the comparative analysis of the educational and methodological activities of the university with alternative forms of education, a number of problems and challenges were identified.

A successful response to these challenges can be obtained only as a result of adapting the educational and methodological support of the university in accordance with the identified requirements: flexibility, the use of analytics, etc.

To implement these requirements, the best practices available in related or similar areas of activity, including in the field of software development, can be successfully applied. And in particular, flexible methodologies, methods for analyzing patterns and antipatterns, and methods for managing documentation can be successfully used.

In the case of appropriate regulatory changes to implement a new model of educational and methodological support, the ideas and tools presented in this article can be used in the educational and methodological activities of universities in accordance with modern requirements and prospects for their development in the context of digitalization of education.

Acknowledgements The research was carried out with the financial support of the Russian Foundation for Basic Research (RFBR) in the framework of the research project 19-013-00690 «Economics of educational and methodological activities in higher education»

References

1. Chen X, Barnett D, Stephens C (2013) Fad or future: the advantages and challenges of massive open online courses (MOOCs). Presented at the research-to practice conference in adult and higher education. Lindenwood University, St. Charles, MO, pp 20–21
2. Stepanova EV (2020) Project training—a way to increase students' motivation. Nauka i obrazovanie: opyt, problemy, perspektivy razvitiya. In: Materials of the international scientific and practical conference, pp 127–130
3. Aruchidi NA, Kalugyan KKh, Shcherbakov SM (2021) Typical mistakes (antipatterns) of educational and methodological activity in higher education. Int Res J 2–2(104):13–18
4. Cardoso A, Mackenzie IF, Kirner C, Tori R (2017) Development of educational resources with virtual and augmented reality: challenges and perspectives. In: XLIII Latin American computer conference (CLEI). IEEE, pp 20–25
5. Yavorsky VV, Chvanova AO, Baidikova NV (2018) Possibilities of using the Github service in the educational process. Modern education: improving the professional competence of university teachers—a guarantee of ensuring the quality of education. In: Materials of the international scientific and methodological conference, pp 159–160
6. Semenova TV, Vilkova KA (2017) Types of integration of mass open online courses in the educational process of universities. Univ Manage Pract Anal 21:114–126

7. Chaubey A, Bhattacharya B (2015) Learning management system in higher education. Int J Sci Technol Eng 2(3):158–162
8. Ablaeva LN, Abdullayeva UB (2019) Foreign and domestic experience of using Github in universities. Inf Comput Technol Econ Educ Soc Sphere 3(25):65–72
9. Clow D (2013) An overview of learning analytics. Teach High Educ 18(6):683–695
10. Kizilcec RF, Brooks C (2017) Diverse big data and randomized field experiments in MOOCs. In: Lang C, Siemens G, Wise A, Gašević D (eds) Handbook of learning analytics, [s. l.], Society for learning analytics research, pp 211–222
11. Shcherbakov SM, Miroshnichenko II, Aruchidi NA, Kurbesov AV (2020) Educational and methodological activity of the university in the digitalization context. In: European proceedings of social and behavioural sciences EpSBS. Krasnoyarsk Science and Technology City Hall, Krasnoyarsk, pp 787–793
12. Mansurov O (2019) Competition "Digital breakthrough" what it will be and what will give the participants an interview with the project leader Oleg Mansurov. Autom IT Power Eng 7(120):68–70
13. Kurbesov AV, Miroshnichenko II, Shcherbakov SM (2020) Agile methodology in the educational and methodological activities of the university. Comput Sci Educ 10:41–46
14. Green D, Stillman E (2018) Comprehending Agile: values, principles, methodologies. Mann, Ivanov and Ferber, Moscow
15. Karwin B (2010) SQL antipatterns: avoiding the pitfalls of database. Pragmatic Bookshelf, Dallas Texas

Psychological Predictors of Socio-psychological Adaptation of Educational Migrants in the Context of Infodemia

Anastasia Grishina and **Victor Klimenko**

Abstract The migration flows of students are growing every year all over the world, which means that the study of the psychological characteristics of this group of young people, as well as the difficulties they face, is extremely relevant for psychological science. The current situation of a pandemic and forced self-isolation throughout the world led to even more serious psychological consequences for educational migrants, because many of them could not return to their homeland, or come to their place of study, which increased anxiety for their future and career, for the inability to help their loved ones, etc. The conducted empirical research was devoted to the study of the level of psychological well-being, socio-psychological adaptation, as well as indicators of the personal potential of migrant students. The empirical base of the study was made up of 380 people—students of higher educational institutions of Yekaterinburg and Tomsk at the age of 18–22 with different migration experience. The empirical study was carried out using the following psychodiagnostic techniques: Test "Hardiness" (S. Maddi, adapted by D. A. Leontiev, E. I. Rasskazova); "Tolerance to Uncertainty" scale (D. MacClain, adapted by E. G. Lukovitskaya); Self-organization questionnaire by E. Yu. Mandrikova; "Socio-psychological adaptation" (K. Rogers, R. Diamond, adapted by A. K. Osnitsky); Questionnaire "Scale of psychological well-being" (K. Riff, adapted by N. N. Lepeshinsky). All students who took part in our study demonstrate the corresponding normative indicators of various personal potential parameters, of socio-psychological adaptation and psychological well-being, but respondents with different migration experience differ among themselves in the degree of expression of the studied parameters.

Keywords Educational migration · Infodemia · Hardiness · Tolerance to uncertainty · Self-organization · Socio-psychological adaptation

---

A. Grishina (✉)
Don State Technical University, Sq. Gagarina, 1, Rostov-on-Don 344003, Russia

V. Klimenko
National Research Tomsk State University, Lenin Ave., 79, Tomsk 634050, Russia

A. Beskopylny and M. Shamtsyan (eds.), *XIV International Scientific Conference "INTERAGROMASH 2021"*, Lecture Notes in Networks and Systems 247,
https://doi.org/10.1007/978-3-030-80946-1_76

1 Introduction

The problem of social and psychological adaptation of educational migrants is of particular importance in the modern world for many reasons. Migrant students are often more active, ready to make decisions and change something in their lives. In addition, the successful adaptation of educational migrants in the host society significantly increases the probability that upon graduation they will move into the category of qualified labor migrants and will be competitive participants in the labor market of the host region. For these and a number of other reasons, universities are responsible for organizing the process of adaptation and integration of students from other regions [1, 2]. At the moment, particular difficulties arise in connection with the conditions of the pandemic and the resulting infodemic. The concept of infodemic was introduced by the World Health Organization during the COVID-19 pandemic to determine the oversupply of information that occurs during an epidemic, which can be either accurate or inaccurate [3, 4]. At the same time, the emphasis is on the complexity of controlling the reliability of the disseminated information.

The situation of migration in general and educational migration in particular poses a number of serious tasks for the individual. How a young person copes with them most often depends not so much on the objective factors of the new educational environment as on the psychological characteristics of the migrant student himself, his personal potential. According to D. A. Leontyev, personal potential is the ability of an individual to self-regulation and self-government [5]. The following parameters of personal potential can be attributed to the main components that determine the success of the socio-psychological adaptation of educational migrants:

- hardiness
- tolerance to uncertainty
- level of self-organization
- level of psychological well-being

Let us briefly describe each of these parameters.

Hardiness (the author of the term is Salvatore Maddi, the author of the term in the Russian-speaking environment is D. A. Leontiev) is understood as the ability of a person to transform the unfavorable circumstances of his life for himself. Hardiness can act both from the point of view of ensuring the stability of the individual under stress and difficult life situations, and from the standpoint of self-regulatory processes, allowing the subject to make decisions and act more consciously and responsibly in relation to his life, relying on his personal attitudes and beliefs, regardless of the general level of stress and the apparent pressure of the environment [6, 7].

Tolerance to uncertainty has received much attention from scientists in the last 20 years. It has become a significant value for a modern man, helping to adapt to the constantly changing world. Many scientists understand intolerance to ambiguity as a tendency to interpret uncertain situations as a source of threat, as a readiness of a person to accept the situation or avoid it. In the situation of the Covid-19 pandemic, this psychological feature has taken on special importance, becoming a competence that allows a person to interact effectively with a constantly changing environment and being under the influence of uncontrolled flows of information [8].

Self-organization of activity is an indicator that, acting as one of the parameters of the socio-psychological adaptation of a person, also allows one to study empirically the features of the psychological time structuring [9].

The construct of psychological well-being deserves special mention, which, being developed by K. Riff, is understood as an integral indicator of the subject's orientation towards the implementation of various components of his positive functioning and assessment of the degree of their implementation from the point of view of the subject's "peak of potential opportunities" [7]. Currently, although there are discussions about whether psychological well-being is the initial personal resource for overcoming a difficult situation or arises as a result of its successful resolution, we believe that, speaking in the form of some kind of attitude, even if it is dynamic, psychological well-being should be considered as part of the search for predictors of personal resources actualization in difficult life situations. This opinion is confirmed by modern works devoted to finding personal resources to overcome stress in the face of the global pandemic in 2020. These studies actively analyze the relationship between psychological well-being and the ability to overcome stress with: the specificity of psychological stability and resilience in relation to the Eastern European samples [10], the level of awareness and the specificity of work with the value-semantic sphere [11–14], as well as with Big Five indicators and the opportunity to gain positive experience from not the most favorable situations [10, 15].

2 Methods

In our opinion, an empirical study of these psychological characteristics among educational migrants will expand the understanding of the psychological consequences of migration and predictors of successful adaptation of migrants in the host society, as well as consider the problem of migration in the context of a situation of uncertainty and infodemic.

Research hypothesis: Students with different migration experience may differ in the level of psychological well-being, socio-psychological adaptation, as well as in indicators of personal potential.

To confirm the hypothesis put forward, the following psychodiagnostic techniques were used:

1. Test "Hardiness" by Salvatore Maddi (original—"Hardiness Survey", 1997) adapted by D. A. Leontiev and E. I. Rasskazova (2006);
2. Scale "Tolerance to Uncertainty" D. McLain (in the original—Multiple Stimulus Types Ambiguity Tolerance Scale-I, 1993), adapted by E. G. Lukovitskaya (1998);
3. Questionnaire of Self-organization of activity by E. Yu. Mandrikova (2007);

4. Methodology “Socio-psychological adaptation” by K. Rogers and R. Diamond (1954) in the adaptation of A. K. Osnitsky (2004);
5. Questionnaire “Scale of psychological well-being” by K. Riff (original—“The scales of psychological well-being”, 1998) as adapted by N. N. Lepeshinsky (2007);

The experimental base of the study consisted of 380 people. The main research group were educational migrants studying in higher educational institutions of Yekaterinburg and Tomsk. N = 280. Age 18–22. Among educational migrants, we additionally single out the following groups: internal educational migrants studying in Yekaterinburg (90 people), internal educational migrants studying in Tomsk (92 people), and external educational migrants (98 people).

Additionally, a group of non-migrants (N = 100) took part in the study (that is, young people similar to the control group of age and employment, but studying at their place of residence, and often even their place of birth).

3 Results

The study of the factors that determine the success of the socio-psychological adaptation of educational migrants began with the study of such indicators of personal potential as hardiness, self-organization and tolerance to uncertainty.

The results of the study of the parameters of the personal potential of educational migrants using the above psychodiagnostic questionnaires show that the respondents who took part in the study demonstrate close to the normative indicators of personal potential. However, it is necessary to note some tendencies towards deviations from the normative indicators. Thus, most of the deviations from the normative indicators of the expression of personal potential are among young people of the “non-migrant” category—that is, those who study at a higher educational institution at their place of residence. Such respondents are less able to control themselves and the situation as a whole. Often such people do not really understand the limits of their own capabilities and the specifics of their potential, which could help them cope with an emergency situation and save or even increase resources.

They do not have well-developed planning skills, and the plans themselves are subject to frequent changes. Quite often, they prefer not to think about their future, and in the process of setting goals, they are guided not so much by their opinion and position as by external sources (parents, teachers, environment). At the same time, they excessively focus on what is happening to him at the moment (a tendency to overestimate the indicator on the scale of “orientation to the present”), to some extent inclined not to plan far ahead, and in some cases even demonstrate hedonistic tendencies. They are rather wary of uncertain situations in their lives, perceiving them quite menacingly for their peace of mind. In fact, they are worried about the need to solve problems that do not have ready-made solutions. This situation is quite understandable from the point of view of the living conditions in which the "non-migrants" live: the home environment, the closeness of the parents and their financial, material and moral support.

Often, while students who came to study from other regions are faced with the need to organize themselves, learn to live independently, organize everyday life and financial planning, students studying at their place of residence are relieved of most of the listed difficulties, and in some cases they were even relieved of the need to build a long-term prospect of one's own life, succumbing to the opinion of the closest circle and significant adults about the choice of a specialty, future employment, etc.

It should be noted that, just in case, we made sure of the similarity of the indicators of the personal potential in a group of non-migrants from Yekaterinburg and a similar one from Tomsk. Analysis of variance between the two subgroups showed no significant differences between their data (Table 1).

Turning to the data of descriptive statistics on samples of educational migrants, we note that these groups of our respondents have sufficient resources to withstand difficult life situations and uncertainty, are able to see new opportunities in life's difficulties and are quite effective in self-organizing their activities, skillfully using various methods and techniques of self-organization. In general, all indicators of personal potential in these samples are in standard values. However, the group of educational migrants studying at the universities of Yekaterinburg slightly overestimated the indicator on the risk scale in the Hardiness test. This indicates the readiness of this category of young people to take any action without guaranteeing future success and the tendency to perceive life events as a personal challenge, allowing them to reveal themselves and learn more about themselves.

To check the significance of differences between the study samples, we used the Student's T-test for independent samples. The check was carried out both between the group of non-migrants and the general sample of educational migrants, and between groups of educational migrants studying in different regions.

In the course of checking for the reliability of differences in the severity of personal potential indicators on samples of internal educational migrants and non-migrants among students, significant differences were identified only on the scales of the Hardness test: Control, Risk and the total indicator of Hardiness ($p = 0.001$).

This allows us to assert that students studying in higher educational institutions at the place of their original residence or place of birth perceive themselves to a lesser extent as being able to control the events taking place in their lives, to show high motivation to look for ways to influence their lives, as well as to act outside guarantees of future success.

In the course of checking for the reliability of differences in the expression of personal potential indicators on samples of internal educational migrants from different regions, only one significant difference was revealed: in the tendency to self-organization of their activities, students of Tomsk universities more often and, possibly, more effectively use various external means and planning tools and self-organization than educational migrants studying at universities in Yekaterinburg ($p \leq 0.05$). Perhaps this is due to the general culture of the "student city" and the presence of a sufficiently large number of resources that are promoted and promoted

Table 1 Indicators of the personal potential of representatives of different groups of educational migrants and non-migrants

	Non-migrants		Educational migrants from Ekaterinburg		Educational migrants from Tomsk		External educational migrants	
	Mean	Std. Dev.	Mean	Std. Dev.	Mean	Std. Dev.	Mean	Std. Dev.
Commitment	30.0769	9.43806	32.9612	9.95078	33.4810	8.15298	**37.1000**	7.40483
Control	**17.3846**	3.09673	29.7864	8.46405	31.6504	7.91584	31.8000	6.95550
Challenge	12.0769	2.13937	**19.1748**	5.73504	18.3716	5.96017	**24.0500**	5.23626
Hardiness	59.5385	9.91308	81.9223	15.16167	83.6824	12.48694	92.9500	12.78352
Orderliness	**14.3846**	6.30527	16.1262	5.37000	16.9514	5.84425	16.3000	5.34199
Purposefulness	34.1538	6.09434	33.9612	6.74162	33.5000	6.29536	34.1500	5.26433
Persistence	20.6154	6.14462	21.8350	5.70190	**23.2044**	5.05340	19.8000	4.96938
Fixation	18.3077	4.90552	20.8738	5.58829	20.9037	4.38778	20.3500	4.17102
Self-organization	11.1538	5.44436	8.7864	4.21168	11.0098	4.32861	10.9500	3.77631
Orientation to the present	**10.7638**	2.16025	9.3301	2.84704	9.7740	2.97135	**8.5000**	2.68524
General self-organization	108.6154	20.89872	110.9126	17.90066	115.3264	14.92825	110.4500	15.30213
Tolerance to uncertainty	**83.0769**	24.49647	94.4757	26.98977	100.1524	22.33306	**103.5000**	18.97228

Source: Bold represents significant differences when p level is ≤ 0.05

in higher educational institutions of Tomsk. In particular, only planning teaching projects for Tomsk students as part of the development of the general digitalization of the Tomsk region and universities located on its territory provide students with regular materials, tips and life hacks on this topic.

This minimal difference in the indicators of personal potential between groups of educational migrants studying at universities in Siberia and the Urals suggests that the phenomenon of educational migration itself imposes, in a sense, its requirements on the personal resources of an individual, manifested in the similarity of the parameters of the personal potential of different groups of educational migrants.

Comparison of indicators of personal potential in a sample of non-migrants and external educational migrants showed the presence of significant differences in the indicators of Hardiness "Control" and "Risk" ($p = 0.001$), indicators of "Engagement" and tolerance to uncertainty ($p \leq 0.05$). For example, students who went to study in another country are much more inclined to exercise control over the events of their lives and to be involved in them, are ready to take risks and take action in situations with unclear conditions.

Analysis of variance also showed significant differences in the parameters of the personal potential of foreign students from internal educational migrants studying in the Ural region, in terms of the propensity to take risks and the total indicator of resilience, as well as the propensity to use external elements of planning and self-organization of the OSD Questionnaire by E. Yu. Mandrikova.

At the same time, there were no significant differences in the group of educational migrants in Tomsk. We believe that this may be due to the educational culture itself prevailing in Tomsk, trends of internationalization and attitude to the city as a stage for further more strategic development, which often manifests itself in the migration moods of migrants studying in Tomsk [16, 17].

The results obtained on the basis of the data of the Rogers-Diamond social-psychological adaptation questionnaire and the K. Riff scale of psychological well-being were processed, systematized and subjected to statistical processing. Table 2 presents descriptive statistics summarized for different samples of migrants.

In the process of analyzing descriptive statistics using the method of socio-psychological adaptation, we found that, in general, most of the indicators of our respondents are within the normal range.

Thus, representatives of different groups of educational migrants, as well as students studying at their place of residence, are generally quite adapted to the conditions of their life. They are able to accept the necessary norms and rules, although they have some difficulties in achieving goals, including (and, probably, in the first place) associated with the need to correlate external social requirements and their own needs.

In particular, in all groups of respondents, a predominance of mainly external control was found. That is, in their opinion, many events have external causes and are determined by the environment, others, and chance. To be honest, we were somewhat confused by this result, since we expected rather the prevalence of the internal locus of control and the respondents' feeling of control over what is happening. However, a deeper analysis and appeal to the results of focus groups and surveys to assess

Table 2 Descriptive statistics of indicators of socio-psychological adaptation in all groups of respondents

Variable	Non-migrants		Educational migrants from Ekaterinburg		Educational migrants from Tomsk		External educational migrants	
	Mean	Std. Dev.	Mean	Std. Dev.	Mean	Std. Dev.	Mean	Std. Dev.
Adaptability	98.3077	30.68763	98.9252	28.09677	93.2642	26.26832	87.9000	18.91783
Maladjustment	107.9231	40.71130	91.6355	23.98167	94.7170	25.36655	85.8500	24.18628
Integral adaptation	48.1348	10.70961	51.7683	10.17306	49.5556	10.28993	51.0516	10.00866
Emotional comfort	22.0769	4.46209	20.3178	4.56731	20.0283	5.99437	19.8000	5.15650
Emotional discomfort	23.0022	5.59128	22.1941	6.94504	23.5050	6.75683	21.9979	5.92380
Integral emotional comfort	49.2308	8.00967	47.6903	9.05561	45.8760	11.17285	47.5714	9.64571
Internal control	42.6154	12.31895	45.6542	13.66175	40.3679	12.67568	39.0000	11.08342
External control	**40.7692**	14.48651	33.6075	8.87944	34.8962	8.13333	33.3500	8.91642
Integral internality	43.3454	10.81244	49.0304	10.71151	44.9363	10.99775	45.7233	10.91264
Domination	8.1538	3.67074	9.0467	3.13947	9.1792	4.07267	9.0000	3.76969
Weakness	22.3077	4.53477	18.8411	5.19824	18.7075	6.39787	19.3500	6.22622
Integral domination	43.2268	15.68253	48.4044	11.13302	48.9239	15.68814	47.5920	14.41291
Escapism	14.6056	3.06109	15.8588	5.83066	17.2483	7.19257	15.7034	4.69165
Positive relation	60.6923	5.55855	62.9643	8.31420	60.6923	8.35508	62.7000	6.85258
Autonomy	60.4615	6.41113	59.5536	8.14621	58.0577	9.00682	56.1000	10.21299

(continued)

Table 2 (continued)

Variable	Non-migrants		Educational migrants from Ekaterinburg		Educational migrants from Tomsk		External educational migrants	
	Mean	Std. Dev.	Mean	Std. Dev.	Mean	Std. Dev.	Mean	Std. Dev.
Environment management	51.8462	8.28499	58.5714	8.35759	58.2308	8.63075	55.4000	8.86982
Personal growth	59.5385	5.15777	64.7500	7.13570	66.7692	6.46309	64.0000	6.35776
Goals in life	56.3077	7.05200	67.1607	7.66047	65.7115	7.70786	63.0500	6.08255
Self-acceptance	55.6154	5.05863	60.4286	7.95866	59.0000	7.20302	56.4500	6.96212
Psychological well-being	344.4615	17.22699	373.4107	26.95705	368.5192	26.53226	358.1000	24.77669

Source: Bold represents significant differences when p level is ≤ 0.05

difficult life situations, suggested that such a result is associated with the conditions of the study: requirements of self-isolation; permanent restrictions and new rules broadcast from the external environment; a general feeling of the lack of the ability to exercise control and influence on what is happening.

Representatives from the non-migrant group have a particularly pronounced external locus of control. We assume that, in addition to the listed signs, this group was still under the conditions of external influence from the immediate environment, which in adolescence can be experienced quite sharply, which is demonstrated by the test results. At the same time, our respondents, on the whole, have quite a positive attitude towards the events and are able to maintain sufficient confidence in conditions of uncertainty, showing a tendency not to avoid the difficulties that have arisen, but nevertheless to solve them.

In addition, according to the Dominance and Weakness scale indicators, some of our respondents tend to show a desire for leadership, expressing their needs for leadership, while the other part of the respondents is more likely to submit to the demands and influence of the social environment.

When working with the indicators of the questionnaire "Scales of psychological well-being" in the adaptation of N. N. Lepeshinsky, we will mainly use "raw scores", because this will make it possible to achieve greater sensitivity in identifying correlations and performing other forms of statistical data processing.

However, at the stage of descriptive statistics, we will present the data in accordance with the recommendations of the author of the adaptation—N. N. Lepeshinskiy [18] (Table 3).

Indicators of all groups of respondents fit within the standard values. Only in some cases approaching the upper (7 walls) or lower (4 walls) border of the norm. Thus, it can be argued that all of our respondents are able to generally create harmonious relationships with others and moderately show independence in thinking and behavior, consider themselves in general able to manage everyday events, have no pronounced problems with the presence of goals and self-acceptance, in general, they are ready to change their perception of the world and self-develop.

Further, the obtained data were subjected to analysis of variance.

We checked the indicators of all respondent groups in pairs to identify differences between the average indicators of socio-psychological adaptation and psychological well-being.

The obtained results show that the groups of students studying at the place of residence (and birth), in comparison with the combined group of internal educational migrants, are reliably more inclined to external control ($p \leq 0.05$), to a greater extent maladjusted and inclined to obey external requirements and instructions (more aware) in comparison ($p \leq 0.05$), they have less pronounced indicators of "environmental management" ($p \leq 0.01$), "personal growth" and "goals in life" ($p \leq 0.001$), and also the total indicator of psychological well-being ($p \leq 0.001$).

In addition, non-migrants, in comparison with Tomsk educational migrants, show self-acceptance to a lesser extent ($p \leq 0.05$). Whereas, with a sample of internal educational migrants studying in Yekaterinburg, significant differences

Table 3 Descriptive statistics of psychological well-being indicators for all groups of respondents

	Non-migrants		Educational migrants from Ekaterinburg		Educational migrants from Tomsk		External educational migrants	
	Raw scores	Stans	Raw scores	Stans	Raw scores	Stans	Raw scores	Stans
Positive relation	60.6923	5	62.9643	6	60.6923	5	62.7000	6
Autonomy	60.4615	7	59.5536	7	58.0577	6	56.1000	6
Environment management	51.8462	4	58.5714	5	58.2308	5	55.4000	5
Personal growth	59.5385	5	64.75	6	66.7692	6	64.0000	6
Goals in life	56.3077	4	67.1607	6	65.7115	6	63.0500	6
Self-acceptance	55.6154	5	60.4286	6	59.0000	6	56.4500	5
Psychologi-cal well-being	344.4615	5	373.4107	6	368.5192	6	358.1000	6

were recorded only in terms of externality and individual scales of psychological well-being.

Students who study at the place of residence and birth reliably differ from external educational ones only in the presence of goals in life ($p \leq 0.05$) and the desire for self-development ($p \leq 0.01$).

The group of students from foreign countries additionally has significant differences in indicators of socio-psychological adaptation and psychological well-being only with the group of internal educational migrants studying in higher educational institutions of Tomsk. Thus, students from other regions studying in Tomsk are reliably more internal, psychologically well and have more formulated and meaningful goals in life ($p \leq 0.05$) compared to external educational migrants.

4 Discussion

The studies of psychological peciliarities of educational migrants compared to native students are widely spread in psychological science. Pan, J. Y., and Wong, D. F. compared the level of negative affect and acculturative stressors between Chinese international students in Australia and Mainland Chinese students in Hong Kong. The predictive effects of acculturative stressors and acculturative strategies on negative affect were also compared between the two groups. It was found that Chinese international students in Australia encounter more acculturative stressors and experience a higher level of negative affect than their counterparts in Hong Kong. The acculturative stressor of academic work and a marginalization strategy significantly predicted negative affect in both groups. The acculturative stressor of cultural difference predicted negative affect in the Hong Kong sample, and assimilation strategy predicted negative affect in the Australian sample only. The conclusion of the study is of a great interest for our study: it is more difficult for Chinese international students to adapt to a host society with greater cultural distance [19].

S. Rodríguez studied the differences between native and immigrant students in mathematics and science skills, and in well-being indicators evaluated in PISA 2018, considering the academic performance to be an indicator of their adaptation to the education system. It was shown that the native students demonstrated higher levels of mathematics and science skills than the two groups of immigrant students, and had significantly higher means in positive affect, self-efficacy-resilience, and feeling of belonging at school. Although life satisfaction was no different between the immigrant and native groups [20].

Bhowmik, M. K., et al, provides an in-depth analysis of acculturative stress and coping behaviors experienced by Mainland Chinese university students in Hong Kong. Findings suggested that Mainland Chinese university students faced acculturative stress stemming from "language barriers," "prejudice and discrimination," "cultural differences," and "transport, food, and accommodation." The study also found that the participants used a number of adaptive and maladaptive coping strategies in the face of acculturative stress [21].

Psychological well-being of student in pandemic situation has also been an object of many research. H. Li et al., revealed substantial adverse effects resulting in varying levels of stress, symptoms of depression, and specific discomfort in the case. Among COVID-19 stressors, financial instability, unpredictability toward future/career, and media exposure have been described as common factors that cause poor psychological well-being and weaken economic sustainability. COVID-19, quarantine, self-isolation, and onerous interventions primarily weaken university students' mental health [22].

All mentioned studies didn't compare the psychological characteristics inside the group of educational migrants mistaking them for heterogeneous groups.

5 Conclusions

Thus, all students who took part in our study demonstrate the corresponding normative indicators of the severity of various parameters of personal potential, socio-psychological adaptation and psychological well-being. That is, in general, they are able to adapt to the surrounding conditions, have sufficient personal resources for effective life fulfillment, and show a sufficient level of psychological well-being. Only the fact of a certain prevalence of external control according to the method of socio-psychological adaptation in all groups of respondents requires a separate mention. That is, in their opinion, many events have external causes and are determined by the environment, by the case that, in our opinion, can be explained by the period of the study and its characteristic requirements for self-isolation, constant restrictions and new rules broadcast from the external environment; a general feeling of the lack of the ability to exercise control and influence what is happening.

Groups of respondents with different migration experience differ among themselves in the degree of expression of the parameters of personal potential, psychological well-being, and socio-psychological adaptation.

References

1. Belousova AK, Barsukova OV, Tushnova YA, Vyshkvyrkina MA, Krischenko EP, Mozgovaya NN, Pavlova TV, Mochalova JA (2015) The scope of values and limits of the personal psychological space of students-emigrants. Mediterr J Soc Sci 6(4):238–241. https://doi.org/10.5901/mjss.2015.v6n4s4p
2. Ryumshina L, Belousova A, Pishchik V (2018) Characteristics of socio-psychological adaptation of foreign students to Russian education. In: EDULEARN18 Proceedings, pp 1787–1791. https://doi.org/10.21125/edulearn.2018
3. Grishina AV, Abakumova IV (2020) Informational behavior in the COVID-19 pandemic: psychological predictors. Int J Cogn Res Sci Eng Educ (IJCRSEE) 8(Special issue of Current Research and Trends in Cognitive Sciences):59–67. https://doi.org/10.23947/2334-8496-2020-8-SI-59-67

4. Freeston M, Tiplady A, Mawn L, Bottesi G, Thwaites S (2020) Towards a model of uncertainty distress in the context of Coronavirus (Covid-19). Cogn Behav Therapist 13:e31. https://doi.org/10.1017/S1754470X2000029X
5. Leontiev DA (2011) Personal potential: structure and diagnostics. Smysl, Moscow
6. Maddi S, Khoshaba D, Harvey R, Fazel M, Resurreccion N (2011) The personality construct of hardiness: relationships with the construction of existential meaning in life. J Humanistic Psychol 51:369–388. https://doi.org/10.1177/0022167810388941
7. Leontiev D (2019) Happiness and well-being: toward the construction of the conceptual field. Monit Publ Opin Econ Soc Changes 1:14–37. https://doi.org/10.14515/monitoring.2020.1.02
8. Abakumova I, Zvezdina G, Grishina A, Zvezdina E, Dyakova E (2020) University students' attitude to distance learning in situation of uncertainty. In: E3S Web of conferences, vol 210, 18017. https://doi.org/10.1051/e3sconf/202021018017
9. Suroedova E, Uvarova G, Shevkieva N (2020) Structural organization of personal resources of coping behavior of students with different levels of self-organization of activity. In: E3S Web of conferences, vol 210, 19005. https://doi.org/10.1051/e3sconf/202021019005
10. Zager Kocjan G, Kavčič T, Avsec A (2021) Resilience matters: explaining the association between personality and psychological functioning during the COVID-19 pandemic. Int J Clin Health Psychol IJCHP 21(1):100198. https://doi.org/10.1016/j.ijchp.2020.08.002
11. Krok D, Zarzycka B (2020) Risk perception of COVID-19, meaning-based resources and psychological well-being amongst healthcare personnel: the mediating role of coping. J Clin Med 9(10):3225. https://doi.org/10.3390/jcm9103225
12. Krok D, Zarzycka B, Telka E (2021) Risk of contracting COVID-19, personal resources and subjective well-being among healthcare workers: the mediating role of stress and meaning-making. J Clin Med 10(1):132. https://doi.org/10.3390/jcm10010132
13. Shiri S, Wexler I, Marmor A, Meiner Z, Schwartz I, Levzion Korach O, Azoulay D (2020) Hospice care: hope and meaning in life mediate subjective well-being of staff. Am J Hospice Palliat Care 37(10):785–790. https://doi.org/10.1177/1049909120905261
14. Park C (2016) Meaning making in the context of disasters. J Clin Psychol 72:1234–1246. https://doi.org/10.1002/jclp.22270
15. Anglim J, Horwood S, Smillie LD, Marrero RJ, Wood JK (2020) Predicting psychological and subjective well-being from personality: a meta-analysis. Psychol Bull 146(4):279–323. https://doi.org/10.1037/bul0000226
16. Meshcheryakova EI (2020) Personal and age characteristics of Chinese and Indian students with positive heterostereotypes in relation to Russia. Modern Sci Actual Probl Theory Pract Ser Cogn 2:58–63
17. Poletaev DV (2018) Development of educational migration to secondary professional institutions in Russia (on the example of Moscow, Tomsk, Yekaterinburg, St. Petersburg). Bull Sci Siberia 1(28):114–127
18. Lepeshinsky NN (2007) Adaptation of the questionnaire "Scales of psychological well-being" K. Riff. Psychol J 3:24–37
19. Pan JY, Wong DF (2011) Acculturative stressors and acculturative strategies as predictors of negative affect among Chinese international students in Australia and Hong Kong: a cross-cultural comparative study. Acad Psychiatry J Am Assoc Direct Psychiatr Residency Training Assoc Acad Psychiatry 35(6):376–381. https://doi.org/10.1176/appi.ap.35.6.376
20. Rodríguez S, Valle A, Gironelli LM, Guerrero E, Regueiro B, Estévez I (2020) Performance and well-being of native and immigrant students. Comparative analysis based on PISA 2018. J Adolesc 85:96–105. https://doi.org/10.1016/j.adolescence.2020.10.001.
21. Bhowmik MK, Cheung R, Hue MT (2018) Acculturative stress and coping strategies among Mainland Chinese university students in Hong Kong: a qualitative inquiry. Am J Orthopsychiatry 88(5):550–562. https://doi.org/10.1037/ort0000338
22. Li H, Hafeez H, Zaheer MA (2021) COVID-19 and pretentious psychological well-being of students: a threat to educational sustainability. Front Psychol 11:628003. https://doi.org/10.3389/fpsyg.2020.628003

Stress Resistance of Future Teachers-Psychologists at the Stage of Training at the University

Oksana Nikolenko and Lyudmila Zheldochenko

Abstract The article presents an analysis of the phenomenon of stress and stress resistance in the scientific psychological and pedagogical literature. The problem of stress tolerance in the professional activity of a teacher-psychologist is considered. It is noted that the stress resistance of students in educational activities is a complex property of the individual, which is characterized by the necessary adaptation of the student to the influence of external and internal factors in the course of educational activities. The factors influencing the development of stress tolerance and a positive attitude to learning activities are considered. The article presents the results of an empirical study of stress resistance in future teachers-psychologists at the stage of training in higher education. The features of the perceptual assessment of students' stress resistance at the final stage of training are shown. It is established that graduates are able to clearly define the goals of their educational and professional activities and choose the best ways to achieve them. They are distinguished by the desire to cope with difficulties independently, the ability to analyze the difficulties encountered and overcome them, unlike first-year students. The study involved students of the 1st and 4th courses of the psychological and pedagogical direction, in the number of 128 people. The following methods were used: self-assessment of the stress resistance of S. Cowhen and G. Willianson, the questionnaire of the perceptual assessment of the type of stress resistance of N. P. Fetiskina, methods of mathematical and statistical processing of the obtained data: descriptive statistics, determination of the reliability of differences: the Fisher criterion. It is reliably established that graduates have a moderate type of stress resistance, and 1st-year students at the initial stage of training are characterized by moderate instability to stress.

Keywords Stress · Stress tolerance · Adaptation · Coping strategies · Coping behavior

O. Nikolenko (✉)
Don State Technical University, Rostov-on-Don, Russia

L. Zheldochenko
Southern Federal University, Rostov-on-Don, Russia

A. Beskopylny and M. Shamtsyan (eds.), *XIV International Scientific Conference "INTERAGROMASH 2021"*, Lecture Notes in Networks and Systems 247,
https://doi.org/10.1007/978-3-030-80946-1_77

1 Introduction

The relevance of the studied problem is due to the fact that at the stage of training at the university, it is important for future teachers-psychologists to have a conscious idea of the specifics of their future professional activities, which are characterized by regular stressful situations that require immediate resolution and responsibility for the decisions made. The professional activity of a teacher-psychologist is characterized by intensive psycho-emotional interaction with the subjects of labor. Therefore, it is important to develop students' skills in managing emotions, mobilizing personal resources, and developing stress tolerance as a professionally important quality during their studies at the university [1]. Studies of both domestic and foreign authors are devoted to the problem of stress and stress resistance. Scientists have identified and described various types of human behavior, depending on the presence of psychological defense mechanisms and coping strategies present in certain stressful situations (V. A. Bodrov, B. H. Vardanyan, J. P. Grebennikov, E. S. Romanova, S. Hobfall, V. Fletcher, S. Folkman, R. Lazarus, et al.). Some studies are devoted to stress resistance within the framework of problems of pedagogical psychology, but they are mainly aimed at studying and developing resistance to stress in teachers (A. A. Baranov, J. B. Karapetyan, A. A. Naprimerov, A. A. Rean, S. V. Subbotin, etc.) [2–4]. The formation of stress resistance in the educational activities of students, today, is a little-studied problem that requires further study. The specifics of professional activity leave their mark on the psycho-emotional state of the teacher-psychologist. The presence of daily stressful factors in the form of frequently occurring problematic pedagogical situations, high dynamism of activity in conditions of lack of time, role uncertainty and frequent changes in the positions of the teacher when interacting with the subjects of the educational space, the constant influence of social assessment are the causes of huge emotional overload for each representative of this profession. According to A. K. Markov, L. M. Mitin, A. I. Shcherbakov et al., the teaching profession makes special demands on such an integral characteristic of the individual as stress resistance, since it justifiably refers to one of the most significant professional qualities of a teacher [5]. Stress resistance of the individual is manifested in the preservation of the ability to social adaptation and maintain significant interpersonal relationships, in the desire for successful self-realization, achievement of life goals, the ability to maintain working capacity, physical and mental health. In the studies of A. A. Baranov, who studied the stress resistance of the teacher, it was empirically established that the level of development of the stress resistance of the teacher significantly increases with the predominance of internal motivation in the motivational complex, a high and adequate assessment of oneself as a professional, as well as formation of professional and pedagogical skills. The researcher associates stress resistance in pedagogical activity with the provision of variable coping strategies due to a different combination of coping resources of the teacher as a person and subject of activity. The author notes that resistance to mental stress is a professionally significant quality of the subject of the educational process, which determines not only the success of mastering pedagogical activity, increasing the productivity of training and education, but also the property that contributes to the self-realization of the teacher as a person [6]. According to G. Selye, stress or adaptation syndrome is "a non-specific reaction of the body in response to a variety of damaging effects" [7]. The author developed a three-component structure of stress: he identified three stages of this process: the stage of anxiety; the stage of resistance (adaptation) and the stage of exhaustion. According to V. V. Suvorova, stress is "a functional state of the body that occurs as a result of an external negative impact on its mental functions, nervous processes or the activity of peripheral organs" [6]. B. H. Vardanyan defines stress resistance as a special interaction of all components of mental activity, including emotional ones. He writes that stress tolerance "… can be more specifically defined as a personality property that provides a harmonious relationship between all components of mental activity in an emotionogenic situation and, thereby, contributes to the successful performance of the activity" [2]. According to most scientists, "stress tolerance" includes such particular components as emotional stability, psychological resistance to stress, stress resistance, and frustration tolerance. According to the empirical research of A. A. Andreeva, the factors influencing the development of stress tolerance and a positive attitude to educational activities are: external factors—pedagogical influence, interpersonal interaction, the system of activities in the university, stimulation; and internal-psychophysiological, socio-psychological, psychological and pedagogical features of the individual. The author notes that the stress resistance of students in educational activities is a complex property of the individual, which is characterized by the necessary adaptation of the student to the influence of external and internal factors in the course of educational activities [8]. Scientists have found that increasing stress tolerance in university students not only contributes to the creation of an emotionally stable, psychologically healthy personality, capable of self-regulation, successful adaptation and socialization in society, but also contributes to the development of a positive attitude to educational activities, which increases the effectiveness of its indicators. Lozgacheva, in her work on the formation of stress resistance at the stage of professionalization, understands stress resistance as a "complex property of a person", which is characterized by the necessary degree of adaptation of an individual to the effects of extreme or cumulative external and internal factors in the course of life, due to the level of activation of the resources of the body and the individual's psyche, manifested in indicators of its functional state and performance [9]. When studying the professional activity of a teacher-psychologist, it is important to take into account the influence of professional stress on the personality of a specialist in the process of performing professional functions. Professional stress is considered by scientists as a process of interaction between the individual and the professional environment, in which the requirements of the professional environment significantly exceed the personal and professional resource of the subject of professional activity and threaten his professional development. Samoukina notes that "Professional stress is a tense state of an employee that occurs under the influence of emotionally negative and extreme factors associated with the professional activity performed" [10, 11]. Researchers have established empirically the leading stressful factors in pedagogical activity. These include:

problematic psychological and pedagogical situations and tasks that generate professional difficulties in the mental structures of the teacher. Pedagogical difficulties are determined, on the one hand, by the complexity of the pedagogical situation, and, on the other hand, by the operational and personal resources of the teacher. From the point of view of overcoming stressful situations, the issues related to increasing the stress tolerance of teachers, expanding the repertoire of constructive strategies for overcoming stressful situations in pedagogical activity, and increasing the coping competence of teachers become important. Most stress theories describe the process of coping with stress as a reactive process that occurs immediately after exposure to a stressor. At the same time, it is important to note that often, not the events themselves, but a subjectively high estimate of the probability of occurrence of certain events can become a serious stressful factor. The threat may actually be insignificant or unlikely, but in humans it creates serious stress due to the prediction of resource loss. S. Hobfall in his theory of resource conservation considers stress in terms of loss or threat of loss of resources. According to the author, resources are values that individuals or communities seek to acquire, preserve, multiply and accumulate. Such resource potential for a person largely determines the quality of life and acts as predictors of well-being. A number of personal traits contribute to successful coping with stress. These include: emotional stability and maturity, self-confidence, the need for self-actualization, the ability to manage your actions and actions, to be responsible to yourself for everything that happens—a high level of subjective control. Intrapersonal mechanisms of overcoming stress states are represented by mechanisms of psychological protection and coping mechanisms. Resistance to psychological stress provides a dynamic balance between all the components of mental activity in a difficult situation and thus contributes to the successful implementation of the teacher's professional activities. In modern Russian psychology, A. G. Maklakov develops the concept of personal adaptive potential—a personal property of a person that provides resistance to various stress factors [12]. Adaptive abilities are determined by the characteristics of the individual, according to A. G. Maklakov, creating the possibility of adequate regulation of the psychophysiological state of a person. The more they are developed, the higher the probability of maintaining normal working capacity and effective activity of the teacher under the negative effects of external conditions, the wider the range of stress factors to which he, as a professional, can adapt. The development of a person's stress tolerance, which helps in overcoming the negative consequences of a stressful situation, is associated with the search, preservation and adequate use of resources. In the works of domestic researchers, many components of stress resistance or qualities associated with stress resistance are identified: tolerance, assertiveness, initiative, desire for self-development, communication. S. K. Bondyreva, A. A. Derkach, for example, point to tolerance as a significant professional quality in the prevention of stress resistance. Vodopyanova identifies such characteristics of teachers-psychologists, based on stress resistance, as the ability to social adaptation, the preservation of significant interpersonal relationships, ensuring successful self-realization, achieving goals [13]. Romanova points out such professional characteristics of a teacher-psychologist associated with his resistance to stress as communication skills, psychological and emotional stability, tolerance and non-evaluative attitude to people, the desire for self-knowledge and self-development, the ability to change the ways of solving problems arising in the course of activity in accordance with changing environmental conditions. It is also important to take into account the factor of readiness and awareness of professional choice, which undoubtedly plays an important role in the formation of the adaptation mechanism [14–16]. The individual style of self-regulation is formed due to the interaction of a person with the surrounding world, and its effectiveness is provided by the mechanism of regulation of the level of activation (energy resource) and the variety of forms of behavior (personal and psychological resource). In connection with the study of the mechanisms of voluntary and effective management of the teacher-psychologist's own psychophysiological states, actions, actions, the problem of preserving coping resources for his professional activity becomes important. The resources used by educational psychologists form the potential for coping with unfavorable situations in professional activity, which provides an adaptive function: it gives confidence, supports individuality, and strengthens self-esteem. However, stressful situations require additional resources. Thus, personal maturity contributes to the formation of holistic professional ideas, building a realistic career trajectory and the use of a wide range of constructive ways to overcome stressful situations [17–21]. Therefore, the issues related to the successful functioning of teachers-psychologists, adaptation and development of effective strategies for overcoming stressful situations associated with uncertainty and depending on the activity of the life position of a particular individual, its potential abilities for self-realization are of particular relevance. It is obvious that the achievement of positive results in professional activity is not possible without the study of each teacher-psychologist of the mechanisms of arbitrary and effective management of their own psychological and physiological states, actions, actions. Mastering the skills of highly effective professional overcoming behavior, high stress resistance helps to prevent the occurrence of symptoms of professional burnout and contributes to the preservation of the health and effectiveness of the professional's personality.

2 Research Materials and Methods

The problem of stress has traditionally been of great interest to scientists and practitioners, and links many fundamental and applied studies. Extensive scientific experience on the problem of studying and dealing with stress has been accumulated abroad (V. Wundt, W. James; R. Yerkes, J. Dodson; W. Cannon; T. Cox, K. McKay; R. Lazarus; R. B. Maimo; D. Mechanik; G. Selye; Z. Freeman, etc.). There is a constantly growing interest in the study of the topic of stress and stress resistance in Russian science. Such prominent psychologists as: V. A. Ababkov, V. V. Bodrov, A. V. Waldman, M. M. Kozlovskaya, O. S. Medvedev, N. E. Vodopyanova, G. I. Kositsky, V. M. Smirnov, L. A. Kitaev-Smyk, A. B. Leonova, I. G. Malkina-Pykh, V. L. Marishchuk, K. I. Pogodaev, Yu. V. Shcherbatykh, etc. However, despite the presence of a significant number of works that reveal the psychological problems of stress and stress resistance of the teacher (K. A. Abulkhanova-Slavskaya, V. A. Bodrov, A. M. Bokovikov, P. G. Zilberman, V. Zhuravkova, V. L. Marishchuk, L. M. Mitina, N. A. Podymov and others), in the scientific literature the problem of stress resistance is presented within the framework of personality theories (G. Allport, H. Eysench, R. Cattell, C. Rogers et al.), "theories of achievement motivation" (D. Mc Clelland et al.), theories of homeostasis (U. Kennon) and stress (G. Selye), theories of temperament and properties of the nervous system (I. P. Pavlov, V. D. Nebylitsyn, V. S. Merlin, etc.), theories of activity reliability (B. F. Lomov, etc.), concepts of professional self-awareness (L. M. Mitina). The professional activity of a modern teacher-psychologist requires constant and maximum mobilization of his own resources, making demands on his stress resistance and adaptability. The ability to successfully cope with a state of stress and minimize its negative impact is an important skill for a teacher-psychologist.

Special attention was paid to the value-semantic orientations of the individual (K. A. Abulkhanova-Slavskaya, L. I. Antsyferova, B. S. Bratus, F. E. Vasilyuk, M. Yu. Denisov, V. S. Nikolaeva, A. V. Poletaeva, A. O. Prokhorov, I. A. Saparova, V. Frankl, etc.), as well as the role of cognitive assessment in the emergence and development of a stress reaction (S. L. Belykh, R. Lazarus, R. P. Milrud, U. Neisser, J. Piaget, H. Remschmidt). Studies were conducted on the influence of gender, age, cultural, and professional characteristics of the individual on the choice and implementation of coping strategies. Argentova, T. Y. Bilgildeeva, A. M. Bogomolov, V. A. Bodrov, V. M. Voynenko, M. A. Kotik, B. F. Lomov, N. A. Sirota, E. V. Topolova, V. M. Yaltonsky, J. Birkimer, F. Blanhard-Fields, G. Seck, L. Sulsky, G. Tome, V. Cleaver).

The analysis of empirical psychological and pedagogical studies devoted to various aspects of professional stress and personal stress resistance has shown that, despite the breadth of research issues, research on the problem of stress resistance of future teachers-psychologists at the stage of training in higher education and the formation of stress resistance as a professionally important quality of a teacher-psychologist is currently insufficiently presented in science.

The purpose of this study was to study the stress resistance of future teachers-psychologists at different stages of training at the university, as well as to identify the features of stress resistance of graduate students in comparison with students of the initial stage of training in the first year.

In our study, to confirm the hypothesis that the indicators of stress resistance in future teachers-psychologists at different stages of training may differ, the following methods were used: the study of self-assessment of stress resistance using the method of S. Cohen and G. Willianson, the questionnaire of perceptual assessment of the type of stress resistance of N. P. Fetiskin, methods of mathematical and statistical processing of the data obtained: descriptive statistics, determination of the reliability of differences: the Fisher criterion.

3 Results and Discussions

The presented empirical study involved 128 subjects, who were future teachers-psychologists of the first and fourth courses of the Southern Federal University. Among them, 9% of boys and 91% of girls, the age range of the subjects is from 18 to 27 years. The study was conducted in several stages. At the first stage, a theoretical analysis of the problem of professional stress and stress tolerance, as a professionally important quality of future teachers-psychologists, was carried out. The selection of methodological tools reflecting the research task is carried out. Then, in two stages, an empirical study was conducted. First, the self-assessment of the respondents' stress resistance was studied according to the method of S. Cowhen and G. Willianson, then the perceptual assessment of the type of stress resistance according to N. P. Fetiskin was studied. The obtained results were subjected to mathematical and statistical processing, the analysis of the obtained results was carried out, and conclusions were drawn.

According to the results of the study of self-assessment of stress resistance of future teachers-psychologists of the first and fourth courses (Fig. 1), it was revealed that almost a third of 1st-year students assess their stress resistance as good (41.2%). A fifth of respondents-18.2%—consider their stress resistance excellent, not much less than a third (27.0%) of respondents have satisfied stress resistance, 10.3%—1st year students rated their stress resistance as poor and 3.3% say about very poor stress resistance.

In the group of 4th year students, the prevailing majority (36.0%) consider their stress resistance excellent, 30.7%—good. Half of the respondents (25.0%) consider their stress tolerance satisfactory, only 6.3%—bad and 3.3%—very bad.

Therefore, it can be assumed that the 4th year students (graduates) are distinguished by a greater confidence in their ability to withstand stress and confidence in their adaptive capabilities in comparison with the 1st year students. To verify this assumption, statistical processing of the obtained data was carried out using the

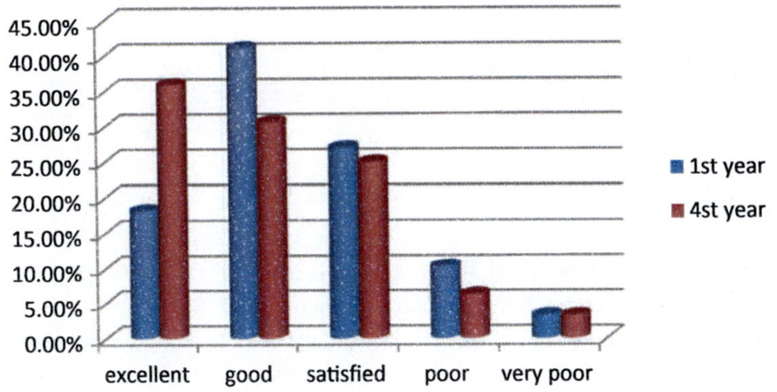

Fig. 1 Self-assessment of stress resistance of 1st and 4th year students in %

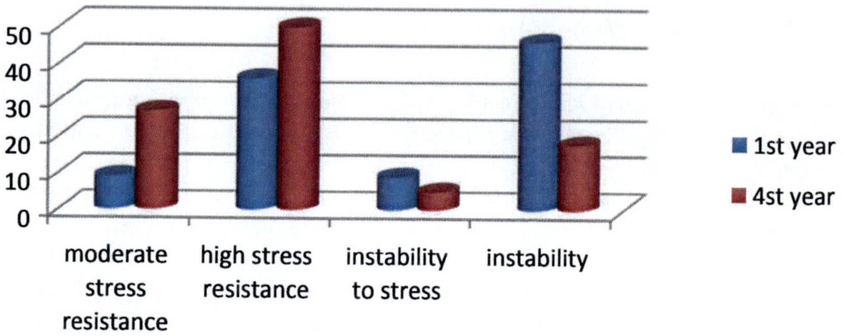

Fig. 2 Perceptual assessment of the type of stress resistance (according to N. P. Fetiskin) of 1st and 4th year students (in %)

Fisher criterion. Significant differences were found in the frequency of occurrence of such assessments as "satisfactory" stress resistance and "very poor" stress resistance. According to other estimates, no significant differences were found. Therefore, we can say that 1st year students are more likely to assess their stress resistance as satisfactory and very poor, they are more likely than 4th year students to feel insecure in their ability to cope with difficulties in personal and educational and professional activities, often unable to control irritation and cope with unforeseen situations that require immediate resolution and responsibility. However, it should be noted that students, both first and fourth years, indicate that they are generally able to cope with difficult, stressful situations in their educational and professional activities and personal life.

Further, according to the results of the study of the perceptual assessment of stress resistance according to the method of N. P. Fetiskin (Fig. 2), it was revealed that half of the 4th year students (50%) have moderate stress resistance, high stress resistance is noted in 27% of graduates, moderate instability to stress was shown by 18% of 4th year students and 5% of graduates are characterized by instability to stress. In the 1st year students, moderate instability to stress prevails—46% of respondents, moderate stress resistance in 39% of first-year students, high stress resistance was observed by 9% of 1st year students and 6% instability to stress.

4 Conclusion

Therefore, it can be assumed that graduates, for the most part, clearly define the goals of their educational and professional activities and choose the best ways to achieve them. They are distinguished by the desire to cope with difficulties independently, the ability to analyze the difficulties encountered and overcome them. Students of the 4th year can work for a long time with great effort, they are able and strive to rationally allocate time. To verify this assumption, statistical processing of the obtained data was carried out using the Fisher criterion. Significant differences were found in the indicators of high stress resistance of 1st and 4th year students and moderate instability to stress. Indicators of moderate stress resistance are in the zone of uncertainty. Therefore, we can conclude that the 4th year students are characterized by a moderate type of stress resistance, and the 1st year students demonstrate a moderate instability to stress.

Thus, the hypothesis of the study that the indicators of stress resistance in future teachers-psychologists indicators of stress resistance differ is confirmed. At the initial stage of training, first-year students showed lower values of stress resistance, in contrast to graduates who complete their studies at the university. Lower indicators of stress tolerance of first-year students can be associated with various factors, both objective and subjective, such as: a change of residence and separation from the parent's home, a different organization of the educational process, in comparison with studying at school; insufficiently formed skills of independent organization of educational and professional activities and life, in general, etc. Higher indicators of stress resistance of graduates can be due to the formed skills of constructive overcoming of stressful situations, skills of psycho-emotional self-regulation, as well as the acquired knowledge and skills, fixed in practical activity, will certainly contribute to high indicators of stress resistance.

References

1. Rogov EI, Zheldochenko LD (2019) Risks of pedagogical profession: the determination of deformations by the sphere of teacher's activity. In: Advances in social science, education and humanities research, vol 321, pp 175–179
2. Vardanyan BKh (2013) Mechanisms of self-regulation of emotional stability. Categories, principles and methods of psychology. Psychological processes
3. Hobfoll SE (1998) Stress, culture, and community. New York
4. Lasarus R (2006) Emotions and interpersonal relationships: toward a person-centered conceptualization of emotions and coping. J Pers 74(1):9–43
5. Shcherbatykh YV (2006) Psikhologiyastressaimetodykorrektsii. St. Petersburg
6. Suvorova VV (1985) Psychophysiology of stress. Pedagogika, Moscow
7. Selye G (2012) At the level of the whole organism. M. Nauka
8. Andreeva EA (2007) Development of stress resistance of students in the conditions of intellectual tests. Diss. cand. psikhol. Nauk. M.
9. Zheldochenko L, Epritskaya N (2019) Representations about the subject of future activities by students in the classroom profile training. In: SHS Web of conferences, vol 70, 08047. https://doi.org/10.1051/shsconf/20197008047
10. Rasskazova EI, Gordeeva TO (2011) Coping strategies in the psychology of stress: approaches, methods and perspectives. Psychol Res 3:17–28
11. Zheldochenko LD (2015) Psychological features of professional deformation of a teacher of a social shelter: Diss. kend. psikhol. Nauk. Rostov-on-Don
12. Maklakov AG (2001) Personal adaptive potential: ego. Mobilization and forecasting in extreme conditions. Psychol J 22(1):16–24
13. Zhuravlev AL, Kryukova TE, Sergienko A (2008) Coping behavior: the current state and prospects in order. M. Publishing House "Institute of Psychology of the Russian Academy of Sciences"

14. Zheldochenko L, Nikolenko O (2020) Relationship of representations about the object of activity and time perspectives for teachers of different ages. In: E3s Web of conferences, vol 210, 18079. https://doi.org/10.1051/e3sconf/202021018079
15. Zheldochenko L, Nikolenko O (2020) The problem of professional identity in the professional development of the individual. In: E3s Web of conferences, vol 210, 22001. https://doi.org/10.1051/e3sconf/202021022001
16. Korochentseva A, Suroedova E, Khachaturyan N, Nikolenko O (2019) Pupil's representation about the socio-psychological qualities of teachers in the process of effective meaning transmission. In: SHS Web of conferences, vol 70, 08020. https://doi.org/10.1051/shsconf/20197008020
17. Rashupkina Ju, Nikolenko O et al (2017) Specific features of students' personal maturity and personal immaturity depending on their social status. In: Gómez Chova L, López Martínez A, Candel Torres I (eds) INTED2017. 11th International technology, education and development conference proceedings. IATED Academy, pp 6300–6303
18. Sheveleva AM, Naumenko MV, Pankratova IA, Rogov EI, Zheldochenko LD (2019) Career orientations and motivation of top and middle-level managers. Opción, Año 35, Especial 21:1590–1607. ISSN 1012-1587, ISSNe: 2477-9385
19. Nikolenko O, Zheldochenko L, Lomova N (2020) Psychological and pedagogical conditions for the formation of value-semantic sphere of students of technical specialties. In: E3s Web of conferences, vol 175, 15029. https://doi.org/10.1051/e3sconf/202017515029
20. Ryumshina L, Pishchik V, Belousova A (2018) Characteristics of socio-psychological adaptation of foreign students to Russian education. In: EDULEARN18 Proceedings. 10th international conference on education and new learning technologies, Palma, Spain, pp 1787–1791. https://doi.org/10.21125/edulearn.2018.0525
21. Kozhukhar G, Belousova A (2017) Psychological resources of reducing the stress factors of defense mechanisms of adult students. In: ICERI2017 Proceedings. 10th International conference of education, research and innovation Seville, Spain, pp 4383–4388. ISBN: 978-84-697-6957-7. Publisher: IATED

Information Literacy and Digital Competence of Teachers in the Age of Digital Transformation

Konstantin Zakharov, Aleksandra Komarova, Tatiana Baranova, and Elena Gulk

Abstract The study discusses the specifics of digital literacy and digital competence of a big community of teachers in one of the regions of St. Petersburg. It contains the calculations of digital literacy indices and digital competency index of the teachers that were made using the range of diagnostic tools of NAFI research center. It was found that today teachers are most advanced in content and assessment, while digital resource technology and management is the least developed field of pedagogical activity where digital technologies are applied. Two levels of digital transformation are introduced in the teaching and learning process—replacement and improvement, to ensure that the zone of traditional teaching is operative. For teachers to be able to transfer successfully to the zone of pedagogical engineering, programs have to be developed to provide psychological and pedagogical support as they improve their ICT competence.

Keywords Information literacy · ICT Competence · Digital transformation · Inventions in pedagogy · Digital competence · Digital literacy index · Digital competence index · Pedagogical engineering

1 Introduction

The history of pedagogy cannot boast of many discoveries that transformed the approach to teaching younger generation in a cardinal way. However, some can be mentioned. The first one is the appearance of the teacher, a person who helped a boy or a girl to learn about the culture of his or her society and become its competent member. The second is the appearance of a written language. That is when literacy was mentioned for the first time. The primary task of a teacher is to teach the learner how to read and write. The third invention is book-printing and mass spread of books that contributed to the cultural, scientific and technological revolution.

K. Zakharov (✉) · A. Komarova · T. Baranova · E. Gulk
Peter the Great St. Petersburg Polytechnic University, 29, Polytechnicheskaya Str., 195251 St. Petersburg, Russia

A. Beskopylny and M. Shamtsyan (eds.), *XIV International Scientific Conference "INTERAGROMASH 2021"*, Lecture Notes in Networks and Systems 247,
https://doi.org/10.1007/978-3-030-80946-1_78

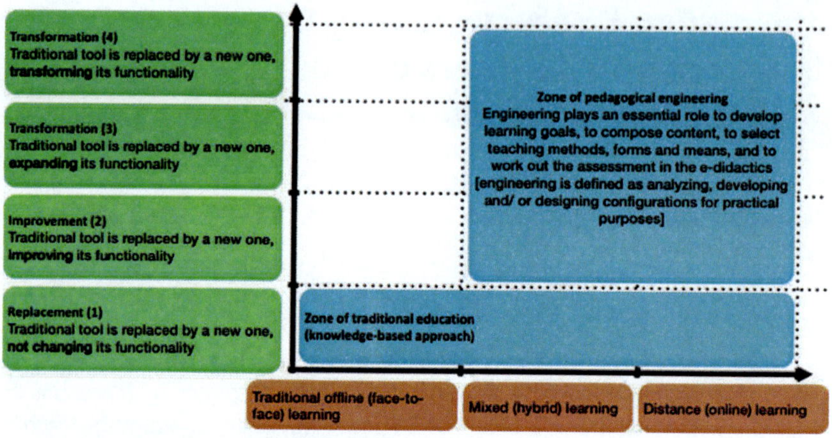

Fig. 1 Four change levels of pedagogy thanks to digital transformation

In the sixteenth–seventeenth century Europe saw the fourth invention—a subject-classroom-lesson system of education, which quickly became traditional and has not lost its significance to this day. Finally, today we all witness the appearance of the fifth invention—the birth of the Internet, digital environment, which leads to rapid digital transformation of society and, among other things, affects education. We understand digital transformation in education as an update of all the structural–functional components of pedagogical systems for improving the learning outcomes of every learner (Uvarov A.) [1]. Following Choshanov [2], Uvarov [1], digital transformation in education can be represented in the following diagram showing the change levels of pedagogy (see Fig. 1).

The ordinate is the level of digital transformation in the educational process, which is known in English as the SAMR model (The Substitution Augmentation Modification Redefinition Model, Puentedura, R.) [3], and the abscissa is the learning format from traditional offline learning (face to face) through hybrid, blended learning to online distance learning. It is seen from the Figure that the zone of traditional teaching occurs at the substitution level, while entering the zone of pedagogical engineering, the zone of electronic didactics (e-didactics) [4] requires that the teacher should have new digital competencies that will allow him or her not only to improve, but also change and transform digital educational environment.

The situation is getting even more complicated because society has been experiencing systemic changes, which affect the education system among other things. In the middle of the previous century, change in technology matched the change of generations and has been accelerating ever since. It is believed that today technologies change every 5–7 years, while generations change every 25. This phenomenon is called the law of time. The education system has faced a challenge. It has to teach the younger generation how to learn, acquire knowledge by themselves, quickly adapt to changes, deal with uncertainty, be capable of multitask, which means that the teaching goal is changing. The world community responds to this challenge with some UN documents [5], and in the Incheon Declaration for the implementation of Sustainable Development Goal 4, UNESCO explicitly declares the promotion of lifelong learning opportunities for everyone [6].

The ordinate is the level of digital transformation in the educational process, which is known in English as the SAMR model (The Substitution Augmentation Modification Redefinition Model, Puentedura, R.) [3], and the abscissa is the learning format from traditional offline learning (face to face) through hybrid, blended learning to online distance learning. It is seen from the Figure that the zone of traditional teaching occurs at the substitution level, while entering the zone of pedagogical engineering, the zone of electronic didactics (e-didactics) [4] requires that the teacher should have new digital competencies that will allow him or her not only to improve, but also change and transform digital educational environment.

The question is whether teachers, whose job is to prepare students for a successful life in this rapidly changing world, are ready for these challenges? The teachers of the older generation remember the time when there were no computers or the Internet, while those who graduate from school today cannot imagine their life without being able to access social media on their phone.

Researchers should ask themselves a question: do digital literacy and digital competence of teachers correspond to the level of digital literacy and digital competence of students? This hypothesis has already been tested and it turned out that teachers' competence not only corresponds to that of their students, but they are ahead of them. The studies were carried out by a group of scientists including Soldatova et al. [7] and experts from the NAFI research center, such as Aimaletdinov et al. [8]. Their conclusions gave us an idea to study digital literacy and competence of teachers in one of St. Petersburg's districts.

The term "digital literacy" was used by P. Gilster as early as in 1997 and has been undergoing changes since then. Basically, it is about the ability to perceive information from various Internet sources. But recently, the emphasis in defining digital literacy has shifted to the cognitive skills of an individual. It is the person who should be in the center of attention of researchers [9]. Consequently, today multidimensional approaches to the concept of "digital literacy" have been adopted and presented in the work by Kewman et al. [10].

In Russia, digital literacy and competence of teachers and educators was researched by G. U. Soldatova (Moscow State University) [11], and S. G. Davydov (HSE). Sharikov [12] proposed a "four-component digital literacy model", according to which the digital literacy index can be measured.

To measure "digital literacy" this concept has to be operationalized, i.e. scales and a diagnostic methodology have to be developed. The attempts to do so were first made in 2005–2006 within the DigEuLit project [13], then in 2011 [14], and later the approach was improved by a group of experts at the G20 Summit held in Berlin in April 2017 [15]. The NAFI research center took it to develop a diagnostic tool for determining digital literacy and competence index, which we used [8]. Thus, teachers' digital literacy index is a generalized indicator showing the level of teachers' readiness for digital economy and based on the assessment of five indicators of information, computer, communication and media literacy, as well as on the attitude to technological innovation.

Digital transformation of the entire society finds it way in education setting new requirements for teachers on the part of the administration of educational institutions, parents, and students. Both positive and negative tendencies of this process should be considered [16]. Since 2015, the digital literacy index (DIC) has been measured at the government level by the regional public organization called "Center for Internet Technologies!" (ROCIT) together with the All-Russian Center for the Study of Public Opinion (FCIOM) [17].

2 Methodology and Methods

The study, conducted by the authors in September–December 2020, used the approach of the NAFI research center. To calculate the ICT Competence Index, which measures the readiness of teachers to actively use ICT in the educational process, the European Digital Competence Framework of Educators (DigCompEdu) was used. It details 22 competencies, grouped into six blocks including Professional Engagement, Digital Resources, Teaching and Learning, Assessment, Empowering Learners, Facilitating Learners' digital competence [18]. Competencies correspond to six levels of experience: Newcomer, Explorer, Integrator, Expert, Leader, and Pioneer.

The goal is to study the digital competence of teachers in secondary schools. The tasks are to assess the general level of digital competence of teachers in secondary schools; identify social factors that contribute to digital competence of teachers working in educational organizations.

The sample consisted of 365 teachers working in secondary schools of St. Petersburg, Russia.

The sample included the following number of teachers by age groups: 100 people aged from 23 to 34 years old, 114 people from 35 to 49 years old, 141 people aged 50 to 64 years old and 10 people over 65 years old.

349 women and 16 men took part in the study.

115 people are teachers at primary schools, 234 people work in secondary schools, 16 people work with all categories of students.

As for the years of teaching experience, the numerical distribution was approximately the same, with the exception of the largest group whose work experience exceeded 30 years (89 people).

As for the experience in using digital technologies, the following groups were the most numerous: 3–5 years (73 people), 6–10 years (110 people) and 11–15 years (74 people), the rest being approximately equal.

238 teachers received advanced training, 127 teachers did not.

The largest group of teachers include those who teach the subjects of the humanitarian cycle (114 people), the groups of teachers who teach other subjects are more or less equal in number.

Methodological tools. The survey methods included a biographical questionnaire and the questionnaire "ICT Competence of Teachers", created by the NAFI Analytical Center for the Russian educational system on the basis of the European Digital Competence Framework 2.0, 2018. These were presented for filling to the respondents in a Google form.

The methods of mathematical and statistical data processing included determining differences: the Mann–Whitney U test, the Kruskal–Wallis test, and Pearson's chi-squared test.

The empirical data were processed using the SPSS software. Statistics. Version 23.

3 Results

The digital competence assessment showed that teachers in general education schools have an average level of ICT competence equal to 44 out of 88 points (50%). At the same time, there is a differentiation in terms of the qualitative content of the acquired experience (see Fig. 2). The results obtained showed that 48.54% of teachers belong to the "Integrator" group, 26.3% to the "Expert" group, 16.9% to the "Explorer" group, 5.0% to the "Leader" group, 3.2% to the "Newcomer" group. No teacher is experienced enough to be in the "Pioneer" group.

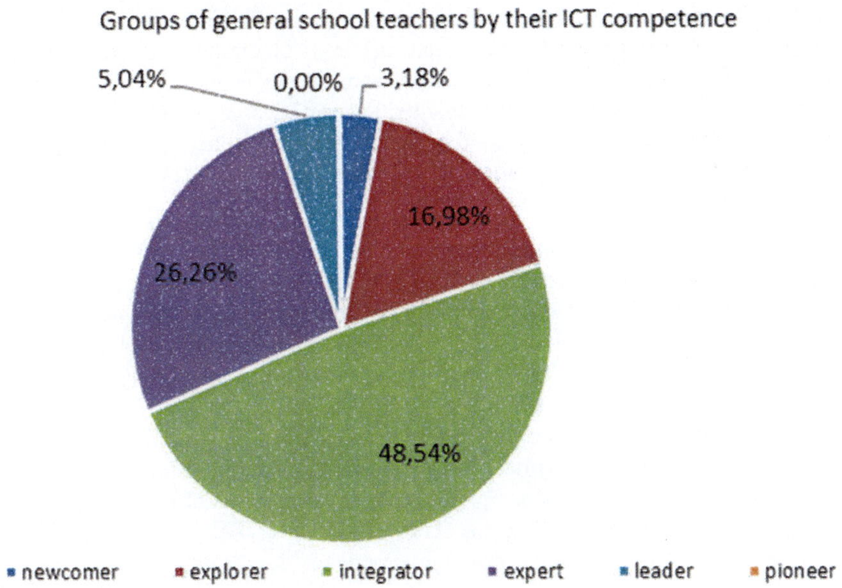

Fig. 2 Distribution of teachers by groups according to the level of ICT competence

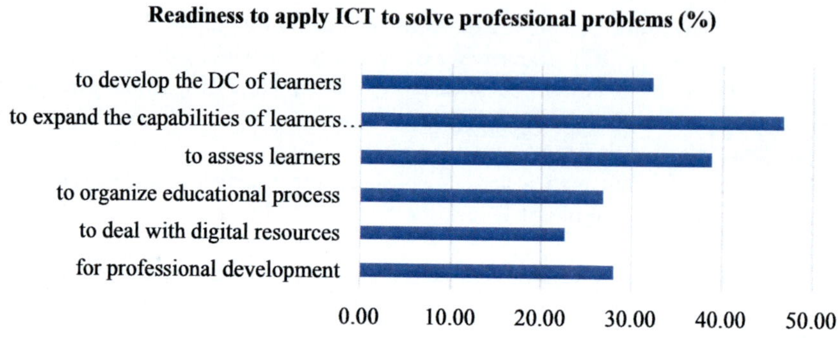

Fig. 3 The degree to which teachers working in general educational organizations are ready to use ICT to solve professional problems

It was also found that the teachers' willingness to use ICT to deal with professional problems is different depending on the type of tasks (see Fig. 3).

The respondents' willingness to use ICT so as to empower students to be independent in the educational process took the first place (noted by 46.8% of the teachers). In the second place (38.6%) teachers indicated their readiness to use ICT for student assessment, the third place (32.3%) was about the wish to develop students' digital literacy, in the fourth place (26.9%) was the organizing of the educational process, the fifth cause (28.1%) was for the teachers' own professional development.

In the last place (indicated by 22.6% of the teachers) was their willingness to work with digital resources in order to create, adapt, manage digital resources, etc.

To identify the factors that determine the level of digital competence of the teachers, the sample was differentiated by age, gender, work experience, experience in using ICT, students' age category, and participation in advanced training programs. What is more, the secondary school teachers were divided into groups depending on the subject they taught.

The following results were obtained:

1. The ICT Competence Index depends on their age (the general Kruskal–Wallis criterion at $p = 0.03$). Teachers with the highest ICT Competence Index (53.8%) are aged 35–49. Pairwise comparison of different age groups showed that the teachers of this age group are ahead of the teachers from other groups (the group aged under 35 (significance level $p = 0.004$) and the group aged 50–65 (significance level $p = 0.026$)).
2. There were no differences in the ICT Competence Index between female and male teachers (Mann–Whitney test at $p = 0.431$). The conclusion made suggests just a trend, because the samples are not equal in terms of composition (349 female teachers and 16 male teachers).
3. Teachers having dissimilar teaching experience do not differ in terms of the digital competence index (Kruskal–Wallis criterion = 8.891 ($p = 0.261$), and

in terms of the levels of experience in using digital technologies (Pearson's chi-squared test = 35.46 (at $p = 0.157$)).

4. The teachers' levels of proficiency in digital technologies and the general ICT Competence Index are predetermined by their experience in using ICT (Pearson's chi-squared test = 107.2 (at $p = 0.0001$)). The highest indicator of the ICT Competence Index is in the group of teachers who have been using digital technologies for 16 years or more. Of all the teachers in this group, 20.9% are at the "Leader" level, 27.9% are at the "Expert" level, 46.5% are at the "Integrator" level.
5. It was found that the teachers who work with different categories of students are at different levels of mastering digital technology (Pearson's chi squared test = 18.59 (at $p = 0.017$). The ICT Competency Index is higher for those teachers who teach adolescents (Kruskal–Wallis test = 20.12 ($p = 0.001$)). The teachers working in primary schools are distributed by their level of proficiency in digital technologies as follows: "Integrator" (56.5%), "Explorer" (21.7%), "Expert" (15.7%), "Newcomer" (4.3%) and "Leader" (1.8%). Secondary school teachers have the following distribution of the level of ICT competence: "Integrator" (45.7%), "Expert" (30.8%), "Explorer" (15%), "Leader" (6.8%), "Newcomer" (1.7%) (see Fig. 4). As noted above, the "Pioneer" group is not represented in this sample at all.
6. The ICT Competence Index is higher (at a reliably significant level, established using the U-Mann–Whitney criterion ($p = 0.021$)) among the teachers who have been improving their qualifications in the field of digital technologies over the past 2 years.
7. It was found that the general digital competence index of the teachers teaching various subjects does not differ (the general Kruskal–Walliss criterion = 3.19

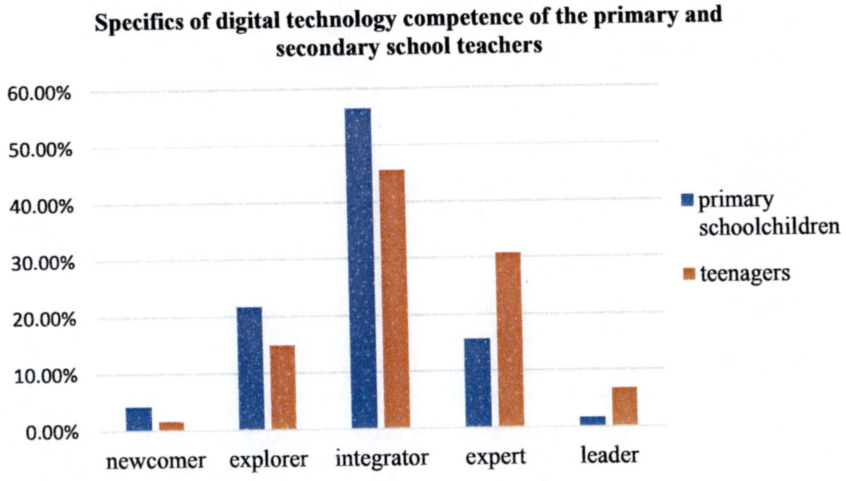

Fig. 4 Groups of primary and secondary school teachers by their level of digital competence

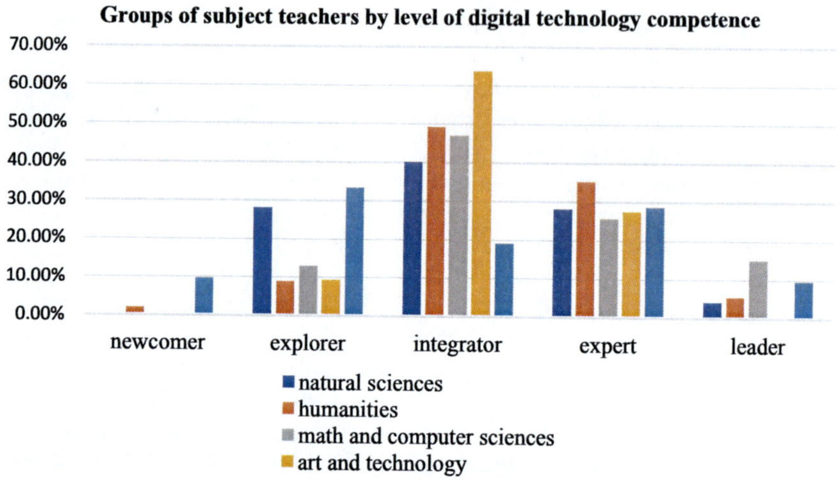

Fig. 5 Groups of subject teachers by their level of competence in digital technologies

at $p = 0.538$). Nevertheless, a significant difference was identified in the distribution of subject teachers by groups depending on their level of experience in digital technologies (Pearson's chi squared test = 33.93 (at $p = 0.006$)) (see Fig. 5).

The math and computer science teachers have the greatest experience and a high level of readiness to apply digital technologies (14.9% of teachers are at the "Leader" level, 25.5% are at the "Expert" level, 46.8% are Integrators, 12.8% are Explorers). The second place in using ICT is occupied by the teachers of the humanitarian cycle of disciplines. Their digital experience is mostly associated with the functions of an Integrator (49.1%) and an Expert (35.1%). The data on the teachers of the natural science cycle correspond to the laws of normal distribution (the predominant function is "Integrator" (40%), "Explorer" and "Expert" take up 28% each, "Leader" is 4%, "Newcomer" is 0%), which gives us an idea of the typicality of the results obtained for the group of subject teachers being considered.

Art and technology teachers are represented mainly by two levels of ICT competence: "Integrator" (63.6%) and "Expert" (27.3%), with the dominating function of searching for opportunities when digital technologies are used by the teachers in their own pedagogical practice.

The group of teachers who teach physical education and the basics of health and safety (H&S) turned out to be heterogeneous in their experience in mastering digital technologies. In this group, the teachers are differentiated as those with a low level of ICT competence ("Newcomers"—9.5% and "Explorers"—33.3%) and with a relatively high level ("Experts" amount for 28.6% and "Leaders" amount for 9.5%). To identify the causes of differentiation, further research is needed to find out how effective psychological and pedagogical support can be provided to the teachers in this group.

4 Discussion

The results obtained in the study are consistent with the data obtained by T. A. Aimaletdinov, L. R. Baimuratova, O. A. Zaitseva, G. R. Imaeva, L. V. Spiridonov and others, but also have a number of specific features associated, as we think, with the period of the study, when digital technologies have been used massively and there has been a sharp increase in getting this experience.

The teachers are mostly ready to use digital technologies to solve the following problems: to expand the degree of students' independence in the educational process, to monitor and assess students. Much less readiness has been demonstrated by the teachers to use digital technologies to resolve the problems of students' development and their own development. The greatest difficulty is caused by the problems associated with the use of digital resources, i.e., their adaptation, creation, management, etc.

About half of the teachers (46.8%) are ready to integrate digital technologies into the educational process. This figure is higher than in the studies conducted by other authors, which is due to the experience that the teachers gained when they organized distance education during the pandemic. In addition, according to this study the number of teachers having an expert level of ICT competence is significantly lower (26.26%), which is caused by the lack of demand for a wide range of digital technologies in the educational institutions and the limited number of problems to be solved. The level of a "leader teacher" in the field of digital technologies is extremely low and there no innovation level has been registered.

Since the results obtained correlate with the SAMR model presented above, we can conclude about the first two levels of digital transformation implemented in the educational process, such as "substitution" and "improvement". These two levels ensure that the zone of traditional education will keep on functioning and only prepare us for the transition to the zone of pedagogical engineering.

Since the sample of the teachers was not homogeneous, we identified some factors that affect the levels of teachers' readiness to use digital technologies and be involved in pedagogical engineering as a whole. These factors include the teachers' age, their experience in using ICT, the educational tier they work with, the specifics and needs of the subject they teach, as well as the teachers' own advancement in the field of information technology.

We believe that a higher level of ICT competence among the teachers aged 35–49 is associated both with a higher level of professional competence in general and with the level of development of the motivational and conceptual sphere of the personality of the teachers in this age group. The study of the motivational and conceptual features of using digital technologies seems to be a promising area of psychological and pedagogical research.

It is quite predictable to say that with growing experience in working with digital technologies, digital competence grows, but at the same time if there is no connection between digital competence and pedagogical activity, the teachers develop their ICT competence outside the field of their main professional activities. In addition, the identified positive effect from taking part in advanced training courses in digital technologies indicates that pedagogical and digital practices can be combined so as to make a step towards entering the zone of pedagogical engineering.

Secondary school teachers are twice as likely as primary school teachers to demonstrate their willingness to use digital technologies in a creative, critical, and consistent manner, and there are fewer Newcomers in this field. These facts are consistent with the demand for experience in the application of digital technologies at the secondary level, and also explain the differences in the levels of ICT competence depending on the subject taught.

The math and computer science teachers are ahead of their peers who teach other subjects in terms of their experience in using digital technologies ("Explorer", "Expert" and "Leader" levels prevail) and are more prepared to apply them in a comprehensive and consistent way. The second and third places are occupied by teachers of humanities and natural sciences. The results of those teaching art and technology show that their competence in digital technologies at the intermediate level, with the "Integrator" level being prevailing. The results of physical education and L&S teachers are ambiguous and indicate the heterogeneity of the digital competence of these subject teachers. A more detailed study has to be carried out to consider not just groups—subjects, but rather individual subjects.

5 Conclusions

Teachers of general educational organizations have an average level of ICT competence. At the same time, they are more open to solving the problems of integrating digital technologies into pedagogical practice and expanding their professional experience. To a lesser extent they are ready to do it comprehensively and consistently, in particular, to train other teachers or to be involved in innovative activities using sophisticated digital technologies.

The area of pedagogical activity most involved in the use of digital technologies is the content, assessment and development area, the least involved is technology and digital resource management.

The digital competence of teachers depends on:

their age (it is higher for teachers aged 35–49);

their experience in using ICT (it is higher for teachers who have experience in using ICT for 16 years or more);

the educational level of students they work with (secondary school teachers are more prepared for creative, critical and consistent use of digital technologies);

the specifics and requirements of the subject they teach (math and computer science teachers have a higher level of ICT competence, the average level of teachers of the humanities, natural science cycle and art and technology teachers, the group of physical education and H&S teachers is heterogeneous in terms of digital competence);

the teachers' own efforts in the field of information technology (the ICT Competency Index is higher if the teachers have undergone advanced training over the past two years).

No relationships were found between the level of digital competence, the gender or teaching experience of the teachers working in general educational organizations, as well as the experience of teaching.

When correlating the results of the study of the teachers' ICT competence with the SMAR model, it can be noted that there are two levels of implementation of digital transformation in the educational process—"substitution" and "improvement". These two levels ensure that the zone of traditional education will keep on functioning and only prepare us for the transition to the zone of pedagogical engineering.

Further research can be dedicated to studying the teachers' motivational and conceptual readiness for digital transformation.

Also, based on the results obtained and the conclusions drawn, programs can be developed to provide psychological and pedagogical support to teachers so that they can transfer to the zone of pedagogical engineering.

References

1. Uvarov A (2019) Difficulties and prospects of digital transformation of education. Higher School of Economics
2. Choshanov M (2011) Didactics and engineering. BINOM. Knowledge Lab, Moscow
3. Puentedura R (2012) Thinking about change in learning and technology. Presentation given 25 Sept 2012 at the 1st Global mobile learning conference, Al Ain, UAE
4. D'Angelo G (2007) From didactics to e-didactics: e-learning paradigms, models and techniques. Liguori, Napoli
5. The United Nations Organization (2015) Transforming our world: the 2030 agenda for sustainable development. https://sustainabledevelopment.un.org/post2015/transformingourworld. Date of retrieval 03.03.2021
6. UNESCO Institute for Lifelong Learning (2010) CONFINTEA VI: the six international conference on adult education: final report. http://unesdoc.unesco.org/images/0018/001877/187790e.pdf. Date of retrieval 03.03.2021
7. Soldatova GU, Nestik TA, Rasskazova EI, Zotova EYu (2013) Digital competence of adolescents and their parents. Internet Development Fund, Moscow
8. Aimaletdinov TA, Baimuratova LR, Zaitseva OA, Imaeva GR, Spiridonova LV (2019) Digital literacy of Russian teachers. NAFI Publishing House, Moscow
9. Chan BS, Churchill D, Chiu TK (2017) Digital literacy learning in higher education through digital storytelling approach. J Int Educ Res (HER) 13(1):2
10. Kewman ML, Sohee P, Bang GJ, Byeong-Young C (2019) Multidimensional approaches to examining digital literacies in the contemporary global. Soc Media Commun 7(2):37
11. Soldatova GU, Shlyapnikov VN (2015) Digital competence of Russian educators. Psychol Sci Educ 20(4):5–18
12. Sharikov A (2018) The concepts of digital literacy: experience from Russia. Commun Media Design 3(3):112
13. Grudziecki M (2006) DigEuLit: concepts and tools for digital literacy development

14. Wilson C, Grizzle A, Tuazon R, Akyempong K, Cheung C-K (2011) Media and information literacy: curriculum for teachers. UNESCO
15. Chetty K, Wenwei L, Josie J, Shenglin B (2017) Bridging the digital divide: measuring digital literacy
16. Sokolova IV, Mudrakova OA (2013) IT penetration in society and education: modern theory and practice, Moscow
17. Digital Literacy Index [Electronic resource]. Moscow 2015. https://wciom.ru/fileadmin/file/reports_conferences/2015/2015-12-21-rocit.pdf. Date of retrieval 04.03.2021
18. The Digital Competence Framework [Electronic resource] (2018). https://ec.europa.eu/jrc/en/digcomp/digital-competence-framework. Date of retrieval: 04.04.2021

Provision and Management of Educational Activities in the Conditions of Distance Education at the North-Eastern Federal University

Tatyana Tretyakova, Elizabeth Barakhsanova, Tuyara Alexeeva, and Inna Bogushevich

Abstract This paper presents the experience and problems of forming electronic information educational environment in the North-Eastern Federal University, identifies the importance and urgency of developing management systems in a digital format and the impact of territorial, climatic, demographic and national characteristics of the region on the development of distance education. The peculiarities of the organization of electronic information educational environment in conditions of the identity of the requirements of federal state educational standards in educational institutions of higher education were disclosed. It is shown how the organization of work in digital format in conditions of self-isolation during the spread of a new coronavirus infection has shown itself in practice. This study predetermines the need for further theoretical analysis and practical use of academic disciplines implemented in online format, the development of scientific and methodological foundations for the organization of the educational process, the organization of different types of practices in the distance mode. All this requires improving and updating the systems of electronic information educational environment to provide individual educational trajectories, development of digital academic mobility of students to participate in educational courses and programs of leading domestic and foreign universities, creating their own online courses to develop inbound digital mobility.

Keywords Digital format · Distance education · Higher education · Electronic information educational environment

1 Introduction

The educational policy of the country through the introduction of the National Project "Education", the Strategy for Development of Information Society in the Russian

T. Tretyakova · E. Barakhsanova (✉) · T. Alexeeva · I. Bogushevich
M. K. Ammosov North-Eastern Federal University, Yakutsk, Republic of Sakha (Yakutia) 677000, Russia

A. Beskopylny and M. Shamtsyan (eds.), *XIV International Scientific Conference "INTERAGROMASH 2021"*, Lecture Notes in Networks and Systems 247,
https://doi.org/10.1007/978-3-030-80946-1_79

Federation for 2017–2030 and the priority project "Modern Digital Educational Environment in the Russian Federation" involves the consistent development and implementation of digital technologies. In turn, this requires the provision of material computer park and the development of software products of educational organizations, training of teaching staff (according to these projects only in 2020 it was planned to train at least 6 million teachers and educators in Russia as a whole).

In this study, we introduce the concept of "Digital Educational Ecosystem". In our opinion, it is a complex social system, which includes a set of the following components: information and telecommunication system, information systems, software applications, mobile devices, participants in the educational process (students and university staff), employers, which in general form the infrastructure of information ecosystem and represent a full range of information support of the educational process. Such ecosystem is in demand in the republic with difficult climatic conditions, transport and information communications, the great remoteness of the regions and the high cost of travel to the centers where professional educational organizations operate. These and other problems significantly reduce the possibility of providing accessible high-quality vocational education to citizens from remote settlements.

The digital educational ecosystem proposed by the NEFU Pedagogical Institute as an integral socially significant multilevel system whose participants are students, teachers of the Institute, teachers and employees of the republic's general education system is represented by educational organizations of all levels. It assumes maintenance of interaction between participants of educational process at online level through realization of the educational programs actual for a customer and improvement of quality of rendering of educational services important for management structures in system of general education, the educational organizations and a pedagogical university. This project provides a full cycle of teacher training, which begins with professional orientation and strategies for self-determination of a schoolchild as a future teacher in school, determining the individual trajectory of professional education in high school, successful employment and ends with the provision of opportunities for further continuous professional growth [1].

The purpose of the study-based on the analysis of the availability and use of digital technologies and distance learning technologies, the practice of implementation of these technologies in the university to show the need and ability of the educational process and its management in the digital format, to formulate recommendations and suggestions for optimizing the educational process of the university.

2 Materials and Methods of Research

To achieve the goal of the study and to solve its tasks, the following research methods were used: theoretical (analysis of pedagogical works and educational and methodological publications; study of FSES HE, legislative and regulatory documents, etc.), comprehensive and systematic study of the object of research; comparison of different objects of research with each other; generalization of theoretical and practical experience, etc.

Theoretical foundations for the study in the aspect of providing and managing educational activities in the conditions of distance education were the works of Russian and foreign authors published over the past 5 years in foreign and domestic publications. The articles of E. Z. Vlasova (Herzen Russian State Pedagogical University, St. Petersburg) substantiate the need for special ICT-training of teachers in the North to implement e-learning in their professional activities [2]. The works of E. A. Barakhsanova (M. K. Ammossov NEFU) reflect the interest in the problems of developing online educational models related to the training of future school teachers in an electronic information and educational environment that takes into account the regional characteristics of the North [3, 4]. M. S. Prokopyev's research is devoted to the study of the issues of teaching methodology of specific disciplines in online ways [5]. A collective article by M. A. Sorochinsky with teachers of other higher education institutions in the country shows the relationship between social and professional in the self-determination of an individual, concludes that professional self-determination is inextricably linked with the self-realization of an individual in other important spheres of life [6]. In the analyzed works the results of the study showed that e-learning makes the process of knowledge exchange more intensive, actualizes the process of knowledge extraction by students themselves, which is important for solving the problem of providing and managing educational activities in distance education in remote rural schools of the region.

The control and assessment activity, both at the level of higher education and at other stages of youth education, is of particular importance for improving the quality of education. In foreign practice much attention is paid to ensuring the reliability of educational achievements assessment based on the method of reasoning and Evidence-Centered Design assessment [7–9]. This technology is quite complex and requires the development of several models to design a comprehensive assessment tool. Among them the important ones are the student model (construct of assessed characteristics) and the model of a complex practice-oriented task, preferably on an interdisciplinary content basis. Various forms of data collection and methods of statistical processing of results are used to obtain authentic assessment. According to Messick, S. properly created constructs contribute to the development of relevant assignments, as well as the preparation of an appropriate scoring system and assessment rubrics [10]. When building an internal system of university assessment we paid special attention to the content of assessment tools, taking into account domestic and foreign experience in this area.

T. V. Tretiakova's research on the formation of education quality assessment systems emphasizes the urgent need to use automated remote systems to record students' educational achievements, based on longitudinal studies of quality through the theory and practice of pedagogical measurement [11]. Accordingly, the formation of funds of assessment tools in a systemic form includes codifiers of the controlled area, databases of assessment tools with assessment criteria, assessment materials and tools, a set of competence-oriented tasks, measurement scales, software and tools and databases, appropriate methodological materials [12]. This led to the substantive content of the electronic information educational system of the university to obtain reliable assessments. It should be kept in mind that the same data can be good evidence for one conclusion, but bad evidence for another [13]. The better the assignments reflect the goals of verification, the more effective the process is in the context of obtaining a reliable assessment.

3 Results

The need to develop distance education in the northern territories, in our opinion, is especially urgent. This is primarily due to the territorial, and indirectly also to the climatic, demographic and national peculiarities of the region. In particular, the peculiarity of ways to develop distance education in Sakha Yakutia is due to the fact that the region currently has a predominance of rural settlements (89% of the total number), located at a fairly large distance from each other. The average distance between ulus centers is 173 km, low population density is 0.1 persons per km^2. Lack of sufficient communications in the uluses, weak Internet, interruption of transport communication between settlements during the period of ice and mudflows, often lasting up to 5 months. Citizens of more than 120 nationalities live in the Republic, and their different mentalities form different stereotypes of educational behavior. All this creates certain problems in the organization of the educational process in real mode and provision of quality education, the main of which are: lack or weak Internet, lack or shortage of qualified personnel, especially in the uluses belonging to the Arctic zone of the Russian Federation, presence of economic problems with formation of sufficient class occupancy, lack of social communication in conditions of small number of classes. The introduction of distance education, implemented in a digital format, could solve these problems.

A special place is occupied by the organization of training and professional competencies of the students of vocational education organizations. The accessibility of learning materials and the organization of the learning process in all aspects, including the organization of internships and the preparation of graduate qualification work, the timeliness of curriculum implementation, the possibility of communication to ensure constant communication with teachers and classmates, the ability to organize group work in digital format—all these are problems whose solution is possible only with appropriate organization of electronic information and educational environment of the professional educational organization. Problems requiring practical solutions are mastering the methods and ways of independent work and self-assessment in self-isolation, organization of remote control and evaluation of educational achievements, etc. [14]. A reliable assessment has a positive effect on the educational motivation of students, increases interest in mastering the subject content, in turn, it requires motivation of students during assessment [15, 16].

Our study is devoted to practical options for solving the above-mentioned problems. First, a bit of statistics showing the number of implemented educational programs and the contingent of students. At present, 12 institutes, 6 faculties, and 2 colleges of NEFU implement 438 basic professional educational programs, including 124 master's programs, 53 postgraduate programs, and 165 additional professional education programs. The urgency of the problem of distance education organization is also dictated by the fact that more than 18,000 students from 42 regions of the Russian Federation and 38 foreign countries are studying at the university today. Different political, social and personal realities require the use of temporary or permanent distance education modes for students.

In accordance with the requirements of the Federal State Educational Standards (FSES), the NEFU EIOS provides students in basic vocational educational programs of secondary and higher education with access to curricula, working programs of disciplines (modules), practicum programs, publications of electronic library systems and electronic educational resources specified in the working programs of disciplines (modules), practicum programs. The formation of an electronic portfolio, including the preservation of student work, reviews and evaluations of the work by the participants of the educational process is provided. There is a record of the educational process, the results of interim certification and the results of the development of the basic professional educational program; all types of classes, procedures for assessing the results of learning, the implementation of which is provided by using e-learning, distance learning technologies; interaction between participants in the educational process, including synchronous and (or) asynchronous interaction through the "Internet".

Electronic information educational environment (EIE) of educational institutions is formed in accordance with the relevant requirements of the federal state educational standards (FSES), and although these requirements for all areas of training, specialties of higher education are formulated identically, but universities have different approaches to the content and functioning of intra-university EIEs. This article shows the possibilities of using electronic information environment to organize distance education in digital format.

Since the NEFU EIOS users are students and employees of NEFU, they are divided into two main groups according to the level of access to the information posted: authorized and unauthorized users. Unauthorized users have access to the elements of the NEFU EIOS, which are in the public domain. Authorized users have access to the SDFU EIOS sections in accordance with the access level.

Below are the structural components of the NEFU EIOS information system, functioning both separately and in integration with other systems:

- "Personal profiles of employees" are correlated with the staff schedule of the university. On the page of the employee there is free access to the following information: contact information, position, education, academic rank and degree, duties, biography, achievements and awards, information about professional development, scientific interests, participation in conferences, length of service, publication activity. When logged in, authorized users, the employee has access to profile editing, the reporting system, student profiles, schedule publication, access to the NEFU electronic library, abstracts of disciplines, layout of the program complex for formation of basic professional educational programs (BEP) of NEFU;

- The NEFU Student Personal Account is available on the NEFU student portal. Authorized users have the opportunity to keep a portfolio, view class schedules and information about instructors through the student's personal account. In addition, they have access to the results of interim certification, educational and methodological documentation for the basic vocational training program. There are opportunities to take questionnaires, use the services of the NEFU Scientific Library, Moodle e-learning system, receive anonymous psychological assistance, apply for an increased state academic scholarship, register for university-wide elective courses, view the news of their educational unit, etc.;
- "Management of VET" provides the formation of relationships and accompanying materials for the main professional educational programs of secondary vocational and higher education;
- The "Integrated Student Personnel Management Information System" is used to record student progress and generate statistical reports;
- "Questionnaires in NEFU" is used to conduct any type of questionnaire, both personalized and anonymous;
- "Electronic schedule" displays information in the personal office of the student and personal office of the employee, it is posted on the general website of NEFU. When logging in, authorized users can see the list of group members, and the student can see the contact information of the instructor;
- "Electives" organizes enrollment in university-wide electives, shows the composition of elective groups and general information about the elective;
- "Electronic sheets" is used to record the results of interim certification, the formation of scholarship protocols for the purpose of assigning state academic scholarship to bachelors, graduate students, residents, to assign increased state academic scholarship (in terms of accounting for the results of the point-rating system);
- "Naukometriya" contains information for accounting the results of research activities, the assignment of increased state academic scholarship for achievements in research activities;
- "Curriculum" is used to form a relationship between the "E-List" and "E-Schedule" to record the results of the educational program. This system is important for the organization of the educational process and directly displays the data in the personal office of the student;
- Organization of all types of classes using e-learning and distance learning technologies is carried out using the electronic distance learning system "Moodle", which contains educational and methodological material on the courses of disciplines, allows for current monitoring of progress and interim certification. "Moodle" is practically introduced in the activities of all schools of the republic and is recommended for use by potential applicants, so this platform is the main product for the organization of the distance learning process in digital format;
- Electronic Library of NEFU provides access to the electronic information systems of the library, which allows remote receipt of any literature designated in the work programs of the disciplines;

- Subsystem "Reception of applications for increased scholarship" serves to submit documents for increased state academic scholarship in 5 areas of activity, is used by local commissions to process applications submitted and the appointment of scholarship;
- "Monitoring the employment of graduates" helps to keep records of graduates' employment, the formation of reporting materials;
- The "News of NEFU academic units" subsystem allows you to receive timely information on the life of the university and provides opportunities for personal communication through the appropriate services;
- EHostel, an information system that contains open information about living in the dormitories, supports the process of settling students in the dormitories;
- "Psychological help service" allows you to receive anonymous psychological help.

The main system for distance learning, including for interim certification, is the system of electronic and distance learning Moodle (hereinafter—Moodle EDMS). It is in Moodle EDMS that teachers place course materials for students and conduct the current progress control. NEFU has a point-rating system, according to which students earn points by completing assignments during the semester. If a student has completed a full course of the discipline and all assignments in a timely and high-quality manner, then at the end of the semester the accumulated amount of points will allow him/her to qualify for an automatic examination, if such an opportunity is provided for in this discipline.

Based on the indicators of educational activities, which are reflected in EDMS Moodle, the Department for Quality Assurance in Education has developed an algorithm to manage the educational process by tracking classes in real time, to obtain information about changes in the training schedule, to mark the attendance of students, etc. This is especially important to ensure control measures by the administrations of educational departments of the university and take appropriate management decisions.

The created system passed the practical test when the university was forced to switch completely to the distance learning form of students and remote mode of teachers' activity in the situation connected with the pandemic of coronavirus infection since March 17, 2020. In general, the educational process was organized in the normal mode, while the educational units were tasked with adaptive management of the quality of education, solving new problems associated with the organization of practices, volunteer help to the population and conducting career guidance work by students with schoolchildren who are in rural areas. A variety of forms of distance learning organization was determined due to the great remoteness of territories, weak Internet in some areas. The interested participation of local administrations of settlements, taking into account the possibilities up to providing students with communication opportunities from the premises of the administration was provided. In general, no student, forced to stay at the place of permanent residence, was not left without attention. As a result, the level of dropouts for various reasons did not exceed the usual statistics.

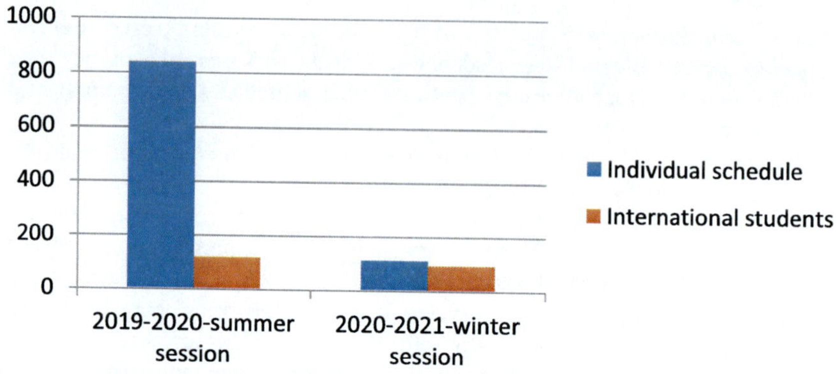

Fig. 1 Statistics on the use of individual educational trajectories of students

Due to the emerging problems associated with the organization of distance education, individual schedules for the session and individual study plans were prepared. During the summer session of the 2019–2020 academic year, individual session schedules were prepared for 837 students. The current 2020–2021 academic year, due to the difficult epidemiological situation, is also in a distance learning format, with 118 international students currently studying on an individual schedule. For the winter session of 2020–2021 academic year to date 111 students have completed individual session schedules, including 93 international students (Fig. 1).

The state final certification of graduates is organized in accordance with the "Temporary procedure for organizing and conducting state final certification using e-learning and distance learning technologies in 2020" approved by NEFU. An information system, the "State Final Assessment" module, has been developed to provide information and methodological support for the State Final Assessment, which proved to be the next step in the practical use of digital technologies and was successfully tested in the summer of 2020. Access to the module "State Final Assessment" is provided to the heads of educational units through a personal account of the employee, the deputy head of the educational unit for academic work, heads of graduate departments, secretaries of state examination commissions (hereinafter—SEC), responsible for technical support of SEC meetings, the Vice Rector for Educational Activities, employees of the Department of Education Quality Assurance.

At present, the information system module "State Final Assessment" includes all basic professional educational programs being implemented at NEFU, for which the period of study ended in 2020 and at the beginning of 2021. Information has been posted, including the number of graduates by form of study, the names of groups, the schedule of State Examination Commissions (SEC), and contact information for the SEC secretary and the person responsible for technical support of SEC meetings. The module contains documents, divided into the following 4 blocks: documents of the State Examination (SE); documents of the defense of the GIA; general documents of the GIA; normative and local documents for the GIA. The documents in the blocks are selected taking into account the levels of education of the corresponding main professional educational programs.

Meetings of the stipend commissions for awarding state academic stipends to students and state stipends to graduate students and residents studying on a full-time basis at the expense of budgetary allocations from the federal budget after completion of midterm evaluations were held in a remote format. The orders for the assignment of scholarships, the movement of students and the issuance of diplomas were made through the electronic document management system Directum.

Management of the educational process is carried out by the Department of Educational Quality Assurance in conjunction with the academic units through weekly monitoring of classes according to the approved schedule. The Department of Education Quality Monitoring DOKO systematically conducted surveys of students on satisfaction with distance learning. Thus, from April 7 to 9, 2020 full-time students took the first survey in the system of online questionnaires through their personal accounts. In general, the students noted the accessibility of the educational material in the NEFU EIOS. According to the results of the survey, most students expressed a desire to continue learning in a blended learning format. The students took a second survey from November 6 to 16, 2020. In general, the survey yielded the same positive results, the students noted the possibility of full access to learning materials and as a positive point noted the possibility to study at their own pace. Students of pedagogical specialties noted the possibility of more immersion and obtaining relevant competencies during internships in schools and kindergartens. At the same time, in many questionnaires students expressed a desire to study in a traditional format, because they were tired of being in self-isolation mode and experienced a lack of contact with teachers and classmates.

The year 2020 brought additional challenges and at the same time motivation for the active use of digital technologies for NEFU, as it became the year of preparation for state accreditation of the university. Preparation for this procedure began in 2019, due to changes in the regulatory documents accompanying the state accreditation procedure, the fragmentation of areas and profiles of training by educational units of the university. At the beginning of 2020, the "State Accreditation" module was introduced, allowing almost every employee to work remotely on the university server simultaneously. This development helped a lot in organizational issues in the preparation of the university for accreditation examination, especially with the introduction of isolated work mode during the pandemic period from March 17, 2020, when all university employees, including teaching staff, work remotely. In total, materials were prepared for 680 SPE and HE Bachelor's degree programs (Table 1).

4 Discussion

The emerged educational situation associated with coronavirus infection and the forced prolonged self-isolation regime activated the participation of teachers,

Table 1 Educational programs introduced in digital format

NPC levels	Head university	MPTI	NTI	BSF	Total OPOP by year of recruitment
Bachelor	280	18	36	10	341
Master's program	141	0	0	0	142
Specialty	41	25	25	0	93
Residency program	36	0	0	0	36
Post-graduate school	56	0	0	0	56
SPO	12	0	0	0	12
Total	**566**	**43**	**61**	**10**	**680**

students and undergraduates in numerous webinars and advanced training courses on the organization of distance learning, methodological and didactic support of the educational process in the online mode. In a fairly short period of time the university has acquired invaluable experience of forming a digital educational environment in which the educational process can be organized in different formats (traditional, distance, mixed):

1. A precedent has been set for a complete immersion in the problem of organizing the educational process in an online format.
2. Provided state final evaluations, interim evaluations, training sessions in the format of video conferencing in accordance with the training schedule.
3. Forced period of self-isolation, learning activities in a distance mode allowed to focus on the social and psychological component, especially students of pedagogical and medical training, to form the competence of an adequate and independent response to emerging pedagogical situations.

Any initiatives of the students to volunteer in the community were supported, if possible corrected by the teachers as part of the educational process and credited as a completed educational practice. A total of 136 students of the Pedagogical Institute of NEFU underwent pedagogical practice in the course of voluntary volunteer activities.

5 Conclusions

The experience of forming a digital educational environment at North-Eastern Federal University, which positively proved itself in the force majeure situation that arose in 2020, can be useful to other universities in a similar situation. At present, management approaches and software have been adapted, a lot of work has been done on the formation of educational and methodological material, correction of educational documentation, and ensuring the completion of the academic year in a remote mode. There has been a productive practical experience in the use of distance (online) learning opportunities, which only have to be comprehended, analyzed and envisaged for use in educational practice. We see the need for further action on the organization and management of the educational process:

- first of all, it is necessary to develop scientific and methodological foundations for the organization of the educational process and all the mechanisms and procedures associated with it;
- taking into account the experience gained, it is necessary to develop proposals for the optimization of the educational process in terms of transferring a number of academic disciplines to the online format, educational programs in a mixed format of training;
- this should inevitably lead to the optimization of the number of basic educational programs and the corresponding activities of departments in the mixed and distance format;
- there is a lot of work to be done on the formation of training documentation to ensure the implementation of educational programs in the online format;
- one of the popular areas of study is the possibility of taking some types of internships in distance mode;
- waiting to analyze the accumulated experience in providing feedback to students and evaluating learning achievements, as well as taking into account the independent work of students;
- a large layer of scientific and pedagogical work is the provision in the electronic information educational system of individual educational trajectories, the development of digital academic mobility of students to participate in educational courses and programs of leading domestic and foreign universities, creating their own online courses to develop inbound digital mobility;
- continuous and necessary readjustment and updating of EIOS systems to meet emerging practical tasks and system challenges.

In the Russian Federation as a whole, we believe that the issue of introducing adjustments to the content of federal state educational standards (FSES) related to the implementation of educational programs in a distance mode requires reflection. The education system of the country within the framework of modernization and optimization of the educational process on the basis of digital technologies is waiting for solutions for the mass organization of measures to improve the qualifications of teachers aimed at achieving a sufficient level of digital competencies, automation of the placement of educational programs on the relevant sites and their modernization on the basis of modular structuring.

References

1. Tretyakova T, Ignatiev V, Barakhsanova E, Varlamova L (2017) Quality of education as a guarantee of the effectiveness of the university. Modern Probl Sci Educ 2. https://science-education.ru/ru/article/view?id=26227

2. Vlasova EZ, Goncharova SV, Barakhsanov VP, Ivanova EA, Karpova NA, Iljina TS, Sysoeva AS (2019) Digital transformation of the pedagogical education in the Russian federation. Dilemas contemporáneos: Educación, Política y Valores. Mexico 7(S10):52
3. Barakhsanova EA, Varlamova V, Vlasova E, Nikitina E, Prokopyev M, Myreeva A (2018) Vocational training of school teachers in Yakutia's universities through the principle of regionalization (case study of the methodology of teaching natural science to prospective to prospective teachers). Espacios. Sociacion de Profesionales y Tecnicos del CONICIT. Venezuela 39(20):35
4. Barakhsanova EA, Barakhsanov VP, Olesov NP, Malgarov II, Neustroev AA (2019) Current trends in digital education development in the Republic of Sakha (Yakutia) Espacios. Sociacion de Profesionales y Tecnicos del CONICIT. Venezuela 40(9):18
5. Prokopiev MS, Vlasova EZ, Tretyakova TV, Sorochinsky MA, Solovieva RA (2020) Development of a programming course for students of a teacher training higher education institution using the programming language Python. Peru 8(3):15. https://doi.org/10.20511/pyr2020.v8n3.484
6. Sergushina E, Kabanov O, Godunov I, Garkusha N, Krutykh E, Rustamova I, Khrameshin S, Sorochinsky M, Nikolskaya E (2020) Relationship of professional and social self-determination of personality in modern conditions. Palarch's J Archaeol Egypt/Egyptol 17(7)
7. Mislevy RJ (2013) Evidence-centered design for simulation-based assessment. Mil Med 178(10):107–114. https://doi.org/10.7205/MILMED-D-13-00213
8. Efremova NF (2010) Approaches to the assessment of competencies of first-year students, embarking on the basic educational programs. Vestnik of Don State Technical University. T. 10. 5(48):774–782
9. Toulmin SE (1958) The uses of argument. Cambridge University Press, Cambridge. van der Linden W, Glas CAW (2010) Elements of adaptive testing. Springer, New York
10. Messick S (1994) The interplay of evidence and consequences in the validation of performance assessments. Educ Res 32(2):13–23
11. Tretyakova TV (2010) The system of education quality assessment and its construction in the regions with territorial and national features (on the materials of the system of complete secondary education in the Republic of Sakha (Yakutia)). D. thesis for the degree of Doctor of pedagogical sciences. Publishing House of SVFU, Yakutsk
12. Schum (1994) Problems of the development of modern science: theory and practice. Cartero Publishing House, Madrid
13. Siegesmund A (2017) Using self-assessment to develop metacognition and self-regulated learners. FEMS Microbiol Lett 364(11):1–4
14. Tretyakova TV, Prokopiev MS, Goncharova SV, Karpova NA, Iljina TS (2020) Distance pedagogical education in the conditions of the coronavirus. Univ san ignacio loyola, ave la fontana, s-n, la molina, lima, peru, propositos y representaciones 8(S3):e718
15. Efremova NF, Huseynova AA (2020) Psychological and pedagogical aspects of increasing educational motivation. In: E3S Web of conferences, vol 175, 15015. INTERAGROMASH 202. https://doi.org/10.1051/e3sconf/202017515015
16. Efremova NF, Huseynova AA, Shvedova S (2019) The influence of assessment on learning motivation. Trends in the development of psychological and pedagogical education in a transitive society. In: SHS Web conference, 04003. ICTDPP-2019. https://doi.org/10.1051/shsconf/20197004003

Interdisciplinary Effects in Modeling Technical and Technological Programs of the University

Sergey Zolotarev, Ivan Gulyak, Ludmila Filimonyuk, Valentina Ivashova, and Nikolay Guzynin

Abstract The article presents the views of employers and students about the demand for interdisciplinary knowledge in technical and technological programs of higher education. The regional survey was attended by 29 experts from among the chief and leading engineers of agricultural enterprises and 47 graduates of engineering programs in 2019 of the Stavropol Territory; 176 people from among students of 2nd–4th courses of faculties of economics, biotechnology. The survey database was processed in SPSS Statistics (version 21) Using the principal component extraction method with Varimax Rotation and Kaiser Normalization, models of engineering education were generated, and we determined the boundaries of the use of an interdisciplinary approach in the training of engineering personnel. An algorithm for mathematical modeling of engineering educational programs is proposed, taking into account the views of stakeholders and an interdisciplinary approach in the formation of graduates' competencies.

Keywords Interdisciplinarity · Modeling · Technical and technological programs

1 Introduction

Interdisciplinarity as a modern approach in modeling technical and technological programs of the university is playing an ever-increasing role. As a working concept of research "interdisciplinary approach" we will understand learning in the framework of more general topics or areas, the integration of educational content in various subject areas, profiles, and areas of training. We are interested in the modeling of programs with technical and technological content, which must be integrated with the humanitarian block, justifying its volume and content.

Discussion of the integration of humanitarian knowledge with engineering and technological modules of educational programs occupies an important place in

S. Zolotarev · I. Gulyak · V. Ivashova (✉) · N. Guzynin
Stavropol State Agrarian University, 12, Zootekhnichesky Lane, Stavropol 355017, Russia

L. Filimonyuk
North Caucasus Federal University, 1, Pushkin Street, Stavropol 355017, Russia

A. Beskopylny and M. Shamtsyan (eds.), *XIV International Scientific Conference "INTERAGROMASH 2021"*, Lecture Notes in Networks and Systems 247,
https://doi.org/10.1007/978-3-030-80946-1_80

modern scientific and educational discourse. This is evidenced by analytical reviews in articles by Russian scientists and publications by foreign researchers.

The activities of engineers and technologists are always associated with active transformations in the material sphere. Often, such transformations affect the interests of many parties. That is why it is important to equip this category of workers with knowledge in the field of effective communications, psychology and sociology. We see this approach in the work of J. E. Masson, I. Soustre-Gacougnolle, M. Perrin, F. Cousin, R. Lassablière, which emphasize the need to include stakeholders in the decision-making process in the field of sustainable development of the grape industry [1].

The solution of complex socio-economic and engineering-technological problems is associated with the work of large professional teams. The issues of interaction in the implementation of the set production tasks require not only a high level of knowledge in the field of engineering and technological processes. They are related to the subject areas of humanitarian knowledge, as they affect the competence of the optimal building of status-role relationships in a team, innovative thinking, pedagogical skills in broadcasting professional experience, and personality psychology. The authors of the publication E. Graeff, N. Maranzana, A. A. Aoussat write about the implementation of the approaches of team interaction and synergetic thinking of engineers and biologists [2].

The emphasis on the interdisciplinarity of training engineering personnel is made in their research work by Mazzoleni et al. [3]. In their opinion, disciplinary diversity is a prerequisite for the development of interdisciplinarity. It is an important factor in a vibrant academic environment with great potential for advances in research and technology. Many discoveries lie at the intersection of sciences and subject areas. Effective innovation and research activities are ensured, among other things, by high-quality interpersonal communications. Knowledge in this area is provided by an interdisciplinary approach to modeling engineering and technological programs of the university.

The brief review of publications shows the growing importance of sociological, philosophical and psychological knowledge in the modern world and the relevance of the quality of teaching not only for students of humanitarian programs, but also for the natural science and engineering cycle. Generalization of socio-psychological practices based on the concept of interdisciplinarity helps to overcome critical gaps in the management of socio-technological processes in production teams and society.

2 Materials and Methods

To develop and test a working algorithm for modeling university programs of a technical and technological profile with interdisciplinary integration of the humanitarian block, we used the following groups of methods: theoretical analysis, empirical research, and design. This is the traditional analysis of documentary sources and the method of factor analysis of the database of sociological surveys.

The questionnaire is developed on the basis of the ideal CDIO Syllabus model, which defines and details the requirements for the learning outcomes and preparation of bachelors for complex engineering activities. To bring the content and effectiveness of engineering educational programs in line with the level of development of modern technologies and the expectations of employers and graduates of educational programs, we carried out a study of the opinions of stakeholders. In total, 29 experts from among the chief and leading engineers of agricultural enterprises and 47 graduates of engineering programs in 2019 took part in the regional survey.

In addition, for a comparative analysis of the importance of an interdisciplinary approach in the implementation of technological and economic programs, in February 2021, Stavropol State Agrarian University conducted a survey "Opinion of University Students on the Quality of Training, Employment Prospects and Career Growth" using Google Forms. In total, 176 people from among the 2nd–4th year students of the Faculties of Economics and Biotechnology took part in the survey. The bulk of the survey participants are enrolled in undergraduate programs. The survey database is processed with SPSS Statistics (version 21).

Using the principal component extraction method with Varimax Rotation and Kaiser Normalization, we generated models of engineering education and determined the boundaries of the use of an interdisciplinary approach in the training of engineering personnel. We suggested an algorithm for mathematical modeling of engineering educational programs, taking into account the views of stakeholders and an interdisciplinary approach in the formation of graduates' competencies.

3 Results

Students rated the quality of undergraduate programs, work programs and specialist programs most highly. According to survey participants, the lowest level of quality of vocational education is observed in Master's and SSE (specialized secondary education) programs.

According to students, the quality of vocational education suffers mainly due to the lack of demand from employers for graduates and the low level of interaction of educational organizations with employers (noted by 48.6% of survey participants, and for economic programs this problem is most acute in comparison with biotechnology programs). Next comes the problem, which is a consequence of the above-mentioned—the inconsistency of the proposed list of specialties with the modern needs of the labor market (noted by 29.1% of survey participants). The following named problems of the quality of vocational education noted by students at the level of 25.6–25.7% refer to the inconsistency of the content of educational programs with the qualification requirements of employees and the lack (or low degree) of student involvement in research and practical activities in the profile of the main curriculum. At the level of relevance of 20.2–20.3% are the problems of the absence or insufficient number of proposals for targeted training of students and the lack of practical experience of teachers.

The rest of the problems that were named in the course of the survey gained less than 15% of relevance:

- the lack (limitation) of the student's ability to individually form his curriculum, to use the resources of different organizations (inter-faculty and inter-university courses, remote access to courses by leading teachers and scientists) (noted by 14.9% of survey participants);
- the lack of effective "social elevator" to support gifted, highly motivated students (noted by 14.2% of survey participants);
- lack of qualified teachers (noted by 12.3% of survey participants);
- low salaries of teachers (noted by 11.5% of survey participants);
- lack of modern material and technical base, poor condition of buildings (noted by 9.7% of survey participants).

The level of criticality of these problems can be reduced due to multidisciplinary training and the quality of the content of educational programs, the involvement of students in research and project activities, the development of leadership competencies, teamwork, technological entrepreneurship, solving non-standard work tasks, increasing labor responsibility and obtaining additional professional skills in related fields of activity.

Based on the results of a survey of 29 experts (chief and leading engineers of agricultural enterprises) and 47 graduates of engineering programs in 2019, we formed a database was. Also we carried out a factor analysis of 24 characteristics of the professional profile of graduates of engineering programs. The total variance explained is 88.6% and is determined by 6 components.

The listed 24 characteristics of a graduate of engineering programs, the significance of which was assessed during the survey, as a result of the factor analysis performed by the Rotation Method: Varimax with Kaiser Normalization (Rotation converged in 13 iterations). They were grouped into 6 factors that, in the opinion of stakeholders, ensure the success of professional activities in the future (Table 1).

According to the content of the grouped characteristics, we can say that the first factor is determined by a set of variables: the ability to organize the production process and ensure compliance with safety and sanitation requirements (load factor 0.917); in-depth knowledge of the basics of engineering, methods and tools (load factor 0.881); the ability to organize a team to perform production tasks efficiently and on time (load factor 0.867); personnel training in the use of devices, mechanisms, technologies, models, systems (load factor 0.821); the engineer has entrepreneurial and business initiative (load factor 0.769); the ability to analyze the technological process, evaluate the results of work, plan the activities of the unit (load factor 0.761); ability to conduct testing, inspection, certification procedures and production facilities (load factor 0.738); key knowledge of the basics of engineering (load factor 0.703). A meaningful interpretation of the first factor can be formulated as the proactive role of a graduate of an engineering program in organizing the production process of a team of workers with comprehensive safety, sanitation, personnel training based on engineering knowledge and testing, verification, certification procedures.

Table 1 A matrix of rotated components of the engineering graduates' professional profile

	Component					
	1	2	3	4	5	6
1. Basic knowledge of Mathematics, Physics	0.334	**0.775**	0.044	−0.012	0.005	−0.058
2. Key knowledge of the basics of engineering	**0.703**	0.359	−0.028	0.184	−0.554	0.089
3. In-depth knowledge of engineering fundamentals, methods and tools	**0.881**	0.106	0.054	0.144	−0.313	0.059
4. Analytical rationale and problem solving	0.310	**0.696**	0.226	0.223	0.119	0.016
5. Ability to conduct experiments, research	−0.098	0.015	0.084	0.160	**0.871**	0.023
6. Systems thinking	−0.029	−0.009	**0.883**	0.162	0.248	−0.013
7. Ethics, justice and other types of responsibility	−0.166	0.605	**0.724**	0.065	0.022	0.008
8. Ability to carry out effective interaction with people	−0.097	**0.631**	0.586	0.070	−0.275	0.344
9. Ability to manage a team	0.083	0.615	0.199	−0.354	**0.641**	0.113
10. Ability to carry out effective communication in foreign languages	0.024	0.152	0.251	0.188	−0.086	**0.875**
11. Understanding the role and responsibility of the engineer	0.413	0.198	**0.839**	−0.094	0.052	0.175
12. The engineer has entrepreneurial and business initiative	**0.769**	0.465	0.099	0.148	0.075	0.137
13. Ability to analyze the technological process, evaluate the results of work, plan the activities of the unit	**0.761**	0.494	0.156	−0.188	0.230	−0.014
14. Willingness to participate in the design of technical means and technological processes and production systems	0.430	0.494	0.020	−0.177	**0.507**	0.464
15. Ability to organize the production process and ensure compliance with safety and sanitation requirements	**0.917**	0.266	−0.042	0.144	−0.107	– 0.004
16. Readiness for operation of production machines, technological equipment and systems	0.500	0.342	**0.595**	−0.085	−0.304	0.356

(continued)

Table 1 (continued)

	Component					
	1	2	3	4	5	6
17. Ability to organize a team to perform production tasks efficiently and on time	**0.867**	0.209	0.339	−0.085	0.040	0.223
18. Ability to conduct testing, inspection, certification procedures and production facilities	**0.738**	0.135	−0.004	0.477	−0.224	0.165
19. Ability to use standard technologies of maintenance, repair and restoration of worn-out machine parts	0.367	**0.710**	0.228	−0.036	0.063	0.470
20. Personnel training in the use of devices, mechanisms, technologies, models, systems	**0.821**	−0.213	−0.064	−0.158	0.236	0.042
21. Ability to complete the life cycle and recycle products, waste	0.397	−0.132	−0.147	0.486	0.165	**0.651**
22. Ability to use modern methods of installation, adjustment of machines and installations	0.101	−0.072	−0.050	**0.876**	−0.026	0.111
23. Implementation of innovations—from concept, design, production, to the launch of new products and services on the market	0.027	0.082	0.202	**0.883**	0.077	0.177
24. Demonstrate skills in engineering entrepreneurship	−0.091	0.236	0.028	0.603	**0.611**	−0.265

The second factor is determined by a set of variables: basic knowledge of mathematics, physics (0.775); the ability to use standard technologies for maintenance, repair and restoration of worn-out machine parts (0.710); analytical reasoning and problem solving (0.696); the ability to effectively interact with people around you (0.631). Thus, the second factor can be interpreted as the qualities of systems engineering thinking, taking into account factors outside the engineering system.

The third factor is determined by a set of variables: systems thinking (0.883); understanding the role and responsibility of an engineer (0.839); ethics, fairness and other types of responsibility (0.724); readiness for operation of production machines, technological equipment and systems (0.595). Thus, the third factor can be interpreted as the responsible operation of production machines, technological equipment and systems based on systems thinking.

The fourth factor is determined by a set of variables: the introduction of innovations—from concept, design, production, to the launch of new goods and services on the market (0.883); the ability to use modern methods of installation, adjustment of machines and installations (0.876). Thus, the fourth factor can be interpreted as promoting innovation in engineering production.

The fifth factor is determined by a set of variables: the ability to conduct experiments, research (0.871); ability to manage a team (0.641); demonstration of engineering entrepreneurship skills (0.611); willingness to participate in the design of technical means and technological processes and production systems (0.507). Thus, the fifth factor can be interpreted as engineering and technological entrepreneurship to achieve economic benefits.

The sixth factor is determined by a set of variables: the ability to carry out effective communications in foreign languages (0.875); the ability to complete the life cycle and recycle products, waste (0.651). Thus, the sixth factor can be interpreted as effective communications to ensure the life cycle of engineering production.

4 Discussion

In the educational programs of the humanitarian orientation of foreign universities, an important place is occupied by the block of social sciences. Doctor of Historical Sciences, Professor of Russian State University for the Humanities I. V. Karapetyants in the article “On Teaching Sociology in Foreign Universities” emphasizes the importance of studying foreign experience in the implementation of these programs. Such experience allows us to see the variety of approaches to the organization and implementation of training courses through which professional training of specialists in the field of social sciences is carried out, to take into account foreign experience in this area.

Sociology, as one of the main components in the block of social sciences, shows trends in the rapidly changing world of social relations based on the analysis of qualitative and quantitative indicators of social processes in modern society. Professor Karapetyants [4] emphasizes the ability of sociology in its development to resort to systemic self-construction, to create on the basis of this interdisciplinary education, which is an obligatory and objective condition for the formation of a disciplinary line that makes up modern programs of sociological education.

An important conclusion made by I. V. Karapetyats in the article, is that a significant proportion of social sciences are updated by the results of sociological research. Therefore, in many universities, specialization in the field of sociology is an important addition to the general professional and special training of both humanities specialists and students of technical and technological educational programs.

Rousseau [5] in the article “The Sociological Imagination, Neoliberalism, and Higher Education” shows the importance of the system of ideological coordinates and sociological imagination of teachers of sociology. It is emphasized that the theoretical and methodological foundations of the research practice of sociology teachers significantly affect the educational component of university education in humanitarian and technical educational programs.

The issue of teaching sociology abroad is interesting not so much from the point of view of the process of transferring sociological knowledge in the educational process of the university as it is, but from the point of view of the interdisciplinary multiplier effect of sociological approaches in understanding the humanities and natural science cycles.

Sociological knowledge helps in teaching various disciplines at the university, as it allows you to better understand the motivation of students to acquire knowledge. Eddy [6] in the article "Recent Research in Science Teaching and Learning" offers an overview of a number of publications that help teachers actualize the socio-psychological portrait of the modern student and thereby ensure better interaction between teachers and students.

Davies [7] writes about the importance of the sociology of education in the university educational process and the promotion of E. Durkheim's concept in his article "Durkheim and the Sociology of Education in Britain". Currently, the sociology of education in Great Britain especially needs to rethink approaches to the system analysis of the functioning of the education system, conceptual consistency and modeling, taking into account modern empirical markers.

Sociology helps in understanding the current challenges of our time, such as, for example, the COVID-19 pandemic. Researcher M. J. Reiss in the article "Science Education in the Light of COVID-19: The Contribution of History, Philosophy and Sociology of Science" talks about the need to acquaint students with the methods of sociological analysis of social practices of overcoming the life problems of different people associated with the pandemic. The author notes that sociological knowledge has great potential in expanding the interdisciplinarity of natural science university courses and, in general, increases the scientific literacy of students [8].

5 Conclusion

Thus, the conducted theoretical analysis of publications, touching upon the issues of relevance and mechanisms for the implementation of an interdisciplinary approach in the training of engineering and technological personnel, confirms the need to search for practical tools for modeling engineering educational programs [9–12]. In our opinion, the proposed mathematical modeling algorithm, taking into account the views of stakeholders, will help in practice to implement an interdisciplinary approach in the formation of graduates' competencies.

The conducted empirical research shows the importance of interdisciplinary training of modern engineers and technologists. In the process of modeling, 6 models were identified:

- the proactive role of a graduate of an engineering program in organizing the production process of a team with comprehensive safety, sanitation, personnel

training based on engineering knowledge and testing, verification, certification procedures;
- quality of systems engineering thinking, taking into account factors outside the engineering system;
- responsible operation of production machines, technological equipment and systems based on systems thinking;
- promotion of innovations in engineering production;
- engineering and technological entrepreneurship to achieve economic benefits;
- effective communications to ensure the life cycle of engineering production.

Each of these models contains an interdisciplinary component of the modern engineering and technological graduate's profile, which determine the success of further production activities.

References

1. Masson JE, Soustre-Gacougnolle I, Perrin M, Cousin F, Lassablière R (2021) Transdisciplinary participatory-action-research from questions to actionable knowledge for sustainable viticulture development. Humanit Soc Sci Commun 8(1):24
2. Graeff E, Maranzana N, Aoussat AA (2021) Shared framework of reference, a first step toward engineers' and biologists' synergic reasoning in biomimetic design teams. J Mech Design Trans ASME 143(4):041402
3. Mazzoleni S, Russo L, Giannino F, Toraldo G, Siettos C (2021) Mathematical modelling and numerical bifurcation analysis of inbreeding and interdisciplinarity dynamics in academia. J Comput Appl Math 385:113194
4. Karapetyats IV (2000) Teaching sociology at foreign universities. Sociol Stud 9:123–125
5. Rousseau N (2020) The sociological imagination, neoliberalism, and higher education. Soc Curr 7(5):395–401
6. Eddy SL (2020) Recent research in science teaching and learning. CBE Life Sci Educ 19(3):124–131
7. Davies B (2020) Durkheim and the sociology of education in Britain. Sport Educ Soc 25(7):712–734
8. Reiss MJ (2020) Science education in the light of COVID-19: the contribution of history, philosophy and sociology of science. Sci Educ 29(4):1079–1092
9. Rudoy E, Petukhova M, Ryumkin S (2019) Key challenges and opportunities of scientific and technological development of Russia's crop industry. IOP Conf Ser Earth Environ Sci 274(1):012071
10. Kenzhin Z, Bayandin M, Primbetova S, Tlesova A, Bogdashkina I (2016) Streamlining human resource management at enterprises operating within Kazakhstan's present-day agro-industrial complex. J Internet Bank Commer 21(6):14
11. Altaeva OA, Imeskenova EG, Chibikova OM (2019) Sustainable development of organic agriculture in the Baikal region, as a cross-border area. IOP Conf Ser Earth Environ Sci 395(1):012006
12. Goncharov VN, Erokhin AM, Ivashova VA, Tronina LA, Kamalova ON (2020) Social responsibility and professional competence of safeguard specialists for the quality and safety of food products. IOP Conf Ser Earth Environ Sci 422(1):012124

Modified Gypsum Binder for Interior Systems

Gulzar Utegenova, Marat Asamatdinov, Bachauatdin Kalbaev, Andrey Medvedev, and Aleksey Zhukov

Abstract Clay gypsum as a natural material of sedimentary origin is widespread both in Russia and in the states formed in the post-Soviet space. In terms of energy intensity and manufacturability, the processing of raw materials into a clay-gypsum binder does not differ from the conditions of traditional processing of natural gypsum. Compared to lime or cement mortars, mortars based on gypsum binder have greater elasticity and plasticity. Such properties associated with the manufacturability of the application, such as workability and thixotropy, as well as the interval for maintaining the pot life of the mixture, are controlled by the introduction of modifying additives, the evaluation of the formulation of which was the purpose of the research, the results of which are presented in the article. The studies carried out have established that varying the recipe parameters make it possible to regulate both the strength and performance characteristics of mixtures based on gypsum plaster, as well as the manufacturability of their application.

Keywords Modified binder · Gypsum plaster · Plaster · Additives · Gypsum binder

G. Utegenova · M. Asamatdinov · B. Kalbaev
Karakalpak State University named after Berdakh, Academician Ch. Abdirov 1, Nukus, Uzbekistan, Karakalpakstan 230100

A. Medvedev (✉) · A. Zhukov
Moscow State University of Civil Engineering, Yaroslavskoye Av. 26, 129337 Moscow, Russia

A. Medvedev
Russian State Geological Prospecting University, Miklukcho-Maklaya, 23, 117997 Moscow, Russia

A. Zhukov
Research Institute of Building Physics of the Russian Academy of Architecture and Building, Lokomotivny Pr. 21, 127238 Moscow, Russia

A. Beskopylny and M. Shamtsyan (eds.), *XIV International Scientific Conference "INTERAGROMASH 2021"*, Lecture Notes in Networks and Systems 247,
https://doi.org/10.1007/978-3-030-80946-1_81

1 Introduction

Improving the quality of plastering works and optimizing their cost is associated with the development of modified compositions based on gypsum, justified use of waste and local building materials, which include clay-gypsum binder.

Complete systems have already become one of the traditional ways of introducing interior finishing works. Regardless of the type of cladding used, in interiors, less often lime, and more often gypsum plaster coatings are used, the main operational requirement for which is good adhesion to the base and the absence of negative effects on the internal environment of the premises [1–3]. According to the criteria for the organization of work, the manufacturability of the application is important. Determined by such properties of plaster mixes as workability, thixotropy, the pot life of the mix [4–6].

The formed coating based on gypsum-containing plaster systems has a capillary-porous structure. Such a structure allows, due to the processes of sorption of water vapor and the accompanying condensation of moisture in the form of thin films, allows you to regulate the humidity in the room. Thus, excess moisture and, along with it, molecules of harmful substances are absorbed from the atmosphere of the room. When airing the premises, the reverse process takes place, and the absorbed moisture evaporates and is removed from the premises. Another important feature of gypsum, as well as of its derivatives and modifications, is the ability to release chemically bound moisture when heated. That is, non-combustible gypsum-containing facings and plaster coatings in the event of a fire perform the function of a fire-retardant barrier, and at temperatures exceeding 180° C, dihydrate gypsum dehydrates, water evaporates, lowering the temperature of the fire effect and forming a protective vapor barrier [7–9].

Clay gypsum (ganch, drywall) has been used as a building material in the Central Asian region and Transcaucasia since ancient times. Interest in clay gypsum as a local binder and an analogue of gypsum manifested itself in the 50–70 s of the last century, but the lack of a scientifically substantiated concept of its application made it impossible to effectively use the results obtained. Interest in this material is now returning. Its use is supported by low production costs, large reserves, non-energy-intensive processing technology, and properties that meet regulatory requirements [10–12].

2 Methods

Clay-gypsum raw material belongs to the group of sedimentary rocks. This rock is a mechanical mixture of gypsum dihydrate and montmorillonite or kaolinite clays. This rock may contain finely dispersed calcium carbonate, silicon oxide and water-soluble salts. The impurities change the properties of the gypsum binder, which is the rationale for the general direction of research: the development of recipes and technological methods for obtaining the gypsum binder that meets the regulatory requirements for gypsum-based binders [13–15].

The properties of plaster mixes and plaster coatings depend on many factors. First of all, it is the material and chemical composition of the main raw material, which becomes especially important when using gypsum-containing wastes of chemical industries (for example, phosphogypsum) or local his-containing raw materials, which may include marls, sand fractions, and clays. Also important are the type, activity and concentration of modifying additives, used to correct the technological and operational characteristics of plaster mixes [16, 17].

The study of the properties and formulations of clay gypsum was carried out by the method of mathematical planning of the experiment. Optimization of the obtained results was carried out by the method of mathematical planning and processing of the experimental results. The dependence of the properties of plaster mixes depending on the recipe factors was carried out in the process of implementing a three-factor experiment.

The study of the obtained dependencies (regression equations, or response functions) was carried out using the general methodology of analytical optimization developed at NRU MSBU.

The analytical optimization method is based on the application of methods of mathematical analysis to the study of response functions as a complete or incomplete quadratic equation and a mathematical function of several variables. In particular, one of the criteria for the extremum of a function of several variables is used for optimization: the equality of their first partial derivatives to zero.

3 Experiments and Results

The purpose of the experiments, the results of which are presented in the article, was the development of a method for selecting the composition of a modified gypsum binder based on a comprehensive assessment of the effect of additives on the properties of this binder.

The study of scientific literature and analysis of the results of a preliminary series of experiments made it possible to identify a group of factors that influence the result to the greatest extent and to establish the intervals of their variation. Experimental conditions, factors, their average established values and variation intervals are shown in Table 1.

Taking into account the requirements of the technology for applying plaster mixtures, the following response functions are determined as the time of the plasticizer consumption (X1) and the time of the Portland cement consumption (X2) of the gypsum binder. Using these parameters, the pot life interval of the modified clay-gypsum binder is determined, that is, the time interval of its defect-free application to the plastered surfaces. The compressive strength of the citric acid consumption (X3) is an optimization characteristic of its properties. The gypsum binder hardens quickly, and the measurement age of 7 days is set based on regulatory requirements.

Table 1 Experimental conditions

Factor name	Math symbol, X_i	Average factor value, $\overline{X}_i$	Variation interval, ΔX_i	Meaning at levels	
				– 1	+ 1
Plasticizer consumption, %	X_1	1.0	0.4	0.6	1.4
Portland cement consumption, %	X_2	4.6	1.2	3.4	5.8
Citric acid consumption, %	X_3	0.05	0.01	0.04	0.06

According to the results of the experiment, after assessing the significance of the factors, regression equations were obtained for the beginning of setting (Y1), the end of setting (Y2) and compressive strength (Y3):

$$\widehat{Y}_1 = 22 + 10X_1 + 5X_2 + 4X_3 - 4X_1X_2 - 2X_2X_3 - 3X_1^2 - 1X_2^2$$
$$\widehat{Y}_2 = 34 + 11X_1 + 6X_2 + 5X_3 - 4X_1X_2 - 3X_1^2 - 2X_2^2$$
$$\widehat{Y}_3 = 8.1 + 2.2X_1 + 1.6X_2 + 1.2X_3 + 0.8X_2X_3 - 1.2X_2^2$$

Confidence intervals were determined for each response function separately based on the convergence of parallel experiments and the value of the t-test. Accordingly, the confidence intervals are: $\Delta b1 = 0.8$; $\Delta b2 = 1.0$; $\Delta b3 = 0.4$. The coefficients of the regression equations that are less in absolute value than the certain confidence intervals were considered insignificant and equated to zero.

4 Discussion

The analysis of the equations shows that, on the areas for determining the variable factors established according to the experimental conditions (Table 1), the plasticizer consumption and, to a lesser extent, the consumption of Portland cement (coefficients at X1 and X2) have the greatest influence on the setting time.

With an increase in the consumption of the plasticizer, the timing of both the beginning and the end of setting increases significantly (coefficients at X12), and the increase in the effect of the consumption of Portland cement on increasing the beginning of setting gradually slows down (negative coefficients at X22). The presence of antagonistic effects of the combined effect of the plasticizer consumption and the consumption of Portland cement, as well as the consumption of Portland cement and citric acid on the setting time (negative coefficients at X1X2 and X2X3) is due to the nature of the chemical interaction of these components.

The greatest influence on the compressive strength is exerted by the consumption of the plasticizer (coefficient at X1), and the influence of the consumption of Portland cement and citric acid is manifested to a lesser extent. But it is also very significant (coefficients at X2 and X3). There is also a synergistic effect of the combined effect of the consumption of Portland cement and citric acid on the strength of the plaster coating (positive coefficient at X2X3), due to the nature of the interaction of free calcium hydroxide and polar radicals during the formation of the structure of the material.

The significant value of the coefficient at X22 allows the use of an analytical method to optimize this dependence, that is, functions of three variables Y3 = f (X1, X2, X3).

Based on the general methodology of analytical optimization, we study the function $Y_3 = f(X_1, X_2, X_3)$ by methods of mathematical analysis: we determine its optimum in the partial derivative with respect to the variable X_2:

$$\partial Y_3/\partial X_2 = 1.6 - 0.8X_3 - 2.4X_2 = 0$$
$$X_2 = 0.67 - 0.33X_3$$

An analytical study of the strength function Y3 = f (X1, X2, X3) shows that the optimal consumption of Portland cement (X2) is a function of the consumption of citric acid (X3) introduced into the mortar to regulate the setting time. The transition from coded values of the factor to natural (using the data in Table 1) allows us to conclude that the optimal consumption of Portland cement is from 4.8 to 5.8%

We solve the mathematical polynomials Y_1 and Y_2 by substituting the optimization function and taking into account the confidence intervals $\Delta b_1 = 0.8$; $\Delta b_2 = 1.0$. As a result, we get the dependences optimized for X_2:

$$Y_1 = 25 + 10X_1 + 1.7X_3$$
$$Y_2 = 37 + 8X_1 + 4X_3 + 1.2X_1X_3 - 3X_1^2$$
$$Y_3 = 8.6 + 2.2X_1 + 0.6X_3$$

The graphic interpretation of the obtained dependencies made it possible to construct nomograms (Fig. 1), with the help of which it is possible to determine the optimal values of the consumption of plasticizer, Portland cement and citric acid and to estimate the setting time of the modified gypsum binder (sectors I and II of the nomogram) and the values strength of the hardened plaster coating (sector III of the nomogram).

Using the nomogram allows you to establish preliminary results, and, first of all, to estimate the costs of the main components of the modified clay-gypsum binder, set based on the requirements for the strength and pot life of the mixture. Further, control mixes are carried out and the actual values of the setting time of the clay-gypsum test and the strength of the plaster coating samples are determined. The control tests carried out showed an error in the results. Not exceeding 8%.

Optimization equations (as well as basic polynomials) confirm that in the ranges of factors used in the experiment (Table 1), the consumption of superplasticizer C-3 has the greatest influence on the result.

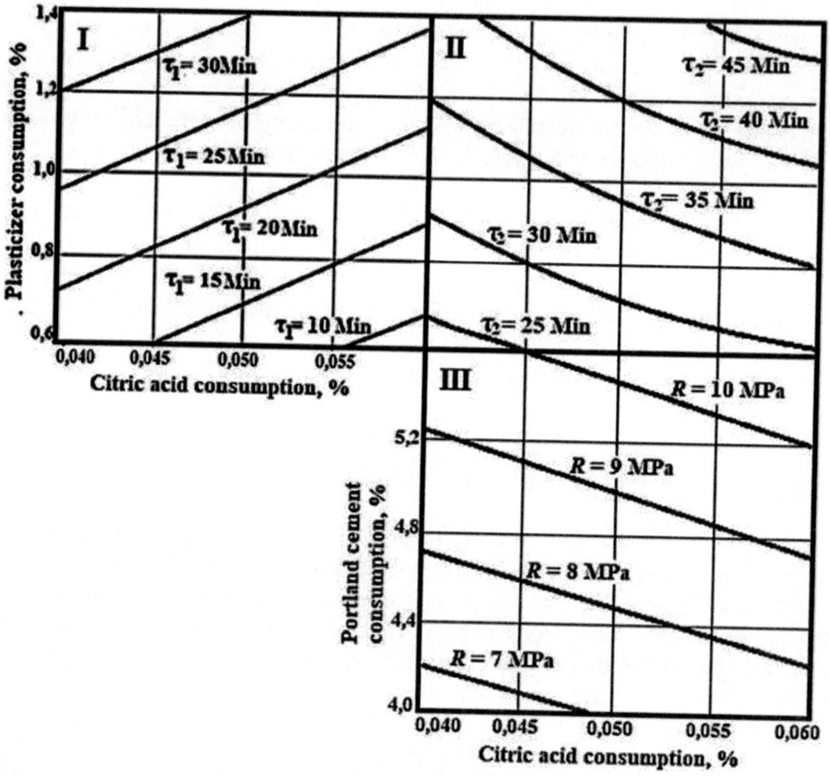

Fig. 1 Nomograms for determining the characteristics of the modified clay-gypsum binder: I—the beginning of setting (τ1, min.); II—end of setting (τ2, min.); III—compressive strength (R, MPa)

The use of gypsum plasters, including those based on modified clay gypsum binder, makes it possible to form high-quality plastered surfaces of light gray color and to minimize the filling work. The final decorative coating (decorative plaster, wallpaper, etc.) can be applied directly to the plaster layer.

5 Conclusions

The solution of technological problems with the use of experimental planning, the implementation of field studies in the laboratory and the use of modern methods of processing and optimization of experimental data significantly increases the efficiency of research, and ensures obtaining reliable results that adequately correspond to real processes.

The developed technique for selecting the composition of the modified clay-gypsum binder is based on the use of nomograms obtained as a result of the experiment, processing of mathematical polynomials, and analytical optimization of the results.

The use of a modified clay-gypsum binder and products based on it, including interior plaster mixes, makes it possible to expand the possibilities of the raw material base and reduce the labor intensity of applying plaster coating. The use of local raw material bases allows to optimize logistics, reduce transportation costs and ensure the stability of the supply of both raw materials and finished products.

References

1. Kodzoev M-BKh, Isachenko SL, Kosarev SA, Basova AV, Skvortzov AV, Asamatdinov MO, Zhukov AD (2018) Modified gypsum binder. In: MATEC Web of conferences, vol 170, 03022. Published online: 13 June 2018. https://doi.org/10.1051/matecconf/201817003022
2. Rumyantsev BM, Zhukov AD, Smirnova TV (2014) Energy efficiency and methodology for creating heat-insulating materials. Internet Bull VolgSABU 3(23):3
3. Asamatdinov M, Shokodko E, Ushakov A, Bessonov I, Kozlov S, Zhukov A (2019) Composition and rheological characteristics of plaster mixtures 02006. In: E3S Web of conference, vol 97. FORM-2019. Published online: 29 May 2019. https://doi.org/10.1051/e3sconf/201997 02006
4. Korovyakov VF (2002) Prospects for the use of waterproof gypsum binders in modern construction. Materials of the All-Russian seminar "Increasing the efficiency of production and use of gypsum materials and products" M.: RAASN, Apr 2002, pp 51–56
5. Bessonov IV (2002) Gypsum materials of a new generation for finishing the facades of buildings. Materials of the All-Russian seminar "Increasing the efficiency of production and use of gypsum materials and products" M.: RAASN, Apr 2002, pp 82–87
6. Zhukov AD, Orlova AM, Naumova TA, Nikushkina TP, Mayorova AA (2015) Environmental aspects of the formation of the insulation shell of buildings. Sci Rev 7:209–212
7. Telichenko VI, Oreshkin DV (2015) Materials science aspects of geoecological and ecological safety in construction. Ecol Urbanized Territ 2:31–33
8. Asamatdinov M, Zhukov A, Medvedtv A, Mukhametzianov V (2020) Fire protection systems using clay-gypsum plaster in agriculture. In: E3S Web of conferences, vol 175, 11008. INTERAGROMASH 2020. https://doi.org/10.1051/e3sconf/202017511008
9. Semenov VS, Bessonov IV, Ter-Zakaryan KA, Zhukov AD, Mednikova EA (2020) Energy-saving seamless insulation systems for frame buildings using foamed polyethylene. regional energy problems (Problemele energeticii regionale), electronic edn. Kishinev, p 4. https://doi.org/10.5281/zenodo.4018999. UDC: 691.1752./6./8.
10. Caetano DE et al (2017) Development of plaster mixtures formulations for additive manufacturing. materials design and applications. Springer, Cham, pp 257–277
11. Asamatdinov MO, Medvedev AA, Zhukov AD, Zarmanyan EV, Poserenin AI (2018) Modeling of the composition of ecologically safe clay-gypsum binder. In: MATEC Web of conferences. https://doi.org/10.1051/matecconf/201819303045
12. Pyataev ER, Medvedev AA, Poserenin AI, Burtseva MA, Mednikova EA, Mukhametzyanov VM (2018) Theoretical principles of creation of cellular concrete with the use of secondary raw materials and dispersed reinforcement. In: IPICSE. https://doi.org/10.1051/matecconf/201825101012
13. Zhukov A, Shokodko E (2020) Mathematical methods for optimizing the technologies of building materials. In: Popovic Z, Manakov A, Breskich V (eds) VIII International scientific

Siberian transport forum, TransSiberia 2019. Advances in intelligent systems and computing, vol 1116. Springer, Cham. First Online 31 Jan 2020, pp 413–421. https://doi.org/10.1007/978-3-030-37919-3_40
14. Zhukov AD, Bobrova EYu, Bessonov IV, Medvedev AA, Demissi BA (2020) Application of statistical methods for solving problems of building materials science. Nanotechnol Constr 12(6):313–319. https://doi.org/10.15828/2075-8545-2020-12-6-313-319
15. Zhukov AD, Korovyakov VF, Asamatdinov MO, Chkunin AS (2016) Modified gypsum marl binders. Sci Rev 7:86–90
16. Zhukov AD, Korovyakov VF, Asamatdinov MO (2017) Kinetics of heat release of clay-gypsum binder 1(20):104–113
17. Kondratieva N, Barre M, Goutenoire F, Sanytsky M et al (2017) Study of modified gypsum binder. Constr Build Mater 149:535–542

Gypsum Polymer Materials in Construction

Igor Bessonov, Aleksey Zhukov, Boris Efimov, Elina Gorbunova, and Ilya Govryakov

Abstract The modern level of technological development involves the use of traditional materials modified with additives of various types and functional purposes, as well as composite materials allowing to obtain a product with improved properties. Expanding the area of application of products based on gypsum for facade systems involves the creation of weather-resistant, and, first of all, waterproof materials based on gypsum polymers. The purpose of the experiment, the results of which are presented in the article, was to assess the possibility of using polycondensation polymers as a component of gypsum polymer, to model the properties of the material and to evaluate its characteristics as a result of climatic and humidity influences. The modeling and optimization of gypsum polymer properties were based on statistical methods as well as methods of mathematical analysis of functions of several variables. The assessment of the water resistance of gypsum polymer samples was carried out under test conditions in an open reservoir with an almost unlimited reaction capacity of the medium. The weather resistance was checked according to the results of tests in a climatic chamber. Experiments have shown that the strength of samples with 20% modified melamine-formaldehyde resin in compression and in bending for 80 days of storage in air increases by 30% and 25%, respectively. The compressive strength is 60 MPa, and the flexural strength is 12 MPa. Gypsum polymer has high frost resistance up to 150 cycles of alternate freezing and thawing. The result of the research was the confirmation of the possibility of using polycondensation resins and the foundations of the method for selecting the composition of the gypsum polymer were developed. The results obtained can be used in the development of the technology of gypsum polymer products, and, in particular, piece products (building cladding tiles).

Keywords Gypsum binder · Thermosetting resins · Mathematical planning · Water resistance · Gypsum polymer

I. Bessonov (✉) · A. Zhukov · E. Gorbunova · I. Govryakov
Research Institute of Building Physics of the Russian Academy of Architecture and Building, Lokomotivny Pr. 21, 127238 Moscow, Russia

A. Zhukov · B. Efimov · E. Gorbunova · I. Govryakov
Moscow State University of Civil Engineering, Yaroslavskoye Av. 26, 129337 Moscow, Russia

A. Beskopylny and M. Shamtsyan (eds.), *XIV International Scientific Conference "INTERAGROMASH 2021"*, Lecture Notes in Networks and Systems 247,
https://doi.org/10.1007/978-3-030-80946-1_82

1 Introduction

The traditional application area of gypsum products is indoor work. For the construction of facings and partitions, frame systems or foam gypsum blocks are used; decorative and acoustic materials based on gypsum are widely used in suspended ceilings; when decorating premises, gypsum plaster compositions are widely used. Gypsum materials not only make it possible to create interiors, but also contribute to the formation of the microclimate of the premises as a whole, i.e. create comfortable conditions for people [1–3]. Gypsum plasterboard or gypsum fiber sheets, stucco decorations, self-leveling floors, various dry plaster, putty and glue gypsum mixtures. Suspended ceiling panels and etc. are used in the interior decoration.

Studying the historical buildings of big cities, estates, buildings of the seventeenth–nineteenth centuries, you can see gypsum-containing materials not only in the interiors, but also on the facades of buildings. These decorative stucco architectural decorations were made mainly of gypsum-lime mortar, treated with lye or vitriol and repeated impregnation with hot linseed oil.

Many researchers have proposed a method for increasing the water resistance of gypsum by introducing additives that have a common ion with calcium sulfate (for example lime), which reduces its solubility [4–6], or blast furnace slags.

Later gypsum-cement-puzzolanic (GCP) and gypsum-slag-cement-pozzolanic (GSCP) binders were created and comprehensively studied, new technologies were developed for obtaining waterproof gypsum binders. CGB—composite gypsum binders and WGLW—waterproof gypsum binders of low water demand [7–9].

The use of secondary products from other industries, and, in particular, modified phosphogypsum, finely dispersed compositions and clay gypsum also make it possible to obtain products with increased water resistance [10–14].

Hydrophobization can be superficial and volumetric. Superficial treatment with water repellents tends to lose its effeciency over time, and for a longer use of the sheets, it must be repeated periodically. The frequency of treatment with water repellents depends on the operational conditions of the products. Taking into account the experience in the research, a method was adopted for modifying the gypsum binder with various polymers in the form of aqueous solutions or emulsions. The influence of the type of polymer and its consumption on the properties of the composite was studied, as well as the influence of temperature and humidity conditions on these properties.

Note that the use of products based on a modified gypsum binder in facade systems is directly related to the development of a method for assessing the weather resistance of this material, and, first of all, water resistance and frost resistance.

2 Methods

In the course of the work, the effect of various polymers in the form of aqueous solutions or emulsions on the properties of the composite was studied. The best results were obtained when using polycondensation aminoaldehyde resins with non-linear monomers (melamine, resorcinol, benzoguanamine).

The experiment to assess the influence of the gypsum polymer composition on its properties was carried out according to the general methodology for planning the experiment and analytical optimization of its results [15–17], based on the matrix of a full quadratic three-factor experiment. The resulting regression equations were tested against all statistical hypotheses. On the fact of comparison with confidence intervals (Δbj), only significant factors were left, and as a result of verification by Fisher's criterion, a conclusion was made about the adequacy (or inadequacy) of the obtained models.

In the study of the water resistance and atmosphere resistance of gypsum polymer samples, the conditions of exposure to the medium samples in an open reservoir were simulated with an almost unlimited reaction capacity of the medium. The atmosphere resistance of the products was assessed based on the results of testing their properties after processing in a climatic chamber.

Along with conducting frost resistance tests on cube samples, the consequences of cyclic climatic influences on the products themselves were studied. The research was carried out in a cooling and sprinkling installation «Thermal insulation KhDU-0.2», equipped with a swivel holder, into which the investigated element of the facade cladding is installed. The control is carried out both visually and with the use of the «Onyx-2.3» device intended for the control of surface strength by the non-destructive shock-pulse method.

3 Results

The experiment was built on the basis of mathematical planning methods and statistical processing of its results. The following factors are accepted as variable factors: the strength of the gypsum binder (X_1), the consumption of melamine-formaldehyde resin (X_2) and the consumption of the structuring additive (X_3). The strength of the gypsum polymer after 7 days of specimen hardening (Y_1) was taken as a response function, and the softening coefficient of the gypsum polymer specimens was taken as the optimization parameter according to the results of climatic tests ($Y_2 = Y_1/R_w$). Experimental conditions are shown in Table 1.

Mathematical processing of the results of the experiment made it possible to obtain the regression equations for the compressive strength of gypsum polymer samples (Y1) and its softening coefficient (Y2). The significance of the coefficient was checked by confidence intervals; accordingly, the confidence interval for strength was $\Delta b1 = 1$ MPa, and the softening coefficient was $\Delta b2 = 0.04$.

Table 1 Intervals of variation of factors

Factor name	Symbol X_i	Average value of the factor, $\overline{X}_i$	Variation interval, ΔX_i	Factor values at levels	
				– 1	+ 1
Gypsum strength, MPa	X_1	6	1	5	7
Resin consumption, %	X_2	17	4	13	21
Consumption of structuring additive, %	X_3	3	1	2	4

The resulting models were tested for adequacy by Fisher's criterion. The calculated values of Fisher's criteria are equal for the model of compressive strength F1 = 16.2 and for the model for the softening coefficient F2 = 15.9 and less than the table values, therefore the models are considered adequate.

The following mathematical models (polynomials) are obtained:

- for compressive strength

$$Y_1 = 54 + 15X_1 + 7X_2 + 6X_3 + 3X_1X_2 + 2X_1X_3 \quad (1)$$

- for the softening coefficient:

$$Y_2 = 0.82 + 0.08X_1 + 0.14X_2 + 0.09X_3 - 0.04X_1X_2 - 0.05X_3^2 \quad (2)$$

4 Discussion

Analysis of the coefficients of the equation Y1 = f1 (X1, X2, X3) shows that the strength of the gypsum polymer increases with increasing values of all factors (in the areas of their definition). The strength of the gypsum binder (its brand) has the greatest effect on strength. The softening factor is more dependent on the resin consumption.

Analytical optimization is based on the fact that the functions for strength and density Y1 = f1 (X1, X2, X3) and Y2 = f2 (X1, X2, X3) are mathematical and methods of mathematical analysis can be applied to them, provided that there is no the condition of adequacy is violated. In the case under consideration, the following scheme is adopted:

- the equation $Y_2 = f_2$ (X_1, X_2, X_3) is differentiated by X_3 and equated to zero, determining the extremum of the function Y_2 by X_3;
- solve the functions $Y_1 = f_1$ (X_1, X_2, X_3) and $Y_2 = f_2$ (X_1, X_2, X_3) with X_3 = opt and carry out local optimization.

Analytical optimization includes the following sequence of actions:

1. Determine the value of the local extremum of the function $Y_2 = f_1(X_1, X_2, X_3)$ by X_3:

$$\frac{\partial Y_2}{\partial X_3} = 0.09 - 0.10X_3 = 0 \rightarrow X_3 = \frac{0.09}{0.1} = 0.9 \quad (3)$$

2. We calculate the value of the natural value of the plasticizer consumption (corresponding to the possible obtaining of the maximum compressive strength of polymer concrete), using the deciding factor formula and the experimental conditions (Table 1):

$$\widetilde{X}_3 = \overline{X}_3 + \Delta X_3 \cdot 0.9 = 3 + 1 \cdot 0.9 = 3.9 \text{ kg/m}^3 \quad (4)$$

3. We calculate mathematical models (polynomials) for the optimized value of the factor $X_3 = 0.9$:

for compressive strength:

$$Y_1 = 59 + 17X_1 + 7X_2 + 3X_1X_2 \quad (5)$$

for the softening coefficient:

$$Y_2 = 0.86 + 0.08X_1 + 0.14X_2 - 0.04X_1X_2 \quad (6)$$

The graphic interpretation of the obtained models is shown in Figs. 1 and 2 and can be used when designing a gypsum polymer composition.

In the course of research, it was determined that the strength of samples with 20% modified melamine-formaldehyde resin in compression and in bending for 80 days of storage in air increases by 30% and 25%, respectively. The compressive strength is 60 MPa, and the flexural strength is 12 MPa. Gypsum polymer has a fairly high frost

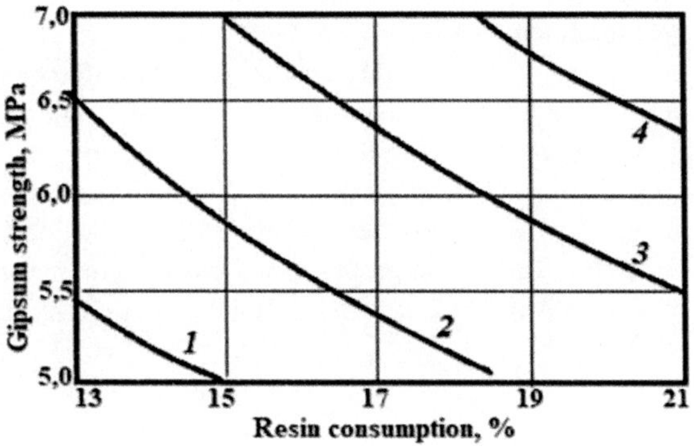

Fig. 1 Dependence of compressive strength on variable factors at an optimized consumption of a structuring additive of 3.9 kg/m^3. Compressive strength: 1—45 MPa; 2—50 MPa; 3—55 MPa; 4—60 MPa

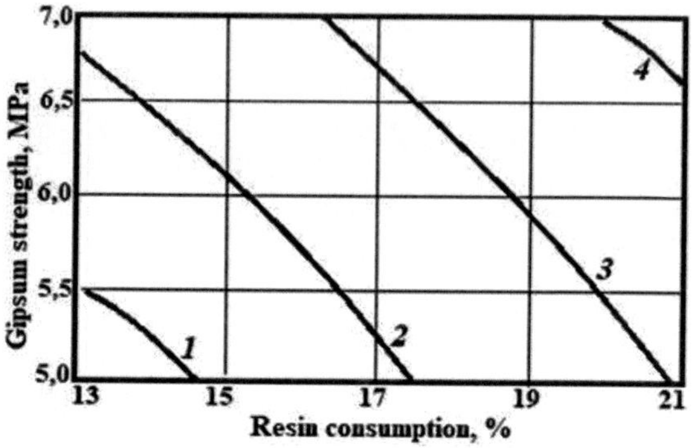

Fig. 2 Dependence of the softening coefficient on variable factors at an optimized consumption of a structuring additive of 3.9 kg/m^3. Softening factor: 1—0.7; 2—0.8; 3—0.9; 4—1.0

resistance. Samples with 20% melamine-formaldehyde resin withstand 150 cycles of alternate freezing and thawing.

In the study of the water resistance and atmosphere resistance of gypsum polymer samples, the conditions of exposure to the medium samples in an open reservoir were simulated with an almost unlimited reaction capacity of the medium.

It has been established that the depth of destruction of gypsum polymer samples in distilled water significantly decreases with an increase in the density of the hardened stone due to an increase in the content of the polymer component and a decrease in the water-gypsum ratio. The best results were obtained using modified melamine-formaldehyde resin.

With constant immersion of samples in distilled water, the strength of gypsum polymer with 20% modified melamine-formaldehyde resin in 8 months of testing decreased by only 20%, and for control gypsum samples during the same time—by 70%. Under the conditions of alternating moistening and drying during the same test time, the strength of gypsum polymer samples practically did not change, and for gypsum samples it decreased by 70%.

With increasing age of the samples, no significant changes in the characteristics of the composite occur. The increase in strength over time can be explained by the continued polymerization of the resin. The degree of polymerization of the resin in the presence of a hardener under natural conditions is practically the same as during heat treatment.

The vapor permeability of the gypsum polymer is 0.092 mg/(m.h. Pa), which determines a favorable humidity regime for walls made of brickwork with facing from this material. The durability of the exterior decoration with modified plaster products is confirmed by practice. Currently, architectural and decorative products made of gypsum polymer decorate the facades of buildings in Moscow, Nizhny Novgorod, Perm. The properties of gypsum polymer make it possible to use it from small plastic products with a fine relief to volumetric landscape gardening sculptures.

5 Conclusions

The studies carried out made it possible to determine the optimal consumption of poly-condensation resin and finely ground mineral additives, and also allowed the development of important elements for the formation of a method for selecting a gypsum polymer composition. The possibility of using polycondensation resins in a composition with a gypsum binder has been confirmed.

The strength of samples with 20% modified melamine-formaldehyde resin in compression and in bending for 80 days of storage in air increases by 30% and 25%, respectively. The compressive strength is 60 MPa, and the flexural strength is 12 MPa. Gypsum polymer has a fairly high frost resistance. Samples with 20% melamine-formaldehyde resin withstand 150 cycles of alternate freezing and thawing.

The results obtained can be used in the development of the technology of gypsum polymer products, and, in particular, piece products (tiles) for facing the facades of buildings. The vapor permeability of the gypsum polymer is 0.092 mg/(m.h. Pa), which determines the favorable humidity regime of the walls made of brickwork with facing from this material.

References

1. Korolev EV, Danilov AM, Garkina IA (2006) Building materials as systems. Build Mater 7:55–57
2. Buryanov AF (2005) Gypsum, its research and application from P.P. Budnikov to the present day. Build Mater 9:46–48
3. Panchenko AI, Buryanov AF, Kozlov NV, Soloviev VG, Pashkevich SA (2014) Comprehensive assessment of the effectiveness of the use of gypsum binder with increased water resistance. Build Mater 12:72–74
4. Bessonov IV (1998) Gypsum of increased water resistance. In: 3 Scientific-practical. conf. "Problems of construction thermal physics and energy saving in buildings". Sat. report - M.: NIISF, pp 112–117
5. Garkavi MS, Buryanov AF, Fisher HB (2015) Features of crystallization of dihydrate gypsum during artificial aging of gypsum binder. Stroitel'nye Materialy 12:73–75
6. Bessonov IV, Yalunina OV (2004) Environmental aspects of the use of gypsum building materials. Constr Mater Mag 4:10–12
7. Ferronskaya AV, Korovyakov VF, Melnichenko SV, Chumakov LD (1992) Water-resistant gypsum binders of low water demand for winter concreting. Build Mater 5:34–35
8. Pustovgar AP, Burianov AF, Vasilik PG (2010) Features of the use of hyperplasticizers in dry building mixtures. Stroitel'nye Materialy 12:61–64
9. Asamatdinov MO, Medvedev AA, Zhukov AD, Zarmanyan EV, Poserenin AI (2018) Modeling of the composition of ecologically safe clay-gypsum binder. In: Web of conferences. Published online: 20 Aug 2018. https://doi.org/10.1051/matecconf/201819303045
10. Khaev TE, Tkach EV, Oreshkin DV (2017) Modified lightweight gypsum material with hollow glass microspheres for restoration work. Build Mater 10:45–51
11. Kodzoev M-BKh, Isachenkov SL, Kosarev SA, Basova AV, Skvortzov AV, Asamatdinov MO, Zhukov AD (2018) Modified gypsum binder. In: MATEC Web of conferences, vol 170, 03022. Published online: 13 June 2018. https://doi.org/10.1051/matecconf/201817003022
12. Meshheryakov YuG, Fedorov SV (2008) Industrielle verarbeitung von Phosphogips and Phospho Energiesparende technologien. Int Kongress Fachmesse Euro ECO. 141 Hannover

13. Yakovlev G, Polyanskikh I, Fedorova G, Gordina A, Buryanov A (2015) Anhydrite and gypsum compositions modified with ultrafine man-made admixtures. In: Procedia engineering "7th Scientific technical conference material problems in civil engineering", vol 108, pp 13–21
14. Bekmanssurov M, Gordina A, Poljanskich I, Jakovlev G, Fisher H-B (2015) Gipsverbundwerkstoffe, modifiziert mit Technogenultradispers-zusatzmitteln. 19 Internationale Baustofftagung, 16–18 september. Weimar: Bauhaus. Band 2, pp 645–651
15. Bessonov I, Zhukov A, Shokodko E, Chernov A (2020) Optimization of the technology for the production of foam glass aggregate. In: TPACEE 2019, E3S Web of conferences, vol 164, 14016. Published online: 05 May 2020. https://doi.org/10.1051/e3sconf/202016414016
16. Zhukov A, Shokodko E, Bobrova E, Bessonov I, Dosanova G, Ushakov N (2019) Interior acoustic materials and systems. In: EMMFT-2018. Advances in intelligent systems and computing, vol 983. Springer, Cham, pp 740–747. https://doi.org/10.1007/978-3-030-19868-8_72
17. Zhukov A, Shokodko E (2020) Mathematical methods for optimizing the technologies of building materials. In: VIII International scientific Siberian transport forum. TransSiberia 2019. Advances in intelligent systems and computing, vol 1116. Springer, Cham, pp 413–421. https://doi.org/10.1007/978-3-030-37919-3_40

Insulation Systems for Structures on Pile Supports

Aleksey Zhukov, Igor Bessonov, Andrey Medvedev, Ekaterina Zinovieva, and Elizaveta Mednikova

Abstract Construction on problem soils or in permafrost conditions involves using of pile foundations with a ventilated space under the floor structure. In this case, additional thermal insulation is required under the first-floor structure (above the ventilated space). This problem is compounded by construction in cold regions. The goal of research was to develop insulation systems for buildings on pile foundations for different climatic zones, including conditions of the arctic region and other regions with a predominance of ever-frozen ground. With the help of the THERM computer program, the conditions of bidimensional heat interchange in the enclosing structures of a building with pile foundation were simulated. The construction of such buildings is practiced on problem soils. The resulting models were analyzed in terms of the thermophysical characteristics of the structures. As a result, the optimal version of the insulation system was chosen, effective both in the climatic conditions of the midland and in the especially cold conditions of Yakutia and Trans-Polar region (The Subarctic). This system included insulation with mineral wool slabs along the facade walls, with extruded polystyrene (XPS) foams along the basement part and the floor structure, and with roll polyethylene (PE) foams (with the formation of a seamless insulation shell) along the ventilated space under the floor structure and above, on top of insulation boards.

Keywords Insulation shell · Seamless joint · Extruded polystyrene foam (XPS) · Polyethylene foam (PE) · Thermal bridge · Thermal resistance · Calculation program

A. Zhukov (✉) · A. Medvedev · E. Zinovieva · E. Mednikova
Moscow State University of Civil Engineering, 26, Yaroslavskoe Shosse, 129337 Moscow, Russia

A. Zhukov · I. Bessonov
Research Institute of Building Physics of the Russian Academy of Architecture and Building, Lokomotivny Pr. 21, 127238 Moscow, Russia

A. Zhukov · E. Mednikova
Higher School of Economics University, 20 Myasnitskaya Ulitsa, 101000 Moscow, Russia

A. Medvedev
Russian State Geological Prospecting University, Miklukcho-Maklaya, 23, 117997 Moscow, Russia

A. Beskopylny and M. Shamtsyan (eds.), *XIV International Scientific Conference "INTERAGROMASH 2021"*, Lecture Notes in Networks and Systems 247,
https://doi.org/10.1007/978-3-030-80946-1_83

1 Introduction

The minimum permissible heat transfer resistance of walls and coverings of buildings for various purposes and different climatic conditions is regulated by SP 50.13330-2012 and it is determined by the operating conditions of the insulation systems, including atmospheric manifestations [1–3]. During the reconstruction of walls and coverings, the thickness of the additional thermal insulation layer was determined based on the difference between the required and existing heat transfer resistances.

The effective use of thermal insulation is based on the following conditions. Firstly, the thermal insulation material is used only dry or at the equilibrium moisture content. Any excessive moistening of the structure leads to an increase in the thermal conductivity of the insulating layer and a sharp drop in the real thermal resistance. Secondly, the installation of insulation boards must be close to each other and to the structural element. The appearance of a gap between the boards is not allowed due to leading to the formation of thermal bridge, which will increase heat loss through the enclosing structure [4–6].

Moisture can enter building structures in various ways. The movement of the vapour-air mixture occurs constantly in the construction of the basement part of the wall and foundation. Moisture condensation is possible under certain conditions, for example, when the temperature in the material reaches the dew point. Capillary rise of ground moisture through the foundation structure and through the capillaries of the wall material can occur if the insulation system is incorrectly performed. All these reasons make it advisable to use seamless heat-vapor-waterproofing shells [7–9] in building insulation systems.

2 Materials and Methods

Insulation systems for residential buildings for the Arctic region have a number of significant features, the main of which are installations allowing to preserve permafrost. Preservation of the soil in a frozen state implies either the use of floating foundations (like the "Swedish slab"), which is not effective for residential buildings (heat escapes through the floor, the soil thaws and loses its bearing capacity), or construction on pile foundations, which is used everywhere [10, 11].

The insulation diagram of a residential building with a pile foundation is shown in Fig. 1. The insulating wall cladding is made according to the technology of insulated plaster facades (EIFS), and the floor structure above the ventilated space is insulated with several layers of XPS boards laid with offset joints.

Such an insulation system has two negative factors, these are cold bridges at the joints of the slabs (displacement during installation allows to reduce heat losses, but not completely) and, most importantly, a serious channel of heat transfer between the joint of thermal insulation plates and enclosing structures (see Fig. 1a). The joint

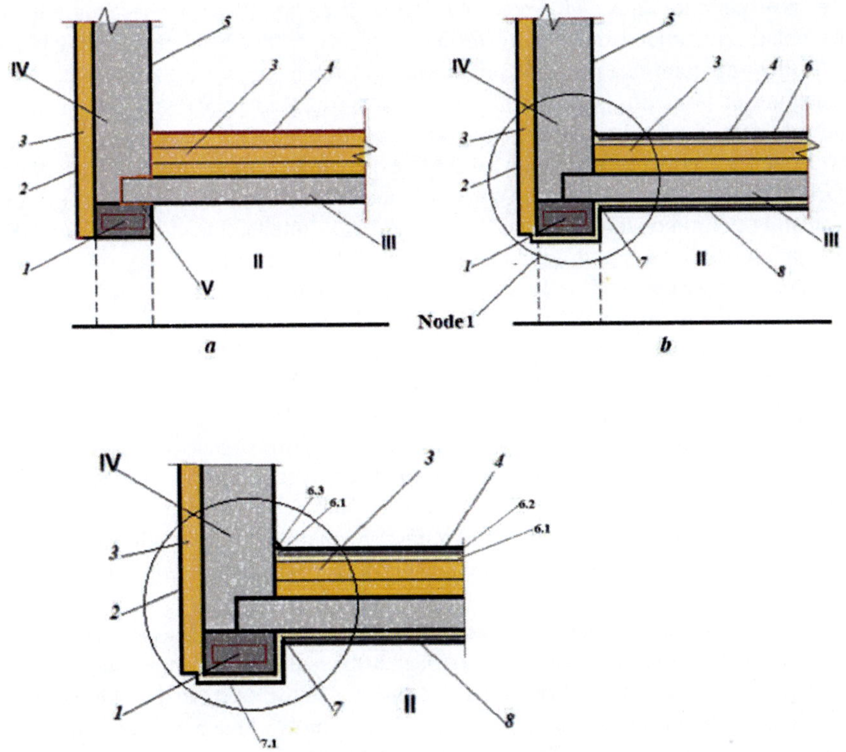

Fig. 1 The floor structure of residential building diagram (section 2—wall between columns and detailing node 1): **a** without insulation of the floor structure over the ventilated space; **b** with insulation of the floor structure over the ventilated space; I—column; II—ventilated space; III—the floor structure over the ventilated space; IV—bearing wall; V—area of increased heat transfer and cold air infiltration; 1—binding of columns; 2—facade insulation system; 3—XPS thermal insulation board; 4—top coat of floor; 5—interior wall cladding; 6—floating floor system (dry build), includes: 6.1—EPE foam layer 20 mm thick, installed on the wall; 6.2—chrysotile cement sheet 10 mm thick (or fiber cement sheet), then the top coat of floor; 6.3—floor molding; 7—insulation above the ventilated space; 7.1—insulation before EIFS (completely blocks the infiltration flow at the junction); 8—protective lining

is not only an area of active conductive heat transfer, but also a path for infiltration of cold air through leaks in the placement of slabs and their abutment to structures.

At significant negative outside temperatures (from − 30 °C and below), the perimeter of the building in its lower basement part remains in the area of negative temperatures, which is extremely undesirable, both from the point of view of the formation of a comfortable microclimate in the premises, and from the point of view of the durability of the structure, freezing to the full thickness. At the same time, the thickness of the thermal insulation laid along the overlap can reach 500 mm.

The experience of using foamed polyethylene in central Russia made it possible to formulate recommendations for the formation of an effective insulating coating [12–14]. It is important that the basic principles of the technology based on mechanical fastening of insulating sheets (rolls) and welding of their joints with hot air have already been implemented at dozens of facilities throughout the country.

Rolled polyethylene foam with a reflective coating (foil or with a metallized film) forms a protective layer along the outer perimeter of the structure, and is the basis of a floating floor laid along XPS insulating boards. With a thickness of XPS-thermal insulation of 200 mm and a thickness of UPF (PE)-insulating coating of 20 mm, the thermal resistance of the insulating circuit is not less than 4.0 m^2 °C/W, at the same time, the thermal uniformity of the structure along the surface is significantly increases and the paths of infiltration of cold air are completely excluded.

The program for calculating temperature fields for the THERM computer, which meets the requirements of set of rules SR 50.13330.2012, contributes to the study of heat transfer processes. Using the program makes it possible to simulate 2D heat transfer in building components such as windows, walls, floors, roofs and doors. The analysis of heat transfer using the program allows you to evaluate the energy efficiency of the structure and the local temperatures of the sample, allowing you to solve problems with condensation, moisture content of the structure material and its tightness.

The program version includes several additional technical characteristics and user interface features. The most important of them is the radiative heat transfer algorithm taking into account mutual irradiation and shading of the calculated surfaces. This ability increases the computational accuracy in situations where non-planar surfaces are analyzed that have different temperatures and exchange energy through radiative heat transfer.

3 Results

The insulating cladding of the walls is made according to the technology of insulated plaster facades (SFTC), and the ceiling over the ventilated space is insulated with several layers of XPS-plates, laid with offset joints. Insulation made of rolled polyethylene foam is mounted along the entire base (under the overlap) and installed on the wall up to the basement part of the SFTC. Thus, rolled polyethylene foam with a metallized coating completely blocks the infiltration flow at the junction of the wall ceiling. Separate rolls of polyethylene foam are mounted in the lock, followed by welding the joint with hot air using a construction hair dryer. Considering that polyethylene foam has not only low thermal conductivity, but also very low vapor, moisture and wind permeability, we can say that a seamless insulating shell is formed.

The implementation of computer calculations was aimed at assessing the distribution of temperature fields and thermal resistance of insulation systems operating in different temperature conditions.

The calculations used the following input data:

The strapping of the columns was carried out with reinforced concrete half-timbered timber (density 2400 kg/m^3, thickness 200 mm, thermal conductivity under operating conditions B (λ_B)—1.86 W/(m °C)).

Facade insulation system (SFTC) along the surface of the walls: mineral wool slab with a density of 80 kg/m^3, thickness of 100 mm, λ_B—0.46 W/(m °C). The basement part is made of extruded polystyrene foam (XPS-plates) 100 mm thick at a height of 1 m from the bottom edge of the wall above the ventilated space. XPS-boards with a density of 30 kg/m^3, with a thickness of 50 mm each layer λ_B 0.031 W/(m °C) XPS-plates over cold.

The insulation of the overlap was carried out over the ventilated space made of rolled polyethylene foam with a metallized coating 20 mm thick, with a density of 30 kg/m^3, λ_B 0.032 W/(m °C). On top of the overlap, two layers of XPS-plates were laid (in a spaced manner with displaced joints) and a layer of rolled polyethylene foam, which was put of the wall (100 mm) with mechanical fastening and welded into a seamless insulating shell.

With the help of the THERM computer program, the conditions of two-dimensional heat transfer in the enclosing structures of a pile building were simulated. The construction of such buildings is practiced on problem soils. Structures were taken as the basis for the calculation, the nodes of which are shown in Figs. 2 and 3.

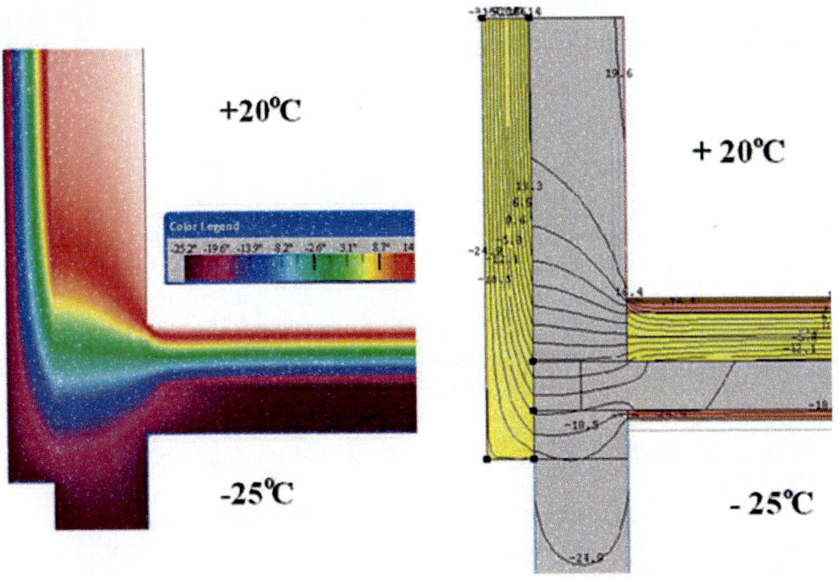

a *b*

Fig. 2 The structure of the formation of the temperature field in a structure partially insulated above the ventilated space: **a** visualization of the temperature field; **b** graphic interpretation of the temperature distribution (Moscow, Moscow region. Leaning on a column)

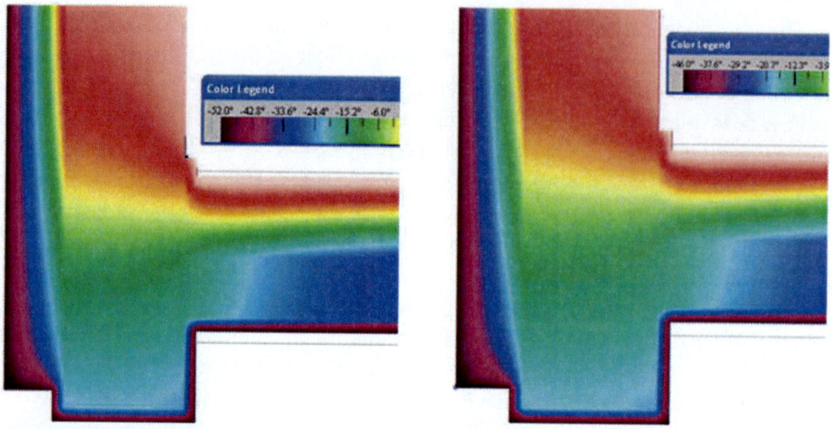

a *b*

Fig. 3 The structure of the formation of the temperature field in the structure isolated above the ventilated space (between the support columns): **a** visualization of the temperature field for Yakutsk (cold—52 °C); **b** visualization of the temperature field for Norilsk (cold—46 °C)

The resulting models were analyzed in terms of the thermophysical characteristics of the structures. As a result, the optimal version of the insulation system was chosen, which is effective both in the climatic conditions of the middle zone and in the especially cold conditions of Yakutia and the Arctic.

4 Discussions

Modeling shows that in the places where the structure rests on the column, in view of the high thermal and thermal diffusivity of the supporting structures, a "temperature bridge" is preserved, passing through the contact points of structural elements ("column–strapping–load-bearing wall"). At the same time, the placement of rolled polyethylene foam on the outer insulated surface (above the cold space under the structure) and as an element of the floating floor inside the room, allows you to completely block possible paths of heat loss both due to the "temperature bridge" and due to the blocking of direct infiltration of cold air along joints of structures (Figs. 1 and 2).

The effect of a higher level is manifested in the area between the columns, where the use of combined insulation (panel extrusion polystyrene (XPS) or polyisocyanurate foam (PIR) with a contour protection from rolled polyethylene foam) allows a completely seamless insulation shell to be formed (Fig. 3).

Thus, the proposed design solutions based on the use of a combined insulation system make it possible to form an insulating shell that meets the requirements of both heat conservation and energy efficiency.

The effect of using this system solution increases if structures operating in conditions of significant negative temperatures are insulated (Fig. 3). It should be noted that the studies carried out at the NRU MSCU [15–17] showed that all insulating materials can operate at temperatures up to minus 80 °C without loss of properties and thermal insulation ability.

The operating conditions of thermal insulation based on extruded polystyrene foam, polyisocyanurate foam and polyethylene foam are quite consistent with the studied ranges of operation of these materials and their operational resistance, which assumes the possibility of long-term operation of the insulation system. The developed insulation systems for building structures allow achieving the standard thermal resistance, minimizing possible heat transfer bridges.

In construction, the following types of building systems are implemented: a composite facade thermal insulation system (SFTC), a ventilated facade system (SVF), a layered masonry system (SLM), a translucent facade system (SFT), frame structures. A group of building systems, implemented in pile structures, which are used especially widely on soft soils or in permafrost conditions, is located separately.

The development of the northern territories is relevant for the development of all countries bordering the Arctic regions. For Russia, these regions are of particular importance in view of the expediency of developing the Northern Sea Route, increasing the country's defense capability, exploration and production of oil and gas on the shelf of the northern seas, as well as the extraction of other minerals. The development of the northern territories, from the standpoint of construction, involves the implementation of the following tasks:

- construction of buildings and structures that provide heat conservation, comfortable indoor conditions and the possibility of implementing technological processes (which also requires positive temperatures);
- development of road infrastructure taking into account the preservation of permafrost and taking into account the presence (or formation during thawing) of weak soils.

Materials for insulation systems of buildings and structures for northern latitudes must have not only low thermal conductivity, vapor and water permeability, but also resistance to aggressive environments and resistance to groundwater.

Foamed plastics are most suitable for the structures under consideration, namely products based on extruded polystyrene foam (XPS-plates) and polyethylene foam (PE-products). Thermal insulation products based on these polymers have low thermal conductivity and vapor permeability, high resistance to moisture and the whole complex of frost factors. XPS-boards have greater rigidity and resistance to mechanical stress and are used in the construction of road foundations and facade systems with plastering (SFTC). PE-products (mats or rolls) are elastic material that allows insulating surfaces of complex configuration and allowing a seamless welded joint.

5 Conclusion

The operating conditions of thermal insulation based on extruded polystyrene foam, polyisocyanurate foam and polyethylene foam are quite consistent with the studied ranges of operation of these materials and their operational resistance, which assumes the possibility of long-term operation of the insulation system. Rolled polyethylene foam with a metallized coating, forms a seamless insulating shell, completely blocks the possible paths of cold air infiltration at the joints of the structural elements, thereby reducing possible heat losses.

Using the program makes it possible to simulate 2D heat transfer in building components such as windows, walls, floors, roofs and doors. The analysis of heat transfer using the program allows you to evaluate the energy efficiency of the structure and the local temperatures of the sample, allowing you to solve problems with condensation, moisture content of the structure material and its tightness.

Calculation of temperature fields using the THERM computer program showed that the developed insulation systems for building structures allow achieving the standard thermal resistance, minimizing possible heat transfer bridges.

References

1. Zhukov AD, Ter-Zakaryan KA, Tuchaev DU, Petrovsky ES (2018) Energy-efficient insulation of food warehouses and vegetable stores. Int Agric J 1(361):65–67
2. Zinovieva EA, Zhukov AD, Ter-Zakaryan AK, Bessonov IV (2019) Dome house vegetarian. Housing 7:35–40. https://doi.org/10.31659/0044-4472-2019-7-35-40
3. Zhukov AD, Ter-Zakaryan KA, Semenov VS, Kozlov SD, Zinovieva EA, Fomina ED (2018) Insulation systems for buildings and structures based on polyethylene foam. In: MSCEU. IPICSE-2018. https://doi.org/10.1051/matecconf/201825101014
4. Zhukov AD, Ter-Zakaryan KA, Semenov VS (2018) Insulation systems with the expanded polyethylene application. ScienceDirect IFAC PaperOnLine 51(30):803–807. https://doi.org/10.1016/j.ifacol.2018.11.191
5. Fedyuk R, Mochalov AV, Simonov VA (2012) Trends in the development of norms for thermal protection of buildings in Russia. Bull Eng Sch FEFU 2(11):39–44
6. Umnyakova NP, Tsygankov VM, Kuzmin VA (2018) Experimental heat engineering studies for rational design of wall structures with reflecting heat insulation. Zhilishchnoe Stroitel'stvo 1–2:38–42
7. Gnip IYa, Kerchulis VI, Vaitkus SY (2012) Confidence intervals forecasting creep deformation of foam polystyrene. Stroitel'nye Materialy 12:40–44
8. Umnyakova NP, Bessonov IV, Zhukov AD, Zinoveva EA (2019) The effectiveness of facade systems. In: MATEC Web of conferences (ICMTMTE 2019), 00013. Published online: 18 Nov 2019. https://doi.org/10.1051/matecconf/201929800013
9. Gudkov P, Kagan P, Pilipenko A, Zhukova EYu, Zinovieva EA, Ushakov NA (2019) Usage of thermal isolation systems for low-rise buildings as a component of information models, 01039. Published online: 29 May 2019. https://doi.org/10.1051/e3sconf/20199701039
10. Zhuk PM, Zhukov AD (2018) Normative legal base of environmental assessment of building materials: prospects for improvement. Ecol Ind Russ 4:52–57
11. Zhukov AD, Ter-Zakaryan KA, Bessonov IV, Semenov VS, Starostin AV (2018) Building insulation systems using polyethylene foam. Build Mater 9:58–61

12. Zhukov AD, Ter-Zakaryan KA, Bessonov IV, Lobanov VA, Starostin AV (2019) Energy efficiency of seamless insulating shells. Build Mater 6:49–55. https://doi.org/10.31659/0585-430X-2019-771-6-49-55
13. Kozlov S, Efimov B, Bobrova E, Zinovieva E, Zhukova E (2019) Optimization of foamed plastic technology, 06010. Published online: 29 May 2019. https://doi.org/10.1051/e3sconf/20199706010
14. Semenov VS, Bessonov IV, Ter-Zakaryan KA, Zhukov AD, Mednikova EA (2020) Energy-saving seamless insulation systems for frame buildings using foamed polyethylene. In: Regional energy problems (Problemele energeticii regionale), electronic edn. Kishinev, p 4. https://doi.org/10.5281/zenodo.4018999 UDC: 691.1752./6./8
15. Zhukov A, Dovydenko T, Kozlov S, Ter-Zakaryan K, Bobrova E (2019) Innovative technologies for low-rise construction. In: TPACEE 2018, 02032. Published online: 02 Apr 2019. https://doi.org/10.1051/e3sconf/20199102032
16. Zhukov AD, Ter-Zakaryan KA, Bobrova EYu (2018) Innovative technologies of rural construction. Moscow Econ J (QJE.SU) 5
17. Bobrova E, Pilipenko A, Zhukov A (2019) Insulating sheath system and energy efficiency of buildings. In: TPACEE 2018, 02019. Published online: 02 Apr 2019. https://doi.org/10.1051/e3sconf/20199102019

Estimates for the Approval of the Project of Engineering Networks in the Organization Responsible for Connecting Consumers to the Electric Grids

Tatyana Kyzmina and Sergey Kourkin

Abstract This paper raises the issue of the lack of regulatory documentation in the field of approval of the project of engineering networks at the stage of "working design", which leads to repeated appeals to the balance-holding organizations due to the absence of specific requirements in preparing a project for submission to the investigated authority, which, as a rule, leads to an increase in the terms of providing the agreed documentation to the technical customer, which are also not regulated. The purpose of the study is to determine the average time indicators for the approval of the developed project in the organization responsible for the provision of services in the field of electricity distribution and technological connection to power grids, as well as the degree of their mutual influence. The study is based on a static method, in which the actual terms for the development and removal of comments of the developed project of already existing residential objects, which are typical in terms of space-planning and structural solutions, are taken as the factors under consideration. The static method of correlation and regression analysis makes it possible to find out whether there is a relationship between the accepted factors based on the determined correlation coefficients. The obtained averaged time indicators, as well as the degree of their mutual influence, will allow design organizations to predict the time frame required to obtain approval from the organization responsible for connecting consumers to the electric grids in order to competently distribute labor resources and avoid violation of the terms approved in the contract for the implementation of design work.

Keywords Approval · Working design · Electric grids · Averaged calculated indicators · Correlation and regression analysis

1 Introduction

In order to provide the developed documentation to the technical customer in accordance with the signed and legalized contract for the execution of design work on

T. Kyzmina · S. Kourkin (✉)
Moscow State University of Civil Engineering, 26, Yaroslavskoe Sh, 129337 Moscow, Russia

A. Beskopylny and M. Shamtsyan (eds.), *XIV International Scientific Conference "INTERAGROMASH 2021"*, Lecture Notes in Networks and Systems 247,
https://doi.org/10.1007/978-3-030-80946-1_84

time for further transfer of the documentation to the construction site with the stamp "approved for construction", it is necessary to approve the developed project of engineering networks at the stage of working documentation with all organizations whose area of activity is affected by its preparation [1]. One of such organizations is the authority responsible for connecting consumers to power grids, since the functioning of the projected facility, as well as the construction site, is impossible without providing electricity.

Based on the developed working documentation, the contractor will carry out construction and installation work of the projected facility, track the location of the routes of the projected networks and the points of their connection. Besides, in accordance with the Urban Planning Code of the Russian Federation, without an approved working documentation, the work manufacturer does not have the right to issue an order for construction work.

The working project is considered approved in the case when the balance-holding organization has agreed on the submitted documentation. Approval implies substantiation of the need for the developed documentation, confirmation of its financial and legal reliability, as well as technological and methodological relevance.

Agreements are divided into two stages: internal and external. At the stage of internal approval, working documentation is compared with the current regulatory requirements, adopted technical solutions developed at the stage of design documentation, and, if necessary, supplemented with detailed elements or auxiliary drawings. At the stage of external approval, the developed project is approved in all organizations whose area of activity has been affected, as well as with other interested authorities.

Each organization that performs the coordination functions of the provided organization posts on its official website all the necessary information on the terms and requirements for the approval of the provided project of engineering networks in the public domain. All terms are indicated in business days. In case of incompleteness of the documentation and the admission of a number of errors, the documentation is sent for revision with subsequent re-appeal. If everything is developed in accordance with the current standards, and also meets the requirements provided by the coordinating organization, the design organization is notified of the need to obtain a paper copy of the reviewed working documentation.

2 Investigation Objects

As a result of the lack of regulatory documents governing the timing of instructions carried out by the design organization in order to obtain approval from the organization responsible for connecting consumers to power grids, with the exception of the Order of the Ministry of Construction, Housing and Utilities of the Russian Federation dated May 15, 2020 No. 264/pr, we will select ten objects from the archive of one of the largest development organizations in order to analyze and process the labor input data available for their projects in order to determine the averaged time indicators that will be used as assumptions in the design of such objects in the future. It should be noted that the Order of the Ministry of Construction, Housing and Utilities of the Russian Federation gives only generalized terms necessary for the implementation of architectural and construction design in order to predict the term for a lease of a land plot, which does not give any understanding about the timing of approval of a project of engineering networks at the stage of "working design" and also its limits [2].

As the objects under study, we will take civil residential buildings of the same type in terms of space-planning and design solutions, which are located on the territory of Moscow. The design of each studied object was approved by the Moscow State Expertise, was agreed with all interested organizations and transferred to the construction site with the stamp "approved for construction".

Among the coordinating organizations, the choice fell on the authority responsible for connecting consumers to the electric grids, since it is one of the most labor-intensive ones in terms of obtaining the approval of the documentation being developed, based on expert estimate [3], which is based on the many years of experience of employees in the field of design of standard objects, considered by the developer company, as well as studied Russian and foreign scientific works and technical literature [4–7].

3 Average Indicators

Nowadays, the Resolution of the Ministry of Construction, Housing and Utilities of the Russian Federation on the development, coordination, approval and composition of the developed documentation has not been registered, and the previous normative document SNiP 11-01-95 was canceled early. As a result of the decisions made, for this period of time, there is no normative document regulating these stages of project preparation.

In addition to the fact that the regulatory documentation does not control the development, approval and the time frame in which it is necessary to prepare and approve the working documentation with various organizations, there is also no document that takes into account the possibility of repetitions of appeals caused by various factors, instructing the developer on his further actions.

Based on the shortcomings in the field of regulatory documentation regarding the correctness of the design of the engineering network project, as well as the time limits for this process, we will derive the average time indicators for the organization responsible for connecting consumers to power grids to obtain approval, based on the actual data of the objects under study [8].

The averaged time indicators for the development of the volume, internal approval (verification by the chief specialist of the prepared documentation), as well as the removal of comments after the initial appeal to the coordinating organization will be determined by the formula of the arithmetic mean numbers taken from the sample of the objects under study, based on the company's calendar planner:

$$\bar{x} = \frac{1}{n}\sum_{i=1}^{n} x_i = \frac{1}{n}(x_1 + \cdots + x_n) \quad (1)$$

The complexity of the instructions required for the submission of documentation to the organization responsible for connecting consumers to the electric grids (for each of the facilities) were taken on the basis of their working documentation and are presented in Table 1.

Formula 1 will allow you to determine the average indicators for each of the required orders:

$$\bar{T}_{development} = \frac{(6+3+3+4+5+5+2+4+5+3)}{10} = 4 \text{ man-hour};$$

$$\bar{T}_{check} = \frac{(4+1+1+2+3+3+1+2+2+1)}{10} = 2 \text{ man-hour};$$

Table 1 Actual labor intensity of obtaining project approval

No.	Research object name	Actual labor intensity of development, man-hours	Actual complexity of the check, man-hours	Actual complexity of correcting comments, man-hours
1	Residential building No. 1	6	4	5
2	Residential building No. 2	3	1	11
3	Residential building No. 3	3	1	11
4	Residential building No. 4	4	2	6
5	Residential building No. 5	5	3	5
6	Residential building No. 6	5	3	6
7	Residential building No. 7	2	1	12
8	Residential building No. 8	4	2	7
9	Residential building No. 9	5	2	6
10	Residential building No. 10	3	1	11

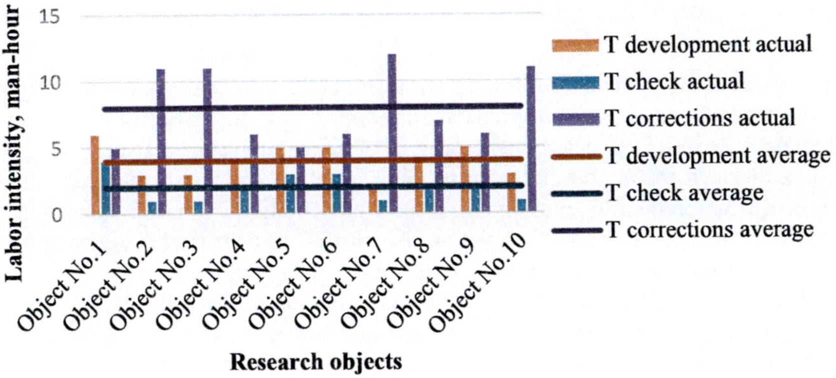

Fig. 1 Labor intensity diagram

$$\bar{T}_{corrections} = \frac{(5 + 11 + 11 + 6 + 5 + 6 + 12 + 7 + 6 + 11)}{10} = 8 \text{ man-hour.}$$

The obtained results will be presented on the diagram in a graphical form with the aim of the most visual display (Fig. 1).

In a number of cases, it is impossible to obtain approval from the balance-holding organization from the first time, which leads to the correction of the shortcomings issued by them and repeated appeal, which significantly extends the time frames that are not always available.

Complaints can arise for a number of reasons: differences in the official and unofficial requirements of organizations, mistakes made by designers in the development of documentation, contradictions between coordinating organizations, and so on.

By analogy with the average time indicators of instructions required to obtain approval in the instance in question, we will calculate the average number of repeated appeals to this organization caused for some reason, based on the data of the development company, according to Formula 1.

The number of repeated appeals is presented in Table 2.

The average number of repeated appeals is determined by Formula 1:

$$\bar{N} = \frac{(2 + 1 + 0 + 0 + 0 + 1 + 0 + 0 + 1 + 0)}{10} = 0.50 \approx 1 \text{ time.}$$

Table 2 Number of repeated appeals for project approval

	Residential building									
Object name	No. 1	No. 2	No .3	No. 4	No. 5	No. 6	No. 7	No. 8	No. 9	No. 10
Number of appeals	2	1	0	0	0	1	0	0	1	0

4 Research Methods

To predict the timing and the most rational distribution of labor resources in order to predict the time required for the preparation and approval of the project of engineering networks, as well as to determine the degree of mutual influence between the labor costs allocated for the development of the project and the removal of comments provided by the coordinating organization, we will build communication models based on the static method of correlation and regression analysis of a sample of data borrowed from the archived data of the studied civil buildings.

Correlation-regression analysis is based on a method for calculating correlation coefficients, based on which an assumption is made about the existence of a relationship between the variables under consideration.

The correlation coefficient is a quantitative measure of the probabilistic relationship, i.e. joint variability of two variables. The correlation coefficient value ranges from -1 to $+1$.

The linearity of the correlation is manifested in the fact that the standard deviations of the analyzed points on the scatter plot originate from a straight line.

Regression analysis will consist in building a prediction line (trend line), each point of which deviates minimally from each point of the scatter plot (i.e. the sum of the squares of the deviations between the points under consideration will tend to zero), as well as determining the coefficient of determination with the subsequent identification of the significance of the constructed model and the probability of making an error in our judgment about the significance of the model based on the Fisher criterion [9, 10].

In order to determine whether an increase in labor costs at the stage of document development (more detailed preparation of a project) affects a decrease in labor costs spent on the removal of claims provided by an approving organization, on the total duration of approval of the entire project, taking into account repeated appeals, it is necessary to establish the degree of mutual influence of the factors in question. The labor input for the development of documentation is taken as a variable for the analysis.

The labor input taken to determine the degree of mutual influence was adopted in accordance with the documentation of the objects of the development company under study and their ranking (see Table 3).

The determination of the design parameters of the model, as well as its construction, in order to analyze the obtained data was carried out in Excel and are presented in Table 4 and Fig. 2.

The analysis of the compiled model gives the following idea of the mutual influence of the considered values: the relationship is linear, inverse and very high. The true value of the correlation coefficient is located in the range from -0.73 to -0.98. In 87%, changing one parameter will change another. The judgment about the significance of the model is confirmed, and the probability of making a mistake tends to zero.

Table 3 Actual labor intensity and its ranking

No.	Object name	Actual labor intensity of development, man-hours	Actual complexity of correcting comments, man-hours	Ranking along the y-axis, ry	Ranking along the x-axis, rx
1	Residential building No. 1	6	5	1.5	10
2	Residential building No. 2	3	11	8	3
3	Residential building No. 3	3	11	8	3
4	Residential building No. 4	4	6	4	5.5
5	Residential building No. 5	5	5	1.5	8
6	Residential building No. 6	5	6	4	8
7	Residential building No. 7	2	12	10	1
8	Residential building No. 8	4	7	6	5.5
9	Residential building No. 9	5	6	4	8
10	Residential building No. 10	3	11	8	3

Table 4 Design parameters of the model

Pearson coef.	Spearman coef.	Left border of distribution	Right border of distribution	Coef. of determination	Model relevance	Error probability
−0.93	− 0.85	−0.98	−0.73	0.87	52.94 > 5.32	0.000086 < 0.05

5 Reducing Deadlines

Since repeated appeals for the purpose of approving the project being developed are already an integral part of acquiring project experience and maintaining an archive of comments from the coordinating organization, an analysis of the existing claims for the accepted ten objects is carried out, and the most frequent ones are identified in order to take them into account in the initial preparation of the project of engineering networks for increase the likelihood of agreeing on the project at the first contact. The list of comments is presented in Table 5.

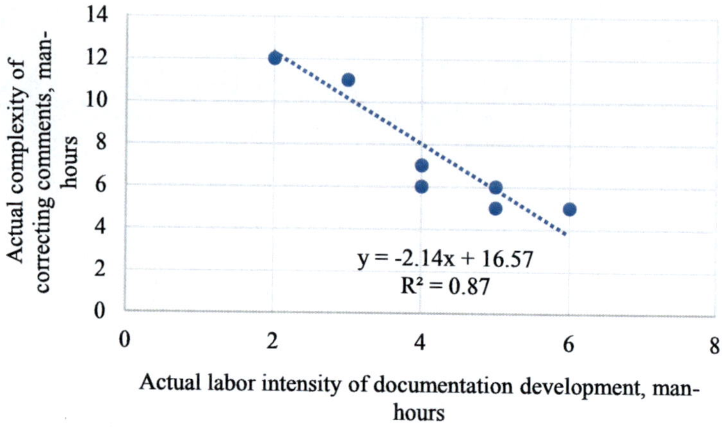

Fig. 2 Scatter diagram

Table 5 List of typical comments

No.	Comments
1	The plan should be provided on a scale of 1:500 with an exact reference of the land plot
2	Provide sections in the places where temporary roads overlap with the projected networks
3	The graphic representation of the cross-sections of the projected conductors must correspond to the calculated values specified in the project

In addition to comments in the field of design, the organization accepts documentation only in a specially prepared form, which must also be taken into account at the development stage. It is necessary to provide a bound project in one copy, as well as an electronic version in *pdf format—for archives.

6 Conclusion

In the course of the study, the averaged time indicators for the development of the volume, internal approval (checking by the chief specialist of the prepared documentation) were determined, as well as the removal of comments after the initial appeal to the coordinating organization responsible for connecting consumers to power grids, which will be laid as assumptions in the design of such objects in the future. In addition, the average number of repeated appeals and the degree of mutual influence of the labor inputs spent on the preparation of volumes and the removal of the provided comments were determined in order to predict and reduce the time indicators for the removal of comments and repeated appeals.

To reduce the time for approval, an analysis of the comments issued by the approving organization was carried out to identify the most typical ones in order to take them into account in the initial preparation of the project, as well as the documents that the organization requests in addition to the official requirements.

In order to identify how much the approval process will be reduced in the instance responsible for connecting consumers to power grids, it is necessary to build network diagrams for the actual and expected options with a visual presentation of time indicators and make a comparison to analyze the effectiveness of the implemented proposals.

References

1. Bolshakova PV (2020) Functions of a technical customer (developer) in the implementation of an investment project and factors affecting their implementation. Constr Prod 1:27–32
2. Abramov IL, Lapidus AA (2018) Systemic integrated method for assessing factors affecting construction timelines. In: MATEC Web of Conferences, 05033
3. Zagorskaya AV, Lapidus AA (2020) Application of expert assessment methods in scientific research. The required number of experts. Constr Prod 3:21–34
4. Kuzmina TK, Kourkin SV (2021) Study on the design indicators of the approval of the project of engineering networks in JSC "Mosvodokanal". Eng J Don 4
5. Riddell T (2018) RFI process in the field: 10 tips you can start using today. http://esub.com/10-tips-managing-rfi-process-field
6. Rodriguez JA (2019) Critical final step of a construction project. http://thebalancesmb.com/expert-tips-when-recording-asbuiltdrawings-844876
7. Kuzmina TK, Bolshakova PV (2020) Identification and systematization of factors in the preparation of objects for construction by a technical customer (developer). Constr Prod 4:38–43
8. Oleynik PP, Kuzmina TK (2018) Modeling the activities of a technical customer at the stage of pre-design study and preparation for construction. Technol Organ Constr Prod 4:7–10
9. Nasledov AD (2012) Mathematical research methods. Analysis and interpretation of data. SPb.: Rech, p 392
10. Nasledov AD (2011) SPSS 19. Professional statistical data analysis. SPb.: Piter, p 400.

Development of a Simulation Automated System for Address Sorting of Correspondence

Sergey Morozov, Konstantin Kuzmin, and Vladimir Vershinin

Abstract The paper discusses a model of an automated system for sorting correspondence, developed in the VisSim environment. A simulation model of information coding has been developed. The possibility of optimizing the operation of similar lines is shown. A generalized algorithm for managing the sorting and accumulation of correspondence has been developed.

1 Introduction

The use of technical means for sorting and processing correspondence makes it possible to increase labor productivity and, on this basis, to achieve a higher efficiency of the enterprise, first of all, to accelerate the processing and delivery of correspondence, to create more favorable working conditions in the production activity of the enterprise.

To solve these problems, a wide variety of equipment is needed: lifting and transporting, special post-processing machines, means of small mechanization.

2 Methods

Processing of incoming letters (and all incoming correspondence) begins from the moment the document arrives at the enterprise [1]. To start to describe this process of work, one should analyze the factors that have a direct impact on its organization.

S. Morozov · K. Kuzmin (✉)
Smolensk Regional Cossack Institute of Industrial Technology and Business (Branch), Moscow State University of Technology and Management Named After K.G. Rasumovsky (1St Cossack University), 77, Lenin Str., 215100 Vyazma, Russia

V. Vershinin
Moscow State University of Civil Engineering, 26, Yaroslavskoye Shosse, Moscow 129337, Russia

A. Beskopylny and M. Shamtsyan (eds.), *XIV International Scientific Conference "INTERAGROMASH 2021"*, Lecture Notes in Networks and Systems 247,
https://doi.org/10.1007/978-3-030-80946-1_85

These factors include: the method of delivery of correspondence; the content of the document; author of the document; deadlines for the execution of the document.

The second important factor affecting the processing of incoming correspondence is the authorship of the document. The significance and speed of reaction to an incoming letter depends on who is its author [2]. A letter from the control authorities requires an immediate response, while a letter from a dubious source, in most cases, will not even be considered.

The content of the document is also significant. Modern clerical processes involve the development of special routes for certain groups of documents that are characteristic of documents of a certain content [3].

If we carefully read the content of incoming letters, it is possible to safely classify them according to the following criteria:

- letters containing any demand, request, proposal, complaint, i.e. letters of inquiry requiring any action or decision-making;
- letters of response to the initiative documents sent by you;
- reference letters.

It is clear from the name "letters of request" that this type of documents requires careful study, and work with a document of this group may include all stages of processing incoming correspondence, from registration to control of execution. Letters of this type can also include the so-called circular letters sent from higher organizations to subordinate ones. In these letters, the organization is assigned any tasks that require mandatory and sometimes urgent execution. Therefore, for this category of service letters, two factors are important: the need for execution and, as a consequence, the terms of execution.

Letters of response are response documents to outgoing requests, inquiries, proposals, demands and other initiative letters sent to other organizations in order to solve any of your problems. Most often, letters of response do not require further action, except for their transfer to the employees who initiated the consideration of this issue.

Due to the large volume of incoming and outgoing correspondence and the need to sort it, it is advisable to use automated control systems for sorting and processing correspondence.

3 Results and Discussions

The model of the automated system for sorting the correspondence was developed in the VisSim environment. The model consists of two autonomous submodels [4].

Without losing the generality of the solution of the problem, this work contains 4 tracks for recording the coding and decoding of information out of 12 actually used in practice of sorting correspondence.

Figure 1 shows the coding model of randomly distributed correspondence arriving to the conveyor.

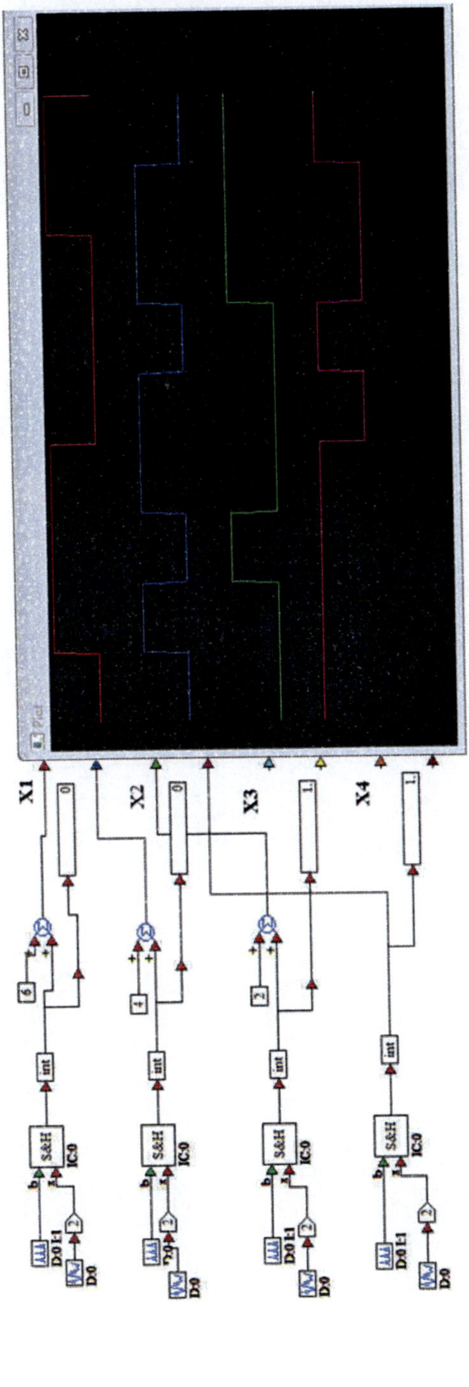

Fig. 1 Model of coding of randomly distributed correspondences arriving to the conveyor

The processing time of information in the form of a postal address, taking into account the speed of the conveyor, does not exceed 1 s. Therefore, the pulse repetition rate from the clock generator is set to 1 Hz. The indication of the information signal at the current time on the tracks (X1, X2, X3 and X4) is recorded by one-bit digital indicators in the form of logical “0” or “1” [5].

The results of encoding randomly received objects of sorting (correspondence) on the tracks (X1, X2, X3 and X4) are shown on the oscilloscope graphs. In Fig. 1, the following designations are adopted:

 random number generator with uniform distribution in the range of numbers from 0 to 1

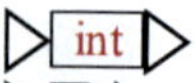

block for separating an integer part from a real number.

random number multiplier.

 single pulse clock generator with a frequency of 1 Hz.

block that fixes the value of a random variable with a delay of 1 s.

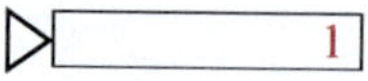

one-bit digital indicator of the latching of the bit for binary coding.

Using the tracks shown in Fig. 1, it is possible to encode 16 addresses, which is shown in Fig. 2.

The developed information coding model is the first part of a complete model of an automated system for sorting correspondence and operates in real time [6]. This

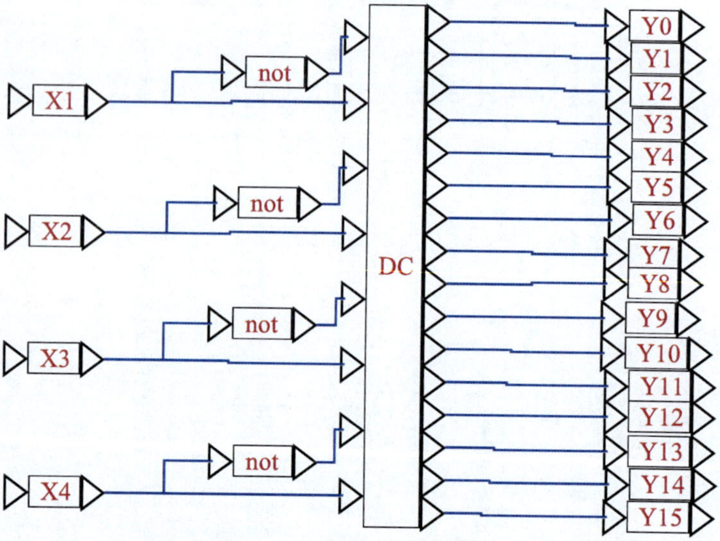

Fig. 2 Block of coding information from 16 addresses

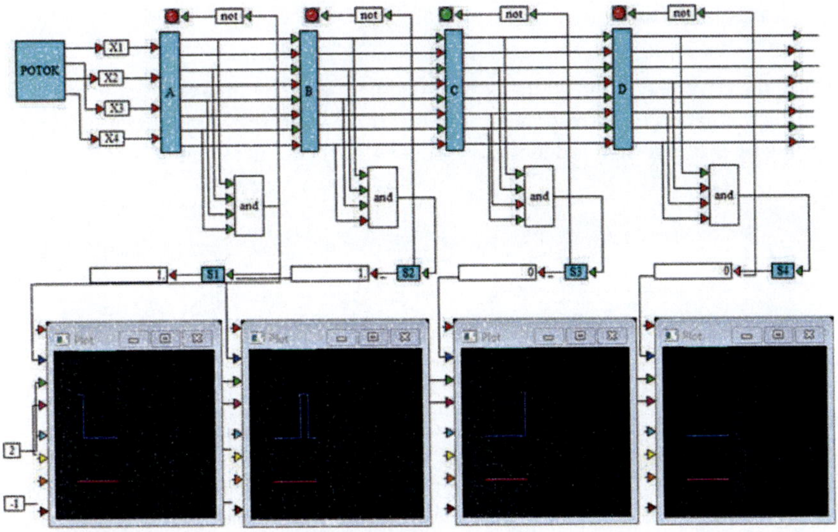

Fig. 3 Simulation model of an automated system for sorting correspondence

model is simulated, since the input parameters (number of tracks, clock frequency, etc.) and the speed of the conveyor can be set depending on the load of the post office.

For a clear illustration of the process of encoding address information, the graphics displaying the state of the tracks are displaced vertically and are presented in Fig. 2.

A simulation model of an automated system for sorting correspondence is shown in Fig. 3. This model works in the VisSim environment in real time [7].

The basic element of the decoder of coded forwarding addresses in the proposed model is a conjuncture logic element ("AND") with four inputs.

Each such element performs the function of selecting and forwarding the corresponding postal items to a separate storage device with a real corresponding code address.

A light alarm is provided for each storage device, which allows one to visually determine the state of the current gateway channel:

- "Red light"—corresponds to closed channels, "green light"—open ones.

In each channel, there is also a counter of incoming correspondence in the accumulator [8–18].

4 Conclusions

Thus, the developed model allows reproducing the technological process of sorting and accumulation of correspondence in real time: to carry out optimization for mail sorting lines of different productivity by changing the input parameters of the system.

References

1. Besekerskij VA, Popov EP (2017) Teoriya sistem avtomaticheskogo upravleniya. 4-e izd., pererab. i dop. – SPb.: Professiya. 747
2. Dzhekson RG (2016) Novejshie datchiki: uchebnik – monografiya / per. s angl. – 2-e izd., dop. – M.: Tekhnosfera. 397
3. Prangishvili IV, Ambarcumyan AA (2015) Osnovy postroeniya ASU slozhnymi tekhnologicheskimi processami. M.: Energoatom-izdat
4. Morozov SM, Kuzmin KA, Kochetkova LI, Balmashnova EV (2019) Razrabotka iskhodnyh koncepcij metrologicheskogo obespecheniya izmeritel'nyh i raschetnyh operacij pri avtomatizacii izmerenij.- Agrarnyj nauchnyj zhurnal. 4, 87–89
5. Morozov S, Kuzmin K, Makarov G (2018) Neural network principle of implementation of digital filters. In: MATEC web of conferences (sm. v knigah), vol 193, p 102
6. Kuzmin KA, Morozov SM (2018) Opredelenie optimal'nyh rezhimov raboty svch-ustanovki dlya sushki i obzhiga stroitel'nogo kirpicha.- Inzhenernyj vestnik Dona. 3, 75
7. Morozov S, Kuzmin K, Pogodin D, Kochetkova L, Rogozhina A (2019) Modeling of the management of the microwave grain drying process. In: E3S web of conferences. 2018 International science conference on business technologies for sustainable urban development, SPbWOSCE 2018. 02142
8. Belostotsky AA (2020) Analysis of the correspondence of the established and achieved target indicators in national projects. Vestnik Inst World Civil 11(2–27):102–107
9. Bobkov VN (2019) Theoretical and methodological approaches to resolving the contradictions between national goals and national projects. Sci Works Free Econ Soc Russia 217(3):77–99
10. Bukhvald EM (2020) Financial and budgetary aspects of the implementation of national projects in the Russian Federation. Theory Pract Soc Dev 3(145):41–44
11. Valishvili MA (2020) Financing of priority national projects: interim results and prospects for implementation. Entrepr Guide 13(4):46–54
12. Vdovin IA, Venglinsky DR (2020) The problem of attracting off-budget investments in national projects. Bus Soc Power 2–3(36–37):74–85
13. Gorokhova SS (2020) National projects of the Russian Federation as a tool for achieving strategic goals and objectives of the state. Russian Justice 6:41–44
14. Ivanenko LV, Taseev VB (2020) Problems of the implementation of national projects and determining the options for their solution. Fundam Econ Manag Law 6(25):43–47
15. Korostelkina IA, Chebneva ON (2020) Achievement of national development goals of Russia during a pandemic: features of monitoring of national projects by the Accounts Chamber of the Russian Federation. Econ Dev Region Manag Innov Training Personnel 7:196–200
16. Polyanskaya EE (2020) The role of interaction between the population and authorities in the transition to sustainable development of the territory. Collection of abstracts "ECONOMIC AGENDA of the 2020s". Annual Scientific Conference "Lomonosov Readings", 20–24Oct 2020, Moscow
17. Kirillova A, Zotov V, Musinova N (2018) Engineering solutions in the construction of production facilities for waste processing. In: 2018 international scientific conference environmental science for construction industry, ESCI 2018 Ho Chi Minh City, 02–05 Mar 2018

18. Kirillova AN, Musinova NN (2017) Structuring problems, technologies and mechanisms for implementing strategies in the field of waste management. Econ Entrepr 10–1(87–1):910–914

Slippage and Wear in Rolling Bearings of Machines

Aleksandr Dykha, Dmytro Marchenko, Viktor Artiukh, and Elena Solovyeva

Abstract Rolling bearings are used as support units in almost all modern transport and technological machines. Bearing units largely determine the durability of the machine as a whole. The main cause of bearing failure is wear of bearing parts due to slippage. The quantity of slip is needed to calculate the friction path in predicting wear. The paper deals with the problem of experimental determination of the slip value and predicting wear in rolling bearings of machines. A general methodology for solving the problem is proposed. The design of an experimental setup for determining slip is proposed. Experimental studies of the effect of load, speed and lubrication in rolling bearings have been carried out. For the calculated prediction of the wear of the raceway, the wear-contact problem is solved when the ball and the plane interact. The wear model is shown in dimensionless form. For the numerical determination of the parameters of the law of wear, a modified calculation and experimental technique is proposed. The proposed approaches make it possible to effectively predict the durability of rolling bearings based on the wear criterion for technological and transport machines.

Keywords Bearing · Slip value · Laboratory test · Raceway wear · Contact pressure

A. Dykha
Khmelnytskyi National University, Khmelnytskyi, Ukraine

D. Marchenko
Mykolayiv National Agrarian University, Mykolaiv, Ukraine

V. Artiukh (✉)
Peter the Great St. Petersburg Polytechnic University, Saint Petersburg, Russia

E. Solovyeva
K.G. Razumovsky Moscow State University of Technologies and Management (the First Cossack University), Moscow, Russia

A. Beskopylny and M. Shamtsyan (eds.), *XIV International Scientific Conference "INTERAGROMASH 2021"*, Lecture Notes in Networks and Systems 247,
https://doi.org/10.1007/978-3-030-80946-1_86

1 Introduction

The basic bearing units of the machines are rolling bearings. These units are integral parts in rotating elements of machines and mechanisms operating under various technological conditions with the coupling action of heavy axial loads and torques [1–4], high temperatures [5, 6], humid or abrasive environments [7, 8], etc. Precision accuracy is most often also included in the list of requirements for bearing assemblies, which, however, does not eliminate their negative slippage effect [1, 9, 10]. The main causes of bearing failure are [11, 12]: fatigue spalling of raceways; wear of raceways; plastic deformation and creep of raceways. At present, the process of contact fatigue failure of bearings is well studied. Methods for calculating the contact strength based on the solution of the contact problem according to Hertz [13–15], including for the different shapes of the indenters in the plastic region of deformations [16–18], have been developed. Test methods and definitions of damage accumulation models have been developed [19–21].

At the same time, many bearings fail due to wear rather than fatigue. This primarily applies to bearings in abrasive and high temperature conditions in which different methods of materials hardening and extending of bearing elements life are used [22–26]. There are practically no methods for calculating and testing bearings for wear, which complicates their design. Methodology for calculations and tests for wear and reliability of bearings are the following stages [1, 15]:

(a) The features of the design and technology of the friction unit are studied. The loads on the rolling elements are determined.
(b) The contact pressures between the body and the raceway are determined.
(c) The sliding path of the ball or roller along the raceway is determined.
(d) Selection of a wear model and experimental determination of its parameters.
(e) Solving wear-contact problems for bearing parts.
(f) Calculations for bearing wear reliability using Gaussian normal distribution.

Thus, one of the main values for calculating and analyzing the process of wear of rolling friction units is the sliding friction path between the contacting bodies.

The roller bearing model was developed to evaluate cage slip, roller slip, film thickness and cage forces for a given bearing geometry and operating conditions. The model takes into account the friction of the cage guide surface, the friction of the roller pocket, and the cage unbalance. The description of the lubricating film thickness, adhesion forces and pressure is based on the solution of the problem of contact surface mechanics [27].

The article [28] proposes a numerical method for determining the sliding of a cage in high-speed ball bearings with an axial load. The model agrees well with the experimental results for sliding ball bearings. The model is recommended for studying the causes of sliding in loaded ball bearings.

The influence of operating parameters on the slip of the cage in cylindrical roller bearings is considered in paper [29]. The cylindrical roller bearing test bench is designed to measure the movement of bearing elements under various operating conditions. The influence of operating parameters, namely, shaft rotation speed, radial load, lubricating oil viscosity, number of rollers and bearing temperature, on the cage sliding is obtained experimentally.

The slip of the cage and balls of ball bearings used in paper [30] for the main shafts of jet engines is evaluated. A new method has been developed for assessing slip, taking into account the increase in oil temperature caused by slip in the contacts between the ball and the raceway. The analytical results of the study were compared with experimental data.

In [31], the influence of different models of a radially loaded cylindrical roller bearing is investigated using a dynamic model of a bearing assembly. The cage rotation speed, depending on the force between the roller and the raceways, is designed for a wide range of speeds and loads. Comparison of the simulation results with the results of experimental measurements shows that the model is able to predict the operation of the rolling elements under various operating conditions.

Triboacoustic control is one of the effective methods for controlling the processes of destruction and slipping in bearings [32]. Mechanical and mathematical models of damage accumulation in tribosystems and forms of wear laws with identification of their parameters are discussed in detail in works [17, 18, 33, 34]. Further modeling of sliding in rolling bearings from various points of view was also considered in papers [35–38].

2 Methodology of Research

2.1 *Slip in Ball Bearings*

It is known from experiments that the length of the path of a ball of a ball during rolling is not equal to the length of a circle made up of points of contact. This means that rolling occurs with slippage. In this paper, slippage is necessary to determine the sliding friction path for wear calculations. Determination of the sliding path in rolling with slip in most of the known studies was done by theoretical methods. Determination of slippage by theoretical methods leads to complex mathematical problems, the solution of which again requires simplifications and approximations. Most often, it is difficult to assess the accuracy of these solutions.

It is more rational to experimentally determine the value of the slip coefficient of bearings, taking into account various factors and apply them to other types of bearings.

Consider the basic relations for the rolling of a ball on a plane with slippage.

The rolling friction force is related to the normal load, ball radius and rolling friction coefficient by the ratio:

$$F_r = \mu_r \frac{N}{R} \tag{1}$$

When the ball is rotated through an angle 2π or a full turn in the absence of slippage, the length of the contact line on the plane will be equal to the circumference of the ball:

$$l = 2\pi R \tag{2}$$

In the presence of slippage, the rolling friction path will differ from the circumference by an amount Δl.

Attitude:

$$S_c = \frac{\Delta l}{l} \tag{3}$$

called the slip coefficient. Experimental determination of the slip for a single ball is not difficult.

Let us assume that rolling resistance occurs due to the slipping of the ball along the plane at the points of contact. Thus, the rolling resistance due to surface deformations is neglected.

Slip resistance frictional force:

$$F_s = \mu_s N \tag{4}$$

We assume that the work of rolling friction forces on the rolling path is equal to the work of sliding friction forces on the sliding path Δl:

$$W_r = W_s \tag{5}$$

or

$$F_r l = F_s \Delta l \tag{6}$$

Substituting @@(1) and (4) into (6), we obtain:

$$\mu_r \frac{N}{R} l = \mu_s N \Delta l \tag{7}$$

Or, taking into account (3), we obtain:

$$S_c = \frac{\mu_r}{\mu_s R} \tag{8}$$

Therefore, the coefficient of rolling slip is equal to the coefficient of rolling friction divided by the coefficient of sliding friction.

An example of a ball rolling $d = 12.3$ mm on hardened steel.

The rolling friction coefficient is assumed to be $\mu_r = 0.01$ mm. Sliding friction coefficient of steel on steel without lubrication $\mu_s = 0.1 \ldots 0.3$. Then:

$$S_c = \frac{\mu_r}{\mu_s R} = \frac{0.01}{(0.1\ldots0.3)\cdot 6.15} = 0.016\ldots0.005.$$

With a cage diameter in the bearing of $d_c = 51$ mm, the sliding friction path for the balls along the raceway per revolution is $D_s = S_c\pi d_c = (0.016\ldots0.005)\pi \cdot 51 = 2.56\ldots0.80$ mm.

2.2 *The Basis of the Experimental Method*

The mechanics of the motion of balls in rolling bearings of machines with a slip is a complex problem in contact mechanics. The complexity is explained by the variety of ball movements: around the rotor axis; rotation around its own axis; spinning; gyroscopic motion; slippage in all contact zones.

Slip mechanics are complicated by the proximity of the ball radii and the radius of the ring track. At the same time, the calculation of rolling bearings for wear requires an accurate assessment of the sliding friction path in contact between the ball and the raceways.

A computational and experimental method is proposed for determining the amount of slip in ball bearings. The basis of the approach is as follows. From the kinematics of the bearings, precise relationships are obtained for the movements of the balls and rings without slipping. On the other hand, experimentally, you can simply measure the number of revolutions of the separator. The difference between the calculated and actual numbers of revolutions of the cage allows you to determine the average value of the ball sliding along the raceway.

The kinematics of rings and a ball in a bearing is considered under the following conditions: the inner ring rotates; the outer ring is stationary; the ball rolls over the rings without slipping. Kinematics are considered to determine the sliding friction path of a ball along the raceways.

For a bearing with a fixed outer ring, the geometric ratio is adopted:

$$V_c = \frac{1}{2}V_1 \tag{9}$$

where V_c, V_1 is the linear speeds of movement of points of the separator and the inner ring V_1.

The linear speed of the ring is expressed in terms of the number of revolutions by the ratio:

$$V_1 = \frac{\pi d_1 n_1}{60} \tag{10}$$

Similarly, for the linear speed of the separator:

$$V_c = \frac{\pi d_c n_c}{60} = (d_1 + d_0)\frac{\pi n_c}{60} \tag{11}$$

where d_c is the cage average diameter;
d_1 is the diameter of the inner ring;
n_1 is the number of revolutions of the inner ring;
n_c is the number of revolutions of the separator;
d_0 is the ball diameter.
After mutual substitutions and transformations, we get:

$$n_c = \frac{n_1}{2(1 + d_0/d_1)} \tag{12}$$

The experimental speed of the separator n_c^* differs from the theoretical speed n_c. The difference between experimental and theoretical rpm is explained by the slippage of the balls along the raceways.

With this in mind, the coefficient of slip in the rolling bearing is determined by the relationship:

$$S_c = \frac{n_c^* - n_c}{n_c^*} = 1 - \frac{n_c}{n_c^*} \tag{13}$$

3 Results and Analysis

3.1 Installation, Test Procedure and Results

An installation for testing rolling bearings for slip is shown in Fig. 1.

Carrier 1 transmits rotation with frequency n through pins 3 to the upper disk 4 of the working head. The load Q on the bearing assembly is transmitted through ball 2. Ball 3 is fixed in a cylindrical cage 11. A radial bearing 5 is needed to center the upper disc and the lower head housing 9. A thrust ball bearing is tested, consisting of an upper ring 6, a lower ring 8 and a cage 7. The working head is attached to the table of the test bench with a nut 10. During wear tests, the test specimen is installed in place of the lower ring of the thrust bearing. When testing for slippage when the upper ring rotates at a speed n_1, the separator 7 is simultaneously rotated at a speed n_c.

The number of revolutions of the ring and cage was measured using a DT2234B electronic digital tachometer. Foil pieces 5 × 10 mm in size were attached to the upper ring and to the separator. The beam from the tachometer periodically hits the foil and records the number of revolutions per minute accurate to the first decimal place. The measurement results are shown in Tables 1 and 2.

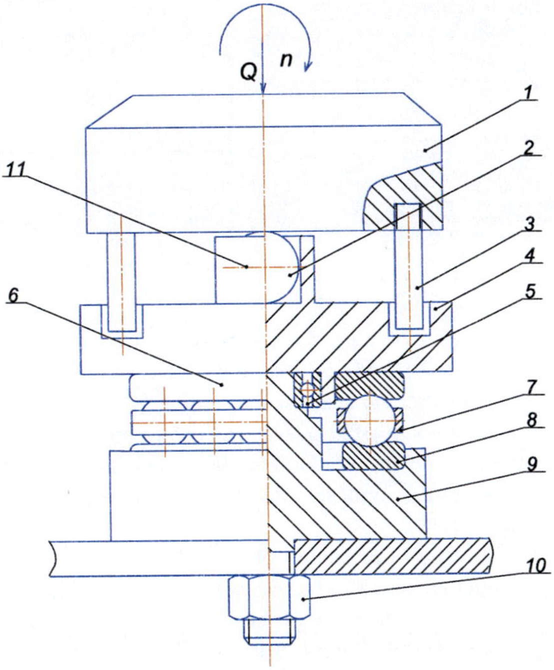

Fig. 1 Installation test machine for rolling bearings

Table 1 Test results without lubrication

Speed, rpm	Q, N						n_0, rpm
	100	S_c	250	S_c	500	S_c	
n_1	250.2	0.0312	250.9	0.0322	250.4	0.0309	250
n_c	129.1		129.6		129.2		
n_1	497.4	0.0293	496.6	0.0312	495.9	0.0315	500
n_c	256.2		256.3		256.0		
n_1	1350	0.0321	1349	0.0320	1348	0.0317	1350
n_c	697.4		696.8		696.1		

Table 2 Test results with Fiol-3 grease

Speed, rpm	Q, N						n_0, rpm
	100	S_c	250	S_c	500	S_c	
n_1	251.0	0.0331	250.5	0.0342	250.8	0.0350	250
n_c	129.8		129.7		129.9		
n_1	499.4	0.0323	498.5	0.0343	497.9	0.0335	500
n_c	258.0		258.1		257.6		
n_1	1351	0.0352	1349	0.0343	1349	0.0362	1350
n_c	700.1		698.5		699.8		

Fig. 2 Scheme of wear tests

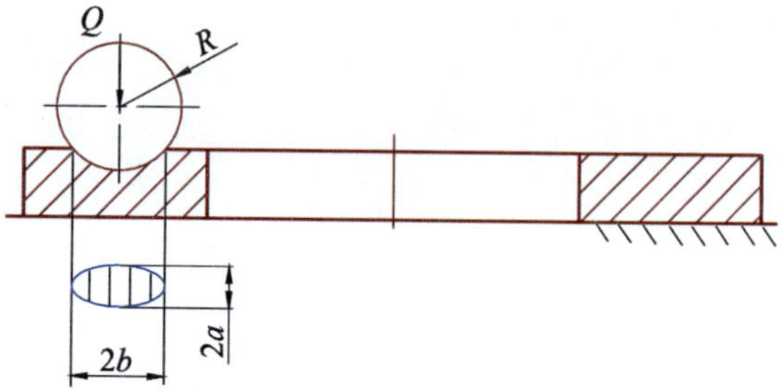

Slip tests were carried out at loads Q = 100; 250; 500 N for bearing 8204. The revolutions of the upper ring were set in steps: n1 = 250; 500; 1350 rpm.

The theoretical speed of rotation of the separator was calculated by the Formula (9), the amount of slippage was determined by the Formula (12). Slip tests were carried out under two lubrication modes: in the absence of lubrication and lubrication with Fiol-3 grease. The results of measurements and calculations are shown in Tables 1 and 2.

Analysis of the results obtained shows that the slip coefficient is practically independent of the load and rotation speed. Slip fluctuations due to these factors are rather random. Average value of the slip coefficient for bearing 8204 without lubrication $S_c = 0, 0313$, and when lubricated with Fiol-3 grease $S_c = 0, 0342$. Thus, the use of lubricant increases slip by 8%.

3.2 *Modeling of Rolling Bearing Wear*

The calculation of wear is based on the mechanics of the contact interaction of a ball and a plane with wear. The sliding of a ball on a ring (Fig. 2) under a load Q is considered.

The ring is stationary, the surface of the ring wears out. During the wear process, a groove with a width of 2b is formed, which increases with wear. We assume that the ball does not wear out.

Ring wear model in differential form:

$$\frac{dU_w}{ds} = k_w\left(\frac{\sigma}{HB}\right)^m\left(\frac{\upsilon R}{\nu}\right); \tag{14}$$

or in the form:

$$\frac{dU_w}{ds} = k_w\sigma^m\Pi, \quad \Pi = (HB)^{-m}\left(\frac{\upsilon R}{\nu}\right),$$

where σ is the contact pressure;

k_w, m are the parameters of the wear model;
U_w is wear;
s is the friction path for ring;
HB is Brinell ring material hardness;
υ is the relative sliding speed;
R is the radius of the ball;
ν is the kinematic viscosity of the lubricant.

In the process of wear, a platform is formed with dimensions: 2a along the friction path and 2b across the friction path. We assume that the contact pressure is evenly distributed over the contact area:

$$\sigma = \frac{Q}{\pi ab} \tag{15}$$

The depth of the wear groove U_w is related to the width of the groove and the radius of the ball in the ratio:

$$U_w = \frac{b^2(s)}{2R} \tag{16}$$

The friction path s for ring points is related to the size of the contact area in the sliding direction by the ratio:

$$s = 2azntS_c \tag{17}$$

where z is the number of balls passing through the contact area, per revolution;
n is the number of revolutions per minute;;
t is the duration of the bearing;
S_c is the slip value.

The size a of the contact area can be determined by the Hertz formula:

$$a = 1.09\sqrt[3]{\left(\frac{QR}{E}\right)} \tag{18}$$

where E is the modulus of elasticity of the ring material.

The main assumption is that the size of the contact area in the direction of motion of the ball is taken to be equal to the initial size a according to (18).

Differentiating Eq. (16) and substituting in (14) after transformations, we obtain the differential equation of the problem:

$$\frac{Q}{\pi ab} = \left(\frac{b}{\Pi k_w R}\frac{db}{ds}\right)\frac{1}{m} \tag{19}$$

The solution of the differential Eq. (20) with a zero initial area $b = 0$ gives an expression for determining the size of the wear area of the bearing ring $b(s)$:

$$b = \left((m + 2) \left(\frac{Q}{\pi a} \right)^m \Pi k_w RS \right)^{\frac{1}{m+2}} \quad (20)$$

The inverse problem consists in determining the parameters of the bearing ring wear model K_w and *m* for the function *b(s)* known from the experiment. Based on the test results, this function can be represented as a power-law approximation:

$$b(s) = cs^{\beta} \quad (21)$$

Then Eq. (20) can be represented in the form:

$$\frac{c^{m+2} S^{\beta m + 2\beta}}{(m + 2)} = \left(\frac{Q}{\pi a} \right)^m \Pi k_w RS. \quad (22)$$

From the condition that this equation is satisfied, we obtain:

$$\beta m + 2\beta = 1. \quad (23)$$

or

$$m = \frac{1 - 2\beta}{\beta} \quad (24)$$

Taking into account (23), from (22) it follows:

$$k_w = \frac{c^{m+2}}{(m + 2)} \left(\frac{\pi a}{Q} \right)^m \Pi R \quad (25)$$

Thus, a generalized solution of the contact problem of the interaction of a ball and a bearing ring during ring wear is proposed. At the same time, an approximate solution of the problem of ring wear by a ball is carried out under the assumption of a slight change in the size of the contact area in the direction of the ball movement.

4 Conclusions

1. The main bearing unit of modern machines is rolling bearings. There are practically no methods for calculating and testing bearings for wear, which complicates their design.
2. A computational and experimental method for determining the amount of slippage in ball bearings is proposed. From the kinematics of bearings, relations for the movements of balls and rings without slipping are known; on the other

hand, the number of revolutions of the separator can be measured experimentally. The difference between the calculated and actual numbers of revolutions of the separator allows you to determine the amount of slip.

3. Analysis of the results obtained shows that the slip coefficient does not depend on the load and rotation speed. Slip fluctuations due to these factors are random. The use of lubricant increases slip by 8%.
4. The solution of the wear contact problem for the ball and bearing ring in case of wear of the ring has been carried out. An approximate solution of the problem of ring wear by a ball is carried out under the assumption that the contact area remains unchanged in the direction of the ball movement.

Acknowledgements The reported study was funded by RFBR according to the research project №19-08-01241a. The authors declare that there is no conflict of interest regarding the publication of this paper.

References

1. Zaretsky EV, Poplavski JV, Miller CR (2013) Rolling bearing life prediction-past, present, and future. NASA Glenn Research Center, Cleveland, OH, United States, March 1, p 15
2. Artiukh V, Belyaev M, Ignatovich I, Miloradova N (2017) Depreciation of bearing blocks of rollers of roller conveyers of rolling mills. IOP Conf Ser Earth Environ Sci 9(1):012228. https://doi.org/10.1088/1755-1315/90/1/012228
3. Artiukh V, Mazur V, Adamtsevich A (2017) Priority influence of horizontal forces at rolling on operation of main sheet rolling equipment. MATEC Web Conf 106:04001. https://doi.org/10.1051/matecconf/201710604001
4. Puzyr R, Kukhar V, Maslov A, Shchipkovskyi Y (2018) The development of the method for the calculation of the shaping force in the production of vehicle wheel rims. Int J Eng Technol (UAE) 7(4.3):30–34. https://doi.org/10.14419/ijet.v7i4.3.20128
5. Zaretsky EV, Anderson WJ, Bumberger EN (1969) Rolling-element bearing life from 400° to 600° F. NASA Technical Note. NASA, Washington D.C., 1969. https://ntrs.nasa.gov/api/citations/19690007035/downloads/19690007035.pdf
6. Kukhar V, Kurpe O, Klimov E, Balalayeva E, Dragobetskii V (2018) Improvement of the method for calculation the metal temperature loss on a coilbox unit at the rolling on hot strip mills. Int J Eng Technol (UAE) 7(4.3):35–39. https://doi.org/10.14419/ijet.v7i4.3.19548
7. Anishchenko AS, Sosnovskij NY (1993) Rolling machines for washing machines bodies working. Kuznechno-Shtampovochnoe Proizvodstvo 11:27–28
8. Tanaka H, Tsushima N (2002) Estimation of rolling bearing life under contaminated lubrication. In: Beswick J (ed) Bearing steel technology, ASTM International, West Conshohocken, PA, pp 213–225. https://doi.org/10.1520/STP10856S
9. Bostan I, Mazuru S, Casian M, Slătineanu L, Merticaru V et al (2018) Method of axial adjustment for precessional transmissions. MATEC Web Conf 178:06024. https://doi.org/10.1051/matecconf/201817806024
10. Chen G, Wen J (2015) Effects of size and raceway hardness on the fatigue life of large rolling bearing. J Mech Sci Technol 29:3873–3883. https://doi.org/10.1007/s12206-015-0833-3
11. Branch NA, Arakere NK, Svendsen V, Forster NH (2010) Stress field evolution in a ball bearing raceway fatigue spall. Bearing steel technology, 8: Developments in rolling bearing steels and testing, ed. J. Beswick (ASTM International, West Conshohocken, PA), pp 57–80. https://doi.org/10.1520/STP49123S

12. Vasilchenko T, Yavtushenko G, Bondarenko Y, Belokon Y (2015) Calculation of planetary drive of mechanical press. Metallur Mining Ind 7(12):178–182
13. Anoopnath PR, Suresh Babu V, Vishwanath AK (2018) Hertz Contact Stress of Deep Groove Ball Bearing. Mater Today Proc 5(2):3283–3288. https://doi.org/10.1016/j.matpr.2017.11.570
14. You H, Zhu C, Li W (2012) Contact analysis on large negative clearance four-point contact ball bearing. Procedia Eng 37:174–178. https://doi.org/10.1016/j.proeng.2012.04.222
15. Houpert L, Chevalier F (2012) Rolling bearing stress based Life—Part I: calculation model. J Tribol 134(2):021103. https://doi.org/10.1115/1.4006135
16. Artiukh V, Kukhar V, Balalayeva E (2018) Refinement issue of displaced volume at upsetting of cylindrical workpiece by radial dies. MATEC Web Conf 224:01036. https://doi.org/10.1051/matecconf/201822401036
17. Li P, Zhou L, Cui F, Wang Q, Guo M, Zhai Y, Li Y, Hua D (2018) Elasto-plastic analysis of the contact region between a rigid ellipsoid and a semi-flat surface. Adv Mech Eng 10(9):168781401879739. https://doi.org/10.1177/1687814018797397
18. Ghaednia H, Wang X, Saha S, Xu Y, Sharma A, Jackson RL (2017) A review of elastic-plastic contact mechanics. Appl Mech Rev 69(6):060804. https://doi.org/10.1115/1.4038187
19. Gao H, Huang H, Zhu S, Li Y, Yuan R (2014) A modified nonlinear damage accumulation model for fatigue life prediction considering load interaction effects. Sci World J 2014(1), 164378. https://doi.org/10.1155/2014/164378
20. Kukhar VV, Grushko AV, Vishtak IV (2018) Shape indexes for dieless forming of elongated forgings with sharpened end by tensile drawing with rupture. Solid State Phenom 284:408–415. https://doi.org/10.4028/www.scientific.net/SSP.284.408
21. Markov OE, Gerasimenko OV, Shapoval AA, Abdulov OR, Zhytnikov RU (2019) Computerized simulation of shortened ingots with a controlled crystallization for manufacturing of high-quality forgings. Int J Adv Manuf Technol 103(5–8):3057–3065. https://doi.org/10.1007/s00170-019-03749-4
22. Eckels M, Kotzalas MN, Doll GL (2013) Attaining high levels of bearing performance with a nanocomposite diamond-like carbon coating. Tribol Trans 56(3):410–416. https://doi.org/10.1080/10402004.2012.756567
23. Malinov LS, Malysheva IE, Klimov ES, Kukhar VV, Balalayeva EY (2019) Effect of particular combinations of quenching, tempering and carburization on abrasive wear of low-carbon manganese steels with metastable austenite. Mater Sci Forum 945:574–578. https://doi.org/10.4028/www.scientific.net/MSF.945.574
24. Anishchenko AS (1996) Heat treatment effect on properties of deformed alloy type 36N. Metallur Heat Treat Metals 4:31–32
25. Ishchenko A, Radionenko A, Ischenko E (2017) Tribotechnical research into friction surfaces based on polymeric composite materials. Eastern-Eur J Enterpr Technol 6(12–90):12–19. https://doi.org/10.15587/1729-4061.2017.114367
26. Dragobetskii V, Zagirnyak V, Shlyk S, Shapoval A, Naumova O (2019) Application of explosion treatment methods for production items of powder materials. Przeglad Elektrotechniczny 95(5):39–42. https://doi.org/10.15199/48.2019.05.10
27. Dykha A, Makovkin O (2019) Physical basis of contact mechanics of surfaces. J Phys Conf Ser 1172:012003. https://doi.org/10.1088/1742-6596/1172/1/012003
28. Dominy J (1986) The nature of slip in high-speed axially loaded ball bearings. Proc Inst Mech Eng C J Mech Eng Sci 200(5):359–365. https://doi.org/10.1243/PIME_PROC_1986_200_139_02
29. Selvaraj A, Marappan R (2011) Experimental analysis of factors influencing the cage slip in cylindrical roller bearing. Int J Adv Manuf Technol 53:635–644. https://doi.org/10.1007/s00170-010-2854-5
30. Yoshida T, Hirota M, Yoshikawa K (2013) Analysis of cage slip in ball bearings considering non-newtonian behavior and temperature rise of lubricating oil. Tribol Online 8(3):210–218. https://doi.org/10.2474/trol.8.210
31. Takabi J, Khonsari MM (2014) On the Influence of traction coefficient on the cage angular velocity in roller bearings. Tribol Trans 57(5):793–805. https://doi.org/10.1080/10402004.2014.913327

32. Dykha AV, Zaspa YP, Slashchuk VO (2018) Triboacoustic control of fretting. J Frict Wear 39(2):169–172. https://doi.org/10.3103/S1068366618020046
33. Sorokatyi R, Chernets M, Dykha A, Mikosyanchyk O (2019) Phenomenological model of accumulation of fatigue tribological damage in the surface layer of materials. Mech Mach Sci 73:3761–3769. https://doi.org/10.1007/978-3-030-20131-9_371
34. Dykha A, Marchenko D (2018) Prediction the wear of sliding bearings. Int J Eng Technol (UAE) 7(2.23):4–8. https://doi.org/10.14419/ijet.v7i2.23.11872
35. Cavallaro G, Nelias D (2005) Analysis of high-speed inter-shaft cylindrical roller bearing with flexible rings. Tribol Trans 48(2):154–164
36. Hu X, Luo GH, Gao DP (2006) Quasi-static analysis of cylindrical roller inter-shaft bearing. J Aerosp Power 21(6):1069–1074
37. Liu X, Deng S (2011) Dynamic stability analysis of cages in highspeed oil-lubricated angular contact ball bearings. Trans Tianjin Univ 17:20–27
38. Deng SE, Jia QY, Xue JX (2014) Design principle of rolling bearings, 2nd edn. Standards Press of China, Beijing

Modeling of Loading and Topological Optimization of Metal Structures of Portal Cranes

Yurii Sahirov, Viktor Artiukh, Sergey Gridnev, and Elena Solovyeva

Abstract Clarification of methods for calculating strength of portal cranes metal structures, development of new methods and methodologies are important scientific and technical tasks which solutions will improve safety of crane operation in general. Purpose of this paper is modeling of loading of metal structures of 'Sokol' portal cranes during their operation using finite element method (FEM) and analysis of actual stress–strain state of their elements. Object of research is regularity of stress distribution in the metal structures elements. The paper proposes methodology for modeling and analysis of stress–strain state of metal structures of 'Sokol' portal crane, which is based on classical methods of permissible stress strength calculations, method of computer mathematical modeling, FEM and topological optimization. Loading scheme and solid-deformable model were developed, acting forces were determined, and the calculation was performed by FEM in CAD/CAM/CAE system. Results of calculations and analysis are presented. Obtained regularity of stress distribution in the elements of metal structures clearly showed that stress is distributed unevenly; the stress–strain state of metal structure elements is rather complicated and requires additional research.

Keywords Portal crane · Metal structure · Topological optimization · Solid-deformable model · Loading scheme · Finite element method · Criterion

Y. Sahirov
Pryazovskyi State Technical University, Universytets'ka, 7, Mariupol 87500, Ukraine

V. Artiukh (✉)
Peter the Great St. Petersburg Polytechnic University, Polytechnicheskaya, 29, St. Petersburg 195251, Russia

S. Gridnev
Voronezh State Technical University, Moscow Avenue, 14, Voronezh 394026, Russia

E. Solovyeva
K.G. Razumovsky Moscow State University of Technologies and Management (the First Cossack University), Moscow, Russia

A. Beskopylny and M. Shamtsyan (eds.), *XIV International Scientific Conference "INTERAGROMASH 2021"*, Lecture Notes in Networks and Systems 247,
https://doi.org/10.1007/978-3-030-80946-1_87

1 Introduction

Portal cranes are the most common loading and unloading machines of port transshipment complexes. Number of portal cranes in Ukraine is more than 1000 items. Designs of portals, columns, etc. are quite diverse that can be explained by variety of cranes requirements, experience of crane-building enterprises as well as insufficient knowledge of rational use limits of structures and parameters of metal structures elements. Portal cranes of such manufacturers as 'KIROW ARDELT' GmbH (former known as 'VEB Kranbau Eberswalde', Germany), 'Liebherr', 'Takraff' GmbH (Germany), 'Ganz Danubius HUTI' Ltd. (Hungary), PJSC 'Azovmash' (Ukraine) (refer to Fig. 1), LLC 'Plant of lifting and handling equipment named after S.M. Kirov' (Russia), LLC 'Siberian heavy machine building plant' (Russia), etc. are the most popular cranes in Ukraine. There is almost no renewal of cranes despite significant workload during their operation [1–6].

Fig. 1 Portal cranes produced by **a** 'KIROW ARDELT' GmbH; **b, c** 'Liebherr'; **d** 'Takraff' GmbH; **e** 'Ganz Danubius HUTI' Ltd.; **f** PJSC 'Azovmash'

Cost, weight, operational reliability of hoisting-and-transport machines are largely determined by their metal structures. Cost of material in the total cost of metal structure is around 70% and more [7–12].

Traditional strength calculations of cranes steel structures and their elements are based on assumption of flat scheme of their operation while their elements work as spatial systems. This approach is traditional due to the fact that calculation methods have been developed by manual methods of calculation using simple calculating devices. In complicated cases it is difficult to assess accuracy of such calculations without experimental verification on computer 3D models and products [13–16].

At present, need for breaking metal structures into flat elements is gradually disappearing due to development of computer technology and special programs for calculating spatial structures. Almost all modern strength calculations are carried out using computer solid-deformable models and finite element method (FEM). In recent decades FEM has taken leading position and become widespread. Thus, use of traditional engineering methods of computer modeling and FEM for studying stress–strain states and optimizing metal structures of 'Sokol' portal cranes is actual research [17–19].

2 Formulation of Task

Durability and reliability of cranes mechanisms usually depends on properties of their metal structures. Weight of payload of portal cranes is up to 11% of weight of their metal structures while, for example, in overhead cranes this figure reaches around 50%. Creation of rational design schemes with favorable values of their geometric parameters and sizes of individual elements, improvement of design methods using modern CAD/CAM/CAE systems, introduction of new developments into practice of design and technological enterprises producing portal cranes in order to reduce their material consumption and increase energy efficiency are important resource-saving tasks solution of which can be achieved with optimal design.

Loaded metal structures of portal cranes determine their final cost. In addition, during design process much attention is paid to loading modes and compliance with requirements for rigidity, fatigue and static strengths, etc. Usage of topological optimization in design process of loaded metal structures is very important, because even small weight reduction, optimization of stress distribution in metal structures provide significant savings in material, labor financial resources that makes it possible to increase operational reliability.

Modern CAD/CAM/CAE systems even at initial stages make it possible to optimize design of metal structures that helps to find the best structural schemes and during operation to apply the most rational options for repairs and reinforcement. It is known that improvement of complex FEM models is actual task and significant problem. At the same time, usage of topological optimization during design is a relatively new component. There are three methods for calculating metal structures of loading and unloading machines:

1. by permissible stress;
2. by system of probabilistic calculations;
3. by limit states.

Some authors also highlight FEM which is advisable to apply to analyze stress–strain states of spatial structures that include many elements. FEM is modern and quite common numerical method for solving differential equations. It is widely used for solving problems of fracture mechanics and deformable solids, calculations and analysis of stress–strain states of metal structures of loading and unloading machines.

It is also often used to solve problems of optimizing designs of machines parts, elements of metal structures of loading and unloading machines. FEM has become widespread due to significant development of CAD/CAM/CAE systems.

At department 'Loading and unloading machines and machine details' of 'Priazovskyi State Technical University' (Ukraine, Mariupol city), many important problems have been solved during last years:

- design of elements of metal structures and equipment of portal cranes;
- reducing tension level of elements of metal structures;
- forces and power calculations;
- structural parametric synthesis.

FEM was used using modern CAD/CAM/CAE systems during execution of described tasks. Nowadays, in design of metal structures, in particular, in their strength analysis software 'NX', 'Ansys', 'Nastran', 'SCAD', 'SolidWorks', Lira, etc. are used. All of them basically implement classical methods of structural analysis based on FEM. They got the greatest application due to development of computer technology. Enterprises also use CALS systems that make it possible to automate processes associated with product life cycle which includes stages of design, preparation and production itself, operation of product and its further disposal at the end of operation.

It is impossible to imagine search of the most loaded section or unit of metal structure (which determines its reliability) without qualitative analysis of stress–strain state of entire system and clear understanding of physical processes that occur during operation of the structure. At this time, it becomes possible to search for certain structure as a result of which optimal shape and geometric parameters of each of its elements should be obtained within given level of tension.

Thus, study of stress–strain state of the metal structure was carried out on three-dimensional model using personal computer and modern CAD/CAM/CAE systems. Geometric parameters of the model (geometric similarity) fully correspond to full-scale sample and are made according to drawings of the crane manufacturer at scale of 1:1. Simulation of different crane loading options was carried out.

Object of the research is the metal structures of 'Sokol' portal cranes. It is necessary to have general picture of their stress–strain states to analyze loading of elements of the crane metal structures. Obtaining this data is possible by creating 3D calculating model. Solid-deformable model (scale is 1:1) of the crane steel structures was developed (refer to Fig. 2) using modern CAD/CAM/CAE software consid-

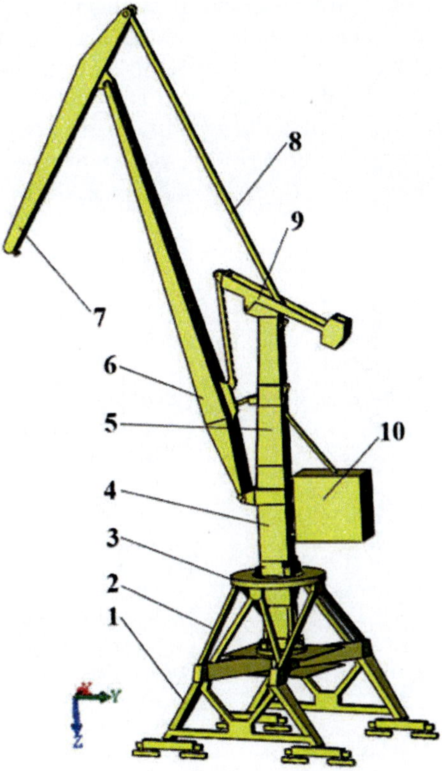

Fig. 2 General view of developed 3D model of 'Sokol' portal cranes metal structures

ering information on detected defects in the metal structures and their repairs during operation.

Analysis of information about revealed defects during operation allowed to establish following information (refer to Tables 1 and 2).

Analysis of the detected defects and performed repairs allows concluding that most of the defects are in the elements of the metal structure; no changes (quantity and geometrical parameters) were made to the loaded elements of the metal structures of both 'Sokol' cranes because repairs were carried out to restore their performance.

2.1 *Material of Metal Structures and Values of Permissible Stress*

Material written on original drawings is steel grades St38 b-2 and St38-3 which formerly were widely used in Germany. An analogue of steel grades St38 b-2 and St38-3 in Germany is steel grade St37-2 mechanical properties of which are given in Table 3. The closest analogues of steel grade St37-2 is analyzed in Table 4.

It is possible to specify steel grade S235JR for the model (which was adopted in all load cases) in CAD system materials database. Yield stress for this steel grade (taking into account metal thickness) is equal to σy = 225 MPa. Safety factor for the considered metal structures is taken to be equal to nII = 1.4. Permissible stress is

$$[\sigma] = \frac{\sigma_y}{n_{II}} = \frac{225}{1.4} = 161 \text{ MPa}.$$

Table 1 Defects identified for 'Sokol' crane # 1

#	Defect location	Defect description
1	Rope-block system	Axial gaps of two inner end blocks of jib are around d = 2.0 … 2.5 mm
2	Platforms, ladders and railings	Corrosion damage is up to 8% of initial thickness of walls metal of bracket, fastening of platform to upper flange of right beam of rocker arm of movable counterweight
3	Metal structures	Cracks in: • welded seams $L_{1,2}$ = 100 … 350 mm • welding of repair sleeves in outer left and inner right lugs of boom of hinge joint 'boom-column'
4		Unsatisfactory condition of hinges for fastening machine room to column: • radial gap is d = 4.0 … 6.0 mm • axial gap is d = 8.0 … 10 mm

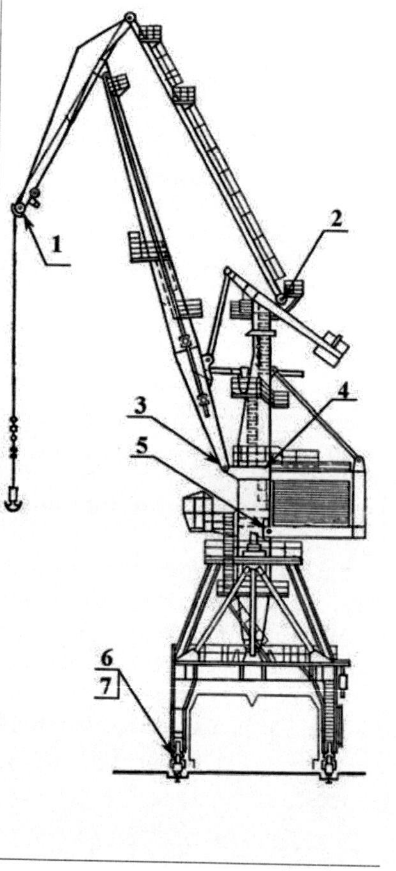

(continued)

Table 1 (continued)

5		Corrosion damage of the column in the cavity of bolted joints packages of its parts above roof of the machine room are: • right wall is up to 7% of initial metal thickness • left wall is up to 10% of initial metal thickness • back and front walls is up to 6% of initial metal thickness • horizontal diaphragm is up to 7% of initial metal thickness	
6	Movement mechanism	Weakened fit of open gear drive shaft-pinion in hollow shaft of mechanism gearbox drive on supports 'A' and 'D'	
7		Radial gaps in bearings of axle-boxes of drive shaft-pinion of gearbox of mechanism drive on supports 'A' and 'D' are equal to 1.5 … 2.0 mm	

3 Objectives of This Paper

Purpose of this paper is to study stress–strain state of 'Sokol' portal cranes metal structures in order to identify the most loaded areas with subsequent development of structural solutions for their strengthening.

Portal cranes 'Sokol' with lifting capacity of 32 tons are installed and operated in 'State enterprise Mariupol commercial sea port'. Main elements of the crane metal structures are portal, column and hinge joint boom system. All of them are made as welded steel sheets. In developed solid model (refer to Fig. 2) implemented main loaded metal structures are hingedly connected to each other that corresponds to real object.

Table 2 Defects identified for 'Sokol' crane # 2

#	Defect location	Defect description
1	Metal structures	Deformation of handrail
2		Deformation of handrail, bars and entrance ladder
3		Cracks: • L = 60 mm of welded seam of connecting flange on support 'B'; • L = 70 mm of welded seam of connecting flange on support 'A'

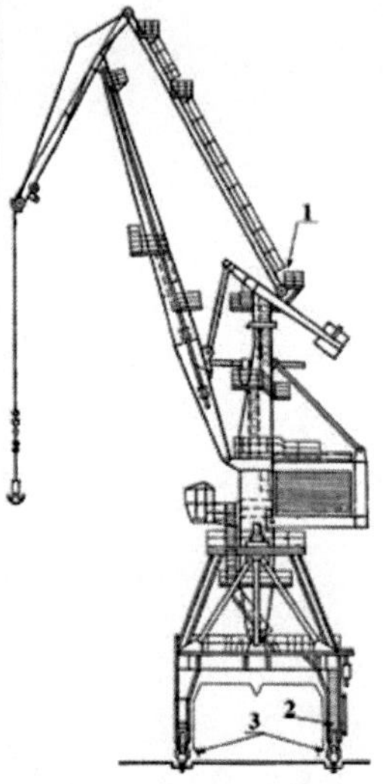

Table 3 Mechanical properties of steel grade St37-2

Nominal thickness (mm)	Minimum yield strength, R_eH (MPa)	Ultimate strength, R_m (MPa)
<3	235	360–510
$\geq 3 \leq 16$	235	340–470
$> 16 \leq 40$	225	340–470
$> 40 \leq 63$	215	340–470
$> 63 \leq 80$	205	340–470
$> 80 \leq 100$	195	340–470

Table 4 The closest equivalents (analogs) of steel grade St37-2

The USA (ASME SA/EN 10025-2)	S235JR, 1.0038
The RSA (SANS 50025-2)	S235JR, 1.0038
France (AFNOR A 35-501)	E 24-2
The United Kingdom (BS 4360)	40B
Spain (UNE 36080)	AE 235 B
Italy (UNI 7070)	Fe 360 B
Japan (JIS)	SS400
International (ISO 630)	Fe 360-B
Russian analogues (GOST)	Ст2пс, Ст2сп, Ст3пс, Ст3сп

The portal consists of box beams and I-beams. Lower part pos. 1 includes four supports interconnected in pairs along the crane runways by beams, crosspiece on central part of which column thrust bearing is attached. Supports that tighten the beam structure are additionally connected with braces to ensure sufficient rigidity of it. Upper part pos. 2 is girder. There is support circle pos. 3 on the upper part on which rail of support-rotary device and ring gear are attached. Such installation of the support circle provides ability to move it that is necessary when installing the crane. Horizontal loads acting on the column are perceived through the rail of the support-rotary device by the support circle and transferred to the portal.

Column of box-shaped structure consists of lower part pos. 4 and upper part pos. 5 connected to each other. To ensure required rigidity of the column channels are welded longitudinally inside it and diaphragms are installed which are also made in the model. At the bottom of the column there are lugs for the boom and machine room. In the upper part of the column there are lugs for fastening boom braces, counterweight rocker arm and rod that supports the machine room. In the upper part of the column there is an oval hole for gear rail of horizontal movement mechanism.

Movement of the gear rail during crane operation at horizontal movements close to the maximum ones is limited by stops welded to the column inside of the oval hole.

Hingely articulated boom system consists of boom pos. 6 with jib pos. 7, rigid guy-line pos. 8 and rocker arm of movable counterweight pos. 9. All of the listed elements of the metal structure are box-type beams required rigidity of which is provided by diaphragms, channels and angles welded inside of it. The boom has lugs by which it is hingely connected to the jib, rocker arm of the movable counterweight, gear rail and the column. The rigid guy-line is hingely connected to the boom and to the column. All connections of the boom system are hinged. The machine room pos. 10 is a welded sheet steel structure that is attached to the column. There are lifting mechanism and fixed counterweight (mass of which is taken into account in the model) inside machine room of the crane.

4 Materials and Methods

At present, FEM is the main method for strength analysis. It is necessary to prepare finite element model of structure to make its strength calculation by FEM. Complexity of this model depends on complexity of geometric model and accuracy requirements. Different types of calculations can determine different requirements for the model. Qualitative model adequately reflects real physical processes.

Finite element mesh was built on the basis of three-dimensional model (refer to Fig. 3) to implement FEM. Hinge and rigid fastening of the column, rigid fastening of the crane running wheels were used as support for the structure (refer to Fig. 4) for the design models. To simulate loaded state of the metal structure the design loads and their combinations are accepted and calculated for two loading options in accordance with norms for calculating metal structures of portal cranes.

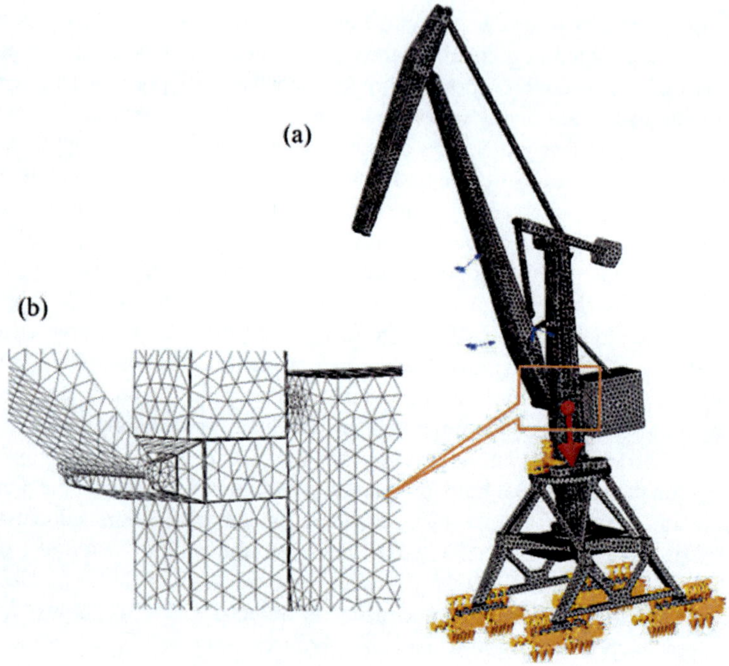

Fig. 3 Finite element mesh (**a**) and its zoomed fragment (**b**)

Design case II is determination of strength of metal structures by yield point by method of permissible stress which takes into account the maximum (ultimate) working loads during operation in the most severe operating conditions which include (in addition to loads from own and nominal weight of load and load gripping device) also the maximum dynamic loads that arise during sudden starts and emergency braking of mechanisms, sudden switching on (off) of electric current and limiting values of wind load of working state W (refer to Fig. 5).

Variant I is when only lifting mechanism works, load combination is II a where:

- Load weight is G;
- Dynamic coefficient is ψ_{II};
- Angle of inclined rise is α_i.

Variant II is when mechanisms for changing horizontal movement and turn work. Combination of loads are II b2 and II c where:

- Load weight is G;
- Angles of ropes deflection from vertical are $a_{II,}\beta_{II}$ (in the boom swing plane and perpendicular to the swing plane respectively);

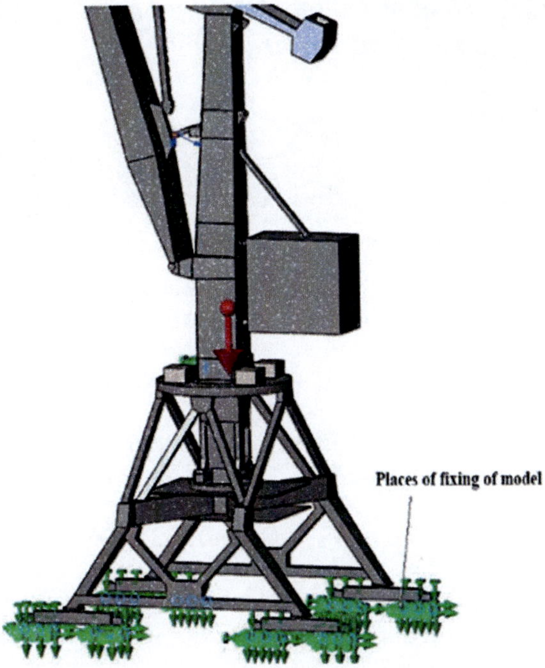

Fig. 4 Model fastening places

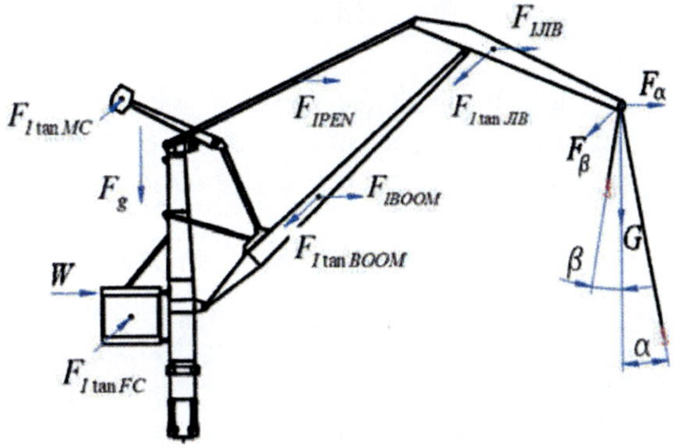

Fig. 5 Calculation scheme

- Tangential forces of inertia of the boom, jib, movable counterweight and fixed counterweight when turning the crane are $F_{I\tan BOOM}$, $F_{I\tan JIB}$, $F_{I\tan MC}$, $F_{I\tan FC}$.
- Forces of inertia of the boom, jib and guy-line when changing horizontal movement are F_{IBOOM}, F_{IJIB}, F_{IPEN}.

Both variants take into account action of gravity F_g on the structure.

4.1 Determination of Acting Forces Under Loading According to Variant I

Weight of the lifted load

$$G = Q \cdot g = 32 \cdot 9.81 \approx 314{,}000 \text{ N},$$

where Q = 32 tons is crane lifting capacity.

- Horizontal movement of the boom is R = 16 m.
- Weight of the lifting load taking into account dynamic coefficient

$$G_{II} = G \cdot \psi_{II} = 314000 \cdot 1.6 = 502400 \text{ N},$$

where ψII = 1.6 is dynamic coefficient during lifting load.

- Horizontal force from deflection of the load

$$F_a = G \cdot tg\alpha_i = 314000 \cdot 0.105 = 33000 \text{ N},$$

where $\alpha_i = 6°$ is angle of ropes deflection from vertical (inclined rise).

4.2 Determination of Acting Forces Under Loading According to Variant II

- Weight of the lifted load is $G = 31{,}4000$ N,
- Horizontal movement of the boom is R = 16 m,
- Horizontal force from deflection of the load

$$F_a = G \cdot tg\alpha_{II} = 31{,}4000 \cdot tg17 = 95{,}975 \text{ N},$$

$$F_\beta = G \cdot tg\beta_{II} = 1314000 \cdot tg15 = 84{,}115 \text{ N}.$$

where $\alpha_{II} = 17°$ is angle of ropes deflection from vertical in the boom swing plane; $\beta_{II} = 15°$ is angle of ropes deflection from vertical perpendicular to the boom swing plane.

- Force of inertia of boom system during change of boom horizontal movement acting in boom plane

$$F_I = \frac{m \cdot V}{t}, \text{kN},$$

where m is mass of moving element (jib, boom and guy-line) of boom system, ton;
V is velocity of moving element of boom system, m/s;
t = 5.0 s is acceleration (braking) time of horizontal movement mechanism.

Horizontal displacements and velocities of moving elements centers of gravity of boom system were determined by graphical-analytical method according to calculation formula

$$V = \frac{l}{l_m} \cdot V_m \text{ , m/s}$$

where l is horizontal displacement of element center of gravity per unit of time, mm (refer to Fig. 6);
$l_m = 5000$ mm is change of horizontal movement per unit of time;
$V_m = 1.05$ m/s is velocity change of horizontal movement.
Graphical definition of horizontal displacement of moving elements center of gravity of boom system is shown on Fig. 6.
Velocity of the jib center of gravity

$$V_J = \frac{3399}{5000} \cdot 1.05 = 0.714 \text{ m/s,}$$

Force of inertia of the jib

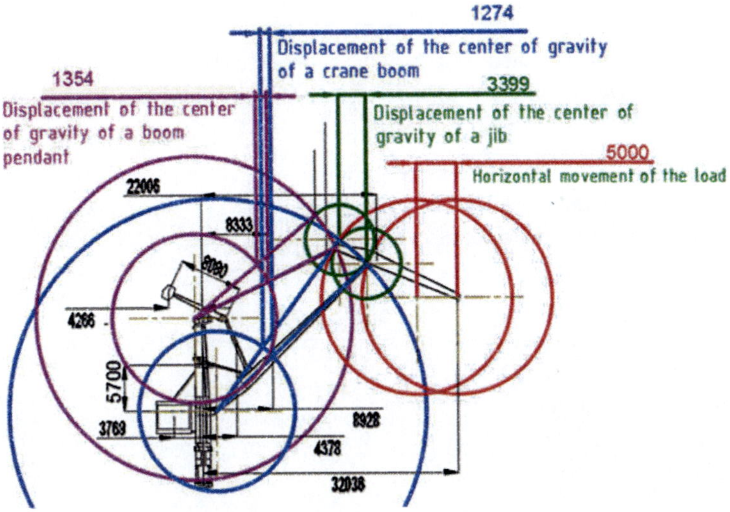

Fig. 6 Determination of horizontal displacements of moving elements centers of gravity of the boom system

$$F_{IJIB} = \frac{8.8 \cdot 0.714}{5.0} = 1.256\ \text{kN}$$

Velocity of the boom center of gravity

$$V_{boom} = \frac{1274}{5000} \cdot 1.05 = 0.286\ \text{m/s},$$

Force of inertia of the boom

$$F_{IBOOM} = \frac{16 \cdot 0.286}{5.0} = 0.856\ \text{kN}$$

Velocity of the guy-line center of gravity

$$V_{pen} = \frac{1353}{5000} \cdot 1.05 = 0.284\ \text{m/s},$$

Force of inertia of the guy-line

$$F_{IPEN} = \frac{2.7 \cdot 0.284}{5.0} = 0.153\ \text{kN}$$

- Tangential inertial forces of boom, jib and counterweight during turning of the crane are determined by formula

$$F_{I\tan} = \frac{m \cdot \pi \cdot r \cdot n}{30 t_R},\ \text{N}$$

where m is mass of element, ton (m of the boom is 16.0 tons; m of the jib is 8.8 tons; m of the movable counterweight is 15.5 tons; m of the fixed counterweight is 14.8 tons);

r is distance from element center of gravity to its axis of rotation, m;
п is rotation frequency, min – 1;
tR is acceleration time of rotation mechanism, s.
Force of inertia of boom

$$F_{I\tan BOOM} = \frac{16 \cdot 3.14 \cdot 8.93 \cdot 1.6}{30 \cdot 8} \approx 3\ \text{kN}.$$

Force of inertia of jib.

$$F_{I\tan JIB} = \frac{8.8 \cdot 3.14 \cdot 8.3 \cdot 1.6}{30 \cdot 8} = 1.5\ \text{kN}.$$

Force of inertia of movable counterweight

$$F_{I\tan MC} = \frac{15.5 \cdot 3.14 \cdot 4.3 \cdot 1.6}{30 \cdot 8} = 1.4\ \text{kN}.$$

Force of inertia of fixed counterweight

$$F_{I\tan FC} = \frac{14.8 \cdot 3.14 \cdot 3.8 \cdot 1.6}{30 \cdot 8} = 1.2\ \text{kN}.$$

- Determination of wind load W. Dynamic pressure of wind is $p_w = 250$ Pa acts on operating state of the crane rotating part.

Obtained values of the forces as well as dynamic pressure of wind of working state under loads according to variants I and II were used to simulate stress–strain state according to the first and second variants.

Self-weight of the structure and action of the previously determined loads was taken into account when calculating the models. General views of models with applied loads under two versions are shown on Fig. 7. Very wide range of results was obtained after the calculation which includes, among other things, stress and displacement diagrams.

Analysis of equivalent stress level on surface of the column from the boom side according to the first and second loading variants showed that the most unfavorable (high stress values) is the first loading variant which was chosen for further analysis of the stress–strain state and identifying the most dangerous areas.

Further analysis of the stress–strain state of metal structures and the identified most dangerous places are given only for the first design variant.

Structures are identified using method of Evolutionary Structural Optimization (ESO) shape and placement of which in the structure are ineffective. ESO method removes (or adds) material using heuristic criteria and it is simple enough to be implemented for solving topological optimization problems. The method is based on FEM determination of stress values in structural elements. It is generally accepted

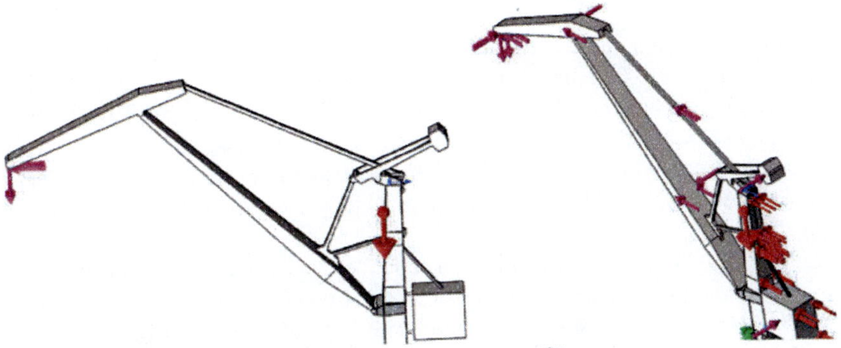

(a) (b)

Fig. 7 Places of forces application: **a** forces application is under the first variant; **b** forces application is under the second variant that optimal location in one or another part of the material structure and optimal shape of elements is low level of stress in them. Ideal design solution is that in which stress values in all elements (sections) of the structure is the same and close in value to permissible ones but at the same time are safe stress values.

Equivalent stresses (according to von Mises σ_e) in structural elements were determined in this paper. It is accepted $[\sigma] = 161$ MPa as admissible stress. It was decided to consider the optimal elements that satisfy condition

$$\frac{\sigma_e}{[\sigma]} \leq R_{Ri},$$

where R_{Ri} is rejection rate at which element is removed.

Analysis of the stress–strain state can consist of several cycles depending on set goals. The rejection rate can be increased by evolution rate H_i in each subsequent cycle

$$R_{R(i+1)} = R_{Ri} + H_i.$$

The main objective of this paper is to identify the most loaded sections of the metal structure for subsequent decision making on their possible strengthening. At the first stage of the analysis R_{Ri} was taken equal to 0.25 in order to identify non-optimal placement of material and shapes of elements.

5 Results

Following diagrams were developed as a result of FEM calculation:

- diagrams of equivalent stress distribution in metal structure (refer to Fig. 8);
- diagrams of equivalent stress distribution in metal structure with parameter 'ISO Limitations' (refer to Fig. 9) taking into account $R_{Ri} = 0.25$.

The stress–strain state of the metal structures of 'Sokol' portal cranes can be analyzed in order to find the most dangerous areas of them.

Analyzing the diagram of equivalent stress distribution (refer to Fig. 9) using 'ISO Limitations' tool (taking into account $R_{Ri} = 0.25$, limiting value of equivalent stress = 40 MPa) places (refer to Table 5) where the maximum stress values are locally concentrated can be highlighted. Non-optimal elements and material placements are 'cut' from the model.

It is accepted that $R_{Ri} \approx 1$ in the next cycle of analysis of the stress–strain state to identify the most dangerous places of metal structures, thereby increasing stress level to 160 MPa when using 'ISO Limitations' tool (refer to Fig. 10).

Analyzing diagram of equivalent stress distribution (refer to Fig. 10) the most dangerous places of metal structures can be highlighted (refer to Table 6) in which the maximum values of stress are close to 160 MPa.

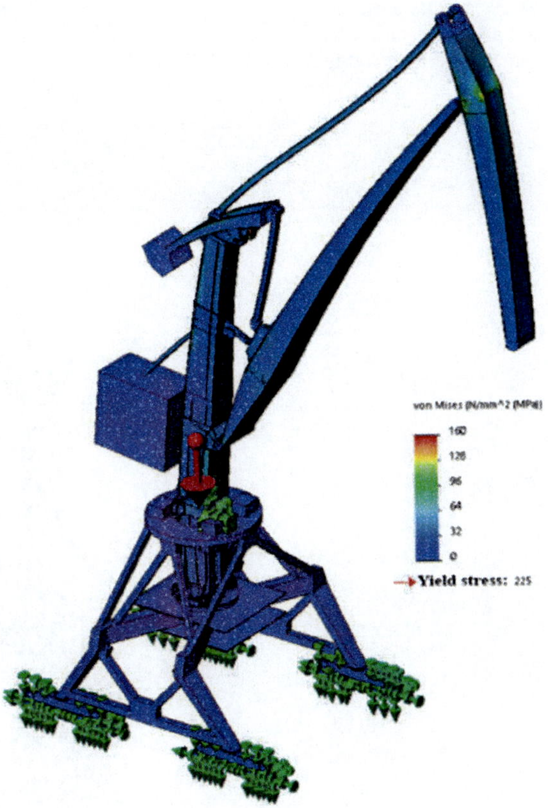

Fig. 8 Diagrams of equivalent stress distribution in the metal structure

Based on the above, following solutions can be recommended to strengthen the most dangerous places in the metal structures of 'Sokol' portal cranes.

The first solution can be aimed at ensuring uniform distribution of stress between elements of the metal structures in the most loaded places that can be achieved by installing additional stiffeners and diaphragms. Concentration of stress occurs primarily with sharp change of thickness of the metal structure (loaded area) and it is source of cracks and destruction. Their concentration is facilitated not only by abrupt changes in cross-sections but also by presence of sharp corners on path of power flow e.g. corner edges at junction of various elements of the metal structures.

The second solution is strengthening of the most loaded sections (elements) of the metal structure by installing pads that will locally reduce the stress level.

For the found most loaded places of the metal structures of 'Sokol' portal cranes:

- It can be recommended to install pad in area of detected maximum stress within lower wall of the jib at place where the hinge is fastened to the boom.

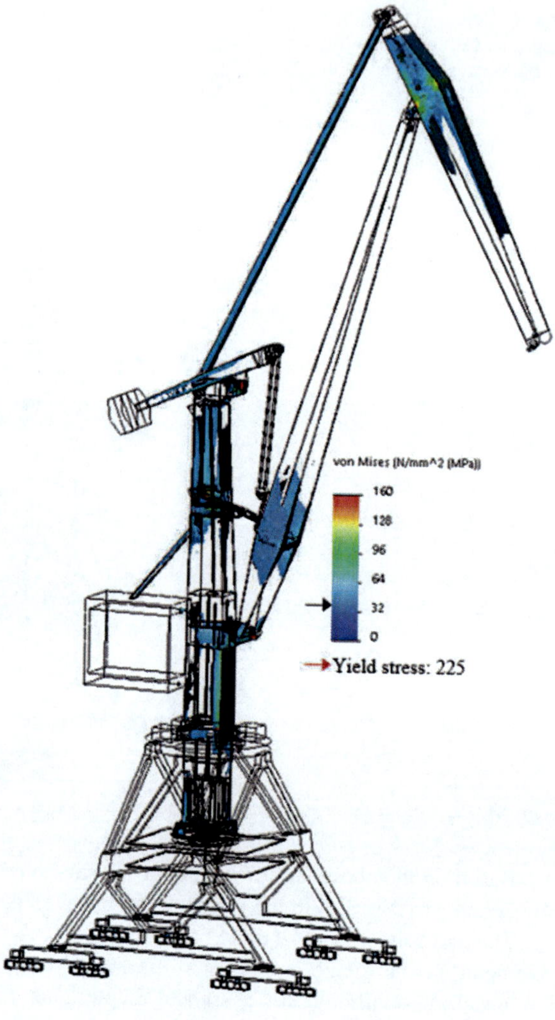

Fig. 9 Loaded areas of metal structures using 'ISO Limitations' tool (40 MPa)

- It is possible to recommend installing additional stiffeners and diaphragms which will ensure more uniform distribution of stress between elements of metal structures within place of fastening counterweight rod on the column, place on the column under the hinge of fastening the boom to the column, the metal structures elements outside and inside at place of fastening the support rollers of the column (from the boom side).

Multiple modeling and further analysis of the stress–strain state of the most loaded places of the metal structures with different reinforcement options will reveal optimal design solutions for their reinforcement.

Table 5 The most loaded areas in the model

#	Area description	General view of the area
1	Jib in place of its fastening hinge to the boom	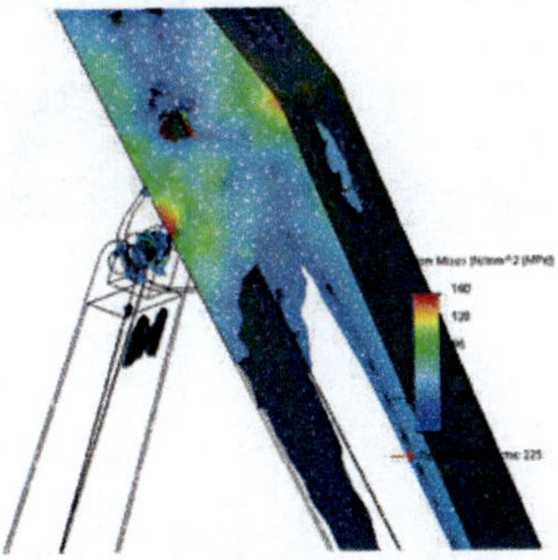
2	Place of counterweight rocker arm hinge fastening	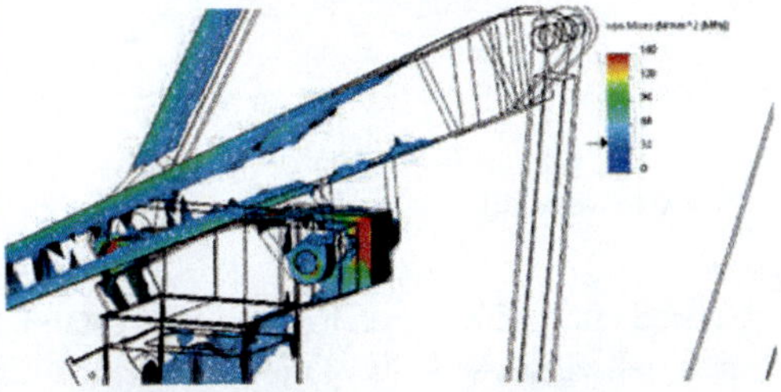
3	Place of horizontal movement mechanism fastening	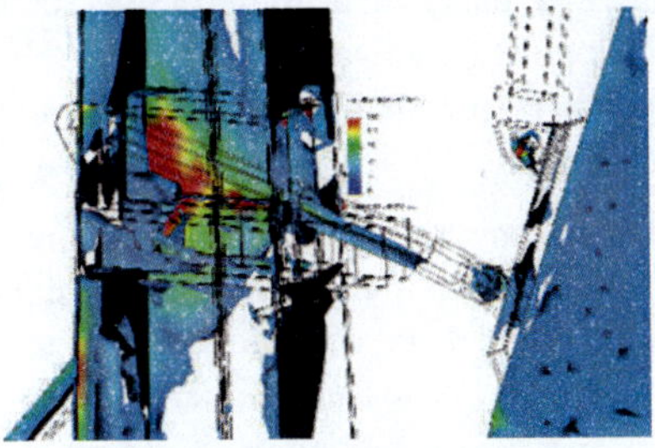
4	Place on the column under hinge of boom fastening and hinge of boom fastening to the column	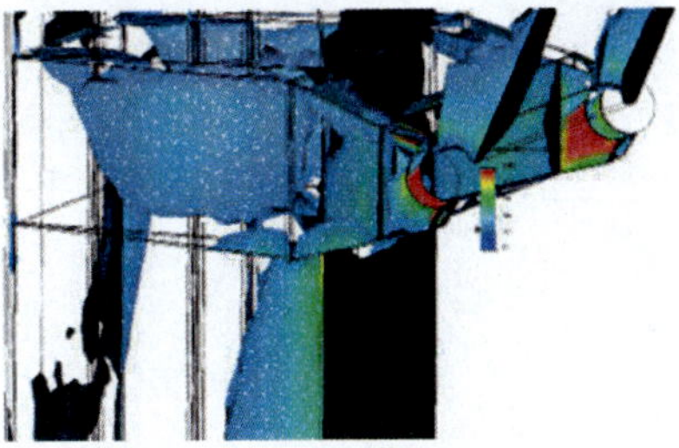

(continued)

Table 5 (continued)

#	Area description	General view of the area
5	Column support rollers fastening points from the boom side	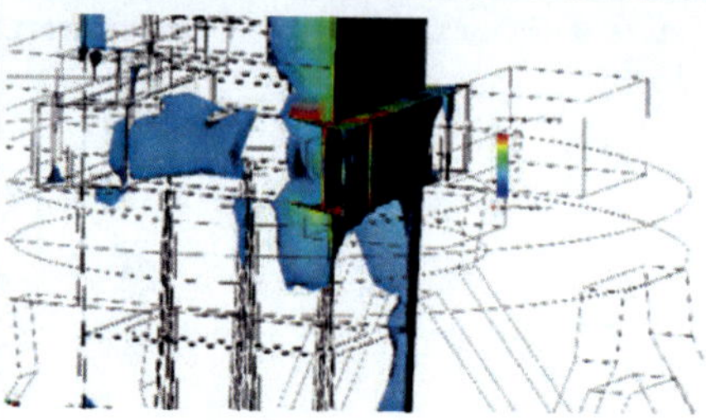
6	Places of fastening points of crane supports	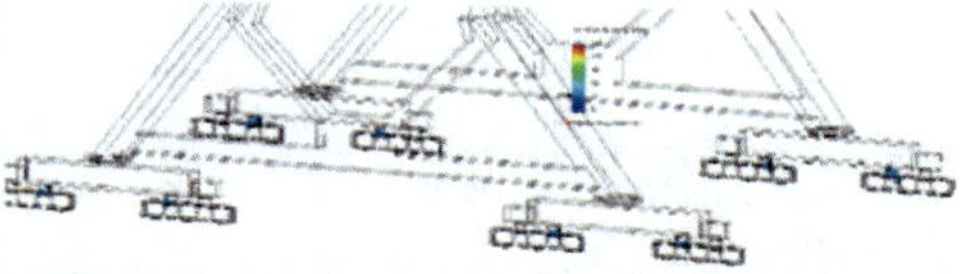

6 Conclusions

- Carried out studies made it possible to establish law of distribution of equivalent stress between elements of the metal structures of 'Sokol' portal cranes.
- Ineffective elements and locations of material of the metal structures are revealed using topological optimization according to ESO method.
- The most loaded places of metal structures of 'Sokol' portal cranes are:

  lower wall of the jib at place of hinge joint fastening with the boom;
  place of fastening on the counterweight rod;
  place on the column under hinge of fastening boom to column;
  elements of the metal structures outside and inside at place of fastening of the column support rollers (from the boom side).

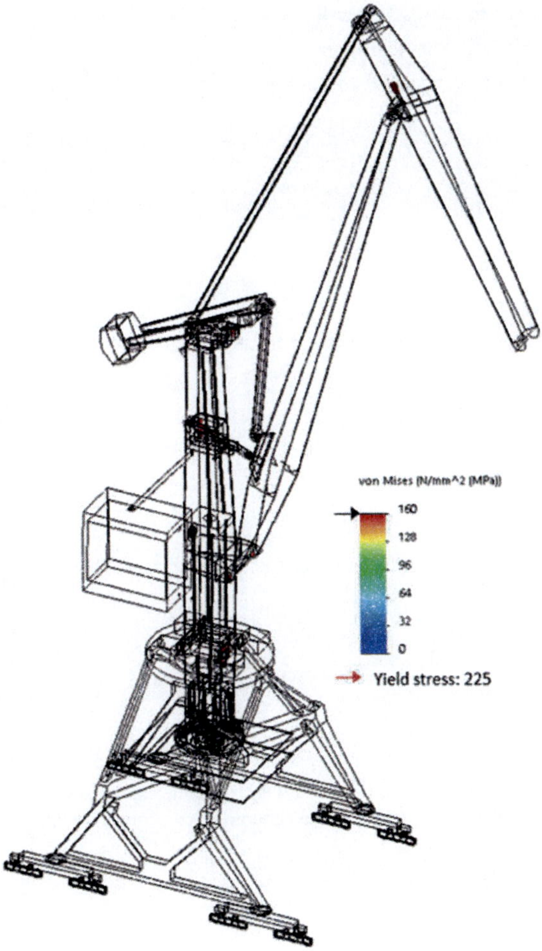

Fig. 10 The most dangerous places of metal structures when using 'ISO Limitations' tool (160 MPa)

Table 6 The most loaded places of the metal structures

#	Area description	General view of the area
1	Lower wall of the jib at place of hinge joint fastening with the boom	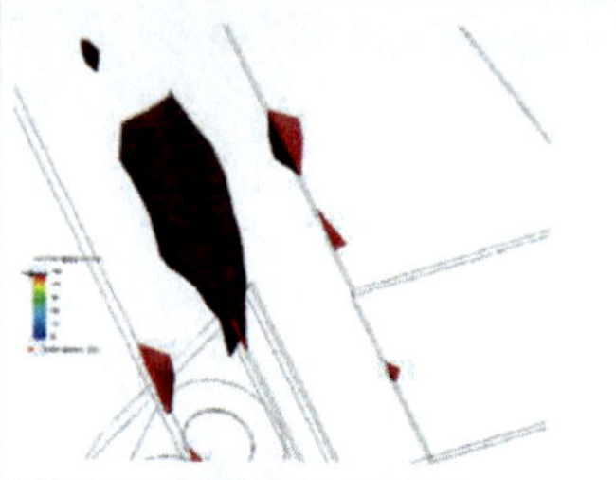
2	Place of fastening on the counterweight rod	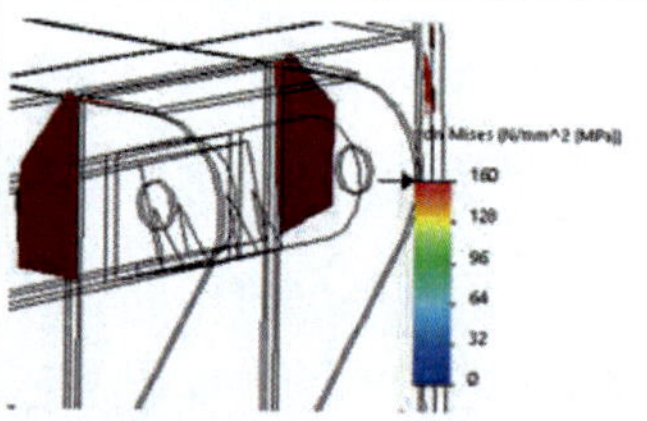
3	Place on the column under hinge of fastening boom to column	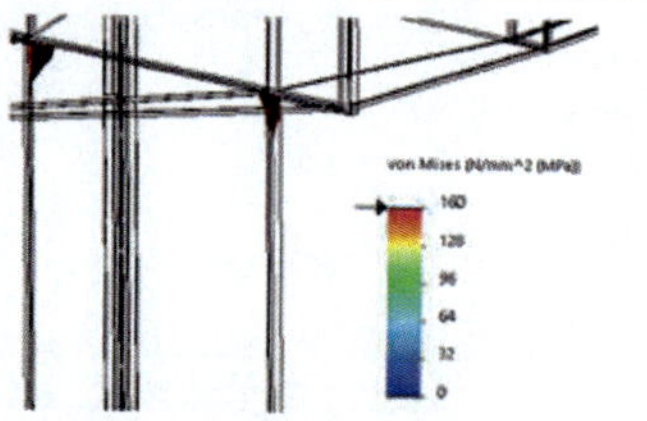
4	Elements of the metal structures outside and inside at place of fastening of the column support rollers (from the boom side)	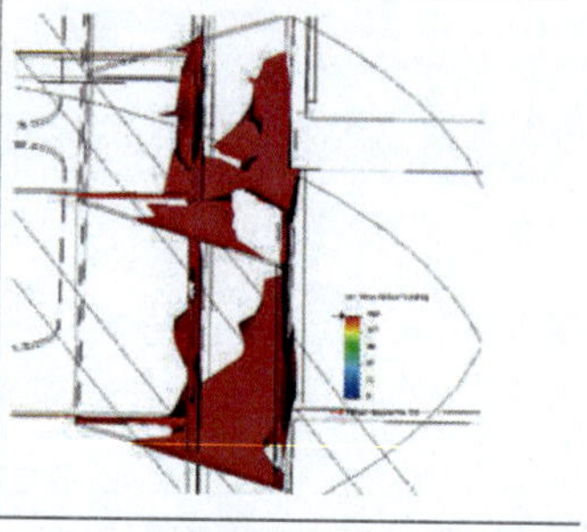

Acknowledgements The reported study was funded by RFBR according to the research project №19-08-01241a. The authors declare that there is no conflict of interest regarding the publication of this paper.

References

1. Artiukh V, Mazur V, Sahirov Y, Kapustina N (2020) Protection of metallurgical machines from breakdowns at iron and steel works. Adv Intell Syst Comput 1115:950–962. https://doi.org/10.1007/978-3-030-37916-2_94
2. Gharaibeh NS, Matarneh MI, Artyukh VG (2014) Loading decrease in metallurgical machines. Res J Appl Sci Eng Technol 8(12):1461–1464
3. Sagirov Y, Artiukh V, Mazur V, Aleksandrovskiy M (2020) Scientific and methodological bases of rational design of hoisting-and-transport machines metal structures. In: E3S Web of Conferences, vol 164, p 03005. https://doi.org/10.1051/e3sconf/202016403005
4. Huang X, Xie YM (2007) Convergent and mesh-independent solutions for the bi-directional evolutionary structural optimization method. Finite Elem Anal Des 43(14):1039–1049
5. Bendsøe MP, Sigmund O (2003) Topology optimization: theory, methods and applications. Springer, Berlin. http://help.solidworks.com/2019/russian/SolidWorks/cworks/c_simp_method_topology.htm?format=P&value=1
6. Zhou X, Xiao H (2010) Research on testing for security evaluation for port crane. In: International conference on computer, mechatronics, control and electronic engineering, CMCE 2010, Changchun, pp 123–125. https://doi.org/10.1109/CMCE.2010.5609600
7. Maruschak PO, Panin SV, Zakiev IM, Poltaranin MA, Sotnikov AL (2016) Scale levels of damage to the raceway of a spherical roller bearing. Eng Fail Anal 59:69–78
8. Maksarov VV, Keksin AI (2018) Forming conditions of complex-geometry profiles in corrosion-resistant materials. In: IOP Conference Series: Earth and Environmental Science, vol 194(6), p 062016. https://doi.org/10.1088/1755-1315/194/6/062016
9. Solomonov KN, Tishchuk LI (2019) Simulation of deformation processes in upsetting. J Phys Conf Ser 1348:012020
10. Osadchii VA, Gorbatyuk SM, Filippov DI, Kuprienko NS (2019) Development of a method for the visualization of the process of forming of curved profiles in the software used for the numerical analyses of the roll passes. Metallurgist 63(5–6):658–664. https://doi.org/10.1007/s11015-019-00872-1
11. Ishchenko A, Golinka S (2014) Peculiarities of equipment repair of crushing and grading complex with the help of composite. Metallur Mining Ind 6(3):45–49
12. Sahirov Y, Artiukh V, Mazur V, Vaynshtok N, Reshetnikova I (2019) Analysis of stress-strain states of casting crane traverse. In: E3S web of conferences, vol 110, p 01049. https://doi.org/10.1051/e3sconf/201911001049
13. Zhang Y, Liu J, Chen L (2007) Approach to safety assessment system and its application to port machineries. Port Oper 3:7–9
14. Kitaeva D, Rudaev Y, Subbotina E (2014) About the volume forming of aluminium details in superplasticity conditions. In: METAL 2014—23rd international conference on metallurgy and materials, conference proceedings, pp 347–352
15. Malinov LS, Malysheva IE, Klimov ES, Kukhar VV, Balalayeva EY (2019) Effect of particular combinations of quenching, tempering and carburization on abrasive wear of low-carbon manganese steels with metastable austenite. Mater Sci Forum 945:574–578. https://doi.org/10.4028/www.scientific.net/MSF.945.574
16. Nemchuk OO, Nesterov OA (2020) In-service Brittle fracture resistance degradation of steel in a ship-to-shore portal crane. Strength Mater 52(2):275–280. https://doi.org/10.1007/s11223-020-00175-w
17. Vodzianskiy V, Artiukh V, Kargin S, Borisova A (2020) Study of some aspects of refining process physics of large-sized details precision surfaces. In: E3S web of conferences, vol 175, p 05049. https://doi.org/10.1051/e3sconf/202017505049

18. Artiukh V, Mazur V, Sahirov Y, Geraskin Y (2020) Approximation of limit amplitude diagram for destructive elements of safety devices. In: E3S web of conferences, vol 164, p 14009. https://doi.org/10.1051/e3sconf/202016414009
19. Shi L-W, Xie H, Wang M, Guan Y, Leng G (2008) The weight and angle of depression detection and control system of a large portal crane. Fourth Int Symp on Precision Mechanical Measurements (2008) 71304N-6. DOI: https://doi.org/10.1117/12.819727

The Impact of External Influences on the Characteristics of Metals of Welded Structures of Construction Machines

Alexander Scherbakov, Anna Babanina, Vera Breskich, and Vadim Klyovan

Abstract Thermal cycling treatment is the most effective treatment for increasing the reliability and durability of metal structures. In this work, an assessment of the effect of the structure of a metal with a different degree of dispersion, obtained during thermal cycling after a different number of processing cycles, on the mechanical properties of steels 08ps, St3 and 10KhSND was made. Additionally, the influence of the degree of cold plastic deformation on the change in the stray magnetic field strength H_p in low-carbon and low-alloy steels is considered. As a result, methods have been developed for assessing the effect of thermal cycling modes on the mechanical properties of structural steels, as well as the effect of acting stresses during cold plastic deformation of steels 08ps, St3 and 10KhSND on the stray magnetic field strength. Graphical dependencies of values are obtained. On the basis of the research results, the relationship between the structure, magnetic and mechanical parameters was revealed, which allows a reliable assessment of the structural state of the metal and the acting stresses in welded joints and elements of welded metal structures using other methods of passive fluxgate control method. The study is devoted to identifying which zones and places of the welded joint are the most dangerous, and which of them must be controlled first.

Keywords Welded metal structures · Construction machines · Mechanical properties of steels · Thermal cycling · Plastic deformation

A. Scherbakov
Saint Petersburg State University of Architecture and Civil Engineering, Vtoraya Krasnoarmeiskaya Street, 4, Saint Petersburg 190005, Russia

A. Babanina · V. Breskich (✉)
Peter the Great St. Petersburg Polytechnic University, Polytechnicheskaya, 29, St. Petersburg 195251, Russia

V. Klyovan
Saint Petersburg State University of Architecture and Civil Engineering (SPbGASU), 2nd Krasnoarmeyskaya St., St. Petersburg, Russia

A. Beskopylny and M. Shamtsyan (eds.), *XIV International Scientific Conference "INTERAGROMASH 2021"*, Lecture Notes in Networks and Systems 247,
https://doi.org/10.1007/978-3-030-80946-1_88

1 Introduction

Heat treatment, which makes it possible to obtain a fine-grained structure in steels, significantly changes their mechanical characteristics. Thus, mechanical tests of carbon steels showed that thermal cycling increases the specific work of destruction by at least 50% due to an increase in toughness, which can increase the workability of steels by pressure even at room temperatures: stamping ability with deep drawing of metals increases [1, 2].

The most structurally sensitive characteristic of the mechanical properties of steels is impact toughness, due to which the search for the optimal mode of thermal cycling treatment is effectively carried out by determining and analyzing the values of impact toughness. When testing carbon steels for impact toughness at various numbers of thermal cycles, it showed that the most optimal is 5–6 cycles, which significantly increases the impact toughness and shifts the cold brittleness threshold to the region of negative temperatures [3–7]. The hardness of steels after thermal cycling remained almost unchanged; it remained at the level of normalized steels. At the same time, an increase in the yield stress and plasticity was noted in comparison with the normalized state, which is a very positive fact from the point of view of the strength of metals.

Obviously, thermal cycling reduces the sensitivity of steel to stress concentration due to its greater plasticity, which is very important when increasing hazardous stress concentration zones of elements of welded metal structures by conducting controlled thermal cycling treatment of these zones with step-by-step passive flux gate control.

The strength and plastic properties of steels can be significantly increased by thermal cycling, which leads to grain refinement and spheroidization of cementite inclusions. Thermal cycling also has a significant effect on the mechanical properties of alloyed structural steels. Alloyed steels differ from carbon steels in that the thermodynamic activity of carbon in them is lower than in simple steels. However, the methods of thermal cycling treatment of carbon steels can be applied to alloyed steels without significant changes in technology, if the total content of alloying elements does not exceed 5%. It should be remembered that alloying elements change the critical points in alloys.

Structural changes during thermal cycling of alloy steels are slower. With an increase in the content of alloying elements in steel, the required number of cycles also increases in comparison with carbon steels. The thermal cycling of alloy steels also has a significant effect on the mechanical properties. Thus, a 5-8-fold thermal cycling of steels led to a sharp refinement of the structure, while the plasticity and impact toughness increased [8–10].

In this work, we will consider the effect of thermal cycling modes on the mechanical properties of structural steels, as well as the effect of the acting stresses during cold plastic deformation of 08ps, St3 and 10KhSND steels on the stray magnetic field strength.

2 Method for Assessing the Effect of Heat Treatment Modes on the Mechanical Properties of Structural Steels

Thermal cycling, in contrast to heat treatment, to a greater extent reveals the positive effect of alloying on the strength and plastic characteristics during thermal cycling. By sharply increasing the strength, ductility and impact toughness, it is possible to obtain previously unattainable values of the fracture work of alloy steels under various types of loading. Consequently, thermal cycling is the most effective treatment for increasing the reliability and durability of metal structures.

In this work, an assessment of the effect of the structure of a metal with a different degree of dispersion, obtained during thermal cycling after a different number of processing cycles, on the mechanical properties of 08ps, St3 and 10KhSND steels was made. In this case, it was assumed to use the effect of increasing the strength properties of structural steels while strengthening the elements of metal structures of construction machines in local zones of stress concentrations during repair. The studied steels were subjected to high-temperature annealing at 1050 °C and subsequent 1–10-fold thermal cycling.

3 Method for Assessing the Effect of Cold Plastic Deformation on the Structure and Properties of Structural Steels

Considering that in the elements of welded metal structures of construction machines and welded joints, there may be dangerous stress concentration zones in which cold plastic deformation has passed or is somehow undergoing, it was of scientific and practical interest to consider the effect of the degree of cold plastic deformation on the change in the strength of the stray magnetic field Hp in mild and low alloy steels. For this purpose, the initial samples of structural steels 08ps and 10KhSND in the as delivered state with a thickness of 2 mm were subjected to fractional rolling at room temperature to the degree of deformation of 8, 15, 22, 30, 40 and 50%. The strength of the stray magnetic field on the surface of the samples during deformation was measured in three zones separated from each other at a distance of 30 mm, with the middle zone in the center of the sample.

4 Results of Assessing the Effect of Heat Treatment Modes on the Mechanical Properties of Structural Steels

The mechanical properties of the steels under study after high-temperature annealing at 1050 °C and subsequent 1–10-fold thermal cycling are presented in Table 1.

Table 1 Results of the study of steels 08 ps, St3, 10KhSND

Grade	Mechanical properties	The number of thermal cycles							
		0	1	2	3	4	5	7	10
08ps	$\sigma_{0.2}$, MPa	166.0	172.0	181.5	188.0	190.5	193.0	193.5	195.0
	σ_t, MPa	281.5	295.5	301.0	309.5	315.0	317.5	320.0	323.5
	δ, %	24.0	23.5	23.0	23.0	23.5	23.0	23.0	23.0
St3	$\sigma_{0.2}$, MPa	233.0	242.5	250.0	255.5	259.0	262.5	263.0	263.5
	σ_t, MPa	418.0	441.5	451.0	468.5	471.5	476.0	477.5	484.5
	δ, %	23.0	22.5	22.0	22.0	22.5	22.0	22.0	21.5
10KhSND	$\sigma_{0.2}$, MPa	385.0	400.5	416.0	423.5	428.5	435.0	434.0	435.0
	σ_t, MPa	503.0	538.0	538.0	563.5	568.5	573.5	568.5	583.5
	δ, %	19.0	18.0	19.0	18.5	19.0	18.5	18.5	18.0

It can be seen that after thermal cycling, a significant increase in strength properties is observed both for low-carbon steels 08ps and St3 and for low-alloy steel 10KhSND. However, with an increase in the number of cycles over three, the increment in strength properties somewhat slows down, which is in good agreement with the experimental data of magnetic control and metallographic analysis. Such an increase in the mechanical properties of steels makes it possible to recommend thermal cycling for metal reinforcement in hazardous local stress concentration zones both in welded joints and in elements of operated metal structures during repairs.

5 Results of Assessing the Effect of Cold Plastic Deformation on the Structure and Properties of Structural Steels

Figure 1 shows the dependences of the stray magnetic field strength Hp on the degree of cold plastic deformation ε for three zones on samples made of steels 08ps and 10KhSND, respectively.

It can be seen (Fig. 1a) that with an increase in the degree of cold plastic deformation for points 1 and 2, the stray magnetic field strength Hp changes its negative sign to positive, and after $\varepsilon = 30\%$, the sign changes again. The values of the magnetic parameter Hp at point 3 remain negative for all degrees of cold plastic deformation. It should be noted that at $\varepsilon = 50\%$, the Hp values of all points approach their minimum values.

All the initial values of the stray magnetic field on the samples made of steel 10KhSND (Fig. 1b), in contrast to steel 08ps, have only a positive sign. They significantly decrease at $\varepsilon = 8\%$ and take negative values at $\varepsilon = 15\%$. It should be emphasized that at $\varepsilon = 50\%$, the spread of Hp values between the extreme points of the sample is much larger than that of steel 08ps. So, if for steel 10KhSND the spread is 18 units, then for steel 08ps—6 units.

Figure 2 shows the dependence of the average values of Hp in zones 1–3 on the degree of deformation ε for steels 08ps and 10KhSND.

It can be seen that, despite a significant scatter of H_p values and different signs in the initial state, this scatter decreases with an increase in the degree of deformation, and at $\varepsilon = 40\%$ or more, all H_p values have a negative sign. At the degree of deformation $\varepsilon = 50\%$, the values of the stray magnetic field for samples made of steels 10KhSND almost coincide. Significant changes in the stray magnetic field H_p during cold plastic deformation and the similarity of the shape of the H_p curves of the steels under study are largely caused by structural changes occurring in the steels.

The structure of steel 10KhSND after cold plastic deformation to different degrees is shown in Fig. 3.

We can see that with an increase in the degree of deformation, the crystallographic orientation of the grains along the rolling direction increases. At the initial stages of rolling, there are undeformed grains, while grains with a significant degree of

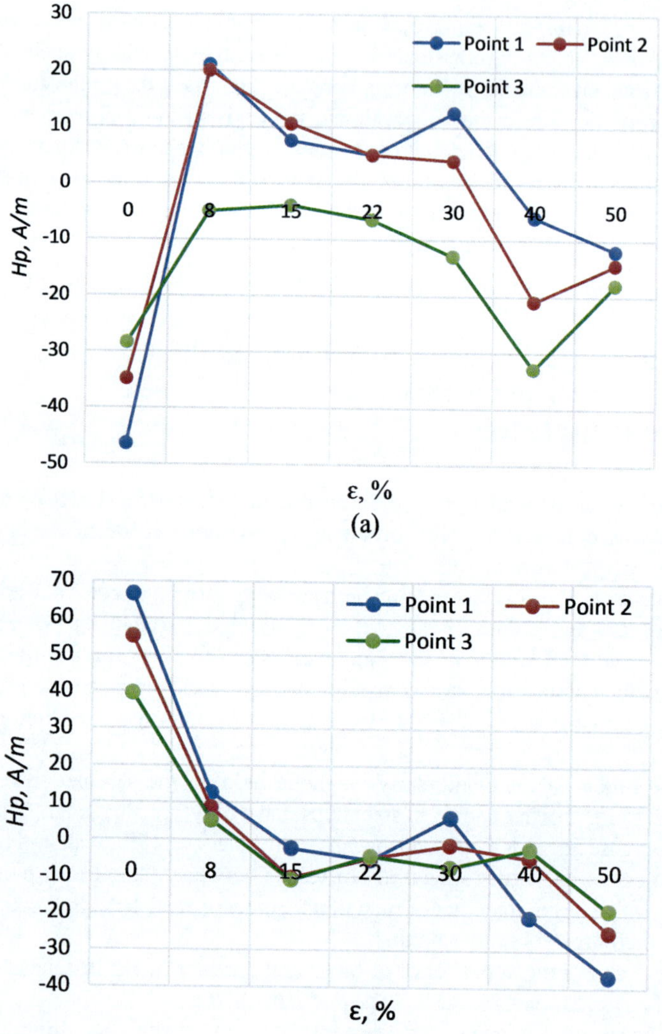

Fig. 1 Dependence of the stray magnetic field strength H_p on the degree of plastic deformation ε for steels 08ps (**a**) and 10KhSND (**b**)

deformation are located nearby. At the degree of deformation ε = 30%, there are almost no grains that have not undergone plastic deformation.

A different character of the rolling texture formation is noted in steel 08ps (Fig. 4), which has a coarser-grained initial microstructure.

The formation of the texture of this steel at the initial stages of rolling has a less pronounced character, as a result of which, even after rolling by ε = 50%, weakly

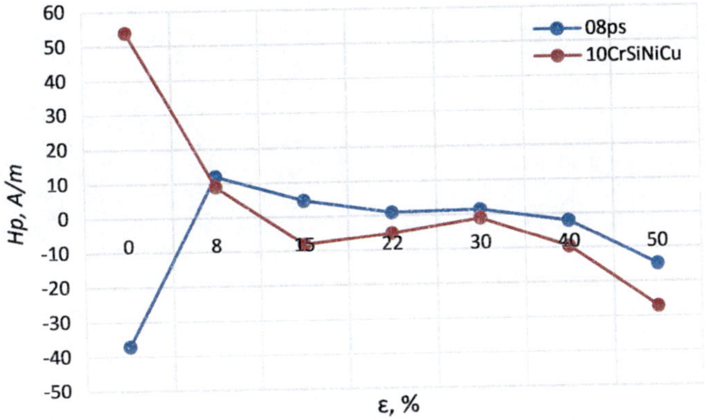

Fig. 2 Dependence of the stray magnetic field strength H_p on the degree of plastic deformation ε for steels 08ps and 10KhSND

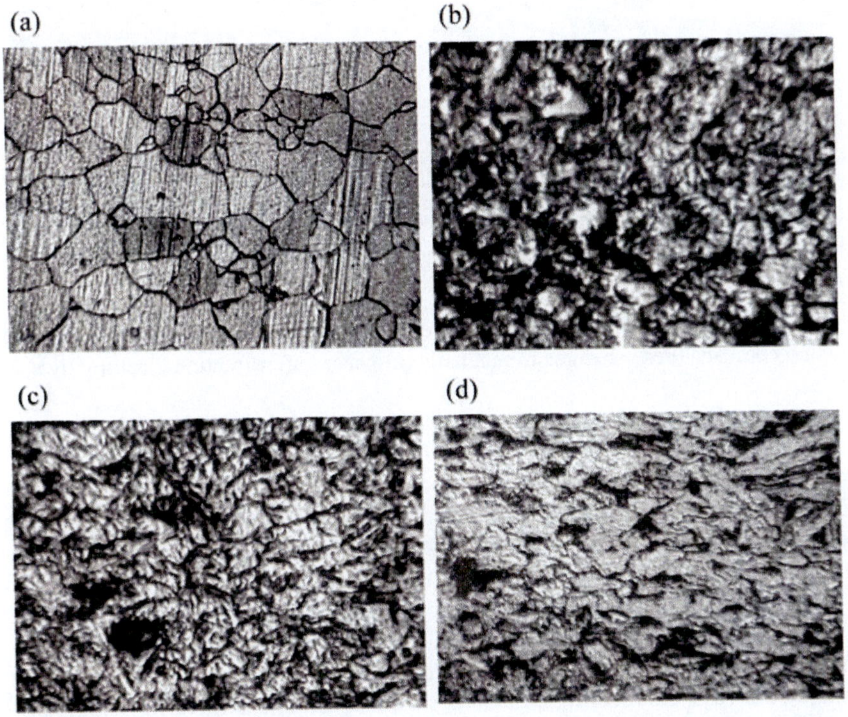

Fig. 3 Structure of steel 10KhSND, ×650: **a** as delivered state, **b–d** after rolling to a degree of deformation of 8, 30 and 50%, respectively

(a) (b) (c) (d)

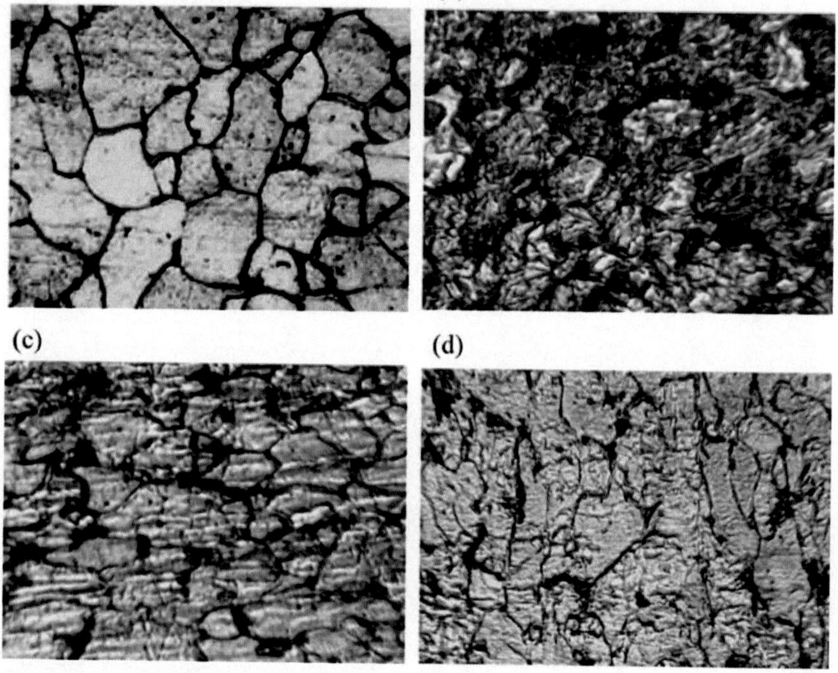

Fig. 4 Structure of steel 08ps, × 650: **a** as delivered state, **b–d** after rolling to a degree of deformation of 15, 30 and 50%, respectively

deformed grains are visible in its structure, in contrast to low-alloy steel 10KhSND, which confirms these changes in magnetic properties after fractional rolling by ε = 50%.

Structural changes during cold plastic deformation are also observed in the fine structure of the metal. If a chaotic distribution of dislocations is observed in the structure before deformation, then after cold rolling to a degree of deformation of 10%, two types of structures are formed: a dislocation forest with a uniform distribution of a large number of dislocations and a coil structure. In this case, the volume fractions of the resulting structures are approximately equal.

The formation of a coil dislocation structure indicates the onset of a spontaneously proceeding process of redistribution of dislocation density and interaction of dislocations, as a result of which a structure is created that is more stable than the previous one. In our case, the ordering of the dislocation structure leads to a convergence of the magnetic field strengths in the steels under study, which can be seen in Fig. 3 with deformation by the degree ε = 8%. With an increase in the degree of deformation to 25%, a cellular structure is formed, which is characterized by wide boundaries with a high density of dislocations and inter-boundary volumes of the metal relatively free from dislocations. A further increase in ε to 40% promotes the formation of a strip structure, which, in contrast to an equiaxial cellular structure, is elongated volumes bounded by rather narrow walls, while a significant number of dislocations are observed in the strip region. One of the features of this structure is the presence of significant misorientation introduced by extended walls.

The change of the positive sign of the magnetic field strength to a negative one at $\varepsilon > 22\%$ and a further increase in the magnetic field strength Hp is probably associated with the formation of cellular and then strip dislocation structures.

Thus, it follows from the above that in the steels under study, there is a correlation between the degree of cold plastic deformation, magnetic field strength, and structural changes in steels, which makes it possible to recommend a passive fluxgate method for monitoring structural changes in metal during cold plastic deformation, which is especially important for assessment of the technical condition of long-term operating metal structures of construction machines.

It should be emphasized that with an increase in the degree of cold plastic deformation, regardless of the sign and the initial value of the magnetic stray field strength Hp, the values of the magnetic parameter take negative values, and their value tends to the values of the Earth's magnetic field, which is of practical importance in assessing the technical condition of long-operated metal structures of construction machines.

6 Conclusions

As a result of the study, a method was developed for assessing the effect of thermal cycling modes on the mechanical properties of structural steels, which allows finding optimal processing modes by changing the magnetic parameter Hp, which makes it possible to recommend the developed method for strengthening metal in local zones of stress concentration in welded joints and elements of operating metal structures of construction machines during repair.

A method has also been developed for assessing the effect of the acting stresses during cold plastic deformation of steels 08ps, St3 and 10KhSND on the stray magnetic field strength and structural changes in the metal, which makes it possible to increase the reliability of the results of diagnosing the elements of metal structures of operating construction machines, especially in areas that have undergone plastic deformation for some degree.

References

1. Scherbakov AP (2020) Experimental studies of heat treatment influence on the welded connections properties in working mechanisms of road building. Russ Automobile Highway Ind J 17(6):664–675
2. Scherbakov A, Monastyreva D, Smirnov V (2019) Passive fluxgate control of structural transformations in structural steels during thermal cycling. In: E3S web of conferences, vol 135, p 03022

3. Scherbakov A, Babanina A, Graboviy K (2021) Acting stresses in structural steels during elastoplastic deformation. Adv Intell Syst Comput 1259:298–311
4. Lupachev AV, Pavlyuk SK (2019) Determination of electrode potentials on local surfaces of metals and welded joints Factory laboratory. Diagnost Mater 8:43–46
5. Vedyakov II, Odessa PD, Gurov SV (2018) Ensuring the strength of welded joints for unique structures from rolled large thicknesses of increased and high strength. Struct Mech Calcul Struct 2(277);68–75 (2018)
6. Lekhov OS, Mikhalev AV, Shevelev MM (2018) The load and stress state of the strikers of the continuous casting and deformation of the strip upon receipt of steel sheets for welded pipes. In: Proceedings of higher educational institutions. Ferrous Metall 61(4):268–273
7. Scherbakov AP (2020) Material and method selection for increasing the wear resistance of construction machines components. Russ Automobile Highway Ind J 17(4):464–475
8. Fireushin AM, Fatkullin MR, Ulyabaev RR, Chernyatyev RR, Kadoshnikova AV (2018) Improving the technology for repairing cases of thick-walled machines and assemblies. Oil Gas Bus 16(3):76–83
9. Plotnikov DG (2015) Methods for assessing the strength of metal structures of hoisting-and-transport machines. Sci Techn Statements SPbGPU 1(214):186–193
10. Scherbakov A, Babanina A, Matusevich A (2021) Passive probe-coil magnetic field test of stress-strain state for welded joints. Adv Intell Syst Comput 1259:312–323

Methodology for Assessing the Quality of Services Based on the Discrepancy Model

Alexandra Kadykova, Andrey Smolyaninov, Alexander Kolosov, and Irina Pocebneva

Abstract An approach to assessing the quality of services is considered, which will improve and timely identify shortcomings, factors of loss of competitiveness and quickly respond to them. To solve this problem, a gap model was considered with the addition of a quality tool—the Deming cycle (PDCA cycle) for more dangerous discrepancies (gap). At the initial stage, during the analysis, it was decided that the most dangerous discrepancies in the model are found at gaps three and four. At the second stage, a Deming cycle was imposed on the selected gaps in order to reduce them and, as a result, improve the service under study and increase the quality. A significant advantage of this model is the continuity of the Deming cycle and its ability to continually improve the selected service if the desired result is not achieved immediately. Interpretation of the obtained result means that the developed model reduces the mismatch between consumer expectations and service perception.

Keywords Service · Deming cycle · Quality of service · Consumer expectations

1 Introduction

Service is the result of direct interaction of the contractor to meet the needs of the consumer [1].

The concept of a service includes two parts—the result and the process. The results are the work performed “service in material form”, “service in immaterial form”.

The second component of the service is the process of providing services—the activity of the service provider. The provision of a service can be subdivided into separate stages (provision of the necessary resources, technological execution process, monitoring and evaluation, service process).

A. Kadykova (✉)
Moscow Aviation Institute, National Research University, Volokolamskoe shosse, 4, 125993 Moscow, Russia

A. Smolyaninov · A. Kolosov · I. Pocebneva
Voronezh State Technical University, 20-letiya Oktyabrya Street, 84, 394006 Voronezh, Russia

A. Beskopylny and M. Shamtsyan (eds.), *XIV International Scientific Conference “INTERAGROMASH 2021”*, Lecture Notes in Networks and Systems 247,
https://doi.org/10.1007/978-3-030-80946-1_89

The characteristic of the service process is the service culture. It is determined by the professionalism and ethics of staff service.

Service conditions—a set of factors affecting the consumer of the service in the process of service.

The revealed differences between the service market and the product market allow us to classify the features of the service market and expand their list by highlighting such a specific feature as the dependence of the formation of this market on the territorial nature of population settlement.

The features of the services market include:

1. high dynamism of market processes. The dynamic nature of demand for services is significantly influenced by the time factor, the dynamics of supply is due to the flexibility of the industry structure of services,
2. territorial segmentation. The forms of service provision, the demand and the conditions for the functioning of service enterprises depend on the characteristics of the territory covered by a particular market,
3. local character. Under the influence of territorial conditions, the service market acquires a clear spatial form, within the framework of which socio-economic characteristics that are different from others, but similar to each other, are formed,
4. high sensitivity to changes in market conditions. The impossibility of storing, storing and transporting services and, at the same time, the temporal and spatial coincidence of their production and consumption,
5. high rate of capital turnover. The service market has a shorter production cycle,
6. the specifics of the organization of production of services. Service providers are usually small and medium-sized enterprises of various profiles, since these enterprises are more mobile and have ample opportunities for a sensitive response to changes in market conditions,
7. the specifics of the service rendering process. Assumes personal contact between the manufacturer and the consumer,
8. a high degree of differentiation of services. The complex structure of demand leads to the emergence of new, non-standard services,
9. uncertainty of the result of the service rendering. Evaluation of the result of activities for the provision of services, as a rule, is possible only after its consumption,
10. dependence on the territorial nature of population settlement. The formation and development of the service sector market, especially at the regional level, is largely due to the nature of the settlement that has developed within the territorial boundaries of a particular market [2, 3].

The service has a number of features:

- obscurity (intangibility)—non-obviousness of properties,
- instability, instability of properties,
- inseparability of production and consumption of services,
- non-persistence in time.

2 Experimental

Gap model—defines the quality of service in terms of meeting customer expectations. In other words, this is a comprehensive analytical study that studies the inconsistencies, the gaps between the current state of the company and the desired one. The analysis also makes it possible to identify problem areas (bottlenecks) that impede development and assess the company's readiness to make the transition from the current state to the desired one. According to the modelers, "The first step in servicing is knowing what the customer expects and what kind of critical situation may arise in the quality of the service" [4].

Simply put, the company needs to know what the customer is expecting and meet their expectations with excellent quality. "The model is closely related to marketing, since it is based on an orientation to consumer needs" [5] (Fig. 1).

The central element of the gap model is the "consumer gap", which consists in the mismatch between consumer expectations and service perception—the key concepts of the service system. Accordingly, the main task of the organization is to reduce

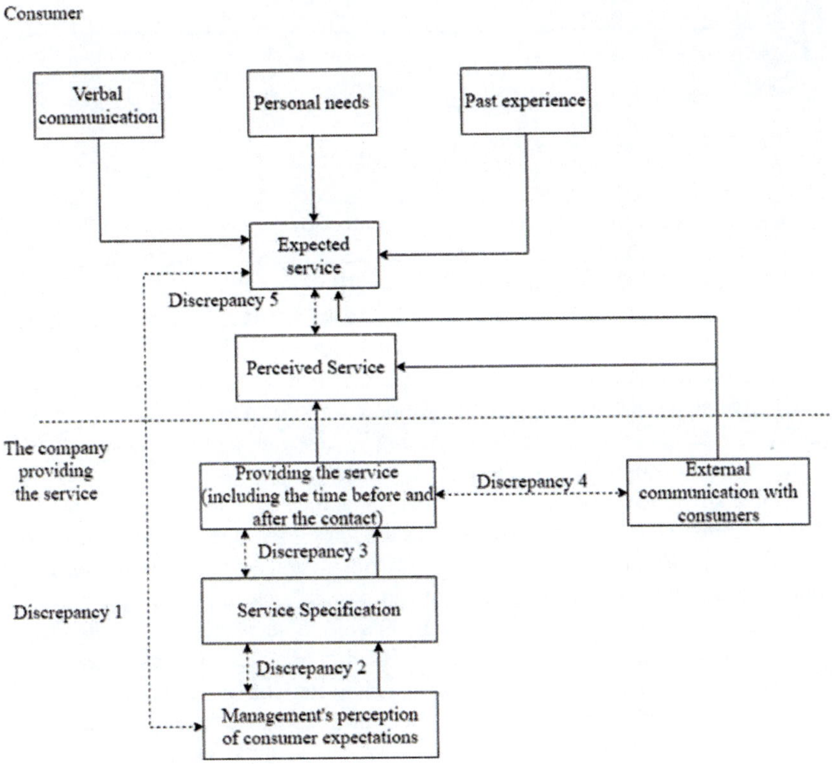

Fig. 1 Discrepancy model (gap model)

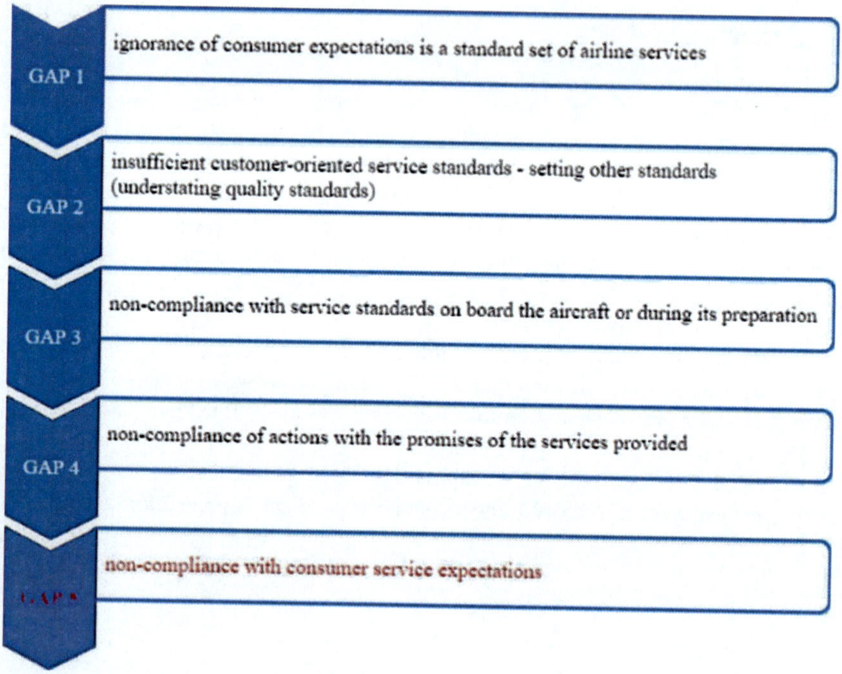

Fig. 2 Divergence model steps (gap model)

this gap in order to meet the needs of consumers of services and build long-term relationships with them. To do this, the organization needs to narrow the remaining "gaps" in the area of corporate governance [6, 7].

For each industry and company, the breaks of this model must be adapted, since the air services market is quite specific. The divergence model has five steps, which are discussed in Fig. 2.

The five-step service quality analysis model provides a proper understanding of quality service delivery. By exploring this model, one can develop an understanding of the potential problems associated with providing quality service. And this will help solve any problems at each of the stages that may exist in our actions [8].

It should be noted that the proposed methodology for assessing the quality of a service system [9] successfully combines functional, process and system approaches, which are well known in quality management and are presented in Fig. 3.

The analysis of the essence of the consumer's requirements for the communication system shows that the characteristics of the "quality of air travel" feature are divided into two groups: technical characteristics and characteristics of service management.

In the process of studying the Gap model, it was revealed that the discontinuities can be tried to be reduced or tried to be minimized using the E. Deming cycle (PDCA Cycle) [10, 11].

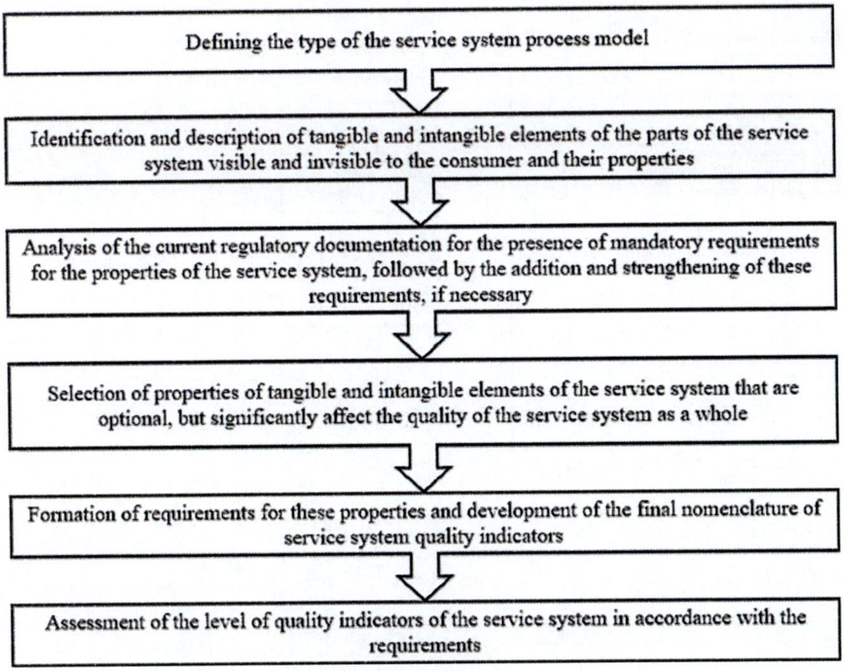

Fig. 3 Determination of the type of process model of the service system

The Deming Cycle is a business process that improves the quality of the airline's services through flexibility, quality improvement and elimination of defects and errors. In practice, the tool helps to improve the (gap model) and improve the quality of services provided by the airline.

It should be noted that it was decided to apply this tool to the (gap) 2 and 3, therefore, minimizing critical situations can help improve the gap model, which is shown in Fig. 4 [12].

3 Evaluation

As an example, let's take the service "transporting an animal on board an aircraft". To begin with, we will determine which gap this service belongs to and perform a full description of the current situation. Information is taken from questionnaires, surveys of users of the service under investigation [13].

Description of the problem: transportation of an animal is a serious problem in the company today. Cases of death of animals during transportation in the luggage compartment of an aircraft have become more frequent.

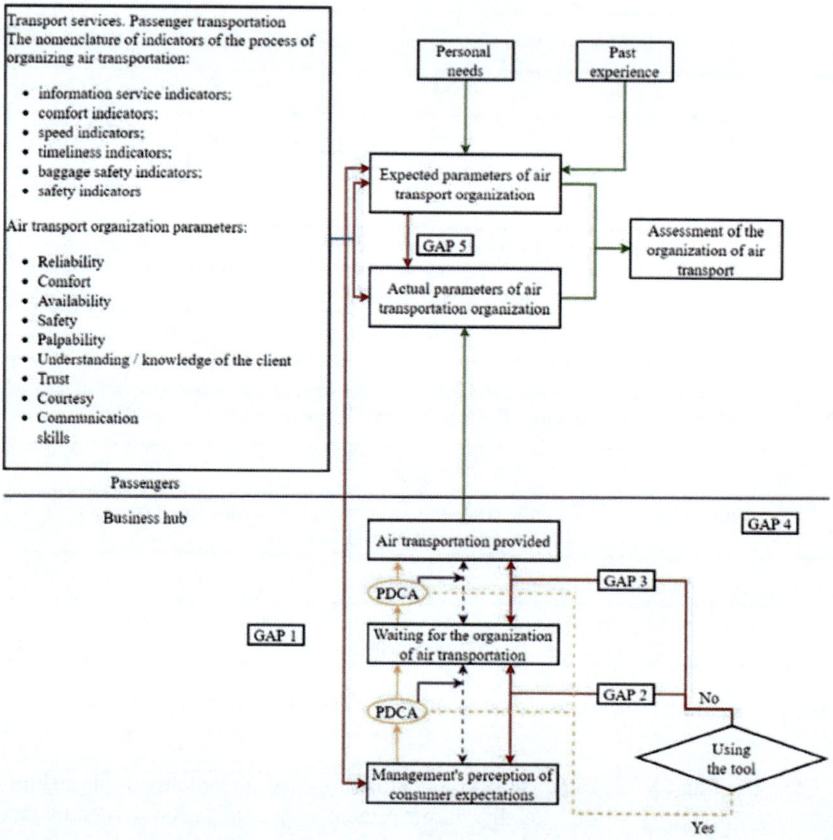

Fig. 4 Improving the discrepancy model using the PDCA cycle

Desired situation: improve the control systems for the transport of the animal. Eliminate the wording "Excess baggage". Increase the survivability of pets during transport [14].

Obstacle: reluctance of the management to amend the internal documentation for the transportation of the animal (weight, shape and size of the cage).

Expected result: increase the rating of the airline in the market of services provided.

To eliminate shortcomings and failures, we will apply the Deming cycle.

P—When checking in on board, the pet is weighed, visually inspected and the cage is inspected [15].

D—At this stage of the check, the airline employees strictly follow the internal documents for the carriage of the animal and make a decision on the place of its carriage (in the cabin and in the luggage compartment). Even if the norm has been exceeded, the company can insignificantly allow the animal to be transported in the cabin, provided that it fits into the cage [16, 17].

When transporting an animal in the hold, you must strictly follow the transport instructions. A person responsible for transportation from the crew is appointed and until the end of the trip this responsibility is assigned to him. The parameters of the life support of the four-footed passenger in the luggage compartment are clearly controlled. At the end of the trip, the responsible person transfers the animal to the owner with an acceptance certificate [18].

C—Analysis of the obtained information using a statistical tool—Graphs/Questionnaires/Animal mortality.

A—Based on the information received, the management of the airline makes a decision to fix the system for transporting animals in internal documents. With a decrease in mortality, it will be possible to talk about the effectiveness of the measures taken by the airline.

Since the Deming cycle has no end, and if the desired result is not achieved, it is possible to repeat it again until the result is achieved [11, 19].

4 Conclusions

The quality programs applied in practice are diverse, each of them has its own advantages and disadvantages, however, experts note that today there is no specific model or methodology adapted for the air transportation market, so the quality of services in this market is very difficult to assess.

Building an effective quality management system based on various programs, including E. Deming's cycle, is, first of all, necessary to find a quantitative measure of the quality of service as the degree of perception of its compliance with consumer expectations. An effective quality system allows:

1. predict and evaluate the actually achieved level of quality;
2. structure the activities of the firm;
3. design service processes, taking into account its main goal—to achieve maximum customer satisfaction.

Having carried out an improved gap analysis using E. Deming's cycle, we can conclude how the company operates, namely how many gaps it has and which ones.

References

1. Kanhva V, Andryunina Y, Belyaeva S, Sonin Y (2021) Construction in the digital economy: prospects and directions of transformation. In: Paper presented at E3S web of conferences, vol 244. https://doi.org/10.1051/e3sconf/202124405008

2. Skachko G, Surkova E, Ermolaeva E, Potsebneva I (2021) Adoption of management decisions based on the risk management model. Paper presented at E3S web of conferences, vol 244. https://doi.org/10.1051/e3sconf/202124411008
3. Surkova E, Klonitskaya A, Ermolaeva E (2020) Modeling of business processes of complex organizational systems. Paper presented at E3S web of conferences, vol 164. https://doi.org/10.1051/e3sconf/202016410040
4. Gumba K, Uvarova S, Belyaeva S, Vlasenko V (2021) Innovations as sustainable competitive advantages in the digital economy: justification and forecasting. Paper presented at E3S web of conferences, vol 244. https://doi.org/10.1051/e3sconf/202124410011
5. Uvarova S, Belyaeva S, Myshovskaya L (2017) Development of a "quality-efficiency" model of construction investment projects for road construction based on sustainable growth. Paper presented at IOP conference series: earth and environment science, vol 90(1). https://doi.org/10.1088/1755-1315/90/1/012169.
6. Nikolaevich LA, Viktorovna DM, Klonitskaya AY, Vyacheslavovna KS, Alekseevna ST, Viktorovna SE, Mikhailovskaya NM, Ermolaeva EN (2017) Problems of investment and budget planning and NWF in Russia. Int. J. Econ. Res. 14(15):371–380
7. Golov R, Alkhimovich I, Kazarnovsky V, Ermolaeva E (2019) Development of a mechanism for assessing the economic sustainability of small enterprises. Paper presented at E3S web of conferences, vol 91. https://doi.org/10.1051/e3sconf/20199108052
8. Viktorovna DM, Vladimirovich TP, Valerievna SE, Nikolaevna TN, Yurievna KA, Nikolaevna EE (2017) Business management in the regions of Russia: threat analysis. Int J Econ Res 14(4):55–67
9. Belyaeva S, Voronov D, Erypalov S (2017) Methodological framework for assessing the competitiveness of the construction industry and development companies. Article presented on MATEC conferences website, vol 106. https://doi.org/10.1051/matecconf/201710608033
10. Desyatirikova EN, Belousov VE, Fedosova SP, Ievleva AA (2017) DSS design for project risk management. Paper presented at the 2017 international conference "Quality management, transport and information security, information technology", IT, Quality Management and IS 2017, pp 492–495. https://doi.org/10.1109/ITMQIS.2017.8085869
11. Barkalov SA, Bekirova ON, Kalinina NY, Moiseev SI (2020) Identity model based on hidden variables. J Phys Conf Ser 1679(5). https://doi.org/10.1088/1742-6596/1679/5/052029
12. Desyatirikov AN, Desyatirikova EN, Sergeeva SI, Fedosova SP, Mager VE (2018) Investment mechanisms for increasing the sustainability of the region's development on the example of the Voronezh region. Report presented in the materials of the International Conference 2018 "Quality management, transport and information security, information technology", IT, quality management and IS 2018, pp 832–835 (2018). https://doi.org/10.1109/ITMQIS.2018.8525113
13. Uvarova SS, Belyaeva SV, Provotorov LA, Myshovskaya LP (2017) The mechanism of system integration of operational control institutions to ensure the effectiveness of investment projects. Bull Higher Educ Inst, Text Ind Technol Ser 1:56–60
14. Korchagin A, Deniskina A, Fateeva I (2019) Lean and energy efficient production based on the Internet of Things (IOT) in the aviation industry. Presented at E3S web of conferences, vol 110. https://doi.org/10.1051/e3sconf/201911002124
15. Burkov V, Burkova I, Barkalov S, Averina T (2020) Project risk management. Report presented at the 2nd international conference on control systems, mathematical modeling, automation and energy efficiency, Proceedings—2020, SUMMA 2020, pp 145–148. https://doi.org/10.1109/SUMMA50634.2020.9280817
16. Barkalov SA, Moiseev SI, Bekirova ON (2018) Expert information processing models based on the theory of hidden variables. Report presented on the materials of the 11th international conference of 2018 "Managing the development of large-scale systems", MLSD. https://doi.org/10.1109/MLSD.2018.8551943
17. Barkalov S, Averina T, Bautina E, Perevalova O (2020) Management of social investment projects for digital transformation of education J Phys Conf Ser 1691(1). https://doi.org/10.1088/1742-6596/1691/1/012083

18. Barkalov S, Kurochka P, Khodunov A, Kalinina N (2020) Model of choice of labor technology based on multi-criteria assessments. E3S web of conferences, vol 164. https://doi.org/10.1051/e3sconf/202016408030
19. Deniskin Y, Deniskina A, Potsebneva I, Revunova S (2020) Application of complex information objects in production management systems. In: E3S web of conferences, vol 164. https://doi.org/10.1051/e3sconf/202016410042

Model of Information Support of the Quality Management System

Sergey Serebryansky, Boris Safoklov, Irina Pocebneva, and Alexander Kolosov

Abstract The article discusses the construction of a model of information support for the process of the quality management system, and its integration into a single information space, which separates all stages of the product life cycle. Also, the article presents an algorithm for determining vector estimates that should be presented to the information and analytical center, in order to ensure the division of the set of alternatives into classes of solutions for a relatively small number of calls. The application of this algorithm will allow to effectively solve the problems of improving the QMS, by means of automatic classification of the decisions made in order to minimize the risk of not achieving the planned quality indicators at one or another level or stage of the product life cycle.

Keywords Quality management system · Unified information space · Information support · Life cycle stages · Product competitiveness · Information security · Functional model

1 Introduction

Quality assurance and control rules at all stages of the product life cycle are a system of interaction of many factors. Requirements for product quality are laid down at the design stage by the developer, in production conditions they are provided with technological processes, and operating organizations are obliged to support these requirements for product quality using the latest equipment, control equipment and tools [1]. The competitiveness of aviation products is a complex concept that requires a multifaceted and continuous assessment of the condition of the product and its components in real time [2]. This is facilitated by the use of digital technologies

S. Serebryansky · B. Safoklov
Moscow Aviation Institute, National Research University, Volokolamskoe shosse, 4, Moscow 125993, Russia

I. Pocebneva (✉) · A. Kolosov
Voronezh State Technical University, 20-letiya Oktyabrya Street, 84, Voronezh 394006, Russia
e-mail: ipocebneva@vgasu.vrn.ru

A. Beskopylny and M. Shamtsyan (eds.), *XIV International Scientific Conference "INTERAGROMASH 2021"*, Lecture Notes in Networks and Systems 247,
https://doi.org/10.1007/978-3-030-80946-1_90

and the transition to flexible product lifecycle management, including the recycling stage.

Control of information about the properties of a product is carried out by a quality management system (QMS), which is a part of the overall management system of an organization, the purpose of which is to ensure the stable quality of services provided and products. The QMS of the enterprise provides for the constant formation of goals in the field of quality and the solution of the assigned tasks [3, 4].

The most important principle for the functioning of the QMS, in accordance with the ISO 9000 series, is the process approach. This assumes that the QMS is a chain (sequence) of cyclically repeating processes, each of which has an input and an output. Moreover, the output of one process is the input of another. These inputs and outputs can be tangible objects. But without exception, all (both material and non-material) inputs and outputs carry information about their properties. It is this information that is used to control the processes. The ability and ability to receive, process, analyze information and make an adequate decision on its basis is one of the conditions for effective product lifecycle management [5].

The presence of the information space of the QMS will allow to combine all stages and processes of the organization's production. It will lay the foundation for integration into the information field of the product life cycle [6]. The formation of databases of quality indicators at the stages of the life cycle will ensure the introduction and application of technologies for automating business processes of an enterprise at a modern level, ensure the availability and intuitive adaptability for all participants in the stages of the product life cycle.

Interaction and mutual understanding during the joint work of geographically distant participants in the same process, i.e. working on a project of a complex technical product from different places and at different times, a fairly common phenomenon in our time. Such work leads to a general increase in the amount of information and its redistribution among the participants.

The most important factor for improving the efficiency and effectiveness of the QMS is the information support of the process of analysis and management decision-making at the stages of the product life cycle. The use of information support for the process (IOP) of the QMS will allow maintaining the quality indicators of the business processes of the enterprise, and will lead to an increase in the competitiveness of products [7, 8].

Consider the consequences of using the IEP QMS:

- information automated analysis of input and output data accompanying various stages of research and development and development of products or services in order to form solutions that ensure optimal performance of the technical task (TOR);
- analysis of costs, at the "pre-production" stage for products or for the provision of services, in order to select the optimal set of technological processes (TP);
- integration into a single information environment of all stages, stages of the product (product) life cycle, will lead to automation of quality control, reduction of production costs;

- information interaction with external agents (contractors, subcontractors, suppliers, transport companies, etc.) will ensure high efficiency of the processes.

2 Experimental

Each stage of the life cycle of a complex technical object is characterized by two layers: information and material. Descriptions of products and processes for various purposes are created in the information layer. It is characterized by the predominance of intellectual work. In the material layer, according to descriptions presented in various forms, the materialization of products and processes is carried out.

From the information environment of the organization in the design, creation, use, analysis and automation of IEP QMS, uniform information standards of the enterprise should be applied [9].

The regulatory and organizational base of the QMS is formed from two parts:

1. Regulatory and methodological support of the QMS, which includes the main set of documents on quality management and organization.
2. Information system for collecting, registering, storing and processing quality data (IOP QMS system), which should be integrated into a single information space.

Preparation for the implementation of these parts is based on the use of functional models of quality management processes [10]. The model is built by the decomposition method: from large composite structures to simpler ones. On the basis of the functional model, a matrix of responsibility of all participants in the process is determined, ensuring the improvement of the quality and competitiveness of the products. The main task of the created model is to reflect the activity at all stages of the product or service life cycle. Thus, the creation of a functional model makes it possible to determine the structure of a documented quality management system that meets the requirements of GOST R ISO 9001-2015 [11, 12].

It is advisable to start the implementation of information support technologies for QMS with technologies for information support of product life cycle processes at the design and production stages (PDM technologies).

Considering the huge choice of PDM-systems, the question of using the most optimal system for a particular enterprise is determined by the requirements for the PDM-system.

Structuring the system, it is advisable to use the IEP QMS model of the organization, which will function on the basis of the information technologies used by the enterprise (Fig. 1). This form is the most effective.

This form:

- complies with modern principles of quality management,
- principles of the process approach to management.

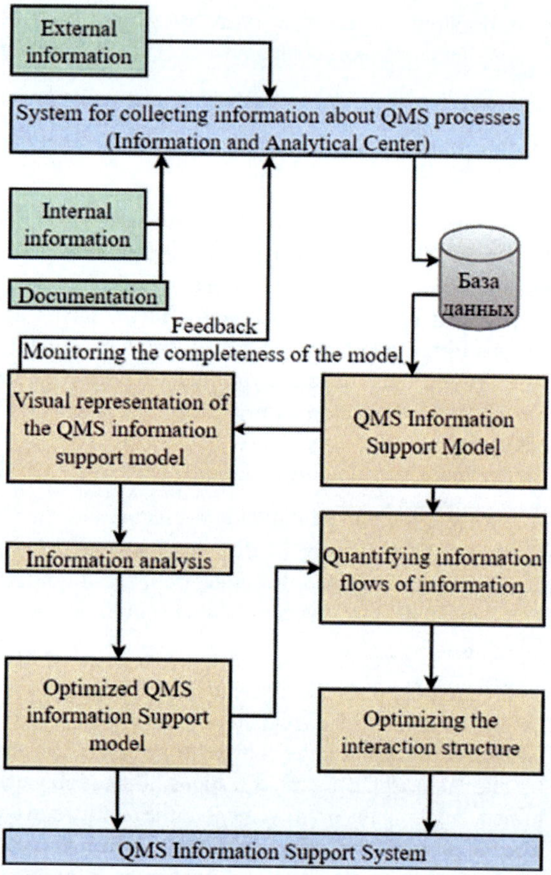

Fig. 1 IOP QMS model

Following the above principles, the organization's divisions are in a single information field through the use of unified software tools [13].

In accordance with this model, the workflow is carried out in a single information space by the participants in the process by means of information support technologies for the QMS.

It is necessary to form a list of information to build a model of information support for the QMS. The list (composition) of information should cover the maximum possible number of processes at the stages of the product life cycle.

It should be formed in such a way that, according to the data received, the structure of the workflow is clearly built not only at the level of departments, but also at the level of employees, i.e. the cycle of each document must be known, from its concept to its placement in the archive and subsequent use [14]. The results of the analysis of information should identify all points of stopping the document (control, sighting, etc.).

Thus, the following requirements for the content of the list of necessary information can be formulated:

- coverage of the maximum number of processes, including processes of information support of the QMS;
- ensuring the ability to trace the cycle of passing the documentation;
- ensuring the formation of a list of applicable regulatory documents;
- ensuring the formation of a list of used materials, purchased parts, as well as a list of their suppliers;
- collection of information on quality assurance processes;
- formation of a list of consumers of the organization and the products they purchase.

The proposed model of information support for the QMS together with IT services and automation departments allows you to create a detailed model of information support for the enterprise QMS. In this model, the mechanism of interaction between departments and the processes of their information support are investigated in real time. The methods of interaction, the level of interaction, the degree of use of computers and the capabilities of the local network during interaction are studied. A list of processes implemented by each participant and officials performing these actions is formed [15].

As a result, a single information space is formed, Fig. 2, which ensures complete digitalization of all stages of the life cycle and their management.

Integration of information flows allows to optimize the interaction between the stages from the standpoint of the quality management system for the conformity of the qualities of the manufactured product and the efficiency of resource use [16, 17]. At all stages of the life cycle, some decisions are made that can have a different effect on the strategic development of the company, while in order to assess the risk of making a particular decision, they must be classified. To automate this process, it is proposed to use the method of statistical tests implemented in the information and analytical center in accordance with the following algorithm.

3 Evaluation

The sequence diagram of the vector estimates of the decisions made to the Center is based on the maximization of the informativeness indicator at each step $\widehat{O}_i$, vector evaluation $y^i \in Y$. At the same time, the actual information obtained when assigning a vector estimate y^i to l—MY class, determined by the indicator g_{li} (i.e., the number of vector estimates whose membership in a certain class becomes known if y^i referred to l—MY class). Let the partition of the set into classes have already been obtained. Consequently, it is known in advance to which class the presented vector estimate will be assigned. Thus, it is possible to determine the minimum number of questions to the Center required to construct a given partition, if we use the previous procedure, but define $\widehat{O}_i$ according to the formula $\widehat{O}_i = g_{li}$, where l—the number of the class

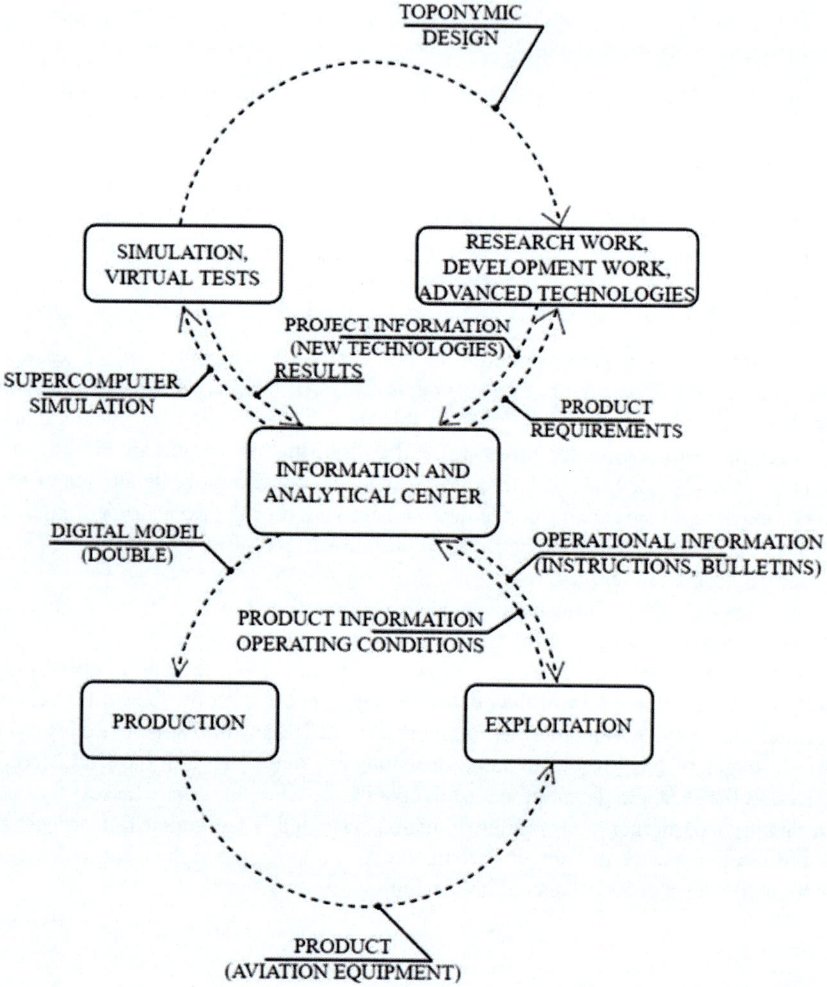

Fig. 2 A single space for information support of the process (IOP) of the QMS

to which the vector evaluation belongs y^i in this partition [1]. Such a procedure was called "reference" (that is, the best one within the proposed approach), since at each step of the procedure a vector estimate is presented, which determines, in accordance with the ratio P^0 belonging to one of the classes of the maximum number of vector estimates [18].

To generate the initial partition of the set of alternatives into classes of solutions, the following procedure is proposed. The researcher sets the number of criteria Q, the number of gradations on the scale of each criterion $\omega_q(q = \overline{1, Q})$ number of classes N. Many alternatives (vector estimates) are being formed Y, representing all possible combinations of scores on the criteria scales. It is known that the vector estimate having the first estimates on the scales of all criteria belongs to the first class, and the vector estimate having the last estimates on the scales of all criteria—to N—my class. Then, in accordance with the proposed survey procedure, the Center determines the vector estimate $y^i \in Y$, which must be presented to the Center in order for it to assign it to one of the classes. The Center's response is modeled using a pseudo-random number generator: the coefficients of proximity of a given vector estimate to different classes determined in accordance with the procedure are considered as the probabilities of assigning the presented vector estimate yi to the appropriate class. The value of the next pseudo-random number R is used to determine the class number to which the presented vector estimate will be assigned: $y^l \in Y_k$,

$$\sum_{t=0}^{k-1} P_{ti} < R \leq \sum_{t=1}^{k} P_{ti} \tag{1}$$

where $P_{0i} = 0$. After that, in accordance with the dominance ratio, the sets of numbers are corrected G^j for each vector estimate $y^j \in Y$, whose class has not yet been determined. Next, the coordinates of the centers of the classes are recalculated, and the procedure is repeated until the belonging to one of the classes is determined [2] of each vector estimate of the set Y. The number of calls to the pseudo-random number generator characterizes the number of questions to the Center for constructing this partition based on the proposed rational procedure for polling the Center.

The scheme of statistical tests of the behavior of the proposed algorithm consists of the following stages [19].

1. Generation of Center responses using a random number generator and determination of the initial partitioning of the set of alternatives into classes of solutions.
2. Determination of a "reference" sequence of questions to the Center (in this case, the Center is a partition of the set of alternatives generated at stage I) for partitioning the original set of alternatives into classes.
3. Repeated repetition of stages I–II and comparison of results.

Stages I and II are carried out in accordance with the described procedures. Stage III consists in multiple repetition of stages I and II for a different number of criteria Q, gradations on the scales of criteria c and solution classes N. The data obtained at each of the stages on the number of presented vector estimates are averaged, and the value of this indicator obtained at stage I is compared (N0) and II ($N_{\hat{y}}$).This attitude characterizes the effectiveness of the proposed rational procedure for polling the Center.

For each variant of the number of criteria, gradations according to the scales of criteria and the number of classes of solutions, about 500 realizations of procedures were carried out, on each of which an estimate of the number of presentations of vector estimates was determined, when the partition was known in advance and when it was not known.

Table 1 Average scores

Q	ω	L	$N_{\acute{y}}$			N_O			$N_O/N_{\acute{y}}$		
			$N=2$	N= 3	N= 4	$N=2$	N= 3	N= 4	$N=2$	N= 3	N= 4
4	3	81	3	8	11	8	13	17	2.67	1.63	1.55
	4	256	4	9	13	10	14	21	2.50	1.67	1.55
5	3	243	4	9	15	10	18	25	2.50	2.0	1.67
	4	1024	5	12	19	14	24	33	2.82	2.0	1.68

The average values of these estimates are given in Table 1 for the case of four and five criteria with three and four gradations according to their scales and two, three and four classes of solutions.

The data presented show that the proposed procedure requires the presentation of no more than 2.8 times more vector estimates than the reference algorithm. Moreover, this ratio decreases with an increase in the number of classes of solutions. The absolute values of the number of presented vector estimates are significantly less than the power Y, which indicates the advisability of using the proposed procedure for polling the Center.

Note that the proposed algorithm for determining vector estimates that should be presented to the Center ensures the construction of a partition of the set Y into classes of solutions for a relatively small number of calls to the Center. At the same time, its application significantly reduces the possibility of checking the correctness of the Center's answers based on the relation P^0, so for the case N = 2 when using the algorithm, the possibility of inconsistencies in the partition is completely excluded Y. On the one hand, this is a positive phenomenon, since we need to build a consistent partition. On the other hand, a random error of the Center in assigning the presented vector estimate to one of the classes can lead to a partition that does not correspond to the real preferences of the Center.

A practical method of decision-making should provide the ability to verify the information received [3]. In this case, the requirement is put forward that each vector estimate of the set not presented to the Center Y was assessed directly or indirectly (based on the relationship P^0) at least twice. Therefore, after constructing the partition using the proposed algorithm, additional presentation of a part of the vector estimates for which this condition was not satisfied is provided.

The following procedure is proposed for an additional survey of the Center. Into subset $Z \subset Y$ vector estimates are distinguished, the belonging of which to a certain class of solutions was determined on the basis of the relation only once. For this subset, a "reference" procedure is used to determine the sequence of presented vector estimates, that is, the indicator of information content $\widehat{O}_i$, for $y^i \in Z$ defined as $\widehat{O}_i = g_{li}$, where g_{li}—number of vector estimates from Z, whose belonging to one of the classes of solutions is established unambiguously if the Center is presented with a vector estimate y^i (in this case, it is known that the vector estimate in the construction of the partition of the set using the proposed algorithm was assigned to the class Y_l).

Thus, the number of vector estimates presented to the Center for constructing a partition of the set Z for classes of solutions will not exceed the number of presented vector estimates indicated in Table 1 for the corresponding parameter values Q, ω and N.

In the resulting partition of the set Y on the classes of solutions, vector estimates of the set Z belong to the classes in which they fell when dividing the set into classes Z.

The thus constructed partition of the set into solution classes even for $N = 2$ does not guarantee the asymmetry of the relationship P^*. In this regard, one should use the procedure for reducing the resulting partition of the set Y to a consistent view. The proposed procedure for polling the Center allows one to obtain the necessary information to solve the initial problem for a relatively small number of questions to the Center [20].

4 Conclusions

Thus, it is possible to single out the priority tasks in the development of information technologies to support the QMS for organizations—developers of standards and software products, as well as industrial enterprises:

- development of a regulatory framework in the field of information technology that meets the requirements of international standards;
- development of a complex of domestic means of information support for the QMS;
- introduction of a regulatory framework and software and hardware solutions for the preparation of electronic operational documentation for products;
- the use of information technologies that improve the information support of the QMS, allows you to develop an optimized model of continuous improvement, which makes it possible to optimize and improve the QMS processes.
- a single information space implements the continuity of improving the QMS processes.

It is also worth noting that the use of information technologies allows you to effectively solve the problems of improving the QMS by automatically classifying the decisions made in order to minimize the risk of not achieving the planned quality indicators at one or another decision level.

References

1. Korchagin A, Deniskina A, Fateeva I (2019) Lean and energy efficient production based on the Internet of Things (IOT) in the aviation industry. In: E3S web of conferences. https://doi.org/10.1051/e3sconf/201911002124

2. Bityukov YI, Deniskin YI ()2018 Quality control of structures made of composite materials. Polym Sci Ser D 11(2):197–201. https://doi.org/10.1134/S1995421218020041
3. Yurin D, Deniskina A, Boytsov B, Karpovich M (2021) Quality 4.0. Time of revolutionary changes in the QMS. In: E3S web of conferences, vol 244, 11010
4. Skachko G, Surkova E, Ermolaeva E, Potsebneva I (2021) Management decision making based on the risk management model. In: E3S web of conferences, vol 244, 11008
5. Smolyaninov A, Potsebneva I, Fateeva I, Singur K (2021) Software implementation of a virtual laboratory stand for distance learning. In: E3S web of conferences, vol 244. https://doi.org/10.1051/e3sconf/202124411009
6. Deniskin Y, Deniskina A, Potsebneva I, Revunova S (2020) Application of complex information objects in industrial control systems. In: E3S web of conferences, vol 164, 10042
7. Barabanov A, Serebryansky S (2020) Modeling optimization of the stages of the product life cycle taking into account the uncertainty of the main parametric variables. In: Proceedings of the 13th international conference on the management of large-scale systems development. MLSD 2020, 9247777
8. Strelets DY, Serebryansky SA, Skurin MV (2019) A digital approach to lifecycle management of aviation products. In: Materials of the 12th international conference 2019, Large-scale systems development management, MLSD 2019, 8911020
9. Serebryansky SA, Barabanov AV. On the question of multicriteria optimization of aircraft components in order to optimize its life cycle. In: Advances in science, technology and engineering systems, pp 408–415
10. Desyatirikova EN, Khodar A, Rechinsky AV, Chernenkaya LV, Alkhayat I (2020) Performance analysis of available service broker algorithms in cloud analyst. Lecture notes in networks and systems, vol 95, pp 449–457
11. Desyatirikova EN, Myshovskaya LP, Lutin VI, Mager VE, Khripunov YV (2020) Algorithmization of evaluating solutions by the Neumann-Pearson criterion. In: Proceedings of the 23rd international conference on soft computing and measurements. SCM 2020, pp 178–181, 9198783
12. Desyatirikova EN, Ivanov SA, Sergeeva SI, Zuev, SA (2020) Creation of digital archives of project documentation. In: Materials of the international conference IEEE 2020 "Quality management, transport and information security, information technologies", IT and QM and IS, pp 357–359, 9322944
13. Averina T, Barkalov S, Fedorova I, Poryadina V (2021) Influence of digital technologies on the company's business model. In: E3S web of conferences, vol 244, 10002
14. Barkalov S, Bekirova O, Sokolova O, Zenkova E (2021) New approaches to multicenter assessment of alternatives in decision making. In: E3S web of conferences, vol 244, 11011
15. Barkalov SA, Lomakin MI, Mistrov LE, Morozov VP (2020) Management of an organization in difficult economic conditions based on an information system. Phys Conf Ser 1679(2):022086
16. Potsebneva I, Belousov V, Fateeva I, Lukinov V, Folomeeva T (2018) Algorithms for hierarchical mixing of experts' opinions on the problems of synthesis of information systems for urban development management 2018. In: MATEC web of conferences, vol 170, 01120. https://doi.org/10.1051/matecconf/201817001120
17. Mager VE, Belousov VE, Desyatirikova EN, Ivanov SA, Potsebneva IV (2017) Algorithm of information processing when constructing the structure of a self-adjusting technical system, optimal in terms of quality parameters. In: 2017 2nd IEEE international conference on control in technical systems, CTS 2017 8109503, pp 114–117. https://doi.org/10.1109/CTSYS.2017.8109503
18. Chernenkaya LV, Desyatirikova EN, Belousov VE, Sergeeva SI, Slinkova NV (2017) Optimal planning of distributed control systems with active elements. In: Proceedings of the 2nd IEEE 2017 international conference on control in technical systems, CTS 2017, pp 37–40, 8109482
19. Khodar A, Chernenkaya LV, Alhayat I, Fadhil A-AKA, Desyatirikova EN (2020) Design model for improving task scheduling in cloud computing based on particle swarm optimization. In: Proceedings of the IEEE 2020 conference in russian. young researchers in electrical and electronics engineering, EIConRus 2020, pp 345–350, 9039501

20. Velikorosov VV, Rechinsky AV, Chernenkaya LV, Filin SA, Chernenky AV (2019) Digital economy as a tool for reducing uncertainty in strategic management decisions. In: Mater international conference. ACM

Improving the Efficiency of Production Processes of Enterprises of the Aviation Industry

Sergey Serebryansky, Boris Safoklov, Irina Pocebneva, and Vitaly Lepeshkin

Abstract The article discusses the application of the mechanism for selecting alternatives from the semantic network of conditional statements when managing product quality parameters at the stages of the life cycle. This will make it possible to implement the timely improvement of quality management processes for aviation products based on the use of information technology to support the product life cycle in real time.

Keywords Product quality management · Standardization · Quality management system · Unified information space · Information support · Life cycle stages · Product competitiveness · Information security · Functional model

1 Introduction

The development and implementation of technologies in the aviation industry have made the development and effective use of information resources, both local and national, global, urgent. Currently, only those participants who will use modern information technologies (IT) in their activities will be able to resist the competition in the aviation market. It is these technologies, along with the advanced technologies of material production, that can significantly increase labor productivity, quality and competitiveness of products with a significant reduction in the terms of launching products that meet the needs and expectations of consumers.

It is assumed that within the framework of the existing object-oriented approach, each entity of the material implementation of a complex technical system (part, assembly, unit, subsystem, system, product) can have a physical and mathematical

S. Serebryansky (✉) · B. Safoklov
Moscow Aviation Institute, National Research University, Volokolamskoe shosse, 4, 125993 Moscow, Russia

I. Pocebneva
Voronezh State Technical University, 20-letiya Oktyabrya Street, 84, 394006 Voronezh, Russia

V. Lepeshkin
Voronezh Institute of High Technologies, Lenina Street, 73A, 394043 Voronezh, Russia

A. Beskopylny and M. Shamtsyan (eds.), *XIV International Scientific Conference "INTERAGROMASH 2021"*, Lecture Notes in Networks and Systems 247,
https://doi.org/10.1007/978-3-030-80946-1_91

description (modeled) in the form of a corresponding information-digital object. For example, a part can be described by a corresponding CAD information model. Accordingly, a product unit, which is a collection of parts or a system that is a collection of aggregates connected by certain physical relations, can also be described by an information model based on the interaction of models of parts or aggregates [1].

The life cycle (LC) is a set of interrelated processes of changing the state of a product and includes a number of stages, from the birth of an idea (concept) of a new product to its disposal at the end of its useful life. These include the stages of marketing research of the aircraft market, the formation of requirements for the product, design, technological preparation of its production, production, certification, after-sales service and operation of the product, disposal. All these stages have their own targets and criteria for the quality of processes. At the same time, participants in the life cycle strive to achieve their goals with maximum efficiency and quality. Quality management, in a broad sense, should be understood as the management of processes aimed at ensuring the quality of their results. This approach is consistent with the principles of Total Quality Management, which are based on product lifecycle management through quality management at its stages [2].

The effectiveness of the decisions made is significantly influenced by the use of modern information technologies. Information is understood as meaningful and processed data that is used to solve management problems. The data reflects events taking place both at the very stage of the life cycle and beyond. To ensure operational management, it is necessary to receive timely data on the processes occurring at various stages of the product life cycle.

Data on products and processes, in this article, must be considered together, since management processes directly affect the quality of the stages of the product life cycle. To manage data, it is also necessary to aim at solving problems that arise in the process of functioning of quality management systems (QMS).

The description of the process can be represented as a set of operations that make up the process, necessary conditions and resources, input and output streams. The set of standardized information models of a product, processes and resources forms a single integrated model that provides information support for tasks performed during the life cycle.

Each stage of the life cycle requires its own amount of data, determined by the content of the tasks being solved (Fig. 1). The totality of this data can be interpreted as contextual information models of the product, processes and resources corresponding to the stages of the product life cycle.

In the course of data exchange within the stages of the life cycle and between them, information flows are formed, the number of paths and the intensity of which from stage to stage increases, and becomes much more than the ways of moving material and financial flows. Information flows are the physical movements of information from one employee of an enterprise to another or from one department to another. Any change in information is not considered as information flows. The information flow can exist in the form of paper and electronic documents (carriers).

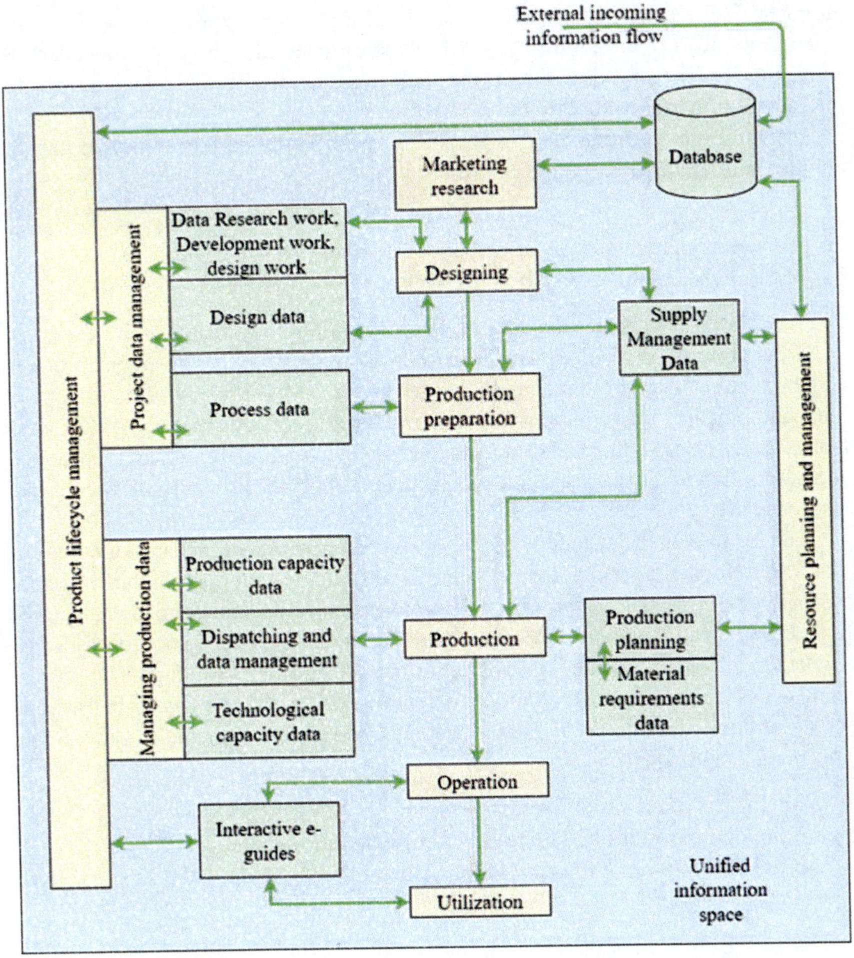

Fig. 1 Formation of information flows of data about the product by stages of the life cycle

Automation and integration of all stages of the production activity of an enterprise is possible only on the basis of a single environment—methodological, organizational and informational, with the latter playing the leading role [3].

2 Experimental

The system of information flows is the set of all physical movements of information about the product and the conditions of its operation, makes it possible to carry out any process and implement any solution. The sum of information flows (Fig. 1), allows an enterprise, a participant in the life cycle of a product, administrative, financial and economic activities. Information flows ensure that the organization operates satisfactorily. The purpose of working with information flows is to maximize the interaction of participants [4].

The information circulating in the information space of the life cycle can be conditionally divided into three classes:

1. product data;
2. data on the processes performed at the stages of the life cycle;
3. data on the resources required for the execution of processes.

Product data constitutes the bulk of the information flow during its life cycle.

At the design and development stage, data about the product, the design process, and the required organizational and other resources are used. An information model of technological preparation of production is interpreted as a description of a process that uses data on a product and resources. The production model can also be presented as a description of the process associated with data about the product and the required resources [5].

The type of data, from stage to stage, can have its own set of tools and methods of work, which forms a "technological" layer of information support—a data management system (or a set of systems) that takes into account the semantics of data, features of their organization and provides a high level of exchange with other processes J C.

Data management technology is a set of methods, information models, rules for using access to data, necessary and sufficient to work with them in solving various problems during the life cycle of a product. You can control the data flow in a single information space as follows:

- changing the direction of the flow;
- limiting the transmission rate to an appropriate receive rate;
- limiting the volume of the flow to the value of the throughput of a separate stage of the life cycle.

During the creation of information support, one should be guided by the averaged and leveled need for special information of managers and specialists. A special place here is occupied by the type of information associated with management, which will reflect the applied techniques and methods of organizing management [6, 7].

The process of forming information support for the life cycle includes several stages:

- a description of the state of the object, i.e. the physical display of the product. Assumes a set of technical and economic indicators and parameters characterizing the control and controlled systems, with the appropriate classification of these indicators;
- modeling of classification links in information arrays with the allocation of cause-and-effect relationships, that is, the formation of private static models;
- reflection in information models of the dynamics of individual elements and processes, i.e., substantiation of trends in quantitative and qualitative changes

in production. At the same time, a quantitative change presupposes the correction of information, and a qualitative change presupposes its partial or complete restructuring;
- an integrated information model of the production process, reflecting the relationship and dynamics of local processes and the entire production [8].

High-quality use of resources at each stage of the life cycle is achieved through the systemic management of information and communications. It should be noted that at each stage of the product life cycle, the task is to select alternatives for organizing production.

The Semantic Web (SS) is a convenient way to graphically describe objects in the domain. In this case, an object can be understood as a process, state, any entity, etc. Define CC as a directed graph with labeled vertices and arcs. In this case, the vertices are identified with the corresponding objects of the subject area, and the arcs—with the relations between them. A wide variety of SS is formed on the basis of the following considerations:

- is the CC summit simple or difficult?
- how many and what relationships are used in the SS?
- what processes need to be modeled on the SS?

Suppose that the CC vertex is simple (ordinary) if it has no internal structure. In other words, simple vertices are identified with terminal (end) objects of the simulated environment. Obviously, numbers, lexemes, pentagrams, etc. can appear as such vertices. Complex nodes can be expanded up to the semantic network, in which only simple nodes participate.

Consider an example of product production management based on the choice of alternatives from the semantic network.

Consider an example of using conditional statements in decision making.

Let be $A = \{a_l, l = \overline{1, D}\}$—many alternatives for organizing production, and each alternative is described by a set of parameters $Y = \left\{Y_j, j = \overline{1, Y}\right\}$ (technical conditions, indicators of raw materials, qualifications of workers, etc.), and it is estimated by a set of indicators $X = \{X_i, i = \overline{1, I}\}$ (consumer properties of manufactured products), the values of which can be calculated based on information on product quality parameters. The relationship between various parameters and indicators can be known only approximately and expressed by a set of statements of the form:

$$g_t : \text{if } L_t, X_i = H \tag{1}$$

where L_t is a logical statement of the form:

$$L_t : Y_{j_1} = G_{j_1 t} \wedge \cdots \wedge Y_{j_m} = G_{j_m t};$$

$$\{Y_{j_1}, ..., Y_{j_m}\} \subseteq Y.$$

Some degree of confidence can be assigned to a statement. $\alpha \in [0, 1]$ in its truth. More complex statements consist of several simple conditional statements (1), connected by a link "differently":

$$\frac{IF\ \ L_1^*}{IF\ L_{n-1}^*}, \frac{\text{then}\ \ X_i = H_i}{\text{then}\ X_i = H_{n-1}} \frac{\text{otherwise}}{\text{otherwise}\ X_i = H_n} \tag{2}$$

Alternatives can be probabilistic when parameters take on values G_{jk} with probability λ_{jk}, moreover λ_{jk} can be numeric, fuzzy, or linguistic.

Preparation of information for solving the problem. We denote $L_t = \overline{L}_1^* \wedge \cdots \wedge \overline{L}_t^*$. Then (2) can be written as:

If L_1, then $X_i = H_1$.

$$\text{If } L_2, \quad \text{then} \quad X_i = H_2, \tag{3}$$

If L_n, then $X_i = H_n$.

Each line (3) is a simple conditional granule. The aggregate of granules is evidence [9]:

$$E = \{g_1, \ldots, g_n\} \tag{4}$$

Based on the information contained in (4), it is possible to formulate the fuzzy relation R between indicator X_i and parameters $Y_j \in Y$, used in the certificate E.

Let be Y^E many parameters used in the certificate E, Y^t—set of parameters, the values of which are set in the granule g_t this testimony; power $|Y^E|$ multitudes Y^E is equal to m_E, $|Y^t| = m_t$,

$$Y^t = \{Y_{j1}^t, \ldots, Y_{jm1}^t\}, \ \ Y^E = \left\{Y_{j1}^E, \ldots, Y_{jm_E}^E\right\}, \ \ Y^E = \bigcup_{t=1}^{n} Y^t.$$

Let us denote by relationship between parameters and exponent induced by one granule g_t. R_t there is a fuzzy set with a membership function [10]

$$\mu_{Rt}(y_{j1}^t, \ldots, y_{jm_t}^t, x) = 1 - \mu_{L_t}(y_{j1}^t, \ldots, y_{jm_t}^t) + \mu_{H_t}(x),$$

$$y_{jk}^t \in Y_{jk}^t, x \in \chi,$$

Y_{jk}^t, χ—sets of clear values for Y_{jk} and X,

$\mu_{L_t}(\cdot)$—vector value membership function $y_t = (y_{j1}^t, \ldots, y_{jm_t}^t)$ to the fuzzy set defined by the expression L_t.

We calculate the cylindrical extensions $C(R)$ relations R_t in $Y_{j1}^E \times Y_{j2}^E \times \cdots \times Y_{jm_E}^E$, equal $C(R_t) = \int_{y_{j1}^E \times \cdots \times y_{jm_E}^E} \mu_{R_1}(y^t)/y^E$.

Then the ratio R corresponding to the evidence E is determined by the membership function [11].

$$\mu_R(y^E, x) = \min_t \mu_{C(R_t)}(y^t, x). \tag{5}$$

Based on the obtained relations, using the compositional inference rule [12], one can calculate the fuzzy value H indicator μ_H for given fuzzy values of parameters:

$$\mu_H(x) = \max_{y^E}(\mu_{C(G_0^E)}(y^E, x), \mu_R(y^E, x)), \tag{6}$$

where $G_0^E = G_{j_1} \wedge \cdots \wedge G_{j_{m_E}}$, $C(G_0^E)$—cylindrical expansion G_0^E in $Y_{j_1}^E \times Y_{j_2}^E \times \cdots \times Y_{j_{m_E}}^E$, and the operation of the limited work is defined as follows:

$$\mu_F(u) \otimes \mu_G(u) = \max(0, \mu_F(u) + \mu_G(u) - 1).$$

As an alternative method for calculating fuzzy values of H, one can use the method of fuzzy linear interpolation [13], but in the multidimensional case it requires large computational costs.

Each evidence E can be reflected in the form of a semantic network (Fig. 2), in which the vertices correspond to the parameters, the calculated quality indicators and the operator of combining the scalar values of the parameters into a vector; unmarked arcs only indicate the direction in which the values of parameters and indicators are transmitted; the marked arcs, in addition, determine the transformation of information in accordance with the fuzzy relation R; the point at the vertex denotes the synapse and shows that to calculate the vector of values, information is needed from all vertices, the arcs from which converge into the synapse.

We will write down the fact that the value of X is deducible based on evidence E in the form:

$$X = E(Y^E) \tag{7}$$

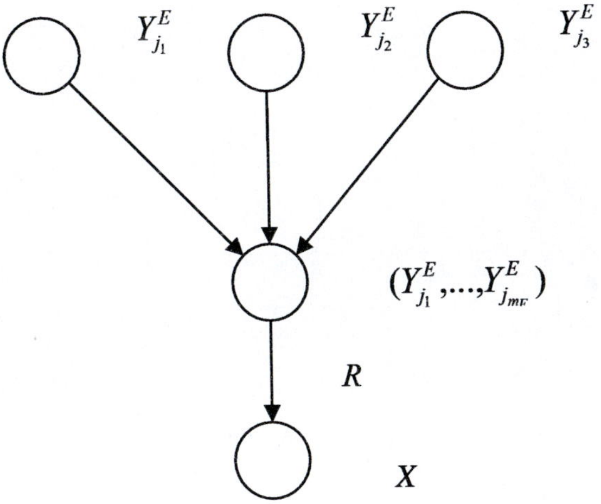

Fig. 2 Representation of an utterance in the form of a network section

As parameters Y^E there can also be indicators calculated on the basis of other evidence, for example:

$$Z = E_1(Y_1, Y_2), X = E_2(Z, Y_3) \tag{8}$$

The network corresponding to the set of evidence (8) is shown in Fig. 3.

Let's consider the process of preliminary identification of inconsistency, incompleteness and redundancy of statements.

Let us select from the general set of statements a subset in which different values of the same parameters and indicators are compared, and consider one of such subsets.

Without loss of generality, let us assume that it includes statements

$$g_1, g_2, \ldots, g_n \quad E = \{g_1, g_2, \ldots, g_n\} \tag{9}$$

For simplicity, consider the case when $g_1, g_2, \ldots, g_n$ determine the dependence of some indicator X_i from one parameter Y_j.

The subjective nature of the statements can lead to their mutual contradiction. The manifestation of the internal inconsistency of evidence is the presence of such g'_t, g''_t, for which:

$$G_{jot'} \subseteq G_{jot''} \wedge H_{t'} \not\subset H_{t''},$$

where $G_1 \subseteq G_2 \Leftrightarrow \mu_{G_1}(y) \leq \mu_{G_2}(y)$.

The degree of incompleteness of the evidence is expressed by the value:

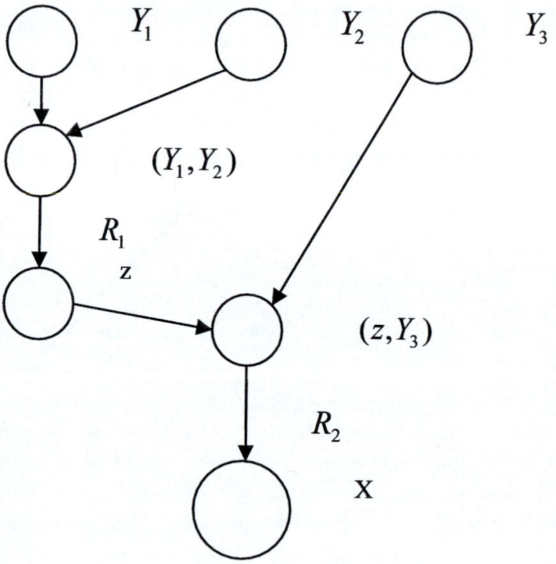

Fig. 3 A section of the network for presenting a set of evidence

$$in\ E = \max_{y_j \in Y_j} \max\left(0, 1 - \sum_{t=1}^{n} \mu_{G_{jt}}(y_j)\right) \tag{10}$$

Some evidence may indicate metric values for the same or similar parameter values. The degree of redundancy of evidence can be determined by the value:

$$red E = \max_{y_j \in Y_j} \max\left(0, \sum_{t=1}^{n} \mu_{G_{jt}}(y_j) - 1\right) \tag{11}$$

Similar indicators can be calculated for evidence, in which each granule determines the dependence of the indicator on a subset Y^E parameters. For this, in the Formulas (10) and (11) it is enough to replace $\mu_{G_{jt}}(y_j)$ on $\mu_{G_t^E}(y^E)$. In this case, information in the form of conditional statements IF …, THEN … is set by experts. If the number of statements is large, some of them turn out to be contradictory. If the inconsistency could not be eliminated at the stage of collecting statements, then it is advisable to assign a degree of reliability to the evidence $\alpha \in [0, 1]$. If the evidence E_h It has $\alpha_h \neq 1$, then when writing it in the form (7) on the right we will assign α_h

$$X_i = E_h(Y^{E_h}),\ \alpha_h.$$

If a certain indicator X should be calculated taking into account several conflicting evidence, then in the network representation this fact will be represented by a square pseudo-synapse at the vertex-indicator (Fig. 4). The purpose of the pseudo-synapse is to smooth out the inconsistency of evidence. Confidence levels are assigned to evidence heuristically or determined using statistical methods in the process of training the system [14, 15]. We will assume that they are given.

Let there be evidence E_1 and E_2:

$X = E_1(Y_1, Y_2), \alpha_1; X = E_2(Y_3, Y_4), \alpha_2.$

In general, the values H_1, H_2 evidence-based E_1 and E_2, differ. It is proposed to use H as the resulting estimate:

$$\mu_H(x) = \min(\mu_{H_1}(x), \mu_{H_2}(x)) \tag{12}$$

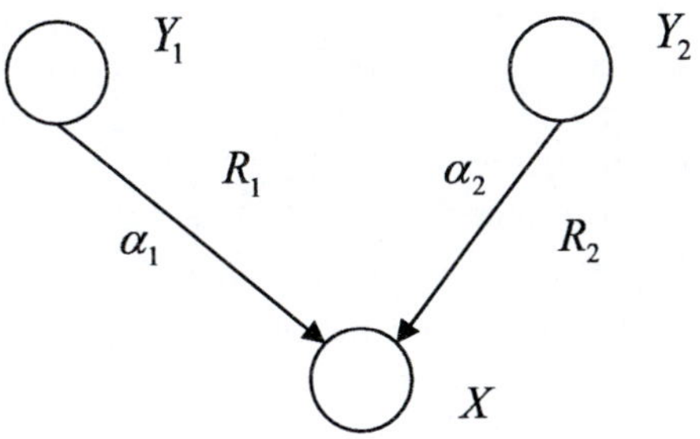

Fig. 4 Section of the network for two conflicting statements

or taking into account the degrees of confidence in the evidence:

$$\mu_H(x) = \min(\mu_{H_1}^{\alpha_1}(x), \mu_{H_2}^{\alpha_2}(x)) \tag{13}$$

Using the product instead of min and carrying out the subsequent normalization, we get:

$$\mu_H(x) = \frac{\mu_{H_1}(x) \cdot \mu_{H_2}(x)}{\max\limits_x \mu_{H_1}(x) \cdot \mu_{H_2}(x)}. \tag{14}$$

When deriving Formulas (12), (13) and (14), Dempster's approach for combining evidence was used as a basis. But, as noted, if the mutual inconsistency of the evidence is great, then it is more appropriate to use some other method of combination, for example, the summation of "confidence masses". In the case of indistinct evidence, this corresponds to the calculation of H by the formula:

$$\mu_H(x) = \frac{\alpha_1}{\alpha_1 + \alpha_2}\mu_{H_1}(x) + \frac{\alpha_2}{\alpha_1 + \alpha_2}\mu_{H_2}(x).$$

The analysis of the advantages and disadvantages of various methods of combination allows us to propose a generalized formula, which, depending on the degree of mutual contradiction and the degree of reliability of individual evidence, gives a result close to one of those listed.

Let M be the set of numbers of evidence for which the values of the parameters are known or can be calculated and the use of which makes it possible to calculate the value of some indicator X. Let us denote by the set of all non-empty subsets of M. The membership function of the combined estimate is defined as follows [16]:

$$\mu_H(x) = \bigoplus_{m \in M_1} ((1 - \max_{i \in M \backslash m} \alpha_i) \cdot \min(\min_{i \in m} \mu_i(x), \min_{j \in M \backslash m} (1 - \mu_j(x)))$$

$$\oplus (a_1, \ldots, a_n) = \min(1, a_1 + \cdots + a_n). \tag{15}$$

Next, we will calculate the arterial estimates of the alternatives.

Indicator value X_i can be calculated by the waveform sequential vertex excitation algorithm. The values of the initial parameters required for calculating the values of the indicators must be preset. The initial vertices corresponding to these parameters are considered excited. The essence of the algorithm is the sequential excitation of vertices that have at least one synapse, all input arcs of which originate from the excited vertices.

Algorithm for calculating the values of indicators.

1. A multitude of M_1 target vertices, i.e. vertices corresponding to the indicators, the values of which need to be calculated.
2. A multitude of M_2 all vertices of the network.
3. The top is sought $m \in M_2$, from which you can get to any vertex $S \in M_1$ in one step. If there is no such m, then go to step 5.

4. Inclusion of vertices m in the set $M_1 : M_1 = M_1 \cup m$. Eliminating vertices m of the multitude $M_2 : M_2 = M_2 \backslash m$. Go to step 3.
5. A multitude of S_1 excited network vertices that belong M.
6. A multitude of $S_2 = M_1 \backslash S_1$.
7. For each $s \in S_2$ the possibility of excitation of one of the synapses is checked based on information about the excitation of the vertices S_1. If the synapse can be excited, then, taking into account the ratios indicated on the input arcs, the value of the index corresponding to the vertex is calculated, the latter is entered into the list of excited vertices: $S_1 = S_1 \cup s$, $S_2 = S_2 \backslash s$. If s corresponds to the indicator of interest to us (is a terminal vertex), then the required value is found and the algorithm is successfully completed.
8. If at step 7 at least one vertex has been added to S_1, then go to step 7. If at step 7 no vertex added to S_1, then the desired value cannot be calculated, the algorithm ends in vain.

Next, consider the non-deterministic parameter values.

Let it be known that for some alternative the parameters Y_j, used in the certificate E, take on fuzzy values G_{jk_j} with probability P_{jk_j}, $k_j = \overline{1, n_j}$, $j \in M^E$, M^E—a variety of indexes of the parameters of the certificate. The algorithm for calculating the estimates remains the same, with the only difference that the excitation of the network vertex is accompanied by the assignment of more than one fuzzy value to it H, and some set of values $\{H_k\}$ each of which has a probability p_k, $k = \overline{1, n}$, $n = n_1 \cdot \ldots \cdot n_{m_E}$ n. Calculation method p_k is determined by the nature of the relationship between the parameters Y_j. If the parameters are mutually independent, then

$$p_k = \prod_{j \in M^E} p_{jk_j}. \tag{16}$$

If the degree of mutual dependence is unknown, then p_k can be estimated at intervals:

$$p_k = [0, \min_{j \in M^E} p_{jk_j}]. \tag{17}$$

If the probabilities of different parameter values are fuzzy and equal λ_{jk_j}, then for independent parameters:

$$\lambda_k = \prod_{j \in M^E} \lambda_{jk_j}, \tag{18}$$

and for parameters with an unknown degree of mutual dependence:

$$\lambda_k = \overline{\min}(\delta, \overline{\min}\lambda_{jk_j}), \tag{19}$$

where $\overline{\min}$—extended minimum operation, and δ -fuzzy number with membership function:

$$\mu_\delta(\eta) = \begin{cases} 1, 0 \le \eta \le 1 \\ 0, \text{otherwise} \end{cases}$$

Next, consider the following conditions of the system: inconsistency, incompleteness, redundancy in the description of alternatives.

The unwanted qualities of the information contained in the evidence have analogues in the description of alternatives. The description of an alternative is considered to be contradictory, in which such subjective probabilities of fuzzy values of at least one parameter are specified, which cannot take place for any objective distribution of probabilities for the set of basic values of this parameter, i.e. under no conceivable probability distribution $f(y_j)$ it is impossible to satisfy the conditions:

$$\int\limits_{y_j \in Y_j} f(y_j)dy_j = 1; \;\; P_{jk_j} = \int\limits_{y_j \in Y_j} \mu_{G_{jk_j}}(y_j) f(y_j)dy_j; \;\; k_j = \overline{1, n_j} \tag{20}$$

where p_{jk_j} there is a given clear value of the probability for G_{jk_j} or a subset p_j clear base values fuzzy meaning λ_{jk_j} c $\mu_{\lambda_{jk_j}}(p_j) = 1$.

The degree of inconsistency can be calculated as

$$contr A = \min_{f \in F}\left(1 - \int\limits_{y_j \in Y_j} f(y_j)dy_j\right)$$

subject to constraints (20), where F—set of possible probability distributions.

The incompleteness of the contradictory description of the alternative can be calculated as:

$$nf A = \min\left(0, 1 - \max_{\mu_\sigma(p)=1} p\right), \text{ where } \sigma = \sum_{k_j=1}^{n_j} \lambda_{jk_j}$$

and redundancy:

$$red A = \min(0, \max_{\mu_\sigma(p)=1} p - 1).$$

3 Evaluation

Let's make a choice of alternatives.

We will consider the best alternative, which allows achieving the ideal values of the criteria.

It is required to determine the degree of reliability of statements $X_i = Q$, where Q—some linguistic value of the indicator X_i, which corresponds to the fuzzy set Q with membership function $\mu_Q(x)$.

The degree of fulfillment of the statement $X_i = Q$ will be determined by two values: the expected certainty *EC(Q)*, which is equal [17]

$$\mathrm{EC(Q)} = \sum_{k=1}^{n} \lambda_k(\sup H_k - \sup(H_k \cap \overline{Q})), \tag{21}$$

and the expected opportunity, which is [18]

$$\mathrm{E\Pi(Q)} = \sum_{k=1}^{n} \lambda_k \sup(H_k \cap Q), \tag{22}$$

where $\sup A = \max\limits_x \mu_A(x)$, $\mu_{\overline{Q}}(x) = 1 - \mu_Q(x)$, but H_k calculated by Formulas (12) and (15). If a more complex statement of the form acts as a hypothetical one:

$$X_{i_1} = Q_{i_1} \wedge \cdots \vee \cdots \wedge X_{i_m} = Q_{i_m},$$

then the assessment of its reliability is carried out in the same way. Let, for example, the requirements for the values of indicators are expressed as follows:

$$X_1 = Q_1 \wedge X_2 = Q_2 \vee X_3 = Q_3$$

.

Then in expressions (21) and (22) instead of H_k and Q fuzzy sets should be considered accordingly $(H_1 \cap H_2)_k$ and $Q_1 \cap Q_2 \cup Q_3$ with membership functions:

$$\mu_{(H_1 \cap H_2)_H}(x_1, x_2) = \min(\mu_{H_{1k}}(x_1), \mu_{H_{2k}}(x_2));$$

$$\mu_{Q_1 \cap Q_2 \cup Q_3}(x_1, x_2, x_3) = \max(\min(\mu_{Q_1}(x_1), \mu_{Q_2}(x_2)), \mu_{Q_3}(x_3))$$

As already noted, to calculate the value of a certain indicator, it is not necessary to know the values of all parameters. The set of parameters, the values of which are necessary to display the desired value of the indicator, are determined at steps 1–4 of the algorithm for calculating the values of the indicators. On the other hand, if in the process of outputting a vertex corresponding to the primary parameter is included in the list of vertices to be excited, and the parameter is not assigned any value, then the output procedure should provide an opportunity to request information from the database or (in the interactive mode) to the expert. If estimates are calculated on the network $EC_l(Q)$, $E\Pi_l(Q)$ for each alternative, then the best one can be considered the one for which these estimates are maximum [19, 20].

4 Conclusions

The life cycle (LC) is a set of interrelated processes of changing the state of a product and includes a number of stages, from the birth of an idea (concept) of a new product to its disposal at the end of its useful life. These include the stages of marketing research of the aircraft market, the formation of requirements for the product, design, technological preparation of its production, production, certification, after-sales service and operation of the product, disposal. All these stages have their own targets and criteria for the quality of processes. At the same time, participants in the life cycle strive to achieve their goals with maximum efficiency and quality. Quality management, in a broad sense, should be understood as the management of processes aimed at ensuring the quality of their results. Along with those described, other approaches to assessing the quality of alternatives are possible. The advantage of the proposed method for choosing an alternative on the semantic network of conditionally-investigative statements is that it allows you to process information about non-deterministic fuzzy values of quality parameters and linguistic probabilities of these values, allows you to take into account the degree of reliability of various statements, makes it possible to calculate fuzzy values of indicators.

References

1. Deniskin Y, Deniskina A, Potsebneva I, Revunova S (2020) Application of complex information objects in industrial control systems. In: E3S web of conferences, vol 164, p 10042
2. Barabanov A, Serebryansky S (2020) Modeling the optimization of the stages of the product life cycle taking into account the uncertainty of the main parametric variables. In: Proceedings of the 13th international conference on the management of large-scale systems development, MLSD 2020, 9247777
3. Strelets DY, Serebryansky SA, Skurin MV (2019) A digital approach to lifecycle management of aviation products. In: Materials of the 12th International Conference of Large-scale systems development management, MLSD 2019, 8911020
4. Serebryansky SA, Barabanov AV. On the question of multicriteria optimization of aircraft components in order to optimize its life cycle. In: Advances in science, technology and engineering systems, pp 408–415
5. Korchagin A, Deniskina A, Fateeva I (2019) Lean and energy efficient production based on the internet of things (IOT) in the aviation industry, In: E3S web of conferences. https://doi.org/10.1051/e3sconf/201911002124
6. Bityukov YI, Deniskin YI (2018) Quality control of structures made of composite materials. Polym Sci Ser D 11(2):197–201. https://doi.org/10.1134/S1995421218020041

7. Desyatirikova EN, Khodar A, Rechinsky AV, Chernenkaya LV, Alkhayat I (2020) Performance analysis of available service broker algorithms in cloud analyst. In: Lecture notes in networks and systems, vol 95, pp 449–457
8. Desyatirikova EN, Myshovskaya LP, Lutin VI, Mager VE, Khripunov YV (2020) Algorithmization of evaluating solutions by the Neumann-Pearson criterion. In: Proceedings of the 23rd international conference on soft computing and measurements, SCM 2020, pp 178–181, 9198783
9. Yurin D, Deniskina A, Boytsov B, Karpovich M (2021) Quality 4.0. Time of revolutionary changes in the QMS. In: E3S web of conferences, vol 244, p 11010
10. Skachko G, Surkova E, Ermolaeva E, Potsebneva I (2021) Management decision-making based on the risk management model. In: E3S web of conferences, vol 244, p 11008
11. Barkalov SA, Lomakin MI, Mistrov LE, Morozov VP (2020) Management of an organization in difficult economic conditions based on an information system. Phys Conf Ser 1679(2):022086.
12. Smolyaninov A., Potsebneva I, Fateeva I, Singur K (2021) Software implementation of a virtual laboratory stand for distance learning. In: E3S web of conferences, vol 244. https://doi.org/10.1051/e3sconf/202124411009
13. Desyatirikova EN, Ivanov SA, Sergeeva SI, Zuev SA (2020) Creation of digital archives of project documentation. In: Materials of the international conference IEEE 2020 "Quality management, transport and information security, information technology", IT and QM and IS 2020, pp 357–359, 932294
14. Averina T, Barkalov S, Fedorova I, Poryadina V (2021) Influence of digital technologies on the company's business model. In: E3S web of conferences, vol 244, p 10002
15. Khodar A, Chernenkaya LV, Alhayat I, Fadhil Al-Afare KA, Desyatirikova EN (2020) Design model for improving task scheduling in cloud computing based on particle swarm optimization. In: Proceedings of the IEEE 2020 conference in Russian. Young researchers in electrical and electronics engineering, EIConRus 2020, pp 345–350, 9039501
16. Velikorosov VV, Rechinsky AV, Chernenkaya LV, Filin SA, Chernenky AV (2019) Digital economy as a tool for reducing uncertainty in strategic management decisions. In: Materials of the international conference, ACM
17. Barkalov S, Bekirova O, Sokolova O, Zenkova E (2021) New approaches to multicenter assessment of alternatives in decision making. In: E3S web of conferences, vol 244, p 11011
18. Mager VE, Belousov VE, Desyatirikova EN, Ivanov SA, Potsebneva IV (2017) Algorithm of information processing when constructing the structure of a self-adjusting technical system, optimal in terms of quality parameters. In: 2017 2nd IEEE international conference on control in technical systems, CTS 2017 8109503, pp 114–117. https://doi.org/10.1109/CTSYS.2017.8109503
19. Potsebneva I, Belousov V, Fateeva I, Lukinov V, Folomeeva T (2018) Algorithms for hierarchical mixing of experts' opinions on the problems of synthesis of information systems for urban development management. In: MATEC web of conferences, vol 170, p 01120. https://doi.org/10.1051/matecconf/201817001120
20. Chernenkaya LV, Desyatirikova EN, Belousov VE, Sergeeva SI, Slinkova NV (2017) Optimal planning of distributed control systems with active elements. In: Proceedings of the 2nd international conference IEEE 2017 on control in technical systems, CTS 2017, pp 37–40, 8109482

Algorithm for Scheduling Construction and Installation Work Using Bulk Construction

Andrey Smolyaninov, Antonina Deniskina, Natalia Mikhailovskaya, and Olga Malyutina

Abstract An algorithm for scheduling construction and installation works by the OBS method, taking into account the schedule of movement of teams between objects, is considered, characterized by the ability to efficiently calculate the values of minimizing the time of work by the parameters of organizational and technological reliability through the use of the branch and bound method.

Keywords Minimax algorithm · Schedule of construction and installation works · Schedule of movement of teams · Minimization of work time

1 Introduction

Currently, three-dimensional housing construction is widely used. Volumetric blocks are large structural elements of a volumetric form, in the hollow space of which a certain functional fragment of the building is enclosed. Volumetric blocks can contain a room, a staircase, or serve as a spatial boundary between the premises of the building and the external environment. The structures of volumetric blocks were developed in the USSR in the 1950s and, after being tested in experimental construction, were introduced into mass production in the late 1960s and early 1970s [1].

The block layout of buildings is the most industrial, since volumetric elements with full finishing and installed sanitary and electrical equipment are brought to the construction site, which allows maximum use of the possibilities of factory production due to the transfer of 75–80% of labor processes, while with large-panel buildings this figure does not exceed 50%. As a result, the number of assembled elements is reduced by 5–6 times, the construction process from volumetric blocks is reduced to the installation of elements, welding of their embedded parts, sealing of seams and

A. Smolyaninov (✉)
Voronezh State Technical University, 20-Letiya Oktyabrya Street, 84, 394006 Voronezh, Russia
e-mail: u00781@vgasu.vrn.ru

A. Deniskina · N. Mikhailovskaya · O. Malyutina
Moscow Aviation Institute, National Research University, Volokolamskoe shosse, 4, 125993 Moscow, Russia

A. Beskopylny and M. Shamtsyan (eds.), *XIV International Scientific Conference "INTERAGROMASH 2021"*, Lecture Notes in Networks and Systems 247,
https://doi.org/10.1007/978-3-030-80946-1_92

connection of plumbing communications. Carrying out of the volumetric block of communications and all embedded parts allows them to be mounted on the construction site without going inside the block that comes to it already locked with a key. As a result of these features, the method of volume-block construction allows 2–3 times to reduce the construction time of buildings in comparison with large-panel construction and reduce the total construction costs by 70%. Thus, the new technology of block construction of apartment buildings makes it possible to turn construction production into a highly mechanized process of assembling and erecting buildings at a construction site and transferring the main technological processes to factory conditions, which contributes to an increase in the quality of work. At the same time, the effectiveness of construction and installation works is often determined by their organization, and therefore it seems appropriate to develop an algorithm for scheduling construction and installation works used in the bulk-block construction method when performing work on several objects, taking into account the minimization of downtime associated with the movement of construction and assembly teams [2, 3].

2 Experimental

Let there be one brigade and n objects, where she works. Duration j-th work on the object is determined by the expression:

$$t_{ij} = \tau_{ij} + \tau_j,$$

where is τ_{ij}—time the team moves in the transition from service *i*-th object to *j*-th, τ_j—running time *j*-th work on site.

You want to determine the sequence of work performed at the facilities, at which the maximum time of work at each individual object is minimal [4].

The problem is reduced to a well-known task about the salesman in the minimax staging of it.

Mathematically, the task can be presented in the following form.

You want to define:

$$t = \min_{i;j=\overline{0;n},i\neq j}\max t_{ij}x_{ij} \tag{1}$$

Restrictions:

$$\sum_{i=0}^{n} x_{ij} = 1, \sum_{j=0}^{n} x_{ij} = 1, \quad i, j = \overline{0; n}, i \neq j, \tag{2}$$

$$U_i - U_j + (n+1)x_{ij} \leq n, \quad i, j = \overline{0, n}, i \neq j, \tag{3}$$

$$x_{ij} \in \{0, 1\}; \quad i, j = \overline{0, n}, i \neq j, \tag{4}$$

where is: $U_i \geq 0$; $U_j \geq 0$—additional variables,

t_{oj}—working time j-ом object, if he's the first $(j = \overline{1,n})$,

t_{io} —time the brigade moves to its original state if the work on the i-ом the object is the last $(i = \overline{1,n})$.

Restrictions (1)–(4) are typical for a salesman task. Condition (1) determines the minimax nature of the task in question [5, 6].

Practical tasks that boil down to (1) and (4) can have a very different physical meaning. For example, (1)–(4) the task of minimizing the flow of the system resource when controlling the parameters is reduced.

The branch and boundary method can be used to solve the problem (1) and (4). In this case, you need to determine how to evaluate the lower boundary and the conditions for building a tree of possible options.

Denote: U—lots of variables x_{ij},$S_l = \{x_{ij}; x_{ij} = 1\}$—a lot of fixed variables that are part of separate tree branches of possible;

$l = \overline{0, n}$—number of variables x_{ij}, fixed per unit and included in many S_l (at p and $l = 0$, $S_0 \neq \varphi$), $E_{si} = \{x_{ij}, x_{ij} = 0\}$—lots of variables x_{ij}, fixed at zero, the introduction of which in a set of S_l violates restrictions (7)–(9) or sub-optimal result, $G_{sl} = U \backslash (S_l \cup E_{sl})$—a lot of free variables, from which the choice is made at the next step of the solution to be included in the set S_l [7];

T_{sl}—bottom boundary for the target function of the tree branch of possible variants, made up of variables $x_{ij} \in S_l$, $T_{sl}(\overline{x}_{ij})$—bottom line for the target function $\overline{x}_{ij} = 0$.

Let's say that $l = 0$ and $S_0 \neq \varphi$. We will find in each line and each column of the original matrix $|||t_{ij}||s_o$ Minimal items $t_{ih}(i = \overline{0, n})$ and $t_{Rj}(j = \overline{0, n})$. Then the lower boundary of the solution can be determined by the expression:

$$T_{So} = \max_{ij}(t_{ih}; t_{Rj})h; \quad k = \overline{0, n}. \tag{5}$$

To increase the probability of cutting off unpromising branches into many S_0, it is advisable to introduce a variable x_{ih} or $x_{Rj}(i; j = \overline{0; n}, i \neq j)$, which results in a maximum increase in the lower boundary, provided that the variable $x_{ih} = 0$ $\overline{x}_{Rj} = 0$. Substituting in the matrix $||t_{ij}||s_o t_{ih} = \infty$, Determine the value $T_{So}(\overline{x}_{ih}) = t_{iq} = \min_{j=\overline{0,n}} t_{ij}$, which characterizes the lower boundary for the target function if the variable $x_{ih}(\overline{x}_{ih} = 0)$ set S_0, Not introduced.

Using the condition:

$$T_{So}(\overline{x}'_{ij}) = \max_{i;j}\left[T_{So}(\overline{x}_{ih}); T_{So}(\overline{x}_{Rj})\right] \tag{6}$$

identify the variable x'_{ij}, which is introduced into a set S_0. To exclude from G_{S_0}, variables $x_{ij} \in E_{S1}$ in the matrix $||t_{ij}||_{S_o}$ cross out the linei' and column j', corresponding to the variable x'_{ij}, and we're putting it up $t_{j'i'} = \infty$.

In all subsequent steps of the computing process, the lower limit for the target function $T_{Sl}(l = \overline{1, n})$ and choosing a variable to be included in a set S_l similar expressions (5), (6). Given that the conditions must be met to determine the lower boundary and select the variable x'_{ij} you can use the following expressions:

$$T_{S_l}(\overline{x}'_{ij}) = \max_{i,j-\overline{0,n}}\left[T_{S_l}(x'_{ih}); T_{S_l}(\overline{x}_{Rj}); T_{S_l}\right], i \neq p, \quad j \neq q, \; x_{pq} \in S_l, \; l = \overline{1, n}, \quad (7)$$

$$T_{S_l} = \max_{i,j=\overline{0,n}}\left(t_{ih}; t_{Rj}; T_{S_{l-1}}\right), \quad i \neq p, \; j \neq q, \; x_{pq} \in S_l, \; 1 = \overline{1, n.} \quad (8)$$

Consistently using conditions (7) and (8), we build a branch of a tree of possible options, which includes variables $x_{ij} \in S_l$. At $L = n$ get the first record-breaking solution $t^0 = T_{Sn}$, which is used to cut off unpromising branches by checking inequality:

$$T_{S_l}(x'_{ij}) < t^0, \quad 1 = n - 1, n - 2, \ldots, L, \ldots, 0. \quad (9)$$

Let's say that inequality (9) is not met $1 = n - 1, n - 2, \ldots, L + 1$ and is done when $l = L$. Since all branches of the options tree for which inequality (9) is not met are unpromising, then to cut them off in the matrix instead of an element that corresponds to the variable x'_{ij}, put into a set S_l, let's set it up ∞. Using conditions (8) and (9) define the lower boundary for the target function and select a new variable x'_{ij} to be included in the set S_l. When building a new branch of the tree of possible options, check the implementation of inequality:

$$T_{Sl} < t^0, \quad 1 = L, L + 1, \ldots, n. \quad (10)$$

If inequality (10) is met with $l = n$, we get a new record-breaking solution T_{Sn}, which is used in the future to check inequality (9) and (10). The computing process ends if the condition (9) is not performed $l = n - 1, n - 2, \ldots, 0$. In this case, the last record-breaking decision t^0 and the corresponding set of variables S_n is the best [8, 9].

3 Evaluation

Let's look at the solution (1)–(4) on a specific example, the raw data for which is presented in the Table 1. Using expression (10), we find the lower boundary of the target function $T_{S0} = 5$. In Table 1 t_{iq} and the string T_{pj} determine that the condition (11) corresponds to a variable $x'_{0.3}$, which is introduced into a set S_0. At the same

Table 1 The matrix 1

i	j						
	0	1	2	3	4	t_{ij}	t_{ij}
0	∞	7	8	5	9	5	7
1	4	∞	6	2	7	2	4
2	2	7	∞	4	5	2	4
3	1	5	4	∞	6	1	4
4	3	6	8	7	∞	3	6
t_{ij}	1	5	4	2	5	–	–
t_{ij}	2	6	6	4	6	–	–

time, the value of $T_{S0}(\overline{x'}_{0.3}) = 7$. Striking out in the Table 2 first line and fourth column and substituting $t_{3.0} = \infty$, we get a matrix $||t_{ij}||_{S1}$, which is brought to the Table 3. Looking at this table t_{Rj} and column t_{ih}, Find $T_{S1} = 5$ [10, 11].

Condition (10) is satisfied with variables $x_{1.0}, x_{4.0}, x_{3.1}, x_{2.4}$, for whom $T_{S1}(\overline{x}_{ij}) = 6$. Choose any of these variables, such as $x'_{1,0}$, which we include in the set S_1. All further calculations are done in the same way. In Tables 3 and 4 given matrix $||t_{ij}||_{S1}$ ($l = 2, 3$) and marked elements t'_{ij}, that correspond to variables x'_{ij}, introduced into the set S_l [12].

Table 2 The matrix 2

i	j					
	0	1	2	4	t_{ij}	t_{ij}
1	4	∞	6	7	4	6
2	2	7	∞	5	2	5
3	∞	5	4	6	4	5
4	3	6	8	∞	3	6
t_{ij}	2	5	4	5	5	–
t_{ij}	3	6	6	6	–	–

Table 3 The matrix 3

I	j				
	1	2	4	t_{ij}	t_{ij}
2	7	∞	5	5	7
3	∞	4	6	4	6
4	6	8	∞	6	8
t_{ij}	6	4	5	–	–
t_{ij}	7	8	6	–	–

Table 4 The matrix 4

I	j			
	2	4	t_{ij}	t_{ij}
2	∞	5	5	∞
3	4	∞	4	∞
t_{ij}	4	5	–	–
t_{ij}	∞	∞	–	–

On rice. 1 presented a tree of possible variants, near each top of which is indicated an estimate of the lower boundary of the target function. The process of forming a set S_l ends with $l = 5$. First record-breaking decision $t^0 = 6$ variables $x_{0.3} = x_{3.2} = x_{2.4} = x_{4.1} = x_{1.0} = 1$, which are part of a plethora of S_5 [13, 14].

Because the condition (10) is performed $l = 4, 3, 7, \ldots, 0$, the resulting solution is the best solution (Fig. 1).

The evaluation of the effectiveness of the method of determining the lower boundary of the solution was cited through computational experiments [15].

As the original data of the matrix $||t_{ij}||$ random numbers were taken, evenly distributed in the range of 1 to 100. Time to solve a problem n = 20, 30, 40, 50 accordingly, it is 0.5; 2.5; 7.5; 27 min. Thus, the proposed way of assessing the lower boundary of the solution is quite accurate and allows to effectively solve the minimax problem of scheduling construction and installation work [16, 17].

Example 1. Let the number of objects equal to 4. The original data on the time (in minutes) of moving between objects are in Table 5.

Fig. 1 A tree of possible solutions

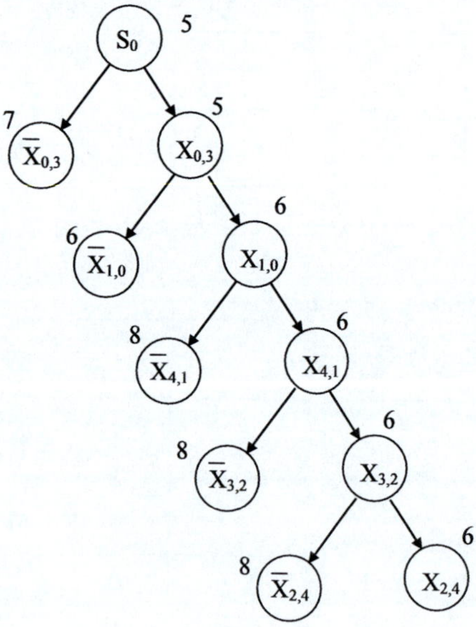

Table 5 Time baseline data

An object	1	2	3	4
1	M	50	110	90
2	100	M	80	70
3	70	140	M	80
4	120	60	150	M

The travel time from an object to the same object is designated by the letter M.

1. Find the minimum value in each row (di) and write it out in a separate column:

An object	1	2	3	4	di
1	M	50	110	90	50
2	100	M	80	70	70
3	70	140	M	80	70
4	120	60	150	M	60

We edit the lines—from each item in the line we subtract the corresponding value of the found minimum (di).

An object	1	2	3	4	di
1	M	0	60	40	50
2	30	M	10	0	70
3	0	70	M	10	70
4	60	0	90	M	60

2. Next, we find the minimum value in each column (dj).
3. Next, we find the minimum value in each column (dj). We enter these minimums on a separate line.

An object	1	2	3	4	di
1	M	0	60	40	50
2	30	M	10	0	70
3	0	70	M	10	70
4	60	0	90	M	60
dj	0	0	10	0	–

4. We edit columns—subtract from each element of the matrix corresponding to it dj.

An object	1	2	3	4	di
1	M	0	50	40	50
2	30	M	0	0	70
3	0	70	M	10	70
4	60	0	80	M	60
dj	0	0	10	0	–

5. For each zero cell of the resulting converted matrix we find an "assessment." It will be the sum of the minimum item by line and the minimum element on the column in which this zero cell is placed. It itself is not taken into account. The previously found di and dj are not counted [18, 19].

An object	1	2	3	4
1	M	0(40)	50	40
2	30	M	0(50)	0(10)
3	0(40)	70	M	10
4	60	0(60)	80	M

6. Choose the zero cell with the highest score. We're replacing it with A M. One of the segments of the path was found: from the 4th to the 2nd [20].

That line and that column where two "M" were formed completely cross out.

An object	1	2	3	4
1	M	0(40)	50	40
2	30	M	0(50)	0(10)
3	0(40)	70	M	10
4	60	0(60)	80	M

Back to p. 1. and again look for lows on rows and columns, carry out their reduction, consider estimates of zero cells.

An object	1	3	4
1	M	50	40
2	30	0	0
3	0	M	10

Finding all the stretches of the way, we get the optimal route 4 → 2 → 3 → 1 → 4. Total travel time: L = 300 min (5 h).

Thus, the algorithm of scheduling construction and installation work by the OBS method, taking into account the schedule of the movement of brigades between objects, differs from the ability to effectively calculate the values of minimizing the time of work on the parameters of organizational and technological reliability by using the method of branches and boundaries.

4 Conclusions

Due to the fact that the effectiveness of construction and installation work is determined by their organization, the article presents an algorithm for scheduling construction and installation work for the construction of a building using the volume-block construction method, which is characterized by the ability to efficiently calculate the values of minimizing the time of work according to the parameters of organizational and technological reliability due to using the branch and bound method. This algorithm makes it possible to organize the execution of work at several sites, taking into account the minimization of downtime associated with the movement of construction and installation teams.

References

1. Pocebneva I, Belousov V, Fateeva I, Lukinov V, Folomeeva T (2018) Models of resource planning during formation of calendar construction plans for erection of high-rise buildings. E3S Web Conf 33:03032. https://doi.org/10.1051/e3sconf/20183303032
2. Mager VE, Belousov VE, Desyatirikova EN, Ivanov SA, Pocebneva IV (2017) Information processing algorithm at creation of optimum structure of the self-adjusted technical system in quality parameters 2017. In: Proceedings of 2017 IEEE 2nd international conference on control in technical systems (CTS 2017), vol 8109503. pp 114–117. https://doi.org/10.1109/CTSYS.2017.8109503
3. Zherlykina M, Smorodina E, Kalabukhov G, Smolyaninov A (2020) Modeling of air handling processes in systems of conditioning premises. E3S Web Conf 175:11017. https://doi.org/10.1051/e3sconf/202017511017
4. Makovetskaya E, Deniskina A, Krylov E, Urumova F (2019) Organizational optimization of construction processes by virtue of robotization. E3S Web Conf 91:02036. https://doi.org/10.1051/e3sconf/20199102036
5. Deniskin Y, Deniskina A, Pocebneva I, Revunova S (2020) Application of complex information objects in industry management systems. E3S Web Conf 164:10042. https://doi.org/10.1051/e3sconf/202016410042
6. Chernenkaya LV, Desyatirikova EN, Fedosova SP, Ievleva AA, Vertakova AY (2018) Proceedings of the 2018 IEEE conference of Russian young researchers in electrical and electronic engineering. ElConRus. https://doi.org/10.1109/EIConRus.2018.8317321
7. Algazinov EK, Garshina VV, Stepantsov VA, Desyatirikova EN (2019) Experimental study reliability and functional stability of the social graph. J Phys: Conf Ser 1202(1):12009. https://doi.org/10.1088/1742-6596/1202/1/012009
8. Belyaeva S, Makeeva T, Chugunov A, Andreeva P (2018) The substantiation of methodical instrumentation to increase the tempo of high-rise construction in region. E3S Web Conf 33:03023. https://doi.org/10.1051/e3sconf/20183303023
9. Uvarova S, Belyaeva S, Myshovskaya L (2017) Development of the quality-efficiency model of construction investment projects for road construction on the basis of sustainable growth. IOP Conf Ser: Earth Environ Sci 90(1):012169. https://doi.org/10.1088/1755-1315/90/1/012169
10. Uvarova S, Gumba K, Kiseleva E, Sonin Y (2020) Digitalization of design and construction of agricultural facilities as a basis for ensuring their energy efficiency. E3S Web Conf 175:11004. https://doi.org/10.1051/e3sconf/202017511004
11. Dobrosotskikh MG, Mishchenko VY, Mishchenko AV (2020) Dynamic correction of the construction projects scheduling implementation under stochastic influences. IOP Conf Ser: Mater Sci Eng 972(1):012050. https://doi.org/10.1088/1757-899X/972/1/012050

12. Dobrosotskikh MG, Mishchenko VY, Preobrazhenskii MA (2019) Improvement the scheduling based on the reducing of the dimensionality of the system. IOP Conf Ser: Mater Sci Eng 481(1):012029. https://doi.org/10.1088/1757-899X/481/1/012029
13. Mishchenko VY, Dobrosotskikh MG (2016) NP solvable task of scheduling of construction, reconstruction and repair of objects. Izvestiya Vysshikh Uchebnykh Zavedenii, Seriya Teknologiya Tekstil'noi Promyshlennosti, 6 Jan 2016, pp 13–20
14. Emelyanov DI, Ponyavina NA, Chesnokova EA (2017) The solution of the problem of planning of difficult productions at the enterprise on the basis of methods of network planning. Izvestiya Vysshikh Uchebnykh Zavedenii, Seriya Teknologiya Tekstil'noi Promyshlennosti 371(5):28–32
15. Averina TA, Barkalov SA, Smolyanova OL (2019) Resource allocation models in a crisis management programme. In: Proceedings of 2019 12th international conference & management of large-scale system development & MLSD, 8911048. https://doi.org/10.1109/MLSD.2019.8911048
16. Barkalov SA, Kurochka PN (2017) Model for determining the term of execution of sub-conflicting works. In: Proceedings of 2017 10th international conference management of large-scale system development (MLSD 2017), 8109598. https://doi.org/10.1109/MLSD.2017.8109598
17. Barkalov SA, Kurochka PN (2017) Modeling of production activity of a construction enterprise under analyzed law of distribution. In: Proceedings of 2017 10th international conference management of large-scale system development (MLSD 2017), 8109599. https://doi.org/10.1109/MLSD.2017.8109599
18. Barkalov S, Kurochka P, Nasonova T (2018) Optimal placement of maintenance facilities. MATEC Web Conf 170:01124. https://doi.org/10.1051/matecconf/201817001124
19. Barkalov SA, Kurochka PN, Pinaeva MA (2019) Model of formation plans for the urban areas development. IOP Conf Ser: Mater Sci Eng 537(6):062035. https://doi.org/10.1088/1757-899X/537/6/062035
20. Skachko G, Surkova E, Ermolaeva E, Pocebneva I (2021) Adoption of management decisions on the basis of the risk management model. E3S Web Conf 244:11008. https://doi.org/10.1051/e3sconf/202124411008

Formation of Nanosized Films of Chromium Silicides on Silicon Surface

Makhmudhodzha Isaev, Abdumalik Gaibov, Abdigani Eshkulov, and Pulat Saidachmetov

Abstract Chromium in silicon has an increased reactivity: no metallic chromium phases are found on the surface of the doped crystal, but monosilicides, disilicides, and higher chromium silicides are formed in the case of diffusion alloying of silicon from the gas phase. The amorphous phase was observed in a near-surface layer with a thickness of 3 nm–10 μm. At a depth of 10–30 μm, chromium mono and disilicides were mainly observed. No silicide inclusions were found at depths greater than 30 μm. An analysis of the results obtained shows that during the formation of silicides, silicides with rich chromium atoms are first formed, and then chromium monosilicides, in the case of diffusion alloying of silicon from the gas phase. And in the case of silicon sprayed with chromium, silicide of chromium enriched in the latter is formed at the interface between silicon and chromium, after thermal annealing. The spectrum contains oxygen lines during thermal annealing, which is a Cr_2O_3 compound. Analysis of the results showed that the surface of the deposited chromium in the course of thermal annealing is covered with a thin layer of Cr_2O_3, which plays the role of a diffusion barrier for oxygen from the annealing atmosphere. The theory of nucleation and electron microscopic observations make it possible to distinguish the sequence of the following stages of film growth: Formation of nuclei and island structure, coalescence of silicide islands, formation of channels, formation of a continuous film.

Keywords Silicide · Film · Alloying · Annealing · Quenching · Nucleation · Cluster

M. Isaev · A. Gaibov · A. Eshkulov (✉)
Tashkent State Technical University Named After Islam Karimov, University St. 2A, Tashkent 100095, Republic of Uzbekistan

P. Saidachmetov
South Kazakhstan State University Named After M. Auezov, Shymkent, Kazakhstan

A. Beskopylny and M. Shamtsyan (eds.), *XIV International Scientific Conference "INTERAGROMASH 2021"*, Lecture Notes in Networks and Systems 247,
https://doi.org/10.1007/978-3-030-80946-1_93

1 Introduction

The properties of the thin film may be very different from those of the massive material, especially if the thickness of the films is very small. These differences are due to the specificity of the film structure, which is due to the processes occurring during film formation.

Currently, nanoscale structures created on the surface of semiconductors and dielectric films have great prospects in the development of unique devices for micro-, opto-, and nanoelectronics. Due to the high thermal stability and resistance to chemical treatments, the technological compatibility of silicides with silicon and the possibility of choosing semiconductors with a band gap width of 0.07–2 eV, it makes them promising for the manufacture of almost all semiconductor devices.

At the same time, nanofilms of $Si/Mn_{1,75}$, $Si/CoSi_2$, Si/LIF_2, Si/CaF_2, type are of particular interest, which can be used in instrument structures for forming ohmic and barrier contacts, in optoelectronic devices, memory circuits, transistors with permeable and metal bases [1, 2]. Calculations show that such transistors can have a boundary gain frequency greater than 100 GHz. Due to their unique properties, metal silicides are now the subject of study by the largest world research centers [3–8].

There are the following conventional and relatively novel methods for producing thin slicidal films:

- application of metal to silicon by evaporation, spraying or electrolytic deposition and subsequent heat treatment,
- chemical deposition of metal from the gas phase, from solutions,
- driving metal atoms from the metal film deposited on the surface of Si with Si^+ ions or inert gases,
- molecular beam epitaxy,
- ion implantation, ion synthesis, etc.

Using scanning tunnel microscopy, the main growth patterns of ultrafine epitaxial films $CoSi_2$ (111) obtained by sequential deposition of Co and Si atoms in stoichiometric ratio on the surface of Co–Si (111) at room temperature and subsequent annealing at 600–700 °C were studied. The formation of solid epitaxial films $CoSi_2$ which include 3–4 triple layers of Si–Co–Si, occurs under the condition of accurate control of deposition. At a temperature above 700 °C, the multilevel film $CoSi_2$ with point punctures grows due to vertical growth due to the difference in free energies of the surfaces $CoSi_2$ and Si (111) [3].

At the same time, the use of metal silicides in electronics is so far limited by experiments on their use in photovoltaic, thermoelectric converters, and SBIS. A deterrent is the lack of knowledge of their properties. In this regard, the purpose of our research was to study the formation of films of chromium silicides on the surface of silicon.

2 Experiment Procedure

Chromium has a large diffusion coefficient in silicon, so we used the diffusion doping method. This method has a number of other advantages:

- Relative simplicity of technology;
- Possibility of investigating the effect of annealing temperature on the initial crystal parameters;
- The possibility of controlling the concentration of electrically active chromium atoms by changing the temperature.

For the doping of chromium silicon by us, ingots of silicon monocrystal of both p-type grade SDB-10 and n-type grade ESPh-20 grown by the Chokhralsky method were used. Their initial parameters are as follows: for n-type silicon, the resistivity $\rho = 20\ \Omega$ cm, electron mobility $\mu\rho = 600$ cm_2/V s, hole concentration $p = 1.5 \times 10^{15}$ cm^{-3}, oxygen concentration 1.1017 cm^{-3}.

Samples in the form of a parallelepiped with a size of (2–3) × (3–5) × (10–30) mm^3 were cut from ingots of silicon monocrystal with a diamond disk. Samples ground with use of micropowder of carbide of M-14 silicon. In order to remove the surface layer disturbed during grinding, the samples were degreased in toluene at a temperature of 40–50 °C and subjected to chemical etching in $1HF:5HNO_3$ solution for 1–2 min, washed in deionized water and dried at a temperature of not more than 1000 °C. The silicon samples were placed in an amount of 3 pcs in quartz vials previously washed in solution HNO_3:3HCl and boiled in distilled water.

The mass of the alloying atoms of metal was determined from Mendeleyev-Klapeyrona's equation of $pV = mRT \cdot \mu^{-1}$. From there is $m = pV\mu\ R^{-1} \cdot T^{-1}$, where metal p-vapor pressure in an ampoule which is equal to 1×10^{-2}–2×10^{-5} mm Hg. at T = 1000–1250 °C. The weight of the metal thus determined was 4 mg. The mass of the metal is not limited upwards, i.e. at least 4 mg is used. A purity metal powder of 99.99% was placed in an ampoule in an amount of 4–5 mg. Ampoules with samples and a diffuser were pumped to a vacuum of ~10^{-3} wave 10^{-4} mm Hg. Art. and sealed [9]. The ampoules were then placed in a horizontal furnace and annealed at a temperature of 1050–1070 °C for 20–120 min. Temperature fluctuations in the working zone of the furnace did not exceed ± 3 °C. After annealing, the samples were quenched by cooling at a rate of 100–150 K/s by dropping the vials into water and held at a crystal temperature of T = 15–20 °C. The water vessel was rotated at 30 turns/mines.

After opening the ampoules, the surface of the samples had p-type conductivity. In order to assess the effect of heat treatment on electrophysical properties, control samples without a diffuser were annealed under the same conditions (temperature, time, cooling rate).

To remove the conductivity distribution profile over the thickness of the samples, we used a four-probe resistance measurement method [10]. The probes were made of tungsten sharpened electrolytically. The DC source was the B5-48 device, the ampervoltomometer Ф-30 was used as a current device, and B7-30 was used to measure the potential difference on internal probes.

A small current meter was passed through the external probes from the current generator, and the internal probes were used to measure the corresponding voltage drop. Based on the measured values of potential and current difference, the surface resistance of the sample was determined by the formula

$$R_S = 4.53\frac{V}{I} \tag{1}$$

where, RS—specific surface resistance (Ω/cm^2), V—voltage drop at internal probes (V), I—current flowing through external probes (A).

The found values of specific surface resistances and two layers differing by thickness allow calculating the surface and volume resistivity of the removed layer:

$$\overline{R_S} = \frac{R_{S_1} \cdot R_{S_2}}{R_{S_1} + R_{S_2}} \quad \overline{\rho} = \overline{R_S} \cdot \Delta t \tag{2}$$

$\overline{\rho}$ where is the average volume resistivity, Δt is the thickness of the layer to be removed.

The crystal thickness was controlled by a micrometer with an accuracy of ± 1 μm.

Measurements showed that after doping in a single crystal, a wide spectrum of conductivity is obtained, varying by ten orders of magnitude as it moves from surface to volume in the case of samples of the original silicon type p and by six orders of magnitude in the case of samples of the original silicon type n. When measuring the specific electrical conductivity and mobility of current carriers, a layer-by-layer chemical release of the doped sample in a 1HF:8HNO_3 solution with fixed layer removal was carried out. After removal of each layer, conductivity by thickness of samples was measured by a four-probe method, while correction was made for the possibility of non-uniform distribution of impurity over the surface.

Measurements carried out at the Hall installation made it possible to determine the near-surface concentration of current carriers and mobility. Measurement of superficial resistance after the removed layer was shown that it changes from 10–25 Ω/cm^2 up to 120–130 Ω/cm^2 in all alloyed crystals in near-surface area.

The low resistivity of Si <Cr> is due to the presence of two d-electrons in the outer shell. Miller et al. [11], Iwati et al. [12] suggest the existence of d-s-p-hybridization involving d-electrons Cr and Si and α electrons Si.

3 Results of Experiments and Their Discussion

A typical dependence of electrical conductivity on the thickness of the layer to be removed at T = 300 K is shown in Fig. 1 for Si <Cr> samples. As can be seen

Fig. 1 The electrical conductivity depends on the thickness of the layer to be removed

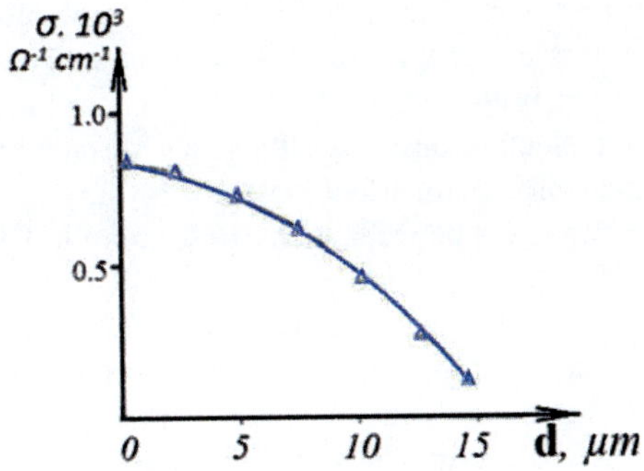

from the figure, the electrical conductivity decreases monotonically with increasing thickness of the layer to be removed. Changes in electrical conductivity at low and high temperatures are small. The relative decrease in electrical conductivity with an increase in the thickness of the layer to be removed is mainly due to a decrease in the concentration of current carriers in the sample. Figure 2 shows the dependence of the mobility of current carriers on the thickness of the removed layer for silicon doped with chromium. As can be seen from the figure, with an increase in the thickness of the layer to be removed for all crystals, a rapid increase in the mobility of current carriers is observed. This is due to a decrease in the concentration of holes at an increase in d, leading to a decrease in scattering during collisions with crystal lattice atoms and impurity atoms.

To explain the observed phenomena, it was necessary to study the structural and phase composition of the near-surface layer of the sample. A raster electron microscope (REM-200) made a visual observation of the image of the surface structure. At that, thin electron beam with specified parameters is formed and controlled by its

Fig. 2 Dependence of current carriers mobility on thickness of layer to be removed

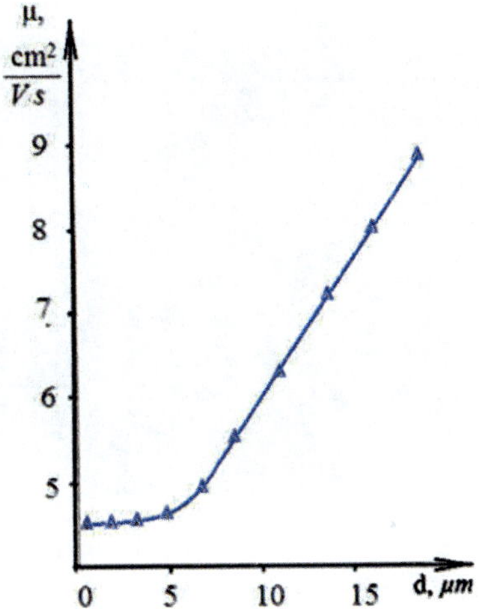

movement. The electron beam, moving along a sample of diffusion-doped silicon, forms a raster on the display screen. The signal from the sample modulates the beam brightness in the electron beam tube. This beam scans synchronously with the electron beam, resulting in a surface image on the screen. Studies have shown that block structures exist on the surface of the sample and there are interfacial boundaries between them. Resolution in REM reached ~10 nm. An estimate of the cross-sectional size between the blocks showed that it was of the order of 100 nm.

Samples were studied on infrared (IR) microscopes of type IM-1 and IM-4. It has been found that continuous structures opaque to IR rays are formed on the surface of the doped sample. When removing a layer with a thickness of ~3 μm from the surface, a grid structure is observed.

For the purpose of clarification of the chemical composition of observed block structures, qualitative and quantitative analyses on the Cameca MS-46 microanalyzer were carried out. The analysis was carried out at an acceleration voltage of 20 kV, current through a sample of 20 nA. An electron probe of a microassay with a focused diameter of up to 3 μm interacts with the sample and the element composition of Si <Cr> was studied from characteristic X-rays (Fig. 3).

The presence of Au, Ag, Ni, Co, Mn, Fe elements was not detected within the sensitivity of the installation. Against the background of the signal from silicon, the presence of chromium clusters was recorded, the dimensions of which reached 30–40 μm. With successive removal from the surface of the layers, the signal intensity due to impurity clusters was reduced and individual grains of impurity inclusions were detected. When more than 25–30 μm is removed from the surface, the signal caused by impurity atoms disappeared against the background of the signal from silicon. These studies showed that impurity clusters are the second phase. The chromium

Fig. 3 Microphotography of the surface of the doped crystal. (1400× magnification), i.e. alternating transparent and opaque regions for IR rays

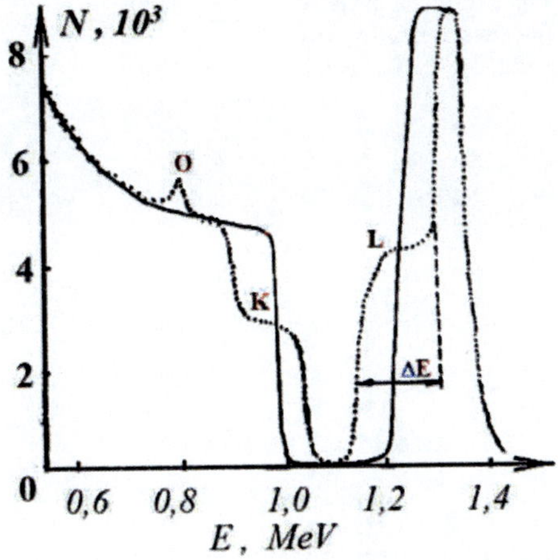

Fig. 4 Energy spectrum from surface Si <Cr>, Solid line—after annealing, dotted line—before annealing

content varies not only between but also within the impurity clusters. The dimensions of the clusters depend on the removal from the outer surface of the sample.

To clarify the kinetics of the growth of chromium silicide, a layer of chromium with a thickness of 3000 A was sprayed on n-type silicon ($n = 10^{16}$ cm^{-3}). At the same time, the silicon crystal was heated to a temperature of 200 °C (Fig. 4). The energy spectrum from the silicon surface is given, having a chromium layer on the surface in cases before and after annealing at a temperature of 450 °C for 60 min. The spectrum taken after annealing shows that the step L corresponding to the chromium protrudes in the reverse direction, and the step K corresponding to silicon protrudes in the forward direction. This shows that a mixture of chromium and silicon occurs at 450 °C. From the average height of these protrusions, the Cr-Si ratio in the mixed layer, which was 2:1, is determined.

This corresponds to chromium $CrSi_2$ disilicide. In this figure, peak O corresponds to the presence of oxygen observed after annealing the crystal.

Examination of these crystals in an electron microscope shows that the surface of the sample before annealing is smooth, and after annealing is not smooth (step). This phenomenon is intensified at the borders of chromium with disilicide of chromium and silicon with $CrSi_2$ (Fig. 5).

The growth rate of the silicidal phase was determined by the width of the ΔE plate at the peak of chromium, which is directly proportional to the thickness of the silicide layer [Figure at annealing temperature of crystal with chromium silicide above 450 °C, silicide growth has linear dependence on annealing time (Fig. 6)].

Silicide thickness is determined by formula:

$$W = \left[B \exp(-E_a/kT)\right]t, \tag{3}$$

where, B is a constant, E_a is the average activation energy of the growth process, the Boltzmann k-constant, T is the absolute temperature, t is the annealing time. The

Fig. 5 Silicide layer thickness versus energy width ΔE for $CrSi_2$

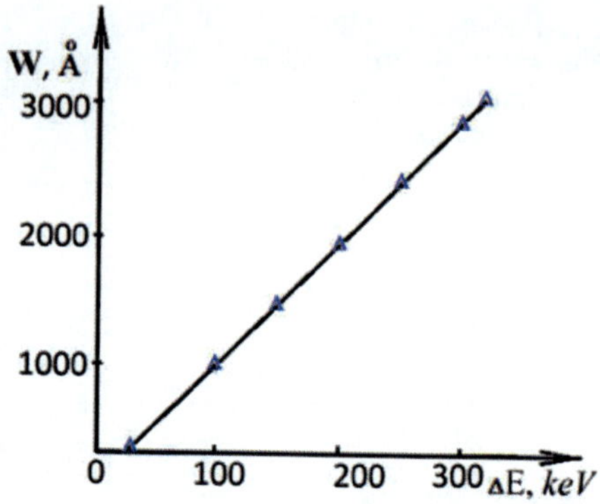

Fig. 6 Silicide layer thickness versus annealing time at 450 °C for $CrSi_2$

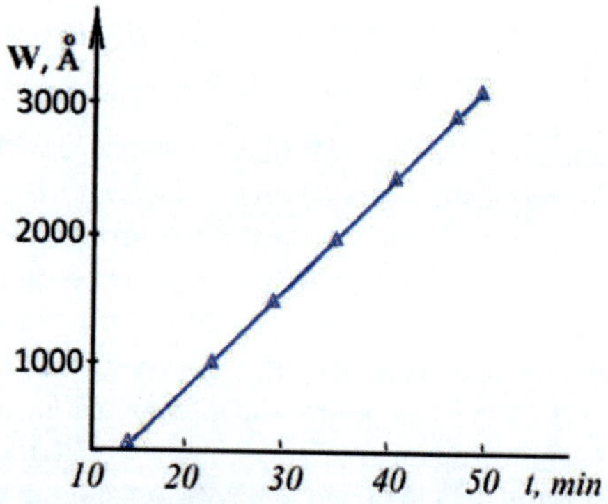

value of the diffusion coefficient D can be calculated from the slope of the time lines W. The activation energy of the phase formation process can be determined from the dependence of the diffusion coefficient on the inverse temperature, i.e., according to Arrhenius graphs, which was $E_a = (1.5 \pm 0.1)$ eV.

The growth kinetics of the silicide layer can be studied by identifying the predominant particles in diffusion in this process. Throughout the temperature range studied, silicon is a rapidly diffusing particle during the growth of chromium disilicide.

Figure 7 shows the dependence of the size of the silicidal inclusion at the three points of the crystal on the thickness of the layer to be removed from the crystal surface. From the figure, it can be seen that the dimensions of silicidal inclusions are linearly reduced deep into the crystal.

Figure 8 shows the relationship between the concentration of silicidal inclusions and the distance (from the surface deep into the crystal). Figure 8 shows that on a surface with a thickness of ~3 nm, silicidal inclusions have saturation, that is, a continuous film is formed. At depths from 3 nm to 30 μm, the concentration of inclusions decreases smoothly.

Images of the surface of the sample after layer-by-layer removal were obtained using an electron microscope REM-200, where the kinetics of reducing the size of

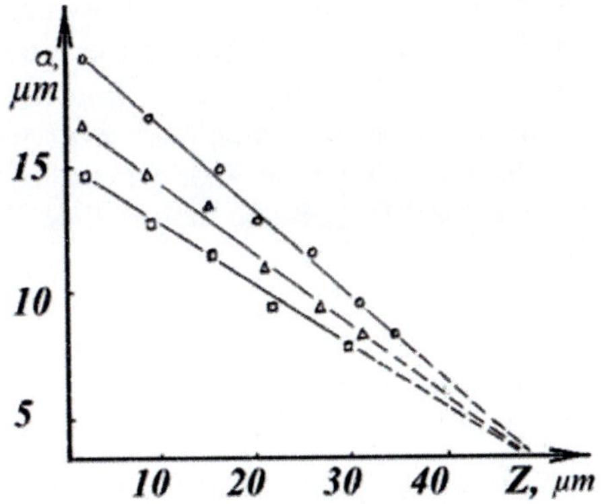

Fig. 7 The dependence of silicidal inclusions at three points of the crystal on the thickness of the layer to be removed

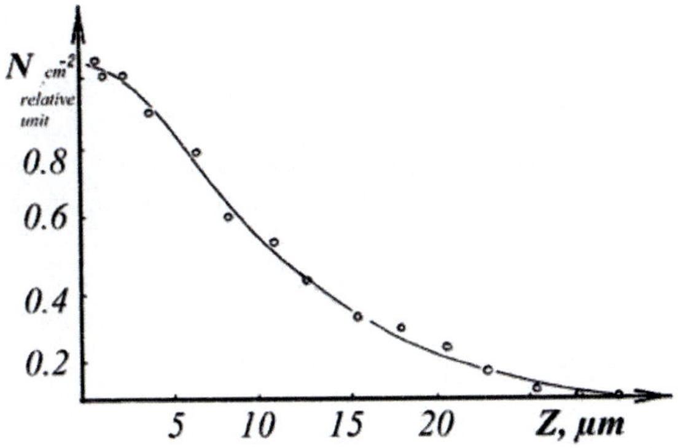

Fig. 8 Concentration of silicidal inclusions versus distance

the metal silicide in the high-resistance matrix of doped silicon are quite clearly visible. IR microscopy studies showed that prior to removal of the near-surface layer, samples were completely opaque, as the layers were removed, formed silicidal films [13–15] were detected, and after removal of the layer 30–40 μm, the crystal became completely transparent. Thus, it can be considered that the current transfer along the layer is carried out by passing the carriers through relatively high-ohmic areas (silicon compensated by chromium atoms) and formed thin nanoscale films of chromium silicides.

It has been found that the concentration of silicidal inclusions in the near-surface region of the crystal depends on the type of substrate, the amount of diffuser, the diffusion temperature and the quenching rate. The obtained experimental results show that chromium in silicon has an increased reactivity: metal phases of chromium are not detected on the surface of the doped crystal, but mono-, di- and higher silicides of chromium are formed in the case of diffusion doping of silicon from the gas phase. An amorphous phase was observed in a near-surface layer with a thickness of 3 nm–10 mm. At a depth of 10–30 μm, mainly mono and chromium disilicides were observed. No silicidal inclusions were detected at a depth of more than 30 μm.

Analysis of the obtained results shows that when silicides are formed, silicides are first formed with rich chromium atoms, and then chromium monosilicides, in the case of diffusion doping of silicon from the gas phase. And in the case of chromium-sputtered silicon, chromium silicide enriched by the latter is formed on the border of silicon and chromium, after thermal annealing.

It should be noted that oxygen lines are observed in the spectrum during the heat annealing process, which is a compound of Cr_2O_3. Analysis of the results showed that the surface of the applied chromium during temperature annealing is covered with a thin layer of Cr_2O_3. which acts as a diffusion barrier for oxygen from the annealing atmosphere. Nucleation theory and electron microscopic observations make it possible to distinguish the sequence of the following stages of film growth:

- Formation of embryos and island structure;
- Coalescence of silicidal islets;
- Formation of channels;
- Forming a continuous film.

The absence of formation of chromium silicides at a temperature below ~400 °C is explained by the increasing difficulty in removing silicon atoms from its covalent bond. At temperatures above 400 °C, phonons can provide the necessary energy to release the silicon atom. Therefore, at low temperatures (less than 400 °C), chromium silicides are not formed.

4 Conclusion

Thus, we have established that when diffusion doping silicon with chromium atoms on the crystal surface, a silicidal film with a thickness of 2–3 nm is formed, at a depth of more than 30–40 μm, a high-ohm region is formed (in the case of the original p-type silicon), and an intermediate amorphous layer with a thickness of~30 mm is formed between the film and the high-ohm region.

The activation energy of the silicidal phase formation process was determined to be 1.5 eV.

References

1. Murarka S (1986) Silicides for CBIS. M: World
2. Umirzakov BE, Ashurov RK, Donaev SB (2019) Morphology and electronic properties of Si nanoscale structures created on the surface CaF_2. J Tech Phys 89(vyp2):264–267
3. Alekseev AA, Olyanich DA, Utas TV, Kotlyar VG, Zotov AV, Saranin AA (2015) STM-observation of ultra-thin epitaxial films $CoSi_2$ (111), painted at high temperature. J Tech Phys 85(10):94–100

4. Varnakov SN, Gomoyunova MV, Grebenyuk GS, Zabluda VN, Ovchinikov SG, Pronin II (2014) Initial growth stages of manganese films on the Si (100) 2x1 surface. Phys Solid State 56(2):380–384
5. Filonov AB, Ivanenko LI, Migas DB, Shaposhnikov VL, Krivosheeva AV, Krivosheeva AE, Borisenko VE (2004) Semiconductor silicides: properties and prospects of application. Rep BCID 3:168–177
6. Galkin NG, Goroshko DL, Chusovitin EA, Galkin KN, Dotsenko SA (2013) Silicon-silicide quazi-zero dimensional heterostructures for silicon-based photonics, opto- and thermoelectronics. Phys Stat Sol (c) 10(12):1670–1676
7. Wang D, Zou Q (2009) Formation of manganese silicide nanowires on Si (111) surfaces by the reactive epitaxy method. Nanotechnol 20(27):5607–5612
8. Li B, Liu J (2009) $CoSi_2$-coated Si nanocrystal memory. J Appl phys 105, 084905
9. Kassymov AK, Mukhitdinov DZ, Isaev MS (1994) Method of making semiconductor photovoltaic device. Patent №. 2014672. M, pp 1–10
10. Zi S (1984) Fizik semiconductor devices. M World
11. Miller JN, Shwartz SA, Lindau I, Spicer WE, Debilishelis B, Abbati I, Braincovich LJ (1980) Vac Sci Thechnol 17:920
12. Iwati M, Okuno K, Kamel S, Ito T, Hiraki A (1980) J Electrochem Soc 127:1542–1546
13. Kamilov TS, Kabilov DK, Samiyev IS, Khushnutdinova KK (2005) On the possibility developing terboelectric sensors based on multielement higher manganese silicide film structures. Tech Phys 50(10):1370–1373
14. Ivanova LD (2017) Thermoelectric materials for different temperature levels. Phys Semicond Technol 51(7):948–951
15. Varnakov SN, Gomoyunova MV, Grebenyuk GS, Zabluda VN, Ovchinnikov SG, Pronin II (2014) Initial stages of growth of surface manganese films Si <100> −2x1. Solid State Phys 56(2):375–379

Modern Gondola with Lightweight Body

Farida Galimova, Yaxyo Khurmatov, Mirzo Abdulloev, Bobur Jumabekov, Doston Sultonaliev, and Dilorom Ergeshova

Abstract The article describes the need for the development of international ways, the renewal of transport routes and the improvement of the transport infrastructure of Uzbekistan. The introduction of modern energy-saving and resource-saving technologies at the enterprises of Uzbekistan will ensure the economical usage of raw materials, materials and energy resources in the construction of rolling stock units. To increase the efficiency rolling stock usage, it is necessary to make changes in their design or, to use more advanced steel grades in their construction. The gondola car was chosen as the object of research. By mastering the introduction of high-strength steels of grades 16G2AF and 18G2AFps, it will be possible to create cars that will have a lighter tare weight and increased carrying capacity. In the course of the study, the technology of digital prototyping was used in the environment of modern engineering programs in two stages. Numerical experiments were carried out to optimize the thicknesses of the gondola car body bearing parts and the gondola car with an optimized design was tested for strength. In further calculations, the safety margin of the most loaded parts with the largest flow of failures in operation was increased. In the article, on the basis of the research carried out, specific technical proposals have been developed for the use of high-strength steels in the construction of gondolas developed and manufactured by the Subsidiary "Casting and Mechanical Plant".

Keywords Rolling stock · Gondola car · Body · High-strength steel · Reliability · Stress · Enterprise · Optimization · Engineering program

1 Introduction

Uzbekistan attaches great importance to the development of international ways going through the territory of the republic, to the renewal of transport routes, as well as the improvement of the transport infrastructure of Uzbekistan [1–3]. For an integrated solution of the issues of increasing the international cargo transportation volume and

F. Galimova · Y. Khurmatov · M. Abdulloev · B. Jumabekov · D. Sultonaliev (✉) · D. Ergeshova
Department "Wagons and Wagon Facilities", Tashkent State Transport University, Adilkhodjaev St. 1, Tashkent, Uzbekistan 100067

A. Beskopylny and M. Shamtsyan (eds.), *XIV International Scientific Conference "INTERAGROMASH 2021"*, Lecture Notes in Networks and Systems 247,
https://doi.org/10.1007/978-3-030-80946-1_94

attracting transit cargo traffic through the territory of the republic, active work is constantly being carried out to improve the quality of services provided by transport and logistics companies of the republic [4]. In this relation, one of the priorities of the republic's economic development has become the introduction of modern energy-saving and resource-saving technologies at the enterprises of Uzbekistan, ensuring the economical use of raw materials, materials and energy resources in the construction of rolling stock units (cars and locomotives) [5–7].

2 Relevance of the Research

There are several main ways to improve the design of newly built the same type rolling stock of railways for increasing their usage efficiency: increasing the carrying capacity and decreasing the tare-load ratio of a railway car, increasing the body volume and reducing operating costs [8–12].

It is possible to increase the carrying capacity, for example, of a universal gondola car model 12–9922 with an axle load of 23.5 tf and a body volume of 92 m^3 produced in the Republic of Uzbekistan [13–15] by reducing the tare weight (Fig. 1). Leading foreign car-building enterprises are actively introducing high-strength steel grades, which make it possible to reduce the tare-load of cars without sacrificing safety factors [16]. It is remarkable, that a slight decrease in the tare of one car leads to an increase in the carrying capacity on the railways as a whole, since the net mass of trains, consisting of wagons with small tare ratios, is increasing [17, 18]. Maintaining the overall dimensions of cars and reducing their tare provides a reduction in the rolling

Fig. 1 General view of the gondola body of a model 12-9922, developed and manufactured by subsidiary "casting and mechanical plant"

stock and the need for the construction of new tracks to increase the carrying capacity [19]. In addition, metal for the construction of wagons and locomotives, electricity and fuel consumed by locomotives are saved, and the cost of transportation is reduced.

Detailed planning and optimization of all types' cars construction, subject to the maximum use of the permitted overall space, lead to a significant increase in the profitability of freight traffic [19].

In today's competitive environment, rolling stock with improved parameters is of greatest interest to the consumer. Note that in the market of the Republic of Uzbekistan at the Subsidiary "Casting and Mechanical Plant", the maintenance center of which was updated in 2012, and at the Subsidiary "Andijan Mechanical Plant", the load-bearing parts of the body and frame of a four-axle gondola car with gate arrangements are made of steels below 390 MPa. There is an unoccupied niche in the introduction of gondola car parts made of high-strength steels of 16G2AF and 18G2AFps grades of 440 MPa strength class [5, 6], the development of which will allow us to take a leading position in the market of universal freight cars. Therefore, the development of specific technical proposals for the use of high-strength steels in the construction of gondolas manufactured in the Republic of Uzbekistan, in order to reduce the tare ratios and increase the reliability in operation, is relevant.

3 Research to Improve the Design of a Gondola Car

The specific effect and feasibility of using high-strength steels in the bearing parts of a model 12-9922 gondola car were determined using digital prototyping technology in the environment of modern engineering programs.

At the first stage, a gondola car-analogue of model 12-9922 was calculated for strength according to two main design modes, zones of increased stress concentration were determined. Then, at the second stage, a numerical experiment was carried out to optimize the thicknesses of the load-bearing body and frame parts made of high-strength steels of the 440 MPa strength class, and the gondola car of an alternative design was tested for strength. The main technical parameters of the gondola car model 12-9922 are shown in Table 1.

As a result of strength analysis of an analog four-axle general-purpose gondola car of model 12-9922, manufactured by Subsidiary "Casting and Mechanical Plant" [9, 10], the following results were obtained:

- With design mode I, the maximum equivalent stresses in the body elements are: in the buffer beam (upon impact) 309 MPa (95.1% of the permissible stresses).
- With design mode III, the maximum equivalent stresses in the body elements are: in the end wall belt (upon impact) 189 MPa (92.2% of the permissible stresses).

So, according to the I and III design modes, there is a high loading of the side walls racks, the upper strapping, the center beam, the sheets of the pivot and end beams. In addition, according to mode I, the current maximum stresses are close to those allowed in the frame of the end wall, the trim of the manhole cover.

Table 1 Main technical parameters and overall dimensions of a general-purpose gondola car model 12-9922 manufactured by subsidiary "CMP"

Name of parameter	Parameter standards
Carrying capacity (t)	70
Gondola car weight (tare) (t)	23.5 ± 0.2
Cubic capacity (m^3)	92
Wheel pair axle load on the rails [kN (tf)]	230 (23.5)
Length of the gondola car along the coupler pulling face axes (mm)	$13{,}920^{+20}$
Specific volume (m^3/t)	1.295
Specific material consumption (ratio of tare weight to volume) (t/m^3)	0.244
Design speed (km/h)	120
Rail gage (mm)	1520
Dimensions as per GOST standards 9238	I-BM
Frame length along the front beams (mm)	12,700
Gondola car base (mm)	8650
Operating life (years)	22

Based on the results of the calculations, it can be concluded that the strength of a general-purpose gondola-analogue of model 12-9922 with gate arrangement and an axial load of 23.5 tf, developed and manufactured by the Subsidiary "Casting and Mechanical Plant" under the main loads according to I and III design modes, meets the standard requirements [6].

The most loaded parts of an analogue gondola car identified in the calculations have the highest flow of failures in operation. Therefore, with the further optimization of the thicknesses of these parts and the replacement of the steel grade with high-strength one, their safety factors should be increased.

The use of high-strength steels of grades 16G2AF and 18G2AFps with a yield strength of 440 MPa in the bearing parts of an analogue gondola car will lead to the appearance of a strength reserve, with constant boundary conditions. Reducing the thickness of the parts will eliminate the resulting reserve and will reduce the high-strength steel gondola tare, without compromising its strength.

For this purpose, at the second stage, a comparative numerical experiment was carried out and the strength of the gondola car was calculated from the action of the main loads according to the I and III design modes, the body elements of which are made of high-strength steel with a minimum yield strength of 440 MPa [5, 6].

During the optimization of the design of a gondola car-analogue of model 12-9922 produced by the Subsidiary "Casting and Mechanical Plant", the boundary conditions in the design model, the mesh, the geometry of the nodes and parts of the gondola car, remained unchanged, and only the thicknesses of the bearing parts changed.

In the Ansys Workbench engineering program, more than ten variants of a gondola car with different thicknesses of load-bearing elements made of steel of a strength class of 440 MPa were calculated, and the stress state from the action of a set of basic loads according to I and III design modes was determined. The best option was considered to be the one at which the maximum stresses in all elements did not exceed the corresponding permissible stresses for steel with a strength class of 440 MPa. The thicknesses of the load-bearing parts of the gondola car body made of steels of strength class 440 MPa, obtained as a result of optimization, are shown in Table 2.

Table 2 Thickness of bearing units of a gondola body model 12-9922

Name of the detail	Thickness of a gondola-analogue unit of the model 12-9922, mm	Thickness of a part made of steel of strength class 440 MPa, mm
Sidewall		
Shell plate	4	3.6
Corner support	8	7
Stake	9/13	8/12
Bottom rail	12	8
Centre sill		
H-beam	9	8
Middle beam		
Crown plate	8	5
Web plate	8	6
Bottom plate	10	6
Side wall		
Vertical post	10	6
End pillar	4.8/7.8	4/6
Moulding	8	6
Cover plate	4.5	4
Buffer beam		
End plate	7	5
Top plate	8	6
Web plate	6	4
Centre bearer		
Web plate	8	5
Top plate	10	6
Bottom plate	12	8
Manhole cover		
Rear reinforcement	4	2
Manhole cover sheet	5	4
Front plate/ right/ left	5/5/5	4/4/4

The steel grade and thickness of the modernized gondola car parts, which were not included in Table 2, remained unchanged. The introduction of high-strength steels with a strength class of 440 MPa in these elements is inexpedient from a technical and economic point of view.

In the nodes of the optimized gondola car from the action of the main loads according to the I and III design modes, the bearing parts of which are made of high-strength steel of the strength class 440 MPa, and the thicknesses correspond to Table 2, the following maximum equivalent stresses were obtained according to the Mises theory:

1. In the design mode I, the maximum equivalent stresses in the body elements are: on the center beam in the zone of embedment in the pivot girder of the Z-shaped profile (during compression) 382 MPa (96.5% of the permissible stresses) and in the zone of embedding in the pivot girder of the I-section (in compression) 368 MPa (92.9% of the permissible stress).
2. In design mode III, the maximum equivalent stresses in the body elements are: on the side wall in the intermediate post (under tension) 259 MPa (95.9% of the permissible stresses), in the corner post (under tension) 254 MPa (94.1% of the permissible stresses), on an intermediate beam in a vertical sheet (under compression) 257 MPa (95.2% of the permissible stresses).

As can be seen from the results, a decrease in the thicknesses of the main load-bearing elements of the body and frame of a gondola made of high-strength steel did not lead to an excess of the permissible stresses in I and III design modes.

4 Conclusion

Based on the studies carried out, it was found that in order to improve the structure of the gondola car body, seven subassemblies of the gondola car (side and end walls; center, pivot, end, intermediate beams; manhole cover), the most vulnerable in operation, must be made entirely of high-strength steels with a yield point 440 MPa. Reducing the thickness of more than 20 parts in these nodes by 12–38% did not damage the strength of the structure.

Thus, due to the introduction of high-strength steels of grades 16G2AF and 18G2AFps with a yield point of 440 MPa and a decrease in the thickness of metal-consuming body parts (Table 2), it was possible to reduce the tare weight of the optimized gondola car to 21.5 t and increase its carrying capacity by 2 t, to 72 t.

References

1. Rasulov MK, Ibragimov UN, Rahimov RV (2013) Problems of increasing the competitiveness of domestic railway corridors. In: Proceeding scientific works of the Republican scientific and

technical conference with the participation of foreign scientists resource-saving technologies in railway transport. Tashkent Institute of Railway Engineering, Tashkent, pp 14–17
2. Rahimov RV, Galimova FS (2015) Progressive technologies for the development of interstate railway transportation. Int Sci Inst Educ 8(15):86–88
3. Rahimov RV, Zairova DN (2017) The ways to increase the transit potential of the Republic of Uzbekistan proc. Bull Tashkent Inst Railway Eng 1:120–127
4. Rahimov RV, Galimova FS (2015) Introduction of modern technologies in the enterprise SJSRC Uzbekistan railways. Bull Tashkent State Tech Univ 3:159–164
5. Ergashev ZZ, Rahimov RV, Galimova FS (2014) Implementation of a digital prototype at the railway car-building enterprises of Uzbekistan. Bull Tashkent Inst Railway Eng 2(3):43–46
6. Rahimov RV, Zairova DN, Galimova FS (2015) Introduction of digital prototype technology in the construction of new cars at the enterprises of JSC O'zbekiston temir yo'llari. Bull Tashkent Inst Railway Eng 3(4):54–60
7. Ruzmetov YO, Rahimov RV (2019) Prospects for the development of car building in the Republic of Uzbekistan. In: Proceeding collection of scientific papers of the VIII all-Russian scientific-practical conference problems and prospects for the development of carriage building, BSTU, Bryansk, pp 147–150
8. Rahimov RV (2018) State and prospects for the development of the wagon fleet of railways in Uzbekistan. In: Proceeding materials of the XIII international scientific and technical conference rolling stock of the XXI century: ideas, requirements, projects. Emperor Alexander I St. Petersburg State Transport University, Saint Petersburg pp 124–128
9. Rahimov RV (2018) Choice of directions for the development of the wagon fleet of Uzbekistan railways proc. Transp Russ Federation (Tashkent) 1(74):71–74
10. Rahimov RV, Ruzmetov YO (2018) Analysis of the state and prospects of the development of the freight wagon fleet of the Republic of Uzbekistan. Non-Ferrous Met 44(1):7–11
11. Rahimov RV (2017) Analysis of the state and development prospects of the freight car fleet of JSC Uzbekistan railways. In: Proceeding scientific works of the Republican scientific and technical conference with the participation of foreign scientists resource-saving technologies in railway transport. Tashkent Institute of Railway Engineering, Tashkent, pp 91–94
12. Rahimov RV (2014) Researches of the stressed—deformed state of the open wagon body model 12–9922, produced in Uzbekistan. In: Proceedings VI international scientific conference transport problems 2014. Silesian University of Technology Faculty of Transport, Katowice, pp 614–621
13. Rahimov RV (2012) Calculation of the stress-strain state of the metal structure of the body of a new gondola car model 12-9922 manufactured in the Republic of Uzbekistan. In: Proceeding scientific works of the Republican scientific and technical conference with the participation of foreign scientists resource-saving technologies in railway transport. Tashkent Institute of Railway Engineering, Tashkent, pp 18–23
14. Rahimov RV, Zairova DN (2012) Investigation of the calculation of the effectiveness of the gondola car brake. In: Proceeding scientific works of the Republican scientific and technical conference with the participation of foreign scientists resource-saving technologies in railway transport. Tashkent Institute of Railway Engineering, Tashkent, pp 100–102
15. Boronenko YP, Filippova IO (2015) The use of high-strength steels in car building. Transp Russ Federation 3(58):16–19
16. Rahimov RV (2019) Assessment of the stress-strain behavior of structural elements of the railway superstructures in the Republic of Uzbekistan during the operation of rolling stock with increased axle loads. Bull Sci Res Results 3:67–88
17. Rahimov RV (2019) Estimated determination of indicators of the impact of rolling stock with increased axial loads on the track in the conditions of the railways of the Republic of Uzbekistan. Bull Transp Povolzhya 5(77):23–33
18. Rahimov RV, Ruzmetov YO (2019) Assessment of the impact of the rolling stock with increased axial loads on a way and setting the conditions of their circulation on the railways of the Republic of Uzbekistan. Railway Transp: Topical Issues Innov 1–2:5–13

19. Rahimov RV (2020) On the possibility of plying freight cars with increased axle loads on the railway infrastructure of the Republic of Uzbekistan. Sci Educ Transp 1:79–81

Mathematical Models of Asymmetric Modes in High-Speed Traffic

Kamolbek Turdibekov, Mirjalil Yakubov, Absaid Sulliev, and Amangul Sanbetova

Abstract Considered asymmetric modes caused by the high-speed electric vehicles symmetrical components circuit direct, inverse and zero sequence, mathematical models of the network elements relative to the phase variables, pocenie mathematical models will provide calculations in different productions of design problems and identification of active power losses in conditions of asymmetric modes of rail transport.

Keywords Asymmetric modes · Mathematical models · Traction loads · Three-phase lines · Current and voltage matrices

1 Introduction

The energy consumption of high-speed trains reaches significant values. Significant currents occur in power lines (power lines) and the contact network. An AC electric locomotive is a purely single-phase load of high power. Therefore, the loads created by electric traction in a three-phase network are not symmetrical, i.e. different currents flow in its individual phases. Different currents in phases. Power lines with their identical resistances lead to different losses and voltage levels in them, i.e., to an asymmetry of the voltage supplied to three-phase electric energy receivers. Electrified single-phase railways are powered by power systems and create an uneven load of three-phase lines [1–3].

Electromotive forces, voltages, and currents in a three-phase circuit form a three-phase system of EMF, voltages, and currents, respectively. If all EMF, voltages or currents are equal and shifted relative to each other by 120°, then such systems of EMF, voltages or currents are symmetric. In the case when the modules of these values are not equal, or the angles between these values are not equal to 120°, or both occur simultaneously, such systems are asymmetric. If single-phase consumers

K. Turdibekov · M. Yakubov · A. Sulliev (✉)
Tashkent State Transport University, Tashkent, Uzbekistan

A. Sanbetova
Tashkent Institute of Irrigation and Agricultural Engineering Engineers, Tashkent, Uzbekistan

A. Beskopylny and M. Shamtsyan (eds.), *XIV International Scientific Conference "INTERAGROMASH 2021"*, Lecture Notes in Networks and Systems 247,
https://doi.org/10.1007/978-3-030-80946-1_95

also receive power from a three-phase system, then the load is distributed unevenly across the phases, which leads to worse use of all elements of the three-phase circuit. The asymmetric load of power systems causes voltage losses in its elements and as a result leads to the appearance of voltage asymmetry in three-phase consumers. Three-phase asynchronous motors are the main non-traction consumers powered by the power supply system of electrified railways.

2 Methods

Unbalanced voltage at the terminals of three-phase motors leads to unequal loading of their phases and hence to greater heating of the most loaded phases or to the need to reduce the load on the motors. In the case of unbalanced voltage on some devices, the voltage may be higher than nominal, and the other below, which leads to reduction of service life first and reduce the light flux of the second. In non-symmetric chains, the symmetric components of one sequence generally affect the symmetric components of the other sequence. The unbalanced system of resistances of a three-phase circuit can be decomposed into symmetrical components of various sequences, just as it was done above for voltages. As a result, symmetric components of the resistances of the forward, reverse and zero sequences will be obtained. In general, the currents of each of the sequences will cause.

Mathematical models of the corresponding network elements with respect to phase variables are necessary for the analysis of asymmetric steady-state modes in high-speed mode that feed the traction load. The network that feeds the traction load can be attributed to an extended and multi-wire network. At voltages of 110–220 kV, and more recently at higher rated voltages, it is often possible to find two-chain lines with two lines located on common supports. It is known that a three-phase system has different resistances for currents of different sequences of industrial frequency, if we take into account the variety of geometry of the location of wires on the supports of power lines [1–4].

3 Results and Discussion

The mathematical model of an unbalanced mode should take into account [2–4]:—unsymmetric mode of the electrical network;—distribution and asymmetry of line parameters;—ability to calculate voltages and currents in any cross-section of the line;—ability to calculate active and reactive components of power flows in any cross-section of the line. In three-phase lines (consisting of m wires), the voltages and currents in the line are connected by known telegraph equations. The equations of a three-phase symmetric line have the form

$$\left.\begin{aligned} -\frac{\partial u}{\partial x} &= L\frac{\partial i}{\partial t} + Ri; \\ -\frac{\partial i}{\partial x} &= C\frac{\partial u}{\partial t} + Gu, \end{aligned}\right\} \tag{1}$$

where L, C, R, and G is a square matrix of the third order: L—matrix whose elements are the self and mutual inductance of the wires of the line; C is the matrix whose elements are the self and mutual capacitance coefficients; R, G—matrix of resistances and conductances; u, i—matrix-columns of the voltages and currents [3–5]. Applying Eq. (2) for the steady state and introducing complex voltages, currents, resistances and conductivities, the line equations can be represented as the conductivity of a line of unit length; U, I are matrices of voltage and current images.

The elements of the matrix G are almost zero due to the small leakage of currents through the insulators, By simple transformations it is possible to obtain

$$\left.\begin{aligned} -\frac{\partial \underline{U}}{\partial x} &= (R_0 + j\omega L_0)\underline{I} = \underline{Z}_0\underline{I}; \\ -\frac{\partial I}{\partial x} &= (G_0 + j\omega C_0)\underline{U} = \underline{Y}_0\underline{U}, \end{aligned}\right\} \tag{2}$$

where

$$Z_0 = \begin{vmatrix} r+r_3 & r_3 & r_3 \\ r_3 & r+r_3 & r_3 \\ r_3 & r_3 & r+r_3 \end{vmatrix} + j\omega \begin{vmatrix} L_{aa} & L_{ab} & L_{ac} \\ L_{ba} & L_{bb} & L_{bc} \\ L_{ca} & L_{cb} & L_{cc} \end{vmatrix} = \begin{vmatrix} Z_{aa} & Z_{ab} & Z_{ac} \\ Z_{ba} & Z_{bb} & Z_{bc} \\ Z_{ca} & Z_{cb} & Z_{cc} \end{vmatrix}$$

$$\text{the } Y_0 = \begin{vmatrix} Y_{aa} & Y_{ab} & Y_{ac} \\ Y_{ba} & Y_{bb} & Y_{bc} \\ Y_{ca} & Y_{cb} & Y_{cc} \end{vmatrix}$$

this is the complex resistance and complex conductivity of a line of unit length; U, I are matrices of voltage and current images. For overhead lines, the elements of the matrix G are practically zero due to the small leakage of currents through the insulators and by simple transformations it is possible to obtain

$$\left.\begin{aligned} -\frac{\partial^2 \underline{U}}{\partial x^2} &= \underline{Z}_0\underline{Y}_0\underline{U}; \\ -\frac{\partial^2 I}{\partial x^2} &= \underline{Y}_0\underline{Z}_0\underline{I}. \end{aligned}\right\} \tag{3}$$

For an unbalanced line consisting of m wires, these equations are obtained after solving Eq. (3), in the form of column matrices as follows

$$U = e^{-\gamma_u x} \cdot A + e^{\gamma_u x} \cdot B;$$
$$I = e^{-\gamma_i x} \cdot C + e^{\gamma_i x} \cdot D, \tag{4}$$

where $e^{\gamma x}$ is the matrix exponent; A, B, C, D are constant column vectors of $m \times 1$ dimension calculated from boundary conditions; γ—matrix of propagation coefficients.

It follows from (4) that

$$\left.\begin{aligned} \underline{\gamma}_u &= \sqrt{\underline{Z}_0\underline{Y}_0} \\ \underline{\gamma}_i &= \sqrt{\underline{Y}_0\underline{Z}_0} \end{aligned}\right\} \tag{5}$$

In the case of a symmetric three-phase line, $\gamma_u = \gamma_i = \gamma$ due to the symmetry of the matrices Z_o and Y_o. y_u and y_i are complex square matrices of dimension $m \times m$, representing the corresponding functions of the matrices of proper and mutual linear resistances and conductivities of wires. Using (5), we can obtain the relation of the matrices A, B, C, D with the matrices of currents and voltages at the beginning and end of the line I(0), U(0), I(l), U(l). We have

$$\left.\begin{aligned} \left.\frac{\partial U}{\partial x}\right|_{x=l} &= Z_0 I(l) = -\gamma e^{-\gamma_u x} \cdot A + \gamma e^{\gamma_u x} \cdot B; \\ \left.\frac{\partial I}{\partial x}\right|_{x=l} &= Y_0 U(l) = -\gamma e^{-\gamma_i x} \cdot C + \gamma e^{\gamma_i x} \cdot D, \end{aligned}\right\} \tag{6}$$

where from

$$\left.\begin{aligned} \frac{\partial U_x}{\partial x} &= -Z \cdot I_x = -\gamma_u e^{-\gamma_u x} \cdot A + \gamma_u e^{\gamma_u x} \cdot B; \\ \frac{\partial I_x}{\partial x} &= -Y \cdot U_x = -\gamma_i e^{-\gamma_i x} \cdot C + \gamma_i e^{\gamma_i x} \cdot D, \end{aligned}\right\} \tag{7}$$

where U_x and I_x are vectors-columns of complex voltages and currents of dimension $m \times 1$ in the cross section at a distance x from the beginning of the line; at the beginning of the line

$$\left.\begin{aligned} -Z \cdot I_H &= -\gamma_u \cdot A + \gamma_u \cdot B; \\ -Y \cdot U_H &= -\gamma_i \cdot C + \gamma_i \cdot D, \end{aligned}\right\} \tag{8}$$

at the end of the line

$$\left.\begin{aligned} Z \cdot I_k &= -\gamma_u e^{-\gamma_u x} \cdot A + \gamma_u e^{\gamma_u x} \cdot B; \\ -Y \cdot U_k &= -\gamma_i e^{-\gamma_i x} \cdot C + \gamma_i e^{\gamma_i x} \cdot D, \end{aligned}\right\} \tag{9}$$

The original model was originally developed for the calculation and analysis of transients in lines with distributed parameters. At the same time, the problems formulated above belong to the class of steady-state problems.

For a single (one or two-chain) line, when mathematically modeling its modes, different problem statements are possible with different composition of the specified (independent) parameters and calculated (dependent) parameters. To formalize the mathematical model (5, 9), we convert it to a form that is convenient for dividing the mode parameters into dependent and independent mode parameters. This procedure includes several steps [5, 6]:

1. The column vectors A, B are determined by solving a system of matrix equations written on the basis of the first equation of the system (9) for the corresponding beginning and end of the line

$$\begin{aligned} U_H &= A + B; \\ U_k &= e^{-\gamma_u l} \cdot A + e^{\gamma_u l} \cdot B, \end{aligned} \tag{10}$$

where l is the length of the line, and the indices h and k denote its beginning and end. The solution of this system when using the block form of matrix recording has the following form

$$\begin{vmatrix} A \\ B \end{vmatrix} = \begin{vmatrix} E & E \\ e^{-\gamma_u l} & e^{-\gamma_u l} \end{vmatrix}^{-1} \times \begin{vmatrix} U_H \\ U_k \end{vmatrix} = \begin{vmatrix} H_{11} & H_{12} \\ H_{21} & H_{22} \end{vmatrix} \times \begin{vmatrix} U_H \\ U_k \end{vmatrix}, \tag{11}$$

where $E = \begin{vmatrix} 1 & 0 & 0 \\ 0 & 1 & 0 \\ 0 & 0 & 1 \end{vmatrix}$—a unit complex matrix of dimension m × m;

$\underline{\gamma}_u = \sqrt{\underline{Z}_0 \underline{Y}_0} = \left(\begin{vmatrix} Z_{aa} & Z_{ab} & Z_{ac} \\ Z_{ba} & Z_{bb} & Z_{bc} \\ Z_{ca} & Z_{cb} & Z_{cc} \end{vmatrix} x \begin{vmatrix} Y_{aa} & Y_{ab} & Y_{ac} \\ Y_{ba} & Y_{bb} & Y_{bc} \\ Y_{ca} & Y_{cb} & Y_{cc} \end{vmatrix} \right)^{\frac{1}{2}}$—matrix of propagation coefficients.

$U_H = \begin{vmatrix} U_{a.H} \\ U_{b.H} \\ U_{c.H} \end{vmatrix}$; $U_k = \begin{vmatrix} U_{a.k} \\ U_{b.k} \\ U_{c.k} \end{vmatrix}$ phase voltages of the beginning and end of the line.

Where from

$$\begin{aligned} A &= H_{11} U_H + H_{12} U_k \\ B &= H_{21} U_H + H_{22} U_k \end{aligned} \tag{12}$$

2. Column vectors C, D are determined by solving a system of matrix equations written on the basis of the second equation of the system (12) for the

corresponding start and end of the line:

$$\begin{aligned} I_H &= C + D; \\ I_k &= e^{-\gamma_u l} \cdot C + e^{\gamma_u l} \cdot D \end{aligned} \tag{13}$$

The solution of this system when using the block form of matrix recording has the following form

$$\begin{vmatrix} C \\ D \end{vmatrix} = \begin{vmatrix} E & E \\ e^{-\gamma_i^l} & e^{-\gamma_i^l} \end{vmatrix}^{-1} \times \begin{vmatrix} I_H \\ I_k \end{vmatrix} = \begin{vmatrix} W_{11} & W_{12} \\ W_{21} & W_{22} \end{vmatrix} \times \begin{vmatrix} I_H \\ I_k \end{vmatrix} \tag{14}$$

where $\underline{\gamma}_i = \sqrt{\underline{Y}_0 \underline{Z}_0} = \left(\begin{vmatrix} Y_{aa} & Y_{ab} & Y_{ac} \\ Y_{ba} & Y_{bb} & Y_{bc} \\ Y_{ca} & Y_{cb} & Y_{cc} \end{vmatrix} \times \begin{vmatrix} Z_{aa} & Z_{ab} & Z_{ac} \\ Z_{ba} & Z_{bb} & Z_{bc} \\ Z_{ca} & Z_{cb} & Z_{cc} \end{vmatrix} \right)^{\frac{1}{2}}$ matrix of propagation coefficients.

$I_H = \begin{vmatrix} I_{a.H} \\ I_{b.H} \\ I_{c.H} \end{vmatrix}$; $I_k = \begin{vmatrix} I_{a.k} \\ I_{b.k} \\ I_{c.k} \end{vmatrix}$—phase currents of the beginning and end of the line.

Where from

$$\begin{aligned} C &= W_{11} \cdot I_H + W_{12} \cdot I_k \\ D &= W_{21} \cdot I_H + W_{22} \cdot I_k \end{aligned} \tag{15}$$

3. The column vectors of voltages and currents at the beginning and end of the line are determined using the equations of the system (13, 15), taking into account certain values of the vectors-integration constants

$$\begin{aligned} &Y \cdot U_H - \gamma_i \cdot (W_{11} \cdot I_H + W_{12} \cdot I_k) + \gamma_i \cdot (W_{21} \cdot I_H + W_{22} \cdot I_k) = 0 \\ &Y \cdot U_k - \gamma_i e^{-\gamma_i l} \cdot (W_{11} \cdot I_H + W_{12} \cdot I_k) + \gamma_i e^{\gamma_i l} \cdot (W_{21} \cdot I_H + W_{22} \cdot I_k) = 0 \\ &Z \cdot I_H - \gamma_u \cdot (H_{11} U_H + H_{12} U_k) + \gamma_u \cdot (H_{21} U_H + H_{22} U_k) = 0 \\ &Z \cdot I_k - \gamma_u e^{-\gamma_u l} \cdot (H_{11} U_H + H_{12} U_k) + \gamma_u e^{\gamma_u l} \cdot (H_{21} U_H + H_{22} U_k) = 0 \end{aligned}$$

The resulting matrix equations, combined in pairs into systems, form mathematical models of modes and provide calculations in different formulations of computational problems. So, when specifying voltages and currents at the beginning of a line, the voltages and currents at the end of the line can be determined, and vice versa. It is possible to set the operating parameters at the beginning and end of the line, and it is also possible to set the phase voltages at the beginning of the line and the phase currents at the end of the line. Such a statement is appropriate, for example, in the study of idle modes of the line. Another formulation of the calculation problem is the setting of voltages at the beginning and end of the line, for example, when studying short-circuit modes at the end of the line [7, 8].

There is a certain pattern in the formation of Y_{ij} blocks:—on the main diagonal there are Y_{ii} blocks representing the sums of matrices defined by matrix expressions of the type $-Z^{ij-1} \cdot (-\gamma_u^{ij} \cdot H_{11}^{ij} + \gamma_u^{ij} \cdot H_{21}^{ij})$ provided that the branches lying between nodes i and j are adjacent to the node by their origin and matrix expressions of the type.

$-Z^{ji-1} \cdot (-\gamma_u^{ji} \cdot e^{-\gamma_u^{ji} l^{ji}} \cdot H_{12}^{ji} + \gamma_u^{ji} \cdot e^{-\gamma_u^{ji} l^{ji}} \cdot H_{22}^{ji})$ provided that the branches lying between nodes i and j are adjacent to node i with their end—the off-diagonal blocks of the matrix Y_{ij} is computed expressions of the type $-Z^{ij-1} \cdot (-\gamma_u^{ij} \cdot H_{11}^{ij} + \gamma_u^{ij} \cdot H_{21}^{ij})$ if the indices i and j coincide with the node numbers of the beginning and end of the branch and matrix expressions of the type $-Z^{ji-1} \cdot (-\gamma_u^{ji} \cdot e^{-\gamma_u^{ji} l^{ji}} \cdot H_{11}^{ji} + \gamma_u^{ji} \cdot e^{-\gamma_u^{ji} l^{ji}} \cdot H_{21}^{ji})$ when changing the order of the indexes and node numbers of the beginning and end of the branch.

In cases where the load currents are unknown in advance and are determined in the calculation process depending on the nodal voltages using appropriate mathematical models [8, 9], the calculation should be performed iteratively.

$$\begin{vmatrix} Y_{11} & Y_{12} \\ Y_{21} & Y_{22} \end{vmatrix} \mathrm{x} \begin{vmatrix} U_1 \\ U_2 \end{vmatrix}^{k+1} = \begin{vmatrix} J_1(u_1^k) \\ J_2(u_2^k) \end{vmatrix}^{k+1} - U_6 \mathrm{x} \begin{vmatrix} U_{16} \\ U_{26} \end{vmatrix} \tag{16}$$

These equations allow us to study modes with nonlinear loads. In addition, on the basis of these equations, it is possible to study the frequency properties of complex areas and determine the possibility of resonant modes at higher harmonic frequencies. When setting the voltage of the base node, power and energy losses, as well as their structure, are calculated based on the results of measurements in a complex electrical network [10]. The most difficult procedure for performing calculations based on the above equations is the procedure for calculating functions from matrices of the type:

$$\underline{\gamma}_u = \sqrt{\underline{Z}_0 \underline{Y}_0}, \quad \underline{\gamma}_i = \sqrt{\underline{Y}_0 \underline{Z}_0} \text{ the } e^{-\gamma_u l}, \; e^{\gamma_u l}, \; e^{-\gamma_i l}, \; e^{\gamma_i l} \tag{17}$$

Calculations of these functions are performed using the Cayley-Hamilton theorem, according to which the matrix function is defined by the formula:

$$f(A) = \frac{1}{\Delta} \sum_{k=1}^{n} \Delta_{n-k} A^{n-k} \tag{18}$$

where Δ is the Vandermonde determinant $\det[\lambda_i^{k-1}]$, where λ_i are the eigenvalues of the matrix A, a Δ_j—and a is the determinant obtained if Δ is substituted $\lambda_1^j, \lambda_2^j, \ldots, \lambda_n^j$ for the values of the functions $f(\lambda_1), f(\lambda_2), \ldots, f(\lambda_n)$.

4 Conclusion

The presented equations allow us to study the modes of lines with different numbers of wires and lightning protection cables. On the basis of Eq. (18), plots of the distribution of voltages and currents are calculated. Calculation of current distribution plots provides determination of active power losses under conditions of asymmetry.

References

1. Zakaryukin VP, Kryukov AV (2005) Complex-symmetric modes of electrical systems. Publishing House of Irkutsk State University, Irkutsk, p 273
2. Zakaryukin VP, Kryukov AV (2005) Methods of joint modeling of traction and external power supply systems of alternating current railways. Publishing House of Irkutsk State University, Irkutsk, p 305
3. Grinkrug MS (2011) Unsymmetric modes of operation of electric networks. M. Lambert Academic Publishing, p 124
4. Ded AV (2014) Estimated estimation of additional power losses in the elements of electric networks. Mod High-Tech Technol 10:21–25
5. Naumov IV (2006) Experimental study of the parameters of asymmetry in an unsymmetric system of power supply voltage. Uspekhi 195 sovremennogo estestvoznaniya 11:64–65
6. Turdibekov KK, Kurbanov IB (2018) Voltage symmetry in the power supply system of high-speed railway transport. J TASHSTU Prob Energy Resour Saving—Tashkent 3–4:156–159
7. Kholiddinov IK (2015) Algorithm for determining the level of stress asymmetry//Power engineering: management, quality and efficiency of energy resources use. In: International scientific and technical conference, 27–29 May. Blagoveshchensk, Amur Region, pp 196–201
8. Rustamov DS, Zhumabaev SK (2017) Questions of current asymmetry in three-phase traction networks and a device for measuring it. Resource-saving technologies in railway transport. In: Proceeding Doklady reports of the scientific and technical conference with the participation of foreign scientists, Tashkent, pp 170–172
9. Amirov SF, Jumaboev SH, Umarov MG (2018) Electromagnetic Converter unbalance current to voltage with flat measuring winding. Actual problems of modern science, Moscow, vol 6, pp 246–248
10. Amirov SF, Safarov AM, Jumaboev SH (2019) Measuring device for the detection of asymmetry of three-phase currents in the traction power supply. Prob Inf Power Eng—Toshkent 4:63–67